$136.95 per copy (in United States).
Price subject to change without prior notice.

0037

SO-AXN-822

RSMeans

Electrical Cost Data

30th Annual Edition

RSMeans
Construction Publishers & Consultants
63 Smiths Lane
Kingston, MA 02364-0800
(781) 422-5000

Senior Editor
John H. Chiang, PE

Contributing Editors
Christopher Babbitt
Ted Baker
Barbara Balboni
Robert A. Bastoni
Gary W. Christensen
Cheryl Elsmore
Robert J. Kuchta
Robert C. McNichols
Robert W. Mewis, CCC
Melville J. Mossman, PE
Jeannene D. Murphy
Stephen C. Plotner
Eugene R. Spencer
Marshall J. Stetson
Phillip R. Waier, PE

**Senior Engineering
Operations Manager**
John H. Ferguson, PE

**Senior Vice President
& General Manager**
John Ware

**Vice President
of Direct Response**
John M. Shea

**Director of Product
Development**
Thomas J. Dion

Production Manager
Michael Kokernak

Production Coordinator
Wayne D. Anderson

Technical Support
Jonathan Forgit
Mary Lou Geary
Jill Goodman
Gary L. Hoitt
Genevieve Medeiros
Paula Reale-Camelio
Kathryn S. Rodriguez
Sheryl A. Rose

Book & Cover Design
Norman R. Forgit

 This book is recyclable.

 This book is printed on recycled stock.

 Reed Construction Data®

First Printing

Foreword

Reed Construction Data's portfolio of project information products and services includes national, regional, and local construction data, project leads, and project plans, specifications, and addenda available online or in print. Reed Bulletin (www.reedbulletin.com) and Reed CONNECT™ (www.reedconnect.com) deliver the most comprehensive, timely, and reliable project information to support contractors, distributers, and building product manufacturers in identifying, bidding, and tracking projects. The Reed First Source (www.reedfirstsource.com) suite of products used by design professionals for the search, selection, and specification of nationally available building projects consists of the First Source™ annual, SPEC-DATA™, MANU-SPEC™, First Source CAD, and manufacturer catalogs. Reed Design Registry, a database of more than 30,000 U.S. architectural firms, is also published by Reed Construction Data. RSMeans (www.rsmeans.com) provides construction cost data, training, and consulting services in print, CD-ROM, and online. Associated Construction Publications (www.reedpubs.com) reports on heavy, highway, and non-residential construction through a network of 14 regional construction magazines. Reed Construction Data (www.reedconstructiondata.com), headquartered in Atlanta, is a subsidiary of Reed Business Information (www.reedbusinessinformation.com), North America's largest business-to-business information provider. With more than 80 market-leading publications and 55 websites, Reed Business Information's wide range of services also includes research, business development, direct marketing lists, training and development programs, and technology solutions. Reed Business Information is a member of the Reed Elsevier plc group (NYSE: RUK and ENL)—a leading provider of global information-driven services and solutions in the science and medical, legal, education, and business-to-business industry sectors.

Our Mission

Since 1942, RSMeans has been actively engaged in construction cost publishing and consulting throughout North America.

Today, over 60 years after RSMeans began, our primary objective remains the same: to provide you, the construction and facilities professional, with the most current and comprehensive construction cost data possible.

Whether you are a contractor, an owner, an architect, an engineer, a facilities manager, or anyone else who needs a reliable construction cost estimate, you'll find this publication to be a highly useful and necessary tool.

Today, with the constant flow of new construction methods and materials, it's difficult to find the time to look at and evaluate all the different construction cost possibilities. In addition, because labor and material costs keep changing, last year's cost information is not a reliable basis for today's estimate or budget.

That's why so many construction professionals turn to RSMeans. We keep track of the costs for you, along with a wide range of other key information, from city cost indexes . . . to productivity rates . . . to crew composition . . . to contractor's overhead and profit rates.

RSMeans performs these functions by collecting data from all facets of the industry and organizing it in a format that is instantly accessible to you. From the preliminary budget to the detailed unit price estimate, you'll find the data in this book useful for all phases of construction cost determination.

The Staff, the Organization, and Our Services

When you purchase one of RSMeans' publications, you are, in effect, hiring the services of a full-time staff of construction and engineering professionals.

Our thoroughly experienced and highly qualified staff works daily at collecting, analyzing, and disseminating comprehensive cost information for your needs. These staff members have years of practical construction experience and engineering training prior to joining the firm. As a result, you can count on them not only for the cost figures, but also for additional background reference information that will help you create a realistic estimate.

The RSMeans organization is always prepared to help you solve construction problems through its five major divisions: Construction and Cost Data Publishing, Electronic Products and Services, Consulting and Research Services, Insurance Services, and Professional Development Services.

Besides a full array of construction cost estimating books, RSMeans also publishes a number of other reference works for the construction industry. Subjects include construction estimating and project and business management; special topics such as HVAC, roofing, plumbing, and hazardous waste remediation; and a library of facility management references.

In addition, you can access all of our construction cost data electronically using *Means CostWorks*® CD or on the Web.

What's more, you can increase your knowledge and improve your construction estimating and management performance with an RSMeans Construction Seminar or In-House Training Program. These two-day seminar programs offer unparalleled opportunities for everyone in your organization to get updated on a wide variety of construction-related issues.

RSMeans is also a worldwide provider of construction cost management and analysis services for commercial and government owners, and of claims and valuation services for insurers.

In short, RSMeans can provide you with the tools and expertise for constructing accurate and dependable construction estimates and budgets in a variety of ways.

Robert Snow Means Established a Tradition of Quality That Continues Today

Robert Snow Means spent years building RSMeans, making certain he always delivered a quality product.

Today, at RSMeans, we do more than talk about the quality of our data and the usefulness of our books. We stand behind all of our data, from historical cost indexes to construction materials and techniques to current costs.

If you have any questions about our products or services, please call us toll-free at 1-800-334-3509. Our customer service representatives will be happy to assist you. You can also visit our Web site at www.rsmeans.com.

Table of Contents

Related RSMeans Products and Services

John H. Chiang, P.E., Senior Editor of this cost data book, suggests the following RSMeans products and services as companion information resources to *RSMeans Electrical Cost Data:*

Construction Cost Data Books
Electrical Change Order Cost Data 2007
Building Construction Cost Data 2007
Mechanical Cost Data 2007

Reference Books
Electrical Estimating Methods, 3rd Ed.
ADA Compliance Pricing Guide, 2nd Ed.
Building Security: Strategies & Costs
Designing & Building with the IBC, 2nd Ed.
Estimating Building Costs
Estimating Handbook, 2nd Ed.
Green Building: Project Planning & Estimating, 2nd Ed.
How to Estimate with Means Data and CostWorks, 3rd Ed.
Plan Reading & Material Takeoff
Project Scheduling and Management for Construction

Seminars and In-House Training
Mechanical & Electrical Estimating
Means CostWorks® Training
Means Data for Job Order Contracting (JOC)
Plan Reading & Material Takeoff
Scheduling & Project Management
Unit Price Estimating

RSMeans on the Internet
Visit RSMeans at **www.rsmeans.com.** The site contains useful interactive cost and reference material. Request or download **FREE** estimating software demos. Visit our bookstore for convenient ordering and to learn more about new publications and companion products.

RSMeans Electronic Data
Get the information found in RSMeans cost books electronically on *Means CostWorks®* CD or on the Web.

RSMeans Business Solutions
Engineers and Analysts offer research studies, benchmark analysis, predictive cost modeling, analytics, job order contracting, and real property management consultation, as well as custom-designed, web-based dashboards and calculators that apply RCD/RSMeans extensive databases. Clients include federal government agencies, architects, construction management firms, and institutional organizations such as school systems, health care facilities, associations, and corporations.

New! RSMeans for Job Order Contracting (JOC)
Best practice JOC tools for cost estimating and project management to help streamline delivery processes for renovation projects. Renovation is a $147 billion market in the U.S., and includes projects in school districts, municipalities, health care facilities, colleges and universities, and corporations.
- RSMeans Engineers consult in contracting methods and conduct JOC Facility Audits
- JOCWorks™ Software (Basic, Advanced, PRO levels)
- RSMeans Job Order Contracting Cost Data for the entire U.S.

Construction Costs for Software Applications
Over 25 unit price and assemblies cost databases are available through a number of leading estimating and facilities management software providers (listed below). For more information see the "Other RSMeans Products" pages at the back of this publication.

MeansData™ is also available to federal, state, and local government agencies as multi-year, multi-seat licenses.

- 3D International
- 4Clicks-Solutions, LLC
- Aepco, Inc.
- Applied Flow Technology
- ArenaSoft Estimating
- Ares Corporation
- Beck
- BSD – Building Systems Design, Inc.
- CMS – Construction Management Software
- Corecon Technologies, Inc.
- CorVet Systems
- Earth Tech
- Estimating Systems, Inc.
- HCSS
- Maximus Asset Solutions
- MC^2 – Management Computer Controls
- Sage Timberline Office
- Shaw Beneco Enterprises, Inc.
- US Cost, Inc.
- VFA – Vanderweil Facility Advisers
- WinEstimator, Inc.

How the Book Is Built: An Overview

A Powerful Construction Tool

You have in your hands one of the most powerful construction tools available today. A successful project is built on the foundation of an accurate and dependable estimate. This book will enable you to construct just such an estimate.

For the casual user the book is designed to be:

- quickly and easily understood so you can get right to your estimate.
- filled with valuable information so you can understand the necessary factors that go into the cost estimate.

For the regular user, the book is designed to be:

- a handy desk reference that can be quickly referred to for key costs.
- a comprehensive, fully reliable source of current construction costs and productivity rates so you'll be prepared to estimate any project.
- a source book for preliminary project cost, product selections, and alternate materials and methods.

To meet all of these requirements we have organized the book into the following clearly defined sections.

How To Use the Book: The Details

This section contains an in-depth explanation of how the book is arranged ... and how you can use it to determine a reliable construction cost estimate. It includes information about how we develop our cost figures and how to completely prepare your estimate.

Unit Price Section

All unit price cost data has been divided into the 50 divisions according to the MasterFormat system of classification and numbering. For a listing of these divisions and an outline of their subdivisions, see the Unit Price Section Table of Contents.

Estimating tips are included at the beginning of each division.

Assemblies Section

The cost data in this section has been organized in an "Assemblies" format. These assemblies are the functional elements of a building and are arranged according to the 7 divisions of the UNIFORMAT II classification system. For a complete explanation of a typical "Assemblies" page, see "How To Use the Assemblies Cost Tables."

Reference Section

This section includes information on Equipment Rental Costs, Crew Listings, Historical Cost Indexes, City Cost Indexes, Location Factors, Reference Tables, Square Foot Costs, and a listing of Abbreviations.

Equipment Rental Costs: This section contains the average costs to rent and operate hundreds of pieces of construction equipment.

Crew Listings: This section lists all the crews referenced in the book. For the purposes of this book, a crew is composed of more than one trade classification and/or the addition of power equipment to any trade classification. Power equipment is included in the cost of the crew. Costs are shown both with the bare labor rates and with the installing contractor's overhead and profit added. For each, the total crew cost per eight-hour day and the composite cost per labor-hour are listed.

Historical Cost Indexes: These indexes provide you with data to adjust construction costs over time.

City Cost Indexes: All costs in this book are U.S. national averages. Costs vary because of the regional economy. You can adjust costs by CSI Division to over 316 locations throughout the U.S. and Canada by using the data in this section.

Location Factors: You can adjust total project costs to over 900 locations throughout the U.S. and Canada by using the data in this section.

Reference Tables: At the beginning of selected major classifications in the Unit Price and Assemblies section are "reference numbers" shown in a shaded box. These numbers refer you to related information in the Reference Section. In this section, you'll find reference tables, explanations, estimating information that support how we develop the unit price data, technical data, and estimating procedures.

Square Foot Costs: This section contains costs for 59 different building types that allow you to make a rough estimate for the overall cost of a project or its major components.

Abbreviations: A listing of abbreviations used throughout this book, along with the terms they represent, is included in this section.

Index

A comprehensive listing of all terms and subjects in this book will help you quickly find what you need when you are not sure where it falls in MasterFormat.

The Scope of This Book

This book is designed to be as comprehensive and as easy to use as possible. To that end we have made certain assumptions and limited its scope in two key ways:

1. We have established material prices based on a national average.
2. We have computed labor costs based on a 30-city national average of union wage rates.

For a more detailed explanation of how the cost data is developed, see "How To Use the Book:The Details."

Project Size

This book is aimed primarily at commercial and industrial projects costing $1,000,000 and up, or large multi-family housing projects. Costs are primarily for new construction or major renovation of buildings rather than repairs or minor alterations.

With reasonable exercise of judgment the figures can be used for any building work. However, for civil engineering structures such as bridges, dams, highways, or the like, please refer to *RSMeans Heavy Construction Cost Data.*

How to Use the Book: The Details

What's Behind the Numbers? The Development of Cost Data

The staff at RSMeans continuously monitors developments in the construction industry in order to ensure reliable, thorough and up-to-date cost information.

While *overall* construction costs may vary relative to general economic conditions, price fluctuations within the industry are dependent upon many factors. Individual price variations may, in fact, be opposite to overall economic trends. Therefore, costs are continually monitored and complete updates are published yearly. Also, new items are frequently added in response to changes in materials and methods.

Costs — $ (U.S.)

All costs represent U.S. national averages and are given in U.S. dollars. The RSMeans City Cost Indexes can be used to adjust costs to a particular location. The City Cost Indexes for Canada can be used to adjust U.S. national averages to local costs in Canadian dollars. No exchange rate conversion is necessary.

Material Costs

The RSMeans staff contacts manufacturers, dealers, distributors, and contractors all across the U.S. and Canada to determine national average material costs. If you have access to current material costs for your specific location, you may wish to make adjustments to reflect differences from the national average. Included within material costs are fasteners for a normal installation. RSMeans engineers use manufacturers' recommendations, written specifications, and/or standard construction practice for size and spacing of fasteners. Adjustments to material costs may be required for your specific application or location. Material costs do not include sales tax.

Labor Costs

Labor costs are based on the average of wage rates from 30 major U.S. cities. Rates are determined from labor union agreements or prevailing wages for construction trades for the current year. Rates, along with overhead and profit markups, are listed on the inside back cover of this book.

- If wage rates in your area vary from those used in this book, or if rate increases are expected within a given year, labor costs should be adjusted accordingly.

Labor costs reflect productivity based on actual working conditions. These figures include time spent during a normal workday on tasks other than actual installation, such as material receiving and handling, mobilization at site, site movement, breaks, and cleanup.

Productivity data is developed over an extended period so as not to be influenced by abnormal variations and reflects a typical average.

Equipment Costs

Equipment costs include not only rental, but also operating costs for equipment under normal use. The operating costs include parts and labor for routine servicing such as repair and replacement of pumps, filters, and worn lines. Normal operating expendables, such as fuel, lubricants, tires, and electricity (where applicable), are also included. Extraordinary operating expendables with highly variable wear patterns, such as diamond bits and blades, are excluded. These costs are included under materials. Equipment rental rates are obtained from industry sources throughout North America—contractors, suppliers, dealers, manufacturers, and distributors.

Crew Equipment Cost/Day—The power equipment required for each crew is included in the crew cost. The daily cost for crew equipment is based on dividing the weekly bare rental rate by 5 (number of working days per week) and then adding the hourly operating cost times 8 (hours per day). This "Crew Equipment Cost/Day" is listed in the Reference Section.

Factors Affecting Costs

Costs can vary depending upon a number of variables. Here's how we have handled the main factors affecting costs.

Quality—The prices for materials and the workmanship upon which productivity is based represent sound construction work. They are also in line with U.S. government specifications.

Overtime—We have made no allowance for overtime. If you anticipate premium time or work beyond normal working hours, be sure to make an appropriate adjustment to your labor costs.

Productivity—The productivity, daily output, and labor-hour figures for each line item are based on working an eight-hour day in daylight hours in moderate temperatures. For work that extends beyond normal work hours or is performed under adverse conditions, productivity may decrease. (See the section in "How To Use the Unit Price Pages" for more on productivity.)

Size of Project—The size, scope of work, and type of construction project will have a significant impact on cost. Economies of scale can reduce costs for large projects. Unit costs can often run higher for small projects. Costs in this book are intended for the size and type of project as previously described in "How the Book Is Built: An Overview." Costs for projects of a significantly different size or type should be adjusted accordingly.

Location—Material prices in this book are for metropolitan areas. However, in dense urban areas, traffic and site storage limitations may increase costs. Beyond a 20-mile radius of large cities, extra trucking or transportation charges may also increase the material costs slightly. On the other hand, lower wage rates may be in effect. Be sure to consider both of these factors when preparing an estimate, particularly if the job site is located in a central city or remote rural location.

In addition, highly specialized subcontract items may require travel and per-diem expenses for mechanics.

Other Factors—

- season of year
- contractor management
- weather conditions
- local union restrictions
- building code requirements
- availability of:
  - adequate energy
  - skilled labor
  - building materials
- owner's special requirements/restrictions
- safety requirements
- environmental considerations

Unpredictable Factors—General business conditions influence "in-place" costs of all items. Substitute materials and construction methods may have to be employed. These may affect the installed cost and/or life cycle costs. Such factors may be difficult to evaluate and cannot necessarily be predicted on the basis of the job's location in a particular section of the country. Thus, where these factors apply, you may find significant but unavoidable cost variations for which you will have to apply a measure of judgment to your estimate.

Rounding of Costs

In general, all unit prices in excess of $5.00 have been rounded to make them easier to use and still maintain adequate precision of the results. The rounding rules we have chosen are in the following table.

Prices from ...	Rounded to the nearest ...
$.01 to $5.00	$.01
$5.01 to $20.00	$.05
$20.01 to $100.00	$.50
$100.01 to $300.00	$1.00
$300.01 to $1,000.00	$5.00
$1,000.01 to $10,000.00	$25.00
$10,000.01 to $50,000.00	$100.00
$50,000.01 and above	$500.00

Estimating Labor-Hours

The labor-hours expressed in this publication are based on Average Installation time, using an efficiency level of approximately 60-65% (see item 7 below), which has been found reasonable and acceptable by many contractors.

The book uses this National Efficiency Average to establish a consistent benchmark.

For bid situations, adjustments to this efficiency level should be the responsibility of the contractor bidding the project.

The unit labor-hour is divided in the following manner. A typical day for a crew might be:

1.	Study Plans	3%	14.4 min.
2.	Material Procurement	3%	14.4 min.
3.	Receiving and Storing	3%	14.4 min
4.	Mobilization	5%	24.0 min.
5.	Site Movement	5%	24.0 min.
6.	Layout and Marking	8%	38.4 min.
7.	Actual Installation	64%	307.2 min.
8.	Cleanup	3%	14.4 min.
9.	Breaks, Non-Productive	6%	28.8 min.
		100%	480.0 min.

If any of the percentages expressed in this breakdown do not apply to the particular work or project situation, then that percentage or a portion of it may be deducted from or added to labor-hours.

Overhead, Profit & Contingencies

General—Prices given in this book are of two kinds: (1) BARE COSTS and (2) TOTAL COST INCLUDING INSTALLING CONTRACTOR'S OVERHEAD & PROFIT.

General Conditions—Cost data in this book is presented in two ways: Bare Costs and Total Cost including O&P (Overhead and Profit). General Conditions, when applicable, should also be added to the Total Cost including O&P. The costs for General Conditions are listed in Division 1 of the Unit Price Section and the Reference Section of this book. General Conditions for the *Installing Contractor* may range from 0% to 10% of the Total Cost including O&P. For the *General* or *Prime Contractor*, costs for General Conditions may range from 5% to 15% of the Total Cost including O&P, with a figure of 10% as the most typical allowance.

Overhead and Profit—Total Cost including O&P for the *Installing Contractor* is shown in the last column on both the Unit Price and the Assemblies pages of this book. This figure is the sum of the bare material cost plus 10% for profit, the base labor cost plus total overhead and profit, and the bare equipment cost plus 10% for profit. Details for the calculation of Overhead and Profit on labor are shown on the inside back cover and in the Reference Section of this book. (See the "How To Use the Unit Price Pages" for an example of this calculation.)

Subcontractors—Usually a considerable portion of all large jobs is subcontracted. In fact, the percentage done by subs is constantly increasing and may run over 90%. Since the workers employed by these companies do nothing else but install their particular product, they soon become experts in that line. The result: installation by these firms is accomplished so efficiently that the total in-place cost, even with the subcontractor's overhead and profit, is no more, and often less, than if the principal contractor had handled the installation himself or herself. Also, the quality of the work may be higher.

Contingencies—The allowance for contingencies generally is to provide for indefinable construction difficulties. On alterations or repair jobs, 20% is not too much. If drawings are final and only field contingencies are being considered, 2% or 3% is probably sufficient, and often nothing need be added. As far as the contract is concerned, future changes in plans will be covered by extras. The contractor should allow for inflationary price trends and possible material shortages during the course of the job. If drawings are not complete or approved, or a budget cost is needed, it is wise to add 5% to 10%. Contingencies, then, are a matter of judgment. Additional allowances are shown in Division 01 21 16.50 for contingencies and job conditions and Reference Table R012153-10 for factors to convert prices for repair and remodeling jobs.

Final Checklist

Estimating can be a straightforward process provided you remember the basics. Here's a checklist of some of the steps you should remember to complete before finalizing your estimate.

Did you remember to . . .

- factor in the City Cost Index for your locale?
- take into consideration which items have been marked up and by how much?
- mark up the entire estimate sufficiently for your purposes?
- read the background information on techniques and technical matters that could impact your project time span and cost?
- include all components of your project in the final estimate?
- double check your figures for accuracy?
- call RSMeans if you have any questions about your estimate or the data you've found in our publications?

Remember, RSMeans stands behind its publications. If you have any questions about your estimate . . .about the costs you've used from our books . . .or even about the technical aspects of the job that may affect your estimate, feel free to call the RSMeans editors at 1-800-334-3509.

Unit Price Section

Table of Contents

How to Use the Unit Price Pages

The following is a detailed explanation of a sample entry in the Unit Price Section. Next to each bold number below is the described item with the appropriate component of the sample entry following in parentheses. Some prices are listed as bare costs; others as costs that include overhead and profit of the installing contractor. In most cases, if the work is to be subcontracted, the general contractor will need to add an additional markup (RSMeans suggests using 10%) to the figures in the column "Total Incl. O&P."

1 Division Number/Title (03 30/Cast-In-Place Concrete)

Use the Unit Price Section Table of Contents to locate specific items. The sections are classified according to the CSI MasterFormat 2004 system.

2 Line Numbers (03 30 53.40 3920)

Each unit price line item has been assigned a unique 12-digit code based on the CSI MasterFormat classification.

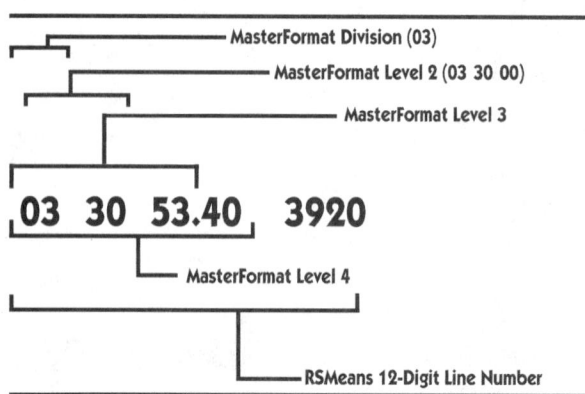

MasterFormat Division (03)
MasterFormat Level 2 (03 30 00)
MasterFormat Level 3

03 30 53.40 3920

MasterFormat Level 4
RSMeans 12-Digit Line Number

3 Description (Concrete In Place, etc.)

Each line item is described in detail. Sub-items and additional sizes are indented beneath the appropriate line items. The first line or two after the main item (in boldface) may contain descriptive information that pertains to all line items beneath this boldface listing.

4 Reference Number Information

R033053 -50 You'll see reference numbers shown in bold boxes at the beginning of some sections. These refer to related items in the Reference Section, visually identified by a vertical gray bar on the page edges.

The relation may be an estimating procedure that should be read before estimating, or technical information.

The "R" designates the Reference Section. The numbers refer to the MasterFormat 2004 classification system.

It is strongly recommended that you review all reference numbers that appear within the section in which you are working.

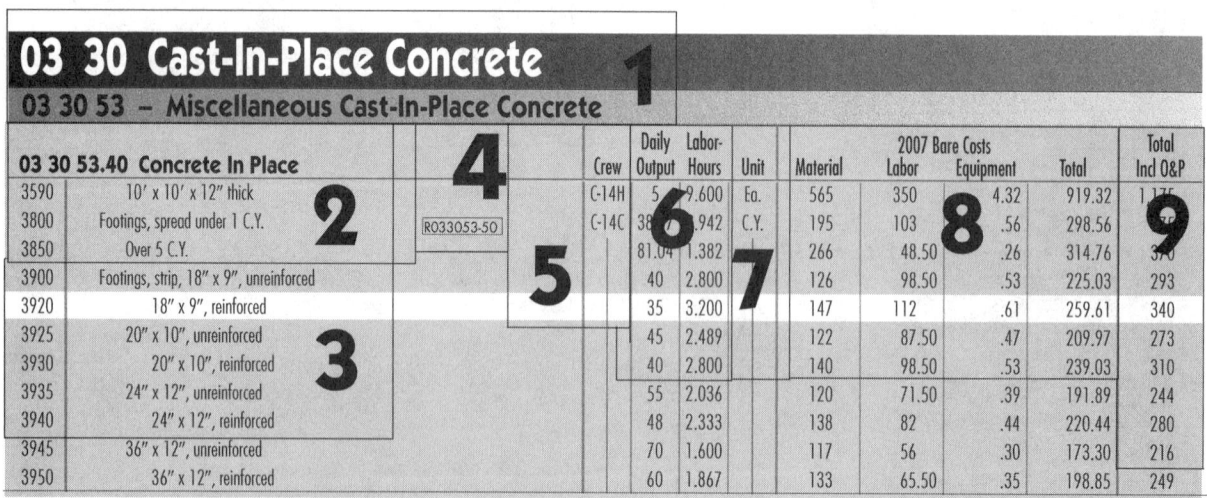

03 30 Cast-In-Place Concrete   1

03 30 53 – Miscellaneous Cast-In-Place Concrete

03 30 53.40 Concrete In Place		Crew	Daily Output	Labor-Hours	Unit	Material	2007 Bare Costs Labor	Equipment	Total	Total Incl O&P
3590	10' x 10' x 12" thick	C-14H	5	9.600	Ea.	565	350	4.32	919.32	1,175
3800	Footings, spread under 1 C.Y.	C-14C	38	.942	C.Y.	195	103	.56	298.56	370
3850	Over 5 C.Y.			81.04	1.382		266	48.50	.26	314.76	370
3900	Footings, strip, 18" x 9", unreinforced			40	2.800		126	98.50	.53	225.03	293
3920	18" x 9", reinforced			35	3.200		147	112	.61	259.61	340
3925	20" x 10", unreinforced			45	2.489		122	87.50	.47	209.97	273
3930	20" x 10", reinforced			40	2.800		140	98.50	.53	239.03	310
3935	24" x 12", unreinforced			55	2.036		120	71.50	.39	191.89	244
3940	24" x 12", reinforced			48	2.333		138	82	.44	220.44	280
3945	36" x 12", unreinforced			70	1.600		117	56	.30	173.30	216
3950	36" x 12", reinforced			60	1.867		133	65.50	.35	198.85	249

Crew (C-14C)

The "Crew" column designates the typical trade or crew used to install the item. If an installation can be accomplished by one trade and requires no power equipment, that trade and the number of workers are listed (for example, "2 Carpenters"). If an installation requires a composite crew, a crew code designation is listed (for example, "C-14C"). You'll find full details on all composite crews in the Crew Listings.

- For a complete list of all trades utilized in this book and their abbreviations, see the inside back cover.

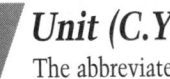

Crews

Crew No.	Bare Costs		Incl. Subs O & P		Cost Per Labor-Hour	
Crew C-14C	Hr.	Daily	Hr.	Daily	Bare Costs	Incl. O&P
1 Carpenter Foreman (out)	$38.70	$309.60	$60.25	$482.00	$35.15	$54.99
6 Carpenters	36.70	1761.60	57.15	2743.20		
2 Rodmen (reinf.)	41.30	660.80	67.75	1084.00		
4 Laborers	28.75	920.00	44.75	1432.00		
1 Cement Finisher	35.55	284.40	52.20	417.60		
1 Gas Engine Vibrator		21.40		23.54	.19	.21
112 L.H., Daily Totals		$3957.80		$6182.34	$35.34	$55.20

Productivity: Daily Output (35.0)/ Labor-Hours (3.20)

The "Daily Output" represents the typical number of units the designated crew will install in a normal 8-hour day. To find out the number of days the given crew would require to complete the installation, divide your quantity by the daily output. For example:

Quantity	÷	Daily Output	=	Duration
100 C.Y.	÷	35.0/ Crew Day	=	2.86 Crew Days

The "Labor-Hours" figure represents the number of labor-hours required to install one unit of work. To find out the number of labor-hours required for your particular task, multiply the quantity of the item times the number of labor-hours shown. For example:

Quantity	x	Productivity Rate	=	Duration
100 C.Y.	x	3.20 Labor-Hours/ C.Y.	=	320 Labor-Hours

Unit (C.Y.)

The abbreviated designation indicates the unit of measure upon which the price, production, and crew are based (C.Y. = Cubic Yard). For a complete listing of abbreviations, refer to the Abbreviations Listing in the Reference Section of this book.

Bare Costs:
Mat. (Bare Material Cost) (147)

The unit material cost is the "bare" material cost with no overhead or profit included. Costs shown reflect national average material prices for January of the current year and include delivery to the job site. No sales taxes are included.

Labor (112)

The unit labor cost is derived by multiplying bare labor-hour costs for Crew C-14C by labor-hour units. The bare labor-hour cost is found in the Crew Section under C-14C. (If a trade is listed, the hourly labor cost—the wage rate—is found on the inside back cover.)

Labor-Hour Cost Crew C-14C	x	Labor-Hour Units	=	Labor
$35.15	x	3.20	=	$112.00

Equip. (Equipment) (.61)

Equipment costs for each crew are listed in the description of each crew. Tools or equipment whose value justifies purchase or ownership by a contractor are considered overhead as shown on the inside back cover. The unit equipment cost is derived by multiplying the bare equipment hourly cost by the labor-hour units.

Equipment Cost Crew C-14C	x	Labor-Hour Units	=	Equip.
$.19	x	3.20	=	$.61

Total (259.61)

The total of the bare costs is the arithmetic total of the three previous columns: mat., labor, and equip.

Material	+	Labor	+	Equip.	=	Total
$147	+	$112	+	$.61	=	$259.61

Total Costs Including O &P

This figure is the sum of the bare material cost plus 10% for profit; the bare labor cost plus total overhead and profit (per the inside back cover or, if a crew is listed, from the crew listings); and the bare equipment cost plus 10% for profit.

Material is Bare Material Cost + 10% = 147 + 14.70	=	$161.70
Labor for Crew C-14C = Labor-Hour Cost (54.99) x Labor-Hour Units (3.20)	=	$175.97
Equip. is Bare Equip. Cost + 10% = .61 + .06	=	$ .67
Total (Rounded)	=	$340

Estimating Tips

01 20 00 Price & Payment Procedures

When estimating historic preservation projects (depending on the condition of the existing structure and the owner's requirements), a 15%-20% contingency or allowance is recommended, regardless of the stage of the drawings.

01 30 00 Administrative Requirements

- Before determining a final cost estimate, it is a good practice to review all the items listed in Subdivision 01 03 00 to make final adjustments for items that may need customizing to specific job conditions.

- Requirements for initial and periodic submittals can represent a significant cost to the General Requirements of a job. Thoroughly check the submittal specifications when estimating a project to determine any costs that should be included.

01 40 00 Quality Requirements

- All projects will require some degree of Quality Control. This cost is not included in the unit cost of construction listed in each division. Depending upon the terms of the contract, the various costs of inspection and testing can be the responsibility of either the owner or the contractor. Be sure to include the required costs in your estimate.

01 50 00 Temporary Facilities and Controls

- Barricades, access roads, safety nets, scaffolding, security, and many more requirements for the execution of a safe project are elements of direct cost. These costs can easily be overlooked when preparing an estimate. When looking through the major classifications of this subdivision, determine which items apply to each division in your estimate.

01 70 00 Execution and Closeout Requirements

- When preparing an estimate, thoroughly read the specifications to determine the requirements for Contract Closeout. Final cleaning, record documentation, operation and maintenance data, warranties and bonds, and spare parts and maintenance materials can all be elements of cost for the completion of a contract. Do not overlook these in your estimate.

Reference Numbers

Reference numbers are shown in shaded boxes at the beginning of some major classifications. These numbers refer to related items in the Reference Section. The reference information may be an estimating procedure, an alternate pricing method, or technical information.

Note: Not all subdivisions listed here necessarily appear in this publication.

Division 1 - General Requirements

01 11 Summary of Work

01 11 31 – Professional Consultants

01 11 31.20 Construction Management Fees		Crew	Daily Output	Labor-Hours	Unit	Material	2007 Bare Costs Labor	Equipment	Total	Total Incl O&P
0010	**CONSTRUCTION MANAGEMENT FEES**									
0020	$1,000,000 job, minimum				Project					4.50%
0050	Maximum									7.50%
0060	For work to $10,000									10%
0300	$5,000,000 job, minimum									2.50%
0350	Maximum				↓					4%

01 11 31.30 Engineering Fees

01 11 31.30 Engineering Fees		Crew	Daily Output	Labor-Hours	Unit	Material	2007 Bare Costs Labor	Equipment	Total	Total Incl O&P
0010	**ENGINEERING FEES** R011110-30									
0020	Educational planning consultant, minimum				Project					.50%
0100	Maximum				"					2.50%
0200	Electrical, minimum				Contrct					4.10%
0300	Maximum									10.10%
0400	Elevator & conveying systems, minimum									2.50%
0500	Maximum									5%
0600	Food service & kitchen equipment, minimum									8%
0700	Maximum									12%
1000	Mechanical (plumbing & HVAC), minimum									4.10%
1100	Maximum				↓					10.10%

01 21 Allowances

01 21 16 – Contingency Allowances

01 21 16.50 Contingencies		Crew	Daily Output	Labor-Hours	Unit	Material	2007 Bare Costs Labor	Equipment	Total	Total Incl O&P
0010	**CONTINGENCIES**									
0020	For estimate at conceptual stage				Project					20%
0050	Schematic stage									15%
0100	Preliminary working drawing stage (Design Dev.)									10%
0150	Final working drawing stage				↓					3%

01 21 53 – Factors Allowance

01 21 53.50 Factors		Crew	Daily Output	Labor-Hours	Unit	Material	2007 Bare Costs Labor	Equipment	Total	Total Incl O&P
0010	**FACTORS** Cost adjustments									
0100	Add to construction costs for particular job requirements									
0500	Cut & patch to match existing construction, add, minimum				Costs	2%	3%			
0550	Maximum					5%	9%			
0800	Dust protection, add, minimum					1%	2%			
0850	Maximum					4%	11%			
1100	Equipment usage curtailment, add, minimum					1%	1%			
1150	Maximum					3%	10%			
1400	Material handling & storage limitation, add, minimum					1%	1%			
1450	Maximum					6%	7%			
1700	Protection of existing work, add, minimum					2%	2%			
1750	Maximum					5%	7%			
2000	Shift work requirements, add, minimum						5%			
2050	Maximum						30%			
2300	Temporary shoring and bracing, add, minimum					2%	5%			
2350	Maximum					5%	12%			
2400	Work inside prisons and high security areas, add, minimum						30%			
2450	Maximum				↓		50%			

01 21 55 – Job Conditions Allowance

01 21 55.50 Job Conditions		Crew	Daily Output	Labor-Hours	Unit	Material	2007 Bare Costs Labor	Equipment	Total	Total Incl O&P
0010	**JOB CONDITIONS** Modifications to total									

01 21 Allowances

01 21 55 – Job Conditions Allowance

01 21 55.50 Job Conditions

		Crew	Daily Output	Labor-Hours	Unit	Material	2007 Bare Costs Labor	Equipment	Total	Total Incl O&P
0020	project cost summaries									
0100	Economic conditions, favorable, deduct				Project				2%	2%
0200	Unfavorable, add								5%	5%
0300	Hoisting conditions, favorable, deduct								2%	2%
0400	Unfavorable, add								5%	5%
0700	Labor availability, surplus, deduct								1%	1%
0800	Shortage, add								10%	10%
0900	Material storage area, available, deduct								1%	1%
1000	Not available, add								2%	2%
1100	Subcontractor availability, surplus, deduct								5%	5%
1200	Shortage, add								12%	12%
1300	Work space, available, deduct								2%	2%
1400	Not available, add								5%	5%

01 21 57 – Overtime Allowance

01 21 57.50 Overtime

0010	**OVERTIME** for early completion of projects or where	R012909-90								
0020	labor shortages exist, add to usual labor, up to				Costs		100%			

01 21 61 – Cost Indexes

01 21 61.10 Construction Cost Index

0010	**CONSTRUCTION COST INDEX** (Reference) over 930 zip code locations in									
0020	The U.S. and Canada, total bldg cost, min. (Clarksdale, MS)				%					67.20%
0050	Average									100%
0100	Maximum (New York, NY)									130.90%

01 21 61.20 Historical Cost Indexes

0010	**HISTORICAL COST INDEXES** (See Reference section)									

01 21 61.30 Labor Index

0010	**LABOR INDEX** (Reference) For over 930 zip code locations in									
0020	the U.S. and Canada, minimum (Clarksdale, MS)				%		29.90%			
0050	Average						100%			
0100	Maximum (New York, NY)						164.50%			

01 21 61.50 Material Index

0011	**MATERIAL INDEX** For over 930 zip code locations in									
0020	the U.S. and Canada, minimum (Elizabethtown, KY)				%	90.70%				
0040	Average					100%				
0060	Maximum (Ketchikan, AK)					141.60%				

01 21 63 – Taxes

01 21 63.10 Taxes

0010	**TAXES**	R012909-80								
0020	Sales tax, State, average				%	4.84%				
0050	Maximum	R012909-85				7.25%				
0300	Unemployment, combined Federal and State, minimum						.80%			
0350	Average						6.20%			
0400	Maximum						11.76%			

01 31 Project Management and Coordination

01 31 13 – Project Coordination

01 31 13.30 Insurance

	01 31 13.30 Insurance		Crew	Daily Output	Labor-Hours	Unit	Material	2007 Bare Costs Labor	Equipment	Total	Total Incl O&P
0010	**INSURANCE**	R013113-40									
0020	Builders risk, standard, minimum					Job					.24%
0050	Maximum	R013113-60									.64%
0200	All-risk type, minimum										.25%
0250	Maximum					↓					.62%
0400	Contractor's equipment floater, minimum					Value					.50%
0450	Maximum					"					1.50%
0600	Public liability, average					Job					2.02%
0800	Workers' compensation & employer's liability, average										
0850	by trade, carpentry, general					Payroll		18.39%			
1000	Electrical							6.59%			
1150	Insulation							15.81%			
1450	Plumbing							8.10%			
1550	Sheet metal work (HVAC)					↓		11.89%			

01 31 13.50 Mark-Up

	01 31 13.50 Mark-Up		Crew	Daily Output	Labor-Hours	Unit	Material	Labor	Equipment	Total	Total Incl O&P
0010	**MARK-UP** For General Contractors for change										
0100	of scope of job as bid										
0200	Extra work, by subcontractors, add					%					10%
0250	By General Contractor, add										15%
0400	Omitted work, by subcontractors, deduct all but										5%
0450	By General Contractor, deduct all but										7.50%
0600	Overtime work, by subcontractors, add										15%
0650	By General Contractor, add										10%
1000	Installing contractors, on his own labor, minimum							46.90%			
1100	Maximum					↓		87.30%			

01 31 13.60 Overhead

	01 31 13.60 Overhead		Crew	Daily Output	Labor-Hours	Unit	Material	Labor	Equipment	Total	Total Incl O&P
0010	**OVERHEAD**										
0020	As percent of direct costs, minimum					%				5%	
0050	Average									13%	
0100	Maximum					↓				30%	

01 31 13.80 Overhead and Profit

	01 31 13.80 Overhead and Profit		Crew	Daily Output	Labor-Hours	Unit	Material	Labor	Equipment	Total	Total Incl O&P
0010	**OVERHEAD & PROFIT** Allowance to add to items in this										
0020	book that do not include Subs O&P, average					%				25%	
0100	Allowance to add to items in this book that										
0110	do include Subs O&P, minimum					%					5%
0150	Average										10%
0200	Maximum										15%
0300	Typical, by size of project, under $100,000									30%	
0350	$500,000 project									25%	
0400	$2,000,000 project									20%	
0450	Over $10,000,000 project					↓				15%	

01 31 13.90 Performance Bond

	01 31 13.90 Performance Bond		Crew	Daily Output	Labor-Hours	Unit	Material	Labor	Equipment	Total	Total Incl O&P
0010	**PERFORMANCE BOND**	R013113-80									
0020	For buildings, minimum					Job					.60%
0100	Maximum					"					2.50%

01 32 Construction Progress Documentation

01 32 33 – Photographic Documentation

01 32 33.50 Photographs		Crew	Daily Output	Labor-Hours	Unit	Material	2007 Bare Costs Labor	Equipment	Total	Total Incl O&P
0010	**PHOTOGRAPHS**									
0020	8" x 10", 4 shots, 2 prints ea., std. mounting				Set	450			450	495
0100	Hinged linen mounts					530			530	580
0200	8" x 10", 4 shots, 2 prints each, in color					455			455	505
0300	For I.D. slugs, add to all above					5.30			5.30	5.85
1500	Time lapse equipment, camera and projector, buy					3,775			3,775	4,175
1550	Rent per month					565			565	620
1700	Cameraman and film, including processing, B.&W.				Day	1,375			1,375	1,525
1720	Color				"	1,375			1,375	1,525

01 41 Regulatory Requirements

01 41 26 – Permits

01 41 26.50 Permits		Crew	Daily Output	Labor-Hours	Unit	Material	2007 Bare Costs Labor	Equipment	Total	Total Incl O&P
0010	**PERMITS**									
0020	Rule of thumb, most cities, minimum				Job					.50%
0100	Maximum				"					2%

01 51 Temporary Utilities

01 51 13 – Temporary Electricity

01 51 13.50 Temporary Power Equip (Pro-Rated Per Job)		Crew	Daily Output	Labor-Hours	Unit	Material	2007 Bare Costs Labor	Equipment	Total	Total Incl O&P
0010	**TEMPORARY POWER EQUIP (PRO-RATED PER JOB)** R015113-65									
0020	Service, overhead feed, 3 use									
0030	100 Amp	1 Elec	1.25	6.400	Ea.	600	281		881	1,075
0040	200 Amp		1	8		790	350		1,140	1,400
0050	400 Amp		.75	10.667		1,500	470		1,970	2,325
0060	600 Amp		.50	16		2,175	700		2,875	3,450
0100	Underground feed, 3 use									
0110	100 Amp	1 Elec	2	4	Ea.	565	176		741	885
0120	200 Amp		1.15	6.957		775	305		1,080	1,300
0130	400 Amp		1	8		1,475	350		1,825	2,150
0140	600 Amp		.75	10.667		1,850	470		2,320	2,725
0150	800 Amp		.50	16		2,825	700		3,525	4,150
0160	1000 Amp		.35	22.857		3,150	1,000		4,150	4,950
0170	1200 Amp		.25	32		3,525	1,400		4,925	5,975
0180	2000 Amp		.20	40		4,300	1,750		6,050	7,350
0200	Transformers, 3 use									
0210	30 KVA	1 Elec	1	8	Ea.	935	350		1,285	1,550
0220	45 KVA		.75	10.667		1,075	470		1,545	1,875
0230	75 KVA		.50	16		1,625	700		2,325	2,850
0240	112.5 KVA		.40	20		2,225	880		3,105	3,725
0250	Feeder, PVC, CU wire									
0260	60 Amp w/trench	1 Elec	96	.083	L.F.	4.59	3.66		8.25	10.50
0270	100 Amp w/trench		85	.094		9.05	4.13		13.18	16.15
0280	200 Amp w/trench		59	.136		19.15	5.95		25.10	30
0290	400 Amp w/trench		42	.190		50	8.35		58.35	68
0300	Feeder, PVC, aluminum wire									
0310	60 Amp w/trench	1 Elec	96	.083	L.F.	3.79	3.66		7.45	9.60
0320	100 Amp w/trench		85	.094		4.17	4.13		8.30	10.75
0330	200 Amp w/trench		59	.136		8.90	5.95		14.85	18.65

01 51 Temporary Utilities

01 51 13 – Temporary Electricity

01 51 13.50 Temporary Power Equip (Pro-Rated Per Job)	Crew	Daily Output	Labor-Hours	Unit	Material	2007 Bare Costs Labor	2007 Bare Costs Equipment	Total	Total Incl O&P
0340	400 Amp w/trench	1 Elec	42	.190	L.F.	18.85	8.35		27.20	33.50
0350	Feeder, EMT, CU wire									
0360	60 Amp	1 Elec	90	.089	L.F.	4.04	3.90		7.94	10.25
0370	100 Amp		80	.100		9.20	4.39		13.59	16.65
0380	200 Amp		60	.133		18.35	5.85		24.20	28.50
0390	400 Amp	↓	35	.229	↓	52.50	10.05		62.55	72.50
0400	Feeder, EMT, Al wire									
0410	60 Amp	1 Elec	90	.089	L.F.	4.22	3.90		8.12	10.45
0420	100 Amp		80	.100		5.30	4.39		9.69	12.35
0430	200 Amp		60	.133		13.15	5.85		19	23
0440	400 Amp	↓	35	.229	↓	24.50	10.05		34.55	42
0500	Equipment, 3 use									
0510	Spider box 50 Amp	1 Elec	8	1	Ea.	246	44		290	335
0520	Lighting cord 100'		8	1		39	44		83	109
0530	Light stanchion	↓	8	1	↓	78	44		122	152
0540	Temporary cords, 100', 3 use									
0550	Feeder cord, 50 Amp	1 Elec	16	.500	Ea.	160	22		182	209
0560	Feeder cord, 100 Amp		12	.667		460	29.50		489.50	550
0570	Tap cord, 50 Amp		12	.667		440	29.50		469.50	530
0580	Tap cord, 100 Amp	↓	6	1.333	↓	630	58.50		688.50	780
0590	Temporary cords, 50', 3 use									
0600	Feeder cord, 50 Amp	1 Elec	16	.500	Ea.	91	22		113	133
0610	Feeder cord, 100 Amp		12	.667		225	29.50		254.50	292
0620	Tap cord, 50 Amp		12	.667		196	29.50		225.50	260
0630	Tap cord, 100 Amp	↓	6	1.333		325	58.50		383.50	445
0700	Connections									
0710	Compressor or pump									
0720	30 Amp	1 Elec	7	1.143	Ea.	19.80	50		69.80	96.50
0730	60 Amp		5.30	1.509		45	66.50		111.50	148
0740	100 Amp	↓	4	2	↓	58.50	88		146.50	196
0750	Tower crane									
0760	60 Amp	1 Elec	4.50	1.778	Ea.	45	78		123	166
0770	100 Amp	"	3	2.667	"	58.50	117		175.50	239
0780	Manlift									
0790	Single	1 Elec	3	2.667	Ea.	46.50	117		163.50	225
0800	Double	"	2	4	"	60	176		236	325
0810	Welder									
0820	50 Amp w/disconnect	1 Elec	5	1.600	Ea.	292	70		362	425
0830	100 Amp w/disconnect		3.80	2.105		445	92.50		537.50	630
0840	200 Amp w/disconnect		2.50	3.200		710	140		850	990
0850	400 Amp w/disconnect	↓	1	8	↓	2,275	350		2,625	3,025
0860	Office trailer									
0870	60 Amp	1 Elec	4.50	1.778	Ea.	91.50	78		169.50	217
0880	100 Amp		3	2.667		123	117		240	310
0890	200 Amp	↓	2	4	↓	540	176		716	855
0900	Lamping, add per floor				Total				625	690
0910	Maintenance, total temp. power cost				Job				5%	

01 51 13.80 Temporary Utilities

		Crew	Daily Output	Labor-Hours	Unit	Material	Labor	Equipment	Total	Total Incl O&P
0010	**TEMPORARY UTILITIES**									
0100	Heat, incl. fuel and operation, per week, 12 hrs. per day	1 Skwk	100	.080	CSF Flr	10.35	3.04		13.39	16.10
0200	24 hrs. per day	"	60	.133		19.95	5.05		25	30
0350	Lighting, incl. service lamps, wiring & outlets, minimum	1 Elec	34	.235	↓	2.63	10.35		12.98	18.30

01 51 Temporary Utilities

01 51 13 – Temporary Electricity

01 51 13.80 Temporary Utilities

		Crew	Daily Output	Labor-Hours	Unit	Material	2007 Bare Costs Labor	Equipment	Total	Total Incl O&P
0360	Maximum	1 Elec	17	.471	CSF Flr	5.70	20.50		26.20	37.50
0400	Power for temp lighting only, per month, min/month 6.6 KWH								.75	.83
0450	Maximum/month 23.6 KWH								2.85	3.14
0600	Power for job duration incl. elevator, etc., minimum								47	51.70
0650	Maximum								110	121
1000	Toilet, portable, see equipment 01 54 33.40 in reference section									

01 52 Construction Facilities

01 52 13 – Field Offices and Sheds

01 52 13.20 Office and Storage Space

		Crew	Daily Output	Labor-Hours	Unit	Material	2007 Bare Costs Labor	Equipment	Total	Total Incl O&P
0010	**OFFICE AND STORAGE SPACE**									
0020	Trailer, furnished, no hookups, 20' x 8', buy	2 Skwk	1	16	Ea.	7,975	610		8,585	9,700
0250	Rent per month					201			201	221
0300	32' x 8', buy	2 Skwk	.70	22.857		11,900	870		12,770	14,400
0350	Rent per month					241			241	265
0400	50' x 10', buy	2 Skwk	.60	26.667		20,500	1,025		21,525	24,100
0450	Rent per month					330			330	365
0500	50' x 12', buy	2 Skwk	.50	32		25,600	1,225		26,825	30,100
0550	Rent per month					375			375	410
0700	For air conditioning, rent per month, add					41			41	45
0800	For delivery, add per mile				Mile	4.50			4.50	4.95
1000	Portable buildings, prefab, on skids, economy, 8' x 8'	2 Carp	265	.060	S.F.	85	2.22		87.22	97
1100	Deluxe, 8' x 12'	"	150	.107	"	95	3.91		98.91	111
1200	Storage boxes, 20' x 8', buy	2 Skwk	1.80	8.889	Ea.	4,175	340		4,515	5,100
1250	Rent per month					76			76	83.50
1300	40' x 8', buy	2 Skwk	1.40	11.429		6,175	435		6,610	7,450
1350	Rent per month					101			101	111

01 52 13.40 Field Office Expense

		Crew	Daily Output	Labor-Hours	Unit	Material	2007 Bare Costs Labor	Equipment	Total	Total Incl O&P
0010	**FIELD OFFICE EXPENSE**									
0100	Field office expense, office equipment rental average				Month	150			150	165
0120	Office supplies, average				"	95			95	105
0125	Office trailer rental, see division 01 52 13.20									
0140	Telephone bill; avg. bill/month incl. long dist.				Month	210			210	231
0160	Field office lights & HVAC				"	110			110	121

01 54 Construction Aids

01 54 23 – Temporary Scaffolding and Platforms

01 54 23.70 Scaffolding

			Crew	Daily Output	Labor-Hours	Unit	Material	2007 Bare Costs Labor	Equipment	Total	Total Incl O&P
0010	**SCAFFOLDING**	R015423-10									
0015	Steel tube, regular, no plank, labor only to erect & dismantle										
0090	Building exterior, wall face, 1 to 5 stories, 6'-4" x 5' frames		3 Carp	8	3	C.S.F.		110		110	171
0200	6 to 12 stories		4 Carp	8	4			147		147	229
0310	13 to 20 stories		5 Carp	8	5			184		184	286
0460	Building interior, wall face area, up to 16' high		3 Carp	12	2			73.50		73.50	114
0560	16' to 40' high			10	2.400			88		88	137
0800	Building interior floor area, up to 30' high			150	.160	C.C.F.		5.85		5.85	9.15
0900	Over 30' high		4 Carp	160	.200	"		7.35		7.35	11.45
0906	Complete system for face of walls, no plank, material only rent/mo					C.S.F.	34.50			34.50	37.50
0908	Interior spaces, no plank, material only rent/mo					"	3.30			3.30	3.63

01 54 Construction Aids

01 54 23 – Temporary Scaffolding and Platforms

01 54 23.70 Scaffolding		Crew	Daily Output	Labor-Hours	Unit	Material	2007 Bare Costs Labor	Equipment	Total	Total Incl O&P
0910	Steel tubular, heavy duty shoring, buy									
0920	Frames 5' high 2' wide				Ea.	82.50			82.50	91
0925	5' high 4' wide					93.50			93.50	103
0930	6' high 2' wide					94.50			94.50	104
0935	6' high 4' wide					110			110	121
0940	Accessories									
0945	Cross braces				Ea.	16			16	17.60
0950	U-head, 8" x 8"					19.30			19.30	21
0955	J-head, 4" x 8"					14.10			14.10	15.50
0960	Base plate, 8" x 8"					15.70			15.70	17.25
0965	Leveling jack					33.50			33.50	37
1000	Steel tubular, regular, buy									
1100	Frames 3' high 5' wide				Ea.	64			64	70
1150	5' high 5' wide					73.50			73.50	81
1200	6'-4" high 5' wide					92.50			92.50	102
1350	7'-6" high 6' wide					160			160	175
1500	Accessories cross braces					18.15			18.15	19.95
1550	Guardrail post					16.50			16.50	18.15
1600	Guardrail 7' section					8.80			8.80	9.70
1650	Screw jacks & plates					24			24	26.50
1700	Sidearm brackets					31			31	34
1750	8" casters					33			33	36.50
1800	Plank 2" x 10" x 16'-0"					44			44	48.50
1900	Stairway section					270			270	296
1910	Stairway starter bar					32			32	35
1920	Stairway inside handrail					58.50			58.50	64
1930	Stairway outside handrail					80.50			80.50	88.50
1940	Walk-thru frame guardrail					40.50			40.50	44.50
2000	Steel tubular, regular, rent/mo.									
2100	Frames 3' high 5' wide				Ea.	5			5	5.50
2150	5' high 5' wide					5			5	5.50
2200	6'-4" high 5' wide					4.50			4.50	4.95
2250	7'-6" high 6' wide					7			7	7.70
2500	Accessories, cross braces					1			1	1.10
2550	Guardrail post					1			1	1.10
2600	Guardrail 7' section					1			1	1.10
2650	Screw jacks & plates					2			2	2.20
2700	Sidearm brackets					2			2	2.20
2750	8" casters					8			8	8.80
2800	Outrigger for rolling tower					3			3	3.30
2850	Plank 2" x 10" x 16'-0"					6			6	6.60
2900	Stairway section					40			40	44
2940	Walk-thru frame guardrail					2.50			2.50	2.75
3000	Steel tubular, heavy duty shoring, rent/mo.									
3250	5' high 2' & 4' wide				Ea.	5			5	5.50
3300	6' high 2' & 4' wide					5			5	5.50
3500	Accessories, cross braces					1			1	1.10
3600	U - head, 8" x 8"					1			1	1.10
3650	J - head, 4" x 8"					1			1	1.10
3700	Base plate, 8" x 8"					1			1	1.10
3750	Leveling jack					2			2	2.20
5700	Planks, 2x10x16'-0", labor only to erect & remove to 50' H	3 Carp	72	.333			12.25		12.25	19.05
5800	Over 50' high	4 Carp	80	.400			14.70		14.70	23

01 54 Construction Aids

01 54 23 – Temporary Scaffolding and Platforms

01 54 23.75 Scaffolding Specialties		Crew	Daily Output	Labor-Hours	Unit	Material	2007 Bare Costs Labor	Equipment	Total	Total Incl O&P
0010	**SCAFFOLDING SPECIALTIES**									
1500	Sidewalk bridge using tubular steel									
1510	scaffold frames, including planking	3 Carp	45	.533	L.F.	5.60	19.55		25.15	36.50
1600	For 2 uses per month, deduct from all above					50%				
1700	For 1 use every 2 months, add to all above					100%				
1900	Catwalks, 20" wide, no guardrails, 7' span, buy				Ea.	121			121	133
2000	10' span, buy					170			170	187
3720	Putlog, standard, 8' span, with hangers, buy					67			67	74
3750	12' span, buy					101			101	111
3760	Trussed type, 16' span, buy					232			232	255
3790	22' span, buy					277			277	305
3795	Rent per month					30			30	33
3800	Rolling ladders with handrails, 30" wide, buy, 2 step					196			196	215
4000	7 step					590			590	645
4050	10 step					820			820	900
4100	Rolling towers, buy, 5' wide, 7' long, 10' high					1,175			1,175	1,275
4200	For 5' high added sections, to buy, add					184			184	202
4300	Complete incl. wheels, railings, outriggers,									
4350	21' high, to buy				Ea.	1,975			1,975	2,150
4400	Rent/month = 5% of purchase cost				"	600			600	660

01 54 26 – Temporary Swing Staging

01 54 26.50 Swing Staging		Crew	Daily Output	Labor-Hours	Unit	Material	2007 Bare Costs Labor	Equipment	Total	Total Incl O&P
0010	**SWING STAGING**, 500 lb cap., 2' wide to 24' long, hand operat									
0020	steel cable type, with 60' cables, buy				Ea.	5,100			5,100	5,625
0030	Rent per month				"	510			510	560
2200	Move swing staging (setup and remove)	E-4	2	16	Move		670	57.50	727.50	1,300

01 54 36 – Equipment Mobilization

01 54 36.50 Mobilization or Demob.		Crew	Daily Output	Labor-Hours	Unit	Material	2007 Bare Costs Labor	Equipment	Total	Total Incl O&P
0010	**MOBILIZATION OR DEMOB.** (One or the other, unless noted) R015433-10									
0015	Up to 25 mi haul dist (50 mi RT for mob/demob crew)									
0020	Dozer, loader, backhoe, excav., grader, paver, roller, 70 to 150 H.P.	B-34N	4	2	Ea.		59	124	183	228
0100	Above 150 HP	B-34K	3	2.667			79	192	271	335
0900	Shovel or dragline, 3/4 C.Y.	"	3.60	2.222			65.50	160	225.50	277
1100	Small equipment, placed in rear of, or towed by pickup truck	A-3A	8	1			28.50	11.40	39.90	56.50
1150	Equip up to 70 HP, on flatbed trailer behind pickup truck	A-3D	4	2			57	46	103	139
2000	Mob & demob truck-mounted crane up to 75 ton, driver only	1 Eqhv	3.60	2.222			88.50		88.50	133
2100	Crane, truck-mounted, over 75 ton	A-3E	2.50	6.400			222	36.50	258.50	380
2200	Crawler-mounted, up to 75 ton	A-3F	2	8			277	305	582	760
2300	Over 75 ton	A-3G	1.50	10.667			370	445	815	1,050
2500	For each additional 5 miles haul distance, add						10%	10%		
3000	For large pieces of equipment, allow for assembly/knockdown									
3100	For mob/demob of micro-tunneling equip, see section 33 05 23.19									

01 54 39 – Construction Equipment

01 54 39.70 Small Tools		Crew	Daily Output	Labor-Hours	Unit	Material	2007 Bare Costs Labor	Equipment	Total	Total Incl O&P
0010	**SMALL TOOLS**									
0020	As % of contractor's work, minimum				Total				.45%	.50%
0100	Maximum				"				1.82%	2%

01 55 Vehicular Access and Parking

01 55 23 – Temporary Roads

01 55 23.50 Roads and Sidewalks	Crew	Daily Output	Labor-Hours	Unit	Material	2007 Bare Costs Labor	Equipment	Total	Total Incl O&P
0010 **ROADS AND SIDEWALKS** Temporary									
0050 Roads, gravel fill, no surfacing, 4" gravel depth	B-14	715	.067	S.Y.	3.87	2.04	.34	6.25	7.80
0100 8" gravel depth	"	615	.078	"	7.75	2.38	.40	10.53	12.60
1000 Ramp, 3/4" plywood on 2" x 6" joists, 16" O.C.	2 Carp	300	.053	S.F.	1.28	1.96		3.24	4.46
1100 On 2" x 10" joists, 16" O.C.	"	275	.058	"	1.89	2.14		4.03	5.40

01 56 Temporary Barriers and Enclosures

01 56 13 – Temporary Air Barriers

01 56 13.90 Winter Protection	Crew	Daily Output	Labor-Hours	Unit	Material	Labor	Equipment	Total	Total Incl O&P
0010 **WINTER PROTECTION** Reinforced plastic on wood									
0100 Framing to close openings	2 Clab	750	.021	S.F.	.39	.61		1	1.38
0200 Tarpaulins hung over scaffolding, 8 uses, not incl. scaffolding		1500	.011		.25	.31		.56	.76
0300 Prefab fiberglass panels, steel frame, 8 uses		1200	.013		.85	.38		1.23	1.54

01 56 23 – Temporary Barricades

01 56 23.10 Barricades	Crew	Daily Output	Labor-Hours	Unit	Material	Labor	Equipment	Total	Total Incl O&P
0010 **BARRICADES**									
0020 5' high, 3 rail @ 2" x 8", fixed	2 Carp	20	.800	L.F.	5.65	29.50		35.15	51.50
0150 Movable		30	.533		4.72	19.55		24.27	35.50
1000 Guardrail, wooden, 3' high, 1" x 6", on 2" x 4" posts		200	.080		1.14	2.94		4.08	5.85
1100 2" x 6", on 4" x 4" posts		165	.097		2.33	3.56		5.89	8.10
1200 Portable metal with base pads, buy					13.15			13.15	14.50
1250 Typical installation, assume 10 reuses	2 Carp	600	.027		1.31	.98		2.29	2.96
1300 Barricade tape, polyethelyne, 7 mil, 3" wide x 500' long roll				Ea.	25			25	27.50

01 56 26 – Temporary Fencing

01 56 26.50 Temporary Fencing	Crew	Daily Output	Labor-Hours	Unit	Material	Labor	Equipment	Total	Total Incl O&P
0010 **TEMPORARY FENCING**									
0020 Chain link, 11 ga, 5' high	2 Clab	400	.040	L.F.	6	1.15		7.15	8.40
0100 6' high		300	.053		6.50	1.53		8.03	9.55
0200 Rented chain link, 6' high, to 1000' (up to 12 mo.)		400	.040		2.85	1.15		4	4.93
0250 Over 1000' (up to 12 mo.)		300	.053		2.65	1.53		4.18	5.30
0350 Plywood, painted, 2" x 4" frame, 4' high	A-4	135	.178		5.20	6.30		11.50	15.45
0400 4" x 4" frame, 8' high	"	110	.218		10.35	7.70		18.05	23.50
0500 Wire mesh on 4" x 4" posts, 4' high	2 Carp	100	.160		10.05	5.85		15.90	20
0550 8' high	"	80	.200		15.15	7.35		22.50	28

01 56 29 – Temporary Protective Walkways

01 56 29.50 Protection	Crew	Daily Output	Labor-Hours	Unit	Material	Labor	Equipment	Total	Total Incl O&P
0010 **PROTECTION**									
0020 Stair tread, 2" x 12" planks, 1 use	1 Carp	75	.107	Tread	4.29	3.91		8.20	10.80
0100 Exterior plywood, 1/2" thick, 1 use		65	.123		1.42	4.52		5.94	8.60
0200 3/4" thick, 1 use		60	.133		2.12	4.89		7.01	9.95
2200 Sidewalks, 2" x 12" planks, 2 uses		350	.023	S.F.	.72	.84		1.56	2.10
2300 Exterior plywood, 2 uses, 1/2" thick		750	.011		.24	.39		.63	.87
2400 5/8" thick		650	.012		.30	.45		.75	1.03
2500 3/4" thick		600	.013		.35	.49		.84	1.15

01 58 Project Identification

01 58 13 – Temporary Project Signage

01 58 13.50 Signs	Crew	Daily Output	Labor-Hours	Unit	Material	2007 Bare Costs Labor	Equipment	Total	Total Incl O&P
0010 **SIGNS**									
0020 High intensity reflectorized, no posts, buy				S.F.	16.55			16.55	18.20

01 74 Cleaning and Waste Management

01 74 13 – Progress Cleaning

01 74 13.20 Cleaning Up

01 74 13.20 Cleaning Up	Crew	Daily Output	Labor-Hours	Unit	Material	2007 Bare Costs Labor	Equipment	Total	Total Incl O&P
0010 **CLEANING UP**									
0020 After job completion, allow, minimum				Job					.30%
0040 Maximum				"					1%
0050 Cleanup of floor area, continuous, per day, during const.	A-5	24	.750	M.S.F.	1.70	21.50	1.47	24.67	37
0100 Final by GC at end of job	"	11.50	1.565	"	2.71	45	3.07	50.78	76.50

01 91 Commissioning

01 91 13 – General Commissioning Requirements

01 91 13.50 Commissioning

01 91 13.50 Commissioning	Crew	Daily Output	Labor-Hours	Unit	Material	2007 Bare Costs Labor	Equipment	Total	Total Incl O&P
0010 **COMMISSIONING** Including documentation of design intent									
0100 performance verification, O&M, training, min				Project					.50%
0150 Maximum				"					.75%

01 93 Facility Maintenance

01 93 13 – Facility Maintenance Procedures

01 93 13.16 Electrical Facilities Maintenance

01 93 13.16 Electrical Facilities Maintenance	Crew	Daily Output	Labor-Hours	Unit	Material	2007 Bare Costs Labor	Equipment	Total	Total Incl O&P
0010 **ELECTRICAL FACILITIES MAINTENANCE**									
0700 Cathodic protection systems									
0720 Check and adjust reading on rectifier	1 Elec	20	.400	Ea.		17.55		17.55	26
0730 Check pipe to soil potential		20	.400			17.55		17.55	26
0740 Replace lead connection		4	2			88		88	131
0800 Control device, install		5.70	1.404			61.50		61.50	91.50
0810 Disassemble, clean and reinstall		7	1.143			50		50	74.50
0820 Replace		10.70	.748			33		33	49
0830 Trouble shoot		10	.800			35		35	52.50
0900 Demolition, for electrical demolition see Division 26 05 05.10									
1000 Distribution systems and equipment install or repair a breaker									
1010 In power panels up to 200 amps	1 Elec	7	1.143	Ea.		50		50	74.50
1020 Over 200 amps		2	4			176		176	261
1030 Reset breaker or replace fuse		20	.400			17.55		17.55	26
1100 Megger test MCC (each stack)		4	2			88		88	131
1110 MCC vacuum and clean (each stack)		5.30	1.509			66.50		66.50	98.50
2500 Remove/replace or maint. road fixture & lamp		3	2.667		760	117		877	1,000
2510 Fluorescent fixture		7	1.143		72	50		122	154
2515 Relamp (fluor.) facility area each tube		60	.133		5.60	5.85		11.45	14.85
2518 Fluorescent fixture, clean (area)		44	.182		.08	8		8.08	12
2520 Incandescent fixture		11	.727		53	32		85	106
2530 Lamp (incadescent or fluorescent)		60	.133		3.15	5.85		9	12.15
2535 Replace cord in socket lamp		13	.615		2.20	27		29.20	42.50
2540 Ballast electronic type for two tubes		8	1		35	44		79	104
2541 Starter		30	.267		1.70	11.70		13.40	19.30

15

01 93 13.16 Electrical Facilities Maintenance	Crew	Daily Output	Labor-Hours	Unit	Material	2007 Bare Costs Labor	Equipment	Total	Total Incl O&P
2545	Replace other lighting parts	1 Elec	11	.727	Ea.	16.60	32		48.60	66
2550	Switch		11	.727		7.40	32		39.40	55.50
2555	Receptacle		11	.727		9.40	32		41.40	58
2560	Floodlight		4	2		281	88		369	440
2570	Christmas lighting, indoor, per string		16	.500			22		22	32.50
2580	Outdoor		13	.615			27		27	40
2590	Test battery operated emergency lights		40	.200			8.80		8.80	13.05
2600	Repair/replace component in communication system		6	1.333		53.50	58.50		112	146
2700	Repair misc. appliances (incl. clocks, vent fan, blower, etc.)		6	1.333			58.50		58.50	87
2710	Reset clocks & timers		50	.160			7		7	10.45
2720	Adjust time delay relays		16	.500			22		22	32.50
2730	Test specific gravity of lead-acid batteries	↓	80	.100	↓		4.39		4.39	6.55
3000	Motors and generators									
3020	Disassemble, clean and reinstall motor, up to 1/4 HP	1 Elec	4	2	Ea.		88		88	131
3030	Up to 3/4 HP		3	2.667			117		117	174
3040	Up to 10 HP		2	4			176		176	261
3050	Replace part, up to 1/4 HP		6	1.333			58.50		58.50	87
3060	Up to 3/4 HP		4	2			88		88	131
3070	Up to 10 HP		3	2.667			117		117	174
3080	Megger test motor windings		5.33	1.501			66		66	98
3082	Motor vibration check		16	.500			22		22	32.50
3084	Oil motor bearings		25	.320			14.05		14.05	21
3086	Run test emergency generator for 30 minutes		11	.727			32		32	47.50
3090	Rewind motor, up to 1/4 HP		3	2.667			117		117	174
3100	Up to 3/4 HP		2	4			176		176	261
3110	Up to 10 HP		1.50	5.333			234		234	350
3150	Generator, repair or replace part		4	2			88		88	131
3160	Repair DC generator		2	4			176		176	261
4000	Stub pole, install or remove		3	2.667			117		117	174
4500	Transformer maintenance up to 15 kVA	↓	2.70	2.963	↓		130		130	194

Estimating Tips

02 20 00 Assessment

- If possible, visit the site and take an inventory of the type, quantity, and size of the trees. Certain trees may have a landscape resale value or firewood value. Stump disposal can be very expensive, particularly if they cannot be buried at the site. Consider using a bulldozer in lieu of hand-cutting trees.

- Estimators should visit the site to determine the need for haul road access, storage of materials, and security considerations. When estimating for access roads on unstable soil, consider using a geotextile stabilization fabric. It can greatly reduce the quantity of crushed stone or gravel. Sites of limited size and access can cause cost overruns due to lost productivity. Theft and damage is another consideration if the location is isolated. A temporary fence or security guards may be required. Investigate the site thoroughly.

02 30 00 Subsurface Investigation

In preparing estimates on structures involving earthwork or foundations, all information concerning soil characteristics should be obtained. Look particularly for hazardous waste, evidence of prior dumping of debris, and previous stream beds.

02 40 00 Demolition and Structure Moving

The costs shown for selective demolition do not include rubbish handling or disposal. These items should be estimated separately using RSMeans data or other sources.

- Historic preservation often requires that the contractor remove materials from the existing structure, rehab them, and replace them. The estimator must be aware of any related measures and precautions that must be taken when doing selective demolition and cutting and patching. Requirements may include special handling and storage, as well as security.

- In addition to Section 02 41 00, you can find selective demolition items in each division. Example: Roofing demolition is in Division 7.

Reference Numbers

Reference numbers are shown in shaded boxes at the beginning of some major classifications. These numbers refer to related items in the Reference Section. The reference information may be an estimating procedure, an alternate pricing method, or technical information.

Note: Not all subdivisions listed here necessarily appear in this publication.

02 41 Demolition

02 41 13 – Selective Site Demolition

02 41 13.17 Demolish, Remove Pavement and Curb		Crew	Daily Output	Labor-Hours	Unit	Material	2007 Bare Costs Labor	Equipment	Total	Total Incl O&P
0010	**DEMOLISH, REMOVE PAVEMENT AND CURB** R024119-10									
5010	Pavement removal, bituminous roads, 3" thick	B-38	690	.058	S.Y.		1.90	1.30	3.20	4.35
5050	4" to 6" thick		420	.095			3.11	2.14	5.25	7.15
5100	Bituminous driveways		640	.063			2.04	1.41	3.45	4.69
5200	Concrete to 6" thick, hydraulic hammer, mesh reinforced		255	.157			5.15	3.53	8.68	11.75
5300	Rod reinforced		200	.200			6.55	4.50	11.05	15
5400	Concrete, 7" to 24" thick, plain		33	1.212	C.Y.		39.50	27.50	67	91
5500	Reinforced		24	1.667	"		54.50	37.50	92	125
5600	With hand held air equipment, bituminous, to 6" thick	B-39	1900	.025	S.F.		.77	.09	.86	1.29
5700	Concrete to 6" thick, no reinforcing		1600	.030			.91	.11	1.02	1.53
5800	Mesh reinforced		1400	.034			1.04	.13	1.17	1.75
5900	Rod reinforced		765	.063			1.91	.24	2.15	3.21

02 41 13.33 Minor Site Demolition

		Crew	Daily Output	Labor-Hours	Unit	Material	Labor	Equipment	Total	Total Incl O&P
0010	**MINOR SITE DEMOLITION** R024119-10									
0015	No hauling, abandon catch basin or manhole	B-6	7	3.429	Ea.		108	35	143	205
0020	Remove existing catch basin or manhole, masonry		4	6			189	61	250	355
0030	Catch basin or manhole frames and covers, stored		13	1.846			58	18.70	76.70	110
0040	Remove and reset		7	3.429			108	35	143	205
4000	Sidewalk removal, bituminous, 2-1/2" thick		325	.074	S.Y.		2.32	.75	3.07	4.39
4050	Brick, set in mortar		185	.130			4.08	1.32	5.40	7.70
4100	Concrete, plain, 4"		160	.150			4.72	1.52	6.24	8.90
4200	Mesh reinforced		150	.160			5.05	1.62	6.67	9.55

02 41 19 – Selective Structure Demolition

02 41 19.13 Selective Building Demolition

		Crew	Daily Output	Labor-Hours	Unit	Material	Labor	Equipment	Total	Total Incl O&P
0010	**SELECTIVE BUILDING DEMOLITION**									
0020	Costs related to selective demolition of specific building components									
0025	are included under Common Work Results (XX 05 00)									
0030	in the component's appropriate division.									

02 41 19.16 Selective Demolition, Cutout

		Crew	Daily Output	Labor-Hours	Unit	Material	Labor	Equipment	Total	Total Incl O&P
0010	**SELECTIVE DEMOLITION, CUTOUT** R024119-10									
0020	Concrete, elev. slab, light reinforcement, under 6 CF	B-9C	65	.615	C.F.		17.95	2.77	20.72	31
0050	Light reinforcing, over 6 C.F.	"	75	.533	"		15.55	2.40	17.95	26.50
6000	Walls, interior, not including re-framing,									
6010	openings to 5 S.F.									
6100	Drywall to 5/8" thick	1 Clab	24	.333	Ea.		9.60		9.60	14.90
6200	Paneling to 3/4" thick		20	.400			11.50		11.50	17.90
6300	Plaster, on gypsum lath		20	.400			11.50		11.50	17.90
6340	On wire lath		14	.571			16.45		16.45	25.50

02 41 19.19 Selective Demolition, Dump Charges

		Crew	Daily Output	Labor-Hours	Unit	Material	Labor	Equipment	Total	Total Incl O&P
0010	**SELECTIVE DEMOLITION, DUMP CHARGES** R024119-10									
0020	Dump charges, typical urban city, tipping fees only									
0100	Building construction materials					Ton	90		90	90
0200	Trees, brush, lumber						65		65	65
0300	Rubbish only						78		78	78
0500	Reclamation station, usual charge						95		95	95

02 41 19.23 Selective Demolition, Rubbish Handling

		Crew	Daily Output	Labor-Hours	Unit	Material	Labor	Equipment	Total	Total Incl O&P
0010	**SELECTIVE DEMOLITION, RUBBISH HANDLING** R024119-10									
0020	The following are to be added to the demolition prices									
0400	Chute, circular, prefabricated steel, 18" diameter	B-1	40	.600	L.F.	37	17.65		54.65	68
0440	30" diameter	"	30	.800	"	44	23.50		67.50	85
0725	Dumpster, weekly rental, 1 dump/week, 20 C.Y. capacity (8 Tons)					Week	690		690	759

02 41 Demolition

02 41 19 – Selective Structure Demolition

02 41 19.23 Selective Demolition, Rubbish Handling

		Crew	Daily Output	Labor-Hours	Unit	Material	2007 Bare Costs Labor	Equipment	Total	Total Incl O&P
0800	30 C.Y. capacity (10 Tons)				Week	900			900	990
0840	40 C.Y. capacity (13 Tons)				↓	1,160			1,160	1,276
1000	Dust partition, 6 mil polyethylene, 1" x 3" frame	2 Carp	2000	.008	S.F.	.28	.29		.57	.77
1080	2" x 4" frame	"	2000	.008	"	.33	.29		.62	.82
2000	Load, haul, and dump, 50' haul	2 Clab	24	.667	C.Y.		19.15		19.15	30
2040	100' haul		16.50	.970			28		28	43.50
2080	Over 100' haul, add per 100 L.F.		35.50	.451			12.95		12.95	20
2120	In elevators, per 10 floors, add	↓	140	.114			3.29		3.29	5.10
3000	Loading & trucking, including 2 mile haul, chute loaded	B-16	45	.711			21	11.80	32.80	45.50
3040	Hand loading truck, 50' haul	"	48	.667			19.65	11.05	30.70	42.50
3080	Machine loading truck	B-17	120	.267			8.25	5.10	13.35	18.30
5000	Haul, per mile, up to 8 C.Y. truck	B-34B	1165	.007			.20	.46	.66	.81
5100	Over 8 C.Y. truck	"	1550	.005	↓		.15	.34	.49	.62

02 41 19.27 Selective Demolition, Torch Cutting

		Crew	Daily Output	Labor-Hours	Unit	Material	2007 Bare Costs Labor	Equipment	Total	Total Incl O&P
0010	**SELECTIVE DEMOLITION, TORCH CUTTING** R024119-10									
0020	Steel, 1" thick plate	1 Clab	360	.022	L.F.	.20	.64		.84	1.21
0040	1" diameter bar	"	210	.038	Ea.		1.10		1.10	1.71
1000	Oxygen lance cutting, reinforced concrete walls									
1040	12" to 16" thick walls	1 Clab	10	.800	L.F.		23		23	36
1080	24" thick walls	"	6	1.333	"		38.50		38.50	59.50

02 82 Asbestos Remediation

02 82 13 – Asbestos Abatement

02 82 13.47 Asbestos Waste Packaging, Handling, and Disposal

		Crew	Daily Output	Labor-Hours	Unit	Material	2007 Bare Costs Labor	Equipment	Total	Total Incl O&P
0010	**ASBESTOS WASTE PACKAGING, HANDLING, AND DISPOSAL**									
0100	Collect and bag bulk material, 3 C.F. bags, by hand	A-9	400	.160	Ea.	1.24	6.65		7.89	11.85
0200	Large production vacuum loader	A-12	880	.073		.80	3.02	.75	4.57	6.50
1000	Double bag and decontaminate	A-9	960	.067		1.24	2.77		4.01	5.75
2000	Containerize bagged material in drums, per 3 C.F. drum	"	800	.080		6.50	3.32		9.82	12.40
3000	Cart bags 50' to dumpster	2 Asbe	400	.040	↓		1.66		1.66	2.62
5000	Disposal charges, not including haul, minimum				C.Y.				45.45	50
5020	Maximum				"				159.09	175
5100	Remove refrigerant from system	1 Plum	40	.200	Lb.		8.95		8.95	13.50
9000	For type C (supplied air) respirator equipment, add				%					10%

Division Notes

	CREW	DAILY OUTPUT	LABOR-HOURS	UNIT	2007 BARE COSTS				TOTAL INCL O&P
					MAT.	LABOR	EQUIP.	TOTAL	

Estimating Tips

General

- Carefully check all the plans and specifications. Concrete often appears on drawings other than structural drawings, including mechanical and electrical drawings for equipment pads. The cost of cutting and patching is often difficult to estimate. See Subdivisions 02 41 19 and 03 01 05 for demolition costs.

- Always obtain concrete prices from suppliers near the job site. A volume discount can often be negotiated depending upon competition in the area. Remember to add for waste, particularly for slabs and footings on grade.

03 10 00 Concrete Forming and Accessories

- A primary cost for concrete construction is forming. Most jobs today are constructed with prefabricated forms. The selection of the forms best suited for the job and the total square feet of forms required for efficient concrete forming and placing are key elements in estimating concrete construction. Enough forms must be available for erection to make efficient use of the concrete placing equipment and crew.

- Concrete accessories for forming and placing depend upon the systems used. Study the plans and specifications to assure that all special accessory requirements have been included in the cost estimate, such as anchor bolts, inserts, and hangers.

03 20 00 Concrete Reinforcing

- Ascertain that the reinforcing steel supplier has included all accessories, cutting, bending, and an allowance for lapping, splicing, and waste. A good rule of thumb is 10% for lapping, splicing, and waste. Also, 10% waste should be allowed for welded wire fabric.

- The unit price items in the section for Reinforcing In Place include the labor to install accessories such as beam and slab bolsters, high chairs, and bar ties and tie wire. The material cost for these accessories is not included; they may be obtained from the Accessories section.

03 30 00 Cast-in-Place Concrete

- When estimating structural concrete, pay particular attention to requirements for concrete additives, curing methods, and surface treatments. Special consideration for climate, hot or cold, must be included in your estimate. Be sure to include requirements for concrete placing equipment and concrete finishing.

03 40 00 Precast Concrete
03 50 00 Cast Decks and Underlayment

- The cost of hauling precast concrete structural members is often an important factor. For this reason, it is important to get a quote from the nearest supplier. It may become economically feasible to set up precasting beds on the site if the hauling costs are prohibitive.

Reference Numbers

Reference numbers are shown in shaded boxes at the beginning of some major classifications. These numbers refer to related items in the Reference Section. The reference information may be an estimating procedure, an alternate pricing method, or technical information.

Note: Not all subdivisions listed here necessarily appear in this publication.

03 01 Maintenance of Concrete

03 01 30 – Maintenance of Cast-In-Place Concrete

03 01 30.62 Concrete Patching

		Crew	Daily Output	Labor-Hours	Unit	Material	2007 Bare Costs Labor	Equipment	Total	Total Incl O&P
0010	**CONCRETE PATCHING**									
0100	Floors, 1/4" thick, small areas, regular grout	1 Cefi	170	.047	S.F.	1	1.67		2.67	3.56
0150	Epoxy grout	"	100	.080	"	7.40	2.84		10.24	12.35
2000	Walls, including chipping, cleaning and epoxy grout									
2100	1/4" deep	1 Cefi	65	.123	S.F.	7.45	4.38		11.83	14.60
2150	1/2" deep		50	.160		14.90	5.70		20.60	25
2200	3/4" deep	↓	40	.200	↓	22.50	7.10		29.60	35

03 11 Concrete Forming

03 11 13 – Structural Cast-In-Place Concrete Forming

03 11 13.40 Forms In Place, Equipment Foundations

		Crew	Daily Output	Labor-Hours	Unit	Material	2007 Bare Costs Labor	Equipment	Total	Total Incl O&P
0010	**FORMS IN PLACE, EQUIPMENT FOUNDATIONS**									
0020	1 use	C-2	160	.300	SFCA	2.59	10.70		13.29	19.55
0050	2 use		190	.253		1.42	9		10.42	15.60
0100	3 use		200	.240		1.04	8.55		9.59	14.50
0150	4 use	↓	205	.234	↓	.84	8.35		9.19	13.95

03 11 13.45 Forms In Place, Footings

		Crew	Daily Output	Labor-Hours	Unit	Material	2007 Bare Costs Labor	Equipment	Total	Total Incl O&P
0010	**FORMS IN PLACE, FOOTINGS**									
0020	Continuous wall, plywood, 1 use	C-1	375	.085	SFCA	2.64	2.96		5.60	7.50
0050	2 use		440	.073		1.45	2.52		3.97	5.55
0100	3 use		470	.068		1.06	2.36		3.42	4.84
0150	4 use		485	.066		.86	2.29		3.15	4.52
5000	Spread footings, job-built lumber, 1 use		305	.105		1.81	3.64		5.45	7.65
5050	2 use		371	.086		1	2.99		3.99	5.75
5100	3 use		401	.080		.72	2.77		3.49	5.10
5150	4 use	↓	414	.077	↓	.59	2.68		3.27	4.83

03 11 13.65 Forms In Place, Slab On Grade

		Crew	Daily Output	Labor-Hours	Unit	Material	2007 Bare Costs Labor	Equipment	Total	Total Incl O&P
0010	**FORMS IN PLACE, SLAB ON GRADE**									
3000	Edge forms, wood, 4 use, on grade, to 6" high	C-1	600	.053	L.F.	.29	1.85		2.14	3.20
6000	Trench forms in floor, wood, 1 use		160	.200	SFCA	2.51	6.95		9.46	13.55
6050	2 use		175	.183		1.21	6.35		7.56	11.25
6100	3 use		180	.178		.88	6.15		7.03	10.55
6150	4 use	↓	185	.173	↓	.71	6		6.71	10.15

03 15 Concrete Accessories

03 15 05 – Concrete Forming Accessories

03 15 05.75 Sleeves and Chases

		Crew	Daily Output	Labor-Hours	Unit	Material	2007 Bare Costs Labor	Equipment	Total	Total Incl O&P
0010	**SLEEVES AND CHASES**									
0100	Plastic, 1 use, 9" long, 2" diameter	1 Carp	100	.080	Ea.	1.55	2.94		4.49	6.30
0150	4" diameter		90	.089		4.14	3.26		7.40	9.65
0200	6" diameter		75	.107		7.30	3.91		11.21	14.15
0250	12" diameter		60	.133		21	4.89		25.89	30.50
5000	Sheet metal, 2" diameter		100	.080		.97	2.94		3.91	5.65
5100	4" diameter		90	.089		1.21	3.26		4.47	6.45
5150	6" diameter		75	.107		1.76	3.91		5.67	8.05
5200	12" diameter		60	.133		3.52	4.89		8.41	11.45
6000	Steel pipe, 2" diameter		100	.080		5.65	2.94		8.59	10.75
6100	4" diameter		90	.089		16.35	3.26		19.61	23
6150	6" diameter	↓	75	.107	↓	30	3.91		33.91	39

03 15 Concrete Accessories

03 15 05 – Concrete Forming Accessories

03 15 05.75 Sleeves and Chases	Crew	Daily Output	Labor-Hours	Unit	Material	2007 Bare Costs Labor	Equipment	Total	Total Incl O&P
6200	12" diameter	1 Carp	60	.133	Ea.	75	4.89		79.89	90

03 21 Reinforcing Steel

03 21 10 – Uncoated Reinforcing Steel

03 21 10.60 Reinforcing In Place

		Crew	Daily Output	Labor-Hours	Unit	Material	2007 Bare Costs Labor	Equipment	Total	Total Incl O&P
0015	**REINFORCING IN PLACE** A615 Grade 60, incl. access. labor									
0502	Footings, #4 to #7	4 Rodm	4200	.008	Lb.	.47	.31		.78	1.03
0552	#8 to #18	↓	7200	.004	↓	.47	.18		.65	.81
0602	Slab on grade, #3 to #7	↓	4200	.008	↓	.45	.31		.76	1.01

03 22 Welded Wire Fabric Reinforcing

03 22 05 – Uncoated Welded Wire Fabric

03 22 05.50 Welded Wire Fabric

		Crew	Daily Output	Labor-Hours	Unit	Material	2007 Bare Costs Labor	Equipment	Total	Total Incl O&P
0010	**WELDED WIRE FABRIC** ASTM A185									
0040	Reinforcing sheets, 6x6-W1.4xW1.4				S.F.	.28			.28	.31

03 30 Cast-In-Place Concrete

03 30 53 – Miscellaneous Cast-In-Place Concrete

03 30 53.40 Concrete In Place

		Crew	Daily Output	Labor-Hours	Unit	Material	2007 Bare Costs Labor	Equipment	Total	Total Incl O&P
0010	**CONCRETE IN PLACE**									
0020	Including forms (4 uses), concrete, placement, reinforcing									
0050	steel and finishing unless otherwise indicated									
3540	Equipment pad, 3' x 3' x 6" thick	C-14H	45	1.067	Ea.	42.50	38.50	.48	81.48	108
3550	4' x 4' x 6" thick		30	1.600		64.50	58	.72	123.22	162
3560	5' x 5' x 8" thick		18	2.667		115	97	1.20	213.20	279
3570	6' x 6' x 8" thick		14	3.429		157	124	1.54	282.54	370
3580	8' x 8' x 10" thick		8	6		330	218	2.70	550.70	705
3590	10' x 10' x 12" thick	↓	5	9.600	↓	565	350	4.32	919.32	1,175
3800	Footings, spread under 1 C.Y.	C-14C	38.07	2.942	C.Y.	195	103	.56	298.56	375
3850	Over 5 C.Y.		81.04	1.382		266	48.50	.26	314.76	370
3900	Footings, strip, 18" x 9", unreinforced		40	2.800		126	98.50	.53	225.03	293
3920	18" x 9", reinforced		35	3.200		147	112	.61	259.61	340
3925	20" x 10", unreinforced		45	2.489		122	87.50	.47	209.97	273
3930	20" x 10", reinforced		40	2.800		140	98.50	.53	239.03	310
3935	24" x 12", unreinforced		55	2.036		120	71.50	.39	191.89	244
3940	24" x 12", reinforced		48	2.333		138	82	.44	220.44	280
3945	36" x 12", unreinforced		70	1.600		117	56	.30	173.30	216
3950	36" x 12", reinforced		60	1.867		133	65.50	.35	198.85	249
4000	Foundation mat, under 10 C.Y.		38.67	2.896		197	102	.55	299.55	375
4050	Over 20 C.Y.	↓	56.40	1.986		174	70	.38	244.38	300
4650	Slab on grade, not including finish, 4" thick	C-14E	60.75	1.449		124	52.50	.35	176.85	219
4700	6" thick	"	92	.957	↓	120	34.50	.23	154.73	187

03 31 Structural Concrete

03 31 05 – Normal Weight Structural Concrete

03 31 05.35 Normal Weight Concrete, Ready Mix	Crew	Daily Output	Labor-Hours	Unit	Material	2007 Bare Costs Labor	Equipment	Total	Total Incl O&P
0010	**NORMAL WEIGHT CONCRETE, READY MIX**									
0012	Includes local aggregate, sand, portland cement, and water									
0015	Excludes all additives and treatments									
0020	2000 psi				C.Y.	99.50			99.50	110
0100	2500 psi					101			101	111
0150	3000 psi					104			104	114
0200	3500 psi					106			106	116
0300	4000 psi					108			108	119
1000	For high early strength cement, add					10%				
2000	For all lightweight aggregate, add					45%				

03 31 05.70 Placing Concrete

		Crew	Daily Output	Labor-Hours	Unit	Material	Labor	Equipment	Total	Total Incl O&P
0010	**PLACING CONCRETE**									
0020	Includes labor and equipment to place and vibrate									
1900	Footings, continuous, shallow, direct chute	C-6	120	.400	C.Y.		12.10	.36	12.46	19
1950	Pumped	C-20	150	.427			13.25	5	18.25	26
2000	With crane and bucket	C-7	90	.800			25	12.45	37.45	52
2100	Footings, continuous, deep, direct chute	C-6	140	.343			10.35	.31	10.66	16.30
2150	Pumped	C-20	160	.400			12.40	4.70	17.10	24.50
2200	With crane and bucket	C-7	110	.655			20.50	10.20	30.70	42.50
2400	Footings, spread, under 1 C.Y., direct chute	C-6	55	.873			26.50	.78	27.28	41.50
2450	Pumped	C-20	65	.985			30.50	11.55	42.05	59.50
2500	With crane and bucket	C-7	45	1.600			50	25	75	105
2600	Over 5 C.Y., direct chute	C-6	120	.400			12.10	.36	12.46	19
2650	Pumped	C-20	150	.427			13.25	5	18.25	26
2700	With crane and bucket	C-7	100	.720			22.50	11.20	33.70	47
2900	Foundation mats, over 20 C.Y., direct chute	C-6	350	.137			4.14	.12	4.26	6.55
2950	Pumped	C-20	400	.160			4.97	1.88	6.85	9.70
3000	With crane and bucket	C-7	300	.240			7.55	3.73	11.28	15.65

03 35 Concrete Finishing

03 35 29 – Tooled Concrete Finishing

03 35 29.30 Finishing Floors

		Crew	Daily Output	Labor-Hours	Unit	Material	Labor	Equipment	Total	Total Incl O&P
0010	**FINISHING FLOORS**									
0020	Monolithic, screed finish	1 Cefi	900	.009	S.F.		.32		.32	.46
0100	Screed and bull float (darby) finish		725	.011			.39		.39	.58
0150	Screed, float, and broom finish		630	.013			.45		.45	.66
0200	Screed, float, and hand trowel		600	.013			.47		.47	.70
0250	Machine trowel		550	.015			.52		.52	.76

03 35 29.35 Control Joints, Saw Cut

		Crew	Daily Output	Labor-Hours	Unit	Material	Labor	Equipment	Total	Total Incl O&P
0010	**CONTROL JOINTS, SAW CUT**									
0100	Sawcut in green concrete									
0120	1" depth	C-27	2000	.008	L.F.	.09	.28	.06	.43	.59
0140	1-1/2" depth		1800	.009		.14	.32	.07	.53	.68
0160	2" depth		1600	.010		.18	.36	.07	.61	.80
0200	Clean out control joint of debris	C-28	6000	.001			.05		.05	.07
0300	Joint sealant									
0320	Backer rod, polyethylene, 1/4" diameter	1 Cefi	460	.017	L.F.	.04	.62		.66	.96
0340	Sealant, polyurethane									
0360	1/4" x 1/4" (308 LF/Gal)	1 Cefi	270	.030	L.F.	.17	1.05		1.22	1.74
0380	1/4" x 1/2" (154 LF/Gal)	"	255	.031	"	.34	1.12		1.46	2.01

03 54 Cast Underlayment

03 54 16 – Hydraulic Cement Underlayment

03 54 16.50 Cement Underlayment		Crew	Daily Output	Labor-Hours	Unit	Material	2007 Bare Costs Labor	Equipment	Total	Total Incl O&P
0010	**CEMENT UNDERLAYMENT**									
2510	Underlayment, P.C based self-leveling, 4100 psi, pumped, 1/4"	C-8	20000	.003	S.F.	1.46	.09	.04	1.59	1.79
2520	1/2"		19000	.003		2.92	.10	.04	3.06	3.40
2530	3/4"		18000	.003		4.38	.10	.04	4.52	5
2540	1"		17000	.003		5.85	.11	.04	6	6.60
2550	1-1/2"		15000	.004		8.75	.12	.05	8.92	9.90
2560	Hand mix, 1/2"	C-18	4000	.002		2.92	.07	.01	3	3.32
2610	Topping, P.C. based self-level/dry 6100 psi, pumped, 1/4"	C-8	20000	.003		2.28	.09	.04	2.41	2.69
2620	1/2"		19000	.003		4.56	.10	.04	4.70	5.20
2630	3/4"		18000	.003		6.85	.10	.04	6.99	7.70
2660	1"		17000	.003		9.10	.11	.04	9.25	10.25
2670	1-1/2"		15000	.004		13.70	.12	.05	13.87	15.30
2680	Hand mix, 1/2"	C-18	4000	.002		4.56	.07	.01	4.64	5.10

03 63 Epoxy Grouting

03 63 05 – Grouting of Dowels and Fasteners

03 63 05.10 Epoxy Only

03 63 05.10 Epoxy Only		Crew	Daily Output	Labor-Hours	Unit	Material	2007 Bare Costs Labor	Equipment	Total	Total Incl O&P
0010	**EPOXY ONLY**									
1500	Chemical anchoring, epoxy cartridge, excludes layout, drilling, fastener									
1530	For fastener 3/4" dia x 6" embedment	B-89A	27	.593	Ea.	5.55	19.80	3.89	29.24	41.50
1535	1" dia x 8" embedment		24	.667		8.35	22.50	4.37	35.22	48.50
1540	1-1/4" dia x 10" embedment		21	.762		16.65	25.50	5	47.15	63.50
1545	1-3/4" dia x 12" embedment		20	.800		28	26.50	5.25	59.75	78
1550	14" embedment		17	.941		33.50	31.50	6.15	71.15	92.50
1555	2" dia x 12" embedment		16	1		44.50	33.50	6.55	84.55	108
1560	18" embedment		15	1.067		55.50	35.50	7	98	124

03 82 Concrete Boring

03 82 13 – Concrete Core Drilling

03 82 13.10 Core Drilling

03 82 13.10 Core Drilling		Crew	Daily Output	Labor-Hours	Unit	Material	2007 Bare Costs Labor	Equipment	Total	Total Incl O&P
0010	**CORE DRILLING**									
0020	Reinf. conc slab, up to 6" thick, incl. bit, layout & set up									
0100	1" diameter core	B-89A	28	.571	Ea.	2.80	19.05	3.75	25.60	36.50
0150	Each added inch thick, add		300	.053		.50	1.78	.35	2.63	3.71
0300	3" diameter core		23	.696		6.20	23	4.56	33.76	48
0350	Each added inch thick, add		186	.086		1.12	2.87	.56	4.55	6.30
0500	4" diameter core		19	.842		6.20	28	5.50	39.70	57
0550	Each added inch thick, add		170	.094		1.41	3.14	.62	5.17	7.10
0700	6" diameter core		14	1.143		10.25	38	7.50	55.75	79
0750	Each added inch thick, add		140	.114		1.74	3.82	.75	6.31	8.70
0900	8" diameter core		11	1.455		14	48.50	9.55	72.05	101
0950	Each added inch thick, add		95	.168		2.35	5.60	1.10	9.05	12.55
1100	10" diameter core		10	1.600		18.70	53.50	10.50	82.70	115
1150	Each added inch thick, add		80	.200		3.09	6.70	1.31	11.10	15.25
1300	12" diameter core		9	1.778		22.50	59.50	11.65	93.65	130
1350	Each added inch thick, add		68	.235		3.70	7.85	1.54	13.09	17.95
1500	14" diameter core		7	2.286		27	76.50	15	118.50	166
1550	Each added inch thick, add		55	.291		4.71	9.70	1.91	16.32	22.50
1700	18" diameter core		4	4		35.50	134	26	195.50	276

03 82 Concrete Boring

03 82 13 – Concrete Core Drilling

03 82 13.10 Core Drilling

		Crew	Daily Output	Labor-Hours	Unit	Material	2007 Bare Costs Labor	Equipment	Total	Total Incl O&P
1750	Each added inch thick, add	B-89A	28	.571	Ea.	6.20	19.05	3.75	29	40.50
1760	For horizontal holes, add to above				▼				30%	30%
1770	Prestressed hollow core plank, 6" thick									
1780	1" diameter core	B-89A	52	.308	Ea.	1.86	10.25	2.02	14.13	20.50
1790	Each added inch thick, add		350	.046		.32	1.53	.30	2.15	3.05
1800	3" diameter core		50	.320		4.09	10.70	2.10	16.89	23.50
1810	Each added inch thick, add		240	.067		.68	2.23	.44	3.35	4.69
1820	4" diameter core		48	.333		5.45	11.15	2.19	18.79	25.50
1830	Each added inch thick, add		216	.074		.94	2.47	.49	3.90	5.40
1840	6" diameter core		44	.364		6.75	12.15	2.39	21.29	29
1850	Each added inch thick, add		175	.091		1.12	3.05	.60	4.77	6.65
1860	8" diameter core		32	.500		9.05	16.70	3.28	29.03	39.50
1870	Each added inch thick, add		118	.136		1.57	4.53	.89	6.99	9.75
1880	10" diameter core		28	.571		12.20	19.05	3.75	35	47
1890	Each added inch thick, add		99	.162		1.68	5.40	1.06	8.14	11.40
1900	12" diameter core		22	.727		14.85	24.50	4.77	44.12	59.50
1910	Each added inch thick, add		85	.188	▼	2.47	6.30	1.23	10	13.90
1950	Minimum charge for above, 3" diameter core		7	2.286	Total		76.50	15	91.50	136
2000	4" diameter core		6.80	2.353			78.50	15.45	93.95	139
2050	6" diameter core		6	2.667			89	17.50	106.50	158
2100	8" diameter core		5.50	2.909			97	19.10	116.10	172
2150	10" diameter core		4.75	3.368			112	22	134	200
2200	12" diameter core		3.90	4.103			137	27	164	243
2250	14" diameter core		3.38	4.734			158	31	189	280
2300	18" diameter core	▼	3.15	5.079	▼		170	33.50	203.50	300

03 82 16 – Concrete Drilling

03 82 16.10 Concrete Drilling

		Crew	Daily Output	Labor-Hours	Unit	Material	2007 Bare Costs Labor	Equipment	Total	Total Incl O&P
0010	**CONCRETE DRILLING**									
0050	Up to 4" deep in conc/brick floor/wall, incl. bit & layout, no anchor									
0100	Holes, 1/4" diameter	1 Carp	75	.107	Ea.	.10	3.91		4.01	6.20
0150	For each additional inch of depth, add		430	.019		.02	.68		.70	1.09
0200	3/8" diameter		63	.127		.09	4.66		4.75	7.35
0250	For each additional inch of depth, add		340	.024		.02	.86		.88	1.36
0300	1/2" diameter		50	.160		.09	5.85		5.94	9.25
0350	For each additional inch of depth, add		250	.032		.02	1.17		1.19	1.86
0400	5/8" diameter		48	.167		.17	6.10		6.27	9.75
0450	For each additional inch of depth, add		240	.033		.04	1.22		1.26	1.95
0500	3/4" diameter		45	.178		.20	6.50		6.70	10.35
0550	For each additional inch of depth, add		220	.036		.05	1.33		1.38	2.14
0600	7/8" diameter		43	.186		.25	6.85		7.10	10.90
0650	For each additional inch of depth, add		210	.038		.06	1.40		1.46	2.25
0700	1" diameter		40	.200		.28	7.35		7.63	11.75
0750	For each additional inch of depth, add		190	.042		.07	1.55		1.62	2.49
0800	1-1/4" diameter		38	.211		.40	7.75		8.15	12.50
0850	For each additional inch of depth, add		180	.044		.10	1.63		1.73	2.65
0900	1-1/2" diameter		35	.229		.61	8.40		9.01	13.70
0950	For each additional inch of depth, add	▼	165	.048	▼	.15	1.78		1.93	2.94
1000	For ceiling installations, add						40%			

Estimating Tips

05 05 00 Common Work Results for Metals

- Nuts, bolts, washers, connection angles, and plates can add a significant amount to both the tonnage of a structural steel job and the estimated cost. As a rule of thumb, add 10% to the total weight to account for these accessories.

- Type 2 steel construction, commonly referred to as "simple construction," consists generally of field-bolted connections with lateral bracing supplied by other elements of the building, such as masonry walls or x-bracing. The estimator should be aware, however, that shop connections may be accomplished by welding or bolting. The method may be particular to the fabrication shop and may have an impact on the estimated cost.

05 20 00 Metal Joists

- In any given project the total weight of open web steel joists is determined by the loads to be supported and the design. However, economies can be realized in minimizing the amount of labor used to place the joists. This is done by maximizing the joist spacing, and therefore minimizing the number of joists required to be installed on the job. Certain spacings and locations may be required by the design, but in other cases maximizing the spacing and keeping it as uniform as possible will keep the costs down.

05 30 00 Metal Decking

- The takeoff and estimating of metal deck involves more than simply the area of the floor or roof and the type of deck specified or shown on the drawings. Many different sizes and types of openings may exist. Small openings for individual pipes or conduits may be drilled after the floor/roof is installed, but larger openings may require special deck lengths as well as reinforcing or structural support. The estimator should determine who will be supplying this reinforcing. Additionally, some deck terminations are part of the deck package, such as screed angles and pour stops, and others will be part of the steel contract, such as angles attached to structural members and cast-in-place angles and plates. The estimator must ensure that all pieces are accounted for in the complete estimate.

05 50 00 Metal Fabrications

- The most economical steel stairs are those that use common materials, standard details, and most importantly, a uniform and relatively simple method of field assembly. Commonly available A36 channels and plates are very good choices for the main stringers of the stairs, as are angles and tees for the carrier members. Risers and treads are usually made by specialty shops, and it is most economical to use a typical detail in as many places as possible. The stairs should be pre-assembled and shipped directly to the site. The field connections should be simple and straightforward to be accomplished efficiently, and with minimum equipment and labor.

Reference Numbers

Reference numbers are shown in shaded boxes at the beginning of some major classifications. These numbers refer to related items in the Reference Section. The reference information may be an estimating procedure, an alternate pricing method, or technical information.

Note: Not all subdivisions listed here necessarily appear in this publication.

05 05 Common Work Results for Metals

05 05 21 – Fastening Methods for Metal

05 05 21.15 Drilling Steel		Crew	Daily Output	Labor-Hours	Unit	Material	2007 Bare Costs Labor	Equipment	Total	Total Incl O&P
0010	**DRILLING STEEL**									
1910	Drilling & layout for steel, up to 1/4" deep, no anchor									
1920	Holes, 1/4" diameter	1 Sswk	112	.071	Ea.	.13	2.95		3.08	5.50
1925	For each additional 1/4" depth, add		336	.024		.13	.98		1.11	1.92
1930	3/8" diameter		104	.077		.15	3.18		3.33	5.90
1935	For each additional 1/4" depth, add		312	.026		.15	1.06		1.21	2.08
1940	1/2" diameter		96	.083		.17	3.45		3.62	6.45
1945	For each additional 1/4" depth, add		288	.028		.17	1.15		1.32	2.26
1950	5/8" diameter		88	.091		.27	3.76		4.03	7.10
1955	For each additional 1/4" depth, add		264	.030		.27	1.25		1.52	2.57
1960	3/4" diameter		80	.100		.30	4.14		4.44	7.85
1965	For each additional 1/4" depth, add		240	.033		.30	1.38		1.68	2.83
1970	7/8" diameter		72	.111		.36	4.59		4.95	8.70
1975	For each additional 1/4" depth, add		216	.037		.36	1.53		1.89	3.17
1980	1" diameter		64	.125		.41	5.15		5.56	9.80
1985	For each additional 1/4" depth, add		192	.042		.41	1.72		2.13	3.57
1990	For drilling up, add						40%			

05 05 23 – Metal Fastenings

05 05 23.10 Bolts and Hex Nuts		Crew	Daily Output	Labor-Hours	Unit	Material	2007 Bare Costs Labor	Equipment	Total	Total Incl O&P
0010	**BOLTS & HEX NUTS**, Steel, A307									
0100	1/4" diameter, 1/2" long	1 Sswk	140	.057	Ea.	.07	2.36		2.43	4.36
0200	1" long		140	.057		.08	2.36		2.44	4.37
0300	2" long		130	.062		.11	2.54		2.65	4.73
0400	3" long		130	.062		.16	2.54		2.70	4.79
0500	4" long		120	.067		.18	2.76		2.94	5.20
0600	3/8" diameter, 1" long		130	.062		.12	2.54		2.66	4.74
0700	2" long		130	.062		.15	2.54		2.69	4.78
0800	3" long		120	.067		.20	2.76		2.96	5.20
0900	4" long		120	.067		.25	2.76		3.01	5.25
1000	5" long		115	.070		.31	2.88		3.19	5.55
1100	1/2" diameter, 1-1/2" long		120	.067		.24	2.76		3	5.25
1200	2" long		120	.067		.27	2.76		3.03	5.30
1300	4" long		115	.070		.41	2.88		3.29	5.65
1400	6" long		110	.073		.56	3.01		3.57	6.05
1500	8" long		105	.076		.73	3.15		3.88	6.50
1600	5/8" diameter, 1-1/2" long		120	.067		.47	2.76		3.23	5.50
1700	2" long		120	.067		.51	2.76		3.27	5.55
1800	4" long		115	.070		.71	2.88		3.59	6
1900	6" long		110	.073		.89	3.01		3.90	6.45
2000	8" long		105	.076		1.29	3.15		4.44	7.10
2100	10" long		100	.080		1.60	3.31		4.91	7.75
2200	3/4" diameter, 2" long		120	.067		.74	2.76		3.50	5.80
2300	4" long		110	.073		1.02	3.01		4.03	6.60
2400	6" long		105	.076		1.30	3.15		4.45	7.15
2500	8" long		95	.084		1.92	3.48		5.40	8.40
2600	10" long		85	.094		2.50	3.89		6.39	9.80
2700	12" long		80	.100		2.91	4.14		7.05	10.70
2800	1" diameter, 3" long		105	.076		1.90	3.15		5.05	7.80
2900	6" long		90	.089		2.93	3.68		6.61	9.85
3000	12" long		75	.107		5.50	4.41		9.91	14.05
3100	For galvanized, add					75%				
3200	For stainless, add					350%				

05 05 Common Work Results for Metals

05 05 23 – Metal Fastenings

05 05 23.15 Chemical Anchors

		Crew	Daily Output	Labor-Hours	Unit	Material	2007 Bare Costs Labor	2007 Bare Costs Equipment	Total	Total Incl O&P
0010	**CHEMICAL ANCHORS**									
0020	Includes layout & drilling									
1430	Chemical anchor, w/rod & epoxy cartridge, 3/4" diam. x 9-1/2" long	B-89A	27	.593	Ea.	13.70	19.80	3.89	37.39	50.50
1435	1" diameter x 11-3/4" long		24	.667		26.50	22.50	4.37	53.37	68.50
1440	1-1/4" diameter x 14" long		21	.762		50.50	25.50	5	81	101
1445	1-3/4" diameter x 15" long		20	.800		95	26.50	5.25	126.75	152
1450	18" long		17	.941		114	31.50	6.15	151.65	182
1455	2" diameter x 18" long		16	1		145	33.50	6.55	185.05	219
1460	24" long		15	1.067		190	35.50	7	232.50	272

05 05 23.20 Expansion Anchors

		Crew	Daily Output	Labor-Hours	Unit	Material	2007 Bare Costs Labor	2007 Bare Costs Equipment	Total	Total Incl O&P
0010	**EXPANSION ANCHORS**									
0100	Anchors for concrete, brick or stone, no layout and drilling									
0200	Expansion shields, zinc, 1/4" diameter, 1-5/16" long, single	1 Carp	90	.089	Ea.	1.11	3.26		4.37	6.30
0300	1-3/8" long, double		85	.094		1.22	3.45		4.67	6.75
0400	3/8" diameter, 1-1/2" long, single		85	.094		1.83	3.45		5.28	7.40
0500	2" long, double		80	.100		2.26	3.67		5.93	8.20
0600	1/2" diameter, 2-1/16" long, single		80	.100		3.03	3.67		6.70	9.05
0700	2-1/2" long, double		75	.107		2.92	3.91		6.83	9.30
0800	5/8" diameter, 2-5/8" long, single		75	.107		4.33	3.91		8.24	10.85
0900	2-3/4" long, double		70	.114		4.33	4.19		8.52	11.30
1000	3/4" diameter, 2-3/4" long, single		70	.114		6.45	4.19		10.64	13.60
1100	3-15/16" long, double		65	.123		8.60	4.52		13.12	16.50
1500	Self drilling anchor, snap-off, for 1/4" diameter bolt		26	.308		.95	11.30		12.25	18.65
1600	3/8" diameter bolt		23	.348		1.39	12.75		14.14	21.50
1700	1/2" diameter bolt		20	.400		2.13	14.70		16.83	25.50
1800	5/8" diameter bolt		18	.444		3.56	16.30		19.86	29.50
1900	3/4" diameter bolt		16	.500		6	18.35		24.35	35
2100	Hollow wall anchors for gypsum wall board, plaster or tile									
2300	1/8" diameter, short	1 Carp	160	.050	Ea.	.30	1.84		2.14	3.19
2400	Long		150	.053		.36	1.96		2.32	3.45
2500	3/16" diameter, short		150	.053		.64	1.96		2.60	3.75
2600	Long		140	.057		.69	2.10		2.79	4.03
2700	1/4" diameter, short		140	.057		.78	2.10		2.88	4.13
2800	Long		130	.062		.88	2.26		3.14	4.49
3000	Toggle bolts, bright steel, 1/8" diameter, 2" long		85	.094		.28	3.45		3.73	5.70
3100	4" long		80	.100		.43	3.67		4.10	6.15
3400	1/4" diameter, 3" long		75	.107		.54	3.91		4.45	6.70
3500	6" long		70	.114		.76	4.19		4.95	7.40
3600	3/8" diameter, 3" long		70	.114		1.04	4.19		5.23	7.70
3700	6" long		60	.133		1.82	4.89		6.71	9.60
3800	1/2" diameter, 4" long		60	.133		2.70	4.89		7.59	10.55
3900	6" long		50	.160		4.45	5.85		10.30	14.05
4000	Nailing anchors									
4100	Nylon nailing anchor, 1/4" diameter, 1" long	1 Carp	3.20	2.500	C	21	92		113	166
4200	1-1/2" long		2.80	2.857		26.50	105		131.50	193
4300	2" long		2.40	3.333		44.50	122		166.50	240
4400	Metal nailing anchor, 1/4" diameter, 1" long		3.20	2.500		32	92		124	178
4500	1-1/2" long		2.80	2.857		43.50	105		148.50	211
4600	2" long		2.40	3.333		55.50	122		177.50	252
5000	Screw anchors for concrete, masonry,									
5100	stone & tile, no layout or drilling included									
5200	Jute fiber, #6, #8, & #10, 1" long	1 Carp	240	.033	Ea.	.26	1.22		1.48	2.19

05 05 23 – Metal Fastenings

05 05 23.20 Expansion Anchors		Crew	Daily Output	Labor-Hours	Unit	Material	2007 Bare Costs Labor	Equipment	Total	Total Incl O&P
5400	#14, 2" long	1 Carp	160	.050	Ea.	.60	1.84		2.44	3.52
5500	#16, 2" long		150	.053		.63	1.96		2.59	3.74
5600	#20, 2" long		140	.057		1.01	2.10		3.11	4.38
5700	Lag screw shields, 1/4" diameter, short		90	.089		.48	3.26		3.74	5.65
5900	3/8" diameter, short		85	.094		.88	3.45		4.33	6.35
6100	1/2" diameter, short		80	.100		1.22	3.67		4.89	7.05
6300	3/4" diameter, short		70	.114		3.44	4.19		7.63	10.35
6600	Lead, #6 & #8, 3/4" long		260	.031		.17	1.13		1.30	1.95
6700	#10 - #14, 1-1/2" long		200	.040		.25	1.47		1.72	2.57
6800	#16 & #18, 1-1/2" long		160	.050		.34	1.84		2.18	3.23
6900	Plastic, #6 & #8, 3/4" long		260	.031		.12	1.13		1.25	1.89
7100	#10 & #12, 1" long	↓	220	.036	↓	.16	1.33		1.49	2.26
8000	Wedge anchors, not including layout or drilling									
8050	Carbon steel, 1/4" diameter, 1-3/4" long	1 Carp	150	.053	Ea.	.48	1.96		2.44	3.58
8200	5" long		140	.057		1.28	2.10		3.38	4.68
8250	1/2" diameter, 2-3/4" long		140	.057		1.11	2.10		3.21	4.49
8300	7" long		125	.064		1.89	2.35		4.24	5.75
8350	5/8" diameter, 3-1/2" long		130	.062		2.19	2.26		4.45	5.95
8400	8-1/2" long		115	.070		4.67	2.55		7.22	9.15
8450	3/4" diameter, 4-1/4" long		115	.070		2.65	2.55		5.20	6.90
8500	10" long		95	.084		6	3.09		9.09	11.40
8550	1" diameter, 6" long		100	.080		8.90	2.94		11.84	14.30
8575	9" long		85	.094		11.55	3.45		15	18.10
8600	12" long		75	.107		12.45	3.91		16.36	19.80
8950	Self-drilling concrete screw, hex washer head, 3/16" dia x 1-3/4" long		300	.027		.37	.98		1.35	1.93
8960	2-1/4" long		250	.032		.56	1.17		1.73	2.45
8970	Phillips flat head, 3/16" dia x 1-3/4" long		300	.027		.38	.98		1.36	1.94
8980	2-1/4" long	↓	250	.032	↓	.56	1.17		1.73	2.45

05 05 23.30 Lag Screws

		Crew	Daily Output	Labor-Hours	Unit	Material	Labor	Equipment	Total	Total Incl O&P
0010	**LAG SCREWS**									
0020	Steel, 1/4" diameter, 2" long	1 Carp	200	.040	Ea.	.10	1.47		1.57	2.40
0100	3/8" diameter, 3" long		150	.053		.27	1.96		2.23	3.35
0200	1/2" diameter, 3" long		130	.062		.44	2.26		2.70	4
0300	5/8" diameter, 3" long	↓	120	.067	↓	.86	2.45		3.31	4.76

05 05 23.35 Machine Screws

		Crew	Daily Output	Labor-Hours	Unit	Material	Labor	Equipment	Total	Total Incl O&P
0010	**MACHINE SCREWS**									
0020	Steel, round head, #8 x 1" long	1 Carp	4.80	1.667	C	2.63	61		63.63	98.50
0110	#8 x 2" long		2.40	3.333		5.75	122		127.75	197
0200	#10 x 1" long		4	2		3.75	73.50		77.25	118
0300	#10 x 2" long	↓	2	4	↓	7	147		154	237

05 05 23.40 Machinery Anchors

		Crew	Daily Output	Labor-Hours	Unit	Material	Labor	Equipment	Total	Total Incl O&P
0010	**MACHINERY ANCHORS**, heavy duty, incl. sleeve, floating base nut,									
0020	Lower stud & coupling nut, fiber plug, connecting stud, washer & nut.									
0030	For flush mounted embedment in poured concrete heavy equip. pads.									
0200	Stud & bolt, 1/2" diameter	E-16	40	.400	Ea.	61	16.95	2.88	80.83	101
0300	5/8" diameter		35	.457		67.50	19.35	3.29	90.14	113
0500	3/4" diameter		30	.533		78	22.50	3.84	104.34	131
0600	7/8" diameter		25	.640		85	27	4.61	116.61	148
0800	1" diameter		20	.800		89.50	34	5.75	129.25	166
0900	1-1/4" diameter	↓	15	1.067	↓	119	45	7.70	171.70	220

05 05 23.50 Powder Actuated Tools and Fasteners

0010	**POWDER ACTUATED TOOLS & FASTENERS**									

05 05 Common Work Results for Metals

05 05 23 – Metal Fastenings

05 05 23.50 Powder Actuated Tools and Fasteners	Crew	Daily Output	Labor-Hours	Unit	Material	2007 Bare Costs Labor	Equipment	Total	Total Incl O&P
0020	Stud driver, .22 caliber, buy, minimum				Ea.	355			355	390
0100	Maximum				"	575			575	630
0300	Powder charges for above, low velocity				C	18.30			18.30	20
0400	Standard velocity					26			26	28.50
0600	Drive pins & studs, 1/4" & 3/8" diam., to 3" long, minimum	1 Carp	4.80	1.667		13.70	61		74.70	111
0700	Maximum	"	4	2		53.50	73.50		127	173
0800	Pneumatic stud driver for 1/8" diameter studs				Ea.	2,475			2,475	2,725
0900	Drive pins for above, 1/2" to 3/4" long	1 Carp	1	8	M	565	294		859	1,075

05 05 23.55 Rivets

		Crew	Daily Output	Labor-Hours	Unit	Material	Labor	Equipment	Total	Total Incl O&P
0010	**RIVETS**									
0100	Aluminum rivet & mandrel, 1/2" grip length x 1/8" diameter	1 Carp	4.80	1.667	C	5.50	61		66.50	102

05 12 Structural Steel Framing

05 12 23 – Structural Steel for Buildings

05 12 23.60 Pipe Support Framing

		Crew	Daily Output	Labor-Hours	Unit	Material	Labor	Equipment	Total	Total Incl O&P
0010	**PIPE SUPPORT FRAMING**									
0020	Under 10#/L.F.	E-4	3900	.008	Lb.	1.37	.34	.03	1.74	2.16
0200	10.1 to 15#/L.F.		4300	.007		1.35	.31	.03	1.69	2.08
0400	15.1 to 20#/L.F.		4800	.007		1.33	.28	.02	1.63	2.01
0600	Over 20#/L.F.		5400	.006		1.31	.25	.02	1.58	1.91

05 35 Raceway Decking Assemblies

05 35 13 – Cellular Decking

05 35 13.50 Cellular Decking

		Crew	Daily Output	Labor-Hours	Unit	Material	Labor	Equipment	Total	Total Incl O&P
0010	**CELLULAR DECKING**									
0200	Cellular units, galv, 2" deep, 20-20 gauge, over 15 squares	E-4	1460	.022	S.F.	6.15	.92	.08	7.15	8.50
0250	18-20 gauge		1420	.023		7	.94	.08	8.02	9.50
0300	18-18 gauge		1390	.023		7.15	.96	.08	8.19	9.75
0320	16-18 gauge		1360	.024		8.55	.98	.08	9.61	11.25
0340	16-16 gauge		1330	.024		9.50	1.01	.09	10.60	12.35
0400	3" deep, galvanized, 20-20 gauge		1375	.023		6.75	.97	.08	7.80	9.30
0500	18-20 gauge		1350	.024		8.15	.99	.09	9.23	10.90
0600	18-18 gauge		1290	.025		8.15	1.04	.09	9.28	10.95
0700	16-18 gauge		1230	.026		9.20	1.09	.09	10.38	12.15
0800	16-16 gauge		1150	.028		10	1.16	.10	11.26	13.20
1000	4-1/2" deep, galvanized, 20-18 gauge		1100	.029		9.40	1.22	.10	10.72	12.70
1100	18-18 gauge		1040	.031		9.35	1.29	.11	10.75	12.75
1200	16-18 gauge		980	.033		10.55	1.37	.12	12.04	14.20
1300	16-16 gauge		935	.034		11.50	1.43	.12	13.05	15.40
1900	For multi-story or congested site, add						50%			

Estimating Tips

06 05 00 Common Work Results for Wood, Plastics, and Composites

- Common to any wood-framed structure are the accessory connector items such as screws, nails, adhesives, hangers, connector plates, straps, angles, and hold-downs. For typical wood-framed buildings, such as residential projects, the aggregate total for these items can be significant, especially in areas where seismic loading is a concern. For floor and wall framing, the material cost is based on 10 to 25 lbs. per MBF. Hold-downs, hangers, and other connectors should be taken off by the piece.

06 10 00 Carpentry

- Lumber is a traded commodity and therefore sensitive to supply and demand in the marketplace. Even in "budgetary" estimating of wood-framed projects, it is advisable to call local suppliers for the latest market pricing.
- Common quantity units for wood-framed projects are "thousand board feet" (MBF). A board foot is a volume of wood, 1" x 1' x 1', or 144 cubic inches. Board-foot quantities are generally calculated using nominal material dimensions—dressed sizes are ignored. Board foot per lineal foot of any stick of lumber can be calculated by dividing the nominal cross-sectional area by 12. As an example, 2,000 lineal feet of 2 x 12 equates to 4 MBF by dividing the nominal area, 2 x 12, by 12, which equals 2, and multiplying by 2,000 to give 4,000 board feet. This simple rule applies to all nominal dimensioned lumber.
- Waste is an issue of concern at the quantity takeoff for any area of construction. Framing lumber is sold in even foot lengths, i.e., 10', 12', 14', 16', and depending on spans, wall heights and the grade of lumber, waste is inevitable. A rule of thumb for lumber waste is 5% to 10% depending on material quality and the complexity of the framing.
- Wood in various forms and shapes is used in many projects, even where the main structural framing is steel, concrete, or masonry. Plywood as a back-up partition material and 2x boards used as blocking and cant strips around roof edges are two common examples. The estimator should ensure that the costs of all wood materials are included in the final estimate.

06 20 00 Finish Carpentry

- It is necessary to consider the grade of workmanship when estimating labor costs for erecting millwork and interior finish. In practice, there are three grades: premium, custom, and economy. The RSMeans daily output for base and case moldings is in the range of 200 to 250 L.F. per carpenter per day. This is appropriate for most average custom-grade projects. For premium projects, an adjustment to productivity of 25% to 50% should be made depending on the complexity of the job.

Reference Numbers

Reference numbers are shown in shaded boxes at the beginning of some major classifications. These numbers refer to related items in the Reference Section. The reference information may be an estimating procedure, an alternate pricing method, or technical information.

Note: Not all subdivisions listed here necessarily appear in this publication.

06 05 23 – Wood, Plastic, and Composite Fastenings

06 05 23.10 Nails		Crew	Daily Output	Labor-Hours	Unit	Material	2007 Bare Costs Labor	Equipment	Total	Total Incl O&P
0010	**NAILS**									
0020	Copper nails, plain				Lb.	6.60			6.60	7.25
0400	Stainless steel, plain					5.95			5.95	6.55
0500	Box, 3d to 20d, bright					.76			.76	.84
0520	Galvanized					.88			.88	.97
0600	Common, 3d to 60d, plain					.82			.82	.90
0700	Galvanized					1.06			1.06	1.17
0800	Aluminum					4.45			4.45	4.90
1000	Annular or spiral thread, 4d to 60d, plain					1.88			1.88	2.07
1200	Galvanized					1.95			1.95	2.15
1400	Drywall nails, plain					.79			.79	.87
1600	Galvanized					1.59			1.59	1.75
1800	Finish nails, 4d to 10d, plain					.91			.91	1
2000	Galvanized					1.51			1.51	1.66
2100	Aluminum					4.10			4.10	4.51
2300	Flooring nails, hardened steel, 2d to 10d, plain					1.61			1.61	1.77
2400	Galvanized					2.25			2.25	2.48
2500	Gypsum lath nails, 1-1/8", 13 ga. flathead, blued					1.54			1.54	1.69
2600	Masonry nails, hardened steel, 3/4" to 3" long, plain					1.61			1.61	1.77
2700	Galvanized					2			2	2.20
5000	Add to prices above for cement coating					.10			.10	.11
5200	Zinc or tin plating					.13			.13	.14
5500	Vinyl coated sinkers, 8d to 16d					.57			.57	.63

06 05 23.20 Pneumatic Nails		Crew	Daily Output	Labor-Hours	Unit	Material	2007 Bare Costs Labor	Equipment	Total	Total Incl O&P
0010	**PNEUMATIC NAILS**									
0020	Framing, per carton of 5000, 2"				Ea.	37			37	41
0100	2-3/8"					42.50			42.50	46.50
0200	Per carton of 4000, 3"					37.50			37.50	41.50
0300	3-1/4"					40			40	44
0400	Per carton of 5000, 2-3/8", galv.					57.50			57.50	63.50
0500	Per carton of 4000, 3", galv.					65			65	71.50
0600	3-1/4", galv.					80.50			80.50	88.50
0700	Roofing, per carton of 7200, 1"					34			34	37
0800	1-1/4"					31.50			31.50	34.50
0900	1-1/2"					36.50			36.50	40
1000	1-3/4"					45.50			45.50	50

06 05 23.40 Sheet Metal Screws		Crew	Daily Output	Labor-Hours	Unit	Material	2007 Bare Costs Labor	Equipment	Total	Total Incl O&P
0010	**SHEET METAL SCREWS**									
0020	Steel, standard, #8 x 3/4", plain				C	3.29			3.29	3.62
0100	Galvanized					3.44			3.44	3.78
0300	#10 x 1", plain					4.49			4.49	4.94
0400	Galvanized					4.60			4.60	5.05
0600	With washers, #14 x 1", plain					10.20			10.20	11.20
0700	Galvanized					10.50			10.50	11.55
0900	#14 x 2", plain					18			18	19.80
1000	Galvanized					18.15			18.15	19.95
1500	Self-drilling, with washers, (pinch point) #8 x 3/4", plain					7.10			7.10	7.80
1600	Galvanized					7.20			7.20	7.90
1800	#10 x 3/4", plain					9.30			9.30	10.20
1900	Galvanized					9.40			9.40	10.35
3000	Stainless steel w/aluminum or neoprene washers, #14 x 1", plain					18.35			18.35	20
3100	#14 x 2", plain					25			25	27.50

06 05 Common Work Results for Wood, Plastics and Composites

06 05 23 – Wood, Plastic, and Composite Fastenings

06 05 23.50 Wood Screws	Crew	Daily Output	Labor-Hours	Unit	Material	2007 Bare Costs Labor	Equipment	Total	Total Incl O&P
0010 **WOOD SCREWS**									
0020 Steel, #8 x 1" long				C	3.29			3.29	3.62
0100 Brass					11.50			11.50	12.65
0200 #8, 2" long, steel					5.60			5.60	6.15
0300 Brass					19.50			19.50	21.50
0400 #10, 1" long, steel					5.90			5.90	6.50
0500 Brass					14			14	15.40
0600 #10, 2" long, steel					6.30			6.30	6.90
0700 Brass					23			23	25.50
0800 #10, 3" long, steel					10.50			10.50	11.55
1000 #12, 2" long, steel					7.50			7.50	8.25
1100 Brass					30			30	33
1500 #12, 3" long, steel					11.50			11.50	12.65
2000 #12, 4" long, steel					43			43	47.50

06 16 Sheathing

06 16 36 – Wood Panel Product Sheathing

06 16 36.10 Sheathing	Crew	Daily Output	Labor-Hours	Unit	Material	2007 Bare Costs Labor	Equipment	Total	Total Incl O&P
0010 **SHEATHING**, plywood on roofs									
0012 Plywood on roofs, CDX									
0030 5/16" thick	2 Carp	1600	.010	S.F.	.62	.37		.99	1.25
0035 Pneumatic nailed		1952	.008		.62	.30		.92	1.15
0050 3/8" thick		1525	.010		.44	.39		.83	1.08
0055 Pneumatic nailed		1860	.009		.44	.32		.76	.97
0100 1/2" thick		1400	.011		.47	.42		.89	1.17
0105 Pneumatic nailed		1708	.009		.47	.34		.81	1.06
0200 5/8" thick		1300	.012		.59	.45		1.04	1.35
0205 Pneumatic nailed		1586	.010		.59	.37		.96	1.23
0300 3/4" thick		1200	.013		.71	.49		1.20	1.54
0305 Pneumatic nailed		1464	.011		.71	.40		1.11	1.40
0500 Plywood on walls with exterior CDX, 3/8" thick		1200	.013		.44	.49		.93	1.24
0505 Pneumatic nailed		1488	.011		.44	.39		.83	1.09
0600 1/2" thick		1125	.014		.47	.52		.99	1.33
0605 Pneumatic nailed		1395	.011		.47	.42		.89	1.18
0700 5/8" thick		1050	.015		.59	.56		1.15	1.52
0705 Pneumatic nailed		1302	.012		.59	.45		1.04	1.35
0800 3/4" thick		975	.016		.71	.60		1.31	1.72
0805 Pneumatic nailed		1209	.013		.71	.49		1.20	1.54

Division Notes

	CREW	DAILY OUTPUT	LABOR-HOURS	UNIT	2007 BARE COSTS				TOTAL INCL O&P
					MAT.	LABOR	EQUIP.	TOTAL	

Estimating Tips

07 10 00 Dampproofing and Waterproofing

- Be sure of the job specifications before pricing this subdivision. The difference in cost between waterproofing and dampproofing can be great. Waterproofing will hold back standing water. Dampproofing prevents the transmission of water vapor. Also included in this section are vapor retarding membranes.

07 20 00 Thermal Protection

- Insulation and fireproofing products are measured by area, thickness, volume or R value. Specifications may give only what the specific R value should be in a certain situation. The estimator may need to choose the type of insulation to meet that R value.

07 30 00 Steep Slope Roofing
07 40 00 Roofing and Siding Panels

- Many roofing and siding products are bought and sold by the square. One square is equal to an area that measures 100 square feet.

This simple change in unit of measure could create a large error if the estimator is not observant. Accessories necessary for a complete installation must be figured into any calculations for both material and labor.

07 50 00 Membrane Roofing
07 60 00 Flashing and Sheet Metal
07 70 00 Roofing and Wall Specialties and Accessories

- The items in these subdivisions compose a roofing system. No one component completes the installation, and all must be estimated. Built-up or single-ply membrane roofing systems are made up of many products and installation trades. Wood blocking at roof perimeters or penetrations, parapet coverings, reglets, roof drains, gutters, downspouts, sheet metal flashing, skylights, smoke vents, and roof hatches all need to be considered along with the roofing material. Several different installation trades will need to work together on the roofing system. Inherent difficulties in the scheduling and coordination of various trades must be accounted for when estimating labor costs.

07 90 00 Joint Protection

- To complete the weather-tight shell, the sealants and caulkings must be estimated. Where different materials meet—at expansion joints, at flashing penetrations, and at hundreds of other locations throughout a construction project—they provide another line of defense against water penetration. Often, an entire system is based on the proper location and placement of caulking or sealants. The detailed drawings that are included as part of a set of architectural plans show typical locations for these materials. When caulking or sealants are shown at typical locations, this means the estimator must include them for all the locations where this detail is applicable. Be careful to keep different types of sealants separate, and remember to consider backer rods and primers if necessary.

Reference Numbers

Reference numbers are shown in shaded boxes at the beginning of some major classifications. These numbers refer to related items in the Reference Section. The reference information may be an estimating procedure, an alternate pricing method, or technical information.

Note: Not all subdivisions listed here necessarily appear in this publication.

07 84 13 – Penetration Firestopping

07 84 13.10 Firestopping		Crew	Daily Output	Labor-Hours	Unit	Material	2007 Bare Costs Labor	Equipment	Total	Total Incl O&P
0010	**FIRESTOPPING**	R078413-30								
0100	Metallic piping, non insulated									
0110	Through walls, 2" diameter	1 Carp	16	.500	Ea.	10.75	18.35		29.10	40.50
0120	4" diameter		14	.571		16.45	21		37.45	50.50
0130	6" diameter		12	.667		22	24.50		46.50	62.50
0140	12" diameter		10	.800		39	29.50		68.50	88.50
0150	Through floors, 2" diameter		32	.250		6.55	9.20		15.75	21.50
0160	4" diameter		28	.286		9.40	10.50		19.90	26.50
0170	6" diameter		24	.333		12.35	12.25		24.60	32.50
0180	12" diameter		20	.400		21	14.70		35.70	46
0190	Metallic piping, insulated									
0200	Through walls, 2" diameter	1 Carp	16	.500	Ea.	15.25	18.35		33.60	45.50
0210	4" diameter		14	.571		21	21		42	55.50
0220	6" diameter		12	.667		26.50	24.50		51	67.50
0230	12" diameter		10	.800		43.50	29.50		73	93.50
0240	Through floors, 2" diameter		32	.250		11.05	9.20		20.25	26.50
0250	4" diameter		28	.286		13.90	10.50		24.40	31.50
0260	6" diameter		24	.333		16.85	12.25		29.10	37.50
0270	12" diameter		20	.400		21	14.70		35.70	46
0280	Non metallic piping, non insulated									
0290	Through walls, 2" diameter	1 Carp	12	.667	Ea.	44.50	24.50		69	87
0300	4" diameter		10	.800		56	29.50		85.50	107
0310	6" diameter		8	1		77.50	36.50		114	143
0330	Through floors, 2" diameter		16	.500		34.50	18.35		52.85	66.50
0340	4" diameter		6	1.333		43	49		92	124
0350	6" diameter		6	1.333		51.50	49		100.50	133
0370	Ductwork, insulated & non insulated, round									
0380	Through walls, 6" diameter	1 Carp	12	.667	Ea.	22.50	24.50		47	62.50
0390	12" diameter		10	.800		45	29.50		74.50	95
0400	18" diameter		8	1		73	36.50		109.50	138
0410	Through floors, 6" diameter		16	.500		12.35	18.35		30.70	42
0420	12" diameter		14	.571		22.50	21		43.50	57
0430	18" diameter		12	.667		39.50	24.50		64	81.50
0440	Ductwork, insulated & non insulated, rectangular									
0450	With stiffener/closure angle, through walls, 6" x 12"	1 Carp	8	1	Ea.	18.75	36.50		55.25	77.50
0460	12" x 24"		6	1.333		25	49		74	104
0470	24" x 48"		4	2		71	73.50		144.50	192
0480	With stiffener/closure angle, through floors, 6" x 12"		10	.800		10.10	29.50		39.60	56.50
0490	12" x 24"		8	1		18.20	36.50		54.70	77
0500	24" x 48"		6	1.333		36	49		85	116
0510	Multi trade openings									
0520	Through walls, 6" x 12"	1 Carp	2	4	Ea.	39.50	147		186.50	273
0530	12" x 24"	"	1	8		158	294		452	630
0540	24" x 48"	2 Carp	1	16		635	585		1,220	1,600
0550	48" x 96"	"	.75	21.333		2,550	785		3,335	4,025
0560	Through floors, 6" x 12"	1 Carp	2	4		39.50	147		186.50	273
0570	12" x 24"	"	1	8		158	294		452	630
0580	24" x 48"	2 Carp	.75	21.333		635	785		1,420	1,925
0590	48" x 96"	"	.50	32		2,550	1,175		3,725	4,625
0600	Structural penetrations, through walls									
0610	Steel beams, W8 x 10	1 Carp	8	1	Ea.	24.50	36.50		61	84
0620	W12 x 14		6	1.333		39.50	49		88.50	120
0630	W21 x 44		5	1.600		78.50	58.50		137	178

07 84 Firestopping

07 84 13 – Penetration Firestopping

07 84 13.10 Firestopping

		Crew	Daily Output	Labor-Hours	Unit	Material	2007 Bare Costs Labor	Equipment	Total	Total Incl O&P
0640	W36 x 135	1 Carp	3	2.667	Ea.	191	98		289	360
0650	Bar joists, 18" deep		6	1.333		36	49		85	116
0660	24" deep		6	1.333		45	49		94	126
0670	36" deep		5	1.600		67.50	58.50		126	166
0680	48" deep		4	2		78.50	73.50		152	201
0690	Construction joints, floor slab at exterior wall									
0700	Precast, brick, block or drywall exterior									
0710	2" wide joint	1 Carp	125	.064	L.F.	5.60	2.35		7.95	9.85
0720	4" wide joint	"	75	.107	"	11.25	3.91		15.16	18.45
0730	Metal panel, glass or curtain wall exterior									
0740	2" wide joint	1 Carp	40	.200	L.F.	13.30	7.35		20.65	26
0750	4" wide joint	"	25	.320	"	18.15	11.75		29.90	38.50
0760	Floor slab to drywall partition									
0770	Flat joint	1 Carp	100	.080	L.F.	5.50	2.94		8.44	10.60
0780	Fluted joint		50	.160		11.25	5.85		17.10	21.50
0790	Etched fluted joint		75	.107		7.30	3.91		11.21	14.15
0800	Floor slab to concrete/masonry partition									
0810	Flat joint	1 Carp	75	.107	L.F.	12.35	3.91		16.26	19.70
0820	Fluted joint	"	50	.160	"	14.60	5.85		20.45	25
0830	Concrete/CMU wall joints									
0840	1" wide	1 Carp	100	.080	L.F.	6.75	2.94		9.69	11.95
0850	2" wide		75	.107		12.35	3.91		16.26	19.70
0860	4" wide		50	.160		23.50	5.85		29.35	35
0870	Concrete/CMU floor joints									
0880	1" wide	1 Carp	200	.040	L.F.	3.37	1.47		4.84	6
0890	2" wide		150	.053		6.20	1.96		8.16	9.85
0900	4" wide		100	.080		11.80	2.94		14.74	17.50

07 92 Joint Sealants

07 92 10 – Caulking and Sealants

07 92 10.10 Caulking and Sealants

		Crew	Daily Output	Labor-Hours	Unit	Material	2007 Bare Costs Labor	Equipment	Total	Total Incl O&P
0010	**CAULKING AND SEALANTS**									
3200	Polyurethane, 1 or 2 component				Gal.	52.50			52.50	57.50
3300	Cartridges				"	49			49	54
3500	Bulk, in place, 1/4" x 1/4"	1 Bric	150	.053	L.F.	.17	2.03		2.20	3.28
3600	1/2" x 1/4"		145	.055		.34	2.10		2.44	3.56
3800	3/4" x 3/8", 68 L.F./gal.		130	.062		.77	2.34		3.11	4.41
3900	1" x 1/2"		110	.073		1.36	2.77		4.13	5.70

Division Notes

	CREW	DAILY OUTPUT	LABOR-HOURS	UNIT	2007 BARE COSTS				TOTAL INCL O&P
					MAT.	LABOR	EQUIP.	TOTAL	

Estimating Tips

08 10 00 Doors and Frames

- Most metal doors and frames look alike, but there may be significant differences among them. When estimating these items be sure to choose the line item that most closely compares to the specification or door schedule requirements regarding:
  - type of metal
  - metal gauge
  - door core material
  - fire rating
  - finish
- Wood and plastic doors vary considerably in price. The primary determinant is the veneer material. Lauan, birch, and oak are the most common veneers. Other variables include the following:
  - hollow or solid core
  - fire rating
  - flush or raised panel
  - finish

08 30 00 Specialty Doors and Frames

- There are many varieties of special doors, and they are usually priced per each. Add frames, hardware, or operators required for a complete installation.

08 40 00 Entrances, Storefronts, and Curtain Walls

- Glazed curtain walls consist of the metal tube framing and the glazing material. The cost data in this subdivision is presented for the metal tube framing alone or the composite wall. If your estimate requires a detailed takeoff of the framing, be sure to add the glazing cost.

08 50 00 Windows

- Most metal windows are delivered preglazed. However, some metal windows are priced without glass. Refer to 08 80 00 Glazing for glass pricing. The grade C indicates commercial grade windows, usually ASTM C-35.
- All wood windows are priced preglazed. The two glazing options priced are single pane float glass and insulating glass 1/2" thick. Add the cost of screens and grills if required.

08 70 00 Hardware

- Hardware costs add considerably to the cost of a door. The most efficient method to determine the hardware requirements for a project is to review the door schedule.

- Door hinges are priced by the pair, with most doors requiring 1-1/2 pairs per door. The hinge prices do not include installation labor because it is included in door installation. Hinges are classified according to the frequency of use.

08 80 00 Glazing

- Different openings require different types of glass. The three most common types are:
  - float
  - tempered
  - insulating
- Most exterior windows are glazed with insulating glass. Entrance doors and window walls, where the glass is less than 18" from the floor, are generally glazed with tempered glass. Interior windows and some residential windows are glazed with float glass.

Reference Numbers

Reference numbers are shown in shaded boxes at the beginning of some major classifications. These numbers refer to related items in the Reference Section. The reference information may be an estimating procedure, an alternate pricing method, or technical information.

Note: Not all subdivisions listed here necessarily appear in this publication.

08 74 Access Control Hardware

08 74 19 – Biometric Identity Access Control Hardware

08 74 19.50 Biometric Identity Access	Crew	Daily Output	Labor-Hours	Unit	Material	2007 Bare Costs Labor	Equipment	Total	Total Incl O&P
0010 **BIOMETRIC IDENTITY ACCESS**									
0220 Hand geometry scanner, mem of 512 users, excl striker/powr	1 Elec	3	2.667	Ea.	1,725	117		1,842	2,075
0230 Memory upgrade for, adds 9,700 user profiles		8	1		225	44		269	315
0240 Adds 32,500 user profiles		8	1		525	44		569	645
0250 Prison type, memory of 256 users, excl striker, power		3	2.667		2,175	117		2,292	2,550
0260 Memory upgrade for, adds 3,300 user profiles		8	1		180	44		224	264
0270 Adds 9,700 user profiles		8	1		360	44		404	460
0280 Adds 27,900 user profiles		8	1		505	44		549	620
0290 All weather, mem of 512 users, excl striker/pwr		3	2.667		3,225	117		3,342	3,725
0300 Facial & fingerprint scanner, combination unit, excl striker/power		3	2.667		4,200	117		4,317	4,800
0310 Access for, for initial setup, excl striker/power		3	2.667		1,000	117		1,117	1,275

Estimating Tips

General

- Room Finish Schedule: A complete set of plans should contain a room finish schedule. If one is not available, it would be well worth the time and effort to obtain one.

09 20 00 Plaster and Gypsum Board

- Lath is estimated by the square yard plus a 5% allowance for waste. Furring, channels, and accessories are measured by the linear foot. An extra foot should be allowed for each accessory miter or stop.

- Plaster is also estimated by the square yard. Deductions for openings vary by preference, from zero deduction to 50% of all openings over 2 feet in width. The estimator should allow one extra square foot for each linear foot of horizontal interior or exterior angle located below the ceiling level. Also, double the areas of small radius work.

- Drywall accessories, studs, track, and acoustical caulking are all measured by the linear foot. Drywall taping is figured by the square foot. Gypsum wallboard is estimated by the square foot. No material deductions should be made for door or window openings under 32 S.F.

09 60 00 Flooring

- Tile and terrazzo areas are taken off on a square foot basis. Trim and base materials are measured by the linear foot. Accent tiles are listed per each. Two basic methods of installation are used. Mud set is approximately 30% more expensive than thin set. In terrazzo work, be sure to include the linear footage of embedded decorative strips, grounds, machine rubbing, and power cleanup.

- Wood flooring is available in strip, parquet, or block configuration. The latter two types are set in adhesives with quantities estimated by the square foot. The laying pattern will influence labor costs and material waste. In addition to the material and labor for laying wood floors, the estimator must make allowances for sanding and finishing these areas unless the flooring is prefinished.

- Sheet flooring is measured by the square yard. Roll widths vary, so consideration should be given to use the most economical width, as waste must be figured into the total quantity. Consider also the installation methods available, direct glue down or stretched.

09 70 00 Wall Finishes

- Wall coverings are estimated by the square foot. The area to be covered is measured, length by height of wall above baseboards, to calculate the square footage of each wall. This figure is divided by the number of square feet in the single roll which is being used. Deduct, in full, the areas of openings such as doors and windows. Where a pattern match is required allow 25%-30% waste.

09 80 00 Acoustic Treatment

- Acoustical systems fall into several categories. The takeoff of these materials should be by the square foot of area with a 5% allowance for waste. Do not forget about scaffolding, if applicable, when estimating these systems.

09 90 00 Painting and Coating

- A major portion of the work in painting involves surface preparation. Be sure to include cleaning, sanding, filling, and masking costs in the estimate.

- Protection of adjacent surfaces is not included in painting costs. When considering the method of paint application, an important factor is the amount of protection and masking required. These must be estimated separately and may be the determining factor in choosing the method of application.

Reference Numbers

Reference numbers are shown in shaded boxes at the beginning of some major classifications. These numbers refer to related items in the Reference Section. The reference information may be an estimating procedure, an alternate pricing method, or technical information.

Note: Not all subdivisions listed here necessarily appear in this publication.

09 22 Supports for Plaster and Gypsum Board

09 22 03 – Fastening Methods for Finishes

09 22 03.20 Drilling Plaster/Drywall	Crew	Daily Output	Labor-Hours	Unit	Material	2007 Bare Costs Labor	Equipment	Total	Total Incl O&P
0010 **DRILLING PLASTER/DRYWALL**									
1100 Drilling & layout for drywall/plaster walls, up to 1" deep, no anchor									
1200 Holes, 1/4" diameter	1 Carp	150	.053	Ea. *	.01	1.96		1.97	3.06
1300 3/8" diameter		140	.057		.01	2.10		2.11	3.28
1400 1/2" diameter		130	.062		.01	2.26		2.27	3.53
1500 3/4" diameter		120	.067		.03	2.45		2.48	3.84
1600 1" diameter		110	.073		.04	2.67		2.71	4.20
1700 1-1/4" diameter		100	.080		.05	2.94		2.99	4.62
1800 1-1/2" diameter		90	.089		.08	3.26		3.34	5.20
1900 For ceiling installations, add						40%			

09 69 Access Flooring

09 69 13 – Rigid-Grid Access Flooring

09 69 13.10 Access Floors

09 69 13.10 Access Floors	Crew	Daily Output	Labor-Hours	Unit	Material	2007 Bare Costs Labor	Equipment	Total	Total Incl O&P
0010 **ACCESS FLOORS**									
0015 System pricing including panels, pedestals, stringers, and laminate cover									
0100 Computer room, greater than 6,000 S.F.	4 Carp	750	.043	S.F.	8.15	1.57		9.72	11.40
0110 Less than 6,000 S.F.	2 Carp	375	.043		9.40	1.57		10.97	12.80
0120 Office greater than 6,000 S.F.	4 Carp	1050	.030		4.30	1.12		5.42	6.45
0250 Panels, particle board or steel, 1250# load, no covering, under 6,000 S.F.	2 Carp	600	.027		4	.98		4.98	5.90
0300 Over 6,000 S.F.		640	.025		3.50	.92		4.42	5.30
0400 Aluminum, 24" panels		500	.032		30	1.17		31.17	35.50
0600 For carpet covering, add					8.30			8.30	9.10
0700 For vinyl floor covering, add					6.45			6.45	7.10
0900 For high pressure laminate covering, add					5.25			5.25	5.80
0910 For stringer system, add	2 Carp	1000	16		1.40	.59		10.36	11.70
0950 Office applications, steel or concrete panels,									
0960 no covering, over 6,000 S.F.	2 Carp	960	.017	S.F.	9.75	.61		10.36	11.70
1000 Machine cutouts after initial installation	1 Carp	50	.160	Ea.	4.87	5.85		10.72	14.50
1050 Pedestals, 6" to 12"	2 Carp	85	.188		7.70	6.90		14.60	19.20
1100 Air conditioning grilles, 4" x 12"	1 Carp	17	.471		64	17.25		81.25	97.50
1150 4" x 18"	"	14	.571		87.50	21		108.50	129
1200 Approach ramps, minimum	2 Carp	60	.267	S.F.	24.50	9.80		34.30	42.50
1300 Maximum	"	40	.400	"	32.50	14.70		47.20	58.50
1500 Handrail, 2 rail, aluminum	1 Carp	15	.533	L.F.	96.50	19.55		116.05	137

Estimating Tips

General

- The items in this division are usually priced per square foot or each.
- Many items in Division 10 require some type of support system or special anchors that are not usually furnished with the item. The required anchors must be added to the estimate in the appropriate division.
- Some items in Division 10, such as lockers, may require assembly before installation. Verify the amount of assembly required. Assembly can often exceed installation time.

10 20 00 Interior Specialties

- Support angles and blocking are not included in the installation of toilet compartments, shower/dressing compartments, or cubicles. Appropriate line items from Divisions 5 or 6 may need to be added to support the installations.
- Toilet partitions are priced by the stall. A stall consists of a side wall, pilaster, and door with hardware. Toilet tissue holders and grab bars are extra.
- The required acoustical rating of a folding partition can have a significant impact on costs. Verify the sound transmission coefficient rating of the panel priced to the specification requirements.

- Grab bar installation does not include supplemental blocking or backing to support the required load. When grab bars are installed at an existing facility, provisions must be made to attach the grab bars to solid structure.

Reference Numbers

Reference numbers are shown in shaded boxes at the beginning of some major classifications. These numbers refer to related items in the Reference Section. The reference information may be an estimating procedure, an alternate pricing method, or technical information.

Note: Not all subdivisions listed here necessarily appear in this publication.

Division 10 - Specialties

10 28 Toilet, Bath, and Laundry Accessories

10 28 16 – Bath Accessories

10 28 16.20 Medicine Cabinets	Crew	Daily Output	Labor-Hours	Unit	Material	2007 Bare Costs Labor	Equipment	Total	Total Incl O&P
0010 **MEDICINE CABINETS**									
0020 With mirror, st. st. frame, 16" x 22", unlighted	1 Carp	14	.571	Ea.	76.50	21		97.50	117
0100 Wood frame		14	.571		106	21		127	150
0300 Sliding mirror doors, 20" x 16" x 4-3/4", unlighted		7	1.143		95	42		137	171
0400 24" x 19" x 8-1/2", lighted		5	1.600		150	58.50		208.50	257
0600 Triple door, 30" x 32", unlighted, plywood body		7	1.143		225	42		267	315
0700 Steel body		7	1.143		296	42		338	390
0900 Oak door, wood body, beveled mirror, single door		7	1.143		145	42		187	225
1000 Double door		6	1.333		350	49		399	460
1200 Hotel cabinets, stainless, with lower shelf, unlighted		10	.800		190	29.50		219.50	255
1300 Lighted		5	1.600		282	58.50		340.50	400

Estimating Tips

General

- The items in this division are usually priced per square foot or each. Many of these items are purchased by the owner for installation by the contractor. Check the specifications for responsibilities and include time for receiving, storage, installation, and mechanical and electrical hook-ups in the appropriate divisions.

- Many items in Division 11 require some type of support system that is not usually furnished with the item. Examples of these systems include blocking for the attachment of casework and support angles for ceiling-hung projection screens. The required blocking or supports must be added to the estimate in the appropriate division.

- Some items in Division 11 may require assembly or electrical hookups. Verify the amount of assembly required or the need for a hard electrical connection and add the appropriate costs.

Reference Numbers

Reference numbers are shown in shaded boxes at the beginning of some major classifications. These numbers refer to related items in the Reference Section. The reference information may be an estimating procedure, an alternate pricing method, or technical information.

Note: Not all subdivisions listed here necessarily appear in this publication.

11 12 Parking Control Equipment

11 12 13 – Parking Key and Card Control Units

11 12 13.10 Parking Control Units

		Crew	Daily Output	Labor-Hours	Unit	Material	2007 Bare Costs Labor	Equipment	Total	Total Incl O&P
0010	**PARKING CONTROL UNITS**									
5100	Card reader	1 Elec	2	4	Ea.	1,800	176		1,976	2,225
5120	Proximity with customer display	2 Elec	1	16		5,500	700		6,200	7,100
6000	Parking control software, minimum	1 Elec	.50	16		21,900	700		22,600	25,200
6020	Maximum	"	.20	40		91,000	1,750		92,750	103,000

11 12 16 – Parking Ticket Dispensers

11 12 16.10 Ticket Dispensers

		Crew	Daily Output	Labor-Hours	Unit	Material	Labor	Equipment	Total	Total Incl O&P
0010	**TICKET DISPENSERS**									
5900	Ticket spitter with time/date stamp, standard	2 Elec	2	8	Ea.	6,275	350		6,625	7,425
5920	Mag stripe encoding	"	2	8	"	18,600	350		18,950	20,900

11 12 26 – Parking Fee Collection Equipment

11 12 26.13 Parking Fee Coin Collection Equipment

		Crew	Daily Output	Labor-Hours	Unit	Material	Labor	Equipment	Total	Total Incl O&P
0010	**PARKING FEE COIN COLLECTION EQUIPMENT**									
5200	Cashier booth, average	B-22	1	30	Ea.	9,450	1,025	245	10,720	12,200
5300	Collector station, pay on foot	2 Elec	.20	80		109,500	3,500		113,000	125,500
5320	Credit card only	"	.50	32		20,200	1,400		21,600	24,400

11 12 26.23 Fee Equipment

		Crew	Daily Output	Labor-Hours	Unit	Material	Labor	Equipment	Total	Total Incl O&P
0010	**FEE EQUIPMENT**									
5600	Fee computer	1 Elec	1.50	5.333	Ea.	13,200	234		13,434	14,900

11 12 33 – Parking Gates

11 12 33.13 Parking Gates

		Crew	Daily Output	Labor-Hours	Unit	Material	Labor	Equipment	Total	Total Incl O&P
0010	**PARKING GATES**									
5000	Barrier gate with programmable controller	2 Elec	3	5.333	Ea.	3,275	234		3,509	3,950
5020	Industrial		3	5.333		4,425	234		4,659	5,225
5500	Exit verifier		1	16		17,400	700		18,100	20,300
5700	Full sign, 4" letters	1 Elec	2	4		1,225	176		1,401	1,575
5800	Inductive loop	2 Elec	4	4		169	176		345	445
5950	Vehicle detector, microprocessor based	1 Elec	3	2.667		385	117		502	600

11 23 Commercial Laundry and Dry Cleaning Equipment

11 23 13 – Dry Cleaning Equipment

11 23 13.13 Dry Cleaning Equipment

		Crew	Daily Output	Labor-Hours	Unit	Material	Labor	Equipment	Total	Total Incl O&P
0010	**DRY CLEANING EQUIPMENT** Not incl. rough-in									
2000	Dry cleaners, electric, 20 lb. capacity	L-1	.20	80	Ea.	32,800	3,550		36,350	41,400
2050	25 lb. capacity		.17	94.118		44,200	4,175		48,375	55,000
2100	30 lb. capacity		.15	106		46,600	4,725		51,325	58,500
2150	60 lb. capacity		.09	177		72,000	7,875		79,875	91,000

11 23 19 – Finishing Equipment

11 23 19.13 Folders and Spreaders

		Crew	Daily Output	Labor-Hours	Unit	Material	Labor	Equipment	Total	Total Incl O&P
0010	**FOLDERS AND SPREADERS**									
3500	Folders, blankets & sheets, minimum	1 Elec	.17	47.059	Ea.	29,500	2,075		31,575	35,500
3700	King size with automatic stacker		.10	80		53,500	3,500		57,000	64,000
3800	For conveyor delivery, add		.45	17.778		6,000	780		6,780	7,750

11 23 23 – Commercial Ironing Equipment

11 23 23.13 Irons and Pressers

		Crew	Daily Output	Labor-Hours	Unit	Material	Labor	Equipment	Total	Total Incl O&P
0010	**IRONS AND PRESSERS**									
4500	Ironers, institutional, 110", single roll	1 Elec	.20	40	Ea.	29,100	1,750		30,850	34,600

11 23 Commercial Laundry and Dry Cleaning Equipment

11 23 26 – Commercial Washers and Extractors

11 23 26.13 Washers and Extractors

		Crew	Daily Output	Labor-Hours	Unit	Material	2007 Bare Costs Labor	Equipment	Total	Total Incl O&P
0010	**WASHERS AND EXTRACTORS**, not including rough-in									
6000	Combination washer/extractor, 20 lb. capacity	L-6	1.50	8	Ea.	5,100	355		5,455	6,125
6100	30 lb. capacity		.80	15		8,275	670		8,945	10,100
6200	50 lb. capacity		.68	17.647		9,700	785		10,485	11,900
6300	75 lb. capacity		.30	40		18,200	1,775		19,975	22,700
6350	125 lb. capacity		.16	75		24,400	3,350		27,750	31,800

11 23 33 – Coin-Operated Laundry Equipment

11 23 33.13 Coin Operated Laundry Equipment

		Crew	Daily Output	Labor-Hours	Unit	Material	Labor	Equipment	Total	Total Incl O&P
0010	**COIN OPERATED LAUNDRY EQUIPMENT**									
5290	Clothes washer									
5300	Commercial, coin operated, average	1 Plum	3	2.667	Ea.	1,100	119		1,219	1,400

11 26 Unit Kitchens

11 26 13 – Metal Unit Kitchens

11 26 13.10 Unit Kitchens

		Crew	Daily Output	Labor-Hours	Unit	Material	Labor	Equipment	Total	Total Incl O&P
0010	**UNIT KITCHENS**									
1500	Combination range, refrigerator and sink, 30" wide, minimum	L-1	2	8	Ea.	840	355		1,195	1,450
1550	Maximum		1	16		2,675	710		3,385	4,000
1570	60" wide, average		1.40	11.429		2,225	505		2,730	3,175
1590	72" wide, average		1.20	13.333		3,100	590		3,690	4,300
1600	Office model, 48" wide		2	8		2,250	355		2,605	3,000
1620	Refrigerator and sink only		2.40	6.667		2,950	296		3,246	3,700
1640	Combination range, refrigerator, sink, microwave									
1660	oven and ice maker	L-1	.80	20	Ea.	3,450	885		4,335	5,125

11 31 Residential Appliances

11 31 13 – Residential Kitchen Appliances

11 31 13.13 Cooking Equipment

		Crew	Daily Output	Labor-Hours	Unit	Material	Labor	Equipment	Total	Total Incl O&P
0010	**COOKING EQUIPMENT**									
0020	Cooking range, 30" free standing, 1 oven, minimum	2 Clab	10	1.600	Ea.	270	46		316	370
0050	Maximum		4	4		1,550	115		1,665	1,875
0700	Free-standing, 1 oven, 21" wide range, minimum		10	1.600		310	46		356	410
0750	21" wide, maximum		4	4		325	115		440	540
0900	Counter top cook tops, 4 burner, standard, minimum	1 Elec	6	1.333		215	58.50		273.50	325
0950	Maximum		3	2.667		585	117		702	820
1050	As above, but with grille and griddle attachment, minimum		6	1.333		520	58.50		578.50	655
1100	Maximum		3	2.667		870	117		987	1,125
1200	Induction cooktop, 30" wide		3	2.667		590	117		707	820
1250	Microwave oven, minimum		4	2		85.50	88		173.50	225
1300	Maximum		2	4		420	176		596	720

11 31 13.33 Kitchen Cleaning Equipment

		Crew	Daily Output	Labor-Hours	Unit	Material	Labor	Equipment	Total	Total Incl O&P
0010	**KITCHEN CLEANING EQUIPMENT**									
2750	Dishwasher, built-in, 2 cycles, minimum	L-1	4	4	Ea.	268	177		445	560
2800	Maximum		2	8		310	355		665	870
2950	4 or more cycles, minimum		4	4		285	177		462	580
2960	Average		4	4		380	177		557	680
3000	Maximum		2	8		530	355		885	1,125

11 31 Residential Appliances

11 31 13 – Residential Kitchen Appliances

11 31 13.43 Waste Disposal Equipment

		Crew	Daily Output	Labor-Hours	Unit	Material	2007 Bare Costs Labor	Equipment	Total	Total Incl O&P
0010	**WASTE DISPOSAL EQUIPMENT**									
3300	Garbage disposal, sink type, minimum	L-1	10	1.600	Ea.	50	71		121	161
3350	Maximum	"	10	1.600	"	159	71		230	281

11 31 13.53 Kitchen Ventilation Equipment

		Crew	Daily Output	Labor-Hours	Unit	Material	2007 Bare Costs Labor	Equipment	Total	Total Incl O&P
0010	**KITCHEN VENTILATION EQUIPMENT**									
4150	Hood for range, 2 speed, vented, 30" wide, minimum	L-3	5	3.200	Ea.	40.50	129		169.50	242
4200	Maximum		3	5.333		595	214		809	980
4300	42" wide, minimum		5	3.200		242	129		371	465
4330	Custom		5	3.200		670	129		799	930
4350	Maximum		3	5.333		820	214		1,034	1,225
4500	For ventless hood, 2 speed, add					16.15			16.15	17.75
4650	For vented 1 speed, deduct from maximum					42			42	46

11 31 23 – Residential Laundry Appliances

11 31 23.13 Washers

		Crew	Daily Output	Labor-Hours	Unit	Material	2007 Bare Costs Labor	Equipment	Total	Total Incl O&P
0010	**WASHERS**									
5000	Washers, residential, 4 cycle, average	1 Plum	3	2.667	Ea.	740	119		859	995

11 31 33 – Miscellaneous Residential Appliances

11 31 33.23 Water Heaters

		Crew	Daily Output	Labor-Hours	Unit	Material	2007 Bare Costs Labor	Equipment	Total	Total Incl O&P
0010	**WATER HEATERS**									
6900	Water heater, electric, glass lined, 30 gallon, minimum	L-1	5	3.200	Ea.	345	142		487	590
6950	Maximum		3	5.333		480	237		717	885
7100	80 gallon, minimum		2	8		630	355		985	1,225
7150	Maximum		1	16		875	710		1,585	2,000

11 31 33.43 Air Quality

		Crew	Daily Output	Labor-Hours	Unit	Material	2007 Bare Costs Labor	Equipment	Total	Total Incl O&P
0010	**AIR QUALITY**									
2450	Dehumidifier, portable, automatic, 15 pint				Ea.	159			159	175
2550	40 pint					194			194	213
3550	Heater, electric, built-in, 1250 watt, ceiling type, minimum	1 Elec	4	2		74.50	88		162.50	213
3600	Maximum		3	2.667		122	117		239	310
3700	Wall type, minimum		4	2		106	88		194	247
3750	Maximum		3	2.667		140	117		257	330
3900	1500 watt wall type, with blower		4	2		131	88		219	275
3950	3000 watt		3	2.667		267	117		384	470
4850	Humidifier, portable, 8 gallons per day					160			160	176
5000	15 gallons per day					193			193	212

11 41 Food Storage Equipment

11 41 13 – Refrigerated Food Storage Cases

11 41 13.20 Refrigerated Food Storage Equipment

		Crew	Daily Output	Labor-Hours	Unit	Material	2007 Bare Costs Labor	Equipment	Total	Total Incl O&P
0010	**REFRIGERATED FOOD STORAGE EQUIPMENT**									
2350	Cooler, reach-in, beverage, 6' long	Q-1	6	2.667	Ea.	3,775	108		3,883	4,300
8310	With glass doors, 68 C.F.	"	4	4	"	5,775	161		5,936	6,600

11 42 Food Preparation Equipment

11 42 10 - Food Preparation Equipment

11 42 10.10 Food Preparation Equipment

		Crew	Daily Output	Labor-Hours	Unit	Material	2007 Bare Costs Labor	Equipment	Total	Total Incl O&P
0010	**FOOD PREPARATION EQUIPMENT**									
1850	Coffee urn, twin 6 gallon urns	1 Plum	2	4	Ea.	2,650	179		2,829	3,175
3800	Food mixers, 20 quarts	L-7	7	4		3,350	142		3,492	3,925
3900	60 quarts	"	5	5.600	↓	11,000	199		11,199	12,400

11 44 Food Cooking Equipment

11 44 13 - Commercial Ranges

11 44 13.10 Cooking Equipment

		Crew	Daily Output	Labor-Hours	Unit	Material	2007 Bare Costs Labor	Equipment	Total	Total Incl O&P
0010	**COOKING EQUIPMENT**									
0020	Bake oven, gas, one section	Q-1	8	2	Ea.	4,425	80.50		4,505.50	5,000
1300	Broiler, without oven, standard	"	8	2		2,950	80.50		3,030.50	3,375
1550	Infra-red	L-7	4	7		6,750	248		6,998	7,800
6350	Kettle, w/steam jacket, tilting, w/positive lock, SS, 20 gallons		7	4		5,350	142		5,492	6,125
6600	60 gallons		6	4.667		7,550	165		7,715	8,550
8850	Steamer, electric 27 KW		7	4		9,425	142		9,567	10,600
9100	Electric, 10 KW or gas 100,000 BTU	↓	5	5.600		4,500	199		4,699	5,250
9150	Toaster, conveyor type, 16-22 slices per minute				↓	1,125			1,125	1,225
9200	For deluxe models of above equipment, add					75%				
9400	Rule of thumb: Equipment cost based									
9410	on kitchen work area									
9420	Office buildings, minimum	L-7	77	.364	S.F.	64.50	12.90		77.40	91
9450	Maximum		58	.483		109	17.10		126.10	147
9550	Public eating facilities, minimum		77	.364		84.50	12.90		97.40	113
9600	Maximum		46	.609		137	21.50		158.50	185
9750	Hospitals, minimum		58	.483		87	17.10		104.10	122
9800	Maximum	↓	39	.718	↓	145	25.50		170.50	200

11 46 Food Dispensing Equipment

11 46 16 - Service Line Equipment

11 46 16.10 Food Dispensing Equipment

		Crew	Daily Output	Labor-Hours	Unit	Material	2007 Bare Costs Labor	Equipment	Total	Total Incl O&P
0010	**FOOD DISPENSING EQUIPMENT**									
3300	Food warmer, counter, 1.2 KW				Ea.	545			545	600
3550	1.6 KW				"	1,600			1,600	1,775

11 48 Cleaning and Disposal Equipment

11 48 13 - Commercial Dishwashers

11 48 13.10 Dishwashers

		Crew	Daily Output	Labor-Hours	Unit	Material	2007 Bare Costs Labor	Equipment	Total	Total Incl O&P
0010	**DISHWASHERS**									
2700	Dishwasher, commercial, rack type									
2720	10 to 12 racks per hour	Q-1	3.20	5	Ea.	3,625	202		3,827	4,275
2750	Semi-automatic 38 to 50 racks per hour	"	1.30	12.308	"	7,150	495		7,645	8,625

11 52 Audio-Visual Equipment

11 52 16 – Projectors

11 52 16.10 Movie Equipment	Crew	Daily Output	Labor-Hours	Unit	Material	2007 Bare Costs Labor	Equipment	Total	Total Incl O&P
0010 **MOVIE EQUIPMENT**									
0020 Changeover, minimum				Ea.	430			430	475
0100 Maximum					835			835	920
0800 Lamphouses, incl. rectifiers, xenon, 1,000 watt	1 Elec	2	4		6,150	176		6,326	7,000
0900 1,600 watt		2	4		6,575	176		6,751	7,475
1000 2,000 watt		1.50	5.333		7,050	234		7,284	8,100
1100 4,000 watt		1.50	5.333		8,725	234		8,959	9,925
3700 Sound systems, incl. amplifier, mono, minimum		.90	8.889		3,050	390		3,440	3,925
3800 Dolby/Super Sound, maximum		.40	20		16,700	880		17,580	19,700
4100 Dual system, 2 channel, front surround, minimum		.70	11.429		4,275	500		4,775	5,450
4200 Dolby/Super Sound, 4 channel, maximum		.40	20		15,300	880		16,180	18,100
5300 Speakers, recessed behind screen, minimum		2	4		980	176		1,156	1,325
5400 Maximum		1	8		2,850	350		3,200	3,675
7000 For automation, varying sophistication, minimum		1	8	System	2,200	350		2,550	2,950
7100 Maximum	2 Elec	.30	53.333	"	5,150	2,350		7,500	9,125

11 53 Laboratory Equipment

11 53 19 – Laboratory Sterilizers

11 53 19.13 Sterilizers

	Crew	Daily Output	Labor-Hours	Unit	Material	Labor	Equipment	Total	Total Incl O&P
0010 **STERILIZERS**									
0700 Glassware washer, undercounter, minimum	L-1	1.80	8.889	Ea.	6,100	395		6,495	7,300
0710 Maximum	"	1	16	"	8,950	710		9,660	10,900

11 53 33 – Emergency Safety Appliances

11 53 33.13 Emergency Equipment

	Crew	Daily Output	Labor-Hours	Unit	Material	Labor	Equipment	Total	Total Incl O&P
0010 **EMERGENCY EQUIPMENT**									
1400 Safety equipment, eye wash, hand held				Ea.	405			405	445
1450 Deluge shower				"	640			640	705

11 53 43 – Service Fittings and Accessories

11 53 43.13 Fittings

	Crew	Daily Output	Labor-Hours	Unit	Material	Labor	Equipment	Total	Total Incl O&P
0010 **FITTINGS**									
8000 Alternate pricing method: as percent of lab furniture									
8050 Installation, not incl. plumbing & duct work				% Furn.				20%	22%
8100 Plumbing, final connections, simple system								9.09%	10%
8110 Moderately complex system								13.64%	15%
8120 Complex system								18.18%	20%
8150 Electrical, simple system								9.09%	10%
8160 Moderately complex system								18.18%	20%
8170 Complex system								31.80%	35%

11 61 Theater and Stage Equipment

11 61 33 – Rigging Systems and Controls

11 61 33.10 Controls

		Crew	Daily Output	Labor-Hours	Unit	Material	2007 Bare Costs Labor	Equipment	Total	Total Incl O&P
0010	**CONTROLS**									
0050	Control boards with dimmers and breakers, minimum	1 Elec	1	8	Ea.	12,100	350		12,450	13,800
0100	Average		.50	16		33,600	700		34,300	38,100
0150	Maximum		.20	40		111,000	1,750		112,750	125,000

11 66 Athletic Equipment

11 66 43 – Interior Scoreboards

11 66 43.10 Scoreboards

		Crew	Daily Output	Labor-Hours	Unit	Material	2007 Bare Costs Labor	Equipment	Total	Total Incl O&P
0010	**SCOREBOARDS**									
7000	Scoreboards, baseball, minimum	R-3	1.30	15.385	Ea.	3,050	665	126	3,841	4,475
7200	Maximum		.05	400		14,800	17,300	3,275	35,375	45,700
7300	Football, minimum		.86	23.256		3,775	1,000	190	4,965	5,850
7400	Maximum		.20	100		12,000	4,325	815	17,140	20,600
7500	Basketball (one side), minimum		2.07	9.662		2,125	420	79	2,624	3,025
7600	Maximum		.30	66.667		5,275	2,875	545	8,695	10,700
7700	Hockey-basketball (four sides), minimum		.25	80		5,300	3,450	655	9,405	11,700
7800	Maximum		.15	133		5,375	5,775	1,100	12,250	15,700

11 71 Medical Sterilizing Equipment

11 71 10 – Medical Sterilizing Equipment

11 71 10.10 Medical Sterilizing Equipment

		Crew	Daily Output	Labor-Hours	Unit	Material	2007 Bare Costs Labor	Equipment	Total	Total Incl O&P
0010	**MEDICAL STERILIZING EQUIPMENT**									
6200	Steam generators, electric 10 KW to 180 KW, freestanding									
6250	Minimum	1 Elec	3	2.667	Ea.	7,400	117		7,517	8,325
6300	Maximum	"	.70	11.429	"	26,600	500		27,100	30,000

11 72 Examination and Treatment Equipment

11 72 53 – Treatment Equipment

11 72 53.13 Treatment Equipment

		Crew	Daily Output	Labor-Hours	Unit	Material	2007 Bare Costs Labor	Equipment	Total	Total Incl O&P
0010	**TREATMENT EQUIPMENT**									
6700	Surgical lights, doctor's office, single arm	2 Elec	2	8	Ea.	1,200	350		1,550	1,825
6750	Dual arm	"	1	16	"	5,375	700		6,075	6,950

11 76 Operating Room Equipment

11 76 10 – Operating Room Equipment

11 76 10.10 Operating Room Equipment

		Crew	Daily Output	Labor-Hours	Unit	Material	2007 Bare Costs Labor	Equipment	Total	Total Incl O&P
0010	**OPERATING ROOM EQUIPMENT**									
6800	Surgical lights, major operating room, dual head, minimum	2 Elec	1	16	Ea.	16,100	700		16,800	18,900
6850	Maximum	"	1	16	"	26,900	700		27,600	30,700

11 78 Mortuary Equipment

11 78 13 – Mortuary Refrigerators

11 78 13.10 Mortuary Equipment	Crew	Daily Output	Labor-Hours	Unit	Material	2007 Bare Costs Labor	Equipment	Total	Total Incl O&P
0010 **MORTUARY EQUIPMENT**									
0015 Autopsy table, standard	1 Plum	1	8	Ea.	8,200	360		8,560	9,575

11 78 16 – Crematorium Equipment

11 78 16.10 Crematory

0010 **CREMATORY**									
1500 Crematory, not including building, 1 place	Q-3	.20	160	Ea.	55,000	6,825		61,825	71,000
1750 2 place	"	.10	320	"	78,500	13,700		92,200	107,000

11 82 Solid Waste Handling Equipment

11 82 26 – Waste Compactors and Destructors

11 82 26.10 Compactors

	Crew	Daily Output	Labor-Hours	Unit	Material	Labor	Equipment	Total	Total Incl O&P
0010 **COMPACTORS**									
0020 Compactors, 115 volt, 250#/hr., chute fed	L-4	1	24	Ea.	10,100	825		10,925	12,400
1000 Heavy duty industrial compactor, 0.5 C.Y. capacity		1	24		6,575	825		7,400	8,500
1050 1.0 C.Y. capacity		1	24		10,000	825		10,825	12,300
1100 3 C.Y. capacity		.50	48		13,700	1,650		15,350	17,600
1150 5.0 C.Y. capacity		.50	48		16,900	1,650		18,550	21,200
1200 Combination shredder/compactor (5,000 lbs./hr.)		.50	48		33,400	1,650		35,050	39,300
1400 For handling hazardous waste materials, 55 gallon drum packer, std.					15,500			15,500	17,100
1410 55 gallon drum packer w/HEPA filter					19,400			19,400	21,300
1420 55 gallon drum packer w/charcoal & HEPA filter					25,900			25,900	28,400
1430 All of the above made explosion proof, add					11,800			11,800	13,000
5800 Shredder, industrial, minimum					20,000			20,000	22,000
5850 Maximum					107,500			107,500	118,000
5900 Baler, industrial, minimum					8,025			8,025	8,825
5950 Maximum					468,500			468,500	515,500

Estimating Tips

General

- The items in this division are usually priced per square foot or each. Most of these items are purchased by the owner and placed by the supplier. Do not assume the items in Division 12 will be purchased and installed by the supplier. Check the specifications for responsibilities and include receiving, storage, installation, and mechanical and electrical hookups in the appropriate divisions.

- Some items in this division require some type of support system that is not usually furnished with the item. Examples of these systems include blocking for the attachment of casework and heavy drapery rods. The required blocking must be added to the estimate in the appropriate division.

Reference Numbers

Reference numbers are shown in shaded boxes at the beginning of some major classifications. These numbers refer to related items in the Reference Section. The reference information may be an estimating procedure, an alternate pricing method, or technical information.

Note: Not all subdivisions listed here necessarily appear in this publication.

12 46 Furnishing Accessories

12 46 19 – Clocks

	12 46 19.50 Clocks	Crew	Daily Output	Labor-Hours	Unit	Material	2007 Bare Costs Labor	Equipment	Total	Total Incl O&P
0010	**CLOCKS**									
0080	12" diameter, single face	1 Elec	8	1	Ea.	75.50	44		119.50	149
0100	Double face	"	6.20	1.290	"	144	56.50		200.50	243

Estimating Tips

General

- The items and systems in this division are usually estimated, purchased, supplied, and installed as a unit by one or more subcontractors. The estimator must ensure that all parties are operating from the same set of specifications and assumptions, and that all necessary items are estimated and will be provided. Many times the complex items and systems are covered, but the more common ones, such as excavation or a crane, are overlooked for the very reason that everyone assumes nobody could miss them. The estimator should be the central focus and be able to ensure that all systems are complete.

- Another area where problems can develop in this division is at the interface between systems. The estimator must ensure, for instance, that anchor bolts, nuts, and washers are estimated and included for the air-supported structures and pre-engineered buildings to be bolted to their foundations. Utility supply is a common area where essential items or pieces of equipment can be missed or overlooked due to the fact that each subcontractor may feel it is another's responsibility. The estimator should also be aware of certain items which may be supplied as part of a package but installed by others, and ensure that the installing contractor's estimate includes the cost of installation. Conversely, the estimator must also ensure that items are not costed by two different subcontractors, resulting in an inflated overall estimate.

13 30 00 Special Structures

- The foundations and floor slab, as well as rough mechanical and electrical, should be estimated, as this work is required for the assembly and erection of the structure. Generally, as noted in the book, the pre-engineered building comes as a shell and additional features, such as windows and doors, must be included by the estimator. Here again, the estimator must have a clear understanding of the scope of each portion of the work and all the necessary interfaces.

Reference Numbers

Reference numbers are shown in shaded boxes at the beginning of some major classifications. These numbers refer to related items in the Reference Section. The reference information may be an estimating procedure, an alternate pricing method, or technical information.

Note: Not all subdivisions listed here necessarily appear in this publication.

Division 13 - Special Construction

13 11 Swimming Pools

13 11 13 – Below-Grade Swimming Pools

13 11 13.50 Swimming Pools

		Crew	Daily Output	Labor-Hours	Unit	Material	2007 Bare Costs Labor	Equipment	Total	Total Incl O&P
0010	**SWIMMING POOLS** Residential in-ground, vinyl lined, concrete									
0020	Sides including equipment, sand bottom	B-52	300	.187	SF Surf	12.25	6.30	1.39	19.94	25
0100	Metal or polystyrene sides	B-14	410	.117		10.25	3.56	.59	14.40	17.40
0200	Add for vermiculite bottom					.79			.79	.87
0500	Gunite bottom and sides, white plaster finish									
0600	12' x 30' pool	B-52	145	.386	SF Surf	23	13.10	2.88	38.98	48
0720	16' x 32' pool		155	.361		20.50	12.25	2.69	35.44	44.50
0750	20' x 40' pool		250	.224		18.35	7.60	1.67	27.62	33.50
0810	Concrete bottom and sides, tile finish									
0820	12' x 30' pool	B-52	80	.700	SF Surf	23	23.50	5.20	51.70	68
0830	16' x 32' pool		95	.589		19.05	19.95	4.39	43.39	57
0840	20' x 40' pool		130	.431		15.15	14.60	3.21	32.96	42.50
1100	Motel, gunite with plaster finish, incl. medium									
1150	capacity filtration & chlorination	B-52	115	.487	SF Surf	28	16.50	3.63	48.13	60.50
1200	Municipal, gunite with plaster finish, incl. high									
1250	capacity filtration & chlorination	B-52	100	.560	SF Surf	36.50	18.95	4.17	59.62	74
1350	Add for formed gutters				L.F.	53.50			53.50	59
1360	Add for stainless steel gutters				"	158			158	174
1700	Filtration and deck equipment only, as % of total				Total				20%	20%
1800	Deck equipment, rule of thumb, 20' x 40' pool				SF Pool				1.18	1.30
1900	5000 S.F. pool				"				1.73	1.90

13 11 46 – Swimming Pool Accessories

13 11 46.50 Swimming Pool Equipment

		Crew	Daily Output	Labor-Hours	Unit	Material	2007 Bare Costs Labor	Equipment	Total	Total Incl O&P
0010	**SWIMMING POOL EQUIPMENT**									
0020	Diving stand, stainless steel, 3 meter	2 Carp	.40	40	Ea.	6,150	1,475		7,625	9,050
2100	Lights, underwater, 12 volt, with transformer, 300 watt	1 Elec	.40	20		175	880		1,055	1,500
2200	110 volt, 500 watt, standard		.40	20		122	880		1,002	1,425
2400	Low water cutoff type		.40	20		156	880		1,036	1,475
2800	Heaters, see division 23 52 28.10									

13 21 Controlled Environment Rooms

13 21 13 – Clean Rooms

13 21 13.50 Clean Rooms

		Crew	Daily Output	Labor-Hours	Unit	Material	2007 Bare Costs Labor	Equipment	Total	Total Incl O&P
0010	**CLEAN ROOMS**									
1100	Clean room, soft wall, 12' x 12', Class 100	1 Carp	.18	44.444	Ea.	13,600	1,625		15,225	17,600
1110	Class 1,000		.18	44.444		10,600	1,625		12,225	14,200
1120	Class 10,000		.21	38.095		9,000	1,400		10,400	12,100
1130	Class 100,000		.21	38.095		8,325	1,400		9,725	11,300
2800	Ceiling grid support, slotted channel struts 4'-0" O.C., ea. way				S.F.				5.91	6.50
3000	Ceiling panel, vinyl coated foil on mineral substrate									
3020	Sealed, non-perforated				S.F.				1.27	1.40
4000	Ceiling panel seal, silicone sealant, 150 L.F./gal.	1 Carp	150	.053	L.F.	.24	1.96		2.20	3.31
4100	Two sided adhesive tape	"	240	.033	"	.11	1.22		1.33	2.02
4200	Clips, one per panel				Ea.	.95			.95	1.05
6000	HEPA filter,2'x4',99.97% eff.,3" dp beveled frame (silicone seal)					290			290	320
6040	6" deep skirted frame (channel seal)					335			335	370
6100	99.99% efficient, 3" deep beveled frame (silicone seal)					315			315	345
6140	6" deep skirted frame (channel seal)					355			355	395
6200	99.999% efficient, 3" deep beveled frame (silicone seal)					355			355	395
6240	6" deep skirted frame (channel seal)					400			400	440

13 21 Controlled Environment Rooms

13 21 13 – Clean Rooms

13 21 13.50 Clean Rooms	Crew	Daily Output	Labor-Hours	Unit	Material	2007 Bare Costs Labor	Equipment	Total	Total Incl O&P
7000	Wall panel systems, including channel strut framing									
7020	Polyester coated aluminum, particle board				S.F.				18.18	20
7100	Porcelain coated aluminum, particle board								31.82	35
7400	Wall panel support, slotted channel struts, to 12' high				↓				16.36	18

13 24 Special Activity Rooms

13 24 16 – Saunas

13 24 16.50 Saunas

		Crew	Daily Output	Labor-Hours	Unit	Material	2007 Bare Costs Labor	Equipment	Total	Total Incl O&P
0010	**SAUNAS**									
0020	Prefabricated, incl. heater & controls, 7' high, 6' x 4', C/C	L-7	2.20	12.727	Ea.	3,875	450		4,325	4,950
0050	6' x 4', C/P		2	14		3,575	495		4,070	4,725
0400	6' x 5', C/C		2	14		4,325	495		4,820	5,525
0450	6' x 5', C/P		2	14		4,050	495		4,545	5,225
0600	6' x 6', C/C		1.80	15.556		4,625	550		5,175	5,925
0650	6' x 6', C/P		1.80	15.556		4,300	550		4,850	5,575
0800	6' x 9', C/C		1.60	17.500		5,775	620		6,395	7,300
0850	6' x 9', C/P		1.60	17.500		5,475	620		6,095	6,975
1000	8' x 12', C/C		1.10	25.455		8,950	905		9,855	11,200
1050	8' x 12', C/P		1.10	25.455		8,200	905		9,105	10,400
1200	8' x 8', C/C		1.40	20		6,800	710		7,510	8,600
1250	8' x 8', C/P		1.40	20		6,400	710		7,110	8,125
1400	8' x 10', C/C		1.20	23.333		7,575	825		8,400	9,600
1450	8' x 10', C/P		1.20	23.333		7,025	825		7,850	9,000
1600	10' x 12', C/C		1	28		9,475	995		10,470	11,900
1650	10' x 12', C/P	↓	1	28		8,575	995		9,570	11,000
2500	Heaters only (incl. above), wall mounted, to 200 C.F.					500			500	550
2750	To 300 C.F.					615			615	675
3000	Floor standing, to 720 C.F., 10,000 watts, w/controls	1 Elec	3	2.667		1,575	117		1,692	1,900
3250	To 1,000 C.F., 16,000 watts	"	3	2.667	↓	1,750	117		1,867	2,100

13 24 26 – Steam Baths

13 24 26.50 Steam Baths

		Crew	Daily Output	Labor-Hours	Unit	Material	2007 Bare Costs Labor	Equipment	Total	Total Incl O&P
0010	**STEAM BATHS**									
0020	Heater, timer & head, single, to 140 C.F.	1 Plum	1.20	6.667	Ea.	1,125	299		1,424	1,700
0500	To 300 C.F.		1.10	7.273		1,275	325		1,600	1,900
1000	Commercial size, with blow-down assembly, to 800 C.F.		.90	8.889		4,375	400		4,775	5,425
1500	To 2500 C.F.	↓	.80	10		6,900	450		7,350	8,275
2000	Multiple, motels, apts., 2 baths, w/ blow-down assm., 500 C.F.	Q-1	1.30	12.308		4,325	495		4,820	5,500
2500	4 baths	"	.70	22.857	↓	4,725	920		5,645	6,575

13 49 Radiation Protection

13 49 13 – Lead Sheet

13 49 13.50 Lead Sheets

		Crew	Daily Output	Labor-Hours	Unit	Material	2007 Bare Costs Labor	Equipment	Total	Total Incl O&P
0010	**LEAD SHEETS**									
0300	Lead sheets, 1/16" thick	2 Lath	135	.119	S.F.	5.85	3.99		9.84	12.35
0400	1/8" thick		120	.133		11.90	4.49		16.39	19.75
0500	Lead shielding, 1/4" thick		135	.119		25	3.99		28.99	33.50
0550	1/2" thick	↓	120	.133	↓	44.50	4.49		48.99	55.50
0950	Lead headed nails (average 1 lb. per sheet)				Lb.	6.20			6.20	6.80
1000	Butt joints in 1/8" lead or thicker, 2" batten strip x 7' long	2 Lath	240	.067	Ea.	15.35	2.25		17.60	20.50
1200	X-ray protection, average radiography or fluoroscopy									
1210	room, up to 300 S.F. floor, 1/16" lead, minimum	2 Lath	.25	64	Total	5,125	2,150		7,275	8,850
1500	Maximum, 7'-0" walls	"	.15	106	"	6,225	3,600		9,825	12,200
1600	Deep therapy X-ray room, 250 KV capacity,									
1800	up to 300 S.F. floor, 1/4" lead, minimum	2 Lath	.08	200	Total	18,800	6,750		25,550	30,700
1900	Maximum, 7'-0" walls	"	.06	266	"	24,300	8,975		33,275	40,100

13 49 19 – Lead-Lined Materials

13 49 19.50 Shielding Lead

		Crew	Daily Output	Labor-Hours	Unit	Material	2007 Bare Costs Labor	Equipment	Total	Total Incl O&P
0010	**SHIELDING LEAD**									
0100	Laminated lead in wood doors, 1/16" thick, no hardware				S.F.	37			37	41
0200	Lead lined door frame, not incl. hardware,									
0210	1/16" thick lead, butt prepared for hardware	1 Lath	2.40	3.333	Ea.	495	112		607	710
0850	Window frame with 1/16" lead and voice passage, 36" x 60"	2 Glaz	2	8		925	288		1,213	1,450
0870	24" x 36" frame		8	2	↓	750	72		822	935
0900	Lead gypsum board, 5/8" thick with 1/16" lead		160	.100	S.F.	5.60	3.61		9.21	11.60
0910	1/8" lead	↓	140	.114		11.35	4.12		15.47	18.75
0930	1/32" lead	2 Lath	200	.080	↓	3.64	2.70		6.34	8

13 49 21 – Lead Glazing

13 49 21.50 Lead Glazing

		Crew	Daily Output	Labor-Hours	Unit	Material	2007 Bare Costs Labor	Equipment	Total	Total Incl O&P
0010	**LEAD GLAZING**									
0600	Lead glass, 1/4" thick, 2.0 mm LE, 12" x 16"	2 Glaz	13	1.231	Ea.	240	44.50		284.50	330
0700	24" x 36"		8	2		950	72		1,022	1,150
0800	36" x 60"	↓	2	8	↓	3,225	288		3,513	3,975
2000	X-ray viewing panels, clear lead plastic									
2010	7 mm thick, 0.3 mm LE, 2.3 lbs/S.F.	H-3	139	.115	S.F.	119	3.65		122.65	137
2020	12 mm thick, 0.5 mm LE, 3.9 lbs/S.F.		82	.195		161	6.20		167.20	187
2030	18 mm thick, 0.8mm LE, 5.9 lbs/S.F.		54	.296		175	9.40		184.40	207
2040	22 mm thick, 1.0 mm LE, 7.2 lbs/S.F.		44	.364		179	11.55		190.55	215
2050	35 mm thick, 1.5 mm LE, 11.5 lbs/S.F.		28	.571		201	18.10		219.10	249
2060	46 mm thick, 2.0 mm LE, 15.0 lbs/S.F.	↓	21	.762	↓	264	24		288	330
2090	For panels 12 S.F. to 48 S.F., add crating charge				Ea.					50

13 49 23 – Modular Shielding Partitions

13 49 23.50 Modular Shielding Partitions

		Crew	Daily Output	Labor-Hours	Unit	Material	2007 Bare Costs Labor	Equipment	Total	Total Incl O&P
0010	**MODULAR SHIELDING PARTITIONS**									
4000	X-ray barriers, modular, panels mounted within framework for									
4002	attaching to floor, wall or ceiling, upper portion is clear lead									
4005	plastic window panels 48"H, lower portion is opaque leaded									
4008	steel panels 36"H, structural supports not incl.									
4010	1-section barrier, 36"W x 84"H overall									
4020	0.5 mm LE panels	H-3	6.40	2.500	Ea.	2,700	79.50		2,779.50	3,075
4030	0.8 mm LE panels		6.40	2.500		2,875	79.50		2,954.50	3,275
4040	1.0 mm LE panels		5.33	3.002		2,925	95		3,020	3,375
4050	1.5 mm LE panels	↓	5.33	3.002		3,125	95		3,220	3,575
4060	2-section barrier, 72"W x 84"H overall									

13 49 Radiation Protection

13 49 23 – Modular Shielding Partitions

13 49 23.50 Modular Shielding Partitions	Crew	Daily Output	Labor-Hours	Unit	Material	2007 Bare Costs Labor	Equipment	Total	Total Incl O&P
4070	0.5 mm LE panels	H-3	4	4	Ea.	5,500	127		5,627	6,250
4080	0.8 mm LE panels		4	4		5,825	127		5,952	6,600
4090	1.0 mm LE panels		3.56	4.494		5,950	142		6,092	6,750
5000	1.5 mm LE panels		3.20	5		6,750	159		6,909	7,675
5010	3-section barrier, 108"W x 84"H overall									
5020	0.5 mm LE panels	H-3	3.20	5	Ea.	8,275	159		8,434	9,350
5030	0.8 mm LE panels		3.20	5		8,725	159		8,884	9,850
5040	1.0 mm LE panels		2.67	5.993		8,875	190		9,065	10,100
5050	1.5 mm LE panels		2.46	6.504		9,650	206		9,856	10,900
7000	X-ray barriers, mobile, mounted within framework w/casters on									
7005	bottom, clear lead plastic window panels on upper portion,									
7010	opaque on lower, 30"W x 75"H overall, incl. framework									
7020	24"H upper w/0.5 mm LE, 48"H lower w/0.8 mm LE	1 Carp	16	.500	Ea.	2,125	18.35		2,143.35	2,350
7030	48"W x 75"H overall, incl. framework									
7040	36"H upper w/0.5 mm LE, 36"H lower w/0.8 mm LE	1 Carp	16	.500	Ea.	3,750	18.35		3,768.35	4,150
7050	36"H upper w/1.0 mm LE, 36"H lower w/1.5 mm LE	"	16	.500	"	4,575	18.35		4,593.35	5,050
7060	72"W x 75"H overall, incl. framework									
7070	36"H upper w/0.5 mm LE, 36"H lower w/0.8 mm LE	1 Carp	16	.500	Ea.	4,450	18.35		4,468.35	4,925
7080	36"H upper w/1.0 mm LE, 36"H lower w/1.5 mm LE	"	16	.500	"	5,625	18.35		5,643.35	6,200

13 49 33 – Radio Frequency Shielding

13 49 33.50 Shielding, Radio Frequency	Crew	Daily Output	Labor-Hours	Unit	Material	2007 Bare Costs Labor	Equipment	Total	Total Incl O&P
0010	**SHIELDING, RADIO FREQUENCY**									
0020	Prefabricated or screen-type copper or steel, minimum	2 Carp	180	.089	SF Surf	25	3.26		28.26	32.50
0100	Average		155	.103		27.50	3.79		31.29	36
0150	Maximum		145	.110		32.50	4.05		36.55	42.50

Division Notes

		CREW	DAILY OUTPUT	LABOR-HOURS	UNIT	2007 BARE COSTS				TOTAL INCL O&P
						MAT.	LABOR	EQUIP.	TOTAL	

Estimating Tips

Pipe for fire protection and all uses is located in Section 22 22 13.

When installing piping above 10' review Section 22 02 02.20 for percentage adds due to the elevated work.

Many, but not all, areas require backflow protection in the fire system. It is advisable to check local building codes for specific requirements.

For your reference, the following is a list of the most applicable Fire Codes and Standards which may be purchased from the NFPA, 1 Batterymarch Park, Quincy, MA 02169-7471.

NFPA 1: Uniform Fire Code
NFPA 10: Portable Fire Extinguishers
NFPA 11: Low-, Medium-, and High-Expansion Foam
NFPA 12: Carbon Dioxide Extinguishing Systems (Also companion 12A)
NFPA 13: Installation of Sprinkler Systems (Also companion 13D, 13E, and 13R)
NFPA 14: Installation of Standpipe and Hose Systems
NFPA 15: Water Spray Fixed Systems for Fire Protection
NFPA 16: Installation of Foam-Water Sprinkler and Foam-Water Spray Systems
NFPA 17: Dry Chemical Extinguishing Systems (Also companion 17A)
NFPA 18: Wetting Agents

NFPA 20: Installation of Stationary Pumps for Fire Protection
NFPA 22: Water Tanks for Private Fire Protection
NFPA 24: Installation of Private Fire Service Mains and their Appurtenances
NFPA 25: Inspection, Testing and Maintenance of Water-Based Fire Protection

Reference Numbers

Reference numbers are shown in shaded boxes at the beginning of some major classifications. These numbers refer to related items in the Reference Section. The reference information may be an estimating procedure, an alternate pricing method, or technical information.

Note: Not all subdivisions listed here necessarily appear in this publication.

Note: **i2 Trade Service,** *in part, has been used as a reference source for some of the material prices used in Division 21.*

21 21 Carbon-Dioxide Fire-Extinguishing Systems

21 21 16 – Carbon-Dioxide Fire-Extinguishing Equipment

21 21 16.50 CO2 Fire Extinguishing System	Crew	Daily Output	Labor-Hours	Unit	Material	2007 Bare Costs Labor	Equipment	Total	Total Incl O&P
0010 **CO₂ FIRE EXTINGUISHING SYSTEM**									
0100 Control panel, single zone with batteries (2 zones det., 1 suppr.)	1 Elec	1	8	Ea.	1,350	350		1,700	2,000
0150 Multizone (4) with batteries (8 zones det., 4 suppr.)	"	.50	16		2,575	700		3,275	3,875
1000 Dispersion nozzle, CO₂, 3" x 5"	1 Plum	18	.444		52.50	19.90		72.40	88
2000 Extinguisher, CO₂ system, high pressure, 75 lb. cylinder	Q-1	6	2.667		1,000	108		1,108	1,250
2100 100 lb. cylinder	"	5	3.200		1,025	129		1,154	1,325
3000 Electro/mechanical release	L-1	4	4		131	177		308	410
3400 Manual pull station	1 Plum	6	1.333		47.50	59.50		107	142
4000 Pneumatic damper release	"	8	1	↓	175	45		220	261

21 22 Clean-Agent Fire-Extinguishing Systems

21 22 16 – Clean-Agent Fire-Extinguishing Equipment

21 22 16.50 FM200 Fire Extinguishing System	Crew	Daily Output	Labor-Hours	Unit	Material	2007 Bare Costs Labor	Equipment	Total	Total Incl O&P
0010 **FM200 FIRE EXTINGUISHING SYSTEM**									
1100 Dispersion nozzle FM200, 1-1/2"	1 Plum	14	.571	Ea.	52.50	25.50		78	96.50
2400 Extinguisher, FM 200 system, filled, with mounting bracket									
2460 26 lb. container	Q-1	8	2	Ea.	1,800	80.50		1,880.50	2,100
2480 44 lb. container		7	2.286		2,400	92		2,492	2,800
2500 63 lb. container		6	2.667		2,800	108		2,908	3,225
2520 101 lb. container		5	3.200		3,750	129		3,879	4,325
2540 196 lb. container	↓	4	4	↓	6,100	161		6,261	6,950
6000 Average FM200 system, minimum				C.F.	1.38			1.38	1.52
6020 Maximum				"	2.75			2.75	3.03

Estimating Tips

22 10 00 Plumbing Piping and Pumps

This subdivision is primarily basic pipe and related materials. The pipe may be used by any of the mechanical disciplines, i.e., plumbing, fire protection, heating, and air conditioning.

- The piping section lists the add to labor for elevated pipe installation. These adds apply to all elevated pipe, fittings, valves, insulation, etc., that are placed above 10' high. CAUTION: the correct percentage may vary for the same pipe. For example, the percentage add for the basic pipe installation should be based on the maximum height that the craftsman must install for that particular section. If the pipe is to be located 14' above the floor but it is suspended on threaded rod from beams, the bottom flange of which is 18' high (4' rods), then the height is actually 18' and the add is 20%. The pipe coverer, however, does not have to go above the 14', and so his or her add should be 10%.

- Most pipe is priced first as straight pipe with a joint (coupling, weld, etc.) every 10' and a hanger usually every 10'.

There are exceptions with hanger spacing such as for cast iron pipe (5') and plastic pipe (3 per 10'). Following each type of pipe there are several lines listing sizes and the amount to be subtracted to delete couplings and hangers. This is for pipe that is to be buried or supported together on trapeze hangers. The reason that the couplings are deleted is that these runs are usually long, and frequently longer lengths of pipe are used. By deleting the couplings, the estimator is expected to look up and add back the correct reduced number of couplings.

- When preparing an estimate it may be necessary to approximate the fittings. Fittings usually run between 25% and 50% of the cost of the pipe. The lower percentage is for simpler runs, and the higher number is for complex areas such as mechanical rooms.

- For historic restoration projects, the systems must be as invisible as possible, and pathways must be sought for pipes, conduit, and ductwork. While installations in accessible spaces (such as basements and attics) are relatively straightforward to estimate, labor costs may be more difficult to determine when delivery systems must be concealed.

22 40 00 Plumbing Fixtures

- Plumbing fixture costs usually require two lines: the fixture itself and its "rough-in, supply and waste."

- In the Assemblies Section (Plumbing D2010) for the desired fixture, the System Components Group at the center of the page shows the fixture on the first line. The rest of the list (fittings, pipe, tubing, etc.) will total up to what we refer to in the Unit Price section as "Rough-in, supply, waste and vent." Note that for most fixtures we allow a nominal 5' of tubing to reach from the fixture to a main or riser.

- Remember that gas- and oil-fired units need venting.

Reference Numbers

Reference numbers are shown in shaded boxes at the beginning of some major classifications. These numbers refer to related items in the Reference Section. The reference information may be an estimating procedure, an alternate pricing method, or technical information.
Note: Not all subdivisions listed here necessarily appear in this publication.

Note: **i2 Trade Service,** *in part, has been used as a reference source for some of the material prices used in Division 22.*

22 01 Operation and Maintenance of Plumbing

22 01 02 – General Plumbing

22 01 02.10 Boilers, General	Crew	Daily Output	Labor-Hours	Unit	Material	2007 Bare Costs Labor	Equipment	Total	Total Incl O&P
0010	**BOILERS, GENERAL**, Prices do not include flue piping, elec. wiring,								
0020	gas or oil piping, boiler base, pad, or tankless unless noted								
0100	Boiler H.P.: 10 KW = 34 lbs/steam/hr = 33,475 BTU/hr.								
0120									
0150	To convert SFR to BTU rating: Hot water, 150 x SFR;								
0160	Forced hot water, 180 x SFR; steam, 240 x SFR								

22 05 Common Work Results for Plumbing

22 05 29 – Hangers and Supports for Plumbing Piping and Equipment

22 05 29.10 Hangers & Supp. for Plumb'g/HVAC Pipe/Equip.

22 05 29.10 Hangers & Supp. for Plumb'g/HVAC Pipe/Equip.	Crew	Daily Output	Labor-Hours	Unit	Material	2007 Bare Costs Labor	Equipment	Total	Total Incl O&P
0010	**HANGERS AND SUPPORTS FOR PLUMB'G/HVAC PIPE/EQUIP.**								
0011	TYPE numbers per MSS-SP58								
0050	Brackets								
0060	Beam side or wall, malleable iron, TYPE 34								
0070	3/8" threaded rod size	1 Plum	48	.167	Ea.	1.77	7.45		9.22	13.20
0080	1/2" threaded rod size		48	.167		3.41	7.45		10.86	15
0090	5/8" threaded rod size		48	.167		4.54	7.45		11.99	16.25
0100	3/4" threaded rod size		48	.167		5.45	7.45		12.90	17.25
0110	7/8" threaded rod size		48	.167		6.35	7.45		13.80	18.25
0120	For concrete installation, add						30%			
0150	Wall, welded steel, medium, TYPE 32								
0160	0 size, 12" wide, 18" deep	1 Plum	34	.235	Ea.	133	10.55		143.55	162
0170	1 size, 18" wide 24" deep		34	.235		158	10.55		168.55	190
0180	2 size, 24" wide, 30" deep		34	.235		209	10.55		219.55	246
0300	Clamps								
0310	C-clamp, for mounting on steel beam flange, w/locknut, TYPE 23								
0320	3/8" threaded rod size	1 Plum	160	.050	Ea.	1.28	2.24		3.52	4.78
0330	1/2" threaded rod size		160	.050		1.63	2.24		3.87	5.15
0340	5/8" threaded rod size		160	.050		2.70	2.24		4.94	6.35
0350	3/4" threaded rod size		160	.050		3.72	2.24		5.96	7.45
0750	Riser or extension pipe, carbon steel, TYPE 8								
0760	3/4" pipe size	1 Plum	48	.167	Ea.	2.17	7.45		9.62	13.65
0770	1" pipe size		47	.170		2.19	7.65		9.84	13.85
0780	1-1/4" pipe size		46	.174		2.72	7.80		10.52	14.70
0790	1-1/2" pipe size		45	.178		2.90	7.95		10.85	15.20
0800	2" pipe size		43	.186		3.02	8.35		11.37	15.85
0810	2-1/2" pipe size		41	.195		3.18	8.75		11.93	16.65
0820	3" pipe size		40	.200		3.37	8.95		12.32	17.20
0830	3-1/2" pipe size		39	.205		7.40	9.20		16.60	22
0840	4" pipe size		38	.211		4.25	9.45		13.70	18.90
0850	5" pipe size		37	.216		6.35	9.70		16.05	21.50
0860	6" pipe size		36	.222		7.25	9.95		17.20	23
1150	Insert, concrete								
1160	Wedge type, carbon steel body, malleable iron nut								
1170	1/4" threaded rod size	1 Plum	96	.083	Ea.	1.01	3.73		4.74	6.70
1180	3/8" threaded rod size		96	.083		1.03	3.73		4.76	6.75
1190	1/2" threaded rod size		96	.083		1.09	3.73		4.82	6.80
1200	5/8" threaded rod size		96	.083		1.11	3.73		4.84	6.80
1210	3/4" threaded rod size		96	.083		1.31	3.73		5.04	7.05
1220	7/8" threaded rod size		96	.083		1.65	3.73		5.38	7.40
1230	For galvanized, add					.49			.49	.54

22 05 29 – Hangers and Supports for Plumbing Piping and Equipment

22 05 29.10 Hangers & Supp. for Plumb'g/HVAC Pipe/Equip.	Crew	Daily Output	Labor-Hours	Unit	Material	2007 Bare Costs Labor	Equipment	Total	Total Incl O&P
2650	Rods, carbon steel									
2660	Continuous thread									
2670	1/4" thread size	1 Plum	144	.056	L.F.	.17	2.49		2.66	3.93
2680	3/8" thread size		144	.056		.24	2.49		2.73	4
2690	1/2" thread size		144	.056		.47	2.49		2.96	4.26
2700	5/8" thread size		144	.056		.65	2.49		3.14	4.46
2710	3/4" thread size		144	.056		1.10	2.49		3.59	4.95
2720	7/8" thread size		144	.056		1.42	2.49		3.91	5.30
2725	1/4" thread size, bright finish		144	.056		1.13	2.49		3.62	4.98
2726	1/2" thread size, bright finish		144	.056		1.94	2.49		4.43	5.85
2730	For galvanized, add					40%				
2750	Both ends machine threaded 18" length									
2760	3/8" thread size	1 Plum	240	.033	Ea.	2.16	1.49		3.65	4.63
2770	1/2" thread size		240	.033		3.47	1.49		4.96	6.05
2780	5/8" thread size		240	.033		4.99	1.49		6.48	7.75
2790	3/4" thread size		240	.033		7.65	1.49		9.14	10.65
2800	7/8" thread size		240	.033		9.30	1.49		10.79	12.45
2810	1" thread size		240	.033		12.75	1.49		14.24	16.25
4400	U-bolt, carbon steel									
4410	Standard, with nuts, TYPE 42									
4420	1/2" pipe size	1 Plum	160	.050	Ea.	.91	2.24		3.15	4.37
4430	3/4" pipe size		158	.051		.91	2.27		3.18	4.41
4450	1" pipe size		152	.053		1	2.36		3.36	4.65
4460	1-1/4" pipe size		148	.054		1.17	2.42		3.59	4.93
4470	1-1/2" pipe size		143	.056		1.28	2.51		3.79	5.20
4480	2" pipe size		139	.058		1.31	2.58		3.89	5.30
4490	2-1/2" pipe size		134	.060		2.17	2.67		4.84	6.40
4500	3" pipe size		128	.063		2.28	2.80		5.08	6.70
4510	3-1/2" pipe size		122	.066		2.38	2.94		5.32	7.05
4520	4" pipe size		117	.068		2.41	3.06		5.47	7.25
4530	5" pipe size		114	.070		2.63	3.14		5.77	7.60
4540	6" pipe size		111	.072		4.57	3.23		7.80	9.90
4580	For plastic coating on 1/2" thru 6" size, add					150%				
8800	Wire cable support system									
8810	Cable with hook terminal and locking device									
8830	2 mm, (.079") dia cable, (100 lb. cap.)									
8840	1 m, (3.3') length, with hook	1 Shee	96	.083	Ea.	3.20	3.63		6.83	9.10
8850	2 m, (6.6') length, with hook		84	.095		3.69	4.15		7.84	10.45
8860	3 m, (9.9') length, with hook		72	.111		4.19	4.84		9.03	12.05
8870	5 m, (16.4') length, with hook	Q-9	60	.267		5.25	10.45		15.70	22
8880	10 m, (32.8') length, with hook	"	30	.533		7.65	21		28.65	40.50
8900	3mm, (.118") dia cable, (200 lb. cap.)									
8910	1 m, (3.3') length, with hook	1 Shee	96	.083	Ea.	4.10	3.63		7.73	10.10
8920	2 m, (6.6') length, with hook		84	.095		4.61	4.15		8.76	11.45
8930	3 m, (9.9') length, with hook		72	.111		5.10	4.84		9.94	13.05
8940	5 m, (16.4') length, with hook	Q-9	60	.267		6.30	10.45		16.75	23
8950	10 m, (32.8') length, with hook	"	30	.533		9.05	21		30.05	42
9000	Cable system accessories									
9010	Anchor bolt, 3/8", with nut	1 Shee	140	.057	Ea.	1.17	2.49		3.66	5.15
9020	Air duct corner protector		160	.050		.55	2.18		2.73	3.97
9030	Air duct support attachment		140	.057		.20	2.49		2.69	4.06
9040	Flange clip, hammer-on style									
9044	For flange thickness 3/32" - 9/64", 160 lb. cap.	1 Shee	180	.044	Ea.	.28	1.94		2.22	3.29

22 05 Common Work Results for Plumbing

22 05 29 – Hangers and Supports for Plumbing Piping and Equipment

22 05 29.10 Hangers & Supp. for Plumb'g/HVAC Pipe/Equip.		Crew	Daily Output	Labor-Hours	Unit	Material	2007 Bare Costs Labor	2007 Bare Costs Equipment	Total	Total Incl O&P
9048	For flange thickness 1/8" - 1/4", 200 lb. cap.	1 Shee	160	.050	Ea.	.28	2.18		2.46	3.67
9052	For flange thickness 5/16" - 1/2", 200 lb. cap.		150	.053		.59	2.32		2.91	4.23
9056	For flange thickness 9/16" - 3/4", 200 lb. cap.		140	.057		.79	2.49		3.28	4.71
9060	Wire insulation protection tube		180	.044	L.F.	.35	1.94		2.29	3.37
9070	Wire cutter				Ea.	34			34	37.50

22 33 Electric Domestic Water Heaters

22 33 13 – Instantaneous Electric Domestic Water Heaters

22 33 13.10 Hot Water Dispensers

		Crew	Daily Output	Labor-Hours	Unit	Material	Labor	Equipment	Total	Total Incl O&P
0010	**HOT WATER DISPENSERS**									
0160	Commercial, 100 cup, 11.3 amp	1 Plum	14	.571	Ea.	345	25.50		370.50	420
3180	Household, 60 cup	"	14	.571	"	173	25.50		198.50	230

22 33 30 – Residential, Electric Domestic Water Heaters

22 33 30.13 Residential, Small-Capacity Electric Domestic Water Heaters

		Crew	Daily Output	Labor-Hours	Unit	Material	Labor	Equipment	Total	Total Incl O&P
0010	**RESIDENTIAL, SMALL-CAPACITY ELECTRIC DOMESTIC WATER HEATERS**									
1000	Residential, electric, glass lined tank, 5 yr, 10 gal., single element [R2240000-10] [R2240000-20]	1 Plum	2.30	3.478	Ea.	273	156		429	535
1040	20 gallon, single element		2.20	3.636		345	163		508	625
1060	30 gallon, double element		2.20	3.636		385	163		548	670
1080	40 gallon, double element		2	4		410	179		589	720
1100	52 gallon, double element		2	4		460	179		639	775
1120	66 gallon, double element		1.80	4.444		625	199		824	990
1140	80 gallon, double element		1.60	5		700	224		924	1,100
1180	120 gallon, double element		1.40	5.714		975	256		1,231	1,450

22 33 33 – Light-Commercial Electric Domestic Water Heaters

22 33 33.10 Commercial Electric Water Heaters

		Crew	Daily Output	Labor-Hours	Unit	Material	Labor	Equipment	Total	Total Incl O&P
0010	**COMMERCIAL ELECTRIC WATER HEATERS**									
4000	Commercial, 100° rise. NOTE: for each size tank, a range of									
4010	heaters between the ones shown are available									
4020	Electric									
4100	5 gal., 3 kW, 12 GPH, 208V	1 Plum	2	4	Ea.	2,550	179		2,729	3,075
4120	10 gal., 6 kW, 25 GPH, 208V		2	4		2,825	179		3,004	3,375
4140	50 gal., 9 kW, 37 GPH, 208V		1.80	4.444		3,850	199		4,049	4,550
4160	50 gal., 36 kW, 148 GPH, 208V		1.80	4.444		5,900	199		6,099	6,800
4180	80 gal., 12 kW, 49 GPH, 208V		1.50	5.333		4,775	239		5,014	5,600
4200	80 gal., 36 kW, 148 GPH, 208V		1.50	5.333		6,575	239		6,814	7,600
4220	100 gal., 36 kW, 148 GPH, 208V		1.20	6.667		6,875	299		7,174	8,000
4240	120 gal., 36 kW, 148 GPH, 208V		1.20	6.667		7,150	299		7,449	8,325
4260	150 gal., 15 kW , 61 GPH, 480V		1	8		17,000	360		17,360	19,200
4280	150 gal., 120 kW, 490 GPH, 480V		1	8		23,800	360		24,160	26,700
4300	200 gal., 15 kW, 61 GPH, 480V	Q-1	1.70	9.412		18,500	380		18,880	21,000
4320	200 gal., 120 kW , 490 GPH, 480V		1.70	9.412		25,300	380		25,680	28,400
4340	250 gal., 15 kW, 61 GPH, 480V		1.50	10.667		19,100	430		19,530	21,600
4360	250 gal., 150 kW, 615 GPH, 480V		1.50	10.667		27,800	430		28,230	31,200
4380	300 gal., 30 kW, 123 GPH, 480V		1.30	12.308		20,600	495		21,095	23,400
4400	300 gal., 180 kW, 738 GPH, 480V		1.30	12.308		29,800	495		30,295	33,500
4420	350 gal., 30 kW, 123 GPH, 480V		1.10	14.545		22,600	585		23,185	25,800
4440	350 gal., 180 kW, 738 GPH, 480V		1.10	14.545		30,700	585		31,285	34,700
4460	400 gal., 30 kW, 123 GPH, 480V		1	16		25,500	645		26,145	29,100
4480	400 gal., 210 kW, 860 GPH, 480V		1	16		35,500	645		36,145	40,000

22 33 Electric Domestic Water Heaters

22 33 33 – Light-Commercial Electric Domestic Water Heaters

22 33 33.10 Commercial Electric Water Heaters

		Crew	Daily Output	Labor-Hours	Unit	Material	2007 Bare Costs Labor	Equipment	Total	Total Incl O&P
4500	500 gal., 30 kW, 123 GPH, 480V	Q-1	.80	20	Ea.	29,400	805		30,205	33,600
4520	500 gal., 240 kW, 984 GPH, 480V		.80	20		42,400	805		43,205	47,800
4540	600 gal., 30 kW, 123 GPH, 480V	Q-2	1.20	20		34,000	835		34,835	38,700
4560	600 gal., 300 kW, 1230 GPH, 480V		1.20	20		51,000	835		51,835	57,500
4580	700 gal., 30 kW, 123 GPH, 480V		1	24		36,800	1,000		37,800	42,000
4600	700 gal., 300 kW, 1230 GPH, 480V		1	24		52,500	1,000		53,500	59,000
4620	800 gal., 60 kW, 245 GPH, 480V		.90	26.667		38,200	1,125		39,325	43,700
4640	800 gal., 300 kW, 1230 GPH, 480V		.90	26.667		54,500	1,125		55,625	61,000
4660	1000 gal., 60 kW, 245 GPH, 480V		.70	34.286		41,900	1,425		43,325	48,200
4680	1000 gal., 480 kW, 1970 GPH, 480V		.70	34.286		69,500	1,425		70,925	78,500
4700	1250 gal., 60 kW, 245 GPH, 480V		.60	40		47,300	1,675		48,975	54,500
4720	1250 gal., 480 kW, 1970 GPH, 480V		.60	40		72,000	1,675		73,675	81,500
4740	1500 gal., 60 kW, 245 GPH, 480V		.50	48		62,500	2,000		64,500	71,500
4760	1500 gal., 480 kW, 1970 GPH, 480V		.50	48		100,000	2,000		102,000	113,000
5400	Modulating step control, 2-5 steps	1 Elec	5.30	1.509		1,850	66.50		1,916.50	2,125
5440	6-10 steps		3.20	2.500		2,375	110		2,485	2,800
5460	11-15 steps		2.70	2.963		2,875	130		3,005	3,350
5480	16-20 steps		1.60	5		3,350	220		3,570	4,000

22 34 Fuel-Fired Domestic Water Heaters

22 34 30 – Residential Gas Domestic Water Heaters

22 34 30.13 Residential, Atmospheric, Gas Domestic Water Heaters

		Crew	Daily Output	Labor-Hours	Unit	Material	2007 Bare Costs Labor	Equipment	Total	Total Incl O&P
0010	**RESIDENTIAL, ATMOSPHERIC, GAS DOMESTIC WATER HEATERS**									
2000	Gas fired, foam lined tank, 10 yr, vent not incl.,									
2040	30 gallon	1 Plum	2	4	Ea.	625	179		804	960
2060	40 gallon		1.90	4.211		635	189		824	980
2080	50 gallon		1.80	4.444		665	199		864	1,025
2100	75 gallon		1.50	5.333		880	239		1,119	1,325
2120	100 gallon		1.30	6.154		1,350	276		1,626	1,925

22 34 46 – Oil-Fired Domestic Water Heaters

22 34 46.10 Residential Oil-Fired Water Heaters

		Crew	Daily Output	Labor-Hours	Unit	Material	2007 Bare Costs Labor	Equipment	Total	Total Incl O&P
0010	**RESIDENTIAL OIL-FIRED WATER HEATERS**									
3000	Oil fired, glass lined tank, 5 yr, vent not included, 30 gallon	1 Plum	2	4	Ea.	665	179		844	1,000
3040	50 gallon		1.80	4.444		1,375	199		1,574	1,800
3060	70 gallon		1.50	5.333		1,725	239		1,964	2,250
3080	85 gallon		1.40	5.714		4,750	256		5,006	5,600

Division Notes

	CREW	DAILY OUTPUT	LABOR-HOURS	UNIT	2007 BARE COSTS				TOTAL INCL O&P
					MAT.	LABOR	EQUIP.	TOTAL	

Estimating Tips

23 10 00 Facility Fuel Systems

- The prices in this subdivision for above- and below-ground storage tanks do not include foundations or hold-down slabs. The estimator should refer to Divisions 3 and 31 for foundation system pricing. In addition to the foundations, required tank accessories, such as tank gauges, leak detection devices, and additional manholes and piping, must be added to the tank prices.

23 50 00 Central Heating Equipment

- When estimating the cost of an HVAC system, check to see who is responsible for providing and installing the temperature control system. It is possible to overlook controls, assuming that they would be included in the electrical estimate.
- When looking up a boiler, be careful on specified capacity. Some manufacturers rate their products on output while others use input.
- Include HVAC insulation for pipe, boiler, and duct (wrap and liner).
- Be careful when looking up mechanical items to get the correct pressure rating and connection type (thread, weld, flange).

23 70 00 Central HVAC Equipment

- Combination heating and cooling units are sized by the air conditioning requirements. (See Reference No. R236000-20 for preliminary sizing guide.)
- A ton of air conditioning is nominally 400 CFM.
- Rectangular duct is taken off by the linear foot for each size, but its cost is usually estimated by the pound. Remember that SMACNA standards now base duct on internal pressure.
- Prefabricated duct is estimated and purchased like pipe: straight sections and fittings.
- Note that cranes or other lifting equipment are not included on any lines in Division 15. For example, if a crane is required to lift a heavy piece of pipe into place high above a gym floor, or to put a rooftop unit on the roof of a four-story building, etc., it must be added. Due to the potential for extreme variation—from nothing additional required to a major crane or helicopter—we feel that including a nominal amount for "lifting contingency" would be useless and detract from the accuracy of the estimate. When using equipment rental from RSMeans do not forget to include the cost of the operator(s).

Reference Numbers

Reference numbers are shown in shaded boxes at the beginning of some major classifications. These numbers refer to related items in the Reference Section. The reference information may be an estimating procedure, an alternate pricing method, or technical information.

Note: Not all subdivisions listed here necessarily appear in this publication.

Note: **i2 Trade Service,** *in part, has been used as a reference source for some of the material prices used in Division 23.*

23 09 23 – Direct-Digital Control System for HVAC

23 09 23.10 Control Components/DDC Systems	Crew	Daily Output	Labor-Hours	Unit	Material	2007 Bare Costs Labor	Equipment	Total	Total Incl O&P
0010 **CONTROL COMPONENTS/DDC SYSTEMS** (Sub's quote incl. M & L)									
0100 Analog inputs									
0110 Sensors (avg. 50' run in 1/2" EMT)									
0120 Duct temperature				Ea.				321.57	353.73
0130 Space temperature								227	250
0140 Duct humidity, +/- 3%								605.32	665.85
0150 Space humidity, +/- 2%								926.90	1,019.59
0160 Duct static pressure								491.82	541
0170 C.F.M./Transducer								662.07	728.28
0180 K.W./Transducer								1,182.27	1,300.50
0182 K.W.H. totalization (not incl. elec. meter pulse xmtr.)								543.38	597.72
0190 Space static pressure				↓				927.27	1,020
1000 Analog outputs (avg. 50' run in 1/2" EMT)									
1010 P/I Transducer				Ea.				548.57	603.43
1020 Analog output, matl. in MUX								264.83	291.31
1030 Pneumatic (not incl. control device)								558	613.83
1040 Electric (not incl control device)				↓				331	364.14
2000 Status (Alarms)									
2100 Digital inputs (avg. 50' run in 1/2" EMT)									
2110 Freeze				Ea.				378.33	416.16
2120 Fire								340.49	374.54
2130 Differential pressure, (air)								520.20	572.22
2140 Differential pressure, (water)								737.74	811.51
2150 Current sensor								378.33	416.16
2160 Duct high temperature thermostat								496.55	546.21
2170 Duct smoke detector				↓				614.78	676.26
2200 Digital output (avg. 50' run in 1/2" EMT)									
2210 Start/stop				Ea.				296.73	326.40
2220 On/off (maintained contact)				"				510	561
3000 Controller M.U.X. panel, incl. function boards									
3100 48 point				Ea.				4,587.22	5,045.94
3110 128 point				"				6,290	6,918.66
3200 D.D.C. controller (avg. 50' run in conduit)									
3210 Mechanical room									
3214 16 point controller (incl. 120v/1ph power supply)				Ea.				1,937.45	3,121.20
3229 32 point controller (incl. 120v/1ph power supply)				"				4,729	5,202
3230 Includes software programming and checkout									
3260 Space									
3266 V.A.V. terminal box (incl. space temp. sensor)				Ea.				733	806.31
3280 Host computer (avg. 50' run in conduit)									
3281 Package complete with PC, keyboard,									
3282 printer, color CRT, modem, basic software				Ea.				8,512.36	9,363.60
4000 Front end costs									
4100 Computer (P.C.)/software program				Ea.				5,675	6,242.40
4200 Color graphics software								3,405	3,745.44
4300 Color graphics slides								426	468.18
4350 Additional dot matrix printer				↓				851	936.36
4400 Communications trunk cable				L.F.				3.31	3.64
4500 Engineering labor, (not incl. dftg.)				Point				71.88	79.07
4600 Calibration labor								71.88	79.07
4700 Start-up, checkout labor				↓				108.76	119.64
4800 Drafting labor, as req'd									
5000 Communications bus (data transmission cable)									

23 09 23 – Direct-Digital Control System for HVAC

23 09 23.10 Control Components/DDC Systems	Crew	Daily Output	Labor-Hours	Unit	Material	2007 Bare Costs Labor	Equipment	Total	Total Incl O&P
5010	#18 twisted shielded pair in 1/2" EMT conduit				C.L.F.				331.04	364.14
8000	Applications software									
8050	Basic maintenance manager software (not incl. data base entry)				Ea.				1,702	1,872.72
8100	Time program				Point				5.95	6.55
8120	Duty cycle								11.86	13.05
8140	Optimum start/stop								35.94	39.53
8160	Demand limiting								17.78	19.56
8180	Enthalpy program				↓				35.94	39.53
8200	Boiler optimization				Ea.				1,064	1,170.45
8220	Chiller optimization				"				1,419	1,560.60
8240	Custom applications									
8260	Cost varies with complexity									

23 09 43 – Pneumatic Control System for HVAC

23 09 43.10 Pneumatic Control Systems

		Crew	Daily Output	Labor-Hours	Unit	Material	2007 Bare Costs Labor	Equipment	Total	Total Incl O&P
0010	**PNEUMATIC CONTROL SYSTEMS** (Sub's quote incl. mat. & labor)									
0011	Including a nominal 50 Ft. of tubing. Add control panelboard if req'd.									
0100	Heating and Ventilating, split system									
0200	Mixed air control, economizer cycle,panel readout,tubing									
0220	Up to 10 tons	Q-19	.68	35.294	Ea.	3,150	1,475		4,625	5,675
0240	For 10 to 20 tons		.63	37.915		3,375	1,575		4,950	6,075
0260	For over 20 tons		.58	41.096		3,650	1,725		5,375	6,600
0270	Enthalpy cycle, up to 10 tons		.50	48.387		3,475	2,025		5,500	6,850
0280	For 10 to 20 tons		.46	52.174		3,750	2,175		5,925	7,400
0290	For over 20 tons	↓	.42	56.604	↓	4,075	2,375		6,450	8,025
0300	Heating coil, hot water, 3 way valve,									
0320	Freezestat, limit control on discharge, readout	Q-5	.69	23.088	Ea.	2,350	940		3,290	4,000
0500	Cooling coil, chilled water, room									
0520	Thermostat, 3 way valve	Q-5	2	8	Ea.	1,050	325		1,375	1,650
0600	Cooling tower, fan cycle, damper control,									
0620	Control system including water readout in/out at panel	Q-19	.67	35.821	Ea.	4,150	1,500		5,650	6,800
1000	Unit ventilator, day/night operation,									
1100	freezestat, ASHRAE, cycle 2	Q-19	.91	26.374	Ea.	2,300	1,100		3,400	4,175
2000	Compensated hot water from boiler, valve control,									
2100	readout and reset at panel, up to 60 GPM	Q-19	.55	43.956	Ea.	4,300	1,825		6,125	7,475
2120	For 120 GPM		.51	47.059		4,600	1,975		6,575	8,000
2140	For 240 GPM		.49	49.180		4,800	2,050		6,850	8,375
3000	Boiler room combustion air, damper to 5 SF, controls		1.37	17.582		2,075	735		2,810	3,375
3500	Fan coil, heating and cooling valves, 4 pipe control system		3	8		935	335		1,270	1,525
3600	Heat exchanger system controls	↓	.86	27.907	↓	2,000	1,175		3,175	3,975
3900	Multizone control (one per zone), includes thermostat, damper									
3910	motor and reset of discharge temperature	Q-5	.51	31.373	Ea.	2,075	1,275		3,350	4,200
4000	Pneumatic thermostat, including controlling room radiator valve	"	2.43	6.593	↓	625	268		893	1,100
4040	Program energy saving optimizer	Q-19	1.21	19.786		5,175	825		6,000	6,950
4060	Pump control system	"	3	8		960	335		1,295	1,550
4080	Reheat coil control system, not incl coil	Q-5	2.43	6.593	↓	810	268		1,078	1,300
4500	Air supply for pneumatic control system									
4600	Tank mounted duplex compressor, starter, alternator,									
4620	piping, dryer, PRV station and filter									
4630	1/2 HP	Q-19	.68	35.139	Ea.	7,700	1,475		9,175	10,700
4640	3/4 HP		.64	37.383		8,075	1,550		9,625	11,200
4650	1 HP		.61	39.539		8,825	1,650		10,475	12,200
4660	1-1/2 HP	↓	.58	41.739	↓	9,400	1,750		11,150	12,900

23 09 Instrumentation and Control for HVAC

23 09 43 – Pneumatic Control System for HVAC

23 09 43.10 Pneumatic Control Systems		Crew	Daily Output	Labor-Hours	Unit	Material	2007 Bare Costs Labor	2007 Bare Costs Equipment	Total	Total Incl O&P
4680	3 HP	Q-19	.55	43.956	Ea.	12,800	1,825		14,625	16,900
4690	5 HP	↓	.42	57.143	↓	22,400	2,375		24,775	28,200
4800	Main air supply, includes 3/8" copper main and labor	Q-5	1.82	8.791	C.L.F.	259	360		619	825
4810	If poly tubing used, deduct									30%
7000	Static pressure control for air handling unit, includes pressure									
7010	sensor, receiver controller, readout and damper motors	Q-19	.64	37.383	Ea.	6,125	1,550		7,675	9,100
7020	If return air fan requires control, add									70%
8600	VAV boxes, incl. thermostat, damper motor, reheat coil & tubing	Q-5	1.46	10.989		960	445		1,405	1,725
8610	If no reheat coil, deduct		↓							204

23 09 53 – Pneumatic and Electric Control System for HVAC

23 09 53.10 Control Components

		Crew	Daily Output	Labor-Hours	Unit	Material	2007 Bare Costs Labor	2007 Bare Costs Equipment	Total	Total Incl O&P
0010	**CONTROL COMPONENTS**									
0700	Controller, receiver									
0850	Electric, single snap switch	1 Elec	4	2	Ea.	360	88		448	525
0860	Dual snap switches	"	3	2.667	"	450	117		567	670
3590	Sensor, electric operated									
3620	Humidity	1 Elec	8	1	Ea.	54.50	44		98.50	125
3650	Pressure		8	1		1,150	44		1,194	1,350
3680	Temperature	↓	10	.800	↓	83	35		118	144
5000	Thermostats									
5200	24 hour, automatic, clock	1 Shee	8	1	Ea.	106	43.50		149.50	184
5220	Electric, low voltage, 2 wire	1 Elec	13	.615		18	27		45	60
5230	3 wire		10	.800		22	35		57	77
5420	Electric operated, humidity		8	1		50.50	44		94.50	121
5430	DPST	↓	8	1	↓	71	44		115	144
7090	Valves, motor controlled, including actuator									
7100	Electric motor actuated									
7200	Brass, two way, screwed									
7210	1/2" pipe size	L-6	36	.333	Ea.	162	14.85		176.85	200
7220	3/4" pipe size		30	.400		209	17.80		226.80	256
7230	1" pipe size		28	.429		245	19.05		264.05	299
7240	1-1/2" pipe size		19	.632		310	28		338	380
7250	2" pipe size	↓	16	.750	↓	530	33.50		563.50	630
7350	Brass, three way, screwed									
7360	1/2" pipe size	L-6	33	.364	Ea.	188	16.20		204.20	231
7370	3/4" pipe size		27	.444		267	19.80		286.80	325
7380	1" pipe size		25.50	.471		237	21		258	292
7384	1-1/4" pipe size		21	.571		345	25.50		370.50	420
7390	1-1/2" pipe size		17	.706		340	31.50		371.50	420
7400	2" pipe size	↓	14	.857	↓	470	38		508	570
7550	Iron body, two way, flanged									
7560	2-1/2" pipe size	L-6	4	3	Ea.	470	134		604	715
7570	3" pipe size		3	4		555	178		733	875
7580	4" pipe size	↓	2	6	↓	925	267		1,192	1,425
7850	Iron body, three way, flanged									
7860	2-1/2" pipe size	L-6	3	4	Ea.	915	178		1,093	1,275
7870	3" pipe size		2.50	4.800		1,025	214		1,239	1,450
7880	4" pipe size	↓	2	6	↓	1,300	267		1,567	1,825

23 34 HVAC Fans

23 34 13 – Axial HVAC Fans

23 34 13.10 Axial HVAC Fans		Crew	Daily Output	Labor-Hours	Unit	Material	2007 Bare Costs Labor	Equipment	Total	Total Incl O&P
0010	**AXIAL HVAC FANS**									
0020	Air conditioning and process air handling									
0030	Axial flow, compact, low sound, 2.5" S.P.									
0050	3,800 CFM, 5 HP	Q-20	3.40	5.882	Ea.	4,025	236		4,261	4,800
0080	6,400 CFM, 5 HP		2.80	7.143		4,500	287		4,787	5,400
0100	10,500 CFM, 7-1/2 HP		2.40	8.333		5,625	335		5,960	6,675
0120	15,600 CFM, 10 HP		1.60	12.500		7,075	500		7,575	8,550
0140	23,000 CFM, 15 HP		.70	28.571		10,900	1,150		12,050	13,700
0160	28,000 CFM, 20 HP		.40	50		12,100	2,000		14,100	16,400
1500	Vaneaxial, low pressure, 2000 CFM, 1/2 HP		3.60	5.556		1,375	223		1,598	1,875
1520	4,000 CFM, 1 HP		3.20	6.250		1,575	251		1,826	2,125
1540	8,000 CFM, 2 HP		2.80	7.143		2,050	287		2,337	2,700
1560	16,000 CFM, 5 HP		2.40	8.333		2,950	335		3,285	3,725

23 34 14 – Blower HVAC Fans

23 34 14.10 Blower HVAC Fans		Crew	Daily Output	Labor-Hours	Unit	Material	2007 Bare Costs Labor	Equipment	Total	Total Incl O&P
0010	**BLOWER HVAC FANS**									
2000	Blowers, direct drive with motor, complete									
2020	1045 CFM @ .5" S.P., 1/5 HP	Q-20	18	1.111	Ea.	156	44.50		200.50	241
2040	1385 CFM @ .5" S.P., 1/4 HP		18	1.111		159	44.50		203.50	244
2060	1640 CFM @ .5" S.P., 1/3 HP		18	1.111		189	44.50		233.50	276
2080	1760 CFM @ .5" S.P., 1/2 HP		18	1.111		204	44.50		248.50	294
2090	4 speed									
2100	1164 to 1739 CFM @ .5" S.P., 1/3 HP	Q-20	16	1.250	Ea.	251	50		301	355
2120	1467 to 2218 CFM @ 1.0" S.P., 3/4 HP	"	14	1.429	"	325	57.50		382.50	450
2500	Ceiling fan, right angle, extra quiet, 0.10" S.P.									
2520	95 CFM	Q-20	20	1	Ea.	173	40		213	252
2540	210 CFM		19	1.053		204	42.50		246.50	289
2560	385 CFM		18	1.111		259	44.50		303.50	355
2580	885 CFM		16	1.250		510	50		560	635
2600	1,650 CFM		13	1.538		705	62		767	870
2620	2,960 CFM		11	1.818		940	73		1,013	1,125
2680	For speed control switch, add	1 Elec	16	.500		94	22		116	137
7500	Utility set, steel construction, pedestal, 1/4" S.P.									
7520	Direct drive, 150 CFM, 1/8 HP	Q-20	6.40	3.125	Ea.	680	125		805	935
7540	485 CFM, 1/6 HP		5.80	3.448		855	138		993	1,150
7560	1950 CFM, 1/2 HP		4.80	4.167		1,000	167		1,167	1,350
7580	2410 CFM, 3/4 HP		4.40	4.545		1,850	182		2,032	2,300
7600	3328 CFM, 1-1/2 HP		3	6.667		2,050	268		2,318	2,650
7680	V-belt drive, drive cover, 3 phase									
7700	800 CFM, 1/4 HP	Q-20	6	3.333	Ea.	560	134		694	820
7720	1,300 CFM, 1/3 HP		5	4		585	161		746	890
7740	2,000 CFM, 1 HP		4.60	4.348		695	175		870	1,025
7760	2,900 CFM, 3/4 HP		4.20	4.762		935	191		1,126	1,325
7780	3,600 CFM, 3/4 HP		4	5		1,150	201		1,351	1,575
7800	4,800 CFM, 1 HP		3.50	5.714		1,350	229		1,579	1,850
7820	6,700 CFM, 1-1/2 HP		3	6.667		1,675	268		1,943	2,250
7830	7,500 CFM, 2 HP		2.50	8		2,275	320		2,595	3,000
7840	11,000 CFM, 3 HP		2	10		3,050	400		3,450	3,975
7860	13,000 CFM, 3 HP		1.60	12.500		3,100	500		3,600	4,175
7880	15,000 CFM, 5 HP		1	20		3,200	805		4,005	4,750
7900	17,000 CFM, 7-1/2 HP		.80	25		3,425	1,000		4,425	5,300
7920	20,000 CFM, 7-1/2 HP		.80	25		4,075	1,000		5,075	6,025

23 34 HVAC Fans

23 34 16 – Centrifugal HVAC Fans

23 34 16.10 Centrifugal HVAC Fans	Crew	Daily Output	Labor-Hours	Unit	Material	2007 Bare Costs Labor	Equipment	Total	Total Incl O&P
0010 **CENTRIFUGAL HVAC FANS**									
0200 In-line centrifugal, supply/exhaust booster									
0220 aluminum wheel/hub, disconnect switch, 1/4" S.P.									
0240 500 CFM, 10" diameter connection	Q-20	3	6.667	Ea.	985	268		1,253	1,475
0260 1,380 CFM, 12" diameter connection		2	10		1,050	400		1,450	1,775
0280 1,520 CFM, 16" diameter connection		2	10		1,150	400		1,550	1,875
0300 2,560 CFM, 18" diameter connection		1	20		1,225	805		2,030	2,575
0320 3,480 CFM, 20" diameter connection		.80	25		1,475	1,000		2,475	3,150
0326 5,080 CFM, 20" diameter connection		.75	26.667		1,600	1,075		2,675	3,400
3500 Centrifugal, airfoil, motor and drive, complete									
3520 1000 CFM, 1/2 HP	Q-20	2.50	8	Ea.	1,175	320		1,495	1,775
3540 2,000 CFM, 1 HP		2	10		1,300	400		1,700	2,050
3560 4,000 CFM, 3 HP		1.80	11.111		1,575	445		2,020	2,400
3580 8,000 CFM, 7-1/2 HP		1.40	14.286		2,475	575		3,050	3,600
3600 12,000 CFM, 10 HP		1	20		3,025	805		3,830	4,575
4500 Corrosive fume resistant, plastic									
4600 roof ventilators, centrifugal, V belt drive, motor									
4620 1/4" S.P., 250 CFM, 1/4 HP	Q-20	6	3.333	Ea.	2,525	134		2,659	3,000
4640 895 CFM, 1/3 HP		5	4		2,750	161		2,911	3,275
4660 1630 CFM, 1/2 HP		4	5		3,250	201		3,451	3,875
4680 2240 CFM, 1 HP		3	6.667		3,400	268		3,668	4,125
4700 3810 CFM, 2 HP		2	10		3,775	400		4,175	4,775
4710 5000 CFM, 2 HP		1.80	11.111		7,925	445		8,370	9,400
4715 8000 CFM, 5 HP		1.40	14.286		8,825	575		9,400	10,600
4720 11760 CFM, 5 HP		1	20		8,975	805		9,780	11,100
4740 18810 CFM, 10 HP		.70	28.571		9,300	1,150		10,450	12,000
4800 For intermediate capacity, motors may be varied									
4810 For explosion proof motor, add				Ea.	15%				
5000 Utility set, centrifugal, V belt drive, motor									
5020 1/4" S.P., 1200 CFM, 1/4 HP	Q-20	6	3.333	Ea.	3,075	134		3,209	3,600
5040 1520 CFM, 1/3 HP		5	4		3,075	161		3,236	3,650
5060 1850 CFM, 1/2 HP		4	5		3,100	201		3,301	3,700
5080 2180 CFM, 3/4 HP		3	6.667		3,125	268		3,393	3,850
5100 1/2" S.P., 3600 CFM, 1 HP		2	10		4,525	400		4,925	5,625
5120 4250 CFM, 1-1/2 HP		1.60	12.500		4,600	500		5,100	5,825
5140 4800 CFM, 2 HP		1.40	14.286		4,650	575		5,225	6,000
5160 6920 CFM, 5 HP		1.30	15.385		4,800	620		5,420	6,225
5180 7700 CFM, 7-1/2 HP		1.20	16.667		4,950	670		5,620	6,475
5200 For explosion proof motor, add					15%				
5500 Fans, industrial exhauster, for air which may contain granular matl.									
5520 1000 CFM, 1-1/2 HP	Q-20	2.50	8	Ea.	1,725	320		2,045	2,400
5540 2000 CFM, 3 HP		2	10		2,125	400		2,525	2,975
5560 4000 CFM, 7-1/2 HP		1.80	11.111		2,925	445		3,370	3,875
5580 8000 CFM, 15 HP		1.40	14.286		3,925	575		4,500	5,200
5600 12,000 CFM, 30 HP		1	20		6,325	805		7,130	8,175
7000 Roof exhauster, centrifugal, aluminum housing, 12" galvanized									
7020 curb, bird screen, back draft damper, 1/4" S.P.									
7100 Direct drive, 320 CFM, 11" sq. damper	Q-20	7	2.857	Ea.	385	115		500	600
7120 600 CFM, 11" sq. damper		6	3.333		390	134		524	635
7140 815 CFM, 13" sq. damper		5	4		390	161		551	675
7160 1450 CFM, 13" sq. damper		4.20	4.762		505	191		696	850
7180 2050 CFM, 16" sq. damper		4	5		500	201		701	860

23 34 HVAC Fans

23 34 16 – Centrifugal HVAC Fans

23 34 16.10 Centrifugal HVAC Fans

		Crew	Daily Output	Labor-Hours	Unit	Material	2007 Bare Costs Labor	Equipment	Total	Total Incl O&P
7200	V-belt drive, 1650 CFM, 12" sq. damper	Q-20	6	3.333	Ea.	835	134		969	1,125
7220	2750 CFM, 21" sq. damper		5	4		965	161		1,126	1,300
7230	3500 CFM, 21" sq. damper		4.50	4.444		1,075	178		1,253	1,450
7240	4910 CFM, 23" sq. damper		4	5		1,325	201		1,526	1,750
7260	8525 CFM, 28" sq. damper		3	6.667		1,650	268		1,918	2,225
7280	13,760 CFM, 35" sq. damper		2	10		2,275	400		2,675	3,125
7300	20,558 CFM, 43" sq. damper		1	20		4,825	805		5,630	6,550
7320	For 2 speed winding, add					15%				
7340	For explosionproof motor, add					340			340	375
7360	For belt driven, top discharge, add					15%				
8500	Wall exhausters, centrifugal, auto damper, 1/8" S.P.									
8520	Direct drive, 610 CFM, 1/20 HP	Q-20	14	1.429	Ea.	251	57.50		308.50	365
8540	796 CFM, 1/12 HP		13	1.538		259	62		321	380
8560	822 CFM, 1/6 HP		12	1.667		410	67		477	550
8580	1,320 CFM, 1/4 HP		12	1.667		415	67		482	555
8600	1756 CFM, 1/4 HP		11	1.818		480	73		553	640
8620	1983 CFM, 1/4 HP		10	2		575	80.50		655.50	760
8640	2900 CFM, 1/2 HP		9	2.222		700	89		789	905
8660	3307 CFM, 3/4 HP		8	2.500		765	100		865	1,000
9500	V-belt drive, 3 phase									
9520	2,800 CFM, 1/4 HP	Q-20	9	2.222	Ea.	1,150	89		1,239	1,375
9540	3,740 CFM, 1/2 HP		8	2.500		1,175	100		1,275	1,450
9560	4400 CFM, 3/4 HP		7	2.857		1,200	115		1,315	1,500
9580	5700 CFM, 1-1/2 HP		6	3.333		1,250	134		1,384	1,575

23 34 23 – HVAC Power Ventilators

23 34 23.10 HVAC Power Ventilators

		Crew	Daily Output	Labor-Hours	Unit	Material	2007 Bare Costs Labor	Equipment	Total	Total Incl O&P
0010	**HVAC POWER VENTILATORS**									
3000	Paddle blade air circulator, 3 speed switch									
3020	42", 5,000 CFM high, 3000 CFM low	1 Elec	2.40	3.333	Ea.	80	146		226	305
3040	52", 6,500 CFM high, 4000 CFM low	"	2.20	3.636	"	85.50	160		245.50	330
3100	For antique white motor, same cost									
3200	For brass plated motor, same cost									
3300	For light adaptor kit, add				Ea.	28.50			28.50	31.50
6000	Propeller exhaust, wall shutter, 1/4" S.P.									
6020	Direct drive, two speed									
6100	375 CFM, 1/10 HP	Q-20	10	2	Ea.	310	80.50		390.50	465
6120	730 CFM, 1/7 HP		9	2.222		345	89		434	515
6140	1000 CFM, 1/8 HP		8	2.500		480	100		580	685
6160	1890 CFM, 1/4 HP		7	2.857		490	115		605	715
6180	3275 CFM, 1/2 HP		6	3.333		500	134		634	750
6200	4720 CFM, 1 HP		5	4		770	161		931	1,100
6300	V-belt drive, 3 phase									
6320	6175 CFM, 3/4 HP	Q-20	5	4	Ea.	645	161		806	955
6340	7500 CFM, 3/4 HP		5	4		680	161		841	995
6360	10,100 CFM, 1 HP		4.50	4.444		830	178		1,008	1,200
6380	14,300 CFM, 1-1/2 HP		4	5		975	201		1,176	1,375
6400	19,800 CFM, 2 HP		3	6.667		1,125	268		1,393	1,625
6420	26,250 CFM, 3 HP		2.60	7.692		1,350	310		1,660	1,950
6440	38,500 CFM, 5 HP		2.20	9.091		1,575	365		1,940	2,300
6460	46,000 CFM, 7-1/2 HP		2	10		1,700	400		2,100	2,500
6480	51,500 CFM, 10 HP		1.80	11.111		1,750	445		2,195	2,625
6650	Residential, bath exhaust, grille, back draft damper									

23 34 HVAC Fans

23 34 23 – HVAC Power Ventilators

23 34 23.10 HVAC Power Ventilators	Crew	Daily Output	Labor-Hours	Unit	Material	2007 Bare Costs Labor	Equipment	Total	Total Incl O&P
6660	50 CFM	Q-20	24	.833	Ea.	39.50	33.50		73	94
6670	110 CFM		22	.909		54.50	36.50		91	116
6680	Light combination, squirrel cage, 100 watt, 70 CFM	↓	24	.833	↓	63	33.50		96.50	120
6700	Light/heater combination, ceiling mounted									
6710	70 CFM, 1450 watt	Q-20	24	.833	Ea.	76	33.50		109.50	135
6800	Heater combination, recessed, 70 CFM		24	.833		36	33.50		69.50	90.50
6820	With 2 infrared bulbs		23	.870		54	35		89	113
6900	Kitchen exhaust, grille, complete, 160 CFM		22	.909		63.50	36.50		100	126
6910	180 CFM		20	1		54	40		94	121
6920	270 CFM		18	1.111		97.50	44.50		142	176
6930	350 CFM	↓	16	1.250	↓	76	50		126	161
6940	Residential roof jacks and wall caps									
6944	Wall cap with back draft damper									
6946	3" & 4" dia. round duct	1 Shee	11	.727	Ea.	13.80	31.50		45.30	64
6948	6" dia. round duct	"	11	.727	"	33.50	31.50		65	85.50
6958	Roof jack with bird screen and back draft damper									
6960	3" & 4" dia. round duct	1 Shee	11	.727	Ea.	13.30	31.50		44.80	63.50
6962	3-1/4" x 10" rectangular duct	"	10	.800	"	24.50	35		59.50	80.50
6980	Transition									
6982	3-1/4" x 10" to 6" dia. round	1 Shee	20	.400	Ea.	14.15	17.40		31.55	42.50
8020	Attic, roof type									
8030	Aluminum dome, damper & curb									
8040	6" diameter, 300 CFM	1 Elec	16	.500	Ea.	288	22		310	350
8050	7" diameter, 450 CFM		15	.533		315	23.50		338.50	380
8060	9" diameter, 900 CFM		14	.571		505	25		530	595
8080	12" diameter, 1000 CFM (gravity)		10	.800		355	35		390	445
8090	16" diameter, 1500 CFM (gravity)		9	.889		430	39		469	535
8100	20" diameter, 2500 CFM (gravity)		8	1		525	44		569	645
8110	26" diameter, 4000 CFM (gravity)		7	1.143		640	50		690	780
8120	32" diameter, 6500 CFM (gravity)		6	1.333		875	58.50		933.50	1,050
8130	38" diameter, 8000 CFM (gravity)		5	1.600		1,300	70		1,370	1,525
8140	50" diameter, 13,000 CFM (gravity)	↓	4	2	↓	1,875	88		1,963	2,200
8160	Plastic, ABS dome									
8180	1050 CFM	1 Elec	14	.571	Ea.	105	25		130	153
8200	1600 CFM	"	12	.667	"	157	29.50		186.50	217
8240	Attic, wall type, with shutter, one speed									
8250	12" diameter, 1000 CFM	1 Elec	14	.571	Ea.	227	25		252	287
8260	14" diameter, 1500 CFM		12	.667		245	29.50		274.50	315
8270	16" diameter, 2000 CFM	↓	9	.889	↓	278	39		317	365
8290	Whole house, wall type, with shutter, one speed									
8300	30" diameter, 4800 CFM	1 Elec	7	1.143	Ea.	595	50		645	730
8310	36" diameter, 7000 CFM		6	1.333		645	58.50		703.50	795
8320	42" diameter, 10,000 CFM		5	1.600		725	70		795	900
8330	48" diameter, 16,000 CFM	↓	4	2		900	88		988	1,125
8340	For two speed, add				↓	54			54	59.50
8350	Whole house, lay-down type, with shutter, one speed									
8360	30" diameter, 4500 CFM	1 Elec	8	1	Ea.	635	44		679	760
8370	36" diameter, 6500 CFM		7	1.143		680	50		730	825
8380	42" diameter, 9000 CFM		6	1.333		750	58.50		808.50	905
8390	48" diameter, 12,000 CFM		5	1.600		850	70		920	1,050
8440	For two speed, add					40.50			40.50	45
8450	For 12 hour timer switch, add	1 Elec	32	.250	↓	40.50	11		51.50	61.50

23 52 Heating Boilers

23 52 13 – Electric Boilers

23 52 13.10 Electric Boilers, ASME	Crew	Daily Output	Labor-Hours	Unit	Material	2007 Bare Costs Labor	Equipment	Total	Total Incl O&P
0010 **ELECTRIC BOILERS, ASME**, Standard controls and trim.									
1000 Steam, 6 KW, 20.5 MBH	Q-19	1.20	20	Ea.	3,175	835		4,010	4,725
1040 9 KW, 30.7 MBH		1.20	20		3,125	835		3,960	4,675
1060 18 KW, 61.4 MBH		1.20	20		3,225	835		4,060	4,775
1080 24 KW, 81.8 MBH		1.10	21.818		3,725	910		4,635	5,475
1120 36 KW, 123 MBH		1.10	21.818		4,125	910		5,035	5,900
1160 60 KW, 205 MBH		1	24		5,150	1,000		6,150	7,175
1220 112 KW, 382 MBH		.75	32		7,075	1,325		8,400	9,775
1240 148 KW, 505 MBH		.65	36.923		7,450	1,550		9,000	10,500
1260 168 KW, 573 MBH		.60	40		10,200	1,675		11,875	13,800
1280 222 KW, 758 MBH		.55	43.636		11,700	1,825		13,525	15,600
1300 296 KW, 1010 MBH		.45	53.333		11,800	2,225		14,025	16,300
1320 300 KW, 1023 MBH		.40	60		15,200	2,500		17,700	20,500
1340 370 KW, 1263 MBH		.35	68.571		16,300	2,875		19,175	22,200
1360 444 KW, 1515 MBH		.30	80		17,500	3,350		20,850	24,200
1380 518 KW, 1768 MBH	Q-21	.36	88.889		19,800	3,800		23,600	27,400
1400 592 KW, 2020 MBH		.34	94.118		22,300	4,000		26,300	30,600
1420 666 KW, 2273 MBH		.32	100		23,500	4,250		27,750	32,300
1460 740 KW, 2526 MBH		.28	114		24,700	4,875		29,575	34,400
1480 814 KW, 2778 MBH		.25	128		27,700	5,450		33,150	38,600
1500 962 KW, 3283 MBH		.22	145		28,400	6,200		34,600	40,500
1520 1036 KW, 3536 MBH		.20	160		29,800	6,825		36,625	43,000
1540 1110 KW, 3788 MBH		.19	168		31,300	7,175		38,475	45,200
1560 2070 KW, 7063 MBH		.18	177		43,800	7,575		51,375	59,500
1580 2250 KW, 7677 MBH		.17	188		51,000	8,025		59,025	68,000
1600 2,340 KW, 7984 MBH		.16	200		56,500	8,525		65,025	75,000
2000 Hot water, 7.5 KW, 25.6 MBH	Q-19	1.30	18.462		3,225	770		3,995	4,675
2020 15 KW, 51.2 MBH		1.30	18.462		3,300	770		4,070	4,775
2040 30 KW, 102 MBH		1.20	20		3,425	835		4,260	5,000
2060 45 KW, 164 MBH		1.20	20		3,850	835		4,685	5,475
2070 60 KW, 205 MBH		1.20	20		4,050	835		4,885	5,700
2080 75 KW, 256 MBH		1.10	21.818		4,350	910		5,260	6,175
2100 90 KW, 307 MBH		1.10	21.818		4,700	910		5,610	6,550
2120 105 KW, 358 MBH		1	24		5,025	1,000		6,025	7,025
2140 120 KW, 410 MBH		.90	26.667		5,275	1,125		6,400	7,475
2160 135 KW, 461 MBH		.75	32		5,700	1,325		7,025	8,275
2180 150 KW, 512 MBH		.65	36.923		5,875	1,550		7,425	8,775
2200 165 KW, 563 MBH		.60	40		6,175	1,675		7,850	9,275
2220 296 KW, 1010 MBH		.55	43.636		10,300	1,825		12,125	14,100
2280 370 KW, 1263 MBH		.40	60		11,400	2,500		13,900	16,400
2300 444 KW, 1515 MBH		.35	68.571		13,900	2,875		16,775	19,600
2340 518 KW, 1768 MBH	Q-21	.44	72.727		15,300	3,100		18,400	21,600
2360 592 KW, 2020 MBH		.43	74.419		16,900	3,175		20,075	23,400
2400 666 KW, 2273 MBH		.40	80		18,100	3,400		21,500	25,000
2420 740 KW, 2526 MBH		.39	82.051		18,500	3,500		22,000	25,700
2440 814 KW, 2778 MBH		.38	84.211		19,100	3,600		22,700	26,500
2460 888 KW, 3031 MBH		.37	86.486		20,200	3,675		23,875	27,700
2480 962 KW, 3283 MBH		.36	88.889		22,600	3,800		26,400	30,600
2500 1,036 KW, 3536 MBH		.34	94.118		24,000	4,000		28,000	32,400
2520 1110 KW, 3788 MBH		.33	96.970		26,500	4,125		30,625	35,400
2540 1440 KW, 4915 MBH		.32	100		30,500	4,250		34,750	39,900
2560 1560 KW, 5323 MBH		.31	103		34,200	4,400		38,600	44,300
2580 1680 KW, 5733 MBH		.30	106		36,600	4,550		41,150	47,000

23 52 Heating Boilers

23 52 13 – Electric Boilers

23 52 13.10 Electric Boilers, ASME		Crew	Daily Output	Labor-Hours	Unit	Material	2007 Bare Costs Labor	Equipment	Total	Total Incl O&P
2600	1800 KW, 6143 MBH	Q-21	.29	110	Ea.	38,500	4,700		43,200	49,400
2620	1980 KW, 6757 MBH		.28	114		41,100	4,875		45,975	52,500
2640	2100 KW, 7167 MBH		.27	118		45,900	5,050		50,950	58,000
2660	2220 KW, 7576 MBH		.26	123		47,400	5,250		52,650	60,000
2680	2,400 KW, 8191 MBH		.25	128		50,000	5,450		55,450	63,000
2700	2610 KW, 8905 MBH		.24	133		54,000	5,675		59,675	67,500
2720	2790 KW, 9519 MBH		.23	139		56,000	5,925		61,925	71,000
2740	2970 KW, 10133 MBH		.21	152		57,500	6,500		64,000	73,500
2760	3150 KW, 10748 MBH		.19	168		63,000	7,175		70,175	80,500
2780	3240 KW, 11055 MBH		.18	177		64,000	7,575		71,575	82,000
2800	3420 KW, 11669 MBH		.17	188		68,000	8,025		76,025	87,000
2820	3,600 KW, 12,283 MBH		.16	200		70,500	8,525		79,025	90,500

23 52 28 – Swimming Pool Boilers

23 52 28.10 Swimming Pool Heaters

		Crew	Daily Output	Labor-Hours	Unit	Material	2007 Bare Costs Labor	Equipment	Total	Total Incl O&P
0010	**SWIMMING POOL HEATERS**, Not including wiring, external									
0020	piping, base or pad,									
2000	Electric, 12 KW, 4,800 gallon pool	Q-19	3	8	Ea.	2,050	335		2,385	2,750
2020	15 KW, 7,200 gallon pool		2.80	8.571		2,075	360		2,435	2,800
2040	24 KW, 9,600 gallon pool		2.40	10		2,775	420		3,195	3,675
2060	30 KW, 12,000 gallon pool		2	12		2,875	500		3,375	3,925
2080	35 KW, 14,400 gallon pool		1.60	15		2,900	625		3,525	4,125
2100	55 KW, 24,000 gallon pool		1.20	20		3,950	835		4,785	5,600
9000	To select pool heater: 12 BTUH x S.F. pool area									
9010	X temperature differential =required output									
9050	For electric, KW = gallons x 2.5 divided by 1000									
9100	For family home type pool, double the									
9110	Rated gallon capacity = 1/2°F rise per hour									

23 54 Furnaces

23 54 13 – Electric-Resistance Furnaces

23 54 13.10 Electric-Resistance Furnaces

		Crew	Daily Output	Labor-Hours	Unit	Material	2007 Bare Costs Labor	Equipment	Total	Total Incl O&P
0010	**ELECTRIC-RESISTANCE FURNACES**, Hot air, blowers, std. controls.									
1000	Electric, UL listed									
1020	10.2 MBH	Q-20	5	4	Ea.	325	161		486	605
1040	17.1 MBH		4.80	4.167		340	167		507	630
1060	27.3 MBH		4.60	4.348		410	175		585	720
1100	34.1 MBH		4.40	4.545		425	182		607	745
1120	51.6 MBH		4.20	4.762		525	191		716	870
1140	68.3 MBH		4	5		635	201		836	1,000
1160	85.3 MBH		3.80	5.263		695	211		906	1,100

23 82 Convection Heating and Cooling Units

23 82 16 – Air Coils

23 82 16.20 Duct Heaters	Crew	Daily Output	Labor-Hours	Unit	Material	2007 Bare Costs Labor	Equipment	Total	Total Incl O&P
0010 **DUCT HEATERS**, Electric, 480 V, 3 Ph.									
0020 Finned tubular insert, 500 °F									
0100 8" wide x 6" high, 4.0 kW	Q-20	16	1.250	Ea.	640	50		690	775
0120 12" high, 8.0 kW		15	1.333		1,050	53.50		1,103.50	1,250
0140 18" high, 12.0 kW		14	1.429		1,475	57.50		1,532.50	1,725
0160 24" high, 16.0 kW		13	1.538		1,900	62		1,962	2,200
0180 30" high, 20.0 kW		12	1.667		2,325	67		2,392	2,675
0300 12" wide x 6" high, 6.7 kW		15	1.333		675	53.50		728.50	825
0320 12" high, 13.3 kW		14	1.429		1,100	57.50		1,157.50	1,300
0340 18" high, 20.0 kW		13	1.538		1,525	62		1,587	1,775
0360 24" high, 26.7 kW		12	1.667		1,975	67		2,042	2,275
0380 30" high, 33.3 kW		11	1.818		2,400	73		2,473	2,750
0500 18" wide x 6" high, 13.3 kW		14	1.429		695	57.50		752.50	855
0520 12" high, 26.7 kW		13	1.538		1,250	62		1,312	1,475
0540 18" high, 40.0 kW		12	1.667		1,675	67		1,742	1,925
0560 24" high, 53.3 kW		11	1.818		2,225	73		2,298	2,525
0580 30" high, 66.7 kW		10	2		2,775	80.50		2,855.50	3,175
0700 24" wide x 6" high, 17.8 kW		13	1.538		770	62		832	940
0720 12" high, 35.6 kW		12	1.667		1,350	67		1,417	1,600
0740 18" high, 53.3 kW		11	1.818		1,825	73		1,898	2,100
0760 24" high, 71.1 kW		10	2		2,450	80.50		2,530.50	2,825
0780 30" high, 88.9 kW		9	2.222		3,025	89		3,114	3,450
0900 30" wide x 6" high, 22.2 kW		12	1.667		810	67		877	995
0920 12" high, 44.4 kW		11	1.818		1,425	73		1,498	1,675
0940 18" high, 66.7 kW		10	2		1,925	80.50		2,005.50	2,250
0960 24" high, 88.9 kW		9	2.222		2,500	89		2,589	2,875
0980 30" high, 111.0 kW	▼	8	2.500	▼	3,200	100		3,300	3,650
1400 Note decreased kW available for									
1410 each duct size at same cost									
1420 See line 5000 for modifications and accessories									
2000 Finned tubular flange with insulated									
2020 terminal box, 500 °F									
2100 12" wide x 36" high, 54 kW	Q-20	10	2	Ea.	3,150	80.50		3,230.50	3,575
2120 40" high, 60 kW		9	2.222		3,600	89		3,689	4,100
2200 24" wide x 36" high, 118.8 kW		9	2.222		3,575	89		3,664	4,050
2220 40" high, 132 kW		8	2.500		2,725	100		2,825	3,150
2400 36" wide x 8" high, 40 kW		11	1.818		1,550	73		1,623	1,800
2420 16" high, 80 kW		10	2		2,125	80.50		2,205.50	2,475
2440 24" high, 120 kW		9	2.222		2,800	89		2,889	3,200
2460 32" high, 160 kW		8	2.500		3,575	100		3,675	4,075
2480 36" high, 180 kW		7	2.857		4,400	115		4,515	5,025
2500 40" high, 200 kW		6	3.333		4,875	134		5,009	5,575
2600 40" wide x 8" high, 45 kW		11	1.818		1,675	73		1,748	1,925
2620 16" high, 90 kW		10	2		2,300	80.50		2,380.50	2,650
2640 24" high, 135 kW		9	2.222		2,925	89		3,014	3,325
2660 32" high, 180 kW		8	2.500		3,875	100		3,975	4,425
2680 36" high, 202.5 kW		7	2.857		4,475	115		4,590	5,100
2700 40" high, 225 kW		6	3.333		5,100	134		5,234	5,800
2800 48" wide x 8" high, 54.8 kW		10	2		1,750	80.50		1,830.50	2,050
2820 16" high, 109.8 kW		9	2.222		2,425	89		2,514	2,800
2840 24" high, 164.4 kW		8	2.500		3,075	100		3,175	3,550
2860 32" high, 219.2 kW		7	2.857		4,125	115		4,240	4,700
2880 36" high, 246.6 kW	▼	6	3.333		4,725	134		4,859	5,400

23 82 Convection Heating and Cooling Units

23 82 16 – Air Coils

23 82 16.20 Duct Heaters	Crew	Daily Output	Labor-Hours	Unit	Material	2007 Bare Costs Labor	Equipment	Total	Total Incl O&P
2900	40" high, 274 kW	Q-20	5	4	Ea.	5,350	161		5,511	6,150
3000	56" wide x 8" high, 64 kW		9	2.222		2,000	89		2,089	2,325
3020	16" high, 128 kW		8	2.500		2,750	100		2,850	3,175
3040	24" high, 192 kW		7	2.857		3,350	115		3,465	3,850
3060	32" high, 256 kW		6	3.333		4,675	134		4,809	5,350
3080	36" high, 288kW		5	4		5,350	161		5,511	6,150
3100	40" high, 320kW		4	5		5,925	201		6,126	6,825
3200	64" wide x 8" high, 74kW		8	2.500		2,050	100		2,150	2,400
3220	16" high, 148kW		7	2.857		2,850	115		2,965	3,300
3240	24" high, 222kW		6	3.333		3,625	134		3,759	4,175
3260	32" high, 296kW		5	4		4,850	161		5,011	5,575
3280	36" high, 333kW		4	5		5,775	201		5,976	6,650
3300	40" high, 370kW	▼	3	6.667	▼	6,375	268		6,643	7,400
3800	Note decreased kW available for									
3820	each duct size at same cost									
5000	Duct heater modifications and accessories									
5120	T.C.O. limit auto or manual reset	Q-20	42	.476	Ea.	104	19.10		123.10	145
5140	Thermostat		28	.714		445	28.50		473.50	535
5160	Overheat thermocouple (removable)		7	2.857		630	115		745	870
5180	Fan interlock relay		18	1.111		152	44.50		196.50	237
5200	Air flow switch		20	1		132	40		172	207
5220	Split terminal box cover	▼	100	.200	▼	43.50	8.05		51.55	60.50
8000	To obtain BTU multiply kW by 3413									

23 82 27 – Infrared Units

23 82 27.10 Infrared Units

		Crew	Daily Output	Labor-Hours	Unit	Material	2007 Bare Costs Labor	Equipment	Total	Total Incl O&P
0010	**INFRARED UNITS**									
2000	Electric, single or three phase									
2050	6 kW, 20,478 BTU	1 Elec	2.30	3.478	Ea.	380	153		533	645
2100	13.5 KW, 40,956 BTU		2.20	3.636		575	160		735	875
2150	24 KW, 81,912 BTU	▼	2	4		940	176		1,116	1,275

23 83 Radiant Heating Units

23 83 33 – Electric Radiant Heaters

23 83 33.10 Electric Heating

		Crew	Daily Output	Labor-Hours	Unit	Material	2007 Bare Costs Labor	Equipment	Total	Total Incl O&P
0010	**ELECTRIC HEATING**, Not incl. conduit or feed wiring.									
1100	Rule of thumb: Baseboard units, including control	1 Elec	4.40	1.818	kW	84	80		164	212
1300	Baseboard heaters, 2' long, 375 watt		8	1	Ea.	35	44		79	104
1400	3' long, 500 watt		8	1		39	44		83	109
1600	4' long, 750 watt		6.70	1.194		49	52.50		101.50	132
1800	5' long, 935 watt		5.70	1.404		58	61.50		119.50	156
2000	6' long, 1125 watt		5	1.600		64.50	70		134.50	176
2200	7' long, 1310 watt		4.40	1.818		70.50	80		150.50	197
2400	8' long, 1500 watt		4	2		81.50	88		169.50	221
2600	9' long, 1680 watt		3.60	2.222		90.50	97.50		188	245
2800	10' long, 1875 watt	▼	3.30	2.424	▼	101	106		207	269
2950	Wall heaters with fan, 120 to 277 volt									
3160	Recessed, residential, 750 watt	1 Elec	6	1.333	Ea.	94	58.50		152.50	191
3170	1000 watt		6	1.333		96.50	58.50		155	193
3180	1250 watt		5	1.600		106	70		176	221
3190	1500 watt	▼	4	2		106	88		194	247

23 83 Radiant Heating Units

23 83 33 – Electric Radiant Heaters

23 83 33.10 Electric Heating		Crew	Daily Output	Labor-Hours	Unit	Material	2007 Bare Costs Labor	Equipment	Total	Total Incl O&P
3210	2000 watt	1 Elec	4	2	Ea.	112	88		200	254
3230	2500 watt		3.50	2.286		240	100		340	415
3240	3000 watt		3	2.667		360	117		477	575
3250	4000 watt		2.70	2.963		365	130		495	595
3260	Commercial, 750 watt		6	1.333		153	58.50		211.50	255
3270	1000 watt		6	1.333		153	58.50		211.50	255
3280	1250 watt		5	1.600		153	70		223	273
3290	1500 watt		4	2		153	88		241	299
3300	2000 watt		4	2		153	88		241	299
3310	2500 watt		3.50	2.286		153	100		253	315
3320	3000 watt		3	2.667		270	117		387	470
3330	4000 watt		2.70	2.963		270	130		400	490
3600	Thermostats, integral		16	.500		19.50	22		41.50	54
3800	Line voltage, 1 pole		8	1		23.50	44		67.50	91.50
3810	2 pole		8	1		30.50	44		74.50	99
3820	Low voltage, 1 pole		8	1		26	44		70	94
4000	Heat trace system, 400 degree R238313-10									
4020	115V, 2.5 watts per L.F.	1 Elec	530	.015	L.F.	6.70	.66		7.36	8.35
4030	5 watts per L.F. R238313-20		530	.015		6.70	.66		7.36	8.35
4050	10 watts per L.F.		530	.015		6.70	.66		7.36	8.35
4060	208V, 5 watts per L.F.		530	.015		6.70	.66		7.36	8.35
4080	480V, 8 watts per L.F.		530	.015		6.70	.66		7.36	8.35
4200	Heater raceway									
4260	Heat transfer cement									
4280	1 gallon				Ea.	57.50			57.50	63
4300	5 gallon				"	235			235	258
4320	Snap band, clamp									
4340	3/4" pipe size	1 Elec	470	.017	Ea.		.75		.75	1.11
4360	1" pipe size		444	.018			.79		.79	1.18
4380	1-1/4" pipe size		400	.020			.88		.88	1.31
4400	1-1/2" pipe size		355	.023			.99		.99	1.47
4420	2" pipe size		320	.025			1.10		1.10	1.63
4440	3" pipe size		160	.050			2.20		2.20	3.27
4460	4" pipe size		100	.080			3.51		3.51	5.25
4480	Thermostat NEMA 3R, 22 amp, 0-150 Deg, 10' cap.		8	1		196	44		240	281
4500	Thermostat NEMA 4X, 25 amp, 40 Deg, 5-1/2' cap.		7	1.143		196	50		246	290
4520	Thermostat NEMA 4X, 22 amp, 25-325 Deg, 10' cap.		7	1.143		530	50		580	655
4540	Thermostat NEMA 4X, 22 amp, 15-140 Deg,		6	1.333		445	58.50		503.50	575
4580	Thermostat NEMA 4,7,9, 22 amp, 25-325 Deg, 10' cap.		3.60	2.222		650	97.50		747.50	860
4600	Thermostat NEMA 4,7,9, 22 amp, 15-140 Deg,		3	2.667		645	117		762	885
4720	Fiberglass application tape, 36 yard roll		11	.727		59.50	32		91.50	113
5000	Radiant heating ceiling panels, 2' x 4', 500 watt		16	.500		203	22		225	256
5050	750 watt		16	.500		223	22		245	279
5200	For recessed plaster frame, add		32	.250		42.50	11		53.50	63
5300	Infrared quartz heaters, 120 volts, 1000 watts		6.70	1.194		135	52.50		187.50	227
5350	1500 watt		5	1.600		135	70		205	254
5400	240 volts, 1500 watt		5	1.600		135	70		205	254
5450	2000 watt		4	2		135	88		223	280
5500	3000 watt		3	2.667		155	117		272	345
5550	4000 watt		2.60	3.077		155	135		290	370
5570	Modulating control		.80	10		78	440		518	740
5600	Unit heaters, heavy duty, with fan & mounting bracket									
5650	Single phase, 208-240-277 volt, 3 kW	1 Elec	3.20	2.500	Ea.	345	110		455	545

23 83 33 – Electric Radiant Heaters

23 83 33.10 Electric Heating		Crew	Daily Output	Labor-Hours	Unit	Material	2007 Bare Costs Labor	Equipment	Total	Total Incl O&P
5750	5 kW	1 Elec	2.40	3.333	Ea.	360	146		506	615
5800	7 kW		1.90	4.211		550	185		735	885
5850	10 kW		1.30	6.154		630	270		900	1,100
5950	15 kW		.90	8.889		1,025	390		1,415	1,700
6000	480 volt, 3 kW		3.30	2.424		380	106		486	580
6020	4 kW		3	2.667		390	117		507	600
6040	5 kW		2.60	3.077		455	135		590	700
6060	7 kW		2	4		595	176		771	915
6080	10 kW		1.40	5.714		650	251		901	1,100
6100	13 kW		1.10	7.273		1,025	320		1,345	1,600
6120	15 kW		1	8		1,025	350		1,375	1,650
6140	20 kW		.90	8.889		1,350	390		1,740	2,050
6300	3 phase, 208-240 volt, 5 kW		2.40	3.333		340	146		486	590
6320	7 kW		1.90	4.211		520	185		705	850
6340	10 kW		1.30	6.154		565	270		835	1,025
6360	15 kW		.90	8.889		945	390		1,335	1,625
6380	20 kW		.70	11.429		1,350	500		1,850	2,225
6400	25 kW		.50	16		1,575	700		2,275	2,800
6500	480 volt, 5 kW		2.60	3.077		475	135		610	725
6520	7 kW		2	4		490	176		666	800
6540	10 kW		1.40	5.714		630	251		881	1,075
6560	13 kW		1.10	7.273		1,025	320		1,345	1,600
6580	15 kW		1	8		1,025	350		1,375	1,650
6600	20 kW		.90	8.889		1,325	390		1,715	2,025
6620	25 kW		.60	13.333		1,575	585		2,160	2,625
6630	30 kW		.70	11.429		1,825	500		2,325	2,775
6640	40 kW		.60	13.333		2,350	585		2,935	3,450
6650	50 kW	▼	.50	16	▼	2,825	700		3,525	4,150
6800	Vertical discharge heaters, with fan									
6820	Single phase, 208-240-277 volt, 10 kW	1 Elec	1.30	6.154	Ea.	585	270		855	1,050
6840	15 kW		.90	8.889		980	390		1,370	1,650
6900	3 phase, 208-240 volt, 10 kW		1.30	6.154		565	270		835	1,025
6920	15 kW		.90	8.889		945	390		1,335	1,625
6940	20 kW		.70	11.429		1,350	500		1,850	2,225
6960	25 kW		.50	16		1,575	700		2,275	2,800
6980	30 kW		.40	20		1,825	880		2,705	3,325
7000	40 kW		.36	22.222		2,350	975		3,325	4,025
7020	50 kW		.32	25		2,825	1,100		3,925	4,725
7100	480 volt, 10 kW		1.40	5.714		630	251		881	1,075
7120	15 kW		1	8		1,025	350		1,375	1,650
7140	20 kW		.90	8.889		1,575	390		1,965	2,325
7160	25 kW		.60	13.333		1,575	585		2,160	2,625
7180	30 kW		.50	16		1,825	700		2,525	3,075
7200	40 kW		.40	20		2,350	880		3,230	3,875
7220	50 kW		.35	22.857		2,825	1,000		3,825	4,600
7410	Sill height convector heaters, 5" high x 2' long, 500 watt		6.70	1.194		242	52.50		294.50	345
7420	3' long, 750 watt		6.50	1.231		285	54		339	395
7430	4' long, 1000 watt		6.20	1.290		330	56.50		386.50	450
7440	5' long, 1250 watt		5.50	1.455		375	64		439	505
7450	6' long, 1500 watt		4.80	1.667		425	73		498	575
7460	8' long, 2000 watt		3.60	2.222		635	97.50		732.50	845
7470	10' long, 2500 watt	▼	3	2.667	▼	715	117		832	965
7900	Cabinet convector heaters, 240 volt									

23 83 33 – Electric Radiant Heaters

23 83 33.10 Electric Heating		Crew	Daily Output	Labor-Hours	Unit	Material	2007 Bare Costs Labor	Equipment	Total	Total Incl O&P
7920	3' long, 2000 watt	1 Elec	5.30	1.509	Ea.	1,700	66.50		1,766.50	1,975
7940	3000 watt		5.30	1.509		1,775	66.50		1,841.50	2,050
7960	4000 watt		5.30	1.509		1,825	66.50		1,891.50	2,100
7980	6000 watt		4.60	1.739		1,900	76.50		1,976.50	2,200
8000	8000 watt		4.60	1.739		1,975	76.50		2,051.50	2,300
8020	4' long, 4000 watt		4.60	1.739		1,850	76.50		1,926.50	2,150
8040	6000 watt		4	2		1,925	88		2,013	2,225
8060	8000 watt		4	2		2,000	88		2,088	2,325
8080	10,000 watt		4	2		2,000	88		2,088	2,325
8100	Available also in 208 or 277 volt									
8200	Cabinet unit heaters, 120 to 277 volt, 1 pole,									
8220	wall mounted, 2 kW	1 Elec	4.60	1.739	Ea.	1,600	76.50		1,676.50	1,875
8230	3 kW		4.60	1.739		1,675	76.50		1,751.50	1,950
8240	4 kW		4.40	1.818		1,725	80		1,805	2,025
8250	5 kW		4.40	1.818		1,775	80		1,855	2,075
8260	6 kW		4.20	1.905		1,800	83.50		1,883.50	2,100
8270	8 kW		4	2		1,850	88		1,938	2,175
8280	10 kW		3.80	2.105		1,900	92.50		1,992.50	2,225
8290	12 kW		3.50	2.286		1,925	100		2,025	2,275
8300	13.5 kW		2.90	2.759		2,100	121		2,221	2,475
8310	16 kW		2.70	2.963		2,150	130		2,280	2,550
8320	20 kW		2.30	3.478		3,100	153		3,253	3,625
8330	24 kW		1.90	4.211		3,225	185		3,410	3,825
8350	Recessed, 2 kW		4.40	1.818		1,600	80		1,680	1,875
8370	3 kW		4.40	1.818		1,675	80		1,755	1,950
8380	4 kW		4.20	1.905		1,725	83.50		1,808.50	2,025
8390	5 kW		4.20	1.905		1,775	83.50		1,858.50	2,075
8400	6 kW		4	2		1,800	88		1,888	2,100
8410	8 kW		3.80	2.105		1,850	92.50		1,942.50	2,200
8420	10 kW		3.50	2.286		1,900	100		2,000	2,225
8430	12 kW		2.90	2.759		1,925	121		2,046	2,300
8440	13.5 kW		2.70	2.963		2,075	130		2,205	2,475
8450	16 kW		2.30	3.478		2,450	153		2,603	2,900
8460	20 kW		1.90	4.211		2,950	185		3,135	3,500
8470	24 kW		1.60	5		3,075	220		3,295	3,700
8490	Ceiling mounted, 2 kW		3.20	2.500		1,600	110		1,710	1,925
8510	3 kW		3.20	2.500		1,675	110		1,785	2,000
8520	4 kW		3	2.667		1,725	117		1,842	2,075
8530	5 kW		3	2.667		1,775	117		1,892	2,125
8540	6 kW		2.80	2.857		1,800	125		1,925	2,150
8550	8 kW		2.40	3.333		1,850	146		1,996	2,275
8560	10 kW		2.20	3.636		1,900	160		2,060	2,325
8570	12 kW		2	4		2,075	176		2,251	2,525
8580	13.5 kW		1.50	5.333		2,100	234		2,334	2,650
8590	16 kW		1.30	6.154		2,150	270		2,420	2,750
8600	20 kW		.90	8.889		3,025	390		3,415	3,900
8610	24 kW		.60	13.333		3,075	585		3,660	4,250
8630	208 to 480 V, 3 pole									
8650	Wall mounted, 2 kW	1 Elec	4.60	1.739	Ea.	1,800	76.50		1,876.50	2,100
8670	3 kW		4.60	1.739		1,875	76.50		1,951.50	2,175
8680	4 kW		4.40	1.818		1,900	80		1,980	2,225
8690	5 kW		4.40	1.818		1,975	80		2,055	2,300
8700	6 kW		4.20	1.905		1,975	83.50		2,058.50	2,300

23 83 33.10 Electric Heating		Crew	Daily Output	Labor- Hours	Unit	Material	2007 Bare Costs Labor	Equipment	Total	Total Incl O&P
8710	8 kW	1 Elec	4	2	Ea.	2,050	88		2,138	2,375
8720	10 kW		3.80	2.105		2,100	92.50		2,192.50	2,450
8730	12 kW		3.50	2.286		2,125	100		2,225	2,500
8740	13.5 kW		2.90	2.759		2,150	121		2,271	2,550
8750	16 kW		2.70	2.963		2,200	130		2,330	2,625
8760	20 kW		2.30	3.478		3,025	153		3,178	3,550
8770	24 kW		1.90	4.211		3,075	185		3,260	3,650
8790	Recessed, 2 kW		4.40	1.818		1,800	80		1,880	2,100
8810	3 kW		4.40	1.818		1,875	80		1,955	2,175
8820	4 kW		4.20	1.905		1,925	83.50		2,008.50	2,225
8830	5 kW		4.20	1.905		1,975	83.50		2,058.50	2,300
8840	6 kW		4	2		1,975	88		2,063	2,300
8850	8 kW		3.80	2.105		2,050	92.50		2,142.50	2,400
8860	10 kW		3.50	2.286		2,100	100		2,200	2,450
8870	12 kW		2.90	2.759		2,125	121		2,246	2,525
8880	13.5 kW		2.70	2.963		2,150	130		2,280	2,575
8890	16 kW		2.30	3.478		2,200	153		2,353	2,650
8900	20 kW		1.90	4.211		3,025	185		3,210	3,600
8920	24 kW		1.60	5		3,075	220		3,295	3,700
8940	Ceiling mount, 2 kW		3.20	2.500		1,800	110		1,910	2,150
8950	3 kW		3.20	2.500		1,850	110		1,960	2,225
8960	4 kW		3	2.667		1,900	117		2,017	2,275
8970	5 kW		3	2.667		1,975	117		2,092	2,350
8980	6 kW		2.80	2.857		1,975	125		2,100	2,350
8990	8 kW		2.40	3.333		2,050	146		2,196	2,475
9000	10 kW		2.20	3.636		2,100	160		2,260	2,550
9020	13.5 kW		1.50	5.333		2,150	234		2,384	2,725
9030	16 kW		1.30	6.154		2,200	270		2,470	2,825
9040	20 kW		.90	8.889		3,025	390		3,415	3,900
9060	24 kW		.60	13.333		3,050	585		3,635	4,250

Estimating Tips

26 05 00 Common Work Results for Electrical

- Conduit should be taken off in three main categories–power distribution, branch power, and branch lighting–so the estimator can concentrate on systems and components, therefore making it easier to ensure all items have been accounted for.

- For cost modifications for elevated conduit installation, add the percentages to labor according to the height of installation, and only to the quantities exceeding the different height levels, not to the total conduit quantities.

- Remember that aluminum wiring of equal ampacity is larger in diameter than copper and may require larger conduit.

- If more than three wires at a time are being pulled, deduct percentages from the labor hours of that grouping of wires.

- The estimator should take the weights of materials into consideration when completing a takeoff. Topics to consider include: How will the materials be supported? What methods of support are available? How high will the support structure have to reach? Will the final support structure be able to withstand the total burden? Is the support material included or separate from the fixture, equipment, and material specified?

- Do not overlook the costs for equipment used in the installation. If scaffolding or highlifts are available in the field, contractors may use them in lieu of the proposed ladders and rolling staging.

26 20 00 Low-Voltage Electrical Transmission

- Supports and concrete pads may be shown on drawings for the larger equipment, or the support system may be only a piece of plywood for the back of a panelboard. In either case, it must be included in the costs.

26 40 00 Electrical and Cathodic Protection

- When taking off grounding system, identify separately the type and size of wire, and list each unique type of ground connection.

26 50 00 Lighting

- Fixtures should be taken off room by room, using the fixture schedule, specifications, and the ceiling plan. For large concentrations of lighting fixtures in the same area, deduct the percentages from labor hours.

Reference Numbers

Reference numbers are shown in shaded boxes at the beginning of some major classifications. These numbers refer to related items in the Reference Section. The reference information may be an estimating procedure, an alternate pricing method, or technical information.

Note: Not all subdivisions listed here necessarily appear in this publication.

Note: **i2 Trade Service,** *in part, has been used as a reference source for some of the material prices used in Division 26.*

26 01 Operation and Maintenance of Electrical Systems

26 01 40 – Operation and Maintenance of Electrical and Cathodic Protection Systems

26 01 40.51 Operation and Maintenance of Electrical Systems	Crew	Daily Output	Labor-Hours	Unit	Material	2007 Bare Costs Labor	Equipment	Total	Total Incl O&P
0010 **OPERATION AND MAINTENANCE OF ELECTRICAL SYSTEMS**									
3000 Remove and replace (reinstall), switch cover	1 Elec	60	.133	Ea.		5.85		5.85	8.70
3020 Outlet cover	"	60	.133	"		5.85		5.85	8.70

26 01 50 – Operation and Maintenance of Lighting

26 01 50.81 Luminaire Replacement

	Crew	Daily Output	Labor-Hours	Unit	Material	Labor	Equipment	Total	Total Incl O&P
0010 **LUMINAIRE REPLACEMENT**									
3200 Remove and replace (reinstall), lighting fixture	1 Elec	4	2	Ea.		88		88	131

26 05 Common Work Results for Electrical

26 05 05 – Selective Electrical Demolition

26 05 05.10 Electrical Demolition

		Crew	Daily Output	Labor-Hours	Unit	Material	Labor	Equipment	Total	Total Incl O&P
0010	**ELECTRICAL DEMOLITION** R260105-30									
0020	Conduit to 15' high, including fittings & hangers									
0100	Rigid galvanized steel, 1/2" to 1" diameter	1 Elec	242	.033	L.F.		1.45		1.45	2.16
0120	1-1/4" to 2"	"	200	.040			1.76		1.76	2.61
0140	2-1/2" to 3-1/2"	2 Elec	302	.053			2.33		2.33	3.46
0160	4" to 6"	"	160	.100			4.39		4.39	6.55
0170	PVC #40, 1/2" to 1"	1 Elec	410	.020			.86		.86	1.28
0172	1-1/4" to 2"		350	.023			1		1	1.49
0174	2-1/2"		250	.032			1.40		1.40	2.09
0176	3" to 3-1/2"	2 Elec	340	.047			2.07		2.07	3.08
0178	4" to 6"	"	230	.070			3.05		3.05	4.55
0200	Electric metallic tubing (EMT), 1/2" to 1"	1 Elec	394	.020			.89		.89	1.33
0220	1-1/4" to 1-1/2"		326	.025			1.08		1.08	1.60
0240	2" to 3"		236	.034			1.49		1.49	2.22
0260	3-1/2" to 4"	2 Elec	310	.052			2.27		2.27	3.37
0270	Armored cable, (BX) avg. 50' runs									
0280	#14, 2 wire	1 Elec	690	.012	L.F.		.51		.51	.76
0290	#14, 3 wire		571	.014			.62		.62	.92
0300	#12, 2 wire		605	.013			.58		.58	.86
0310	#12, 3 wire		514	.016			.68		.68	1.02
0320	#10, 2 wire		514	.016			.68		.68	1.02
0330	#10, 3 wire		425	.019			.83		.83	1.23
0340	#8, 3 wire		342	.023			1.03		1.03	1.53
0350	Non metallic sheathed cable (Romex)									
0360	#14, 2 wire	1 Elec	720	.011	L.F.		.49		.49	.73
0370	#14, 3 wire		657	.012			.53		.53	.80
0380	#12, 2 wire		629	.013			.56		.56	.83
0390	#10, 3 wire		450	.018			.78		.78	1.16
0400	Wiremold raceway, including fittings & hangers									
0420	No. 3000	1 Elec	250	.032	L.F.		1.40		1.40	2.09
0440	No. 4000		217	.037			1.62		1.62	2.41
0460	No. 6000		166	.048			2.12		2.12	3.15
0465	Telephone/power pole		12	.667	Ea.		29.50		29.50	43.50
0470	Non-metallic, straight section		480	.017	L.F.		.73		.73	1.09
0500	Channels, steel, including fittings & hangers									
0520	3/4" x 1-1/2"	1 Elec	308	.026	L.F.		1.14		1.14	1.70
0540	1-1/2" x 1-1/2"		269	.030			1.31		1.31	1.94
0560	1-1/2" x 1-7/8"		229	.035			1.53		1.53	2.28
0600	Copper bus duct, indoor, 3 phase									

26 05 05 – Selective Electrical Demolition

26 05 05.10 Electrical Demolition	Crew	Daily Output	Labor-Hours	Unit	Material	2007 Bare Costs Labor	Equipment	Total	Total Incl O&P
0610 Including hangers & supports									
0620 225 amp	2 Elec	135	.119	L.F.		5.20		5.20	7.75
0640 400 amp		106	.151			6.65		6.65	9.85
0660 600 amp		86	.186			8.15		8.15	12.15
0680 1000 amp		60	.267			11.70		11.70	17.45
0700 1600 amp		40	.400			17.55		17.55	26
0720 3000 amp		10	1.600			70		70	105
0800 Plug-in switches, 600V 3 ph, incl. disconnecting									
0820 wire, conduit terminations, 30 amp	1 Elec	15.50	.516	Ea.		22.50		22.50	33.50
0840 60 amp		13.90	.576			25.50		25.50	37.50
0850 100 amp		10.40	.769			34		34	50.50
0860 200 amp		6.20	1.290			56.50		56.50	84.50
0880 400 amp	2 Elec	5.40	2.963			130		130	194
0900 600 amp		3.40	4.706			207		207	310
0920 800 amp		2.60	6.154			270		270	400
0940 1200 amp		2	8			350		350	525
0960 1600 amp		1.70	9.412			415		415	615
1010 Safety switches, 250 or 600V, incl. disconnection									
1050 of wire & conduit terminations									
1100 30 amp	1 Elec	12.30	.650	Ea.		28.50		28.50	42.50
1120 60 amp		8.80	.909			40		40	59.50
1140 100 amp		7.30	1.096			48		48	71.50
1160 200 amp		5	1.600			70		70	105
1180 400 amp	2 Elec	6.80	2.353			103		103	154
1200 600 amp	"	4.60	3.478			153		153	227
1210 Panel boards, incl. removal of all breakers,									
1220 conduit terminations & wire connections									
1230 3 wire, 120/240V, 100A, to 20 circuits	1 Elec	2.60	3.077	Ea.		135		135	201
1240 200 amps, to 42 circuits	2 Elec	2.60	6.154			270		270	400
1250 400 amps, to 42 circuits	"	2.20	7.273			320		320	475
1260 4 wire, 120/208V, 125A, to 20 circuits	1 Elec	2.40	3.333			146		146	218
1270 200 amps, to 42 circuits	2 Elec	2.40	6.667			293		293	435
1280 400 amps, to 42 circuits	"	1.92	8.333			365		365	545
1300 Transformer, dry type, 1 ph, incl. removal of									
1320 supports, wire & conduit terminations									
1340 1 kVA	1 Elec	7.70	1.039	Ea.		45.50		45.50	68
1360 5 kVA		4.70	1.702			74.50		74.50	111
1380 10 kVA		3.60	2.222			97.50		97.50	145
1400 37.5 kVA	2 Elec	3	5.333			234		234	350
1420 75 kVA	"	2.50	6.400			281		281	420
1440 3 Phase to 600V, primary									
1460 3 kVA	1 Elec	3.85	2.078	Ea.		91		91	136
1480 15 kVA	2 Elec	4.20	3.810			167		167	249
1500 30 kVA		3.50	4.571			201		201	299
1510 45 kVA		3.10	5.161			227		227	335
1520 75 kVA		2.70	5.926			260		260	385
1530 112.5 kVA	R-3	2.90	6.897			298	56.50	354.50	505
1540 150 kVA		2.70	7.407			320	60.50	380.50	545
1550 300 kVA		1.80	11.111			480	91	571	820
1560 500 kVA		1.40	14.286			620	117	737	1,050
1570 750 kVA		1.10	18.182			785	149	934	1,350
1600 Pull boxes & cabinets, sheet metal, incl. removal									
1620 of supports and conduit terminations									

26 05 05.10 Electrical Demolition		Crew	Daily Output	Labor-Hours	Unit	Material	2007 Bare Costs Labor	Equipment	Total	Total Incl O&P
1640	6" x 6" x 4"	1 Elec	31.10	.257	Ea.		11.30		11.30	16.80
1660	12" x 12" x 4"		23.30	.343			15.05		15.05	22.50
1680	24" x 24" x 6"		12.30	.650			28.50		28.50	42.50
1700	36" x 36" x 8"		7.70	1.039			45.50		45.50	68
1720	Junction boxes, 4" sq. & oct.		80	.100			4.39		4.39	6.55
1740	Handy box		107	.075			3.28		3.28	4.89
1760	Switch box		107	.075			3.28		3.28	4.89
1780	Receptacle & switch plates		257	.031			1.37		1.37	2.03
1790	Receptacles & switches, 15 to 30 amp		135	.059			2.60		2.60	3.87
1800	Wire, THW-THWN-THHN, removed from									
1810	in place conduit, to 15' high									
1830	#14	1 Elec	65	.123	C.L.F.		5.40		5.40	8.05
1840	#12		55	.145			6.40		6.40	9.50
1850	#10		45.50	.176			7.70		7.70	11.50
1860	#8		40.40	.198			8.70		8.70	12.95
1870	#6		32.60	.245			10.75		10.75	16.05
1880	#4	2 Elec	53	.302			13.25		13.25	19.75
1890	#3		50	.320			14.05		14.05	21
1900	#2		44.60	.359			15.75		15.75	23.50
1910	1/0		33.20	.482			21		21	31.50
1920	2/0		29.20	.548			24		24	36
1930	3/0		25	.640			28		28	42
1940	4/0		22	.727			32		32	47.50
1950	250 kcmil		20	.800			35		35	52.50
1960	300 kcmil		19	.842			37		37	55
1970	350 kcmil		18	.889			39		39	58
1980	400 kcmil		17	.941			41.50		41.50	61.50
1990	500 kcmil		16.20	.988			43.50		43.50	64.50
2000	Interior fluorescent fixtures, incl. supports									
2010	& whips, to 15' high									
2100	Recessed drop-in 2' x 2', 2 lamp	2 Elec	35	.457	Ea.		20		20	30
2120	2' x 4', 2 lamp		33	.485			21.50		21.50	31.50
2140	2' x 4', 4 lamp		30	.533			23.50		23.50	35
2160	4' x 4', 4 lamp		20	.800			35		35	52.50
2180	Surface mount, acrylic lens & hinged frame									
2200	1' x 4', 2 lamp	2 Elec	44	.364	Ea.		15.95		15.95	24
2220	2' x 2', 2 lamp		44	.364			15.95		15.95	24
2260	2' x 4', 4 lamp		33	.485			21.50		21.50	31.50
2280	4' x 4', 4 lamp		23	.696			30.50		30.50	45.50
2300	Strip fixtures, surface mount									
2320	4' long, 1 lamp	2 Elec	53	.302	Ea.		13.25		13.25	19.75
2340	4' long, 2 lamp		50	.320			14.05		14.05	21
2360	8' long, 1 lamp		42	.381			16.70		16.70	25
2380	8' long, 2 lamp		40	.400			17.55		17.55	26
2400	Pendant mount, industrial, incl. removal									
2410	of chain or rod hangers, to 15' high									
2420	4' long, 2 lamp	2 Elec	35	.457	Ea.		20		20	30
2440	8' long, 2 lamp	"	27	.593	"		26		26	38.50
2460	Interior incandescent, surface, ceiling									
2470	or wall mount, to 12' high									
2480	Metal cylinder type, 75 Watt	2 Elec	62	.258	Ea.		11.35		11.35	16.85
2500	150 Watt	"	62	.258	"		11.35		11.35	16.85
2520	Metal halide, high bay									

26 05 Common Work Results for Electrical

26 05 05 – Selective Electrical Demolition

26 05 05.10 Electrical Demolition

		Crew	Daily Output	Labor-Hours	Unit	Material	2007 Bare Costs Labor	2007 Bare Costs Equipment	Total	Total Incl O&P
2540	400 Watt	2 Elec	15	1.067	Ea.		47		47	69.50
2560	1000 Watt		12	1.333			58.50		58.50	87
2580	150 Watt, low bay	↓	20	.800	↓		35		35	52.50
2600	Exterior fixtures, incandescent, wall mount									
2620	100 Watt	2 Elec	50	.320	Ea.		14.05		14.05	21
2640	Quartz, 500 Watt		33	.485			21.50		21.50	31.50
2660	1500 Watt	↓	27	.593	↓		26		26	38.50
2680	Wall pack, mercury vapor									
2700	175 Watt	2 Elec	25	.640	Ea.		28		28	42
2720	250 Watt	"	25	.640			28		28	42
7000	Weatherhead/mast, 2"	1 Elec	16	.500			22		22	32.50
7002	3"		10	.800			35		35	52.50
7004	3-1/2"		8.50	.941			41.50		41.50	61.50
7006	4"		8	1	↓		44		44	65.50
7100	Service entry cable, #6, +#6 neutral		420	.019	L.F.		.84		.84	1.24
7102	#4, +#4 neutral		360	.022			.98		.98	1.45
7104	#2, +#4 neutral		345	.023			1.02		1.02	1.52
7106	#2, +#2 neutral		330	.024	↓		1.06		1.06	1.58
9000	Minimum labor/equipment charge	↓	4	2	Job		88		88	131

26 05 05.15 Electrical Demolition, Grounding

		Crew	Daily Output	Labor-Hours	Unit	Material	2007 Bare Costs Labor	2007 Bare Costs Equipment	Total	Total Incl O&P
0010	**ELECTRICAL DEMOLITION, GROUNDING** Addition									
0100	Ground clamp, bronze	1 Elec	64	.125	Ea.		5.50		5.50	8.15
0140	Water pipe ground clamp, bronze, heavy duty	"	24	.333	"		14.65		14.65	22
0200	Ground wire, bare armored	2 Elec	900	.018	L.F.		.78		.78	1.16
0240	Bare copper or aluminum	"	2800	.006	"		.25		.25	.37

26 05 05.20 Electrical Demolition, Wiring Methods

			Crew	Daily Output	Labor-Hours	Unit	Material	2007 Bare Costs Labor	2007 Bare Costs Equipment	Total	Total Incl O&P
0010	**ELECTRICAL DEMOLITION, WIRING METHODS** Addition										
0100	Armored cable, w/ PVC jacket, in cable tray										
0110	#6	R024119-10	1 Elec	9.30	.860	C.L.F.		38		38	56
0120	#4		2 Elec	16.20	.988			43.50		43.50	64.50
0130	#2			13.80	1.159			51		51	76
0140	#1			12	1.333			58.50		58.50	87
0150	1/0			10.80	1.481			65		65	97
0160	2/0			10.20	1.569			69		69	103
0170	3/0			9.60	1.667			73		73	109
0180	4/0		↓	9	1.778			78		78	116
0190	250 kcmil		3 Elec	10.80	2.222			97.50		97.50	145
0210	350 kcmil			9.90	2.424			106		106	158
0230	500 kcmil			9	2.667			117		117	174
0240	750 kcmil		↓	8.10	2.963	↓		130		130	194
1100	Control cable, 600 V or less										
1110	3 wires		1 Elec	24	.333	C.L.F.		14.65		14.65	22
1120	5 wires			20	.400			17.55		17.55	26
1130	7 wires			16.50	.485			21.50		21.50	31.50
1140	9 wires		↓	15	.533			23.50		23.50	35
1150	12 wires		2 Elec	26	.615			27		27	40
1160	15 wires			22	.727			32		32	47.50
1170	19 wires			19	.842			37		37	55
1180	25 wires		↓	16	1	↓		44		44	65.50
1500	Mineral insulated (MI) cable, 600 V										
1510	#10		1 Elec	4.80	1.667	C.L.F.		73		73	109
1520	#8			4.50	1.778	↓		78		78	116

26 05 05.20 Electrical Demolition, Wiring Methods		Crew	Daily Output	Labor-Hours	Unit	Material	2007 Bare Costs Labor	2007 Bare Costs Equipment	Total	Total Incl O&P
1530	#6	1 Elec	4.20	1.905	C.L.F.		83.50		83.50	124
1540	#4	2 Elec	7.20	2.222			97.50		97.50	145
1550	#2		6.60	2.424			106		106	158
1560	#1		6.30	2.540			111		111	166
1570	1/0		6	2.667			117		117	174
1580	2/0		5.70	2.807			123		123	183
1590	3/0		5.40	2.963			130		130	194
1600	4/0		4.80	3.333			146		146	218
1610	250 kcmil	3 Elec	7.20	3.333			146		146	218
1620	500 kcmil	"	5.90	4.068			179		179	266
2000	Shielded cable, XLP shielding, to 35 kV									
2010	#4	2 Elec	13.20	1.212	C.L.F.		53		53	79
2020	#1		12	1.333			58.50		58.50	87
2030	1/0		11.40	1.404			61.50		61.50	91.50
2040	2/0		10.80	1.481			65		65	97
2050	4/0		9.60	1.667			73		73	109
2060	250 kcmil	3 Elec	13.50	1.778			78		78	116
2070	350 kcmil		11.70	2.051			90		90	134
2080	500 kcmil		11.20	2.143			94		94	140
2090	750 kcmil		10.80	2.222			97.50		97.50	145
3000	Modular flexible wiring									
3010	Cable set	1 Elec	120	.067	Ea.		2.93		2.93	4.36
3020	Conversion module		48	.167			7.30		7.30	10.90
3030	Switching assembly		96	.083			3.66		3.66	5.45
3200	Undercarpet									
3210	Power or telephone, flat cable	1 Elec	1320	.006	L.F.		.27		.27	.40
3220	Transition block assemblies w/ fitting		75	.107	Ea.		4.68		4.68	6.95
3230	Floor box with fitting		60	.133			5.85		5.85	8.70
3300	Data system, cable with connection		50	.160			7		7	10.45
4000	Cable tray, including fitting & support									
4010	Galvanized steel, 6" wide	2 Elec	310	.052	L.F.		2.27		2.27	3.37
4020	9" wide		295	.054			2.38		2.38	3.54
4030	12" wide		285	.056			2.46		2.46	3.67
4040	18" wide		270	.059			2.60		2.60	3.87
4050	24" wide		260	.062			2.70		2.70	4.02
4060	30" wide		240	.067			2.93		2.93	4.36
4070	36" wide		220	.073			3.19		3.19	4.75
4110	Aluminum, 6" wide		420	.038			1.67		1.67	2.49
4120	9" wide		415	.039			1.69		1.69	2.52
4130	12" wide		390	.041			1.80		1.80	2.68
4140	18" wide		370	.043			1.90		1.90	2.83
4150	24" wide		350	.046			2.01		2.01	2.99
4160	30" wide		325	.049			2.16		2.16	3.22
4170	36" wide		300	.053			2.34		2.34	3.49
4310	Cable channel, aluminum, 4" wide straight	1 Elec	240	.033			1.46		1.46	2.18
5000	Conduit nipples, with locknuts and bushings									
5020	1/2"	1 Elec	108	.074	Ea.		3.25		3.25	4.84
5040	3/4"		96	.083			3.66		3.66	5.45
5060	1"		81	.099			4.34		4.34	6.45
5080	1-1/4"		69	.116			5.10		5.10	7.60
5100	1-1/2"		60	.133			5.85		5.85	8.70
5120	2"		54	.148			6.50		6.50	9.70
5140	2-1/2"		45	.178			7.80		7.80	11.60

26 05 05 – Selective Electrical Demolition

26 05 05.20 Electrical Demolition, Wiring Methods	Crew	Daily Output	Labor-Hours	Unit	Material	2007 Bare Costs Labor	Equipment	Total	Total Incl O&P
5160	3"	1 Elec	36	.222	Ea.		9.75		9.75	14.50
5180	3-1/2"		33	.242			10.65		10.65	15.85
5200	4"		27	.296			13		13	19.35
5220	5"		21	.381			16.70		16.70	25
5240	6"		18	.444	↓		19.50		19.50	29
5500	Electric nonmetallic tubing (ENT), flexible, 1/2" - 1" diameter		690	.012	L.F.		.51		.51	.76
5510	1-1/4" - 2" diameter		300	.027			1.17		1.17	1.74
5600	Flexible metallic tubing, steel, 3/8" - 3/4" diameter		600	.013			.59		.59	.87
5610	1" - 1-1/4" diameter		260	.031			1.35		1.35	2.01
5620	1-1/2" - 2" diameter		140	.057			2.51		2.51	3.73
5630	2-1/2" diameter		100	.080			3.51		3.51	5.25
5640	3" - 3-1/2" diameter	2 Elec	150	.107			4.68		4.68	6.95
5650	4" diameter	"	100	.160			7		7	10.45
5700	Sealtite flexible conduit, 3/8" - 3/4" diameter	1 Elec	420	.019			.84		.84	1.24
5710	1" - 1-1/4" diameter		180	.044			1.95		1.95	2.90
5720	1-1/2" - 2" diameter		120	.067			2.93		2.93	4.36
5730	2-1/2" diameter		80	.100			4.39		4.39	6.55
5740	3" diameter	2 Elec	150	.107			4.68		4.68	6.95
5750	4" diameter	"	100	.160			7		7	10.45
5800	Wiring duct, plastic, 1-1/2" - 2-1/2" wide	1 Elec	360	.022			.98		.98	1.45
5810	3" wide		330	.024			1.06		1.06	1.58
5820	4" wide	↓	300	.027	↓		1.17		1.17	1.74
6300	Wireway, with fittings and supports, to 15' high									
6310	2-1/2" x 2-1/2"	1 Elec	180	.044	L.F.		1.95		1.95	2.90
6320	4" x 4"	"	160	.050			2.20		2.20	3.27
6330	6" x 6"	2 Elec	240	.067			2.93		2.93	4.36
6340	8" x 8"		160	.100			4.39		4.39	6.55
6350	10" x 10"		120	.133			5.85		5.85	8.70
6360	12" x 12"	↓	80	.200	↓		8.80		8.80	13.05

26 05 05.25 Electrical Demolition, Electrical Power

26 05 05.25		Crew	Daily Output	Labor-Hours	Unit	Material	Labor	Equipment	Total	Total Incl O&P
0010	**ELECTRICAL DEMOLITION, ELECTRICAL POWER** R024119-10									
0100	Meter centers and sockets									
0120	Meter socket, 4 terminal R260105-30	1 Elec	8.40	.952	Ea.		42		42	62
0140	Trans-socket, 13 terminal, 400 A	"	3.60	2.222			97.50		97.50	145
0160	800 A	2 Elec	4.40	3.636			160		160	238
0200	Meter center, 400 A		5.80	2.759			121		121	180
0210	600 A		4	4			176		176	261
0220	800 A		3.30	4.848			213		213	315
0230	1200 A		2.80	5.714			251		251	375
0240	1600 A		2.50	6.400			281		281	420
0300	Base meter devices, 3 meter		3.60	4.444			195		195	290
0310	4 meter		3.30	4.848			213		213	315
0320	5 meter		2.90	5.517			242		242	360
0330	6 meter		2.20	7.273			320		320	475
0340	7 meter		2	8			350		350	525
0350	8 meter		1.90	8.421	↓		370		370	550
0400	Branch meter devices									
0410	Socket w/ circuit breaker 200 A, 2 meter	2 Elec	3.30	4.848	Ea.		213		213	315
0420	3 meter		2.90	5.517			242		242	360
0430	4 meter		2.60	6.154			270		270	400
0450	Main circuit breaker, 400 A		5.80	2.759			121		121	180
0460	600 A	↓	4	4	↓		176		176	261

26 05 05.25 Electrical Demolition, Electrical Power	Crew	Daily Output	Labor-Hours	Unit	Material	2007 Bare Costs Labor	Equipment	Total	Total Incl O&P
0470	800 A	2 Elec	3.30	4.848	Ea.		213		213	315
0480	1200 A		2.80	5.714			251		251	375
0490	1600 A		2.50	6.400			281		281	420
0500	Main lug terminal box, 800 A		3.40	4.706			207		207	310
0510	1200 A		2.60	6.154			270		270	400
1000	Motors, 230/460 V, 60 Hz, 3/4 HP	1 Elec	10.70	.748			33		33	49
1010	5 HP		9	.889			39		39	58
1020	10 HP		8	1			44		44	65.50
1030	15 HP		6.40	1.250			55		55	81.50
1040	20 HP	2 Elec	10.40	1.538			67.50		67.50	101
1050	50 HP		9.60	1.667			73		73	109
1060	75 HP		5.60	2.857			125		125	187
1070	100 HP	3 Elec	5.40	4.444			195		195	290
1080	150 HP		3.60	6.667			293		293	435
1090	200 HP		3	8			350		350	525
1200	Variable frequency drive, 460 V, for 5 HP motor size	1 Elec	3.20	2.500			110		110	163
1210	10 HP motor size	"	2.70	2.963			130		130	194
1220	20 HP motor size	2 Elec	3.60	4.444			195		195	290
1230	50 HP motor size	"	2.10	7.619			335		335	500
1240	75 HP motor size	R-3	2.20	9.091			395	74.50	469.50	665
1250	100 HP motor size		2	10			435	81.50	516.50	735
1260	150 HP motor size		2	10			435	81.50	516.50	735
1270	200 HP motor size		1.70	11.765			510	96	606	865
2000	Generator set w/ accessories, 3 ph 4 wire, 277/480 V									
2010	7.5 kW	3 Elec	2	12	Ea.		525		525	785
2020	20 kW		1.70	14.118			620		620	925
2040	30 kW		1.33	18.045			790		790	1,175
2060	50 kW		1	24			1,050		1,050	1,575
2080	100 kW		.75	32			1,400		1,400	2,100
2100	150 kW		.63	38.095			1,675		1,675	2,500
2120	250 kW		.59	40.678			1,775		1,775	2,650
2140	400 kW		.50	48			2,100		2,100	3,125
2160	500 kW		.44	54.545			2,400		2,400	3,575
2180	750 kW	4 Elec	.56	57.143			2,500		2,500	3,725
2200	1000 kW	"	.46	69.565			3,050		3,050	4,550
3000	Uninterruptible power supply system (UPS)									
3010	Single phase, 120 V, 1 kVA	1 Elec	3.20	2.500	Ea.		110		110	163
3020	2 kVA	2 Elec	3.60	4.444			195		195	290
3030	5 kVA	3 Elec	2.50	9.600			420		420	625
3040	10 kVA		2.30	10.435			460		460	680
3050	15 kVA		1.80	13.333			585		585	870
4000	Transformer, incl support, wire & conduit termination									
4010	Buck-boost, single phase, 120/240 V, 0.1 kVA	1 Elec	25	.320	Ea.		14.05		14.05	21
4020	0.5 kVA		12.50	.640			28		28	42
4030	1 kVA		6.30	1.270			56		56	83
4040	5 kVA		3.80	2.105			92.50		92.50	138
4800	5 kV or 15 kV primary, 277/480 V second, 112.5 kVA	R-3	3.50	5.714			247	46.50	293.50	420
4810	150 kVA		2.70	7.407			320	60.50	380.50	545
4820	225 kVA		2.30	8.696			375	71	446	640
4830	500 kVA		1.45	13.793			595	113	708	1,025
4840	750 kVA		1.33	15.038			650	123	773	1,100
4850	1000 kVA		1.25	16			690	131	821	1,175
4860	2000 kVA		1.05	19.048			825	156	981	1,400

26 05 Common Work Results for Electrical

26 05 05 – Selective Electrical Demolition

26 05 05.25 Electrical Demolition, Electrical Power

		Crew	Daily Output	Labor-Hours	Unit	Material	2007 Bare Costs Labor	Equipment	Total	Total Incl O&P
4870	3000 kVA	R-3	.75	26.667	Ea.		1,150	218	1,368	1,975
6010	Isolation panel, 3 kVA	1 Elec	2.30	3.478			153		153	227
6020	5 kVA		2.20	3.636			160		160	238
6030	7.5 kVA		2.10	3.810			167		167	249
6040	10 kVA		1.80	4.444			195		195	290
6050	15 kVA		1.40	5.714			251		251	375
7000	Power filters & conditioners									
7100	Automatic voltage regulator	2 Elec	6	2.667	Ea.		117		117	174
7200	Capacitor, 1 kVAR	1 Elec	8.60	.930			41		41	61
7210	5 kVAR		5.80	1.379			60.50		60.50	90
7220	10 kVAR		4.80	1.667			73		73	109
7230	15 kVAR		4.20	1.905			83.50		83.50	124
7240	20 kVAR		3.50	2.286			100		100	149
7250	30 kVAR		3.40	2.353			103		103	154
7260	50 kVAR		3.20	2.500			110		110	163
7400	Computer isolator transformer									
7410	Single phase, 120/240 V, 0.5 kVAR	1 Elec	12.80	.625	Ea.		27.50		27.50	41
7420	1 kVAR		8.50	.941			41.50		41.50	61.50
7430	2.5 kVAR		6.40	1.250			55		55	81.50
7440	5 kVAR		3.70	2.162			95		95	141
7500	Computer regulator transformer									
7510	Single phase, 240 V, 0.5 kVAR	1 Elec	8.50	.941	Ea.		41.50		41.50	61.50
7520	1 kVAR		6.40	1.250			55		55	81.50
7530	2 kVAR		3.20	2.500			110		110	163
7540	Single phase, plug-in unit 120 V, 0.5 kVAR		26	.308			13.50		13.50	20
7550	1 kVAR		17	.471			20.50		20.50	31
7600	Power conditioner transformer									
7610	Single phase 115 V - 240 V, 3 kVA	2 Elec	5.10	3.137	Ea.		138		138	205
7620	5 kVA	"	3.70	4.324			190		190	283
7630	7.5 kVA	3 Elec	4.80	5			220		220	325
7640	10 kVA	"	4.30	5.581			245		245	365
7700	Transient suppressor/voltage regulator									
7710	Single phase 115 or 220 V, 1 kVA	1 Elec	8.50	.941	Ea.		41.50		41.50	61.50
7720	2 kVA		7.30	1.096			48		48	71.50
7730	4 kVA		6.80	1.176			51.50		51.50	77
7800	Transient voltage suppressor transformer									
7810	Single phase, 115 or 220 V, 3.6 kVA	1 Elec	12.80	.625	Ea.		27.50		27.50	41
7820	7.2 kVA		11.50	.696			30.50		30.50	45.50
7830	14.4 kVA		10.20	.784			34.50		34.50	51.50
7840	Single phase, plug-in, 120 V, 1.8 kVA		26	.308			13.50		13.50	20
8000	Power measurement & control									
8010	Switchboard instruments, 3 phase 4 wire, indicating unit	1 Elec	25	.320	Ea.		14.05		14.05	21
8020	Recording unit		12	.667			29.50		29.50	43.50
8100	3 current transformers, 3 phase 4 wire, 5 to 800 A		6.40	1.250			55		55	81.50
8110	1000 to 1500 A		4.20	1.905			83.50		83.50	124
8120	2000 to 4000 A		3.20	2.500			110		110	163

26 05 05.30 Electrical Demolition, Transmission and Distr.

		Crew	Daily Output	Labor-Hours	Unit	Material	2007 Bare Costs Labor	Equipment	Total	Total Incl O&P
0010	**ELECTRICAL DEMOLITION, TRANSMISSION & DISTR.**									
0100	Load interrupter switch, 600 A, NEMA 1, 4.8 kV	R-3	1.33	15.038	Ea.		650	123	773	1,100
0120	13.8 kV	R024119-10	"	1.27	15.748			680	129	809	1,175
0200	Lightning arrester, 4.8 kV	1 Elec	9	.889			39		39	58
0220	13.8 kV	R260105-30		6.67	1.199			52.50		52.50	78.50

95

26 05 05 – Selective Electrical Demolition

26 05 05.30 Electrical Demolition, Transmission and Distr.		Crew	Daily Output	Labor-Hours	Unit	Material	2007 Bare Costs Labor	Equipment	Total	Total Incl O&P
0300	Alarm or option items	1 Elec	3.33	2.402	Ea.		105		105	157

26 05 05.35 Electrical Demolition, L.V. Distribution

		Crew	Daily Output	Labor-Hours	Unit	Material	Labor	Equipment	Total	Total Incl O&P
0010	**ELECTRICAL DEMOLITION, L.V. DISTRIBUTION** Addition R024119-10									
0100	Circuit breakers in enclosure									
0120	Enclosed (NEMA 1), 600 V, 3 pole, 30 A R260105-30	1 Elec	12.30	.650	Ea.		28.50		28.50	42.50
0140	60 A		8.80	.909			40		40	59.50
0160	100 A		7.30	1.096			48		48	71.50
0180	225 A		5	1.600			70		70	105
0200	400 A	2 Elec	6.80	2.353			103		103	154
0220	600 A		4.60	3.478			153		153	227
0240	800 A		3.60	4.444			195		195	290
0260	1200 A		3.10	5.161			227		227	335
0280	1600 A		2.80	5.714			251		251	375
0300	2000 A		2.50	6.400			281		281	420
0400	Enclosed (NEMA 7), 600 V, 3 pole, 50 A	1 Elec	8.80	.909			40		40	59.50
0410	100 A		5.80	1.379			60.50		60.50	90
0420	150 A		3.80	2.105			92.50		92.50	138
0430	250 A	2 Elec	6.20	2.581			113		113	169
0440	400 A	"	4.60	3.478			153		153	227
0460	Manual motor starter, NEMA 1	1 Elec	24	.333			14.65		14.65	22
0480	NEMA 4 or NEMA 7		15	.533			23.50		23.50	35
0500	Time switches, single pole single throw		15.40	.519			23		23	34
0520	Photo cell		30	.267			11.70		11.70	17.45
0540	Load management device, 4 loads		7.70	1.039			45.50		45.50	68
0550	8 loads		3.80	2.105			92.50		92.50	138
0600	Transfer switches, enclosed, 30 A		12	.667			29.50		29.50	43.50
0610	60 A		9.50	.842			37		37	55
0620	100 A		6.50	1.231			54		54	80.50
0630	150 A	2 Elec	12	1.333			58.50		58.50	87
0640	260 A		10	1.600			70		70	105
0660	400 A		8	2			88		88	131
0670	600 A		5	3.200			140		140	209
0680	800 A		4	4			176		176	261
0690	1200 A		3.50	4.571			201		201	299
0700	1600 A		3	5.333			234		234	350
0710	2000 A		2.50	6.400			281		281	420
1000	Enclosed controller, NEMA 1, 30 A	1 Elec	13.80	.580			25.50		25.50	38
1020	60 A		11.50	.696			30.50		30.50	45.50
1040	100 A		9.60	.833			36.50		36.50	54.50
1060	150 A		7.70	1.039			45.50		45.50	68
1080	200 A		5.40	1.481			65		65	97
1100	400 A	2 Elec	6.90	2.319			102		102	152
1120	600 A		4.60	3.478			153		153	227
1140	800 A		3.80	4.211			185		185	275
1160	1200 A		3.10	5.161			227		227	335
1200	Control station, NEMA 1	1 Elec	30	.267			11.70		11.70	17.45
1220	NEMA 7		23	.348			15.25		15.25	22.50
1230	Control switches, push button		69	.116			5.10		5.10	7.60
1240	Indicating light unit		120	.067			2.93		2.93	4.36
1250	Relay		15	.533			23.50		23.50	35
2000	Motor control center components									
2020	Starter, NEMA 1, size 1	1 Elec	9	.889	Ea.		39		39	58

26 05 Common Work Results for Electrical

26 05 05 – Selective Electrical Demolition

26 05 05.35 Electrical Demolition, L.V. Distribution		Crew	Daily Output	Labor-Hours	Unit	Material	2007 Bare Costs Labor	Equipment	Total	Total Incl O&P
2040	Size 2	2 Elec	13.30	1.203	Ea.		53		53	78.50
2060	Size 3		6.70	2.388			105		105	156
2080	Size 4		5.30	3.019			133		133	197
2100	Size 5		3.30	4.848			213		213	315
2120	NEMA 7, size 1	1 Elec	8.70	.920			40.50		40.50	60
2140	Size 2	2 Elec	12.70	1.260			55.50		55.50	82.50
2160	Size 3		6.30	2.540			111		111	166
2180	Size 4		5	3.200			140		140	209
2200	Size 5		3.20	5			220		220	325
2300	Fuse, light contactor, NEMA 1, 30 A	1 Elec	9	.889			39		39	58
2310	60 A		6.70	1.194			52.50		52.50	78
2320	100 A		3.30	2.424			106		106	158
2330	200 A		2.70	2.963			130		130	194
2350	Motor control center, incoming section	2 Elec	4	4			176		176	261
2400	Starter & structure, 10 HP	1 Elec	9	.889			39		39	58
2410	25 HP	2 Elec	13.30	1.203			53		53	78.50
2420	50 HP		6.70	2.388			105		105	156
2430	75 HP		5.30	3.019			133		133	197
2440	100 HP		4.70	3.404			149		149	222
2450	200 HP		3.30	4.848			213		213	315
2460	400 HP		2.70	5.926			260		260	385
2500	Motor starter & control									
2510	Motor starter, NEMA 1, 5 HP	1 Elec	8.90	.899	Ea.		39.50		39.50	58.50
2520	10 HP	"	6.20	1.290			56.50		56.50	84.50
2530	25 HP	2 Elec	8.40	1.905			83.50		83.50	124
2540	50 HP		6.90	2.319			102		102	152
2550	100 HP		4.60	3.478			153		153	227
2560	200 HP		3.50	4.571			201		201	299
2570	400 HP		3.10	5.161			227		227	335
2610	Motor starter, NEMA 7, 5 HP	1 Elec	6.20	1.290			56.50		56.50	84.50
2620	10 HP	"	4.20	1.905			83.50		83.50	124
2630	25 HP	2 Elec	6.90	2.319			102		102	152
2640	50 HP		4.60	3.478			153		153	227
2650	100 HP		3.50	4.571			201		201	299
2660	200 HP		1.90	8.421			370		370	550
2710	Combination control unit, NEMA 1, 5 HP	1 Elec	6.90	1.159			51		51	76
2720	10 HP	"	5	1.600			70		70	105
2730	25 HP	2 Elec	7.70	2.078			91		91	136
2740	50 HP		5.10	3.137			138		138	205
2750	100 HP		3.10	5.161			227		227	335
2810	NEMA 7, 5 HP	1 Elec	5	1.600			70		70	105
2820	10 HP	"	3.90	2.051			90		90	134
2830	25 HP	2 Elec	5.10	3.137			138		138	205
2840	50 HP		3.10	5.161			227		227	335
2850	100 HP		2.30	6.957			305		305	455
2860	200 HP		1.50	10.667			470		470	695
3000	Panelboard or load center circuit breaker									
3010	Bolt-on or plug in, 15 A - 50 A	1 Elec	20	.400	Ea.		17.55		17.55	26
3020	60 A - 70 A		16	.500			22		22	32.50
3030	Bolt-on, 80 A - 100 A		14	.571			25		25	37.50
3040	Up to 250 A		7.30	1.096			48		48	71.50
3050	Motor operated, 30 A		13	.615			27		27	40
3060	60 A		10	.800			35		35	52.50

26 05 05 – Selective Electrical Demolition

26 05 05.35 Electrical Demolition, L.V. Distribution

26 05 05.35 Electrical Demolition, L.V. Distribution		Crew	Daily Output	Labor-Hours	Unit	Material	2007 Bare Costs Labor	Equipment	Total	Total Incl O&P
3070	100 A	1 Elec	8	1	Ea.		44		44	65.50
3200	Switchboard circuit breaker									
3210	15 A - 60 A	1 Elec	19	.421	Ea.		18.50		18.50	27.50
3220	70 A - 100 A		14	.571			25		25	37.50
3230	125 A - 400 A		10	.800			35		35	52.50
3240	450 A - 600 A		5.30	1.509			66.50		66.50	98.50
3250	700 A - 800 A		4.30	1.860			81.50		81.50	122
3260	1000 A		3.30	2.424			106		106	158
3270	1200 A		2.70	2.963			130		130	194
3500	Switchboard, incoming section, 400 A	2 Elec	3.70	4.324			190		190	283
3510	600 A		3.30	4.848			213		213	315
3520	800 A		2.90	5.517			242		242	360
3530	1200 A		2.40	6.667			293		293	435
3540	1600 A		2.20	7.273			320		320	475
3550	2000 A		2.10	7.619			335		335	500
3560	3000 A		1.90	8.421			370		370	550
3570	4000 A		1.70	9.412			415		415	615
3610	Distribution section, 600 A		4	4			176		176	261
3620	800 A		3.60	4.444			195		195	290
3630	1200 A		3.10	5.161			227		227	335
3640	1600 A		2.90	5.517			242		242	360
3650	2000 A		2.70	5.926			260		260	385
3710	Transition section, 600 A		3.80	4.211			185		185	275
3720	800 A		3.30	4.848			213		213	315
3730	1200 A		2.70	5.926			260		260	385
3740	1600 A		2.40	6.667			293		293	435
3750	2000 A		2.20	7.273			320		320	475
3760	2500 A		2.10	7.619			335		335	500
3770	3000 A		1.90	8.421			370		370	550
4000	Bus duct, aluminum or copper, 30 A	1 Elec	200	.040	L.F.		1.76		1.76	2.61
4020	60 A		160	.050			2.20		2.20	3.27
4040	100 A		140	.057			2.51		2.51	3.73
5000	Feedrail, trolley busway, up to 60 A		160	.050			2.20		2.20	3.27
5020	100 A		120	.067			2.93		2.93	4.36
5040	Busway, 50 A		200	.040			1.76		1.76	2.61
6000	Fuse, 30 A		133	.060			2.64		2.64	3.93
6010	60 A		133	.060			2.64		2.64	3.93
6020	100 A		105	.076			3.34		3.34	4.98
6030	200 A		95	.084			3.70		3.70	5.50
6040	400 A		80	.100			4.39		4.39	6.55
6050	600 A		53	.151			6.65		6.65	9.85
6060	601 A - 1200 A		42	.190			8.35		8.35	12.45
6070	1500 A - 1600 A		34	.235			10.35		10.35	15.40
6080	Up to 2500 A		26	.308			13.50		13.50	20
6090	4000 A		21	.381			16.70		16.70	25
6100	4500 A - 5000 A		18	.444			19.50		19.50	29
6120	6000 A		15	.533			23.50		23.50	35
6200	Fuse plug or fustat		100	.080			3.51		3.51	5.25

26 05 05.50 Electrical Demolition, Lighting

26 05 05.50 Electrical Demolition, Lighting			Crew	Daily Output	Labor-Hours	Unit	Material	Labor	Equipment	Total	Total Incl O&P
0010	**ELECTRICAL DEMOLITION, LIGHTING** Addition	R024119-10									
0100	Fixture hanger, flexible, 1/2" diameter		1 Elec	36	.222	Ea.		9.75		9.75	14.50
0120	3/4" diameter	R260105-30	"	30	.267	"		11.70		11.70	17.45

26 05 Common Work Results for Electrical

26 05 05 – Selective Electrical Demolition

26 05 05.50 Electrical Demolition, Lighting		Crew	Daily Output	Labor-Hours	Unit	Material	2007 Bare Costs Labor	Equipment	Total	Total Incl O&P
3000	Light pole, anchor base, excl concrete bases									
3010	Metal light pole, 10'	2 Elec	24	.667	Ea.		29.50		29.50	43.50
3020	16'	"	18	.889			39		39	58
3030	20'	R-3	8.70	2.299			99.50	18.80	118.30	169
3040	40'	"	6	3.333			144	27	171	245
3100	Wood light pole, 10'	2 Elec	36	.444			19.50		19.50	29
3120	20'	"	24	.667			29.50		29.50	43.50
3140	Bollard light, 42"	1 Elec	9	.889			39		39	58
3160	Walkway luminaire	"	8.10	.988			43.50		43.50	64.50
4000	Explosionproof									
4010	Metal halide, 175 W	1 Elec	8.70	.920	Ea.		40.50		40.50	60
4020	250 W	"	8.10	.988			43.50		43.50	64.50
4030	400 W	2 Elec	14.40	1.111			49		49	72.50
4050	High pressure sodium, 70 W	1 Elec	9	.889			39		39	58
4060	100 W		9	.889			39		39	58
4070	150 W		8.10	.988			43.50		43.50	64.50
4100	Incandescent		8.70	.920			40.50		40.50	60
4200	Flurorescent		8.10	.988			43.50		43.50	64.50
5000	Ballast, flurorescent fixture		24	.333			14.65		14.65	22
5040	High intensity discharge fixture		24	.333			14.65		14.65	22
5300	Exit and emergency lighting									
5310	Exit light	1 Elec	24	.333	Ea.		14.65		14.65	22
5320	Emergency battery pack lighting unit		12	.667			29.50		29.50	43.50
5330	Remote lamp only		80	.100			4.39		4.39	6.55
5340	Self-contained fluorescent lamp pack		30	.267			11.70		11.70	17.45
5800	Energy saving devices									
5810	Occupancy sensor	1 Elec	21	.381	Ea.		16.70		16.70	25
5820	Automatic wall switch		72	.111			4.88		4.88	7.25
5830	Remote power pack		30	.267			11.70		11.70	17.45
5840	Photoelectric control		24	.333			14.65		14.65	22
6000	Lamps									
6010	Fluorescent	1 Elec	200	.040	Ea.		1.76		1.76	2.61
6030	High intensity discharge lamp, up to 400 W		68	.118			5.15		5.15	7.70
6040	Up to 1000 W		45	.178			7.80		7.80	11.60
6050	Quartz		90	.089			3.90		3.90	5.80
6070	Incandescent		360	.022			.98		.98	1.45
6080	Exterior, PAR		290	.028			1.21		1.21	1.80
6090	Guards for fluorescent lamp		24	.333			14.65		14.65	22

26 05 13 – Medium-Voltage Cables

26 05 13.10 Cable Terminations

26 05 13.10 Cable Terminations		Crew	Daily Output	Labor-Hours	Unit	Material	2007 Bare Costs Labor	Equipment	Total	Total Incl O&P
0010	**CABLE TERMINATIONS,** 5 kV to 35 kV									
0100	Indoor, insulation diameter range .525" to 1.025"									
0300	Padmount, 5 kV	1 Elec	8	1	Ea.	60.50	44		104.50	132
0400	15 kV		6.40	1.250		109	55		164	202
0500	25 kV		6	1.333		127	58.50		185.50	226
0600	35 kV		5.60	1.429		162	62.50		224.50	272
0700	insulation diameter range .975" to 1.570"									
0800	Padmount, 5 kV	1 Elec	8	1	Ea.	73.50	44		117.50	147
0900	15 kV		6	1.333		145	58.50		203.50	247
1000	25 kV		5.60	1.429		167	62.50		229.50	278
1100	35 kV		5.30	1.509		200	66.50		266.50	320
1200	insulation diameter range 1.540" to 1.900"									

26 05 Common Work Results for Electrical

26 05 13 – Medium-Voltage Cables

26 05 13.10 Cable Terminations

		Crew	Daily Output	Labor-Hours	Unit	Material	2007 Bare Costs Labor	Equipment	Total	Total Incl O&P
1300	Padmount, 5 KV	1 Elec	7.40	1.081	Ea.	111	47.50		158.50	193
1400	15 kV		5.60	1.429		184	62.50		246.50	296
1500	25 kV		5.30	1.509		234	66.50		300.50	355
1600	35 kV		5	1.600		262	70		332	395
1700	Outdoor systems, #4 stranded to 1/0 stranded									
1800	5 kV	1 Elec	7.40	1.081	Ea.	120	47.50		167.50	203
1900	15 kV		5.30	1.509		149	66.50		215.50	262
2000	25 kV		5	1.600		167	70		237	289
2100	35 kV		4.80	1.667		195	73		268	325
2200	#1 solid to 4/0 stranded, 5 kV		6.90	1.159		134	51		185	224
2300	15 kV		5	1.600		149	70		219	268
2400	25 kV		4.80	1.667		200	73		273	330
2500	35 kV		4.60	1.739		200	76.50		276.50	335
2600	3/0 solid to 350 kcmil stranded, 5 kV		6.40	1.250		167	55		222	266
2700	15 kV		4.80	1.667		167	73		240	293
2800	25 kV		4.60	1.739		234	76.50		310.50	370
2900	35 kV		4.40	1.818		234	80		314	375
3000	400 kcmil compact to 750 kcmil stranded, 5 kV		6	1.333		217	58.50		275.50	325
3100	15 kV		4.60	1.739		217	76.50		293.50	350
3200	25 kV		4.40	1.818		262	80		342	405
3300	35 kV		4.20	1.905		262	83.50		345.50	410
3400	1000 kcmil, 5 kV		5.60	1.429		234	62.50		296.50	350
3500	15 kV		4.40	1.818		251	80		331	395
3600	25 kV		4.20	1.905		262	83.50		345.50	410
3700	35 kV		4	2		262	88		350	420

26 05 13.16 Medium-Voltage, Single Cable

		Crew	Daily Output	Labor-Hours	Unit	Material	2007 Bare Costs Labor	Equipment	Total	Total Incl O&P
0010	**MEDIUM-VOLTAGE, SINGLE CABLE** Splicing & terminations not included									
0040	Copper, XLP shielding, 5 kV, #6	2 Elec	4.40	3.636	C.L.F.	206	160		366	465
0050	#4		4.40	3.636		267	160		427	530
0100	#2		4	4		315	176		491	610
0200	#1		4	4		365	176		541	660
0400	1/0		3.80	4.211		400	185		585	715
0600	2/0		3.60	4.444		500	195		695	840
0800	4/0		3.20	5		655	220		875	1,050
1000	250 kcmil	3 Elec	4.50	5.333		810	234		1,044	1,250
1200	350 kcmil		3.90	6.154		1,050	270		1,320	1,550
1400	500 kcmil		3.60	6.667		1,275	293		1,568	1,825
1600	15 kV, ungrounded neutral, #1	2 Elec	4	4		440	176		616	745
1800	1/0		3.80	4.211		525	185		710	855
2000	2/0		3.60	4.444		600	195		795	950
2200	4/0		3.20	5		795	220		1,015	1,200
2400	250 kcmil	3 Elec	4.50	5.333		880	234		1,114	1,325
2600	350 kcmil		3.90	6.154		1,125	270		1,395	1,625
2800	500 kcmil		3.60	6.667		1,375	293		1,668	1,925
3000	25 kV, grounded neutral, #1/0	2 Elec	3.60	4.444		730	195		925	1,100
3200	2/0		3.40	4.706		800	207		1,007	1,200
3400	4/0		3	5.333		1,000	234		1,234	1,450
3600	250 kcmil	3 Elec	4.20	5.714		1,250	251		1,501	1,750
3800	350 kcmil		3.60	6.667		1,475	293		1,768	2,025
3900	500 kcmil		3.30	7.273		1,725	320		2,045	2,350
4000	35 kV, grounded neutral, #1/0	2 Elec	3.40	4.706		770	207		977	1,150
4200	2/0		3.20	5		905	220		1,125	1,325

26 05 Common Work Results for Electrical

26 05 13 – Medium-Voltage Cables

26 05 13.16 Medium-Voltage, Single Cable

		Crew	Daily Output	Labor-Hours	Unit	Material	2007 Bare Costs Labor	Equipment	Total	Total Incl O&P
4400	4/0	2 Elec	2.80	5.714	C.L.F.	1,150	251		1,401	1,625
4600	250 kcmil	3 Elec	3.90	6.154		1,325	270		1,595	1,875
4800	350 kcmil		3.30	7.273		1,600	320		1,920	2,250
5000	500 kcmil		3	8		1,875	350		2,225	2,600
5050	Aluminum, XLP shielding, 5 kV, #2	2 Elec	5	3.200		152	140		292	375
5070	#1		4.40	3.636		157	160		317	410
5090	1/0		4	4		182	176		358	460
5100	2/0		3.80	4.211		204	185		389	500
5150	4/0		3.60	4.444		242	195		437	555
5200	250 kcmil	3 Elec	4.80	5		293	220		513	645
5220	350 kcmil		4.50	5.333		345	234		579	730
5240	500 kcmil		3.90	6.154		440	270		710	885
5260	750 kcmil		3.60	6.667		585	293		878	1,075
5300	15 kV aluminum, XLP, #1	2 Elec	4.40	3.636		195	160		355	455
5320	1/0		4	4		202	176		378	485
5340	2/0		3.80	4.211		241	185		426	540
5360	4/0		3.60	4.444		267	195		462	585
5380	250 kcmil	3 Elec	4.80	5		320	220		540	675
5400	350 kcmil		4.50	5.333		360	234		594	745
5420	500 kcmil		3.90	6.154		495	270		765	945
5440	750 kcmil		3.60	6.667		680	293		973	1,175

26 05 19 – Low-Voltage Electrical Power Conductors and Cables

26 05 19.13 Undercarpet Electrical Power Cables

		Crew	Daily Output	Labor-Hours	Unit	Material	2007 Bare Costs Labor	Equipment	Total	Total Incl O&P
0010	**UNDERCARPET ELECTRICAL POWER CABLES** R260519-80									
0020	Power System									
0100	Cable flat, 3 conductor, #12, w/attached bottom shield	1 Elec	982	.008	L.F.	4.42	.36		4.78	5.40
0200	Shield, top, steel		1768	.005	"	4.80	.20		5	5.60
0250	Splice, 3 conductor		48	.167	Ea.	14.65	7.30		21.95	27
0300	Top shield		96	.083		1.30	3.66		4.96	6.90
0350	Tap		40	.200		18.80	8.80		27.60	33.50
0400	Insulating patch, splice, tap, & end		48	.167		46.50	7.30		53.80	62
0450	Fold		230	.035			1.53		1.53	2.27
0500	Top shield, tap & fold		96	.083		1.30	3.66		4.96	6.90
0700	Transition, block assembly		77	.104		67.50	4.56		72.06	81.50
0750	Receptacle frame & base		32	.250		37	11		48	57
0800	Cover receptacle		120	.067		3.15	2.93		6.08	7.85
0850	Cover blank		160	.050		3.70	2.20		5.90	7.35
0860	Receptacle, direct connected, single		25	.320		79.50	14.05		93.55	108
0870	Dual		16	.500		130	22		152	176
0880	Combination Hi & Lo, tension		21	.381		96	16.70		112.70	130
0900	Box, floor with cover		20	.400		80.50	17.55		98.05	115
0920	Floor service w/barrier		4	2		227	88		315	380
1000	Wall, surface, with cover		20	.400		53	17.55		70.55	84
1100	Wall, flush, with cover		20	.400		37	17.55		54.55	66.50
1450	Cable flat, 5 conductor #12, w/attached bottom shield		800	.010	L.F.	7.25	.44		7.69	8.60
1550	Shield, top, steel		1768	.005	"	7.20	.20		7.40	8.20
1600	Splice, 5 conductor		48	.167	Ea.	23.50	7.30		30.80	37
1650	Top shield		96	.083		1.30	3.66		4.96	6.90
1700	Tap		48	.167		31	7.30		38.30	45
1750	Insulating patch, splice tap, & end		83	.096		45.50	4.23		49.73	56.50
1800	Transition, block assembly		77	.104		48.50	4.56		53.06	60.50
1850	Box, wall, flush with cover		20	.400		49.50	17.55		67.05	80

26 05 19 – Low-Voltage Electrical Power Conductors and Cables

26 05 19.13 Undercarpet Electrical Power Cables		Crew	Daily Output	Labor-Hours	Unit	Material	2007 Bare Costs Labor	Equipment	Total	Total Incl O&P
1900	Cable flat, 4 conductor, #12	1 Elec	933	.009	L.F.	5.95	.38		6.33	7.10
1950	3 conductor #10		982	.008		5.15	.36		5.51	6.20
1960	4 conductor #10		933	.009		6.75	.38		7.13	8
1970	5 conductor #10		884	.009		8.25	.40		8.65	9.70
2500	Telephone System									
2510	Transition fitting wall box, surface	1 Elec	24	.333	Ea.	37.50	14.65		52.15	63
2520	Flush		24	.333		37.50	14.65		52.15	63
2530	Flush, for PC board		24	.333		37.50	14.65		52.15	63
2540	Floor service box		4	2		206	88		294	360
2550	Cover, surface					12.65			12.65	13.90
2560	Flush					12.65			12.65	13.90
2570	Flush for PC board					12.65			12.65	13.90
2700	Floor fitting w/duplex jack & cover	1 Elec	21	.381		39	16.70		55.70	68
2720	Low profile		53	.151		13.60	6.65		20.25	25
2740	Miniature w/duplex jack		53	.151		21	6.65		27.65	33
2760	25 pair kit		21	.381		41	16.70		57.70	70
2780	Low profile		53	.151		13.90	6.65		20.55	25
2800	Call director kit for 5 cable		19	.421		65.50	18.50		84	99.50
2820	4 pair kit		19	.421		78	18.50		96.50	113
2840	3 pair kit		19	.421		82.50	18.50		101	118
2860	Comb. 25 pair & 3 cond power		21	.381		65.50	16.70		82.20	97
2880	5 cond power		21	.381		74.50	16.70		91.20	107
2900	PC board, 8-3 pair		161	.050		56.50	2.18		58.68	66
2920	6-4 pair		161	.050		56.50	2.18		58.68	66
2940	3 pair adapter		161	.050		51.50	2.18		53.68	60
2950	Plug		77	.104		2.40	4.56		6.96	9.45
2960	Couplers		321	.025		6.75	1.09		7.84	9.10
3000	Bottom shield for 25 pr. cable		4420	.002	L.F.	.67	.08		.75	.86
3020	4 pair		4420	.002		.32	.08		.40	.47
3040	Top shield for 25 pr. cable		4420	.002		.67	.08		.75	.86
3100	Cable assembly, double-end, 50', 25 pr.		11.80	.678	Ea.	199	30		229	264
3110	3 pair		23.60	.339		58.50	14.90		73.40	86.50
3120	4 pair		23.60	.339		65.50	14.90		80.40	94
3140	Bulk 3 pair		1473	.005	L.F.	1	.24		1.24	1.45
3160	4 pair		1473	.005	"	1.25	.24		1.49	1.73
3500	Data System									
3520	Cable 25 conductor w/conn. 40', 75 ohm	1 Elec	14.50	.552	Ea.	59	24		83	101
3530	Single lead		22	.364		161	15.95		176.95	201
3540	Dual lead		22	.364		200	15.95		215.95	244
3560	Shields same for 25 cond. as 25 pair tele.									
3570	Single & dual, none req'd.									
3590	BNC coax connectors, Plug	1 Elec	40	.200	Ea.	8.55	8.80		17.35	22.50
3600	TNC coax connectors, Plug	"	40	.200	"	11	8.80		19.80	25
3700	Cable-bulk									
3710	Single lead	1 Elec	1473	.005	L.F.	2.15	.24		2.39	2.72
3720	Dual lead	"	1473	.005	"	3.10	.24		3.34	3.76
3730	Hand tool crimp				Ea.	430			430	470
3740	Hand tool notch				"	15.35			15.35	16.90
3750	Boxes & floor fitting same as telephone									
3790	Data cable notching, 90°	1 Elec	97	.082	Ea.		3.62		3.62	5.40
3800	180°		60	.133			5.85		5.85	8.70
8100	Drill floor		160	.050		1.70	2.20		3.90	5.15
8200	Marking floor		1600	.005	L.F.		.22		.22	.33

26 05 Common Work Results for Electrical

26 05 19 – Low-Voltage Electrical Power Conductors and Cables

26 05 19.13 Undercarpet Electrical Power Cables

		Crew	Daily Output	Labor-Hours	Unit	Material	2007 Bare Costs Labor	Equipment	Total	Total Incl O&P
8300	Tape, hold down	1 Elec	6400	.001	L.F.	.13	.05		.18	.22
8350	Tape primer, 500 ft. per can	↓	96	.083	Ea.	26	3.66		29.66	34
8400	Tool, splicing				"	207			207	228

26 05 19.20 Armored Cable

		Crew	Daily Output	Labor-Hours	Unit	Material	2007 Bare Costs Labor	Equipment	Total	Total Incl O&P
0010	**ARMORED CABLE** R260519-20									
0050	600 volt, copper (BX), #14, 2 conductor, solid	1 Elec	2.40	3.333	C.L.F.	74.50	146		220.50	300
0100	3 conductor, solid		2.20	3.636		118	160		278	370
0120	4 conductor, solid		2	4		171	176		347	450
0150	#12, 2 conductor, solid		2.30	3.478		75.50	153		228.50	310
0200	3 conductor, solid		2	4		121	176		297	395
0220	4 conductor, solid		1.80	4.444		175	195		370	480
0240	#12, 19 conductor, stranded		1.10	7.273		890	320		1,210	1,450
0250	#10, 2 conductor, solid		2	4		138	176		314	410
0300	3 conductor, solid		1.60	5		190	220		410	535
0320	4 conductor, solid		1.40	5.714		293	251		544	695
0350	#8, 3 conductor, solid		1.30	6.154		355	270		625	790
0370	4 conductor, stranded		1.10	7.273		510	320		830	1,025
0380	#6, 2 conductor, stranded		1.30	6.154		370	270		640	810
0400	3 conductor with PVC jacket, in cable tray, #6	↓	3.10	2.581		405	113		518	615
0450	#4	2 Elec	5.40	2.963		520	130		650	770
0500	#2		4.60	3.478		695	153		848	985
0550	#1		4	4		940	176		1,116	1,275
0600	1/0		3.60	4.444		1,100	195		1,295	1,525
0650	2/0		3.40	4.706		1,350	207		1,557	1,775
0700	3/0		3.20	5		1,575	220		1,795	2,050
0750	4/0	↓	3	5.333		1,800	234		2,034	2,325
0800	250 kcmil	3 Elec	3.60	6.667		2,025	293		2,318	2,650
0850	350 kcmil		3.30	7.273		2,700	320		3,020	3,450
0900	500 kcmil	↓	3	8		3,500	350		3,850	4,375
0910	4 conductor with PVC jacket, in cable tray, #6	1 Elec	2.70	2.963		515	130		645	760
0920	#4	2 Elec	4.60	3.478		660	153		813	950
0930	#2		4	4		820	176		996	1,150
0940	#1		3.60	4.444		1,200	195		1,395	1,625
0950	1/0		3.40	4.706		1,375	207		1,582	1,800
0960	2/0		3.20	5		1,550	220		1,770	2,025
0970	3/0		3	5.333		1,850	234		2,084	2,375
0980	4/0	↓	2.40	6.667		2,375	293		2,668	3,025
0990	250 kcmil	3 Elec	3.30	7.273		2,825	320		3,145	3,575
1000	350 kcmil		3	8		3,325	350		3,675	4,200
1010	500 kcmil	↓	2.70	8.889	↓	4,575	390		4,965	5,600
1050	5 kV, copper, 3 conductor with PVC jacket,									
1060	non-shielded, in cable tray, #4	2 Elec	380	.042	L.F.	8	1.85		9.85	11.55
1100	#2		360	.044		10.35	1.95		12.30	14.30
1200	#1		300	.053		13.20	2.34		15.54	18
1400	1/0		290	.055		15.25	2.42		17.67	20.50
1600	2/0		260	.062		17.60	2.70		20.30	23.50
2000	4/0	↓	240	.067		23.50	2.93		26.43	30.50
2100	250 kcmil	3 Elec	330	.073		32	3.19		35.19	40.50
2150	350 kcmil		315	.076		39.50	3.34		42.84	48.50
2200	500 kcmil	↓	270	.089	↓	53	3.90		56.90	64
2400	15 kV, copper, 3 conductor with PVC jacket galv steel armored									
2500	grounded neutral, in cable tray, #2	2 Elec	300	.053	L.F.	16.75	2.34		19.09	22

26 05 19.20 Armored Cable		Crew	Daily Output	Labor-Hours	Unit	Material	2007 Bare Costs Labor	Equipment	Total	Total Incl O&P
2600	#1	2 Elec	280	.057	L.F.	17.85	2.51		20.36	23.50
2800	1/0		260	.062		20.50	2.70		23.20	26.50
2900	2/0		220	.073		27	3.19		30.19	35
3000	4/0		190	.084		30.50	3.70		34.20	39.50
3100	250 kcmil	3 Elec	270	.089		34.50	3.90		38.40	43.50
3150	350 kcmil		240	.100		40.50	4.39		44.89	51
3200	500 kcmil		210	.114		54	5		59	67
3400	15 kV, copper, 3 conductor with PVC jacket,									
3450	ungrounded neutral, in cable tray, #2	2 Elec	260	.062	L.F.	18.05	2.70		20.75	24
3500	#1		230	.070		20	3.05		23.05	26.50
3600	1/0		200	.080		23	3.51		26.51	30.50
3700	2/0		190	.084		28	3.70		31.70	36.50
3800	4/0		160	.100		34	4.39		38.39	44
4000	250 kcmil	3 Elec	210	.114		39.50	5		44.50	51
4050	350 kcmil		195	.123		52	5.40		57.40	65.50
4100	500 kcmil		180	.133		63.50	5.85		69.35	78
4200	600 volt, aluminum, 3 conductor in cable tray with PVC jacket									
4300	#2	2 Elec	540	.030	L.F.	3.99	1.30		5.29	6.35
4400	#1		460	.035		4.42	1.53		5.95	7.15
4500	#1/0		400	.040		5.50	1.76		7.26	8.65
4600	#2/0		360	.044		5.60	1.95		7.55	9.05
4700	#3/0		340	.047		6.55	2.07		8.62	10.30
4800	#4/0		320	.050		7.90	2.20		10.10	11.95
4900	250 kcmil	3 Elec	450	.053		9.50	2.34		11.84	13.95
5000	350 kcmil		360	.067		11.30	2.93		14.23	16.80
5200	500 kcmil		330	.073		14.05	3.19		17.24	20.50
5300	750 kcmil		285	.084		18.15	3.70		21.85	25.50
5400	600 volt, aluminum, 4 conductor in cable tray with PVC jacket									
5410	#2	2 Elec	520	.031	L.F.	4.57	1.35		5.92	7.05
5430	#1		440	.036		5.60	1.60		7.20	8.55
5450	1/0		380	.042		6.55	1.85		8.40	9.95
5470	2/0		340	.047		6.65	2.07		8.72	10.45
5480	3/0		320	.050		7.85	2.20		10.05	11.85
5500	4/0		300	.053		9.20	2.34		11.54	13.65
5520	250 kcmil	3 Elec	420	.057		9.85	2.51		12.36	14.60
5540	350 kcmil		330	.073		12.75	3.19		15.94	18.80
5560	500 kcmil		300	.080		15.95	3.51		19.46	23
5580	750 kcmil		270	.089		22	3.90		25.90	30
5600	5 kV, aluminum, unshielded in cable tray, #2 with PVC jacket	2 Elec	380	.042		5.60	1.85		7.45	8.90
5700	#1 with PVC jacket		360	.044		6.25	1.95		8.20	9.75
5800	1/0 with PVC jacket		300	.053		6.40	2.34		8.74	10.50
6000	2/0 with PVC jacket		290	.055		6.55	2.42		8.97	10.80
6200	3/0 with PVC jacket		260	.062		7.85	2.70		10.55	12.60
6300	4/0 with PVC jacket		240	.067		9.20	2.93		12.13	14.50
6400	250 kcmil with PVC jacket	3 Elec	330	.073		10.10	3.19		13.29	15.85
6500	350 kcmil with PVC jacket		315	.076		11.80	3.34		15.14	18
6600	500 kcmil with PVC jacket		300	.080		14	3.51		17.51	20.50
6800	750 kcmil with PVC jacket		270	.089		17.25	3.90		21.15	25
6900	15 KV, aluminum, shielded-grounded, #2 with PVC jacket	2 Elec	320	.050		12.70	2.20		14.90	17.20
7000	#1 with PVC jacket		300	.053		13.05	2.34		15.39	17.85
7200	1/0 with PVC jacket		280	.057		14.05	2.51		16.56	19.25
7300	2/0 with PVC jacket		260	.062		14.30	2.70		17	19.70
7400	3/0 with PVC jacket		240	.067		16	2.93		18.93	22

26 05 19.20 Armored Cable	Crew	Daily Output	Labor-Hours	Unit	Material	2007 Bare Costs Labor	2007 Bare Costs Equipment	Total	Total Incl O&P
7500	4/0 with PVC jacket	2 Elec	220	.073	L.F.	16.45	3.19		19.64	23
7600	250 kcmil with PVC jacket	3 Elec	300	.080		18.05	3.51		21.56	25
7700	350 kcmil with PVC jacket		270	.089		21	3.90		24.90	29.50
7800	500 kcmil with PVC jacket		240	.100		25.50	4.39		29.89	34.50
8000	750 kcmil with PVC jacket		204	.118		30	5.15		35.15	40.50
8200	15 kV, aluminum, shielded-ungrounded, #1 with PVC jacket	2 Elec	250	.064		15.90	2.81		18.71	21.50
8300	1/0 with PVC jacket		230	.070		16.45	3.05		19.50	22.50
8400	2/0 with PVC jacket		210	.076		18.05	3.34		21.39	25
8500	3/0 with PVC jacket		200	.080		18.35	3.51		21.86	25.50
8600	4/0 with PVC jacket		190	.084		19.95	3.70		23.65	27.50
8700	250 kcmil with PVC jacket	3 Elec	270	.089		21.50	3.90		25.40	29.50
8800	350 kcmil with PVC jacket		240	.100		24.50	4.39		28.89	33.50
8900	500 kcmil with PVC jacket		210	.114		29.50	5		34.50	40
8950	750 kcmil with PVC jacket		174	.138		36.50	6.05		42.55	49.50
9010	600 volt, copper (MC) steel clad, #14, 2 wire	1 Elec	2.40	3.333	C.L.F.	75.50	146		221.50	300
9020	3 wire		2.20	3.636		117	160		277	365
9030	4 wire		2	4		158	176		334	435
9040	#12, 2 wire		2.30	3.478		77.50	153		230.50	310
9050	3 wire		2	4		119	176		295	390
9060	4 wire		1.80	4.444		161	195		356	465
9070	#10, 2 wire		2	4		139	176		315	415
9080	3 wire		1.60	5		216	220		436	565
9090	4 wire		1.40	5.714		340	251		591	750
9091	For Health Care Facilities cable,add,minimum					5%				
9092	Maximum					20%				
9100	#8, 2 wire, stranded	1 Elec	1.80	4.444	C.L.F.	269	195		464	585
9110	3 wire, stranded		1.30	6.154		415	270		685	855
9120	4 wire, stranded		1.10	7.273		610	320		930	1,150
9130	#6, 2 wire, stranded		1.30	6.154		420	270		690	860
9200	600 volt, copper (MC) aluminum clad, #14, 2 wire		2.65	3.019		75.50	133		208.50	280
9210	3 wire		2.45	3.265		117	143		260	340
9220	4 wire		2.20	3.636		158	160		318	410
9230	#12, 2 wire		2.55	3.137		77.50	138		215.50	290
9240	3 wire		2.20	3.636		119	160		279	370
9250	4 wire		2	4		161	176		337	440
9260	#10, 2 wire		2.20	3.636		139	160		299	390
9270	3 wire		1.80	4.444		216	195		411	530
9280	4 wire		1.55	5.161		340	227		567	710
9600	Alum (MC) aluminum clad, #6, 3 conductor w/#6 grnd		1.67	4.790		242	210		452	580
9610	4 conductor w/#6 grnd		1.64	4.878		262	214		476	610
9620	#4, 3 conductor w/#6 grnd	2 Elec	2.86	5.594		259	246		505	650
9630	4 conductor w/#6 grnd		2.82	5.674		325	249		574	725
9640	#2, 3 conductor w/#4 grnd		2.50	6.400		345	281		626	800
9650	4 conductor w/#4 grnd		2.47	6.478		390	284		674	855
9660	#1, 3 conductor w/#4 grnd		2	8		415	350		765	980
9670	4 conductor w/#4 grnd		1.98	8.081		500	355		855	1,075
9680	1/0, 3 conductor w/#4 grnd		1.82	8.791		485	385		870	1,100
9690	4 conductor w/#4 grnd		1.79	8.939		620	390		1,010	1,275
9700	2/0, 3 conductor w/#4 grnd		1.75	9.143		535	400		935	1,175
9710	4 conductor w/#4 grnd		1.71	9.357		685	410		1,095	1,375
9720	3/0, 3 conductor w/#4 grnd		1.71	9.357		675	410		1,085	1,350
9730	4 conductor w/#4 grnd		1.67	9.581		790	420		1,210	1,500
9740	4/0, 3 conductor w/#2 grnd		1.67	9.581		715	420		1,135	1,400

26 05 19.20 Armored Cable

		Crew	Daily Output	Labor-Hours	Unit	Material	2007 Bare Costs Labor	Equipment	Total	Total Incl O&P
9750	4 conductor w/#2 grnd	2 Elec	1.61	9.938	C.L.F.	950	435		1,385	1,700
9760	250 kcmil, 3 conductor w/#1 grnd	3 Elec	2.42	9.917		845	435		1,280	1,575
9770	4 conductor w/#1 grnd		2.36	10.169		1,225	445		1,670	2,025
9775	300 kcmil, 4 conductor w/ #1 grnd		2.22	10.811		1,300	475		1,775	2,125
9780	350 kcmil, 3 conductor w/ 1/0 grnd		2.19	10.959		1,275	480		1,755	2,125
9790	4 conductor w/ 1/0 grnd		2.10	11.429		1,625	500		2,125	2,550
9800	500 kcmil, 3 conductor w/#1 grnd		2.10	11.429		1,625	500		2,125	2,525
9810	4 conductor w/ 2/0 grnd		2	12		1,950	525		2,475	2,925
9840	750 kcmil, 3 conductor w/ 1/0 grnd		1.95	12.308		2,350	540		2,890	3,375
9850	4 conductor w/ 3/0 grnd		1.89	12.698		2,775	555		3,330	3,875

26 05 19.25 Cable Connectors

		Crew	Daily Output	Labor-Hours	Unit	Material	2007 Bare Costs Labor	Equipment	Total	Total Incl O&P
0010	**CABLE CONNECTORS**									
0100	600 volt, nonmetallic, #14-2 wire	1 Elec	160	.050	Ea.	.96	2.20		3.16	4.33
0200	#14-3 wire to #12-2 wire		133	.060		.96	2.64		3.60	4.99
0300	#12-3 wire to #10-2 wire		114	.070		.96	3.08		4.04	5.65
0400	#10-3 wire to #14-4 and #12-4 wire		100	.080		.96	3.51		4.47	6.30
0500	#8-3 wire to #10-4 wire		80	.100		1.84	4.39		6.23	8.55
0600	#6-3 wire		40	.200		2.46	8.80		11.26	15.75
0800	SER aluminum, 3 #8 insulated + 1 #8 ground		32	.250		2.80	11		13.80	19.45
0900	3 #6 + 1 #6 ground		24	.333		2.80	14.65		17.45	25
1000	3 #4 + 1 #6 ground		22	.364		4.10	15.95		20.05	28.50
1100	3 #2 + 1 #4 ground		20	.400		7.30	17.55		24.85	34
1200	3 1/0 + 1 #2 ground		18	.444		18.75	19.50		38.25	49.50
1400	3 2/0 + 1 #1 ground		16	.500		18.75	22		40.75	53
1600	3 4/0 + 1 #2/0 ground		14	.571		22.50	25		47.50	62.50
1800	600 volt, armored, #14-2 wire		80	.100		.59	4.39		4.98	7.20
2200	#14-4, #12-3 and #10-2 wire		40	.200		.59	8.80		9.39	13.70
2400	#12-4, #10-3 and #8-2 wire		32	.250		1.39	11		12.39	17.90
2600	#8-3 and #10-4 wire		26	.308		2.47	13.50		15.97	22.50
2650	#8-4 wire		22	.364		4.29	15.95		20.24	28.50
2700	PVC jacket connector, #6-3 wire, #6-4 wire		16	.500		10.95	22		32.95	44.50
2800	#4-3 wire, #4-4 wire		16	.500		10.95	22		32.95	44.50
2900	#2-3 wire		12	.667		10.95	29.50		40.45	55.50
3000	#1-3 wire, #2-4 wire		12	.667		15.40	29.50		44.90	60.50
3200	1/0-3 wire		11	.727		15.40	32		47.40	64.50
3400	2/0-3 wire, 1/0-4 wire		10	.800		15.40	35		50.40	69.50
3500	3/0-3 wire, 2/0-4 wire		9	.889		31.50	39		70.50	92.50
3600	4/0-3 wire, 3/0-4 wire		7	1.143		31.50	50		81.50	109
3800	250 kcmil-3 wire, 4/0-4 wire		6	1.333		43.50	58.50		102	135
4000	350 kcmil-3 wire, 250 kcmil-4 wire		5	1.600		43.50	70		113.50	153
4100	350 kcmil-4 wire		4	2		170	88		258	320
4200	500 kcmil-3 wire		4	2		170	88		258	320
4250	500 kcmil-4 wire, 750 kcmil-3 wire		3.50	2.286		223	100		323	395
4300	750 kcmil-4 wire		3	2.667		223	117		340	420
4400	5 kV, armored, #4		8	1		61	44		105	133
4600	#2		8	1		61	44		105	133
4800	#1		8	1		61	44		105	133
5000	1/0		6.40	1.250		75.50	55		130.50	165
5200	2/0		5.30	1.509		75.50	66.50		142	182
5500	4/0		4	2		99.50	88		187.50	241
5600	250 kcmil		3.60	2.222		122	97.50		219.50	280
5650	350 kcmil		3.20	2.500		144	110		254	320

26 05 Common Work Results for Electrical

26 05 19 – Low-Voltage Electrical Power Conductors and Cables

26 05 19.25 Cable Connectors

		Crew	Daily Output	Labor-Hours	Unit	Material	2007 Bare Costs Labor	Equipment	Total	Total Incl O&P
5700	500 kcmil	1 Elec	2.50	3.200	Ea.	144	140		284	365
5720	750 kcmil		2.20	3.636		210	160		370	470
5750	1000 kcmil		2	4		281	176		457	570
5800	15 kV, armored, #1		4	2		96	88		184	237
5900	1/0		4	2		96	88		184	237
6000	3/0		3.60	2.222		124	97.50		221.50	281
6100	4/0		3.40	2.353		124	103		227	290
6200	250 kcmil		3.20	2.500		137	110		247	315
6300	350 kcmil		2.70	2.963		158	130		288	370
6400	500 kcmil		2	4		158	176		334	435

26 05 19.30 Cable Splicing

		Crew	Daily Output	Labor-Hours	Unit	Material	2007 Bare Costs Labor	Equipment	Total	Total Incl O&P
0010	**CABLE SPLICING** URD or similar, ideal conditions									
0100	#6 stranded to #1 stranded, 5 kV	1 Elec	4	2	Ea.	114	88		202	256
0120	15 kV		3.60	2.222		114	97.50		211.50	270
0140	25 kV		3.20	2.500		114	110		224	288
0200	#1 stranded to 4/0 stranded, 5 kV		3.60	2.222		119	97.50		216.50	276
0210	15 kV		3.20	2.500		119	110		229	294
0220	25 kV		2.80	2.857		119	125		244	320
0300	4/0 stranded to 500 kcmil stranded, 5 kV		3.30	2.424		268	106		374	450
0310	15 kV		2.90	2.759		268	121		389	475
0320	25 kV		2.50	3.200		268	140		408	505
0400	500 kcmil, 5 kV		3.20	2.500		268	110		378	455
0410	15 kV		2.80	2.857		268	125		393	480
0420	25 kV		2.30	3.478		268	153		421	520
0500	600 kcmil, 5 kV		2.90	2.759		268	121		389	475
0510	15 kV		2.40	3.333		268	146		414	510
0520	25 kV		2	4		268	176		444	555
0600	750 kcmil, 5 kV		2.60	3.077		315	135		450	545
0610	15 kV		2.20	3.636		315	160		475	585
0620	25 kV		1.90	4.211		315	185		500	620
0700	1000 kcmil, 5 kV		2.30	3.478		340	153		493	600
0710	15 kV		1.90	4.211		340	185		525	650
0720	25 kV		1.60	5		340	220		560	700

26 05 19.35 Cable Terminations

		Crew	Daily Output	Labor-Hours	Unit	Material	2007 Bare Costs Labor	Equipment	Total	Total Incl O&P
0010	**CABLE TERMINATIONS**									
0015	Wire connectors, screw type, #22 to #14	1 Elec	260	.031	Ea.	.08	1.35		1.43	2.10
0020	#18 to #12		240	.033		.09	1.46		1.55	2.28
0025	#18 to #10		240	.033		.14	1.46		1.60	2.33
0030	Screw-on connectors, insulated, #18 to #12		240	.033		.09	1.46		1.55	2.28
0035	#16 to #10		230	.035		.12	1.53		1.65	2.40
0040	#14 to #8		210	.038		.26	1.67		1.93	2.78
0045	#12 to #6		180	.044		.44	1.95		2.39	3.38
0050	Terminal lugs, solderless, #16 to #10		50	.160		.48	7		7.48	11
0100	#8 to #4		30	.267		.74	11.70		12.44	18.25
0150	#2 to #1		22	.364		1	15.95		16.95	25
0200	1/0 to 2/0		16	.500		2.22	22		24.22	35
0250	3/0		12	.667		3.40	29.50		32.90	47
0300	4/0		11	.727		3.40	32		35.40	51
0350	250 kcmil		9	.889		3.40	39		42.40	61.50
0400	350 kcmil		7	1.143		4.40	50		54.40	79.50
0450	500 kcmil		6	1.333		8.60	58.50		67.10	96.50
0500	600 kcmil		5.80	1.379		9.20	60.50		69.70	100

26 05 19.35 Cable Terminations		Crew	Daily Output	Labor-Hours	Unit	Material	2007 Bare Costs Labor	Equipment	Total	Total Incl O&P
0550	750 kcmil	1 Elec	5.20	1.538	Ea.	10.10	67.50		77.60	112
0600	Split bolt connectors, tapped, #6		16	.500		3.85	22		25.85	36.50
0650	#4		14	.571		4.65	25		29.65	42.50
0700	#2		12	.667		7	29.50		36.50	51
0750	#1		11	.727		8.85	32		40.85	57.50
0800	1/0		10	.800		8.85	35		43.85	62.50
0850	2/0		9	.889		14.35	39		53.35	74
0900	3/0		7.20	1.111		20	49		69	94.50
1000	4/0		6.40	1.250		24.50	55		79.50	109
1100	250 kcmil		5.70	1.404		24.50	61.50		86	119
1200	300 kcmil		5.30	1.509		44.50	66.50		111	148
1400	350 kcmil		4.60	1.739		44.50	76.50		121	163
1500	500 kcmil		4	2		58.50	88		146.50	195
1600	Crimp 1 hole lugs, copper or aluminum, 600 volt									
1620	#14	1 Elec	60	.133	Ea.	.34	5.85		6.19	9.05
1630	#12		50	.160		.47	7		7.47	10.95
1640	#10		45	.178		.47	7.80		8.27	12.10
1780	#8		36	.222		1.81	9.75		11.56	16.50
1800	#6		30	.267		2.92	11.70		14.62	20.50
2000	#4		27	.296		3.15	13		16.15	23
2200	#2		24	.333		4.30	14.65		18.95	26.50
2400	#1		20	.400		5.15	17.55		22.70	31.50
2500	1/0		17.50	.457		5.80	20		25.80	36.50
2600	2/0		15	.533		6.95	23.50		30.45	42.50
2800	3/0		12	.667		8.60	29.50		38.10	53
3000	4/0		11	.727		9.35	32		41.35	58
3200	250 kcmil		9	.889		11.15	39		50.15	70.50
3400	300 kcmil		8	1		12.50	44		56.50	79.50
3500	350 kcmil		7	1.143		12.95	50		62.95	89
3600	400 kcmil		6.50	1.231		16.10	54		70.10	98
3800	500 kcmil		6	1.333		19.50	58.50		78	109
4000	600 kcmil		5.80	1.379		31	60.50		91.50	124
4200	700 kcmil		5.50	1.455		33.50	64		97.50	132
4400	750 kcmil		5.20	1.538		33.50	67.50		101	138
4500	Crimp 2-way connectors, copper or alum., 600 volt,									
4510	#14	1 Elec	60	.133	Ea.	.54	5.85		6.39	9.30
4520	#12		50	.160		.85	7		7.85	11.40
4530	#10		45	.178		.85	7.80		8.65	12.55
4540	#8		27	.296		2	13		15	21.50
4600	#6		25	.320		4.10	14.05		18.15	25.50
4800	#4		23	.348		4.40	15.25		19.65	27.50
5000	#2		20	.400		6.35	17.55		23.90	33
5200	#1		16	.500		9.15	22		31.15	42.50
5400	1/0		13	.615		10.10	27		37.10	51
5420	2/0		12	.667		10.30	29.50		39.80	55
5440	3/0		11	.727		12.40	32		44.40	61
5460	4/0		10	.800		12.90	35		47.90	66.50
5480	250 kcmil		9	.889		14.70	39		53.70	74
5500	300 kcmil		8.50	.941		16.50	41.50		58	79.50
5520	350 kcmil		8	1		17.10	44		61.10	84.50
5540	400 kcmil		7.30	1.096		22	48		70	95.50
5560	500 kcmil		6.20	1.290		27	56.50		83.50	115
5580	600 kcmil		5.50	1.455		40.50	64		104.50	140

26 05 19.35 Cable Terminations

		Crew	Daily Output	Labor-Hours	Unit	Material	2007 Bare Costs Labor	Equipment	Total	Total Incl O&P
5600	700 kcmil	1 Elec	4.50	1.778	Ea.	41.50	78		119.50	162
5620	750 kcmil		4	2		41.50	88		129.50	177
7000	Compression equipment adapter, aluminum wire, #6		30	.267		8.30	11.70		20	26.50
7020	#4		27	.296		8.75	13		21.75	29
7040	#2		24	.333		9.20	14.65		23.85	32
7060	#1		20	.400		10.60	17.55		28.15	37.50
7080	1/0		18	.444		11.10	19.50		30.60	41
7100	2/0		15	.533		21	23.50		44.50	58
7140	4/0		11	.727		21	32		53	70.50
7160	250 kcmil		9	.889		21	39		60	81
7180	300 kcmil		8	1		25	44		69	93
7200	350 kcmil		7	1.143		25	50		75	102
7220	400 kcmil		6.50	1.231		32.50	54		86.50	116
7240	500 kcmil		6	1.333		32.50	58.50		91	123
7260	600 kcmil		5.80	1.379		44	60.50		104.50	139
7280	750 kcmil	▼	5.20	1.538		44	67.50		111.50	150
8000	Compression tool, hand					1,225			1,225	1,325
8100	Hydraulic					2,525			2,525	2,775
8500	Hydraulic dies				▼	380			380	420

26 05 19.50 Mineral Insulated Cable

		Crew	Daily Output	Labor-Hours	Unit	Material	2007 Bare Costs Labor	Equipment	Total	Total Incl O&P
0010	**MINERAL INSULATED CABLE** 600 volt									
0100	1 conductor, #12	1 Elec	1.60	5	C.L.F.	253	220		473	605
0200	#10		1.60	5		330	220		550	690
0400	#8		1.50	5.333		390	234		624	775
0500	#6	▼	1.40	5.714		435	251		686	855
0600	#4	2 Elec	2.40	6.667		605	293		898	1,100
0800	#2		2.20	7.273		805	320		1,125	1,350
0900	#1		2.10	7.619		920	335		1,255	1,500
1000	1/0		2	8		1,050	350		1,400	1,700
1100	2/0		1.90	8.421		1,275	370		1,645	1,950
1200	3/0		1.80	8.889		1,500	390		1,890	2,225
1400	4/0	▼	1.60	10		1,725	440		2,165	2,550
1410	250 kcmil	3 Elec	2.40	10		1,950	440		2,390	2,800
1420	350 kcmil		1.95	12.308		2,200	540		2,740	3,225
1430	500 kcmil	▼	1.95	12.308		2,775	540		3,315	3,850
1500	2 conductor, #12	1 Elec	1.40	5.714		525	251		776	950
1600	#10		1.20	6.667		635	293		928	1,125
1800	#8		1.10	7.273		790	320		1,110	1,350
2000	#6	▼	1.05	7.619		1,000	335		1,335	1,600
2100	#4	2 Elec	2	8		1,325	350		1,675	1,975
2200	3 conductor, #12	1 Elec	1.20	6.667		660	293		953	1,150
2400	#10		1.10	7.273		760	320		1,080	1,300
2600	#8		1.05	7.619		920	335		1,255	1,525
2800	#6	▼	1	8		1,200	350		1,550	1,850
3000	#4	2 Elec	1.80	8.889		1,475	390		1,865	2,175
3100	4 conductor, #12	1 Elec	1.20	6.667		700	293		993	1,200
3200	#10		1.10	7.273		830	320		1,150	1,400
3400	#8		1	8		1,075	350		1,425	1,725
3600	#6		.90	8.889		1,350	390		1,740	2,075
3620	7 conductor, #12		1.10	7.273		885	320		1,205	1,450
3640	#10		1	8	▼	1,100	350		1,450	1,750
3800	M.I. terminations, 600 volt, 1 conductor, #12	▼	8	1	Ea.	12.45	44		56.45	79

26 05 Common Work Results for Electrical

26 05 19 – Low-Voltage Electrical Power Conductors and Cables

26 05 19.50 Mineral Insulated Cable		Crew	Daily Output	Labor-Hours	Unit	Material	2007 Bare Costs Labor	Equipment	Total	Total Incl O&P
4000	#10	1 Elec	7.60	1.053	Ea.	12.45	46		58.45	82.50
4100	#8		7.30	1.096		12.45	48		60.45	85
4200	#6		6.70	1.194		12.45	52.50		64.95	91.50
4400	#4		6.20	1.290		12.45	56.50		68.95	98
4600	#2		5.70	1.404		18.70	61.50		80.20	112
4800	#1		5.30	1.509		18.70	66.50		85.20	119
5000	1/0		5	1.600		18.70	70		88.70	126
5100	2/0		4.70	1.702		18.70	74.50		93.20	132
5200	3/0		4.30	1.860		18.70	81.50		100.20	143
5400	4/0		4	2		41.50	88		129.50	177
5410	250 kcmil		4	2		41.50	88		129.50	177
5420	350 kcmil		4	2		72.50	88		160.50	211
5430	500 kcmil		4	2		72.50	88		160.50	211
5500	2 conductor, #12		6.70	1.194		12.45	52.50		64.95	91.50
5600	#10		6.40	1.250		18.70	55		73.70	102
5800	#8		6.20	1.290		18.70	56.50		75.20	105
6000	#6		5.70	1.404		18.70	61.50		80.20	112
6200	#4		5.30	1.509		41.50	66.50		108	144
6400	3 conductor, #12		5.70	1.404		18.70	61.50		80.20	112
6500	#10		5.50	1.455		18.70	64		82.70	116
6600	#8		5.20	1.538		18.70	67.50		86.20	122
6800	#6		4.80	1.667		18.70	73		91.70	130
7200	#4		4.60	1.739		41.50	76.50		118	160
7400	4 conductor, #12		4.60	1.739		21	76.50		97.50	137
7500	#10		4.40	1.818		21	80		101	142
7600	#8		4.20	1.905		21	83.50		104.50	147
8400	#6		4	2		43.50	88		131.50	179
8500	7 conductor, #12		3.50	2.286		21	100		121	172
8600	#10	▼	3	2.667		45	117		162	224
8800	Crimping tool, plier type					45.50			45.50	50
9000	Stripping tool					171			171	188
9200	Hand vise				▼	51			51	56

26 05 19.55 Non-Metallic Sheathed Cable

		Crew	Daily Output	Labor-Hours	Unit	Material	Labor	Equipment	Total	Total Incl O&P
0010	**NON-METALLIC SHEATHED CABLE** 600 volt									
0100	Copper with ground wire, (Romex)									
0150	#14, 2 conductor	1 Elec	2.70	2.963	C.L.F.	38.50	130		168.50	236
0200	3 conductor		2.40	3.333		53.50	146		199.50	277
0220	4 conductor		2.20	3.636		89	160		249	335
0250	#12, 2 conductor		2.50	3.200		58	140		198	273
0300	3 conductor		2.20	3.636		87.50	160		247.50	335
0320	4 conductor		2	4		136	176		312	410
0350	#10, 2 conductor		2.20	3.636		96.50	160		256.50	345
0400	3 conductor		1.80	4.444		138	195		333	440
0420	4 conductor		1.60	5		185	220		405	530
0450	#8, 3 conductor		1.50	5.333		224	234		458	595
0480	4 conductor		1.40	5.714		340	251		591	750
0500	#6, 3 conductor	▼	1.40	5.714		355	251		606	770
0520	#4, 3 conductor	2 Elec	2.40	6.667		645	293		938	1,150
0540	#2, 3 conductor	"	2.20	7.273	▼	965	320		1,285	1,525
0550	SE type SER aluminum cable, 3 RHW and									
0600	1 bare neutral, 3 #8 & 1 #8	1 Elec	1.60	5	C.L.F.	133	220		353	470
0650	3 #6 & 1 #6	"	1.40	5.714	▼	150	251		401	540

26 05 Common Work Results for Electrical

26 05 19 – Low-Voltage Electrical Power Conductors and Cables

26 05 19.55 Non-Metallic Sheathed Cable

		Crew	Daily Output	Labor-Hours	Unit	Material	2007 Bare Costs Labor	Equipment	Total	Total Incl O&P
0700	3 #4 & 1 #6	2 Elec	2.40	6.667	C.L.F.	168	293		461	620
0750	3 #2 & 1 #4		2.20	7.273		248	320		568	745
0800	3 #1/0 & 1 #2		2	8		375	350		725	935
0850	3 #2/0 & 1 #1		1.80	8.889		440	390		830	1,075
0900	3 #4/0 & 1 #2/0		1.60	10		630	440		1,070	1,350
1450	UF underground feeder cable, copper with ground, #14, 2 conductor	1 Elec	4	2		42	88		130	177
1500	#12, 2 conductor		3.50	2.286		63.50	100		163.50	219
1550	#10, 2 conductor		3	2.667		101	117		218	285
1600	#14, 3 conductor		3.50	2.286		57	100		157	212
1650	#12, 3 conductor		3	2.667		84.50	117		201.50	267
1700	#10, 3 conductor		2.50	3.200		133	140		273	355
2400	SEU service entrance cable, copper 2 conductors, #8 + #8 neut.		1.50	5.333		200	234		434	570
2600	#6 + #8 neutral		1.30	6.154		276	270		546	705
2800	#6 + #6 neutral		1.30	6.154		305	270		575	740
3000	#4 + #6 neutral	2 Elec	2.20	7.273		405	320		725	925
3200	#4 + #4 neutral		2.20	7.273		455	320		775	980
3400	#3 + #5 neutral		2.10	7.619		535	335		870	1,100
3600	#3 + #3 neutral		2.10	7.619		575	335		910	1,125
3800	#2 + #4 neutral		2	8		625	350		975	1,225
4000	#1 + #1 neutral		1.90	8.421		915	370		1,285	1,550
4200	1/0 + 1/0 neutral		1.80	8.889		1,050	390		1,440	1,725
4400	2/0 + 2/0 neutral		1.70	9.412		1,275	415		1,690	2,025
4600	3/0 + 3/0 neutral		1.60	10		1,575	440		2,015	2,400
4800	Aluminum 2 conductors, #8 + #8 neutral	1 Elec	1.60	5		109	220		329	445
5000	#6 + #6 neutral	"	1.40	5.714		110	251		361	495
5100	#4 + #6 neutral	2 Elec	2.50	6.400		141	281		422	575
5200	#4 + #4 neutral		2.40	6.667		141	293		434	590
5300	#2 + #4 neutral		2.30	6.957		188	305		493	660
5400	#2 + #2 neutral		2.20	7.273		188	320		508	680
5450	1/0 + #2 neutral		2.10	7.619		287	335		622	815
5500	1/0 + 1/0 neutral		2	8		287	350		637	840
5550	2/0 + #1 neutral		1.90	8.421		330	370		700	910
5600	2/0 + 2/0 neutral		1.80	8.889		330	390		720	940
5800	3/0 + 1/0 neutral		1.70	9.412		420	415		835	1,075
6000	3/0 + 3/0 neutral		1.70	9.412		420	415		835	1,075
6200	4/0 + 2/0 neutral		1.60	10		460	440		900	1,150
6400	4/0 + 4/0 neutral		1.60	10		460	440		900	1,150
6500	Service entrance cap for copper SEU									
6600	100 amp	1 Elec	12	.667	Ea.	12.15	29.50		41.65	57
6700	150 amp		10	.800		22	35		57	77
6800	200 amp		8	1		33.50	44		77.50	103

26 05 19.70 Portable Cord

		Crew	Daily Output	Labor-Hours	Unit	Material	2007 Bare Costs Labor	Equipment	Total	Total Incl O&P
0010	**PORTABLE CORD** 600 volt									
0100	Type SO, #18, 2 conductor	1 Elec	980	.008	L.F.	.29	.36		.65	.85
0110	3 conductor		980	.008		.44	.36		.80	1.01
0120	#16, 2 conductor		840	.010		.31	.42		.73	.96
0130	3 conductor		840	.010		.45	.42		.87	1.12
0140	4 conductor		840	.010		.64	.42		1.06	1.32
0240	#14, 2 conductor		840	.010		.46	.42		.88	1.13
0250	3 conductor		840	.010		.68	.42		1.10	1.37
0260	4 conductor		840	.010		.89	.42		1.31	1.60
0280	#12, 2 conductor		840	.010		.92	.42		1.34	1.63

26 05 Common Work Results for Electrical

26 05 19 – Low-Voltage Electrical Power Conductors and Cables

26 05 19.70 Portable Cord		Crew	Daily Output	Labor-Hours	Unit	Material	2007 Bare Costs Labor	2007 Bare Costs Equipment	Total	Total Incl O&P
0290	3 conductor	1 Elec	840	.010	L.F.	.98	.42		1.40	1.70
0320	#10, 2 conductor		765	.010		.95	.46		1.41	1.73
0330	3 conductor		765	.010		1.22	.46		1.68	2.02
0340	4 conductor		765	.010		1.96	.46		2.42	2.84
0360	#8, 2 conductor		555	.014		2.24	.63		2.87	3.40
0370	3 conductor		540	.015		2.82	.65		3.47	4.07
0380	4 conductor		525	.015		3.75	.67		4.42	5.15
0400	#6, 2 conductor		525	.015		1.98	.67		2.65	3.18
0410	3 conductor		490	.016		3.81	.72		4.53	5.25
0420	4 conductor		415	.019		5.35	.85		6.20	7.10
0440	#4, 2 conductor	2 Elec	830	.019		3.15	.85		4	4.73
0450	3 conductor		700	.023		4.77	1		5.77	6.75
0460	4 conductor		660	.024		5.65	1.06		6.71	7.85
0480	#2, 2 conductor		450	.036		3.94	1.56		5.50	6.65
0490	3 conductor		350	.046		5.70	2.01		7.71	9.30
0500	4 conductor		280	.057		7	2.51		9.51	11.45

26 05 19.75 Modular Flexible Wiring System		Crew	Daily Output	Labor-Hours	Unit	Material	2007 Bare Costs Labor	2007 Bare Costs Equipment	Total	Total Incl O&P
0010	**MODULAR FLEXIBLE WIRING SYSTEM**									
0050	Standard selector cable, 3 conductor, 15' long	1 Elec	40	.200	Ea.	54	8.80		62.80	72.50
0100	Conversion Module		16	.500		19.60	22		41.60	54
0150	Cable set, 3 conductor, 15' long		24	.333		21	14.65		35.65	45
0200	Switching assembly, 3 conductor, 10' long		32	.250		41.50	11		52.50	62

26 05 19.90 Wire			Crew	Daily Output	Labor-Hours	Unit	Material	2007 Bare Costs Labor	2007 Bare Costs Equipment	Total	Total Incl O&P
0010	**WIRE**	R260519-90									
0020	600 volt type THW, copper, solid, #14		1 Elec	13	.615	C.L.F.	10	27		37	51
0030	#12			11	.727		15.20	32		47.20	64
0040	#10			10	.800		23	35		58	78
0050	Stranded, #14	R260519-91		13	.615		11.60	27		38.60	53
0100	#12			11	.727		17.60	32		49.60	67
0120	#10	R260519-92		10	.800		27	35		62	82
0140	#8			8	1		45	44		89	115
0160	#6	R260519-93		6.50	1.231		75	54		129	163
0180	#4		2 Elec	10.60	1.509		118	66.50		184.50	229
0200	#3	R260519-94		10	1.600		148	70		218	267
0220	#2			9	1.778		186	78		264	320
0240	#1	R260533-22		8	2		235	88		323	390
0260	1/0			6.60	2.424		266	106		372	450
0280	2/0			5.80	2.759		335	121		456	545
0300	3/0			5	3.200		415	140		555	670
0350	4/0			4.40	3.636		520	160		680	815
0400	250 kcmil		3 Elec	6	4		600	176		776	925
0420	300 kcmil			5.70	4.211		715	185		900	1,075
0450	350 kcmil			5.40	4.444		840	195		1,035	1,225
0480	400 kcmil			5.10	4.706		960	207		1,167	1,350
0490	500 kcmil			4.80	5		1,200	220		1,420	1,625
0500	600 kcmil			3.90	6.154		1,425	270		1,695	1,975
0510	750 kcmil			3.30	7.273		2,075	320		2,395	2,750
0520	1000 kcmil			2.70	8.889		2,750	390		3,140	3,600
0540	Aluminum, stranded, #6		1 Elec	8	1		25.50	44		69.50	93.50
0560	#4		2 Elec	13	1.231		31.50	54		85.50	115
0580	#2			10.60	1.509		42.50	66.50		109	145
0600	#1			9	1.778		62	78		140	185

26 05 Common Work Results for Electrical

26 05 19 – Low-Voltage Electrical Power Conductors and Cables

26 05 19.90 Wire		Crew	Daily Output	Labor-Hours	Unit	Material	2007 Bare Costs Labor	Equipment	Total	Total Incl O&P
0620	1/0	2 Elec	8	2	C.L.F.	75	88		163	214
0640	2/0		7.20	2.222		88	97.50		185.50	242
0680	3/0		6.60	2.424		109	106		215	278
0700	4/0		6.20	2.581		122	113		235	305
0720	250 kcmil	3 Elec	8.70	2.759		149	121		270	345
0740	300 kcmil		8.10	2.963		205	130		335	420
0760	350 kcmil		7.50	3.200		208	140		348	440
0780	400 kcmil		6.90	3.478		244	153		397	495
0800	500 kcmil		6	4		269	176		445	555
0850	600 kcmil		5.70	4.211		340	185		525	650
0880	700 kcmil		5.10	4.706		395	207		602	745
0900	750 kcmil		4.80	5		400	220		620	760
0920	Type THWN-THHN, copper, solid, #14	1 Elec	13	.615		10	27		37	51
0940	#12		11	.727		15.20	32		47.20	64
0960	#10		10	.800		23	35		58	78
1000	Stranded, #14		13	.615		11.60	27		38.60	53
1200	#12		11	.727		17.60	32		49.60	67
1250	#10		10	.800		27	35		62	82
1300	#8		8	1		45	44		89	115
1350	#6		6.50	1.231		75	54		129	163
1400	#4	2 Elec	10.60	1.509		118	66.50		184.50	229
1450	#3		10	1.600		148	70		218	267
1500	#2		9	1.778		186	78		264	320
1550	#1		8	2		235	88		323	390
1600	1/0		6.60	2.424		266	106		372	450
1650	2/0		5.80	2.759		335	121		456	545
1700	3/0		5	3.200		415	140		555	670
2000	4/0		4.40	3.636		520	160		680	815
2200	250 kcmil	3 Elec	6	4		600	176		776	925
2400	300 kcmil		5.70	4.211		715	185		900	1,075
2600	350 kcmil		5.40	4.444		840	195		1,035	1,225
2700	400 kcmil		5.10	4.706		960	207		1,167	1,350
2800	500 kcmil		4.80	5		1,200	220		1,420	1,625
2900	600 volt, copper type XHHW, solid, #14	1 Elec	13	.615		16.40	27		43.40	58
2920	#12		11	.727		22.50	32		54.50	72
2940	#10		10	.800		30.50	35		65.50	86
3000	Stranded, #14		13	.615		17.60	27		44.60	59.50
3020	#12		11	.727		25	32		57	75
3040	#10		10	.800		36.50	35		71.50	92.50
3060	#8		8	1		55.50	44		99.50	127
3080	#6		6.50	1.231		89.50	54		143.50	179
3100	#4	2 Elec	10.60	1.509		139	66.50		205.50	252
3120	#2		9	1.778		216	78		294	355
3140	#1		8	2		273	88		361	430
3160	1/0		6.60	2.424		300	106		406	490
3180	2/0		5.80	2.759		375	121		496	595
3200	3/0		5	3.200		465	140		605	725
3220	4/0		4.40	3.636		585	160		745	885
3240	250 kcmil	3 Elec	6	4		715	176		891	1,050
3260	300 kcmil		5.70	4.211		850	185		1,035	1,200
3280	350 kcmil		5.40	4.444		990	195		1,185	1,400
3300	400 kcmil		5.10	4.706		1,125	207		1,332	1,550
3320	500 kcmil		4.80	5		1,300	220		1,520	1,750

26 05 19.90 Wire		Crew	Daily Output	Labor-Hours	Unit	Material	2007 Bare Costs Labor	Equipment	Total	Total Incl O&P
3340	600 kcmil	3 Elec	3.90	6.154	C.L.F.	1,550	270		1,820	2,100
3360	750 kcmil		3.30	7.273		2,525	320		2,845	3,250
3380	1000 kcmil	↓	2.40	10		3,325	440		3,765	4,300
5020	600 volt, aluminum type XHHW, stranded, #6	1 Elec	8	1		25.50	44		69.50	93.50
5040	#4	2 Elec	13	1.231		31.50	54		85.50	115
5060	#2		10.60	1.509		42.50	66.50		109	145
5080	#1		9	1.778		62	78		140	185
5100	1/0		8	2		75	88		163	214
5120	2/0		7.20	2.222		88	97.50		185.50	242
5140	3/0		6.60	2.424		109	106		215	278
5160	4/0	↓	6.20	2.581		122	113		235	305
5180	250 kcmil	3 Elec	8.70	2.759		149	121		270	345
5200	300 kcmil		8.10	2.963		205	130		335	420
5220	350 kcmil		7.50	3.200		208	140		348	440
5240	400 kcmil		6.90	3.478		244	153		397	495
5260	500 kcmil		6	4		269	176		445	555
5280	600 kcmil		5.70	4.211		340	185		525	650
5300	700 kcmil		5.40	4.444		395	195		590	725
5320	750 kcmil		5.10	4.706		400	207		607	745
5340	1000 kcmil	↓	3.60	6.667		590	293		883	1,075
5400	600 volt, copper type XLPE-USE(RHW), solid, #12	1 Elec	11	.727		24	32		56	74
5420	#10		10	.800		29.50	35		64.50	85
5440	Stranded, #14		13	.615		25	27		52	67.50
5460	#12		11	.727		28	32		60	78.50
5480	#10		10	.800		39	35		74	95
5500	#8		8	1		67.50	44		111.50	140
5520	#6	↓	6.50	1.231		101	54		155	192
5540	#4	2 Elec	10.60	1.509		150	66.50		216.50	264
5560	#2		9	1.778		224	78		302	360
5580	#1		8	2		298	88		386	460
5600	1/0		6.60	2.424		340	106		446	535
5620	2/0		5.80	2.759		420	121		541	640
5640	3/0		5	3.200		505	140		645	765
5660	4/0	↓	4.40	3.636		630	160		790	930
5680	250 kcmil	3 Elec	6	4		770	176		946	1,100
5700	300 kcmil		5.70	4.211		915	185		1,100	1,275
5720	350 kcmil		5.40	4.444		1,025	195		1,220	1,450
5740	400 kcmil		5.10	4.706		1,200	207		1,407	1,625
5760	500 kcmil		4.80	5		1,425	220		1,645	1,900
5780	600 kcmil		3.90	6.154		1,700	270		1,970	2,275
5800	750 kcmil		3.30	7.273		2,550	320		2,870	3,300
5820	1000 kcmil	↓	2.70	8.889		3,375	390		3,765	4,275
5840	600 volt, aluminum type XLPE-USE (RHW), stranded, #6	1 Elec	8	1		31	44		75	99.50
5860	#4	2 Elec	13	1.231		35	54		89	119
5880	#2		10.60	1.509		49	66.50		115.50	153
5900	#1		9	1.778		68	78		146	191
5920	1/0		8	2		84	88		172	224
5940	2/0		7.20	2.222		98.50	97.50		196	253
5960	3/0		6.60	2.424		116	106		222	286
5980	4/0	↓	6.20	2.581		129	113		242	310
6000	250 kcmil	3 Elec	8.70	2.759		175	121		296	370
6020	300 kcmil		8.10	2.963		228	130		358	445
6040	350 kcmil	↓	7.50	3.200	↓	233	140		373	465

26 05 19 – Low-Voltage Electrical Power Conductors and Cables

26 05 19.90 Wire		Crew	Daily Output	Labor-Hours	Unit	Material	2007 Bare Costs Labor	Equipment	Total	Total Incl O&P
6060	400 kcmil	3 Elec	6.90	3.478	C.L.F.	284	153		437	540
6080	500 kcmil		6	4		310	176		486	605
6100	600 kcmil		5.70	4.211		400	185		585	715
6110	700 kcmil		5.40	4.444		460	195		655	795
6120	750 kcmil		5.10	4.706		465	207		672	825

26 05 23 – Control-Voltage Electrical Power Cables

26 05 23.10 Control Cable

		Crew	Daily Output	Labor-Hours	Unit	Material	2007 Bare Costs Labor	Equipment	Total	Total Incl O&P
0010	**CONTROL CABLE**									
0020	600 volt, copper, #14 THWN wire with PVC jacket, 2 wires	1 Elec	9	.889	C.L.F.	26.50	39		65.50	87
0030	3 wires		8	1		37.50	44		81.50	107
0100	4 wires		7	1.143		46.50	50		96.50	126
0150	5 wires		6.50	1.231		56	54		110	142
0200	6 wires		6	1.333		83	58.50		141.50	179
0300	8 wires		5.30	1.509		102	66.50		168.50	211
0400	10 wires		4.80	1.667		121	73		194	242
0500	12 wires		4.30	1.860		139	81.50		220.50	275
0600	14 wires		3.80	2.105		167	92.50		259.50	320
0700	16 wires		3.50	2.286		170	100		270	335
0800	18 wires		3.30	2.424		191	106		297	370
0810	19 wires		3.10	2.581		195	113		308	385
0900	20 wires		3	2.667		221	117		338	415
1000	22 wires		2.80	2.857		230	125		355	440

26 05 23.20 Special Wires and Fittings

		Crew	Daily Output	Labor-Hours	Unit	Material	2007 Bare Costs Labor	Equipment	Total	Total Incl O&P
0010	**SPECIAL WIRES & FITTINGS**									
0100	Fixture TFFN 600 volt 90°C stranded, #18	1 Elec	13	.615	C.L.F.	9.60	27		36.60	50.50
0150	#16		13	.615		13.20	27		40.20	54.50
0500	Thermostat, no jacket, twisted, #18-2 conductor		8	1		8.90	44		52.90	75.50
0550	#18-3 conductor		7	1.143		14	50		64	90
0600	#18-4 conductor		6.50	1.231		19.90	54		73.90	103
0650	#18-5 conductor		6	1.333		23.50	58.50		82	113
0700	#18-6 conductor		5.50	1.455		27	64		91	125
0750	#18-7 conductor		5	1.600		41.50	70		111.50	151
0800	#18-8 conductor		4.80	1.667		54	73		127	169
1500	Fire alarm FEP teflon 150 volt to 200°C									
1550	#22, 1 pair	1 Elec	10	.800	C.L.F.	94.50	35		129.50	157
1600	2 pair		8	1		157	44		201	239
1650	4 pair		7	1.143		241	50		291	340
1700	6 pair		6	1.333		315	58.50		373.50	430
1750	8 pair		5.50	1.455		395	64		459	530
1800	10 pair		5	1.600		475	70		545	630
1850	#18, 1 pair		8	1		108	44		152	185
1900	2 pair		6.50	1.231		207	54		261	310
1950	4 pair		4.80	1.667		315	73		388	455
2000	6 pair		4	2		410	88		498	585
2050	8 pair		3.50	2.286		540	100		640	740
2100	10 pair		3	2.667		615	117		732	855
2460	Tray cable, type TC, copper, #16-2 conductor		9.40	.851		19.50	37.50		57	77
2464	#16-3 conductor		8.40	.952		26	42		68	90.50
2468	#16-4 conductor		7.30	1.096		32.50	48		80.50	108
2472	#16-5 conductor		6.70	1.194		39	52.50		91.50	121
2476	#16-7 conductor		5.50	1.455		53.50	64		117.50	154
2480	#16-9 conductor		5	1.600		63	70		133	175

26 05 23 – Control-Voltage Electrical Power Cables

26 05 23.20 Special Wires and Fittings		Crew	Daily Output	Labor-Hours	Unit	Material	2007 Bare Costs Labor	Equipment	Total	Total Incl O&P
2484	#16-12 conductor	2 Elec	8.80	1.818	C.L.F.	87	80		167	215
2488	#16-15 conductor		7.20	2.222		114	97.50		211.50	270
2492	#16-19 conductor		6.60	2.424		138	106		244	310
2496	#16-25 conductor		6.40	2.500		201	110		311	385
2500	#14-2 conductor	1 Elec	9	.889		26.50	39		65.50	87
2520	#14-3 conductor		8	1		37.50	44		81.50	107
2540	#14-4 conductor		7	1.143		46.50	50		96.50	126
2560	#14-5 conductor		6.50	1.231		56	54		110	142
2564	#14-7 conductor		5.30	1.509		83	66.50		149.50	190
2568	#14-9 conductor	2 Elec	9.60	1.667		102	73		175	221
2572	#14-12 conductor		8.60	1.860		139	81.50		220.50	275
2576	#14-15 conductor		7.20	2.222		167	97.50		264.50	330
2578	#14-19 conductor		6.20	2.581		211	113		324	400
2582	#14-25 conductor		5.30	3.019		278	133		411	500
2590	#12-2 conductor	1 Elec	8.40	.952		38	42		80	104
2592	#12-3 conductor		7.60	1.053		55.50	46		101.50	130
2594	#12-4 conductor		6.60	1.212		72	53		125	158
2596	#12-5 conductor		6.20	1.290		85	56.50		141.50	178
2598	#12-7 conductor	2 Elec	10.40	1.538		119	67.50		186.50	232
2602	#12-9 conductor		8.80	1.818		156	80		236	291
2604	#12-12 conductor		7.80	2.051		200	90		290	355
2606	#12-15 conductor		6.80	2.353		237	103		340	415
2608	#12-19 conductor		6	2.667		310	117		427	520
2610	#12-25 conductor	3 Elec	7.70	3.117		390	137		527	635
2618	#10-2 conductor	1 Elec	8	1		57	44		101	128
2622	#10-3 conductor	"	7.30	1.096		86	48		134	166
2624	#10-4 conductor	2 Elec	12.80	1.250		110	55		165	203
2626	#10-5 conductor		11.80	1.356		137	59.50		196.50	240
2628	#10-7 conductor		9.40	1.702		187	74.50		261.50	315
2630	#10-9 conductor		8.40	1.905		240	83.50		323.50	390
2632	#10-12 conductor		7.20	2.222		310	97.50		407.50	485
2640	300 V, copper braided shield, PVC jacket									
2650	2 conductor #18 stranded	1 Elec	7	1.143	C.L.F.	58.50	50		108.50	139
2660	3 conductor #18	"	6	1.333	"	82.50	58.50		141	178
3000	Strain relief grip for cable									
3050	Cord, top, #12-3	1 Elec	40	.200	Ea.	14.10	8.80		22.90	28.50
3060	#12-4		40	.200		14.10	8.80		22.90	28.50
3070	#12-5		39	.205		16.40	9		25.40	31.50
3100	#10-3		39	.205		16.40	9		25.40	31.50
3110	#10-4		38	.211		16.40	9.25		25.65	32
3120	#10-5		38	.211		17.90	9.25		27.15	33.50
3200	Bottom, #12-3		40	.200		36	8.80		44.80	52.50
3210	#12-4		40	.200		36	8.80		44.80	52.50
3220	#12-5		39	.205		36.50	9		45.50	54
3230	#10-3		39	.205		39	9		48	56.50
3300	#10-4		38	.211		42	9.25		51.25	60
3310	#10-5		38	.211		46	9.25		55.25	64.50
3400	Cable ties, standard, 4" length		190	.042		.18	1.85		2.03	2.95
3410	7" length		160	.050		.28	2.20		2.48	3.58
3420	14.5" length		90	.089		.61	3.90		4.51	6.45
3430	Heavy, 14.5" length		80	.100		.89	4.39		5.28	7.55

26 05 Common Work Results for Electrical

26 05 26 - Grounding and Bonding for Electrical Systems

26 05 26.80 Grounding

		Crew	Daily Output	Labor-Hours	Unit	Material	2007 Bare Costs Labor	Equipment	Total	Total Incl O&P
0010	**GROUNDING** R260526-80									
0030	Rod, copper clad, 8' long, 1/2" diameter	1 Elec	5.50	1.455	Ea.	14.05	64		78.05	110
0040	5/8" diameter		5.50	1.455		15.40	64		79.40	112
0050	3/4" diameter		5.30	1.509		29.50	66.50		96	131
0080	10' long, 1/2" diameter		4.80	1.667		18.10	73		91.10	129
0090	5/8" diameter		4.60	1.739		22.50	76.50		99	139
0100	3/4" diameter		4.40	1.818		32.50	80		112.50	155
0130	15' long, 3/4" diameter		4	2		87	88		175	227
0150	Coupling, bronze, 1/2" diameter					3.40			3.40	3.74
0160	5/8" diameter					5			5	5.50
0170	3/4" diameter					11.30			11.30	12.45
0190	Drive studs, 1/2" diameter					6.70			6.70	7.35
0210	5/8" diameter					8.20			8.20	9
0220	3/4" diameter					9.25			9.25	10.20
0230	Clamp, bronze, 1/2" diameter	1 Elec	32	.250		4.09	11		15.09	21
0240	5/8" diameter		32	.250		4.52	11		15.52	21.50
0250	3/4" diameter		32	.250		5.75	11		16.75	22.50
0260	Wire ground bare armored, #8-1 conductor		2	4	C.L.F.	109	176		285	380
0270	#6-1 conductor		1.80	4.444		131	195		326	435
0280	#4-1 conductor		1.60	5		175	220		395	520
0320	Bare copper wire, #14 solid		14	.571		9.40	25		34.40	48
0330	#12		13	.615		14.30	27		41.30	55.50
0340	#10		12	.667		21.50	29.50		51	67
0350	#8		11	.727		37.50	32		69.50	88.50
0360	#6		10	.800		71	35		106	131
0370	#4		8	1		114	44		158	191
0380	#2		5	1.600		160	70		230	281
0390	Bare copper wire, stranded, #8		11	.727		40	32		72	91.50
0400	#6		10	.800		73	35		108	133
0450	#4	2 Elec	16	1		114	44		158	191
0600	#2		10	1.600		167	70		237	289
0650	#1		9	1.778		224	78		302	365
0700	1/0		8	2		250	88		338	405
0750	2/0		7.20	2.222		295	97.50		392.50	470
0800	3/0		6.60	2.424		375	106		481	570
1000	4/0		5.70	2.807		470	123		593	700
1200	250 kcmil	3 Elec	7.20	3.333		555	146		701	830
1210	300 kcmil		6.60	3.636		620	160		780	925
1220	350 kcmil		6	4		780	176		956	1,125
1230	400 kcmil		5.70	4.211		890	185		1,075	1,250
1240	500 kcmil		5.10	4.706		1,100	207		1,307	1,500
1260	750 kcmil		3.60	6.667		1,725	293		2,018	2,325
1270	1000 kcmil		3	8		2,300	350		2,650	3,050
1360	Bare aluminum, stranded, #6	1 Elec	9	.889		18.15	39		57.15	78
1370	#4	2 Elec	16	1		21	44		65	88.50
1380	#2		13	1.231		33	54		87	117
1390	#1		10.60	1.509		40.50	66.50		107	143
1400	1/0		9	1.778		46.50	78		124.50	167
1410	2/0		8	2		58.50	88		146.50	196
1420	3/0		7.20	2.222		72	97.50		169.50	224
1430	4/0		6.60	2.424		86	106		192	253
1440	250 kcmil	3 Elec	9.30	2.581		97.50	113		210.50	276

117

26 05 26 – Grounding and Bonding for Electrical Systems

26 05 26.80 Grounding		Crew	Daily Output	Labor-Hours	Unit	Material	2007 Bare Costs Labor	Equipment	Total	Total Incl O&P
1450	300 kcmil	3 Elec	8.70	2.759	C.L.F.	148	121		269	345
1460	400 kcmil		7.50	3.200		159	140		299	385
1470	500 kcmil		6.90	3.478		201	153		354	450
1480	600 kcmil		6	4		242	176		418	525
1490	700 kcmil		5.70	4.211		278	185		463	580
1500	750 kcmil		5.10	4.706		310	207		517	655
1510	1000 kcmil		4.80	5		430	220		650	800
1800	Water pipe ground clamps, heavy duty									
2000	Bronze, 1/2" to 1" diameter	1 Elec	8	1	Ea.	21	44		65	88.50
2100	1-1/4" to 2" diameter		8	1		27.50	44		71.50	96
2200	2-1/2" to 3" diameter		6	1.333		46.50	58.50		105	138
2730	Exothermic weld, 4/0 wire to 1" ground rod		7	1.143		9.75	50		59.75	85
2740	4/0 wire to building steel		7	1.143		7.75	50		57.75	83
2750	4/0 wire to motor frame		7	1.143		7.75	50		57.75	83
2760	4/0 wire to 4/0 wire		7	1.143		6.75	50		56.75	82
2770	4/0 wire to #4 wire		7	1.143		6.75	50		56.75	82
2780	4/0 wire to #8 wire		7	1.143		6.75	50		56.75	82
2790	Mold, reusable, for above					111			111	123
2800	Brazed connections, #6 wire	1 Elec	12	.667		12.90	29.50		42.40	57.50
3000	#2 wire		10	.800		17.30	35		52.30	71.50
3100	3/0 wire		8	1		26	44		70	94
3200	4/0 wire		7	1.143		29.50	50		79.50	107
3400	250 kcmil wire		5	1.600		34.50	70		104.50	143
3600	500 kcmil wire		4	2		42.50	88		130.50	178
3700	Insulated ground wire, copper #14		13	.615	C.L.F.	11.60	27		38.60	53
3710	#12		11	.727		17.60	32		49.60	67
3720	#10		10	.800		27	35		62	82
3730	#8		8	1		45	44		89	115
3740	#6		6.50	1.231		75	54		129	163
3750	#4	2 Elec	10.60	1.509		118	66.50		184.50	229
3770	#2		9	1.778		186	78		264	320
3780	#1		8	2		235	88		323	390
3790	1/0		6.60	2.424		266	106		372	450
3800	2/0		5.80	2.759		335	121		456	545
3810	3/0		5	3.200		415	140		555	670
3820	4/0		4.40	3.636		520	160		680	815
3830	250 kcmil	3 Elec	6	4		600	176		776	925
3840	300 kcmil		5.70	4.211		715	185		900	1,075
3850	350 kcmil		5.40	4.444		840	195		1,035	1,225
3860	400 kcmil		5.10	4.706		960	207		1,167	1,350
3870	500 kcmil		4.80	5		1,200	220		1,420	1,625
3880	600 kcmil		3.90	6.154		1,425	270		1,695	1,975
3890	750 kcmil		3.30	7.273		2,075	320		2,395	2,750
3900	1000 kcmil		2.70	8.889		2,750	390		3,140	3,600
3960	Insulated ground wire, aluminum, #6	1 Elec	8	1		25.50	44		69.50	93.50
3970	#4	2 Elec	13	1.231		31.50	54		85.50	115
3980	#2		10.60	1.509		42.50	66.50		109	145
3990	#1		9	1.778		62	78		140	185
4000	1/0		8	2		75	88		163	214
4010	2/0		7.20	2.222		88	97.50		185.50	242
4020	3/0		6.60	2.424		109	106		215	278
4030	4/0		6.20	2.581		122	113		235	305
4040	250 kcmil	3 Elec	8.70	2.759		149	121		270	345

26 05 Common Work Results for Electrical

26 05 26 – Grounding and Bonding for Electrical Systems

26 05 26.80 Grounding

		Crew	Daily Output	Labor-Hours	Unit	Material	2007 Bare Costs Labor	Equipment	Total	Total Incl O&P
4050	300 kcmil	3 Elec	8.10	2.963	C.L.F.	205	130		335	420
4060	350 kcmil		7.50	3.200		208	140		348	440
4070	400 kcmil		6.90	3.478		244	153		397	495
4080	500 kcmil		6	4		269	176		445	555
4090	600 kcmil		5.70	4.211		340	185		525	650
4100	700 kcmil		5.10	4.706		395	207		602	745
4110	750 kcmil		4.80	5		400	220		620	760
5000	Copper Electrolytic ground rod system									
5010	Includes augering hole, mixing bentonite clay,									
5020	Installing rod, and terminating ground wire									
5100	Straight Vertical type, 2" Dia.									
5120	8.5' long, Clamp Connection	1 Elec	2.67	2.996	Ea.	685	132		817	945
5130	With exothermic weld Connection		1.95	4.103		685	180		865	1,025
5140	10' long		2.35	3.404		760	149		909	1,050
5150	With exothermic weld Connection		1.78	4.494		760	197		957	1,125
5160	12' long		2.16	3.704		920	163		1,083	1,250
5170	With exothermic weld Connection		1.67	4.790		920	210		1,130	1,325
5180	20' long		1.74	4.598		1,400	202		1,602	1,850
5190	With exothermic weld Connection		1.40	5.714		1,450	251		1,701	1,975
5195	40' long with exothermic weld connection	2 Elec	2	8		2,700	350		3,050	3,475
5200	L-Shaped, 2" Dia.									
5220	4' Vert. x 10' Horz., Clamp Connection	1 Elec	5.33	1.501	Ea.	1,100	66		1,166	1,300
5230	With exothermic weld Connection	"	3.08	2.597	"	1,100	114		1,214	1,375
5300	Protective Box at grade level, with breather slots									
5320	Round 12" long, Plastic	1 Elec	32	.250	Ea.	45	11		56	66
5330	Concrete	"	16	.500		75	22		97	115
5400	Bentonite Clay, 50# bag, 1 per 10' of rod					38			38	42

26 05 29 – Hangers and Supports for Electrical Systems

26 05 29.20 Hangers

		Crew	Daily Output	Labor-Hours	Unit	Material	2007 Bare Costs Labor	Equipment	Total	Total Incl O&P
0010	**HANGERS** R260533-20									
0030	Conduit supports									
0050	Strap w/2 holes, rigid steel conduit									
0100	1/2" diameter	1 Elec	470	.017	Ea.	.21	.75		.96	1.34
0150	3/4" diameter		440	.018		.25	.80		1.05	1.47
0200	1" diameter		400	.020		.41	.88		1.29	1.76
0300	1-1/4" diameter		355	.023		.65	.99		1.64	2.19
0350	1-1/2" diameter		320	.025		.76	1.10		1.86	2.47
0400	2" diameter		266	.030		.88	1.32		2.20	2.94
0500	2-1/2" diameter		160	.050		1.83	2.20		4.03	5.30
0550	3" diameter		133	.060		2.27	2.64		4.91	6.45
0600	3-1/2" diameter		100	.080		2.97	3.51		6.48	8.50
0650	4" diameter		80	.100		3.21	4.39		7.60	10.10
0700	EMT, 1/2" diameter		470	.017		.17	.75		.92	1.30
0800	3/4" diameter		440	.018		.21	.80		1.01	1.42
0850	1" diameter		400	.020		.39	.88		1.27	1.74
0900	1-1/4" diameter		355	.023		.63	.99		1.62	2.16
0950	1-1/2" diameter		320	.025		.66	1.10		1.76	2.36
1000	2" diameter		266	.030		.96	1.32		2.28	3.03
1100	2-1/2" diameter		160	.050		1.63	2.20		3.83	5.05
1150	3" diameter		133	.060		2.19	2.64		4.83	6.35
1200	3-1/2" diameter		100	.080		2.47	3.51		5.98	7.95
1250	4" diameter		80	.100		2.80	4.39		7.19	9.65

119

26 05 29.20 Hangers		Crew	Daily Output	Labor-Hours	Unit	Material	2007 Bare Costs Labor	Equipment	Total	Total Incl O&P
1400	Hanger, with bolt, 1/2" diameter	1 Elec	200	.040	Ea.	.51	1.76		2.27	3.17
1450	3/4" diameter		190	.042		.53	1.85		2.38	3.33
1500	1" diameter		176	.045		.87	2		2.87	3.93
1550	1-1/4" diameter		160	.050		1.23	2.20		3.43	4.62
1600	1-1/2" diameter		140	.057		1.59	2.51		4.10	5.50
1650	2" diameter		130	.062		1.75	2.70		4.45	5.95
1700	2-1/2" diameter		100	.080		2.12	3.51		5.63	7.60
1750	3" diameter		64	.125		2.49	5.50		7.99	10.90
1800	3-1/2" diameter		50	.160		3.67	7		10.67	14.50
1850	4" diameter		40	.200		8.35	8.80		17.15	22.50
1900	Riser clamps, conduit, 1/2" diameter		40	.200		8.70	8.80		17.50	22.50
1950	3/4" diameter		36	.222		8.70	9.75		18.45	24
2000	1" diameter		30	.267		8.80	11.70		20.50	27
2100	1-1/4" diameter		27	.296		11.70	13		24.70	32
2150	1-1/2" diameter		27	.296		12.35	13		25.35	33
2200	2" diameter		20	.400		12.95	17.55		30.50	40.50
2250	2-1/2" diameter		20	.400		13.55	17.55		31.10	41
2300	3" diameter		18	.444		14.60	19.50		34.10	45
2350	3-1/2" diameter		18	.444		18.60	19.50		38.10	49.50
2400	4" diameter		14	.571	▼	19.25	25		44.25	58.50
2500	Threaded rod, painted, 1/4" diameter		260	.031	L.F.	1.79	1.35		3.14	3.98
2600	3/8" diameter		200	.040		2.35	1.76		4.11	5.20
2700	1/2" diameter		140	.057		3.81	2.51		6.32	7.90
2800	5/8" diameter		100	.080		5.25	3.51		8.76	11.05
2900	3/4" diameter	▼	60	.133	▼	6.60	5.85		12.45	15.95
2940	Couplings painted, 1/4" diameter				C	216			216	238
2960	3/8" diameter					294			294	325
2970	1/2" diameter					445			445	490
2980	5/8" diameter					640			640	700
2990	3/4" diameter					945			945	1,050
3000	Nuts, galvanized, 1/4" diameter					11.95			11.95	13.10
3050	3/8" diameter					17.95			17.95	19.75
3100	1/2" diameter					40			40	44
3150	5/8" diameter					60.50			60.50	66.50
3200	3/4" diameter					90			90	99
3250	Washers, galvanized, 1/4" diameter					6.40			6.40	7.05
3300	3/8" diameter					13.40			13.40	14.75
3350	1/2" diameter					16.25			16.25	17.85
3400	5/8" diameter					50			50	55
3450	3/4" diameter					72.50			72.50	80
3500	Lock washers, galvanized, 1/4" diameter					5.40			5.40	5.95
3550	3/8" diameter					11.70			11.70	12.85
3600	1/2" diameter					16.10			16.10	17.70
3650	5/8" diameter					29.50			29.50	32.50
3700	3/4" diameter				▼	46			46	50.50
3800	Channels, steel, 3/4" x 1-1/2"	1 Elec	80	.100	L.F.	5.30	4.39		9.69	12.35
3900	1-1/2" x 1-1/2"		70	.114		7.05	5		12.05	15.20
4000	1-7/8" x 1-1/2"		60	.133		17.05	5.85		22.90	27.50
4100	3" x 1-1/2"		50	.160	▼	19.70	7		26.70	32
4200	Spring nuts, long, 1/4"		120	.067	Ea.	1.59	2.93		4.52	6.10
4250	3/8"		100	.080		2.18	3.51		5.69	7.65
4300	1/2"		80	.100		2.34	4.39		6.73	9.10
4350	Spring nuts, short, 1/4"	▼	120	.067	▼	1.96	2.93		4.89	6.50

26 05 29.20 Hangers		Crew	Daily Output	Labor-Hours	Unit	Material	2007 Bare Costs Labor	Equipment	Total	Total Incl O&P
4400	3/8"	1 Elec	100	.080	Ea.	2.05	3.51		5.56	7.50
4450	1/2"		80	.100	↓	2.53	4.39		6.92	9.35
4500	Closure strip		200	.040	L.F.	2.96	1.76		4.72	5.85
4550	End cap		60	.133	Ea.	1.20	5.85		7.05	10
4600	End connector 3/4" conduit		40	.200		3.99	8.80		12.79	17.45
4650	Junction box, 1 channel		16	.500		24.50	22		46.50	59.50
4700	2 channel		14	.571		30	25		55	70.50
4750	3 channel		12	.667		34	29.50		63.50	81
4800	4 channel		10	.800		38	35		73	94
4850	Spliceplate		40	.200		9.30	8.80		18.10	23.50
4900	Continuous concrete insert, 1-1/2" deep, 1' long		16	.500		12.85	22		34.85	46.50
4950	2' long		14	.571		16.05	25		41.05	55
5000	3' long		12	.667		19.25	29.50		48.75	64.50
5050	4' long		10	.800		22.50	35		57.50	77
5100	6' long		8	1		33.50	44		77.50	102
5150	3/4" deep, 1' long		16	.500		10.40	22		32.40	44
5200	2' long		14	.571		12.85	25		37.85	51.50
5250	3' long		12	.667		15.45	29.50		44.95	60.50
5300	4' long		10	.800		17.80	35		52.80	72
5350	6' long		8	1		26.50	44		70.50	94.50
5400	90° angle fitting 2-1/8" x 2-1/8"		60	.133		2.20	5.85		8.05	11.10
5450	Supports, suspension rod type, small		60	.133		10.60	5.85		16.45	20.50
5500	Large		40	.200		11.70	8.80		20.50	26
5550	Beam clamp, small		60	.133		4.90	5.85		10.75	14.10
5600	Large		40	.200		6.35	8.80		15.15	20
5650	U-support, small		60	.133		3.44	5.85		9.29	12.50
5700	Large		40	.200		9.15	8.80		17.95	23
5750	Concrete insert, cast, for up to 1/2" threaded rod		16	.500		2.88	22		24.88	35.50
5800	Beam clamp, 1/4" clamp, for 1/4" threaded drop rod		32	.250		2.58	11		13.58	19.20
5900	3/8" clamp, for 3/8" threaded drop rod		32	.250		5.80	11		16.80	23
6000	Strap, rigid conduit, 1/2" diameter		540	.015		1.12	.65		1.77	2.20
6050	3/4" diameter		440	.018		1.25	.80		2.05	2.57
6100	1" diameter		420	.019		1.34	.84		2.18	2.71
6150	1-1/4" diameter		400	.020		1.55	.88		2.43	3.02
6200	1-1/2" diameter		400	.020		1.76	.88		2.64	3.25
6250	2" diameter		267	.030		1.99	1.32		3.31	4.15
6300	2-1/2" diameter		267	.030		2.16	1.32		3.48	4.34
6350	3" diameter		160	.050		2.39	2.20		4.59	5.90
6400	3-1/2" diameter		133	.060		2.92	2.64		5.56	7.15
6450	4" diameter		100	.080		3.25	3.51		6.76	8.85
6500	5" diameter		80	.100		8.35	4.39		12.74	15.75
6550	6" diameter		60	.133		9.45	5.85		15.30	19.10
6600	EMT, 1/2" diameter		540	.015		1.12	.65		1.77	2.20
6650	3/4" diameter		440	.018		1.19	.80		1.99	2.50
6700	1" diameter		420	.019		1.34	.84		2.18	2.71
6750	1-1/4" diameter		400	.020		1.50	.88		2.38	2.96
6800	1-1/2" diameter		400	.020		1.82	.88		2.70	3.31
6850	2" diameter		267	.030		1.96	1.32		3.28	4.12
6900	2-1/2" diameter		267	.030		2.48	1.32		3.80	4.69
6950	3" diameter		160	.050		2.61	2.20		4.81	6.15
6970	3-1/2" diameter		133	.060		2.98	2.64		5.62	7.20
6990	4" diameter		100	.080		3.42	3.51		6.93	9
7000	Clip, 1 hole for rigid conduit, 1/2" diameter	↓	500	.016	↓	.64	.70		1.34	1.75

26 05 29.20 Hangers		Crew	Daily Output	Labor- Hours	Unit	Material	2007 Bare Costs Labor	Equipment	Total	Total Incl O&P
7050	3/4" diameter	1 Elec	470	.017	Ea.	.94	.75		1.69	2.14
7100	1" diameter		440	.018		1.17	.80		1.97	2.48
7150	1-1/4" diameter		400	.020		2.52	.88		3.40	4.08
7200	1-1/2" diameter		355	.023		2.90	.99		3.89	4.66
7250	2" diameter		320	.025		5.65	1.10		6.75	7.85
7300	2-1/2" diameter		266	.030		12.60	1.32		13.92	15.80
7350	3" diameter		160	.050		18.10	2.20		20.30	23
7400	3-1/2" diameter		133	.060		25.50	2.64		28.14	32.50
7450	4" diameter		100	.080		57.50	3.51		61.01	69
7500	5" diameter		80	.100		180	4.39		184.39	205
7550	6" diameter		60	.133		194	5.85		199.85	223
7820	Conduit hangers, with bolt & 12" rod, 1/2" diameter		150	.053		2.86	2.34		5.20	6.65
7830	3/4" diameter		145	.055		2.88	2.42		5.30	6.80
7840	1" diameter		135	.059		3.22	2.60		5.82	7.40
7850	1-1/4" diameter		120	.067		3.58	2.93		6.51	8.30
7860	1-1/2" diameter		110	.073		3.94	3.19		7.13	9.10
7870	2" diameter		100	.080		5.55	3.51		9.06	11.35
7880	2-1/2" diameter		80	.100		5.95	4.39		10.34	13.05
7890	3" diameter		60	.133		6.30	5.85		12.15	15.65
7900	3-1/2" diameter		45	.178		7.50	7.80		15.30	19.85
7910	4" diameter		35	.229		12.20	10.05		22.25	28.50
7920	5" diameter		30	.267		13.60	11.70		25.30	32.50
7930	6" diameter		25	.320		25	14.05		39.05	48.50
7950	Jay clamp, 1/2" diameter		32	.250		4.76	11		15.76	21.50
7960	3/4" diameter		32	.250		4.76	11		15.76	21.50
7970	1" diameter		32	.250		4.76	11		15.76	21.50
7980	1-1/4" diameter		30	.267		6.90	11.70		18.60	25
7990	1-1/2" diameter		30	.267		6.90	11.70		18.60	25
8000	2" diameter		30	.267		8.60	11.70		20.30	27
8010	2-1/2" diameter		28	.286		12.30	12.55		24.85	32
8020	3" diameter		28	.286		16.70	12.55		29.25	37
8030	3-1/2" diameter		25	.320		21.50	14.05		35.55	44.50
8040	4" diameter		25	.320		37.50	14.05		51.55	62.50
8050	5" diameter		20	.400		95	17.55		112.55	131
8060	6" diameter		16	.500		178	22		200	229
8070	Channels, 3/4" x 1-1/2" w/12" rods for 1/2" to 1" conduit		30	.267		10.25	11.70		21.95	29
8080	1-1/2" x 1-1/2" w/12" rods for 1-1/4" to 2" conduit		28	.286		11.15	12.55		23.70	31
8090	1-1/2" x 1-1/2" w/12" rods for 2-1/2" to 4" conduit		26	.308		14.05	13.50		27.55	35.50
8100	1-1/2" x 1-7/8" w/12" rods for 5" to 6" conduit		24	.333		40.50	14.65		55.15	67
8110	Beam clamp, conduit, plastic coated steel, 1/2" diam.		30	.267		17.35	11.70		29.05	36.50
8120	3/4" diameter		30	.267		18.25	11.70		29.95	37.50
8130	1" diameter		30	.267		18.45	11.70		30.15	38
8140	1-1/4" diameter		28	.286		25	12.55		37.55	46
8150	1-1/2" diameter		28	.286		30	12.55		42.55	51.50
8160	2" diameter		28	.286		40	12.55		52.55	62.50
8170	2-1/2" diameter		26	.308		44	13.50		57.50	68.50
8180	3" diameter		26	.308		49	13.50		62.50	74
8190	3-1/2" diameter		23	.348		51	15.25		66.25	78.50
8200	4" diameter		23	.348		55.50	15.25		70.75	83.50
8210	5" diameter		18	.444		163	19.50		182.50	209
8220	Channels, plastic coated									
8250	3/4" x 1-1/2", w/12" rods for 1/2" to 1" conduit	1 Elec	28	.286	Ea.	26	12.55		38.55	47.50
8260	1-1/2" x 1-1/2", w/12" rods for 1-1/4" to 2" conduit		26	.308		29.50	13.50		43	52.50

26 05 29 – Hangers and Supports for Electrical Systems

26 05 29.20 Hangers		Crew	Daily Output	Labor-Hours	Unit	Material	2007 Bare Costs Labor	Equipment	Total	Total Incl O&P
8270	1-1/2" x 1-1/2", w/12" rods for 2-1/2" to 3-1/2" conduit	1 Elec	24	.333	Ea.	32	14.65		46.65	57.50
8280	1-1/2" x 1-7/8", w/12" rods for 4" to 5" conduit		22	.364		53.50	15.95		69.45	82.50
8290	1-1/2" x 1-7/8", w/12" rods for 6" conduit		20	.400		58.50	17.55		76.05	90
8320	Conduit hangers, plastic coated steel, with bolt & 12" rod, 1/2" diam.		140	.057		15.10	2.51		17.61	20.50
8330	3/4" diameter		135	.059		15.45	2.60		18.05	21
8340	1" diameter		125	.064		15.80	2.81		18.61	21.50
8350	1-1/4" diameter		110	.073		16.75	3.19		19.94	23
8360	1-1/2" diameter		100	.080		18.80	3.51		22.31	26
8370	2" diameter		90	.089		20.50	3.90		24.40	29
8380	2-1/2" diameter		70	.114		24.50	5		29.50	34.50
8390	3" diameter		50	.160		30	7		37	43.50
8400	3-1/2" diameter		35	.229		31.50	10.05		41.55	50
8410	4" diameter		25	.320		44.50	14.05		58.55	69.50
8420	5" diameter		20	.400		48.50	17.55		66.05	79.50
9000	Parallel type, conduit beam clamp, 1/2"		32	.250		4.15	11		15.15	21
9010	3/4"		32	.250		4.55	11		15.55	21.50
9020	1"		32	.250		4.85	11		15.85	21.50
9030	1-1/4"		30	.267		6.70	11.70		18.40	25
9040	1-1/2"		30	.267		7.60	11.70		19.30	26
9050	2"		30	.267		9.85	11.70		21.55	28.50
9060	2-1/2"		28	.286		13.10	12.55		25.65	33
9070	3"		28	.286		16.65	12.55		29.20	37
9090	4"		25	.320		21.50	14.05		35.55	44.50
9110	Right angle, conduit beam clamp, 1/2"		32	.250		2.75	11		13.75	19.40
9120	3/4"		32	.250		2.80	11		13.80	19.45
9130	1"		32	.250		3.15	11		14.15	19.80
9140	1-1/4"		30	.267		3.85	11.70		15.55	21.50
9150	1-1/2"		30	.267		4.35	11.70		16.05	22
9160	2"		30	.267		6.30	11.70		18	24.50
9170	2-1/2"		28	.286		7.80	12.55		20.35	27.50
9180	3"		28	.286		9	12.55		21.55	28.50
9190	3-1/2"		25	.320		11.30	14.05		25.35	33.50
9200	4"		25	.320		12.20	14.05		26.25	34.50
9230	Adjustable, conduit hanger, 1/2"		32	.250		3.55	11		14.55	20.50
9240	3/4"		32	.250		3.30	11		14.30	20
9250	1"		32	.250		4.10	11		15.10	21
9260	1-1/4"		30	.267		4.90	11.70		16.60	23
9270	1-1/2"		30	.267		5.60	11.70		17.30	23.50
9280	2"		30	.267		6.05	11.70		17.75	24
9290	2-1/2"		28	.286		7.55	12.55		20.10	27
9300	3"		28	.286		7.95	12.55		20.50	27.50
9310	3-1/2"		25	.320		10.30	14.05		24.35	32.50
9320	4"		25	.320		18	14.05		32.05	41
9330	5"		20	.400		23.50	17.55		41.05	52
9340	6"		16	.500		29	22		51	64.50
9350	Combination conduit hanger, 3/8"		32	.250		8.20	11		19.20	25.50
9360	Adjustable flange 3/8"		32	.250		9.60	11		20.60	27

26 05 33 – Raceway and Boxes for Electrical Systems

26 05 33.05 Conduit

	26 05 33.05 Conduit	Crew	Daily Output	Labor-Hours	Unit	Material	2007 Bare Costs Labor	Equipment	Total	Total Incl O&P
0010	**CONDUIT** To 15' high, includes 2 terminations, 2 elbows,	R260533-20								
0020	11 beam clamps, and 11 couplings per 100 L.F.									
0300	Aluminum, 1/2" diameter	1 Elec	100	.080	L.F.	2.08	3.51		5.59	7.55

26 05 33 – Raceway and Boxes for Electrical Systems

26 05 33.05 Conduit		Crew	Daily Output	Labor-Hours	Unit	Material	2007 Bare Costs Labor	Equipment	Total	Total Incl O&P
0500	3/4" diameter R260533-21	1 Elec	90	.089	L.F.	2.81	3.90		6.71	8.90
0700	1" diameter		80	.100		3.72	4.39		8.11	10.65
1000	1-1/4" diameter		70	.114		5	5		10	12.95
1030	1-1/2" diameter		65	.123		6	5.40		11.40	14.65
1050	2" diameter		60	.133		8.20	5.85		14.05	17.70
1070	2-1/2" diameter		50	.160		13.30	7		20.30	25
1100	3" diameter	2 Elec	90	.178		18.15	7.80		25.95	31.50
1130	3-1/2" diameter		80	.200		24	8.80		32.80	39.50
1140	4" diameter		70	.229		29	10.05		39.05	47
1150	5" diameter		50	.320		61.50	14.05		75.55	89
1160	6" diameter		40	.400		91	17.55		108.55	126
1161	Field bends, 45° to 90°, 1/2" diameter	1 Elec	53	.151	Ea.		6.65		6.65	9.85
1162	3/4" diameter		47	.170			7.45		7.45	11.10
1163	1" diameter		44	.182			8		8	11.90
1164	1-1/4" diameter		23	.348			15.25		15.25	22.50
1165	1-1/2" diameter		21	.381			16.70		16.70	25
1166	2" diameter		16	.500			22		22	32.50
1170	Elbows, 1/2" diameter		40	.200		11	8.80		19.80	25
1200	3/4" diameter		32	.250		14.90	11		25.90	33
1230	1" diameter		28	.286		21	12.55		33.55	41.50
1250	1-1/4" diameter		24	.333		33	14.65		47.65	58.50
1270	1-1/2" diameter		20	.400		44	17.55		61.55	74.50
1300	2" diameter		16	.500		64.50	22		86.50	104
1330	2-1/2" diameter		12	.667		109	29.50		138.50	164
1350	3" diameter		8	1		169	44		213	252
1370	3-1/2" diameter		6	1.333		263	58.50		321.50	375
1400	4" diameter		5	1.600		310	70		380	450
1410	5" diameter		4	2		920	88		1,008	1,150
1420	6" diameter		2.50	3.200		1,275	140		1,415	1,600
1430	Couplings, 1/2" diameter					3.64			3.64	4
1450	3/4" diameter					5.45			5.45	6
1470	1" diameter					7.20			7.20	7.95
1500	1-1/4" diameter					8.80			8.80	9.70
1530	1-1/2" diameter					10.20			10.20	11.25
1550	2" diameter					14.50			14.50	15.95
1570	2-1/2" diameter					32.50			32.50	36
1600	3" diameter					42.50			42.50	47
1630	3-1/2" diameter					58.50			58.50	64.50
1650	4" diameter					70.50			70.50	78
1670	5" diameter					207			207	227
1690	6" diameter					325			325	355
1750	Rigid galvanized steel, 1/2" diameter	1 Elec	90	.089	L.F.	2.48	3.90		6.38	8.55
1770	3/4" diameter		80	.100		2.83	4.39		7.22	9.65
1800	1" diameter		65	.123		3.89	5.40		9.29	12.35
1830	1-1/4" diameter		60	.133		5.40	5.85		11.25	14.65
1850	1-1/2" diameter		55	.145		6.25	6.40		12.65	16.35
1870	2" diameter		45	.178		8	7.80		15.80	20.50
1900	2-1/2" diameter		35	.229		15.05	10.05		25.10	31.50
1930	3" diameter	2 Elec	50	.320		18.40	14.05		32.45	41
1950	3-1/2" diameter		44	.364		23	15.95		38.95	49
1970	4" diameter		40	.400		26	17.55		43.55	54.50
1980	5" diameter		30	.533		54	23.50		77.50	94.50
1990	6" diameter		20	.800		80.50	35		115.50	141

26 05 Common Work Results for Electrical

26 05 33 – Raceway and Boxes for Electrical Systems

26 05 33.05 Conduit		Crew	Daily Output	Labor-Hours	Unit	Material	2007 Bare Costs Labor	Equipment	Total	Total Incl O&P
1991	Field bends, 45° to 90°, 1/2" diameter	1 Elec	44	.182	Ea.		8		8	11.90
1992	3/4" diameter		40	.200			8.80		8.80	13.05
1993	1" diameter		36	.222			9.75		9.75	14.50
1994	1-1/4" diameter		19	.421			18.50		18.50	27.50
1995	1-1/2" diameter		18	.444			19.50		19.50	29
1996	2" diameter		13	.615			27		27	40
2000	Elbows, 1/2" diameter		32	.250		10.60	11		21.60	28
2030	3/4" diameter		28	.286		11.60	12.55		24.15	31.50
2050	1" diameter		24	.333		17	14.65		31.65	40.50
2070	1-1/4" diameter		18	.444		23.50	19.50		43	54.50
2100	1-1/2" diameter		16	.500		29	22		51	64
2130	2" diameter		12	.667		41.50	29.50		71	89
2150	2-1/2" diameter		8	1		78.50	44		122.50	152
2170	3" diameter		6	1.333		108	58.50		166.50	206
2200	3-1/2" diameter		4.20	1.905		171	83.50		254.50	310
2220	4" diameter		4	2		194	88		282	345
2230	5" diameter		3.50	2.286		535	100		635	740
2240	6" diameter		2	4		805	176		981	1,150
2250	Couplings, 1/2" diameter					2.47			2.47	2.72
2270	3/4" diameter					3.02			3.02	3.32
2300	1" diameter					4.46			4.46	4.91
2330	1-1/4" diameter					5.60			5.60	6.15
2350	1-1/2" diameter					7.05			7.05	7.75
2370	2" diameter					9.35			9.35	10.30
2400	2-1/2" diameter					23			23	25.50
2430	3" diameter					30			30	32.50
2450	3-1/2" diameter					40			40	44
2470	4" diameter					40.50			40.50	44.50
2480	5" diameter					92.50			92.50	102
2490	6" diameter					129			129	142
2500	Steel, intermediate conduit (IMC), 1/2" diameter	1 Elec	100	.080	L.F.	1.92	3.51		5.43	7.35
2530	3/4" diameter		90	.089		2.37	3.90		6.27	8.40
2550	1" diameter		70	.114		3.32	5		8.32	11.10
2570	1-1/4" diameter		65	.123		4.42	5.40		9.82	12.90
2600	1-1/2" diameter		60	.133		5.20	5.85		11.05	14.40
2630	2" diameter		50	.160		6.75	7		13.75	17.85
2650	2-1/2" diameter		40	.200		12.80	8.80		21.60	27
2670	3" diameter	2 Elec	60	.267		16.80	11.70		28.50	36
2700	3-1/2" diameter		54	.296		21	13		34	42.50
2730	4" diameter		50	.320		24	14.05		38.05	47.50
2731	Field bends, 45° to 90°, 1/2" diameter	1 Elec	44	.182	Ea.		8		8	11.90
2732	3/4" diameter		40	.200			8.80		8.80	13.05
2733	1" diameter		36	.222			9.75		9.75	14.50
2734	1-1/4" diameter		19	.421			18.50		18.50	27.50
2735	1-1/2" diameter		18	.444			19.50		19.50	29
2736	2" diameter		13	.615			27		27	40
2750	Elbows, 1/2" diameter		32	.250		9.45	11		20.45	27
2770	3/4" diameter		28	.286		11.55	12.55		24.10	31.50
2800	1" diameter		24	.333		16.65	14.65		31.30	40.50
2830	1-1/4" diameter		18	.444		27	19.50		46.50	58.50
2850	1-1/2" diameter		16	.500		30.50	22		52.50	66
2870	2" diameter		12	.667		43.50	29.50		73	91.50
2900	2-1/2" diameter		8	1		77.50	44		121.50	151

125

26 05 33.05 Conduit		Crew	Daily Output	Labor-Hours	Unit	Material	2007 Bare Costs Labor	Equipment	Total	Total Incl O&P
2930	3" diameter	1 Elec	6	1.333	Ea.	118	58.50		176.50	217
2950	3-1/2" diameter		4.20	1.905		180	83.50		263.50	320
2970	4" diameter	↓	4	2		210	88		298	360
3000	Couplings, 1/2" diameter					2.21			2.21	2.43
3030	3/4" diameter					2.72			2.72	2.99
3050	1" diameter					4			4	4.40
3070	1-1/4" diameter					5			5	5.50
3100	1-1/2" diameter					6.30			6.30	6.90
3130	2" diameter					8.35			8.35	9.15
3150	2-1/2" diameter					20.50			20.50	22.50
3170	3" diameter					26.50			26.50	29
3200	3-1/2" diameter					35.50			35.50	39.50
3230	4" diameter				↓	36			36	39.50
4100	Rigid steel, plastic coated, 40 mil. thick									
4130	1/2" diameter	1 Elec	80	.100	L.F.	5.25	4.39		9.64	12.30
4150	3/4" diameter		70	.114		5.90	5		10.90	13.95
4170	1" diameter		55	.145		7.45	6.40		13.85	17.70
4200	1-1/4" diameter		50	.160		9.45	7		16.45	21
4230	1-1/2" diameter		45	.178		11.30	7.80		19.10	24
4250	2" diameter		35	.229		14.95	10.05		25	31.50
4270	2-1/2" diameter		25	.320		24	14.05		38.05	47.50
4300	3" diameter	2 Elec	44	.364		30.50	15.95		46.45	57.50
4330	3-1/2" diameter		40	.400		42	17.55		59.55	72
4350	4" diameter		36	.444		44.50	19.50		64	78
4370	5" diameter	↓	30	.533	↓	89	23.50		112.50	133
4400	Elbows, 1/2" diameter	1 Elec	28	.286	Ea.	15.90	12.55		28.45	36
4430	3/4" diameter		24	.333		16.60	14.65		31.25	40.50
4450	1" diameter		18	.444		19	19.50		38.50	50
4470	1-1/4" diameter		16	.500		23.50	22		45.50	58
4500	1-1/2" diameter		12	.667		28.50	29.50		58	75
4530	2" diameter		8	1		40	44		84	110
4550	2-1/2" diameter		6	1.333		75.50	58.50		134	170
4570	3" diameter		4.20	1.905		120	83.50		203.50	256
4600	3-1/2" diameter		4	2		158	88		246	305
4630	4" diameter		3.80	2.105		173	92.50		265.50	330
4650	5" diameter	↓	3.50	2.286		415	100		515	605
4680	Couplings, 1/2" diameter					4.56			4.56	5
4700	3/4" diameter					4.79			4.79	5.25
4730	1" diameter					6.20			6.20	6.80
4750	1-1/4" diameter					7.25			7.25	7.95
4770	1-1/2" diameter					8.55			8.55	9.45
4800	2" diameter					12.60			12.60	13.85
4830	2-1/2" diameter					31			31	34
4850	3" diameter					37.50			37.50	41.50
4870	3-1/2" diameter					48.50			48.50	53.50
4900	4" diameter					57			57	62.50
4950	5" diameter				↓	185			185	203
5000	Electric metallic tubing (EMT), 1/2" diameter	1 Elec	170	.047	L.F.	.60	2.07		2.67	3.74
5020	3/4" diameter		130	.062		.99	2.70		3.69	5.10
5040	1" diameter		115	.070		1.74	3.05		4.79	6.45
5060	1-1/4" diameter		100	.080		2.76	3.51		6.27	8.30
5080	1-1/2" diameter		90	.089		3.50	3.90		7.40	9.65
5100	2" diameter	↓	80	.100	↓	4.41	4.39		8.80	11.40

26 05 33.05 Conduit		Crew	Daily Output	Labor-Hours	Unit	Material	2007 Bare Costs Labor	2007 Bare Costs Equipment	Total	Total Incl O&P
5120	2-1/2" diameter	1 Elec	60	.133	L.F.	9	5.85		14.85	18.55
5140	3" diameter	2 Elec	100	.160		11.75	7		18.75	23.50
5160	3-1/2" diameter		90	.178		15.20	7.80		23	28.50
5180	4" diameter	▼	80	.200	▼	16.85	8.80		25.65	31.50
5200	Field bends, 45° to 90°, 1/2" diameter	1 Elec	89	.090	Ea.		3.95		3.95	5.85
5220	3/4" diameter		80	.100			4.39		4.39	6.55
5240	1" diameter		73	.110			4.81		4.81	7.15
5260	1-1/4" diameter		38	.211			9.25		9.25	13.75
5280	1-1/2" diameter		36	.222			9.75		9.75	14.50
5300	2" diameter		26	.308			13.50		13.50	20
5320	Offsets, 1/2" diameter		65	.123			5.40		5.40	8.05
5340	3/4" diameter		62	.129			5.65		5.65	8.45
5360	1" diameter		53	.151			6.65		6.65	9.85
5380	1-1/4" diameter		30	.267			11.70		11.70	17.45
5400	1-1/2" diameter		28	.286			12.55		12.55	18.65
5420	2" diameter		20	.400			17.55		17.55	26
5700	Elbows, 1" diameter		40	.200		7.05	8.80		15.85	21
5720	1-1/4" diameter		32	.250		9.35	11		20.35	26.50
5740	1-1/2" diameter		24	.333		10.90	14.65		25.55	34
5760	2" diameter		20	.400		15.95	17.55		33.50	43.50
5780	2-1/2" diameter		12	.667		39	29.50		68.50	86.50
5800	3" diameter		9	.889		57.50	39		96.50	122
5820	3-1/2" diameter		7	1.143		77.50	50		127.50	160
5840	4" diameter	▼	6	1.333		91	58.50		149.50	187
6200	Couplings, set screw, steel, 1/2" diameter					1.34			1.34	1.47
6220	3/4" diameter					2.04			2.04	2.24
6240	1" diameter					3.36			3.36	3.70
6260	1-1/4" diameter					7.05			7.05	7.75
6280	1-1/2" diameter					9.75			9.75	10.75
6300	2" diameter					13.15			13.15	14.50
6320	2-1/2" diameter					32			32	35.50
6340	3" diameter					42.50			42.50	47
6360	3-1/2" diameter					47.50			47.50	52.50
6380	4" diameter					53			53	58.50
6500	Box connectors, set screw, steel, 1/2" diameter	1 Elec	120	.067		.89	2.93		3.82	5.35
6520	3/4" diameter		110	.073		1.43	3.19		4.62	6.30
6540	1" diameter		90	.089		2.71	3.90		6.61	8.80
6560	1-1/4" diameter		70	.114		5.30	5		10.30	13.25
6580	1-1/2" diameter		60	.133		7.65	5.85		13.50	17.10
6600	2" diameter		50	.160		10.55	7		17.55	22
6620	2-1/2" diameter		36	.222		37	9.75		46.75	55
6640	3" diameter		27	.296		44	13		57	68
6680	3-1/2" diameter		21	.381		61.50	16.70		78.20	93
6700	4" diameter		16	.500		68.50	22		90.50	108
6740	Insulated box connectors, set screw, steel, 1/2" diameter		120	.067		1.21	2.93		4.14	5.70
6760	3/4" diameter		110	.073		1.92	3.19		5.11	6.85
6780	1" diameter		90	.089		3.51	3.90		7.41	9.65
6800	1-1/4" diameter		70	.114		6.40	5		11.40	14.50
6820	1-1/2" diameter		60	.133		9.05	5.85		14.90	18.65
6840	2" diameter		50	.160		13.10	7		20.10	25
6860	2-1/2" diameter		36	.222		62.50	9.75		72.25	83
6880	3" diameter		27	.296		73.50	13		86.50	100
6900	3-1/2" diameter		21	.381	▼	103	16.70		119.70	138

26 05 33.05 Conduit	Crew	Daily Output	Labor-Hours	Unit	Material	2007 Bare Costs Labor	Equipment	Total	Total Incl O&P
6920 4" diameter	1 Elec	16	.500	Ea.	112	22		134	156
7000 EMT to conduit adapters, 1/2" diameter (compression)		70	.114		4.23	5		9.23	12.10
7020 3/4" diameter		60	.133		6.05	5.85		11.90	15.35
7040 1" diameter		50	.160		9.20	7		16.20	20.50
7060 1-1/4" diameter		40	.200		15.95	8.80		24.75	30.50
7080 1-1/2" diameter		30	.267		19.65	11.70		31.35	39
7100 2" diameter		25	.320		28.50	14.05		42.55	52
7200 EMT to Greenfield adapters, 1/2" to 3/8" diameter (compression)		90	.089		3.06	3.90		6.96	9.15
7220 1/2" diameter		90	.089		5.75	3.90		9.65	12.15
7240 3/4" diameter		80	.100		7.40	4.39		11.79	14.65
7260 1" diameter		70	.114		19.60	5		24.60	29
7270 1-1/4" diameter		60	.133		24.50	5.85		30.35	35.50
7280 1-1/2" diameter		50	.160		27.50	7		34.50	41
7290 2" diameter		40	.200		41.50	8.80		50.30	58.50
7400 EMT, LB, LR or LL fittings with covers, 1/2" dia, set screw		24	.333		10.60	14.65		25.25	33.50
7420 3/4" diameter		20	.400		13.05	17.55		30.60	40.50
7440 1" diameter		16	.500		20	22		42	54.50
7450 1-1/4" diameter		13	.615		29.50	27		56.50	72.50
7460 1-1/2" diameter		11	.727		35.50	32		67.50	86.50
7470 2" diameter		9	.889		56	39		95	120
7600 EMT, "T" fittings with covers, 1/2" diameter, set screw		16	.500		13.70	22		35.70	47.50
7620 3/4" diameter		15	.533		17.35	23.50		40.85	54
7640 1" diameter		12	.667		25	29.50		54.50	71
7650 1-1/4" diameter		11	.727		36	32		68	87
7660 1-1/2" diameter		10	.800		44.50	35		79.50	102
7670 2" diameter		8	1		59	44		103	130
8000 EMT, expansion fittings, no jumper, 1/2" diameter		24	.333		42.50	14.65		57.15	69
8020 3/4" diameter		20	.400		50	17.55		67.55	81
8040 1" diameter		16	.500		58.50	22		80.50	97
8060 1-1/4" diameter		13	.615		79	27		106	127
8080 1-1/2" diameter		11	.727		109	32		141	168
8100 2" diameter		9	.889		162	39		201	236
8110 2-1/2" diameter		7	1.143		255	50		305	355
8120 3" diameter		6	1.333		315	58.50		373.50	430
8140 4" diameter		5	1.600		550	70		620	710
8200 Split adapter, 1/2" diameter		110	.073		3.20	3.19		6.39	8.25
8210 3/4" diameter		90	.089		2.90	3.90		6.80	9
8220 1" diameter		70	.114		3.95	5		8.95	11.80
8230 1-1/4" diameter		60	.133		6.35	5.85		12.20	15.70
8240 1-1/2" diameter		50	.160		9.80	7		16.80	21.50
8250 2" diameter		36	.222		29	9.75		38.75	46
8300 1 hole clips, 1/2" diameter		500	.016		.35	.70		1.05	1.44
8320 3/4" diameter		470	.017		.45	.75		1.20	1.61
8340 1" diameter		444	.018		.84	.79		1.63	2.10
8360 1-1/4" diameter		400	.020		1.12	.88		2	2.54
8380 1-1/2" diameter		355	.023		1.71	.99		2.70	3.35
8400 2" diameter		320	.025		2.68	1.10		3.78	4.58
8420 2-1/2" diameter		266	.030		4.98	1.32		6.30	7.45
8440 3" diameter		160	.050		6.25	2.20		8.45	10.15
8460 3-1/2" diameter		133	.060		9.85	2.64		12.49	14.80
8480 4" diameter		100	.080		12.75	3.51		16.26	19.25
8500 Clamp back spacers, 1/2" diameter		500	.016		1.04	.70		1.74	2.19
8510 3/4" diameter		470	.017		1.24	.75		1.99	2.47

26 05 33.05 Conduit		Crew	Daily Output	Labor-Hours	Unit	Material	2007 Bare Costs Labor	2007 Bare Costs Equipment	Total	Total Incl O&P
8520	1" diameter	1 Elec	444	.018	Ea.	2.18	.79		2.97	3.58
8530	1-1/4" diameter		400	.020		3.89	.88		4.77	5.60
8540	1-1/2" diameter		355	.023		4.36	.99		5.35	6.25
8550	2" diameter		320	.025		7.35	1.10		8.45	9.70
8560	2-1/2" diameter		266	.030		14.10	1.32		15.42	17.45
8570	3" diameter		160	.050		20	2.20		22.20	25.50
8580	3-1/2" diameter		133	.060		27.50	2.64		30.14	34
8590	4" diameter		100	.080		59.50	3.51		63.01	71
8600	Offset connectors, 1/2" diameter		40	.200		5.05	8.80		13.85	18.60
8610	3/4" diameter		32	.250		7.25	11		18.25	24.50
8620	1" diameter		24	.333		10	14.65		24.65	33
8650	90° pulling elbows, female, 1/2" diameter, with gasket		24	.333		4.95	14.65		19.60	27.50
8660	3/4" diameter		20	.400		7.65	17.55		25.20	34.50
8700	Couplings, compression, 1/2" diameter, steel					2.76			2.76	3.04
8710	3/4" diameter					3.84			3.84	4.22
8720	1" diameter					6.60			6.60	7.30
8730	1-1/4" diameter					12.55			12.55	13.80
8740	1-1/2" diameter					18.30			18.30	20
8750	2" diameter					25			25	27.50
8760	2-1/2" diameter					111			111	122
8770	3" diameter					139			139	153
8780	3-1/2" diameter					226			226	249
8790	4" diameter					231			231	254
8800	Box connectors, compression, 1/2" diam., steel	1 Elec	120	.067		2.70	2.93		5.63	7.35
8810	3/4" diameter		110	.073		3.95	3.19		7.14	9.10
8820	1" diameter		90	.089		6.45	3.90		10.35	12.90
8830	1-1/4" diameter		70	.114		12.75	5		17.75	21.50
8840	1-1/2" diameter		60	.133		18.60	5.85		24.45	29
8850	2" diameter		50	.160		26.50	7		33.50	39.50
8860	2-1/2" diameter		36	.222		83.50	9.75		93.25	107
8870	3" diameter		27	.296		105	13		118	135
8880	3-1/2" diameter		21	.381		159	16.70		175.70	200
8890	4" diameter		16	.500		176	22		198	226
8900	Box connectors, insulated compression, 1/2" diam., steel		120	.067		3.48	2.93		6.41	8.20
8910	3/4" diameter		110	.073		4.78	3.19		7.97	10
8920	1" diameter		90	.089		7.95	3.90		11.85	14.55
8930	1-1/4" diameter		70	.114		16.55	5		21.55	25.50
8940	1-1/2" diameter		60	.133		25	5.85		30.85	36
8950	2" diameter		50	.160		34.50	7		41.50	48.50
8960	2-1/2" diameter		36	.222		103	9.75		112.75	129
8970	3" diameter		27	.296		141	13		154	174
8980	3-1/2" diameter		21	.381		232	16.70		248.70	281
8990	4" diameter		16	.500		209	22		231	262
9100	PVC, #40, 1/2" diameter		190	.042	L.F.	1.02	1.85		2.87	3.87
9110	3/4" diameter		145	.055		1.23	2.42		3.65	4.96
9120	1" diameter		125	.064		2.01	2.81		4.82	6.40
9130	1-1/4" diameter		110	.073		2.61	3.19		5.80	7.65
9140	1-1/2" diameter		100	.080		2.99	3.51		6.50	8.55
9150	2" diameter		90	.089		3.79	3.90		7.69	9.95
9160	2-1/2" diameter		65	.123		6.30	5.40		11.70	15
9170	3" diameter	2 Elec	110	.145		7.10	6.40		13.50	17.30
9180	3-1/2" diameter		100	.160		9.65	7		16.65	21
9190	4" diameter		90	.178		10.35	7.80		18.15	23

26 05 33.05 Conduit		Crew	Daily Output	Labor-Hours	Unit	Material	2007 Bare Costs Labor	Equipment	Total	Total Incl O&P
9200	5" diameter	2 Elec	70	.229	L.F.	14.70	10.05		24.75	31
9210	6" diameter	↓	60	.267	↓	20	11.70		31.70	39.50
9220	Elbows, 1/2" diameter	1 Elec	50	.160	Ea.	1.58	7		8.58	12.20
9230	3/4" diameter		42	.190		1.59	8.35		9.94	14.20
9240	1" diameter		35	.229		2.69	10.05		12.74	17.90
9250	1-1/4" diameter		28	.286		3.83	12.55		16.38	23
9260	1-1/2" diameter		20	.400		5.20	17.55		22.75	31.50
9270	2" diameter		16	.500		7.55	22		29.55	41
9280	2-1/2" diameter		11	.727		13.15	32		45.15	62
9290	3" diameter		9	.889		23	39		62	83.50
9300	3-1/2" diameter		7	1.143		32	50		82	110
9310	4" diameter		6	1.333		40	58.50		98.50	131
9320	5" diameter		4	2		70	88		158	208
9330	6" diameter		3	2.667		119	117		236	305
9340	Field bends, 45° & 90°, 1/2" diameter		45	.178			7.80		7.80	11.60
9350	3/4" diameter		40	.200			8.80		8.80	13.05
9360	1" diameter		35	.229			10.05		10.05	14.95
9370	1-1/4" diameter		32	.250			11		11	16.35
9380	1-1/2" diameter		27	.296			13		13	19.35
9390	2" diameter		20	.400			17.55		17.55	26
9400	2-1/2" diameter		16	.500			22		22	32.50
9410	3" diameter		13	.615			27		27	40
9420	3-1/2" diameter		12	.667			29.50		29.50	43.50
9430	4" diameter		10	.800			35		35	52.50
9440	5" diameter		9	.889			39		39	58
9450	6" diameter		8	1			44		44	65.50
9460	PVC adapters, 1/2" diameter		50	.160		.51	7		7.51	11
9470	3/4" diameter		42	.190		.94	8.35		9.29	13.50
9480	1" diameter		38	.211		1.19	9.25		10.44	15.05
9490	1-1/4" diameter		35	.229		1.59	10.05		11.64	16.70
9500	1-1/2" diameter		32	.250		1.85	11		12.85	18.40
9510	2" diameter		27	.296		2.66	13		15.66	22.50
9520	2-1/2" diameter		23	.348		4.57	15.25		19.82	27.50
9530	3" diameter		18	.444		6.75	19.50		26.25	36.50
9540	3-1/2" diameter		13	.615		8.70	27		35.70	49.50
9550	4" diameter		11	.727		11.80	32		43.80	60.50
9560	5" diameter		8	1		23	44		67	91
9570	6" diameter	↓	6	1.333	↓	28	58.50		86.50	118
9580	PVC-LB, LR or LL fittings & covers									
9590	1/2" diameter	1 Elec	20	.400	Ea.	6.55	17.55		24.10	33.50
9600	3/4" diameter		16	.500		8.45	22		30.45	42
9610	1" diameter		12	.667		9.30	29.50		38.80	54
9620	1-1/4" diameter		9	.889		14.15	39		53.15	73.50
9630	1-1/2" diameter		7	1.143		17	50		67	93
9640	2" diameter		6	1.333		30	58.50		88.50	120
9650	2-1/2" diameter		6	1.333		109	58.50		167.50	207
9660	3" diameter		5	1.600		113	70		183	229
9670	3-1/2" diameter		4	2		121	88		209	264
9680	4" diameter	↓	3	2.667	↓	124	117		241	310
9690	PVC-tee fitting & cover									
9700	1/2"	1 Elec	14	.571	Ea.	8.50	25		33.50	47
9710	3/4"		13	.615		10.40	27		37.40	51.50
9720	1"	↓	10	.800	↓	10.85	35		45.85	64.50

26 05 33 – Raceway and Boxes for Electrical Systems

26 05 33.05 Conduit		Crew	Daily Output	Labor-Hours	Unit	Material	2007 Bare Costs Labor	Equipment	Total	Total Incl O&P
9730	1-1/4"	1 Elec	9	.889	Ea.	17.70	39		56.70	77.50
9740	1-1/2"		8	1		23.50	44		67.50	91
9750	2"		7	1.143		33	50		83	111
9760	PVC-reducers, 3/4" x 1/2" diameter					1.55			1.55	1.71
9770	1" x 1/2" diameter					3.47			3.47	3.82
9780	1" x 3/4" diameter					3.68			3.68	4.05
9790	1-1/4" x 3/4" diameter					4.83			4.83	5.30
9800	1-1/4" x 1" diameter					4.83			4.83	5.30
9810	1-1/2" x 1-1/4" diameter					4.85			4.85	5.35
9820	2" x 1-1/4" diameter					6.05			6.05	6.65
9830	2-1/2" x 2" diameter					19.60			19.60	21.50
9840	3" x 2" diameter					21			21	23
9850	4" x 3" diameter					26			26	28.50
9860	Cement, quart					23			23	25
9870	Gallon					97.50			97.50	107
9880	Heat bender, to 6" diameter					1,225			1,225	1,350
9900	Add to labor for higher elevated installation									
9910	15' to 20' high, add						10%			
9920	20' to 25' high, add						20%			
9930	25' to 30' high, add						25%			
9940	30' to 35' high, add						30%			
9950	35' to 40' high, add						35%			
9960	Over 40' high, add						40%			
9980	Allow. for cond. ftngs., 5% min.-20% max.									

26 05 33.10 Conduit

		Crew	Daily Output	Labor-Hours	Unit	Material	2007 Bare Costs Labor	Equipment	Total	Total Incl O&P
0010	**CONDUIT** To 15' high, includes couplings only R260533-23									
0200	Electric metallic tubing, 1/2"diameter	1 Elec	435	.018	L.F.	.44	.81		1.25	1.68
0220	3/4" diameter		253	.032		.80	1.39		2.19	2.95
0240	1" diameter		207	.039		1.38	1.70		3.08	4.05
0260	1-1/4"diameter		173	.046		2.29	2.03		4.32	5.55
0280	1-1/2" diameter		153	.052		2.95	2.30		5.25	6.65
0300	2" diameter		130	.062		3.67	2.70		6.37	8.05
0320	2-1/2" diameter		92	.087		7.05	3.82		10.87	13.45
0340	3" diameter	2 Elec	148	.108		9.30	4.75		14.05	17.30
0360	3-1/2" diameter		134	.119		11.95	5.25		17.20	21
0380	4" diameter		114	.140		12.95	6.15		19.10	23.50
0500	Steel rigid galvanized, 1/2" diameter R260533-24	1 Elec	146	.055		1.73	2.41		4.14	5.50
0520	3/4" diameter		125	.064		2.05	2.81		4.86	6.45
0540	1" diameter		93	.086		2.99	3.78		6.77	8.90
0560	1-1/4" diameter		88	.091		4.14	3.99		8.13	10.50
0580	1-1/2" diameter		80	.100		4.84	4.39		9.23	11.85
0600	2" diameter		65	.123		6.10	5.40		11.50	14.75
0620	2-1/2" diameter		48	.167		11.85	7.30		19.15	24
0640	3" diameter	2 Elec	64	.250		14.50	11		25.50	32.50
0660	3-1/2" diameter		60	.267		17.95	11.70		29.65	37
0680	4" diameter		52	.308		19.30	13.50		32.80	41
0700	5" diameter		50	.320		39.50	14.05		53.55	64.50
0720	6" diameter		48	.333		56.50	14.65		71.15	84.50
1000	Steel intermediate conduit (IMC), 1/2 diameter R260533-25	1 Elec	155	.052		1.29	2.27		3.56	4.79
1010	3/4" diameter		130	.062		1.63	2.70		4.33	5.80
1020	1" diameter		100	.080		2.46	3.51		5.97	7.95
1030	1-1/4" diameter		93	.086		3.12	3.78		6.90	9.05

26 05 33 – Raceway and Boxes for Electrical Systems

26 05 33.10 Conduit		Crew	Daily Output	Labor-Hours	Unit	Material	2007 Bare Costs Labor	Equipment	Total	Total Incl O&P
1040	1-1/2" diameter	1 Elec	85	.094	L.F.	3.82	4.13		7.95	10.35
1050	2" diameter		70	.114		5.10	5		10.10	13.05
1060	2-1/2" diameter		53	.151		10.40	6.65		17.05	21.50
1070	3" diameter	2 Elec	80	.200		13.15	8.80		21.95	27.50
1080	3-1/2" diameter		70	.229		15.85	10.05		25.90	32.50
1090	4" diameter		60	.267		17.60	11.70		29.30	37

26 05 33.20 Conduit Nipples		Crew	Daily Output	Labor-Hours	Unit	Material	2007 Bare Costs Labor	Equipment	Total	Total Incl O&P
0010	**CONDUIT NIPPLES** With locknuts and bushings									
0100	Aluminum, 1/2" diameter, close	1 Elec	36	.222	Ea.	5.40	9.75		15.15	20.50
0120	1-1/2" long		36	.222		5.50	9.75		15.25	20.50
0140	2" long		36	.222		5.80	9.75		15.55	21
0160	2-1/2" long		36	.222		6.55	9.75		16.30	21.50
0180	3" long		36	.222		6.80	9.75		16.55	22
0200	3-1/2" long		36	.222		7.20	9.75		16.95	22.50
0220	4" long		36	.222		7.60	9.75		17.35	23
0240	5" long		36	.222		8.45	9.75		18.20	24
0260	6" long		36	.222		8.80	9.75		18.55	24
0280	8" long		36	.222		11.20	9.75		20.95	27
0300	10" long		36	.222		13.20	9.75		22.95	29
0320	12" long		36	.222		15	9.75		24.75	31
0340	3/4" diameter, close		32	.250		7.60	11		18.60	24.50
0360	1-1/2" long		32	.250		7.80	11		18.80	25
0380	2" long		32	.250		8	11		19	25
0400	2-1/2" long		32	.250		8.35	11		19.35	25.50
0420	3" long		32	.250		8.80	11		19.80	26
0440	3-1/2" long		32	.250		9	11		20	26.50
0460	4" long		32	.250		9.35	11		20.35	26.50
0480	5" long		32	.250		10.65	11		21.65	28
0500	6" long		32	.250		11.75	11		22.75	29.50
0520	8" long		32	.250		14.60	11		25.60	32.50
0540	10" long		32	.250		16.60	11		27.60	34.50
0560	12" long		32	.250		16.60	11		27.60	34.50
0580	1" diameter, close		27	.296		10.75	13		23.75	31
0600	2" long		27	.296		11.40	13		24.40	32
0620	2-1/2" long		27	.296		12	13		25	32.50
0640	3" long		27	.296		12.55	13		25.55	33
0660	3-1/2" long		27	.296		13.60	13		26.60	34.50
0680	4" long		27	.296		14.40	13		27.40	35
0700	5" long		27	.296		16.15	13		29.15	37
0720	6" long		27	.296		18.20	13		31.20	39.50
0740	8" long		27	.296		21.50	13		34.50	43
0760	10" long		27	.296		25.50	13		38.50	48
0780	12" long		27	.296		29.50	13		42.50	52
0800	1-1/4" diameter, close		23	.348		14.75	15.25		30	38.50
0820	2" long		23	.348		14.90	15.25		30.15	39
0840	2-1/2" long		23	.348		15.70	15.25		30.95	40
0860	3" long		23	.348		16.90	15.25		32.15	41
0880	3-1/2" long		23	.348		18.30	15.25		33.55	42.50
0900	4" long		23	.348		19.45	15.25		34.70	44
0920	5" long		23	.348		21.50	15.25		36.75	46.50
0940	6" long		23	.348		24	15.25		39.25	49
0960	8" long		23	.348		29	15.25		44.25	54

26 05 33.20 Conduit Nipples		Crew	Daily Output	Labor-Hours	Unit	Material	2007 Bare Costs Labor	Equipment	Total	Total Incl O&P
0980	10" long	1 Elec	23	.348	Ea.	33.50	15.25		48.75	59.50
1000	12" long		23	.348		38.50	15.25		53.75	64.50
1020	1-1/2" diameter, close		20	.400		20	17.55		37.55	48
1040	2" long		20	.400		20.50	17.55		38.05	48.50
1060	2-1/2" long		20	.400		21	17.55		38.55	49
1080	3" long		20	.400		22.50	17.55		40.05	51
1100	3-1/2" long		20	.400		25.50	17.55		43.05	54
1120	4" long		20	.400		26	17.55		43.55	54.50
1140	5" long		20	.400		28	17.55		45.55	57
1160	6" long		20	.400		31	17.55		48.55	60
1180	8" long		20	.400		36.50	17.55		54.05	66.50
1200	10" long		20	.400		42.50	17.55		60.05	73
1220	12" long		20	.400		48	17.55		65.55	79
1240	2" diameter, close		18	.444		28	19.50		47.50	59.50
1260	2-1/2" long		18	.444		29	19.50		48.50	60.50
1280	3" long		18	.444		30.50	19.50		50	62.50
1300	3-1/2" long		18	.444		33	19.50		52.50	65
1320	4" long		18	.444		34	19.50		53.50	66.50
1340	5" long		18	.444		37.50	19.50		57	70.50
1360	6" long		18	.444		41	19.50		60.50	74
1380	8" long		18	.444		48.50	19.50		68	82
1400	10" long		18	.444		55.50	19.50		75	90
1420	12" long		18	.444		62	19.50		81.50	97.50
1440	2-1/2" diameter, close		15	.533		59	23.50		82.50	100
1460	3" long		15	.533		59.50	23.50		83	101
1480	3-1/2" long		15	.533		62.50	23.50		86	104
1500	4" long		15	.533		64.50	23.50		88	106
1520	5" long		15	.533		68.50	23.50		92	111
1540	6" long		15	.533		72	23.50		95.50	115
1560	8" long		15	.533		83	23.50		106.50	126
1580	10" long		15	.533		93.50	23.50		117	138
1600	12" long		15	.533		104	23.50		127.50	149
1620	3" diameter, close		12	.667		67.50	29.50		97	118
1640	3" long		12	.667		71	29.50		100.50	122
1660	3-1/2" long		12	.667		77	29.50		106.50	129
1680	4" long		12	.667		78.50	29.50		108	130
1700	5" long		12	.667		84.50	29.50		114	136
1720	6" long		12	.667		91	29.50		120.50	144
1740	8" long		12	.667		105	29.50		134.50	159
1760	10" long		12	.667		119	29.50		148.50	175
1780	12" long		12	.667		133	29.50		162.50	190
1800	3-1/2" diameter, close		11	.727		122	32		154	182
1820	4" long		11	.727		134	32		166	195
1840	5" long		11	.727		140	32		172	202
1860	6" long		11	.727		148	32		180	211
1880	8" long		11	.727		164	32		196	228
1900	10" long		11	.727		182	32		214	248
1920	12" long		11	.727		198	32		230	266
1940	4" diameter, close		9	.889		146	39		185	219
1960	4" long		9	.889		157	39		196	231
1980	5" long		9	.889		166	39		205	240
2000	6" long		9	.889		176	39		215	251
2020	8" long		9	.889		196	39		235	274

26 05 33.20 Conduit Nipples		Crew	Daily Output	Labor-Hours	Unit	Material	2007 Bare Costs Labor	Equipment	Total	Total Incl O&P
2040	10" long	1 Elec	9	.889	Ea.	216	39		255	296
2060	12" long		9	.889		236	39		275	320
2080	5" diameter, close		7	1.143		320	50		370	430
2100	5" long		7	1.143		340	50		390	450
2120	6" long		7	1.143		345	50		395	455
2140	8" long		7	1.143		380	50		430	495
2160	10" long		7	1.143		415	50		465	530
2180	12" long		7	1.143		445	50		495	560
2200	6" diameter, close		6	1.333		525	58.50		583.50	660
2220	5" long		6	1.333		555	58.50		613.50	695
2240	6" long		6	1.333		570	58.50		628.50	710
2260	8" long		6	1.333		615	58.50		673.50	760
2280	10" long		6	1.333		660	58.50		718.50	810
2300	12" long		6	1.333		685	58.50		743.50	840
2320	Rigid galvanized steel, 1/2" diameter, close		32	.250		3.31	11		14.31	20
2340	1-1/2" long		32	.250		3.66	11		14.66	20.50
2360	2" long		32	.250		3.95	11		14.95	20.50
2380	2-1/2" long		32	.250		4.15	11		15.15	21
2400	3" long		32	.250		4.38	11		15.38	21
2420	3-1/2" long		32	.250		4.66	11		15.66	21.50
2440	4" long		32	.250		4.93	11		15.93	22
2460	5" long		32	.250		5.35	11		16.35	22
2480	6" long		32	.250		6.10	11		17.10	23
2500	8" long		32	.250		9.50	11		20.50	27
2520	10" long		32	.250		10.70	11		21.70	28
2540	12" long		32	.250		12.10	11		23.10	29.50
2560	3/4" diameter, close		27	.296		4.48	13		17.48	24.50
2580	2" long		27	.296		4.84	13		17.84	24.50
2600	2-1/2" long		27	.296		4.90	13		17.90	25
2620	3" long		27	.296		5.20	13		18.20	25
2640	3-1/2" long		27	.296		5.50	13		18.50	25.50
2660	4" long		27	.296		6.15	13		19.15	26
2680	5" long		27	.296		6.75	13		19.75	27
2700	6" long		27	.296		7.50	13		20.50	27.50
2720	8" long		27	.296		11	13		24	31.50
2740	10" long		27	.296		12.80	13		25.80	33.50
2760	12" long		27	.296		14.15	13		27.15	35
2780	1" diameter, close		23	.348		6.80	15.25		22.05	30
2800	2" long		23	.348		7.15	15.25		22.40	30.50
2820	2-1/2" long		23	.348		7.50	15.25		22.75	31
2840	3" long		23	.348		8	15.25		23.25	31.50
2860	3-1/2" long		23	.348		8.65	15.25		23.90	32
2880	4" long		23	.348		9	15.25		24.25	32.50
2900	5" long		23	.348		9.80	15.25		25.05	33.50
2920	6" long		23	.348		10.40	15.25		25.65	34
2940	8" long		23	.348		14.55	15.25		29.80	38.50
2960	10" long		23	.348		17.80	15.25		33.05	42
2980	12" long		23	.348		19.80	15.25		35.05	44.50
3000	1-1/4" diameter, close		20	.400		9.10	17.55		26.65	36
3020	2" long		20	.400		9.45	17.55		27	36.50
3040	3" long		20	.400		10.35	17.55		27.90	37.50
3060	3-1/2" long		20	.400		11.15	17.55		28.70	38.50
3080	4" long		20	.400		11.50	17.55		29.05	38.50

26 05 Common Work Results for Electrical

26 05 33 – Raceway and Boxes for Electrical Systems

26 05 33.20 Conduit Nipples		Crew	Daily Output	Labor-Hours	Unit	Material	2007 Bare Costs Labor	Equipment	Total	Total Incl O&P
3100	5" long	1 Elec	20	.400	Ea.	12.65	17.55		30.20	40
3120	6" long		20	.400		13.65	17.55		31.20	41
3140	8" long		20	.400		19.55	17.55		37.10	47.50
3160	10" long		20	.400		23.50	17.55		41.05	52
3180	12" long		20	.400		26.50	17.55		44.05	55.50
3200	1-1/2" diameter, close		18	.444		12.35	19.50		31.85	42.50
3220	2" long		18	.444		12.65	19.50		32.15	43
3240	2-1/2" long		18	.444		13.40	19.50		32.90	44
3260	3" long		18	.444		13.90	19.50		33.40	44.50
3280	3-1/2" long		18	.444		15	19.50		34.50	45.50
3300	4" long		18	.444		15.70	19.50		35.20	46.50
3320	5" long		18	.444		16.80	19.50		36.30	47.50
3340	6" long		18	.444		19.10	19.50		38.60	50
3360	8" long		18	.444		25.50	19.50		45	57.50
3380	10" long		18	.444		29.50	19.50		49	61.50
3400	12" long		18	.444		31.50	19.50		51	63.50
3420	2" diameter, close		16	.500		17.10	22		39.10	51.50
3440	2-1/2" long		16	.500		18.30	22		40.30	52.50
3460	3" long		16	.500		19.55	22		41.55	54
3480	3-1/2" long		16	.500		21	22		43	55.50
3500	4" long		16	.500		22	22		44	56.50
3520	5" long		16	.500		24	22		46	59
3540	6" long		16	.500		26	22		48	61
3560	8" long		16	.500		33	22		55	68.50
3580	10" long		16	.500		37	22		59	73.50
3600	12" long		16	.500		41	22		63	77.50
3620	2-1/2" diameter, close		13	.615		42.50	27		69.50	87
3640	3" long		13	.615		43	27		70	87
3660	3-1/2" long		13	.615		46.50	27		73.50	91
3680	4" long		13	.615		47.50	27		74.50	92
3700	5" long		13	.615		52	27		79	97
3720	6" long		13	.615		55.50	27		82.50	101
3740	8" long		13	.615		65.50	27		92.50	113
3760	10" long		13	.615		72.50	27		99.50	120
3780	12" long		13	.615		80.50	27		107.50	129
3800	3" diameter, close		12	.667		52	29.50		81.50	101
3820	3" long		12	.667		56.50	29.50		86	106
3900	3-1/2" long		12	.667		56.50	29.50		86	106
3920	4" long		12	.667		58.50	29.50		88	108
3940	5" long		12	.667		63.50	29.50		93	113
3960	6" long		12	.667		68	29.50		97.50	119
3980	8" long		12	.667		79.50	29.50		109	131
4000	10" long		12	.667		88.50	29.50		118	141
4020	12" long		12	.667		101	29.50		130.50	156
4040	3-1/2" diameter, close		10	.800		83.50	35		118.50	145
4060	4" long		10	.800		91	35		126	153
4080	5" long		10	.800		96	35		131	158
4100	6" long		10	.800		102	35		137	165
4120	8" long		10	.800		113	35		148	177
4140	10" long		10	.800		125	35		160	191
4160	12" long		10	.800		137	35		172	203
4180	4" diameter, close		8	1		102	44		146	178
4200	4" long		8	1		109	44		153	186

135

26 05 33.20 Conduit Nipples		Crew	Daily Output	Labor-Hours	Unit	Material	2007 Bare Costs Labor	Equipment	Total	Total Incl O&P
4220	5" long	1 Elec	8	1	Ea.	116	44		160	193
4240	6" long		8	1		121	44		165	199
4260	8" long		8	1		134	44		178	213
4280	10" long		8	1		149	44		193	230
4300	12" long		8	1		164	44		208	246
4320	5" diameter, close		6	1.333		204	58.50		262.50	310
4340	5" long		6	1.333		232	58.50		290.50	340
4360	6" long		6	1.333		240	58.50		298.50	350
4380	8" long		6	1.333		258	58.50		316.50	370
4400	10" long		6	1.333		278	58.50		336.50	390
4420	12" long		6	1.333		305	58.50		363.50	425
4440	6" diameter, close		5	1.600		297	70		367	430
4460	5" long		5	1.600		330	70		400	465
4480	6" long		5	1.600		340	70		410	480
4500	8" long		5	1.600		360	70		430	500
4520	10" long		5	1.600		395	70		465	540
4540	12" long		5	1.600		415	70		485	565
4560	Plastic coated, 40 mil thick, 1/2" diameter, 2" long		32	.250		13.60	11		24.60	31.50
4580	2-1/2" long		32	.250		15.10	11		26.10	33
4600	3" long		32	.250		15.30	11		26.30	33
4680	3-1/2" long		32	.250		16.85	11		27.85	35
4700	4" long		32	.250		16.95	11		27.95	35
4720	5" long		32	.250		17.15	11		28.15	35
4740	6" long		32	.250		17.60	11		28.60	35.50
4760	8" long		32	.250		18.20	11		29.20	36.50
4780	10" long		32	.250		18.95	11		29.95	37.50
4800	12" long		32	.250		19.60	11		30.60	38
4820	3/4" diameter, 2" long		26	.308		14.50	13.50		28	36
4840	2-1/2" long		26	.308		16.10	13.50		29.60	38
4860	3" long		26	.308		16.20	13.50		29.70	38
4880	3-1/2" long		26	.308		17.80	13.50		31.30	39.50
4900	4" long		26	.308		18.05	13.50		31.55	40
4920	5" long		26	.308		18.35	13.50		31.85	40
4940	6" long		26	.308		18.60	13.50		32.10	40.50
4960	8" long		26	.308		19.35	13.50		32.85	41.50
4980	10" long		26	.308		20	13.50		33.50	42
5000	12" long		26	.308		20.50	13.50		34	43
5020	1" diameter, 2" long		22	.364		16.35	15.95		32.30	42
5040	2-1/2" long		22	.364		17.90	15.95		33.85	43.50
5060	3" long		22	.364		18.15	15.95		34.10	44
5080	3-1/2" long		22	.364		19.85	15.95		35.80	46
5100	4" long		22	.364		20	15.95		35.95	46
5120	5" long		22	.364		20.50	15.95		36.45	46.50
5140	6" long		22	.364		21	15.95		36.95	47
5160	8" long		22	.364		22	15.95		37.95	48
5180	10" long		22	.364		22.50	15.95		38.45	49
5200	12" long		22	.364		25	15.95		40.95	52
5220	1-1/4" diameter, 2" long		18	.444		20.50	19.50		40	51.50
5240	2-1/2" long		18	.444		22	19.50		41.50	53.50
5260	3" long		18	.444		22.50	19.50		42	53.50
5280	3-1/2" long		18	.444		22.50	19.50		42	53.50
5300	4" long		18	.444		24.50	19.50		44	56
5320	5" long		18	.444		25	19.50		44.50	56.50

Reed Construction Data ®

Since its founding in 1975, Reed Construction Data has developed an online and print portfolio of innovative products and services for the construction, design and manufacturing community. Our products and services are designed specifically to help construction industry professionals advance their businesses with timely, accurate and actionable project, product and cost data. Reed Construction Data is your all-inclusive source of construction information encompassing all phases of the construction process.

Reed Bulletin and Reed Connect™ deliver the most comprehensive, timely and reliable project information to support contractors, distributors and building product manufacturers in identifying, bidding and tracking projects – private and public, general building and civil. Reed Construction Data also offers in-depth construction activity statistics and forecasts covering major project categories, many at the county and metropolitan level.

Reed Bulletin

www.reedbulletin.com

- Project leads targeted by geographic region and formatted by construction stage – available online or in print.

- Locate those hard-to-find jobs that are more profitable to your business.

- Optional automatic e-mail updates sent whenever project details change.

- Download plans and specs online or order print copies.

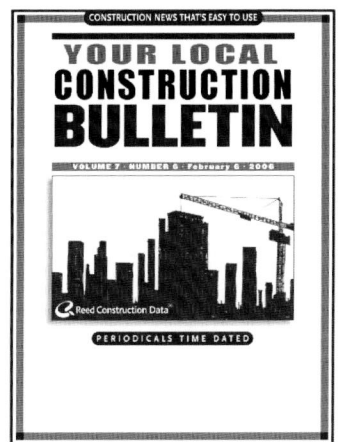

Reed Research & Analytics

www.buildingteamforecast.com

Reed Construction Forecast

- Delivers timely construction industry activity combining historical data, current year projections and forecasts.

- Modeled at the individual MSA-level to capture changing local market conditions.

- Covers 21 major project categories.

Reed Construction Starts

- Available in a monthly report or as an interactive database.

- Data provided in square footage and dollar value.

- Highly effective and efficient business planning tool.

Market Fundamentals

- Metropolitan area-specific reporting.

- Five-year forecast of industry performance including major projects in development and underway.

- Property types include office, retail, hotel, warehouse and apartment.

Reed Connect

www.reedconnect.com

- Customized web-based project lead delivery service featuring advanced search capabilities.

- Manage and track actionable leads from planning to quote through winning the job.

- Competitive analysis tool to analyze lost sales opportunities.

- Potential integration with your CRM application.

For more information about Reed Construction Data, please call 877-REED411, visit our website at www.reedconstructiondata.com, or E-mail: marketing@reedbusiness.com

Reed Construction Data
CONNECT • FIRST SOURCE • BULLETIN
RSMEANS • ACP • RESEARCH • PLANSDIRECT

Reed Construction Data ®

The design community utilizes the Reed First Source® suite of products to search, select and specify nationally available building products during the formative stages of project design, as well as during other stages of product selection and specification. Reed Design Registry is a detailed database of architecture firms. This tool features sophisticated search and sort capabilities to support the architect selection process and ensure locating the best manufacturers to partner with on your next project.

Reed First Source - The Leading Product Directory to "Find It, Choose It, Use It"

www.reedfirstsource.com

- Comprehensive directory of over 11,000 commercial building product manufacturers classified by MasterFormat™ 2004 categories.

- SPEC-DATA's 10-part format provides performance data along with technical and installation information.

- MANU-SPEC delivers manufacturer guide specifications in the CSI 3-part SectionFormat.

- Search for products, download CAD, research building codes and view catalogs online.

Reed Design Registry – The Premier Source of Architecture Firms

www.reedregistry.com

- Comprehensive website directory contains over 30,000 architect firms.

- Profiles list firm size, firm's specialties and more.

- Official database of AIA member-owned firms.

- Allows architects to share project information with the building community.

- Coming soon – Reed Registry to include engineers and landscape architects.

 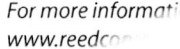

26 05 33.20 Conduit Nipples		Crew	Daily Output	Labor-Hours	Unit	Material	2007 Bare Costs Labor	Equipment	Total	Total Incl O&P
5340	6" long	1 Elec	18	.444	Ea.	25.50	19.50		45	57.50
5360	8" long		18	.444		26	19.50		45.50	57.50
5380	10" long		18	.444		29	19.50		48.50	61
5400	12" long		18	.444		34	19.50		53.50	66
5420	1-1/2" diameter, 2" long		16	.500		23.50	22		45.50	58
5440	2-1/2" long		16	.500		25	22		47	60
5460	3" long		16	.500		25.50	22		47.50	60.50
5480	3-1/2" long		16	.500		26.50	22		48.50	61.50
5500	4" long		16	.500		27.50	22		49.50	63
5520	5" long		16	.500		28	22		50	63.50
5540	6" long		16	.500		30.50	22		52.50	66
5560	8" long		16	.500		32.50	22		54.50	68
5580	10" long		16	.500		38.50	22		60.50	75
5600	12" long		16	.500		44	22		66	81
5620	2" diameter, 2-1/2" long		14	.571		31	25		56	72
5640	3" long		14	.571		31.50	25		56.50	72
5660	3-1/2" long		14	.571		33.50	25		58.50	74.50
5680	4" long		14	.571		37	25		62	78.50
5700	5" long		14	.571		37	25		62	78.50
5720	6" long		14	.571		38.50	25		63.50	79.50
5740	8" long		14	.571		43	25		68	84.50
5760	10" long		14	.571		51.50	25		76.50	94
5780	12" long		14	.571		57.50	25		82.50	101
5800	2-1/2" diameter, 3-1/2" long		12	.667		60.50	29.50		90	110
5820	4" long		12	.667		62	29.50		91.50	112
5840	5" long		12	.667		71.50	29.50		101	122
5860	6" long		12	.667		73	29.50		102.50	124
5880	8" long		12	.667		82.50	29.50		112	134
5900	10" long		12	.667		89.50	29.50		119	142
5920	12" long		12	.667		101	29.50		130.50	155
5940	3" diameter, 3-1/2" long		11	.727		72	32		104	127
5960	4" long		11	.727		75.50	32		107.50	131
5980	5" long		11	.727		80.50	32		112.50	136
6000	6" long		11	.727		87.50	32		119.50	144
6020	8" long		11	.727		97.50	32		129.50	155
6040	10" long		11	.727		111	32		143	170
6060	12" long		11	.727		128	32		160	189
6080	3-1/2" diameter, 4" long		9	.889		113	39		152	182
6100	5" long		9	.889		114	39		153	184
6120	6" long		9	.889		120	39		159	191
6140	8" long		9	.889		136	39		175	208
6160	10" long		9	.889		153	39		192	226
6180	12" long		9	.889		163	39		202	238
6200	4" diameter, 4" long		7.50	1.067		130	47		177	213
6220	5" long		7.50	1.067		133	47		180	217
6240	6" long		7.50	1.067		144	47		191	228
6260	8" long		7.50	1.067		160	47		207	246
6280	10" long		7.50	1.067		188	47		235	277
6300	12" long		7.50	1.067		196	47		243	286
6320	5" diameter, 5" long		5.50	1.455		242	64		306	360
6340	6" long		5.50	1.455		257	64		321	380
6360	8" long		5.50	1.455		268	64		332	390
6380	10" long		5.50	1.455		290	64		354	415

26 05 Common Work Results for Electrical

26 05 33 – Raceway and Boxes for Electrical Systems

26 05 33.20 Conduit Nipples		Crew	Daily Output	Labor-Hours	Unit	Material	2007 Bare Costs Labor	Equipment	Total	Total Incl O&P
6400	12" long	1 Elec	5.50	1.455	Ea.	293	64		357	420
6420	6" diameter, 5" long		4.50	1.778		345	78		423	495
6440	6" long		4.50	1.778		360	78		438	510
6460	8" long		4.50	1.778		375	78		453	525
6480	10" long		4.50	1.778		395	78		473	550
6500	12" long		4.50	1.778		420	78		498	575

26 05 33.25 Conduit Fittings for Rigid Galvanized Steel

		Crew	Daily Output	Labor-Hours	Unit	Material	2007 Bare Costs Labor	Equipment	Total	Total Incl O&P
0010	**CONDUIT FITTINGS FOR RIGID GALVANIZED STEEL**									
0050	Standard, locknuts, 1/2" diameter				Ea.	.22			.22	.24
0100	3/4" diameter					.37			.37	.41
0300	1" diameter					.60			.60	.66
0500	1-1/4" diameter					.82			.82	.90
0700	1-1/2" diameter					1.35			1.35	1.49
1000	2" diameter					1.97			1.97	2.17
1030	2-1/2" diameter					5.50			5.50	6.05
1050	3" diameter					7.05			7.05	7.75
1070	3-1/2" diameter					11.90			11.90	13.10
1100	4" diameter					14.85			14.85	16.35
1110	5" diameter					32			32	35
1120	6" diameter					54.50			54.50	59.50
1130	Bushings, plastic, 1/2" diameter	1 Elec	40	.200		.21	8.80		9.01	13.30
1150	3/4" diameter		32	.250		.35	11		11.35	16.75
1170	1" diameter		28	.286		.58	12.55		13.13	19.30
1200	1-1/4" diameter		24	.333		.88	14.65		15.53	23
1230	1-1/2" diameter		18	.444		1.19	19.50		20.69	30.50
1250	2" diameter		15	.533		2.22	23.50		25.72	37.50
1270	2-1/2" diameter		13	.615		5.15	27		32.15	45.50
1300	3" diameter		12	.667		5.30	29.50		34.80	49.50
1330	3-1/2" diameter		11	.727		7	32		39	55
1350	4" diameter		9	.889		8.60	39		47.60	67.50
1360	5" diameter		7	1.143		18.60	50		68.60	95
1370	6" diameter		5	1.600		35.50	70		105.50	144
1390	Steel, 1/2" diameter		40	.200		.51	8.80		9.31	13.60
1400	3/4" diameter		32	.250		.68	11		11.68	17.10
1430	1" diameter		28	.286		1.02	12.55		13.57	19.75
1450	Steel insulated, 1-1/4" diameter		24	.333		6.20	14.65		20.85	29
1470	1-1/2" diameter		18	.444		7.90	19.50		27.40	37.50
1500	2" diameter		15	.533		11.35	23.50		34.85	47.50
1530	2-1/2" diameter		13	.615		21	27		48	63
1550	3" diameter		12	.667		28	29.50		57.50	74.50
1570	3-1/2" diameter		11	.727		36	32		68	87
1600	4" diameter		9	.889		45.50	39		84.50	109
1610	5" diameter		7	1.143		99.50	50		149.50	184
1620	6" diameter		5	1.600		153	70		223	273
1630	Sealing locknuts, 1/2" diameter		40	.200		1.60	8.80		10.40	14.80
1650	3/4" diameter		32	.250		1.90	11		12.90	18.45
1670	1" diameter		28	.286		2.80	12.55		15.35	21.50
1700	1-1/4" diameter		24	.333		4.70	14.65		19.35	27
1730	1-1/2" diameter		18	.444		6	19.50		25.50	35.50
1750	2" diameter		15	.533		7.50	23.50		31	43.50
1760	Grounding bushing, insulated, 1/2" diameter		32	.250		4.75	11		15.75	21.50
1770	3/4" diameter		28	.286		6.05	12.55		18.60	25.50

26 05 33.25 Conduit Fittings for Rigid Galvanized Steel		Crew	Daily Output	Labor-Hours	Unit	Material	2007 Bare Costs Labor	Equipment	Total	Total Incl O&P
1780	1" diameter	1 Elec	20	.400	Ea.	6.80	17.55		24.35	33.50
1800	1-1/4" diameter		18	.444		8.60	19.50		28.10	38.50
1830	1-1/2" diameter		16	.500		9.25	22		31.25	42.50
1850	2" diameter		13	.615		11.25	27		38.25	52.50
1870	2-1/2" diameter		12	.667		20.50	29.50		50	66
1900	3" diameter		11	.727		26	32		58	76
1930	3-1/2" diameter		9	.889		34.50	39		73.50	95.50
1950	4" diameter		8	1		36.50	44		80.50	106
1960	5" diameter		6	1.333		93	58.50		151.50	189
1970	6" diameter		4	2		142	88		230	287
1990	Coupling, with set screw, 1/2" diameter		50	.160		4.80	7		11.80	15.75
2000	3/4" diameter		40	.200		6.05	8.80		14.85	19.70
2030	1" diameter		35	.229		9.90	10.05		19.95	26
2050	1-1/4" diameter		28	.286		16.90	12.55		29.45	37.50
2070	1-1/2" diameter		23	.348		22	15.25		37.25	46.50
2090	2" diameter		20	.400		48.50	17.55		66.05	79.50
2100	2-1/2" diameter		18	.444		105	19.50		124.50	144
2110	3" diameter		15	.533		124	23.50		147.50	172
2120	3-1/2" diameter		12	.667		177	29.50		206.50	238
2130	4" diameter		10	.800		230	35		265	305
2140	5" diameter		9	.889		375	39		414	470
2150	6" diameter		8	1		485	44		529	595
2160	Box connector with set screw, plain, 1/2" diameter		70	.114		3.10	5		8.10	10.85
2170	3/4" diameter		60	.133		4.45	5.85		10.30	13.60
2180	1" diameter		50	.160		7	7		14	18.15
2190	Insulated, 1-1/4" diameter		40	.200		14.50	8.80		23.30	29
2200	1-1/2" diameter		30	.267		21	11.70		32.70	40.50
2210	2" diameter		20	.400		41.50	17.55		59.05	71.50
2220	2-1/2" diameter		18	.444		100	19.50		119.50	139
2230	3" diameter		15	.533		133	23.50		156.50	181
2240	3-1/2" diameter		12	.667		187	29.50		216.50	250
2250	4" diameter		10	.800		247	35		282	325
2260	5" diameter		9	.889		298	39		337	385
2270	6" diameter		8	1		380	44		424	485
2280	LB, LR or LL fittings & covers, 1/2" diameter		16	.500		10.25	22		32.25	44
2290	3/4" diameter		13	.615		12.40	27		39.40	53.50
2300	1" diameter		11	.727		18.30	32		50.30	67.50
2330	1-1/4" diameter		8	1		27.50	44		71.50	96
2350	1-1/2" diameter		6	1.333		34	58.50		92.50	125
2370	2" diameter		5	1.600		57	70		127	168
2380	2-1/2" diameter		4	2		118	88		206	261
2390	3" diameter		3.50	2.286		153	100		253	315
2400	3-1/2" diameter		3	2.667		247	117		364	445
2410	4" diameter		2.50	3.200		276	140		416	515
2420	T fittings, with cover, 1/2" diameter		12	.667		12.40	29.50		41.90	57
2430	3/4" diameter		11	.727		14.80	32		46.80	64
2440	1" diameter		9	.889		22	39		61	82
2450	1-1/4" diameter		6	1.333		31	58.50		89.50	121
2470	1-1/2" diameter		5	1.600		40	70		110	149
2500	2" diameter		4	2		62.50	88		150.50	200
2510	2-1/2" diameter		3.50	2.286		124	100		224	285
2520	3" diameter		3	2.667		174	117		291	365
2530	3-1/2" diameter		2.50	3.200		281	140		421	520

26 05 33.25 Conduit Fittings for Rigid Galvanized Steel	Crew	Daily Output	Labor-Hours	Unit	Material	2007 Bare Costs Labor	Equipment	Total	Total Incl O&P
2540	4" diameter	1 Elec	2	4	Ea.	296	176		472	585
2550	Nipples chase, plain, 1/2" diameter		40	.200		.94	8.80		9.74	14.10
2560	3/4" diameter		32	.250		1.17	11		12.17	17.65
2570	1" diameter		28	.286		2.38	12.55		14.93	21.50
2600	Insulated, 1-1/4" diameter		24	.333		8.90	14.65		23.55	32
2630	1-1/2" diameter		18	.444		11.70	19.50		31.20	42
2650	2" diameter		15	.533		17.65	23.50		41.15	54.50
2660	2-1/2" diameter		12	.667		42	29.50		71.50	90
2670	3" diameter		10	.800		45.50	35		80.50	103
2680	3-1/2" diameter		9	.889		64	39		103	128
2690	4" diameter		8	1		97	44		141	173
2700	5" diameter		7	1.143		286	50		336	390
2710	6" diameter		6	1.333		445	58.50		503.50	570
2720	Nipples offset, plain, 1/2" diameter		40	.200		5.60	8.80		14.40	19.20
2730	3/4" diameter		32	.250		5.65	11		16.65	22.50
2740	1" diameter		24	.333		7.55	14.65		22.20	30.50
2750	Insulated, 1-1/4" diameter		20	.400		33.50	17.55		51.05	62.50
2760	1-1/2" diameter		18	.444		41	19.50		60.50	74
2770	2" diameter		16	.500		64.50	22		86.50	104
2780	3" diameter		14	.571		132	25		157	183
2850	Coupling, expansion, 1/2" diameter		12	.667		42.50	29.50		72	90.50
2880	3/4" diameter		10	.800		50	35		85	108
2900	1" diameter		8	1		58.50	44		102.50	130
2920	1-1/4" diameter		6.40	1.250		79	55		134	169
2940	1-1/2" diameter		5.30	1.509		109	66.50		175.50	219
2960	2" diameter		4.60	1.739		162	76.50		238.50	292
2980	2-1/2" diameter		3.60	2.222		255	97.50		352.50	425
3000	3" diameter		3	2.667		315	117		432	520
3020	3-1/2" diameter		2.80	2.857		465	125		590	695
3040	4" diameter		2.40	3.333		550	146		696	825
3060	5" diameter		2	4		1,150	176		1,326	1,500
3080	6" diameter		1.80	4.444		1,850	195		2,045	2,350
3100	Expansion deflection, 1/2" diameter		12	.667		152	29.50		181.50	211
3120	3/4" diameter		12	.667		162	29.50		191.50	222
3140	1" diameter		10	.800		205	35		240	279
3160	1-1/4" diameter		6.40	1.250		255	55		310	365
3180	1-1/2" diameter		5.30	1.509		270	66.50		336.50	395
3200	2" diameter		4.60	1.739		350	76.50		426.50	500
3220	2-1/2" diameter		3.60	2.222		490	97.50		587.50	685
3240	3" diameter		3	2.667		620	117		737	860
3260	3-1/2" diameter		2.80	2.857		740	125		865	1,000
3280	4" diameter		2.40	3.333		825	146		971	1,125
3300	5" diameter		2	4		1,325	176		1,501	1,700
3320	6" diameter		1.80	4.444		2,250	195		2,445	2,800
3340	Ericson, 1/2" diameter		16	.500		4.40	22		26.40	37.50
3360	3/4" diameter		14	.571		5.65	25		30.65	43.50
3380	1" diameter		11	.727		10.40	32		42.40	59
3400	1-1/4" diameter		8	1		22	44		66	90
3420	1-1/2" diameter		7	1.143		26.50	50		76.50	104
3440	2" diameter		5	1.600		53	70		123	164
3460	2-1/2" diameter		4	2		129	88		217	273
3480	3" diameter		3.50	2.286		196	100		296	365
3500	3-1/2" diameter		3	2.667		325	117		442	535

26 05 33.25 Conduit Fittings for Rigid Galvanized Steel	Crew	Daily Output	Labor-Hours	Unit	Material	2007 Bare Costs Labor	Equipment	Total	Total Incl O&P
3520	4" diameter	1 Elec	2.70	2.963	Ea.	385	130		515	620
3540	5" diameter		2.50	3.200		710	140		850	990
3560	6" diameter		2.30	3.478		950	153		1,103	1,275
3580	Split, 1/2" diameter		32	.250		4.50	11		15.50	21.50
3600	3/4" diameter		27	.296		5.60	13		18.60	25.50
3620	1" diameter		20	.400		10.30	17.55		27.85	37.50
3640	1-1/4" diameter		16	.500		14.90	22		36.90	49
3660	1-1/2" diameter		14	.571		18.65	25		43.65	58
3680	2" diameter		12	.667		37.50	29.50		67	84.50
3700	2-1/2" diameter		10	.800		86.50	35		121.50	148
3720	3" diameter		9	.889		130	39		169	201
3740	3-1/2" diameter		8	1		209	44		253	296
3760	4" diameter		7	1.143		246	50		296	345
3780	5" diameter		6	1.333		425	58.50		483.50	555
3800	6" diameter		5	1.600		570	70		640	730
4600	Reducing bushings, 3/4" to 1/2" diameter		54	.148		1.29	6.50		7.79	11.10
4620	1" to 3/4" diameter		46	.174		1.99	7.65		9.64	13.55
4640	1-1/4" to 1" diameter		40	.200		4.62	8.80		13.42	18.15
4660	1-1/2" to 1-1/4" diameter		36	.222		6.65	9.75		16.40	22
4680	2" to 1-1/2" diameter		32	.250		15.05	11		26.05	33
4740	2-1/2" to 2" diameter		30	.267		17.20	11.70		28.90	36.50
4760	3" to 2-1/2" diameter		28	.286		22.50	12.55		35.05	43
4800	Through-wall seal, 1/2" diameter		8	1		162	44		206	244
4820	3/4" diameter		7.50	1.067		162	47		209	248
4840	1" diameter		6.50	1.231		162	54		216	259
4860	1-1/4" diameter		5.50	1.455		224	64		288	340
4880	1-1/2" diameter		5	1.600		224	70		294	350
4900	2" diameter		4.20	1.905		243	83.50		326.50	390
4920	2-1/2" diameter		3.50	2.286		259	100		359	435
4940	3" diameter		3	2.667		259	117		376	460
4960	3-1/2" diameter		2.50	3.200		405	140		545	655
4980	4" diameter		2	4		465	176		641	770
5000	5" diameter		1.50	5.333		590	234		824	1,000
5020	6" diameter	▼	1	8	▼	590	350		940	1,175
5100	Cable supports, 2 or more wires									
5120	1-1/2" diameter	1 Elec	8	1	Ea.	64	44		108	136
5140	2" diameter		6	1.333		94.50	58.50		153	191
5160	2-1/2" diameter		4	2		105	88		193	247
5180	3" diameter		3.50	2.286		143	100		243	305
5200	3-1/2" diameter		2.60	3.077		189	135		324	410
5220	4" diameter		2	4		235	176		411	520
5240	5" diameter		1.50	5.333		440	234		674	830
5260	6" diameter		1	8		740	350		1,090	1,325
5280	Service entrance cap, 1/2" diameter		16	.500		7.30	22		29.30	40.50
5300	3/4" diameter		13	.615		8.50	27		35.50	49.50
5320	1" diameter		10	.800		10.90	35		45.90	64.50
5340	1-1/4" diameter		8	1		14.80	44		58.80	82
5360	1-1/2" diameter		6.50	1.231		31.50	54		85.50	115
5380	2" diameter		5.50	1.455		40.50	64		104.50	140
5400	2-1/2" diameter		4	2		127	88		215	271
5420	3" diameter		3.40	2.353		193	103		296	365
5440	3-1/2" diameter		3	2.667		236	117		353	435
5460	4" diameter	▼	2.70	2.963	▼	360	130		490	590

26 05 33.25 Conduit Fittings for Rigid Galvanized Steel	Crew	Daily Output	Labor-Hours	Unit	Material	Labor	Equipment	Total	Total Incl O&P
5600	Fire stop fittings, to 3/4" diameter	1 Elec	24	.333	Ea.	112	14.65		126.65	145
5610	1" diameter		22	.364		137	15.95		152.95	175
5620	1-1/2" diameter		20	.400		172	17.55		189.55	215
5640	2" diameter		16	.500		266	22		288	325
5660	3" diameter		12	.667		335	29.50		364.50	410
5680	4" diameter		10	.800		445	35		480	540
5700	6" diameter		8	1		765	44		809	910
5750	90° pull elbows steel, female, 1/2" diameter		16	.500		8.90	22		30.90	42.50
5760	3/4" diameter		13	.615		10.35	27		37.35	51.50
5780	1" diameter		11	.727		17.50	32		49.50	67
5800	1-1/4" diameter		8	1		26	44		70	94
5820	1-1/2" diameter		6	1.333		37	58.50		95.50	128
5840	2" diameter		5	1.600		59.50	70		129.50	171
6000	Explosion proof, flexible coupling									
6010	1/2" diameter, 4" long	1 Elec	12	.667	Ea.	99	29.50		128.50	153
6020	6" long		12	.667		110	29.50		139.50	165
6050	12" long		12	.667		141	29.50		170.50	199
6070	18" long		12	.667		172	29.50		201.50	233
6090	24" long		12	.667		203	29.50		232.50	267
6110	30" long		12	.667		235	29.50		264.50	305
6130	36" long		12	.667		266	29.50		295.50	335
6140	3/4" diameter, 4" long		10	.800		119	35		154	184
6150	6" long		10	.800		134	35		169	200
6180	12" long		10	.800		177	35		212	248
6200	18" long		10	.800		220	35		255	295
6220	24" long		10	.800		266	35		301	345
6240	30" long		10	.800		310	35		345	395
6260	36" long		10	.800		355	35		390	445
6270	1" diameter, 6" long		8	1		240	44		284	330
6300	12" long		8	1		305	44		349	400
6320	18" long		8	1		365	44		409	465
6340	24" long		8	1		425	44		469	535
6360	30" long		8	1		580	44		624	705
6380	36" long		8	1		660	44		704	790
6390	1-1/4" diameter, 12" long		6.40	1.250		460	55		515	585
6410	18" long		6.40	1.250		630	55		685	775
6430	24" long		6.40	1.250		750	55		805	905
6450	30" long		6.40	1.250		845	55		900	1,000
6470	36" long		6.40	1.250		1,075	55		1,130	1,250
6480	1-1/2" diameter, 12" long		5.30	1.509		615	66.50		681.50	775
6500	18" long		5.30	1.509		715	66.50		781.50	890
6520	24" long		5.30	1.509		815	66.50		881.50	995
6540	30" long		5.30	1.509		1,100	66.50		1,166.50	1,300
6560	36" long		5.30	1.509		1,250	66.50		1,316.50	1,475
6570	2" diameter, 12" long		4.60	1.739		940	76.50		1,016.50	1,150
6590	18" long		4.60	1.739		1,175	76.50		1,251.50	1,400
6610	24" long		4.60	1.739		1,275	76.50		1,351.50	1,525
6630	30" long		4.60	1.739		1,400	76.50		1,476.50	1,650
6650	36" long		4.60	1.739		1,700	76.50		1,776.50	2,000
7000	Close up plug, 1/2" diameter, explosion proof		40	.200		2.18	8.80		10.98	15.45
7010	3/4" diameter		32	.250		2.45	11		13.45	19.05
7020	1" diameter		28	.286		3	12.55		15.55	22
7030	1-1/4" diameter		24	.333		3.30	14.65		17.95	25.50

26 05 33 – Raceway and Boxes for Electrical Systems

26 05 33.25 Conduit Fittings for Rigid Galvanized Steel	Crew	Daily Output	Labor-Hours	Unit	Material	2007 Bare Costs Labor	Equipment	Total	Total Incl O&P
7040	1-1/2" diameter	1 Elec	18	.444	Ea.	4.65	19.50		24.15	34
7050	2" diameter		15	.533		7.90	23.50		31.40	43.50
7060	2-1/2" diameter		13	.615		12.30	27		39.30	53.50
7070	3" diameter		12	.667		17.90	29.50		47.40	63
7080	3-1/2" diameter		11	.727		23	32		55	73
7090	4" diameter		9	.889		31	39		70	92
7091	Elbow female, 45°, 1/2"		16	.500		9.20	22		31.20	42.50
7092	3/4"		13	.615		9.95	27		36.95	51
7093	1"		11	.727		13.45	32		45.45	62.50
7094	1-1/4"		8	1		18.90	44		62.90	86.50
7095	1-1/2"		6	1.333		19.50	58.50		78	109
7096	2"		5	1.600		24	70		94	132
7097	2-1/2"		4.50	1.778		65	78		143	188
7098	3"		4.20	1.905		70.50	83.50		154	202
7099	3-1/2"		4	2		106	88		194	248
7100	4"		3.80	2.105		131	92.50		223.50	282
7101	90°, 1/2"		16	.500		8.70	22		30.70	42
7102	3/4"		13	.615		9.50	27		36.50	50.50
7103	1"		11	.727		12.85	32		44.85	61.50
7104	1-1/4"		8	1		20	44		64	87.50
7105	1-1/2"		6	1.333		29	58.50		87.50	119
7106	2"		5	1.600		46	70		116	156
7107	2-1/2"		4.50	1.778		83.50	78		161.50	208
7110	Elbows 90°, long male & female, 1/2" diameter, explosion proof		16	.500		12.85	22		34.85	46.50
7120	3/4" diameter		13	.615		14.40	27		41.40	56
7130	1" diameter		11	.727		20	32		52	69.50
7140	1-1/4" diameter		8	1		23	44		67	91
7150	1-1/2" diameter		6	1.333		34.50	58.50		93	125
7160	2" diameter		5	1.600		52	70		122	162
7170	Capped elbow, 1/2" diameter, explosion proof		11	.727		12	32		44	60.50
7180	3/4" diameter		8	1		13.40	44		57.40	80.50
7190	1" diameter		6	1.333		16.60	58.50		75.10	105
7200	1-1/4" diameter		5	1.600		39	70		109	148
7210	Pulling elbow, 1/2" diameter, explosion proof		11	.727		58	32		90	111
7220	3/4" diameter		8	1		60.50	44		104.50	132
7230	1" diameter		6	1.333		158	58.50		216.50	261
7240	1-1/4" diameter		5	1.600		163	70		233	284
7250	1-1/2" diameter		5	1.600		241	70		311	370
7260	2" diameter		4	2		250	88		338	405
7270	2-1/2" diameter		3.50	2.286		550	100		650	755
7280	3" diameter		3	2.667		530	117		647	760
7290	3-1/2" diameter		2.50	3.200		990	140		1,130	1,300
7300	4" diameter		2.20	3.636		970	160		1,130	1,325
7310	LB conduit body, 1/2" diameter		11	.727		37	32		69	88.50
7320	3/4" diameter		8	1		42.50	44		86.50	112
7330	T conduit body, 1/2" diameter		9	.889		39	39		78	101
7340	3/4" diameter		6	1.333		45.50	58.50		104	138
7350	Explosion proof, round box w/cover, 3 threaded hubs, 1/2" diameter		8	1		35.50	44		79.50	105
7351	3/4" diameter		8	1		38.50	44		82.50	108
7352	1" diameter		7.50	1.067		48	47		95	122
7353	1-1/4" diameter		7	1.143		87.50	50		137.50	171
7354	1-1/2" diameter		7	1.143		159	50		209	250
7355	2" diameter		6	1.333		165	58.50		223.50	269

26 05 33.25 Conduit Fittings for Rigid Galvanized Steel		Crew	Daily Output	Labor-Hours	Unit	Material	2007 Bare Costs Labor	Equipment	Total	Total Incl O&P
7356	Round box w/cover & mtng flange, 3 threaded hubs, 1/2" diameter	1 Elec	8	1	Ea.	53.50	44		97.50	125
7357	3/4" diameter		8	1		56	44		100	127
7358	4 threaded hubs, 1" diameter		7	1.143		54	50		104	134
7400	Unions, 1/2" diameter		20	.400		9.60	17.55		27.15	36.50
7410	3/4" - 1/2" diameter		16	.500		14	22		36	48
7420	3/4" diameter		16	.500		13.20	22		35.20	47
7430	1" diameter		14	.571		24	25		49	64
7440	1-1/4" diameter		12	.667		36	29.50		65.50	83
7450	1-1/2" diameter		10	.800		46	35		81	104
7460	2" diameter		8.50	.941		59.50	41.50		101	127
7480	2-1/2" diameter		8	1		91	44		135	166
7490	3" diameter		7	1.143		123	50		173	210
7500	3-1/2" diameter		6	1.333		193	58.50		251.50	299
7510	4" diameter		5	1.600		214	70		284	340
7680	Reducer, 3/4" to 1/2"		54	.148		1.94	6.50		8.44	11.85
7690	1" to 1/2"		46	.174		2.76	7.65		10.41	14.40
7700	1" to 3/4"		46	.174		2.76	7.65		10.41	14.40
7710	1-1/4" to 3/4"		40	.200		3.97	8.80		12.77	17.40
7720	1-1/4" to 1"		40	.200		3.97	8.80		12.77	17.40
7730	1-1/2" to 1"		36	.222		6.50	9.75		16.25	21.50
7740	1-1/2" to 1-1/4"		36	.222		6.50	9.75		16.25	21.50
7750	2" to 3/4"		32	.250		12.60	11		23.60	30
7760	2" to 1-1/4"		32	.250		8.80	11		19.80	26
7770	2" to 1-1/2"		32	.250		9.35	11		20.35	26.50
7780	2-1/2" to 1-1/2"		30	.267		15.05	11.70		26.75	34
7790	3" to 2"		30	.267		17.70	11.70		29.40	37
7800	3-1/2" to 2-1/2"		28	.286		34.50	12.55		47.05	56.50
7810	4" to 3"		28	.286		40.50	12.55		53.05	63
7820	Sealing fitting, vertical/horizontal, 1/2" diameter		14.50	.552		13.85	24		37.85	51.50
7830	3/4" diameter		13.30	.602		16.25	26.50		42.75	57.50
7840	1" diameter		11.40	.702		21	31		52	69
7850	1-1/4" diameter		10	.800		25.50	35		60.50	80.50
7860	1-1/2" diameter		8.80	.909		38.50	40		78.50	102
7870	2" diameter		8	1		50	44		94	121
7880	2-1/2" diameter		6.70	1.194		70	52.50		122.50	155
7890	3" diameter		5.70	1.404		86	61.50		147.50	187
7900	3-1/2" diameter		4.70	1.702		230	74.50		304.50	365
7910	4" diameter		4	2		360	88		448	525
7920	Sealing hubs, 1" by 1-1/2"		12	.667		26	29.50		55.50	72.50
7930	1-1/4" by 2"		10	.800		29.50	35		64.50	85
7940	1-1/2" by 2"		9	.889		56	39		95	120
7950	2" by 2-1/2"		8	1		74	44		118	147
7960	3" by 4"		7	1.143		127	50		177	215
7970	4" by 5"		6	1.333		264	58.50		322.50	375
7980	Drain, 1/2"		32	.250		43	11		54	64
7990	Breather, 1/2"		32	.250		43	11		54	64
8000	Plastic coated 40 mil thick									
8010	LB, LR or LL conduit body w/cover, 1/2" diameter	1 Elec	13	.615	Ea.	44.50	27		71.50	88.50
8020	3/4" diameter		11	.727		49	32		81	102
8030	1" diameter		8	1		66	44		110	138
8040	1-1/4" diameter		6	1.333		95.50	58.50		154	192
8050	1-1/2" diameter		5	1.600		117	70		187	233
8060	2" diameter		4.50	1.778		172	78		250	305

26 05 33.25 Conduit Fittings for Rigid Galvanized Steel	Crew	Daily Output	Labor-Hours	Unit	Material	2007 Bare Costs Labor	Equipment	Total	Total Incl O&P
8070	2-1/2" diameter	1 Elec	4	2	Ea.	315	88		403	475
8080	3" diameter		3.50	2.286		395	100		495	585
8090	3-1/2" diameter		3	2.667		585	117		702	820
8100	4" diameter		2.50	3.200		655	140		795	930
8150	T conduit body with cover, 1/2" diameter		11	.727		51	32		83	104
8160	3/4" diameter		9	.889		64	39		103	129
8170	1" diameter		6	1.333		76.50	58.50		135	171
8180	1-1/4" diameter		5	1.600		108	70		178	224
8190	1-1/2" diameter		4.50	1.778		135	78		213	264
8200	2" diameter		4	2		196	88		284	345
8210	2-1/2" diameter		3.50	2.286		330	100		430	515
8220	3" diameter		3	2.667		440	117		557	660
8230	3-1/2" diameter		2.50	3.200		585	140		725	855
8240	4" diameter		2	4		635	176		811	960
8300	FS conduit body, 1 gang, 3/4" diameter		11	.727		50.50	32		82.50	103
8310	1" diameter		10	.800		55	35		90	113
8350	2 gang, 3/4" diameter		9	.889		87	39		126	154
8360	1" diameter		8	1		104	44		148	180
8400	Duplex receptacle cover		64	.125		37	5.50		42.50	49
8410	Switch cover		64	.125		50.50	5.50		56	63.50
8420	Switch, vaportight cover		53	.151		116	6.65		122.65	137
8430	Blank, cover		64	.125		33.50	5.50		39	45
8520	FSC conduit body, 1 gang, 3/4" diameter		10	.800		59.50	35		94.50	118
8530	1" diameter		9	.889		67	39		106	132
8550	2 gang, 3/4" diameter		8	1		101	44		145	177
8560	1" diameter		7	1.143		111	50		161	197
8590	Conduit hubs, 1/2" diameter		18	.444		33.50	19.50		53	66
8600	3/4" diameter		16	.500		38	22		60	74.50
8610	1" diameter		14	.571		47.50	25		72.50	89.50
8620	1-1/4" diameter		12	.667		54.50	29.50		84	104
8630	1-1/2" diameter		10	.800		62.50	35		97.50	121
8640	2" diameter		8.80	.909		90	40		130	158
8650	2-1/2" diameter		8.50	.941		140	41.50		181.50	216
8660	3" diameter		8	1		193	44		237	278
8670	3-1/2" diameter		7.50	1.067		250	47		297	345
8680	4" diameter		7	1.143		305	50		355	410
8690	5" diameter		6	1.333		365	58.50		423.50	490
8700	Plastic coated 40 mil thick									
8710	Pipe strap, stamped 1 hole, 1/2" diameter	1 Elec	470	.017	Ea.	8.40	.75		9.15	10.35
8720	3/4" diameter		440	.018		8.40	.80		9.20	10.45
8730	1" diameter		400	.020		10.80	.88		11.68	13.20
8740	1-1/4" diameter		355	.023		13.35	.99		14.34	16.15
8750	1-1/2" diameter		320	.025		16.35	1.10		17.45	19.65
8760	2" diameter		266	.030		16.90	1.32		18.22	20.50
8770	2-1/2" diameter		200	.040		20.50	1.76		22.26	25
8780	3" diameter		133	.060		21.50	2.64		24.14	27.50
8790	3-1/2" diameter		110	.073		23	3.19		26.19	30.50
8800	4" diameter		90	.089		28.50	3.90		32.40	37
8810	5" diameter		70	.114		116	5		121	134
8840	Clamp back spacers, 3/4" diameter		440	.018		14.25	.80		15.05	16.90
8850	1" diameter		400	.020		18.55	.88		19.43	22
8860	1-1/4" diameter		355	.023		18.85	.99		19.84	22
8870	1-1/2" diameter		320	.025		26	1.10		27.10	30

26 05 33 – Raceway and Boxes for Electrical Systems

26 05 33.25 Conduit Fittings for Rigid Galvanized Steel

		Crew	Daily Output	Labor-Hours	Unit	Material	2007 Bare Costs Labor	Equipment	Total	Total Incl O&P
8880	2" diameter	1 Elec	266	.030	Ea.	40.50	1.32		41.82	47
8900	3" diameter		133	.060		92	2.64		94.64	105
8920	4" diameter	↓	90	.089		134	3.90		137.90	153
8950	Touch-up plastic coating, spray, 12 oz.					38			38	41.50
8960	Sealing fittings, 1/2" diameter	1 Elec	11	.727		55	32		87	108
8970	3/4" diameter		9	.889		56.50	39		95.50	120
8980	1" diameter		7.50	1.067		67.50	47		114.50	144
8990	1-1/4" diameter		6.50	1.231		81.50	54		135.50	170
9000	1-1/2" diameter		5.50	1.455		112	64		176	218
9010	2" diameter		4.80	1.667		135	73		208	258
9020	2-1/2" diameter		4	2		196	88		284	345
9030	3" diameter		3.50	2.286		247	100		347	420
9040	3-1/2" diameter		3	2.667		620	117		737	855
9050	4" diameter		2.50	3.200		1,000	140		1,140	1,300
9060	5" diameter		1.70	4.706		1,300	207		1,507	1,750
9070	Unions, 1/2" diameter		18	.444		46	19.50		65.50	79.50
9080	3/4" diameter		15	.533		47	23.50		70.50	86.50
9090	1" diameter		13	.615		62.50	27		89.50	109
9100	1-1/4" diameter		11	.727		101	32		133	159
9110	1-1/2" diameter		9.50	.842		122	37		159	189
9120	2" diameter		8	1		163	44		207	245
9130	2-1/2" diameter		7.50	1.067		246	47		293	340
9140	3" diameter		6.80	1.176		340	51.50		391.50	445
9150	3-1/2" diameter		5.80	1.379		420	60.50		480.50	550
9160	4" diameter		4.80	1.667		540	73		613	705
9170	5" diameter	↓	4	2	↓	755	88		843	965

26 05 33.30 Electrical Nonmetallic Tubing (ENT)

		Crew	Daily Output	Labor-Hours	Unit	Material	2007 Bare Costs Labor	Equipment	Total	Total Incl O&P
0010	**ELECTRICAL NONMETALLIC TUBING (ENT)**									
0050	Flexible, 1/2" diameter	1 Elec	270	.030	L.F.	.56	1.30		1.86	2.56
0100	3/4" diameter		230	.035		.77	1.53		2.30	3.12
0200	1" diameter		145	.055		1.25	2.42		3.67	4.99
0210	1-1/4" diameter		125	.064		1.71	2.81		4.52	6.05
0220	1-1/2" diameter		100	.080		2.33	3.51		5.84	7.80
0230	2" diameter		75	.107	↓	2.70	4.68		7.38	9.90
0300	Connectors, to outlet box, 1/2" diameter		230	.035	Ea.	1.10	1.53		2.63	3.48
0310	3/4" diameter		210	.038		2.05	1.67		3.72	4.75
0320	1" diameter		200	.040		3.30	1.76		5.06	6.25
0400	Couplings, to conduit, 1/2" diameter		145	.055		1.26	2.42		3.68	5
0410	3/4" diameter		130	.062		2.40	2.70		5.10	6.65
0420	1" diameter	↓	125	.064	↓	3.65	2.81		6.46	8.20

26 05 33.35 Flexible Metallic Conduit

		Crew	Daily Output	Labor-Hours	Unit	Material	2007 Bare Costs Labor	Equipment	Total	Total Incl O&P
0010	**FLEXIBLE METALLIC CONDUIT**									
0050	Steel, 3/8" diameter	1 Elec	200	.040	L.F.	.34	1.76		2.10	2.98
0100	1/2" diameter		200	.040		.37	1.76		2.13	3.02
0200	3/4" diameter		160	.050		.51	2.20		2.71	3.83
0250	1" diameter		100	.080		.94	3.51		4.45	6.30
0300	1-1/4" diameter		70	.114		1.23	5		6.23	8.80
0350	1-1/2" diameter		50	.160		2.16	7		9.16	12.85
0370	2" diameter		40	.200		2.65	8.80		11.45	15.95
0380	2-1/2" diameter	↓	30	.267		3.20	11.70		14.90	21
0390	3" diameter	2 Elec	50	.320		4.73	14.05		18.78	26
0400	3-1/2" diameter	↓	40	.400	↓	5.95	17.55		23.50	32.50

26 05 33.35 Flexible Metallic Conduit		Crew	Daily Output	Labor-Hours	Unit	Material	2007 Bare Costs Labor	Equipment	Total	Total Incl O&P
0410	4" diameter	2 Elec	30	.533	L.F.	8.10	23.50		31.60	44
0420	Connectors, plain, 3/8" diameter	1 Elec	100	.080	Ea.	2.88	3.51		6.39	8.40
0430	1/2" diameter		80	.100		3.36	4.39		7.75	10.25
0440	3/4" diameter		70	.114		3.82	5		8.82	11.65
0450	1" diameter		50	.160		8.05	7		15.05	19.35
0452	1-1/4" diameter		45	.178		9.90	7.80		17.70	22.50
0454	1-1/2" diameter		40	.200		14.65	8.80		23.45	29
0456	2" diameter		28	.286		22	12.55		34.55	42.50
0458	2-1/2" diameter		25	.320		37.50	14.05		51.55	62
0460	3" diameter		20	.400		54	17.55		71.55	85
0462	3-1/2" diameter		16	.500		156	22		178	205
0464	4" diameter		13	.615		200	27		227	260
0490	Insulated, 1" diameter		40	.200		8.10	8.80		16.90	22
0500	1-1/4" diameter		40	.200		14.50	8.80		23.30	29
0550	1-1/2" diameter		32	.250		21	11		32	39.50
0600	2" diameter		23	.348		37.50	15.25		52.75	63.50
0610	2-1/2" diameter		20	.400		94.50	17.55		112.05	130
0620	3" diameter		17	.471		126	20.50		146.50	169
0630	3-1/2" diameter		13	.615		219	27		246	281
0640	4" diameter		10	.800		289	35		324	375
0650	Connectors 90°, plain, 3/8" diameter		80	.100		1.95	4.39		6.34	8.70
0660	1/2" diameter		60	.133		3.30	5.85		9.15	12.35
0700	3/4" diameter		50	.160		5.95	7		12.95	17
0750	1" diameter		40	.200		10.35	8.80		19.15	24.50
0790	Insulated, 1" diameter		40	.200		15.15	8.80		23.95	30
0800	1-1/4" diameter		30	.267		32.50	11.70		44.20	53
0850	1-1/2" diameter		23	.348		56.50	15.25		71.75	85
0900	2" diameter		18	.444		75	19.50		94.50	112
0910	2-1/2" diameter		16	.500		161	22		183	210
0920	3" diameter		14	.571		201	25		226	259
0930	3-1/2" diameter		11	.727		585	32		617	690
0940	4" diameter		8	1		880	44		924	1,025
0960	Couplings, to flexible conduit, 1/2" diameter		50	.160		1.70	7		8.70	12.30
0970	3/4" diameter		40	.200		2.79	8.80		11.59	16.10
0980	1" diameter		35	.229		4.86	10.05		14.91	20.50
0990	1-1/4" diameter		28	.286		10.70	12.55		23.25	30.50
1000	1-1/2" diameter		23	.348		13.85	15.25		29.10	38
1010	2" diameter		20	.400		28	17.55		45.55	57
1020	2-1/2" diameter		18	.444		38.50	19.50		58	71
1030	3" diameter		15	.533		96	23.50		119.50	141
1070	Sealtite, 3/8" diameter		140	.057	L.F.	1.67	2.51		4.18	5.55
1080	1/2" diameter		140	.057		1.72	2.51		4.23	5.60
1090	3/4" diameter		100	.080		2.41	3.51		5.92	7.90
1100	1" diameter		70	.114		3.64	5		8.64	11.45
1200	1-1/4" diameter		50	.160		4.98	7		11.98	15.95
1300	1-1/2" diameter		40	.200		5.80	8.80		14.60	19.40
1400	2" diameter		30	.267		7.20	11.70		18.90	25.50
1410	2-1/2" diameter		27	.296		13	13		26	33.50
1420	3" diameter	2 Elec	50	.320		17.95	14.05		32	41
1440	4" diameter	"	30	.533		28	23.50		51.50	65.50
1490	Connectors, plain, 3/8" diameter	1 Elec	70	.114	Ea.	4.06	5		9.06	11.90
1500	1/2" diameter		70	.114		4	5		9	11.85
1700	3/4" diameter		50	.160		5.85	7		12.85	16.90

26 05 33.35 Flexible Metallic Conduit		Crew	Daily Output	Labor-Hours	Unit	Material	2007 Bare Costs Labor	Equipment	Total	Total Incl O&P
1900	1" diameter	1 Elec	40	.200	Ea.	10	8.80		18.80	24
1910	Insulated, 1" diameter		40	.200		12.95	8.80		21.75	27.50
2000	1-1/4" diameter		32	.250		20.50	11		31.50	39
2100	1-1/2" diameter		27	.296		26.50	13		39.50	48.50
2200	2" diameter		20	.400		47.50	17.55		65.05	78
2210	2-1/2" diameter		15	.533		280	23.50		303.50	345
2220	3" diameter		12	.667		310	29.50		339.50	390
2240	4" diameter		8	1		385	44		429	490
2290	Connectors, 90°, 3/8" diameter		70	.114		5.85	5		10.85	13.85
2300	1/2" diameter		70	.114		5.65	5		10.65	13.65
2400	3/4" diameter		50	.160		8.75	7		15.75	20
2600	1" diameter		40	.200		17.95	8.80		26.75	33
2790	Insulated, 1" diameter		40	.200		23.50	8.80		32.30	38.50
2800	1-1/4" diameter		32	.250		38	11		49	58.50
3000	1-1/2" diameter		27	.296		44	13		57	68
3100	2" diameter		20	.400		61.50	17.55		79.05	93.50
3110	2-1/2" diameter		14	.571		287	25		312	355
3120	3" diameter		11	.727		330	32		362	415
3140	4" diameter		7	1.143		430	50		480	550
4300	Coupling sealtite to rigid, 1/2" diameter		20	.400		4.90	17.55		22.45	31.50
4500	3/4" diameter		18	.444		6.95	19.50		26.45	36.50
4800	1" diameter		14	.571		9.55	25		34.55	48
4900	1-1/4" diameter		12	.667		15.95	29.50		45.45	61
5000	1-1/2" diameter		11	.727		29	32		61	79
5100	2" diameter		10	.800		49.50	35		84.50	107
5110	2-1/2" diameter		9.50	.842		190	37		227	264
5120	3" diameter		9	.889		203	39		242	281
5130	3-1/2" diameter		9	.889		250	39		289	335
5140	4" diameter		8.50	.941		250	41.50		291.50	335

26 05 33.45 Wireway

		Crew	Daily Output	Labor-Hours	Unit	Material	2007 Bare Costs Labor	Equipment	Total	Total Incl O&P
0010	**WIREWAY** to 15' high R260533-60									
0020	For higher elevations, see 26 05 36.40 9900									
0100	Screw cover, NEMA 1 w/ fittings and supports, 2-1/2" x 2-1/2"	1 Elec	45	.178	L.F.	11.45	7.80		19.25	24
0200	4" x 4"	"	40	.200		12.65	8.80		21.45	27
0400	6" x 6"	2 Elec	60	.267		19.70	11.70		31.40	39
0600	8" x 8"		40	.400		33	17.55		50.55	62.50
0620	10" x 10"		30	.533		42	23.50		65.50	81
0640	12" x 12"		20	.800		60	35		95	119
0800	Elbows, 90°, 2-1/2"	1 Elec	24	.333	Ea.	35.50	14.65		50.15	61
1000	4"		20	.400		40.50	17.55		58.05	70.50
1200	6"		18	.444		46	19.50		65.50	79.50
1400	8"		16	.500		73.50	22		95.50	113
1420	10"		12	.667		81.50	29.50		111	134
1440	12"		10	.800		132	35		167	198
1500	Elbows, 45°, 2-1/2"		24	.333		33.50	14.65		48.15	59
1510	4"		20	.400		43.50	17.55		61.05	74
1520	6"		18	.444		45.50	19.50		65	79
1530	8"		16	.500		73.50	22		95.50	113
1540	10"		12	.667		80	29.50		109.50	132
1550	12"		10	.800		185	35		220	257
1600	"T" box, 2-1/2"		18	.444		41	19.50		60.50	74
1800	4"		16	.500		49	22		71	86.50

26 05 33.45	Wireway	Crew	Daily Output	Labor-Hours	Unit	Material	2007 Bare Costs Labor	2007 Bare Costs Equipment	Total	Total Incl O&P
2000	6"	1 Elec	14	.571	Ea.	55	25		80	98
2200	8"		12	.667		102	29.50		131.50	157
2220	10"		10	.800		118	35		153	183
2240	12"		8	1		196	44		240	282
2300	Cross, 2-1/2"		16	.500		45.50	22		67.50	82.50
2310	4"		14	.571		55.50	25		80.50	98.50
2320	6"		12	.667		68.50	29.50		98	119
2400	Panel adapter, 2-1/2"		24	.333		15.20	14.65		29.85	38.50
2600	4"		20	.400		17.20	17.55		34.75	45
2800	6"		18	.444		22.50	19.50		42	53.50
3000	8"		16	.500		32	22		54	67.50
3020	10"		14	.571		37	25		62	78
3040	12"		12	.667		51.50	29.50		81	100
3200	Reducer, 4" to 2-1/2"		24	.333		24	14.65		38.65	48
3400	6" to 4"		20	.400		45.50	17.55		63.05	76
3600	8" to 6"		18	.444		53	19.50		72.50	87
3620	10" to 8"		16	.500		60.50	22		82.50	99
3640	12" to 10"		14	.571		75.50	25		100.50	121
3780	End cap, 2-1/2"		24	.333		5.10	14.65		19.75	27.50
3800	4"		20	.400		6.25	17.55		23.80	33
4000	6"		18	.444		7.50	19.50		27	37.50
4200	8"		16	.500		10.20	22		32.20	43.50
4220	10"		14	.571		13.20	25		38.20	52
4240	12"		12	.667		23	29.50		52.50	69
4300	U-connector, 2-1/2"		200	.040		5.10	1.76		6.86	8.20
4320	4"		200	.040		6.25	1.76		8.01	9.50
4340	6"		180	.044		7.50	1.95		9.45	11.15
4360	8"		170	.047		12.20	2.07		14.27	16.50
4380	10"		150	.053		14.85	2.34		17.19	19.85
4400	12"		130	.062		30.50	2.70		33.20	37.50
4420	Hanger, 2-1/2"		100	.080		8.90	3.51		12.41	15.05
4430	4"		100	.080		11.30	3.51		14.81	17.70
4440	6"		80	.100		20.50	4.39		24.89	29
4450	8"		65	.123		27	5.40		32.40	37.50
4460	10"		50	.160		35	7		42	49
4470	12"		40	.200		71.50	8.80		80.30	92
4475	Screw cover, NEMA 3R w/ fittings and supports, 4" x 4"		36	.222	L.F.	27	9.75		36.75	44
4480	6" x 6"	2 Elec	55	.291		34	12.75		46.75	56.50
4485	8" x 8"		36	.444		55	19.50		74.50	89.50
4490	12" x 12"		18	.889		77	39		116	143
4500	Hinged cover, with fittings and supports, 2-1/2" x 2-1/2"	1 Elec	60	.133		11.45	5.85		17.30	21.50
4520	4" x 4"	"	45	.178		12.65	7.80		20.45	25.50
4540	6" x 6"	2 Elec	80	.200		20	8.80		28.80	35
4560	8" x 8"		60	.267		33	11.70		44.70	54
4580	10" x 10"		50	.320		42	14.05		56.05	67
4600	12" x 12"		24	.667		60	29.50		89.50	110
4700	Elbows 90°, hinged cover, 2-1/2" x 2-1/2"	1 Elec	32	.250	Ea.	35.50	11		46.50	55.50
4720	4"		27	.296		40.50	13		53.50	64
4730	6"		23	.348		46	15.25		61.25	73
4740	8"		18	.444		73.50	19.50		93	110
4750	10"		14	.571		81.50	25		106.50	128
4760	12"		12	.667		132	29.50		161.50	189
4800	Tee box, hinged cover, 2-1/2" x 2-1/2"		23	.348		41	15.25		56.25	67.50

26 05 33.45 Wireway

		Crew	Daily Output	Labor-Hours	Unit	Material	2007 Bare Costs Labor	Equipment	Total	Total Incl O&P
4810	4"	1 Elec	20	.400	Ea.	49	17.55		66.55	80
4820	6"		18	.444		55	19.50		74.50	89.50
4830	8"		16	.500		102	22		124	146
4840	10"		12	.667		118	29.50		147.50	174
4860	12"		10	.800		196	35		231	269
4880	Cross box, hinged cover, 2-1/2" x 2-1/2"		18	.444		45.50	19.50		65	79
4900	4"		16	.500		55.50	22		77.50	93.50
4920	6"		13	.615		68.50	27		95.50	116
4940	8"		11	.727		102	32		134	160
4960	10"		10	.800		175	35		210	246
4980	12"		9	.889		191	39		230	268
5000	Flanged, oil tite w/screw cover, 2-1/2" x 2-1/2"		40	.200	L.F.	27	8.80		35.80	42.50
5020	4" x 4"		35	.229		31.50	10.05		41.55	49.50
5040	6" x 6"	2 Elec	60	.267		45	11.70		56.70	67
5060	8" x 8"	"	50	.320		62	14.05		76.05	89
5120	Elbows 90°, flanged, 2-1/2" x 2-1/2"	1 Elec	23	.348	Ea.	57.50	15.25		72.75	85.50
5140	4"		20	.400		73	17.55		90.55	107
5160	6"		18	.444		90	19.50		109.50	128
5180	8"		15	.533		134	23.50		157.50	182
5240	Tee box, flanged, 2-1/2" x 2-1/2"		18	.444		79	19.50		98.50	116
5260	4"		16	.500		90	22		112	132
5280	6"		15	.533		122	23.50		145.50	169
5300	8"		13	.615		176	27		203	234
5360	Cross box, flanged, 2-1/2" x 2-1/2"		15	.533		97.50	23.50		121	142
5380	4"		13	.615		127	27		154	180
5400	6"		12	.667		165	29.50		194.50	226
5420	8"		10	.800		194	35		229	266
5480	Flange gasket, 2-1/2"		160	.050		2.86	2.20		5.06	6.40
5500	4"		80	.100		3.89	4.39		8.28	10.85
5520	6"		53	.151		5.20	6.65		11.85	15.55
5530	8"		40	.200		6.90	8.80		15.70	20.50

26 05 33.50 Outlet Boxes

		Crew	Daily Output	Labor-Hours	Unit	Material	2007 Bare Costs Labor	Equipment	Total	Total Incl O&P
0010	**OUTLET BOXES** R260533-65									
0020	Pressed steel, octagon, 4"	1 Elec	20	.400	Ea.	2.73	17.55		20.28	29
0040	For Romex or BX		20	.400		4.11	17.55		21.66	30.50
0050	For Romex or BX, with bracket		20	.400		6.45	17.55		24	33
0060	Covers, blank		64	.125		1.14	5.50		6.64	9.40
0100	Extension rings		40	.200		4.53	8.80		13.33	18.05
0150	Square, 4"		20	.400		2.35	17.55		19.90	28.50
0160	For Romex or BX		20	.400		7.30	17.55		24.85	34
0170	For Romex or BX, with bracket		20	.400		7.95	17.55		25.50	35
0200	Extension rings		40	.200		4.58	8.80		13.38	18.10
0220	2-1/8" deep, 1" KO		20	.400		6.10	17.55		23.65	32.50
0250	Covers, blank		64	.125		1.29	5.50		6.79	9.55
0260	Raised device		64	.125		2.42	5.50		7.92	10.80
0300	Plaster rings		64	.125		2.51	5.50		8.01	10.90
0350	Square, 4-11/16"		20	.400		6.65	17.55		24.20	33.50
0370	2-1/8" deep, 3/4" to 1-1/4" KO		20	.400		7.70	17.55		25.25	34.50
0400	Extension rings		40	.200		10.30	8.80		19.10	24.50
0450	Covers, blank		53	.151		2.32	6.65		8.97	12.40
0460	Raised device		53	.151		6.90	6.65		13.55	17.45
0500	Plaster rings		53	.151		6.85	6.65		13.50	17.40

26 05 33.50 Outlet Boxes		Crew	Daily Output	Labor-Hours	Unit	Material	2007 Bare Costs Labor	2007 Bare Costs Equipment	Total	Total Incl O&P
0550	Handy box	1 Elec	27	.296	Ea.	2.92	13		15.92	22.50
0560	Covers, device		64	.125		1.06	5.50		6.56	9.30
0600	Extension rings		54	.148		3.61	6.50		10.11	13.65
0650	Switchbox		27	.296		4.38	13		17.38	24
0660	Romex or BX		27	.296		6.30	13		19.30	26.50
0670	with bracket		27	.296		7.20	13		20.20	27.50
0680	Partition, metal		27	.296		2.74	13		15.74	22.50
0700	Masonry, 1 gang, 2-1/2" deep		27	.296		8.35	13		21.35	28.50
0710	3-1/2" deep		27	.296		8.40	13		21.40	28.50
0750	2 gang, 2-1/2" deep		20	.400		15.65	17.55		33.20	43.50
0760	3-1/2" deep		20	.400		12.65	17.55		30.20	40
0800	3 gang, 2-1/2" deep		13	.615		18.75	27		45.75	60.50
0850	4 gang, 2-1/2" deep		10	.800		21.50	35		56.50	76.50
0860	5 gang, 2-1/2" deep		9	.889		23	39		62	83.50
0870	6 gang, 2-1/2" deep		8	1		47.50	44		91.50	118
0880	Masonry thru-the-wall, 1 gang, 4" block		16	.500		18.55	22		40.55	53
0890	6" block		16	.500		21.50	22		43.50	56
0900	8" block		16	.500		24	22		46	59
0920	2 gang, 6" block		16	.500		22	22		44	57
0940	Bar hanger with 3/8" stud, for wood and masonry boxes		53	.151		3.92	6.65		10.57	14.15
0950	Concrete, set flush, 4" deep		20	.400		13.05	17.55		30.60	40.50
1000	Plate with 3/8" stud		80	.100		5.30	4.39		9.69	12.40
1100	Concrete, floor, 1 gang		5.30	1.509		73	66.50		139.50	179
1150	2 gang		4	2		115	88		203	258
1200	3 gang		2.70	2.963		163	130		293	375
1250	For duplex receptacle, pedestal mounted, add		24	.333		72.50	14.65		87.15	102
1270	Flush mounted, add		27	.296		17.80	13		30.80	39
1300	For telephone, pedestal mounted, add		30	.267		77	11.70		88.70	102
1350	Carpet flange, 1 gang		53	.151		34.50	6.65		41.15	48
1400	Cast, 1 gang, FS (2" deep), 1/2" hub		12	.667		15.50	29.50		45	60.50
1410	3/4" hub		12	.667		16.30	29.50		45.80	61.50
1420	FD (2-11/16" deep), 1/2" hub		12	.667		17.10	29.50		46.60	62.50
1430	3/4" hub		12	.667		18.60	29.50		48.10	64
1450	2 gang, FS, 1/2" hub		10	.800		28	35		63	83
1460	3/4" hub		10	.800		28.50	35		63.50	84
1470	FD, 1/2" hub		10	.800		32.50	35		67.50	88
1480	3/4" hub		10	.800		33	35		68	89
1500	3 gang, FS, 3/4" hub		9	.889		41.50	39		80.50	104
1510	Switch cover, 1 gang, FS		64	.125		4.10	5.50		9.60	12.65
1520	2 gang		53	.151		6.50	6.65		13.15	17
1530	Duplex receptacle cover, 1 gang, FS		64	.125		4.10	5.50		9.60	12.65
1540	2 gang, FS		53	.151		6.50	6.65		13.15	17
1550	Weatherproof switch cover		64	.125		7.30	5.50		12.80	16.20
1600	Weatherproof receptacle cover		64	.125		7.30	5.50		12.80	16.20
1750	FSC, 1 gang, 1/2" hub		11	.727		17.15	32		49.15	66.50
1760	3/4" hub		11	.727		17.85	32		49.85	67
1770	2 gang, 1/2" hub		9	.889		29.50	39		68.50	90.50
1780	3/4" hub		9	.889		32.50	39		71.50	94
1790	FDC, 1 gang, 1/2" hub		11	.727		20.50	32		52.50	70
1800	3/4" hub		11	.727		22	32		54	71.50
1810	2 gang, 1/2" hub		9	.889		36	39		75	97.50
1820	3/4" hub		9	.889		35	39		74	96.50
2000	Poke-thru fitting, fire rated, for 3-3/4" floor		6.80	1.176		100	51.50		151.50	187

26 05 33 – Raceway and Boxes for Electrical Systems

26 05 33.50 Outlet Boxes		Crew	Daily Output	Labor-Hours	Unit	Material	2007 Bare Costs Labor	Equipment	Total	Total Incl O&P
2040	For 7" floor	1 Elec	6.80	1.176	Ea.	100	51.50		151.50	187
2100	Pedestal, 15 amp, duplex receptacle & blank plate		5.25	1.524		103	67		170	214
2120	Duplex receptacle and telephone plate		5.25	1.524		104	67		171	214
2140	Pedestal, 20 amp, duplex recept. & phone plate		5	1.600		105	70		175	220
2160	Telephone plate, both sides		5.25	1.524		99	67		166	209
2200	Abandonment plate		32	.250		30.50	11		41.50	50

26 05 33.60 Outlet Boxes, Plastic		Crew	Daily Output	Labor-Hours	Unit	Material	2007 Bare Costs Labor	Equipment	Total	Total Incl O&P
0010	**OUTLET BOXES, PLASTIC**									
0050	4" diameter, round with 2 mounting nails	1 Elec	25	.320	Ea.	2.64	14.05		16.69	24
0100	Bar hanger mounted		25	.320		4.79	14.05		18.84	26.50
0200	4", square with 2 mounting nails		25	.320		3.94	14.05		17.99	25.50
0300	Plaster ring		64	.125		1.65	5.50		7.15	9.95
0400	Switch box with 2 mounting nails, 1 gang		30	.267		1.79	11.70		13.49	19.40
0500	2 gang		25	.320		3.66	14.05		17.71	25
0600	3 gang		20	.400		5.75	17.55		23.30	32.50
0700	Old work box		30	.267		3.52	11.70		15.22	21.50
1400	PVC, FSS, 1 gang, 1/2" hub		14	.571		18.40	25		43.40	57.50
1410	3/4" hub		14	.571		19.75	25		44.75	59
1420	FD, 1 gang for variable terminations		14	.571		17.20	25		42.20	56.50
1450	FS, 2 gang for variable terminations		12	.667		23	29.50		52.50	69
1480	Blank cover, FS, 1 gang		64	.125		4.90	5.50		10.40	13.55
1500	2 gang		53	.151		5.55	6.65		12.20	15.95
1510	Switch cover, FS, 1 gang		64	.125		11.20	5.50		16.70	20.50
1520	2 gang		53	.151		18.70	6.65		25.35	30.50
1530	Duplex receptacle cover, FS, 1 gang		64	.125		16.75	5.50		22.25	26.50
1540	2 gang		53	.151		18.30	6.65		24.95	30
1750	FSC, 1 gang, 1/2" hub		13	.615		18.05	27		45.05	60
1760	3/4" hub		13	.615		19.75	27		46.75	61.50
1770	FSC, 2 gang, 1/2" hub		11	.727		23.50	32		55.50	73.50
1780	3/4" hub		11	.727		23.50	32		55.50	73.50
1790	FDC, 1 gang, 1/2" hub		13	.615		17.25	27		44.25	59
1800	3/4" hub		13	.615		18.80	27		45.80	60.50
1810	Weatherproof, T box w/ 3 holes		14	.571		17.70	25		42.70	57
1820	4" diameter round w/ 5 holes		14	.571		26	25		51	66

26 05 33.65 Pull Boxes		Crew	Daily Output	Labor-Hours	Unit	Material	2007 Bare Costs Labor	Equipment	Total	Total Incl O&P
0010	**PULL BOXES** R260533-70									
0100	Sheet metal, pull box, NEMA 1, type SC, 6" W x 6" H x 4" D	1 Elec	8	1	Ea.	11.65	44		55.65	78.50
0180	6" W x 8" H x 4" D		8	1		13.70	44		57.70	80.50
0200	8" W x 8" H x 4" D		8	1		16	44		60	83
0210	10" W x 10" H x 4" D		7	1.143		21	50		71	97.50
0220	12" W x 12" H x 4" D		6.50	1.231		27	54		81	111
0230	15" W x 15" H x 4" D		5.20	1.538		36.50	67.50		104	141
0240	18" W x 18" H x 4" D		4.40	1.818		48.50	80		128.50	173
0250	6" W x 6" H x 6" D		8	1		14.15	44		58.15	81
0260	8" W x 8" H x 6" D		7.50	1.067		19.20	47		66.20	90.50
0270	10" W x 10" H x 6" D		5.50	1.455		25	64		89	123
0300	10" W x 12" H x 6" D		5.30	1.509		28	66.50		94.50	130
0310	12" W x 12" H x 6" D		5.20	1.538		32	67.50		99.50	136
0320	15" W x 15" H x 6" D		4.60	1.739		45.50	76.50		122	164
0330	18" W x 18" H x 6" D		4.20	1.905		56.50	83.50		140	187
0340	24" W x 24" H x 6" D		3.20	2.500		115	110		225	289
0350	12" W x 12" H x 8" D		5	1.600		35.50	70		105.50	144

26 05 33.65 Pull Boxes		Crew	Daily Output	Labor-Hours	Unit	Material	2007 Bare Costs Labor	Equipment	Total	Total Incl O&P
0360	15" W x 15" H x 8" D	1 Elec	4.50	1.778	Ea.	73	78		151	196
0370	18" W x 18" H x 8" D		4	2		112	88		200	254
0380	24" W x 18" H x 6" D		3.70	2.162		89	95		184	239
0400	16" W x 20" H x 8" D		4	2		106	88		194	248
0500	20" W x 24" H x 8" D		3.20	2.500		125	110		235	300
0510	24" W x 24" H x 8" D		3	2.667		127	117		244	315
0600	24" W x 36" H x 8" D		2.70	2.963		176	130		306	390
0610	30" W x 30" H x 8" D		2.70	2.963		235	130		365	455
0620	36" W x 36" H x 8" D		2	4		287	176		463	575
0630	24" W x 24" H x 10" D		2.50	3.200		145	140		285	370
0650	Pull box, hinged , NEMA 1, 6" W x 6" H x 4" D		8	1		11.90	44		55.90	78.50
0660	8" W x 8" H x 4" D		8	1		16.30	44		60.30	83.50
0670	10" W x 10" H x 4" D		7	1.143		21.50	50		71.50	98
0680	12" W x 12" H x 4" D		6	1.333		27.50	58.50		86	118
0690	15" W x 15" H x 4" D		5.20	1.538		42.50	67.50		110	148
0700	18" W x 18" H x 4" D		4.40	1.818		49	80		129	173
0710	6" W x 6" H x 6" D		8	1		14.50	44		58.50	81.50
0720	8" W x 8" H x 6" D		7.50	1.067		19.55	47		66.55	91
0730	10" W x 10" H x 6" D		5.50	1.455		25.50	64		89.50	123
0740	12" W x 12" H x 6" D		5.20	1.538		32.50	67.50		100	137
0800	12" W x 16" H x 6" D		4.70	1.702		39	74.50		113.50	154
0810	15" W x 15" H x 6" D		4.60	1.739		45.50	76.50		122	164
0820	18" W x 18" H x 6" D		4.20	1.905		57.50	83.50		141	188
1000	20" W x 20" H x 6" D		3.60	2.222		78.50	97.50		176	232
1010	24" W x 24" H x 6" D		3.20	2.500		117	110		227	292
1020	12" W x 12" H x 8" D		5	1.600		67	70		137	179
1030	15" W x 15" H x 8" D		4.50	1.778		95	78		173	221
1040	18" W x 18" H x 8" D		4	2		134	88		222	278
1200	20" W x 20" H x 8" D		3.20	2.500		157	110		267	335
1210	24" W x 24" H x 8" D		3	2.667		180	117		297	370
1220	30" W x 30" H x 8" D		2.70	2.963		255	130		385	475
1400	24" W x 36" H x 8" D		2.70	2.963		252	130		382	470
1600	24" W x 42" H x 8" D		2	4		380	176		556	680
1610	36" W x 36" H x 8" D		2	4		350	176		526	645
2100	Pull box, NEMA 3R, type SC, raintight & weatherproof									
2150	6" L x 6" W x 6" D	1 Elec	10	.800	Ea.	20	35		55	74.50
2200	8" L x 6" W x 6" D		8	1		25	44		69	93
2250	10" L x 6" W x 6" D		7	1.143		33	50		83	111
2300	12" L x 12" W x 6" D		5	1.600		47	70		117	157
2350	16" L x 16" W x 6" D		4.50	1.778		93.50	78		171.50	219
2400	20" L x 20" W x 6" D		4	2		128	88		216	272
2450	24" L x 18" W x 8" D		3	2.667		140	117		257	330
2500	24" L x 24" W x 10" D		2.50	3.200		188	.140		328	415
2550	30" L x 24" W x 12" D		2	4		340	176		516	635
2600	36" L x 36" W x 12" D		1.50	5.333		450	234		684	845
2800	Cast iron, pull boxes for surface mounting R260533-75									
3000	NEMA 4, watertight & dust tight									
3050	6" L x 6" W x 6" D	1 Elec	4	2	Ea.	171	88		259	320
3100	8" L x 6" W x 6" D		3.20	2.500		232	110		342	420
3150	10" L x 6" W x 6" D		2.50	3.200		289	140		429	530
3200	12" L x 12" W x 6" D		2.30	3.478		490	153		643	760
3250	16" L x 16" W x 6" D		1.30	6.154		1,000	270		1,270	1,500
3300	20" L x 20" W x 6" D		.80	10		1,850	440		2,290	2,700

26 05 33.65 Pull Boxes		Crew	Daily Output	Labor-Hours	Unit	Material	2007 Bare Costs Labor	Equipment	Total	Total Incl O&P
3350	24" L x 18" W x 8" D	1 Elec	.70	11.429	Ea.	1,950	500		2,450	2,900
3400	24" L x 24" W x 10" D		.50	16		3,000	700		3,700	4,350
3450	30" L x 24" W x 12" D		.40	20		5,025	880		5,905	6,825
3500	36" L x 36" W x 12" D		.20	40		6,200	1,750		7,950	9,450
3510	NEMA 4 clamp cover, 6" L x 6" W x 4" D		4	2		211	88		299	365
3520	8" L x 6" W x 4" D		4	2		248	88		336	405
4000	NEMA 7, explosionproof									
4050	6" L x 6" W x 6" D	1 Elec	2	4	Ea.	640	176		816	960
4100	8" L x 6" W x 6" D		1.80	4.444		730	195		925	1,100
4150	10" L x 6" W x 6" D		1.60	5		895	220		1,115	1,300
4200	12" L x 12" W x 6" D		1	8		1,475	350		1,825	2,150
4250	16" L x 14" W x 6" D		.60	13.333		2,025	585		2,610	3,125
4300	18" L x 18" W x 8" D		.50	16		3,600	700		4,300	5,000
4350	24" L x 18" W x 8" D		.40	20		4,825	880		5,705	6,600
4400	24" L x 24" W x 10" D		.30	26.667		6,250	1,175		7,425	8,625
4450	30" L x 24" W x 12" D		.20	40		9,525	1,750		11,275	13,100
5000	NEMA 9, dust tight 6" L x 6" W x 6" D		3.20	2.500		182	110		292	365
5050	8" L x 6" W x 6" D		2.70	2.963		218	130		348	435
5100	10" L x 6" W x 6" D		2	4		293	176		469	580
5150	12" L x 12" W x 6" D		1.60	5		580	220		800	960
5200	16" L x 16" W x 6" D		1	8		995	350		1,345	1,625
5250	18" L x 18" W x 8" D		.70	11.429		1,575	500		2,075	2,475
5300	24" L x 18" W x 8" D		.60	13.333		2,200	585		2,785	3,300
5350	24" L x 24" W x 10" D		.40	20		2,850	880		3,730	4,425
5400	30" L x 24" W x 12" D		.30	26.667		4,475	1,175		5,650	6,675
6000	J.I.C. wiring boxes, NEMA 12, dust tight & drip tight									
6050	6" L x 8" W x 4" D	1 Elec	10	.800	Ea.	46.50	35		81.50	104
6100	8" L x 10" W x 4" D		8	1		58.50	44		102.50	130
6150	12" L x 14" W x 6" D		5.30	1.509		95	66.50		161.50	203
6200	14" L x 16" W x 6" D		4.70	1.702		113	74.50		187.50	235
6250	16" L x 20" W x 6" D		4.40	1.818		241	80		321	385
6300	24" L x 30" W x 6" D		3.20	2.500		355	110		465	555
6350	24" L x 30" W x 8" D		2.90	2.759		380	121		501	595
6400	24" L x 36" W x 8" D		2.70	2.963		420	130		550	655
6450	24" L x 42" W x 8" D		2.30	3.478		460	153		613	730
6500	24" L x 48" W x 8" D		2	4		500	176		676	810

26 05 33.95 Cutting and Drilling

		Crew	Daily Output	Labor-Hours	Unit	Material	2007 Bare Costs Labor	Equipment	Total	Total Incl O&P
0010	**CUTTING AND DRILLING**									
0100	Hole drilling to 10' high, concrete wall									
0110	8" thick, 1/2" pipe size	R-31	12	.667	Ea.	3.80	29.50	5.95	39.25	54
0120	3/4" pipe size		12	.667		3.80	29.50	5.95	39.25	54
0130	1" pipe size		9.50	.842		7.85	37	7.50	52.35	72
0140	1-1/4" pipe size		9.50	.842		7.85	37	7.50	52.35	72
0150	1-1/2" pipe size		9.50	.842		7.85	37	7.50	52.35	72
0160	2" pipe size		4.40	1.818		8.45	80	16.20	104.65	146
0170	2-1/2" pipe size		4.40	1.818		8.45	80	16.20	104.65	146
0180	3" pipe size		4.40	1.818		8.45	80	16.20	104.65	146
0190	3-1/2" pipe size		3.30	2.424		9.05	106	21.50	136.55	191
0200	4" pipe size		3.30	2.424		9.05	106	21.50	136.55	191
0500	12" thick, 1/2" pipe size		9.40	.851		5.80	37.50	7.55	50.85	70.50
0520	3/4" pipe size		9.40	.851		5.80	37.50	7.55	50.85	70.50
0540	1" pipe size		7.30	1.096		11.10	48	9.75	68.85	94.50

26 05 33.95 Cutting and Drilling	Crew	Daily Output	Labor-Hours	Unit	Material	2007 Bare Costs Labor	Equipment	Total	Total Incl O&P
0560	1-1/4" pipe size	R-31	7.30	1.096	Ea.	11.10	48	9.75	68.85	94.50
0570	1-1/2" pipe size		7.30	1.096		11.10	48	9.75	68.85	94.50
0580	2" pipe size		3.60	2.222		12.95	97.50	19.80	130.25	181
0590	2-1/2" pipe size		3.60	2.222		12.95	97.50	19.80	130.25	181
0600	3" pipe size		3.60	2.222		12.95	97.50	19.80	130.25	181
0610	3-1/2" pipe size		2.80	2.857		14.70	125	25.50	165.20	231
0630	4" pipe size		2.50	3.200		14.70	140	28.50	183.20	257
0650	16" thick, 1/2" pipe size		7.60	1.053		7.80	46	9.35	63.15	88
0670	3/4" pipe size		7	1.143		7.80	50	10.15	67.95	94.50
0690	1" pipe size		6	1.333		14.30	58.50	11.85	84.65	116
0710	1-1/4" pipe size		5.50	1.455		14.30	64	12.95	91.25	125
0730	1-1/2" pipe size		5.50	1.455		14.30	64	12.95	91.25	125
0750	2" pipe size		3	2.667		17.40	117	23.50	157.90	219
0770	2-1/2" pipe size		2.70	2.963		17.40	130	26.50	173.90	242
0790	3" pipe size		2.50	3.200		17.40	140	28.50	185.90	260
0810	3-1/2" pipe size		2.30	3.478		20.50	153	31	204.50	284
0830	4" pipe size		2	4		20.50	176	35.50	232	325
0850	20" thick, 1/2" pipe size		6.40	1.250		9.80	55	11.15	75.95	105
0870	3/4" pipe size		6	1.333		9.80	58.50	11.85	80.15	111
0890	1" pipe size		5	1.600		17.55	70	14.25	101.80	140
0910	1-1/4" pipe size		4.80	1.667		17.55	73	14.85	105.40	145
0930	1-1/2" pipe size		4.60	1.739		17.55	76.50	15.50	109.55	150
0950	2" pipe size		2.70	2.963		22	130	26.50	178.50	247
0970	2-1/2" pipe size		2.40	3.333		22	146	29.50	197.50	275
0990	3" pipe size		2.20	3.636		22	160	32.50	214.50	298
1010	3-1/2" pipe size		2	4		26	176	35.50	237.50	330
1030	4" pipe size		1.70	4.706		26	207	42	275	385
1050	24" thick, 1/2" pipe size		5.50	1.455		11.80	64	12.95	88.75	122
1070	3/4" pipe size		5.10	1.569		11.80	69	13.95	94.75	131
1090	1" pipe size		4.30	1.860		21	81.50	16.55	119.05	163
1110	1-1/4" pipe size		4	2		21	88	17.80	126.80	174
1130	1-1/2" pipe size		4	2		21	88	17.80	126.80	174
1150	2" pipe size		2.40	3.333		26.50	146	29.50	202	280
1170	2-1/2" pipe size		2.20	3.636		26.50	160	32.50	219	305
1190	3" pipe size		2	4		26.50	176	35.50	238	330
1210	3-1/2" pipe size		1.80	4.444		31.50	195	39.50	266	370
1230	4" pipe size		1.50	5.333		31.50	234	47.50	313	435
1500	Brick wall, 8" thick, 1/2" pipe size		18	.444		3.80	19.50	3.96	27.26	37.50
1520	3/4" pipe size		18	.444		3.80	19.50	3.96	27.26	37.50
1540	1" pipe size		13.30	.602		7.85	26.50	5.35	39.70	54
1560	1-1/4" pipe size		13.30	.602		7.85	26.50	5.35	39.70	54
1580	1-1/2" pipe size		13.30	.602		7.85	26.50	5.35	39.70	54
1600	2" pipe size		5.70	1.404		8.45	61.50	12.50	82.45	115
1620	2-1/2" pipe size		5.70	1.404		8.45	61.50	12.50	82.45	115
1640	3" pipe size		5.70	1.404		8.45	61.50	12.50	82.45	115
1660	3-1/2" pipe size		4.40	1.818		9.05	80	16.20	105.25	147
1680	4" pipe size		4	2		9.05	88	17.80	114.85	161
1700	12" thick, 1/2" pipe size		14.50	.552		5.80	24	4.91	34.71	48
1720	3/4" pipe size		14.50	.552		5.80	24	4.91	34.71	48
1740	1" pipe size		11	.727		11.10	32	6.45	49.55	67
1760	1-1/4" pipe size		11	.727		11.10	32	6.45	49.55	67
1780	1-1/2" pipe size		11	.727		11.10	32	6.45	49.55	67
1800	2" pipe size		5	1.600		12.95	70	14.25	97.20	135

26 05 33.95 Cutting and Drilling		Crew	Daily Output	Labor-Hours	Unit	Material	2007 Bare Costs Labor	Equipment	Total	Total Incl O&P
1820	2-1/2" pipe size	R-31	5	1.600	Ea.	12.95	70	14.25	97.20	135
1840	3" pipe size		5	1.600		12.95	70	14.25	97.20	135
1860	3-1/2" pipe size		3.80	2.105		14.70	92.50	18.75	125.95	175
1880	4" pipe size		3.30	2.424		14.70	106	21.50	142.20	198
1900	16" thick, 1/2" pipe size		12.30	.650		7.80	28.50	5.80	42.10	57.50
1920	3/4" pipe size		12.30	.650		7.80	28.50	5.80	42.10	57.50
1940	1" pipe size		9.30	.860		14.30	38	7.65	59.95	80
1960	1-1/4" pipe size		9.30	.860		14.30	38	7.65	59.95	80
1980	1-1/2" pipe size		9.30	.860		14.30	38	7.65	59.95	80
2000	2" pipe size		4.40	1.818		17.40	80	16.20	113.60	156
2010	2-1/2" pipe size		4.40	1.818		17.40	80	16.20	113.60	156
2030	3" pipe size		4.40	1.818		17.40	80	16.20	113.60	156
2050	3-1/2" pipe size		3.30	2.424		20.50	106	21.50	148	204
2070	4" pipe size		3	2.667		20.50	117	23.50	161	223
2090	20" thick, 1/2" pipe size		10.70	.748		9.80	33	6.65	49.45	67
2110	3/4" pipe size		10.70	.748		9.80	33	6.65	49.45	67
2130	1" pipe size		8	1		17.55	44	8.90	70.45	94.50
2150	1-1/4" pipe size		8	1		17.55	44	8.90	70.45	94.50
2170	1-1/2" pipe size		8	1		17.55	44	8.90	70.45	94.50
2190	2" pipe size		4	2		22	88	17.80	127.80	175
2210	2-1/2" pipe size		4	2		22	88	17.80	127.80	175
2230	3" pipe size		4	2		22	88	17.80	127.80	175
2250	3-1/2" pipe size		3	2.667		26	117	23.50	166.50	229
2270	4" pipe size		2.70	2.963		26	130	26.50	182.50	252
2290	24" thick, 1/2" pipe size		9.40	.851		11.80	37.50	7.55	56.85	77
2310	3/4" pipe size		9.40	.851		11.80	37.50	7.55	56.85	77
2330	1" pipe size		7.10	1.127		21	49.50	10.05	80.55	108
2350	1-1/4" pipe size		7.10	1.127		21	49.50	10.05	80.55	108
2370	1-1/2" pipe size		7.10	1.127		21	49.50	10.05	80.55	108
2390	2" pipe size		3.60	2.222		26.50	97.50	19.80	143.80	196
2410	2-1/2" pipe size		3.60	2.222		26.50	97.50	19.80	143.80	196
2430	3" pipe size		3.60	2.222		26.50	97.50	19.80	143.80	196
2450	3-1/2" pipe size		2.80	2.857		31.50	125	25.50	182	250
2470	4" pipe size		2.50	3.200		31.50	140	28.50	200	276
3000	Knockouts to 8' high, metal boxes & enclosures									
3020	With hole saw, 1/2" pipe size	1 Elec	53	.151	Ea.		6.65		6.65	9.85
3040	3/4" pipe size		47	.170			7.45		7.45	11.10
3050	1" pipe size		40	.200			8.80		8.80	13.05
3060	1-1/4" pipe size		36	.222			9.75		9.75	14.50
3070	1-1/2" pipe size		32	.250			11		11	16.35
3080	2" pipe size		27	.296			13		13	19.35
3090	2-1/2" pipe size		20	.400			17.55		17.55	26
4010	3" pipe size		16	.500			22		22	32.50
4030	3-1/2" pipe size		13	.615			27		27	40
4050	4" pipe size		11	.727			32		32	47.50
4070	With hand punch set, 1/2" pipe size		40	.200			8.80		8.80	13.05
4090	3/4" pipe size		32	.250			11		11	16.35
4110	1" pipe size		30	.267			11.70		11.70	17.45
4130	1-1/4" pipe size		28	.286			12.55		12.55	18.65
4150	1-1/2" pipe size		26	.308			13.50		13.50	20
4170	2" pipe size		20	.400			17.55		17.55	26
4190	2-1/2" pipe size		17	.471			20.50		20.50	31
4200	3" pipe size		15	.533			23.50		23.50	35

26 05 33 – Raceway and Boxes for Electrical Systems

26 05 33.95 Cutting and Drilling		Crew	Daily Output	Labor-Hours	Unit	Material	2007 Bare Costs Labor	Equipment	Total	Total Incl O&P
4220	3-1/2" pipe size	1 Elec	12	.667	Ea.		29.50		29.50	43.50
4240	4" pipe size		10	.800			35		35	52.50
4260	With hydraulic punch, 1/2" pipe size		44	.182			8		8	11.90
4280	3/4" pipe size		38	.211			9.25		9.25	13.75
4300	1" pipe size		38	.211			9.25		9.25	13.75
4320	1-1/4" pipe size		38	.211			9.25		9.25	13.75
4340	1-1/2" pipe size		38	.211			9.25		9.25	13.75
4360	2" pipe size		32	.250			11		11	16.35
4380	2-1/2" pipe size		27	.296			13		13	19.35
4400	3" pipe size		23	.348			15.25		15.25	22.50
4420	3-1/2" pipe size		20	.400			17.55		17.55	26
4440	4" pipe size	▼	18	.444	▼		19.50		19.50	29

26 05 36 – Cable Trays for Electrical Systems

26 05 36.10 Cable Tray Ladder Type		Crew	Daily Output	Labor-Hours	Unit	Material	2007 Bare Costs Labor	Equipment	Total	Total Incl O&P
0010	**CABLE TRAY LADDER TYPE** w/ ftngs. & supports, 4" dp., to 15' elev.									
0100	For higher elevations, see 26 05 36.40 9900									
0160	Galvanized steel tray R260536-10									
0170	4" rung spacing, 6" wide	2 Elec	98	.163	L.F.	10.85	7.15		18	22.50
0180	9" wide R260536-11		92	.174		11.85	7.65		19.50	24.50
0200	12" wide		86	.186		13.05	8.15		21.20	26.50
0400	18" wide		82	.195		15.15	8.55		23.70	29.50
0600	24" wide		78	.205		17.45	9		26.45	32.50
0650	30" wide		68	.235		22	10.35		32.35	40
0700	36" wide		60	.267		24.50	11.70		36.20	44.50
0800	6" rung spacing, 6" wide		100	.160		9.90	7		16.90	21.50
0850	9" wide		94	.170		10.85	7.45		18.30	23
0860	12" wide		88	.182		11.40	8		19.40	24.50
0870	18" wide		84	.190		13	8.35		21.35	27
0880	24" wide		80	.200		14.55	8.80		23.35	29
0890	30" wide		70	.229		17.30	10.05		27.35	34
0900	36" wide		64	.250		18.85	11		29.85	37.50
0910	9" rung spacing, 6" wide		102	.157		9.30	6.90		16.20	20.50
0920	9" wide		98	.163		9.50	7.15		16.65	21
0930	12" wide		94	.170		10.95	7.45		18.40	23
0940	18" wide		90	.178		12.30	7.80		20.10	25
0950	24" wide		86	.186		14.10	8.15		22.25	27.50
0960	30" wide		80	.200		15.40	8.80		24.20	30
0970	36" wide		74	.216		16.40	9.50		25.90	32
0980	12" rung spacing, 6" wide		106	.151		9.15	6.65		15.80	19.90
0990	9" wide		104	.154		9.35	6.75		16.10	20.50
1000	12" wide		100	.160		9.90	7		16.90	21.50
1010	18" wide		96	.167		10.60	7.30		17.90	22.50
1020	24" wide		94	.170		11.40	7.45		18.85	23.50
1030	30" wide		88	.182		12.95	8		20.95	26
1040	36" wide		84	.190		13.65	8.35		22	27.50
1041	18" rung spacing, 6" wide		108	.148		9.05	6.50		15.55	19.65
1042	9" wide		106	.151		9.20	6.65		15.85	20
1043	12" wide		102	.157		9.80	6.90		16.70	21
1044	18" wide		98	.163		10.30	7.15		17.45	22
1045	24" wide		96	.167		10.65	7.30		17.95	22.50
1046	30" wide		90	.178		11.20	7.80		19	24
1047	36" wide	▼	86	.186	▼	11.50	8.15		19.65	25

26 05 36 – Cable Trays for Electrical Systems

26 05 36.10 Cable Tray Ladder Type	Crew	Daily Output	Labor-Hours	Unit	Material	2007 Bare Costs Labor	Equipment	Total	Total Incl O&P
1050	Elbows horiz. 9" rung spacing, 90°, 12" radius, 6" wide	2 Elec	9.60	1.667	Ea.	42	73		115	156
1060	9" wide		8.40	1.905		45	83.50		128.50	174
1070	12" wide		7.60	2.105		48.50	92.50		141	192
1080	18" wide		6.20	2.581		64	113		177	240
1090	24" wide		5.40	2.963		71.50	130		201.50	273
1100	30" wide		4.80	3.333		94.50	146		240.50	320
1110	36" wide		4.20	3.810		110	167		277	370
1120	90° 24" radius, 6" wide		9.20	1.739		87	76.50		163.50	210
1130	9" wide		8	2		89.50	88		177.50	230
1140	12" wide		7.20	2.222		92	97.50		189.50	246
1150	18" wide		5.80	2.759		100	121		221	290
1160	24" wide		5	3.200		108	140		248	325
1170	30" wide		4.40	3.636		116	160		276	365
1180	36" wide		3.80	4.211		138	185		323	425
1190	90° 36" radius, 6" wide		8.80	1.818		96	80		176	225
1200	9" wide		7.60	2.105		101	92.50		193.50	249
1210	12" wide		6.80	2.353		106	103		209	271
1220	18" wide		5.40	2.963		122	130		252	330
1230	24" wide		4.60	3.478		133	153		286	375
1240	30" wide		4	4		160	176		336	435
1250	36" wide		3.40	4.706		174	207		381	500
1260	45° 12" radius, 6" wide		13.20	1.212		32.50	53		85.50	115
1270	9" wide		11	1.455		34	64		98	133
1280	12" wide		9.60	1.667		35.50	73		108.50	148
1290	18" wide		7.60	2.105		37.50	92.50		130	179
1300	24" wide		6.20	2.581		48	113		161	222
1310	30" wide		5.40	2.963		61.50	130		191.50	262
1320	36" wide		4.60	3.478		64	153		217	298
1330	45° 24" radius, 6" wide		12.80	1.250		41.50	55		96.50	127
1340	9" wide		10.60	1.509		44	66.50		110.50	147
1350	12" wide		9.20	1.739		49.50	76.50		126	168
1360	18" wide		7.20	2.222		50.50	97.50		148	201
1370	24" wide		5.80	2.759		61	121		182	248
1380	30" wide		5	3.200		73	140		213	290
1390	36" wide		4.20	3.810		84	167		251	340
1400	45° 36" radius, 6" wide		12.40	1.290		59.50	56.50		116	150
1410	9" wide		10.20	1.569		62.50	69		131.50	172
1420	12" wide		8.80	1.818		64	80		144	190
1430	18" wide		6.80	2.353		69	103		172	230
1440	24" wide		5.40	2.963		73	130		203	275
1450	30" wide		4.60	3.478		94.50	153		247.50	330
1460	36" wide		3.80	4.211		105	185		290	390
1470	Elbows horizontal, 4" rung spacing, use 9" rung x 1.50									
1480	6" rung spacing use 9" rung x 1.20									
1490	12" rung spacing use 9" rung x .93									
1500	Elbows vert. 9" rung spacing, 90°, 12" radius, 6" wide	2 Elec	9.60	1.667	Ea.	58	73		131	173
1510	9" wide		8.40	1.905		59.50	83.50		143	189
1520	12" wide		7.60	2.105		61	92.50		153.50	206
1530	18" wide		6.20	2.581		64	113		177	240
1540	24" wide		5.40	2.963		66.50	130		196.50	267
1550	30" wide		4.80	3.333		71.50	146		217.50	297
1560	36" wide		4.20	3.810		74	167		241	330
1570	24" radius, 6" wide		9.20	1.739		82	76.50		158.50	204

26 05 36 – Cable Trays for Electrical Systems

26 05 36.10 Cable Tray Ladder Type	Crew	Daily Output	Labor-Hours	Unit	Material	2007 Bare Costs Labor	Equipment	Total	Total Incl O&P
1580	9" wide	2 Elec	8	2	Ea.	84.50	88		172.50	224
1590	12" wide		7.20	2.222		87	97.50		184.50	241
1600	18" wide		5.80	2.759		92	121		213	281
1610	24" wide		5	3.200		97.50	140		237.50	315
1620	30" wide		4.40	3.636		101	160		261	350
1630	36" wide		3.80	4.211		114	185		299	400
1640	36" radius, 6" wide		8.80	1.818		110	80		190	240
1650	9" wide		7.60	2.105		114	92.50		206.50	263
1660	12" wide		6.80	2.353		115	103		218	281
1670	18" wide		5.40	2.963		125	130		255	330
1680	24" wide		4.60	3.478		127	153		280	365
1690	30" wide		4	4		143	176		319	420
1700	36" wide		3.40	4.706		154	207		361	480
1710	Elbows vertical, 4" rung spacing, use 9" rung x 1.25									
1720	6" rung spacing, use 9" rung x 1.15									
1730	12" rung spacing, use 9" rung x .90									
1740	Tee horizontal, 9" rung spacing, 12" radius, 6" wide	2 Elec	5	3.200	Ea.	89.50	140		229.50	310
1750	9" wide		4.60	3.478		96	153		249	335
1760	12" wide		4.40	3.636		101	160		261	350
1770	18" wide		4	4		115	176		291	390
1780	24" wide		3.60	4.444		136	195		331	440
1790	30" wide		3.40	4.706		163	207		370	490
1800	36" wide		3	5.333		184	234		418	555
1810	24" radius, 6" wide		4.60	3.478		140	153		293	380
1820	9" wide		4.20	3.810		146	167		313	410
1830	12" wide		4	4		151	176		327	425
1840	18" wide		3.60	4.444		164	195		359	470
1850	24" wide		3.20	5		178	220		398	520
1860	30" wide		3	5.333		193	234		427	565
1870	36" wide		2.60	6.154		248	270		518	675
1880	36" radius, 6" wide		4.20	3.810		209	167		376	480
1890	9" wide		3.80	4.211		218	185		403	515
1900	12" wide		3.60	4.444		232	195		427	545
1910	18" wide		3.20	5		248	220		468	600
1920	24" wide		2.80	5.714		280	251		531	685
1930	30" wide		2.60	6.154		305	270		575	735
1940	36" wide		2.20	7.273		330	320		650	835
1980	Tee vertical, 9" rung spacing, 12" radius, 6" wide		5.40	2.963		152	130		282	360
1990	9" wide		5.20	3.077		154	135		289	370
2000	12" wide		5	3.200		155	140		295	380
2010	18" wide		4.60	3.478		156	153		309	400
2020	24" wide		4.40	3.636		160	160		320	415
2030	30" wide		4	4		170	176		346	450
2040	36" wide		3.60	4.444		177	195		372	485
2050	24" radius, 6" wide		5	3.200		274	140		414	510
2060	9" wide		4.80	3.333		278	146		424	525
2070	12" wide		4.60	3.478		282	153		435	535
2080	18" wide		4.20	3.810		291	167		458	570
2090	24" wide		4	4		298	176		474	590
2100	30" wide		3.60	4.444		305	195		500	625
2110	36" wide		3.20	5		320	220		540	675
2120	36" radius, 6" wide		4.60	3.478		510	153		663	790
2130	9" wide		4.40	3.636		520	160		680	810

26 05 36 – Cable Trays for Electrical Systems

26 05 36.10 Cable Tray Ladder Type		Crew	Daily Output	Labor-Hours	Unit	Material	2007 Bare Costs Labor	Equipment	Total	Total Incl O&P
2140	12" wide	2 Elec	4.20	3.810	Ea.	525	167		692	830
2150	18" wide		3.80	4.211		530	185		715	860
2160	24" wide		3.60	4.444		540	195		735	885
2170	30" wide		3.20	5		560	220		780	940
2180	36" wide		2.80	5.714		565	251		816	995
2190	Tee, 4" rung spacing, use 9" rung x 1.30									
2200	6" rung spacing, use 9" rung x 1.20									
2210	12" rung spacing, use 9" rung x .90									
2220	Cross horizontal, 9" rung spacing, 12" radius, 6" wide	2 Elec	4	4	Ea.	110	176		286	385
2230	9" wide		3.80	4.211		116	185		301	400
2240	12" wide		3.60	4.444		127	195		322	430
2250	18" wide		3.40	4.706		138	207		345	460
2260	24" wide		3	5.333		154	234		388	520
2270	30" wide		2.80	5.714		186	251		437	580
2280	36" wide		2.60	6.154		206	270		476	625
2290	24" radius, 6" wide		3.60	4.444		215	195		410	525
2300	9" wide		3.40	4.706		221	207		428	555
2310	12" wide		3.20	5		228	220		448	575
2320	18" wide		3	5.333		243	234		477	620
2330	24" wide		2.60	6.154		259	270		529	685
2340	30" wide		2.40	6.667		275	293		568	740
2350	36" wide		2.20	7.273		320	320		640	825
2360	36" radius, 6" wide		3.20	5		300	220		520	655
2370	9" wide		3	5.333		310	234		544	695
2380	12" wide		2.80	5.714		325	251		576	735
2390	18" wide		2.60	6.154		360	270		630	795
2400	24" wide		2.20	7.273		400	320		720	915
2410	30" wide		2	8		435	350		785	1,000
2420	36" wide		1.80	8.889		465	390		855	1,100
2430	Cross horizontal, 4" rung spacing, use 9" rung x 1.30									
2440	6" rung spacing, use 9" rung x 1.20									
2450	12" rung spacing, use 9" rung x .90									
2460	Reducer, 9" to 6" wide tray	2 Elec	13	1.231	Ea.	61.50	54		115.50	148
2470	12" to 9" wide tray		12	1.333		62.50	58.50		121	156
2480	18" to 12" wide tray		10.40	1.538		62.50	67.50		130	170
2490	24" to 18" wide tray		9	1.778		64	78		142	187
2500	30" to 24" wide tray		8	2		65.50	88		153.50	203
2510	36" to 30" wide tray		7	2.286		73	100		173	230
2511	Reducer, 18" to 6" wide tray		10.40	1.538		65.50	67.50		133	173
2512	24" to 12" wide tray		9	1.778		66.50	78		144.50	189
2513	30" to 18" wide tray		8	2		73	88		161	212
2514	30" to 12" wide tray		8	2		70.50	88		158.50	209
2515	36" to 24" wide tray		7	2.286		74	100		174	231
2516	36" to 18" wide tray		7	2.286		74	100		174	231
2517	36" to 12" wide tray		7	2.286		74	100		174	231
2520	Dropout or end plate, 6" wide		32	.500		6.85	22		28.85	40
2530	9" wide		28	.571		8.05	25		33.05	46.50
2540	12" wide		26	.615		8.85	27		35.85	49.50
2550	18" wide		22	.727		10.35	32		42.35	59
2560	24" wide		20	.800		12.55	35		47.55	66.50
2570	30" wide		18	.889		13.70	39		52.70	73
2580	36" wide		16	1		15.75	44		59.75	83
2590	Tray connector		48	.333		12.15	14.65		26.80	35.50

26 05 36 – Cable Trays for Electrical Systems

26 05 36.10 Cable Tray Ladder Type	Crew	Daily Output	Labor-Hours	Unit	Material	2007 Bare Costs Labor	Equipment	Total	Total Incl O&P
3200	Aluminum tray, 4" deep, 6" rung spacing, 6" wide	2 Elec	134	.119	L.F.	13.65	5.25		18.90	23
3210	9" wide		128	.125		14.45	5.50		19.95	24
3220	12" wide		124	.129		15.30	5.65		20.95	25.50
3230	18" wide		114	.140		17	6.15		23.15	28
3240	24" wide		106	.151		19.75	6.65		26.40	32
3250	30" wide		100	.160		22	7		29	34.50
3260	36" wide		94	.170		23.50	7.45		30.95	37
3270	9" rung spacing, 6" wide		140	.114		11.10	5		16.10	19.65
3280	9" wide		134	.119		11.60	5.25		16.85	20.50
3290	12" wide		130	.123		11.85	5.40		17.25	21
3300	18" wide		122	.131		13.80	5.75		19.55	24
3310	24" wide		116	.138		16.05	6.05		22.10	26.50
3320	30" wide		108	.148		17.75	6.50		24.25	29
3330	36" wide		100	.160		18.80	7		25.80	31
3340	12" rung spacing, 6" wide		146	.110		11.75	4.81		16.56	20
3350	9" wide		140	.114		11.90	5		16.90	20.50
3360	12" wide		134	.119		12.30	5.25		17.55	21.50
3370	18" wide		128	.125		13	5.50		18.50	22.50
3380	24" wide		124	.129		14.45	5.65		20.10	24.50
3390	30" wide		114	.140		15.45	6.15		21.60	26
3400	36" wide		106	.151		16.25	6.65		22.90	28
3401	18" rung spacing, 6" wide		150	.107		11.70	4.68		16.38	19.80
3402	9" wide tray		144	.111		11.85	4.88		16.73	20.50
3403	12" wide tray		140	.114		12.30	5		17.30	21
3404	18" wide tray		134	.119		12.90	5.25		18.15	22
3405	24" wide tray		130	.123		13.90	5.40		19.30	23.50
3406	30" wide tray		120	.133		14.40	5.85		20.25	24.50
3407	36" wide tray		110	.145		15.05	6.40		21.45	26
3410	Elbows horiz. 9" rung spacing, 90° 12" radius, 6" wide		9.60	1.667	Ea.	52.50	73		125.50	167
3420	9" wide		8.40	1.905		54	83.50		137.50	184
3430	12" wide		7.60	2.105		64	92.50		156.50	209
3440	18" wide		6.20	2.581		72	113		185	248
3450	24" wide		5.40	2.963		89	130		219	292
3460	30" wide		4.80	3.333		94.50	146		240.50	320
3470	36" wide		4.20	3.810		114	167		281	375
3480	24" radius, 6" wide		9.20	1.739		93.50	76.50		170	217
3490	9" wide		8	2		97	88		185	238
3500	12" wide		7.20	2.222		102	97.50		199.50	257
3510	18" wide		5.80	2.759		111	121		232	300
3520	24" wide		5	3.200		122	140		262	345
3530	30" wide		4.40	3.636		132	160		292	385
3540	36" wide		3.80	4.211		150	185		335	440
3550	90°, 36" radius, 6" wide		8.80	1.818		102	80		182	232
3560	9" wide		7.60	2.105		105	92.50		197.50	253
3570	12" wide		6.80	2.353		118	103		221	284
3580	18" wide		5.40	2.963		132	130		262	340
3590	24" wide		4.60	3.478		148	153		301	390
3600	30" wide		4	4		164	176		340	440
3610	36" wide		3.40	4.706		188	207		395	515
3620	45° 12" radius, 6" wide		13.20	1.212		34.50	53		87.50	117
3630	9" wide		11	1.455		36	64		100	135
3640	12" wide		9.60	1.667		37	73		110	150
3650	18" wide		7.60	2.105		44	92.50		136.50	187

26 05 36 – Cable Trays for Electrical Systems

26 05 36.10 Cable Tray Ladder Type		Crew	Daily Output	Labor-Hours	Unit	Material	2007 Bare Costs Labor	Equipment	Total	Total Incl O&P
3660	24" wide	2 Elec	6.20	2.581	Ea.	54.50	113		167.50	229
3670	30" wide		5.40	2.963		62.50	130		192.50	263
3680	36" wide		4.60	3.478		65.50	153		218.50	299
3690	45° 24" radius, 6" wide		12.80	1.250		46	55		101	132
3700	9" wide		10.60	1.509		52.50	66.50		119	156
3710	12" wide		9.20	1.739		54	76.50		130.50	173
3720	18" wide		7.20	2.222		57.50	97.50		155	209
3730	24" wide		5.80	2.759		66.50	121		187.50	253
3740	30" wide		5	3.200		79.50	140		219.50	297
3750	36" wide		4.20	3.810		84.50	167		251.50	340
3760	45° 36" radius, 6" wide		12.40	1.290		62.50	56.50		119	154
3770	9" wide		10.20	1.569		64	69		133	174
3780	12" wide		8.80	1.818		65.50	80		145.50	191
3790	18" wide		6.80	2.353		71.50	103		174.50	233
3800	24" wide		5.40	2.963		88.50	130		218.50	291
3810	30" wide		4.60	3.478		93.50	153		246.50	330
3820	36" wide		3.80	4.211		102	185		287	390
3830	Elbows horizontal, 4" rung spacing, use 9" rung x 1.50									
3840	6" rung spacing, use 9" rung x 1.20									
3850	12" rung spacing, use 9" rung x .93									
3860	Elbows vertical 9" rung spacing, 90° 12" radius, 6" wide	2 Elec	9.60	1.667	Ea.	73	73		146	190
3870	9" wide		8.40	1.905		77	83.50		160.50	209
3880	12" wide		7.60	2.105		78	92.50		170.50	224
3890	18" wide		6.20	2.581		80.50	113		193.50	258
3900	24" wide		5.40	2.963		86	130		216	289
3910	30" wide		4.80	3.333		89.50	146		235.50	315
3920	36" wide		4.20	3.810		92	167		259	350
3930	24" radius, 6" wide		9.20	1.739		79.50	76.50		156	202
3940	9" wide		8	2		82	88		170	221
3950	12" wide		7.20	2.222		84.50	97.50		182	238
3960	18" wide		5.80	2.759		89.50	121		210.50	279
3970	24" wide		5	3.200		94.50	140		234.50	315
3980	30" wide		4.40	3.636		100	160		260	350
3990	36" wide		3.80	4.211		106	185		291	390
4000	36" radius, 6" wide		8.80	1.818		123	80		203	254
4010	9" wide		7.60	2.105		125	92.50		217.50	276
4020	12" wide		6.80	2.353		133	103		236	300
4030	18" wide		5.40	2.963		138	130		268	345
4040	24" wide		4.60	3.478		143	153		296	385
4050	30" wide		4	4		155	176		331	430
4060	36" wide		3.40	4.706		161	207		368	485
4070	Elbows vertical, 4" rung spacing, use 9" rung x 1.25									
4080	6" rung spacing, use 9" rung x 1.15									
4090	12" rung spacing, use 9" rung x .90									
4100	Tee horizontal 9" rung spacing, 12" radius, 6" wide	2 Elec	5	3.200	Ea.	81	140		221	299
4110	9" wide		4.60	3.478		87	153		240	325
4120	12" wide		4.40	3.636		88.50	160		248.50	335
4130	18" wide		4.20	3.810		105	167		272	365
4140	24" wide		4	4		115	176		291	390
4150	30" wide		3.60	4.444		134	195		329	440
4160	36" wide		3.40	4.706		161	207		368	485
4170	24" radius, 6" wide		4.60	3.478		138	153		291	380
4180	9" wide		4.20	3.810		146	167		313	410

26 05 Common Work Results for Electrical

26 05 36 – Cable Trays for Electrical Systems

26 05 36.10 Cable Tray Ladder Type		Crew	Daily Output	Labor-Hours	Unit	Material	2007 Bare Costs Labor	Equipment	Total	Total Incl O&P
4190	12" wide	2 Elec	4	4	Ea.	150	176		326	425
4200	18" wide		3.80	4.211		164	185		349	455
4210	24" wide		3.60	4.444		179	195		374	485
4220	30" wide		3.20	5		195	220		415	540
4230	36" wide		3	5.333		300	234		534	680
4240	36" radius, 6" wide		4.20	3.810		196	167		363	465
4250	9" wide		3.80	4.211		201	185		386	495
4260	12" wide		3.60	4.444		220	195		415	530
4270	18" wide		3.40	4.706		232	207		439	565
4280	24" wide		3.20	5		256	220		476	605
4290	30" wide		2.80	5.714		325	251		576	730
4300	36" wide		2.60	6.154		360	270		630	795
4310	Tee vertical 9" rung spacing, 12" radius, 6" wide		5.40	2.963		182	130		312	395
4320	9" wide		5.20	3.077		188	135		323	410
4330	12" wide		5	3.200		189	140		329	415
4340	18" wide		4.60	3.478		193	153		346	440
4350	24" wide		4.40	3.636		209	160		369	470
4360	30" wide		4.20	3.810		212	167		379	485
4370	36" wide		4	4		220	176		396	505
4380	24" radius, 6" wide		5	3.200		310	140		450	550
4390	9" wide		4.80	3.333		310	146		456	565
4400	12" wide		4.60	3.478		315	153		468	575
4410	18" wide		4.20	3.810		325	167		492	610
4420	24" wide		4	4		335	176		511	625
4430	30" wide		3.80	4.211		345	185		530	650
4440	36" wide		3.60	4.444		370	195		565	700
4450	36" radius, 6" wide		4.60	3.478		680	153		833	970
4460	9" wide		4.40	3.636		690	160		850	1,000
4470	12" wide		4.20	3.810		720	167		887	1,050
4480	18" wide		3.80	4.211		725	185		910	1,075
4490	24" wide		3.60	4.444		745	195		940	1,100
4500	30" wide		3.40	4.706		755	207		962	1,150
4510	36" wide		3.20	5		765	220		985	1,175
4520	Tees, 4" rung spacing, use 9" rung x 1.30									
4530	6" rung spacing, use 9" rung x 1.20									
4540	12" rung spacing, use 9" rung x .90									
4550	Cross horizontal 9" rung spacing, 12" radius, 6" wide	2 Elec	4.40	3.636	Ea.	113	160		273	360
4560	9" wide		4.20	3.810		125	167		292	385
4570	12" wide		4	4		127	176		303	400
4580	18" wide		3.60	4.444		154	195		349	460
4590	24" wide		3.40	4.706		159	207		366	485
4600	30" wide		3	5.333		187	234		421	555
4610	36" wide		2.80	5.714		233	251		484	630
4620	24" radius, 6" wide		4	4		216	176		392	500
4630	9" wide		3.80	4.211		224	185		409	520
4640	12" wide		3.60	4.444		230	195		425	545
4650	18" wide		3.20	5		242	220		462	590
4660	24" wide		3	5.333		264	234		498	640
4670	30" wide		2.60	6.154		282	270		552	710
4680	36" wide		2.40	6.667		320	293		613	785
4690	36" radius, 6" wide		3.60	4.444		256	195		451	570
4700	9" wide		3.40	4.706		266	207		473	605
4710	12" wide		3.20	5		282	220		502	635

26 05 36.10 Cable Tray Ladder Type		Crew	Daily Output	Labor-Hours	Unit	Material	2007 Bare Costs Labor	Equipment	Total	Total Incl O&P
4720	18" wide	2 Elec	2.80	5.714	Ea.	300	251		551	705
4730	24" wide		2.60	6.154		370	270		640	810
4740	30" wide		2.20	7.273		410	320		730	930
4750	36" wide		2	8		460	350		810	1,025
4760	Cross horizontal, 4" rung spacing, use 9" rung x 1.30									
4770	6" rung spacing, use 9" rung x 1.20									
4780	12" rung spacing, use 9" rung x .90									
4790	Reducer, 9" to 6" wide tray	2 Elec	16	1	Ea.	60	44		104	132
4800	12" to 9" wide tray		14	1.143		62.50	50		112.50	144
4810	18" to 12" wide tray		12.40	1.290		62.50	56.50		119	154
4820	24" to 18" wide tray		10.60	1.509		64	66.50		130.50	169
4830	30" to 24" wide tray		9.20	1.739		65.50	76.50		142	186
4840	36" to 30" wide tray		8	2		69	88		157	207
4841	Reducer, 18" to 6" wide tray		12.40	1.290		62.50	56.50		119	154
4842	24" to 12" wide tray		10.60	1.509		64	66.50		130.50	169
4843	30" to 18" wide tray		9.20	1.739		65.50	76.50		142	186
4844	30" to 12" wide tray		9.20	1.739		65.50	76.50		142	186
4845	36" to 24" wide tray		8	2		69	88		157	207
4846	36" to 18" wide tray		8	2		69	88		157	207
4847	36" to 12" wide tray		8	2		69	88		157	207
4850	Dropout or end plate, 6" wide		32	.500		7.95	22		29.95	41.50
4860	9" wide tray		28	.571		8.45	25		33.45	47
4870	12" wide tray		26	.615		9.20	27		36.20	50
4880	18" wide tray		22	.727		12.30	32		44.30	61
4890	24" wide tray		20	.800		14.45	35		49.45	68.50
4900	30" wide tray		18	.889		17.15	39		56.15	77
4910	36" wide tray		16	1		19.05	44		63.05	86.50
4920	Tray connector		48	.333		11.50	14.65		26.15	34.50
8000	Elbow 36" radius horiz., 60°, 6" wide tray		10.60	1.509		87	66.50		153.50	194
8010	9" wide tray		9	1.778		94.50	78		172.50	220
8020	12" wide tray		7.80	2.051		97.50	90		187.50	241
8030	18" wide tray		6.20	2.581		104	113		217	283
8040	24" wide tray		5	3.200		122	140		262	345
8050	30" wide tray		4.40	3.636		124	160		284	375
8060	30°, 6" wide tray		14	1.143		66.50	50		116.50	148
8070	9" wide tray		11.40	1.404		70.50	61.50		132	169
8080	12" wide tray		9.80	1.633		71.50	71.50		143	186
8090	18" wide tray		7.40	2.162		75.50	95		170.50	224
8100	24" wide tray		5.80	2.759		92	121		213	281
8110	30" wide tray		4.80	3.333		94.50	146		240.50	320
8120	Adjustable, 6" wide tray		12.40	1.290		88.50	56.50		145	182
8130	9" wide tray		10.20	1.569		93.50	69		162.50	206
8140	12" wide tray		8.80	1.818		94.50	80		174.50	223
8150	18" wide tray		6.80	2.353		104	103		207	268
8160	24" wide tray		5.40	2.963		110	130		240	315
8170	30" wide tray		4.60	3.478		123	153		276	360
8180	Wye 36" radius horiz., 45°, 6" wide tray		4.60	3.478		96	153		249	335
8190	9" wide tray		4.40	3.636		101	160		261	350
8200	12" wide tray		4.20	3.810		105	167		272	365
8210	18" wide tray		3.80	4.211		119	185		304	405
8220	24" wide tray		3.60	4.444		133	195		328	435
8230	30" wide tray		3.40	4.706		148	207		355	475
8240	Elbow 36" radius vert. in/outside, 60°, 6" wide tray		10.60	1.509		96	66.50		162.50	205

26 05 36.10 Cable Tray Ladder Type		Crew	Daily Output	Labor-Hours	Unit	Material	2007 Bare Costs Labor	Equipment	Total	Total Incl O&P
8250	9" wide tray	2 Elec	9	1.778	Ea.	100	78		178	226
8260	12" wide tray		7.80	2.051		101	90		191	245
8270	18" wide tray		6.20	2.581		105	113		218	284
8280	24" wide tray		5	3.200		111	140		251	330
8290	30" wide tray		4.40	3.636		119	160		279	370
8300	45°, 6" wide tray		12.40	1.290		86.50	56.50		143	180
8310	9" wide tray		10.20	1.569		92	69		161	204
8320	12" wide tray		8.80	1.818		93	80		173	222
8330	18" wide tray		6.80	2.353		96	103		199	260
8340	24" wide tray		5.40	2.963		101	130		231	305
8350	30" wide tray		4.60	3.478		104	153		257	340
8360	30°, 6" wide tray		14	1.143		76	50		126	158
8370	9" wide tray		11.40	1.404		77	61.50		138.50	177
8380	12" wide tray		9.80	1.633		82.50	71.50		154	198
8390	18" wide tray		7.40	2.162		84	95		179	233
8400	24" wide tray		5.80	2.759		86	121		207	275
8410	30" wide tray		4.80	3.333		90.50	146		236.50	320
8660	Adjustable, 6" wide tray		12.40	1.290		86.50	56.50		143	180
8670	9" wide tray		10.20	1.569		92	69		161	204
8680	12" wide tray		8.80	1.818		93.50	80		173.50	222
8690	18" wide tray		6.80	2.353		104	103		207	268
8700	24" wide tray		5.40	2.963		110	130		240	315
8710	30" wide tray		4.60	3.478		122	153		275	360
8720	Cross, vertical, 24" radius, 6" wide tray		3.60	4.444		600	195		795	950
8730	9" wide tray		3.40	4.706		610	207		817	980
8740	12" wide tray		3.20	5		625	220		845	1,025
8750	18" wide tray		2.80	5.714		655	251		906	1,100
8760	24" wide tray		2.60	6.154		680	270		950	1,150
8770	30" wide tray		2.20	7.273		720	320		1,040	1,275
9200	Splice plate	1 Elec	48	.167	Pr.	12.05	7.30		19.35	24
9210	Expansion joint		48	.167		14.45	7.30		21.75	27
9220	Horizontal hinged		48	.167		12.05	7.30		19.35	24
9230	Vertical hinged		48	.167		14.45	7.30		21.75	27
9240	Ladder hanger, vertical		28	.286	Ea.	3.58	12.55		16.13	22.50
9250	Ladder to channel connector		24	.333		42	14.65		56.65	68
9260	Ladder to box connector 30" wide		19	.421		42	18.50		60.50	73.50
9270	24" wide		20	.400		42	17.55		59.55	72
9280	18" wide		21	.381		40.50	16.70		57.20	69.50
9290	12" wide		22	.364		38	15.95		53.95	65.50
9300	9" wide		23	.348		34	15.25		49.25	60
9310	6" wide		24	.333		34	14.65		48.65	59.50
9320	Ladder floor flange		24	.333		27.50	14.65		42.15	52
9330	Cable roller for tray 30" wide		10	.800		215	35		250	289
9340	24" wide		11	.727		182	32		214	249
9350	18" wide		12	.667		174	29.50		203.50	235
9360	12" wide		13	.615		137	27		164	190
9370	9" wide		14	.571		121	25		146	171
9380	6" wide		15	.533		98.50	23.50		122	143
9390	Pulley, single wheel		12	.667		224	29.50		253.50	291
9400	Triple wheel		10	.800		450	35		485	545
9440	Nylon cable tie, 14" long		80	.100		.44	4.39		4.83	7.05
9450	Ladder, hold down clamp		60	.133		7.70	5.85		13.55	17.15
9460	Cable clamp		60	.133		7.80	5.85		13.65	17.30

26 05 36 – Cable Trays for Electrical Systems

		Crew	Daily Output	Labor-Hours	Unit	Material	2007 Bare Costs Labor	2007 Bare Costs Equipment	Total	Total Incl O&P
26 05 36.10 Cable Tray Ladder Type										
9470	Wall bracket, 30" wide tray	1 Elec	19	.421	Ea.	39.50	18.50		58	71
9480	24" wide tray		20	.400		31	17.55		48.55	60
9490	18" wide tray		21	.381		27.50	16.70		44.20	55.50
9500	12" wide tray		22	.364		26	15.95		41.95	52.50
9510	9" wide tray		23	.348		22.50	15.25		37.75	47.50
9520	6" wide tray		24	.333		22	14.65		36.65	46.50
26 05 36.20 Cable Tray Solid Bottom										
0010	**CABLE TRAY SOLID BOTTOM** w/ ftngs. & supports, 3" dp, to 15' high									
0200	For higher elevations, see 26 05 36.40 9900									
0220	Galvanized steel, tray, 6" wide R260536-10	2 Elec	120	.133	L.F.	8.75	5.85		14.60	18.35
0240	12" wide		100	.160		11.25	7		18.25	23
0260	18" wide		70	.229		13.65	10.05		23.70	30
0280	24" wide R260536-11		60	.267		16.05	11.70		27.75	35
0300	30" wide		50	.320		19.05	14.05		33.10	42
0320	36" wide		44	.364		27.50	15.95		43.45	54
0340	Elbow horizontal 90°, 12" radius, 6" wide		9.60	1.667	Ea.	66.50	73		139.50	182
0360	12" wide		6.80	2.353		74.50	103		177.50	236
0370	18" wide		5.40	2.963		85.50	130		215.50	288
0380	24" wide		4.40	3.636		104	160		264	350
0390	30" wide		3.80	4.211		124	185		309	410
0400	36" wide		3.40	4.706		141	207		348	465
0420	24" radius, 6" wide		9.20	1.739		94.50	76.50		171	218
0440	12" wide		6.40	2.500		110	110		220	284
0450	18" wide		5	3.200		128	140		268	350
0460	24" wide		4	4		148	176		324	425
0470	30" wide		3.40	4.706		174	207		381	500
0480	36" wide		3	5.333		196	234		430	565
0500	36" radius, 6" wide		8.80	1.818		138	80		218	271
0520	12" wide		6	2.667		156	117		273	345
0530	18" wide		4.60	3.478		191	153		344	435
0540	24" wide		3.60	4.444		206	195		401	515
0550	30" wide		3	5.333		250	234		484	625
0560	36" wide		2.60	6.154		280	270		550	710
0580	Elbow vertical 90°, 12" radius, 6" wide		9.60	1.667		80.50	73		153.50	198
0600	12" wide		6.80	2.353		86	103		189	249
0610	18" wide		5.40	2.963		87	130		217	290
0620	24" wide		4.40	3.636		101	160		261	350
0630	30" wide		3.80	4.211		108	185		293	395
0640	36" wide		3.40	4.706		111	207		318	435
0670	24" radius, 6" wide		9.20	1.739		114	76.50		190.50	239
0690	12" wide		6.40	2.500		125	110		235	300
0700	18" wide		5	3.200		136	140		276	360
0710	24" wide		4	4		147	176		323	425
0720	30" wide		3.40	4.706		156	207		363	480
0730	36" wide		3	5.333		170	234		404	535
0750	36" radius, 6" wide		8.80	1.818		156	80		236	291
0770	12" wide		6.60	2.424		174	106		280	350
0780	18" wide		4.60	3.478		191	153		344	435
0790	24" wide		3.60	4.444		209	195		404	520
0800	30" wide		3	5.333		228	234		462	600
0810	36" wide		2.60	6.154		247	270		517	670
0840	Tee horizontal, 12" radius, 6" wide		5	3.200		93.50	140		233.50	310

26 05 36.20 Cable Tray Solid Bottom		Crew	Daily Output	Labor-Hours	Unit	Material	2007 Bare Costs Labor	Equipment	Total	Total Incl O&P
0860	12" wide	2 Elec	4	4	Ea.	101	176		277	370
0870	18" wide		3.40	4.706		120	207		327	440
0880	24" wide		2.80	5.714		134	251		385	525
0890	30" wide		2.60	6.154		154	270		424	570
0900	36" wide		2.20	7.273		175	320		495	670
0940	24" radius, 6" wide		4.60	3.478		145	153		298	385
0960	12" wide		3.60	4.444		169	195		364	475
0970	18" wide		3	5.333		188	234		422	555
0980	24" wide		2.40	6.667		255	293		548	715
0990	30" wide		2.20	7.273		276	320		596	780
1000	36" wide		1.80	8.889		300	390		690	910
1020	36" radius, 6" wide		4.20	3.810		228	167		395	500
1040	12" wide		3.20	5		255	220		475	605
1050	18" wide		2.60	6.154		276	270		546	705
1060	24" wide		2.20	7.273		355	320		675	865
1070	30" wide		2	8		385	350		735	950
1080	36" wide		1.60	10		400	440		840	1,100
1100	Tee vertical, 12" radius, 6" wide		5	3.200		154	140		294	380
1120	12" wide		4	4		160	176		336	435
1130	18" wide		3.60	4.444		163	195		358	470
1140	24" wide		3.40	4.706		178	207		385	505
1150	30" wide		3	5.333		191	234		425	560
1160	36" wide		2.60	6.154		197	270		467	615
1180	24" radius, 6" wide		4.60	3.478		227	153		380	475
1200	12" wide		3.60	4.444		237	195		432	550
1210	18" wide		3.20	5		248	220		468	600
1220	24" wide		3	5.333		261	234		495	635
1230	30" wide		2.60	6.154		287	270		557	715
1240	36" wide		2.20	7.273		300	320		620	805
1260	36" radius, 6" wide		4.20	3.810		355	167		522	640
1280	12" wide		3.20	5		360	220		580	720
1290	18" wide		2.80	5.714		385	251		636	800
1300	24" wide		2.60	6.154		400	270		670	840
1310	30" wide		2.20	7.273		465	320		785	990
1320	36" wide		2	8		495	350		845	1,075
1340	Cross horizontal, 12" radius, 6" wide		4	4		111	176		287	385
1360	12" wide		3.40	4.706		122	207		329	445
1370	18" wide		2.80	5.714		142	251		393	530
1380	24" wide		2.40	6.667		160	293		453	610
1390	30" wide		2	8		178	350		528	720
1400	36" wide		1.80	8.889		196	390		586	795
1420	24" radius, 6" wide		3.60	4.444		204	195		399	515
1440	12" wide		3	5.333		228	234		462	600
1450	18" wide		2.40	6.667		255	293		548	715
1460	24" wide		2	8		325	350		675	885
1470	30" wide		1.80	8.889		355	390		745	970
1480	36" wide		1.60	10		370	440		810	1,075
1500	36" radius, 6" wide		3.20	5		335	220		555	690
1520	12" wide		2.60	6.154		365	270		635	805
1530	18" wide		2	8		400	350		750	965
1540	24" wide		1.80	8.889		470	390		860	1,100
1550	30" wide		1.60	10		510	440		950	1,225
1560	36" wide		1.40	11.429		545	500		1,045	1,350

26 05 36.20 Cable Tray Solid Bottom		Crew	Daily Output	Labor-Hours	Unit	Material	2007 Bare Costs Labor	Equipment	Total	Total Incl O&P
1580	Drop out or end plate, 6" wide	2 Elec	32	.500	Ea.	13.55	22		35.55	47.50
1600	12" wide		26	.615		16.90	27		43.90	58.50
1610	18" wide		22	.727		18.10	32		50.10	67.50
1620	24" wide		20	.800		21.50	35		56.50	76
1630	30" wide		18	.889		24	39		63	84.50
1640	36" wide		16	1		25.50	44		69.50	93.50
1660	Reducer, 12" to 6" wide		12	1.333		65.50	58.50		124	159
1680	18" to 12" wide		10.60	1.509		66.50	66.50		133	172
1700	18" to 6" wide		10.60	1.509		66.50	66.50		133	172
1720	24" to 18" wide		9.20	1.739		69	76.50		145.50	190
1740	24" to 12" wide		9.20	1.739		69	76.50		145.50	190
1760	30" to 24" wide		8	2		74	88		162	213
1780	30" to 18" wide		8	2		74	88		162	213
1800	30" to 12" wide		8	2		74	88		162	213
1820	36" to 30" wide		7.20	2.222		77	97.50		174.50	230
1840	36" to 24" wide		7.20	2.222		77	97.50		174.50	230
1860	36" to 18" wide		7.20	2.222		77	97.50		174.50	230
1880	36" to 12" wide		7.20	2.222		77	97.50		174.50	230
2000	Aluminum tray, 6" wide		150	.107	L.F.	10.65	4.68		15.33	18.65
2020	12" wide		130	.123		13.90	5.40		19.30	23.50
2030	18" wide		100	.160		17.45	7		24.45	29.50
2040	24" wide		90	.178		22	7.80		29.80	35.50
2050	30" wide		70	.229		26	10.05		36.05	43.50
2060	36" wide		64	.250		32.50	11		43.50	52.50
2080	Elbow horizontal 90°, 12" radius, 6" wide		9.60	1.667	Ea.	85.50	73		158.50	203
2100	12" wide		7.60	2.105		96	92.50		188.50	243
2110	18" wide		6.80	2.353		118	103		221	284
2120	24" wide		5.80	2.759		134	121		255	330
2130	30" wide		5	3.200		164	140		304	390
2140	36" wide		4.40	3.636		184	160		344	440
2160	24" radius, 6" wide		9.20	1.739		122	76.50		198.50	248
2180	12" wide		7.20	2.222		143	97.50		240.50	305
2190	18" wide		6.40	2.500		164	110		274	345
2200	24" wide		5.40	2.963		202	130		332	415
2210	30" wide		4.60	3.478		232	153		385	480
2220	36" wide		4	4		266	176		442	555
2240	36" radius, 6" wide		8.80	1.818		179	80		259	315
2260	12" wide		6.80	2.353		215	103		318	390
2270	18" wide		6	2.667		256	117		373	455
2280	24" wide		5	3.200		274	140		414	510
2290	30" wide		4.20	3.810		310	167		477	590
2300	36" wide		3.60	4.444		365	195		560	690
2320	Elbow vertical 90°, 12" radius, 6" wide		9.60	1.667		101	73		174	220
2340	12" wide		7.60	2.105		105	92.50		197.50	253
2350	18" wide		6.80	2.353		119	103		222	285
2360	24" wide		5.80	2.759		125	121		246	320
2370	30" wide		5	3.200		133	140		273	355
2380	36" wide		4.40	3.636		137	160		297	390
2400	24" radius, 6" wide		9.20	1.739		143	76.50		219.50	272
2420	12" wide		7.20	2.222		156	97.50		253.50	315
2430	18" wide		6.40	2.500		168	110		278	345
2440	24" wide		5.40	2.963		182	130		312	395
2450	30" wide		4.60	3.478		193	153		346	440

26 05 36.20 Cable Tray Solid Bottom		Crew	Daily Output	Labor-Hours	Unit	Material	2007 Bare Costs Labor	Equipment	Total	Total Incl O&P
2460	36" wide	2 Elec	4	4	Ea.	212	176		388	495
2480	36" radius, 6" wide		8.80	1.818		189	80		269	325
2500	12" wide		6.80	2.353		212	103		315	390
2510	18" wide		6	2.667		232	117		349	430
2520	24" wide		5	3.200		241	140		381	475
2530	30" wide		4.20	3.810		266	167		433	540
2540	36" wide		3.60	4.444		276	195		471	595
2560	Tee horizontal, 12" radius, 6" wide		5	3.200		132	140		272	355
2580	12" wide		4.40	3.636		160	160		320	415
2590	18" wide		4	4		179	176		355	460
2600	24" wide		3.60	4.444		201	195		396	510
2610	30" wide		3	5.333		232	234		466	605
2620	36" wide		2.40	6.667		266	293		559	730
2640	24" radius, 6" wide		4.60	3.478		218	153		371	465
2660	12" wide		4	4		244	176		420	530
2670	18" wide		3.60	4.444		276	195		471	595
2680	24" wide		3	5.333		340	234		574	725
2690	30" wide		2.40	6.667		385	293		678	860
2700	36" wide		2.20	7.273		435	320		755	950
2720	36" radius, 6" wide		4.20	3.810		355	167		522	640
2740	12" wide		3.60	4.444		385	195		580	715
2750	18" wide		3.20	5		445	220		665	815
2760	24" wide		2.60	6.154		520	270		790	970
2770	30" wide		2	8		570	350		920	1,150
2780	36" wide		1.80	8.889		640	390		1,030	1,275
2800	Tee vertical, 12" radius, 6" wide		5	3.200		188	140		328	415
2820	12" wide		4.40	3.636		192	160		352	450
2830	18" wide		4.20	3.810		200	167		367	470
2840	24" wide		4	4		212	176		388	495
2850	30" wide		3.60	4.444		232	195		427	545
2860	36" wide		3	5.333		244	234		478	620
2880	24" radius, 6" wide		4.60	3.478		276	153		429	530
2900	12" wide		4	4		293	176		469	580
2910	18" wide		3.80	4.211		310	185		495	620
2920	24" wide		3.60	4.444		340	195		535	665
2930	30" wide		3.20	5		370	220		590	735
2940	36" wide		2.60	6.154		395	270		665	830
2960	36" radius, 6" wide		4.20	3.810		435	167		602	725
2980	12" wide		3.40	4.706		455	207		662	810
2990	18" wide		3.40	4.706		480	207		687	835
3000	24" wide		3.20	5		495	220		715	865
3010	30" wide		2.80	5.714		530	251		781	960
3020	36" wide		2.20	7.273		590	320		910	1,125
3040	Cross horizontal, 12" radius, 6" wide		4.40	3.636		168	160		328	420
3060	12" wide		4	4		189	176		365	470
3070	18" wide		3.40	4.706		215	207		422	545
3080	24" wide		2.80	5.714		244	251		495	645
3090	30" wide		2.60	6.154		276	270		546	705
3100	36" wide		2.20	7.273		305	320		625	810
3120	24" radius, 6" wide		4	4		293	176		469	580
3140	12" wide		3.60	4.444		340	195		535	665
3150	18" wide		3	5.333		370	234		604	755
3160	24" wide		2.40	6.667		445	293		738	925

26 05 36.20 Cable Tray Solid Bottom

		Crew	Daily Output	Labor-Hours	Unit	Material	2007 Bare Costs Labor	2007 Bare Costs Equipment	Total	Total Incl O&P
3170	30" wide	2 Elec	2.20	7.273	Ea.	495	320		815	1,025
3180	36" wide		1.80	8.889		510	390		900	1,150
3200	36" radius, 6" wide		3.60	4.444		510	195		705	855
3220	12" wide		3.20	5		540	220		760	920
3230	18" wide		2.60	6.154		600	270		870	1,050
3240	24" wide		2	8		655	350		1,005	1,250
3250	30" wide		1.80	8.889		705	390		1,095	1,350
3260	36" wide		1.60	10		825	440		1,265	1,575
3280	Dropout, or end plate, 6" wide		32	.500		16.90	22		38.90	51
3300	12" wide		26	.615		19.05	27		46.05	61
3310	18" wide		22	.727		23	32		55	72.50
3320	24" wide		20	.800		25	35		60	80
3330	30" wide		18	.889		29.50	39		68.50	90
3340	36" wide		16	1		34	44		78	103
3380	Reducer, 12" to 6" wide		14	1.143		82.50	50		132.50	166
3400	18" to 12" wide		12	1.333		85.50	58.50		144	181
3420	18" to 6" wide		12	1.333		85.50	58.50		144	181
3440	24" to 18" wide		10.60	1.509		90.50	66.50		157	198
3460	24" to 12" wide		10.60	1.509		90.50	66.50		157	198
3480	30" to 24" wide		9.20	1.739		98.50	76.50		175	222
3500	30" to 18" wide		9.20	1.739		98.50	76.50		175	222
3520	30" to 12" wide		9.20	1.739		98.50	76.50		175	222
3540	36" to 30" wide		8	2		104	88		192	245
3560	36" to 24" wide		8	2		105	88		193	246
3580	36" to 18" wide		8	2		105	88		193	246
3600	36" to 12" wide		8	2		108	88		196	249

26 05 36.30 Cable Tray Trough

		Crew	Daily Output	Labor-Hours	Unit	Material	2007 Bare Costs Labor	2007 Bare Costs Equipment	Total	Total Incl O&P
0010	**CABLE TRAY TROUGH** vented, w/ ftngs & supports, 6" dp, to 15' high									
0020	For higher elevations, see 26 05 36.40 9900									
0200	Galvanized steel, tray, 6" wide _(R260536-10)_	2 Elec	90	.178	L.F.	11.55	7.80		19.35	24.50
0240	12" wide		80	.200		14.10	8.80		22.90	28.50
0260	18" wide		70	.229		17	10.05		27.05	33.50
0280	24" wide _(R260536-11)_		60	.267		19.45	11.70		31.15	39
0300	30" wide		50	.320		28	14.05		42.05	51.50
0320	36" wide		40	.400		31	17.55		48.55	60
0340	Elbow horizontal 90°, 12" radius, 6" wide		7.60	2.105	Ea.	71	92.50		163.50	217
0360	12" wide		5.60	2.857		84.50	125		209.50	280
0370	18" wide		4.40	3.636		97.50	160		257.50	345
0380	24" wide		3.60	4.444		120	195		315	420
0390	30" wide		3.20	5		136	220		356	475
0400	36" wide		2.80	5.714		157	251		408	550
0420	24" radius, 6" wide		7.20	2.222		105	97.50		202.50	260
0440	12" wide		5.20	3.077		122	135		257	335
0450	18" wide		4	4		146	176		322	420
0460	24" wide		3.20	5		165	220		385	505
0470	30" wide		2.80	5.714		191	251		442	585
0480	36" wide		2.40	6.667		221	293		514	680
0500	36" radius, 6" wide		6.80	2.353		157	103		260	325
0520	12" wide		4.80	3.333		179	146		325	415
0530	18" wide		3.60	4.444		215	195		410	525
0540	24" wide		2.80	5.714		228	251		479	625
0550	30" wide		2.40	6.667		256	293		549	715

26 05 36.30 Cable Tray Trough		Crew	Daily Output	Labor-Hours	Unit	Material	2007 Bare Costs Labor	Equipment	Total	Total Incl O&P
0560	36" wide	2 Elec	2	8	Ea.	285	350		635	840
0580	Elbow vertical 90°, 12" radius, 6" wide		7.60	2.105		92	92.50		184.50	239
0600	12" wide		5.60	2.857		102	125		227	300
0610	18" wide		4.40	3.636		104	160		264	350
0620	24" wide		3.60	4.444		119	195		314	420
0630	30" wide		3.20	5		122	220		342	460
0640	36" wide		2.80	5.714		128	251		379	515
0660	24" radius, 6" wide		7.20	2.222		132	97.50		229.50	290
0680	12" wide		5.20	3.077		140	135		275	355
0690	18" wide		4	4		155	176		331	430
0700	24" wide		3.20	5		160	220		380	500
0710	30" wide		2.80	5.714		179	251		430	570
0720	36" wide		2.40	6.667		188	293		481	640
0740	36" radius, 6" wide		6.80	2.353		178	103		281	350
0760	12" wide		4.80	3.333		189	146		335	425
0770	18" wide		3.60	4.444		210	195		405	520
0780	24" wide		2.80	5.714		225	251		476	625
0790	30" wide		2.40	6.667		230	293		523	690
0800	36" wide		2	8		256	350		606	805
0820	Tee horizontal, 12" radius, 6" wide		4	4		108	176		284	380
0840	12" wide		3.20	5		120	220		340	455
0850	18" wide		2.80	5.714		136	251		387	525
0860	24" wide		2.40	6.667		157	293		450	610
0870	30" wide		2.20	7.273		179	320		499	670
0880	36" wide		2	8		197	350		547	740
0900	24" radius, 6" wide		3.60	4.444		178	195		373	485
0920	12" wide		2.80	5.714		195	251		446	590
0930	18" wide		2.40	6.667		218	293		511	675
0940	24" wide		2	8		282	350		632	835
0950	30" wide		1.80	8.889		305	390		695	915
0960	36" wide		1.60	10		340	440		780	1,025
0980	36" radius, 6" wide		3.20	5		256	220		476	605
1000	12" wide		2.40	6.667		292	293		585	755
1010	18" wide		2	8		320	350		670	875
1020	24" wide		1.60	10		385	440		825	1,075
1030	30" wide		1.40	11.429		405	500		905	1,200
1040	36" wide		1.20	13.333		480	585		1,065	1,400
1060	Tee vertical, 12" radius, 6" wide		4	4		179	176		355	460
1080	12" wide		3.20	5		182	220		402	525
1090	18" wide		3	5.333		187	234		421	555
1100	24" wide		2.80	5.714		197	251		448	590
1110	30" wide		2.60	6.154		209	270		479	630
1120	36" wide		2.20	7.273		212	320		532	710
1140	24" radius, 6" wide		3.60	4.444		241	195		436	555
1160	12" wide		2.80	5.714		253	251		504	655
1170	18" wide		2.60	6.154		265	270		535	690
1180	24" wide		2.40	6.667		285	293		578	750
1190	30" wide		2.20	7.273		300	320		620	805
1200	36" wide		1.80	8.889		325	390		715	940
1220	36" radius, 6" wide		3.20	5		385	220		605	745
1240	12" wide		2.40	6.667		395	293		688	865
1250	18" wide		2.20	7.273		405	320		725	925
1260	24" wide		2	8		430	350		780	1,000

26 05 36.30 Cable Tray Trough		Crew	Daily Output	Labor-Hours	Unit	Material	2007 Bare Costs Labor	Equipment	Total	Total Incl O&P
1270	30" wide	2 Elec	1.80	8.889	Ea.	500	390		890	1,125
1280	36" wide		1.40	11.429		505	500		1,005	1,300
1300	Cross horizontal, 12" radius, 6" wide		3.20	5		138	220		358	475
1320	12" wide		2.80	5.714		140	251		391	530
1330	18" wide		2.40	6.667		156	293		449	605
1340	24" wide		2	8		164	350		514	705
1350	30" wide		1.80	8.889		184	390		574	785
1360	36" wide		1.60	10		206	440		646	880
1380	24" radius, 6" wide		2.80	5.714		218	251		469	615
1400	12" wide		2.40	6.667		228	293		521	685
1410	18" wide		2	8		251	350		601	800
1420	24" wide		1.60	10		320	440		760	1,000
1430	30" wide		1.40	11.429		350	500		850	1,125
1440	36" wide		1.20	13.333		370	585		955	1,275
1460	36" radius, 6" wide		2.40	6.667		370	293		663	845
1480	12" wide		2	8		385	350		735	950
1490	18" wide		1.60	10		395	440		835	1,075
1500	24" wide		1.20	13.333		470	585		1,055	1,400
1510	30" wide		1	16		520	700		1,220	1,625
1520	36" wide		.80	20		560	880		1,440	1,925
1540	Dropout or end plate, 6" wide		26	.615		16	27		43	57.50
1560	12" wide		22	.727		19.60	32		51.60	69
1580	18" wide		20	.800		21.50	35		56.50	76
1600	24" wide		18	.889		24.50	39		63.50	85
1620	30" wide		16	1		26.50	44		70.50	94.50
1640	36" wide		13.40	1.194		29.50	52.50		82	110
1660	Reducer, 12" to 6" wide		9.40	1.702		67	74.50		141.50	185
1680	18" to 12" wide		8.40	1.905		70	83.50		153.50	201
1700	18" to 6" wide		8.40	1.905		70	83.50		153.50	201
1720	24" to 18" wide		7.20	2.222		74.50	97.50		172	227
1740	24" to 12" wide		7.20	2.222		74.50	97.50		172	227
1760	30" to 24" wide		6.40	2.500		77	110		187	248
1780	30" to 18" wide		6.40	2.500		77	110		187	248
1800	30" to 12" wide		6.40	2.500		77	110		187	248
1820	36" to 30" wide		5.80	2.759		82.50	121		203.50	271
1840	36" to 24" wide		5.80	2.759		82.50	121		203.50	271
1860	36" to 18" wide		5.80	2.759		84	121		205	272
1880	36" to 12" wide		5.80	2.759		84	121		205	272
2000	Aluminum, tray, vented, 6" wide		120	.133	L.F.	13.90	5.85		19.75	24
2010	9" wide		110	.145		15.70	6.40		22.10	27
2020	12" wide		100	.160		17.55	7		24.55	30
2030	18" wide		90	.178		21	7.80		28.80	34.50
2040	24" wide		80	.200		25	8.80		33.80	40.50
2050	30" wide		70	.229		33	10.05		43.05	51.50
2060	36" wide		60	.267		36	11.70		47.70	57.50
2080	Elbow horiz. 90°, 12" radius, 6" wide		7.60	2.105	Ea.	92	92.50		184.50	239
2090	9" wide		7	2.286		100	100		200	259
2100	12" wide		6.20	2.581		106	113		219	286
2110	18" wide		5.60	2.857		124	125		249	325
2120	24" wide		4.60	3.478		143	153		296	385
2130	30" wide		4	4		177	176		353	455
2140	36" wide		3.60	4.444		193	195		388	505
2160	24" radius, 6" wide		7.20	2.222		133	97.50		230.50	291

26 05 36 – Cable Trays for Electrical Systems

26 05 36.30 Cable Tray Trough		Crew	Daily Output	Labor-Hours	Unit	Material	2007 Bare Costs Labor	Equipment	Total	Total Incl O&P
2180	12" wide	2 Elec	5.80	2.759	Ea.	156	121		277	350
2190	18" wide		5.20	3.077		179	135		314	400
2200	24" wide		4.20	3.810		205	167		372	475
2210	30" wide		3.60	4.444		234	195		429	550
2220	36" wide		3.20	5		253	220		473	605
2240	36" radius, 6" wide		6.80	2.353		188	103		291	360
2260	12" wide		5.40	2.963		214	130		344	430
2270	18" wide		4.80	3.333		243	146		389	485
2280	24" wide		3.80	4.211		266	185		451	570
2290	30" wide		3.40	4.706		310	207		517	655
2300	36" wide		2.80	5.714		340	251		591	750
2320	Elbow vertical 90°, 12" radius, 6" wide		7.60	2.105		111	92.50		203.50	261
2330	9" wide		7	2.286		119	100		219	280
2340	12" wide		6.20	2.581		120	113		233	300
2350	18" wide		5.60	2.857		124	125		249	325
2360	24" wide		4.60	3.478		137	153		290	380
2370	30" wide		4	4		143	176		319	420
2380	36" wide		3.60	4.444		145	195		340	450
2400	24" radius, 6" wide		7.20	2.222		151	97.50		248.50	310
2420	12" wide		5.80	2.759		164	121		285	360
2430	18" wide		5.20	3.077		178	135		313	395
2440	24" wide		4.20	3.810		180	167		347	450
2450	30" wide		3.60	4.444		191	195		386	500
2460	36" wide		3.20	5		200	220		420	545
2480	36" radius, 6" wide		6.80	2.353		187	103		290	360
2500	12" wide		5.40	2.963		200	130		330	415
2510	18" wide		4.80	3.333		224	146		370	465
2520	24" wide		3.80	4.211		242	185		427	540
2530	30" wide		3.40	4.706		266	207		473	605
2540	36" wide		2.80	5.714		274	251		525	675
2560	Tee horizontal, 12" radius, 6" wide		4	4		140	176		316	415
2570	9" wide		3.80	4.211		143	185		328	435
2580	12" wide		3.60	4.444		156	195		351	460
2590	18" wide		3.20	5		180	220		400	525
2600	24" wide		2.80	5.714		207	251		458	605
2610	30" wide		2.40	6.667		223	293		516	680
2620	36" wide		2.20	7.273		253	320		573	755
2640	24" radius, 6" wide		3.60	4.444		212	195		407	525
2660	12" wide		3.20	5		242	220		462	590
2670	18" wide		2.80	5.714		266	251		517	670
2680	24" wide		2.40	6.667		335	293		628	800
2690	30" wide		2	8		360	350		710	920
2700	36" wide		1.80	8.889		420	390		810	1,050
2720	36" radius, 6" wide		3.20	5		340	220		560	700
2740	12" wide		2.80	5.714		395	251		646	805
2750	18" wide		2.40	6.667		445	293		738	925
2760	24" wide		2	8		520	350		870	1,100
2770	30" wide		1.60	10		555	440		995	1,275
2780	36" wide		1.40	11.429		640	500		1,140	1,450
2800	Tee vertical, 12" radius, 6" wide		4	4		192	176		368	470
2810	9" wide		3.80	4.211		193	185		378	490
2820	12" wide		3.60	4.444		207	195		402	520
2830	18" wide		3.40	4.706		209	207		416	540

26 05 36.30 Cable Tray Trough		Crew	Daily Output	Labor-Hours	Unit	Material	2007 Bare Costs Labor	Equipment	Total	Total Incl O&P
2840	24" wide	2 Elec	3.20	5	Ea.	214	220		434	560
2850	30" wide		3	5.333		224	234		458	595
2860	36" wide		2.60	6.154		233	270		503	655
2880	24" radius, 6" wide		3.60	4.444		271	195		466	590
2900	12" wide		3.20	5		293	220		513	645
2910	18" wide		3	5.333		310	234		544	695
2920	24" wide		2.80	5.714		335	251		586	740
2930	30" wide		2.60	6.154		360	270		630	795
2940	36" wide		2.20	7.273		365	320		685	880
2960	36" radius, 6" wide		3.20	5		420	220		640	785
2980	12" wide		2.80	5.714		445	251		696	865
2990	18" wide		2.60	6.154		460	270		730	905
3000	24" wide		2.40	6.667		495	293		788	975
3010	30" wide		2.20	7.273		530	320		850	1,050
3020	36" wide		1.80	8.889		560	390		950	1,200
3040	Cross horizontal, 12" radius, 6" wide		3.60	4.444		174	195		369	480
3050	9" wide		3.40	4.706		187	207		394	515
3060	12" wide		3.20	5		192	220		412	535
3070	18" wide		2.80	5.714		201	251		452	595
3080	24" wide		2.40	6.667		224	293		517	680
3090	30" wide		2.20	7.273		271	320		591	775
3100	36" wide		1.80	8.889		330	390		720	940
3120	24" radius, 6" wide		3.20	5		310	220		530	670
3140	12" wide		2.80	5.714		340	251		591	750
3150	18" wide		2.40	6.667		370	293		663	840
3160	24" wide		2	8		425	350		775	995
3170	30" wide		1.80	8.889		460	390		850	1,075
3180	36" wide		1.40	11.429		520	500		1,020	1,325
3200	36" radius, 6" wide		2.80	5.714		520	251		771	945
3220	12" wide		2.40	6.667		545	293		838	1,025
3230	18" wide		2	8		595	350		945	1,175
3240	24" wide		1.60	10		685	440		1,125	1,400
3250	30" wide		1.40	11.429		745	500		1,245	1,575
3260	36" wide		1.20	13.333		865	585		1,450	1,825
3280	Dropout, or end plate, 6" wide		26	.615		17.40	27		44.40	59
3300	12" wide		22	.727		20.50	32		52.50	70
3310	18" wide		20	.800		25	35		60	80
3320	24" wide		18	.889		30	39		69	91
3330	30" wide		16	1		32	44		76	101
3340	36" wide		14	1.143		36.50	50		86.50	115
3370	Reducer, 9" to 6" wide		12	1.333		85.50	58.50		144	181
3380	12" to 6" wide		11.40	1.404		89.50	61.50		151	190
3390	12" to 9" wide		11.40	1.404		89.50	61.50		151	190
3400	18" to 12" wide		9.60	1.667		96	73		169	215
3420	18" to 6" wide		9.60	1.667		96	73		169	215
3430	18" to 9" wide		9.60	1.667		96	73		169	215
3440	24" to 18" wide		8.40	1.905		102	83.50		185.50	237
3460	24" to 12" wide		8.40	1.905		102	83.50		185.50	237
3470	24" to 9" wide		8.40	1.905		104	83.50		187.50	238
3475	24" to 6" wide		8.40	1.905		104	83.50		187.50	238
3480	30" to 24" wide		7.20	2.222		105	97.50		202.50	260
3500	30" to 18" wide		7.20	2.222		105	97.50		202.50	260
3520	30" to 12" wide		7.20	2.222		108	97.50		205.50	263

26 05 36 – Cable Trays for Electrical Systems

26 05 36.30 Cable Tray Trough		Crew	Daily Output	Labor-Hours	Unit	Material	2007 Bare Costs Labor	2007 Bare Costs Equipment	Total	Total Incl O&P
3540	36" to 30" wide	2 Elec	6.40	2.500	Ea.	109	110		219	283
3560	36" to 24" wide		6.40	2.500		109	110		219	283
3580	36" to 18" wide		6.40	2.500		109	110		219	283
3600	36" to 12" wide		6.40	2.500		109	110		219	283
3610	Elbow horizontal 60°, 12" radius, 6" wide		7.80	2.051		74	90		164	216
3620	9" wide		7.20	2.222		81	97.50		178.50	235
3630	12" wide		6.40	2.500		89.50	110		199.50	262
3640	18" wide		5.80	2.759		96	121		217	286
3650	24" wide		4.80	3.333		118	146		264	350
3680	Elbow horizontal 45°, 12" radius, 6" wide		8	2		64	88		152	202
3690	9" wide		7.40	2.162		68	95		163	216
3700	12" wide		6.60	2.424		71.50	106		177.50	237
3710	18" wide		6	2.667		80.50	117		197.50	263
3720	24" wide		5	3.200		93.50	140		233.50	310
3750	Elbow horizontal, 30° 12" radius, 6" wide		8.20	1.951		56	85.50		141.50	190
3760	9" wide		7.60	2.105		58.50	92.50		151	203
3770	12" wide		6.80	2.353		62.50	103		165.50	223
3780	18" wide		6.20	2.581		68	113		181	244
3790	24" wide		5.20	3.077		74	135		209	283
3820	Elbow vertical 60° in/outside, 12" radius, 6" wide		7.80	2.051		92	90		182	235
3830	9" wide		7.20	2.222		93.50	97.50		191	248
3840	12" wide		6.40	2.500		94.50	110		204.50	267
3850	18" wide		5.80	2.759		98.50	121		219.50	288
3860	24" wide		4.80	3.333		102	146		248	330
3890	Elbow vertical 45° in/outside, 12" radius, 6" wide		8	2		74	88		162	213
3900	9" wide		7.40	2.162		78.50	95		173.50	228
3910	12" wide		6.60	2.424		81	106		187	248
3920	18" wide		6	2.667		84	117		201	266
3930	24" wide		5	3.200		92	140		232	310
3960	Elbow vertical 30° in/outside, 12" radius, 6" wide		8.20	1.951		64	85.50		149.50	199
3970	9" wide		7.60	2.105		68	92.50		160.50	213
3980	12" wide		6.80	2.353		69	103		172	230
3990	18" wide		6.20	2.581		72	113		185	248
4000	24" wide		5.20	3.077		73	135		208	282
4250	Reducer, left or right hand, 24" to 18" wide		8.40	1.905		97.50	83.50		181	231
4260	24" to 12" wide		8.40	1.905		97.50	83.50		181	231
4270	24" to 9" wide		8.40	1.905		97.50	83.50		181	231
4280	24" to 6" wide		8.40	1.905		98.50	83.50		182	232
4290	18" to 12" wide		9.60	1.667		91	73		164	209
4300	18" to 9" wide		9.60	1.667		91	73		164	209
4310	18" to 6" wide		9.60	1.667		91	73		164	209
4320	12" to 9" wide		11.40	1.404		87	61.50		148.50	187
4330	12" to 6" wide		11.40	1.404		87	61.50		148.50	187
4340	9" to 6" wide		12	1.333		84.50	58.50		143	180
4350	Splice plate	1 Elec	48	.167		8.05	7.30		15.35	19.75
4360	Splice plate, expansion joint		48	.167		8.25	7.30		15.55	20
4370	Splice plate, hinged, horizontal		48	.167		6.65	7.30		13.95	18.25
4380	Vertical		48	.167		9.45	7.30		16.75	21.50
4390	Trough, hanger, vertical		28	.286		26.50	12.55		39.05	48
4400	Box connector, 24" wide		20	.400		36	17.55		53.55	65.50
4410	18" wide		21	.381		30	16.70		46.70	58
4420	12" wide		22	.364		28.50	15.95		44.45	55.50
4430	9" wide		23	.348		27.50	15.25		42.75	52.50

26 05 36.30 Cable Tray Trough		Crew	Daily Output	Labor-Hours	Unit	Material	2007 Bare Costs Labor	Equipment	Total	Total Incl O&P
4440	6" wide	1 Elec	24	.333	Ea.	26	14.65		40.65	50.50
4450	Floor flange		24	.333		27.50	14.65		42.15	52
4460	Hold down clamp		60	.133		3.14	5.85		8.99	12.15
4520	Wall bracket, 24" wide tray		20	.400		26	17.55		43.55	54.50
4530	18" wide tray		21	.381		25	16.70		41.70	52.50
4540	12" wide tray		22	.364		13.45	15.95		29.40	39
4550	9" wide tray		23	.348		12	15.25		27.25	35.50
4560	6" wide tray		24	.333		10.90	14.65		25.55	34
5000	Cable channel aluminum, vented, 1-1/4" deep, 4" wide, straight		80	.100	L.F.	10.25	4.39		14.64	17.85
5010	Elbow horizontal, 36" radius, 90°		5	1.600	Ea.	185	70		255	310
5020	60°		5.50	1.455		141	64		205	251
5030	45°		6	1.333		115	58.50		173.50	214
5040	30°		6.50	1.231		99	54		153	190
5050	Adjustable		6	1.333		94.50	58.50		153	191
5060	Elbow vertical, 36" radius, 90°		5	1.600		198	70		268	325
5070	60°		5.50	1.455		155	64		219	266
5080	45°		6	1.333		128	58.50		186.50	227
5090	30°		6.50	1.231		113	54		167	205
5100	Adjustable		6	1.333		94.50	58.50		153	191
5110	Splice plate, hinged, horizontal		48	.167		7.05	7.30		14.35	18.70
5120	Splice plate, hinged, vertical		48	.167		10.05	7.30		17.35	22
5130	Hanger, vertical		28	.286		11.05	12.55		23.60	31
5140	Single		28	.286		16.80	12.55		29.35	37
5150	Double		20	.400		17.20	17.55		34.75	45
5160	Channel to box connector		24	.333		22.50	14.65		37.15	47
5170	Hold down clip		80	.100		3.19	4.39		7.58	10.05
5180	Wall bracket, single		28	.286		8.70	12.55		21.25	28
5190	Double		20	.400		11.15	17.55		28.70	38.50
5200	Cable roller		16	.500		137	22		159	183
5210	Splice plate		48	.167		4.20	7.30		11.50	15.50

26 05 36.40 Cable Tray, Covers and Dividers

26 05 36.40		Crew	Daily Output	Labor-Hours	Unit	Material	2007 Bare Costs Labor	Equipment	Total	Total Incl O&P
0010	**CABLE TRAY, COVERS AND DIVIDERS** To 15' high R260536-10									
0011	For higher elevations, see 26 05 36.40 9900									
0100	Covers, ventilated galv. steel, straight, 6" wide tray size	2 Elec	520	.031	L.F.	4.22	1.35		5.57	6.65
0200	9" wide tray size		460	.035		5.20	1.53		6.73	7.95
0300	12" wide tray size R260536-11		400	.040		6.20	1.76		7.96	9.45
0400	18" wide tray size		300	.053		8.45	2.34		10.79	12.80
0500	24" wide tray size		220	.073		10.35	3.19		13.54	16.15
0600	30" wide tray size		180	.089		12.55	3.90		16.45	19.60
0700	36" wide tray size		160	.100		14.35	4.39		18.74	22.50
1000	Elbow horizontal 90°, 12" radius, 6" wide tray size		150	.107	Ea.	31	4.68		35.68	41.50
1020	9" wide tray size		128	.125		34.50	5.50		40	46
1040	12" wide tray size		108	.148		36.50	6.50		43	50
1060	18" wide tray size		84	.190		50.50	8.35		58.85	68
1080	24" wide tray size		66	.242		61	10.65		71.65	83.50
1100	30" wide tray size		60	.267		76	11.70		87.70	101
1120	36" wide tray size		50	.320		92	14.05		106.05	122
1160	24" radius, 6" wide tray size		136	.118		52	5.15		57.15	64.50
1180	9" wide tray size		116	.138		53.50	6.05		59.55	67.50
1200	12" wide tray size		96	.167		59.50	7.30		66.80	76
1220	18" wide tray size		76	.211		73	9.25		82.25	94.50
1240	24" wide tray size		60	.267		89	11.70		100.70	115

26 05 36.40 Cable Tray, Covers and Dividers		Crew	Daily Output	Labor-Hours	Unit	Material	2007 Bare Costs Labor	Equipment	Total	Total Incl O&P
1260	30" wide tray size	2 Elec	52	.308	Ea.	117	13.50		130.50	149
1280	36" wide tray size		44	.364		136	15.95		151.95	173
1320	36" radius, 6" wide tray size		120	.133		76	5.85		81.85	92
1340	9" wide tray size		104	.154		84	6.75		90.75	102
1360	12" wide tray size		84	.190		89.50	8.35		97.85	111
1380	18" wide tray size		72	.222		114	9.75		123.75	140
1400	24" wide tray size		52	.308		136	13.50		149.50	169
1420	30" wide tray size		46	.348		161	15.25		176.25	200
1440	36" wide tray size		40	.400		189	17.55		206.55	234
1480	Elbow horizontal 45°, 12" radius, 6" wide tray size		150	.107		22.50	4.68		27.18	31.50
1500	9" wide tray size		128	.125		26.50	5.50		32	37.50
1520	12" wide tray size		108	.148		29	6.50		35.50	41.50
1540	18" wide tray size		88	.182		35.50	8		43.50	51
1560	24" wide tray size		76	.211		41	9.25		50.25	59.50
1580	30" wide tray size		66	.242		48.50	10.65		59.15	69.50
1600	36" wide tray size		60	.267		54.50	11.70		66.20	77.50
1640	24" radius, 6" wide tray size		136	.118		32.50	5.15		37.65	43.50
1660	9" wide tray size		116	.138		36.50	6.05		42.55	49.50
1680	12" wide tray size		96	.167		41	7.30		48.30	56.50
1700	18" wide tray size		80	.200		47	8.80		55.80	65
1720	24" wide tray size		70	.229		54.50	10.05		64.55	75
1740	30" wide tray size		60	.267		69.50	11.70		81.20	93.50
1760	36" wide tray size		52	.308		76	13.50		89.50	104
1800	36" radius, 6" wide tray size		120	.133		47	5.85		52.85	60.50
1820	9" wide tray size		104	.154		53.50	6.75		60.25	68.50
1840	12" wide tray size		84	.190		58.50	8.35		66.85	77
1860	18" wide tray size		76	.211		67	9.25		76.25	88
1880	24" wide tray size		62	.258		82	11.35		93.35	107
1900	30" wide tray size		52	.308		89	13.50		102.50	118
1920	36" wide tray size		48	.333		108	14.65		122.65	140
1960	Elbow vertical 90°, 12" radius, 6" wide tray size		150	.107		26	4.68		30.68	35.50
1980	9" wide tray size		128	.125		26.50	5.50		32	37.50
2000	12" wide tray size		108	.148		28.50	6.50		35	41
2020	18" wide tray size		88	.182		32.50	8		40.50	48
2040	24" wide tray size		68	.235		33.50	10.35		43.85	52
2060	30" wide tray size		60	.267		36.50	11.70		48.20	58
2080	36" wide tray size		50	.320		47	14.05		61.05	73
2120	24" radius, 6" wide tray size		136	.118		32	5.15		37.15	42.50
2140	9" wide tray size		116	.138		35.50	6.05		41.55	48
2160	12" wide tray size		96	.167		36.50	7.30		43.80	51.50
2180	18" wide tray size		80	.200		48.50	8.80		57.30	66.50
2200	24" wide tray size		62	.258		54.50	11.35		65.85	77
2220	30" wide tray size		52	.308		62	13.50		75.50	88
2240	36" wide tray size		44	.364		70.50	15.95		86.45	102
2280	36" radius, 6" wide tray size		120	.133		37.50	5.85		43.35	49.50
2300	9" wide tray size		104	.154		47	6.75		53.75	62
2320	12" wide tray size		84	.190		53.50	8.35		61.85	71
2340	18" wide tray size		76	.211		67	9.25		76.25	88
2350	24" wide tray size		54	.296		74.50	13		87.50	101
2360	30" wide tray size		46	.348		89	15.25		104.25	121
2370	36" wide tray size		40	.400		105	17.55		122.55	142
2400	Tee horizontal, 12" radius, 6" wide tray size		92	.174		47.50	7.65		55.15	63.50
2410	9" wide tray size		80	.200		48.50	8.80		57.30	66.50

26 05 36.40 Cable Tray, Covers and Dividers		Crew	Daily Output	Labor- Hours	Unit	Material	2007 Bare Costs Labor	Equipment	Total	Total Incl O&P
2420	12" wide tray size	2 Elec	68	.235	Ea.	55	10.35		65.35	76
2430	18" wide tray size		60	.267		66.50	11.70		78.20	90.50
2440	24" wide tray size		52	.308		84	13.50		97.50	112
2460	30" wide tray size		36	.444		98.50	19.50		118	137
2470	36" wide tray size		30	.533		119	23.50		142.50	166
2500	24" radius, 6" wide tray size		88	.182		74	8		82	93.50
2510	9" wide tray size		76	.211		85.50	9.25		94.75	108
2520	12" wide tray size		64	.250		88	11		99	113
2530	18" wide tray size		56	.286		113	12.55		125.55	143
2540	24" wide tray size		48	.333		173	14.65		187.65	212
2560	30" wide tray size		32	.500		197	22		219	250
2570	36" wide tray size		26	.615		218	27		245	279
2600	36" radius, 6" wide tray size		84	.190		134	8.35		142.35	160
2610	9" wide tray size		72	.222		137	9.75		146.75	166
2620	12" wide tray size		60	.267		152	11.70		163.70	185
2630	18" wide tray size		52	.308		178	13.50		191.50	216
2640	24" wide tray size		44	.364		234	15.95		249.95	282
2660	30" wide tray size		28	.571		255	25		280	320
2670	36" wide tray size		22	.727		289	32		321	370
2700	Cross horizontal, 12" radius, 6" wide tray size		68	.235		70.50	10.35		80.85	93
2710	9" wide tray size		64	.250		74.50	11		85.50	98.50
2720	12" wide tray size		60	.267		83	11.70		94.70	109
2730	18" wide tray size		52	.308		98.50	13.50		112	128
2740	24" wide tray size		36	.444		120	19.50		139.50	161
2760	30" wide tray size		30	.533		137	23.50		160.50	186
2770	36" wide tray size		28	.571		160	25		185	214
2800	24" radius, 6" wide tray size		64	.250		134	11		145	164
2810	9" wide tray size		60	.267		146	11.70		157.70	178
2820	12" wide tray size		56	.286		157	12.55		169.55	192
2830	18" wide tray size		48	.333		188	14.65		202.65	229
2840	24" wide tray size		32	.500		229	22		251	285
2860	30" wide tray size		26	.615		259	27		286	325
2870	36" wide tray size		24	.667		291	29.50		320.50	365
2900	36" radius, 6" wide tray size		60	.267		225	11.70		236.70	265
2910	9" wide tray size		56	.286		233	12.55		245.55	275
2920	12" wide tray size		52	.308		244	13.50		257.50	289
2930	18" wide tray size		44	.364		282	15.95		297.95	335
2940	24" wide tray size		28	.571		360	25		385	435
2960	30" wide tray size		22	.727		385	32		417	475
2970	36" wide tray size		20	.800		410	35		445	510
3000	Reducer, 9" to 6" wide tray size		128	.125		30	5.50		35.50	41
3010	12" to 6" wide tray size		108	.148		31	6.50		37.50	44
3020	12" to 9" wide tray size		108	.148		31	6.50		37.50	44
3030	18" to 12" wide tray size		88	.182		33.50	8		41.50	48.50
3050	18" to 6" wide tray size		88	.182		33.50	8		41.50	48.50
3060	24" to 18" wide tray size		80	.200		46.50	8.80		55.30	64.50
3070	24" to 12" wide tray size		80	.200		42.50	8.80		51.30	60
3090	30" to 24" wide tray size		70	.229		49.50	10.05		59.55	69
3100	30" to 18" wide tray size		70	.229		49.50	10.05		59.55	69
3110	30" to 12" wide tray size		70	.229		42.50	10.05		52.55	62
3140	36" to 30" wide tray size		64	.250		54	11		65	75.50
3150	36" to 24" wide tray size		64	.250		54	11		65	75.50
3160	36" to 18" wide tray size		64	.250		54	11		65	75.50

26 05 36.40 Cable Tray, Covers and Dividers		Crew	Daily Output	Labor-Hours	Unit	Material	2007 Bare Costs Labor	Equipment	Total	Total Incl O&P
3170	36" to 12" wide tray size	2 Elec	64	.250	Ea.	54	11		65	75.50
3250	Covers, aluminum, straight, 6" wide tray size		520	.031	L.F.	4.10	1.35		5.45	6.50
3270	9" wide tray size		460	.035		4.93	1.53		6.46	7.65
3290	12" wide tray size		400	.040		5.90	1.76		7.66	9.10
3310	18" wide tray size		320	.050		7.80	2.20		10	11.85
3330	24" wide tray size		260	.062		9.75	2.70		12.45	14.70
3350	30" wide tray size		200	.080		10.75	3.51		14.26	17.10
3370	36" wide tray size		180	.089		11.40	3.90		15.30	18.35
3400	Elbow horizontal 90°, 12" radius, 6" wide tray size		150	.107	Ea.	30.50	4.68		35.18	40.50
3410	9" wide tray size		128	.125		32	5.50		37.50	43
3420	12" wide tray size		108	.148		34.50	6.50		41	47.50
3430	18" wide tray size		88	.182		45.50	8		53.50	62
3440	24" wide tray size		70	.229		57	10.05		67.05	78
3460	30" wide tray size		64	.250		68.50	11		79.50	92
3470	36" wide tray size		54	.296		84	13		97	111
3500	24" radius, 6" wide tray size		136	.118		42	5.15		47.15	53.50
3510	9" wide tray size		116	.138		52	6.05		58.05	66
3520	12" wide tray size		96	.167		57	7.30		64.30	74
3530	18" wide tray size		80	.200		68	8.80		76.80	87.50
3540	24" wide tray size		64	.250		84	11		95	108
3560	30" wide tray size		56	.286		102	12.55		114.55	132
3570	36" wide tray size		48	.333		127	14.65		141.65	161
3600	36" radius, 6" wide tray size		120	.133		72	5.85		77.85	87.50
3610	9" wide tray size		104	.154		78.50	6.75		85.25	96.50
3620	12" wide tray size		84	.190		89	8.35		97.35	110
3630	18" wide tray size		76	.211		106	9.25		115.25	131
3640	24" wide tray size		56	.286		129	12.55		141.55	161
3660	30" wide tray size		50	.320		148	14.05		162.05	184
3670	36" wide tray size		44	.364		174	15.95		189.95	215
3700	Elbow horizontal 45°, 12" radius, 6" wide tray size		150	.107		22.50	4.68		27.18	32
3710	9" wide tray size		128	.125		23.50	5.50		29	33.50
3720	12" wide tray size		108	.148		26	6.50		32.50	38
3730	18" wide tray size		88	.182		30	8		38	45
3740	24" wide tray size		80	.200		33.50	8.80		42.30	49.50
3760	30" wide tray size		70	.229		42	10.05		52.05	61
3770	36" wide tray size		64	.250		50.50	11		61.50	72
3800	24" radius, 6" wide tray size		136	.118		26	5.15		31.15	36.50
3810	9" wide tray size		116	.138		33.50	6.05		39.55	45.50
3820	12" wide tray size		96	.167		34.50	7.30		41.80	49
3830	18" wide tray size		80	.200		42	8.80		50.80	59
3840	24" wide tray size		72	.222		52	9.75		61.75	71.50
3860	30" wide tray size		64	.250		60	11		71	82.50
3870	36" wide tray size		56	.286		69.50	12.55		82.05	94.50
3900	36" radius, 6" wide tray size		120	.133		45.50	5.85		51.35	58.50
3910	9" wide tray size		104	.154		49.50	6.75		56.25	64
3920	12" wide tray size		84	.190		52	8.35		60.35	69.50
3930	18" wide tray size		76	.211		62.50	9.25		71.75	83
3940	24" wide tray size		64	.250		76	11		87	100
3960	30" wide tray size		56	.286		86.50	12.55		99.05	114
3970	36" wide tray size		50	.320		98.50	14.05		112.55	129
4000	Elbow vertical 90°, 12" radius, 6" wide tray size		150	.107		26	4.68		30.68	35.50
4010	9" wide tray size		128	.125		26	5.50		31.50	37
4020	12" wide tray size		108	.148		28.50	6.50		35	41

26 05 36.40 Cable Tray, Covers and Dividers	Crew	Daily Output	Labor-Hours	Unit	Material	2007 Bare Costs Labor	Equipment	Total	Total Incl O&P
4030	18" wide tray size	2 Elec	88	.182	Ea.	33.50	8		41.50	48.50
4040	24" wide tray size		70	.229		34.50	10.05		44.55	53
4060	30" wide tray size		64	.250		35.50	11		46.50	55.50
4070	36" wide tray size		54	.296		42.50	13		55.50	66.50
4100	24" radius, 6" wide tray size		136	.118		30	5.15		35.15	40.50
4110	9" wide tray size		116	.138		32.50	6.05		38.55	45
4120	12" wide tray size		96	.167		35.50	7.30		42.80	50
4130	18" wide tray size		80	.200		42.50	8.80		51.30	60
4140	24" wide tray size		64	.250		46.50	11		57.50	68
4160	30" wide tray size		56	.286		60	12.55		72.55	84.50
4170	36" wide tray size		48	.333		65.50	14.65		80.15	94
4200	36" radius, 6" wide tray size		120	.133		34.50	5.85		40.35	46.50
4210	9" wide tray size		104	.154		42.50	6.75		49.25	57
4220	12" wide tray size		84	.190		49.50	8.35		57.85	66.50
4230	18" wide tray size		76	.211		60	9.25		69.25	80
4240	24" wide tray size		56	.286		70.50	12.55		83.05	96
4260	30" wide tray size		50	.320		89	14.05		103.05	119
4270	36" wide tray size		44	.364		96	15.95		111.95	129
4300	Tee horizontal, 12" radius, 6" wide tray size		108	.148		42.50	6.50		49	56.50
4310	9" wide tray size		88	.182		45.50	8		53.50	62
4320	12" wide tray size		80	.200		50.50	8.80		59.30	68.50
4330	18" wide tray size		68	.235		60	10.35		70.35	81.50
4340	24" wide tray size		56	.286		76	12.55		88.55	102
4360	30" wide tray size		44	.364		89	15.95		104.95	122
4370	36" wide tray size		36	.444		108	19.50		127.50	148
4400	24" radius, 6" wide tray size		96	.167		69.50	7.30		76.80	87
4410	9" wide tray size		80	.200		77	8.80		85.80	98
4420	12" wide tray size		72	.222		86.50	9.75		96.25	110
4430	18" wide tray size		60	.267		101	11.70		112.70	128
4440	24" wide tray size		48	.333		161	14.65		175.65	199
4460	30" wide tray size		40	.400		178	17.55		195.55	222
4470	36" wide tray size		32	.500		198	22		220	251
4500	36" radius, 6" wide tray size		88	.182		127	8		135	151
4510	9" wide tray size		72	.222		129	9.75		138.75	157
4520	12" wide tray size		64	.250		142	11		153	172
4530	18" wide tray size		56	.286		161	12.55		173.55	196
4540	24" wide tray size		44	.364		205	15.95		220.95	249
4560	30" wide tray size		36	.444		230	19.50		249.50	282
4570	36" wide tray size		28	.571		260	25		285	325
4600	Cross horizontal, 12" radius, 6" wide tray size		80	.200		65.50	8.80		74.30	85
4610	9" wide tray size		72	.222		69	9.75		78.75	90.50
4620	12" wide tray size		64	.250		77	11		88	101
4630	18" wide tray size		56	.286		93.50	12.55		106.05	122
4640	24" wide tray size		48	.333		108	14.65		122.65	140
4660	30" wide tray size		40	.400		128	17.55		145.55	167
4670	36" wide tray size		32	.500		147	22		169	195
4700	24" radius, 6" wide tray size		72	.222		128	9.75		137.75	156
4710	9" wide tray size		64	.250		137	11		148	167
4720	12" wide tray size		56	.286		147	12.55		159.55	181
4730	18" wide tray size		48	.333		172	14.65		186.65	211
4740	24" wide tray size		40	.400		205	17.55		222.55	251
4760	30" wide tray size		32	.500		239	22		261	296
4770	36" wide tray size		24	.667		260	29.50		289.50	330

26 05 36 – Cable Trays for Electrical Systems

26 05 36.40 Cable Tray, Covers and Dividers	Crew	Daily Output	Labor-Hours	Unit	Material	2007 Bare Costs Labor	Equipment	Total	Total Incl O&P
4800 36" radius, 6" wide tray size	2 Elec	64	.250	Ea.	205	11		216	241
4810 9" wide tray size		56	.286		214	12.55		226.55	254
4820 12" wide tray size		50	.320		230	14.05		244.05	274
4830 18" wide tray size		44	.364		253	15.95		268.95	305
4840 24" wide tray size		36	.444		310	19.50		329.50	375
4860 30" wide tray size		28	.571		360	25		385	435
4870 36" wide tray size		22	.727		395	32		427	480
4900 Reducer, 9" to 6" wide tray size		128	.125		32.50	5.50		38	44
4910 12" to 6" wide tray size		108	.148		34	6.50		40.50	47
4920 12" to 9" wide tray size		108	.148		34	6.50		40.50	47
4930 18" to 12" wide tray size		88	.182		37.50	8		45.50	53
4950 18" to 6" wide tray size		88	.182		37.50	8		45.50	53
4960 24" to 18" wide tray size		80	.200		48	8.80		56.80	65.50
4970 24" to 12" wide tray size		80	.200		41	8.80		49.80	58.50
4990 30" to 24" wide tray size		70	.229		50	10.05		60.05	70
5000 30" to 18" wide tray size		70	.229		50	10.05		60.05	70
5010 30" to 12" wide tray size		70	.229		50	10.05		60.05	70
5040 36" to 30" wide tray size		64	.250		55.50	11		66.50	77.50
5050 36" to 24" wide tray size		64	.250		55.50	11		66.50	77.50
5060 36" to 18" wide tray size		64	.250		55.50	11		66.50	77.50
5070 36" to 12" wide tray size		64	.250		55.50	11		66.50	77.50
5710 Tray cover hold down clamp	1 Elec	60	.133		8.85	5.85		14.70	18.40
8000 Divider strip, straight, galvanized, 3" deep		200	.040	L.F.	3.71	1.76		5.47	6.70
8020 4" deep		180	.044		4.54	1.95		6.49	7.90
8040 6" deep		160	.050		5.95	2.20		8.15	9.80
8060 Aluminum, straight, 3" deep		210	.038		3.71	1.67		5.38	6.55
8080 4" deep		190	.042		4.61	1.85		6.46	7.80
8100 6" deep		170	.047		5.90	2.07		7.97	9.60
8110 Divider strip vertical fitting 3" deep									
8120 12" radius, galvanized, 30°	1 Elec	28	.286	Ea.	18	12.55		30.55	38.50
8140 45°		27	.296		22.50	13		35.50	44.50
8160 60°		26	.308		24	13.50		37.50	46.50
8180 90°		25	.320		30	14.05		44.05	54
8200 Aluminum, 30°		29	.276		15.25	12.10		27.35	35
8220 45°		28	.286		18	12.55		30.55	38.50
8240 60°		27	.296		21	13		34	42.50
8260 90°		26	.308		26.50	13.50		40	49.50
8280 24" radius, galvanized, 30°		25	.320		27.50	14.05		41.55	51
8300 45°		24	.333		30.50	14.65		45.15	55.50
8320 60°		23	.348		38.50	15.25		53.75	65
8340 90°		22	.364		52.50	15.95		68.45	82
8360 Aluminum, 30°		26	.308		25.50	13.50		39	48
8380 45°		25	.320		29.50	14.05		43.55	53
8400 60°		24	.333		36.50	14.65		51.15	62.50
8420 90°		23	.348		50	15.25		65.25	77.50
8440 36" radius, galvanized, 30°		22	.364		36.50	15.95		52.45	64.50
8460 45°		21	.381		42.50	16.70		59.20	72
8480 60°		20	.400		50	17.55		67.55	81
8500 90°		19	.421		69.50	18.50		88	104
8520 Aluminum, 30°		23	.348		38.50	15.25		53.75	65
8540 45°		22	.364		50	15.95		65.95	79
8560 60°		21	.381		64.50	16.70		81.20	96
8570 90°		20	.400		84.50	17.55		102.05	119

26 05 36 – Cable Trays for Electrical Systems

26 05 36.40 Cable Tray, Covers and Dividers	Crew	Daily Output	Labor-Hours	Unit	Material	2007 Bare Costs Labor	Equipment	Total	Total Incl O&P
8590 Divider strip vertical fitting 4" deep									
8600 12" radius, galvanized, 30°	1 Elec	27	.296	Ea.	23	13		36	44.50
8610 45°		26	.308		26.50	13.50		40	49.50
8620 60°		25	.320		30	14.05		44.05	54
8630 90°		24	.333		36.50	14.65		51.15	62.50
8640 Aluminum, 30°		28	.286		21.50	12.55		34.05	42
8650 45°		27	.296		25	13		38	47
8660 60°		26	.308		28.50	13.50		42	51.50
8670 90°		25	.320		34	14.05		48.05	58.50
8680 24" radius, galvanized, 30°		24	.333		36.50	14.65		51.15	62.50
8690 45°		23	.348		46	15.25		61.25	73
8700 60°		22	.364		52	15.95		67.95	81
8710 90°		21	.381		69.50	16.70		86.20	101
8720 Aluminum, 30°		25	.320		34	14.05		48.05	58.50
8730 45°		24	.333		41	14.65		55.65	67.50
8740 60°		23	.348		48.50	15.25		63.75	76
8750 90°		22	.364		66.50	15.95		82.45	97
8760 36" radius, galvanized, 30°		23	.348		42.50	15.25		57.75	69.50
8770 45°		22	.364		48.50	15.95		64.45	77.50
8780 60°		21	.381		61	16.70		77.70	92.50
8790 90°		20	.400		83	17.55		100.55	118
8800 Aluminum, 30°		24	.333		52	14.65		66.65	79
8810 45°		23	.348		66.50	15.25		81.75	95.50
8820 60°		22	.364		81	15.95		96.95	114
8830 90°		21	.381		102	16.70		118.70	138
8840 Divider strip vertical fitting 6" deep									
8850 12" radius, galvanized, 30°	1 Elec	24	.333	Ea.	25	14.65		39.65	49.50
8860 45°		23	.348		28	15.25		43.25	53
8870 60°		22	.364		32	15.95		47.95	59
8880 90°		21	.381		40	16.70		56.70	69
8890 Aluminum, 30°		25	.320		23.50	14.05		37.55	47
8900 45°		24	.333		28	14.65		42.65	52.50
8910 60°		23	.348		29.50	15.25		44.75	54.50
8920 90°		22	.364		34.50	15.95		50.45	62
8930 24" radius, galvanized, 30°		23	.348		36.50	15.25		51.75	63
8940 45°		22	.364		46	15.95		61.95	74.50
8950 60°		21	.381		52.50	16.70		69.20	83
8960 90°		20	.400		69.50	17.55		87.05	102
8970 Aluminum, 30°		24	.333		34.50	14.65		49.15	60
8980 45°		23	.348		47	15.25		62.25	74.50
8990 60°		22	.364		52	15.95		67.95	81
9000 90°		21	.381		70.50	16.70		87.20	103
9010 36" radius, galvanized, 30°		22	.364		42.50	15.95		58.45	71
9020 45°		21	.381		52.50	16.70		69.20	83
9030 60°		20	.400		69.50	17.55		87.05	102
9040 90°		19	.421		89	18.50		107.50	126
9050 Aluminum, 30°		23	.348		52.50	15.25		67.75	80.50
9060 45°		22	.364		70.50	15.95		86.45	102
9070 60°		21	.381		83	16.70		99.70	117
9080 90°		20	.400		100	17.55		117.55	136
9120 Divider strip, horizontal fitting, galvanized, 3" deep		33	.242		25.50	10.65		36.15	44
9130 4" deep		30	.267		28.50	11.70		40.20	49
9140 6" deep		27	.296		36.50	13		49.50	60

26 05 Common Work Results for Electrical

26 05 36 – Cable Trays for Electrical Systems

26 05 36.40 Cable Tray, Covers and Dividers

		Crew	Daily Output	Labor-Hours	Unit	Material	2007 Bare Costs Labor	Equipment	Total	Total Incl O&P
9150	Aluminum, 3" deep	1 Elec	35	.229	Ea.	24	10.05		34.05	41.50
9160	4" deep		32	.250		26.50	11		37.50	46
9170	6" deep		29	.276		35.50	12.10		47.60	57
9300	Divider strip protector		300	.027	L.F.	2.62	1.17		3.79	4.62
9310	Fastener, ladder tray				Ea.	.47			.47	.52
9320	Trough or solid bottom tray				"	.32			.32	.35
9899										
9900	Add to labor for higher elevated installation									
9910	15' to 20' high add						10%			
9920	20' to 25' high add						20%			
9930	25' to 30' high add						25%			
9940	30' to 35' high add						30%			
9960	Over 40' high add						40%			

26 05 39 – Underfloor Raceways for Electrical Systems

26 05 39.30 Conduit In Concrete Slab

		Crew	Daily Output	Labor-Hours	Unit	Material	2007 Bare Costs Labor	Equipment	Total	Total Incl O&P
0010	**CONDUIT IN CONCRETE SLAB** Including terminations,									
0020	fittings and supports									
3230	PVC, schedule 40, 1/2" diameter	1 Elec	270	.030	L.F.	.73	1.30		2.03	2.75
3250	3/4" diameter		230	.035		.88	1.53		2.41	3.24
3270	1" diameter		200	.040		1.20	1.76		2.96	3.93
3300	1-1/4" diameter		170	.047		1.68	2.07		3.75	4.93
3330	1-1/2" diameter		140	.057		2.02	2.51		4.53	5.95
3350	2" diameter		120	.067		2.60	2.93		5.53	7.20
3370	2-1/2" diameter		90	.089		4.35	3.90		8.25	10.60
3400	3" diameter	2 Elec	160	.100		5.40	4.39		9.79	12.50
3430	3-1/2" diameter		120	.133		7.10	5.85		12.95	16.55
3440	4" diameter		100	.160		7.90	7		14.90	19.15
3450	5" diameter		80	.200		11.80	8.80		20.60	26
3460	6" diameter		60	.267		16.55	11.70		28.25	35.50
3530	Sweeps, 1" diameter, 30" radius	1 Elec	32	.250	Ea.	21.50	11		32.50	40
3550	1-1/4" diameter		24	.333		27	14.65		41.65	51.50
3570	1-1/2" diameter		21	.381		28	16.70		44.70	56
3600	2" diameter		18	.444		29	19.50		48.50	61
3630	2-1/2" diameter		14	.571		50	25		75	92.50
3650	3" diameter		10	.800		56.50	35		91.50	115
3670	3-1/2" diameter		8	1		85	44		129	159
3700	4" diameter		7	1.143		81	50		131	164
3710	5" diameter		6	1.333		124	58.50		182.50	224
3730	Couplings, 1/2" diameter					.41			.41	.45
3750	3/4" diameter					.50			.50	.55
3770	1" diameter					.77			.77	.85
3800	1-1/4" diameter					1.02			1.02	1.12
3830	1-1/2" diameter					1.40			1.40	1.54
3850	2" diameter					1.86			1.86	2.05
3870	2-1/2" diameter					3.30			3.30	3.63
3900	3" diameter					5.40			5.40	5.95
3930	3-1/2" diameter					6			6	6.60
3950	4" diameter					8.35			8.35	9.20
3960	5" diameter					21			21	23.50
3970	6" diameter					27			27	30
4030	End bells 1" diameter, PVC	1 Elec	60	.133		5.30	5.85		11.15	14.50
4050	1-1/4" diameter		53	.151		6.60	6.65		13.25	17.10

R260539-30

26 05 Common Work Results for Electrical

26 05 39 – Underfloor Raceways for Electrical Systems

26 05 39.30 Conduit In Concrete Slab

		Crew	Daily Output	Labor-Hours	Unit	Material	2007 Bare Costs Labor	Equipment	Total	Total Incl O&P
4100	1-1/2" diameter	1 Elec	48	.167	Ea.	6.60	7.30		13.90	18.15
4150	2" diameter		34	.235		9.70	10.35		20.05	26
4170	2-1/2" diameter		27	.296		10.75	13		23.75	31
4200	3" diameter		20	.400		11.35	17.55		28.90	38.50
4250	3-1/2" diameter		16	.500		12.45	22		34.45	46
4300	4" diameter		14	.571		13.55	25		38.55	52.50
4310	5" diameter		12	.667		21	29.50		50.50	67
4320	6" diameter		9	.889	↓	23.50	39		62.50	83.50
4350	Rigid galvanized steel, 1/2" diameter		200	.040	L.F.	2.26	1.76		4.02	5.10
4400	3/4" diameter		170	.047		2.61	2.07		4.68	5.95
4450	1" diameter		130	.062		3.67	2.70		6.37	8.05
4500	1-1/4" diameter		110	.073		4.96	3.19		8.15	10.20
4600	1-1/2" diameter		100	.080		5.80	3.51		9.31	11.60
4800	2" diameter	↓	90	.089	↓	7.35	3.90		11.25	13.90

26 05 39.40 Conduit In Trench

		Crew	Daily Output	Labor-Hours	Unit	Material	2007 Bare Costs Labor	Equipment	Total	Total Incl O&P
0010	**CONDUIT IN TRENCH** Includes terminations and fittings R260539-40									
0020	Does not include excavation or backfill, see div. 31 23 16.00									
0200	Rigid galvanized steel, 2" diameter	1 Elec	150	.053	L.F.	7.05	2.34		9.39	11.25
0400	2-1/2" diameter	"	100	.080		13.75	3.51		17.26	20.50
0600	3" diameter	2 Elec	160	.100		17.05	4.39		21.44	25.50
0800	3-1/2" diameter		140	.114		22	5		27	31.50
1000	4" diameter		100	.160		24	7		31	37
1200	5" diameter		80	.200		52	8.80		60.80	70
1400	6" diameter	↓	60	.267	↓	76	11.70		87.70	101

26 05 43 – Underground Ducts and Raceways for Electrical Systems

26 05 43.10 Trench Duct

		Crew	Daily Output	Labor-Hours	Unit	Material	2007 Bare Costs Labor	Equipment	Total	Total Incl O&P
0010	**TRENCH DUCT** Steel with cover									
0020	Standard adjustable, depths to 4"									
0100	Straight, single compartment, 9" wide	2 Elec	40	.400	L.F.	82	17.55		99.55	116
0200	12" wide		32	.500		93	22		115	135
0400	18" wide		26	.615		124	27		151	176
0600	24" wide		22	.727		159	32		191	223
0700	27" wide		21	.762		171	33.50		204.50	238
0800	30" wide		20	.800		192	35		227	264
1000	36" wide		16	1		231	44		275	320
1020	Two compartment, 9" wide		38	.421		93	18.50		111.50	130
1030	12" wide		30	.533		106	23.50		129.50	152
1040	18" wide		24	.667		130	29.50		159.50	187
1050	24" wide		20	.800		170	35		205	240
1060	30" wide		18	.889		207	39		246	286
1070	36" wide		14	1.143		241	50		291	340
1090	Three compartment, 9" wide		36	.444		106	19.50		125.50	146
1100	12" wide		28	.571		117	25		142	167
1110	18" wide		22	.727		147	32		179	210
1120	24" wide		18	.889		185	39		224	262
1130	30" wide		16	1		223	44		267	310
1140	36" wide	↓	12	1.333	↓	261	58.50		319.50	375
1200	Horizontal elbow, 9" wide		5.40	2.963	Ea.	305	130		435	530
1400	12" wide		4.60	3.478		350	153		503	610
1600	18" wide		4	4		450	176		626	755
1800	24" wide		3.20	5		630	220		850	1,025
1900	27" wide		3	5.333		715	234		949	1,125

26 05 Common Work Results for Electrical

26 05 43 – Underground Ducts and Raceways for Electrical Systems

26 05 43.10 Trench Duct		Crew	Daily Output	Labor-Hours	Unit	Material	2007 Bare Costs Labor	Equipment	Total	Total Incl O&P
2000	30" wide	2 Elec	2.60	6.154	Ea.	840	270		1,110	1,325
2200	36" wide		2.40	6.667		1,100	293		1,393	1,625
2220	Two compartment, 9" wide		3.80	4.211		490	185		675	815
2230	12" wide		3	5.333		540	234		774	945
2240	18" wide		2.40	6.667		650	293		943	1,150
2250	24" wide		2	8		825	350		1,175	1,425
2260	30" wide		1.80	8.889		1,100	390		1,490	1,800
2270	36" wide		1.60	10		1,350	440		1,790	2,125
2290	Three compartment, 9" wide		3.60	4.444		515	195		710	855
2300	12" wide		2.80	5.714		575	251		826	1,000
2310	18" wide		2.20	7.273		700	320		1,020	1,250
2320	24" wide		1.80	8.889		885	390		1,275	1,550
2330	30" wide		1.60	10		1,150	440		1,590	1,925
2350	36" wide		1.40	11.429		1,400	500		1,900	2,300
2400	Vertical elbow, 9" wide		5.40	2.963		106	130		236	310
2600	12" wide		4.60	3.478		115	153		268	355
2800	18" wide		4	4		132	176		308	405
3000	24" wide		3.20	5		164	220		384	505
3100	27" wide		3	5.333		171	234		405	540
3200	30" wide		2.60	6.154		181	270		451	600
3400	36" wide		2.40	6.667		199	293		492	655
3600	Cross, 9" wide		4	4		495	176		671	805
3800	12" wide		3.20	5		525	220		745	905
4000	18" wide		2.60	6.154		630	270		900	1,100
4200	24" wide		2.20	7.273		800	320		1,120	1,350
4300	27" wide		2.20	7.273		940	320		1,260	1,500
4400	30" wide		2	8		1,025	350		1,375	1,675
4600	36" wide		1.80	8.889		1,300	390		1,690	2,000
4620	Two compartment, 9" wide		3.80	4.211		520	185		705	845
4630	12" wide		3	5.333		545	234		779	950
4640	18" wide		2.40	6.667		660	293		953	1,150
4650	24" wide		2	8		840	350		1,190	1,450
4660	30" wide		1.80	8.889		1,100	390		1,490	1,800
4670	36" wide		1.60	10		1,350	440		1,790	2,125
4690	Three compartment, 9" wide		3.60	4.444		525	195		720	870
4700	12" wide		2.80	5.714		600	251		851	1,025
4710	18" wide		2.20	7.273		715	320		1,035	1,250
4720	24" wide		1.80	8.889		890	390		1,280	1,550
4730	30" wide		1.60	10		1,175	440		1,615	1,925
4740	36" wide		1.40	11.429		1,425	500		1,925	2,325
4800	End closure, 9" wide		14.40	1.111		31	49		80	107
5000	12" wide		12	1.333		35.50	58.50		94	126
5200	18" wide		10	1.600		54.50	70		124.50	165
5400	24" wide		8	2		72	88		160	210
5500	27" wide		7	2.286		83	100		183	241
5600	30" wide		6.60	2.424		90	106		196	257
5800	36" wide		5.80	2.759		107	121		228	298
6000	Tees, 9" wide		4	4		300	176		476	590
6200	12" wide		3.60	4.444		350	195		545	675
6400	18" wide		3.20	5		450	220		670	820
6600	24" wide		3	5.333		645	234		879	1,050
6700	27" wide		2.80	5.714		715	251		966	1,150
6800	30" wide		2.60	6.154		840	270		1,110	1,325

26 05 43.10 Trench Duct

		Crew	Daily Output	Labor-Hours	Unit	Material	2007 Bare Costs Labor	Equipment	Total	Total Incl O&P
7000	36" wide	2 Elec	2	8	Ea.	1,100	350		1,450	1,725
7020	Two compartment, 9" wide		3.80	4.211		350	185		535	660
7030	12" wide		3.40	4.706		375	207		582	725
7040	18" wide		3	5.333		505	234		739	905
7050	24" wide		2.80	5.714		680	251		931	1,125
7060	30" wide		2.40	6.667		925	293		1,218	1,450
7070	36" wide		1.90	8.421		1,150	370		1,520	1,825
7090	Three compartment, 9" wide		3.60	4.444		400	195		595	730
7100	12" wide		3.20	5		420	220		640	785
7110	18" wide		2.80	5.714		535	251		786	960
7120	24" wide		2.60	6.154		735	270		1,005	1,200
7130	30" wide		2.20	7.273		960	320		1,280	1,525
7140	36" wide		1.80	8.889		1,225	390		1,615	1,925
7200	Riser, and cabinet connector, 9" wide		5.40	2.963		132	130		262	340
7400	12" wide		4.60	3.478		154	153		307	395
7600	18" wide		4	4		189	176		365	470
7800	24" wide		3.20	5		229	220		449	575
7900	27" wide		3	5.333		236	234		470	610
8000	30" wide		2.60	6.154		263	270		533	690
8200	36" wide	▼	2	8		305	350		655	860
8400	Insert assembly, cell to conduit adapter, 1-1/4"	1 Elec	16	.500	▼	52.50	22		74.50	90
8500	Adjustable partition	"	320	.025	L.F.	18.60	1.10		19.70	22
8600	Depth of duct over 4", per 1", add					8.10			8.10	8.90
8700	Support post	1 Elec	240	.033		18.55	1.46		20.01	22.50
8800	Cover double tile trim, 2 sides					28.50			28.50	31.50
8900	4 sides					85			85	93.50
9160	Trench duct 3-1/2" x 4-1/2", add					7.70			7.70	8.45
9170	Trench duct 4" x 5", add					7.70			7.70	8.45
9200	For carpet trim, add					26			26	28.50
9210	For double carpet trim, add				▼	79.50			79.50	87.50

26 05 43.20 Underfloor Duct

		Crew	Daily Output	Labor-Hours	Unit	Material	2007 Bare Costs Labor	Equipment	Total	Total Incl O&P
0010	**UNDERFLOOR DUCT** R260543-50									
0020										
0100	Duct, 1-3/8" x 3-1/8" blank, standard	2 Elec	160	.100	L.F.	10	4.39		14.39	17.55
0200	1-3/8" x 7-1/4" blank, super duct		120	.133		20	5.85		25.85	30.50
0400	7/8" or 1-3/8" insert type, 24" O.C., 1-3/8" x 3-1/8", std.		140	.114		13.40	5		18.40	22
0600	1-3/8" x 7-1/4", super duct	▼	100	.160	▼	23.50	7		30.50	36
0800	Junction box, single duct, 1 level, 3-1/8"	1 Elec	4	2	Ea.	300	88		388	460
0820	3-1/8" x 7-1/4"		4	2		350	88		438	515
0840	2 level, 3-1/8" upper & lower		3.20	2.500		350	110		460	550
0860	3-1/8" upper, 7-1/4" lower		2.70	2.963		350	130		480	580
0880	Carpet pan for above		80	.100		249	4.39		253.39	281
0900	Terrazzo pan for above		67	.119		580	5.25		585.25	650
1000	Junction box, single duct, 1 level, 7-1/4"		2.70	2.963		350	130		480	580
1020	2 level, 7-1/4" upper & lower		2.70	2.963		400	130		530	635
1040	2 duct, two 3-1/8" upper & lower		3.20	2.500		530	110		640	750
1200	1 level, 2 duct, 3-1/8"		3.20	2.500		400	110		510	605
1220	Carpet pan for above boxes		80	.100		249	4.39		253.39	281
1240	Terrazzo pan for above boxes		67	.119		580	5.25		585.25	650
1260	Junction box, 1 level, two 3-1/8" x one 3-1/8" + one 7-1/4"		2.30	3.478		680	153		833	975
1280	2 level, two 3-1/8" upper, one 3-1/8" + one 7-1/4" lower		2	4		750	176		926	1,075
1300	Carpet pan for above boxes	▼	80	.100	▼	249	4.39		253.39	281

26 05 Common Work Results for Electrical

26 05 43 – Underground Ducts and Raceways for Electrical Systems

26 05 43.20 Underfloor Duct

		Crew	Daily Output	Labor-Hours	Unit	Material	2007 Bare Costs Labor	Equipment	Total	Total Incl O&P
1320	Terrazzo pan for above boxes	1 Elec	67	.119	Ea.	580	5.25		585.25	650
1400	Junction box, 1 level, 2 duct, 7-1/4"		2.30	3.478		1,025	153		1,178	1,350
1420	Two 3-1/8" + one 7-1/4"		2	4		1,025	176		1,201	1,375
1440	Carpet pan for above		80	.100		249	4.39		253.39	281
1460	Terrazzo pan for above		67	.119		580	5.25		585.25	650
1580	Junction box, 1 level, one 3-1/8" + one 7-1/4" x same		2.30	3.478		680	153		833	975
1600	Triple duct, 3-1/8"		2.30	3.478		680	153		833	975
1700	Junction box, 1 level, one 3-1/8" + two 7-1/4"		2	4		1,150	176		1,326	1,525
1720	Carpet pan for above		80	.100		249	4.39		253.39	281
1740	Terrazzo pan for above		67	.119		580	5.25		585.25	650
1800	Insert to conduit adapter, 3/4" & 1"		32	.250		24.50	11		35.50	43.50
2000	Support, single cell		27	.296		37	13		50	60
2200	Super duct		16	.500		37	22		59	73
2400	Double cell		16	.500		37	22		59	73
2600	Triple cell		11	.727		37	32		69	88
2800	Vertical elbow, standard duct		10	.800		66.50	35		101.50	126
3000	Super duct		8	1		66.50	44		110.50	139
3200	Cabinet connector, standard duct		32	.250		50	11		61	71.50
3400	Super duct		27	.296		50	13		63	74.50
3600	Conduit adapter, 1" to 1-1/4"		32	.250		50	11		61	71.50
3800	2" to 1-1/4"		27	.296		60	13		73	85.50
4000	Outlet, low tension (tele, computer, etc.)		8	1		70	44		114	143
4200	High tension, receptacle (120 volt)		8	1		70	44		114	143
4300	End closure, standard duct		160	.050		2.80	2.20		5	6.35
4310	Super duct		160	.050		5.30	2.20		7.50	9.10
4350	Elbow, horiz., standard duct		26	.308		167	13.50		180.50	204
4360	Super duct		26	.308		167	13.50		180.50	204
4380	Elbow, offset, standard duct		26	.308		66.50	13.50		80	93
4390	Super duct		26	.308		66.50	13.50		80	93
4400	Marker screw assembly for inserts		50	.160		11.70	7		18.70	23.50
4410	Y take off, standard duct		26	.308		100	13.50		113.50	130
4420	Super duct		26	.308		100	13.50		113.50	130
4430	Box opening plug, standard duct		160	.050		9.35	2.20		11.55	13.55
4440	Super duct		160	.050		9.35	2.20		11.55	13.55
4450	Sleeve coupling, standard duct		160	.050		25	2.20		27.20	31
4460	Super duct		160	.050		31	2.20		33.20	37.50
4470	Conduit adapter, standard duct, 3/4"		32	.250		50	11		61	71.50
4480	1" or 1-1/4"		32	.250		50	11		61	71.50
4500	1-1/2"		32	.250		50	11		61	71.50

26 05 80 – Wiring Connections

26 05 80.10 Motor Connections

		Crew	Daily Output	Labor-Hours	Unit	Material	2007 Bare Costs Labor	Equipment	Total	Total Incl O&P
0010	**MOTOR CONNECTIONS**									
0020	Flexible conduit and fittings, 115 volt, 1 phase, up to 1 HP motor	1 Elec	8	1	Ea.	8.75	44		52.75	75
0050	2 HP motor R260580-75		6.50	1.231		9	54		63	90.50
0100	3 HP motor		5.50	1.455		13.35	64		77.35	110
0120	230 volt, 10 HP motor, 3 phase		4.20	1.905		14.40	83.50		97.90	140
0150	15 HP motor		3.30	2.424		26.50	106		132.50	187
0200	25 HP motor		2.70	2.963		29.50	130		159.50	227
0400	50 HP motor		2.20	3.636		76	160		236	320
0600	100 HP motor		1.50	5.333		183	234		417	550
1500	460 volt, 5 HP motor, 3 phase		8	1		9.50	44		53.50	76
1520	10 HP motor		8	1		9.50	44		53.50	76

26 05 80 – Wiring Connections

26 05 80.10 Motor Connections		Crew	Daily Output	Labor-Hours	Unit	Material	2007 Bare Costs Labor	Equipment	Total	Total Incl O&P
1530	25 HP motor	1 Elec	6	1.333	Ea.	15.85	58.50		74.35	104
1540	30 HP motor		6	1.333		15.85	58.50		74.35	104
1550	40 HP motor		5	1.600		25	70		95	133
1560	50 HP motor		5	1.600		28.50	70		98.50	136
1570	60 HP motor		3.80	2.105		31.50	92.50		124	173
1580	75 HP motor		3.50	2.286		43	100		143	197
1590	100 HP motor		2.50	3.200		63.50	140		203.50	279
1600	125 HP motor		2	4		85.50	176		261.50	355
1610	150 HP motor		1.80	4.444		89.50	195		284.50	390
1620	200 HP motor		1.50	5.333		144	234		378	510
2005	460 Volt, 5 HP motor, 3 Phase, w/sealtite		8	1		12.80	44		56.80	79.50
2010	10 HP motor		8	1		12.80	44		56.80	79.50
2015	25 HP motor		6	1.333		23	58.50		81.50	112
2020	30 HP motor		6	1.333		23	58.50		81.50	112
2025	40 HP motor		5	1.600		43.50	70		113.50	153
2030	50 HP motor		5	1.600		42	70		112	152
2035	60 HP motor		3.80	2.105		64	92.50		156.50	208
2040	75 HP motor		3.50	2.286		69.50	100		169.50	226
2045	100 HP motor		2.50	3.200		93	140		233	310
2055	150 HP motor		1.80	4.444		101	195		296	400
2060	200 HP motor		1.50	5.333		645	234		879	1,050

26 05 90 – Residential Wiring

26 05 90.10 Residential Wiring		Crew	Daily Output	Labor-Hours	Unit	Material	2007 Bare Costs Labor	Equipment	Total	Total Incl O&P
0010	**RESIDENTIAL WIRING**									
0020	20' avg. runs and #14/2 wiring incl. unless otherwise noted									
1000	Service & panel, includes 24' SE-AL cable, service eye, meter,									
1010	Socket, panel board, main bkr., ground rod, 15 or 20 amp									
1020	1-pole circuit breakers, and misc. hardware									
1100	100 amp, with 10 branch breakers	1 Elec	1.19	6.723	Ea.	515	295		810	1,000
1110	With PVC conduit and wire		.92	8.696		575	380		955	1,200
1120	With RGS conduit and wire		.73	10.959		735	480		1,215	1,525
1150	150 amp, with 14 branch breakers		1.03	7.767		800	340		1,140	1,400
1170	With PVC conduit and wire		.82	9.756		925	430		1,355	1,675
1180	With RGS conduit and wire		.67	11.940		1,250	525		1,775	2,125
1200	200 amp, with 18 branch breakers	2 Elec	1.80	8.889		1,050	390		1,440	1,725
1220	With PVC conduit and wire		1.46	10.959		1,175	480		1,655	2,000
1230	With RGS conduit and wire		1.24	12.903		1,600	565		2,165	2,600
1800	Lightning surge suppressor for above services, add	1 Elec	32	.250		46	11		57	67
2000	Switch devices									
2100	Single pole, 15 amp, Ivory, with a 1-gang box, cover plate,									
2110	Type NM (Romex) cable	1 Elec	17.10	.468	Ea.	12.65	20.50		33.15	44.50
2120	Type MC (BX) cable		14.30	.559		26.50	24.50		51	65.50
2130	EMT & wire		5.71	1.401		32	61.50		93.50	127
2150	3-way, #14/3, type NM cable		14.55	.550		17.35	24		41.35	55
2170	Type MC cable		12.31	.650		37	28.50		65.50	83
2180	EMT & wire		5	1.600		35.50	70		105.50	144
2200	4-way, #14/3, type NM cable		14.55	.550		31.50	24		55.50	70.50
2220	Type MC cable		12.31	.650		51	28.50		79.50	98.50
2230	EMT & wire		5	1.600		49.50	70		119.50	160
2250	S.P., 20 amp, #12/2, type NM cable		13.33	.600		21	26.50		47.50	62
2270	Type MC cable		11.43	.700		31	30.50		61.50	80
2280	EMT & wire		4.85	1.649		41.50	72.50		114	154

26 05 90.10 Residential Wiring	Crew	Daily Output	Labor-Hours	Unit	Material	2007 Bare Costs Labor	Equipment	Total	Total Incl O&P
2290	S.P. rotary dimmer, 600W, no wiring	1 Elec	17	.471	Ea.	17.45	20.50		37.95	50
2300	S.P. rotary dimmer, 600W, type NM cable		14.55	.550		25	24		49	63.50
2320	Type MC cable		12.31	.650		39	28.50		67.50	85.50
2330	EMT & wire		5	1.600		46.50	70		116.50	156
2350	3-way rotary dimmer, type NM cable		13.33	.600		23	26.50		49.50	64.50
2370	Type MC cable		11.43	.700		37	30.50		67.50	86
2380	EMT & wire		4.85	1.649		44.50	72.50		117	157
2400	Interval timer wall switch, 20 amp, 1-30 min., #12/2									
2410	Type NM cable	1 Elec	14.55	.550	Ea.	46	24		70	86.50
2420	Type MC cable		12.31	.650		51	28.50		79.50	98.50
2430	EMT & wire		5	1.600		66.50	70		136.50	178
2500	Decorator style									
2510	S.P., 15 amp, type NM cable	1 Elec	17.10	.468	Ea.	16.75	20.50		37.25	49
2520	Type MC cable		14.30	.559		30.50	24.50		55	70
2530	EMT & wire		5.71	1.401		36	61.50		97.50	131
2550	3-way, #14/3, type NM cable		14.55	.550		21.50	24		45.50	59.50
2570	Type MC cable		12.31	.650		41	28.50		69.50	87.50
2580	EMT & wire		5	1.600		39.50	70		109.50	149
2600	4-way, #14/3, type NM cable		14.55	.550		35.50	24		59.50	75
2620	Type MC cable		12.31	.650		55	28.50		83.50	103
2630	EMT & wire		5	1.600		53.50	70		123.50	164
2650	S.P., 20 amp, #12/2, type NM cable		13.33	.600		25	26.50		51.50	66.50
2670	Type MC cable		11.43	.700		35.50	30.50		66	84.50
2680	EMT & wire		4.85	1.649		45.50	72.50		118	158
2700	S.P., slide dimmer, type NM cable		17.10	.468		31	20.50		51.50	65
2720	Type MC cable		14.30	.559		45	24.50		69.50	86
2730	EMT & wire		5.71	1.401		52.50	61.50		114	150
2750	S.P., touch dimmer, type NM cable		17.10	.468		28	20.50		48.50	61.50
2770	Type MC cable		14.30	.559		42	24.50		66.50	82.50
2780	EMT & wire		5.71	1.401		49.50	61.50		111	146
2800	3-way touch dimmer, type NM cable		13.33	.600		48	26.50		74.50	92
2820	Type MC cable		11.43	.700		62	30.50		92.50	114
2830	EMT & wire		4.85	1.649		69.50	72.50		142	185
3000	Combination devices									
3100	S.P. switch/15 amp recpt., Ivory, 1-gang box, plate									
3110	Type NM cable	1 Elec	11.43	.700	Ea.	23	30.50		53.50	71
3120	Type MC cable		10	.800		37	35		72	93
3130	EMT & wire		4.40	1.818		44	80		124	168
3150	S.P. switch/pilot light, type NM cable		11.43	.700		23.50	30.50		54	71.50
3170	Type MC cable		10	.800		37.50	35		72.50	94
3180	EMT & wire		4.43	1.806		45	79.50		124.50	168
3190	2-S.P. switches, 2-#14/2, no wiring		14	.571		7.35	25		32.35	45.50
3200	2-S.P. switches, 2-#14/2, type NM cables		10	.800		29.50	35		64.50	85
3220	Type MC cable		8.89	.900		51	39.50		90.50	115
3230	EMT & wire		4.10	1.951		49.50	85.50		135	183
3250	3-way switch/15 amp recpt., #14/3, type NM cable		10	.800		31.50	35		66.50	87
3270	Type MC cable		8.89	.900		51	39.50		90.50	115
3280	EMT & wire		4.10	1.951		49.50	85.50		135	183
3300	2-3 way switches, 2-#14/3, type NM cables		8.89	.900		44.50	39.50		84	108
3320	Type MC cable		8	1		77	44		121	150
3330	EMT & wire		4	2		58	88		146	195
3350	S.P. switch/20 amp recpt., #12/2, type NM cable		10	.800		35.50	35		70.50	91.50
3370	Type MC cable		8.89	.900		40	39.50		79.50	103

26 05 90.10 Residential Wiring	Crew	Daily Output	Labor-Hours	Unit	Material	2007 Bare Costs Labor	Equipment	Total	Total Incl O&P
3380	EMT & wire	1 Elec	4.10	1.951	Ea.	55.50	85.50		141	190
3400	Decorator style									
3410	S.P. switch/15 amp recpt., type NM cable	1 Elec	11.43	.700	Ea.	27	30.50		57.50	75
3420	Type MC cable		10	.800		41	35		76	97.50
3430	EMT & wire		4.40	1.818		48.50	80		128.50	172
3450	S.P. switch/pilot light, type NM cable		11.43	.700		28	30.50		58.50	76
3470	Type MC cable		10	.800		41.50	35		76.50	98.50
3480	EMT & wire		4.40	1.818		49	80		129	173
3500	2-S.P. switches, 2-#14/2, type NM cables		10	.800		34	35		69	89.50
3520	Type MC cable		8.89	.900		55	39.50		94.50	120
3530	EMT & wire		4.10	1.951		53.50	85.50		139	187
3550	3-way/15 amp recpt., #14/3, type NM cable		10	.800		35.50	35		70.50	91.50
3570	Type MC cable		8.89	.900		55	39.50		94.50	120
3580	EMT & wire		4.10	1.951		53.50	85.50		139	187
3650	2-3 way switches, 2-#14/3, type NM cables		8.89	.900		48.50	39.50		88	113
3670	Type MC cable		8	1		81	44		125	155
3680	EMT & wire		4	2		62.50	88		150.50	200
3700	S.P. switch/20 amp recpt., #12/2, type NM cable		10	.800		39.50	35		74.50	96
3720	Type MC cable		8.89	.900		44	39.50		83.50	108
3730	EMT & wire		4.10	1.951		60	85.50		145.50	194
4000	Receptacle devices									
4010	Duplex outlet, 15 amp recpt., Ivory, 1-gang box, plate									
4015	Type NM cable	1 Elec	14.55	.550	Ea.	11	24		35	48
4020	Type MC cable		12.31	.650		25	28.50		53.50	70
4030	EMT & wire		5.33	1.501		30.50	66		96.50	132
4050	With #12/2, type NM cable		12.31	.650		14.95	28.50		43.45	59
4070	Type MC cable		10.67	.750		25	33		58	76.50
4080	EMT & wire		4.71	1.699		35.50	74.50		110	150
4100	20 amp recpt., #12/2, type NM cable		12.31	.650		23	28.50		51.50	68
4120	Type MC cable		10.67	.750		33	33		66	85.50
4130	EMT & wire		4.71	1.699		43.50	74.50		118	159
4140	For GFI see line 4300 below									
4150	Decorator style, 15 amp recpt., type NM cable	1 Elec	14.55	.550	Ea.	15.10	24		39.10	52.50
4170	Type MC cable		12.31	.650		29	28.50		57.50	74.50
4180	EMT & wire		5.33	1.501		34.50	66		100.50	136
4200	With #12/2, type NM cable		12.31	.650		19.05	28.50		47.55	63.50
4220	Type MC cable		10.67	.750		29	33		62	81
4230	EMT & wire		4.71	1.699		39.50	74.50		114	155
4250	20 amp recpt. #12/2, type NM cable		12.31	.650		27	28.50		55.50	72.50
4270	Type MC cable		10.67	.750		37.50	33		70.50	90
4280	EMT & wire		4.71	1.699		47.50	74.50		122	164
4300	GFI, 15 amp recpt., type NM cable		12.31	.650		39.50	28.50		68	86
4320	Type MC cable		10.67	.750		53.50	33		86.50	108
4330	EMT & wire		4.71	1.699		58.50	74.50		133	176
4350	GFI with #12/2, type NM cable		10.67	.750		43.50	33		76.50	96.50
4370	Type MC cable		9.20	.870		53.50	38		91.50	116
4380	EMT & wire		4.21	1.900		64	83.50		147.50	194
4400	20 amp recpt., #12/2 type NM cable		10.67	.750		45	33		78	98.50
4420	Type MC cable		9.20	.870		55	38		93	118
4430	EMT & wire		4.21	1.900		65.50	83.50		149	196
4500	Weather-proof cover for above receptacles, add		32	.250		4.55	11		15.55	21.50
4550	Air conditioner outlet, 20 amp-240 volt recpt.									
4560	30' of #12/2, 2 pole circuit breaker									

26 05 Common Work Results for Electrical

26 05 90 – Residential Wiring

26 05 90.10 Residential Wiring	Crew	Daily Output	Labor-Hours	Unit	Material	2007 Bare Costs Labor	2007 Bare Costs Equipment	Total	Total Incl O&P
4570	Type NM cable	1 Elec	10	.800	Ea.	59.50	35		94.50	118
4580	Type MC cable		9	.889		71	39		110	137
4590	EMT & wire		4	2		78.50	88		166.50	218
4600	Decorator style, type NM cable		10	.800		63.50	35		98.50	123
4620	Type MC cable		9	.889		75.50	39		114.50	141
4630	EMT & wire	↓	4	2	↓	83	88		171	222
4650	Dryer outlet, 30 amp-240 volt recpt., 20' of #10/3									
4660	2 pole circuit breaker									
4670	Type NM cable	1 Elec	6.41	1.248	Ea.	78.50	55		133.50	168
4680	Type MC cable		5.71	1.401		76	61.50		137.50	175
4690	EMT & wire	↓	3.48	2.299	↓	81.50	101		182.50	240
4700	Range outlet, 50 amp-240 volt recpt., 30' of #8/3									
4710	Type NM cable	1 Elec	4.21	1.900	Ea.	116	83.50		199.50	251
4720	Type MC cable		4	2		160	88		248	305
4730	EMT & wire		2.96	2.703		114	119		233	300
4750	Central vacuum outlet, Type NM cable		6.40	1.250		72	55		127	161
4770	Type MC cable		5.71	1.401		86.50	61.50		148	187
4780	EMT & wire	↓	3.48	2.299	↓	88.50	101		189.50	248
4800	30 amp-110 volt locking recpt., #10/2 circ. bkr.									
4810	Type NM cable	1 Elec	6.20	1.290	Ea.	83	56.50		139.50	176
4820	Type MC cable		5.40	1.481		101	65		166	208
4830	EMT & wire	↓	3.20	2.500	↓	101	110		211	274
4900	Low voltage outlets									
4910	Telephone recpt., 20' of 4/C phone wire	1 Elec	26	.308	Ea.	9.35	13.50		22.85	30.50
4920	TV recpt., 20' of RG59U coax wire, F type connector	"	16	.500	"	16.25	22		38.25	50.50
4950	Door bell chime, transformer, 2 buttons, 60' of bellwire									
4970	Economy model	1 Elec	11.50	.696	Ea.	60	30.50		90.50	112
4980	Custom model		11.50	.696		93.50	30.50		124	149
4990	Luxury model, 3 buttons	↓	9.50	.842	↓	243	37		280	320
6000	Lighting outlets									
6050	Wire only (for fixture), type NM cable	1 Elec	32	.250	Ea.	9.65	11		20.65	27
6070	Type MC cable		24	.333		17.85	14.65		32.50	41.50
6080	EMT & wire		10	.800		22	35		57	76.50
6100	Box (4"), and wire (for fixture), type NM cable		25	.320		17.60	14.05		31.65	40.50
6120	Type MC cable		20	.400		26	17.55		43.55	54.50
6130	EMT & wire	↓	11	.727	↓	29.50	32		61.50	80
6200	Fixtures (use with lines 6050 or 6100 above)									
6210	Canopy style, economy grade	1 Elec	40	.200	Ea.	27	8.80		35.80	42.50
6220	Custom grade		40	.200		50	8.80		58.80	68
6250	Dining room chandelier, economy grade		19	.421		81	18.50		99.50	117
6260	Custom grade		19	.421		239	18.50		257.50	291
6270	Luxury grade		15	.533		530	23.50		553.50	620
6310	Kitchen fixture (fluorescent), economy grade		30	.267		57	11.70		68.70	80
6320	Custom grade		25	.320		165	14.05		179.05	203
6350	Outdoor, wall mounted, economy grade		30	.267		28	11.70		39.70	48.50
6360	Custom grade		30	.267		106	11.70		117.70	134
6370	Luxury grade		25	.320		240	14.05		254.05	285
6410	Outdoor PAR floodlights, 1 lamp, 150 watt		20	.400		27	17.55		44.55	55.50
6420	2 lamp, 150 watt each		20	.400		44	17.55		61.55	74.50
6430	For infrared security sensor, add		32	.250		92	11		103	117
6450	Outdoor, quartz-halogen, 300 watt flood		20	.400		40	17.55		57.55	70
6600	Recessed downlight, round, pre-wired, 50 or 75 watt trim		30	.267		37	11.70		48.70	58
6610	With shower light trim	↓	30	.267	↓	46	11.70		57.70	68

191

26 05 90.10 Residential Wiring	Crew	Daily Output	Labor-Hours	Unit	Material	2007 Bare Costs Labor	Equipment	Total	Total Incl O&P
6620 With wall washer trim	1 Elec	28	.286	Ea.	55	12.55		67.55	79
6630 With eye-ball trim	↓	28	.286		55	12.55		67.55	79
6640 For direct contact with insulation, add					1.85			1.85	2.04
6700 Porcelain lamp holder	1 Elec	40	.200		3.80	8.80		12.60	17.25
6710 With pull switch		40	.200		4.13	8.80		12.93	17.60
6750 Fluorescent strip, 1-20 watt tube, wrap around diffuser, 24"		24	.333		53	14.65		67.65	80.50
6760 1-40 watt tube, 48"		24	.333		65	14.65		79.65	93.50
6770 2-40 watt tubes, 48"		20	.400		79	17.55		96.55	113
6780 With residential ballast		20	.400		89.50	17.55		107.05	125
6800 Bathroom heat lamp, 1-250 watt		28	.286		41	12.55		53.55	63.50
6810 2-250 watt lamps	↓	28	.286	↓	67	12.55		79.55	92
6820 For timer switch, see line 2400									
6900 Outdoor post lamp, incl. post, fixture, 35' of #14/2									
6910 Type NMC cable	1 Elec	3.50	2.286	Ea.	195	100		295	365
6920 Photo-eye, add		27	.296		32	13		45	54.50
6950 Clock dial time switch, 24 hr., w/enclosure, type NM cable		11.43	.700		60	30.50		90.50	112
6970 Type MC cable		11	.727		74	32		106	129
6980 EMT & wire	↓	4.85	1.649	↓	79.50	72.50		152	195
7000 Alarm systems									
7050 Smoke detectors, box, #14/3, type NM cable	1 Elec	14.55	.550	Ea.	35.50	24		59.50	75
7070 Type MC cable		12.31	.650		49.50	28.50		78	97
7080 EMT & wire	↓	5	1.600		48.50	70		118.50	158
7090 For relay output to security system, add				↓	12.90			12.90	14.20
8000 Residential equipment									
8050 Disposal hook-up, incl. switch, outlet box, 3' of flex									
8060 20 amp-1 pole circ. bkr., and 25' of #12/2									
8070 Type NM cable	1 Elec	10	.800	Ea.	33.50	35		68.50	89
8080 Type MC cable		8	1		44.50	44		88.50	115
8090 EMT & wire	↓	5	1.600	↓	56	70		126	167
8100 Trash compactor or dishwasher hook-up, incl. outlet box,									
8110 3' of flex, 15 amp-1 pole circ. bkr., and 25' of #14/2									
8120 Type NM cable	1 Elec	10	.800	Ea.	23.50	35		58.50	78.50
8130 Type MC cable	↓	8	1		39.50	44		83.50	109
8140 EMT & wire	↓	5	1.600	↓	47.50	70		117.50	157
8150 Hot water sink dispensor hook-up, use line 8100									
8200 Vent/exhaust fan hook-up, type NM cable	1 Elec	32	.250	Ea.	9.65	11		20.65	27
8220 Type MC cable		24	.333		17.85	14.65		32.50	41.50
8230 EMT & wire	↓	10	.800	↓	22	35		57	76.50
8250 Bathroom vent fan, 50 CFM (use with above hook-up)									
8260 Economy model	1 Elec	15	.533	Ea.	22.50	23.50		46	60
8270 Low noise model		15	.533		32	23.50		55.50	70
8280 Custom model	↓	12	.667	↓	117	29.50		146.50	173
8300 Bathroom or kitchen vent fan, 110 CFM									
8310 Economy model	1 Elec	15	.533	Ea.	60.50	23.50		84	102
8320 Low noise model	"	15	.533	"	79	23.50		102.50	122
8350 Paddle fan, variable speed (w/o lights)									
8360 Economy model (AC motor)	1 Elec	10	.800	Ea.	105	35		140	169
8370 Custom model (AC motor)		10	.800		182	35		217	253
8380 Luxury model (DC motor)		8	1		360	44		404	460
8390 Remote speed switch for above, add	↓	12	.667	↓	26	29.50		55.50	72
8500 Whole house exhaust fan, ceiling mount, 36", variable speed									
8510 Remote switch, incl. shutters, 20 amp-1 pole circ. bkr.									
8520 30' of #12/2, type NM cable	1 Elec	4	2	Ea.	750	88		838	955

26 05 Common Work Results for Electrical

26 05 90 – Residential Wiring

26 05 90.10 Residential Wiring

		Crew	Daily Output	Labor-Hours	Unit	Material	2007 Bare Costs Labor	Equipment	Total	Total Incl O&P
8530	Type MC cable	1 Elec	3.50	2.286	Ea.	765	100		865	990
8540	EMT & wire	↓	3	2.667	↓	780	117		897	1,025
8600	Whirlpool tub hook-up, incl. timer switch, outlet box									
8610	3' of flex, 20 amp-1 pole GFI circ. bkr.									
8620	30' of #12/2, type NM cable	1 Elec	5	1.600	Ea.	116	70		186	233
8630	Type MC cable		4.20	1.905		121	83.50		204.50	257
8640	EMT & wire	↓	3.40	2.353	↓	131	103		234	298
8650	Hot water heater hook-up, incl. 1-2 pole circ. bkr., box;									
8660	3' of flex, 20' of #10/2, type NM cable	1 Elec	5	1.600	Ea.	37	70		107	146
8670	Type MC cable		4.20	1.905		52	83.50		135.50	181
8680	EMT & wire	↓	3.40	2.353	↓	50	103		153	209
9000	Heating/air conditioning									
9050	Furnace/boiler hook-up, incl. firestat, local on-off switch									
9060	Emergency switch, and 40' of type NM cable	1 Elec	4	2	Ea.	55.50	88		143.50	193
9070	Type MC cable		3.50	2.286		77	100		177	234
9080	EMT & wire	↓	1.50	5.333	↓	87	234		321	445
9100	Air conditioner hook-up, incl. local 60 amp disc. switch									
9110	3' sealtite, 40 amp, 2 pole circuit breaker									
9130	40' of #8/2, type NM cable	1 Elec	3.50	2.286	Ea.	210	100		310	380
9140	Type MC cable		3	2.667		291	117		408	495
9150	EMT & wire	↓	1.30	6.154	↓	236	270		506	660
9200	Heat pump hook-up, 1-40 & 1-100 amp 2 pole circ. bkr.									
9210	Local disconnect switch, 3' sealtite									
9220	40' of #8/2 & 30' of #3/2									
9230	Type NM cable	1 Elec	1.30	6.154	Ea.	610	270		880	1,075
9240	Type MC cable		1.08	7.407		720	325		1,045	1,275
9250	EMT & wire	↓	.94	8.511	↓	645	375		1,020	1,275
9500	Thermostat hook-up, using low voltage wire									
9520	Heating only	1 Elec	24	.333	Ea.	7.10	14.65		21.75	30
9530	Heating/cooling	"	20	.400	"	8.60	17.55		26.15	35.50

26 09 Instrumentation and Control for Electrical Systems

26 09 13 – Electrical Power Monitoring and Control

26 09 13.10 Switchboard Instruments

		Crew	Daily Output	Labor-Hours	Unit	Material	2007 Bare Costs Labor	Equipment	Total	Total Incl O&P
0010	**SWITCHBOARD INSTRUMENTS** 3 phase, 4 wire R260913-80									
0100	AC indicating, ammeter & switch	1 Elec	8	1	Ea.	1,775	44		1,819	2,050
0200	Voltmeter & switch		8	1		1,775	44		1,819	2,050
0300	Wattmeter		8	1		3,525	44		3,569	3,975
0400	AC recording, ammeter		4	2		6,300	88		6,388	7,050
0500	Voltmeter		4	2		6,300	88		6,388	7,050
0600	Ground fault protection, zero sequence		2.70	2.963		5,550	130		5,680	6,325
0700	Ground return path		2.70	2.963		5,550	130		5,680	6,325
0800	3 current transformers, 5 to 800 amp		2	4		2,600	176		2,776	3,100
0900	1000 to 1500 amp		1.30	6.154		3,725	270		3,995	4,500
1200	2000 to 4000 amp		1	8		4,400	350		4,750	5,350
1300	Fused potential transformer, maximum 600 volt	↓	8	1	↓	975	44		1,019	1,150

26 09 13.20 Voltage Monitor Systems

		Crew	Daily Output	Labor-Hours	Unit	Material	2007 Bare Costs Labor	Equipment	Total	Total Incl O&P
0010	**VOLTAGE MONITOR SYSTEMS** (test equipment)									
0100	AC voltage monitor system, 120/240 V, one-channel				Ea.	3,050			3,050	3,375
0110	Modem adapter					385			385	420
0120	Add-on detector only				↓	1,600			1,600	1,775

26 09 Instrumentation and Control for Electrical Systems

26 09 13 - Electrical Power Monitoring and Control

26 09 13.20 Voltage Monitor Systems	Crew	Daily Output	Labor-Hours	Unit	Material	2007 Bare Costs Labor	Equipment	Total	Total Incl O&P
0150	AC voltage remote monitor sys., 3 channel, 120, 230, or 480 V				Ea.	5,550			5,550	6,125
0160	With internal modem					5,875			5,875	6,450
0170	Combination temperature and humidity probe					860			860	950
0180	Add-on detector only					4,025			4,025	4,425
0190	With internal modem					4,375			4,375	4,825

26 12 Medium-Voltage Transformers

26 12 19 - Pad-Mounted, Liquid-Filled, Medium-Voltage Transformers

26 12 19.10 Oil Filled Transformer

	26 12 19.10 Oil Filled Transformer	Crew	Daily Output	Labor-Hours	Unit	Material	2007 Bare Costs Labor	Equipment	Total	Total Incl O&P
0010	**OIL FILLED TRANSFORMER** primary delta or Y, R262213-60									
0050	Pad mounted 5 kV or 15 kV, with taps, 277/480 V secondary, 3 phase									
0100	150 kVA	R-3	.65	30.769	Ea.	8,125	1,325	251	9,701	11,200
0110	225 kVA		.55	36.364		8,950	1,575	297	10,822	12,500
0200	300 kVA		.45	44.444		11,200	1,925	365	13,490	15,600
0300	500 kVA		.40	50		16,200	2,175	410	18,785	21,500
0400	750 kVA		.38	52.632		20,100	2,275	430	22,805	26,000
0500	1000 kVA		.26	76.923		23,800	3,325	630	27,755	31,900
0600	1500 kVA		.23	86.957		28,300	3,775	710	32,785	37,600
0700	2000 kVA		.20	100		35,800	4,325	815	40,940	46,800
0710	2500 kVA		.19	105		43,200	4,550	860	48,610	55,000
0720	3000 kVA		.17	117		52,000	5,100	960	58,060	66,000
0800	3750 kVA		.16	125		67,000	5,400	1,025	73,425	83,000
1990	Pole mounted distribution type, single phase									
2000	13.8 kV primary, 120/240 V secondary, 10 kVA	R-15	7.45	6.443	Ea.	800	276	39	1,115	1,325
2010	50 kVA		3.70	12.973		1,600	555	78.50	2,233.50	2,700
2020	100 kVA		2.75	17.455		2,800	745	105	3,650	4,350
2030	167 kVA		2.15	22.326		3,825	955	135	4,915	5,775
2900	2400 V primary, 120/240 V secondary, 10 kVA		7.45	6.443		825	276	39	1,140	1,375
2910	15 kVA		6.70	7.164		915	305	43.50	1,263.50	1,500
2920	25 kVA		6	8		1,125	340	48.50	1,513.50	1,800
2930	37.5 kVA		4.30	11.163		1,350	480	67.50	1,897.50	2,300
2940	50 kVA		3.70	12.973		1,625	555	78.50	2,258.50	2,725
2950	75 kVA		3	16		2,325	685	96.50	3,106.50	3,700
2960	100 kVA		2.75	17.455		2,850	745	105	3,700	4,375

26 12 19.20 Transformer, Liquid-Filled

	26 12 19.20 Transformer, Liquid-Filled	Crew	Daily Output	Labor-Hours	Unit	Material	2007 Bare Costs Labor	Equipment	Total	Total Incl O&P
0010	**TRANSFORMER, LIQUID-FILLED** Pad mounted									
0020	5 kV or 15 kV primary, 277/480 volt secondary, 3 phase									
0050	225 kVA	R-3	.55	36.364	Ea.	11,200	1,575	297	13,072	15,000
0100	300 kVA		.45	44.444		13,300	1,925	365	15,590	17,900
0200	500 kVA		.40	50		16,800	2,175	410	19,385	22,200
0250	750 kVA		.38	52.632		21,700	2,275	430	24,405	27,700
0300	1000 kVA		.26	76.923		25,200	3,325	630	29,155	33,400
0350	1500 kVA		.23	86.957		29,400	3,775	710	33,885	38,700
0400	2000 kVA		.20	100		36,400	4,325	815	41,540	47,400
0450	2500 kVA		.19	105		43,400	4,550	860	48,810	55,500

26 13 16 – Medium-Voltage Fusible Interrupter Switchgear

26 13 16.10 Switchgear	Crew	Daily Output	Labor-Hours	Unit	Material	2007 Bare Costs Labor	Equipment	Total	Total Incl O&P
0010 **SWITCHGEAR**, Incorporate switch with cable connections, transformer,									
0100 & Low Voltage section									
0200 Load interrupter switch, 600 amp, 2 position									
0300 NEMA 1, 4.8 KV, 300 kVA & below w/CLF fuses	R-3	.40	50	Ea.	17,500	2,175	410	20,085	22,900
0400 400 kVA & above w/CLF fuses		.38	52.632		19,500	2,275	430	22,205	25,300
0500 Non fusible		.41	48.780		14,200	2,100	400	16,700	19,200
0600 13.8 kV, 300 kVA & below		.38	52.632		21,900	2,275	430	24,605	28,000
0700 400 kVA & above		.36	55.556		21,900	2,400	455	24,755	28,200
0800 Non fusible	▼	.40	50		16,600	2,175	410	19,185	22,000
0900 Cable lugs for 2 feeders 4.8 kV or 13.8 kV	1 Elec	8	1		530	44		574	650
1000 Pothead, one 3 conductor or three 1 conductor		4	2		2,550	88		2,638	2,925
1100 Two 3 conductor or six 1 conductor		2	4		5,050	176		5,226	5,800
1200 Key interlocks	▼	8	1	▼	590	44		634	715
1300 Lightning arresters, Distribution class (no charge)									
1400 Intermediate class or line type 4.8 kV	1 Elec	2.70	2.963	Ea.	2,875	130		3,005	3,350
1500 13.8 kV		2	4		3,800	176		3,976	4,425
1600 Station class, 4.8 kV		2.70	2.963		4,900	130		5,030	5,600
1700 13.8 kV	▼	2	4		8,450	176		8,626	9,550
1800 Transformers, 4800 volts to 480/277 volts, 75 kVA	R-3	.68	29.412		14,900	1,275	240	16,415	18,600
1900 112.5 kVA		.65	30.769		18,200	1,325	251	19,776	22,400
2000 150 kVA		.57	35.088		20,800	1,525	287	22,612	25,400
2100 225 kVA		.48	41.667		23,800	1,800	340	25,940	29,300
2200 300 kVA		.41	48.780		26,700	2,100	400	29,200	32,900
2300 500 kVA		.36	55.556		35,100	2,400	455	37,955	42,700
2400 750 kVA		.29	68.966		39,800	2,975	565	43,340	48,900
2500 13,800 volts to 480/277 volts, 75 kVA		.61	32.787		21,000	1,425	268	22,693	25,500
2600 112.5 kVA		.55	36.364		27,900	1,575	297	29,772	33,400
2700 150 kVA		.49	40.816		28,200	1,775	335	30,310	34,000
2800 225 kVA		.41	48.780		32,500	2,100	400	35,000	39,400
2900 300 kVA		.37	54.054		33,200	2,350	440	35,990	40,500
3000 500 kVA		.31	64.516		36,800	2,800	525	40,125	45,200
3100 750 kVA	▼	.42	47.733		40,500	2,075	390	42,965	48,000
3200 Forced air cooling & temperature alarm	1 Elec	1	8	▼	3,250	350		3,600	4,100
3300 Low voltage components									
3400 Maximum panel height 49-1/2", single or twin row									
3500 Breaker heights, type FA or FH, 6"									
3600 type KA or KH, 8"									
3700 type LA, 11"									
3800 type MA, 14"									
3900 Breakers, 2 pole, 15 to 60 amp, type FA	1 Elec	5.60	1.429	Ea.	325	62.50		387.50	450
4000 70 to 100 amp, type FA		4.20	1.905		410	83.50		493.50	575
4100 15 to 60 amp, type FH		5.60	1.429		530	62.50		592.50	680
4200 70 to 100 amp, type FH		4.20	1.905		620	83.50		703.50	810
4300 125 to 225 amp, type KA		3.40	2.353		955	103		1,058	1,200
4400 125 to 225 amp, type KH		3.40	2.353		2,225	103		2,328	2,600
4500 125 to 400 amp, type LA		2.50	3.200		1,750	140		1,890	2,125
4600 125 to 600 amp, type MA		1.80	4.444		2,750	195		2,945	3,325
4700 700 & 800 amp, type MA		1.50	5.333		3,550	234		3,784	4,250
4800 3 pole, 15 to 60 amp, type FA		5.30	1.509		415	66.50		481.50	560
4900 70 to 100 amp, type FA		4	2		515	88		603	695
5000 15 to 60 amp, type FH		5.30	1.509		625	66.50		691.50	790
5100 70 to 100 amp, type FH		4	2		705	88		793	905
5200 125 to 225 amp, type KA	▼	3.20	2.500	▼	1,200	110		1,310	1,500

26 13 Medium-Voltage Switchgear

26 13 16 – Medium-Voltage Fusible Interrupter Switchgear

26 13 16.10 Switchgear		Crew	Daily Output	Labor-Hours	Unit	Material	2007 Bare Costs Labor	Equipment	Total	Total Incl O&P
5300	125 to 225 amp, type KH	1 Elec	3.20	2.500	Ea.	2,675	110		2,785	3,125
5400	125 to 400 amp, type LA		2.30	3.478		2,125	153		2,278	2,550
5500	125 to 600 amp, type MA		1.60	5		3,475	220		3,695	4,150
5600	700 & 800 amp, type MA		1.30	6.154		4,575	270		4,845	5,425

26 22 Low-Voltage Transformers

26 22 13 – Low-Voltage Distribution Transformers

26 22 13.10 Dry Type Transformer

26 22 13.10 Dry Type Transformer			Crew	Daily Output	Labor-Hours	Unit	Material	2007 Bare Costs Labor	Equipment	Total	Total Incl O&P
0010	**DRY TYPE TRANSFORMER**	R262213-10									
0050	Single phase, 240/480 volt primary, 120/240 volt secondary										
0100	1 kVA	R262213-60	1 Elec	2	4	Ea.	235	176		411	520
0300	2 kVA			1.60	5		355	220		575	715
0500	3 kVA			1.40	5.714		435	251		686	855
0700	5 kVA			1.20	6.667		600	293		893	1,100
0900	7.5 kVA		2 Elec	2.20	7.273		835	320		1,155	1,400
1100	10 kVA			1.60	10		1,025	440		1,465	1,775
1300	15 kVA	R262213-65		1.20	13.333		1,400	585		1,985	2,425
1500	25 kVA			1	16		1,900	700		2,600	3,150
1700	37.5 kVA			.80	20		2,475	880		3,355	4,025
1900	50 kVA			.70	22.857		2,950	1,000		3,950	4,725
2100	75 kVA			.65	24.615		3,900	1,075		4,975	5,875
2110	100 kVA		R-3	.90	22.222		5,075	960	182	6,217	7,200
2120	167 kVA		"	.80	25		8,425	1,075	204	9,704	11,100
2190	480V primary 120/240V secondary, nonvent., 15 kVA		2 Elec	1.20	13.333		1,475	585		2,060	2,500
2200	25 kVA			.90	17.778		2,175	780		2,955	3,550
2210	37 kVA			.75	21.333		2,575	935		3,510	4,250
2220	50 kVA			.65	24.615		3,075	1,075		4,150	4,975
2230	75 kVA			.60	26.667		4,050	1,175		5,225	6,200
2240	100 kVA			.50	32		5,275	1,400		6,675	7,900
2250	Low operating temperature(80°C), 25 kVA			1	16		3,950	700		4,650	5,400
2260	37 kVA			.80	20		4,250	880		5,130	5,975
2270	50 kVA			.70	22.857		5,550	1,000		6,550	7,600
2280	75 kVA			.65	24.615		8,850	1,075		9,925	11,300
2290	100 kVA			.55	29.091		9,200	1,275		10,475	12,000
2300	3 phase, 480 volt primary 120/208 volt secondary										
2310	Ventilated, 3 kVA		1 Elec	1	8	Ea.	805	350		1,155	1,400
2700	6 kVA			.80	10		1,100	440		1,540	1,875
2900	9 kVA			.70	11.429		1,250	500		1,750	2,125
3100	15 kVA		2 Elec	1.10	14.545		1,675	640		2,315	2,800
3300	30 kVA			.90	17.778		1,975	780		2,755	3,300
3500	45 kVA			.80	20		2,350	880		3,230	3,900
3700	75 kVA			.70	22.857		3,550	1,000		4,550	5,400
3900	112.5 kVA		R-3	.90	22.222		4,725	960	182	5,867	6,825
4100	150 kVA			.85	23.529		6,150	1,025	192	7,367	8,500
4300	225 kVA			.65	30.769		8,350	1,325	251	9,926	11,500
4500	300 kVA			.55	36.364		10,600	1,575	297	12,472	14,300
4700	500 kVA			.45	44.444		17,500	1,925	365	19,790	22,500
4800	750 kVA			.35	57.143		30,600	2,475	465	33,540	37,900
4820	1000 kVA			.32	62.500		35,100	2,700	510	38,310	43,200
4850	K-4 rated, 15 kVA		2 Elec	1.10	14.545		1,825	640		2,465	2,975
4855	30 kVA			.90	17.778		2,800	780		3,580	4,250

26 22 Low-Voltage Transformers

26 22 13 – Low-Voltage Distribution Transformers

26 22 13.10 Dry Type Transformer	Crew	Daily Output	Labor-Hours	Unit	Material	2007 Bare Costs Labor	Equipment	Total	Total Incl O&P
4860	45 kVA	2 Elec	.80	20	Ea.	3,250	880		4,130	4,875
4865	75 kVA	↓	.70	22.857		4,450	1,000		5,450	6,375
4870	112.5 kVA	R-3	.90	22.222		7,375	960	182	8,517	9,750
4875	150 kVA		.85	23.529		9,900	1,025	192	11,117	12,600
4880	225 kVA		.65	30.769		15,000	1,325	251	16,576	18,800
4885	300 kVA		.55	36.364		20,800	1,575	297	22,672	25,600
4890	500 kVA	↓	.45	44.444		28,800	1,925	365	31,090	35,000
4900	K-13 rated, 15 kVA	2 Elec	1.10	14.545		2,400	640		3,040	3,600
4905	30 kVA		.90	17.778		3,300	780		4,080	4,775
4910	45 kVA		.80	20		3,875	880		4,755	5,575
4915	75 kVA	↓	.70	22.857		5,500	1,000		6,500	7,550
4920	112.5 kVA	R-3	.90	22.222		11,300	960	182	12,442	14,100
4925	150 kVA		.85	23.529		13,800	1,025	192	15,017	16,900
4930	225 kVA		.65	30.769		20,100	1,325	251	21,676	24,400
4935	300 kVA		.55	36.364		28,200	1,575	297	30,072	33,800
4940	500 kVA	↓	.45	44.444	↓	43,100	1,925	365	45,390	50,500
5020	480 volt primary 120/208 volt secondary									
5030	Nonventilated, 15 kVA	2 Elec	1.10	14.545	Ea.	3,400	640		4,040	4,675
5040	30 kVA		.80	20		3,800	880		4,680	5,475
5050	45 kVA		.70	22.857		5,175	1,000		6,175	7,200
5060	75 kVA	↓	.65	24.615		7,550	1,075		8,625	9,900
5070	112.5 kVA	R-3	.85	23.529		10,300	1,025	192	11,517	13,000
5081	150 kVA		.85	23.529		12,400	1,025	192	13,617	15,400
5090	225 kVA		.60	33.333		15,200	1,450	272	16,922	19,200
5100	300 kVA	↓	.50	40		18,200	1,725	325	20,250	22,900
5200	Low operating temperature (80°C), 30 kVA	2 Elec	.90	17.778		3,100	780		3,880	4,550
5210	45 kVA		.80	20		4,650	880		5,530	6,425
5220	75 kVA	↓	.70	22.857		6,200	1,000		7,200	8,325
5230	112.5 kVA	R-3	.90	22.222		11,000	960	182	12,142	13,700
5240	150 kVA		.85	23.529		13,900	1,025	192	15,117	16,900
5250	225 kVA		.65	30.769		19,200	1,325	251	20,776	23,400
5260	300 kVA		.55	36.364		22,900	1,575	297	24,772	27,900
5270	500 kVA	↓	.45	44.444	↓	34,500	1,925	365	36,790	41,300
5380	3 phase, 5 kV primary 277/480 volt secondary									
5400	High voltage, 112.5 kVA	R-3	.85	23.529	Ea.	13,400	1,025	192	14,617	16,400
5410	150 kVA		.65	30.769		14,600	1,325	251	16,176	18,300
5420	225 kVA		.55	36.364		17,200	1,575	297	19,072	21,600
5430	300 kVA		.45	44.444		22,000	1,925	365	24,290	27,500
5440	500 kVA		.35	57.143		28,300	2,475	465	31,240	35,400
5450	750 kVA		.32	62.500		40,000	2,700	510	43,210	48,600
5460	1000 kVA		.30	66.667		46,900	2,875	545	50,320	56,500
5470	1500 kVA		.27	74.074		54,500	3,200	605	58,305	65,500
5480	2000 kVA		.25	80		64,000	3,450	655	68,105	76,500
5490	2500 kVA		.20	100		72,000	4,325	815	77,140	87,000
5500	3000 kVA	↓	.18	111	↓	94,500	4,800	910	100,210	112,000
5590	15 kV primary 277/480 volt secondary									
5600	High voltage, 112.5 kVA	R-3	.85	23.529	Ea.	20,400	1,025	192	21,617	24,200
5610	150 kVA		.65	30.769		23,800	1,325	251	25,376	28,500
5620	225 kVA		.55	36.364		26,300	1,575	297	28,172	31,700
5630	300 kVA		.45	44.444		31,000	1,925	365	33,290	37,400
5640	500 kVA		.35	57.143		39,300	2,475	465	42,240	47,400
5650	750 kVA		.32	62.500		52,000	2,700	510	55,210	61,500
5660	1000 kVA	↓	.30	66.667	↓	59,000	2,875	545	62,420	70,000

26 22 Low-Voltage Transformers

26 22 13 – Low-Voltage Distribution Transformers

26 22 13.10 Dry Type Transformer

		Crew	Daily Output	Labor-Hours	Unit	Material	2007 Bare Costs Labor	Equipment	Total	Total Incl O&P
5670	1500 kVA	R-3	.27	74.074	Ea.	68,000	3,200	605	71,805	80,500
5680	2000 kVA		.25	80		76,000	3,450	655	80,105	89,500
5690	2500 kVA		.20	100		88,000	4,325	815	93,140	104,000
5700	3000 kVA		.18	111		104,500	4,800	910	110,210	123,000
6000	2400V primary, 480V secondary, 300 KVA		.45	44.444		22,000	1,925	365	24,290	27,500
6010	500 KVA		.35	57.143		28,300	2,475	465	31,240	35,400
6020	750 KVA		.32	62.500		37,000	2,700	510	40,210	45,300

26 22 13.20 Isolating Panels

		Crew	Daily Output	Labor-Hours	Unit	Material	2007 Bare Costs Labor	Equipment	Total	Total Incl O&P
0010	**ISOLATING PANELS** used with isolating transformers									
0020	For hospital applications									
0100	Critical care area, 8 circuit, 3 kVA	1 Elec	.58	13.793	Ea.	6,675	605		7,280	8,250
0200	5 kVA		.54	14.815		6,875	650		7,525	8,525
0400	7.5 kVA		.52	15.385		7,000	675		7,675	8,700
0600	10 kVA		.44	18.182		7,225	800		8,025	9,150
0800	Operating room power & lighting, 8 circuit, 3 kVA		.58	13.793		5,575	605		6,180	7,050
1000	5 kVA		.54	14.815		5,900	650		6,550	7,450
1200	7.5 kVA		.52	15.385		6,450	675		7,125	8,075
1400	10 kVA		.44	18.182		6,850	800		7,650	8,750
1600	X-ray systems, 15 kVA, 90 amp		.44	18.182		14,300	800		15,100	16,900
1800	25 kVA, 125 amp		.36	22.222		14,600	975		15,575	17,600

26 22 13.30 Isolating Transformer

		Crew	Daily Output	Labor-Hours	Unit	Material	2007 Bare Costs Labor	Equipment	Total	Total Incl O&P
0010	**ISOLATING TRANSFORMER**									
0100	Single phase, 120/240 volt primary, 120/240 volt secondary									
0200	0.50 kVA R262213-60	1 Elec	4	2	Ea.	310	88		398	470
0400	1 kVA		2	4		385	176		561	685
0600	2 kVA		1.60	5		580	220		800	960
0800	3 kVA		1.40	5.714		660	251		911	1,100
1000	5 kVA		1.20	6.667		870	293		1,163	1,400
1200	7.5 kVA		1.10	7.273		1,100	320		1,420	1,700
1400	10 kVA		.80	10		1,400	440		1,840	2,200
1600	15 kVA		.60	13.333		1,800	585		2,385	2,850
1800	25 kVA		.50	16		2,600	700		3,300	3,900
1810	37.5 kVA	2 Elec	.80	20		4,275	880		5,155	6,000
1820	75 kVA	"	.65	24.615		6,425	1,075		7,500	8,675
1830	3 phase, 120/240 primary, 120/208V secondary, 112.5 kVA	R-3	.90	22.222		8,125	960	182	9,267	10,600
1840	150 kVA		.85	23.529		10,300	1,025	192	11,517	13,100
1850	225 kVA		.65	30.769		14,400	1,325	251	15,976	18,100
1860	300 kVA		.55	36.364		19,100	1,575	297	20,972	23,700
1870	500 kVA		.45	44.444		31,900	1,925	365	34,190	38,400
1880	750 kVA		.35	57.143		32,200	2,475	465	35,140	39,600

26 22 13.90 Transformer Handling

		Crew	Daily Output	Labor-Hours	Unit	Material	2007 Bare Costs Labor	Equipment	Total	Total Incl O&P
0010	**TRANSFORMER HANDLING** Add to normal labor cost in restricted areas									
5000	Transformers									
5150	15 kVA, approximately 200 pounds	2 Elec	2.70	5.926	Ea.		260		260	385
5160	25 kVA, approximately 300 pounds		2.50	6.400			281		281	420
5170	37.5 kVA, approximately 400 pounds		2.30	6.957			305		305	455
5180	50 kVA, approximately 500 pounds		2	8			350		350	525
5190	75 kVA, approximately 600 pounds		1.80	8.889			390		390	580
5200	100 kVA, approximately 700 pounds		1.60	10			440		440	655
5210	112.5 kVA, approximately 800 pounds	3 Elec	2.20	10.909			480		480	715
5220	125 kVA, approximately 900 pounds		2	12			525		525	785
5230	150 kVA, approximately 1000 pounds		1.80	13.333			585		585	870

26 22 Low-Voltage Transformers

26 22 13 – Low-Voltage Distribution Transformers

26 22 13.90 Transformer Handling		Crew	Daily Output	Labor-Hours	Unit	Material	2007 Bare Costs Labor	Equipment	Total	Total Incl O&P
5240	167 kVA, approximately 1200 pounds	3 Elec	1.60	15	Ea.		660		660	980
5250	200 kVA, approximately 1400 pounds		1.40	17.143			755		755	1,125
5260	225 kVA, approximately 1600 pounds		1.30	18.462			810		810	1,200
5270	250 kVA, approximately 1800 pounds		1.10	21.818			960		960	1,425
5280	300 kVA, approximately 2000 pounds		1	24			1,050		1,050	1,575
5290	500 kVA, approximately 3000 pounds		.75	32			1,400		1,400	2,100
5300	600 kVA, approximately 3500 pounds		.67	35.821			1,575		1,575	2,350
5310	750 kVA, approximately 4000 pounds		.60	40			1,750		1,750	2,625
5320	1000 kVA, approximately 5000 pounds		.50	48			2,100		2,100	3,125

26 22 16 – Low-Voltage Buck-Boost Transformers

26 22 16.10 Buck-Boost Transformer

	26 22 16.10 Buck-Boost Transformer	Crew	Daily Output	Labor-Hours	Unit	Material	2007 Bare Costs Labor	Equipment	Total	Total Incl O&P
0010	**BUCK-BOOST TRANSFORMER** R262213-60									
0100	Single phase, 120/240 V primary, 12/24 V secondary									
0200	0.10 kVA	1 Elec	8	1	Ea.	78	44		122	152
0400	0.25 kVA		5.70	1.404		115	61.50		176.50	218
0600	0.50 kVA		4	2		157	88		245	305
0800	0.75 kVA		3.10	2.581		202	113		315	390
1000	1.0 kVA		2	4		251	176		427	535
1200	1.5 kVA		1.80	4.444		310	195		505	630
1400	2.0 kVA		1.60	5		370	220		590	735
1600	3.0 kVA		1.40	5.714		500	251		751	925
1800	5.0 kVA		1.20	6.667		660	293		953	1,150
2000	3 phase, 240 V primary, 208/120 V secondary, 15 kVA	2 Elec	2.40	6.667		2,100	293		2,393	2,750
2200	30 kVA		1.60	10		2,350	440		2,790	3,250
2400	45 kVA		1.40	11.429		2,825	500		3,325	3,875
2600	75 kVA		1.20	13.333		3,425	585		4,010	4,650
2800	112.5 kVA	R-3	1.40	14.286		4,250	620	117	4,987	5,725
3000	150 kVA		1.10	18.182		5,675	785	149	6,609	7,575
3200	225 kVA		1	20		7,400	865	163	8,428	9,600
3400	300 kVA		.90	22.222		10,000	960	182	11,142	12,600

26 24 Switchboards and Panelboards

26 24 13 – Switchboards

26 24 13.10 Switchboards

	26 24 13.10 Switchboards	Crew	Daily Output	Labor-Hours	Unit	Material	2007 Bare Costs Labor	Equipment	Total	Total Incl O&P
0010	**SWITCHBOARDS** Incoming main service section R262419-84									
0100	Aluminum bus bars, not including CT's or PT's									
0200	No main disconnect, includes CT compartment									
0300	120/208 volt, 4 wire, 600 amp	2 Elec	1	16	Ea.	3,650	700		4,350	5,075
0400	800 amp		.88	18.182		3,650	800		4,450	5,225
0500	1000 amp		.80	20		4,400	880		5,280	6,125
0600	1200 amp		.72	22.222		4,400	975		5,375	6,275
0700	1600 amp		.66	24.242		4,400	1,075		5,475	6,400
0800	2000 amp		.62	25.806		4,725	1,125		5,850	6,875
1000	3000 amp		.56	28.571		6,250	1,250		7,500	8,750
1200	277/480 volt, 4 wire, 600 amp		1	16		3,650	700		4,350	5,075
1300	800 amp		.88	18.182		3,650	800		4,450	5,225
1400	1000 amp		.80	20		4,650	880		5,530	6,400
1500	1200 amp		.72	22.222		4,650	975		5,625	6,550
1600	1600 amp		.66	24.242		4,650	1,075		5,725	6,675
1700	2000 amp		.62	25.806		4,725	1,125		5,850	6,875

26 24 Switchboards and Panelboards

26 24 13 – Switchboards

26 24 13.10 Switchboards		Crew	Daily Output	Labor-Hours	Unit	Material	2007 Bare Costs Labor	Equipment	Total	Total Incl O&P
1800	3000 amp	2 Elec	.56	28.571	Ea.	6,250	1,250		7,500	8,750
1900	4000 amp	↓	.52	30.769	↓	7,750	1,350		9,100	10,500
2000	Fused switch & CT compartment									
2100	120/208 volt, 4 wire, 400 amp	2 Elec	1.12	14.286	Ea.	3,775	625		4,400	5,075
2200	600 amp		.94	17.021		4,475	745		5,220	6,000
2300	800 amp		.84	19.048		9,900	835		10,735	12,200
2400	1200 amp		.68	23.529		12,900	1,025		13,925	15,800
2500	277/480 volt, 4 wire, 400 amp		1.14	14.035		3,950	615		4,565	5,275
2600	600 amp		.94	17.021		4,525	745		5,270	6,075
2700	800 amp		.84	19.048		9,900	835		10,735	12,200
2800	1200 amp	↓	.68	23.529	↓	12,900	1,025		13,925	15,800
2900	Pressure switch & CT compartment									
3000	120/208 volt, 4 wire, 800 amp	2 Elec	.80	20	Ea.	8,875	880		9,755	11,100
3100	1200 amp		.66	24.242		17,200	1,075		18,275	20,500
3200	1600 amp		.62	25.806		18,300	1,125		19,425	21,800
3300	2000 amp		.56	28.571		19,400	1,250		20,650	23,300
3310	2500 amp		.50	32		23,900	1,400		25,300	28,400
3320	3000 amp		.44	36.364		32,300	1,600		33,900	37,900
3330	4000 amp		.40	40		41,400	1,750		43,150	48,100
3400	277/480 volt, 4 wire, 800 amp, with ground fault		.80	20		14,700	880		15,580	17,500
3600	1200 amp, with ground fault		.66	24.242		19,000	1,075		20,075	22,500
4000	1600 amp, with ground fault		.62	25.806		20,600	1,125		21,725	24,400
4200	2000 amp, with ground fault	↓	.56	28.571	↓	22,300	1,250		23,550	26,400
4400	Circuit breaker, molded case & CT compartment									
4600	3 pole, 4 wire, 600 amp	2 Elec	.94	17.021	Ea.	7,650	745		8,395	9,525
4800	800 amp		.84	19.048		9,175	835		10,010	11,400
5000	1200 amp	↓	.68	23.529	↓	12,500	1,025		13,525	15,300
5100	Copper bus bars, not incl. CT's or PT's, add, minimum					15%				

26 24 13.20 Switchboards

26 24 13.20 Switchboards		Crew	Daily Output	Labor-Hours	Unit	Material	2007 Bare Costs Labor	Equipment	Total	Total Incl O&P
0010	**SWITCHBOARDS** (in plant distribution)									
0100	Main lugs only, to 600 volt, 3 pole, 3 wire, 200 amp	2 Elec	1.20	13.333	Ea.	1,525	585		2,110	2,550
0110	400 amp		1.20	13.333		1,525	585		2,110	2,550
0120	600 amp		1.20	13.333		1,575	585		2,160	2,600
0130	800 amp		1.08	14.815		1,700	650		2,350	2,850
0140	1200 amp		.92	17.391		2,075	765		2,840	3,400
0150	1600 amp		.86	18.605		2,725	815		3,540	4,225
0160	2000 amp		.82	19.512		3,025	855		3,880	4,600
0250	To 480 volt, 3 pole, 4 wire, 200 amp		1.20	13.333		1,325	585		1,910	2,325
0260	400 amp		1.20	13.333		1,525	585		2,110	2,550
0270	600 amp		1.20	13.333		1,675	585		2,260	2,700
0280	800 amp		1.08	14.815		1,825	650		2,475	2,975
0290	1200 amp		.92	17.391		2,250	765		3,015	3,600
0300	1600 amp		.86	18.605		2,575	815		3,390	4,050
0310	2000 amp		.82	19.512		3,000	855		3,855	4,575
0400	Main circuit breaker, to 600 volt, 3 pole, 3 wire, 200 amp		1.20	13.333		2,800	585		3,385	3,950
0410	400 amp		1.14	14.035		2,800	615		3,415	4,000
0420	600 amp		1.10	14.545		3,550	640		4,190	4,850
0430	800 amp		1.04	15.385		5,925	675		6,600	7,525
0440	1200 amp		.88	18.182		7,750	800		8,550	9,750
0450	1600 amp		.84	19.048		12,500	835		13,335	15,000
0460	2000 amp		.80	20		13,300	880		14,180	15,900
0550	277/480 volt, 3 pole, 4 wire, 200 amp	↓	1.20	13.333	↓	2,950	585		3,535	4,100

26 24 Switchboards and Panelboards

26 24 13 – Switchboards

26 24 13.20 Switchboards		Crew	Daily Output	Labor-Hours	Unit	Material	2007 Bare Costs Labor	Equipment	Total	Total Incl O&P
0560	400 amp	2 Elec	1.14	14.035	Ea.	2,950	615		3,565	4,150
0570	600 amp		1.10	14.545		3,700	640		4,340	5,000
0580	800 amp		1.04	15.385		6,225	675		6,900	7,825
0590	1200 amp		.88	18.182		8,025	800		8,825	10,000
0600	1600 amp		.84	19.048		12,500	835		13,335	15,000
0610	2000 amp		.80	20		13,300	880		14,180	15,900
0700	Main fusible switch w/fuse, 208/240 volt, 3 pole, 3 wire, 200 amp		1.20	13.333		3,075	585		3,660	4,275
0710	400 amp		1.14	14.035		3,075	615		3,690	4,325
0720	600 amp		1.10	14.545		3,750	640		4,390	5,075
0730	800 amp		1.04	15.385		7,675	675		8,350	9,450
0740	1200 amp		.88	18.182		8,925	800		9,725	11,000
0800	120/208, 120/240 volt, 3 pole, 4 wire, 200 amp		1.20	13.333		2,750	585		3,335	3,900
0810	400 amp		1.14	14.035		2,750	615		3,365	3,950
0820	600 amp		1.10	14.545		3,575	640		4,215	4,900
0830	800 amp		1.04	15.385		5,700	675		6,375	7,275
0840	1200 amp		.88	18.182		6,625	800		7,425	8,475
0900	480 or 600 volt, 3 pole, 3 wire, 200 amp		1.20	13.333		2,975	585		3,560	4,150
0910	400 amp		1.14	14.035		2,975	615		3,590	4,200
0920	600 amp		1.10	14.545		3,550	640		4,190	4,850
0930	800 amp		1.04	15.385		5,475	675		6,150	7,025
0940	1200 amp		.88	18.182		6,350	800		7,150	8,200
1000	277 or 480 volt, 3 pole, 4 wire, 200 amp		1.20	13.333		3,075	585		3,660	4,275
1010	400 amp		1.14	14.035		3,075	615		3,690	4,325
1020	600 amp		1.10	14.545		3,725	640		4,365	5,050
1030	800 amp		1.04	15.385		5,700	675		6,375	7,275
1040	1200 amp		.88	18.182		6,625	800		7,425	8,475
1120	1600 amp		.76	21.053		12,100	925		13,025	14,700
1130	2000 amp		.68	23.529		15,900	1,025		16,925	19,100
1150	Pressure switch, bolted, 3 pole, 208/240 volt, 3 wire, 800 amp		.96	16.667		8,950	730		9,680	11,000
1160	1200 amp		.80	20		11,400	880		12,280	13,900
1170	1600 amp		.76	21.053		13,000	925		13,925	15,700
1180	2000 amp		.68	23.529		14,900	1,025		15,925	18,000
1200	120/208 or 120/240 volt, 3 pole, 4 wire, 800 amp		.96	16.667		7,075	730		7,805	8,875
1210	1200 amp		.80	20		8,275	880		9,155	10,400
1220	1600 amp		.76	21.053		13,000	925		13,925	15,700
1230	2000 amp		.68	23.529		14,900	1,025		15,925	18,000
1300	480 or 600 volt, 3 wire, 800 amp		.96	16.667		8,950	730		9,680	11,000
1310	1200 amp		.80	20		12,500	880		13,380	15,100
1320	1600 amp		.76	21.053		14,100	925		15,025	16,900
1330	2000 amp		.68	23.529		15,900	1,025		16,925	19,100
1400	277-480 volt, 4 wire, 800 amp		.96	16.667		8,950	730		9,680	11,000
1410	1200 amp		.80	20		12,500	880		13,380	15,100
1420	1600 amp		.76	21.053		14,100	925		15,025	16,900
1430	2000 amp		.68	23.529		15,900	1,025		16,925	19,100
1500	Main ground fault protector, 1200-2000 amp		5.40	2.963		4,550	130		4,680	5,200
1600	Busway connection, 200 amp		5.40	2.963		400	130		530	635
1610	400 amp		4.60	3.478		400	153		553	665
1620	600 amp		4	4		400	176		576	700
1630	800 amp		3.20	5		400	220		620	765
1640	1200 amp		2.60	6.154		400	270		670	840
1650	1600 amp		2.40	6.667		825	293		1,118	1,350
1660	2000 amp		2	8		825	350		1,175	1,425
1700	Shunt trip for remote operation 200 amp		8	2		905	88		993	1,125

26 24 13 – Switchboards

26 24 13.20 Switchboards

		Crew	Daily Output	Labor-Hours	Unit	Material	2007 Bare Costs Labor	Equipment	Total	Total Incl O&P
1710	400 amp	2 Elec	8	2	Ea.	1,450	88		1,538	1,725
1720	600 amp		8	2		1,700	88		1,788	2,000
1730	800 amp		8	2		2,200	88		2,288	2,550
1740	1200-2000 amp		8	2		4,650	88		4,738	5,225
1800	Motor operated main breaker 200 amp		8	2		2,950	88		3,038	3,375
1810	400 amp		8	2		2,950	88		3,038	3,375
1820	600 amp		8	2		2,375	88		2,463	2,750
1830	800 amp		8	2		2,825	88		2,913	3,225
1840	1200-2000 amp		8	2		4,725	88		4,813	5,300
1900	Current/potential transformer metering compartment 200-800 amp		5.40	2.963		1,925	130		2,055	2,325
1940	1200 amp		5.40	2.963		2,650	130		2,780	3,125
1950	1600-2000 amp		5.40	2.963		2,650	130		2,780	3,125
2000	With watt meter 200-800 amp		4	4		7,550	176		7,726	8,575
2040	1200 amp		4	4		9,350	176		9,526	10,600
2050	1600-2000 amp		4	4		9,350	176		9,526	10,600
2100	Split bus 60-200 amp	1 Elec	5.30	1.509		286	66.50		352.50	415
2130	400 amp	2 Elec	4.60	3.478		495	153		648	770
2140	600 amp		3.60	4.444		605	195		800	955
2150	800 amp		2.60	6.154		770	270		1,040	1,250
2170	1200 amp		2	8		880	350		1,230	1,500
2250	Contactor control 60 amp	1 Elec	2	4		1,875	176		2,051	2,325
2260	100 amp		1.50	5.333		2,100	234		2,334	2,675
2270	200 amp		1	8		3,175	350		3,525	4,025
2280	400 amp	2 Elec	1	16		9,650	700		10,350	11,700
2290	600 amp		.84	19.048		10,800	835		11,635	13,200
2300	800 amp		.72	22.222		12,800	975		13,775	15,600
2500	Modifier, two distribution sections, add		.80	20		3,025	880		3,905	4,625
2520	Three distribution sections, add		.40	40		6,775	1,750		8,525	10,100
2560	Auxiliary pull section, 20", add		2	8		1,300	350		1,650	1,975
2580	24", add		1.80	8.889		1,300	390		1,690	2,025
2600	30", add		1.60	10		1,300	440		1,740	2,100
2620	36", add		1.40	11.429		1,575	500		2,075	2,475
2640	Dog house, 12", add		2.40	6.667		315	293		608	780
2660	18", add		2	8		630	350		980	1,225
3000	Transition section between switchboard and transformer									
3050	or motor control center, 4 wire alum. bus, 600 amp	2 Elec	1.14	14.035	Ea.	2,250	615		2,865	3,400
3100	800 amp		1	16		2,525	700		3,225	3,825
3150	1000 amp		.88	18.182		2,825	800		3,625	4,300
3200	1200 amp		.80	20		3,100	880		3,980	4,725
3250	1600 amp		.72	22.222		3,675	975		4,650	5,500
3300	2000 amp		.66	24.242		4,225	1,075		5,300	6,225
3350	2500 amp		.62	25.806		4,925	1,125		6,050	7,100
3400	3000 amp		.56	28.571		5,675	1,250		6,925	8,125
4000	Weatherproof construction, per vertical section		1.76	9.091		3,075	400		3,475	3,975

26 24 13.30 Switchboards

		Crew	Daily Output	Labor-Hours	Unit	Material	2007 Bare Costs Labor	Equipment	Total	Total Incl O&P
0010	**SWITCHBOARDS** distribution section R262419-80									
0100	Aluminum bus bars, not including breakers									
0160	Subfeed lug-rated at 60 amp	2 Elec	1.30	12.308	Ea.	1,575	540		2,115	2,525
0170	100 amp		1.26	12.698		1,575	555		2,130	2,550
0180	200 amp		1.20	13.333		1,575	585		2,160	2,600
0190	400 amp		1.10	14.545		1,575	640		2,215	2,675
0200	120/208 or 277/480 volt, 4 wire, 600 amp		1	16		1,625	700		2,325	2,825

26 24 Switchboards and Panelboards

26 24 13 – Switchboards

26 24 13.30 Switchboards

26 24 13.30 Switchboards	Crew	Daily Output	Labor-Hours	Unit	Material	2007 Bare Costs Labor	Equipment	Total	Total Incl O&P
0300	800 amp	2 Elec	.88	18.182	Ea.	2,075	800		2,875	3,500
0400	1000 amp		.80	20		2,075	880		2,955	3,600
0500	1200 amp		.72	22.222		2,975	975		3,950	4,725
0600	1600 amp		.66	24.242		3,475	1,075		4,550	5,400
0700	2000 amp		.62	25.806		4,075	1,125		5,200	6,150
0800	2500 amp		.60	26.667		4,600	1,175		5,775	6,800
0900	3000 amp		.56	28.571		5,575	1,250		6,825	8,000
0950	4000 amp		.52	30.769		8,150	1,350		9,500	11,000

26 24 13.40 Switchboards

26 24 13.40 Switchboards	Crew	Daily Output	Labor-Hours	Unit	Material	2007 Bare Costs Labor	Equipment	Total	Total Incl O&P
0010	**SWITCHBOARDS** feeder section group mounted devices R262419-82									
0030	Circuit breakers									
0160	FA frame, 15 to 60 amp, 240 volt, 1 pole	1 Elec	8	1	Ea.	91	44		135	166
0170	2 pole		7	1.143		183	50		233	276
0180	3 pole		5.30	1.509		263	66.50		329.50	390
0210	480 volt, 1 pole		8	1		115	44		159	193
0220	2 pole		7	1.143		300	50		350	405
0230	3 pole		5.30	1.509		385	66.50		451.50	520
0260	600 volt, 2 pole		7	1.143		360	50		410	470
0270	3 pole		5.30	1.509		445	66.50		511.50	590
0280	FA frame, 70 to 100 amp, 240 volt, 1 pole		7	1.143		120	50		170	207
0310	2 pole		5	1.600		284	70		354	415
0320	3 pole		4	2		325	88		413	485
0330	480 volt, 1 pole		7	1.143		144	50		194	233
0360	2 pole		5	1.600		385	70		455	525
0370	3 pole		4	2		460	88		548	635
0380	600 volt, 2 pole		5	1.600		435	70		505	585
0410	3 pole		4	2		535	88		623	720
0420	KA frame, 70 to 225 amp		3.20	2.500		1,000	110		1,110	1,275
0430	LA frame, 125 to 400 amp		2.30	3.478		2,275	153		2,428	2,725
0460	MA frame, 450 to 600 amp		1.60	5		3,775	220		3,995	4,475
0470	700 to 800 amp		1.30	6.154		4,875	270		5,145	5,775
0480	MAL frame, 1000 amp		1	8		5,075	350		5,425	6,100
0490	PA frame, 1200 amp		.80	10		10,300	440		10,740	12,000
0500	Branch circuit, fusible switch, 600 volt, double 30/30 amp		4	2		1,025	88		1,113	1,250
0550	60/60 amp		3.20	2.500		1,025	110		1,135	1,300
0600	100/100 amp		2.70	2.963		1,425	130		1,555	1,775
0650	Single, 30 amp		5.30	1.509		670	66.50		736.50	840
0700	60 amp		4.70	1.702		670	74.50		744.50	850
0750	100 amp		4	2		965	88		1,053	1,175
0800	200 amp		2.70	2.963		1,425	130		1,555	1,775
0850	400 amp		2.30	3.478		2,700	153		2,853	3,200
0900	600 amp		1.80	4.444		3,200	195		3,395	3,825
0950	800 amp		1.30	6.154		5,250	270		5,520	6,175
1000	1200 amp		.80	10		6,250	440		6,690	7,525
1080	Branch circuit, circuit breakers, high interrupting capacity									
1100	60 amp, 240, 480 or 600 volt, 1 pole	1 Elec	8	1	Ea.	234	44		278	325
1120	2 pole		7	1.143		560	50		610	690
1140	3 pole		5.30	1.509		665	66.50		731.50	835
1150	100 amp, 240, 480 or 600 volt, 1 pole		7	1.143		259	50		309	360
1160	2 pole		5	1.600		670	70		740	840
1180	3 pole		4	2		750	88		838	955
1200	225 amp, 240, 480 or 600 volt, 2 pole		3.50	2.286		2,325	100		2,425	2,725

26 24 Switchboards and Panelboards

26 24 13 – Switchboards

26 24 13.40 Switchboards		Crew	Daily Output	Labor-Hours	Unit	Material	2007 Bare Costs Labor	Equipment	Total	Total Incl O&P
1220	3 pole	1 Elec	3.20	2.500	Ea.	2,825	110		2,935	3,300
1240	400 amp, 240, 480 or 600 volt, 2 pole		2.50	3.200		3,125	140		3,265	3,625
1260	3 pole		2.30	3.478		3,750	153		3,903	4,350
1280	600 amp, 240, 480 or 600 volt, 2 pole		1.80	4.444		3,800	195		3,995	4,475
1300	3 pole		1.60	5		4,575	220		4,795	5,350
1320	800 amp, 240, 480 or 600 volt, 2 pole		1.50	5.333		4,650	234		4,884	5,475
1340	3 pole		1.30	6.154		5,825	270		6,095	6,800
1360	1000 amp, 240, 480 or 600 volt, 2 pole		1.10	7.273		8,450	320		8,770	9,750
1380	3 pole		1	8		9,225	350		9,575	10,600
1400	1200 amp, 240, 480 or 600 volt, 2 pole		.90	8.889		6,325	390		6,715	7,525
1420	3 pole		.80	10		9,900	440		10,340	11,600
1700	Fusible switch, 240 V, 60 amp, 2 pole		3.20	2.500		410	110		520	620
1720	3 pole		3	2.667		565	117		682	800
1740	100 amp, 2 pole		2.70	2.963		420	130		550	655
1760	3 pole		2.50	3.200		635	140		775	910
1780	200 amp, 2 pole		2	4		875	176		1,051	1,225
1800	3 pole		1.90	4.211		1,025	185		1,210	1,400
1820	400 amp, 2 pole		1.50	5.333		1,650	234		1,884	2,150
1840	3 pole		1.30	6.154		2,050	270		2,320	2,650
1860	600 amp, 2 pole		1	8		2,325	350		2,675	3,100
1880	3 pole		.90	8.889		2,875	390		3,265	3,725
1900	240-600 V, 800 amp, 2 pole		.70	11.429		5,200	500		5,700	6,475
1920	3 pole		.60	13.333		6,375	585		6,960	7,875
2000	600 V, 60 amp, 2 pole		3.20	2.500		625	110		735	850
2040	100 amp, 2 pole		2.70	2.963		645	130		775	900
2080	200 amp, 2 pole		2	4		1,125	176		1,301	1,475
2120	400 amp, 2 pole		1.50	5.333		2,200	234		2,434	2,775
2160	600 amp, 2 pole		1	8		2,700	350		3,050	3,475
2500	Branch circuit, circuit breakers, 60 amp, 600 volt, 3 pole		5.30	1.509		595	66.50		661.50	755
2520	240, 480 or 600 volt, 1 pole		8	1		96.50	44		140.50	172
2540	240 volt, 2 pole		7	1.143		166	50		216	257
2560	480 or 600 volt, 2 pole		7	1.143		350	50		400	460
2580	240 volt, 3 pole		5.30	1.509		269	66.50		335.50	395
2600	480 volt, 3 pole		5.30	1.509		470	66.50		536.50	615
2620	100 amp, 600 volt, 2 pole		5	1.600		450	70		520	600
2640	3 pole		4	2		565	88		653	755
2660	480 volt, 2 pole		5	1.600		385	70		455	530
2680	240 volt, 2 pole		5	1.600		221	70		291	350
2700	3 pole		4	2		355	88		443	520
2720	480 volt, 3 pole		4	2		520	88		608	700
2740	225 amp, 240, 480 or 600 volt, 2 pole		3.50	2.286		590	100		690	795
2760	3 pole		3.20	2.500		675	110		785	910
2780	400 amp, 240, 480 or 600 volt, 2 pole		2.50	3.200		1,350	140		1,490	1,700
2800	3 pole		2.30	3.478		1,575	153		1,728	1,950
2820	600 amp, 240 or 480 volt, 2 pole		1.80	4.444		2,200	195		2,395	2,725
2840	3 pole		1.60	5		2,700	220		2,920	3,300
2860	800 amp, 240, 480 volt or 600 volt, 2 pole		1.50	5.333		3,325	234		3,559	4,000
2880	3 pole		1.30	6.154		3,875	270		4,145	4,675
2900	1000 amp, 240, 480 or 600 volt, 2 pole		1.10	7.273		4,075	320		4,395	4,950
2920	480 volt, 600 volt, 3 pole		1	8		4,675	350		5,025	5,650
2940	1200 amp, 240, 480 or 600 volt, 2 pole		.90	8.889		5,750	390		6,140	6,900
2960	3 pole		.80	10		6,275	440		6,715	7,550
2980	600 volt, 3 pole		.80	10		6,275	440		6,715	7,550

26 24 Switchboards and Panelboards

26 24 16 – Panelboards

26 24 16.10 Load Centers	Crew	Daily Output	Labor-Hours	Unit	Material	2007 Bare Costs Labor	Equipment	Total	Total Incl O&P
0010 **LOAD CENTERS** (residential type) R262416-50									
0100 3 wire, 120/240V, 1 phase, including 1 pole plug-in breakers									
0200 100 amp main lugs, indoor, 8 circuits	1 Elec	1.40	5.714	Ea.	162	251		413	555
0300 12 circuits		1.20	6.667		226	293		519	685
0400 Rainproof, 8 circuits		1.40	5.714		194	251		445	590
0500 12 circuits		1.20	6.667		274	293		567	735
0600 200 amp main lugs, indoor, 16 circuits	R-1A	1.80	8.889		315	315		630	825
0700 20 circuits		1.50	10.667		395	380		775	1,000
0800 24 circuits		1.30	12.308		515	440		955	1,225
0900 30 circuits		1.20	13.333		595	475		1,070	1,375
1000 42 circuits		.80	20		825	715		1,540	1,975
1200 Rainproof, 16 circuits		1.80	8.889		375	315		690	890
1300 20 circuits		1.50	10.667		450	380		830	1,075
1400 24 circuits		1.30	12.308		670	440		1,110	1,400
1500 30 circuits		1.20	13.333		750	475		1,225	1,550
1600 42 circuits		.80	20		980	715		1,695	2,150
1800 400 amp main lugs, indoor, 42 circuits		.72	22.222		1,150	790		1,940	2,475
1900 Rainproof, 42 circuit		.72	22.222		1,375	790		2,165	2,700
2200 Plug in breakers, 20 amp, 1 pole, 4 wire, 120/208 volts									
2210 125 amp main lugs, indoor, 12 circuits	1 Elec	1.20	6.667	Ea.	296	293		589	760
2300 18 circuits		.80	10		420	440		860	1,125
2400 Rainproof, 12 circuits		1.20	6.667		340	293		633	810
2500 18 circuits		.80	10		455	440		895	1,150
2600 200 amp main lugs, indoor, 24 circuits	R-1A	1.30	12.308		565	440		1,005	1,275
2700 30 circuits		1.20	13.333		645	475		1,120	1,425
2800 36 circuits		1	16		775	570		1,345	1,725
2900 42 circuits		.80	20		855	715		1,570	2,025
3000 Rainproof, 24 circuits		1.30	12.308		625	440		1,065	1,350
3100 30 circuits		1.20	13.333		705	475		1,180	1,500
3200 36 circuits		1	16		975	570		1,545	1,925
3300 42 circuits		.80	20		1,050	715		1,765	2,225
3500 400 amp main lugs, indoor, 42 circuits		.72	22.222		1,300	790		2,090	2,625
3600 Rainproof, 42 circuits		.72	22.222		1,525	790		2,315	2,900
3700 Plug-in breakers, 20 amp, 1 pole, 3 wire, 120/240 volts									
3800 100 amp main breaker, indoor, 12 circuits	1 Elec	1.20	6.667	Ea.	315	293		608	785
3900 18 circuits	"	.80	10		425	440		865	1,125
4000 200 amp main breaker, indoor, 20 circuits	R-1A	1.50	10.667		575	380		955	1,200
4200 24 circuits		1.30	12.308		695	440		1,135	1,425
4300 30 circuits		1.20	13.333		770	475		1,245	1,575
4400 40 circuits		.90	17.778		960	635		1,595	2,000
4500 Rainproof, 20 circuits		1.50	10.667		625	380		1,005	1,250
4600 24 circuits		1.30	12.308		745	440		1,185	1,475
4700 30 circuits		1.20	13.333		855	475		1,330	1,650
4800 40 circuits		.90	17.778		1,050	635		1,685	2,100
5000 400 amp main breaker, indoor, 42 circuits		.72	22.222		2,750	790		3,540	4,225
5100 Rainproof, 42 circuits		.72	22.222		3,075	790		3,865	4,600
5300 Plug in breakers, 20 amp, 1 pole, 4 wire, 120/208 volts									
5400 200 amp main breaker, indoor, 30 circuits	R-1A	1.20	13.333	Ea.	1,175	475		1,650	2,000
5500 42 circuits		.80	20		1,425	715		2,140	2,650
5600 Rainproof, 30 circuits		1.20	13.333		1,250	475		1,725	2,100
5700 42 circuits		.80	20		1,525	715		2,240	2,750

26 24 16 − Panelboards

26 24 16.20 Panelboard and Load Center Circuit Breakers	Crew	Daily Output	Labor-Hours	Unit	Material	2007 Bare Costs Labor	Equipment	Total	Total Incl O&P
0010 **PANELBOARD AND LOAD CENTER CIRCUIT BREAKERS** R262419-82									
0050 Bolt-on, 10,000 amp IC, 120 volt, 1 pole									
0100 15 to 50 amp	1 Elec	10	.800	Ea.	13.80	35		48.80	67.50
0200 60 amp		8	1		13.80	44		57.80	80.50
0300 70 amp	↓	8	1	↓	26	44		70	94.50
0350 240 volt, 2 pole									
0400 15 to 50 amp	1 Elec	8	1	Ea.	30.50	44		74.50	99
0500 60 amp		7.50	1.067		30.50	47		77.50	103
0600 80 to 100 amp		5	1.600		78	70		148	191
0700 3 pole, 15 to 60 amp		6.20	1.290		96	56.50		152.50	191
0800 70 amp		5	1.600		121	70		191	238
0900 80 to 100 amp		3.60	2.222		137	97.50		234.50	296
1000 22,000 amp I.C., 240 volt, 2 pole, 70 - 225 amp		2.70	2.963		590	130		720	845
1100 3 pole, 70 - 225 amp		2.30	3.478		655	153		808	945
1200 14,000 amp I.C., 277 volts, 1 pole, 15 - 30 amp		8	1		36.50	44		80.50	106
1300 22,000 amp I.C., 480 volts, 2 pole, 70 - 225 amp		2.70	2.963		590	130		720	845
1400 3 pole, 70 - 225 amp		2.30	3.478		730	153		883	1,025
2000 Plug-in panel or load center, 120/240 volt, to 60 amp, 1 pole		12	.667		10.70	29.50		40.20	55.50
2010 2 pole		9	.889		24	39		63	84.50
2020 3 pole		7.50	1.067		83.50	47		130.50	162
2030 100 amp, 2 pole		6	1.333		112	58.50		170.50	210
2040 3 pole		4.50	1.778		124	78		202	253
2050 150 to 200 amp, 2 pole		3	2.667		204	117		321	400
2060 Plug-in tandem, 120/240 V, 2-15 A, 1 pole		11	.727		26.50	32		58.50	77
2070 1-15 A & 1-20 A		11	.727		26.50	32		58.50	77
2080 2-20 A		11	.727		26.50	32		58.50	77
2100 High interrupting capacity, 120/240 volt, plug-in, 30 amp, 1 pole		12	.667		21	29.50		50.50	66.50
2110 60 amp, 2 pole		9	.889		49	39		88	112
2120 3 pole		7.50	1.067		127	47		174	210
2130 100 amp, 2 pole		6	1.333		141	58.50		199.50	242
2140 3 pole		4.50	1.778		182	78		260	315
2150 125 amp, 2 pole		3	2.667		375	117		492	590
2200 Bolt-on, 30 amp, 1 pole		10	.800		27	35		62	82
2210 60 amp, 2 pole		7.50	1.067		57	47		104	132
2220 3 pole		6.20	1.290		146	56.50		202.50	246
2230 100 amp, 2 pole		5	1.600		162	70		232	283
2240 3 pole		3.60	2.222		212	97.50		309.50	380
2300 Ground fault, 240 volt, 30 amp, 1 pole		7	1.143		83	50		133	166
2310 2 pole		6	1.333		148	58.50		206.50	250
2350 Key operated, 240 volt, 1 pole, 30 amp		7	1.143		75.50	50		125.50	158
2360 Switched neutral, 240 volt, 30 amp, 2 pole		6	1.333		34.50	58.50		93	125
2370 3 pole		5.50	1.455		52	64		116	152
2400 Shunt trip, for 240 volt breaker, 60 amp, 1 pole		4	2		67.50	88		155.50	206
2410 2 pole		3.50	2.286		67.50	100		167.50	224
2420 3 pole		3	2.667		67.50	117		184.50	249
2430 100 amp, 2 pole		3	2.667		67.50	117		184.50	249
2440 3 pole		2.50	3.200		67.50	140		207.50	284
2450 150 amp, 2 pole		2	4		161	176		337	440
2500 Auxiliary switch, for 240 volt breaker, 60 amp, 1 pole		4	2		61.50	88		149.50	199
2510 2 pole		3.50	2.286		61.50	100		161.50	217
2520 3 pole		3	2.667		61.50	117		178.50	242
2530 100 amp, 2 pole	↓	3	2.667	↓	61.50	117		178.50	242

26 24 Switchboards and Panelboards

26 24 16 – Panelboards

	26 24 16.20 Panelboard and Load Center Circuit Breakers	Crew	Daily Output	Labor-Hours	Unit	Material	2007 Bare Costs Labor	Equipment	Total	Total Incl O&P
2540	3 pole	1 Elec	2.50	3.200	Ea.	61.50	140		201.50	277
2550	150 amp, 2 pole		2	4		102	176		278	375
2600	Panel or load center, 277/480 volt, plug-in, 30 amp, 1 pole		12	.667		49.50	29.50		79	98
2610	60 amp, 2 pole		9	.889		149	39		188	222
2620	3 pole		7.50	1.067		218	47		265	310
2650	Bolt-on, 60 amp, 2 pole		7.50	1.067		149	47		196	234
2660	3 pole		6.20	1.290		218	56.50		274.50	325
2700	I-line, 277/480 volt, 30 amp, 1 pole		8	1		57	44		101	128
2710	60 amp, 2 pole		7.50	1.067		200	47		247	290
2720	3 pole		6.20	1.290		260	56.50		316.50	370
2730	100 amp, 1 pole		7.50	1.067		110	47		157	191
2740	2 pole		5	1.600		260	70		330	390
2750	3 pole		3.50	2.286		305	100		405	490
2800	High interrupting capacity, 277/480 volt, plug-in, 30 amp, 1 pole		12	.667		287	29.50		316.50	360
2810	60 amp, 2 pole		9	.889		445	39		484	545
2820	3 pole		7	1.143		515	50		565	640
2830	Bolt-on, 30 amp, 1 pole		8	1		287	44		331	380
2840	60 amp, 2 pole		7.50	1.067		445	47		492	555
2850	3 pole		6.20	1.290		515	56.50		571.50	650
2900	I-line, 30 amp, 1 pole		8	1		287	44		331	380
2910	60 amp, 2 pole		7.50	1.067		410	47		457	520
2920	3 pole		6.20	1.290		455	56.50		511.50	585
2930	100 amp, 1 pole		7.50	1.067		465	47		512	580
2940	2 pole		5	1.600		465	70		535	615
2950	3 pole		3.60	2.222		510	97.50		607.50	710
2960	Shunt trip, 277/480V breaker, remote oper., 30 amp, 1 pole		4	2		305	88		393	470
2970	60 amp, 2 pole		3.50	2.286		450	100		550	645
2980	3 pole		3	2.667		510	117		627	735
2990	100 amp, 1 pole		3.50	2.286		360	100		460	545
3000	2 pole		3	2.667		510	117		627	735
3010	3 pole		2.50	3.200		560	140		700	825
3050	Under voltage trip, 277/480 volt breaker, 30 amp, 1 pole		4	2		305	88		393	470
3060	60 amp, 2 pole		3.50	2.286		450	100		550	645
3070	3 pole		3	2.667		510	117		627	735
3080	100 amp, 1 pole		3.50	2.286		360	100		460	545
3090	2 pole		3	2.667		510	117		627	735
3100	3 pole		2.50	3.200		560	140		700	825
3150	Motor operated, 277/480 volt breaker, 30 amp, 1 pole		4	2		545	88		633	730
3160	60 amp, 2 pole		3.50	2.286		690	100		790	910
3170	3 pole		3	2.667		750	117		867	1,000
3180	100 amp, 1 pole		3.50	2.286		600	100		700	810
3190	2 pole		3	2.667		750	117		867	1,000
3200	3 pole		2.50	3.200		795	140		935	1,075
3250	Panelboard spacers, per pole		40	.200		4.14	8.80		12.94	17.60

26 24 16.30 Panelboards

		Crew	Daily Output	Labor-Hours	Unit	Material	2007 Bare Costs Labor	Equipment	Total	Total Incl O&P
0010	**PANELBOARDS** (Commercial use) R262416-50									
0050	NQOD, w/20 amp 1 pole bolt-on circuit breakers									
0100	3 wire, 120/240 volts, 100 amp main lugs									
0150	10 circuits	1 Elec	1	8	Ea.	455	350		805	1,025
0200	14 circuits		.88	9.091		535	400		935	1,175
0250	18 circuits		.75	10.667		585	470		1,055	1,325
0300	20 circuits		.65	12.308		655	540		1,195	1,525

26 24 Switchboards and Panelboards

26 24 16 – Panelboards

26 24 16.30 Panelboards	Crew	Daily Output	Labor-Hours	Unit	Material	2007 Bare Costs Labor	Equipment	Total	Total Incl O&P
0350	225 amp main lugs, 24 circuits	2 Elec	1.20	13.333	Ea.	745	585		1,330	1,700
0400	30 circuits		.90	17.778		870	780		1,650	2,100
0450	36 circuits		.80	20		990	880		1,870	2,400
0500	38 circuits		.72	22.222		1,075	975		2,050	2,625
0550	42 circuits		.66	24.242		1,125	1,075		2,200	2,800
0600	4 wire, 120/208 volts, 100 amp main lugs, 12 circuits	1 Elec	1	8		515	350		865	1,100
0650	16 circuits		.75	10.667		595	470		1,065	1,350
0700	20 circuits		.65	12.308		690	540		1,230	1,575
0750	24 circuits		.60	13.333		750	585		1,335	1,700
0800	30 circuits		.53	15.094		865	665		1,530	1,925
0850	225 amp main lugs, 32 circuits	2 Elec	.90	17.778		975	780		1,755	2,225
0900	34 circuits		.84	19.048		995	835		1,830	2,350
0950	36 circuits		.80	20		1,025	880		1,905	2,425
1000	42 circuits		.68	23.529		1,150	1,025		2,175	2,800
1040	225 amp main lugs, NEMA 7, 12 circuits		1	16		2,875	700		3,575	4,225
1100	24 circuits		.40	40		3,600	1,750		5,350	6,575
1200	NEHB,w/20 amp, 1 pole bolt-on circuit breakers									
1250	4 wire, 277/480 volts, 100 amp main lugs, 12 circuits	1 Elec	.88	9.091	Ea.	985	400		1,385	1,675
1300	20 circuits	"	.60	13.333		1,450	585		2,035	2,475
1350	225 amp main lugs, 24 circuits	2 Elec	.90	17.778		1,675	780		2,455	3,000
1400	30 circuits		.80	20		2,025	880		2,905	3,525
1450	36 circuits		.72	22.222		2,350	975		3,325	4,025
1500	42 circuits		.60	26.667		2,675	1,175		3,850	4,700
1510	225 amp main lugs, NEMA 7, 12 circuits		.90	17.778		3,975	780		4,755	5,525
1590	24 circuits		.30	53.333		4,825	2,350		7,175	8,775
1600	NQOD panel, w/20 amp, 1 pole, circuit breakers									
1650	3 wire, 120/240 volt with main circuit breaker									
1700	100 amp main, 12 circuits	1 Elec	.80	10	Ea.	630	440		1,070	1,350
1750	20 circuits	"	.60	13.333		810	585		1,395	1,750
1800	225 amp main, 30 circuits	2 Elec	.68	23.529		1,550	1,025		2,575	3,250
1850	42 circuits		.52	30.769		1,800	1,350		3,150	3,975
1900	400 amp main, 30 circuits		.54	29.630		2,150	1,300		3,450	4,275
1950	42 circuits		.50	32		2,400	1,400		3,800	4,725
2000	4 wire, 120/208 volts with main circuit breaker									
2050	100 amp main, 24 circuits	1 Elec	.47	17.021	Ea.	945	745		1,690	2,150
2100	30 circuits	"	.40	20		1,075	880		1,955	2,475
2200	225 amp main, 32 circuits	2 Elec	.72	22.222		1,800	975		2,775	3,425
2250	42 circuits		.56	28.571		1,975	1,250		3,225	4,050
2300	400 amp main, 42 circuits		.48	33.333		2,675	1,475		4,150	5,100
2350	600 amp main, 42 circuits		.40	40		3,950	1,750		5,700	6,975
2400	NEHB, with 20 amp, 1 pole circuit breaker									
2450	4 wire, 277/480 volts with main circuit breaker									
2500	100 amp main, 24 circuits	1 Elec	.42	19.048	Ea.	1,950	835		2,785	3,375
2550	30 circuits	"	.38	21.053		2,275	925		3,200	3,875
2600	225 amp main, 30 circuits	2 Elec	.72	22.222		2,875	975		3,850	4,600
2650	42 circuits		.56	28.571		3,525	1,250		4,775	5,775
2700	400 amp main, 42 circuits		.46	34.783		4,250	1,525		5,775	6,950
2750	600 amp main, 42 circuits		.38	42.105		5,850	1,850		7,700	9,175
2900	Note: the following line items don't include branch circuit breakers									
2910	For branch circuit breakers information, refer to									
2920	Division 26 24 16.20									
3010	Main lug, no main breaker, 240 volt, 1 pole, 3 wire, 100 amp	1 Elec	2.30	3.478	Ea.	410	153		563	675
3020	225 amp	2 Elec	2.40	6.667		500	293		793	985

26 24 16 – Panelboards

26 24 16.30 Panelboards		Crew	Daily Output	Labor-Hours	Unit	Material	2007 Bare Costs Labor	Equipment	Total	Total Incl O&P
3030	400 amp	2 Elec	1.80	8.889	Ea.	745	390		1,135	1,400
3060	3 pole, 3 wire, 100 amp	1 Elec	2.30	3.478		445	153		598	715
3070	225 amp	2 Elec	2.40	6.667		530	293		823	1,025
3080	400 amp		1.80	8.889		810	390		1,200	1,475
3090	600 amp		1.60	10		945	440		1,385	1,700
3110	3 pole, 4 wire, 100 amp	1 Elec	2.30	3.478		495	153		648	770
3120	225 amp	2 Elec	2.40	6.667		625	293		918	1,125
3130	400 amp		1.80	8.889		810	390		1,200	1,475
3140	600 amp		1.60	10		945	440		1,385	1,700
3160	480 volt, 3 pole, 3 wire, 100 amp	1 Elec	2.30	3.478		595	153		748	880
3170	225 amp	2 Elec	2.40	6.667		730	293		1,023	1,250
3180	400 amp		1.80	8.889		1,025	390		1,415	1,700
3190	600 amp		1.60	10		1,150	440		1,590	1,925
3210	277/480 volt, 3 pole, 4 wire, 100 amp	1 Elec	2.30	3.478		580	153		733	865
3220	225 amp	2 Elec	2.40	6.667		710	293		1,003	1,225
3230	400 amp		1.80	8.889		1,000	390		1,390	1,675
3240	600 amp		1.60	10		1,125	440		1,565	1,875
3260	Main circuit breaker, 240 volt, 1 pole, 3 wire, 100 amp	1 Elec	2	4		560	176		736	880
3270	225 amp	2 Elec	2	8		1,175	350		1,525	1,825
3280	400 amp	"	1.60	10		1,875	440		2,315	2,725
3310	3 pole, 3 wire, 100 amp	1 Elec	2	4		650	176		826	975
3320	225 amp	2 Elec	2	8		1,350	350		1,700	2,025
3330	400 amp		1.60	10		2,150	440		2,590	3,000
3360	120/208 volt, 3 pole, 4 wire, 100 amp		4	4		650	176		826	975
3370	225 amp		2	8		1,350	350		1,700	2,025
3380	400 amp		1.60	10		2,150	440		2,590	3,000
3410	480 volt, 3 pole, 3 wire, 100 amp	1 Elec	2	4		920	176		1,096	1,250
3420	225 amp	2 Elec	2	8		1,600	350		1,950	2,300
3430	400 amp		1.60	10		2,425	440		2,865	3,300
3460	277/480 volt, 3 pole, 4 wire, 100 amp		4	4		895	176		1,071	1,250
3470	225 amp		2	8		1,575	350		1,925	2,250
3480	400 amp		1.60	10		2,425	440		2,865	3,325
3510	Main circuit breaker, HIC, 240 volt, 1 pole, 3 wire, 100 amp	1 Elec	2	4		875	176		1,051	1,225
3520	225 amp	2 Elec	2	8		2,350	350		2,700	3,100
3530	400 amp	"	1.60	10		3,150	440		3,590	4,125
3560	3 pole, 3 wire, 100 amp	1 Elec	2	4		975	176		1,151	1,325
3570	225 amp	2 Elec	2	8		2,650	350		3,000	3,450
3580	400 amp	"	1.60	10		3,525	440		3,965	4,525
3610	120/208 volt, 3 pole, 4 wire, 100 amp	1 Elec	2	4		975	176		1,151	1,325
3620	225 amp	2 Elec	2	8		2,650	350		3,000	3,450
3630	400 amp	"	1.60	10		3,525	440		3,965	4,525
3660	480 volt, 3 pole, 3 wire, 100 amp	1 Elec	2	4		1,475	176		1,651	1,875
3670	225 amp	2 Elec	2	8		2,950	350		3,300	3,775
3680	400 amp	"	1.60	10		3,750	440		4,190	4,775
3710	277/480 volt, 3 pole, 4 wire, 100 amp	1 Elec	2	4		1,400	176		1,576	1,800
3720	225 amp	2 Elec	2	8		2,800	350		3,150	3,600
3730	400 amp	"	1.60	10		3,725	440		4,165	4,750
3760	Main circuit breaker, shunt trip, 100 amp	1 Elec	1.20	6.667		730	293		1,023	1,225
3770	225 amp	2 Elec	1.60	10		1,725	440		2,165	2,550
3780	400 amp	"	1.40	11.429		2,525	500		3,025	3,525

26 24 19.20 Motor Control Center Components		Crew	Daily Output	Labor-Hours	Unit	Material	2007 Bare Costs Labor	Equipment	Total	Total Incl O&P
0010	**MOTOR CONTROL CENTER COMPONENTS** R262419-60									
0100	Starter, size 1, FVNR, NEMA 1, type A, fusible	1 Elec	2.70	2.963	Ea.	1,350	130		1,480	1,700
0120	Circuit breaker		2.70	2.963		1,475	130		1,605	1,825
0140	Type B, fusible		2.70	2.963		1,500	130		1,630	1,850
0160	Circuit breaker		2.70	2.963		1,625	130		1,755	1,975
0180	NEMA 12, type A, fusible		2.60	3.077		1,375	135		1,510	1,725
0200	Circuit breaker		2.60	3.077		1,500	135		1,635	1,850
0220	Type B, fusible		2.60	3.077		1,525	135		1,660	1,875
0240	Circuit breaker		2.60	3.077		1,650	135		1,785	2,000
0300	Starter, size 1, FVR, NEMA 1, type A, fusible		2	4		1,950	176		2,126	2,400
0320	Circuit breaker		2	4		1,950	176		2,126	2,400
0340	Type B, fusible		2	4		2,150	176		2,326	2,600
0360	Circuit breaker		2	4		2,150	176		2,326	2,600
0380	NEMA 12, type A, fusible		1.90	4.211		1,975	185		2,160	2,450
0400	Circuit breaker		1.90	4.211		1,975	185		2,160	2,450
0420	Type B, fusible		1.90	4.211		2,175	185		2,360	2,650
0440	Circuit breaker	▼	1.90	4.211	▼	2,175	185		2,360	2,650
0490	Starter size 1, 2 speed, separate winding									
0500	NEMA 1, type A, fusible	1 Elec	2.60	3.077	Ea.	2,575	135		2,710	3,025
0520	Circuit breaker		2.60	3.077		2,575	135		2,710	3,025
0540	Type B, fusible		2.60	3.077		2,825	135		2,960	3,300
0560	Circuit breaker		2.60	3.077		2,825	135		2,960	3,300
0580	NEMA 12, type A, fusible		2.50	3.200		2,600	140		2,740	3,075
0600	Circuit breaker		2.50	3.200		2,600	140		2,740	3,075
0620	Type B, fusible		2.50	3.200		2,875	140		3,015	3,350
0640	Circuit breaker	▼	2.50	3.200	▼	2,875	140		3,015	3,350
0650	Starter size 1, 2 speed, consequent pole									
0660	NEMA 1, type A, fusible	1 Elec	2.60	3.077	Ea.	2,575	135		2,710	3,025
0680	Circuit breaker		2.60	3.077		2,575	135		2,710	3,025
0700	Type B, fusible		2.60	3.077		2,825	135		2,960	3,300
0720	Circuit breaker		2.60	3.077		2,825	135		2,960	3,300
0740	NEMA 12, type A, fusible		2.50	3.200		2,600	140		2,740	3,075
0760	Circuit breaker		2.50	3.200		2,600	140		2,740	3,075
0780	Type B, fusible		2.50	3.200		2,850	140		2,990	3,325
0800	Circuit breaker	▼	2.50	3.200	▼	2,875	140		3,015	3,350
0810	Starter size 1, 2 speed, space only									
0820	NEMA 1, type A, fusible	1 Elec	16	.500	Ea.	575	22		597	670
0840	Circuit breaker		16	.500		575	22		597	670
0860	Type B, fusible		16	.500		575	22		597	670
0880	Circuit breaker		16	.500		575	22		597	670
0900	NEMA 12, type A, fusible		15	.533		600	23.50		623.50	695
0920	Circuit breaker		15	.533		600	23.50		623.50	695
0940	Type B, fusible		15	.533		600	23.50		623.50	695
0960	Circuit breaker	▼	15	.533		600	23.50		623.50	695
1100	Starter size 2, FVNR, NEMA 1, type A, fusible	2 Elec	4	4		1,525	176		1,701	1,925
1120	Circuit breaker		4	4		1,675	176		1,851	2,100
1140	Type B, fusible		4	4		1,675	176		1,851	2,100
1160	Circuit breaker		4	4		1,850	176		2,026	2,275
1180	NEMA 12, type A, fusible		3.80	4.211		1,550	185		1,735	2,000
1200	Circuit breaker		3.80	4.211		1,700	185		1,885	2,150
1220	Type B, fusible		3.80	4.211		1,700	185		1,885	2,150
1240	Circuit breaker	▼	3.80	4.211	▼	1,850	185		2,035	2,325

26 24 Switchboards and Panelboards

26 24 19 – Motor-Control Centers

26 24 19.20 Motor Control Center Components	Crew	Daily Output	Labor-Hours	Unit	Material	2007 Bare Costs Labor	Equipment	Total	Total Incl O&P
1300 FVR, NEMA 1, type A, fusible	2 Elec	3.20	5	Ea.	2,600	220		2,820	3,175
1320 Circuit breaker		3.20	5		2,600	220		2,820	3,175
1340 Type B, fusible		3.20	5		2,850	220		3,070	3,475
1360 Circuit breaker		3.20	5		2,850	220		3,070	3,475
1380 NEMA type 12, type A, fusible		3	5.333		2,650	234		2,884	3,250
1400 Circuit breaker		3	5.333		2,650	234		2,884	3,250
1420 Type B, fusible		3	5.333		2,900	234		3,134	3,550
1440 Circuit breaker		3	5.333		2,900	234		3,134	3,550
1490 Starter size 2, 2 speed, separate winding									
1500 NEMA 1, type A, fusible	2 Elec	3.80	4.211	Ea.	2,900	185		3,085	3,475
1520 Circuit breaker		3.80	4.211		2,900	185		3,085	3,475
1540 Type B, fusible		3.80	4.211		3,175	185		3,360	3,775
1560 Circuit breaker		3.80	4.211		3,175	185		3,360	3,775
1570 NEMA 12, type A, fusible		3.60	4.444		2,950	195		3,145	3,550
1580 Circuit breaker		3.60	4.444		2,950	195		3,145	3,550
1600 Type B, fusible		3.60	4.444		3,225	195		3,420	3,850
1620 Circuit breaker		3.60	4.444		3,225	195		3,420	3,850
1630 Starter size 2, 2 speed, consequent pole									
1640 NEMA 1, type A, fusible	2 Elec	3.80	4.211	Ea.	3,300	185		3,485	3,900
1660 Circuit breaker		3.80	4.211		3,300	185		3,485	3,900
1680 Type B, fusible		3.80	4.211		3,475	185		3,660	4,100
1700 Circuit breaker		3.80	4.211		3,475	185		3,660	4,100
1720 NEMA 12, type A, fusible		3.80	4.211		3,300	185		3,485	3,900
1740 Circuit breaker		3.60	4.444		3,325	195		3,520	3,975
1760 Type B, fusible		3.60	4.444		3,525	195		3,720	4,175
1780 Circuit breaker		3.60	4.444		3,525	195		3,720	4,175
1830 Starter size 2, autotransformer									
1840 NEMA 1, type A, fusible	2 Elec	3.40	4.706	Ea.	5,125	207		5,332	5,950
1860 Circuit breaker		3.40	4.706		5,250	207		5,457	6,075
1880 Type B, fusible		3.40	4.706		5,600	207		5,807	6,475
1900 Circuit breaker		3.40	4.706		5,600	207		5,807	6,475
1920 NEMA 12, type A, fusible		3.20	5		5,225	220		5,445	6,075
1940 Circuit breaker		3.20	5		5,225	220		5,445	6,075
1960 Type B, fusible		3.20	5		5,700	220		5,920	6,600
1980 Circuit breaker		3.20	5		5,700	220		5,920	6,600
2030 Starter size 2, space only									
2040 NEMA 1, type A, fusible	1 Elec	16	.500	Ea.	575	22		597	670
2060 Circuit breaker		16	.500		575	22		597	670
2080 Type B, fusible		16	.500		575	22		597	670
2100 Circuit breaker		16	.500		575	22		597	670
2120 NEMA 12, type A, fusible		15	.533		600	23.50		623.50	695
2140 Circuit breaker		15	.533		600	23.50		623.50	695
2160 Type B, fusible		15	.533		600	23.50		623.50	695
2180 Circuit breaker		15	.533		600	23.50		623.50	695
2300 Starter size 3, FVNR, NEMA 1, type A, fusible	2 Elec	2	8		2,925	350		3,275	3,750
2320 Circuit breaker		2	8		2,600	350		2,950	3,375
2340 Type B, fusible		2	8		3,225	350		3,575	4,050
2360 Circuit breaker		2	8		2,850	350		3,200	3,675
2380 NEMA 12, type A, fusible		1.90	8.421		2,975	370		3,345	3,825
2400 Circuit breaker		1.90	8.421		2,650	370		3,020	3,450
2420 Type B, fusible		1.90	8.421		3,325	370		3,695	4,200
2440 Circuit breaker		1.90	8.421		2,900	370		3,270	3,750
2500 Starter size 3, FVR, NEMA 1, type A, fusible		1.60	10		3,975	440		4,415	5,025

26 24 19.20 Motor Control Center Components		Crew	Daily Output	Labor-Hours	Unit	Material	2007 Bare Costs Labor	Equipment	Total	Total Incl O&P
2520	Circuit breaker	2 Elec	1.60	10	Ea.	3,800	440		4,240	4,825
2540	Type B, fusible		1.60	10		4,325	440		4,765	5,425
2560	Circuit breaker		1.60	10		4,150	440		4,590	5,225
2580	NEMA 12, type A, fusible		1.50	10.667		4,075	470		4,545	5,175
2600	Circuit breaker		1.50	10.667		3,875	470		4,345	4,975
2620	Type B, fusible		1.50	10.667		4,425	470		4,895	5,550
2640	Circuit breaker		1.50	10.667		4,575	470		5,045	5,750
2690	Starter size 3, 2 speed, separate winding									
2700	NEMA 1, type A, fusible	2 Elec	2	8	Ea.	4,575	350		4,925	5,575
2720	Circuit breaker		2	8		4,050	350		4,400	4,975
2740	Type B, fusible		2	8		5,025	350		5,375	6,050
2760	Circuit breaker		2	8		4,425	350		4,775	5,400
2780	NEMA 12, type A, fusible		1.90	8.421		4,675	370		5,045	5,700
2800	Circuit breaker		1.90	8.421		4,125	370		4,495	5,100
2820	Type B, fusible		1.90	8.421		5,100	370		5,470	6,175
2840	Circuit breaker		1.90	8.421		4,500	370		4,870	5,500
2850	Starter size 3, 2 speed, consequent pole									
2860	NEMA 1, type A, fusible	2 Elec	2	8	Ea.	5,100	350		5,450	6,150
2880	Circuit breaker		2	8		4,575	350		4,925	5,550
2900	Type B, fusible		2	8		5,550	350		5,900	6,625
2920	Circuit breaker		2	8		4,950	350		5,300	5,975
2940	NEMA 12, type A, fusible		1.90	8.421		5,200	370		5,570	6,275
2960	Circuit breaker		1.90	8.421		4,650	370		5,020	5,675
2980	Type B, fusible		1.90	8.421		5,650	370		6,020	6,750
3000	Circuit breaker		1.90	8.421		5,050	370		5,420	6,100
3100	Starter size 3, autotransformer, NEMA 1, type A, fusible		1.60	10		6,350	440		6,790	7,650
3120	Circuit breaker		1.60	10		6,350	440		6,790	7,650
3140	Type B, fusible		1.60	10		6,950	440		7,390	8,300
3160	Circuit breaker		1.60	10		6,950	440		7,390	8,300
3180	NEMA 12, type A, fusible		1.50	10.667		6,475	470		6,945	7,825
3200	Circuit breaker		1.50	10.667		6,475	470		6,945	7,825
3220	Type B, fusible		1.50	10.667		7,075	470		7,545	8,475
3240	Circuit breaker		1.50	10.667		7,075	470		7,545	8,475
3260	Starter size 3, space only, NEMA 1, type A, fusible	1 Elec	15	.533		990	23.50		1,013.50	1,100
3280	Circuit breaker		15	.533		765	23.50		788.50	880
3300	Type B, fusible		15	.533		990	23.50		1,013.50	1,100
3320	Circuit breaker		15	.533		765	23.50		788.50	880
3340	NEMA 12, type A, fusible		14	.571		1,050	25		1,075	1,200
3360	Circuit breaker		14	.571		815	25		840	935
3380	Type B, fusible		14	.571		1,050	25		1,075	1,200
3400	Circuit breaker		14	.571		815	25		840	935
3500	Starter size 4, FVNR, NEMA 1, type A, fusible	2 Elec	1.60	10		3,800	440		4,240	4,825
3520	Circuit breaker		1.60	10		3,450	440		3,890	4,450
3540	Type B, fusible		1.60	10		4,150	440		4,590	5,225
3560	Circuit breaker		1.60	10		3,800	440		4,240	4,825
3580	NEMA 12, type A, fusible		1.50	10.667		3,900	470		4,370	5,000
3600	Circuit breaker		1.50	10.667		3,500	470		3,970	4,550
3620	Type B, fusible		1.50	10.667		4,250	470		4,720	5,375
3640	Circuit breaker		1.50	10.667		3,875	470		4,345	4,950
3700	Starter size 4, FVR, NEMA 1, type A, fusible		1.20	13.333		5,200	585		5,785	6,600
3720	Circuit breaker		1.20	13.333		4,700	585		5,285	6,025
3740	Type B, fusible		1.20	13.333		5,675	585		6,260	7,125
3760	Circuit breaker		1.20	13.333		5,175	585		5,760	6,550

26 24 19.20 Motor Control Center Components	Crew	Daily Output	Labor-Hours	Unit	Material	2007 Bare Costs Labor	2007 Bare Costs Equipment	Total	Total Incl O&P
3780 NEMA 12, type A, fusible	2 Elec	1.16	13.793	Ea.	5,325	605		5,930	6,750
3800 Circuit breaker		1.16	13.793		4,750	605		5,355	6,125
3820 Type B, fusible		1.16	13.793		4,800	605		5,405	6,200
3840 Circuit breaker		1.16	13.793		5,225	605		5,830	6,650
3890 Starter size 4, 2 speed, separate windings									
3900 NEMA 1, type A, fusible	2 Elec	1.60	10	Ea.	6,575	440		7,015	7,875
3920 Circuit breaker		1.60	10		4,925	440		5,365	6,075
3940 Type B, fusible		1.60	10		7,225	440		7,665	8,575
3960 Circuit breaker		1.60	10		5,400	440		5,840	6,600
3980 NEMA 12, type A, fusible		1.50	10.667		6,675	470		7,145	8,050
4000 Circuit breaker		1.50	10.667		5,025	470		5,495	6,225
4020 Type B, fusible		1.50	10.667		7,325	470		7,795	8,775
4040 Circuit breaker		1.50	10.667		5,475	470		5,945	6,725
4050 Starter size 4, 2 speed, consequent pole									
4060 NEMA 1, type A, fusible	2 Elec	1.60	10	Ea.	7,675	440		8,115	9,075
4080 Circuit breaker		1.60	10		5,700	440		6,140	6,925
4100 Type B, fusible		1.60	10		8,425	440		8,865	9,925
4120 Circuit breaker		1.60	10		6,250	440		6,690	7,525
4140 NEMA 12, type A, fusible		1.50	10.667		7,800	470		8,270	9,275
4160 Circuit breaker		1.50	10.667		5,800	470		6,270	7,075
4180 Type B, fusible		1.50	10.667		8,550	470		9,020	10,100
4200 Circuit breaker		1.50	10.667		6,325	470		6,795	7,675
4300 Starter size 4, autotransformer, NEMA 1, type A, fusible		1.30	12.308		7,550	540		8,090	9,100
4320 Circuit breaker		1.30	12.308		7,625	540		8,165	9,175
4340 Type B, fusible		1.30	12.308		8,325	540		8,865	9,950
4360 Circuit breaker		1.30	12.308		8,325	540		8,865	9,950
4380 NEMA 12, type A, fusible		1.24	12.903		7,675	565		8,240	9,300
4400 Circuit breaker		1.24	12.903		7,725	565		8,290	9,350
4420 Type B, fusible		1.24	12.903		8,450	565		9,015	10,100
4440 Circuit breaker		1.24	12.903		8,425	565		8,990	10,100
4500 Starter size 4, space only, NEMA 1, type A, fusible	1 Elec	14	.571		1,375	25		1,400	1,550
4520 Circuit breaker		14	.571		990	25		1,015	1,125
4540 Type B, fusible		14	.571		1,375	25		1,400	1,550
4560 Circuit breaker		14	.571		990	25		1,015	1,125
4580 NEMA 12, type A, fusible		13	.615		1,450	27		1,477	1,650
4600 Circuit breaker		13	.615		1,050	27		1,077	1,200
4620 Type B, fusible		13	.615		1,450	27		1,477	1,650
4640 Circuit breaker		13	.615		1,050	27		1,077	1,200
4800 Starter size 5, FVNR, NEMA 1, type A, fusible	2 Elec	1	16		7,850	700		8,550	9,700
4820 Circuit breaker		1	16		5,575	700		6,275	7,175
4840 Type B, fusible		1	16		8,625	700		9,325	10,500
4860 Circuit breaker		1	16		6,125	700		6,825	7,800
4880 NEMA 12, type A, fusible		.96	16.667		7,950	730		8,680	9,850
4900 Circuit breaker		.96	16.667		5,675	730		6,405	7,350
4920 Type B, fusible		.96	16.667		8,725	730		9,455	10,700
4940 Circuit breaker		.96	16.667		6,250	730		6,980	7,975
5000 Starter size 5, FVR, NEMA 1, type A, fusible		.80	20		11,600	880		12,480	14,000
5020 Circuit breaker		.80	20		9,075	880		9,955	11,300
5040 Type B, fusible		.80	20		12,700	880		13,580	15,300
5060 Circuit breaker		.80	20		9,975	880		10,855	12,300
5080 NEMA 12, type A, fusible		.76	21.053		11,800	925		12,725	14,400
5100 Circuit breaker		.76	21.053		9,225	925		10,150	11,600
5120 Type B, fusible		.76	21.053		12,900	925		13,825	15,600

26 24 19.20 Motor Control Center Components	Crew	Daily Output	Labor-Hours	Unit	Material	2007 Bare Costs Labor	Equipment	Total	Total Incl O&P
5140	Circuit breaker	2 Elec	.76	21.053	Ea.	10,100	925		11,025	12,500
5190	Starter size 5, 2 speed, separate windings									
5200	NEMA 1, type A, fusible	2 Elec	1	16	Ea.	16,000	700		16,700	18,600
5220	Circuit breaker		1	16		11,900	700		12,600	14,200
5240	Type B, fusible		1	16		17,500	700		18,200	20,400
5260	Circuit breaker		1	16		13,100	700		13,800	15,500
5280	NEMA 12, type A, fusible		.96	16.667		16,200	730		16,930	18,900
5300	Circuit breaker		.96	16.667		12,000	730		12,730	14,300
5320	Type B, fusible		.96	16.667		17,700	730		18,430	20,600
5340	Circuit breaker		.96	16.667		13,200	730		13,930	15,600
5400	Starter size 5, autotransformer, NEMA 1, type A, fusible		.70	22.857		13,100	1,000		14,100	15,900
5420	Circuit breaker		.70	22.857		10,900	1,000		11,900	13,400
5440	Type B, fusible		.70	22.857		14,300	1,000		15,300	17,300
5460	Circuit breaker		.70	22.857		11,900	1,000		12,900	14,600
5480	NEMA 12, type A, fusible		.68	23.529		13,200	1,025		14,225	16,100
5500	Circuit breaker		.68	23.529		11,000	1,025		12,025	13,700
5520	Type B, fusible		.68	23.529		14,500	1,025		15,525	17,500
5540	Circuit breakers	↓	.68	23.529		12,000	1,025		13,025	14,800
5600	Starter size 5, space only, NEMA 1, type A, fusible	1 Elec	12	.667		1,750	29.50		1,779.50	1,975
5620	Circuit breaker		12	.667		1,750	29.50		1,779.50	1,975
5640	Type B, fusible		12	.667		1,750	29.50		1,779.50	1,975
5660	Circuit breaker		12	.667		1,150	29.50		1,179.50	1,325
5680	NEMA 12, type A, fusible		11	.727		1,850	32		1,882	2,100
5700	Circuit breaker		11	.727		1,225	32		1,257	1,400
5720	Type B, fusible		11	.727		1,850	32		1,882	2,100
5740	Circuit breaker		11	.727		1,225	32		1,257	1,400
5800	Fuse, light contactor NEMA 1, type A, 30 amp		2.70	2.963		1,350	130		1,480	1,700
5820	60 amp		2	4		1,525	176		1,701	1,925
5840	100 amp		1	8		2,925	350		3,275	3,750
5860	200 amp		.80	10		6,775	440		7,215	8,125
5880	Type B, 30 amp		2.70	2.963		1,475	130		1,605	1,825
5900	60 amp		2	4		1,650	176		1,826	2,075
5920	100 amp		1	8		3,225	350		3,575	4,050
5940	200 amp	2 Elec	1.60	10		7,425	440		7,865	8,825
5960	NEMA 12, type A, 30 amp	1 Elec	2.60	3.077		1,375	135		1,510	1,725
5980	60 amp		1.90	4.211		1,550	185		1,735	2,000
6000	100 amp	↓	.95	8.421		2,975	370		3,345	3,825
6020	200 amp	2 Elec	1.50	10.667		6,900	470		7,370	8,275
6040	Type B, 30 amp	1 Elec	2.60	3.077		1,500	135		1,635	1,850
6060	60 amp		1.90	4.211		1,675	185		1,860	2,125
6080	100 amp	↓	.95	8.421		3,275	370		3,645	4,150
6100	200 amp	2 Elec	1.50	10.667		7,550	470		8,020	9,000
6200	Circuit breaker, light contactor NEMA 1, type A, 30 amp	1 Elec	2.70	2.963		1,475	130		1,605	1,825
6220	60 amp		2	4		1,675	176		1,851	2,100
6240	100 amp	↓	1	8		2,600	350		2,950	3,375
6260	200 amp	2 Elec	1.60	10		5,625	440		6,065	6,850
6280	Type B, 30 amp	1 Elec	2.70	2.963		1,625	130		1,755	1,975
6300	60 amp		2	4		1,850	176		2,026	2,300
6320	100 amp	↓	1	8		2,825	350		3,175	3,650
6340	200 amp	2 Elec	1.60	10		6,125	440		6,565	7,400
6360	NEMA 12, type A, 30 amp	1 Elec	2.60	3.077		1,500	135		1,635	1,850
6380	60 amp		1.90	4.211		1,700	185		1,885	2,150
6400	100 amp	↓	.95	8.421	↓	2,650	370		3,020	3,450

26 24 Switchboards and Panelboards

26 24 19 – Motor-Control Centers

26 24 19.20 Motor Control Center Components	Crew	Daily Output	Labor-Hours	Unit	Material	2007 Bare Costs Labor	Equipment	Total	Total Incl O&P
6420	200 amp	2 Elec	1.50	10.667	Ea.	5,650	470		6,120	6,900
6440	Type B, 30 amp	1 Elec	2.60	3.077		1,775	135		1,910	2,150
6460	60 amp		1.90	4.211		1,875	185		2,060	2,350
6480	100 amp		.95	8.421		2,875	370		3,245	3,725
6500	200 amp	2 Elec	1.50	10.667		6,200	470		6,670	7,525
6600	Fusible switch, NEMA 1, type A, 30 amp	1 Elec	5.30	1.509		900	66.50		966.50	1,100
6620	60 amp		5	1.600		960	70		1,030	1,150
6640	100 amp		4	2		1,050	88		1,138	1,300
6660	200 amp		3.20	2.500		1,825	110		1,935	2,175
6680	400 amp	2 Elec	4.60	3.478		4,300	153		4,453	4,975
6700	600 amp		3.20	5		4,650	220		4,870	5,450
6720	800 amp		2.60	6.154		12,800	270		13,070	14,500
6740	NEMA 12, type A, 30 amp	1 Elec	5.20	1.538		925	67.50		992.50	1,125
6760	60 amp		4.90	1.633		980	71.50		1,051.50	1,175
6780	100 amp		3.90	2.051		1,075	90		1,165	1,325
6800	200 amp		3.10	2.581		1,875	113		1,988	2,250
6820	400 amp	2 Elec	4.40	3.636		4,400	160		4,560	5,100
6840	600 amp		3	5.333		4,800	234		5,034	5,625
6860	800 amp		2.40	6.667		12,900	293		13,193	14,600
6900	Circuit breaker, NEMA 1, type A, 30 amp	1 Elec	5.30	1.509		825	66.50		891.50	1,000
6920	60 amp		5	1.600		825	70		895	1,025
6940	100 amp		4	2		825	88		913	1,050
6960	225 amp		3.20	2.500		1,475	110		1,585	1,800
6980	400 amp	2 Elec	4.60	3.478		2,800	153		2,953	3,300
7000	600 amp		3.20	5		3,275	220		3,495	3,925
7020	800 amp		2.60	6.154		6,850	270		7,120	7,925
7040	NEMA 12, type A, 30 amp	1 Elec	5.20	1.538		850	67.50		917.50	1,025
7060	60 amp		4.90	1.633		850	71.50		921.50	1,050
7080	100 amp		3.90	2.051		850	90		940	1,075
7100	225 amp		3.10	2.581		1,500	113		1,613	1,825
7120	400 amp	2 Elec	4.40	3.636		2,850	160		3,010	3,375
7140	600 amp		3	5.333		3,325	234		3,559	4,000
7160	800 amp		2.40	6.667		6,975	293		7,268	8,100
7300	Incoming line, main lug only, 600 amp, alum, NEMA 1		1.60	10		1,050	440		1,490	1,800
7320	NEMA 12		1.50	10.667		1,075	470		1,545	1,875
7340	Copper, NEMA 1		1.60	10		1,100	440		1,540	1,850
7360	800 amp, alum., NEMA 1		1.50	10.667		2,625	470		3,095	3,575
7380	NEMA 12		1.40	11.429		2,650	500		3,150	3,675
7400	Copper, NEMA 1		1.50	10.667		2,725	470		3,195	3,700
7420	1200 amp, copper, NEMA 1		1.40	11.429		2,825	500		3,325	3,850
7440	Incoming line, fusible switch, 400 amp, alum., NEMA 1		1.20	13.333		3,600	585		4,185	4,825
7460	NEMA 12		1.10	14.545		3,675	640		4,315	5,000
7480	Copper, NEMA 1		1.20	13.333		3,650	585		4,235	4,900
7500	600 amp, alum., NEMA 1		1.10	14.545		4,400	640		5,040	5,800
7520	NEMA 12		1	16		4,475	700		5,175	5,975
7540	Copper, NEMA 1		1.10	14.545		4,450	640		5,090	5,850
7560	Incoming line, circuit breaker, 225 amp, alum., NEMA 1		1.20	13.333		1,875	585		2,460	2,950
7580	NEMA 12		1.10	14.545		1,900	640		2,540	3,050
7600	Copper, NEMA 1		1.20	13.333		1,925	585		2,510	3,000
7620	400 amp, alum., NEMA 1		1.20	13.333		2,800	585		3,385	3,950
7640	NEMA 12		1.10	14.545		2,850	640		3,490	4,075
7660	Copper, NEMA 1		1.20	13.333		2,850	585		3,435	4,025
7680	600 amp, alum., NEMA 1		1.10	14.545		3,275	640		3,915	4,550

215

26 24 19.20 Motor Control Center Components

		Crew	Daily Output	Labor-Hours	Unit	Material	2007 Bare Costs Labor	Equipment	Total	Total Incl O&P
7700	NEMA 12	2 Elec	1	16	Ea.	3,325	700		4,025	4,700
7720	Copper, NEMA 1		1.10	14.545		3,325	640		3,965	4,625
7740	800 amp, copper, NEMA 1	↓	.90	17.778		6,850	780		7,630	8,675
7760	Incoming line, for copper bus, add					103			103	114
7780	For 65000 amp bus bracing, add					153			153	169
7800	For NEMA 3R enclosure, add					4,825			4,825	5,300
7820	For NEMA 12 enclosure, add					133			133	146
7840	For 1/4" x 1" ground bus, add	1 Elec	16	.500		85.50	22		107.50	127
7860	For 1/4" x 2" ground bus, add	"	12	.667		85.50	29.50		115	138
7880	Main rating basic section, alum., NEMA 1, 600 amp	2 Elec	1.60	10			440		440	655
7900	800 amp		1.40	11.429		280	500		780	1,050
7920	1200 amp	↓	1.20	13.333		550	585		1,135	1,475
7940	For copper bus, add					415			415	455
7960	For 65000 amp bus bracing, add					277			277	305
7980	For NEMA 3R enclosure, add					4,825			4,825	5,300
8000	For NEMA 12, enclosure, add					133			133	146
8020	For 1/4" x 1" ground bus, add	1 Elec	16	.500		85.50	22		107.50	127
8040	For 1/4" x 2" ground bus, add		12	.667		85.50	29.50		115	138
8060	Unit devices, pilot light, standard		16	.500		74	22		96	114
8080	Pilot light, push to test		16	.500		103	22		125	147
8100	Pilot light, standard, and push button		12	.667		177	29.50		206.50	239
8120	Pilot light, push to test, and push button		12	.667		207	29.50		236.50	271
8140	Pilot light, standard, and select switch		12	.667		177	29.50		206.50	239
8160	Pilot light, push to test, and select switch	↓	12	.667	↓	207	29.50		236.50	271

26 24 19.30 Motor Control Center

		Crew	Daily Output	Labor-Hours	Unit	Material	2007 Bare Costs Labor	Equipment	Total	Total Incl O&P
0010	**MOTOR CONTROL CENTER** Consists of starters & structures R262419-60									
0050	Starters, class 1, type B, comb. MCP, FVNR, with									
0100	control transformer, 10 HP, size 1, 12" high	1 Elec	2.70	2.963	Ea.	1,475	130		1,605	1,825
0200	25 HP, size 2, 18" high	2 Elec	4	4		1,675	176		1,851	2,100
0300	50 HP, size 3, 24" high		2	8		2,600	350		2,950	3,375
0350	75 HP, size 4, 24" high		1.60	10		3,450	440		3,890	4,450
0400	100 HP, size 4, 30" high		1.40	11.429		4,550	500		5,050	5,775
0500	200 HP, size 5, 48" high		1	16		6,800	700		7,500	8,525
0600	400 HP, size 6, 72" high	↓	.80	20	↓	14,000	880		14,880	16,700
0800	Structures, 600 amp, 22,000 rms, takes any									
0900	combination of starters up to 72" high	2 Elec	1.60	10	Ea.	1,975	440		2,415	2,825
1000	Back to back, 72" front & 66" back	"	1.20	13.333		2,675	585		3,260	3,800
1100	For copper bus add per structure					230			230	253
1200	For NEMA 12, add per structure					133			133	146
1300	For 42,000 rms, add per structure					183			183	201
1400	For 100,000 rms, size 1 & 2, add					590			590	650
1500	Size 3, add					950			950	1,050
1600	Size 4, add					765			765	845
1700	For pilot lights, add per starter	1 Elec	16	.500		103	22		125	147
1800	For push button, add per starter		16	.500		103	22		125	147
1900	For auxilliary contacts, add per starter	↓	16	.500	↓	221	22		243	276

26 24 19.40 Motor Starters and Controls

		Crew	Daily Output	Labor-Hours	Unit	Material	2007 Bare Costs Labor	Equipment	Total	Total Incl O&P
0010	**MOTOR STARTERS AND CONTROLS** R262419-65									
0050	Magnetic, FVNR, with enclosure and heaters, 480 volt									
0080	2 HP, size 00	1 Elec	3.50	2.286	Ea.	191	100		291	360
0100	5 HP, size 0		2.30	3.478		231	153		384	480
0200	10 HP, size 1	↓	1.60	5	↓	259	220		479	610

26 24 19.40 Motor Starters and Controls		Crew	Daily Output	Labor-Hours	Unit	Material	2007 Bare Costs Labor	Equipment	Total	Total Incl O&P
0300	25 HP, size 2	R263413-33	2 Elec	2.20	7.273	Ea.	485	320		805	1,000
0400	50 HP, size 3			1.80	8.889		795	390		1,185	1,450
0500	100 HP, size 4			1.20	13.333		1,775	585		2,360	2,825
0600	200 HP, size 5			.90	17.778		4,125	780		4,905	5,700
0610	400 HP, size 6			.80	20		12,300	880		13,180	14,800
0620	NEMA 7, 5 HP, size 0		1 Elec	1.60	5		1,025	220		1,245	1,450
0630	10 HP, size 1		"	1.10	7.273		1,050	320		1,370	1,650
0640	25 HP, size 2		2 Elec	1.80	8.889		1,700	390		2,090	2,425
0650	50 HP, size 3			1.20	13.333		2,525	585		3,110	3,650
0660	100 HP, size 4			.90	17.778		4,150	780		4,930	5,700
0670	200 HP, size 5			.50	32		9,650	1,400		11,050	12,700
0700	Combination, with motor circuit protectors, 5 HP, size 0		1 Elec	1.80	4.444		750	195		945	1,125
0800	10 HP, size 1		"	1.30	6.154		780	270		1,050	1,250
0900	25 HP, size 2		2 Elec	2	8		1,100	350		1,450	1,725
1000	50 HP, size 3			1.32	12.121		1,575	530		2,105	2,525
1200	100 HP, size 4			.80	20		3,425	880		4,305	5,075
1220	NEMA 7, 5 HP, size 0		1 Elec	1.30	6.154		1,800	270		2,070	2,375
1230	10 HP, size 1		"	1	8		1,825	350		2,175	2,550
1240	25 HP, size 2		2 Elec	1.32	12.121		2,450	530		2,980	3,475
1250	50 HP, size 3			.80	20		4,025	880		4,905	5,725
1260	100 HP, size 4			.60	26.667		6,250	1,175		7,425	8,600
1270	200 HP, size 5			.40	40		13,500	1,750		15,250	17,500
1400	Combination, with fused switch, 5 HP, size 0		1 Elec	1.80	4.444		575	195		770	920
1600	10 HP, size 1		"	1.30	6.154		615	270		885	1,075
1800	25 HP, size 2		2 Elec	2	8		995	350		1,345	1,625
2000	50 HP, size 3			1.32	12.121		1,675	530		2,205	2,650
2200	100 HP, size 4			.80	20		2,925	880		3,805	4,525
2610	NEMA 4, with start-stop pushbutton size 1		1 Elec	1.30	6.154		1,550	270		1,820	2,100
2620	Size 2		2 Elec	2	8		2,050	350		2,400	2,800
2630	Size 3			1.32	12.121		3,275	530		3,805	4,400
2640	Size 4			.80	20		5,025	880		5,905	6,825
2650	NEMA 4, FVNR, including control transformer										
2660	Size 1		2 Elec	2.60	6.154	Ea.	1,425	270		1,695	1,950
2670	Size 2			2	8		2,025	350		2,375	2,750
2680	Size 3			1.32	12.121		3,225	530		3,755	4,350
2690	Size 4			.80	20		4,975	880		5,855	6,775
2710	Magnetic, FVR, control circuit transformer, NEMA 1, size 1		1 Elec	1.30	6.154		835	270		1,105	1,325
2720	Size 2		2 Elec	2	8		1,325	350		1,675	2,000
2730	Size 3			1.32	12.121		2,025	530		2,555	3,025
2740	Size 4			.80	20		4,425	880		5,305	6,150
2760	NEMA 4, size 1		1 Elec	1.10	7.273		1,200	320		1,520	1,800
2770	Size 2		2 Elec	1.60	10		1,925	440		2,365	2,775
2780	Size 3			1.20	13.333		2,900	585		3,485	4,075
2790	Size 4			.70	22.857		5,975	1,000		6,975	8,050
2820	NEMA 12, size 1		1 Elec	1.10	7.273		985	320		1,305	1,550
2830	Size 2		2 Elec	1.60	10		1,550	440		1,990	2,350
2840	Size 3			1.20	13.333		2,450	585		3,035	3,575
2850	Size 4			.70	22.857		5,050	1,000		6,050	7,050
2870	Combination FVR, fused, w/control XFMR & PB, NEMA 1, size 1		1 Elec	1	8		1,450	350		1,800	2,100
2880	Size 2		2 Elec	1.50	10.667		2,050	470		2,520	2,950
2890	Size 3			1.10	14.545		3,050	640		3,690	4,300
2900	Size 4			.70	22.857		6,175	1,000		7,175	8,300
2910	NEMA 4, size 1		1 Elec	.90	8.889		2,125	390		2,515	2,925

26 24 19.40 Motor Starters and Controls		Crew	Daily Output	Labor-Hours	Unit	Material	2007 Bare Costs Labor	Equipment	Total	Total Incl O&P
2920	Size 2	2 Elec	1.40	11.429	Ea.	3,100	500		3,600	4,175
2930	Size 3	1	1	16		4,900	700		5,600	6,450
2940	Size 4		.60	26.667		8,150	1,175		9,325	10,700
2950	NEMA 12, size 1	1 Elec	1	8		1,625	350		1,975	2,325
2960	Size 2	2 Elec	1.40	11.429		2,300	500		2,800	3,275
2970	Size 3		1	16		3,375	700		4,075	4,775
2980	Size 4		.60	26.667		6,700	1,175		7,875	9,125
3010	Manual, single phase, w/pilot, 1 pole 120V NEMA 1	1 Elec	6.40	1.250		60	55		115	148
3020	NEMA 4		4	2		335	88		423	500
3030	2 pole, 120/240 V, NEMA 1		6.40	1.250		65.50	55		120.50	154
3040	NEMA 4		4	2		340	88		428	505
3041	3 phase, 3 pole 600V, NEMA 1		5.50	1.455		234	64		298	350
3042	NEMA 4		3.50	2.286		455	100		555	650
3043	NEMA 12		3.50	2.286		274	100		374	450
3070	Auxiliary contact, normally open					68.50			68.50	75
3500	Magnetic FVNR with NEMA 12, enclosure & heaters, 480 volt									
3600	5 HP, size 0	1 Elec	2.20	3.636	Ea.	218	160		378	475
3700	10 HP, size 1	"	1.50	5.333		330	234		564	710
3800	25 HP, size 2	2 Elec	2	8		615	350		965	1,200
3900	50 HP, size 3		1.60	10		945	440		1,385	1,700
4000	100 HP, size 4		1	16		2,250	700		2,950	3,525
4100	200 HP, size 5		.80	20		5,425	880		6,305	7,250
4200	Combination, with motor circuit protectors, 5 HP, size 0	1 Elec	1.70	4.706		720	207		927	1,100
4300	10 HP, size 1	"	1.20	6.667		750	293		1,043	1,250
4400	25 HP, size 2	2 Elec	1.80	8.889		1,125	390		1,515	1,800
4500	50 HP, size 3		1.20	13.333		1,825	585		2,410	2,875
4600	100 HP, size 4		.74	21.622		4,125	950		5,075	5,950
4700	Combination, with fused switch, 5 HP, size 0	1 Elec	1.70	4.706		700	207		907	1,075
4800	10 HP, size 1	"	1.20	6.667		725	293		1,018	1,225
4900	25 HP, size 2	2 Elec	1.80	8.889		1,100	390		1,490	1,800
5000	50 HP, size 3		1.20	13.333		1,775	585		2,360	2,825
5100	100 HP, size 4		.74	21.622		3,600	950		4,550	5,375
5200	Factory installed controls, adders to size 0 thru 5									
5300	Start-stop push button	1 Elec	32	.250	Ea.	45.50	11		56.50	66.50
5400	Hand-off-auto-selector switch		32	.250		45.50	11		56.50	66.50
5500	Pilot light		32	.250		85.50	11		96.50	110
5600	Start-stop-pilot		32	.250		131	11		142	160
5700	Auxiliary contact, NO or NC		32	.250		62.50	11		73.50	85.50
5800	NO-NC		32	.250		125	11		136	154
5810	Magnetic FVR, NEMA 7, w/heaters, size 1		.66	12.121		1,850	530		2,380	2,850
5830	Size 2	2 Elec	1.10	14.545		3,075	640		3,715	4,325
5840	Size 3		.70	22.857		4,925	1,000		5,925	6,925
5850	Size 4		.60	26.667		7,075	1,175		8,250	9,525
5860	Combination w/circuit breakers, heaters, control XFMR PB, size 1	1 Elec	.60	13.333		1,600	585		2,185	2,625
5870	Size 2	2 Elec	.80	20		2,050	880		2,930	3,550
5880	Size 3		.50	32		2,725	1,400		4,125	5,100
5890	Size 4		.40	40		5,425	1,750		7,175	8,575
5900	Manual, 240 volt, .75 HP motor	1 Elec	4	2		48.50	88		136.50	185
5910	2 HP motor		4	2		131	88		219	275
6000	Magnetic, 240 volt, 1 or 2 pole, .75 HP motor		4	2		191	88		279	340
6020	2 HP motor		4	2		211	88		299	365
6040	5 HP motor		3	2.667		300	117		417	505
6060	10 HP motor		2.30	3.478		750	153		903	1,050

26 24 Switchboards and Panelboards

26 24 19 – Motor-Control Centers

26 24 19.40 Motor Starters and Controls		Crew	Daily Output	Labor-Hours	Unit	Material	2007 Bare Costs Labor	Equipment	Total	Total Incl O&P
6100	3 pole, .75 HP motor	1 Elec	3	2.667	Ea.	191	117		308	385
6120	5 HP motor		2.30	3.478		259	153		412	510
6140	10 HP motor		1.60	5		485	220		705	860
6160	15 HP motor		1.60	5		485	220		705	860
6180	20 HP motor		1.10	7.273		795	320		1,115	1,350
6200	25 HP motor	2 Elec	2.20	7.273		795	320		1,115	1,350
6210	30 HP motor		1.80	8.889		795	390		1,185	1,450
6220	40 HP motor		1.80	8.889		1,775	390		2,165	2,525
6230	50 HP motor		1.80	8.889		1,775	390		2,165	2,525
6240	60 HP motor		1.20	13.333		4,125	585		4,710	5,425
6250	75 HP motor		1.20	13.333		4,125	585		4,710	5,425
6260	100 HP motor		1.20	13.333		4,125	585		4,710	5,425
6270	125 HP motor		.90	17.778		11,600	780		12,380	13,900
6280	150 HP motor		.90	17.778		11,600	780		12,380	13,900
6290	200 HP motor		.90	17.778		11,600	780		12,380	13,900
6400	Starter & nonfused disconnect, 240 volt, 1-2 pole, .75 HP motor	1 Elec	2	4		265	176		441	555
6410	2 HP motor		2	4		285	176		461	575
6420	5 HP motor		1.80	4.444		375	195		570	705
6430	10 HP motor		1.40	5.714		850	251		1,101	1,300
6440	3 pole, .75 HP motor		1.60	5		265	220		485	615
6450	5 HP motor		1.40	5.714		335	251		586	740
6460	10 HP motor		1.10	7.273		585	320		905	1,125
6470	15 HP motor		1	8		585	350		935	1,175
6480	20 HP motor	2 Elec	1.50	10.667		1,025	470		1,495	1,825
6490	25 HP motor		1.50	10.667		1,025	470		1,495	1,825
6500	30 HP motor		1.30	12.308		1,025	540		1,565	1,925
6510	40 HP motor		1.24	12.903		2,175	565		2,740	3,250
6520	50 HP motor		1.12	14.286		2,175	625		2,800	3,325
6530	60 HP motor		.90	17.778		4,550	780		5,330	6,150
6540	75 HP motor		.76	21.053		4,550	925		5,475	6,375
6550	100 HP motor		.70	22.857		4,550	1,000		5,550	6,500
6560	125 HP motor		.60	26.667		12,600	1,175		13,775	15,700
6570	150 HP motor		.52	30.769		12,600	1,350		13,950	15,900
6580	200 HP motor		.50	32		12,600	1,400		14,000	16,000
6600	Starter & fused disconnect, 240 volt, 1-2 pole, .75 HP motor	1 Elec	2	4		283	176		459	570
6610	2 HP motor		2	4		305	176		481	595
6620	5 HP motor		1.80	4.444		395	195		590	725
6630	10 HP motor		1.40	5.714		905	251		1,156	1,375
6640	3 pole, .75 HP motor		1.60	5		283	220		503	635
6650	5 HP motor		1.40	5.714		350	251		601	760
6660	10 HP motor		1.10	7.273		645	320		965	1,175
6690	15 HP motor		1	8		645	350		995	1,225
6700	20 HP motor	2 Elec	1.60	10		1,050	440		1,490	1,825
6710	25 HP motor		1.60	10		1,050	440		1,490	1,825
6720	30 HP motor		1.40	11.429		1,050	500		1,550	1,925
6730	40 HP motor		1.20	13.333		2,350	585		2,935	3,450
6740	50 HP motor		1.20	13.333		2,350	585		2,935	3,450
6750	60 HP motor		.90	17.778		4,700	780		5,480	6,325
6760	75 HP motor		.90	17.778		4,700	780		5,480	6,325
6770	100 HP motor		.70	22.857		4,700	1,000		5,700	6,675
6780	125 HP motor		.54	29.630		13,000	1,300		14,300	16,200
6790	Combination starter & nonfusible disconnect									
6800	240 volt, 1-2 pole, .75 HP motor	1 Elec	2	4	Ea.	545	176		721	860

26 24 19.40 Motor Starters and Controls		Crew	Daily Output	Labor-Hours	Unit	Material	2007 Bare Costs Labor	Equipment	Total	Total Incl O&P
6810	2 HP motor	1 Elec	2	4	Ea.	545	176		721	860
6820	5 HP motor		1.50	5.333		575	234		809	980
6830	10 HP motor		1.20	6.667		890	293		1,183	1,425
6840	3 pole, .75 HP motor		1.80	4.444		545	195		740	890
6850	5 HP motor		1.30	6.154		575	270		845	1,025
6860	10 HP motor		1	8		890	350		1,240	1,500
6870	15 HP motor		1	8		890	350		1,240	1,500
6880	20 HP motor	2 Elec	1.32	12.121		1,475	530		2,005	2,425
6890	25 HP motor		1.32	12.121		1,475	530		2,005	2,425
6900	30 HP motor		1.32	12.121		1,475	530		2,005	2,425
6910	40 HP motor		.80	20		2,800	880		3,680	4,400
6920	50 HP motor		.80	20		2,800	880		3,680	4,400
6930	60 HP motor		.70	22.857		6,275	1,000		7,275	8,400
6940	75 HP motor		.70	22.857		6,275	1,000		7,275	8,400
6950	100 HP motor		.70	22.857		6,275	1,000		7,275	8,400
6960	125 HP motor		.60	26.667		16,500	1,175		17,675	20,000
6970	150 HP motor		.60	26.667		16,500	1,175		17,675	20,000
6980	200 HP motor		.60	26.667		16,500	1,175		17,675	20,000
6990	Combination starter and fused disconnect									
7000	240 volt, 1-2 pole, .75 HP motor	1 Elec	2	4	Ea.	560	176		736	880
7010	2 HP motor		2	4		560	176		736	880
7020	5 HP motor		1.50	5.333		590	234		824	1,000
7030	10 HP motor		1.20	6.667		915	293		1,208	1,425
7040	3 pole, .75 HP motor		1.80	4.444		560	195		755	910
7050	5 HP motor		1.30	6.154		590	270		860	1,050
7060	10 HP motor		1	8		915	350		1,265	1,525
7070	15 HP motor		1	8		915	350		1,265	1,525
7080	20 HP motor	2 Elec	1.32	12.121		1,525	530		2,055	2,475
7090	25 HP motor		1.32	12.121		1,525	530		2,055	2,475
7100	30 HP motor		1.32	12.121		1,525	530		2,055	2,475
7110	40 HP motor		.80	20		2,900	880		3,780	4,500
7120	50 HP motor		.80	20		2,900	880		3,780	4,500
7130	60 HP motor		.80	20		6,475	880		7,355	8,425
7140	75 HP motor		.70	22.857		6,475	1,000		7,475	8,625
7150	100 HP motor		.70	22.857		6,475	1,000		7,475	8,625
7160	125 HP motor		.70	22.857		17,100	1,000		18,100	20,300
7170	150 HP motor		.60	26.667		17,100	1,175		18,275	20,600
7180	200 HP motor		.60	26.667		17,100	1,175		18,275	20,600
7190	Combination starter & circuit breaker disconnect									
7200	240 volt, 1-2 pole, .75 HP motor	1 Elec	2	4	Ea.	565	176		741	885
7210	2 HP motor		2	4		565	176		741	885
7220	5 HP motor		1.50	5.333		595	234		829	1,000
7230	10 HP motor		1.20	6.667		910	293		1,203	1,425
7240	3 pole, .75 HP motor		1.80	4.444		585	195		780	935
7250	5 HP motor		1.30	6.154		615	270		885	1,075
7260	10 HP motor		1	8		925	350		1,275	1,550
7270	15 HP motor		1	8		925	350		1,275	1,550
7280	20 HP motor	2 Elec	1.32	12.121		1,575	530		2,105	2,525
7290	25 HP motor		1.32	12.121		1,575	530		2,105	2,525
7300	30 HP motor		1.32	12.121		1,575	530		2,105	2,525
7310	40 HP motor		.80	20		3,425	880		4,305	5,075
7320	50 HP motor		.80	20		3,425	880		4,305	5,075
7330	60 HP motor		.80	20		7,925	880		8,805	10,000

26 24 Switchboards and Panelboards

26 24 19 – Motor-Control Centers

26 24 19.40 Motor Starters and Controls		Crew	Daily Output	Labor-Hours	Unit	Material	2007 Bare Costs Labor	Equipment	Total	Total Incl O&P
7340	75 HP motor	2 Elec	.70	22.857	Ea.	7,925	1,000		8,925	10,200
7350	100 HP motor		.70	22.857		7,925	1,000		8,925	10,200
7360	125 HP motor		.70	22.857		17,200	1,000		18,200	20,400
7370	150 HP motor		.60	26.667		17,200	1,175		18,375	20,700
7380	200 HP motor		.60	26.667		17,200	1,175		18,375	20,700
7400	Magnetic FVNR with enclosure & heaters, 2 pole,									
7410	230 volt, 1 HP size 00	1 Elec	4	2	Ea.	174	88		262	320
7420	2 HP, size 0		4	2		194	88		282	345
7430	3 HP, size 1		3	2.667		222	117		339	420
7440	5 HP, size 1p		3	2.667		285	117		402	490
7450	115 volt, 1/3 HP, size 00		4	2		174	88		262	320
7460	1 HP, size 0		4	2		194	88		282	345
7470	2 HP, size 1		3	2.667		222	117		339	420
7480	3 HP, size 1P		3	2.667		276	117		393	480
7500	3 pole, 480 volt, 600 HP, size 7	2 Elec	.70	22.857		15,400	1,000		16,400	18,400
7590	Magnetic FVNR with heater, NEMA 1									
7600	600 volt, 3 pole, 5 HP motor	1 Elec	2.30	3.478	Ea.	231	153		384	480
7610	10 HP motor	"	1.60	5		259	220		479	610
7620	25 HP motor	2 Elec	2.20	7.273		485	320		805	1,000
7630	30 HP motor		1.80	8.889		795	390		1,185	1,450
7640	40 HP motor		1.80	8.889		795	390		1,185	1,450
7650	50 HP motor		1.80	8.889		795	390		1,185	1,450
7660	60 HP motor		1.20	13.333		1,775	585		2,360	2,825
7670	75 HP motor		1.20	13.333		1,775	585		2,360	2,825
7680	100 HP motor		1.20	13.333		1,775	585		2,360	2,825
7690	125 HP motor		.90	17.778		4,125	780		4,905	5,700
7700	150 HP motor		.90	17.778		4,125	780		4,905	5,700
7710	200 HP motor		.90	17.778		4,125	780		4,905	5,700
7750	Starter & nonfused disconnect, 600 volt, 3 pole, 5 HP motor	1 Elec	1.40	5.714		360	251		611	775
7760	10 HP motor	"	1.10	7.273		390	320		710	905
7770	25 HP motor	2 Elec	1.50	10.667		620	470		1,090	1,375
7780	30 HP motor		1.30	12.308		1,025	540		1,565	1,925
7790	40 HP motor		1.30	12.308		1,025	540		1,565	1,925
7800	50 HP motor		1.30	12.308		1,025	540		1,565	1,925
7810	60 HP motor		.92	17.391		2,125	765		2,890	3,475
7820	75 HP motor		.92	17.391		2,125	765		2,890	3,475
7830	100 HP motor		.84	19.048		2,125	835		2,960	3,600
7840	125 HP motor		.70	22.857		4,700	1,000		5,700	6,650
7850	150 HP motor		.70	22.857		4,700	1,000		5,700	6,650
7860	200 HP motor		.60	26.667		4,700	1,175		5,875	6,900
7870	Starter & fused disconnect, 600 volt, 3 pole, 5 HP motor	1 Elec	1.40	5.714		480	251		731	905
7880	10 HP motor	"	1.10	7.273		510	320		830	1,025
7890	25 HP motor	2 Elec	1.50	10.667		740	470		1,210	1,500
7900	30 HP motor		1.30	12.308		1,100	540		1,640	2,000
7910	40 HP motor		1.30	12.308		1,100	540		1,640	2,000
7920	50 HP motor		1.30	12.308		1,100	540		1,640	2,000
7930	60 HP motor		.92	17.391		2,325	765		3,090	3,675
7940	75 HP motor		.92	17.391		2,325	765		3,090	3,675
7950	100 HP motor		.84	19.048		2,325	835		3,160	3,800
7960	125 HP motor		.70	22.857		4,925	1,000		5,925	6,925
7970	150 HP motor		.70	22.857		4,925	1,000		5,925	6,925
7980	200 HP motor		.60	26.667		4,925	1,175		6,100	7,175
7990	Combination starter and nonfusible disconnect									

221

26 24 19.40 Motor Starters and Controls		Crew	Daily Output	Labor-Hours	Unit	Material	2007 Bare Costs Labor	2007 Bare Costs Equipment	Total	Total Incl O&P
8000	600 volt, 3 pole, 5 HP motor	1 Elec	1.80	4.444	Ea.	545	195		740	890
8010	10 HP motor	"	1.30	6.154		575	270		845	1,025
8020	25 HP motor	2 Elec	2	8		890	350		1,240	1,500
8030	30 HP motor		1.32	12.121		1,525	530		2,055	2,475
8040	40 HP motor		1.32	12.121		1,525	530		2,055	2,475
8050	50 HP motor		1.32	12.121		1,475	530		2,005	2,425
8060	60 HP motor		.80	20		2,900	880		3,780	4,500
8070	75 HP motor		.80	20		2,900	880		3,780	4,500
8080	100 HP motor		.80	20		2,800	880		3,680	4,400
8090	125 HP motor		.70	22.857		6,475	1,000		7,475	8,625
8100	150 HP motor		.70	22.857		6,475	1,000		7,475	8,625
8110	200 HP motor	▼	.70	22.857	▼	6,475	1,000		7,475	8,625
8140	Combination starter and fused disconnect									
8150	600 volt, 3 pole, 5 HP motor	1 Elec	1.80	4.444	Ea.	560	195		755	910
8160	10 HP motor	"	1.30	6.154		615	270		885	1,075
8170	25 HP motor	2 Elec	2	8		995	350		1,345	1,625
8180	30 HP motor		1.32	12.121		1,550	530		2,080	2,500
8190	40 HP motor		1.32	12.121		1,675	530		2,205	2,650
8200	50 HP motor		1.32	12.121		1,675	530		2,205	2,650
8210	60 HP motor		.80	20		2,925	880		3,805	4,525
8220	75 HP motor		.80	20		2,925	880		3,805	4,525
8230	100 HP motor		.80	20		2,925	880		3,805	4,525
8240	125 HP motor		.70	22.857		6,475	1,000		7,475	8,625
8250	150 HP motor		.70	22.857		6,475	1,000		7,475	8,625
8260	200 HP motor	▼	.70	22.857	▼	6,475	1,000		7,475	8,625
8290	Combination starter & circuit breaker disconnect									
8300	600 volt, 3 pole, 5 HP motor	1 Elec	1.80	4.444	Ea.	750	195		945	1,125
8310	10 HP motor	"	1.30	6.154		780	270		1,050	1,250
8320	25 HP motor	2 Elec	2	8		1,100	350		1,450	1,725
8330	30 HP motor		1.32	12.121		1,575	530		2,105	2,525
8340	40 HP motor		1.32	12.121		1,575	530		2,105	2,525
8350	50 HP motor		1.32	12.121		1,575	530		2,105	2,525
8360	60 HP motor		.80	20		3,425	880		4,305	5,075
8370	75 HP motor		.80	20		3,425	880		4,305	5,075
8380	100 HP motor		.80	20		3,425	880		4,305	5,075
8390	125 HP motor		.70	22.857		7,925	1,000		8,925	10,200
8400	150 HP motor		.70	22.857		7,925	1,000		8,925	10,200
8410	200 HP motor	▼	.70	22.857	▼	7,925	1,000		8,925	10,200
8430	Starter & circuit breaker disconnect									
8440	600 volt, 3 pole, 5 HP motor	1 Elec	1.40	5.714	Ea.	760	251		1,011	1,225
8450	10 HP motor	"	1.10	7.273		790	320		1,110	1,350
8460	25 HP motor	2 Elec	1.50	10.667		1,025	470		1,495	1,825
8470	30 HP motor		1.30	12.308		1,325	540		1,865	2,250
8480	40 HP motor		1.30	12.308		1,325	540		1,865	2,250
8490	50 HP motor		1.30	12.308		1,325	540		1,865	2,250
8500	60 HP motor		.92	17.391		3,175	765		3,940	4,600
8510	75 HP motor		.92	17.391		3,175	765		3,940	4,600
8520	100 HP motor		.84	19.048		3,175	835		4,010	4,725
8530	125 HP motor		.70	22.857		5,525	1,000		6,525	7,575
8540	150 HP motor		.70	22.857		5,525	1,000		6,525	7,575
8550	200 HP motor	▼	.60	26.667		6,525	1,175		7,700	8,925
8900	240 volt, 1-2 pole, .75 HP motor	1 Elec	2	4		720	176		896	1,050
8910	2 HP motor		2	4		740	176		916	1,075

26 24 Switchboards and Panelboards

26 24 19 – Motor-Control Centers

26 24 19.40 Motor Starters and Controls		Crew	Daily Output	Labor-Hours	Unit	Material	2007 Bare Costs Labor	Equipment	Total	Total Incl O&P
8920	5 HP motor	1 Elec	1.80	4.444	Ea.	835	195		1,030	1,200
8930	10 HP motor		1.40	5.714		1,275	251		1,526	1,775
8950	3 pole, .75 HP motor		1.60	5		720	220		940	1,125
8970	5 HP motor		1.40	5.714		790	251		1,041	1,250
8980	10 HP motor		1.10	7.273		1,025	320		1,345	1,600
8990	15 HP motor	↓	1	8		1,025	350		1,375	1,650
9100	20 HP motor	2 Elec	1.50	10.667		1,400	470		1,870	2,250
9110	25 HP motor		1.50	10.667		1,400	470		1,870	2,250
9120	30 HP motor		1.30	12.308		1,400	540		1,940	2,350
9130	40 HP motor		1.24	12.903		3,175	565		3,740	4,325
9140	50 HP motor		1.12	14.286		3,175	625		3,800	4,400
9150	60 HP motor		.90	17.778		6,525	780		7,305	8,325
9160	75 HP motor		.76	21.053		6,525	925		7,450	8,550
9170	100 HP motor		.70	22.857		6,525	1,000		7,525	8,675
9180	125 HP motor		.60	26.667		15,100	1,175		16,275	18,400
9190	150 HP motor		.52	30.769		15,100	1,350		16,450	18,600
9200	200 HP motor	↓	.50	32	↓	15,100	1,400		16,500	18,700

26 25 Enclosed Bus Assemblies

26 25 13 – Enclosed Bus Assemblies

26 25 13.10 Aluminum Bus Duct

		Crew	Daily Output	Labor-Hours	Unit	Material	2007 Bare Costs Labor	Equipment	Total	Total Incl O&P
0010	**ALUMINUM BUS DUCT** 10 ft. long R262513-10									
0050	3 pole 4 wire, plug-in/indoor, straight section, 225 amp	2 Elec	44	.364	L.F.	142	15.95		157.95	181
0100	400 amp		36	.444		142	19.50		161.50	186
0150	600 amp		32	.500		142	22		164	190
0200	800 amp R262513-15		26	.615		164	27		191	221
0250	1000 amp		24	.667		186	29.50		215.50	249
0300	1350 amp		22	.727		285	32		317	365
0310	1600 amp		18	.889		330	39		369	420
0320	2000 amp		16	1		385	44		429	485
0330	2500 amp		14	1.143		495	50		545	615
0340	3000 amp		12	1.333		560	58.50		618.50	700
0350	Feeder, 600 amp		34	.471		131	20.50		151.50	175
0400	800 amp		28	.571		153	25		178	207
0450	1000 amp		26	.615		175	27		202	233
0455	1200 amp		25	.640		175	28		203	235
0500	1350 amp		24	.667		274	29.50		303.50	345
0550	1600 amp		20	.800		315	35		350	405
0600	2000 amp		18	.889		370	39		409	470
0620	2500 amp		14	1.143		480	50		530	605
0630	3000 amp		12	1.333		545	58.50		603.50	685
0640	4000 amp		10	1.600	↓	790	70		860	970
0650	Elbow, 225 amp		4.40	3.636	Ea.	775	160		935	1,100
0700	400 amp		3.80	4.211		775	185		960	1,125
0750	600 amp		3.40	4.706		775	207		982	1,150
0800	800 amp		3	5.333		810	234		1,044	1,250
0850	1000 amp		2.80	5.714		1,050	251		1,301	1,525
0870	1200 amp		2.70	5.926		1,300	260		1,560	1,800
0900	1350 amp		2.60	6.154		1,400	270		1,670	1,925
0950	1600 amp		2.40	6.667		1,650	293		1,943	2,250
1000	2000 amp	↓	2	8	↓	1,825	350		2,175	2,525

26 25 Enclosed Bus Assemblies

26 25 13 – Enclosed Bus Assemblies

26 25 13.10 Aluminum Bus Duct		Crew	Daily Output	Labor-Hours	Unit	Material	2007 Bare Costs Labor	Equipment	Total	Total Incl O&P
1020	2500 amp	2 Elec	1.80	8.889	Ea.	2,150	390		2,540	2,925
1030	3000 amp		1.60	10		2,475	440		2,915	3,375
1040	4000 amp		1.40	11.429		4,000	500		4,500	5,150
1100	Cable tap box end, 225 amp		3.60	4.444		1,075	195		1,270	1,500
1150	400 amp		3.20	5		1,075	220		1,295	1,525
1200	600 amp		2.60	6.154		1,075	270		1,345	1,600
1250	800 amp		2.20	7.273		1,175	320		1,495	1,775
1300	1000 amp		2	8		1,275	350		1,625	1,950
1320	1200 amp		2	8		1,550	350		1,900	2,225
1350	1350 amp		1.60	10		1,675	440		2,115	2,500
1400	1600 amp		1.40	11.429		1,900	500		2,400	2,825
1450	2000 amp		1.20	13.333		2,175	585		2,760	3,275
1460	2500 amp		1	16		2,625	700		3,325	3,950
1470	3000 amp		.80	20		2,900	880		3,780	4,500
1480	4000 amp		.60	26.667		3,875	1,175		5,050	6,000
1500	Switchboard stub, 225 amp		5.80	2.759		955	121		1,076	1,225
1550	400 amp		5.40	2.963		955	130		1,085	1,250
1600	600 amp		4.60	3.478		955	153		1,108	1,275
1650	800 amp		4	4		1,075	176		1,251	1,450
1700	1000 amp		3.20	5		1,200	220		1,420	1,650
1720	1200 amp		3.10	5.161		1,350	227		1,577	1,800
1750	1350 amp		3	5.333		1,500	234		1,734	2,000
1800	1600 amp		2.60	6.154		1,650	270		1,920	2,225
1850	2000 amp		2.40	6.667		1,900	293		2,193	2,525
1860	2500 amp		2.20	7.273		2,275	320		2,595	2,975
1870	3000 amp		2	8		2,650	350		3,000	3,425
1880	4000 amp		1.80	8.889		3,300	390		3,690	4,225
1890	Tee fittings, 225 amp		3.20	5		1,050	220		1,270	1,475
1900	400 amp		2.80	5.714		1,050	251		1,301	1,525
1950	600 amp		2.60	6.154		1,050	270		1,320	1,550
2000	800 amp		2.40	6.667		1,100	293		1,393	1,650
2050	1000 amp		2.20	7.273		1,175	320		1,495	1,750
2070	1200 amp		2.10	7.619		1,375	335		1,710	2,000
2100	1350 amp		2	8		1,900	350		2,250	2,600
2150	1600 amp		1.60	10		2,275	440		2,715	3,150
2200	2000 amp		1.20	13.333		2,525	585		3,110	3,650
2220	2500 amp		1	16		3,025	700		3,725	4,375
2230	3000 amp		.80	20		3,450	880		4,330	5,075
2240	4000 amp		.60	26.667		5,725	1,175		6,900	8,050
2300	Wall flange, 600 amp		20	.800		128	35		163	194
2310	800 amp		16	1		128	44		172	207
2320	1000 amp		13	1.231		128	54		182	222
2325	1200 amp		12	1.333		128	58.50		186.50	228
2330	1350 amp		10.80	1.481		128	65		193	238
2340	1600 amp		9	1.778		128	78		206	257
2350	2000 amp		8	2		128	88		216	272
2360	2500 amp		6.60	2.424		128	106		234	299
2370	3000 amp		5.40	2.963		128	130		258	335
2380	4000 amp		4	4		128	176		304	400
2390	5000 amp		3	5.333		128	234		362	490
2400	Vapor barrier		8	2		300	88		388	460
2420	Roof flange kit		4	4		635	176		811	955
2600	Expansion fitting, 225 amp		10	1.600		1,275	70		1,345	1,500

26 25 13 – Enclosed Bus Assemblies

26 25 13.10 Aluminum Bus Duct		Crew	Daily Output	Labor-Hours	Unit	Material	2007 Bare Costs Labor	Equipment	Total	Total Incl O&P
2610	400 amp	2 Elec	8	2	Ea.	1,275	88		1,363	1,525
2620	600 amp		6	2.667		1,275	117		1,392	1,575
2630	800 amp		4.60	3.478		1,475	153		1,628	1,850
2640	1000 amp		4	4		1,650	176		1,826	2,075
2650	1350 amp		3.60	4.444		2,425	195		2,620	2,950
2660	1600 amp		3.20	5		2,900	220		3,120	3,500
2670	2000 amp		2.80	5.714		3,225	251		3,476	3,925
2680	2500 amp		2.40	6.667		3,900	293		4,193	4,700
2690	3000 amp		2	8		4,475	350		4,825	5,450
2700	4000 amp		1.60	10		5,900	440		6,340	7,150
2800	Reducer nonfused, 400 amp		8	2		1,025	88		1,113	1,250
2810	600 amp		6	2.667		1,025	117		1,142	1,300
2820	800 amp		4.60	3.478		1,225	153		1,378	1,575
2830	1000 amp		4	4		1,425	176		1,601	1,825
2840	1350 amp		3.60	4.444		1,875	195		2,070	2,350
2850	1600 amp		3.20	5		2,525	220		2,745	3,100
2860	2000 amp		2.80	5.714		2,900	251		3,151	3,575
2870	2500 amp		2.40	6.667		3,650	293		3,943	4,450
2880	3000 amp		2	8		4,225	350		4,575	5,175
2890	4000 amp		1.60	10		5,650	440		6,090	6,850
2950	Reducer fuse included, 225 amp		4.40	3.636		2,975	160		3,135	3,525
2960	400 amp		4.20	3.810		3,025	167		3,192	3,600
2970	600 amp		3.60	4.444		3,575	195		3,770	4,250
2980	800 amp		3.20	5		5,700	220		5,920	6,600
2990	1000 amp		3	5.333		6,525	234		6,759	7,525
3000	1200 amp		2.80	5.714		6,525	251		6,776	7,550
3010	1600 amp		2.20	7.273		14,900	320		15,220	16,800
3020	2000 amp		1.80	8.889		16,500	390		16,890	18,800
3100	Reducer circuit breaker, 225 amp		4.40	3.636		2,925	160		3,085	3,475
3110	400 amp		4.20	3.810		3,575	167		3,742	4,175
3120	600 amp		3.60	4.444		5,075	195		5,270	5,875
3130	800 amp		3.20	5		5,950	220		6,170	6,875
3140	1000 amp		3	5.333		6,750	234		6,984	7,775
3150	1200 amp		2.80	5.714		8,125	251		8,376	9,300
3160	1600 amp		2.20	7.273		12,000	320		12,320	13,700
3170	2000 amp		1.80	8.889		13,200	390		13,590	15,100
3250	Reducer circuit breaker, 75,000 AIC, 225 amp		4.40	3.636		4,600	160		4,760	5,300
3260	400 amp		4.20	3.810		4,600	167		4,767	5,300
3270	600 amp		3.60	4.444		6,150	195		6,345	7,050
3280	800 amp		3.20	5		6,750	220		6,970	7,750
3290	1000 amp		3	5.333		10,900	234		11,134	12,300
3300	1200 amp		2.80	5.714		10,900	251		11,151	12,300
3310	1600 amp		2.20	7.273		12,000	320		12,320	13,700
3320	2000 amp		1.80	8.889		13,200	390		13,590	15,100
3400	Reducer circuit breaker CLF 225 amp		4.40	3.636		4,725	160		4,885	5,450
3410	400 amp		4.20	3.810		5,625	167		5,792	6,425
3420	600 amp		3.60	4.444		8,400	195		8,595	9,550
3430	800 amp		3.20	5		8,775	220		8,995	9,975
3440	1000 amp		3	5.333		9,125	234		9,359	10,500
3450	1200 amp		2.80	5.714		11,900	251		12,151	13,500
3460	1600 amp		2.20	7.273		12,000	320		12,320	13,700
3470	2000 amp		1.80	8.889		13,200	390		13,590	15,100
3550	Ground bus added to bus duct, 225 amp		320	.050	L.F.	39	2.20		41.20	46

26 25 13 – Enclosed Bus Assemblies

26 25 13.10 Aluminum Bus Duct		Crew	Daily Output	Labor-Hours	Unit	Material	2007 Bare Costs Labor	Equipment	Total	Total Incl O&P
3560	400 amp	2 Elec	320	.050	L.F.	39	2.20		41.20	46
3570	600 amp		280	.057		39	2.51		41.51	46
3580	800 amp		240	.067		39	2.93		41.93	47
3590	1000 amp		200	.080		39	3.51		42.51	48
3600	1350 amp		180	.089		39	3.90		42.90	48.50
3610	1600 amp		160	.100		39	4.39		43.39	49
3620	2000 amp		160	.100		39	4.39		43.39	49
3630	2500 amp		140	.114		39	5		44	50
3640	3000 amp		120	.133		39	5.85		44.85	51
3650	4000 amp		100	.160		39	7		46	53
3810	High short circuit, 400 amp		36	.444		131	19.50		150.50	173
3820	600 amp		32	.500		131	22		153	177
3830	800 amp		26	.615		153	27		180	209
3840	1000 amp		24	.667		164	29.50		193.50	225
3850	1350 amp		22	.727		219	32		251	289
3860	1600 amp		18	.889		252	39		291	335
3870	2000 amp		16	1		296	44		340	390
3880	2500 amp		14	1.143		495	50		545	615
3890	3000 amp		12	1.333	▼	560	58.50		618.50	700
3920	Cross, 225 amp		5.60	2.857	Ea.	1,550	125		1,675	1,900
3930	400 amp		4.60	3.478		1,550	153		1,703	1,950
3940	600 amp		4	4		1,550	176		1,726	1,975
3950	800 amp		3.40	4.706		1,625	207		1,832	2,100
3960	1000 amp		3	5.333		1,725	234		1,959	2,250
3970	1350 amp		2.80	5.714		2,775	251		3,026	3,450
3980	1600 amp		2.20	7.273		3,300	320		3,620	4,125
3990	2000 amp		1.80	8.889		3,625	390		4,015	4,575
4000	2500 amp		1.60	10		4,300	440		4,740	5,375
4010	3000 amp		1.20	13.333		4,950	585		5,535	6,325
4020	4000 amp		1	16		7,550	700		8,250	9,375
4040	Cable tap box center, 225 amp		3.60	4.444		1,075	195		1,270	1,500
4050	400 amp		3.20	5		1,075	220		1,295	1,525
4060	600 amp		2.60	6.154		1,075	270		1,345	1,600
4070	800 amp		2.20	7.273		1,175	320		1,495	1,775
4080	1000 amp		2	8		1,275	350		1,625	1,950
4090	1350 amp		1.60	10		1,675	440		2,115	2,500
4100	1600 amp		1.40	11.429		1,900	500		2,400	2,825
4110	2000 amp		1.20	13.333		2,175	585		2,760	3,275
4120	2500 amp		1	16		2,625	700		3,325	3,950
4130	3000 amp		.80	20		2,900	880		3,780	4,500
4140	4000 amp		.60	26.667	▼	3,875	1,175		5,050	6,000
4500	Weatherproof, feeder, 600 amp		30	.533	L.F.	157	23.50		180.50	208
4520	800 amp		24	.667		184	29.50		213.50	246
4540	1000 amp		22	.727		210	32		242	279
4550	1200 amp		21	.762		291	33.50		324.50	370
4560	1350 amp		20	.800		330	35		365	415
4580	1600 amp		17	.941		380	41.50		421.50	480
4600	2000 amp		16	1		445	44		489	555
4620	2500 amp		12	1.333		580	58.50		638.50	720
4640	3000 amp		10	1.600		655	70		725	825
4660	4000 amp		8	2		945	88		1,033	1,175
5000	3 pole, 3 wire, feeder, 600 amp		40	.400		120	17.55		137.55	158
5010	800 amp		32	.500	▼	142	22		164	190

26 25 Enclosed Bus Assemblies

26 25 13 – Enclosed Bus Assemblies

26 25 13.10 Aluminum Bus Duct		Crew	Daily Output	Labor-Hours	Unit	Material	2007 Bare Costs Labor	Equipment	Total	Total Incl O&P
5020	1000 amp	2 Elec	30	.533	L.F.	153	23.50		176.50	204
5025	1200 amp		29	.552		153	24		177	205
5030	1350 amp		28	.571		208	25		233	267
5040	1600 amp		24	.667		241	29.50		270.50	310
5050	2000 amp		20	.800		285	35		320	370
5060	2500 amp		16	1		395	44		439	500
5070	3000 amp		14	1.143		470	50		520	595
5080	4000 amp		12	1.333		570	58.50		628.50	710
5200	Plug-in type, 225 amp		50	.320		131	14.05		145.05	165
5210	400 amp		42	.381		131	16.70		147.70	169
5220	600 amp		36	.444		131	19.50		150.50	173
5230	800 amp		30	.533		153	23.50		176.50	204
5240	1000 amp		28	.571		164	25		189	219
5245	1200 amp		27	.593		186	26		212	244
5250	1350 amp		26	.615		219	27		246	281
5260	1600 amp		20	.800		252	35		287	330
5270	2000 amp		18	.889		296	39		335	385
5280	2500 amp		16	1		405	44		449	510
5290	3000 amp		14	1.143		480	50		530	605
5300	4000 amp		12	1.333		580	58.50		638.50	725
5330	High short circuit, 400 amp		42	.381		131	16.70		147.70	169
5340	600 amp		36	.444		131	19.50		150.50	173
5350	800 amp		30	.533		153	23.50		176.50	204
5360	1000 amp		28	.571		164	25		189	219
5370	1350 amp		26	.615		219	27		246	281
5380	1600 amp		20	.800		252	35		287	330
5390	2000 amp		18	.889		296	39		335	385
5400	2500 amp		16	1		465	44		509	575
5410	3000 amp		14	1.143		480	50		530	605
5440	Elbow, 225 amp		5	3.200	Ea.	670	140		810	945
5450	400 amp		4.40	3.636		670	160		830	975
5460	600 amp		4	4		670	176		846	995
5470	800 amp		3.40	4.706		710	207		917	1,100
5480	1000 amp		3.20	5		730	220		950	1,125
5485	1200 amp		3.10	5.161		780	227		1,007	1,200
5490	1350 amp		3	5.333		845	234		1,079	1,275
5500	1600 amp		2.80	5.714		1,300	251		1,551	1,800
5510	2000 amp		2.40	6.667		1,425	293		1,718	2,000
5520	2500 amp		2	8		1,750	350		2,100	2,450
5530	3000 amp		1.80	8.889		2,075	390		2,465	2,875
5540	4000 amp		1.60	10		2,775	440		3,215	3,700
5560	Tee fittings, 225 amp		3.60	4.444		930	195		1,125	1,325
5570	400 amp		3.20	5		930	220		1,150	1,350
5580	600 amp		3	5.333		930	234		1,164	1,375
5590	800 amp		2.80	5.714		995	251		1,246	1,475
5600	1000 amp		2.60	6.154		1,025	270		1,295	1,525
5605	1200 amp		2.50	6.400		1,100	281		1,381	1,625
5610	1350 amp		2.40	6.667		1,525	293		1,818	2,100
5620	1600 amp		1.80	8.889		1,800	390		2,190	2,550
5630	2000 amp		1.40	11.429		2,000	500		2,500	2,950
5640	2500 amp		1.20	13.333		2,500	585		3,085	3,600
5650	3000 amp		1	16		2,925	700		3,625	4,275
5660	4000 amp		.70	22.857		4,275	1,000		5,275	6,200

26 25 Enclosed Bus Assemblies

26 25 13 – Enclosed Bus Assemblies

	26 25 13.10 Aluminum Bus Duct	Crew	Daily Output	Labor-Hours	Unit	Material	2007 Bare Costs Labor	Equipment	Total	Total Incl O&P
5680	Cross, 225 amp	2 Elec	6.40	2.500	Ea.	1,500	110		1,610	1,825
5690	400 amp		5.40	2.963		1,500	130		1,630	1,850
5700	600 amp		4.60	3.478		1,500	153		1,653	1,875
5710	800 amp		4	4		1,600	176		1,776	2,000
5720	1000 amp		3.60	4.444		1,625	195		1,820	2,100
5730	1350 amp		3.20	5		2,450	220		2,670	3,000
5740	1600 amp		2.60	6.154		2,850	270		3,120	3,525
5750	2000 amp		2.20	7.273		3,100	320		3,420	3,900
5760	2500 amp		1.80	8.889		3,750	390		4,140	4,700
5770	3000 amp		1.40	11.429		4,475	500		4,975	5,675
5780	4000 amp		1.20	13.333		6,250	585		6,835	7,750
5800	Expansion fitting, 225 amp		11.60	1.379		1,100	60.50		1,160.50	1,300
5810	400 amp		9.20	1.739		1,100	76.50		1,176.50	1,325
5820	600 amp		7	2.286		1,100	100		1,200	1,350
5830	800 amp		5.20	3.077		1,300	135		1,435	1,625
5840	1000 amp		4.60	3.478		1,425	153		1,578	1,775
5850	1350 amp		4.20	3.810		1,725	167		1,892	2,150
5860	1600 amp		3.60	4.444		2,125	195		2,320	2,650
5870	2000 amp		3.20	5		2,525	220		2,745	3,100
5880	2500 amp		2.80	5.714		2,850	251		3,101	3,500
5890	3000 amp		2.40	6.667		3,325	293		3,618	4,075
5900	4000 amp		1.80	8.889		4,575	390		4,965	5,600
5940	Reducer, nonfused, 400 amp		9.20	1.739		875	76.50		951.50	1,075
5950	600 amp		7	2.286		875	100		975	1,100
5960	800 amp		5.20	3.077		915	135		1,050	1,200
5970	1000 amp		4.60	3.478		1,125	153		1,278	1,475
5980	1350 amp		4.20	3.810		1,675	167		1,842	2,100
5990	1600 amp		3.60	4.444		1,900	195		2,095	2,400
6000	2000 amp		3.20	5		2,225	220		2,445	2,775
6010	2500 amp		2.80	5.714		2,850	251		3,101	3,500
6020	3000 amp		2.20	7.273		3,325	320		3,645	4,125
6030	4000 amp		1.80	8.889		4,575	390		4,965	5,600
6050	Reducer, fuse included, 225 amp		5	3.200		2,300	140		2,440	2,725
6060	400 amp		4.80	3.333		3,025	146		3,171	3,575
6070	600 amp		4.20	3.810		3,900	167		4,067	4,525
6080	800 amp		3.60	4.444		6,025	195		6,220	6,925
6090	1000 amp		3.40	4.706		6,575	207		6,782	7,525
6100	1350 amp		3.20	5		12,000	220		12,220	13,500
6110	1600 amp		2.60	6.154		14,200	270		14,470	16,000
6120	2000 amp		2	8		16,500	350		16,850	18,700
6160	Reducer, circuit breaker, 225 amp		5	3.200		2,800	140		2,940	3,300
6170	400 amp		4.80	3.333		3,450	146		3,596	4,025
6180	600 amp		4.20	3.810		4,950	167		5,117	5,700
6190	800 amp		3.60	4.444		5,775	195		5,970	6,650
6200	1000 amp		3.40	4.706		6,575	207		6,782	7,525
6210	1350 amp		3.20	5		7,925	220		8,145	9,050
6220	1600 amp		2.60	6.154		11,800	270		12,070	13,400
6230	2000 amp		2	8		12,900	350		13,250	14,700
6270	Cable tap box center, 225 amp		4.20	3.810		1,050	167		1,217	1,400
6280	400 amp		3.60	4.444		1,050	195		1,245	1,450
6290	600 amp		3	5.333		1,050	234		1,284	1,500
6300	800 amp		2.60	6.154		1,150	270		1,420	1,650
6310	1000 amp		2.40	6.667		1,225	293		1,518	1,750

26 25 Enclosed Bus Assemblies

26 25 13 – Enclosed Bus Assemblies

26 25 13.10 Aluminum Bus Duct

		Crew	Daily Output	Labor-Hours	Unit	Material	2007 Bare Costs Labor	Equipment	Total	Total Incl O&P
6320	1350 amp	2 Elec	1.80	8.889	Ea.	1,475	390		1,865	2,175
6330	1600 amp		1.60	10		1,650	440		2,090	2,450
6340	2000 amp		1.40	11.429		1,875	500		2,375	2,825
6350	2500 amp		1.20	13.333		2,350	585		2,935	3,450
6360	3000 amp		1	16		2,650	700		3,350	3,975
6370	4000 amp		.70	22.857		3,150	1,000		4,150	4,950
6390	Cable tap box end, 225 amp		4.20	3.810		640	167		807	955
6400	400 amp		3.60	4.444		640	195		835	995
6410	600 amp		3	5.333		640	234		874	1,050
6420	800 amp		2.60	6.154		700	270		970	1,175
6430	1000 amp		2.40	6.667		755	293		1,048	1,275
6435	1200 amp		2.10	7.619		820	335		1,155	1,400
6440	1350 amp		1.80	8.889		910	390		1,300	1,575
6450	1600 amp		1.60	10		1,025	440		1,465	1,775
6460	2000 amp		1.40	11.429		1,175	500		1,675	2,025
6470	2500 amp		1.20	13.333		1,350	585		1,935	2,350
6480	3000 amp		1	16		1,600	700		2,300	2,800
6490	4000 amp		.70	22.857		1,925	1,000		2,925	3,600
7000	Weatherproof, feeder, 600 amp		34	.471	L.F.	155	20.50		175.50	202
7020	800 amp		28	.571		171	25		196	226
7040	1000 amp		26	.615		184	27		211	242
7050	1200 amp		25	.640		215	28		243	278
7060	1350 amp		24	.667		250	29.50		279.50	320
7080	1600 amp		20	.800		289	35		324	370
7100	2000 amp		18	.889		340	39		379	435
7120	2500 amp		14	1.143		475	50		525	595
7140	3000 amp		12	1.333		565	58.50		623.50	705
7160	4000 amp		10	1.600		685	70		755	855

26 25 13.20 Bus Duct

		Crew	Daily Output	Labor-Hours	Unit	Material	2007 Bare Costs Labor	Equipment	Total	Total Incl O&P
0010	**BUS DUCT** 100 amp and less, aluminum or copper, plug-in									
0080	Bus duct, 3 pole 3 wire, 100 amp	1 Elec	42	.190	L.F.	24	8.35		32.35	39
0110	Elbow		4	2	Ea.	61.50	88		149.50	199
0120	Tee		2	4		89.50	176		265.50	360
0130	Wall flange		8	1		7.95	44		51.95	74.50
0140	Ground kit		16	.500		16.90	22		38.90	51
0180	3 pole 4 wire, 100 amp		40	.200	L.F.	27.50	8.80		36.30	43
0200	Cable tap box		3.10	2.581	Ea.	74.50	113		187.50	251
0300	End closure		16	.500		9.95	22		31.95	43.50
0400	Elbow		4	2		61.50	88		149.50	199
0500	Tee		2	4		89.50	176		265.50	360
0600	Hangers		10	.800		5.95	35		40.95	59
0700	Circuit breakers, 15 to 50 amp, 1 pole		8	1		159	44		203	241
0800	15 to 60 amp, 2 pole		6.70	1.194		620	52.50		672.50	765
0900	3 pole		5.30	1.509		620	66.50		686.50	785
1000	60 to 100 amp, 1 pole		6.70	1.194		620	52.50		672.50	765
1100	70 to 100 amp, 2 pole		5.30	1.509		620	66.50		686.50	785
1200	3 pole		4.50	1.778		620	78		698	800
1220	Switch, nonfused, 3 pole, 4 wire		8	1		84.50	44		128.50	159
1240	Fused, 3 fuses, 4 wire, 30 amp		8	1		305	44		349	400
1260	60 amp		5.30	1.509		320	66.50		386.50	455
1280	100 amp		4.50	1.778		535	78		613	705
1300	Plug, fusible, 3 pole 250 volt, 30 amp		5.30	1.509		305	66.50		371.50	435

229

26 25 13.20 Bus Duct		Crew	Daily Output	Labor-Hours	Unit	Material	2007 Bare Costs Labor	Equipment	Total	Total Incl O&P
1310	60 amp	1 Elec	5.30	1.509	Ea.	365	66.50		431.50	500
1320	100 amp		4.50	1.778		535	78		613	705
1330	3 pole 480 volt, 30 amp		5.30	1.509		350	66.50		416.50	485
1340	60 amp		5.30	1.509		380	66.50		446.50	515
1350	100 amp		4.50	1.778		545	78		623	715
1360	Circuit breaker, 3 pole 250 volt, 60 amp		5.30	1.509		620	66.50		686.50	785
1370	3 pole 480 volt, 100 amp		4.50	1.778		620	78		698	800
2000	Bus duct, 2 wire, 250 volt, 30 amp		60	.133	L.F.	2.64	5.85		8.49	11.60
2100	60 amp		50	.160		2.64	7		9.64	13.35
2200	300 volt, 30 amp		60	.133		2.64	5.85		8.49	11.60
2300	60 amp		50	.160		2.64	7		9.64	13.35
2400	3 wire, 250 volt, 30 amp		60	.133		2.64	5.85		8.49	11.60
2500	60 amp		50	.160		2.64	7		9.64	13.35
2600	480/277 volt, 30 amp		60	.133		2.64	5.85		8.49	11.60
2700	60 amp		50	.160		2.64	7		9.64	13.35
2750	End feed, 300 volt 2 wire max. 30 amp		6	1.333	Ea.	51.50	58.50		110	144
2800	60 amp		5.50	1.455		51.50	64		115.50	152
2850	30 amp miniature		6	1.333		51.50	58.50		110	144
2900	3 wire, 30 amp		6	1.333		65	58.50		123.50	159
2950	60 amp		5.50	1.455		65	64		129	167
3000	30 amp miniature		6	1.333		65	58.50		123.50	159
3050	Center feed, 300 volt 2 wire, 30 amp		6	1.333		71.50	58.50		130	166
3100	60 amp		5.50	1.455		71.50	64		135.50	174
3150	3 wire, 30 amp		6	1.333		81.50	58.50		140	177
3200	60 amp		5.50	1.455		81.50	64		145.50	185
3220	Elbow, 30 amp		6	1.333		50	58.50		108.50	142
3240	60 amp		5.50	1.455		50	64		114	150
3260	End cap		40	.200		9.95	8.80		18.75	24
3280	Strength beam, 10 ft.		15	.533		28	23.50		51.50	65.50
3300	Hanger		24	.333		5.95	14.65		20.60	28.50
3320	Tap box, nonfusible		6.30	1.270		84.50	56		140.50	176
3340	Fusible switch 30 amp, 1 fuse		6	1.333		370	58.50		428.50	490
3360	2 fuse		6	1.333		375	58.50		433.50	495
3380	3 fuse		6	1.333		380	58.50		438.50	500
3400	Circuit breaker handle on cover, 1 pole		6	1.333		59.50	58.50		118	153
3420	2 pole		6	1.333		64.50	58.50		123	158
3440	3 pole		6	1.333		84.50	58.50		143	180
3460	Circuit breaker external operhandle, 1 pole		6	1.333		59.50	58.50		118	153
3480	2 pole		6	1.333		64.50	58.50		123	158
3500	3 pole		6	1.333		84.50	58.50		143	180
3520	Terminal plug only		16	.500		10.95	22		32.95	44.50
3540	Terminal with receptacle		16	.500		14.20	22		36.20	48
3560	Fixture plug		16	.500		9.75	22		31.75	43.50
4000	Copper bus duct, lighting, 2 wire 300 volt, 20 amp		70	.114	L.F.	2.74	5		7.74	10.45
4020	35 amp		60	.133		2.74	5.85		8.59	11.70
4040	50 amp		55	.145		2.74	6.40		9.14	12.50
4060	60 amp		50	.160		2.74	7		9.74	13.45
4080	3 wire 300 volt, 20 amp		70	.114		2.99	5		7.99	10.75
4100	35 amp		60	.133		2.99	5.85		8.84	12
4120	50 amp		55	.145		2.99	6.40		9.39	12.80
4140	60 amp		50	.160		2.99	7		9.99	13.75
4160	Feeder in box, end, 1 circuit		6	1.333	Ea.	59.50	58.50		118	153
4180	2 circuit		5.50	1.455		61.50	64		125.50	163

26 25 Enclosed Bus Assemblies

26 25 13 – Enclosed Bus Assemblies

26 25 13.20 Bus Duct

		Crew	Daily Output	Labor-Hours	Unit	Material	2007 Bare Costs Labor	Equipment	Total	Total Incl O&P
4200	Center, 1 circuit	1 Elec	6	1.333	Ea.	81.50	58.50		140	177
4220	2 circuit		5.50	1.455		84.50	64		148.50	188
4240	End cap		40	.200		9.95	8.80		18.75	24
4260	Hanger, surface mount		24	.333		5.95	14.65		20.60	28.50
4280	Coupling	▼	40	.200	▼	7.55	8.80		16.35	21.50

26 25 13.30 Copper Bus Duct

		Crew	Daily Output	Labor-Hours	Unit	Material	2007 Bare Costs Labor	Equipment	Total	Total Incl O&P
0010	**COPPER BUS DUCT** R262513-15									
0100	3 pole 4 wire, weatherproof, feeder duct, 600 amp	2 Elec	24	.667	L.F.	167	29.50		196.50	228
0110	800 amp		18	.889		203	39		242	281
0120	1000 amp		17	.941		227	41.50		268.50	310
0125	1200 amp		16.50	.970		277	42.50		319.50	370
0130	1350 amp		16	1		335	44		379	435
0140	1600 amp		12	1.333		380	58.50		438.50	505
0150	2000 amp		10	1.600		490	70		560	645
0160	2500 amp		7	2.286		610	100		710	820
0170	3000 amp		5	3.200		740	140		880	1,025
0180	4000 amp		3.60	4.444		980	195		1,175	1,375
0200	Plug-in/indoor, bus duct high short circuit, 400 amp		32	.500		159	22		181	208
0210	600 amp		26	.615		159	27		186	215
0220	800 amp		20	.800		189	35		224	261
0230	1000 amp		18	.889		209	39		248	288
0240	1350 amp		16	1		299	44		343	395
0250	1600 amp		12	1.333		340	58.50		398.50	455
0260	2000 amp		10	1.600		430	70		500	575
0270	2500 amp		8	2		525	88		613	710
0280	3000 amp		6	2.667	▼	635	117		752	875
0310	Cross, 225 amp		3	5.333	Ea.	1,875	234		2,109	2,425
0320	400 amp		2.80	5.714		1,875	251		2,126	2,450
0330	600 amp		2.60	6.154		1,875	270		2,145	2,475
0340	800 amp		2.20	7.273		2,050	320		2,370	2,725
0350	1000 amp		2	8		2,275	350		2,625	3,050
0360	1350 amp		1.80	8.889		2,575	390		2,965	3,425
0370	1600 amp		1.70	9.412		2,875	415		3,290	3,775
0380	2000 amp		1.60	10		4,725	440		5,165	5,850
0390	2500 amp		1.40	11.429		5,700	500		6,200	7,025
0400	3000 amp		1.20	13.333		6,575	585		7,160	8,125
0410	4000 amp		1	16		8,575	700		9,275	10,500
0430	Expansion fitting, 225 amp		5.40	2.963		990	130		1,120	1,300
0440	400 amp		4.60	3.478		1,125	153		1,278	1,450
0450	600 amp		4	4		1,400	176		1,576	1,775
0460	800 amp		3.40	4.706		1,625	207		1,832	2,100
0470	1000 amp		3	5.333		1,875	234		2,109	2,425
0480	1350 amp		2.80	5.714		2,350	251		2,601	2,950
0490	1600 amp		2.60	6.154		3,275	270		3,545	4,000
0500	2000 amp		2.20	7.273		3,775	320		4,095	4,625
0510	2500 amp		1.80	8.889		4,600	390		4,990	5,625
0520	3000 amp		1.60	10		5,425	440		5,865	6,625
0530	4000 amp		1.20	13.333		7,000	585		7,585	8,575
0550	Reducer nonfused, 225 amp		5.40	2.963		1,125	130		1,255	1,450
0560	400 amp		4.60	3.478		1,125	153		1,278	1,475
0570	600 amp		4	4		1,125	176		1,301	1,500
0580	800 amp		3.40	4.706		1,375	207		1,582	1,800

231

26 25 13.30 Copper Bus Duct		Crew	Daily Output	Labor-Hours	Unit	Material	2007 Bare Costs Labor	Equipment	Total	Total Incl O&P
0590	1000 amp	2 Elec	3	5.333	Ea.	1,650	234		1,884	2,150
0600	1350 amp		2.80	5.714		2,500	251		2,751	3,125
0610	1600 amp		2.60	6.154		2,875	270		3,145	3,550
0620	2000 amp		2.20	7.273		3,450	320		3,770	4,275
0630	2500 amp		1.80	8.889		4,350	390		4,740	5,350
0640	3000 amp		1.60	10		5,175	440		5,615	6,350
0650	4000 amp		1.20	13.333		6,750	585		7,335	8,300
0670	Reducer fuse included, 225 amp		4.40	3.636		2,425	160		2,585	2,925
0680	400 amp		4.20	3.810		3,025	167		3,192	3,575
0690	600 amp		3.60	4.444		3,725	195		3,920	4,400
0700	800 amp		3.20	5		5,275	220		5,495	6,125
0710	1000 amp		3	5.333		6,600	234		6,834	7,625
0720	1350 amp		2.80	5.714		6,700	251		6,951	7,750
0730	1600 amp		2.20	7.273		14,700	320		15,020	16,600
0740	2000 amp		1.80	8.889		16,500	390		16,890	18,800
0790	Reducer, circuit breaker, 225 amp		4.40	3.636		3,075	160		3,235	3,625
0800	400 amp		4.20	3.810		3,575	167		3,742	4,175
0810	600 amp		3.60	4.444		5,075	195		5,270	5,875
0820	800 amp		3.20	5		5,950	220		6,170	6,875
0830	1000 amp		3	5.333		6,750	234		6,984	7,775
0840	1350 amp		2.80	5.714		8,125	251		8,376	9,300
0850	1600 amp		2.20	7.273		12,000	320		12,320	13,700
0860	2000 amp		1.80	8.889		13,200	390		13,590	15,100
0910	Cable tap box, center, 225 amp		3.20	5		1,150	220		1,370	1,575
0920	400 amp		2.60	6.154		1,150	270		1,420	1,650
0930	600 amp		2.20	7.273		1,150	320		1,470	1,725
0940	800 amp		2	8		1,275	350		1,625	1,925
0950	1000 amp		1.60	10		1,375	440		1,815	2,150
0960	1350 amp		1.40	11.429		1,725	500		2,225	2,650
0970	1600 amp		1.20	13.333		1,925	585		2,510	3,000
0980	2000 amp		1	16		2,325	700		3,025	3,600
1040	2500 amp		.80	20		2,750	880		3,630	4,325
1060	3000 amp		.60	26.667		3,175	1,175		4,350	5,250
1080	4000 amp		.40	40		4,000	1,750		5,750	7,025
1800	3 pole 3 wire, feeder duct, weatherproof, 600 amp		28	.571	L.F.	167	25		192	222
1820	800 amp		22	.727		203	32		235	271
1840	1000 amp		20	.800		227	35		262	305
1850	1200 amp		19	.842		235	37		272	315
1860	1350 amp		18	.889		335	39		374	430
1880	1600 amp		14	1.143		380	50		430	495
1900	2000 amp		12	1.333		490	58.50		548.50	625
1920	2500 amp		8	2		610	88		698	800
1940	3000 amp		6	2.667		740	117		857	990
1960	4000 amp		4	4		980	176		1,156	1,325
2000	Feeder duct/indoor, 600 amp		32	.500		139	22		161	186
2010	800 amp		26	.615		169	27		196	226
2020	1000 amp		24	.667		189	29.50		218.50	252
2025	1200 amp		22	.727		249	32		281	320
2030	1350 amp		20	.800		279	35		314	360
2040	1600 amp		16	1		320	44		364	415
2050	2000 amp		14	1.143		410	50		460	525
2060	2500 amp		10	1.600		505	70		575	665
2070	3000 amp		8	2		615	88		703	810

26 25 Enclosed Bus Assemblies

26 25 13 – Enclosed Bus Assemblies

26 25 13.30 Copper Bus Duct		Crew	Daily Output	Labor-Hours	Unit	Material	2007 Bare Costs Labor	Equipment	Total	Total Incl O&P
2080	4000 amp	2 Elec	6	2.667	L.F.	815	117		932	1,075
2090	5000 amp		5	3.200		995	140		1,135	1,300
2200	Bus duct plug-in/indoor, 225 amp		46	.348		159	15.25		174.25	198
2210	400 amp		36	.444		159	19.50		178.50	204
2220	600 amp		30	.533		159	23.50		182.50	210
2230	800 amp		24	.667		189	29.50		218.50	252
2240	1000 amp		20	.800		209	35		244	283
2250	1350 amp		18	.889		299	39		338	390
2260	1600 amp		14	1.143		340	50		390	445
2270	2000 amp		12	1.333		430	58.50		488.50	555
2280	2500 amp		10	1.600		525	70		595	685
2290	3000 amp		8	2		635	88		723	830
2330	High short circuit, 400 amp		36	.444		159	19.50		178.50	204
2340	600 amp		30	.533		159	23.50		182.50	210
2350	800 amp		24	.667		189	29.50		218.50	252
2360	1000 amp		20	.800		209	35		244	283
2370	1350 amp		18	.889		299	39		338	390
2380	1600 amp		14	1.143		340	50		390	445
2390	2000 amp		12	1.333		430	58.50		488.50	555
2400	2500 amp		10	1.600		525	70		595	685
2410	3000 amp		8	2		635	88		723	830
2440	Elbows, 225 amp		4.60	3.478	Ea.	795	153		948	1,100
2450	400 amp		4.20	3.810		795	167		962	1,125
2460	600 amp		3.60	4.444		795	195		990	1,175
2470	800 amp		3.20	5		855	220		1,075	1,275
2480	1000 amp		3	5.333		895	234		1,129	1,325
2485	1200 amp		2.90	5.517		965	242		1,207	1,400
2490	1350 amp		2.80	5.714		1,075	251		1,326	1,550
2500	1600 amp		2.60	6.154		1,150	270		1,420	1,675
2510	2000 amp		2	8		1,400	350		1,750	2,075
2520	2500 amp		1.80	8.889		2,125	390		2,515	2,925
2530	3000 amp		1.60	10		2,450	440		2,890	3,350
2540	4000 amp		1.40	11.429		3,150	500		3,650	4,200
2560	Tee fittings, 225 amp		2.80	5.714		940	251		1,191	1,400
2570	400 amp		2.40	6.667		940	293		1,233	1,450
2580	600 amp		2	8		940	350		1,290	1,550
2590	800 amp		1.80	8.889		1,025	390		1,415	1,700
2600	1000 amp		1.60	10		1,150	440		1,590	1,900
2605	1200 amp		1.50	10.667		1,225	470		1,695	2,050
2610	1350 amp		1.40	11.429		1,350	500		1,850	2,225
2620	1600 amp		1.20	13.333		1,550	585		2,135	2,600
2630	2000 amp		1	16		2,600	700		3,300	3,900
2640	2500 amp		.70	22.857		3,050	1,000		4,050	4,850
2650	3000 amp		.60	26.667		3,525	1,175		4,700	5,625
2660	4000 amp		.50	32		4,550	1,400		5,950	7,125
2680	Cross, 225 amp		3.60	4.444		1,425	195		1,620	1,875
2690	400 amp		3.20	5		1,425	220		1,645	1,900
2700	600 amp		3	5.333		1,425	234		1,659	1,925
2710	800 amp		2.60	6.154		1,700	270		1,970	2,275
2720	1000 amp		2.40	6.667		1,800	293		2,093	2,400
2730	1350 amp		2.20	7.273		2,150	320		2,470	2,850
2740	1600 amp		2	8		2,300	350		2,650	3,075
2750	2000 amp		1.80	8.889		3,650	390		4,040	4,600

26 25 13 – Enclosed Bus Assemblies

26 25 13.30 Copper Bus Duct		Crew	Daily Output	Labor-Hours	Unit	Material	2007 Bare Costs Labor	Equipment	Total	Total Incl O&P
2760	2500 amp	2 Elec	1.60	10	Ea.	4,250	440		4,690	5,325
2770	3000 amp		1.40	11.429		4,900	500		5,400	6,150
2780	4000 amp		1	16		6,275	700		6,975	7,950
2800	Expansion fitting, 225 amp		6.40	2.500		1,150	110		1,260	1,425
2810	400 amp		5.40	2.963		1,150	130		1,280	1,450
2820	600 amp		4.60	3.478		1,150	153		1,303	1,475
2830	800 amp		4	4		1,375	176		1,551	1,750
2840	1000 amp		3.60	4.444		1,500	195		1,695	1,950
2850	1350 amp		3.20	5		1,875	220		2,095	2,400
2860	1600 amp		3	5.333		2,050	234		2,284	2,600
2870	2000 amp		2.60	6.154		2,475	270		2,745	3,125
2880	2500 amp		2.20	7.273		3,475	320		3,795	4,275
2890	3000 amp		1.80	8.889		4,075	390		4,465	5,075
2900	4000 amp		1.40	11.429		5,225	500		5,725	6,475
2920	Reducer nonfused, 225 amp		6.40	2.500		930	110		1,040	1,200
2930	400 amp		5.40	2.963		930	130		1,060	1,225
2940	600 amp		4.60	3.478		930	153		1,083	1,250
2950	800 amp		4	4		1,075	176		1,251	1,450
2960	1000 amp		3.60	4.444		1,225	195		1,420	1,650
2970	1350 amp		3.20	5		1,625	220		1,845	2,100
2980	1600 amp		3	5.333		1,825	234		2,059	2,350
2990	2000 amp		2.60	6.154		2,200	270		2,470	2,825
3000	2500 amp		2.20	7.273		3,200	320		3,520	4,000
3010	3000 amp		1.80	8.889		3,800	390		4,190	4,750
3020	4000 amp		1.40	11.429		4,925	500		5,425	6,175
3040	Reducer fuse included, 225 amp		5	3.200		2,200	140		2,340	2,625
3050	400 amp		4.80	3.333		2,925	146		3,071	3,450
3060	600 amp		4.20	3.810		3,500	167		3,667	4,075
3070	800 amp		3.60	4.444		4,975	195		5,170	5,775
3080	1000 amp		3.40	4.706		5,875	207		6,082	6,775
3090	1350 amp		3.20	5		6,000	220		6,220	6,925
3100	1600 amp		2.60	6.154		14,000	270		14,270	15,800
3110	2000 amp		2	8		15,800	350		16,150	17,900
3160	Reducer circuit breaker, 225 amp		5	3.200		2,800	140		2,940	3,300
3170	400 amp		4.80	3.333		3,450	146		3,596	4,025
3180	600 amp		4.20	3.810		4,950	167		5,117	5,700
3190	800 amp		3.60	4.444		5,775	195		5,970	6,650
3200	1000 amp		3.40	4.706		6,575	207		6,782	7,525
3210	1350 amp		3.20	5		7,925	220		8,145	9,050
3220	1600 amp		2.60	6.154		11,800	270		12,070	13,400
3230	2000 amp		2	8		12,900	350		13,250	14,700
3280	3 pole, 3 wire, cable tap box center, 225 amp		3.60	4.444		1,300	195		1,495	1,725
3290	400 amp		3	5.333		1,300	234		1,534	1,775
3300	600 amp		2.60	6.154		1,300	270		1,570	1,825
3310	800 amp		2.40	6.667		1,450	293		1,743	2,025
3320	1000 amp		1.80	8.889		1,575	390		1,965	2,300
3330	1350 amp		1.60	10		2,025	440		2,465	2,875
3340	1600 amp		1.40	11.429		2,275	500		2,775	3,250
3350	2000 amp		1.20	13.333		2,750	585		3,335	3,900
3360	2500 amp		1	16		3,275	700		3,975	4,650
3370	3000 amp		.70	22.857		3,800	1,000		4,800	5,700
3380	4000 amp		.50	32		4,825	1,400		6,225	7,425
3400	Cable tap box end, 225 amp		3.60	4.444		705	195		900	1,075

26 25 13 – Enclosed Bus Assemblies

26 25 13.30 Copper Bus Duct		Crew	Daily Output	Labor-Hours	Unit	Material	2007 Bare Costs Labor	Equipment	Total	Total Incl O&P
3410	400 amp	2 Elec	3	5.333	Ea.	705	234		939	1,125
3420	600 amp		2.60	6.154		790	270		1,060	1,275
3430	800 amp		2.40	6.667		790	293		1,083	1,300
3440	1000 amp		1.80	8.889		870	390		1,260	1,550
3445	1200 amp		1.70	9.412		975	415		1,390	1,700
3450	1350 amp		1.60	10		1,075	440		1,515	1,850
3460	1600 amp		1.40	11.429		1,250	500		1,750	2,125
3470	2000 amp		1.20	13.333		1,425	585		2,010	2,450
3480	2500 amp		1	16		1,700	700		2,400	2,925
3490	3000 amp		.70	22.857		1,975	1,000		2,975	3,675
3500	4000 amp		.50	32		2,500	1,400		3,900	4,850
4600	Plug-in, fusible switch w/3 fuses, 3 pole, 250 volt, 30 amp	1 Elec	4	2		276	88		364	435
4610	60 amp		3.60	2.222		360	97.50		457.50	540
4620	100 amp		2.70	2.963		515	130		645	760
4630	200 amp	2 Elec	3.20	5		865	220		1,085	1,275
4640	400 amp		1.40	11.429		2,250	500		2,750	3,225
4650	600 amp		.90	17.778		3,125	780		3,905	4,575
4700	4 pole, 120/208 volt, 30 amp	1 Elec	3.90	2.051		375	90		465	550
4710	60 amp		3.50	2.286		405	100		505	595
4720	100 amp		2.60	3.077		565	135		700	820
4730	200 amp	2 Elec	3	5.333		950	234		1,184	1,400
4740	400 amp		1.30	12.308		2,225	540		2,765	3,250
4750	600 amp		.80	20		3,125	880		4,005	4,725
4800	3 pole, 480 volt, 30 amp	1 Elec	4	2		283	88		371	440
4810	60 amp		3.60	2.222		299	97.50		396.50	475
4820	100 amp		2.70	2.963		505	130		635	750
4830	200 amp	2 Elec	3.20	5		865	220		1,085	1,275
4840	400 amp		1.40	11.429		2,025	500		2,525	2,975
4850	600 amp		.90	17.778		2,875	780		3,655	4,300
4860	800 amp		.66	24.242		8,800	1,075		9,875	11,300
4870	1000 amp		.60	26.667		9,800	1,175		10,975	12,600
4880	1200 amp		.50	32		9,825	1,400		11,225	12,900
4890	1600 amp		.44	36.364		11,400	1,600		13,000	14,900
4900	4 pole, 277/480 volt, 30 amp	1 Elec	3.90	2.051		405	90		495	585
4910	60 amp		3.50	2.286		435	100		535	630
4920	100 amp		2.60	3.077		635	135		770	900
4930	200 amp	2 Elec	3	5.333		1,275	234		1,509	1,750
4940	400 amp		1.30	12.308		2,375	540		2,915	3,425
4950	600 amp		.80	20		3,250	880		4,130	4,875
5050	800 amp		.60	26.667		10,700	1,175		11,875	13,600
5060	1000 amp		.56	28.571		12,300	1,250		13,550	15,500
5070	1200 amp		.48	33.333		12,400	1,475		13,875	15,900
5080	1600 amp		.42	38.095		14,500	1,675		16,175	18,400
5150	Fusible with starter, 3 pole 250 volt, 30 amp	1 Elec	3.50	2.286		1,400	100		1,500	1,675
5160	60 amp		3.20	2.500		1,475	110		1,585	1,800
5170	100 amp		2.50	3.200		1,675	140		1,815	2,025
5180	200 amp	2 Elec	2.80	5.714		2,750	251		3,001	3,400
5200	3 pole 480 volt, 30 amp	1 Elec	3.50	2.286		1,400	100		1,500	1,675
5210	60 amp		3.20	2.500		1,475	110		1,585	1,800
5220	100 amp		2.50	3.200		1,675	140		1,815	2,025
5230	200 amp	2 Elec	2.80	5.714		2,750	251		3,001	3,400
5300	Fusible with contactor, 3 pole 250 volt, 30 amp	1 Elec	3.50	2.286		1,350	100		1,450	1,650
5310	60 amp		3.20	2.500		1,725	110		1,835	2,075

26 25 13 – Enclosed Bus Assemblies

26 25 13.30 Copper Bus Duct		Crew	Daily Output	Labor-Hours	Unit	Material	2007 Bare Costs Labor	Equipment	Total	Total Incl O&P
5320	100 amp	1 Elec	2.50	3.200	Ea.	2,425	140		2,565	2,850
5330	200 amp	2 Elec	2.80	5.714		2,775	251		3,026	3,425
5400	3 pole 480 volt, 30 amp	1 Elec	3.50	2.286		1,450	100		1,550	1,750
5410	60 amp		3.20	2.500		2,050	110		2,160	2,450
5420	100 amp		2.50	3.200		2,825	140		2,965	3,325
5430	200 amp	2 Elec	2.80	5.714		2,900	251		3,151	3,575
5450	Fusible with capacitor, 3 pole 250 volt, 30 amp	1 Elec	3	2.667		3,525	117		3,642	4,050
5460	60 amp		2	4		4,100	176		4,276	4,775
5500	3 pole 480 volt, 30 amp		3	2.667		2,950	117		3,067	3,425
5510	60 amp		2	4		3,700	176		3,876	4,325
5600	Circuit breaker, 3 pole, 250 volt, 60 amp		4.50	1.778		385	78		463	540
5610	100 amp		3.20	2.500		475	110		585	685
5650	4 pole, 120/208 volt, 60 amp		4.40	1.818		435	80		515	600
5660	100 amp		3.10	2.581		520	113		633	740
5700	3 pole, 4 wire 277/480 volt, 60 amp		4.30	1.860		590	81.50		671.50	770
5710	100 amp		3	2.667		655	117		772	895
5720	225 amp	2 Elec	3.20	5		1,450	220		1,670	1,925
5730	400 amp		1.20	13.333		3,025	585		3,610	4,200
5740	600 amp		.96	16.667		4,050	730		4,780	5,550
5750	700 amp		.60	26.667		5,125	1,175		6,300	7,400
5760	800 amp		.60	26.667		5,125	1,175		6,300	7,400
5770	900 amp		.54	29.630		6,750	1,300		8,050	9,350
5780	1000 amp		.54	29.630		6,750	1,300		8,050	9,350
5790	1200 amp		.42	38.095		8,125	1,675		9,800	11,400
5810	Circuit breaker w/HIC fuses, 3 pole 480 volt, 60 amp	1 Elec	4.40	1.818		740	80		820	935
5820	100 amp	"	3.10	2.581		815	113		928	1,075
5830	225 amp	2 Elec	3.40	4.706		2,600	207		2,807	3,150
5840	400 amp		1.40	11.429		4,125	500		4,625	5,275
5850	600 amp		1	16		4,175	700		4,875	5,625
5860	700 amp		.64	25		5,525	1,100		6,625	7,700
5870	800 amp		.64	25		5,525	1,100		6,625	7,700
5880	900 amp		.56	28.571		11,900	1,250		13,150	15,000
5890	1000 amp		.56	28.571		11,900	1,250		13,150	15,000
5950	3 pole 4 wire 277/480 volt, 60 amp	1 Elec	4.30	1.860		740	81.50		821.50	935
5960	100 amp	"	3	2.667		815	117		932	1,075
5970	225 amp	2 Elec	3	5.333		2,600	234		2,834	3,200
5980	400 amp		1.10	14.545		4,125	640		4,765	5,475
5990	600 amp		.94	17.021		4,175	745		4,920	5,675
6000	700 amp		.58	27.586		5,525	1,200		6,725	7,875
6010	800 amp		.58	27.586		5,525	1,200		6,725	7,875
6020	900 amp		.52	30.769		11,900	1,350		13,250	15,100
6030	1000 amp		.52	30.769		11,900	1,350		13,250	15,100
6040	1200 amp		.40	40		11,900	1,750		13,650	15,700
6100	Circuit breaker with starter, 3 pole 250 volt, 60 amp	1 Elec	3.20	2.500		1,050	110		1,160	1,325
6110	100 amp	"	2.50	3.200		1,450	140		1,590	1,800
6120	225 amp	2 Elec	3	5.333		1,875	234		2,109	2,400
6130	3 pole 480 volt, 60 amp	1 Elec	3.20	2.500		1,050	110		1,160	1,325
6140	100 amp	"	2.50	3.200		1,450	140		1,590	1,800
6150	225 amp	2 Elec	3	5.333		1,875	234		2,109	2,400
6200	Circuit breaker with contactor, 3 pole 250 volt, 60 amp	1 Elec	3.20	2.500		995	110		1,105	1,275
6210	100 amp	"	2.50	3.200		1,350	140		1,490	1,675
6220	225 amp	2 Elec	3	5.333		1,750	234		1,984	2,275
6250	3 pole 480 volt, 60 amp	1 Elec	3.20	2.500		995	110		1,105	1,275

26 25 Enclosed Bus Assemblies

26 25 13 – Enclosed Bus Assemblies

26 25 13.30 Copper Bus Duct

		Crew	Daily Output	Labor-Hours	Unit	Material	2007 Bare Costs Labor	2007 Bare Costs Equipment	Total	Total Incl O&P
6260	100 amp	1 Elec	2.50	3.200	Ea.	1,350	140		1,490	1,675
6270	225 amp	2 Elec	3	5.333		1,750	234		1,984	2,275
6300	Circuit breaker with capacitor, 3 pole 250 volt, 60 amp	1 Elec	2	4		4,450	176		4,626	5,125
6310	3 pole 480 volt, 60 amp		2	4		4,125	176		4,301	4,775
6400	Add control transformer with pilot light to above starter		16	.500		320	22		342	385
6410	Switch, fusible, mechanically held contactor optional		16	.500		840	22		862	960
6430	Circuit breaker, mechanically held contactor optional		16	.500		840	22		862	960
6450	Ground neutralizer, 3 pole	↓	16	.500	↓	39	22		61	75

26 25 13.40 Copper Bus Duct

		Crew	Daily Output	Labor-Hours	Unit	Material	2007 Bare Costs Labor	2007 Bare Costs Equipment	Total	Total Incl O&P
0010	**COPPER BUS DUCT** 10 ft. long									
0050	3 pole 4 wire, plug-in/indoor, straight section, 225 amp	2 Elec	40	.400	L.F.	159	17.55		176.55	201
1000	400 amp		32	.500		159	22		181	208
1500	600 amp		26	.615		159	27		186	215
2400	800 amp		20	.800		189	35		224	261
2450	1000 amp		18	.889		209	39		248	288
2470	1200 amp		17	.941		269	41.50		310.50	360
2500	1350 amp		16	1		299	44		343	395
2510	1600 amp		12	1.333		340	58.50		398.50	455
2520	2000 amp		10	1.600		430	70		500	575
2530	2500 amp		8	2		525	88		613	710
2540	3000 amp		6	2.667		635	117		752	875
2550	Feeder, 600 amp		28	.571		139	25		164	191
2600	800 amp		22	.727		169	32		201	234
2700	1000 amp		20	.800		189	35		224	261
2750	1200 amp		19	.842		249	37		286	330
2800	1350 amp		18	.889		279	39		318	365
2900	1600 amp		14	1.143		320	50		370	425
3000	2000 amp		12	1.333		410	58.50		468.50	535
3010	2500 amp		8	2		505	88		593	690
3020	3000 amp		6	2.667		615	117		732	855
3030	4000 amp		4	4		815	176		991	1,150
3040	5000 amp		2	8	↓	995	350		1,345	1,625
3100	Elbows, 225 amp		4	4	Ea.	945	176		1,121	1,300
3200	400 amp		3.60	4.444		945	195		1,140	1,350
3300	600 amp		3.20	5		945	220		1,165	1,375
3400	800 amp		2.80	5.714		1,025	251		1,276	1,500
3500	1000 amp		2.60	6.154		1,150	270		1,420	1,650
3550	1200 amp		2.50	6.400		1,250	281		1,531	1,825
3600	1350 amp		2.40	6.667		1,350	293		1,643	1,900
3700	1600 amp		2.20	7.273		1,450	320		1,770	2,075
3800	2000 amp		1.80	8.889		1,800	390		2,190	2,550
3810	2500 amp		1.60	10		2,850	440		3,290	3,775
3820	3000 amp		1.40	11.429		3,300	500		3,800	4,375
3830	4000 amp		1.20	13.333		4,275	585		4,860	5,575
3840	5000 amp		1	16		6,900	700		7,600	8,625
4000	End box, 225 amp		34	.471		127	20.50		147.50	171
4100	400 amp		32	.500		127	22		149	173
4200	600 amp		28	.571		127	25		152	178
4300	800 amp		26	.615		127	27		154	180
4400	1000 amp		24	.667		127	29.50		156.50	184
4410	1200 amp		23	.696		127	30.50		157.50	186
4500	1350 amp	↓	22	.727	↓	127	32		159	188

26 25 Enclosed Bus Assemblies

26 25 13 – Enclosed Bus Assemblies

26 25 13.40 Copper Bus Duct		Crew	Daily Output	Labor-Hours	Unit	Material	2007 Bare Costs Labor	2007 Bare Costs Equipment	Total	Total Incl O&P
4600	1600 amp	2 Elec	20	.800	Ea.	127	35		162	193
4700	2000 amp		18	.889		156	39		195	230
4710	2500 amp		16	1		156	44		200	238
4720	3000 amp		14	1.143		156	50		206	247
4730	4000 amp		12	1.333		189	58.50		247.50	295
4740	5000 amp		10	1.600		189	70		259	315
4800	Cable tap box end, 225 amp		3.20	5		950	220		1,170	1,375
5000	400 amp		2.60	6.154		950	270		1,220	1,450
5100	600 amp		2.20	7.273		950	320		1,270	1,525
5200	800 amp		2	8		1,075	350		1,425	1,700
5300	1000 amp		1.60	10		1,175	440		1,615	1,950
5350	1200 amp		1.50	10.667		1,325	470		1,795	2,150
5400	1350 amp		1.40	11.429		1,400	500		1,900	2,300
5500	1600 amp		1.20	13.333		1,575	585		2,160	2,600
5600	2000 amp		1	16		1,800	700		2,500	3,025
5610	2500 amp		.80	20		2,150	880		3,030	3,675
5620	3000 amp		.60	26.667		2,425	1,175		3,600	4,425
5630	4000 amp		.40	40		3,075	1,750		4,825	6,025
5640	5000 amp		.20	80		3,600	3,500		7,100	9,200
5700	Switchboard stub, 225 amp		5.40	2.963		955	130		1,085	1,250
5800	400 amp		4.60	3.478		955	153		1,108	1,275
5900	600 amp		4	4		955	176		1,131	1,300
6000	800 amp		3.20	5		1,150	220		1,370	1,600
6100	1000 amp		3	5.333		1,350	234		1,584	1,825
6150	1200 amp		2.80	5.714		1,550	251		1,801	2,075
6200	1350 amp		2.60	6.154		1,725	270		1,995	2,300
6300	1600 amp		2.40	6.667		1,950	293		2,243	2,575
6400	2000 amp		2	8		2,350	350		2,700	3,125
6410	2500 amp		1.80	8.889		2,875	390		3,265	3,725
6420	3000 amp		1.60	10		3,350	440		3,790	4,350
6430	4000 amp		1.40	11.429		4,375	500		4,875	5,550
6440	5000 amp		1.20	13.333		5,350	585		5,935	6,775
6490	Tee fittings, 225 amp		2.40	6.667		1,300	293		1,593	1,850
6500	400 amp		2	8		1,300	350		1,650	1,950
6600	600 amp		1.80	8.889		1,300	390		1,690	2,000
6700	800 amp		1.60	10		1,500	440		1,940	2,300
6750	1000 amp		1.40	11.429		1,725	500		2,225	2,650
6770	1200 amp		1.30	12.308		1,950	540		2,490	2,925
6800	1350 amp		1.20	13.333		2,150	585		2,735	3,250
7000	1600 amp		1	16		2,450	700		3,150	3,750
7100	2000 amp		.80	20		2,900	880		3,780	4,500
7110	2500 amp		.60	26.667		3,575	1,175		4,750	5,675
7120	3000 amp		.50	32		4,150	1,400		5,550	6,675
7130	4000 amp		.40	40		5,375	1,750		7,125	8,525
7140	5000 amp		.20	80		6,350	3,500		9,850	12,200
7200	Plug-in fusible switches w/3 fuses, 600 volt, 3 pole, 30 amp	1 Elec	4	2		565	88		653	755
7300	60 amp		3.60	2.222		635	97.50		732.50	845
7400	100 amp		2.70	2.963		975	130		1,105	1,275
7500	200 amp	2 Elec	3.20	5		1,750	220		1,970	2,225
7600	400 amp		1.40	11.429		5,075	500		5,575	6,350
7700	600 amp		.90	17.778		5,775	780		6,555	7,500
7800	800 amp		.66	24.242		7,775	1,075		8,850	10,100
7900	1200 amp		.50	32		14,600	1,400		16,000	18,100

26 25 Enclosed Bus Assemblies

26 25 13 – Enclosed Bus Assemblies

26 25 13.40 Copper Bus Duct	Crew	Daily Output	Labor-Hours	Unit	Material	2007 Bare Costs Labor	Equipment	Total	Total Incl O&P
7910	1600 amp	2 Elec	.44	36.364	Ea.	14,600	1,600		16,200	18,400
8000	Plug-in circuit breakers, molded case, 15 to 50 amp	1 Elec	4.40	1.818		535	80		615	710
8100	70 to 100 amp	"	3.10	2.581		595	113		708	825
8200	150 to 225 amp	2 Elec	3.40	4.706		1,625	207		1,832	2,075
8300	250 to 400 amp		1.40	11.429		2,825	500		3,325	3,875
8400	500 to 600 amp		1	16		3,825	700		4,525	5,275
8500	700 to 800 amp		.64	25		4,725	1,100		5,825	6,825
8600	900 to 1000 amp		.56	28.571		6,750	1,250		8,000	9,300
8700	1200 amp		.44	36.364		8,125	1,600		9,725	11,300
8720	1400 amp		.40	40		12,000	1,750		13,750	15,800
8730	1600 amp		.40	40		13,200	1,750		14,950	17,100
8750	Circuit breakers, with current limiting fuse, 15 to 50 amp	1 Elec	4.40	1.818		1,075	80		1,155	1,300
8760	70 to 100 amp	"	3.10	2.581		1,275	113		1,388	1,575
8770	150 to 225 amp	2 Elec	3.40	4.706		2,725	207		2,932	3,300
8780	250 to 400 amp		1.40	11.429		4,225	500		4,725	5,400
8790	500 to 600 amp		1	16		4,875	700		5,575	6,400
8800	700 to 800 amp		.64	25		8,000	1,100		9,100	10,400
8810	900 to 1000 amp		.56	28.571		9,125	1,250		10,375	12,000
8850	Combination starter FVNR, fusible switch, NEMA size 0, 30 amp	1 Elec	2	4		1,400	176		1,576	1,775
8860	NEMA size 1, 60 amp		1.80	4.444		1,475	195		1,670	1,925
8870	NEMA size 2, 100 amp		1.30	6.154		1,850	270		2,120	2,450
8880	NEMA size 3, 200 amp	2 Elec	2	8		3,075	350		3,425	3,925
8900	Circuit breaker, NEMA size 0, 30 amp	1 Elec	2	4		1,425	176		1,601	1,825
8910	NEMA size 1, 60 amp		1.80	4.444		1,475	195		1,670	1,925
8920	NEMA size 2, 100 amp		1.30	6.154		2,125	270		2,395	2,750
8930	NEMA size 3, 200 amp	2 Elec	2	8		3,150	350		3,500	4,000
8950	Combination contactor, fusible switch, NEMA size 0, 30 amp	1 Elec	2	4		960	176		1,136	1,300
8960	NEMA size 1, 60 amp		1.80	4.444		990	195		1,185	1,400
8970	NEMA size 2, 100 amp		1.30	6.154		1,475	270		1,745	2,025
8980	NEMA size 3, 200 amp	2 Elec	2	8		1,700	350		2,050	2,400
9000	Circuit breaker, NEMA size 0, 30 amp	1 Elec	2	4		1,100	176		1,276	1,450
9010	NEMA size 1, 60 amp		1.80	4.444		1,125	195		1,320	1,550
9020	NEMA size 2, 100 amp		1.30	6.154		1,750	270		2,020	2,325
9030	NEMA size 3, 200 amp	2 Elec	2	8		2,125	350		2,475	2,850
9050	Control transformer for above, NEMA size 0, 30 amp	1 Elec	8	1		164	44		208	247
9060	NEMA size 1, 60 amp		8	1		164	44		208	247
9070	NEMA size 2, 100 amp		7	1.143		229	50		279	325
9080	NEMA size 3, 200 amp	2 Elec	14	1.143		320	50		370	425
9100	Comb. fusible switch & lighting control, electrically held, 30 amp	1 Elec	2	4		670	176		846	1,000
9110	60 amp		1.80	4.444		965	195		1,160	1,350
9120	100 amp		1.30	6.154		1,250	270		1,520	1,775
9130	200 amp	2 Elec	2	8		3,050	350		3,400	3,900
9150	Mechanically held, 30 amp	1 Elec	2	4		840	176		1,016	1,175
9160	60 amp		1.80	4.444		1,250	195		1,445	1,675
9170	100 amp		1.30	6.154		1,600	270		1,870	2,175
9180	200 amp	2 Elec	2	8		3,275	350		3,625	4,125
9200	Ground bus added to bus duct, 225 amp		320	.050	L.F.	30	2.20		32.20	36.50
9210	400 amp		240	.067		30	2.93		32.93	37.50
9220	600 amp		240	.067		30	2.93		32.93	37.50
9230	800 amp		160	.100		35	4.39		39.39	45
9240	1000 amp		160	.100		40	4.39		44.39	50.50
9250	1350 amp		140	.114		59.50	5		64.50	73
9260	1600 amp		120	.133		64.50	5.85		70.35	79.50

26 25 Enclosed Bus Assemblies

26 25 13 – Enclosed Bus Assemblies

26 25 13.40 Copper Bus Duct		Crew	Daily Output	Labor-Hours	Unit	Material	2007 Bare Costs Labor	Equipment	Total	Total Incl O&P
9270	2000 amp	2 Elec	110	.145	L.F.	84.50	6.40		90.90	103
9280	2500 amp		100	.160		104	7		111	125
9290	3000 amp		90	.178		124	7.80		131.80	149
9300	4000 amp		80	.200		164	8.80		172.80	194
9310	5000 amp		70	.229		199	10.05		209.05	234
9320	High short circuit bracing, add					14.85			14.85	16.35

26 25 13.60 Copper or Aluminum Bus Duct Fittings

26 25 13.60		Crew	Daily Output	Labor-Hours	Unit	Material	2007 Bare Costs Labor	Equipment	Total	Total Incl O&P
0010	**COPPER OR ALUMINUM BUS DUCT FITTINGS**									
0100	Flange, wall, with vapor barrier, 225 amp	2 Elec	6.20	2.581	Ea.	575	113		688	800
0110	400 amp		6	2.667		575	117		692	805
0120	600 amp		5.80	2.759		575	121		696	810
0130	800 amp		5.40	2.963		575	130		705	825
0140	1000 amp		5	3.200		575	140		715	840
0145	1200 amp		4.80	3.333		575	146		721	850
0150	1350 amp		4.60	3.478		575	153		728	855
0160	1600 amp		4.20	3.810		575	167		742	880
0170	2000 amp		4	4		575	176		751	890
0180	2500 amp		3.60	4.444		575	195		770	920
0190	3000 amp		3.20	5		575	220		795	955
0200	4000 amp		2.60	6.154		665	270		935	1,125
0300	Roof, 225 amp		6.20	2.581		635	113		748	865
0310	400 amp		6	2.667		635	117		752	870
0320	600 amp		5.80	2.759		635	121		756	875
0330	800 amp		5.40	2.963		635	130		765	890
0340	1000 amp		5	3.200		635	140		775	905
0345	1200 amp		4.80	3.333		635	146		781	915
0350	1350 amp		4.60	3.478		635	153		788	920
0360	1600 amp		4.20	3.810		635	167		802	945
0370	2000 amp		4	4		635	176		811	955
0380	2500 amp		3.60	4.444		635	195		830	985
0390	3000 amp		3.20	5		635	220		855	1,025
0400	4000 amp		2.60	6.154		635	270		905	1,100
0420	Support, floor mounted, 225 amp		20	.800		128	35		163	194
0430	400 amp		20	.800		128	35		163	194
0440	600 amp		18	.889		128	39		167	199
0450	800 amp		16	1		128	44		172	207
0460	1000 amp		13	1.231		128	54		182	222
0465	1200 amp		11.80	1.356		128	59.50		187.50	230
0470	1350 amp		10.60	1.509		128	66.50		194.50	240
0480	1600 amp		9.20	1.739		128	76.50		204.50	255
0490	2000 amp		8	2		128	88		216	272
0500	2500 amp		6.40	2.500		128	110		238	305
0510	3000 amp		5.40	2.963		128	130		258	335
0520	4000 amp		4	4		128	176		304	400
0540	Weather stop, 225 amp		12	1.333		370	58.50		428.50	495
0550	400 amp		10	1.600		370	70		440	515
0560	600 amp		9	1.778		370	78		448	525
0570	800 amp		8	2		370	88		458	540
0580	1000 amp		6.40	2.500		370	110		480	575
0585	1200 amp		5.90	2.712		370	119		489	585
0590	1350 amp		5.40	2.963		370	130		500	605
0600	1600 amp		4.60	3.478		370	153		523	635

26 25 Enclosed Bus Assemblies

26 25 13 – Enclosed Bus Assemblies

26 25 13.60 Copper or Aluminum Bus Duct Fittings		Crew	Daily Output	Labor-Hours	Unit	Material	2007 Bare Costs Labor	Equipment	Total	Total Incl O&P
0610	2000 amp	2 Elec	4	4	Ea.	370	176		546	670
0620	2500 amp		3.20	5		370	220		590	735
0630	3000 amp		2.60	6.154		370	270		640	810
0640	4000 amp		2	8		370	350		720	935
0660	End closure, 225 amp		34	.471		127	20.50		147.50	171
0670	400 amp		32	.500		127	22		149	173
0680	600 amp		28	.571		127	25		152	178
0690	800 amp		26	.615		127	27		154	180
0700	1000 amp		24	.667		127	29.50		156.50	184
0705	1200 amp		23	.696		127	30.50		157.50	186
0710	1350 amp		22	.727		127	32		159	188
0720	1600 amp		20	.800		127	35		162	193
0730	2000 amp		18	.889		156	39		195	230
0740	2500 amp		16	1		156	44		200	238
0750	3000 amp		14	1.143		156	50		206	247
0760	4000 amp		12	1.333		189	58.50		247.50	295
0780	Switchboard stub, 3 pole 3 wire, 225 amp		6	2.667		765	117		882	1,025
0790	400 amp		5.20	3.077		765	135		900	1,050
0800	600 amp		4.60	3.478		765	153		918	1,075
0810	800 amp		3.60	4.444		925	195		1,120	1,325
0820	1000 amp		3.40	4.706		1,075	207		1,282	1,500
0825	1200 amp		3.20	5		1,075	220		1,295	1,525
0830	1350 amp		3	5.333		1,300	234		1,534	1,775
0840	1600 amp		2.80	5.714		1,525	251		1,776	2,050
0850	2000 amp		2.40	6.667		1,800	293		2,093	2,400
0860	2500 amp		2	8		2,175	350		2,525	2,925
0870	3000 amp		1.80	8.889		2,475	390		2,865	3,300
0880	4000 amp		1.60	10		3,200	440		3,640	4,175
0890	5000 amp		1.40	11.429		3,950	500		4,450	5,100
0900	3 pole 4 wire, 225 amp		5.40	2.963		955	130		1,085	1,250
0910	400 amp		4.60	3.478		955	153		1,108	1,275
0920	600 amp		4	4		955	176		1,131	1,300
0930	800 amp		3.20	5		1,150	220		1,370	1,600
0940	1000 amp		3	5.333		1,350	234		1,584	1,825
0950	1350 amp		2.60	6.154		1,725	270		1,995	2,300
0960	1600 amp		2.40	6.667		1,950	293		2,243	2,575
0970	2000 amp		2	8		2,350	350		2,700	3,125
0980	2500 amp		1.80	8.889		2,875	390		3,265	3,725
0990	3000 amp		1.60	10		3,350	440		3,790	4,350
1000	4000 amp		1.40	11.429		4,350	500		4,850	5,550
1050	Service head, weatherproof, 3 pole 3 wire, 225 amp		3	5.333		1,300	234		1,534	1,775
1060	400 amp		2.80	5.714		1,300	251		1,551	1,800
1070	600 amp		2.60	6.154		1,300	270		1,570	1,825
1080	800 amp		2.40	6.667		1,450	293		1,743	2,025
1090	1000 amp		2	8		1,575	350		1,925	2,275
1100	1350 amp		1.80	8.889		2,050	390		2,440	2,825
1110	1600 amp		1.60	10		2,300	440		2,740	3,175
1120	2000 amp		1.40	11.429		2,800	500		3,300	3,825
1130	2500 amp		1.20	13.333		3,325	585		3,910	4,550
1140	3000 amp		.90	17.778		3,875	780		4,655	5,425
1150	4000 amp		.70	22.857		4,925	1,000		5,925	6,925
1200	3 pole 4 wire, 225 amp		2.60	6.154		1,450	270		1,720	2,000
1210	400 amp		2.40	6.667		1,450	293		1,743	2,025

26 25 Enclosed Bus Assemblies

26 25 13 – Enclosed Bus Assemblies

26 25 13.60 Copper or Aluminum Bus Duct Fittings		Crew	Daily Output	Labor-Hours	Unit	Material	2007 Bare Costs Labor	Equipment	Total	Total Incl O&P
1220	600 amp	2 Elec	2.20	7.273	Ea.	1,450	320		1,770	2,075
1230	800 amp		2	8		1,700	350		2,050	2,375
1240	1000 amp		1.70	9.412		1,950	415		2,365	2,775
1250	1350 amp		1.50	10.667		2,325	470		2,795	3,250
1260	1600 amp		1.40	11.429		2,525	500		3,025	3,525
1270	2000 amp		1.20	13.333		3,375	585		3,960	4,600
1280	2500 amp		1	16		4,150	700		4,850	5,625
1290	3000 amp		.80	20		4,875	880		5,755	6,650
1300	4000 amp		.60	26.667		6,325	1,175		7,500	8,700
1350	Flanged end, 3 pole 3 wire, 225 amp		6	2.667		710	117		827	955
1360	400 amp		5.20	3.077		710	135		845	980
1370	600 amp		4.60	3.478		710	153		863	1,000
1380	800 amp		3.60	4.444		785	195		980	1,150
1390	1000 amp		3.40	4.706		885	207		1,092	1,275
1395	1200 amp		3.20	5		985	220		1,205	1,400
1400	1350 amp		3	5.333		1,050	234		1,284	1,500
1410	1600 amp		2.80	5.714		1,200	251		1,451	1,700
1420	2000 amp		2.40	6.667		1,425	293		1,718	1,975
1430	2500 amp		2	8		1,650	350		2,000	2,350
1440	3000 amp		1.80	8.889		1,875	390		2,265	2,650
1450	4000 amp		1.60	10		2,350	440		2,790	3,225
1500	3 pole 4 wire, 225 amp		5.40	2.963		815	130		945	1,100
1510	400 amp		4.60	3.478		815	153		968	1,125
1520	600 amp		4	4		815	176		991	1,150
1530	800 amp		3.20	5		960	220		1,180	1,375
1540	1000 amp		3	5.333		1,075	234		1,309	1,525
1545	1200 amp		2.80	5.714		1,225	251		1,476	1,700
1550	1350 amp		2.60	6.154		1,300	270		1,570	1,850
1560	1600 amp		2.40	6.667		1,500	293		1,793	2,075
1570	2000 amp		2	8		1,775	350		2,125	2,475
1580	2500 amp		1.80	8.889		2,100	390		2,490	2,900
1590	3000 amp		1.60	10		2,400	440		2,840	3,300
1600	4000 amp		1.40	11.429		3,100	500		3,600	4,175
1650	Hanger, standard, 225 amp		64	.250		14.95	11		25.95	33
1660	400 amp		48	.333		14.95	14.65		29.60	38.50
1670	600 amp		40	.400		14.95	17.55		32.50	42.50
1680	800 amp		32	.500		14.95	22		36.95	49
1690	1000 amp		24	.667		14.95	29.50		44.45	60
1695	1200 amp		22	.727		14.95	32		46.95	64
1700	1350 amp		20	.800		14.95	35		49.95	69
1710	1600 amp		20	.800		14.95	35		49.95	69
1720	2000 amp		18	.889		14.95	39		53.95	74.50
1730	2500 amp		16	1		14.95	44		58.95	82
1740	3000 amp		16	1		14.95	44		58.95	82
1750	4000 amp		16	1		14.95	44		58.95	82
1800	Spring type, 225 amp		16	1		66.50	44		110.50	139
1810	400 amp		14	1.143		66.50	50		116.50	148
1820	600 amp		14	1.143		66.50	50		116.50	148
1830	800 amp		14	1.143		66.50	50		116.50	148
1840	1000 amp		14	1.143		66.50	50		116.50	148
1845	1200 amp		14	1.143		66.50	50		116.50	148
1850	1350 amp		14	1.143		66.50	50		116.50	148
1860	1600 amp		12	1.333		66.50	58.50		125	161

242

26 25 Enclosed Bus Assemblies

26 25 13 – Enclosed Bus Assemblies

26 25 13.60 Copper or Aluminum Bus Duct Fittings	Crew	Daily Output	Labor-Hours	Unit	Material	2007 Bare Costs Labor	2007 Bare Costs Equipment	Total	Total Incl O&P
1870	2000 amp	2 Elec	12	1.333	Ea.	66.50	58.50		125	161
1880	2500 amp		12	1.333		66.50	58.50		125	161
1890	3000 amp		10	1.600		66.50	70		136.50	179
1900	4000 amp		10	1.600		66.50	70		136.50	179

26 25 13.70 Feedrail

		Crew	Daily Output	Labor-Hours	Unit	Material	Labor	Equipment	Total	Total Incl O&P
0010	**FEEDRAIL**, 12 foot mounting									
0050	Trolley busway, 3 pole									
0100	300 volt 60 amp, plain, 10 ft. lengths	1 Elec	50	.160	L.F.	22.50	7		29.50	35.50
0300	Door track		50	.160		25	7		32	38
0500	Curved track		30	.267		289	11.70		300.70	330
0700	Coupling				Ea.	12.95			12.95	14.25
0900	Center feed	1 Elec	5.30	1.509		35	66.50		101.50	137
1100	End feed		5.30	1.509		42	66.50		108.50	145
1300	Hanger set		24	.333		4.98	14.65		19.63	27.50
3000	600 volt 100 amp, plain, 10 ft. lengths		35	.229	L.F.	49	10.05		59.05	68.50
3300	Door track		35	.229	"	56.50	10.05		66.55	77.50
3700	Coupling				Ea.	41			41	45
4000	End cap	1 Elec	40	.200		35	8.80		43.80	51.50
4200	End feed		4	2		213	88		301	365
4500	Trolley, 600 volt, 20 amp		5.30	1.509		275	66.50		341.50	400
4700	30 amp		5.30	1.509		275	66.50		341.50	400
4900	Duplex, 40 amp		4	2		580	88		668	765
5000	60 amp		4	2		580	88		668	765
5300	Fusible, 20 amp		4	2		700	88		788	900
5500	30 amp		4	2		700	88		788	900
5900	300 volt, 20 amp		5.30	1.509		214	66.50		280.50	335
6000	30 amp		5.30	1.509		238	66.50		304.50	360
6300	Fusible, 20 amp		4.70	1.702		330	74.50		404.50	470
6500	30 amp		4.70	1.702		470	74.50		544.50	625
7300	Busway, 250 volt 50 amp, 2 wire		70	.114	L.F.	14.35	5		19.35	23
7330	Coupling				Ea.	28			28	30.50
7340	Center feed	1 Elec	6	1.333		320	58.50		378.50	435
7350	End feed		6	1.333		57.50	58.50		116	151
7360	End cap		40	.200		21	8.80		29.80	36
7370	Hanger set		24	.333		4.98	14.65		19.63	27.50
7400	125/250 volt 50 amp, 3 wire		60	.133	L.F.	15	5.85		20.85	25
7430	Coupling		6	1.333	Ea.	34	58.50		92.50	124
7440	Center feed		6	1.333		340	58.50		398.50	460
7450	End feed		6	1.333		61.50	58.50		120	155
7460	End cap		40	.200		23	8.80		31.80	38
7470	Hanger set		24	.333		4.98	14.65		19.63	27.50
7480	Trolley, 250 volt, 2 pole, 20 amp		6	1.333		35	58.50		93.50	126
7490	30 amp		6	1.333		35	58.50		93.50	126
7500	125/250 volt, 3 pole, 20 amp		6	1.333		37	58.50		95.50	128
7510	30 amp		6	1.333		37	58.50		95.50	128
8000	Cleaning tools, 300 volt, dust remover					64.50			64.50	71
8100	Bus bar cleaner					113			113	125
8300	600 volt, dust remover, 60 amp					240			240	264
8400	100 amp					330			330	360
8600	Bus bar cleaner, 60 amp					292			292	320
8700	100 amp					405			405	445

26 27 13.10 Meter Centers and Sockets		Crew	Daily Output	Labor-Hours	Unit	Material	2007 Bare Costs Labor	Equipment	Total	Total Incl O&P
0010	**METER CENTERS AND SOCKETS**									
0100	Sockets, single position, 4 terminal, 100 amp	1 Elec	3.20	2.500	Ea.	37.50	110		147.50	204
0200	150 amp		2.30	3.478		43.50	153		196.50	275
0300	200 amp		1.90	4.211		56	185		241	335
0400	20 amp		3.20	2.500		97	110		207	270
0500	Double position, 4 terminal, 100 amp		2.80	2.857		153	125		278	355
0600	150 amp		2.10	3.810		180	167		347	445
0700	200 amp		1.70	4.706		385	207		592	735
0800	Trans-socket, 13 terminal, 3 CT mounts, 400 amp		1	8		920	350		1,270	1,525
0900	800 amp	2 Elec	1.20	13.333		1,025	585		1,610	2,000
2000	Meter center, main fusible switch, 1P 3W 120/240 volt									
2030	400 amp	2 Elec	1.60	10	Ea.	1,575	440		2,015	2,375
2040	600 amp		1.10	14.545		2,750	640		3,390	3,975
2050	800 amp		.90	17.778		4,300	780		5,080	5,875
2060	Rainproof 1P 3W 120/240 volt, 400 amp		1.60	10		1,575	440		2,015	2,375
2070	600 amp		1.10	14.545		2,750	640		3,390	3,975
2080	800 amp		.90	17.778		4,300	780		5,080	5,875
2100	3P 4W 120/208V, 400 amp		1.60	10		1,800	440		2,240	2,625
2110	600 amp		1.10	14.545		3,375	640		4,015	4,650
2120	800 amp		.90	17.778		6,250	780		7,030	8,025
2130	Rainproof 3P 4W 120/208V, 400 amp		1.60	10		1,800	440		2,240	2,625
2140	600 amp		1.10	14.545		3,375	640		4,015	4,650
2150	800 amp		.90	17.778		6,250	780		7,030	8,025
2170	Main circuit breaker, 1P 3W 120/240V									
2180	400 amp	2 Elec	1.60	10	Ea.	2,825	440		3,265	3,750
2190	600 amp		1.10	14.545		3,800	640		4,440	5,125
2200	800 amp		.90	17.778		4,425	780		5,205	6,025
2210	1000 amp		.80	20		6,100	880		6,980	8,000
2220	1200 amp		.76	21.053		8,225	925		9,150	10,400
2230	1600 amp		.68	23.529		14,800	1,025		15,825	17,900
2240	Rainproof 1P 3W 120/240V, 400 amp		1.60	10		2,825	440		3,265	3,750
2250	600 amp		1.10	14.545		3,800	640		4,440	5,125
2260	800 amp		.90	17.778		4,425	780		5,205	6,025
2270	1000 amp		.80	20		6,100	880		6,980	8,000
2280	1200 amp		.76	21.053		8,225	925		9,150	10,400
2300	3P 4W 120/208V, 400 amp		1.60	10		3,200	440		3,640	4,175
2310	600 amp		1.10	14.545		4,525	640		5,165	5,925
2320	800 amp		.90	17.778		5,375	780		6,155	7,050
2330	1000 amp		.80	20		7,050	880		7,930	9,050
2340	1200 amp		.76	21.053		9,025	925		9,950	11,300
2350	1600 amp		.68	23.529		14,800	1,025		15,825	17,900
2360	Rainproof 3P 4W 120/208V, 400 amp		1.60	10		3,200	440		3,640	4,175
2370	600 amp		1.10	14.545		4,525	640		5,165	5,925
2380	800 amp		.90	17.778		5,375	780		6,155	7,050
2390	1000 amp		.76	21.053		7,050	925		7,975	9,125
2400	1200 amp		.68	23.529		9,025	1,025		10,050	11,500
2420	Main lugs terminal box, 1P 3W 120/240V									
2430	800 amp	2 Elec	.94	17.021	Ea.	530	745		1,275	1,675
2440	1200 amp		.72	22.222		1,050	975		2,025	2,625
2450	Rainproof 1P 3W 120/240V, 225 amp		2.40	6.667		360	293		653	830
2460	800 amp		.94	17.021		530	745		1,275	1,675
2470	1200 amp		.72	22.222		1,050	975		2,025	2,625
2500	3P 4W 120/208V, 800 amp		.94	17.021		585	745		1,330	1,750

26 27 Low-Voltage Distribution Equipment

26 27 13 – Electricity Metering

26 27 13.10 Meter Centers and Sockets		Crew	Daily Output	Labor-Hours	Unit	Material	2007 Bare Costs Labor	Equipment	Total	Total Incl O&P
2510	1200 amp	2 Elec	.72	22.222	Ea.	1,175	975		2,150	2,750
2520	Rainproof 3P 4W 120/208V, 225 amp		2.40	6.667		360	293		653	830
2530	800 amp		.94	17.021		585	745		1,330	1,750
2540	1200 amp	▼	.72	22.222	▼	1,175	975		2,150	2,750
2590	Basic meter device									
2600	1P 3W 120/240V 4 jaw 125A sockets, 3 meter	2 Elec	1	16	Ea.	605	700		1,305	1,725
2610	4 meter		.90	17.778		725	780		1,505	1,950
2620	5 meter		.80	20		910	880		1,790	2,300
2630	6 meter		.60	26.667		1,050	1,175		2,225	2,900
2640	7 meter		.56	28.571		1,325	1,250		2,575	3,350
2650	8 meter		.52	30.769		1,450	1,350		2,800	3,600
2660	10 meter	▼	.48	33.333	▼	1,825	1,475		3,300	4,175
2680	Rainproof 1P 3W 120/240V 4 jaw 125A sockets									
2690	3 meter	2 Elec	1	16	Ea.	605	700		1,305	1,725
2700	4 meter		.90	17.778		725	780		1,505	1,950
2710	6 meter		.60	26.667		1,050	1,175		2,225	2,900
2720	7 meter		.56	28.571		1,325	1,250		2,575	3,350
2730	8 meter	▼	.52	30.769	▼	1,450	1,350		2,800	3,600
2750	1P 3W 120/240V 4 jaw sockets									
2760	with 125A circuit breaker, 3 meter	2 Elec	1	16	Ea.	1,125	700		1,825	2,300
2770	4 meter		.90	17.778		1,425	780		2,205	2,725
2780	5 meter		.80	20		1,800	880		2,680	3,275
2790	6 meter		.60	26.667		2,100	1,175		3,275	4,050
2800	7 meter		.56	28.571		2,575	1,250		3,825	4,700
2810	8 meter		.52	30.769		2,850	1,350		4,200	5,150
2820	10 meter	▼	.48	33.333	▼	3,575	1,475		5,050	6,100
2830	Rainproof 1P 3W 120/240V 4 jaw sockets									
2840	with 125A circuit breaker, 3 meter	2 Elec	1	16	Ea.	1,125	700		1,825	2,300
2850	4 meter		.90	17.778		1,425	780		2,205	2,725
2870	6 meter		.60	26.667		2,100	1,175		3,275	4,050
2880	7 meter		.56	28.571		2,575	1,250		3,825	4,700
2890	8 meter	▼	.52	30.769	▼	2,850	1,350		4,200	5,150
2920	1P 3W on 3P 4W 120/208V system 5 jaw									
2930	125A sockets, 3 meter	2 Elec	1	16	Ea.	605	700		1,305	1,725
2940	4 meter		.90	17.778		725	780		1,505	1,950
2950	5 meter		.80	20		910	880		1,790	2,300
2960	6 meter		.60	26.667		1,050	1,175		2,225	2,900
2970	7 meter		.56	28.571		1,325	1,250		2,575	3,350
2980	8 meter		.52	30.769		1,450	1,350		2,800	3,600
2990	10 meter	▼	.48	33.333	▼	1,825	1,475		3,300	4,175
3000	Rainproof 1P 3W on 3P 4W 120/208V system									
3020	5 jaw 125A sockets, 3 meter	2 Elec	1	16	Ea.	605	700		1,305	1,725
3030	4 meter		.90	17.778		725	780		1,505	1,950
3050	6 meter		.60	26.667		1,050	1,175		2,225	2,900
3060	7 meter		.56	28.571		1,325	1,250		2,575	3,350
3070	8 meter	▼	.52	30.769	▼	1,450	1,350		2,800	3,600
3090	1P 3W on 3P 4W 120/208V system 5 jaw sockets									
3100	With 125A circuit breaker, 3 meter	2 Elec	1	16	Ea.	1,125	700		1,825	2,300
3110	4 meter		.90	17.778		1,425	780		2,205	2,725
3120	5 meter		.80	20		1,800	880		2,680	3,275
3130	6 meter		.60	26.667		2,100	1,175		3,275	4,050
3140	7 meter		.56	28.571		2,575	1,250		3,825	4,700
3150	8 meter	▼	.52	30.769	▼	2,850	1,350		4,200	5,150

26 27 13 – Electricity Metering

26 27 13.10 Meter Centers and Sockets	Crew	Daily Output	Labor-Hours	Unit	Material	2007 Bare Costs Labor	Equipment	Total	Total Incl O&P
3160 10 meter	2 Elec	.48	33.333	Ea.	3,575	1,475		5,050	6,100
3170 Rainproof 1P 3W on 3P 4W 120/208V system									
3180 5 jaw sockets w/125A circuit breaker, 3 meter	2 Elec	1	16	Ea.	1,125	700		1,825	2,300
3190 4 meter		.90	17.778		1,425	780		2,205	2,725
3210 6 meter		.60	26.667		2,100	1,175		3,275	4,050
3220 7 meter		.56	28.571		2,575	1,250		3,825	4,700
3230 8 meter		.52	30.769		2,850	1,350		4,200	5,150
3250 1P 3W 120/240V 4 jaw sockets									
3260 with 200A circuit breaker, 3 meter	2 Elec	1	16	Ea.	1,700	700		2,400	2,925
3270 4 meter		.90	17.778		2,300	780		3,080	3,675
3290 6 meter		.60	26.667		3,400	1,175		4,575	5,500
3300 7 meter		.56	28.571		4,000	1,250		5,250	6,275
3310 8 meter		.56	28.571		4,600	1,250		5,850	6,925
3330 Rainproof 1P 3W 120/240V 4 jaw sockets									
3350 with 200A circuit breaker, 3 meter	2 Elec	1	16	Ea.	1,700	700		2,400	2,925
3360 4 meter		.90	17.778		2,300	780		3,080	3,675
3380 6 meter		.60	26.667		3,400	1,175		4,575	5,500
3390 7 meter		.56	28.571		4,000	1,250		5,250	6,275
3400 8 meter		.52	30.769		4,600	1,350		5,950	7,050
3420 1P 3W on 3P 4W 120/208V 5 jaw sockets									
3430 with 200A circuit breaker, 3 meter	2 Elec	1	16	Ea.	1,700	700		2,400	2,925
3440 4 meter		.90	17.778		2,300	780		3,080	3,675
3460 6 meter		.60	26.667		3,400	1,175		4,575	5,500
3470 7 meter		.56	28.571		4,000	1,250		5,250	6,275
3480 8 meter		.52	30.769		4,600	1,350		5,950	7,050
3500 Rainproof 1P 3W on 3P 4W 120/208V 5 jaw socket									
3510 with 200A circuit breaker, 3 meter	2 Elec	1	16	Ea.	1,700	700		2,400	2,925
3520 4 meter		.90	17.778		2,300	780		3,080	3,675
3540 6 meter		.60	26.667		3,400	1,175		4,575	5,500
3550 7 meter		.56	28.571		4,000	1,250		5,250	6,275
3560 8 meter		.52	30.769		4,600	1,350		5,950	7,050
3600 Automatic circuit closing, add					65.50			65.50	72.50
3610 Manual circuit closing, add					75			75	82.50
3650 Branch meter device									
3660 3P 4W 208/120 or 240/120 V 7 jaw sockets									
3670 with 200A circuit breaker, 2 meter	2 Elec	.90	17.778	Ea.	2,850	780		3,630	4,275
3680 3 meter		.80	20		4,250	880		5,130	5,975
3690 4 meter		.70	22.857		5,675	1,000		6,675	7,750
3700 Main circuit breaker 42,000 rms, 400 amp		1.60	10		2,475	440		2,915	3,375
3710 600 amp		1.10	14.545		4,000	640		4,640	5,350
3720 800 amp		.90	17.778		5,375	780		6,155	7,050
3730 Rainproof main circ. breaker 42,000 rms, 400 amp		1.60	10		2,475	440		2,915	3,375
3740 600 amp		1.10	14.545		4,000	640		4,640	5,350
3750 800 amp		.90	17.778		5,375	780		6,155	7,050
3760 Main circuit breaker 65,000 rms, 400 amp		1.60	10		3,375	440		3,815	4,350
3770 600 amp		1.10	14.545		4,750	640		5,390	6,175
3780 800 amp		.90	17.778		5,375	780		6,155	7,050
3790 1000 amp		.80	20		7,050	880		7,930	9,050
3800 1200 amp		.76	21.053		9,025	925		9,950	11,300
3810 1600 amp		.68	23.529		14,800	1,025		15,825	17,900
3820 Rainproof main circ. breaker 65,000 rms, 400 amp		1.60	10		3,375	440		3,815	4,350
3830 600 amp		1.10	14.545		4,750	640		5,390	6,175
3840 800 amp		.90	17.778		5,375	780		6,155	7,050

26 27 Low-Voltage Distribution Equipment

26 27 13 – Electricity Metering

26 27 13.10 Meter Centers and Sockets

		Crew	Daily Output	Labor-Hours	Unit	Material	2007 Bare Costs Labor	2007 Bare Costs Equipment	Total	Total Incl O&P
3850	1000 amp	2 Elec	.80	20	Ea.	7,050	880		7,930	9,050
3860	1200 amp		.76	21.053		9,025	925		9,950	11,300
3880	Main circuit breaker 100,000 rms, 400 amp		1.60	10		3,375	440		3,815	4,350
3890	600 amp		1.10	14.545		4,750	640		5,390	6,175
3900	800 amp		.90	17.778		5,575	780		6,355	7,275
3910	Rainproof main circ. breaker 100,000 rms, 400 amp		1.60	10		3,375	440		3,815	4,350
3920	600 amp		1.10	14.545		4,750	640		5,390	6,175
3930	800 amp		.90	17.778		5,575	780		6,355	7,275
3940	Main lugs terminal box, 800 amp		.94	17.021		585	745		1,330	1,750
3950	1600 amp		.72	22.222		1,675	975		2,650	3,300
3960	Rainproof, 800 amp		.94	17.021		585	745		1,330	1,750
3970	1600 amp	▼	.72	22.222	▼	1,675	975		2,650	3,300

26 27 16 – Electrical Cabinets and Enclosures

26 27 16.10 Cabinets

		Crew	Daily Output	Labor-Hours	Unit	Material	2007 Bare Costs Labor	2007 Bare Costs Equipment	Total	Total Incl O&P
0010	**CABINETS** R260533-70									
7000	Cabinets, current transformer									
7050	Single door, 24" H x 24" W x 10" D R262716-40	1 Elec	1.60	5	Ea.	160	220		380	500
7100	30" H x 24" W x 10" D		1.30	6.154		175	270		445	595
7150	36" H x 24" W x 10" D		1.10	7.273		202	320		522	695
7200	30" H x 30" W x 10" D		1	8		210	350		560	755
7250	36" H x 30" W x 10" D		.90	8.889		287	390		677	895
7300	36" H x 36" W x 10" D		.80	10		305	440		745	990
7500	Double door, 48" H x 36" W x 10" D		.60	13.333		570	585		1,155	1,500
7550	24" H x 24" W x 12" D	▼	1	8	▼	248	350		598	800
7600	Telephone with wood backboard									
7620	Single door, 12" H x 12" W x 4" D	1 Elec	5.30	1.509	Ea.	110	66.50		176.50	220
7650	18" H x 12" W x 4" D		4.70	1.702		151	74.50		225.50	277
7700	24" H x 12" W x 4" D		4.20	1.905		171	83.50		254.50	310
7720	18" H x 18" W x 4" D		4.20	1.905		185	83.50		268.50	330
7750	24" H x 18" W x 4" D		4	2		253	88		341	410
7780	36" H x 36" W x 4" D		3.60	2.222		350	97.50		447.50	530
7800	24" H x 24" W x 6" D		3.60	2.222		335	97.50		432.50	510
7820	30" H x 24" W x 6" D		3.20	2.500		430	110		540	635
7850	30" H x 30" W x 6" D		2.70	2.963		495	130		625	740
7880	36" H x 30" W x 6" D		2.50	3.200		530	140		670	795
7900	48" H x 36" W x 6" D		2.20	3.636		935	160		1,095	1,275
7920	Double door, 48" H x 36" W x 6" D	▼	2	4	▼	970	176		1,146	1,325
8000	NEMA 12, double door, floor mounted									
8020	54" H x 42" W x 8" D	2 Elec	6	2.667	Ea.	980	117		1,097	1,250
8040	60" H x 48" W x 8" D		5.40	2.963		1,325	130		1,455	1,675
8060	60" H x 48" W x 10" D		5.40	2.963		1,350	130		1,480	1,675
8080	60" H x 60" W x 10" D		5	3.200		1,550	140		1,690	1,900
8100	72" H x 60" W x 10" D		4	4		1,775	176		1,951	2,200
8120	72" H x 72" W x 10" D		3.40	4.706		1,975	207		2,182	2,450
8140	60" H x 48" W x 12" D		3.40	4.706		1,375	207		1,582	1,825
8160	60" H x 60" W x 12" D		3.20	5		1,550	220		1,770	2,025
8180	72" H x 60" W x 12" D		3	5.333		1,575	234		1,809	2,075
8200	72" H x 72" W x 12" D		3	5.333		1,775	234		2,009	2,300
8220	60" H x 48" W x 16" D		3.20	5		1,475	220		1,695	1,925
8240	72" H x 72" W x 16" D		2.60	6.154		2,075	270		2,345	2,700
8260	60" H x 48" W x 20" D		3	5.333		1,600	234		1,834	2,100
8280	72" H x 72" W x 20" D	▼	2.20	7.273	▼	2,275	320		2,595	2,975

26 27 16 – Electrical Cabinets and Enclosures

26 27 16.10 Cabinets		Crew	Daily Output	Labor-Hours	Unit	Material	2007 Bare Costs Labor	2007 Bare Costs Equipment	Total	Total Incl O&P
8300	60" H x 48" W x 24" D	2 Elec	2.60	6.154	Ea.	1,675	270		1,945	2,250
8320	72" H x 72" W x 24" D	↓	2	8	↓	2,375	350		2,725	3,150
8340	Pushbutton enclosure, oiltight									
8360	3-1/2" H x 3-1/4" W x 2-3/4" D, for 1 P.B.	1 Elec	12	.667	Ea.	44.50	29.50		74	92.50
8380	5-3/4" H x 3-1/4" W x 2-3/4" D, for 2 P.B.		11	.727		44.50	32		76.50	96.50
8400	8" H x 3-1/4" W x 2-3/4" D, for 3 P.B.		10.50	.762		61	33.50		94.50	117
8420	10-1/4" H x 3-1/4" W x 2-3/4" D, for 4 P.B.		10.50	.762		63	33.50		96.50	120
8440	7-1/4" H x 6-1/4" W x 3" D, for 4 P.B.		10	.800		66.50	35		101.50	126
8460	12-1/2" H x 3-1/4" W x 3" D, for 5 P.B.		9	.889		70.50	39		109.50	136
8480	9-1/2" H x 6-1/4" W x 3" D, for 6 P.B.		8.50	.941		80.50	41.50		122	150
8500	9-1/2" H x 8-1/2" W x 3" D, for 9 P.B.		8	1		86.50	44		130.50	161
8510	11-3/4" H x 8-1/2" W x 3" D, for 12 P.B.		7	1.143		95.50	50		145.50	180
8520	11-3/4" H x 10-3/4" W x 3" D, for 16 P.B.		6.50	1.231		125	54		179	218
8540	14" H x 10-3/4" W x 3" D, for 20 P.B.		5	1.600		140	70		210	259
8560	14" H x 13" W x 3" D, for 25 P.B.	↓	4.50	1.778	↓	159	78		237	291
8580	Sloping front pushbutton enclosures									
8600	3-1/2" H x 7-3/4" W x 4-7/8" D, for 3 P.B	1 Elec	10	.800	Ea.	73.50	35		108.50	134
8620	7-1/4" H x 8-1/2" W x 6-3/4" D, for 6 P.B.		8	1		108	44		152	185
8640	9-1/2" H x 8-1/2" W x 7-7/8" D, for 9 P.B.		7	1.143		120	50		170	207
8660	11-1/4" H x 8-1/2" W x 9" D, for 12 P.B.		5	1.600		140	70		210	259
8680	11-3/4" H x 10" W x 9" D, for 16 P.B.		5	1.600		157	70		227	278
8700	11-3/4" H x 13" W x 9" D, for 20 P.B.		5	1.600		183	70		253	305
8720	14" H x 13" W x 10-1/8" D, for 25 P.B.	↓	4.50	1.778	↓	206	78		284	345
8740	Pedestals, not including P.B. enclosure or base									
8760	Straight column 4" x 4"	1 Elec	4.50	1.778	Ea.	187	78		265	320
8780	6" x 6"		4	2		280	88		368	440
8800	Angled column 4" x 4"		4.50	1.778		219	78		297	355
8820	6" x 6"		4	2		330	88		418	495
8840	Pedestal, base 18" x 18"		10	.800		114	35		149	178
8860	24" x 24"	↓	9	.889	↓	257	39		296	340
8900	Electronic rack enclosures									
8920	72" H x 25" W x 24" D	1 Elec	1.50	5.333	Ea.	1,925	234		2,159	2,475
8940	72" H x 30" W x 24" D		1.50	5.333		2,050	234		2,284	2,600
8960	72" H x 25" W x 30" D		1.30	6.154		2,075	270		2,345	2,675
8980	72" H x 25" W x 36" D		1.20	6.667		2,225	293		2,518	2,875
9000	72" H x 30" W x 36" D	↓	1.20	6.667	↓	2,350	293		2,643	3,025
9020	NEMA 12 & 4 enclosure panels									
9040	12" x 24"	1 Elec	20	.400	Ea.	26.50	17.55		44.05	55.50
9060	16" x 12"		20	.400		18.80	17.55		36.35	46.50
9080	20" x 16"		20	.400		27.50	17.55		45.05	56.50
9100	20" x 20"		19	.421		33.50	18.50		52	64.50
9120	24" x 20"		18	.444		43.50	19.50		63	76.50
9140	24" x 24"		17	.471		49.50	20.50		70	85.50
9160	30" x 20"		16	.500		52	22		74	89.50
9180	30" x 24"		16	.500		60	22		82	98.50
9200	36" x 24"		15	.533		72	23.50		95.50	115
9220	36" x 30"		15	.533		96	23.50		119.50	141
9240	42" x 24"		15	.533		82.50	23.50		106	126
9260	42" x 30"		14	.571		110	25		135	159
9280	42" x 36"		14	.571		130	25		155	181
9300	48" x 24"		14	.571		93.50	25		118.50	141
9320	48" x 30"		14	.571		123	25		148	173
9340	48" x 36"	↓	13	.615	↓	146	27		173	201

26 27 16.10 Cabinets

26 27 16.10 Cabinets	Crew	Daily Output	Labor-Hours	Unit	Material	2007 Bare Costs Labor	2007 Bare Costs Equipment	Total	Total Incl O&P
9360	60" x 36"	1 Elec	12	.667	Ea.	179	29.50		208.50	241
9400	Wiring trough steel JIC, clamp cover									
9490	4" x 4", 12" long	1 Elec	12	.667	Ea.	65	29.50		94.50	115
9510	24" long		10	.800		83	35		118	144
9530	36" long		8	1		101	44		145	178
9540	48" long		7	1.143		119	50		169	206
9550	60" long		6	1.333		137	58.50		195.50	238
9560	6" x 6", 12" long		11	.727		85.50	32		117.50	142
9580	24" long		9	.889		112	39		151	181
9600	36" long		7	1.143		140	50		190	229
9610	48" long		6	1.333		168	58.50		226.50	272
9620	60" long		5	1.600		198	70		268	325

26 27 16.20 Cabinets and Enclosures

	26 27 16.20 Cabinets and Enclosures	Crew	Daily Output	Labor-Hours	Unit	Material	2007 Bare Costs Labor	2007 Bare Costs Equipment	Total	Total Incl O&P
0010	**CABINETS AND ENCLOSURES** Nonmetallic									
0080	Enclosures fiberglass NEMA 4X									
0100	Wall mount, quick release latch door, 20"H x 16"W x 6"D	1 Elec	4.80	1.667	Ea.	465	73		538	620
0110	20"H x 20"W x 6"D		4.50	1.778		550	78		628	720
0120	24"H x 20"W x 6"D		4.20	1.905		590	83.50		673.50	775
0130	20"H x 16"W x 8"D		4.50	1.778		525	78		603	690
0140	20"H x 20"W x 8"D		4.20	1.905		580	83.50		663.50	765
0150	24"H x 24"W x 8"D		3.80	2.105		655	92.50		747.50	860
0160	30"H x 24"W x 8"D		3.20	2.500		730	110		840	965
0170	36"H x 30"W x 8"D		3	2.667		1,025	117		1,142	1,300
0180	20"H x 16"W x 10"D		3.50	2.286		600	100		700	810
0190	20"H x 20"W x 10"D		3.20	2.500		645	110		755	875
0200	24"H x 20"W x 10"D		3	2.667		695	117		812	940
0210	30"H x 24"W x 10"D		2.80	2.857		785	125		910	1,050
0220	20"H x 16"W x 12"D		3	2.667		660	117		777	905
0230	20"H x 20"W x 12"D		2.80	2.857		695	125		820	950
0240	24"H x 24"W x 12"D		2.60	3.077		760	135		895	1,025
0250	30"H x 24"W x 12"D		2.40	3.333		840	146		986	1,150
0260	36"H x 30"W x 12"D		2.20	3.636		1,175	160		1,335	1,525
0270	36"H x 36"W x 12"D		2.10	3.810		1,275	167		1,442	1,650
0280	48"H x 36"W x 12"D		2	4		1,500	176		1,676	1,900
0290	60"H x 36"W x 12"D		1.80	4.444		1,675	195		1,870	2,150
0300	30"H x 24"W x 16"D		1.40	5.714		1,000	251		1,251	1,475
0310	48"H x 36"W x 16"D		1.20	6.667		1,675	293		1,968	2,275
0320	60"H x 36"W x 16"D		1	8		1,900	350		2,250	2,600
0480	Freestanding, one door, 72"H x 25"W x 25"D		.80	10		3,250	440		3,690	4,225
0490	Two doors with two panels, 72"H x 49"W x 24"D		.50	16		7,675	700		8,375	9,500
0500	Floor stand kits, for NEMA 4 & 12, 20"W or more		24	.333		95	14.65		109.65	126
0510	6"H x 10"D		24	.333		104	14.65		118.65	136
0520	6"H x 12"D		24	.333		119	14.65		133.65	153
0530	6"H x 18"D		24	.333		140	14.65		154.65	176
0540	12"H x 8"D		22	.364		122	15.95		137.95	158
0550	12"H x 10"D		22	.364		132	15.95		147.95	169
0560	12"H x 12"D		22	.364		141	15.95		156.95	179
0570	12"H x 16"D		22	.364		159	15.95		174.95	199
0580	12"H x 18"D		22	.364		168	15.95		183.95	209
0590	12"H x 20"D		22	.364		177	15.95		192.95	219
0600	18"H x 8"D		20	.400		150	17.55		167.55	191
0610	18"H x 10"D		20	.400		159	17.55		176.55	201

26 27 Low-Voltage Distribution Equipment

26 27 16 – Electrical Cabinets and Enclosures

26 27 16.20 Cabinets and Enclosures	Crew	Daily Output	Labor-Hours	Unit	Material	2007 Bare Costs Labor	Equipment	Total	Total Incl O&P
0620	18"H x 12"D	1 Elec	20	.400	Ea.	168	17.55		185.55	211
0630	18"H x 16"D		20	.400		188	17.55		205.55	233
0640	24"H x 8"D		16	.500		177	22		199	228
0650	24"H x 10"D		16	.500		188	22		210	240
0660	24"H x 12"D		16	.500		197	22		219	250
0670	24"H x 16"D		16	.500		216	22		238	271
0680	Small, screw cover, 5-1/2"H x 4"W x 4-15/16"D		12	.667		57.50	29.50		87	107
0690	7-1/2"H x 4"W x 4-15/16"D		12	.667		60	29.50		89.50	110
0700	7-1/2"H x 6"W x 5-3/16"D		10	.800		69	35		104	129
0710	9-1/2"H x 6"W x 5-11/16"D		10	.800		72	35		107	132
0720	11-1/2"H x 8"W x 6-11/16"D		8	1		105	44		149	182
0730	13-1/2"H x 10"W x 7-3/16"D		7	1.143		129	50		179	217
0740	15-1/2"H x 12"W x 8-3/16"D		6	1.333		171	58.50		229.50	275
0750	17-1/2"H x 14"W x 8-11/16"D		5	1.600		197	70		267	320
0760	Screw cover with window, 6"H x 4"W x 5"D		12	.667		93	29.50		122.50	146
0770	8"H x 4"W x 5"D		11	.727		99.50	32		131.50	157
0780	8"H x 6"W x 5"D		11	.727		140	32		172	202
0790	10"H x 6"W x 6"D		10	.800		147	35		182	215
0800	12"H x 8"W x 7"D		8	1		191	44		235	276
0810	14"H x 10"W x 7"D		7	1.143		227	50		277	325
0820	16"H x 12"W x 8"D		6	1.333		273	58.50		331.50	385
0830	18"H x 14"W x 9"D		5	1.600		325	70		395	465
0840	Quick-release latch cover, 5-1/2"H x 4"W x 5"D		12	.667		74	29.50		103.50	125
0850	7-1/2"H x 4"W x 5"D		12	.667		77	29.50		106.50	128
0860	7-1/2"H x 6"W x 5-1/4"D		10	.800		89	35		124	151
0870	9-1/2"H x 6"W x 5-3/4"D		10	.800		92.50	35		127.50	155
0880	11-1/2"H x 8"W x 6-3/4"D		8	1		135	44		179	215
0890	13-1/2"H x 10"W x 7-1/4"D		7	1.143		167	50		217	259
0900	15-1/2"H x 12"W x 8-1/4"D		6	1.333		215	58.50		273.50	325
0910	17-1/2"H x 14"W x 8-3/4"D		5	1.600		254	70		324	385
0920	Pushbutton, 1 hole 5-1/2"H x 4"W x 4-15/16"D		12	.667		54.50	29.50		84	104
0930	2 hole 7-1/2"H x 4"W x 4-15/16"D		11	.727		61.50	32		93.50	116
0940	4 hole 7-1/2"H x 6"W x 5-3/16"D		10.50	.762		73.50	33.50		107	131
0950	6 hole 9-1/2"H x 6"W x 5-11/16"D		9	.889		89.50	39		128.50	157
0960	8 hole 11-1/2"H x 8"W x 6-11/16"D		8.50	.941		113	41.50		154.50	186
0970	12 hole 13-1/2"H x 10"W x 7-3/16"D		8	1		147	44		191	228
0980	20 hole 15-1/2"H x 12"W x 8-3/16"D		5	1.600		201	70		271	325
0990	30 hole 17-1/2"H x 14"W x 8-11/16"D	↓	4.50	1.778	↓	223	78		301	360
1450	Enclosures polyester NEMA 4X									
1460	Small, screw cover,									
1500	3-15/16"H x 3-15/16"W x 3-1/16"D	1 Elec	12	.667	Ea.	46	29.50		75.50	94
1510	5-3/16"H x 3-5/16"W x 3-1/16"D		12	.667		44.50	29.50		74	92.50
1520	5-7/8"H x 3-7/8"W x 4-3/16"D		12	.667		47.50	29.50		77	95.50
1530	5-7/8"H x 5-7/8"W x 4-3/16"D		12	.667		54	29.50		83.50	103
1540	7-5/8"H x 3-5/16"W x 3-1/16"D		12	.667		49	29.50		78.50	97.50
1550	10-3/16"H x 3-5/16"W x 3-1/16"D		10	.800		59	35		94	117
1560	Clear cover, 3-15/16"H x 3-15/16"W x 2-7/8"D		12	.667		51.50	29.50		81	101
1570	5-3/16"H x 3-5/16"W x 2-7/8"D		12	.667		55.50	29.50		85	105
1580	5-7/8"H x 3-7/8"W x 4"D		12	.667		69	29.50		98.50	119
1590	5-7/8"H x 5-7/8"W x 4"D		12	.667		84.50	29.50		114	137
1600	7-5/8"H x 3-5/16"W x 2-7/8"D		12	.667		61.50	29.50		91	112
1610	10-3/16"H x 3-5/16"W x 2-7/8"D		10	.800		77.50	35		112.50	138
1620	Pushbutton, 1 hole, 5-5/16"H x 3-5/16"W x 3-1/16"D		12	.667		40.50	29.50		70	88

26 27 Low-Voltage Distribution Equipment

26 27 16 – Electrical Cabinets and Enclosures

26 27 16.20 Cabinets and Enclosures

		Crew	Daily Output	Labor-Hours	Unit	Material	2007 Bare Costs Labor	Equipment	Total	Total Incl O&P
1630	2 hole, 7-5/8"H x 3-5/16"W x 3-1/8"D	1 Elec	11	.727	Ea.	45	32		77	97
1640	3 hole, 10-3/16"H x 3-5/16"W x 3-1/16"D		10.50	.762		54	33.50		87.50	110
8000	Wireway fiberglass, straight sect. screwcover, 12" L, 4" W x 4" D		40	.200		138	8.80		146.80	165
8010	6" W x 6" D		30	.267		186	11.70		197.70	222
8020	24" L, 4" W x 4" D		20	.400		176	17.55		193.55	220
8030	6" W x 6" D		15	.533		271	23.50		294.50	335
8040	36" L, 4" W x 4" D		13.30	.602		215	26.50		241.50	277
8050	6" W x 6" D		10	.800		340	35		375	430
8060	48" L, 4" W x 4" D		10	.800		252	35		287	330
8070	6" W x 6" D		7.50	1.067		410	47		457	520
8080	60" L, 4" W x 4" D		8	1		290	44		334	385
8090	6" W x 6" D		6	1.333		480	58.50		538.50	615
8100	Elbow, 90°, 4" W x 4" D		20	.400		137	17.55		154.55	177
8110	6" W x 6" D		18	.444		275	19.50		294.50	335
8120	Elbow, 45°, 4" W x 4" D		20	.400		137	17.55		154.55	177
8130	6" W x 6" D		18	.444		270	19.50		289.50	325
8140	Tee, 4" W x 4" D		16	.500		176	22		198	227
8150	6" W x 6" D		14	.571		320	25		345	390
8160	Cross, 4" W x 4" D		14	.571		247	25		272	310
8170	6" W x 6" D		12	.667		460	29.50		489.50	555
8180	Cut-off fitting, w/flange & adhesive, 4" W x 4" D		18	.444		106	19.50		125.50	146
8190	6" W x 6" D		16	.500		215	22		237	270
8200	Flexible ftng., hvy neoprene coated nylon, 4" W x 4" D		20	.400		176	17.55		193.55	220
8210	6" W x 6" D		18	.444		262	19.50		281.50	315
8220	Closure plate, fiberglass, 4" W x 4" D		20	.400		34	17.55		51.55	63.50
8230	6" W x 6" D		18	.444		38.50	19.50		58	71
8240	Box connector, stainless steel type 304, 4" W x 4" D		20	.400		59.50	17.55		77.05	91.50
8250	6" W x 6" D		18	.444		65	19.50		84.50	101
8260	Hanger, 4" W x 4" D		100	.080		14.80	3.51		18.31	21.50
8270	6" W x 6" D		80	.100		17.70	4.39		22.09	26
8280	Straight tube section fiberglass, 4"W x 4"D, 12" long		40	.200		120	8.80		128.80	145
8290	24" long		20	.400		139	17.55		156.55	179
8300	36" long		13.30	.602		173	26.50		199.50	230
8310	48" long		10	.800		192	35		227	264
8320	60" long		8	1		217	44		261	305
8330	120" long		4	2		330	88		418	495

26 27 19 – Multi-Outlet Assemblies

26 27 19.10 Wiring Duct

		Crew	Daily Output	Labor-Hours	Unit	Material	2007 Bare Costs Labor	Equipment	Total	Total Incl O&P
0010	**WIRING DUCT** Plastic									
1250	PVC, snap-in slots, adhesive backed									
1270	1-1/2"W x 2"H	2 Elec	120	.133	L.F.	5.20	5.85		11.05	14.40
1280	1-1/2"W x 3"H		120	.133		6.35	5.85		12.20	15.70
1290	1-1/2"W x 4"H		120	.133		7.45	5.85		13.30	16.90
1300	2"W x 1"H		120	.133		4.92	5.85		10.77	14.10
1310	2"W x 1-1/2"H		120	.133		5.20	5.85		11.05	14.40
1320	2"W x 2"H		120	.133		5.30	5.85		11.15	14.50
1330	2"W x 2-1/2"H		120	.133		6.50	5.85		12.35	15.85
1340	2"W x 3"H		120	.133		6.70	5.85		12.55	16.10
1350	2"W x 4"H		120	.133		7.90	5.85		13.75	17.40
1360	2-1/2"W x 3"H		120	.133		7.05	5.85		12.90	16.45
1370	3"W x 1"H		110	.145		5.50	6.40		11.90	15.55
1380	3"W x 1-1/4"H		110	.145		6.45	6.40		12.85	16.60

26 27 Low-Voltage Distribution Equipment

26 27 19 – Multi-Outlet Assemblies

26 27 19.10 Wiring Duct

		Crew	Daily Output	Labor-Hours	Unit	Material	2007 Bare Costs Labor	Equipment	Total	Total Incl O&P
1390	3"W x 2"H	2 Elec	110	.145	L.F.	6.50	6.40		12.90	16.65
1400	3"W x 3"H		110	.145		7.65	6.40		14.05	17.90
1410	3"W x 4"H		110	.145		9.40	6.40		15.80	19.80
1420	3"W x 5"H		110	.145		12.05	6.40		18.45	23
1430	4"W x 1-1/2"H		100	.160		6.50	7		13.50	17.60
1440	4"W x 2"H		100	.160		7.25	7		14.25	18.45
1450	4"W x 3"H		100	.160		8.35	7		15.35	19.65
1460	4"W x 4"H		100	.160		10.20	7		17.20	21.50
1470	4"W x 5"H		100	.160		14.10	7		21.10	26
1550	Cover, 1-1/2"W		200	.080		.96	3.51		4.47	6.30
1560	2"W		200	.080		1.27	3.51		4.78	6.65
1570	2-1/2"W		200	.080		1.62	3.51		5.13	7.05
1580	3"W		200	.080		1.93	3.51		5.44	7.35
1590	4"W		200	.080		2.34	3.51		5.85	7.80

26 27 23 – Indoor Service Poles

26 27 23.40 Surface Raceway

		Crew	Daily Output	Labor-Hours	Unit	Material	2007 Bare Costs Labor	Equipment	Total	Total Incl O&P
0010	**SURFACE RACEWAY**									
0090	Metal, straight section									
0100	No. 500	1 Elec	100	.080	L.F.	.95	3.51		4.46	6.30
0110	No. 700		100	.080		1.07	3.51		4.58	6.45
0400	No. 1500, small pancake		90	.089		1.93	3.90		5.83	7.90
0600	No. 2000, base & cover, blank		90	.089		1.88	3.90		5.78	7.85
0610	Receptacle, 6" O.C.		40	.200		10.50	8.80		19.30	24.50
0620	12" O.C.		44	.182		6.85	8		14.85	19.45
0630	18" O.C.		46	.174		7.25	7.65		14.90	19.35
0650	30" O.C.		50	.160		4.97	7		11.97	15.90
0670	No. 2200, base & cover, blank		80	.100		2.75	4.39		7.14	9.60
0700	Receptacle, 18" O.C.		36	.222		6.80	9.75		16.55	22
0720	30" O.C.		40	.200		6.45	8.80		15.25	20
0800	No. 3000, base & cover, blank		75	.107		3.80	4.68		8.48	11.15
0810	Receptacle, 6" O.C.		45	.178		32.50	7.80		40.30	47
0820	12" O.C.		62	.129		18.30	5.65		23.95	28.50
0830	18" O.C.		64	.125		14.75	5.50		20.25	24.50
0840	24" O.C.		66	.121		11.10	5.30		16.40	20
0850	30" O.C.		68	.118		9.95	5.15		15.10	18.65
0860	60" O.C.		70	.114		7.30	5		12.30	15.50
1000	No. 4000, base & cover, blank		65	.123		6.15	5.40		11.55	14.80
1010	Receptacle, 6" O.C.		41	.195		45	8.55		53.55	62.50
1020	12" O.C.		52	.154		27	6.75		33.75	40
1030	18" O.C.		54	.148		22.50	6.50		29	34
1040	24" O.C.		56	.143		18.30	6.25		24.55	29.50
1050	30" O.C.		58	.138		16.30	6.05		22.35	27
1060	60" O.C.		60	.133		12.75	5.85		18.60	23
1200	No. 6000, base & cover, blank		50	.160		10.30	7		17.30	22
1210	Receptacle, 6" O.C.		30	.267		55	11.70		66.70	78
1220	12" O.C.		37	.216		35.50	9.50		45	53
1230	18" O.C.		39	.205		30	9		39	46.50
1240	24" O.C.		41	.195		24.50	8.55		33.05	40
1250	30" O.C.		43	.186		23.50	8.15		31.65	38
1260	60" O.C.		45	.178		18.15	7.80		25.95	31.50
2400	Fittings, elbows, No. 500		40	.200	Ea.	1.72	8.80		10.52	14.95
2800	Elbow cover, No. 2000		40	.200		3.28	8.80		12.08	16.65

26 27 Low-Voltage Distribution Equipment

26 27 23 – Indoor Service Poles

26 27 23.40 Surface Raceway	Crew	Daily Output	Labor-Hours	Unit	Material	2007 Bare Costs Labor	Equipment	Total	Total Incl O&P
2880	Tee, No. 500	1 Elec	42	.190	Ea.	3.30	8.35		11.65	16.10
2900	No. 2000		27	.296		10.30	13		23.30	30.50
3000	Switch box, No. 500		16	.500		11.20	22		33.20	45
3400	Telephone outlet, No. 1500		16	.500		12.40	22		34.40	46
3600	Junction box, No. 1500	▼	16	.500	▼	8.70	22		30.70	42
3800	Plugmold wired sections, No. 2000									
4000	1 circuit, 6 outlets, 3 ft. long	1 Elec	8	1	Ea.	31.50	44		75.50	101
4100	2 circuits, 8 outlets, 6 ft. long		5.30	1.509		52.50	66.50		119	157
4110	Tele-power pole, alum, w/ 2 recept, 10'		4	2		163	88		251	310
4120	12'		3.85	2.078		195	91		286	350
4130	15'		3.70	2.162		256	95		351	425
4140	Steel, w/ 2 recept, 10'		4	2		112	88		200	254
4150	One phone fitting, 10'		4	2		154	88		242	300
4160	Alum, 4 outlets, 10'	▼	3.70	2.162	▼	206	95		301	370
4300	Overhead distribution systems, 125 volt									
4800	No. 2000, entrance end fitting	1 Elec	20	.400	Ea.	4.33	17.55		21.88	31
5000	Blank end fitting		40	.200		1.92	8.80		10.72	15.15
5200	Supporting clip		40	.200		1.08	8.80		9.88	14.25
5800	No. 3000, entrance end fitting		20	.400		8	17.55		25.55	35
6000	Blank end fitting		40	.200		2.36	8.80		11.16	15.65
6020	Internal elbow		20	.400		10.80	17.55		28.35	38
6030	External elbow		20	.400		14.80	17.55		32.35	42.50
6040	Device bracket		53	.151		3.69	6.65		10.34	13.90
6400	Hanger clamp		32	.250	▼	5.20	11		16.20	22
7000	Base		90	.089	L.F.	3.92	3.90		7.82	10.10
7200	Divider		100	.080	"	.76	3.51		4.27	6.10
7400	Entrance end fitting		16	.500	Ea.	18.95	22		40.95	53.50
7600	Blank end fitting		40	.200		5.60	8.80		14.40	19.20
7610	Recp. & tele. cover		53	.151		9.50	6.65		16.15	20.50
7620	External elbow		16	.500		30	22		52	65.50
7630	Coupling		53	.151		4.50	6.65		11.15	14.80
7640	Divider clip & coupling		80	.100		.89	4.39		5.28	7.55
7650	Panel connector		16	.500		18.90	22		40.90	53.50
7800	Take off connector		16	.500		59.50	22		81.50	98
8000	No. 6000, take off connector		16	.500		71.50	22		93.50	112
8100	Take off fitting		16	.500		53.50	22		75.50	91.50
8200	Hanger clamp	▼	32	.250		12.20	11		23.20	30
8230	Coupling					5.60			5.60	6.15
8240	One gang device plate	1 Elec	53	.151		8.80	6.65		15.45	19.55
8250	Two gang device plate		40	.200		8.80	8.80		17.60	23
8260	Blank end fitting		40	.200		7.95	8.80		16.75	22
8270	Combination elbow		14	.571		31	25		56	71.50
8300	Panel connector	▼	16	.500	▼	13.65	22		35.65	47.50
8500	Chan-L-Wire system installed in 1-5/8" x 1-5/8" strut. Strut									
8600	not incl., 30 amp, 4 wire, 3 phase	1 Elec	200	.040	L.F.	4.21	1.76		5.97	7.25
8700	Junction box		8	1	Ea.	28.50	44		72.50	96.50
8800	Insulating end cap		40	.200		7.55	8.80		16.35	21.50
8900	Strut splice plate		40	.200		9.95	8.80		18.75	24
9000	Tap		40	.200		19.80	8.80		28.60	35
9100	Fixture hanger	▼	60	.133		8.55	5.85		14.40	18.10
9200	Pulling tool				▼	76.50			76.50	84
9300	Non-metallic, straight section									
9310	7/16" x 7/8", base & cover, blank	1 Elec	160	.050	L.F.	1.42	2.20		3.62	4.83

26 27 Low-Voltage Distribution Equipment

26 27 23 – Indoor Service Poles

26 27 23.40 Surface Raceway

		Crew	Daily Output	Labor-Hours	Unit	Material	2007 Bare Costs Labor	Equipment	Total	Total Incl O&P
9320	Base & cover w/ adhesive	1 Elec	160	.050	L.F.	1.57	2.20		3.77	5
9340	7/16" x 1-5/16", base & cover, blank		145	.055		1.57	2.42		3.99	5.35
9350	Base & cover w/ adhesive		145	.055		1.83	2.42		4.25	5.60
9370	11/16" x 2-1/4", base & cover, blank		130	.062		2.20	2.70		4.90	6.45
9380	Base & cover w/ adhesive		130	.062		2.54	2.70		5.24	6.80
9400	Fittings, elbows, 7/16" x 7/8"		50	.160	Ea.	1.57	7		8.57	12.20
9410	7/16" x 1-5/16"		45	.178		1.63	7.80		9.43	13.40
9420	11/16" x 2-1/4"		40	.200		1.77	8.80		10.57	15
9430	Tees, 7/16" x 7/8"		35	.229		2.06	10.05		12.11	17.20
9440	7/16" x 1-5/16"		32	.250		2.12	11		13.12	18.70
9450	11/16" x 2-1/4"		30	.267		2.15	11.70		13.85	19.80
9460	Cover clip, 7/16" x 7/8"		80	.100		.41	4.39		4.80	7
9470	7/16" x 1-5/16"		72	.111		.37	4.88		5.25	7.65
9480	11/16" x 2-1/4"		64	.125		.63	5.50		6.13	8.85
9490	Blank end, 7/16" x 7/8"		50	.160		.59	7		7.59	11.10
9510	11/16" x 2-1/4"		40	.200		1	8.80		9.80	14.15
9520	Round fixture box 5.5" dia x 1"		25	.320		9.45	14.05		23.50	31.50
9530	Device box, 1 gang		30	.267		4.18	11.70		15.88	22
9540	2 gang		25	.320		6.20	14.05		20.25	28

26 27 26 – Wiring Devices

26 27 26.10 Low Voltage Switching

		Crew	Daily Output	Labor-Hours	Unit	Material	2007 Bare Costs Labor	Equipment	Total	Total Incl O&P
0010	**LOW VOLTAGE SWITCHING**									
3600	Relays, 120 V or 277 V standard	1 Elec	12	.667	Ea.	28	29.50		57.50	74.50
3800	Flush switch, standard		40	.200		9.75	8.80		18.55	24
4000	Interchangeable		40	.200		12.75	8.80		21.55	27
4100	Surface switch, standard		40	.200		7.15	8.80		15.95	21
4200	Transformer 115 V to 25 V		12	.667		100	29.50		129.50	154
4400	Master control, 12 circuit, manual		4	2		102	88		190	243
4500	25 circuit, motorized		4	2		110	88		198	252
4600	Rectifier, silicon		12	.667		33	29.50		62.50	79.50
4800	Switchplates, 1 gang, 1, 2 or 3 switch, plastic		80	.100		3.24	4.39		7.63	10.10
5000	Stainless steel		80	.100		8.75	4.39		13.14	16.20
5400	2 gang, 3 switch, stainless steel		53	.151		16.90	6.65		23.55	28.50
5500	4 switch, plastic		53	.151		7.25	6.65		13.90	17.80
5600	2 gang, 4 switch, stainless steel		53	.151		17.65	6.65		24.30	29.50
5700	6 switch, stainless steel		53	.151		39	6.65		45.65	53
5800	3 gang, 9 switch, stainless steel		32	.250		54	11		65	76
5900	Receptacle, triple, 1 return, 1 feed		26	.308		34.50	13.50		48	58
6000	2 feed		20	.400		34.50	17.55		52.05	64
6100	Relay gang boxes, flush or surface, 6 gang		5.30	1.509		59.50	66.50		126	164
6200	12 gang		4.70	1.702		62.50	74.50		137	180
6400	18 gang		4	2		78	88		166	217
6500	Frame, to hold up to 6 relays		12	.667		52	29.50		81.50	101
7200	Control wire, 2 conductor		6.30	1.270	C.L.F.	26.50	56		82.50	112
7400	3 conductor		5	1.600		37.50	70		107.50	147
7600	19 conductor		2.50	3.200		310	140		450	555
7800	26 conductor		2	4		390	176		566	690
8000	Weatherproof, 3 conductor		5	1.600		79	70		149	192

26 27 26.20 Wiring Devices

		Crew	Daily Output	Labor-Hours	Unit	Material	2007 Bare Costs Labor	Equipment	Total	Total Incl O&P
0010	**WIRING DEVICES** R262726-90									
0200	Toggle switch, quiet type, single pole, 15 amp	1 Elec	40	.200	Ea.	5.10	8.80		13.90	18.70
0500	20 amp		27	.296		7.40	13		20.40	27.50

26 27 26.20 Wiring Devices	Crew	Daily Output	Labor-Hours	Unit	Material	2007 Bare Costs Labor	Equipment	Total	Total Incl O&P
0510	30 amp	1 Elec	23	.348	Ea.	19.55	15.25		34.80	44
0530	Lock handle, 20 amp		27	.296		23	13		36	45
0540	Security key, 20 amp		26	.308		55	13.50		68.50	80.50
0550	Rocker, 15 amp		40	.200		6	8.80		14.80	19.65
0560	20 amp		27	.296		13.70	13		26.70	34.50
0600	3 way, 15 amp		23	.348		7.65	15.25		22.90	31
0800	20 amp		18	.444		8.35	19.50		27.85	38
0810	30 amp		9	.889		29	39		68	90
0830	Lock handle, 20 amp		18	.444		25.50	19.50		45	57
0840	Security key, 20 amp		17	.471		60	20.50		80.50	97
0850	Rocker, 15 amp		23	.348		8.45	15.25		23.70	32
0860	20 amp		18	.444		19.85	19.50		39.35	51
0900	4 way, 15 amp		15	.533		21.50	23.50		45	59
1000	20 amp		11	.727		36.50	32		68.50	87.50
1020	Lock handle, 20 amp		11	.727		56.50	32		88.50	110
1030	Rocker, 15 amp		15	.533		32.50	23.50		56	71
1040	20 amp		11	.727		49	32		81	101
1100	Toggle switch, quiet type, double pole, 15 amp		15	.533		14.80	23.50		38.30	51.50
1200	20 amp		11	.727		14.50	32		46.50	63.50
1210	30 amp		9	.889		29.50	39		68.50	90.50
1230	Lock handle, 20 amp		11	.727		26.50	32		58.50	76.50
1250	Security key, 20 amp		10	.800		60	35		95	119
1420	Toggle switch quiet type, 1 pole, 2 throw center off, 15 amp		23	.348		49.50	15.25		64.75	77
1440	20 amp		18	.444		54	19.50		73.50	88.50
1460	Lock handle, 20 amp		18	.444		54.50	19.50		74	89
1480	Momentary contact, 15 amp		23	.348		21.50	15.25		36.75	46
1500	20 amp		18	.444		28	19.50		47.50	60
1520	Momentary contact, lock handle, 20 amp		18	.444		37.50	19.50		57	70.50
1650	Dimmer switch, 120 volt, incandescent, 600 watt, 1 pole		16	.500		11.25	22		33.25	45
1700	600 watt, 3 way		12	.667		9.10	29.50		38.60	53.50
1750	1000 watt, 1 pole		16	.500		46	22		68	83
1800	1000 watt, 3 way		12	.667		64	29.50		93.50	114
2000	1500 watt, 1 pole		11	.727		87	32		119	143
2100	2000 watt, 1 pole		8	1		130	44		174	209
2110	Fluorescent, 600 watt		15	.533		71.50	23.50		95	114
2120	1000 watt		15	.533		89.50	23.50		113	134
2130	1500 watt		10	.800		168	35		203	238
2160	Explosionproof, toggle switch, wall, single pole 20 amp		5.30	1.509		155	66.50		221.50	269
2180	Receptacle, single outlet, 20 amp		5.30	1.509		245	66.50		311.50	370
2190	30 amp		4	2		495	88		583	675
2290	60 amp		2.50	3.200		535	140		675	800
2360	Plug, 20 amp		16	.500		115	22		137	159
2370	30 amp		12	.667		204	29.50		233.50	268
2380	60 amp		8	1		264	44		308	355
2410	Furnace, thermal cutoff switch with plate		26	.308		12.65	13.50		26.15	34
2460	Receptacle, duplex, 120 volt, grounded, 15 amp		40	.200		1.23	8.80		10.03	14.40
2470	20 amp		27	.296		9.40	13		22.40	29.50
2480	Ground fault interrupting, 15 amp		27	.296		29.50	13		42.50	52
2482	20 amp		27	.296		31.50	13		44.50	54
2486	Clock receptacle, 15 amp		40	.200		19.05	8.80		27.85	34
2490	Dryer, 30 amp		15	.533		5.85	23.50		29.35	41.50
2500	Range, 50 amp		11	.727		11.95	32		43.95	60.50
2530	Surge suppresser receptacle, duplex, 20 amp		27	.296		44	13		57	68

26 27 26 – Wiring Devices

26 27 26.20 Wiring Devices		Crew	Daily Output	Labor-Hours	Unit	Material	2007 Bare Costs Labor	Equipment	Total	Total Incl O&P
2532	Quad, 20 amp	1 Elec	20	.400	Ea.	130	17.55		147.55	169
2540	Isolated ground receptacle, duplex, 20 amp		27	.296		21.50	13		34.50	43.50
2542	Quad, 20 amp		20	.400		39	17.55		56.55	69
2550	Simplex, 20 amp		27	.296		24.50	13		37.50	46.50
2560	Simplex, 30 amp		15	.533		28.50	23.50		52	66.50
2570	Cable reel w/ receptacle 50' w/3#12, 120 V, 20 A	2 Elec	2.67	5.999		670	263		933	1,125
2600	Wall plates, stainless steel, 1 gang	1 Elec	80	.100		2.06	4.39		6.45	8.80
2800	2 gang		53	.151		3.80	6.65		10.45	14.05
3000	3 gang		32	.250		5.95	11		16.95	23
3100	4 gang		27	.296		9.85	13		22.85	30
3110	Brown plastic, 1 gang		80	.100		.31	4.39		4.70	6.90
3120	2 gang		53	.151		.82	6.65		7.47	10.75
3130	3 gang		32	.250		1.20	11		12.20	17.65
3140	4 gang		27	.296		1.87	13		14.87	21.50
3150	Brushed brass, 1 gang		80	.100		4.40	4.39		8.79	11.40
3160	Anodized aluminum, 1 gang		80	.100		2.55	4.39		6.94	9.35
3170	Switch cover, weatherproof, 1 gang		60	.133		7.90	5.85		13.75	17.40
3180	Vandal proof lock, 1 gang		60	.133		8.55	5.85		14.40	18.10
3200	Lampholder, keyless		26	.308		11.20	13.50		24.70	32.50
3400	Pullchain with receptacle		22	.364		12.20	15.95		28.15	37.50
3500	Pilot light, neon with jewel		27	.296		13.55	13		26.55	34.50
3600	Receptacle, 20 amp, 250 volt, NEMA 6		27	.296		15.45	13		28.45	36.50
3620	277 volt NEMA 7		27	.296		18.35	13		31.35	39.50
3640	125/250 volt NEMA 10		27	.296		19.20	13		32.20	40.50
3680	125/250 volt NEMA 14		25	.320		24.50	14.05		38.55	48
3700	3 pole, 250 volt NEMA 15		25	.320		25.50	14.05		39.55	49
3720	120/208 volt NEMA 18		25	.320		26.50	14.05		40.55	50.50
3740	30 amp, 125 volt NEMA 5		15	.533		17.20	23.50		40.70	54
3760	250 volt NEMA 6		15	.533		19.55	23.50		43.05	56.50
3780	277 volt NEMA 7		15	.533		25	23.50		48.50	62.50
3820	125/250 volt NEMA 14		14	.571		54	25		79	96.50
3840	3 pole, 250 volt NEMA 15		14	.571		50.50	25		75.50	93
3880	50 amp, 125 volt NEMA 5		11	.727		22.50	32		54.50	72.50
3900	250 volt NEMA 6		11	.727		22.50	32		54.50	72
3920	277 volt NEMA 7		11	.727		29	32		61	79.50
3960	125/250 volt NEMA 14		10	.800		69.50	35		104.50	129
3980	3 pole 250 volt NEMA 15		10	.800		68.50	35		103.50	128
4020	60 amp, 125/250 volt, NEMA 14		8	1		77	44		121	151
4040	3 pole, 250 volt NEMA 15		8	1		81	44		125	155
4060	120/208 volt NEMA 18		8	1		75.50	44		119.50	149
4100	Receptacle locking, 20 amp, 125 volt NEMA L5		27	.296		17.30	13		30.30	38.50
4120	250 volt NEMA L6		27	.296		17.30	13		30.30	38.50
4140	277 volt NEMA L7		27	.296		17.30	13		30.30	38.50
4150	3 pole, 250 volt, NEMA L11		27	.296		27.50	13		40.50	49.50
4160	20 amp, 480 volt NEMA L8		27	.296		22	13		35	43.50
4180	600 volt NEMA L9		27	.296		33	13		46	55.50
4200	125/250 volt NEMA L10		27	.296		27.50	13		40.50	49.50
4230	125/250 volt NEMA L14		25	.320		24	14.05		38.05	47.50
4280	250 volt NEMA L15		25	.320		24.50	14.05		38.55	47.50
4300	480 volt NEMA L16		25	.320		28.50	14.05		42.55	52.50
4320	3 phase, 120/208 volt NEMA L18		25	.320		37	14.05		51.05	62
4340	277/480 volt NEMA L19		25	.320		37.50	14.05		51.55	62
4360	347/600 volt NEMA L20		25	.320		37.50	14.05		51.55	62

26 27 26.20 Wiring Devices	Crew	Daily Output	Labor-Hours	Unit	Material	2007 Bare Costs Labor	Equipment	Total	Total Incl O&P
4380	120/208 volt NEMA L21	1 Elec	23	.348	Ea.	29.50	15.25		44.75	55
4400	277/480 volt NEMA L22		23	.348		35.50	15.25		50.75	61.50
4420	347/600 volt NEMA L23		23	.348		35.50	15.25		50.75	61.50
4440	30 amp, 125 volt NEMA L5		15	.533		24.50	23.50		48	62
4460	250 volt NEMA L6		15	.533		26.50	23.50		50	64
4480	277 volt NEMA L7		15	.533		26.50	23.50		50	64
4500	480 volt NEMA L8		15	.533		31.50	23.50		55	70
4520	600 volt NEMA L9		15	.533		31.50	23.50		55	70
4540	125/250 volt NEMA L10		15	.533		32	23.50		55.50	70
4560	3 phase, 250 volt NEMA L11		15	.533		32	23.50		55.50	70
4620	125/250 volt NEMA L14		14	.571		38.50	25		63.50	79.50
4640	250 volt NEMA L15		14	.571		38.50	25		63.50	79.50
4660	480 volt NEMA L16		14	.571		45	25		70	87
4680	600 volt NEMA L17		14	.571		44	25		69	85.50
4700	120/208 volt NEMA L18		14	.571		54.50	25		79.50	97.50
4720	277/480 volt NEMA L19		14	.571		54.50	25		79.50	97.50
4740	347/600 volt NEMA L20		14	.571		55	25		80	98
4760	120/208 volt NEMA L21		13	.615		41	27		68	85
4780	277/480 volt NEMA L22		13	.615		49	27		76	93.50
4800	347/600 volt NEMA L23		13	.615		60.50	27		87.50	107
4840	Receptacle, corrosion resistant, 15 or 20 amp, 125 volt NEMA L5		27	.296		26.50	13		39.50	48.50
4860	250 volt NEMA L6		27	.296		26.50	13		39.50	48.50
4900	Receptacle, cover plate, phenolic plastic, NEMA 5 & 6		80	.100		.49	4.39		4.88	7.10
4910	NEMA 7-23		80	.100		.57	4.39		4.96	7.20
4920	Stainless steel, NEMA 5 & 6		80	.100		1.95	4.39		6.34	8.70
4930	NEMA 7-23		80	.100		3	4.39		7.39	9.85
4940	Brushed brass NEMA 5 & 6		80	.100		4.80	4.39		9.19	11.85
4950	NEMA 7-23		80	.100		5.20	4.39		9.59	12.25
4960	Anodized aluminum, NEMA 5 & 6		80	.100		2.65	4.39		7.04	9.45
4970	NEMA 7-23		80	.100		4.20	4.39		8.59	11.15
4980	Weatherproof NEMA 7-23		60	.133		28	5.85		33.85	39.50
5100	Plug, 20 amp, 250 volt, NEMA 6		30	.267		14.70	11.70		26.40	33.50
5110	277 volt NEMA 7		30	.267		16	11.70		27.70	35
5120	3 pole, 120/250 volt, NEMA 10		26	.308		20.50	13.50		34	42.50
5130	125/250 volt NEMA 14		26	.308		40.50	13.50		54	64.50
5140	250 volt NEMA 15		26	.308		42.50	13.50		56	67
5150	120/208 volt NEMA 8		26	.308		43.50	13.50		57	68
5160	30 amp, 125 volt NEMA 5		13	.615		41.50	27		68.50	86
5170	250 volt NEMA 6		13	.615		43	27		70	87.50
5180	277 volt NEMA 7		13	.615		53	27		80	98.50
5190	125/250 volt NEMA 14		13	.615		47.50	27		74.50	92.50
5200	3 pole, 250 volt NEMA 15		12	.667		50	29.50		79.50	98.50
5210	50 amp, 125 volt NEMA 5		9	.889		52	39		91	115
5220	250 volt NEMA 6		9	.889		54.50	39		93.50	118
5230	277 volt NEMA 7		9	.889		58	39		97	122
5240	125/250 volt NEMA 14		9	.889		58	39		97	122
5250	3 pole, 250 volt NEMA 15		8	1		60.50	44		104.50	132
5260	60 amp, 125/250 volt NEMA 14		7	1.143		67	50		117	148
5270	3 pole, 250 volt NEMA 15		7	1.143		71.50	50		121.50	153
5280	120/208 volt NEMA 18		7	1.143		69.50	50		119.50	151
5300	Plug angle, 20 amp, 250 volt NEMA 6		30	.267		20	11.70		31.70	39.50
5310	30 amp, 125 volt NEMA 5		13	.615		41.50	27		68.50	86
5320	250 volt NEMA 6		13	.615		43	27		70	87.50

26 27 26.20 Wiring Devices	Crew	Daily Output	Labor-Hours	Unit	Material	2007 Bare Costs Labor	Equipment	Total	Total Incl O&P
5330	277 volt NEMA 7	1 Elec	13	.615	Ea.	53	27		80	98.50
5340	125/250 volt NEMA 14		13	.615		53	27		80	98
5350	3 pole, 250 volt NEMA 15		12	.667		55.50	29.50		85	105
5360	50 amp, 125 volt NEMA 5		9	.889		45.50	39		84.50	108
5370	250 volt NEMA 6		9	.889		45	39		84	108
5380	277 volt NEMA 7		9	.889		58	39		97	122
5390	125/250 volt NEMA 14		9	.889		64	39		103	129
5400	3 pole, 250 volt NEMA 15		8	1		67	44		111	140
5410	60 amp, 125/250 volt NEMA 14		7	1.143		74.50	50		124.50	157
5420	3 pole, 250 volt NEMA 15		7	1.143		77.50	50		127.50	160
5430	120/208 volt NEMA 18		7	1.143		78	50		128	160
5500	Plug, locking, 20 amp, 125 volt NEMA L5		30	.267		13.50	11.70		25.20	32.50
5510	250 volt NEMA L6		30	.267		13.50	11.70		25.20	32.50
5520	277 volt NEMA L7		30	.267		13.50	11.70		25.20	32.50
5530	480 volt NEMA L8		30	.267		17	11.70		28.70	36
5540	600 volt NEMA L9		30	.267		20	11.70		31.70	39.50
5550	3 pole, 125/250 volt NEMA L10		26	.308		18.40	13.50		31.90	40
5560	250 volt NEMA L11		26	.308		18.40	13.50		31.90	40
5570	480 volt NEMA L12		26	.308		21.50	13.50		35	43.50
5580	125/250 volt NEMA L14		26	.308		20.50	13.50		34	42.50
5590	250 volt NEMA L15		26	.308		22	13.50		35.50	44
5600	480 volt NEMA L16		26	.308		26	13.50		39.50	48.50
5610	4 pole, 120/208 volt NEMA L18		24	.333		33	14.65		47.65	58
5620	277/480 volt NEMA L19		24	.333		33	14.65		47.65	58
5630	347/600 volt NEMA L20		24	.333		33	14.65		47.65	58
5640	120/208 volt NEMA L21		24	.333		26.50	14.65		41.15	51
5650	277/480 volt NEMA L22		24	.333		31.50	14.65		46.15	57
5660	347/600 volt NEMA L23		24	.333		31.50	14.65		46.15	57
5670	30 amp, 125 volt NEMA L5		13	.615		21	27		48	63
5680	250 volt NEMA L6		13	.615		21.50	27		48.50	63.50
5690	277 volt NEMA L7		13	.615		21	27		48	63
5700	480 volt NEMA L8		13	.615		25	27		52	67.50
5710	600 volt NEMA L9		13	.615		24.50	27		51.50	67
5720	3 pole, 125/250 volt NEMA L10		11	.727		20	32		52	69.50
5730	250 volt NEMA L11		11	.727		20	32		52	69.50
5760	125/250 volt NEMA L14		11	.727		28	32		60	78.50
5770	250 volt NEMA L15		11	.727		28.50	32		60.50	78.50
5780	480 volt NEMA L16		11	.727		33	32		65	84
5790	600 volt NEMA L17		11	.727		33.50	32		65.50	84.50
5800	4 pole, 120/208 volt NEMA L18		10	.800		41.50	35		76.50	98
5810	120/208 volt NEMA L19		10	.800		41.50	35		76.50	98
5820	347/600 volt NEMA L20		10	.800		41.50	35		76.50	98
5830	120/208 volt NEMA L21		10	.800		31.50	35		66.50	87
5840	277/480 volt NEMA L22		10	.800		37.50	35		72.50	94
5850	347/600 volt NEMA L23		10	.800		46.50	35		81.50	104
6000	Connector, 20 amp, 250 volt NEMA 6		30	.267		23	11.70		34.70	43
6010	277 volt NEMA 7		30	.267		26	11.70		37.70	46.50
6020	3 pole, 120/250 volt NEMA 10		26	.308		29.50	13.50		43	52.50
6030	125/250 volt NEMA 14		26	.308		29.50	13.50		43	52.50
6040	250 volt NEMA 15		26	.308		29.50	13.50		43	52.50
6050	120/208 volt NEMA 18		26	.308		32.50	13.50		46	55.50
6060	30 amp, 125 volt NEMA 5		13	.615		51	27		78	96
6070	250 volt NEMA 6		13	.615		51	27		78	96

26 27 Low-Voltage Distribution Equipment

26 27 26 – Wiring Devices

26 27 26.20 Wiring Devices	Crew	Daily Output	Labor-Hours	Unit	Material	2007 Bare Costs Labor	Equipment	Total	Total Incl O&P
6080	277 volt NEMA 7	1 Elec	13	.615	Ea.	51	27		78	96
6110	50 amp, 125 volt NEMA 5		9	.889		70.50	39		109.50	136
6120	250 volt NEMA 6		9	.889		70.50	39		109.50	136
6130	277 volt NEMA 7		9	.889		70.50	39		109.50	136
6200	Connector, locking, 20 amp, 125 volt NEMA L5		30	.267		21	11.70		32.70	40.50
6210	250 volt NEMA L6		30	.267		21	11.70		32.70	40.50
6220	277 volt NEMA L7		30	.267		21	11.70		32.70	40.50
6230	480 volt NEMA L8		30	.267		31.50	11.70		43.20	52
6240	600 volt NEMA L9		30	.267		31.50	11.70		43.20	52
6250	3 pole, 125/250 volt NEMA L10		26	.308		29	13.50		42.50	51.50
6260	250 volt NEMA L11		26	.308		29	13.50		42.50	51.50
6280	125/250 volt NEMA L14		26	.308		29	13.50		42.50	51.50
6290	250 volt NEMA L15		26	.308		29	13.50		42.50	51.50
6300	480 volt NEMA L16		26	.308		34.50	13.50		48	58
6310	4 pole, 120/208 volt NEMA L18		24	.333		45.50	14.65		60.15	72
6320	277/480 volt NEMA L19		24	.333		45.50	14.65		60.15	72
6330	347/600 volt NEMA L20		24	.333		46	14.65		60.65	72.50
6340	120/208 volt NEMA L21		24	.333		43	14.65		57.65	69.50
6350	277/480 volt NEMA L22		24	.333		54.50	14.65		69.15	82
6360	347/600 volt NEMA L23		24	.333		54.50	14.65		69.15	82
6370	30 amp, 125 volt NEMA L5		13	.615		42	27		69	86
6380	250 volt NEMA L6		13	.615		42	27		69	86.50
6390	277 volt NEMA L7		13	.615		42	27		69	86.50
6400	480 volt NEMA L8		13	.615		50	27		77	95
6410	600 volt NEMA L9		13	.615		50	27		77	95
6420	3 pole, 125/250 volt NEMA L10		11	.727		55.50	32		87.50	109
6430	250 volt NEMA L11		11	.727		55.50	32		87.50	109
6460	125/250 volt NEMA L14		11	.727		58	32		90	112
6470	250 volt NEMA L15		11	.727		58	32		90	112
6480	480 volt NEMA L16		11	.727		70	32		102	125
6490	600 volt NEMA L17		11	.727		70	32		102	125
6500	4 pole, 120/208 volt NEMA L18		10	.800		87	35		122	149
6510	120/208 volt NEMA L19		10	.800		87.50	35		122.50	149
6520	347/600 volt NEMA L20		10	.800		87.50	35		122.50	149
6530	120/208 volt NEMA L21		10	.800		57.50	35		92.50	116
6540	277/480 volt NEMA L22		10	.800		69	35		104	129
6550	347/600 volt NEMA L23		10	.800		85.50	35		120.50	147
7000	Receptacle computer, 250 volt, 15 amp, 3 pole 4 wire		8	1		66	44		110	139
7010	20 amp, 2 pole 3 wire		8	1		66	44		110	139
7020	30 amp, 2 pole 3 wire		6.50	1.231		104	54		158	195
7030	30 amp, 3 pole 4 wire		6.50	1.231		111	54		165	203
7040	60 amp, 3 pole 4 wire		4.50	1.778		186	78		264	320
7050	100 amp, 3 pole 4 wire		3	2.667		235	117		352	435
7100	Connector computer, 250 volt, 15 amp, 3 pole 4 wire		27	.296		91.50	13		104.50	120
7110	20 amp, 2 pole 3 wire		27	.296		91.50	13		104.50	120
7120	30 amp, 2 pole 3 wire		15	.533		146	23.50		169.50	195
7130	30 amp, 3 pole 4 wire		15	.533		137	23.50		160.50	186
7140	60 amp, 3 pole 4 wire		8	1		234	44		278	325
7150	100 amp, 3 pole 4 wire		4	2		325	88		413	490
7200	Plug, computer, 250 volt, 15 amp, 3 pole 4 wire		27	.296		83	13		96	110
7210	20 amp, 2 pole, 3 wire		27	.296		83	13		96	110
7220	30 amp, 2 pole, 3 wire		15	.533		122	23.50		145.50	169
7230	30 amp, 3 pole, 4 wire		15	.533		131	23.50		154.50	180

26 27 Low-Voltage Distribution Equipment

26 27 26 – Wiring Devices

26 27 26.20 Wiring Devices

		Crew	Daily Output	Labor-Hours	Unit	Material	2007 Bare Costs Labor	2007 Bare Costs Equipment	Total	Total Incl O&P
7240	60 amp, 3 pole, 4 wire	1 Elec	8	1	Ea.	196	44		240	282
7250	100 amp, 3 pole, 4 wire		4	2		255	88		343	410
7300	Connector adapter to flexible conduit, 1/2"		60	.133		3.95	5.85		9.80	13.05
7310	3/4"		50	.160		5.80	7		12.80	16.85
7320	1-1/4"		30	.267		14.50	11.70		26.20	33.50
7330	1-1/2"		23	.348		21	15.25		36.25	45.50

26 27 73 – Door Chimes

26 27 73.10 Doorbell System

		Crew	Daily Output	Labor-Hours	Unit	Material	2007 Bare Costs Labor	2007 Bare Costs Equipment	Total	Total Incl O&P
0010	**DOORBELL SYSTEM** Incl. transformer, button & signal									
0100	6" bell	1 Elec	4	2	Ea.	89	88		177	229
0200	Buzzer		4	2		70.50	88		158.50	209
1000	Door chimes, 2 notes, minimum		16	.500		23	22		45	58
1020	Maximum		12	.667		122	29.50		151.50	178
1100	Tube type, 3 tube system		12	.667		173	29.50		202.50	235
1180	4 tube system		10	.800		277	35		312	360
1900	For transformer & button, minimum add		5	1.600		13.10	70		83.10	119
1960	Maximum, add		4.50	1.778		39.50	78		117.50	159
3000	For push button only, minimum		24	.333		2.65	14.65		17.30	25
3100	Maximum		20	.400		20.50	17.55		38.05	49
3200	Bell transformer		16	.500		18.50	22		40.50	53

26 28 Low-Voltage Circuit Protective Devices

26 28 13 – Fuses

26 28 13.10 Fuses

		Crew	Daily Output	Labor-Hours	Unit	Material	2007 Bare Costs Labor	2007 Bare Costs Equipment	Total	Total Incl O&P
0010	**FUSES**									
0020	Cartridge, nonrenewable									
0050	250 volt, 30 amp	1 Elec	50	.160	Ea.	1.62	7		8.62	12.25
0100	60 amp		50	.160		2.74	7		9.74	13.45
0150	100 amp		40	.200		11.50	8.80		20.30	25.50
0200	200 amp		36	.222		28	9.75		37.75	45
0250	400 amp		30	.267		63.50	11.70		75.20	87.50
0300	600 amp		24	.333		107	14.65		121.65	140
0400	600 volt, 30 amp		40	.200		8.95	8.80		17.75	23
0450	60 amp		40	.200		12.95	8.80		21.75	27.50
0500	100 amp		36	.222		27	9.75		36.75	44.50
0550	200 amp		30	.267		66	11.70		77.70	90
0600	400 amp		24	.333		136	14.65		150.65	172
0650	600 amp		20	.400		205	17.55		222.55	251
0800	Dual element, time delay, 250 volt, 30 amp		50	.160		4.10	7		11.10	14.95
0850	60 amp		50	.160		7.45	7		14.45	18.65
0900	100 amp		40	.200		16.70	8.80		25.50	31.50
0950	200 amp		36	.222		37	9.75		46.75	55.50
1000	400 amp		30	.267		67	11.70		78.70	91.50
1050	600 amp		24	.333		103	14.65		117.65	135
1300	600 volt, 15 to 30 amp		40	.200		8.95	8.80		17.75	23
1350	35 to 60 amp		40	.200		15.35	8.80		24.15	30
1400	70 to 100 amp		36	.222		32	9.75		41.75	49.50
1450	110 to 200 amp		30	.267		63.50	11.70		75.20	87.50
1500	225 to 400 amp		24	.333		128	14.65		142.65	162
1550	600 amp		20	.400		183	17.55		200.55	227

26 28 Low-Voltage Circuit Protective Devices

26 28 13 – Fuses

26 28 13.10 Fuses

		Crew	Daily Output	Labor-Hours	Unit	Material	2007 Bare Costs Labor	Equipment	Total	Total Incl O&P
1800	Class RK1, high capacity, 250 volt, 30 amp	1 Elec	50	.160	Ea.	5.55	7		12.55	16.55
1850	60 amp		50	.160		10.05	7		17.05	21.50
1900	100 amp		40	.200		22.50	8.80		31.30	38
1950	200 amp		36	.222		49.50	9.75		59.25	69
2000	400 amp		30	.267		89.50	11.70		101.20	116
2050	600 amp		24	.333		137	14.65		151.65	173
2200	600 volt, 30 amp		40	.200		12.55	8.80		21.35	27
2250	60 amp		40	.200		21.50	8.80		30.30	36.50
2300	100 amp		36	.222		44.50	9.75		54.25	63.50
2350	200 amp		30	.267		89	11.70		100.70	115
2400	400 amp		24	.333		179	14.65		193.65	218
2450	600 amp		20	.400		257	17.55		274.55	310
2700	Class J, CLF, 250 or 600 volt, 30 amp		40	.200		15.75	8.80		24.55	30.50
2750	60 amp		40	.200		25.50	8.80		34.30	41.50
2800	100 amp		36	.222		39	9.75		48.75	57.50
2850	200 amp		30	.267		73.50	11.70		85.20	98
2900	400 amp		24	.333		162	14.65		176.65	200
2950	600 amp		20	.400		230	17.55		247.55	279
3100	Class L, 250 or 600 volt, 601 to 1200 amp		16	.500		420	22		442	495
3150	1500-1600 amp		13	.615		540	27		567	635
3200	1800-2000 amp		10	.800		720	35		755	850
3250	2500 amp		10	.800		955	35		990	1,100
3300	3000 amp		8	1		1,100	44		1,144	1,275
3350	3500-4000 amp		8	1		1,525	44		1,569	1,750
3400	4500-5000 amp		6.70	1.194		2,375	52.50		2,427.50	2,675
3450	6000 amp		5.70	1.404		3,925	61.50		3,986.50	4,400
3600	Plug, 120 volt, 1 to 10 amp		50	.160		1.29	7		8.29	11.85
3650	15 to 30 amp		50	.160		1.72	7		8.72	12.35
3700	Dual element 0.3 to 14 amp		50	.160		7.20	7		14.20	18.35
3750	15 to 30 amp		50	.160		2.15	7		9.15	12.80
3800	Fustat, 120 volt, 15 to 30 amp		50	.160		3.05	7		10.05	13.80
3850	0.3 to 14 amp		50	.160		7.40	7		14.40	18.60
3900	Adapters 0.3 to 10 amp		50	.160		3.90	7		10.90	14.75
3950	15 to 30 amp		50	.160		3.33	7		10.33	14.10

26 28 16 – Enclosed Switches and Circuit Breakers

26 28 16.10 Circuit Breakers

		Crew	Daily Output	Labor-Hours	Unit	Material	2007 Bare Costs Labor	Equipment	Total	Total Incl O&P
0010	**CIRCUIT BREAKERS** (in enclosure)									
0100	Enclosed (NEMA 1), 600 volt, 3 pole, 30 amp	1 Elec	3.20	2.500	Ea.	530	110		640	750
0200	60 amp		2.80	2.857		530	125		655	770
0400	100 amp		2.30	3.478		605	153		758	895
0600	225 amp		1.50	5.333		1,400	234		1,634	1,900
0700	400 amp	2 Elec	1.60	10		2,400	440		2,840	3,300
0800	600 amp		1.20	13.333		3,475	585		4,060	4,700
1000	800 amp		.94	17.021		4,525	745		5,270	6,075
1200	1000 amp		.84	19.048		5,725	835		6,560	7,525
1220	1200 amp		.80	20		7,325	880		8,205	9,350
1240	1600 amp		.72	22.222		14,000	975		14,975	16,900
1260	2000 amp		.64	25		15,100	1,100		16,200	18,200
1400	1200 amp with ground fault		.80	20		14,900	880		15,780	17,600
1600	1600 amp with ground fault		.72	22.222		15,800	975		16,775	18,900
1800	2000 amp with ground fault		.64	25		17,000	1,100		18,100	20,300
2000	Disconnect, 240 volt 3 pole, 5 HP motor	1 Elec	3.20	2.500		370	110		480	575

26 28 16.10 Circuit Breakers

		Crew	Daily Output	Labor-Hours	Unit	Material	2007 Bare Costs Labor	Equipment	Total	Total Incl O&P
2020	10 HP motor	1 Elec	3.20	2.500	Ea.	370	110		480	575
2040	15 HP motor		2.80	2.857		370	125		495	595
2060	20 HP motor		2.30	3.478		450	153		603	720
2080	25 HP motor		2.30	3.478		450	153		603	720
2100	30 HP motor		2.30	3.478		450	153		603	720
2120	40 HP motor		2	4		775	176		951	1,100
2140	50 HP motor		1.50	5.333		775	234		1,009	1,200
2160	60 HP motor		1.50	5.333		1,775	234		2,009	2,300
2180	75 HP motor	2 Elec	2	8		1,775	350		2,125	2,475
2200	100 HP motor		1.60	10		1,775	440		2,215	2,600
2220	125 HP motor		1.60	10		1,775	440		2,215	2,600
2240	150 HP motor		1.20	13.333		3,475	585		4,060	4,700
2260	200 HP motor		1.20	13.333		4,525	585		5,110	5,850
2300	Enclosed (NEMA 7), explosion proof, 600 volt 3 pole, 50 amp	1 Elec	2.30	3.478		1,125	153		1,278	1,475
2350	100 amp		1.50	5.333		1,175	234		1,409	1,650
2400	150 amp		1	8		2,825	350		3,175	3,625
2450	250 amp	2 Elec	1.60	10		3,525	440		3,965	4,525
2500	400 amp	"	1.20	13.333		3,900	585		4,485	5,175

26 28 16.20 Safety Switches

		Crew	Daily Output	Labor-Hours	Unit	Material	2007 Bare Costs Labor	Equipment	Total	Total Incl O&P
0010	**SAFETY SWITCHES** R262816-80									
0100	General duty 240 volt, 3 pole NEMA 1, fusible, 30 amp	1 Elec	3.20	2.500	Ea.	92	110		202	264
0200	60 amp		2.30	3.478		155	153		308	400
0300	100 amp		1.90	4.211		267	185		452	570
0400	200 amp		1.30	6.154		575	270		845	1,025
0500	400 amp	2 Elec	1.80	8.889		1,450	390		1,840	2,175
0600	600 amp	"	1.20	13.333		2,700	585		3,285	3,850
0610	Nonfusible, 30 amp	1 Elec	3.20	2.500		74.50	110		184.50	245
0650	60 amp		2.30	3.478		98.50	153		251.50	335
0700	100 amp		1.90	4.211		228	185		413	525
0750	200 amp		1.30	6.154		425	270		695	865
0800	400 amp	2 Elec	1.80	8.889		1,025	390		1,415	1,700
0850	600 amp	"	1.20	13.333		1,950	585		2,535	3,025
1100	Heavy duty, 600 volt, 3 pole NEMA 1 nonfused									
1110	30 amp	1 Elec	3.20	2.500	Ea.	131	110		241	305
1500	60 amp		2.30	3.478		237	153		390	485
1700	100 amp		1.90	4.211		370	185		555	685
1900	200 amp		1.30	6.154		565	270		835	1,025
2100	400 amp	2 Elec	1.80	8.889		1,275	390		1,665	1,975
2300	600 amp		1.20	13.333		2,250	585		2,835	3,350
2500	800 amp		.94	17.021		4,600	745		5,345	6,150
2700	1200 amp		.80	20		6,175	880		7,055	8,100
2900	Heavy duty, 240 volt, 3 pole NEMA 1 fusible									
2910	30 amp	1 Elec	3.20	2.500	Ea.	148	110		258	325
3000	60 amp		2.30	3.478		251	153		404	505
3300	100 amp		1.90	4.211		395	185		580	710
3500	200 amp		1.30	6.154		680	270		950	1,150
3700	400 amp	2 Elec	1.80	8.889		1,750	390		2,140	2,500
3900	600 amp		1.20	13.333		3,000	585		3,585	4,175
4100	800 amp		.94	17.021		7,050	745		7,795	8,850
4300	1200 amp		.80	20		9,250	880		10,130	11,500
4340	2 pole fused, 30 amp	1 Elec	3.50	2.286		112	100		212	272
4350	600 volt, 3 pole, fusible, 30 amp		3.20	2.500		251	110		361	440

26 28 16 – Enclosed Switches and Circuit Breakers

26 28 16.20 Safety Switches	Crew	Daily Output	Labor-Hours	Unit	Material	2007 Bare Costs Labor	Equipment	Total	Total Incl O&P
4380	60 amp	1 Elec	2.30	3.478	Ea.	305	153		458	560
4400	100 amp		1.90	4.211		555	185		740	890
4420	200 amp		1.30	6.154		805	270		1,075	1,275
4440	400 amp	2 Elec	1.80	8.889		2,100	390		2,490	2,875
4450	600 amp		1.20	13.333		3,525	585		4,110	4,750
4460	800 amp		.94	17.021		7,050	745		7,795	8,850
4480	1200 amp		.80	20		9,250	880		10,130	11,500
4500	240 volt 3 pole NEMA 3R (no hubs), fusible									
4510	30 amp	1 Elec	3.10	2.581	Ea.	266	113		379	460
4700	60 amp		2.20	3.636		420	160		580	700
4900	100 amp		1.80	4.444		600	195		795	950
5100	200 amp		1.20	6.667		825	293		1,118	1,350
5300	400 amp	2 Elec	1.60	10		1,825	440		2,265	2,650
5500	600 amp	"	1	16		3,825	700		4,525	5,250
5510	Heavy duty, 600 volt, 3 pole 3ph. NEMA 3R fusible, 30 amp	1 Elec	3.10	2.581		425	113		538	635
5520	60 amp		2.20	3.636		495	160		655	785
5530	100 amp		1.80	4.444		770	195		965	1,150
5540	200 amp		1.20	6.667		1,075	293		1,368	1,600
5550	400 amp	2 Elec	1.60	10		2,500	440		2,940	3,400
5700	600 volt, 3 pole NEMA 3R nonfused									
5710	30 amp	1 Elec	3.10	2.581	Ea.	235	113		348	430
5900	60 amp		2.20	3.636		410	160		570	695
6100	100 amp		1.80	4.444		575	195		770	925
6300	200 amp		1.20	6.667		700	293		993	1,200
6500	400 amp	2 Elec	1.60	10		1,725	440		2,165	2,550
6700	600 amp	"	1	16		3,475	700		4,175	4,875
6900	600 volt, 6 pole NEMA 3R nonfused, 30 amp	1 Elec	2.70	2.963		1,500	130		1,630	1,850
7100	60 amp		2	4		1,725	176		1,901	2,150
7300	100 amp		1.50	5.333		2,125	234		2,359	2,700
7500	200 amp		1.20	6.667		4,800	293		5,093	5,700
7600	600 volt, 3 pole NEMA 7 explosion proof nonfused									
7610	30 amp	1 Elec	2.20	3.636	Ea.	1,150	160		1,310	1,500
7620	60 amp		1.80	4.444		1,375	195		1,570	1,800
7630	100 amp		1.20	6.667		1,675	293		1,968	2,250
7640	200 amp		.80	10		3,450	440		3,890	4,450
7710	600 volt 6 pole, NEMA 3R fusible, 30 amp		2.70	2.963		1,675	130		1,805	2,050
7900	60 amp		2	4		2,025	176		2,201	2,475
8100	100 amp		1.50	5.333		2,450	234		2,684	3,050
8110	240 volt 3 pole, NEMA 12 fusible, 30 amp		3.10	2.581		325	113		438	525
8120	60 amp		2.20	3.636		435	160		595	720
8130	100 amp		1.80	4.444		645	195		840	995
8140	200 amp		1.20	6.667		920	293		1,213	1,450
8150	400 amp	2 Elec	1.60	10		2,125	440		2,565	2,975
8160	600 amp	"	1	16		3,425	700		4,125	4,825
8180	600 volt 3 pole, NEMA 12 fused, 30 amp	1 Elec	3.10	2.581		430	113		543	640
8190	60 amp		2.20	3.636		440	160		600	725
8200	100 amp		1.80	4.444		705	195		900	1,075
8210	200 amp		1.20	6.667		1,100	293		1,393	1,625
8220	400 amp	2 Elec	1.60	10		2,425	440		2,865	3,300
8230	600 amp	"	1	16		4,075	700		4,775	5,525
8240	600 volt 3 pole, NEMA 12 nonfused, 30 amp	1 Elec	3.10	2.581		300	113		413	500
8250	60 amp		2.20	3.636		390	160		550	670
8260	100 amp		1.80	4.444		555	195		750	900

26 28 Low-Voltage Circuit Protective Devices

26 28 16 – Enclosed Switches and Circuit Breakers

26 28 16.20 Safety Switches

		Crew	Daily Output	Labor-Hours	Unit	Material	2007 Bare Costs Labor	Equipment	Total	Total Incl O&P
8270	200 amp	1 Elec	1.20	6.667	Ea.	735	293		1,028	1,250
8280	400 amp	2 Elec	1.60	10		1,825	440		2,265	2,650
8290	600 amp	"	1	16		2,975	700		3,675	4,325
8310	600 volt, 3 pole NEMA 4 fusible, 30 amp	1 Elec	3	2.667		1,175	117		1,292	1,450
8320	60 amp		2.20	3.636		1,300	160		1,460	1,675
8330	100 amp		1.80	4.444		2,875	195		3,070	3,475
8340	200 amp		1.20	6.667		3,650	293		3,943	4,450
8350	400 amp	2 Elec	1.60	10		7,200	440		7,640	8,575
8360	600 volt 3 pole NEMA 4 nonfused, 30 amp	1 Elec	3	2.667		970	117		1,087	1,250
8370	60 amp		2.20	3.636		1,150	160		1,310	1,525
8380	100 amp		1.80	4.444		2,350	195		2,545	2,875
8390	200 amp		1.20	6.667		3,200	293		3,493	3,950
8400	400 amp	2 Elec	1.60	10		6,500	440		6,940	7,800
8490	Motor starters, manual, single phase, NEMA 1	1 Elec	6.40	1.250		59	55		114	147
8500	NEMA 4		4	2		165	88		253	315
8700	NEMA 7		4	2		181	88		269	330
8900	NEMA 1 with pilot		6.40	1.250		76	55		131	165
8920	3 pole, NEMA 1, 230/460 volt, 5 HP, size 0		3.50	2.286		174	100		274	340
8940	10 HP, size 1		2	4		207	176		383	490
9010	Disc. switch, 600V 3 pole fusible, 30 amp, to 10 HP motor		3.20	2.500		268	110		378	460
9050	60 amp, to 30 HP motor		2.30	3.478		615	153		768	905
9070	100 amp, to 60 HP motor		1.90	4.211		615	185		800	955
9100	200 amp, to 125 HP motor		1.30	6.154		925	270		1,195	1,425
9110	400 amp, to 200 HP motor	2 Elec	1.80	8.889		2,325	390		2,715	3,150

26 28 16.40 Time Switches

		Crew	Daily Output	Labor-Hours	Unit	Material	2007 Bare Costs Labor	Equipment	Total	Total Incl O&P
0010	**TIME SWITCHES**									
0100	Single pole, single throw, 24 hour dial	1 Elec	4	2	Ea.	87	88		175	227
0200	24 hour dial with reserve power		3.60	2.222		370	97.50		467.50	550
0300	Astronomic dial		3.60	2.222		149	97.50		246.50	310
0400	Astronomic dial with reserve power		3.30	2.424		480	106		586	690
0500	7 day calendar dial		3.30	2.424		134	106		240	305
0600	7 day calendar dial with reserve power		3.20	2.500		520	110		630	735
0700	Photo cell 2000 watt		8	1		16.05	44		60.05	83
1080	Load management device, 4 loads		2	4		845	176		1,021	1,200
1100	8 loads		1	8		1,375	350		1,725	2,050

26 29 Low-Voltage Controllers

26 29 13 – Enclosed Controllers

26 29 13.10 Contactors, AC

		Crew	Daily Output	Labor-Hours	Unit	Material	2007 Bare Costs Labor	Equipment	Total	Total Incl O&P
0010	**CONTACTORS, AC** Enclosed (NEMA 1)									
0050	Lighting, 600 volt 3 pole, electrically held									
0100	20 amp	1 Elec	4	2	Ea.	253	88		341	410
0200	30 amp		3.60	2.222		268	97.50		365.50	440
0300	60 amp		3	2.667		535	117		652	760
0400	100 amp		2.50	3.200		885	140		1,025	1,175
0500	200 amp		1.40	5.714		2,075	251		2,326	2,675
0600	300 amp	2 Elec	1.60	10		4,425	440		4,865	5,525
0800	600 volt 3 pole, mechanically held, 30 amp	1 Elec	3.60	2.222		405	97.50		502.50	590
0900	60 amp		3	2.667		800	117		917	1,050
1000	75 amp		2.80	2.857		1,150	125		1,275	1,425
1100	100 amp		2.50	3.200		1,150	140		1,290	1,450

26 29 Low-Voltage Controllers

26 29 13 − Enclosed Controllers

26 29 13.10 Contactors, AC

		Crew	Daily Output	Labor-Hours	Unit	Material	2007 Bare Costs Labor	Equipment	Total	Total Incl O&P
1200	150 amp	1 Elec	2	4	Ea.	3,100	176		3,276	3,675
1300	200 amp		1.40	5.714		3,100	251		3,351	3,800
1500	Magnetic with auxiliary contact, size 00, 9 amp		4	2		177	88		265	325
1600	Size 0, 18 amp		4	2		211	88		299	365
1700	Size 1, 27 amp		3.60	2.222		239	97.50		336.50	410
1800	Size 2, 45 amp		3	2.667		445	117		562	665
1900	Size 3, 90 amp		2.50	3.200		720	140		860	1,000
2000	Size 4, 135 amp		2.30	3.478		1,650	153		1,803	2,025
2100	Size 5, 270 amp	2 Elec	1.80	8.889		3,450	390		3,840	4,375
2200	Size 6, 540 amp		1.20	13.333		10,100	585		10,685	12,000
2300	Size 7, 810 amp		1	16		13,600	700		14,300	16,000
2310	Size 8, 1215 amp		.80	20		21,100	880		21,980	24,500
2500	Magnetic, 240 volt, 1-2 pole, .75 HP motor	1 Elec	4	2		143	88		231	288
2520	2 HP motor		3.60	2.222		160	97.50		257.50	320
2540	5 HP motor		2.50	3.200		390	140		530	635
2560	10 HP motor		1.40	5.714		640	251		891	1,075
2600	240 volt or less, 3 pole, .75 HP motor		4	2		143	88		231	288
2620	5 HP motor		3.60	2.222		177	97.50		274.50	340
2640	10 HP motor		3.60	2.222		205	97.50		302.50	370
2660	15 HP motor		2.50	3.200		410	140		550	660
2700	25 HP motor		2.50	3.200		410	140		550	660
2720	30 HP motor	2 Elec	2.80	5.714		685	251		936	1,125
2740	40 HP motor		2.80	5.714		685	251		936	1,125
2760	50 HP motor		1.60	10		685	440		1,125	1,400
2800	75 HP motor		1.60	10		1,600	440		2,040	2,425
2820	100 HP motor		1	16		1,600	700		2,300	2,825
2860	150 HP motor		1	16		3,425	700		4,125	4,800
2880	200 HP motor		1	16		3,425	700		4,125	4,800
3000	600 volt, 3 pole, 5 HP motor	1 Elec	4	2		177	88		265	325
3020	10 HP motor		3.60	2.222		205	97.50		302.50	370
3040	25 HP motor		3	2.667		410	117		527	625
3100	50 HP motor		2.50	3.200		685	140		825	960
3160	100 HP motor	2 Elec	2.80	5.714		1,600	251		1,851	2,150
3220	200 HP motor	"	1.60	10		3,425	440		3,865	4,400

26 29 13.20 Control Stations

		Crew	Daily Output	Labor-Hours	Unit	Material	2007 Bare Costs Labor	Equipment	Total	Total Incl O&P
0010	**CONTROL STATIONS**									
0050	NEMA 1, heavy duty, stop/start	1 Elec	8	1	Ea.	131	44		175	210
0100	Stop/start, pilot light		6.20	1.290		179	56.50		235.50	281
0200	Hand/off/automatic		6.20	1.290		97	56.50		153.50	192
0400	Stop/start/reverse		5.30	1.509		177	66.50		243.50	293
0500	NEMA 7, heavy duty, stop/start		6	1.333		245	58.50		303.50	355
0600	Stop/start, pilot light		4	2		299	88		387	460
0700	NEMA 7 or 9, 1 element		6	1.333		218	58.50		276.50	325
0800	2 element		6	1.333		259	58.50		317.50	370
0900	3 element		4	2		580	88		668	770
0910	Selector switch, 2 position		6	1.333		218	58.50		276.50	325
0920	3 position		4	2		224	88		312	375
0930	Oiltight, 1 element		8	1		91	44		135	166
0940	2 element		6.20	1.290		131	56.50		187.50	229
0950	3 element		5.30	1.509		177	66.50		243.50	293
0960	Selector switch, 2 position		6.20	1.290		97	56.50		153.50	192
0970	3 position		5.30	1.509		97	66.50		163.50	206

26 29 13 – Enclosed Controllers

26 29 13.30 Control Switches	Crew	Daily Output	Labor-Hours	Unit	Material	2007 Bare Costs Labor	Equipment	Total	Total Incl O&P
0010	**CONTROL SWITCHES** Field installed									
6000	Push button 600V 10A, momentary contact									
6150	Standard operator with colored button	1 Elec	34	.235	Ea.	15.40	10.35		25.75	32.50
6160	With single block 1NO 1NC		18	.444		32.50	19.50		52	64.50
6170	With double block 2NO 2NC		15	.533		49.50	23.50		73	89.50
6180	Stnd operator w/mushroom button 1-9/16" diam.		34	.235		32.50	10.35		42.85	51
6190	Stnd operator w/mushroom button 2-1/4" diam.									
6200	With single block 1NO 1NC	1 Elec	18	.444	Ea.	49.50	19.50		69	83.50
6210	With double block 2NO 2NC		15	.533		66.50	23.50		90	109
6500	Maintained contact, selector operator		34	.235		49.50	10.35		59.85	70
6510	With single block 1NO 1NC		18	.444		66.50	19.50		86	103
6520	With double block 2NO 2NC		15	.533		84	23.50		107.50	127
6560	Spring-return selector operator		34	.235		49.50	10.35		59.85	70
6570	With single block 1NO 1NC		18	.444		66.50	19.50		86	103
6580	With double block 2NO 2NC		15	.533		84	23.50		107.50	127
6620	Transformer operator w/illuminated									
6630	button 6V #12 lamp	1 Elec	32	.250	Ea.	84	11		95	108
6640	With single block 1NO 1NC w/guard		16	.500		101	22		123	144
6650	With double block 2NO 2NC w/guard		13	.615		118	27		145	170
6690	Combination operator		34	.235		49.50	10.35		59.85	70
6700	With single block 1NO 1NC		18	.444		66.50	19.50		86	103
6710	With double block 2NO 2NC		15	.533		84	23.50		107.50	127
9000	Indicating light unit, full voltage									
9010	110-125V front mount	1 Elec	32	.250	Ea.	66.50	11		77.50	90
9020	130V resistor type		32	.250		49.50	11		60.50	71
9030	6V transformer type		32	.250		61	11		72	83.50

26 29 13.40 Relays

		Crew	Daily Output	Labor-Hours	Unit	Material	Labor	Equipment	Total	Total Incl O&P
0010	**RELAYS** Enclosed (NEMA 1)									
0050	600 volt AC, 1 pole, 12 amp	1 Elec	5.30	1.509	Ea.	90	66.50		156.50	198
0100	2 pole, 12 amp		5	1.600		90	70		160	204
0200	4 pole, 10 amp		4.50	1.778		120	78		198	248
0500	250 volt DC, 1 pole, 15 amp		5.30	1.509		118	66.50		184.50	229
0600	2 pole, 10 amp		5	1.600		113	70		183	229
0700	4 pole, 4 amp		4.50	1.778		149	78		227	280

26 29 23 – Variable-Frequency Motor Controllers

26 29 23.10 Variable Frequency Drives/Adj. Frequency Drives

		Crew	Daily Output	Labor-Hours	Unit	Material	Labor	Equipment	Total	Total Incl O&P
0010	**VARIABLE FREQUENCY DRIVES/ADJ. FREQUENCY DRIVES**									
0100	Enclosed (NEMA 1), 460 volt, for 3 HP motor size	1 Elec	.80	10	Ea.	1,525	440		1,965	2,350
0110	5 HP motor size		.80	10		1,675	440		2,115	2,475
0120	7.5 HP motor size		.67	11.940		1,975	525		2,500	2,950
0130	10 HP motor size		.67	11.940		1,975	525		2,500	2,950
0140	15 HP motor size	2 Elec	.89	17.978		2,275	790		3,065	3,675
0150	20 HP motor size		.89	17.978		3,375	790		4,165	4,900
0160	25 HP motor size		.67	23.881		3,925	1,050		4,975	5,875
0170	30 HP motor size		.67	23.881		4,775	1,050		5,825	6,825
0180	40 HP motor size		.67	23.881		7,025	1,050		8,075	9,275
0190	50 HP motor size		.53	30.189		7,650	1,325		8,975	10,400
0200	60 HP motor size	R-3	.56	35.714		8,375	1,550	292	10,217	11,800
0210	75 HP motor size		.56	35.714		11,500	1,550	292	13,342	15,200
0220	100 HP motor size		.50	40		11,700	1,725	325	13,750	15,800
0230	125 HP motor size		.50	40		12,900	1,725	325	14,950	17,100

26 29 Low-Voltage Controllers

26 29 23 – Variable-Frequency Motor Controllers

26 29 23.10 Variable Frequency Drives/Adj. Frequency Drives	Crew	Daily Output	Labor-Hours	Unit	Material	2007 Bare Costs Labor	Equipment	Total	Total Incl O&P
0240	150 HP motor size	R-3	.50	40	Ea.	16,500	1,725	325	18,550	21,100
0250	200 HP motor size	↓	.42	47.619		20,500	2,050	390	22,940	26,100
1100	Custom-engineered, 460 volt, for 3 HP motor size	1 Elec	.56	14.286		2,525	625		3,150	3,700
1110	5 HP motor size		.56	14.286		2,650	625		3,275	3,850
1120	7.5 HP motor size		.47	17.021		3,200	745		3,945	4,625
1130	10 HP motor size	↓	.47	17.021		3,200	745		3,945	4,625
1140	15 HP motor size	2 Elec	.62	25.806		3,500	1,125		4,625	5,525
1150	20 HP motor size		.62	25.806		4,775	1,125		5,900	6,925
1160	25 HP motor size		.47	34.043		5,200	1,500		6,700	7,950
1170	30 HP motor size		.47	34.043		7,075	1,500		8,575	10,000
1180	40 HP motor size		.47	34.043		7,900	1,500		9,400	10,900
1190	50 HP motor size	↓	.37	43.243		7,900	1,900		9,800	11,500
1200	60 HP motor size	R-3	.39	51.282		11,800	2,225	420	14,445	16,800
1210	75 HP motor size		.39	51.282		13,600	2,225	420	16,245	18,800
1220	100 HP motor size		.35	57.143		14,200	2,475	465	17,140	19,800
1230	125 HP motor size		.35	57.143		15,000	2,475	465	17,940	20,700
1240	150 HP motor size		.35	57.143		16,800	2,475	465	19,740	22,700
1250	200 HP motor size	↓	.29	68.966	↓	23,300	2,975	565	26,840	30,700
2000	For complex & special design systems to meet specific									
2010	requirements, obtain quote from vendor.									

26 32 Packaged Generator Assemblies

26 32 13 – Engine Generators

26 32 13.13 Diesel-Engine-Driven Generator Sets

		Crew	Daily Output	Labor-Hours	Unit	Material	Labor	Equipment	Total	Total Incl O&P
0010	**DIESEL-ENGINE-DRIVEN GENERATOR SETS**									
2000	Diesel engine, including battery, charger,									
2010	muffler, automatic transfer switch & day tank, 30 kW	R-3	.55	36.364	Ea.	17,000	1,575	297	18,872	21,400
2100	50 kW		.42	47.619		20,900	2,050	390	23,340	26,500
2110	60 kW		.39	51.282		22,600	2,225	420	25,245	28,600
2200	75 kW		.35	57.143		27,200	2,475	465	30,140	34,200
2300	100 kW		.31	64.516		30,200	2,800	525	33,525	38,000
2400	125 kW		.29	68.966		31,800	2,975	565	35,340	40,100
2500	150 kW		.26	76.923		36,500	3,325	630	40,455	45,800
2600	175 kW		.25	80		39,800	3,450	655	43,905	49,600
2700	200 kW		.24	83.333		41,000	3,600	680	45,280	51,000
2800	250 kW		.23	86.957		48,300	3,775	710	52,785	59,500
2850	275 kW		.22	90.909		48,300	3,925	745	52,970	59,500
2900	300 kW		.22	90.909		52,000	3,925	745	56,670	64,000
3000	350 kW		.20	100		59,000	4,325	815	64,140	72,500
3100	400 kW		.19	105		73,000	4,550	860	78,410	87,500
3200	500 kW		.18	111		91,500	4,800	910	97,210	109,000
3220	600 kW	↓	.17	117		120,500	5,100	960	126,560	141,000
3230	650 kW	R-13	.38	110		143,000	4,625	505	148,130	165,000
3240	750 kW		.38	110		151,000	4,625	505	156,130	173,500
3250	800 kW		.36	116		157,000	4,900	530	162,430	180,500
3260	900 kW		.31	135		181,500	5,675	620	187,795	208,500
3270	1000 kW	↓	.27	155	↓	187,000	6,525	710	194,235	216,000

26 32 13.16 Gas-Engine-Driven Generator Sets

0010	**GAS-ENGINE-DRIVEN GENERATOR SETS** R263213-45									
0020	Gas or gasoline operated, includes battery,									
0050	charger, muffler & transfer switch									

26 32 Packaged Generator Assemblies

26 32 13 - Engine Generators

26 32 13.16 Gas-Engine-Driven Generator Sets

		Crew	Daily Output	Labor-Hours	Unit	Material	2007 Bare Costs Labor	Equipment	Total	Total Incl O&P
0200	3 phase 4 wire, 277/480 volt, 7.5 kW	R-3	.83	24.096	Ea.	6,350	1,050	197	7,597	8,775
0300	11.5 kW		.71	28.169		9,000	1,225	230	10,455	12,000
0400	20 kW		.63	31.746		10,600	1,375	259	12,234	14,000
0500	35 kW		.55	36.364		12,700	1,575	297	14,572	16,700
0520	60 kW		.50	40		17,500	1,725	325	19,550	22,100
0600	80 kW		.40	50		21,700	2,175	410	24,285	27,600
0700	100 kW		.33	60.606		23,900	2,625	495	27,020	30,700
0800	125 kW		.28	71.429		48,800	3,100	585	52,485	59,000
0900	185 kW		.25	80		64,500	3,450	655	68,605	77,000

26 33 Battery Equipment

26 33 53 - Static Uninterruptible Power Supply

26 33 53.10 Uninterruptible Power Supply/Conditioner Transformers

			Crew	Daily Output	Labor-Hours	Unit	Material	2007 Bare Costs Labor	Equipment	Total	Total Incl O&P
0010	**UNINTERRUPTIBLE POWER SUPPLY/CONDITIONER TRANSFORMERS**										
0100	Volt. regulating, isolating trans., w/invert. & 10 min. battery pack										
0110	Single-phase, 120 V, 0.35 kVA	R263353-80	1 Elec	2.29	3.493	Ea.	1,000	153		1,153	1,325
0120	0.5 kVA			2	4		1,050	176		1,226	1,400
0130	For additional 55 min. battery, add to .35 kVA			2.29	3.493		590	153		743	880
0140	Add to 0.5 kVA			1.14	7.018		620	310		930	1,150
0150	Single-phase, 120 V, 0.75 kVA			.80	10		1,325	440		1,765	2,100
0160	1.0 kVA			.80	10		1,900	440		2,340	2,750
0170	1.5 kVA		2 Elec	1.14	14.035		3,275	615		3,890	4,525
0180	2 kVA		"	.89	17.978		3,550	790		4,340	5,075
0190	3 kVA		R-3	.63	31.746		4,300	1,375	259	5,934	7,050
0200	5 kVA			.42	47.619		6,225	2,050	390	8,665	10,400
0210	7.5 kVA			.33	60.606		7,950	2,625	495	11,070	13,200
0220	10 kVA			.28	71.429		8,450	3,100	585	12,135	14,500
0230	15 kVA			.22	90.909		11,700	3,925	745	16,370	19,600
0500	For options & accessories add to above, minimum										10%
0520	Maximum										35%
0600	For complex & special design systems to meet specific										
0610	requirements, obtain quote from vendor										

26 35 Power Filters and Conditioners

26 35 13 - Capacitors

26 35 13.10 Capacitors

		Crew	Daily Output	Labor-Hours	Unit	Material	2007 Bare Costs Labor	Equipment	Total	Total Incl O&P
0010	**CAPACITORS** Indoor									
0020	240 volts, single & 3 phase, 0.5 kVAR	1 Elec	2.70	2.963	Ea.	315	130		445	545
0100	1.0 kVAR		2.70	2.963		380	130		510	615
0150	2.5 kVAR		2	4		430	176		606	735
0200	5.0 kVAR		1.80	4.444		515	195		710	855
0250	7.5 kVAR		1.60	5		600	220		820	985
0300	10 kVAR		1.50	5.333		720	234		954	1,150
0350	15 kVAR		1.30	6.154		995	270		1,265	1,500
0400	20 kVAR		1.10	7.273		1,225	320		1,545	1,825
0450	25 kVAR		1	8		1,450	350		1,800	2,125
1000	480 volts, single & 3 phase, 1 kVAR		2.70	2.963		289	130		419	515
1050	2 kVAR		2.70	2.963		330	130		460	560
1100	5 kVAR		2	4		420	176		596	720

26 35 13 – Capacitors

26 35 13.10 Capacitors		Crew	Daily Output	Labor-Hours	Unit	Material	2007 Bare Costs Labor	Equipment	Total	Total Incl O&P
1150	7.5 kVAR	1 Elec	2	4	Ea.	450	176		626	755
1200	10 kVAR		2	4		505	176		681	815
1250	15 kVAR		2	4		625	176		801	945
1300	20 kVAR		1.60	5		695	220		915	1,100
1350	30 kVAR		1.50	5.333		870	234		1,104	1,300
1400	40 kVAR		1.20	6.667		1,100	293		1,393	1,650
1450	50 kVAR		1.10	7.273		1,250	320		1,570	1,850
2000	600 volts, single & 3 phase, 1 kVAR		2.70	2.963		298	130		428	525
2050	2 kVAR		2.70	2.963		340	130		470	570
2100	5 kVAR		2	4		420	176		596	720
2150	7.5 kVAR		2	4		450	176		626	755
2200	10 kVAR		2	4		505	176		681	815
2250	15 kVAR		1.60	5		625	220		845	1,000
2300	20 kVAR		1.60	5		695	220		915	1,100
2350	25 kVAR		1.50	5.333		775	234		1,009	1,200
2400	35 kVAR		1.40	5.714		985	251		1,236	1,450
2450	50 kVAR		1.30	6.154		1,250	270		1,520	1,775

26 35 26 – Harmonic Filters

26 35 26.10 Computer Isolation Transformer

		Crew	Daily Output	Labor-Hours	Unit	Material	Labor	Equipment	Total	Total Incl O&P
0010	**COMPUTER ISOLATION TRANSFORMER**									
0100	Computer grade									
0110	Single-phase, 120/240 V, 0.5 kVA	1 Elec	4	2	Ea.	415	88		503	585
0120	1.0 kVA		2.67	2.996		590	132		722	840
0130	2.5 kVA		2	4		895	176		1,071	1,250
0140	5 kVA		1.14	7.018		1,025	310		1,335	1,575

26 35 26.20 Computer Regulator Transformer

		Crew	Daily Output	Labor-Hours	Unit	Material	Labor	Equipment	Total	Total Incl O&P
0010	**COMPUTER REGULATOR TRANSFORMER**									
0100	Ferro-resonant, constant voltage, variable transformer									
0110	Single-phase, 240 V, 0.5 kVA	1 Elec	2.67	2.996	Ea.	530	132		662	775
0120	1.0 kVA		2	4		725	176		901	1,050
0130	2.0 kVA		1	8		1,250	350		1,600	1,875
0210	Plug-in unit 120 V, 0.14 kVA		8	1		305	44		349	400
0220	0.25 kVA		8	1		355	44		399	455
0230	0.5 kVA		8	1		530	44		574	645
0240	1.0 kVA		5.33	1.501		725	66		791	895
0250	2.0 kVA		4	2		1,250	88		1,338	1,475

26 35 26.30 Power Conditioner Transformer

		Crew	Daily Output	Labor-Hours	Unit	Material	Labor	Equipment	Total	Total Incl O&P
0010	**POWER CONDITIONER TRANSFORMER**									
0100	Electronic solid state, buck-boost, transformer, w/tap switch									
0110	Single-phase, 115 V, 3.0 kVA, + or - 3% accuracy	2 Elec	1.60	10	Ea.	2,850	440		3,290	3,775
0120	208, 220, 230, or 240 V, 5.0 kVA, + or - 1.5% accuracy	3 Elec	1.60	15		3,700	660		4,360	5,025
0130	5.0 kVA, + or - 6% accuracy	2 Elec	1.14	14.035		3,300	615		3,915	4,550
0140	7.5 kVA, + or - 1.5% accuracy	3 Elec	1.50	16		4,675	700		5,375	6,200
0150	7.5 kVA, + or - 6% accuracy		1.60	15		3,900	660		4,560	5,275
0160	10.0 kVA, + or - 1.5% accuracy		1.33	18.045		6,250	790		7,040	8,050
0170	10.0 kVA, + or - 6% accuracy		1.41	17.021		5,300	745		6,045	6,950

26 35 26.40 Transient Voltage Suppressor Transformer

		Crew	Daily Output	Labor-Hours	Unit	Material	Labor	Equipment	Total	Total Incl O&P
0010	**TRANSIENT VOLTAGE SUPPRESSOR TRANSFORMER**									
0110	Single-phase, 120 V, 1.8 kVA	1 Elec	4	2	Ea.	580	88		668	765
0120	3.6 kVA		4	2		1,200	88		1,288	1,450
0130	7.2 kVA		3.20	2.500		1,550	110		1,660	1,875

26 35 Power Filters and Conditioners

26 35 26 – Harmonic Filters

26 35 26.40 Transient Voltage Suppressor Transformer		Crew	Daily Output	Labor-Hours	Unit	Material	2007 Bare Costs Labor	2007 Bare Costs Equipment	Total	Total Incl O&P
0150	240 V, 3.6 kVA	1 Elec	4	2	Ea.	1,200	88		1,288	1,450
0160	7.2 kVA		4	2		1,550	88		1,638	1,825
0170	14.4 kVA		3.20	2.500		2,000	110		2,110	2,375
0210	Plug-in unit, 120 V, 1.8 kVA		8	1		525	44		569	645

26 35 53 – Voltage Regulators

26 35 53.10 Automatic Voltage Regulators

		Crew	Daily Output	Labor-Hours	Unit	Material	2007 Bare Costs Labor	2007 Bare Costs Equipment	Total	Total Incl O&P
0010	**AUTOMATIC VOLTAGE REGULATORS**									
0100	Computer grade, solid state, variable trans. volt. regulator									
0110	Single-phase, 120 V, 8.6 kVA	2 Elec	1.33	12.030	Ea.	5,600	530		6,130	6,950
0120	17.3 kVA		1.14	14.035		6,625	615		7,240	8,225
0130	208/240 V, 7.5/8.6 kVA		1.33	12.030		5,600	530		6,130	6,950
0140	13.5/15.6 kVA		1.33	12.030		6,625	530		7,155	8,075
0150	27.0/31.2 kVA		1.14	14.035		8,375	615		8,990	10,100
0210	Two-phase, single control, 208/240 V, 15.0/17.3 kVA		1.14	14.035		6,625	615		7,240	8,225
0220	Individual phase control, 15.0/17.3 kVA		1.14	14.035		6,625	615		7,240	8,225
0230	30.0/34.6 kVA	3 Elec	1.33	18.045		8,375	790		9,165	10,400
0310	Three-phase single control, 208/240 V, 26/30 kVA	2 Elec	1	16		6,625	700		7,325	8,350
0320	380/480 V, 24/30 kVA	"	1	16		6,625	700		7,325	8,350
0330	43/54 kVA	3 Elec	1.33	18.045		12,000	790		12,790	14,400
0340	Individual phase control, 208 V, 26 kVA	"	1.33	18.045		6,625	790		7,415	8,475
0350	52 kVA	R-3	.91	21.978		8,375	950	180	9,505	10,800
0360	340/480 V, 24/30 kVA	2 Elec	1	16		6,625	700		7,325	8,350
0370	43/54 kVA	"	1	16		8,375	700		9,075	10,300
0380	48/60 kVA	3 Elec	1.33	18.045		12,100	790		12,890	14,500
0390	86/108 kVA	R-3	.91	21.978		13,500	950	180	14,630	16,400
0500	Standard grade, solid state, variable transformer volt. regulator									
0510	Single-phase, 115 V, 2.3 kVA	1 Elec	2.29	3.493	Ea.	1,650	153		1,803	2,050
0520	4.2 kVA		2	4		2,650	176		2,826	3,175
0530	6.6 kVA		1.14	7.018		3,250	310		3,560	4,025
0540	13.0 kVA		1.14	7.018		5,400	310		5,710	6,375
0550	16.6 kVA	2 Elec	1.23	13.008		6,375	570		6,945	7,850
0610	230 V, 8.3 kVA		1.33	12.030		5,400	530		5,930	6,700
0620	21.4 kVA		1.23	13.008		6,375	570		6,945	7,850
0630	29.9 kVA		1.23	13.008		6,375	570		6,945	7,850
0710	460 V, 9.2 kVA		1.33	12.030		5,400	530		5,930	6,700
0720	20.7 kVA		1.23	13.008		6,375	570		6,945	7,850
0810	Three-phase, 230 V, 13.1 kVA	3 Elec	1.41	17.021		5,400	745		6,145	7,025
0820	19.1 kVA		1.41	17.021		6,375	745		7,120	8,100
0830	25.1 kVA		1.23	19.512		6,375	855		7,230	8,275
0840	57.8 kVA	R-3	.95	21.053		11,600	910	172	12,682	14,200
0850	74.9 kVA	"	.91	21.978		11,600	950	180	12,730	14,300
0910	460 V, 14.3 kVA	3 Elec	1.41	17.021		5,400	745		6,145	7,025
0920	19.1 kVA		1.41	17.021		6,375	745		7,120	8,100
0930	27.9 kVA		1.23	19.512		6,375	855		7,230	8,275
0940	59.8 kVA	R-3	1	20		11,600	865	163	12,628	14,200
0950	79.7 kVA		.95	21.053		12,900	910	172	13,982	15,700
0960	118 kVA		.95	21.053		13,600	910	172	14,682	16,500
1000	Laboratory grade, precision, electronic voltage regulator									
1110	Single-phase, 115 V, 0.5 kVA	1 Elec	2.29	3.493	Ea.	1,375	153		1,528	1,750
1120	1.0 kVA		2	4		1,475	176		1,651	1,875
1130	3.0 kVA		.80	10		2,050	440		2,490	2,900
1140	6.0 kVA	2 Elec	1.46	10.959		3,800	480		4,280	4,900

26 35 Power Filters and Conditioners

26 35 53 – Voltage Regulators

26 35 53.10 Automatic Voltage Regulators

		Crew	Daily Output	Labor-Hours	Unit	Material	2007 Bare Costs Labor	Equipment	Total	Total Incl O&P
1150	10.0 kVA	3 Elec	1	24	Ea.	4,925	1,050		5,975	7,000
1160	15.0 kVA	"	1.50	16		5,625	700		6,325	7,250
1210	230 V, 3.0 kVA	1 Elec	.80	10		2,325	440		2,765	3,200
1220	6.0 kVA	2 Elec	1.46	10.959		3,875	480		4,355	5,000
1230	10.0 kVA	3 Elec	1.71	14.035		5,150	615		5,765	6,600
1240	15.0 kVA	"	1.60	15		5,825	660		6,485	7,375

26 35 53.30 Transient Suppressor/Voltage Regulator

		Crew	Daily Output	Labor-Hours	Unit	Material	2007 Bare Costs Labor	Equipment	Total	Total Incl O&P
0010	**TRANSIENT SUPPRESSOR/VOLTAGE REGULATOR** (without isolation)									
0110	Single-phase, 115 V, 1.0 kVA	1 Elec	2.67	2.996	Ea.	985	132		1,117	1,275
0120	2.0 kVA		2.29	3.493		1,350	153		1,503	1,700
0130	4.0 kVA		2.13	3.756		1,650	165		1,815	2,075
0140	220 V, 1.0 kVA		2.67	2.996		985	132		1,117	1,275
0150	2.0 kVA		2.29	3.493		1,350	153		1,503	1,700
0160	4.0 kVA		2.13	3.756		1,750	165		1,915	2,175
0210	Plug-in unit, 120 V, 1.0 kVA		8	1		940	44		984	1,100
0220	2.0 kVA		8	1		1,325	44		1,369	1,525

26 36 Transfer Switches

26 36 13 – Manual Transfer Switches

26 36 13.10 Non-Automatic Transfer Switches

		Crew	Daily Output	Labor-Hours	Unit	Material	2007 Bare Costs Labor	Equipment	Total	Total Incl O&P
0010	**NON-AUTOMATIC TRANSFER SWITCHES** enclosed									
0100	Manual operated, 480 volt 3 pole, 30 amp	1 Elec	2.30	3.478	Ea.	1,550	153		1,703	1,925
0150	60 amp		1.90	4.211		1,550	185		1,735	1,975
0200	100 amp		1.30	6.154		1,550	270		1,820	2,100
0250	200 amp	2 Elec	2	8		3,075	350		3,425	3,900
0300	400 amp		1.60	10		4,475	440		4,915	5,575
0350	600 amp		1	16		6,775	700		7,475	8,500
1000	250 volt 3 pole, 30 amp	1 Elec	2.30	3.478		1,550	153		1,703	1,925
1100	60 amp		1.90	4.211		1,550	185		1,735	1,975
1150	100 amp		1.30	6.154		1,550	270		1,820	2,100
1200	200 amp	2 Elec	2	8		3,075	350		3,425	3,900
1300	600 amp	"	1	16		6,775	700		7,475	8,500
1500	Electrically operated, 480 volt 3 pole, 60 amp	1 Elec	1.90	4.211		2,950	185		3,135	3,525
1600	100 amp	"	1.30	6.154		2,950	270		3,220	3,650
1650	200 amp	2 Elec	2	8		4,850	350		5,200	5,875
1700	400 amp		1.60	10		6,800	440		7,240	8,125
1750	600 amp		1	16		9,800	700		10,500	11,900
2000	250 volt 3 pole, 30 amp	1 Elec	2.30	3.478		2,950	153		3,103	3,475
2050	60 amp	"	1.90	4.211		2,950	185		3,135	3,525
2150	200 amp	2 Elec	2	8		4,850	350		5,200	5,875
2200	400 amp		1.60	10		6,800	440		7,240	8,125
2250	600 amp		1	16		9,800	700		10,500	11,900
2500	NEMA 3R, 480 volt 3 pole, 60 amp	1 Elec	1.80	4.444		3,475	195		3,670	4,125
2550	100 amp	"	1.20	6.667		3,475	293		3,768	4,250
2600	200 amp	2 Elec	1.80	8.889		5,500	390		5,890	6,625
2650	400 amp	"	1.40	11.429		7,450	500		7,950	8,925
2800	NEMA 3R, 250 volt 3 pole solid state, 100 amp	1 Elec	1.20	6.667		3,475	293		3,768	4,250
2850	150 amp	2 Elec	1.80	8.889		4,625	390		5,015	5,650
2900	250 volt 2 pole solid state, 100 amp	1 Elec	1.30	6.154		3,375	270		3,645	4,125
2950	150 amp	2 Elec	2	8		4,475	350		4,825	5,450

26 36 Transfer Switches

26 36 23 – Automatic Transfer Switches

26 36 23.10 Automatic Transfer Switches		Crew	Daily Output	Labor-Hours	Unit	Material	2007 Bare Costs Labor	Equipment	Total	Total Incl O&P
0010	**AUTOMATIC TRANSFER SWITCHES** R263623-60									
0015	Switches, enclosed 120/240 volt, 2 pole, 30 amp	1 Elec	2.40	3.333	Ea.	2,925	146		3,071	3,450
0020	70 amp		2	4		2,925	176		3,101	3,475
0030	100 amp	↓	1.35	5.926		2,925	260		3,185	3,600
0040	225 amp	2 Elec	2.10	7.619		4,325	335		4,660	5,250
0050	400 amp		1.70	9.412		6,575	415		6,990	7,875
0060	600 amp		1.06	15.094		9,675	665		10,340	11,600
0070	800 amp	↓	.84	19.048		11,400	835		12,235	13,800
0100	Switches, enclosed 480 volt, 3 pole, 30 amp	1 Elec	2.30	3.478		3,275	153		3,428	3,825
0200	60 amp		1.90	4.211		3,275	185		3,460	3,875
0300	100 amp	↓	1.30	6.154		3,275	270		3,545	4,000
0400	150 amp	2 Elec	2.40	6.667		4,025	293		4,318	4,850
0500	225 amp		2	8		5,150	350		5,500	6,175
0600	260 amp		2	8		5,950	350		6,300	7,075
0700	400 amp		1.60	10		6,975	440		7,415	8,325
0800	600 amp		1	16		10,700	700		11,400	12,900
0900	800 amp		.80	20		12,600	880		13,480	15,200
1000	1000 amp		.76	21.053		16,600	925		17,525	19,700
1100	1200 amp		.70	22.857		22,800	1,000		23,800	26,600
1200	1600 amp		.60	26.667		26,000	1,175		27,175	30,400
1300	2000 amp	↓	.50	32		29,000	1,400		30,400	34,100
1600	Accessories, time delay on engine starting					255			255	280
1700	Adjustable time delay on retransfer					255			255	280
1800	Shunt trips for customer connections					455			455	500
1900	Maintenance select switch					104			104	114
2000	Auxiliary contact when normal fails					120			120	132
2100	Pilot light-emergency					104			104	114
2200	Pilot light-normal					104			104	114
2300	Auxiliary contact-closed on normal					120			120	132
2400	Auxiliary contact-closed on emergency					120			120	132
2500	Emergency source sensing, frequency relay				↓	525			525	580

26 41 Facility Lightning Protection

26 41 13 – Lightning Protection for Structures

26 41 13.13 Lightning Protection for Buildings

		Crew	Daily Output	Labor-Hours	Unit	Material	2007 Bare Costs Labor	Equipment	Total	Total Incl O&P
0010	**LIGHTNING PROTECTION FOR BUILDINGS**									
0200	Air terminals & base, copper									
0400	3/8" diameter x 10" (to 75' high)	1 Elec	8	1	Ea.	17.70	44		61.70	85
0500	1/2" diameter x 12" (over 75' high)		8	1		20	44		64	87.50
1000	Aluminum, 1/2" diameter x 12" (to 75' high)		8	1		3.05	44		47.05	69
1100	5/8" diameter x 12" (over 75' high)		8	1	↓	4.80	44		48.80	71
2000	Cable, copper, 220 lb. per thousand ft. (to 75' high)		320	.025	L.F.	1.43	1.10		2.53	3.20
2100	375 lb. per thousand ft. (over 75' high)		230	.035		2.26	1.53		3.79	4.76
2500	Aluminum, 101 lb. per thousand ft. (to 75' high)		280	.029	↓	.72	1.25		1.97	2.66
2600	199 lb. per thousand ft. (over 75' high)		240	.033	↓	1.17	1.46		2.63	3.47
3000	Arrester, 175 volt AC to ground		8	1	Ea.	62.50	44		106.50	135
3100	650 volt AC to ground	↓	6.70	1.194	"	69	52.50		121.50	154

26 42 Cathodic Protection

26 42 16 – Passive Cathodic Protection for Underground Storage Tank

26 42 16.50 Cathodic Protection	Crew	Daily Output	Labor-Hours	Unit	Material	2007 Bare Costs Labor	2007 Bare Costs Equipment	Total	Total Incl O&P
0010 **CATHODIC PROTECTION**									
1000 Anodes, magnesium type, 9 #	R-15	18.50	2.595	Ea.	24	111	15.65	150.65	210
1010 17 #		13	3.692		41	158	22.50	221.50	305
1020 32 #		10	4.800		72	205	29	306	415
1030 48 #		7.20	6.667		110	285	40.50	435.50	590
1100 Graphite type w/ epoxy cap, 3" x 60" (32 #)	R-22	8.40	4.438		120	164		284	380
1110 4" x 80" (68 #)		6	6.213		225	229		454	595
1120 6" x 72" (80 #)		5.20	7.169		1,125	264		1,389	1,650
1130 6" x 36" (45 #)		9.60	3.883		570	143		713	840
2000 Rectifiers, silicon type, air cooled, 28 V/10 A	R-19	3.50	5.714		1,950	251		2,201	2,525
2010 20 V/20 A		3.50	5.714		2,100	251		2,351	2,675
2100 Oil immersed, 28 V/10 A		3	6.667		2,750	293		3,043	3,450
2110 20 V/20 A		3	6.667		2,950	293		3,243	3,675
3000 Anode backfill, coke breeze	R-22	3850	.010	Lb.	.15	.36		.51	.71
4000 Cable, OR2, No. 8		2.40	15.533	M.L.F.	233	575		808	1,125
4010 No. 6		2.40	15.533		330	575		905	1,225
4020 No. 4		2.40	15.533		445	575		1,020	1,350
4030 No. 2		2.40	15.533		730	575		1,305	1,675
4040 No. 1		2.20	16.945		910	625		1,535	1,950
4050 No. 1/0		2.20	16.945		1,175	625		1,800	2,250
4060 No. 2/0		2.20	16.945		1,400	625		2,025	2,475
4070 No. 4/0		2	18.640		4,075	685		4,760	5,500
5000 Test station, 7 terminal box, flush curb type w/lockable cover	R-19	12	1.667	Ea.	62.50	73.50		136	178
5010 Reference cell, 2" dia PVC conduit, cplg, plug, set flush	"	4.80	4.167	"	34	183		217	310

26 51 Interior Lighting

26 51 13 – Interior Lighting Fixtures, Lamps, and Ballasts

26 51 13.10 Fixture Hangers

	Crew	Daily Output	Labor-Hours	Unit	Material	2007 Bare Costs Labor	2007 Bare Costs Equipment	Total	Total Incl O&P
0010 **FIXTURE HANGERS**									
0220 Box hub cover	1 Elec	32	.250	Ea.	3.75	11		14.75	20.50
0240 Canopy		12	.667		7.80	29.50		37.30	52
0260 Connecting block		40	.200		3.70	8.80		12.50	17.10
0280 Cushion hanger		16	.500		19.50	22		41.50	54
0300 Box hanger, with mounting strap		8	1		7.20	44		51.20	73.50
0320 Connecting block		40	.200		1.65	8.80		10.45	14.85
0340 Flexible, 1/2" diameter, 4" long		12	.667		9.20	29.50		38.70	53.50
0360 6" long		12	.667		9.90	29.50		39.40	54.50
0380 8" long		12	.667		11.05	29.50		40.55	55.50
0400 10" long		12	.667		11.70	29.50		41.20	56.50
0420 12" long		12	.667		12.70	29.50		42.20	57.50
0440 15" long		12	.667		13.40	29.50		42.90	58.50
0460 18" long		12	.667		15.30	29.50		44.80	60.50
0480 3/4" diameter, 4" long		10	.800		11.10	35		46.10	64.50
0500 6" long		10	.800		12.50	35		47.50	66.50
0520 8" long		10	.800		13.65	35		48.65	67.50
0540 10" long		10	.800		14.45	35		49.45	68.50
0560 12" long		10	.800		15.80	35		50.80	70
0580 15" long		10	.800		17.40	35		52.40	71.50
0600 18" long		10	.800		19.45	35		54.45	74

26 51 13.40 Interior HID Fixtures

	Crew	Daily Output	Labor-Hours	Unit	Material	2007 Bare Costs Labor	2007 Bare Costs Equipment	Total	Total Incl O&P
0010 **INTERIOR HID FIXTURES** Incl. lamps, and mounting hardware									

26 51 Interior Lighting

26 51 13 – Interior Lighting Fixtures, Lamps, and Ballasts

26 51 13.40 Interior HID Fixtures	Crew	Daily Output	Labor-Hours	Unit	Material	2007 Bare Costs Labor	Equipment	Total	Total Incl O&P
0700 High pressure sodium, recessed, round, 70 watt	1 Elec	3.50	2.286	Ea.	390	100		490	580
0720 100 watt		3.50	2.286		415	100		515	605
0740 150 watt		3.20	2.500		465	110		575	675
0760 Square, 70 watt		3.60	2.222		390	97.50		487.50	575
0780 100 watt		3.60	2.222		415	97.50		512.50	600
0820 250 watt		3	2.667		565	117		682	800
0840 1000 watt	2 Elec	4.80	3.333		1,125	146		1,271	1,450
0860 Surface, round, 70 watt	1 Elec	3	2.667		555	117		672	785
0880 100 watt		3	2.667		570	117		687	800
0900 150 watt		2.70	2.963		595	130		725	850
0920 Square, 70 watt		3	2.667		540	117		657	770
0940 100 watt		3	2.667		570	117		687	800
0980 250 watt		2.50	3.200		605	140		745	875
1040 Pendent, round, 70 watt		3	2.667		590	117		707	825
1060 100 watt		3	2.667		565	117		682	800
1080 150 watt		2.70	2.963		580	130		710	830
1100 Square, 70 watt		3	2.667		605	117		722	840
1120 100 watt		3	2.667		615	117		732	850
1140 150 watt		2.70	2.963		630	130		760	890
1160 250 watt		2.50	3.200		850	140		990	1,150
1180 400 watt		2.40	3.333		895	146		1,041	1,200
1220 Wall, round, 70 watt		3	2.667		510	117		627	735
1240 100 watt		3	2.667		520	117		637	745
1260 150 watt		2.70	2.963		530	130		660	780
1300 Square, 70 watt		3	2.667		540	117		657	770
1320 100 watt		3	2.667		565	117		682	800
1340 150 watt		2.70	2.963		600	130		730	855
1360 250 watt		2.50	3.200		645	140		785	920
1380 400 watt	2 Elec	4.80	3.333		795	146		941	1,100
1400 1000 watt	"	3.60	4.444		1,125	195		1,320	1,525
1500 Metal halide, recessed, round, 175 watt	1 Elec	3.40	2.353		390	103		493	585
1520 250 watt	"	3.20	2.500		435	110		545	645
1540 400 watt	2 Elec	5.80	2.759		620	121		741	860
1580 Square, 175 watt	1 Elec	3.40	2.353		360	103		463	550
1640 Surface, round, 175 watt		2.90	2.759		480	121		601	710
1660 250 watt		2.70	2.963		765	130		895	1,025
1680 400 watt	2 Elec	4.80	3.333		850	146		996	1,150
1720 Square, 175 watt	1 Elec	2.90	2.759		525	121		646	755
1800 Pendent, round, 175 watt		2.90	2.759		570	121		691	805
1820 250 watt		2.70	2.963		825	130		955	1,100
1840 400 watt	2 Elec	4.80	3.333		920	146		1,066	1,225
1880 Square, 175 watt	1 Elec	2.90	2.759		555	121		676	790
1900 250 watt	"	2.70	2.963		560	130		690	810
1920 400 watt	2 Elec	4.80	3.333		900	146		1,046	1,200
1980 Wall, round, 175 watt	1 Elec	2.90	2.759		480	121		601	710
2000 250 watt	"	2.70	2.963		760	130		890	1,025
2020 400 watt	2 Elec	4.80	3.333		780	146		926	1,075
2060 Square, 175 watt	1 Elec	2.90	2.759		505	121		626	740
2080 250 watt	"	2.70	2.963		555	130		685	805
2100 400 watt	2 Elec	4.80	3.333		775	146		921	1,075
2800 High pressure sodium, recessed, 70 watt	1 Elec	3.50	2.286		510	100		610	710
2820 100 watt		3.50	2.286		520	100		620	720
2840 150 watt		3.20	2.500		535	110		645	750

26 51 Interior Lighting

26 51 13 – Interior Lighting Fixtures, Lamps, and Ballasts

26 51 13.40 Interior HID Fixtures

		Crew	Daily Output	Labor-Hours	Unit	Material	2007 Bare Costs Labor	Equipment	Total	Total Incl O&P
2900	Surface, 70 watt	1 Elec	3	2.667	Ea.	575	117		692	805
2920	100 watt		3	2.667		595	117		712	830
2940	150 watt		2.70	2.963		625	130		755	880
3000	Pendent, 70 watt		3	2.667		565	117		682	795
3020	100 watt		3	2.667		585	117		702	820
3040	150 watt		2.70	2.963		615	130		745	870
3100	Wall, 70 watt		3	2.667		605	117		722	840
3120	100 watt		3	2.667		635	117		752	875
3140	150 watt		2.70	2.963		660	130		790	920
3200	Metal halide, recessed, 175 watt		3.40	2.353		490	103		593	695
3220	250 watt	▼	3.20	2.500		525	110		635	740
3240	400 watt	2 Elec	5.80	2.759		655	121		776	900
3260	1000 watt	"	4.80	3.333		1,200	146		1,346	1,550
3280	Surface, 175 watt	1 Elec	2.90	2.759		495	121		616	725
3300	250 watt	"	2.70	2.963		765	130		895	1,050
3320	400 watt	2 Elec	4.80	3.333		920	146		1,066	1,225
3340	1000 watt	"	3.60	4.444		1,350	195		1,545	1,775
3360	Pendent, 175 watt	1 Elec	2.90	2.759		495	121		616	725
3380	250 watt	"	2.70	2.963		765	130		895	1,050
3400	400 watt	2 Elec	4.80	3.333		915	146		1,061	1,225
3420	1000 watt	"	3.60	4.444		1,475	195		1,670	1,925
3440	Wall, 175 watt	1 Elec	2.90	2.759		540	121		661	775
3460	250 watt	"	2.70	2.963		810	130		940	1,075
3480	400 watt	2 Elec	4.80	3.333		955	146		1,101	1,275
3500	1000 watt	"	3.60	4.444		1,550	195		1,745	2,000

26 51 13.50 Interior Lighting Fixtures

		Crew	Daily Output	Labor-Hours	Unit	Material	2007 Bare Costs Labor	Equipment	Total	Total Incl O&P
0010	**INTERIOR LIGHTING FIXTURES** Including lamps, mounting R265113-40									
0030	hardware and connections									
0100	Fluorescent, C.W. lamps, troffer, recess mounted in grid, RS									
0200	Acrylic lens, 1'W x 4'L, two 40 watt	1 Elec	5.70	1.404	Ea.	50.50	61.50		112	147
0210	1'W x 4'L, three 40 watt		5.40	1.481		59	65		124	162
0300	2'W x 2'L, two U40 watt		5.70	1.404		54	61.50		115.50	151
0400	2'W x 4'L, two 40 watt		5.30	1.509		54	66.50		120.50	158
0500	2'W x 4'L, three 40 watt		5	1.600		57.50	70		127.50	169
0600	2'W x 4'L, four 40 watt	▼	4.70	1.702		61	74.50		135.50	178
0700	4'W x 4'L, four 40 watt	2 Elec	6.40	2.500		325	110		435	520
0800	4'W x 4'L, six 40 watt		6.20	2.581		335	113		448	535
0900	4'W x 4'L, eight 40 watt	▼	5.80	2.759		345	121		466	560
0910	Acrylic lens, 1'W x 4'L, two 32 watt	1 Elec	5.70	1.404		60	61.50		121.50	158
0930	2'W x 2'L, two U32 watt		5.70	1.404		80	61.50		141.50	180
0940	2'W x 4'L, two 32 watt		5.30	1.509		72	66.50		138.50	178
0950	2'W x 4'L, three 32 watt		5	1.600		76.50	70		146.50	189
0960	2'W x 4'L, four 32 watt	▼	4.70	1.702	▼	78.50	74.50		153	198
1000	Surface mounted, RS									
1030	Acrylic lens with hinged & latched door frame									
1100	1'W x 4'L, two 40 watt	1 Elec	7	1.143	Ea.	73.50	50		123.50	156
1110	1'W x 4'L, three 40 watt		6.70	1.194		75.50	52.50		128	161
1200	2'W x 2'L, two U40 watt		7	1.143		79	50		129	161
1300	2'W x 4'L, two 40 watt		6.20	1.290		90.50	56.50		147	184
1400	2'W x 4'L, three 40 watt		5.70	1.404		91.50	61.50		153	192
1500	2'W x 4'L, four 40 watt	▼	5.30	1.509		93.50	66.50		160	202
1600	4'W x 4'L, four 40 watt	2 Elec	7.20	2.222		445	97.50		542.50	635

26 51 13.50 Interior Lighting Fixtures		Crew	Daily Output	Labor-Hours	Unit	Material	2007 Bare Costs Labor	Equipment	Total	Total Incl O&P
1700	4'W x 4'L, six 40 watt	2 Elec	6.60	2.424	Ea.	485	106		591	695
1800	4'W x 4'L, eight 40 watt		6.20	2.581		505	113		618	725
1900	2'W x 8'L, four 40 watt		6.40	2.500		163	110		273	345
2000	2'W x 8'L, eight 40 watt		6.20	2.581		187	113		300	375
2010	Acrylic wrap around lens									
2020	6"W x 4'L, one 40 watt	1 Elec	8	1	Ea.	71.50	44		115.50	145
2030	6"W x 8'L, two 40 watt	2 Elec	8	2		78.50	88		166.50	217
2040	11"W x 4'L, two 40 watt	1 Elec	7	1.143		47	50		97	126
2050	11"W x 8'L, four 40 watt	2 Elec	6.60	2.424		78.50	106		184.50	244
2060	16"W x 4'L, four 40 watt	1 Elec	5.30	1.509		78.50	66.50		145	185
2070	16"W x 8'L, eight 40 watt	2 Elec	6.40	2.500		170	110		280	350
2080	2'W x 2'L, two U40 watt	1 Elec	7	1.143		96.50	50		146.50	181
2100	Strip fixture									
2130	Surface mounted									
2200	4' long, one 40 watt RS	1 Elec	8.50	.941	Ea.	29.50	41.50		71	94
2300	4' long, two 40 watt RS		8	1		31.50	44		75.50	101
2400	4' long, one 40 watt, SL		8	1		43	44		87	113
2500	4' long, two 40 watt, SL		7	1.143		59	50		109	139
2600	8' long, one 75 watt, SL	2 Elec	13.40	1.194		44	52.50		96.50	127
2700	8' long, two 75 watt, SL	"	12.40	1.290		53	56.50		109.50	143
2800	4' long, two 60 watt, HO	1 Elec	6.70	1.194		86	52.50		138.50	173
2900	8' long, two 110 watt, HO	2 Elec	10.60	1.509		90.50	66.50		157	199
2910	4' long, two 115 watt, VHO	1 Elec	6.50	1.231		120	54		174	213
2920	8' long, two 215 watt, VHO	2 Elec	10.40	1.538		129	67.50		196.50	243
3000	Pendent mounted, industrial, white porcelain enamel									
3100	4' long, two 40 watt, RS	1 Elec	5.70	1.404	Ea.	48.50	61.50		110	145
3200	4' long, two 60 watt, HO	"	5	1.600		77.50	70		147.50	190
3300	8' long, two 75 watt, SL	2 Elec	8.80	1.818		92	80		172	220
3400	8' long, two 110 watt, HO	"	8	2		118	88		206	260
3410	Acrylic finish, 4' long, two 40 watt, RS	1 Elec	5.70	1.404		80.50	61.50		142	180
3420	4' long, two 60 watt, HO		5	1.600		149	70		219	269
3430	4' long, two 115 watt, VHO		4.80	1.667		193	73		266	320
3440	8' long, two 75 watt, SL	2 Elec	8.80	1.818		158	80		238	293
3450	8' long, two 110 watt, HO		8	2		176	88		264	325
3460	8' long, two 215 watt, VHO		7.60	2.105		249	92.50		341.50	410
3470	Troffer, air handling, 2'W x 4'L with four 40 watt, RS	1 Elec	4	2		88.50	88		176.50	229
3480	2'W x 2'L with two U40 watt RS		5.50	1.455		77	64		141	180
3490	Air connector insulated, 5" diameter		20	.400		60.50	17.55		78.05	92.50
3500	6" diameter		20	.400		61.50	17.55		79.05	93.50
3510	Troffer parabolic lay-in, 1'W x 4'L with one F40		5.70	1.404		86	61.50		147.50	187
3520	1'W x 4'L with two F40		5.30	1.509		90.50	66.50		157	199
3530	2'W x 4'L with three F40		5	1.600		125	70		195	243
3535	Downlight, recess mounted		8	1		94.50	44		138.50	170
3540	Wall washer, recess mounted		8	1		94.50	44		138.50	170
3550	Direct/indirect, 4' long, stl., pendent mtd.		5	1.600		145	70		215	264
3560	4' long, alum., pendent mtd.		5	1.600		305	70		375	440
3565	Prefabricated cove, 4' long, stl. continuous row		5	1.600		185	70		255	310
3570	4' long, alum. continuous row		5	1.600		315	70		385	450
4220	Metal halide, integral ballast, ceiling, recess mounted									
4230	prismatic glass lens, floating door									
4240	2'W x 2'L, 250 watt	1 Elec	3.20	2.500	Ea.	305	110		415	500
4250	2'W x 2'L, 400 watt	2 Elec	5.80	2.759		340	121		461	555
4260	Surface mounted, 2'W x 2'L, 250 watt	1 Elec	2.70	2.963		305	130		435	535

26 51 13.50 Interior Lighting Fixtures		Crew	Daily Output	Labor-Hours	Unit	Material	2007 Bare Costs Labor	Equipment	Total	Total Incl O&P
4270	400 watt	2 Elec	4.80	3.333	Ea.	360	146		506	620
4280	High bay, aluminum reflector,									
4290	Single unit, 400 watt	2 Elec	4.60	3.478	Ea.	370	153		523	630
4300	Single unit, 1000 watt		4	4		530	176		706	840
4310	Twin unit, 400 watt		3.20	5		735	220		955	1,125
4320	Low bay, aluminum reflector, 250W DX lamp	1 Elec	3.20	2.500		355	110		465	555
4330	400 watt lamp	2 Elec	5	3.200		515	140		655	775
4340	High pressure sodium integral ballast ceiling, recess mounted									
4350	prismatic glass lens, floating door									
4360	2'W x 2'L, 150 watt lamp	1 Elec	3.20	2.500	Ea.	365	110		475	565
4370	2'W x 2'L, 400 watt lamp	2 Elec	5.80	2.759		435	121		556	660
4380	Surface mounted, 2'W x 2'L, 150 watt lamp	1 Elec	2.70	2.963		420	130		550	660
4390	400 watt lamp	2 Elec	4.80	3.333		470	146		616	740
4400	High bay, aluminum reflector,									
4410	Single unit, 400 watt lamp	2 Elec	4.60	3.478	Ea.	340	153		493	600
4430	Single unit, 1000 watt lamp	"	4	4		490	176		666	800
4440	Low bay, aluminum reflector, 150 watt lamp	1 Elec	3.20	2.500		293	110		403	490
4445	High bay H.I.D. quartz restrike	"	16	.500		146	22		168	194
4450	Incandescent, high hat can, round alzak reflector, prewired									
4470	100 watt	1 Elec	8	1	Ea.	62.50	44		106.50	135
4480	150 watt		8	1		93	44		137	168
4500	300 watt		6.70	1.194		215	52.50		267.50	315
4520	Round with reflector and baffles, 150 watt		8	1		44.50	44		88.50	115
4540	Round with concentric louver, 150 watt PAR		8	1		70	44		114	143
4600	Square glass lens with metal trim, prewired									
4630	100 watt	1 Elec	6.70	1.194	Ea.	47	52.50		99.50	130
4700	200 watt		6.70	1.194		86.50	52.50		139	173
4800	300 watt		5.70	1.404		125	61.50		186.50	230
4810	500 watt		5	1.600		245	70		315	375
4900	Ceiling/wall, surface mounted, metal cylinder, 75 watt		10	.800		48	35		83	106
4920	150 watt		10	.800		69	35		104	129
4930	300 watt		8	1		150	44		194	231
5000	500 watt		6.70	1.194		330	52.50		382.50	445
5010	Square, 100 watt		8	1		96	44		140	172
5020	150 watt		8	1		105	44		149	182
5030	300 watt		7	1.143		310	50		360	415
5040	500 watt		6	1.333		310	58.50		368.50	425
5200	Ceiling, surface mounted, opal glass drum									
5300	8", one 60 watt lamp	1 Elec	10	.800	Ea.	40	35		75	96.50
5400	10", two 60 watt lamps		8	1		44.50	44		88.50	115
5500	12", four 60 watt lamps		6.70	1.194		60	52.50		112.50	144
5510	Pendent, round, 100 watt		8	1		96	44		140	172
5520	150 watt		8	1		106	44		150	182
5530	300 watt		6.70	1.194		147	52.50		199.50	240
5540	500 watt		5.50	1.455		280	64		344	405
5550	Square, 100 watt		6.70	1.194		131	52.50		183.50	222
5560	150 watt		6.70	1.194		136	52.50		188.50	228
5570	300 watt		5.70	1.404		202	61.50		263.50	315
5580	500 watt		5	1.600		273	70		343	405
5600	Wall, round, 100 watt		8	1		56	44		100	127
5620	300 watt		8	1		98	44		142	174
5630	500 watt		6.70	1.194		350	52.50		402.50	465
5640	Square, 100 watt		8	1		91	44		135	166

26 51 13 - Interior Lighting Fixtures, Lamps, and Ballasts

26 51 13.50 Interior Lighting Fixtures		Crew	Daily Output	Labor-Hours	Unit	Material	2007 Bare Costs Labor	Equipment	Total	Total Incl O&P
5650	150 watt	1 Elec	8	1	Ea.	93.50	44		137.50	169
5660	300 watt		7	1.143		143	50		193	232
5670	500 watt		6	1.333		253	58.50		311.50	365
6010	Vapor tight, incandescent, ceiling mounted, 200 watt		6.20	1.290		58.50	56.50		115	149
6020	Recessed, 200 watt		6.70	1.194		94.50	52.50		147	182
6030	Pendent, 200 watt		6.70	1.194		58.50	52.50		111	143
6040	Wall, 200 watt		8	1		66.50	44		110.50	139
6100	Fluorescent, surface mounted, 2 lamps, 4'L, RS, 40 watt		3.20	2.500		104	110		214	277
6110	Industrial, 2 lamps 4' long in tandem, 430 MA		2.20	3.636		190	160		350	445
6130	2 lamps 4' long, 800 MA		1.90	4.211		162	185		347	455
6160	Pendent, indust, 2 lamps 4'L in tandem, 430 MA		1.90	4.211		222	185		407	520
6170	2 lamps 4' long, 430 MA		2.30	3.478		148	153		301	390
6180	2 lamps 4' long, 800 MA		1.70	4.706		187	207		394	515
6850	Vandalproof, surface mounted, fluorescent, two 40 watt		3.20	2.500		220	110		330	405
6860	Incandescent, one 150 watt		8	1		54.50	44		98.50	126
6900	Mirror light, fluorescent, RS, acrylic enclosure, two 40 watt		8	1		87.50	44		131.50	162
6910	One 40 watt		8	1		69.50	44		113.50	142
6920	One 20 watt		12	.667		55	29.50		84.50	104
7000	Low bay, aluminum reflector. 70 watt, high pressure sodium		4	2		256	88		344	415
7010	250 watt		3.20	2.500		345	110		455	545
7020	400 watt	2 Elec	5	3.200		350	140		490	595
7500	Ballast replacement, by weight of ballast, to 15' high									
7520	Indoor fluorescent, less than 2 lbs.	1 Elec	10	.800	Ea.		35		35	52.50
7540	Two 40W, watt reducer, 2 to 5 lbs.		9.40	.851		30	37.50		67.50	88.50
7560	Two F96 slimline, over 5 lbs.		8	1		56.50	44		100.50	128
7580	Vaportite ballast, less than 2 lbs.		9.40	.851			37.50		37.50	55.50
7600	2 lbs. to 5 lbs.		8.90	.899			39.50		39.50	58.50
7620	Over 5 lbs.		7.60	1.053			46		46	69
7630	Electronic ballast for two tubes		8	1		35	44		79	104
7640	Dimmable ballast one lamp		8	1		62.50	44		106.50	135
7650	Dimmable ballast two-lamp		7.60	1.053		102	46		148	181
7690	Emergency ballast (factory installed in fixture)					137			137	151
7990	Decorator									
8000	Pendent RLM in colors, shallow dome, 12" diam. 100 W	1 Elec	8	1	Ea.	70.50	44		114.50	143
8010	Regular dome, 12" diam., 100 watt		8	1		73	44		117	146
8020	16" diam., 200 watt		7	1.143		74	50		124	156
8030	18" diam., 500 watt		6	1.333		89.50	58.50		148	186
8100	Picture framing light, minimum		16	.500		71.50	22		93.50	111
8110	Maximum		16	.500		94.50	22		116.50	137
8150	Miniature low voltage, recessed, pinhole		8	1		145	44		189	225
8160	Star		8	1		120	44		164	198
8170	Adjustable cone		8	1		162	44		206	244
8180	Eyeball		8	1		132	44		176	211
8190	Cone		8	1		116	44		160	194
8200	Coilex baffle		8	1		114	44		158	191
8210	Surface mounted, adjustable cylinder		8	1		117	44		161	195
8250	Chandeliers, incandescent									
8260	24" diam. x 42" high, 6 light candle	1 Elec	6	1.333	Ea.	350	58.50		408.50	470
8270	24" diam. x 42" high, 6 light candle w/glass shade		6	1.333		230	58.50		288.50	340
8280	17" diam. x 12" high, 8 light w/glass panels		8	1		261	44		305	355
8290	32" diam. x 48"H, 10 light bohemian lead crystal		4	2		580	88		668	770
8300	27" diam. x 29"H, 10 light bohemian lead crystal		4	2		540	88		628	725
8310	21" diam. x 9" high 6 light sculptured ice crystal		8	1		420	44		464	525

26 51 Interior Lighting

26 51 13 – Interior Lighting Fixtures, Lamps, and Ballasts

26 51 13.50 Interior Lighting Fixtures		Crew	Daily Output	Labor-Hours	Unit	Material	2007 Bare Costs Labor	Equipment	Total	Total Incl O&P
8500	Accent lights, on floor or edge, 0.5W low volt incandescent									
8520	incl. transformer & fastenings, based on 100' lengths									
8550	Lights in clear tubing, 12" on center	1 Elec	230	.035	L.F.	8	1.53		9.53	11.05
8560	6" on center		160	.050		10.45	2.20		12.65	14.75
8570	4" on center		130	.062		15.95	2.70		18.65	21.50
8580	3" on center		125	.064		17.70	2.81		20.51	23.50
8590	2" on center		100	.080		26	3.51		29.51	34
8600	Carpet, lights both sides 6" OC, in alum. extrusion		270	.030		24	1.30		25.30	28.50
8610	In bronze extrusion		270	.030		27.50	1.30		28.80	32
8620	Carpet-bare floor, lights 18" OC, in alum. extrusion		270	.030		19.25	1.30		20.55	23
8630	In bronze extrusion		270	.030		22.50	1.30		23.80	27
8640	Carpet edge-wall, lights 6" OC in alum. extrusion		270	.030		24	1.30		25.30	28.50
8650	In bronze extrusion		270	.030		27.50	1.30		28.80	32
8660	Bare floor, lights 18" OC, in aluminum extrusion		300	.027		19.25	1.17		20.42	22.50
8670	In bronze extrusion		300	.027		22.50	1.17		23.67	26.50
8680	Bare floor conduit, aluminum extrusion		300	.027		6.40	1.17		7.57	8.75
8690	In bronze extrusion		300	.027	▼	12.75	1.17		13.92	15.80
8700	Step edge to 36", lights 6" OC, in alum. extrusion		100	.080	Ea.	64.50	3.51		68.01	76
8710	In bronze extrusion		100	.080		67	3.51		70.51	79
8720	Step edge to 54", lights 6" OC, in alum. extrusion		100	.080		96.50	3.51		100.01	111
8730	In bronze extrusion		100	.080		102	3.51		105.51	117
8740	Step edge to 72", lights 6" OC, in alum. extrusion		100	.080		129	3.51		132.51	146
8750	In bronze extrusion		100	.080		140	3.51		143.51	159
8760	Connector, male		32	.250		2.38	11		13.38	18.95
8770	Female with pigtail		32	.250		5	11		16	22
8780	Clamps		400	.020		.48	.88		1.36	1.84
8790	Transformers, 50 watt		8	1		67	44		111	139
8800	250 watt		4	2		222	88		310	375
8810	1000 watt	▼	2.70	2.963	▼	425	130		555	665

26 51 13.70 Residential Fixtures

		Crew	Daily Output	Labor-Hours	Unit	Material	Labor	Equipment	Total	Total Incl O&P
0010	**RESIDENTIAL FIXTURES**									
0400	Fluorescent, interior, surface, circline, 32 watt & 40 watt	1 Elec	20	.400	Ea.	77.50	17.55		95.05	112
0500	2' x 2', two U 40 watt		8	1		103	44		147	180
0700	Shallow under cabinet, two 20 watt		16	.500		44	22		66	81
0900	Wall mounted, 4'L, one 40 watt, with baffle		10	.800		119	35		154	184
2000	Incandescent, exterior lantern, wall mounted, 60 watt		16	.500		31.50	22		53.50	67.50
2100	Post light, 150W, with 7' post		4	2		110	88		198	252
2500	Lamp holder, weatherproof with 150W PAR		16	.500		19.50	22		41.50	54
2550	With reflector and guard		12	.667		54.50	29.50		84	103
2600	Interior pendent, globe with shade, 150 watt	▼	20	.400	▼	128	17.55		145.55	167

26 51 13.90 Ballast, Replacement HID

		Crew	Daily Output	Labor-Hours	Unit	Material	Labor	Equipment	Total	Total Incl O&P
0010	**BALLAST, REPLACEMENT HID**									
7510	Multi-tap 120/208/240/277 volt									
7550	High pressure sodium, 70 watt	1 Elec	10	.800	Ea.	150	35		185	218
7560	100 watt		9.40	.851		157	37.50		194.50	229
7570	150 watt		9	.889		170	39		209	245
7580	250 watt		8.50	.941		253	41.50		294.50	340
7590	400 watt		7	1.143		285	50		335	390
7600	1000 watt		6	1.333		395	58.50		453.50	520
7610	Metal halide, 175 watt		8	1		92.50	44		136.50	168
7620	250 watt		8	1		119	44		163	197
7630	400 watt	▼	7	1.143		149	50		199	239

26 51 Interior Lighting

26 51 13 – Interior Lighting Fixtures, Lamps, and Ballasts

	26 51 13.90 Ballast, Replacement HID	Crew	Daily Output	Labor-Hours	Unit	Material	2007 Bare Costs Labor	Equipment	Total	Total Incl O&P
7640	1000 watt	1 Elec	6	1.333	Ea.	256	58.50		314.50	370
7650	1500 watt	↓	5	1.600	↓	320	70		390	455

26 52 Emergency Lighting

26 52 13 – Emergency Lighting

26 52 13.10 Emergency Lighting and Battery Units

		Crew	Daily Output	Labor-Hours	Unit	Material	2007 Bare Costs Labor	Equipment	Total	Total Incl O&P
0010	**EMERGENCY LIGHTING AND BATTERY UNITS**									
0300	Emergency light units, battery operated									
0350	Twin sealed beam light, 25 watt, 6 volt each									
0500	Lead battery operated	1 Elec	4	2	Ea.	122	88		210	265
0700	Nickel cadmium battery operated		4	2		580	88		668	770
0780	Additional remote mount, sealed beam, 25W 6V		26.70	.300		26.50	13.15		39.65	48.50
0790	Twin sealed beam light, 25W 6V each		26.70	.300		46	13.15		59.15	70
0900	Self-contained fluorescent lamp pack	↓	10	.800	↓	135	35		170	202

26 53 Exit Signs

26 53 13 – Exit Lighting

26 53 13.10 Exit Lighting

		Crew	Daily Output	Labor-Hours	Unit	Material	2007 Bare Costs Labor	Equipment	Total	Total Incl O&P
0010	**EXIT LIGHTING**									
0080	Exit light ceiling or wall mount, incandescent, single face	1 Elec	8	1	Ea.	42	44		86	112
0100	Double face		6.70	1.194		48	52.50		100.50	131
0120	Explosion proof		3.80	2.105		440	92.50		532.50	625
0150	Fluorescent, single face		8	1		65	44		109	137
0160	Double face		6.70	1.194		69	52.50		121.50	154
0200	L.E.D. standard, single face		8	1		65	44		109	137
0220	Double face		6.70	1.194		69	52.50		121.50	154
0240	L.E.D. w/battery unit, single face		4.40	1.818		125	80		205	257
0260	Double face	↓	4	2	↓	127	88		215	271

26 54 Classified Location Lighting

26 54 13 – Classified Lighting

26 54 13.20 Explosionproof

		Crew	Daily Output	Labor-Hours	Unit	Material	2007 Bare Costs Labor	Equipment	Total	Total Incl O&P
0010	**EXPLOSIONPROOF**									
6310	Metal halide with ballast, ceiling, surface mounted, 175 watt	1 Elec	2.90	2.759	Ea.	755	121		876	1,025
6320	250 watt	"	2.70	2.963		910	130		1,040	1,200
6330	400 watt	2 Elec	4.80	3.333		975	146		1,121	1,300
6340	Ceiling, pendent mounted, 175 watt	1 Elec	2.60	3.077		720	135		855	990
6350	250 watt	"	2.40	3.333		875	146		1,021	1,175
6360	400 watt	2 Elec	4.20	3.810		935	167		1,102	1,275
6370	Wall, surface mounted, 175 watt	1 Elec	2.90	2.759		815	121		936	1,075
6380	250 watt	"	2.70	2.963		965	130		1,095	1,275
6390	400 watt	2 Elec	4.80	3.333		1,025	146		1,171	1,350
6400	High pressure sodium, ceiling surface mounted, 70 watt	1 Elec	3	2.667		820	117		937	1,075
6410	100 watt		3	2.667		850	117		967	1,100
6420	150 watt		2.70	2.963		875	130		1,005	1,150
6430	Pendent mounted, 70 watt		2.70	2.963		755	130		885	1,025
6440	100 watt	↓	2.70	2.963	↓	810	130		940	1,075

GET THE JOB DONE IN ONE-THIRD THE TIME WITH WIREMOLD® PREWIRED RACEWAY.

Working with your specs, Wiremold will do all the wiring and pre-assembly work for you. You'll get everything you need, clearly labeled and ready for installation, delivered right to the site. All you have to do is mount the raceway on the wall, connect the feed wires, snap on the pre-cut covers, and you're done — no wiring, no cutting. And Wiremold Prewired Raceway is ideal for all kinds of applications and décor. Take less time to get the job done right and be more productive, with Wiremold Prewired Raceway. For details, contact your Wiremold sales representative. Or visit www.wiremold.com.

WIREMOLD

legrand®

Save Time. Be More PRODUCTIVE.

Slash your raceway installation time with Wiremold® Prewired Raceway. Convenient and organized, this product allows you to finish a nine or more hour raceway project in just three hours. Here you see 20 ft. of V4000 Raceway being installed in an "L" pattern 13 ft. on one wall and 7 ft. on the other, with two circuits and devices 36 in. on center.

FIELD-WIRED RACEWAY

Field wiring this job requires 125 parts and pieces — and lots of material handling.

The electrician has to measure, cut, and install base sections individually.

After inserting dividers, wires are pulled, wire clips added, and receptacles are wired up and installed. After more measuring and cutting, covers are installed.

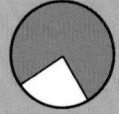 ▪ Total elapsed time 9:43 (8:00 AM to 5:43 PM)
It's a long day!

WIREMOLD PREWIRED RACEWAY

Wiremold prewired raceway arrives labeled so all the components are in the right place. There's less material handling and no going back for missing parts.

The electrician simply installs prewired sections. There's no measuring, no cutting, no wire clips, and no wiring up individual receptacles.

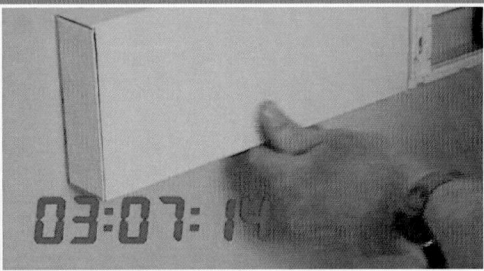

After connecting pigtails, precut covers complete a perfect job — in one-third the time.

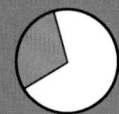 ☐ Total elapsed time 3:30 (8:00 AM to 11:30 AM)
It's not even lunchtime!

Call 800-621-0049 or visit www.wiremold.com

WIREMOLD

▪legrand®

26 54 Classified Location Lighting

26 54 13 – Classified Lighting

26 54 13.20 Explosionproof		Crew	Daily Output	Labor-Hours	Unit	Material	2007 Bare Costs Labor	Equipment	Total	Total Incl O&P
6450	150 watt	1 Elec	2.40	3.333	Ea.	840	146		986	1,150
6460	Wall mounted, 70 watt		3	2.667		875	117		992	1,150
6470	100 watt		3	2.667		905	117		1,022	1,175
6480	150 watt		2.70	2.963		930	130		1,060	1,225
6510	Incandescent, ceiling mounted, 200 watt		4	2		730	88		818	935
6520	Pendent mounted, 200 watt		3.50	2.286		625	100		725	840
6530	Wall mounted, 200 watt		4	2		725	88		813	930
6600	Fluorescent, RS, 4' long, ceiling mounted, two 40 watt		2.70	2.963		1,850	130		1,980	2,225
6610	Three 40 watt		2.20	3.636		2,675	160		2,835	3,175
6620	Four 40 watt		1.90	4.211		3,425	185		3,610	4,050
6630	Pendent mounted, two 40 watt		2.30	3.478		2,175	153		2,328	2,600
6640	Three 40 watt		1.90	4.211		3,075	185		3,260	3,650
6650	Four 40 watt	▼	1.70	4.706	▼	4,050	207		4,257	4,750

26 55 Special Purpose Lighting

26 55 59 – Display Lighting

26 55 59.10 Track Lighting

		Crew	Daily Output	Labor-Hours	Unit	Material	2007 Bare Costs Labor	Equipment	Total	Total Incl O&P
0010	**TRACK LIGHTING**									
0080	Track, 1 circuit, 4' section	1 Elec	6.70	1.194	Ea.	40.50	52.50		93	123
0100	8' section		5.30	1.509		68.50	66.50		135	174
0200	12' section		4.40	1.818		109	80		189	239
0300	3 circuits, 4' section		6.70	1.194		54.50	52.50		107	138
0400	8' section		5.30	1.509		84.50	66.50		151	192
0500	12' section		4.40	1.818		163	80		243	298
1000	Feed kit, surface mounting		16	.500		10.05	22		32.05	43.50
1100	End cover		24	.333		3.50	14.65		18.15	26
1200	Feed kit, stem mounting, 1 circuit		16	.500		29.50	22		51.50	64.50
1300	3 circuit		16	.500		29.50	22		51.50	64.50
2000	Electrical joiner, for continuous runs, 1 circuit		32	.250		14.50	11		25.50	32.50
2100	3 circuit		32	.250		34	11		45	54
2200	Fixtures, spotlight, 75W PAR halogen		16	.500		87	22		109	128
2210	50W MR16 halogen		16	.500		106	22		128	150
3000	Wall washer, 250 watt tungsten halogen		16	.500		101	22		123	144
3100	Low voltage, 25/50 watt, 1 circuit		16	.500		102	22		124	145
3120	3 circuit	▼	16	.500	▼	105	22		127	149

26 55 61 – Theatrical Lighting

26 55 61.10 Lights

		Crew	Daily Output	Labor-Hours	Unit	Material	2007 Bare Costs Labor	Equipment	Total	Total Incl O&P
0010	**LIGHTS**									
2000	Lights, border, quartz, reflector, vented,									
2100	colored or white	1 Elec	20	.400	L.F.	157	17.55		174.55	198
2500	Spotlight, follow spot, with transformer, 2,100 watt	"	4	2	Ea.	1,250	88		1,338	1,500
2600	For no transformer, deduct					825			825	910
3000	Stationary spot, fresnel quartz, 6" lens	1 Elec	4	2		99	88		187	240
3100	8" lens		4	2		198	88		286	350
3500	Ellipsoidal quartz, 1,000W, 6" lens		4	2		299	88		387	460
3600	12" lens		4	2		495	88		583	675
4000	Strobe light, 1 to 15 flashes per second, quartz		3	2.667		680	117		797	925
4500	Color wheel, portable, five hole, motorized	▼	4	2	▼	158	88		246	305

26 56 Exterior Lighting

26 56 13 – Lighting Poles and Standards

26 56 13.10 Lighting Poles		Crew	Daily Output	Labor-Hours	Unit	Material	2007 Bare Costs Labor	Equipment	Total	Total Incl O&P
0010	**LIGHTING POLES**									
2800	Light poles, anchor base									
2820	not including concrete bases									
2840	Aluminum pole, 8' high	1 Elec	4	2	Ea.	610	88		698	800
2850	10' high		4	2		635	88		723	830
2860	12' high		3.80	2.105		665	92.50		757.50	870
2870	14' high		3.40	2.353		690	103		793	915
2880	16' high		3	2.667		760	117		877	1,000
3000	20' high	R-3	2.90	6.897		790	298	56.50	1,144.50	1,375
3200	30' high		2.60	7.692		1,550	335	63	1,948	2,275
3400	35' high		2.30	8.696		1,675	375	71	2,121	2,475
3600	40' high		2	10		1,900	435	81.50	2,416.50	2,825
3800	Bracket arms, 1 arm	1 Elec	8	1		104	44		148	180
4000	2 arms		8	1		209	44		253	296
4200	3 arms		5.30	1.509		315	66.50		381.50	445
4400	4 arms		5.30	1.509		420	66.50		486.50	560
4500	Steel pole, galvanized, 8' high		3.80	2.105		530	92.50		622.50	720
4510	10' high		3.70	2.162		555	95		650	750
4520	12' high		3.40	2.353		600	103		703	815
4530	14' high		3.10	2.581		635	113		748	870
4540	16' high		2.90	2.759		675	121		796	920
4550	18' high		2.70	2.963		710	130		840	980
4600	20' high	R-3	2.60	7.692		940	335	63	1,338	1,600
4800	30' high		2.30	8.696		1,100	375	71	1,546	1,875
5000	35' high		2.20	9.091		1,200	395	74.50	1,669.50	2,000
5200	40' high		1.70	11.765		1,500	510	96	2,106	2,525
5400	Bracket arms, 1 arm	1 Elec	8	1		154	44		198	235
5600	2 arms		8	1		238	44		282	330
5800	3 arms		5.30	1.509		258	66.50		324.50	385
6000	4 arms		5.30	1.509		360	66.50		426.50	495
6100	Fiberglass pole, 1 or 2 fixtures, 20' high	R-3	4	5		485	216	41	742	905
6200	30' high		3.60	5.556		605	240	45.50	890.50	1,075
6300	35' high		3.20	6.250		945	271	51	1,267	1,500
6400	40' high		2.80	7.143		1,100	310	58.50	1,468.50	1,750
6420	Wood pole, 4-1/2" x 5-1/8", 8' high	1 Elec	6	1.333		269	58.50		327.50	385
6430	10' high		6	1.333		305	58.50		363.50	420
6440	12' high		5.70	1.404		385	61.50		446.50	515
6450	15' high		5	1.600		450	70		520	600
6460	20' high		4	2		545	88		633	730
7300	Transformer bases, not including concrete bases									
7320	Maximum pole size, steel, 40' high	1 Elec	2	4	Ea.	1,175	176		1,351	1,525
7340	Cast aluminum, 30' high		3	2.667		630	117		747	865
7350	40' high		2.50	3.200		950	140		1,090	1,250

26 56 19 – Roadway Lighting

26 56 19.20 Roadway Area Luminaire

		Crew	Daily Output	Labor-Hours	Unit	Material	2007 Bare Costs Labor	Equipment	Total	Total Incl O&P
0010	**ROADWAY AREA LUMINAIRE**									
2650	Roadway area luminaire, low pressure sodium, 135 watt	1 Elec	2	4	Ea.	550	176		726	865
2700	180 watt	"	2	4		580	176		756	900
2750	Metal halide, 400 watt	2 Elec	4.40	3.636		460	160		620	745
2760	1000 watt		4	4		520	176		696	830
2780	High pressure sodium, 400 watt		4.40	3.636		475	160		635	765
2790	1000 watt		4	4		545	176		721	860

26 56 Exterior Lighting

26 56 23 – Area Lighting

26 56 23.10 Exterior Fixtures

		Crew	Daily Output	Labor-Hours	Unit	Material	2007 Bare Costs Labor	Equipment	Total	Total Incl O&P
0010	**EXTERIOR FIXTURES** With lamps									
0200	Wall mounted, incandescent, 100 watt	1 Elec	8	1	Ea.	29	44		73	97
0400	Quartz, 500 watt		5.30	1.509		54	66.50		120.50	158
0420	1500 watt		4.20	1.905		107	83.50		190.50	242
1100	Wall pack, low pressure sodium, 35 watt		4	2		220	88		308	375
1150	55 watt		4	2		263	88		351	420
1160	High pressure sodium, 70 watt		4	2		206	88		294	360
1170	150 watt		4	2		237	88		325	390
1180	Metal Halide, 175 watt		4	2		255	88		343	410
1190	250 watt		4	2		273	88		361	430

26 56 26 – Landscape Lighting

26 56 26.20 Landscape Fixtures

		Crew	Daily Output	Labor-Hours	Unit	Material	2007 Bare Costs Labor	Equipment	Total	Total Incl O&P
0010	**LANDSCAPE FIXTURES**									
7380	Landscape recessed uplight, incl. housing, ballast, transformer									
7390	& reflector, not incl. conduit, wire, trench									
7420	Incandescent, 250 watt	1 Elec	5	1.600	Ea.	490	70		560	645
7440	Quartz, 250 watt		5	1.600		465	70		535	620
7460	500 watt		4	2		480	88		568	660

26 56 33 – Walkway Lighting

26 56 33.10 Walkway Luminaire

		Crew	Daily Output	Labor-Hours	Unit	Material	2007 Bare Costs Labor	Equipment	Total	Total Incl O&P
0010	**WALKWAY LUMINAIRE**									
6500	Bollard light, lamp & ballast, 42" high with polycarbonate lens									
6800	Metal halide, 175 watt	1 Elec	3	2.667	Ea.	670	117		787	915
6900	High pressure sodium, 70 watt		3	2.667		690	117		807	935
7000	100 watt		3	2.667		690	117		807	935
7100	150 watt		3	2.667		670	117		787	915
7200	Incandescent, 150 watt		3	2.667		485	117		602	710
7810	Walkway luminaire, square 16", metal halide 250 watt		2.70	2.963		540	130		670	790
7820	High pressure sodium, 70 watt		3	2.667		620	117		737	860
7830	100 watt		3	2.667		635	117		752	870
7840	150 watt		3	2.667		635	117		752	870
7850	200 watt		3	2.667		635	117		752	875
7910	Round 19", metal halide, 250 watt		2.70	2.963		790	130		920	1,075
7920	High pressure sodium, 70 watt		3	2.667		870	117		987	1,125
7930	100 watt		3	2.667		870	117		987	1,125
7940	150 watt		3	2.667		875	117		992	1,150
7950	250 watt		2.70	2.963		915	130		1,045	1,200
8000	Sphere 14" opal, incandescent, 200 watt		4	2		255	88		343	410
8020	Sphere 18" opal, incandescent, 300 watt		3.50	2.286		310	100		410	490
8040	Sphere 16" clear, high pressure sodium, 70 watt		3	2.667		535	117		652	765
8050	100 watt		3	2.667		570	117		687	800
8100	Cube 16" opal, incandescent, 300 watt		3.50	2.286		340	100		440	525
8120	High pressure sodium, 70 watt		3	2.667		500	117		617	725
8130	100 watt		3	2.667		510	117		627	735
8230	Lantern, high pressure sodium, 70 watt		3	2.667		440	117		557	660
8240	100 watt		3	2.667		475	117		592	695
8250	150 watt		3	2.667		445	117		562	665
8260	250 watt		2.70	2.963		620	130		750	880
8270	Incandescent, 300 watt		3.50	2.286		330	100		430	510
8330	Reflector 22" w/globe, high pressure sodium, 70 watt		3	2.667		415	117		532	630
8340	100 watt		3	2.667		420	117		537	640

26 56 Exterior Lighting

26 56 33 – Walkway Lighting

26 56 33.10 Walkway Luminaire

		Crew	Daily Output	Labor-Hours	Unit	Material	2007 Bare Costs Labor	Equipment	Total	Total Incl O&P
8350	150 watt	1 Elec	3	2.667	Ea.	425	117		542	645
8360	250 watt	↓	2.70	2.963	↓	545	130		675	795

26 56 36 – Flood Lighting

26 56 36.20 Floodlights

		Crew	Daily Output	Labor-Hours	Unit	Material	2007 Bare Costs Labor	Equipment	Total	Total Incl O&P
0010	**FLOODLIGHTS** with ballast and lamp,									
1400	pole mounted, pole not included									
1950	Metal halide, 175 watt	1 Elec	2.70	2.963	Ea.	300	130		430	525
2000	400 watt	2 Elec	4.40	3.636		380	160		540	655
2200	1000 watt		4	4		555	176		731	870
2210	1500 watt	↓	3.70	4.324		580	190		770	925
2250	Low pressure sodium, 55 watt	1 Elec	2.70	2.963		500	130		630	745
2270	90 watt		2	4		550	176		726	865
2290	180 watt		2	4		700	176		876	1,025
2340	High pressure sodium, 70 watt		2.70	2.963		216	130		346	430
2360	100 watt		2.70	2.963		222	130		352	440
2380	150 watt	↓	2.70	2.963		260	130		390	480
2400	400 watt	2 Elec	4.40	3.636		365	160		525	640
2600	1000 watt	"	4	4		625	176		801	950
2610	Incandescent, 300 watt	1 Elec	4	2		88.50	88		176.50	229
2620	500 watt	"	4	2		142	88		230	287
2630	1000 watt	2 Elec	6	2.667		152	117		269	340
2640	1500 watt	"	6	2.667	↓	167	117		284	360

26 61 Lighting Systems and Accessories

26 61 13 – Lighting Accessories

26 61 13.10 Energy Saving Lighting Devices

		Crew	Daily Output	Labor-Hours	Unit	Material	2007 Bare Costs Labor	Equipment	Total	Total Incl O&P
0010	**ENERGY SAVING LIGHTING DEVICES**									
0100	Occupancy sensors infrared, ceiling mounted	1 Elec	7	1.143	Ea.	98	50		148	183
0150	Automatic wall switches		24	.333		60	14.65		74.65	88
0200	Remote power pack		10	.800		27	35		62	82
0250	Photoelectric control, S.P.S.T. 120 V		8	1		12.70	44		56.70	79.50
0300	S.P.S.T. 208 V/277 V		8	1		16.10	44		60.10	83
0350	D.P.S.T. 120 V		6	1.333		128	58.50		186.50	228
0400	D.P.S.T. 208 V/277 V		6	1.333		133	58.50		191.50	233
0450	S.P.D.T. 208 V/277 V	↓	6	1.333	↓	158	58.50		216.50	261

26 61 13.30 Fixture Whips

		Crew	Daily Output	Labor-Hours	Unit	Material	2007 Bare Costs Labor	Equipment	Total	Total Incl O&P
0010	**FIXTURE WHIPS**									
0080	3/8" Greenfield, 2 connectors, 6' long									
0100	TFFN wire, three #18	1 Elec	32	.250	Ea.	7.70	11		18.70	25
0150	Four #18		28	.286		7.65	12.55		20.20	27
0200	Three #16		32	.250		7.35	11		18.35	24.50
0250	Four #16		28	.286		8	12.55		20.55	27.50
0300	THHN wire, three #14		32	.250		10	11		21	27.50
0350	Four #14		28	.286		10.80	12.55		23.35	30.50
0360	Three #12	↓	32	.250	↓	12.85	11		23.85	30.50

26 61 23 – Lamps

26 61 23.10 Lamps

		Crew	Daily Output	Labor-Hours	Unit	Material	2007 Bare Costs Labor	Equipment	Total	Total Incl O&P
0010	**LAMPS** R265723-05									
0080	Fluorescent, rapid start, cool white, 2' long, 20 watt	1 Elec	1	8	C	310	350		660	865
0100	4' long, 40 watt	↓	.90	8.889		345	390		735	960

26 61 23 – Lamps

26 61 23.10 Lamps		Crew	Daily Output	Labor-Hours	Unit	Material	2007 Bare Costs Labor	Equipment	Total	Total Incl O&P
0120	3' long, 30 watt	1 Elec	.90	8.889	C	390	390		780	1,000
0125	3' long, 25 watt energy saver R265723-10		.90	8.889		595	390		985	1,225
0150	U-40 watt		.80	10		990	440		1,430	1,750
0155	U-34 watt energy saver		.80	10		990	440		1,430	1,750
0170	4' long, 34 watt energy saver		.90	8.889		305	390		695	920
0176	2' long, T8, 17 W engergy saver R265723-20		1	8		440	350		790	1,000
0178	3' long, T8, 25 W energy saver		.90	8.889		450	390		840	1,075
0180	4' long, T8, 32 watt energy saver		.90	8.889		287	390		677	895
0200	Slimline, 4' long, 40 watt		.90	8.889		735	390		1,125	1,375
0210	4' long, 30 watt energy saver R265723-25		.90	8.889		735	390		1,125	1,375
0300	8' long, 75 watt		.80	10		805	440		1,245	1,550
0350	8' long, 60 watt energy saver		.80	10		510	440		950	1,225
0400	High output, 4' long, 60 watt		.90	8.889		890	390		1,280	1,550
0410	8' long, 95 watt energy saver		.80	10		885	440		1,325	1,625
0500	8' long, 110 watt		.80	10		885	440		1,325	1,625
0512	2' long, T5, 14 watt energy saver		1	8		1,225	350		1,575	1,875
0514	3' long, T5, 21 watt energy saver		.90	8.889		1,225	390		1,615	1,925
0516	4' long, T5, 28 watt energy saver		.90	8.889		1,050	390		1,440	1,725
0520	Very high output, 4' long, 110 watt		.90	8.889		2,350	390		2,740	3,150
0525	8' long, 195 watt energy saver		.70	11.429		2,450	500		2,950	3,450
0550	8' long, 215 watt		.70	11.429		2,400	500		2,900	3,375
0554	Full spectrum, 4' long, 60 watt		.90	8.889		1,750	390		2,140	2,500
0556	6' long, 85 watt		.90	8.889		1,900	390		2,290	2,650
0558	8' long, 110 watt		.80	10		1,625	440		2,065	2,425
0560	Twin tube compact lamp		.90	8.889		555	390		945	1,200
0570	Double twin tube compact lamp		.80	10		1,350	440		1,790	2,125
0600	Mercury vapor, mogul base, deluxe white, 100 watt		.30	26.667		2,825	1,175		4,000	4,850
0650	175 watt		.30	26.667		2,100	1,175		3,275	4,050
0700	250 watt		.30	26.667		3,725	1,175		4,900	5,850
0800	400 watt		.30	26.667		3,000	1,175		4,175	5,050
0900	1000 watt		.20	40		7,000	1,750		8,750	10,300
1000	Metal halide, mogul base, 175 watt		.30	26.667		3,525	1,175		4,700	5,625
1100	250 watt		.30	26.667		4,000	1,175		5,175	6,150
1200	400 watt		.30	26.667		3,800	1,175		4,975	5,925
1300	1000 watt		.20	40		9,150	1,750		10,900	12,700
1320	1000 watt, 125,000 initial lumens		.20	40		14,300	1,750		16,050	18,300
1330	1500 watt		.20	40		13,400	1,750		15,150	17,300
1350	High pressure sodium, 70 watt		.30	26.667		4,100	1,175		5,275	6,275
1360	100 watt		.30	26.667		4,300	1,175		5,475	6,475
1370	150 watt		.30	26.667		4,400	1,175		5,575	6,600
1380	250 watt		.30	26.667		4,675	1,175		5,850	6,900
1400	400 watt		.30	26.667		4,800	1,175		5,975	7,025
1450	1000 watt		.20	40		13,300	1,750		15,050	17,200
1500	Low pressure sodium, 35 watt		.30	26.667		7,175	1,175		8,350	9,625
1550	55 watt		.30	26.667		7,875	1,175		9,050	10,400
1600	90 watt		.30	26.667		9,125	1,175		10,300	11,800
1650	135 watt		.20	40		11,700	1,750		13,450	15,400
1700	180 watt		.20	40		12,800	1,750		14,550	16,700
1750	Quartz line, clear, 500 watt		1.10	7.273		970	320		1,290	1,550
1760	1500 watt		.20	40		3,775	1,750		5,525	6,800
1762	Spot, MR 16, 50 watt		1.30	6.154		815	270		1,085	1,300
1770	Tungsten halogen, T4, 400 watt		1.10	7.273		3,150	320		3,470	3,925
1775	T3, 1200 watt		.30	26.667		3,775	1,175		4,950	5,925

26 61 Lighting Systems and Accessories

26 61 23 – Lamps

26 61 23.10 Lamps	Crew	Daily Output	Labor-Hours	Unit	Material	2007 Bare Costs Labor	Equipment	Total	Total Incl O&P
1778	PAR 30, 50 watt	1 Elec	1.30	6.154	C	875	270		1,145	1,350
1780	PAR 38, 90 watt		1.30	6.154		780	270		1,050	1,250
1800	Incandescent, interior, A21, 100 watt		1.60	5		146	220		366	485
1900	A21, 150 watt		1.60	5		149	220		369	490
2000	A23, 200 watt		1.60	5		209	220		429	555
2200	PS 30, 300 watt		1.60	5		505	220		725	880
2210	PS 35, 500 watt		1.60	5		860	220		1,080	1,275
2230	PS 52, 1000 watt		1.30	6.154		1,825	270		2,095	2,400
2240	PS 52, 1500 watt		1.30	6.154		4,900	270		5,170	5,800
2300	R30, 75 watt		1.30	6.154		540	270		810	990
2400	R40, 100 watt		1.30	6.154		605	270		875	1,075
2500	Exterior, PAR 38, 75 watt		1.30	6.154		1,125	270		1,395	1,650
2600	PAR 38, 150 watt		1.30	6.154		1,250	270		1,520	1,775
2700	PAR 46, 200 watt		1.10	7.273		2,500	320		2,820	3,225
2800	PAR 56, 300 watt		1.10	7.273		3,100	320		3,420	3,900
3000	Guards, fluorescent lamp, 4' long		1	8		960	350		1,310	1,575
3200	8' long	▼	.90	8.889	▼	1,925	390		2,315	2,675

26 71 Motors

26 71 13 – Motors

26 71 13.10 Handling

26 71 13.10 Handling	Crew	Daily Output	Labor-Hours	Unit	Material	Labor	Equipment	Total	Total Incl O&P
0010	**HANDLING** Add to normal labor cost for restricted areas									
5000	Motors									
5100	1/2 HP, 23 pounds	1 Elec	4	2	Ea.		88		88	131
5110	3/4 HP, 28 pounds		4	2			88		88	131
5120	1 HP, 33 pounds		4	2			88		88	131
5130	1-1/2 HP, 44 pounds		3.20	2.500			110		110	163
5140	2 HP, 56 pounds		3	2.667			117		117	174
5150	3 HP, 71 pounds		2.30	3.478			153		153	227
5160	5 HP, 82 pounds		1.90	4.211			185		185	275
5170	7-1/2 HP, 124 pounds		1.50	5.333			234		234	350
5180	10 HP, 144 pounds		1.20	6.667			293		293	435
5190	15 HP, 185 pounds	▼	1	8			350		350	525
5200	20 HP, 214 pounds	2 Elec	1.50	10.667			470		470	695
5210	25 HP, 266 pounds		1.40	11.429			500		500	745
5220	30 HP, 310 pounds		1.20	13.333			585		585	870
5230	40 HP, 400 pounds		1	16			700		700	1,050
5240	50 HP, 450 pounds		.90	17.778			780		780	1,150
5250	75 HP, 680 pounds	▼	.80	20			880		880	1,300
5260	100 HP, 870 pounds	3 Elec	1	24			1,050		1,050	1,575
5270	125 HP, 940 pounds		.80	30			1,325		1,325	1,950
5280	150 HP, 1200 pounds		.70	34.286			1,500		1,500	2,250
5290	175 HP, 1300 pounds		.60	40			1,750		1,750	2,625
5300	200 HP, 1400 pounds	▼	.50	48	▼		2,100		2,100	3,125

26 71 13.20 Motors

26 71 13.20 Motors	Crew	Daily Output	Labor-Hours	Unit	Material	Labor	Equipment	Total	Total Incl O&P
0010	**MOTORS** 230/460 volts, 60 HZ R263413-30									
0050	Dripproof, premium efficiency, 1.15 service factor									
0060	1800 RPM, 1/4 HP	1 Elec	5.33	1.501	Ea.	150	66		216	263
0070	1/3 HP		5.33	1.501		164	66		230	278
0080	1/2 HP		5.33	1.501		193	66		259	310
0090	3/4 HP	▼	5.33	1.501	▼	215	66		281	335

26 71 Motors

26 71 13 – Motors

26 71 13.20 Motors		Crew	Daily Output	Labor-Hours	Unit	Material	2007 Bare Costs Labor	Equipment	Total	Total Incl O&P
0100	1 HP	1 Elec	4.50	1.778	Ea.	226	78		304	365
0150	2 HP		4.50	1.778		239	78		317	380
0200	3 HP		4.50	1.778		241	78		319	380
0250	5 HP		4.50	1.778		345	78		423	495
0300	7.5 HP		4.20	1.905		470	83.50		553.50	645
0350	10 HP		4	2		590	88		678	775
0400	15 HP		3.20	2.500		730	110		840	965
0450	20 HP	2 Elec	5.20	3.077		980	135		1,115	1,275
0500	25 HP		5	3.200		1,050	140		1,190	1,350
0550	30 HP		4.80	3.333		1,325	146		1,471	1,675
0600	40 HP		4	4		1,725	176		1,901	2,150
0650	50 HP		3.20	5		1,925	220		2,145	2,450
0700	60 HP		2.80	5.714		2,600	251		2,851	3,225
0750	75 HP		2.40	6.667		2,825	293		3,118	3,525
0800	100 HP	3 Elec	2.70	8.889		3,750	390		4,140	4,700
0850	125 HP		2.10	11.429		4,350	500		4,850	5,525
0900	150 HP		1.80	13.333		6,025	585		6,610	7,500
0950	200 HP		1.50	16		7,700	700		8,400	9,525
1000	1200 RPM, 1 HP	1 Elec	4.50	1.778		244	78		322	385
1050	2 HP		4.50	1.778		335	78		413	485
1100	3 HP		4.50	1.778		475	78		553	635
1150	5 HP		4.50	1.778		535	78		613	705
1200	3600 RPM, 2 HP		4.50	1.778		284	78		362	425
1250	3 HP		4.50	1.778		310	78		388	460
1300	5 HP		4.50	1.778		335	78		413	485
1350	Totally enclosed, premium efficiency 1.15 service factor									
1360	1800 RPM, 1/4 HP	1 Elec	5.33	1.501	Ea.	152	66		218	265
1370	1/3 HP		5.33	1.501		169	66		235	284
1380	1/2 HP		5.33	1.501		200	66		266	320
1390	3/4 HP		5.33	1.501		215	66		281	335
1400	1 HP		4.50	1.778		246	78		324	385
1450	2 HP		4.50	1.778		315	78		393	460
1500	3 HP		4.50	1.778		380	78		458	535
1550	5 HP		4.50	1.778		430	78		508	590
1600	7.5 HP		4.20	1.905		620	83.50		703.50	810
1650	10 HP		4	2		750	88		838	955
1700	15 HP		3.20	2.500		1,050	110		1,160	1,325
1750	20 HP	2 Elec	5.20	3.077		1,225	135		1,360	1,550
1800	25 HP		5	3.200		1,425	140		1,565	1,750
1850	30 HP		4.80	3.333		1,675	146		1,821	2,075
1900	40 HP		4	4		2,150	176		2,326	2,600
1950	50 HP		3.20	5		2,650	220		2,870	3,225
2000	60 HP		2.80	5.714		3,925	251		4,176	4,700
2050	75 HP		2.40	6.667		5,050	293		5,343	6,000
2100	100 HP	3 Elec	2.70	8.889		7,425	390		7,815	8,725
2150	125 HP		2.10	11.429		9,750	500		10,250	11,400
2200	150 HP		1.80	13.333		11,300	585		11,885	13,400
2250	200 HP		1.50	16		13,800	700		14,500	16,300
2300	1200 RPM, 1 HP	1 Elec	4.50	1.778		240	78		318	380
2350	2 HP		4.50	1.778		275	78		353	420
2400	3 HP		4.50	1.778		375	78		453	525
2450	5 HP		4.50	1.778		520	78		598	685
2500	3600 RPM, 2 HP		4.50	1.778		275	78		353	420

26 71 13.20 Motors		Crew	Daily Output	Labor-Hours	Unit	Material	2007 Bare Costs Labor	Equipment	Total	Total Incl O&P
2550	3 HP	1 Elec	4.50	1.778	Ea.	292	78		370	435
2600	5 HP	↓	4.50	1.778	↓	365	78		443	520

Estimating Tips

27 20 00 Data Communications

27 30 00 Voice Communications

27 40 00 Audio-Video Communications

- When estimating material costs for special systems, it is always prudent to obtain manufacturers' quotations for equipment prices and special installation requirements which will affect the total costs.

Reference Numbers

Reference numbers are shown in shaded boxes at the beginning of some major classifications. These numbers refer to related items in the Reference Section. The reference information may be an estimating procedure, an alternate pricing method, or technical information.

Note: Not all subdivisions listed here necessarily appear in this publication.

Division 27 - Communications

27 01 Operation and Maintenance of Communications Systems

27 01 30 – Operation and Maintenance of Voice Communications

27 01 30.51 Operation and Mainten. of Voice Communic.	Crew	Daily Output	Labor-Hours	Unit	Material	2007 Bare Costs Labor	Equipment	Total	Total Incl O&P
0010	**OPERATION AND MAINTENANCE OF VOICE COMMUNICATIONS**								
3400	Remove and replace (reinstall), speaker	1 Elec	6	1.333	Ea.		58.50		58.50	87

27 05 Common Work Results for Communications

27 05 05 – Selective Communications Demolition

27 05 05.20 Electrical Demolition, Communications

		Crew	Daily Output	Labor-Hours	Unit	Material	2007 Bare Costs Labor	Equipment	Total	Total Incl O&P
0010	**ELECTRICAL DEMOLITION, COMMUNICATIONS** R024119-10									
0100	Fiber optics									
0120	Fiber optic cable R260105-30	1 Elec	2400	.003	L.F.		.15		.15	.22
0160	Multi-channel rack enclosure		6	1.333	Ea.		58.50		58.50	87
0180	Fiber optic patch panel	↓	18	.444	"		19.50		19.50	29
0200	Communication cables & fittings									
0220	Voice/data outlet	1 Elec	140	.057	Ea.		2.51		2.51	3.73
0240	Telephone cable		2800	.003	L.F.		.13		.13	.19
0260	Phone jack		135	.059	Ea.		2.60		2.60	3.87
0300	High performance cable, 2 pair		3000	.003	L.F.		.12		.12	.17
0320	4 pair		2100	.004			.17		.17	.25
0340	25 pair	↓	900	.009	↓		.39		.39	.58
1000	Nurse call system									
1020	Nurse call station	1 Elec	24	.333	Ea.		14.65		14.65	22
1040	Nurse call standard call button		24	.333			14.65		14.65	22
1060	Corridor, dome light or zone indictor	↓	24	.333			14.65		14.65	22
1080	Master control station	2 Elec	2	8	↓		350		350	525

27 05 05.30 Electrical Demolition, Sound and Video

		Crew	Daily Output	Labor-Hours	Unit	Material	2007 Bare Costs Labor	Equipment	Total	Total Incl O&P
0010	**ELECTRICAL DEMOLITION, SOUND & VIDEO** R024119-10									
0100	Sound & video cables									
0120	TV antenna lead-in cable R260105-30	1 Elec	2100	.004	L.F.		.17		.17	.25
0140	Sound cable		2400	.003			.15		.15	.22
0160	Microphone cable		2400	.003			.15		.15	.22
0180	Coaxial cable		2400	.003	↓		.15		.15	.22
0200	Doorbell system		16	.500	Ea.		22		22	32.50
0220	Door chime or devices	↓	36	.222	"		9.75		9.75	14.50
0300	Public address system, not including rough-in									
0320	Conventional office	1 Elec	16	.500	Speaker		22		22	32.50
0340	Conventional industrial	"	8	1	"		44		44	65.50
0400	Sound system, not including rough-in									
0410	Sound components	1 Elec	24	.333	Ea.		14.65		14.65	22
0420	Intercom, master station		6	1.333			58.50		58.50	87
0440	Remote station		24	.333			14.65		14.65	22
0460	Intercom outlets		24	.333			14.65		14.65	22
0480	Handset	↓	12	.667	↓		29.50		29.50	43.50
0500	Emergency call system, not including rough-in									
0520	Annunciator	1 Elec	4	2	Ea.		88		88	131
0540	Devices	"	16	.500			22		22	32.50
0600	Master door, buzzer type unit	2 Elec	1.60	10	↓		440		440	655
0800	TV System, not including rough-in									
0820	Master TV antenna system, per outlet	1 Elec	39	.205	Outlet		9		9	13.40
0840	School & deluxe, per outlet		16	.500	"		22		22	32.50
0860	Amplifier		12	.667	Ea.		29.50		29.50	43.50
0880	Antenna	↓	6	1.333	"		58.50		58.50	87
0900	One camera & one monitor	2 Elec	7.80	2.051	Total		90		90	134

27 05 Common Work Results for Communications

27 05 05 – Selective Communications Demolition

27 05 05.30 Electrical Demolition, Sound and Video

		Crew	Daily Output	Labor-Hours	Unit	Material	2007 Bare Costs Labor	Equipment	Total	Total Incl O&P
0920	One camera	1 Elec	8	1	Ea.		44		44	65.50

27 11 Communications Equipment Room Fittings

27 11 19 – Communications Termination Blocks and Patch Panels

27 11 19.10 Termination Blocks and Patch Panels

		Crew	Daily Output	Labor-Hours	Unit	Material	2007 Bare Costs Labor	Equipment	Total	Total Incl O&P
0010	**TERMINATION BLOCKS AND PATCH PANELS**									
2960	Patch panel, 24 ports, RJ-45/110 type	2 Elec	6	2.667	Ea.	159	117		276	350
3000	48 ports, RJ-45/110 type	3 Elec	6	4		305	176		481	595
3040	96 ports, RJ-45/110 type	"	4	6		505	263		768	945

27 13 Communications Backbone Cabling

27 13 23 – Communications Optical Fiber Backbone Cabling

27 13 23.13 Communications Optical Fiber

			Crew	Daily Output	Labor-Hours	Unit	Material	2007 Bare Costs Labor	Equipment	Total	Total Incl O&P
0010	**COMMUNICATIONS OPTICAL FIBER**										
0020	Fiber optics cable only. Added costs depend on the type of fiber										
0030	special connectors, optical modems, and networking parts.										
0040	Specialized tools & techniques cause installation costs to vary.										
0070	Cable, minimum, bulk simplex	R271323-40	1 Elec	8	1	C.L.F.	23	44		67	91
0080	Cable, maximum, bulk plenum quad		"	2.29	3.493	"	66	153		219	300
0150	Fiber optic jumper					Ea.	55.50			55.50	61
0200	Fiber optic pigtail						30			30	33
0300	Fiber optic connector		1 Elec	24	.333		18.45	14.65		33.10	42.50
0350	Fiber optic finger splice			32	.250		32	11		43	51.50
0400	Transceiver (low cost bi-directional)			8	1		295	44		339	390
0450	Rack housing, 4 rack spaces (12 panels)			2	4		450	176		626	755
0500	Fiber optic patch panel (12 ports)			6	1.333		178	58.50		236.50	283

27 15 Communications Horizontal Cabling

27 15 10 – Special Communications Cabling

27 15 10.23 Sound and Video Cables and Fittings

		Crew	Daily Output	Labor-Hours	Unit	Material	2007 Bare Costs Labor	Equipment	Total	Total Incl O&P
0010	**SOUND AND VIDEO CABLES & FITTINGS**									
0900	TV antenna lead-in, 300 ohm, #20-2 conductor	1 Elec	7	1.143	C.L.F.	26	50		76	103
0950	Coaxial, feeder outlet		7	1.143		32	50		82	110
1000	Coaxial, main riser		6	1.333		45	58.50		103.50	137
1100	Sound, shielded with drain, #22-2 conductor		8	1		22	44		66	89.50
1150	#22-3 conductor		7.50	1.067		29.50	47		76.50	102
1200	#22-4 conductor		6.50	1.231		35	54		89	119
1250	Nonshielded, #22-2 conductor		10	.800		13.15	35		48.15	67
1300	#22-3 conductor		9	.889		18.80	39		57.80	78.50
1350	#22-4 conductor		8	1		24	44		68	92
1400	Microphone cable		8	1		76.50	44		120.50	150

27 15 13 – Communications Copper Horizontal Cabling

27 15 13.13 Communication Cables

		Crew	Daily Output	Labor-Hours	Unit	Material	2007 Bare Costs Labor	Equipment	Total	Total Incl O&P
0010	**COMMUNICATION CABLES**									
2200	Telephone twisted, PVC insulation, #22-2 conductor	1 Elec	10	.800	C.L.F.	8.70	35		43.70	62
2250	#22-3 conductor		9	.889		11.40	39		50.40	70.50
2300	#22-4 conductor		8	1		13.60	44		57.60	80.50

27 15 13.13 Communication Cables		Crew	Daily Output	Labor-Hours	Unit	Material	2007 Bare Costs Labor	Equipment	Total	Total Incl O&P
2350	#18-2 conductor	1 Elec	9	.889	C.L.F.	10.60	39		49.60	69.50
2370	Telephone jack, eight pins	↓	32	.250	Ea.	7.50	11		18.50	24.50
5000	High performance unshielded twisted pair (UTP)									
5100	Category 3, #24, 2 pair solid, PVC jacket R271513-75	1 Elec	10	.800	C.L.F.	6.60	35		41.60	60
5200	4 pair solid		7	1.143		10.50	50		60.50	86
5300	25 pair solid		3	2.667		59	117		176	239
5400	2 pair solid, plenum		10	.800		10.20	35		45.20	63.50
5500	4 pair solid		7	1.143		13.80	50		63.80	89.50
5600	25 pair solid		3	2.667		91	117		208	274
5700	4 pair stranded, PVC jacket		7	1.143		22	50		72	98.50
7000	Category 5, #24, 4 pair solid, PVC jacket		7	1.143		13.40	50		63.40	89.50
7100	4 pair solid, plenum		7	1.143		41.50	50		91.50	120
7200	4 pair stranded, PVC jacket		7	1.143		20	50		70	96.50
7210	Category 5e, #24, 4 pair solid, PVC jacket		7	1.143		11.90	50		61.90	87.50
7212	4 pair solid, plenum		7	1.143		33	50		83	111
7214	4 pair stranded, PVC jacket		7	1.143		24	50		74	101
7240	Category 6, #24, 4 pair solid, PVC jacket		7	1.143		20.50	50		70.50	97
7242	4 pair solid, plenum		7	1.143		74	50		124	156
7244	4 pair stranded, PVC jacket		7	1.143	↓	22	50		72	98.50
7300	Category 5, connector, UTP RJ-45		80	.100	Ea.	1.13	4.39		5.52	7.80
7302	shielded RJ-45		72	.111		3.24	4.88		8.12	10.80
7310	Category 3, jack, UTP RJ-45		72	.111		3.29	4.88		8.17	10.85
7312	Category 5		65	.123		4.27	5.40		9.67	12.75
7314	Category 5e		65	.123		4.27	5.40		9.67	12.75
7316	Category 6		65	.123		4.27	5.40		9.67	12.75
7322	Category 5, jack, shielded RJ-45		60	.133		5.35	5.85		11.20	14.60
7324	Category 5e		60	.133		5.35	5.85		11.20	14.60
7326	Category 6		60	.133		5.35	5.85		11.20	14.60
7400	Category 5e, voice/data expansion module	↓	8	1	↓	37	44		81	106

27 15 33.10 Coaxial Cable and Fittings		Crew	Daily Output	Labor-Hours	Unit	Material	2007 Bare Costs Labor	Equipment	Total	Total Incl O&P
0010	**COAXIAL CABLE & FITTINGS**									
3500	Coaxial connectors, 50 ohm impedance quick disconnect									
3540	BNC plug, for RG A/U #58 cable	1 Elec	42	.190	Ea.	4.58	8.35		12.93	17.50
3550	RG A/U #59 cable		42	.190		4.58	8.35		12.93	17.50
3560	RG A/U #62 cable		42	.190		4.58	8.35		12.93	17.50
3600	BNC jack, for RG A/U #58 cable		42	.190		4.73	8.35		13.08	17.65
3610	RG A/U #59 cable		42	.190		4.73	8.35		13.08	17.65
3620	RG A/U #62 cable		42	.190		4.73	8.35		13.08	17.65
3660	BNC panel jack, for RG A/U #58 cable		40	.200		7.35	8.80		16.15	21
3670	RG A/U #59 cable		40	.200		7.35	8.80		16.15	21
3680	RG A/U #62 cable		40	.200		7.35	8.80		16.15	21
3720	BNC bulkhead jack, for RG A/U #58 cable		40	.200		8.55	8.80		17.35	22.50
3730	RG A/U #59 cable		40	.200		8.55	8.80		17.35	22.50
3740	RG A/U #62 cable		40	.200	↓	8.55	8.80		17.35	22.50
3850	Coaxial cable, RG A/U 58, 50 ohm		8	1	C.L.F.	31.50	44		75.50	100
3860	RG A/U 59, 75 ohm		8	1		29.50	44		73.50	97.50
3870	RG A/U 62, 93 ohm		8	1		31.50	44		75.50	100
3875	RG 6/U, 75 ohm		8	1		26.50	44		70.50	95
3950	RG A/U 58, 50 ohm fire rated		8	1		68.50	44		112.50	141
3960	RG A/U 59, 75 ohm fire rated		8	1		119	44		163	197
3970	RG A/U 62, 93 ohm fire rated	↓	8	1	↓	97	44		141	172

27 15 Communications Horizontal Cabling

27 15 43 – Communications Faceplates and Connectors

27 15 43.13 Communication Outlets

27 15 43.13 Communication Outlets	Crew	Daily Output	Labor-Hours	Unit	Material	2007 Bare Costs Labor	Equipment	Total	Total Incl O&P
0010 **COMMUNICATION OUTLETS**									
0100 Communication outlets, voice/data devises not included									
0120 Voice/Data outlets, single opening	1 Elec	48	.167	Ea.	6.75	7.30		14.05	18.35
0140 Two jack openings		48	.167		2.20	7.30		9.50	13.30
0160 One jack & one 3/4" round opening		48	.167		6.75	7.30		14.05	18.35
0180 One jack & one twinaxial opening		48	.167		6.75	7.30		14.05	18.35
0200 One jack & one connector cabling opening		48	.167		6.75	7.30		14.05	18.35
0220 Two 3/8" coaxial openings		48	.167		6.75	7.30		14.05	18.35
0300 Data outlets, single opening		48	.167		6.75	7.30		14.05	18.35
0320 One 25-pin subminiature opening		48	.167		6.75	7.30		14.05	18.35

27 21 Data Communications Network Equipment

27 21 23 – Data Communications Switches and Hubs

27 21 23.10 Switching and Routing Equipment

	Crew	Daily Output	Labor-Hours	Unit	Material	2007 Bare Costs Labor	Equipment	Total	Total Incl O&P
0010 **SWITCHING AND ROUTING EQUIPMENT**									
1100 Network hub, dual speed, 24 ports, includes cabinet	3 Elec	.66	36.364	Ea.	1,525	1,600		3,125	4,050

27 32 Voice Communications Telephone Sets, Faxes and Modems

27 32 36 – TTY Equipment

27 32 36.10 TTY Telephone Equipment

	Crew	Daily Output	Labor-Hours	Unit	Material	2007 Bare Costs Labor	Equipment	Total	Total Incl O&P
0010 **TTY TELEPHONE EQUIPMENT**									
1620 Telephone, TTY, compact, pocket type				Ea.	289			289	320
1630 Advanced, desk type	2 Elec	20	.800		650	35		685	770
1640 Full-featured public, wall type	"	4	4		995	176		1,171	1,350

27 41 Audio-Video Systems

27 41 19 – Portable Audio-Video Equipment

27 41 19.10 T.V. Systems

	Crew	Daily Output	Labor-Hours	Unit	Material	2007 Bare Costs Labor	Equipment	Total	Total Incl O&P
0010 **T.V. SYSTEMS** not including rough-in wires, cables & conduits									
0100 Master TV antenna system									
0200 VHF reception & distribution, 12 outlets	1 Elec	6	1.333	Outlet	183	58.50		241.50	288
0400 30 outlets		10	.800		120	35		155	185
0600 100 outlets		13	.615		121	27		148	173
0800 VHF & UHF reception & distribution, 12 outlets		6	1.333		182	58.50		240.50	287
1000 30 outlets		10	.800		120	35		155	185
1200 100 outlets		13	.615		123	27		150	175
1400 School and deluxe systems, 12 outlets		2.40	3.333		240	146		386	480
1600 30 outlets		4	2		210	88		298	360
1800 80 outlets		5.30	1.509		202	66.50		268.50	320
1900 Amplifier		4	2	Ea.	600	88		688	790
1910 Antenna		2	4	"	289	176		465	580

27 51 Distributed Audio-Video Communications Systems

27 51 16 – Public Address and Mass Notification Systems

27 51 16.10 Public Address System	Crew	Daily Output	Labor-Hours	Unit	Material	2007 Bare Costs Labor	Equipment	Total	Total Incl O&P
0010	**PUBLIC ADDRESS SYSTEM**									
0100	Conventional, office	1 Elec	5.33	1.501	Speaker	113	66		179	222
0200	Industrial	"	2.70	2.963	"	218	130		348	435
0400	Explosionproof system is 3 times cost of central control									
0600	Installation costs run about 120% of material cost									

27 51 19 – Sound Masking Systems

27 51 19.10 Sound System

27 51 19.10 Sound System	Crew	Daily Output	Labor-Hours	Unit	Material	2007 Bare Costs Labor	Equipment	Total	Total Incl O&P
0010	**SOUND SYSTEM** not including rough-in wires, cables & conduits									
0100	Components, outlet, projector	1 Elec	8	1	Ea.	52	44		96	123
0200	Microphone		4	2		58	88		146	195
0400	Speakers, ceiling or wall		8	1		98.50	44		142.50	174
0600	Trumpets		4	2		183	88		271	335
0800	Privacy switch		8	1		73	44		117	146
1000	Monitor panel		4	2		325	88		413	490
1200	Antenna, AM/FM		4	2		182	88		270	330
1400	Volume control		8	1		73	44		117	146
1600	Amplifier, 250 watts		1	8		1,450	350		1,800	2,125
1800	Cabinets		1	8		710	350		1,060	1,300
2000	Intercom, 25 station capacity, master station	2 Elec	2	8		1,700	350		2,050	2,400
2020	11 station capacity	"	4	4		800	176		976	1,150
2200	Remote station	1 Elec	8	1		137	44		181	217
2400	Intercom outlets		8	1		80.50	44		124.50	154
2600	Handset		4	2		266	88		354	425
2800	Emergency call system, 12 zones, annunciator		1.30	6.154		800	270		1,070	1,275
3000	Bell		5.30	1.509		83	66.50		149.50	190
3200	Light or relay		8	1		41.50	44		85.50	111
3400	Transformer		4	2		182	88		270	330
3600	House telephone, talking station		1.60	5		390	220		610	755
3800	Press to talk, release to listen		5.30	1.509		91	66.50		157.50	199
4000	System-on button					54.50			54.50	60
4200	Door release	1 Elec	4	2		97.50	88		185.50	238
4400	Combination speaker and microphone		8	1		166	44		210	248
4600	Termination box		3.20	2.500		52	110		162	221
4800	Amplifier or power supply		5.30	1.509		600	66.50		666.50	760
5000	Vestibule door unit		16	.500	Name	110	22		132	154
5200	Strip cabinet		27	.296	Ea.	208	13		221	248
5400	Directory		16	.500		98	22		120	141
6000	Master door, button buzzer type, 100 unit	2 Elec	.54	29.630		1,000	1,300		2,300	3,025
6020	200 unit		.30	53.333		1,875	2,350		4,225	5,550
6040	300 unit		.20	80		2,875	3,500		6,375	8,375
6060	Transformer	1 Elec	8	1		25.50	44		69.50	93.50
6080	Door opener		5.30	1.509		36.50	66.50		103	139
6100	Buzzer with door release and plate		4	2		36.50	88		124.50	171
6200	Intercom type, 100 unit	2 Elec	.54	29.630		1,250	1,300		2,550	3,300
6220	200 unit		.30	53.333		2,425	2,350		4,775	6,150
6240	300 unit		.20	80		3,650	3,500		7,150	9,250
6260	Amplifier	1 Elec	2	4		182	176		358	460
6280	Speaker with door release	"	4	2		54.50	88		142.50	191

27 52 Healthcare Communications and Monitoring Systems

27 52 23 – Nurse Call/Code Blue Systems

27 52 23.10 Nurse Call Systems

27 52 23.10 Nurse Call Systems	Crew	Daily Output	Labor-Hours	Unit	Material	2007 Bare Costs Labor	Equipment	Total	Total Incl O&P
0010 **NURSE CALL SYSTEMS**									
0100 Single bedside call station	1 Elec	8	1	Ea.	202	44		246	288
0200 Ceiling speaker station		8	1		59	44		103	130
0400 Emergency call station		8	1		99.50	44		143.50	175
0600 Pillow speaker		8	1		186	44		230	270
0800 Double bedside call station		4	2		191	88		279	340
1000 Duty station		4	2		154	88		242	300
1200 Standard call button		8	1		80.50	44		124.50	154
1400 Lights, corridor, dome or zone indicator		8	1		45	44		89	115
1600 Master control station for 20 stations	2 Elec	.65	24.615	Total	3,575	1,075		4,650	5,550

27 53 Distributed Systems

27 53 13 – Clock Systems

27 53 13.50 Clock Systems

27 53 13.50 Clock Systems	Crew	Daily Output	Labor-Hours	Unit	Material	2007 Bare Costs Labor	Equipment	Total	Total Incl O&P
0010 **CLOCK SYSTEMS**, not including wires & conduits									
0100 Time system components, master controller	1 Elec	.33	24.242	Ea.	1,625	1,075		2,700	3,375
0200 Program bell		8	1		53.50	44		97.50	125
0400 Combination clock & speaker		3.20	2.500		189	110		299	370
0600 Frequency generator		2	4		6,350	176		6,526	7,225
0800 Job time automatic stamp recorder, minimum		4	2		445	88		533	620
1000 Maximum		4	2		680	88		768	880
1200 Time stamp for correspondence, hand operated					340			340	375
1400 Fully automatic					495			495	545
1600 Master time clock system, clocks & bells, 20 room	4 Elec	.20	160		4,075	7,025		11,100	15,000
1800 50 room	"	.08	400		9,400	17,600		27,000	36,400
2000 Time clock, 100 cards in & out, 1 color	1 Elec	3.20	2.500		950	110		1,060	1,225
2200 2 colors		3.20	2.500		1,025	110		1,135	1,300
2400 With 3 circuit program device, minimum		2	4		375	176		551	675
2600 Maximum		2	4		545	176		721	860
2800 Metal rack for 25 cards		7	1.143		60	50		110	141
3000 Watchman's tour station		8	1		63	44		107	135
3200 Annunciator with zone indication		1	8		230	350		580	780
3400 Time clock with tape		1	8		610	350		960	1,200

Division Notes

	CREW	DAILY OUTPUT	LABOR-HOURS	UNIT	2007 BARE COSTS				TOTAL INCL O&P
					MAT.	LABOR	EQUIP.	TOTAL	

Estimating Tips

- When estimating material costs for electronic safety and security systems, it is always prudent to obtain manufacturers' quotations for equipment prices and special installation requirements, which affect the total cost.

- Fire alarm systems consist of control panels, annunciator panels, battery with rack, charger, and fire alarm actuating and indicating devices. Some fire alarm systems include speakers, telephone lines, door closer controls, and other components. Be careful not to overlook the costs related to installation for these items.

Also be aware of costs for integrated automation instrumentation and terminal devices, control equipment, control wiring, and programming.

- Security equipment includes items such as CCTV, access control, and other detection and identification systems to perform alert and alarm functions. Be sure to consider the costs related to installation for this security equipment, such as for integrated automation instrumentation and terminal devices, control equipment, control wiring, and programming.

Reference Numbers

Reference numbers are shown in shaded boxes at the beginning of some major classifications. These numbers refer to related items in the Reference Section. The reference information may be an estimating procedure, an alternate pricing method, or technical information.

Note: Not all subdivisions listed here necessarily appear in this publication.

28 01 Operation and Maint. of Electronic Safety and Security

28 01 30 – Operation and Maint. of Electronic Detection and Alarm

28 01 30.51 Maint. and Admin. of Elec. Detection and Alarm	Crew	Daily Output	Labor-Hours	Unit	Material	2007 Bare Costs Labor	2007 Bare Costs Equipment	Total	Total Incl O&P
0010 **MAINT. AND ADMIN. OF ELEC. DETECTION AND ALARM**									
3300 Remove and replace (reinstall), fire alarm device	1 Elec	5.33	1.501	Ea.		66		66	98

28 13 Access Control

28 13 53 – Security Access Detection

28 13 53.13 Security Access Metal Detectors

	Crew	Daily Output	Labor-Hours	Unit	Material	Labor	Equipment	Total	Total Incl O&P
0010 **SECURITY ACCESS METAL DETECTORS**									
0240 Metal detector, hand-held, wand type, unit only				Ea.				81.82	90
0250 Metal detector, walk through portal type, single zone	1 Elec	2	4		4,000	176		4,176	4,650
0260 Multi zone	"	2	4		5,000	176		5,176	5,750

28 13 53.16 Security Access X-Ray Equipment

	Crew	Daily Output	Labor-Hours	Unit	Material	Labor	Equipment	Total	Total Incl O&P
0010 **SECURITY ACCESS X-RAY EQUIPMENT**									
0290 X-ray machine, desk top, for mail/small packages/letters	1 Elec	4	2	Ea.	3,000	88		3,088	3,425
0300 Conveyor type, incl monitor, minimum		2	4		14,000	176		14,176	15,700
0310 Maximum		2	4		25,000	176		25,176	27,800
0320 X-ray machine, large unit, for airports, incl monitor, min	2 Elec	1	16		35,000	700		35,700	39,600
0330 Maximum	"	.50	32		60,000	1,400		61,400	68,000

28 13 53.23 Security Access Explosive Detection Equipment

	Crew	Daily Output	Labor-Hours	Unit	Material	Labor	Equipment	Total	Total Incl O&P
0010 **SECURITY ACCESS EXPLOSIVE DETECTION EQUIPMENT**									
0270 Explosives detector, walk through portal type	1 Elec	2	4	Ea.	3,500	176		3,676	4,100
0280 Hand-held, battery operated				"				25,455	28,000

28 16 Intrusion Detection

28 16 16 – Intrusion Detection Systems Infrastructure

28 16 16.50 Intrusion Detection

	Crew	Daily Output	Labor-Hours	Unit	Material	Labor	Equipment	Total	Total Incl O&P
0010 **INTRUSION DETECTION**, not including wires & conduits									
0100 Burglar alarm, battery operated, mechanical trigger	1 Elec	4	2	Ea.	254	88		342	410
0200 Electrical trigger		4	2		305	88		393	465
0400 For outside key control, add		8	1		72	44		116	145
0600 For remote signaling circuitry, add		8	1		114	44		158	192
0800 Card reader, flush type, standard		2.70	2.963		850	130		980	1,125
1000 Multi-code		2.70	2.963		1,100	130		1,230	1,400
1010 Card reader, proximity type		2.70	2.963		360	130		490	590
1200 Door switches, hinge switch		5.30	1.509		53.50	66.50		120	158
1400 Magnetic switch		5.30	1.509		63	66.50		129.50	168
1600 Exit control locks, horn alarm		4	2		315	88		403	480
1800 Flashing light alarm		4	2		355	88		443	525
2000 Indicating panels, 1 channel		2.70	2.963		335	130		465	565
2200 10 channel	2 Elec	3.20	5		1,150	220		1,370	1,575
2400 20 channel		2	8		2,250	350		2,600	3,000
2600 40 channel		1.14	14.035		4,075	615		4,690	5,425
2800 Ultrasonic motion detector, 12 volt	1 Elec	2.30	3.478		210	153		363	460
3000 Infrared photoelectric detector		2.30	3.478		173	153		326	420
3200 Passive infrared detector		2.30	3.478		259	153		412	510
3400 Glass break alarm switch		8	1		43.50	44		87.50	113
3420 Switchmats, 30" x 5'		5.30	1.509		77.50	66.50		144	184
3440 30" x 25'		4	2		186	88		274	335
3460 Police connect panel		4	2		223	88		311	375
3480 Telephone dialer		5.30	1.509		350	66.50		416.50	485

28 16 Intrusion Detection

28 16 16 – Intrusion Detection Systems Infrastructure

28 16 16.50 Intrusion Detection		Crew	Daily Output	Labor- Hours	Unit	Material	2007 Bare Costs Labor	Equipment	Total	Total Incl O&P
3500	Alarm bell	1 Elec	4	2	Ea.	71	88		159	209
3520	Siren		4	2		134	88		222	278
3540	Microwave detector, 10' to 200'		2	4		610	176		786	935
3560	10' to 350'	↓	2	4	↓	1,775	176		1,951	2,225

28 23 Video Surveillance

28 23 13 – Video Surveillance Control and Management Systems

28 23 13.10 Closed Circuit Television System

		Crew	Daily Output	Labor- Hours	Unit	Material	2007 Bare Costs Labor	Equipment	Total	Total Incl O&P
0010	**CLOSED CIRCUIT TELEVISION SYSTEM**									
2000	Closed circuit, surveillance, one station (camera & monitor)	2 Elec	2.60	6.154	Total	1,150	270		1,420	1,675
2200	For additional camera stations, add	1 Elec	2.70	2.963	Ea.	650	130		780	910
2400	Industrial quality, one station (camera & monitor)	2 Elec	2.60	6.154	Total	2,400	270		2,670	3,050
2600	For additional camera stations, add	1 Elec	2.70	2.963	Ea.	1,475	130		1,605	1,825
2610	For low light, add		2.70	2.963		1,175	130		1,305	1,500
2620	For very low light, add		2.70	2.963		8,725	130		8,855	9,800
2800	For weatherproof camera station, add		1.30	6.154		910	270		1,180	1,400
3000	For pan and tilt, add		1.30	6.154		2,350	270		2,620	2,975
3200	For zoom lens - remote control, add, minimum		2	4		2,175	176		2,351	2,625
3400	Maximum		2	4		7,925	176		8,101	8,950
3410	For automatic iris for low light, add	↓	2	4	↓	1,900	176		2,076	2,325
3600	Educational T.V. studio, basic 3 camera system, black & white,									
3800	electrical & electronic equip. only, minimum	4 Elec	.80	40	Total	11,300	1,750		13,050	15,000
4000	Maximum (full console)		.28	114		47,900	5,025		52,925	60,000
4100	As above, but color system, minimum		.28	114		63,500	5,025		68,525	77,500
4120	Maximum	↓	.12	266	↓	275,500	11,700		287,200	320,500
4200	For film chain, black & white, add	1 Elec	1	8	Ea.	12,900	350		13,250	14,700
4250	Color, add		.25	32		15,600	1,400		17,000	19,300
4400	For video tape recorders, add, minimum	↓	1	8		2,700	350		3,050	3,500
4600	Maximum	4 Elec	.40	80	↓	22,500	3,500		26,000	30,000

28 23 23 – Video Surveillance Systems Infrastructure

28 23 23.50 Video Surveillance

		Crew	Daily Output	Labor- Hours	Unit	Material	2007 Bare Costs Labor	Equipment	Total	Total Incl O&P
0010	**VIDEO SURVEILLANCE**									
0200	Video cameras, wireless, hidden in exit signs, clocks, etc, incl receiver	1 Elec	3	2.667	Ea.	300	117		417	505
0210	Accessories for, VCR, single camera		3	2.667		500	117		617	725
0220	For multiple cameras		3	2.667		1,500	117		1,617	1,825
0230	Video cameras, wireless, for under vehicle searching, complete	↓	2	4	↓	3,500	176		3,676	4,100

28 31 Fire Detection and Alarm

28 31 23 – Fire Detection and Alarm Annunciation Panels and Fire Stations

28 31 23.50 Alarm Panels and Devices

		Crew	Daily Output	Labor- Hours	Unit	Material	2007 Bare Costs Labor	Equipment	Total	Total Incl O&P
0010	**ALARM PANELS AND DEVICES**									
3594	Fire, alarm control panel									
3600	4 zone	2 Elec	2	8	Ea.	940	350		1,290	1,550
3800	8 zone		1	16		1,400	700		2,100	2,600
4000	12 zone	↓	.67	23.988		1,825	1,050		2,875	3,575
4020	Alarm device	1 Elec	8	1		122	44		166	200
4050	Actuating device	"	8	1		292	44		336	385
4160	Alarm control panel, addressable w/o voice, up to 200 points	2 Elec	1.14	13.998		2,125	615		2,740	3,275
4170	addressable w/ voice, up to 400 points	"	.73	22.008	↓	8,125	965		9,090	10,400

28 31 Fire Detection and Alarm

28 31 23 – Fire Detection and Alarm Annunciation Panels and Fire Stations

28 31 23.50 Alarm Panels and Devices	Crew	Daily Output	Labor-Hours	Unit	Material	2007 Bare Costs Labor	Equipment	Total	Total Incl O&P
4175	Addressable interface device	1 Elec	7.25	1.103	Ea.	205	48.50		253.50	298
4200	Battery and rack		4	2		725	88		813	930
4400	Automatic charger		8	1		445	44		489	555
4600	Signal bell		8	1		52	44		96	123
4800	Trouble buzzer or manual station		8	1		37	44		81	106
5600	Strobe and horn		5.30	1.509		95	66.50		161.50	204
5610	Strobe and horn (ADA type)		5.30	1.509		95	66.50		161.50	204
5620	Visual alarm (ADA type)		6.70	1.194		47.50	52.50		100	131
5800	Fire alarm horn		6.70	1.194		36.50	52.50		89	118
6000	Door holder, electro-magnetic		4	2		77.50	88		165.50	217
6200	Combination holder and closer		3.20	2.500		430	110		540	640
6400	Code transmitter		4	2		690	88		778	890
6600	Drill switch		8	1		86.50	44		130.50	161
6800	Master box		2.70	2.963		3,100	130		3,230	3,600
7000	Break glass station		8	1		50	44		94	121
7010	Break glass station, addressable		7.25	1.103		68	48.50		116.50	147
7800	Remote annunciator, 8 zone lamp		1.80	4.444		201	195		396	510
8000	12 zone lamp	2 Elec	2.60	6.154		345	270		615	780
8200	16 zone lamp	"	2.20	7.273		345	320		665	855

28 31 43 – Fire Detection Sensors

28 31 43.50 Fire and Heat Detectors

0010	FIRE & HEAT DETECTORS									
5000	Detector, rate of rise	1 Elec	8	1	Ea.	35	44		79	104
5010	Detector, heat (addressable type)		7.25	1.103		150	48.50		198.50	237
5100	Fixed temperature		8	1		29.50	44		73.50	98

28 31 46 – Smoke Detection Sensors

28 31 46.50 Smoke Detectors

0010	SMOKE DETECTORS									
5200	Smoke detector, ceiling type	1 Elec	6.20	1.290	Ea.	79	56.50		135.50	171
5240	Smoke detector (addressable type)		6	1.333		150	58.50		208.50	252
5400	Duct type		3.20	2.500		250	110		360	440
5420	Duct (addressable type)		3.20	2.500		285	110		395	480

Estimating Tips

31 05 00 Common Work Results for Earthwork

- Estimating the actual cost of performing earthwork requires careful consideration of the variables involved. This includes items such as type of soil, whether water will be encountered, dewatering, whether banks need bracing, disposal of excavated earth, and length of haul to fill or spoil sites, etc. If the project has large quantities of cut or fill, consider raising or lowering the site to reduce costs, while paying close attention to the effect on site drainage and utilities.

- If the project has large quantities of fill, creating a borrow pit on the site can significantly lower the costs.

- It is very important to consider what time of year the project is scheduled for completion. Bad weather can create large cost overruns from dewatering, site repair, and lost productivity from cold weather.

Reference Numbers

Reference numbers are shown in shaded boxes at the beginning of some major classifications. These numbers refer to related items in the Reference Section. The reference information may be an estimating procedure, an alternate pricing method, or technical information.

Note: Not all subdivisions listed here necessarily appear in this publication.

31 23 Excavation and Fill

31 23 16 – Excavation

31 23 16.14 Excavating, Utility Trench

		Crew	Daily Output	Labor-Hours	Unit	Material	2007 Bare Costs Labor	2007 Bare Costs Equipment	Total	Total Incl O&P
0010	**EXCAVATING, UTILITY TRENCH** Common earth									
0050	Trenching with chain trencher, 12 H.P., operator walking									
0100	4" wide trench, 12" deep	B-53	800	.010	L.F.		.37	.06	.43	.63
0150	18" deep		750	.011			.39	.07	.46	.66
0200	24" deep		700	.011			.42	.07	.49	.71
0300	6" wide trench, 12" deep		650	.012			.45	.08	.53	.77
0350	18" deep		600	.013			.49	.09	.58	.83
0400	24" deep		550	.015			.54	.09	.63	.91
0450	36" deep		450	.018			.66	.11	.77	1.11
0600	8" wide trench, 12" deep		475	.017			.62	.11	.73	1.06
0650	18" deep		400	.020			.74	.13	.87	1.25
0700	24" deep		350	.023			.84	.15	.99	1.43
0750	36" deep	▼	300	.027	▼		.98	.17	1.15	1.67
1000	Backfill by hand including compaction, add									
1050	4" wide trench, 12" deep	A-1G	800	.010	L.F.		.29	.05	.34	.50
1100	18" deep		530	.015			.43	.07	.50	.76
1150	24" deep		400	.020			.58	.10	.68	1.01
1300	6" wide trench, 12" deep		540	.015			.43	.07	.50	.74
1350	18" deep		405	.020			.57	.09	.66	.98
1400	24" deep		270	.030			.85	.14	.99	1.49
1450	36" deep		180	.044			1.28	.21	1.49	2.22
1600	8" wide trench, 12" deep		400	.020			.58	.10	.68	1.01
1650	18" deep		265	.030			.87	.14	1.01	1.51
1700	24" deep		200	.040			1.15	.19	1.34	2
1750	36" deep	▼	135	.059	▼		1.70	.28	1.98	2.96
2000	Chain trencher, 40 H.P. operator riding									
2050	6" wide trench and backfill, 12" deep	B-54	1200	.007	L.F.		.25	.19	.44	.58
2100	18" deep		1000	.008			.29	.23	.52	.69
2150	24" deep		975	.008			.30	.23	.53	.72
2200	36" deep		900	.009			.33	.25	.58	.77
2250	48" deep		750	.011			.39	.30	.69	.92
2300	60" deep		650	.012			.45	.35	.80	1.06
2400	8" wide trench and backfill, 12" deep		1000	.008			.29	.23	.52	.69
2450	18" deep		950	.008			.31	.24	.55	.73
2500	24" deep		900	.009			.33	.25	.58	.77
2550	36" deep		800	.010			.37	.28	.65	.87
2600	48" deep		650	.012			.45	.35	.80	1.06
2700	12" wide trench and backfill, 12" deep		975	.008			.30	.23	.53	.72
2750	18" deep		860	.009			.34	.26	.60	.81
2800	24" deep		800	.010			.37	.28	.65	.87
2850	36" deep		725	.011			.41	.31	.72	.95
3000	16" wide trench and backfill, 12" deep		835	.010			.35	.27	.62	.83
3050	18" deep		750	.011			.39	.30	.69	.92
3100	24" deep	▼	700	.011	▼		.42	.32	.74	.99
3200	Compaction with vibratory plate, add								50%	50%
5100	Hand excavate and trim for pipe bells after trench excavation									
5200	8" pipe	1 Clab	155	.052	L.F.		1.48		1.48	2.31
5300	18" pipe	"	130	.062	"		1.77		1.77	2.75

31 23 16.16 Structural Excavation for Minor Structures

		Crew	Daily Output	Labor-Hours	Unit	Material	2007 Bare Costs Labor	2007 Bare Costs Equipment	Total	Total Incl O&P
0010	**STRUCTURAL EXCAVATION FOR MINOR STRUCTURES**									
0015	Hand, pits to 6' deep, sandy soil	1 Clab	8	1	B.C.Y.		29		29	45
0100	Heavy soil or clay	↓	4	2	↓		57.50		57.50	89.50

31 23 Excavation and Fill

31 23 16 – Excavation

31 23 16.16 Structural Excavation for Minor Structures

		Crew	Daily Output	Labor-Hours	Unit	Material	2007 Bare Costs Labor	2007 Bare Costs Equipment	Total	Total Incl O&P
0300	Pits 6' to 12' deep, sandy soil	1 Clab	5	1.600	B.C.Y.		46		46	71.50
0500	Heavy soil or clay		3	2.667			76.50		76.50	119
0700	Pits 12' to 18' deep, sandy soil		4	2			57.50		57.50	89.50
0900	Heavy soil or clay		2	4			115		115	179
6030	Common earth, hydraulic backhoe, 1/2 C.Y. bucket	B-12E	55	.291			9.95	6.50	16.45	22.50
6035	3/4 C.Y. bucket	B-12F	90	.178			6.10	5.75	11.85	15.60
6040	1 C.Y. bucket	B-12A	108	.148			5.10	5.55	10.65	13.90
6050	1-1/2 C.Y. bucket	B-12B	144	.111			3.81	5.40	9.21	11.70
6060	2 C.Y. bucket	B-12C	200	.080			2.74	4.97	7.71	9.65
6070	Sand and gravel, 3/4 C.Y. bucket	B-12F	100	.160			5.50	5.15	10.65	14.10
6080	1 C.Y. bucket	B-12A	120	.133			4.57	5	9.57	12.50
6090	1-1/2 C.Y. bucket	B-12B	160	.100			3.43	4.85	8.28	10.60
6100	2 C.Y. bucket	B-12C	220	.073			2.49	4.52	7.01	8.80
6110	Clay, till, or blasted rock, 3/4 C.Y. bucket	B-12F	80	.200			6.85	6.45	13.30	17.60
6120	1 C.Y. bucket	B-12A	95	.168			5.75	6.35	12.10	15.75
6130	1-1/2 C.Y. bucket	B-12B	130	.123			4.22	5.95	10.17	13
6140	2 C.Y. bucket	B-12C	175	.091			3.13	5.70	8.83	11.05

31 23 19 – Dewatering

31 23 19.20 Dewatering

		Crew	Daily Output	Labor-Hours	Unit	Material	2007 Bare Costs Labor	2007 Bare Costs Equipment	Total	Total Incl O&P
0010	**DEWATERING**									
0020	Excavate drainage trench, 2' wide, 2' deep	B-11C	90	.178	C.Y.		5.95	2.70	8.65	12.05
0100	2' wide, 3' deep, with backhoe loader	"	135	.119			3.98	1.80	5.78	8.10
0200	Excavate sump pits by hand, light soil	1 Clab	7.10	1.127			32.50		32.50	50.50
0300	Heavy soil	"	3.50	2.286			65.50		65.50	102
0500	Pumping 8 hr., attended 2 hrs. per day, including 20 L.F.									
0550	of suction hose & 100 L.F. discharge hose									
0600	2" diaphragm pump used for 8 hours	B-10H	4	3	Day		106	15.30	121.30	177
0650	4" diaphragm pump used for 8 hours	B-10I	4	3			106	21.50	127.50	184
0800	8 hrs. attended, 2" diaphragm pump	B-10H	1	12			420	61	481	710
0900	3" centrifugal pump	B-10J	1	12			420	70	490	715
1000	4" diaphragm pump	B-10I	1	12			420	86	506	735
1100	6" centrifugal pump	B-10K	1	12			420	300	720	975

31 23 23 – Fill

31 23 23.13 Backfill

		Crew	Daily Output	Labor-Hours	Unit	Material	2007 Bare Costs Labor	2007 Bare Costs Equipment	Total	Total Incl O&P
0010	**BACKFILL**									
0015	By hand, no compaction, light soil	1 Clab	14	.571	L.C.Y.		16.45		16.45	25.50
0100	Heavy soil		11	.727	"		21		21	32.50
0300	Compaction in 6" layers, hand tamp, add to above		20.60	.388	E.C.Y.		11.15		11.15	17.40
0400	Roller compaction operator walking, add	B-10A	100	.120			4.22	1.34	5.56	7.85
0500	Air tamp, add	B-9D	190	.211			6.15	1.13	7.28	10.80
0600	Vibrating plate, add	A-1D	60	.133			3.83	.49	4.32	6.50
0800	Compaction in 12" layers, hand tamp, add to above	1 Clab	34	.235			6.75		6.75	10.55
0900	Roller compaction operator walking, add	B-10A	150	.080			2.81	.89	3.70	5.25
1000	Air tamp, add	B-9	285	.140			4.09	.63	4.72	7.05
1100	Vibrating plate, add	A-1E	90	.089			2.56	.40	2.96	4.42
1300	Dozer backfilling, bulk, up to 300' haul, no compaction	B-10B	1200	.010	L.C.Y.		.35	.82	1.17	1.44
1400	Air tamped, add	B-11B	80	.200	E.C.Y.		6.55	2.92	9.47	13.25
1600	Compacting backfill, 6" to 12" lifts, vibrating roller	B-10C	800	.015			.53	1.67	2.20	2.64
1700	Sheepsfoot roller	B-10D	750	.016			.56	1.82	2.38	2.86
1900	Dozer backfilling, trench, up to 300' haul, no compaction	B-10B	900	.013	L.C.Y.		.47	1.10	1.57	1.92
2000	Air tamped, add	B-11B	80	.200	E.C.Y.		6.55	2.92	9.47	13.25
2200	Compacting backfill, 6" to 12" lifts, vibrating roller	B-10C	700	.017			.60	1.91	2.51	3.02

31 23 Excavation and Fill

31 23 23 – Fill

31 23 23.13 **Backfill**	Crew	Daily Output	Labor-Hours	Unit	Material	2007 Bare Costs Labor	Equipment	Total	Total Incl O&P
2300 Sheepsfoot roller	B-10D	650	.018	E.C.Y.		.65	2.10	2.75	3.30

31 23 23.16 Fill By Borrow and Utility Bedding

	Crew	Daily Output	Labor-Hours	Unit	Material	Labor	Equipment	Total	Total Incl O&P
0010 **FILL BY BORROW AND UTILITY BEDDING**									
0049 Utility bedding, for pipe & conduit, not incl. compaction									
0050 Crushed or screened bank run gravel	B-6	150	.160	L.C.Y.	24	5.05	1.62	30.67	36
0100 Crushed stone 3/4" to 1/2"		150	.160		32	5.05	1.62	38.67	44.50
0200 Sand, dead or bank	↓	150	.160	↓	6.25	5.05	1.62	12.92	16.45
0500 Compacting bedding in trench	A-1D	90	.089	E.C.Y.		2.56	.32	2.88	4.34
0610 See 31 23 23.18 for hauling kilometer add.									

31 23 23.18 Hauling

	Crew	Daily Output	Labor-Hours	Unit	Material	Labor	Equipment	Total	Total Incl O&P
0010 **HAULING**, excavated or borrow, loose cubic yards									
0012 no loading included, highway haulers									
0020 6 C.Y. dump truck, 1/4 mile round trip, 5.0 loads/hr.	B-34A	195	.041	L.C.Y.		1.21	1.89	3.10	3.95
0030 1/2 mile round trip, 4.1 loads/hr.		160	.050			1.48	2.30	3.78	4.81
0040 1 mile round trip, 3.3 loads/hr.		130	.062			1.82	2.83	4.65	5.95
0100 2 mile round trip, 2.6 loads/hr.		100	.080			2.36	3.68	6.04	7.70
0150 3 mile round trip, 2.1 loads/hr.		80	.100			2.96	4.61	7.57	9.60
0200 4 mile round trip, 1.8 loads/hr.	↓	70	.114			3.38	5.25	8.63	11
0310 12 C.Y. dump truck, 1/4 mile round trip 3.7 loads/hr.	B-34B	288	.028			.82	1.84	2.66	3.29
0400 2 mile round trip, 2.2 loads/hr.		180	.044			1.31	2.94	4.25	5.25
0450 3 mile round trip, 1.9 loads/hr.		170	.047			1.39	3.12	4.51	5.60
0500 4 mile round trip, 1.6 loads/hr.	↓	125	.064			1.89	4.24	6.13	7.60
1300 Hauling in medium traffic, add								20%	20%
1400 Heavy traffic, add								30%	30%
1600 Grading at dump, or embankment if required, by dozer	B-10B	1000	.012	↓		.42	.99	1.41	1.73
1800 Spotter at fill or cut, if required	1 Clab	8	1	Hr.		29		29	45

Estimating Tips

33 10 00 Water Utilities
33 30 00 Sanitary Sewerage Utilities
33 40 00 Storm Drainage Utilities

- Never assume that the water, sewer, and drainage lines will go in at the early stages of the project. Consider the site access needs before dividing the site in half with open trenches, loose pipe, and machinery obstructions. Always inspect the site to establish that the site drawings are complete. Check off all existing utilities on your drawings as you locate them. If you find any discrepancies, mark up the site plan for further research. Differing site conditions can be very costly if discovered later in the project.

- See also Section 33 01 00 for restoration of pipe where removal/replacement may be undesirable. Use of new types of piping materials can reduce the overall project cost. Owners/design engineers should consider the installing construction as a valuable source of current information on piping products that could lead to significant utility cost savings.

Reference Numbers

Reference numbers are shown in shaded boxes at the beginning of some major classifications. These numbers refer to related items in the Reference Section. The reference information may be an estimating procedure, an alternate pricing method, or technical information.

Note: Not all subdivisions listed here necessarily appear in this publication.

Note: **i2 Trade Service,** *in part, has been used as a reference source for some of the material prices used in Division 33.*

33 05 Common Work Results for Utilities

33 05 23 – Trenchless Utility Installation

33 05 23.19 Microtunneling

		Crew	Daily Output	Labor-Hours	Unit	Material	2007 Bare Costs Labor	Equipment	Total	Total Incl O&P
0010	**MICROTUNNELING** Not including excavation, backfill, shoring,									
0020	or dewatering, average 50'/day, slurry method									
0100	24" to 48" outside diameter, minimum				L.F.				756	840
0110	Adverse conditions, add				%					50%
1000	Rent microtunneling machine, average monthly lease				Month				89,100	99,000
1010	Operating technician				Day				661.50	735
1100	Mobilization and demobilization, minimum				Job				44,370	49,300
1110	Maximum				"				423,000	470,000

33 05 26 – Utility Line Signs, Markers, and Flags

33 05 26.10 Utility Accessories

		Crew	Daily Output	Labor-Hours	Unit	Material	2007 Bare Costs Labor	Equipment	Total	Total Incl O&P
0010	**UTILITY ACCESSORIES**									
0400	Underground tape, detectable, reinforced, alum. foil core, 2"	1 Clab	150	.053	C.L.F.	2	1.53		3.53	4.59
0500	6"	"	140	.057	"	5	1.64		6.64	8.05

33 12 Water Utility Distribution Equipment

33 12 19 – Water Utility Distribution Fire Hydrants

33 12 19.40 Utility Boxes

		Crew	Daily Output	Labor-Hours	Unit	Material	2007 Bare Costs Labor	Equipment	Total	Total Incl O&P
0010	**UTILITY BOXES** Precast concrete, 6" thick									
0050	5' x 10' x 6' high, I.D.	B-13	2	28	Ea.	1,725	880	370	2,975	3,650
0100	6' x 10' x 6' high, I.D.		2	28		1,775	880	370	3,025	3,700
0150	5' x 12' x 6' high, I.D.		2	28		1,875	880	370	3,125	3,825
0200	6' x 12' x 6' high, I.D.		1.80	31.111		2,100	975	410	3,485	4,275
0250	6' x 13' x 6' high, I.D.		1.50	37.333		2,775	1,175	495	4,445	5,400
0300	8' x 14' x 7' high, I.D.		1	56		3,000	1,750	740	5,490	6,825

33 44 Storm Utility Water Drains

33 44 13 – Utility Area Drains

33 44 13.13 Catch Basin Grates and Frames

		Crew	Daily Output	Labor-Hours	Unit	Material	2007 Bare Costs Labor	Equipment	Total	Total Incl O&P
0010	**CATCH BASIN GRATES AND FRAMES** not including footing, excavation									
1600	Frames & covers, C.I., 24" square, 500 lb.	B-6	7.80	3.077	Ea.	310	97	31	438	525
1700	26" D shape, 600 lb.		7	3.429		535	108	35	678	790
1800	Light traffic, 18" diameter, 100 lb.		10	2.400		174	75.50	24.50	274	335
1900	24" diameter, 300 lb.		8.70	2.759		269	87	28	384	460
2000	36" diameter, 900 lb.		5.80	4.138		540	130	42	712	840
2100	Heavy traffic, 24" diameter, 400 lb.		7.80	3.077		260	97	31	388	470
2200	36" diameter, 1150 lb.		3	8		860	252	81	1,193	1,425
2300	Mass. State standard, 26" diameter, 475 lb.		7	3.429		650	108	35	793	920
2400	30" diameter, 620 lb.		7	3.429		410	108	35	553	660
2500	Watertight, 24" diameter, 350 lb.		7.80	3.077		440	97	31	568	670
2600	26" diameter, 500 lb.		7	3.429		420	108	35	563	665
2700	32" diameter, 575 lb.		6	4		935	126	40.50	1,101.50	1,275
2800	3 piece cover & frame, 10" deep,									
2900	1200 lbs., for heavy equipment	B-6	3	8	Ea.	1,325	252	81	1,658	1,925
3000	Raised for paving 1-1/4" to 2" high,									
3100	4 piece expansion ring									
3200	20" to 26" diameter	1 Clab	3	2.667	Ea.	146	76.50		222.50	279
3300	30" to 36" diameter	"	3	2.667	"	203	76.50		279.50	340
3320	Frames and covers, existing, raised for paving, 2", including									
3340	row of brick, concrete collar, up to 12" wide frame	B-6	18	1.333	Ea.	45.50	42	13.50	101	129

33 44 Storm Utility Water Drains

33 44 13 – Utility Area Drains

33 44 13.13 Catch Basin Grates and Frames

		Crew	Daily Output	Labor-Hours	Unit	Material	2007 Bare Costs Labor	2007 Bare Costs Equipment	Total	Total Incl O&P
3360	20" to 26" wide frame	B-6	11	2.182	Ea.	72	68.50	22	162.50	209
3380	30" to 36" wide frame	↓	9	2.667		89	84	27	200	257
3400	Inverts, single channel brick	D-1	3	5.333		100	178		278	380
3500	Concrete		5	3.200		78	107		185	248
3600	Triple channel, brick		2	8		152	267		419	570
3700	Concrete	↓	3	5.333	↓	134	178		312	420

33 49 Storm Drainage Structures

33 49 13 – Storm Drainage Manholes, Frames, and Covers

33 49 13.10 Storm Drainage Manholes, Frames and Covers

		Crew	Daily Output	Labor-Hours	Unit	Material	2007 Bare Costs Labor	2007 Bare Costs Equipment	Total	Total Incl O&P
0010	**STORM DRAINAGE MANHOLES, FRAMES & COVERS** not including									
0020	footing, excavation, backfill (See line items for frame & cover)									
0050	Brick, 4' inside diameter, 4' deep	D-1	1	16	Ea.	390	535		925	1,250
0100	6' deep		.70	22.857		545	760		1,305	1,750
0150	8' deep		.50	32	↓	695	1,075		1,770	2,400
0200	For depths over 8', add		4	4	V.L.F.	184	133		317	405
0400	Concrete blocks (radial), 4' I.D., 4' deep		1.50	10.667	Ea.	330	355		685	905
0500	6' deep		1	16		435	535		970	1,300
0600	8' deep		.70	22.857	↓	540	760		1,300	1,750
0700	For depths over 8', add	↓	5.50	2.909	V.L.F.	54	97		151	207
0800	Concrete, cast in place, 4' x 4', 8" thick, 4' deep	C-14H	2	24	Ea.	475	870	10.80	1,355.80	1,875
0900	6' deep		1.50	32		685	1,150	14.40	1,849.40	2,575
1000	8' deep		1	48		990	1,750	21.50	2,761.50	3,850
1100	For depths over 8', add	↓	8	6	V.L.F.	110	218	2.70	330.70	465
1110	Precast, 4' I.D., 4' deep	B-22	4.10	7.317	Ea.	850	248	60	1,158	1,375
1120	6' deep		3	10		1,050	340	81.50	1,471.50	1,800
1130	8' deep		2	15	↓	1,275	510	123	1,908	2,325
1140	For depths over 8', add		16	1.875	V.L.F.	174	63.50	15.30	252.80	305
1150	5' I.D., 4' deep	B-6	3	8	Ea.	875	252	81	1,208	1,425
1160	6' deep		2	12		1,175	375	122	1,672	2,025
1170	8' deep		1.50	16	↓	1,475	505	162	2,142	2,575
1180	For depths over 8', add		12	2	V.L.F.	194	63	20.50	277.50	330
1190	6' I.D., 4' deep		2	12	Ea.	1,425	375	122	1,922	2,300
1200	6' deep		1.50	16		1,850	505	162	2,517	3,000
1210	8' deep		1	24	↓	2,300	755	243	3,298	3,950
1220	For depths over 8', add		8	3	V.L.F.	300	94.50	30.50	425	510
1250	Slab tops, precast, 8" thick									
1300	4' diameter manhole	B-6	8	3	Ea.	201	94.50	30.50	326	400
1400	5' diameter manhole		7.50	3.200		395	101	32.50	528.50	625
1500	6' diameter manhole	↓	7	3.429		575	108	35	718	835
3800	Steps, heavyweight cast iron, 7" x 9"	1 Bric	40	.200		16.20	7.60		23.80	29.50
3900	8" x 9"		40	.200		24.50	7.60		32.10	38
3928	12" x 10-1/2"		40	.200		23	7.60		30.60	37
4000	Standard sizes, galvanized steel		40	.200		19.50	7.60		27.10	33
4100	Aluminum	↓	40	.200	↓	26.50	7.60		34.10	40.50

308

33 71 Electrical Utility Transmission and Distribution

33 71 13 – Electrical Utility Towers

33 71 13.23 Steel Electrical Utility Towers

	Crew	Daily Output	Labor-Hours	Unit	Material	2007 Bare Costs Labor	2007 Bare Costs Equipment	Total	Total Incl O&P
0010 **STEEL ELECTRICAL UTILITY TOWERS**									
0100 Excavation and backfill, earth	R-5	135.38	.650	C.Y.		24.50	12.60	37.10	51
0105 Rock		21.46	4.101	"		156	79.50	235.50	320
0200 Steel footings (grillage) in earth		3.91	22.506	Ton	1,550	855	435	2,840	3,475
0205 In rock	↓	3.20	27.500	"	1,550	1,050	535	3,135	3,875
0290 See also Division 33 05 23.19									
0300 Rock anchors	R-5	5.87	14.991	Ea.	400	570	291	1,261	1,625
0400 Concrete foundations	"	12.85	6.848	C.Y.	132	260	133	525	680
0490 See also Division 03 30									
0500 Towers-material handling and spotting	R-7	22.56	2.128	Ton		64	5.15	69.15	104
0540 Steel tower erection	R-5	7.65	11.503		1,500	435	223	2,158	2,550
0550 Lace and box		7.10	12.394	↓	1,500	470	240	2,210	2,625
0560 Painting total structure	↓	1.47	59.864	Ea.	320	2,275	1,150	3,745	5,050
0570 Disposal of surplus material	R-7	20.87	2.300	Mile		69.50	5.55	75.05	113
0600 Special towers-material handling and spotting	"	12.31	3.899	Ton		118	9.45	127.45	191
0640 Special steel structure erection	R-6	6.52	13.497		1,850	510	520	2,880	3,400
0650 Special steel lace and box	"	6.29	13.990	↓	1,850	530	535	2,915	3,450
0670 Disposal of surplus material	R-7	7.87	6.099	Mile		184	14.75	198.75	299

33 71 13.80 Transmission Line Right of Way

	Crew	Daily Output	Labor-Hours	Unit	Material	2007 Bare Costs Labor	2007 Bare Costs Equipment	Total	Total Incl O&P
0010 **TRANSMISSION LINE RIGHT OF WAY**									
0100 Clearing right of way	B-87	6.67	5.997	Acre		219	310	529	670
0200 Restoration & seeding	B-10D	4	3	"	990	106	340	1,436	1,625

33 71 16 – Electrical Utility Poles

33 71 16.20 Line Poles and Fixtures

	Crew	Daily Output	Labor-Hours	Unit	Material	2007 Bare Costs Labor	2007 Bare Costs Equipment	Total	Total Incl O&P
0010 **LINE POLES & FIXTURES** R337116-60									
0100 Digging holes in earth, average	R-5	25.14	3.500	Ea.		133	68	201	275
0105 In rock, average	"	4.51	19.512			740	380	1,120	1,550
0200 Wood poles, material handling and spotting	R-7	6.49	7.396			223	17.90	240.90	365
0220 Erect wood poles & backfill holes, in earth	R-5	6.77	12.999		1,200	495	252	1,947	2,350
0250 In rock	"	5.87	14.991	↓	1,200	570	291	2,061	2,500
0260 Disposal of surplus material	R-7	20.87	2.300	Mile		69.50	5.55	75.05	113
0300 Crossarms for wood pole structure									
0310 Material handling and spotting	R-7	14.55	3.299	Ea.		99.50	8	107.50	162
0320 Install crossarms	R-5	11	8	"	395	305	155	855	1,050
0330 Disposal of surplus material	R-7	40	1.200	Mile		36	2.90	38.90	58.50
0400 Formed plate pole structure									
0410 Material handling and spotting	R-7	2.40	20	Ea.		605	48.50	653.50	980
0420 Erect steel plate pole	R-5	1.95	45.128	↓	7,475	1,700	875	10,050	11,800
0500 Guys, anchors and hardware for pole, in earth		7.04	12.500		455	475	243	1,173	1,475
0510 In rock	↓	17.96	4.900	↓	540	186	95	821	980
0900 Foundations for line poles									
0920 Excavation, in earth	R-5	135.38	.650	C.Y.		24.50	12.60	37.10	51
0940 In rock		20	4.400			167	85.50	252.50	345
0960 Concrete foundations	↓	11	8	↓	132	305	155	592	770

33 71 16.33 Wood Electrical Utility Poles

	Crew	Daily Output	Labor-Hours	Unit	Material	2007 Bare Costs Labor	2007 Bare Costs Equipment	Total	Total Incl O&P
0010 **WOOD ELECTRICAL UTILITY POLES**									
6200 Electric & tel sitework, ps, wd, treatment, see also 26 56 13.10, 20' high	R-3	3.10	6.452	Ea.	251	279	52.50	582.50	750
6400 25' high R337119-30		2.90	6.897		265	298	56.50	619.50	800
6600 30' high		2.60	7.692		292	335	63	690	885
6800 35' high		2.40	8.333		380	360	68	808	1,025
7000 40' high		2.30	8.696		465	375	71	911	1,150
7200 45' high	↓	1.70	11.765	↓	570	510	96	1,176	1,500

33 71 16 – Electrical Utility Poles

33 71 16.33 Wood Electrical Utility Poles	Crew	Daily Output	Labor-Hours	Unit	Material	2007 Bare Costs Labor	Equipment	Total	Total Incl O&P
7400	Cross arms with hardware & insulators									
7600	4' long	1 Elec	2.50	3.200	Ea.	119	140		259	340
7800	5' long		2.40	3.333		138	146		284	370
8000	6' long		2.20	3.636		159	160		319	415

33 71 19 – Electrical Underground Ducts and Manholes

33 71 19.15 Underground Ducts and Manholes

	33 71 19.15 Underground Ducts and Manholes	Crew	Daily Output	Labor-Hours	Unit	Material	2007 Bare Costs Labor	Equipment	Total	Total Incl O&P
0010	**UNDERGROUND DUCTS AND MANHOLES**, In slab or duct bank									
0011	Not including excavation, backfill and cast in place concrete									
1000	Direct burial									
1010	PVC, schedule 40, w/coupling, 1/2" diameter	1 Elec	340	.024	L.F.	.52	1.03		1.55	2.11
1020	3/4" diameter		290	.028		.69	1.21		1.90	2.56
1030	1" diameter		260	.031		1.03	1.35		2.38	3.14
1040	1-1/2" diameter		210	.038		1.68	1.67		3.35	4.34
1050	2" diameter		180	.044		2.20	1.95		4.15	5.30
1060	3" diameter	2 Elec	240	.067		4.19	2.93		7.12	8.95
1070	4" diameter		160	.100		5.90	4.39		10.29	13.05
1080	5" diameter		120	.133		8.55	5.85		14.40	18.10
1090	6" diameter		90	.178		11.30	7.80		19.10	24
1110	Elbows, 1/2" diameter	1 Elec	48	.167	Ea.	1.58	7.30		8.88	12.65
1120	3/4" diameter		38	.211		1.59	9.25		10.84	15.50
1130	1" diameter		32	.250		2.69	11		13.69	19.30
1140	1-1/2" diameter		21	.381		5.20	16.70		21.90	30.50
1150	2" diameter		16	.500		7.55	22		29.55	41
1160	3" diameter		12	.667		23	29.50		52.50	69
1170	4" diameter		9	.889		40	39		79	102
1180	5" diameter		8	1		70	44		114	143
1190	6" diameter		5	1.600		119	70		189	236
1210	Adapters, 1/2" diameter		52	.154		.51	6.75		7.26	10.60
1220	3/4" diameter		43	.186		.94	8.15		9.09	13.20
1230	1" diameter		39	.205		1.19	9		10.19	14.70
1240	1-1/2" diameter		35	.229		1.85	10.05		11.90	17
1250	2" diameter		26	.308		2.66	13.50		16.16	23
1260	3" diameter		20	.400		6.75	17.55		24.30	33.50
1270	4" diameter		14	.571		11.80	25		36.80	50.50
1280	5" diameter		12	.667		23	29.50		52.50	69
1290	6" diameter		9	.889		28	39		67	89
1340	Bell end & cap, 1-1/2" diameter		35	.229		11.95	10.05		22	28
1350	Bell end & plug, 2" diameter		26	.308		12.60	13.50		26.10	34
1360	3" diameter		20	.400		15.65	17.55		33.20	43.50
1370	4" diameter		14	.571		18.45	25		43.45	58
1380	5" diameter		12	.667		28	29.50		57.50	74
1390	6" diameter		9	.889		31.50	39		70.50	92.50
1450	Base spacer, 2" diameter		56	.143		1.90	6.25		8.15	11.45
1460	3" diameter		46	.174		2.11	7.65		9.76	13.65
1470	4" diameter		41	.195		2.22	8.55		10.77	15.20
1480	5" diameter		37	.216		2.73	9.50		12.23	17.15
1490	6" diameter		34	.235		3.83	10.35		14.18	19.60
1550	Intermediate spacer, 2" diameter		60	.133		1.90	5.85		7.75	10.80
1560	3" diameter		46	.174		2.11	7.65		9.76	13.65
1570	4" diameter		41	.195		2.22	8.55		10.77	15.20
1580	5" diameter		37	.216		2.73	9.50		12.23	17.15
1590	6" diameter		34	.235		3.82	10.35		14.17	19.60

33 71 Electrical Utility Transmission and Distribution

33 71 19 – Electrical Underground Ducts and Manholes

33 71 19.15 Underground Ducts and Manholes

		Crew	Daily Output	Labor-Hours	Unit	Material	2007 Bare Costs Labor	Equipment	Total	Total Incl O&P
4010	PVC, schedule 80, w/coupling, 1/2" diameter	1 Elec	215	.037	L.F.	.63	1.63		2.26	3.12
4020	3/4" diameter		180	.044		.85	1.95		2.80	3.84
4030	1" diameter		145	.055		1.21	2.42		3.63	4.94
4040	1-1/2" diameter		120	.067		2.01	2.93		4.94	6.55
4050	2" diameter	▼	100	.080		2.77	3.51		6.28	8.30
4060	3" diameter	2 Elec	130	.123		5.65	5.40		11.05	14.30
4070	4" diameter		90	.178		8.25	7.80		16.05	20.50
4080	5" diameter		70	.229		11.85	10.05		21.90	28
4090	6" diameter	▼	50	.320	▼	16.30	14.05		30.35	39
4110	Elbows, 1/2" diameter	1 Elec	29	.276	Ea.	3.60	12.10		15.70	22
4120	3/4" diameter		23	.348		5.65	15.25		20.90	29
4130	1" diameter		20	.400		7.35	17.55		24.90	34
4140	1-1/2" diameter		16	.500		14.50	22		36.50	48.50
4150	2" diameter		12	.667		19.90	29.50		49.40	65.50
4160	3" diameter		9	.889		67	39		106	132
4170	4" diameter		7	1.143		110	50		160	196
4180	5" diameter		6	1.333		160	58.50		218.50	263
4190	6" diameter		4	2		310	88		398	475
4210	Adapter, 1/2" diameter		39	.205		.51	9		9.51	13.95
4220	3/4" diameter		33	.242		.94	10.65		11.59	16.90
4230	1" diameter		29	.276		1.19	12.10		13.29	19.35
4240	1-1/2" diameter		26	.308		1.85	13.50		15.35	22
4250	2" diameter		23	.348		2.66	15.25		17.91	25.50
4260	3" diameter		18	.444		6.75	19.50		26.25	36.50
4270	4" diameter		13	.615		11.80	27		38.80	53
4280	5" diameter		11	.727		23	32		55	73
4290	6" diameter		8	1		28	44		72	96.50
4310	Bell end & cap, 1-1/2" diameter		26	.308		11.95	13.50		25.45	33
4320	Bell end & plug, 2" diameter		23	.348		12.60	15.25		27.85	36.50
4330	3" diameter		18	.444		15.65	19.50		35.15	46.50
4340	4" diameter		13	.615		18.45	27		45.45	60.50
4350	5" diameter		11	.727		28	32		60	78
4360	6" diameter		8	1		31.50	44		75.50	100
4370	Base spacer, 2" diameter		42	.190		1.90	8.35		10.25	14.55
4380	3" diameter		33	.242		2.11	10.65		12.76	18.15
4390	4" diameter		29	.276		2.22	12.10		14.32	20.50
4400	5" diameter		26	.308		2.73	13.50		16.23	23
4410	6" diameter		25	.320		3.83	14.05		17.88	25
4420	Intermediate spacer, 2" diameter		45	.178		1.90	7.80		9.70	13.70
4430	3" diameter		34	.235		2.11	10.35		12.46	17.70
4440	4" diameter		31	.258		2.22	11.35		13.57	19.30
4450	5" diameter		28	.286		2.73	12.55		15.28	21.50
4460	6" diameter	▼	25	.320	▼	3.82	14.05		17.87	25

33 71 19.17 Electric and Telephone Underground

		Crew	Daily Output	Labor-Hours	Unit	Material	2007 Bare Costs Labor	Equipment	Total	Total Incl O&P
0010	**ELECTRIC AND TELEPHONE UNDERGROUND**, Not including excavation									
0200	backfill and cast in place concrete									
0250	For bedding see div. 31 23 23.16 [R337119-30]									
0400	Hand holes, precast concrete, with concrete cover									
0600	2' x 2' x 3' deep	R-3	2.40	8.333	Ea.	295	360	68	723	940
0800	3' x 3' x 3' deep		1.90	10.526		385	455	86	926	1,200
1000	4' x 4' x 4' deep	▼	1.40	14.286	▼	765	620	117	1,502	1,900
1200	Manholes, precast with iron racks & pulling irons, C.I. frame									

33 71 Electrical Utility Transmission and Distribution

33 71 19 – Electrical Underground Ducts and Manholes

33 71 19.17 Electric and Telephone Underground	Crew	Daily Output	Labor-Hours	Unit	Material	2007 Bare Costs Labor	2007 Bare Costs Equipment	Total	Total Incl O&P
1400	and cover, 4' x 6' x 7' deep	B-13	2	28	Ea.	1,475	880	370	2,725	3,375
1600	6' x 8' x 7' deep		1.90	29.474		1,800	925	390	3,115	3,825
1800	6' x 10' x 7' deep		1.80	31.111		2,025	975	410	3,410	4,175
4200	Underground duct, banks ready for concrete fill, min. of 7.5"									
4400	between conduits, ctr. to ctr.									
4580	PVC, type EB, 1 @ 2" diameter	2 Elec	480	.033	L.F.	.78	1.46		2.24	3.04
4600	2 @ 2" diameter		240	.067		1.56	2.93		4.49	6.10
4800	4 @ 2" diameter		120	.133		3.12	5.85		8.97	12.15
4900	1 @ 3" diameter		400	.040		1.06	1.76		2.82	3.78
5000	2 @ 3" diameter		200	.080		2.12	3.51		5.63	7.60
5200	4 @ 3" diameter		100	.160		4.24	7		11.24	15.10
5300	1 @ 4" diameter		320	.050		1.61	2.20		3.81	5.05
5400	2 @ 4" diameter		160	.100		3.22	4.39		7.61	10.10
5600	4 @ 4" diameter		80	.200		6.45	8.80		15.25	20
5800	6 @ 4" diameter		54	.296		9.65	13		22.65	30
5810	1 @ 5" diameter		260	.062		2.39	2.70		5.09	6.65
5820	2 @ 5" diameter		130	.123		4.79	5.40		10.19	13.30
5840	4 @ 5" diameter		70	.229		9.55	10.05		19.60	25.50
5860	6 @ 5" diameter		50	.320		14.35	14.05		28.40	37
5870	1 @ 6" diameter		200	.080		3.42	3.51		6.93	9
5880	2 @ 6" diameter		100	.160		6.85	7		13.85	18
5900	4 @ 6" diameter		50	.320		13.70	14.05		27.75	36
5920	6 @ 6" diameter		30	.533		20.50	23.50		44	57.50
6200	Rigid galvanized steel, 2 @ 2" diameter		180	.089		14.20	3.90		18.10	21.50
6400	4 @ 2" diameter		90	.178		28.50	7.80		36.30	43
6800	2 @ 3" diameter		100	.160		34	7		41	47.50
7000	4 @ 3" diameter		50	.320		67.50	14.05		81.55	95.50
7200	2 @ 4" diameter		70	.229		47	10.05		57.05	66.50
7400	4 @ 4" diameter		34	.471		93.50	20.50		114	134
7600	6 @ 4" diameter		22	.727		140	32		172	202
7620	2 @ 5" diameter		60	.267		101	11.70		112.70	128
7640	4 @ 5" diameter		30	.533		202	23.50		225.50	257
7660	6 @ 5" diameter		18	.889		300	39		339	395
7680	2 @ 6" diameter		40	.400		147	17.55		164.55	187
7700	4 @ 6" diameter		20	.800		293	35		328	375
7720	6 @ 6" diameter		14	1.143		440	50		490	560
7800	For Cast-in-place Concrete - Add									
7810	Under 1 C.Y.	C-6	16	3	C.Y.	166	90.50	2.67	259.17	325
7820	1 C.Y. - 5 C.Y.		19.20	2.500		151	75.50	2.23	228.73	284
7830	Over 5 C.Y.		24	2		125	60.50	1.78	187.28	232
7850	For Reinforcing Rods - Add									
7860	#4 to #7	2 Rodm	1.10	14.545	Ton	895	600		1,495	1,975
7870	#8 to #14	"	1.50	10.667	"	895	440		1,335	1,700
8000	Fittings, PVC type EB, elbow, 2" diameter	1 Elec	16	.500	Ea.	11	22		33	44.50
8200	3" diameter		14	.571		11.60	25		36.60	50.50
8400	4" diameter		12	.667		16.45	29.50		45.95	61.50
8420	5" diameter		10	.800		39	35		74	95.50
8440	6" diameter		9	.889		74	39		113	140
8500	Coupling, 2" diameter					.79			.79	.87
8600	3" diameter					2.74			2.74	3.01
8700	4" diameter					4.30			4.30	4.73
8720	5" diameter					7.75			7.75	8.55
8740	6" diameter					22.50			22.50	25

33 71 19 – Electrical Underground Ducts and Manholes

33 71 19.17 Electric and Telephone Underground

		Crew	Daily Output	Labor-Hours	Unit	Material	2007 Bare Costs Labor	2007 Bare Costs Equipment	Total	Total Incl O&P
8800	Adapter, 2" diameter	1 Elec	26	.308	Ea.	1.40	13.50		14.90	21.50
9000	3" diameter		20	.400		3.90	17.55		21.45	30.50
9200	4" diameter		16	.500		5.20	22		27.20	38.50
9220	5" diameter		13	.615		13	27		40	54.50
9240	6" diameter		10	.800		17.15	35		52.15	71.50
9400	End bell, 2" diameter		16	.500		6.80	22		28.80	40
9600	3" diameter		14	.571		8.15	25		33.15	46.50
9800	4" diameter		12	.667		9.60	29.50		39.10	54
9810	5" diameter		10	.800		14.50	35		49.50	68.50
9820	6" diameter		8	1		27.50	44		71.50	96
9830	5° angle coupling, 2" diameter		26	.308		11.45	13.50		24.95	32.50
9840	3" diameter		20	.400		14.50	17.55		32.05	42
9850	4" diameter		16	.500		17.15	22		39.15	51.50
9860	5" diameter		13	.615		18.80	27		45.80	60.50
9870	6" diameter		10	.800		19.25	35		54.25	73.50
9880	Expansion joint, 2" diameter		16	.500		33.50	22		55.50	69.50
9890	3" diameter		18	.444		59	19.50		78.50	94
9900	4" diameter		12	.667		85.50	29.50		115	138
9910	5" diameter		10	.800		133	35		168	199
9920	6" diameter	↓	8	1		179	44		223	263
9930	Heat bender, 2" diameter					440			440	485
9940	6" diameter					1,225			1,225	1,350
9950	Cement, quart				↓	23			23	25
9960	Nylon polyethylene pull rope, 1/4"	2 Elec	2000	.008	L.F.	.10	.35		.45	.63

33 71 23 – Insulators and Fittings

33 71 23.16 Post Insulators

		Crew	Daily Output	Labor-Hours	Unit	Material	2007 Bare Costs Labor	2007 Bare Costs Equipment	Total	Total Incl O&P
0010	**POST INSULATORS**									
7400	Insulators, pedestal type	R-11	112	.500	Ea.	20.50	7.50		28	39.50

33 71 26 – Transmission and Distribution Equipment

33 71 26.13 Capacitor Banks

		Crew	Daily Output	Labor-Hours	Unit	Material	2007 Bare Costs Labor	2007 Bare Costs Equipment	Total	Total Incl O&P
0010	**CAPACITOR BANKS**									
1300	Station capacitors									
1350	Synchronous, 13 to 26 kV	R-11	3.11	18.006	MVAR	5,400	740	270	6,410	7,325
1360	46 kV		3.33	16.817		6,875	695	252	7,822	8,900
1370	69 kV		3.81	14.698		6,775	605	220	7,600	8,600
1380	161 kV		6.51	8.602		6,325	355	129	6,809	7,650
1390	500 kV		10.37	5.400		5,500	223	81	5,804	6,475
1450	Static, 13 to 26 kV		3.11	18.006		4,575	740	270	5,585	6,425
1460	46 kV		3.01	18.605		5,775	765	279	6,819	7,800
1470	69 kV		3.81	14.698		5,625	605	220	6,450	7,325
1480	161 kV		6.51	8.602		5,225	355	129	5,709	6,400
1490	500 kV		10.37	5.400	↓	4,750	223	81	5,054	5,650
1600	Voltage regulators, 13 to 26 kV	↓	.75	74.667	Ea.	205,000	3,075	1,125	209,200	231,500

33 71 26.23 Current Transformers

		Crew	Daily Output	Labor-Hours	Unit	Material	2007 Bare Costs Labor	2007 Bare Costs Equipment	Total	Total Incl O&P
0010	**CURRENT TRANSFORMERS**									
4050	Current transformers, 13 to 26 kV	R-11	14	4	Ea.	2,525	165	60	2,750	3,100
4060	46 kV		9.33	6.002		7,350	247	90	7,687	8,575
4070	69 kV		7	8		7,625	330	120	8,075	9,000
4080	161 kV	↓	1.87	29.947	↓	24,700	1,225	450	26,375	29,500

33 71 26.26 Potential Transformers

		Crew	Daily Output	Labor-Hours	Unit	Material	2007 Bare Costs Labor	2007 Bare Costs Equipment	Total	Total Incl O&P
0010	**POTENTIAL TRANSFORMERS**									

33 71 Electrical Utility Transmission and Distribution

33 71 26 – Transmission and Distribution Equipment

33 71 26.26 Potential Transformers		Crew	Daily Output	Labor-Hours	Unit	Material	2007 Bare Costs Labor	Equipment	Total	Total Incl O&P
4100	Potential transformers, 13 to 26 kV	R-11	11.20	5	Ea.	3,600	206	75	3,881	4,350
4110	46 kV		8	7		7,400	289	105	7,794	8,700
4120	69 kV		6.22	9.003		7,850	370	135	8,355	9,325
4130	161 kV		2.24	25		17,000	1,025	375	18,400	20,600
4140	500 kV		1.40	40		50,500	1,650	600	52,750	58,500

33 71 39 – High-Voltage Wiring

33 71 39.13 Overhead High-Voltage Wiring

		Crew	Daily Output	Labor-Hours	Unit	Material	2007 Bare Costs Labor	Equipment	Total	Total Incl O&P
0010	OVERHEAD HIGH-VOLTAGE WIRING R337116-60									
0100	Conductors, primary circuits									
0110	Material handling and spotting	R-5	9.78	8.998	W.Mile		340	175	515	705
0120	For river crossing, add		11	8			305	155	460	625
0150	Conductors, per wire, 210 to 636 kcmil		1.96	44.898		8,300	1,700	870	10,870	12,600
0160	795 to 954 kcmil		1.87	47.059		12,300	1,775	915	14,990	17,200
0170	1000 to 1600 kcmil		1.47	59.864		20,100	2,275	1,150	23,525	26,800
0180	Over 1600 kcmil		1.35	65.185		26,700	2,475	1,275	30,450	34,400
0200	For river crossing, add, 210 to 636 kcmil		1.24	70.968			2,700	1,375	4,075	5,575
0220	795 to 954 kcmil		1.09	80.734			3,050	1,575	4,625	6,325
0230	1000 to 1600 kcmil		.97	90.722			3,450	1,750	5,200	7,100
0240	Over 1600 kcmil		.87	101			3,825	1,950	5,775	7,925
0300	Joints and dead ends	R-8	6	8	Ea.	1,175	310	33.50	1,518.50	1,800
0400	Sagging	R-5	7.33	12.001	W.Mile		455	233	688	940
0500	Clipping, per structure, 69 kV	R-10	9.60	5	Ea.		206	50	256	365
0510	161 kV		5.33	9.006			370	90	460	655
0520	345 to 500 kV		2.53	18.972			780	189	969	1,375
0600	Make and install jumpers, per structure, 69 kV	R-8	3.20	15		315	575	63	953	1,275
0620	161 kV		1.20	40		635	1,550	168	2,353	3,200
0640	345 to 500 kV		.32	150		1,075	5,775	630	7,480	10,500
0700	Spacers	R-10	68.57	.700		63	29	7	99	120
0720	For river crossings, add	"	60	.800			33	8	41	58.50
0800	Installing pulling line (500 kV only)	R-9	1.45	44.138	W.Mile	605	1,575	139	2,319	3,200
0810	Disposal of surplus material	R-7	6.96	6.897	Mile		208	16.70	224.70	340
0820	With trailer mounted reel stands	"	13.71	3.501	"		106	8.45	114.45	171
0900	Insulators and hardware, primary circuits									
0920	Material handling and spotting, 69 kV	R-7	480	.100	Ea.		3.02	.24	3.26	4.91
0930	161 kV		685.71	.070			2.11	.17	2.28	3.43
0950	345 to 500 kV		960	.050			1.51	.12	1.63	2.45
1000	Disk insulators, 69 kV	R-5	880	.100		64.50	3.79	1.94	70.23	79
1020	161 kV		977.78	.090		73.50	3.41	1.75	78.66	88
1040	345 to 500 kV		1100	.080		73.50	3.03	1.55	78.08	87.50
1060	See Div. 33 71 23.16 7400 for pin or pedestal insulator									
1100	Install disk insulator at river crossing, add									
1110	69 kV	R-5	586.67	.150	Ea.		5.70	2.91	8.61	11.75
1120	161 kV		880	.100			3.79	1.94	5.73	7.85
1140	345 to 500 kV		880	.100			3.79	1.94	5.73	7.85
1150	Disposal of surplus material	R-7	41.74	1.150	Mile		34.50	2.78	37.28	56.50
1300	Overhead ground wire installation									
1320	Material handling and spotting	R-7	5.65	8.496	W.Mile		256	20.50	276.50	420
1340	Overhead ground wire	R-5	1.76	50		3,650	1,900	970	6,520	7,925
1350	At river crossing, add		1.17	75.214			2,850	1,450	4,300	5,900
1360	Disposal of surplus material		41.74	2.108	Mile		80	41	121	165
1400	Installing conductors, underbuilt circuits									
1420	Material handling and spotting	R-7	5.65	8.496	W.Mile		256	20.50	276.50	420

33 71 Electrical Utility Transmission and Distribution

33 71 39 – High-Voltage Wiring

33 71 39.13 Overhead High-Voltage Wiring		Crew	Daily Output	Labor-Hours	Unit	Material	2007 Bare Costs Labor	Equipment	Total	Total Incl O&P
1440	Conductors, per wire, 210 to 636 kcmil	R-5	1.96	44.898	W.Mile	8,300	1,700	870	10,870	12,600
1450	795 to 954 kcmil		1.87	47.059		12,300	1,775	915	14,990	17,200
1460	1000 to 1600 kcmil		1.47	59.864		20,100	2,275	1,150	23,525	26,800
1470	Over 1600 kcmil	▼	1.35	65.185	▼	26,700	2,475	1,275	30,450	34,400
1500	Joints and dead ends	R-8	6	8	Ea.	1,175	310	33.50	1,518.50	1,800
1550	Sagging	R-5	8.80	10	W.Mile		380	194	574	785
1600	Clipping, per structure, 69 kV	R-10	9.60	5	Ea.		206	50	256	365
1620	161 kV		5.33	9.006			370	90	460	655
1640	345 to 500 kV	▼	2.53	18.972			780	189	969	1,375
1700	Making and installing jumpers, per structure, 69 kV	R-8	5.87	8.177		315	315	34.50	664.50	865
1720	161 kV		.96	50		635	1,925	210	2,770	3,825
1740	345 to 500 kV	▼	.32	150		1,075	5,775	630	7,480	10,500
1800	Spacers	R-10	96	.500	▼	63	20.50	4.99	88.49	106
1810	Disposal of surplus material	R-7	6.96	6.897	Mile		208	16.70	224.70	340
2000	Insulators and hardware for underbuilt circuits									
2100	Material handling and spotting	R-7	1200	.040	Ea.		1.21	.10	1.31	1.96
2150	Disk insulators, 69 kV	R-8	600	.080		64.50	3.08	.34	67.92	76
2160	161 kV		686	.070		73.50	2.69	.29	76.48	85.50
2170	345 to 500 kV	▼	800	.060	▼	73.50	2.31	.25	76.06	85
2180	Disposal of surplus material	R-7	41.74	1.150	Mile		34.50	2.78	37.28	56.50
2300	Sectionalizing switches, 69 kV	R-5	1.26	69.841	Ea.	16,600	2,650	1,350	20,600	23,800
2310	161 kV		.80	110		18,800	4,175	2,125	25,100	29,300
2500	Protective devices	▼	5.50	16	▼	5,425	605	310	6,340	7,225
2600	Clearance poles, 8 poles per mile									
2650	In earth, 69 kV	R-5	1.16	75.862	Mile	4,475	2,875	1,475	8,825	10,900
2660	161 kV	"	.64	137		7,350	5,225	2,675	15,250	18,900
2670	345 to 500 kV	R-6	.48	183		8,825	6,950	7,025	22,800	27,900
2800	In rock, 69 kV	R-5	.69	127		4,475	4,825	2,475	11,775	14,900
2820	161 kV	"	.35	251		7,350	9,525	4,875	21,750	27,800
2840	345 to 500 kV	R-6	.24	366	▼	8,825	13,900	14,100	36,825	46,100

33 72 Utility Substations

33 72 26 – Substation Bus Assemblies

33 72 26.13 Aluminum Substation Bus Assemblies

		Crew	Daily Output	Labor-Hours	Unit	Material	Labor	Equipment	Total	Total Incl O&P
0010	**ALUMINUM SUBSTATION BUS ASSEMBLIES**									
7300	Bus	R-11	590	.095	Lb.	4.35	3.91	1.42	9.68	12.20

33 72 33 – Control House Equipment

33 72 33.33 Raceway and Boxes for Utility Substations

		Crew	Daily Output	Labor-Hours	Unit	Material	Labor	Equipment	Total	Total Incl O&P
0010	**RACEWAY AND BOXES FOR UTILITY SUBSTATIONS**									
7000	Conduit, conductors, and insulators									
7100	Conduit, metallic	R-11	560	.100	Lb.	2.35	4.12	1.50	7.97	10.45
7110	Non-metallic	"	800	.070	"	2.25	2.89	1.05	6.19	7.95
7190	See also Division 26 05 36.00									
7200	Wire and cable	R-11	700	.080	Lb.	3.55	3.30	1.20	8.05	10.15
7290	See also Division 26 22 00.00									

33 72 33.36 Cable Trays for Utility Substations

		Crew	Daily Output	Labor-Hours	Unit	Material	Labor	Equipment	Total	Total Incl O&P
0010	**CABLE TRAYS FOR UTILITY SUBSTATIONS**									
7700	Cable tray	R-11	40	1.400	L.F.	13.65	57.50	21	92.15	125
7790	See also Division 26 05 36.00									

33 72 Utility Substations

33 72 33 – Control House Equipment

33 72 33.43 Substation Backup Batteries	Crew	Daily Output	Labor-Hours	Unit	Material	2007 Bare Costs Labor	Equipment	Total	Total Incl O&P
0010 **SUBSTATION BACKUP BATTERIES**									
9120 Battery chargers	R-11	11.20	5	Ea.	3,075	206	75	3,356	3,775
9200 Control batteries	"	14	4	K.A.H.	70.50	165	60	295.50	390

33 72 33.46 Substation Converter Stations									
0010 **SUBSTATION CONVERTER STATIONS**									
9000 Station service equipment									
9100 Conversion equipment									
9110 Station service transformers	R-11	5.60	10	Ea.	74,000	410	150	74,560	82,500

33 73 Utility Transformers

33 73 23 – Dry-Type Utility Transformers

33 73 23.20 Primary Transformers

	Crew	Daily Output	Labor-Hours	Unit	Material	2007 Bare Costs Labor	Equipment	Total	Total Incl O&P
0010 **PRIMARY TRANSFORMERS**									
1000 Main conversion equipment									
1050 Power transformers, 13 to 26 kV	R-11	1.72	32.558	MVA	19,400	1,350	490	21,240	23,800
1060 46 kV		3.50	16		18,200	660	240	19,100	21,300
1070 69 kV		3.11	18.006		15,500	740	270	16,510	18,400
1080 110 kV		3.29	17.021		14,700	700	255	15,655	17,500
1090 161 kV		4.31	12.993		13,700	535	195	14,430	16,000
1100 500 kV		7	8		13,600	330	120	14,050	15,500
1200 Grounding transformers		3.11	18.006	Ea.	85,500	740	270	86,510	95,500

33 75 High-Voltage Switchgear and Protection Devices

33 75 13 – Air High-Voltage Circuit Breaker

33 75 13.13 Air High-Voltage Circuit Breaker

	Crew	Daily Output	Labor-Hours	Unit	Material	2007 Bare Costs Labor	Equipment	Total	Total Incl O&P
0010 **AIR HIGH-VOLTAGE CIRCUIT BREAKER**									
2100 Air circuit breakers, 13 to 26 kV	R-11	.56	100	Ea.	47,800	4,125	1,500	53,425	60,500
2110 161 kV	"	.24	233	"	205,500	9,625	3,500	218,625	244,500

33 75 16 – Oil High-Voltage Circuit Breaker

33 75 16.13 Oil Circuit Breaker

	Crew	Daily Output	Labor-Hours	Unit	Material	2007 Bare Costs Labor	Equipment	Total	Total Incl O&P
0010 **OIL CIRCUIT BREAKER**									
2000 Power circuit breakers									
2050 Oil circuit breakers, 13 to 26 kV	R-11	1.12	50	Ea.	45,700	2,050	750	48,500	54,500
2060 46 kV		.75	74.667		66,500	3,075	1,125	70,700	79,500
2070 69 kV		.45	124		155,000	5,125	1,875	162,000	180,000
2080 161 kV		.16	350		237,000	14,400	5,250	256,650	288,000
2090 500 kV		.06	933		887,500	38,500	14,000	940,000	1,049,500

33 75 19 – Gas High-Voltage Circuit Breaker

33 75 19.13 Gas Circuit Breaker

	Crew	Daily Output	Labor-Hours	Unit	Material	2007 Bare Costs Labor	Equipment	Total	Total Incl O&P
0010 **GAS CIRCUIT BREAKER**									
2000 Power circuit breakers									
2150 Gas circuit breakers, 13 to 26 kV	R-11	.56	100	Ea.	179,500	4,125	1,500	185,125	205,500
2160 161 kV		.08	700		236,000	28,900	10,500	275,400	314,500
2170 500 kV		.04	1400		831,000	57,500	21,000	909,500	1,023,500

33 75 23 – Vacuum High-Voltage Circuit Breaker

33 75 23.13 Vacuum Circuit Breaker

0010 **VACUUM CIRCUIT BREAKER**									

33 75 High-Voltage Switchgear and Protection Devices

33 75 23 – Vacuum High-Voltage Circuit Breaker

33 75 23.13 Vacuum Circuit Breaker

	Crew	Daily Output	Labor-Hours	Unit	Material	2007 Bare Costs Labor	2007 Bare Costs Equipment	Total	Total Incl O&P
2000 Power circuit breakers									
2200 Vacuum circuit breakers, 13 to 26 kV	R-11	.56	100	Ea.	39,200	4,125	1,500	44,825	51,000

33 75 36 – High-Voltage Utility Fuses

33 75 36.13 Fuses

	Crew	Daily Output	Labor-Hours	Unit	Material	Labor	Equipment	Total	Total Incl O&P
0010 **FUSES**									
8000 Protective equipment									
8250 Fuses, 13 to 26 kV	R-11	18.67	2.999	Ea.	1,675	124	45	1,844	2,050
8260 46 kV		11.20	5		1,900	206	75	2,181	2,475
8270 69 kV		8	7		1,975	289	105	2,369	2,725
8280 161 kV		4.67	11.991		2,500	495	180	3,175	3,700

33 75 39 – High-Voltage Surge Arresters

33 75 39.13 Surge Arresters

	Crew	Daily Output	Labor-Hours	Unit	Material	Labor	Equipment	Total	Total Incl O&P
0010 **SURGE ARRESTERS**									
8000 Protective equipment									
8050 Lightning arresters, 13 to 26 kV	R-11	18.67	2.999	Ea.	1,225	124	45	1,394	1,575
8060 46 kV		14	4		3,350	165	60	3,575	4,025
8070 69 kV		11.20	5		4,300	206	75	4,581	5,125
8080 161 kV		5.60	10		5,925	410	150	6,485	7,300
8090 500 kV		1.40	40		20,100	1,650	600	22,350	25,200

33 75 43 – Shunt Reactors

33 75 43.13 Reactors

	Crew	Daily Output	Labor-Hours	Unit	Material	Labor	Equipment	Total	Total Incl O&P
0010 **REACTORS**									
8000 Protective equipment									
8150 Reactors and resistors, 13 to 26 kV	R-11	28	2	Ea.	2,800	82.50	30	2,912.50	3,225
8160 46 kV		4.31	12.993		8,450	535	195	9,180	10,300
8170 69 kV		2.80	20		13,800	825	300	14,925	16,800
8180 161 kV		2.24	25		15,700	1,025	375	17,100	19,300
8190 500 kV		.08	700		63,000	28,900	10,500	102,400	124,500

33 75 53 – High-Voltage Switches

33 75 53.13 Switches

	Crew	Daily Output	Labor-Hours	Unit	Material	Labor	Equipment	Total	Total Incl O&P
0010 **SWITCHES**									
3000 Disconnecting switches									
3050 Gang operated switches									
3060 Manual operation, 13 to 26 kV	R-11	1.65	33.939	Ea.	11,000	1,400	510	12,910	14,800
3070 46 kV		1.12	50		18,300	2,050	750	21,100	24,000
3080 69 kV		.80	70		20,500	2,875	1,050	24,425	28,100
3090 161 kV		.56	100		24,700	4,125	1,500	30,325	35,000
3100 500 kV		.14	400		67,500	16,500	6,000	90,000	105,500
3110 Motor operation, 161 kV		.51	109		37,200	4,525	1,650	43,375	49,500
3120 500 kV		.28	200		103,000	8,250	3,000	114,250	128,500
3250 Circuit switches, 161 kV		.41	136		74,000	5,625	2,050	81,675	92,000
3300 Single pole switches									
3350 Disconnecting switches, 13 to 26 kV	R-11	28	2	Ea.	9,675	82.50	30	9,787.50	10,800
3360 46 kV		8	7		17,100	289	105	17,494	19,300
3370 69 kV		5.60	10		18,900	410	150	19,460	21,600
3380 161 kV		2.80	20		72,000	825	300	73,125	80,500
3390 500 kV		.22	254		205,500	10,500	3,825	219,825	246,000
3450 Grounding switches, 46 kV		5.60	10		26,000	410	150	26,560	29,400
3460 69 kV		3.73	15.013		27,000	620	225	27,845	30,900
3470 161 kV		2.24	25		28,600	1,025	375	30,000	33,500

33 75 High-Voltage Switchgear and Protection Devices

33 75 53 – High-Voltage Switches

33 75 53.13 Switches	Crew	Daily Output	Labor-Hours	Unit	Material	2007 Bare Costs Labor	2007 Bare Costs Equipment	Total	Total Incl O&P
3480 500 kV	R-11	.62	90.323	Ea.	35,500	3,725	1,350	40,575	46,200

33 79 Site Grounding

33 79 83 – Site Grounding Conductors

33 79 83.13 Grounding Wire, Bar, and Rod

		Crew	Daily Output	Labor-Hours	Unit	Material	Labor	Equipment	Total	Total Incl O&P
0010	**GROUNDING WIRE, BAR, AND ROD**									
7500	Grounding systems	R-11	280	.200	Lb.	11	8.25	3	22.25	28

33 81 Communications Structures

33 81 13 – Communications Transmission Towers

33 81 13.10 Radio Towers

		Crew	Daily Output	Labor-Hours	Unit	Material	Labor	Equipment	Total	Total Incl O&P
0010	**RADIO TOWERS**									
0020	Guyed, 50'h, 40 lb. sec., 70 MPH basic wind spd.	2 Sswk	1	16	Ea.	2,600	660		3,260	4,050
0100	Wind load 90 MPH basic wind speed	"	1	16		2,600	660		3,260	4,050
0300	190' high, 40 lb. section, wind load 70 MPH basic wind speed	K-2	.33	72.727		7,000	2,750	560	10,310	13,100
0400	200' high, 70 lb. section, wind load 90 MPH basic wind speed		.33	72.727		14,100	2,750	560	17,410	20,900
0600	300' high, 70 lb. section, wind load 70 MPH basic wind speed		.20	120		20,000	4,525	925	25,450	30,900
0700	270' high, 90 lb. section, wind load 90 MPH basic wind speed		.20	120		23,100	4,525	925	28,550	34,300
0800	400' high, 100 lb. section, wind load 70 MPH basic wind speed		.14	171		33,700	6,475	1,325	41,500	49,800
0900	Self-supporting, 60' high, wind load 70 MPH basic wind speed		.80	30		5,400	1,125	232	6,757	8,175
0910	60' high, wind load 90 MPH basic wind speed		.45	53.333		9,725	2,025	410	12,160	14,700
1000	120' high, wind load 70 MPH basic wind speed		.40	60		13,300	2,275	465	16,040	19,100
1200	190' high, wind load 90 MPH basic wind speed		.20	120		32,200	4,525	925	37,650	44,300
2000	For states west of Rocky Mountains, add for shipping					10%				

Estimating Tips

34 11 00 Rail Tracks

This section includes items that may involve either repair of existing, or construction of new, railroad tracks. Additional preparation work, such as the roadbed earthwork, would be found in Division 31. Additional new construction siding and turnouts are found in Section 34 72. Maintenance of railroads is found under 34 01 23 Operation and Maintenance of Railways.

34 41 13 Traffic Signals

This section includes traffic signal systems. Other traffic control devices such as traffic signs are found in Section 10 14 53 Traffic Signage.

34 71 13 Vehicle Barriers

This section includes security vehicle barriers, guide and guard rails, crash barriers, and delineators. The actual maintenance and construction of concrete and asphalt pavement is found in Division 32.

Reference Numbers

Reference numbers are shown in shaded boxes at the beginning of some major classifications. These numbers refer to related items in the Reference Section. The reference information may be an estimating procedure, an alternate pricing method, or technical information.

Note: Not all subdivisions listed here necessarily appear in this publication.

Division 34 - Transportation

34 43 Airfield Signaling and Control Equipment

34 43 13 – Airfield Signals

	34 43 13.10 Airport Lighting	Crew	Daily Output	Labor-Hours	Unit	Material	2007 Bare Costs Labor	Equipment	Total	Total Incl O&P
0010	**AIRPORT LIGHTING**									
0100	Runway centerline, bidir., semi-flush, 200 W, w/shallow insert base	R-22	12.40	3.006	Ea.	1,425	111		1,536	1,750
0120	Flush, 200 W, w/shallow insert base		12.40	3.006		1,425	111		1,536	1,750
0130	for mounting in base housing		18.64	2		755	73.50		828.50	940
0150	Touchdown zone light, unidirectional, 200 W, w/shallow insert base	*	12.40	3.006		1,175	111		1,286	1,475
0160	115 W		12.40	3.006		1,250	111		1,361	1,550
0180	Unidirectional, 200 W, for mounting in base housing		18.60	2.004		575	74		649	740
0190	115 W		18.60	2.004		620	74		694	790
0210	Runway edge & threshold light, bidir., 200 W, for base housing		9.36	3.983		810	147		957	1,100
0240	Threshold & approach light, unidir., 200 W, for base housing		9.36	3.983		475	147		622	740
0260	Runway edge, bi-directional, 2-115 W, for base housing		12.40	3.006		1,250	111		1,361	1,550
0280	Runway threshold & end, bidir., 2-115 W, for base housing		12.40	3.006		1,375	111		1,486	1,700
0370	45 W, flush, for mounting in base housing		18.64	2		680	73.50		753.50	860
0380	115 W	↓	18.64	2	↓	695	73.50		768.50	875

34 43 23 – Weather Observation Equipment

34 43 23.16 Airfield Wind Cones

		Crew	Daily Output	Labor-Hours	Unit	Material	2007 Bare Costs Labor	Equipment	Total	Total Incl O&P
1000	**AIRFIELD WIND CONES**									
1200	Wind cone, 12' lighted assembly, rigid, w/obstruction light	R-21	1.36	24.118	Ea.	7,950	1,050	64.50	9,064.50	10,400
1210	Without obstruction light		1.52	21.579		6,675	945	57.50	7,677.50	8,825
1220	Unlighted assembly, w/obstruction light		1.68	19.524		7,275	855	52	8,182	9,325
1230	Without obstruction light		1.84	17.826		6,675	780	47.50	7,502.50	8,575
1240	Wind cone slip fitter, 2-1/2" pipe		21.84	1.502		87.50	66	4.01	157.51	199
1250	Wind cone sock, 12' x 3', cotton		6.56	5		485	219	13.35	717.35	870
1260	Nylon	R-21	6.56	5	Ea.	490	219	13.35	722.35	880

Assemblies Section

Table of Contents

How to Use the Assemblies Cost Tables

The following is a detailed explanation of a sample Assemblies Cost Table. Most Assembly Tables are separated into three parts: 1) an illustration of the system to be estimated; 2) the components and related costs of a typical system; and 3) the costs for similar systems with dimensional and/or size variations. For costs of the components that comprise these systems, or assemblies, refer to the Unit Price Section. Next to each bold number below is the described item with the appropriate component of the sample entry following in parentheses. In most cases, if the work is to be subcontracted, the general contractor will need to add an additional markup (RSMeans suggests using 10%) to the "Total" figures.

1 System/Line Numbers (D5020 145 0200)

Each Assemblies Cost Line has been assigned a unique identification number based on the UNIFORMAT II classification system.

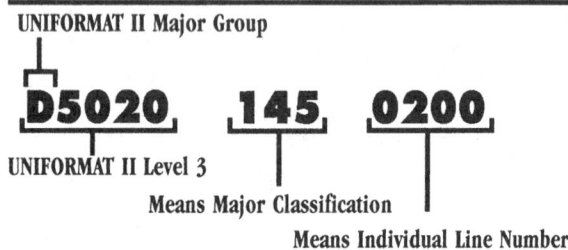

UNIFORMAT II Major Group

D5020   145   0200

UNIFORMAT II Level 3

Means Major Classification

Means Individual Line Number

D50 Electrical

D5020 Lighting and Branch Wiring

2

Power System — Switch — Motor Starter — Switch — Motor Connection — Motor

System D5020 145 installed cost of motor wiring as per Table D5010 170 using 50' of rigid conduit and copper wire. **Cost and setting of motor not included.**

6

System Components	QUANTITY	UNIT	COST EACH		
			MAT.	INST.	TOTAL
SYSTEM D5020 145 0200					
MOTOR INSTALLATION, SINGLE PHASE, 115V, TO AND INCLUDING 1/3 HP MOTOR SIZE					
Wire 600V type THWN-THHN, copper solid #12	1.250	C.L.F.	20.88	59.38	80.26
Steel intermediate conduit, (IMC) 1/2" diam.	50.000	L.F.	105.50	262.50	368
Magnetic FVNR, 115V, 1/3 HP, size 00 starter	1.000	Ea.	191	131	322
Safety switch, fused, heavy duty, 240V 2P 30 amp	1.000	Ea.	123	149	272
Safety switch, non fused, heavy duty, 600V, 3 phase, 30 A	1.000	Ea.	144	163	307
Flexible metallic conduit, Greenfield 1/2" diam.	1.500	L.F.	.62	3.92	4.54
Connectors for flexible metallic conduit Greenfield 1/2" diam.	1.000	Ea.	.70	.55	.25
Coupling for Greenfield to conduit 1/2" diam flexible metalic conduit	1.000	Ea.	1.87	.45	.32
Fuse cartridge nonrenewable, 250V 30 amp	1.000	Ea.	1.78	.45	2.23
TOTAL			592.35	796.25	1,388.60

3 **4** **5** **7** **8** **9**

D5020 145	Motor Installation	COST EACH		
		MAT.	INST.	TOTAL
0200	Motor installation, single phase, 115V, to and including 1/3 HP motor size	590	795	1,385
0240	To and incl. 1 HP motor size	615	795	1,410
0280	To and incl. 2 HP motor size	660	845	1,505
0320	To and incl. 3 HP motor size	750	860	1,610
0360	230V, to and including 1 HP motor size	595	805	1,400
0400	To and incl. 2 HP motor size	620	805	1,425
0440	To and incl. 3 HP motor size	695	865	1,560
0520	Three phase, 200V, to and including 1-1/2 HP motor size	695	885	1,580
0560	To and incl. 3 HP motor size	740	965	1,705
0600	To and incl. 5 HP motor size	795	1,075	1,870
0640	To and incl. 7-1/2 HP motor size	840	1,100	1,940

2 Illustration

At the top of most assembly pages are an illustration, a brief description, and the design criteria used to develop the cost.

3 System Components

The components of a typical system are listed separately to show what has been included in the development of the total system price. The table below contains prices for other similar systems with dimensional and/or size variations.

4 Quantity

This is the number of line item units required for one system unit. For example, we assume that it will take 50 linear feet of Steel Intermediate (IMC) Conduit for each motor to be connected to the other items listed here.

5 Unit of Measure for Each Item

The abbreviated designation indicates the unit of measure, as defined by industry standards, upon which the price of the component is based. For example, wire is priced by C.L.F. (100 linear feet) while conduit is priced by L.F. (linear feet) and the starter and switches are priced by each unit. For a complete listing of abbreviations, see the Reference Section.

6 Unit of Measure for Each System (Cost Each)

Costs shown in the three right-hand columns have been adjusted by the component quantity and unit of measure for the entire system. In this example, "Cost Each" is the unit of measure for this system, or assembly.

7 Materials (590)

This column contains the Materials Cost of each component. These cost figures are bare costs plus 10% for profit.

8 Installation (795)

Installation includes labor and equipment plus the installing contractor's overhead and profit. Equipment costs are the bare rental costs plus 10% for profit. The labor overhead and profit is defined on the inside back cover of this book.

9 Total (1,385)

The figure in this column is the sum of the material and installation costs.

Material Cost	+	Installation Cost	=	Total
$590.00	+	$795.00	=	$1,385.00

D2020 Domestic Water Distribution

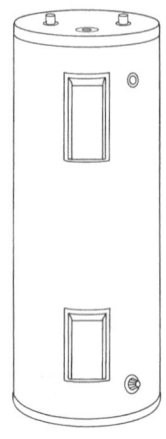

Installation includes piping and fittings within 10' of heater. Electric water heaters do not require venting.

1 Kilowatt hour will raise:			
Gallons of Water	Degrees F	Gallons of Water	Degrees F
4.1	100°	6.8	60°
4.5	90°	8.2	50°
5.1	80°	10.0	40°
5.9	70°		

System Components	QUANTITY	UNIT	COST EACH		
			MAT.	INST.	TOTAL
SYSTEM D2020 210 1780					
ELECTRIC WATER HEATER, RESIDENTIAL, 100° F RISE					
10 GALLON TANK, 7 GPH					
Water heater, residential electric, glass lined tank, 10 gal	1.000	Ea.	300	234	534
Copper tubing, type L, solder joint, hanger 10' OC 1/2" diam	30.000	L.F.	112.80	199.50	312.30
Wrought copper 90° elbow for solder joints 1/2" diam.	4.000	Ea.	3	108	111
Wrought copper Tee for solder joints, 1/2" diam	2.000	Ea.	2.54	83	85.54
Wrought copper union for soldered joints, 1/2" diam	2.000	Ea.	16.10	57	73.10
Valve, gate, bronze, 125 lb, NRS, soldered 1/2" diam	2.000	Ea.	48	45	93
Relief valve, bronze, press & temp, self-close, 3/4" IPS	1.000	Ea.	99	19.25	118.25
Wrought copper adapter, CTS to MPT 3/4" IPS	1.000	Ea.	2.62	31.50	34.12
Copper tubing, type L, solder joints, 3/4" diam	1.000	L.F.	5.70	7.10	12.80
Wrought copper 90° elbow for solder joints 3/4" diam	1.000	Ea.	1.67	28.50	30.17
TOTAL			591.43	812.85	1,404.28

D2020 210	Electric Water Heaters - Residential Systems		COST EACH		
			MAT.	INST.	TOTAL
1760	Electric water heater, residential, 100° F rise				
1780	10 gallon tank, 7 GPH		591.43	812.85	1,405.80
1820	20 gallon tank, 7 GPH	R224000 -20	755	870	1,625
1860	30 gallon tank, 7 GPH		870	910	1,780
1900	40 gallon tank, 8 GPH		1,025	1,000	2,025
1940	52 gallon tank, 10 GPH		1,100	1,000	2,100
1980	66 gallon tank, 13 GPH		1,650	1,150	2,800
2020	80 gallon tank, 16 GPH		1,750	1,200	2,950
2060	120 gallon tank, 23 GPH		2,550	1,400	3,950

D2020 Domestic Water Distribution

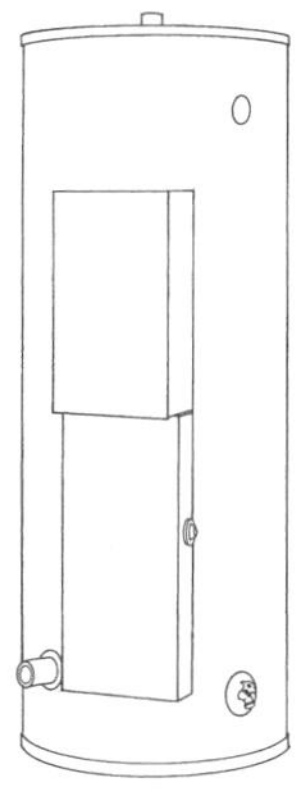

Systems below include piping and fittings within 10′ of heater. Electric water heaters do not require venting.

System Components	QUANTITY	UNIT	COST EACH MAT.	COST EACH INST.	COST EACH TOTAL
SYSTEM D2020 240 1820					
ELECTRIC WATER HEATER, COMMERCIAL, 100° F RISE					
50 GALLON TANK, 9 KW, 37 GPH					
Water heater, commercial, electric, 50 Gal, 9 KW, 37 GPH	1.000	Ea.	4,250	300	4,550
Copper tubing, type L, solder joint, hanger 10′ OC, 3/4″ diam	34.000	L.F.	193.80	241.40	435.20
Wrought copper 90° elbow for solder joints 3/4″ diam	5.000	Ea.	8.35	142.50	150.85
Wrought copper Tee for solder joints, 3/4″ diam	2.000	Ea.	6.14	90	96.14
Wrought copper union for soldered joints, 3/4″ diam	2.000	Ea.	20.20	60	80.20
Valve, gate, bronze, 125 lb, NRS, soldered 3/4″ diam	2.000	Ea.	55	54	109
Relief valve, bronze, press & temp, self-close, 3/4″ IPS	1.000	Ea.	99	19.25	118.25
Wrought copper adapter, copper tubing to male, 3/4″ IPS	1.000	Ea.	2.62	31.50	34.12
TOTAL			4,635.11	938.65	5,573.76

D2020 240	Electric Water Heaters - Commercial Systems		COST EACH MAT.	COST EACH INST.	COST EACH TOTAL
1800	Electric water heater, commercial, 100° F rise				
1820	50 gallon tank, 9 KW 37 GPH		4,625	940	5,565
1860	80 gal, 12 KW 49 GPH	R224000 -10	5,950	1,150	7,100
1900	36 KW 147 GPH		8,125	1,250	9,375
1940	120 gal, 36 KW 147 GPH	R224000 -20	8,750	1,350	10,100
1980	150 gal, 120 KW 490 GPH		27,100	1,450	28,550
2020	200 gal, 120 KW 490 GPH		28,700	1,475	30,175
2060	250 gal, 150 KW 615 GPH		32,000	1,725	33,725
2100	300 gal, 180 KW 738 GPH		34,200	1,825	36,025
2140	350 gal, 30 KW 123 GPH		26,300	1,975	28,275
2180	180 KW 738 GPH		35,200	1,975	37,175
2220	500 gal, 30 KW 123 GPH		33,800	2,325	36,125
2260	240 KW 984 GPH		48,000	2,325	50,325
2300	700 gal, 30 KW 123 GPH		42,100	2,650	44,750
2340	300 KW 1230 GPH		59,000	2,650	61,650
2380	1000 gal, 60 KW 245 GPH		48,800	3,675	52,475
2420	480 KW 1970 GPH		79,500	3,700	83,200
2460	1500 gal, 60 KW 245 GPH		71,500	4,575	76,075
2500	480 KW 1970 GPH		113,000	4,575	117,575

D3020 Heat Generating Systems

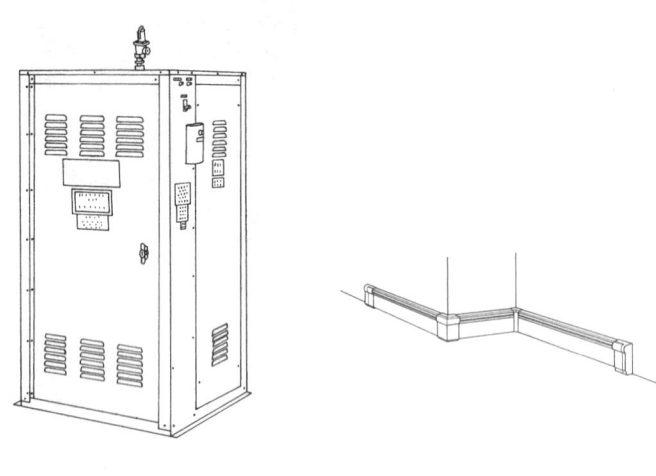

Boiler   **Baseboard Radiation**

Small Electric Boiler
System Considerations:
1. Terminal units are fin tube baseboard radiation rated at 720 BTU/hr with 200° water temperature or 820 BTU/hr steam.
2. Primary use being for residential or smaller supplementary areas, the floor levels are based on 7-1/2′ ceiling heights.
3. All distribution piping is copper for boilers through 205 MBH. All piping for larger systems is steel pipe.

System Components	QUANTITY	UNIT	COST EACH		
			MAT.	INST.	TOTAL
SYSTEM D3020 102 1120					
SMALL HEATING SYSTEM, HYDRONIC, ELECTRIC BOILER					
1,480 S.F., 61 MBH, STEAM, 1 FLOOR					
Boiler, electric steam, std cntrls, trim, ftngs and valves, 18 KW, 61.4 MBH	1.000	Ea.	3,877.50	1,375	5,252.50
Copper tubing type L, solder joint, hanger 10′OC, 1-1/4″ diam	160.000	L.F.	1,832	1,488	3,320
Radiation, 3/4″ copper tube w/alum fin baseboard pkg 7″ high	60.000	L.F.	468	1,014	1,482
Rough in baseboard panel or fin tube with valves & traps	10.000	Set	1,830	5,150	6,980
Pipe covering, calcium silicate w/cover 1″ wall 1-1/4″ diam	160.000	L.F.	337.60	920	1,257.60
Low water cut-off, quick hookup, in gage glass tappings	1.000	Ea.	203	34	237
TOTAL			8,548.10	9,981	18,529.10
COST PER S.F.			5.78	6.74	12.52

D3020 102	Small Heating Systems, Hydronic, Electric Boilers	COST PER S.F.		
		MAT.	INST.	TOTAL
1100	Small heating systems, hydronic, electric boilers			
1120	Steam, 1 floor, 1480 S.F., 61 M.B.H.	5.78	6.74	12.52
1160	3,000 S.F., 123 M.B.H.	4.43	5.90	10.33
1200	5,000 S.F., 205 M.B.H.	3.89	5.45	9.34
1240	2 floors, 12,400 S.F., 512 M.B.H.	2.83	5.40	8.23
1280	3 floors, 24,800 S.F., 1023 M.B.H.	2.96	5.35	8.31
1320	34,750 S.F., 1,433 M.B.H.	2.71	5.20	7.91
1360	Hot water, 1 floor, 1,000 S.F., 41 M.B.H.	8.90	3.74	12.64
1400	2,500 S.F., 103 M.B.H.	6.75	6.75	13.50
1440	2 floors, 4,850 S.F., 205 M.B.H.	6.90	8.10	15
1480	3 floors, 9,700 S.F., 410 M.B.H.	7.70	8.40	16.10

D3020 Heat Generating Systems

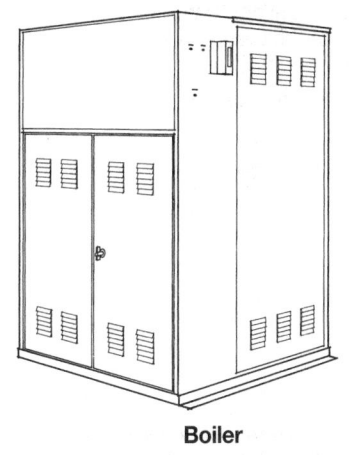

Boiler

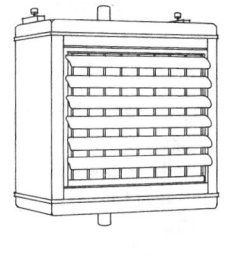

Unit Heater

Large Electric Boiler System Considerations:

1. Terminal units are all unit heaters of the same size. Quantities are varied to accommodate total requirements.
2. All air is circulated through the heaters a minimum of three times per hour.
3. As the capacities are adequate for commercial use, floor levels are based on 10′ ceiling heights.
4. All distribution piping is black steel pipe.

System Components	QUANTITY	UNIT	COST EACH		
			MAT.	INST.	TOTAL
SYSTEM D3020 104 1240					
LARGE HEATING SYSTEM, HYDRONIC, ELECTRIC BOILER					
9,280 S.F., 135 KW, 461 MBH, 1 FLOOR					
Boiler, electric hot water, std ctrls, trim, ftngs, valves, 135 KW, 461 MBH	1.000	Ea.	9,412.50	3,000	12,412.50
Expansion tank, painted steel, 60 Gal capacity ASME	1.000	Ea.	3,325	163	3,488
Circulating pump, CI, close cpld, 50 GPM, 2 HP, 2″ pipe conn	1.000	Ea.	1,900	325	2,225
Unit heater, 1 speed propeller, horizontal, 200° EWT, 72.7 MBH	7.000	Ea.	4,900	1,246	6,146
Unit heater piping hookup with controls	7.000	Set	3,115	8,050	11,165
Pipe, steel, black, schedule 40, welded, 2-1/2″ diam	380.000	L.F.	3,762	8,337.20	12,099.20
Pipe covering, calcium silicate w/cover, 1″ wall, 2-1/2″ diam	380.000	L.F.	995.60	2,242	3,237.60
TOTAL			27,410.10	23,363.20	50,773.30
COST PER S.F.			2.95	2.52	5.47

D3020 104	Large Heating Systems, Hydronic, Electric Boilers	COST PER S.F.		
		MAT.	INST.	TOTAL
1230	Large heating systems, hydronic, electric boilers			
1240	9,280 S.F., 135 K.W., 461 M.B.H., 1 floor	2.95	2.51	5.46
1280	14,900 S.F., 240 K.W., 820 M.B.H., 2 floors	3.68	4.10	7.78
1320	18,600 S.F., 296 K.W., 1,010 M.B.H., 3 floors	3.63	4.45	8.08
1360	26,100 S.F., 420 K.W., 1,432 M.B.H., 4 floors	3.60	4.37	7.97
1400	39,100 S.F., 666 K.W., 2,273 M.B.H., 4 floors	3.07	3.65	6.72
1440	57,700 S.F., 900 K.W., 3,071 M.B.H., 5 floors	2.95	3.60	6.55
1480	111,700 S.F., 1,800 K.W., 6,148 M.B.H., 6 floors	2.68	3.09	5.77
1520	149,000 S.F., 2,400 K.W., 8,191 M.B.H., 8 floors	2.67	3.08	5.75
1560	223,300 S.F., 3,600 K.W., 12,283 M.B.H., 14 floors	2.83	3.50	6.33

D40 Fire Protection

D4090 Other Fire Protection Systems

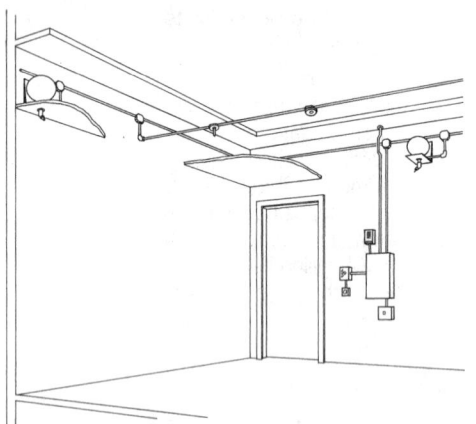

General: Automatic fire protection (suppression) systems other than water sprinklers may be desired for special environments, high risk areas, isolated locations or unusual hazards. Some typical applications would include:

Paint dip tanks
Securities vaults
Electronic data processing
Tape and data storage
Transformer rooms
Spray booths
Petroleum storage
High rack storage

Piping and wiring costs are dependent on the individual application and must be added to the component costs shown below.

All areas are assumed to be open.

D4090 910	Unit Components	COST EACH		
		MAT.	INST.	TOTAL
0020	Detectors with brackets			
0040	Fixed temperature heat detector	32.50	65.50	98
0060	Rate of temperature rise detector	38.50	65.50	104
0080	Ion detector (smoke) detector	86.50	84.50	171
0200	Extinguisher agent			
0240	200 lb FM200, container	6,700	243	6,943
0280	75 lb carbon dioxide cylinder	1,100	162	1,262
0320	Dispersion nozzle			
0340	FM200 1-1/2" dispersion nozzle	58	38.50	96.50
0380	Carbon dioxide 3" x 5" dispersion nozzle	58	30	88
0420	Control station			
0440	Single zone control station with batteries	1,475	525	2,000
0470	Multizone (4) control station with batteries	2,825	1,050	3,875
0490				
0500	Electric mechanical release	144	266	410
0520				
0550	Manual pull station	52	90	142
0570				
0640	Battery standby power 10" x 10" x 17"	800	131	931
0700				
0740	Bell signalling device	57	65.50	122.50

D4090 920	FM200 Systems	COST PER C.F.		
		MAT.	INST.	TOTAL
0820	Average FM200 system, minimum			1.52
0840	Maximum			3.03

330

Figure D5010-111 Typical Overhead Service Entrance

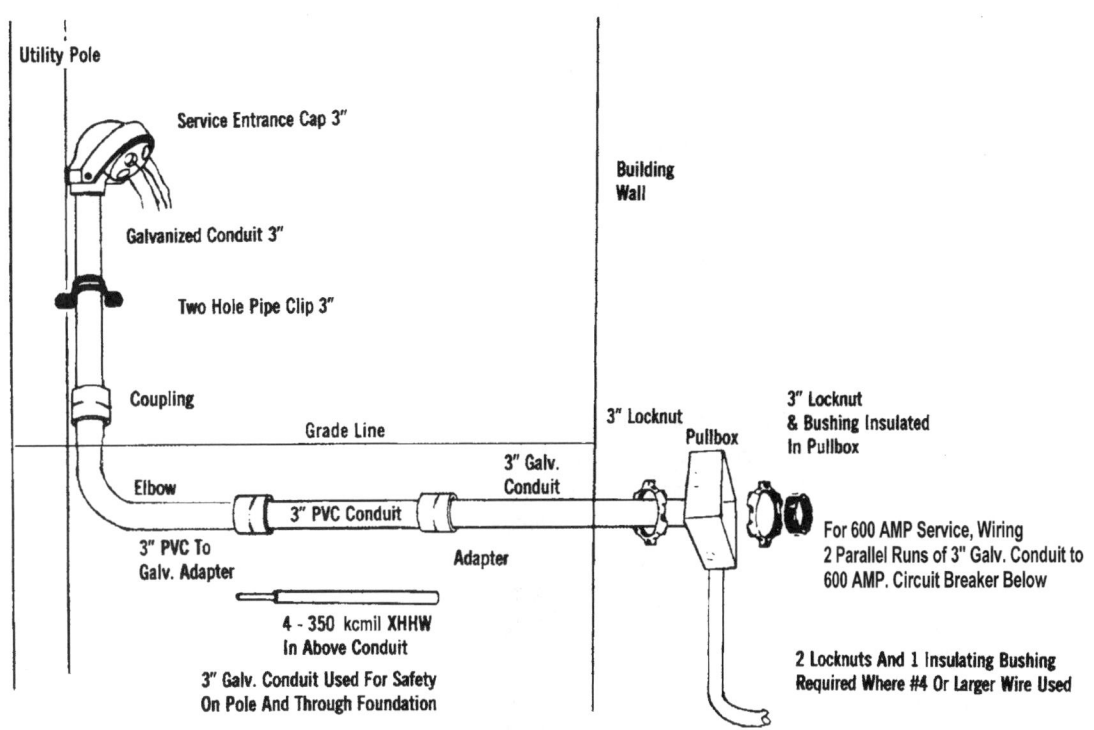

Figure D5010-112 Typical Commercial Electric System

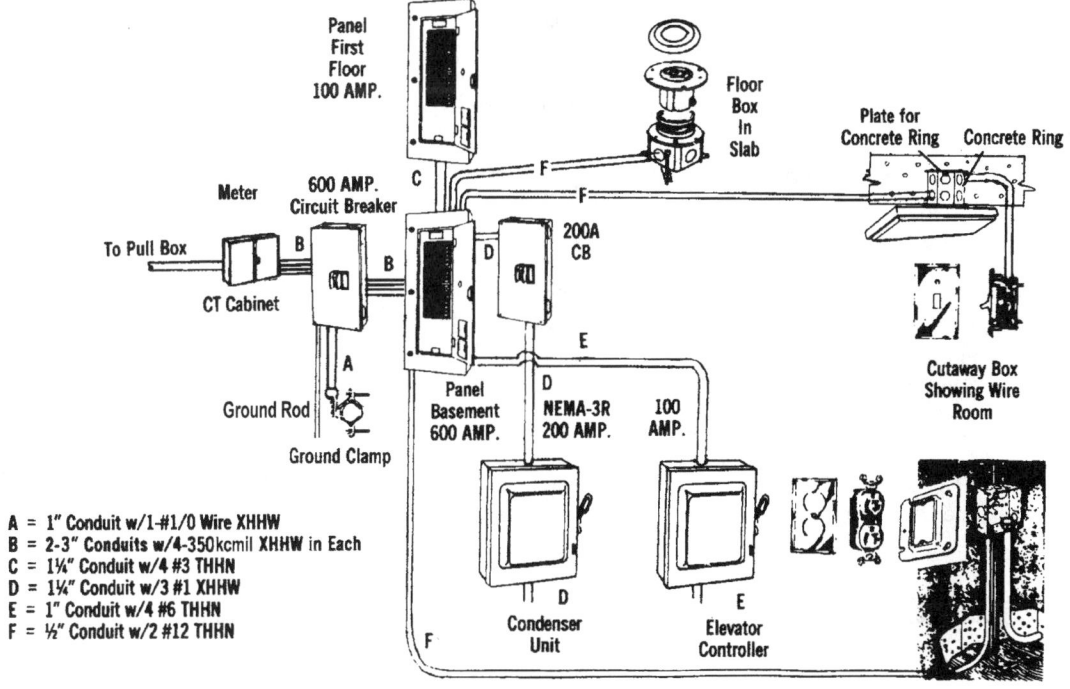

A = 1" Conduit w/1-#1/0 Wire XHHW
B = 2-3" Conduits w/4-350kcmil XHHW in Each
C = 1¼" Conduit w/4 #3 THHN
D = 1¼" Conduit w/3 #1 XHHW
E = 1" Conduit w/4 #6 THHN
F = ½" Conduit w/2 #12 THHN

Figure D5010-113 Preliminary Procedure

1. Determine building size and use.
2. Develop total load in watts
   a. Lighting
   b. Receptacle
   c. Air Conditioning
   d. Elevator
   e. Other power requirements
3. Determine best voltage available from utility company.
4. Determine cost from tables for loads (a) thru (e) above.
5. Determine size of service from formulas (D5010-116).
6. Determine costs for service, panels, and feeders from tables.

Figure D5010-114 Office Building 90' x 210', 3 story, w/garage

Garage Area = 18,900 S.F.
Office Area = 56,700 S.F.
Elevator = 2 @ 125 FPM

Reference Tables	Power Required	Watts
D5010-1151	Garage Lighting .5 Watt/S.F.	9,450
	Office Lighting 3 Watts/S.F.	170,100
D5010-1151	Office Receptacles 2 Watts/S.F.	113,400
D5010-1151, D5020-602	Low Rise Office A.C. 4.3 Watts/S.F.	243,810
D5010-1152, 1153	Elevators - 2 @ 20 HP = 2 @ 17,404 Watts/Ea.	34,808
D5010-1151	Misc. Motors + Power 1.2 Watts/S.F.	68,040
	Total	639,608 Watts

Voltage Available

277/480V, 3 Phase, 4 Wire

Formula

$$\text{D5010-116} \qquad \text{Amperes} = \frac{\text{Watts}}{\text{Volts x Power Factor x 1.73}} = \frac{639,608}{480V \text{ x } .8 \text{ x } 1.73} = 963 \text{ Amps}$$

Use 1200 Amp Service

System	Description	Unit Cost	Unit	Total
D5020 210 0200	Garage Lighting (Interpolated)	$1.15	S.F.	$ 21,641
D5020 210 0280	Office Lighting	6.80	S.F.	385,560
D5020 115 0880	Receptacle - Undercarpet	3.18	S.F.	180,306
D5020 140 0280	Air Conditioning	.56	S.F.	31,752
D5020 135 0320	Misc. Pwr.	.29	S.F.	16,443
D5020 145 2120	Elevators - 2 @ 20HP	2,500.00	Ea.	5,000
D5010 120 0480	Service - 1200 Amp (add 25% for 277/480V)	19,950.00	Ea.	24,938
D5010 230 0480	Feeder - Assume 200 Ft.	333.00	Ft.	66,600
D5010 240 0320	Panels - 1200 Amp (add 20% for 277/480V)	27,250.00	Ea.	32,700
D5030 910 0400	Fire Detection	27,850.00	Ea.	27,850
	Total			$792,790
	or			$792,800

Table D5010-1151 Nominal Watts Per S.F. for Electric Systems for Various Building Types

Type Construction	1. Lighting	2. Devices	3. HVAC	4. Misc.	5. Elevator	Total Watts
Apartment, luxury high rise	2	2.2	3	1		
Apartment, low rise	2	2	3	1		
Auditorium	2.5	1	3.3	.8		
Bank, branch office	3	2.1	5.7	1.4		
Bank, main office	2.5	1.5	5.7	1.4		
Church	1.8	.8	3.3	.8		
College, science building	3	3	5.3	1.3		
College, library	2.5	.8	5.7	1.4		
College, physical education center	2	1	4.5	1.1		
Department store	2.5	.9	4	1		
Dormitory, college	1.5	1.2	4	1		
Drive-in donut shop	3	4	6.8	1.7		
Garage, commercial	.5	.5	0	.5		
Hospital, general	2	4.5	5	1.3		
Hospital, pediatric	3	3.8	5	1.3		
Hotel, airport	2	1	5	1.3		
Housing for the elderly	2	1.2	4	1		
Manufacturing, food processing	3	1	4.5	1.1		
Manufacturing, apparel	2	1	4.5	1.1		
Manufacturing, tools	4	1	4.5	1.1		
Medical clinic	2.5	1.5	3.2	1		
Nursing home	2	1.6	4	1		
Office building, hi rise	3	2	4.7	1.2		
Office building, low rise	3	2	4.3	1.2		
Radio-TV studio	3.8	2.2	7.6	1.9		
Restaurant	2.5	2	6.8	1.7		
Retail store	2.5	.9	5.5	1.4		
School, elementary	3	1.9	5.3	1.3		
School, junior high	3	1.5	5.3	1.3		
School, senior high	2.3	1.7	5.3	1.3		
Supermarket	3	1	4	1		
Telephone exchange	1	.6	4.5	1.1		
Theater	2.5	1	3.3	.8		
Town Hall	2	1.9	5.3	1.3		
U.S. Post Office	3	2	5	1.3		
Warehouse, grocery	1	.6	0	.5		

Rule of Thumb: 1 KVA = 1 HP (Single Phase)

Three Phase:

Watts = 1.73 x Volts x Current x Power Factor x Efficiency

$$\text{Horsepower} = \frac{\text{Volts x Current x 1.73 x Power Factor}}{746 \text{ Watts}}$$

Table D5010-1152 Horsepower Requirements for Elevators with 3 Phase Motors

Type	Maximum Travel Height in Ft.	Travel Speeds in FPM	Capacity of Cars in Lbs.		
			1200	1500	1800
Hydraulic	70	70	10	15	15
		85	15	15	15
		100	15	15	20
		110	20	20	20
		125	20	20	20
		150	25	25	25
		175	25	30	30
		200	30	30	40
Geared Traction	300	200			
		350			
			2000	2500	3000
Hydraulic	70	70	15	20	20
		85	20	20	25
		100	20	25	30
		110	20	25	30
		125	25	30	40
		150	30	40	50
		175	40	50	50
		200	40	50	60
Geared Traction	300	200	10	10	15
		350	15	15	23
			3500	4000	4500
Hydraulic	70	70	20	25	30
		85	25	30	30
		100	30	40	40
		110	40	40	50
		125	40	50	50
		150	50	50	60
		175	60		
		200	60		
Geared Traction	300	200	15		23
		350	23		35

The power factor of electric motors varies from 80% to 90% in larger size motors. The efficiency likewise varies from 80% on a small motor to 90% on a large motor.

Table D5010-1153 Watts per Motor

90% Power Factor & Efficiency @ 200 or 460V			
HP	Watts	HP	Watts
10	9024	30	25784
15	13537	40	33519
20	17404	50	41899
25	21916	60	49634

Table D5010-116 Electrical Formulas

Ohm's Law

Ohm's Law is a method of explaining the relation existing between voltage, current, and resistance in an electrical circuit. It is practically the basis of all electrical calculations. The term "electromotive force" is often used to designate pressure in volts. This formula can be expressed in various forms.

To find the current in amperes:

$$\text{Current} = \frac{\text{Voltage}}{\text{Resistance}} \quad \text{or} \quad \text{Amperes} = \frac{\text{Volts}}{\text{Ohms}} \quad \text{or} \quad I = \frac{E}{R}$$

The flow of current in amperes through any circuit is equal to the voltage or electromotive force divided by the resistance of that circuit.

To find the pressure or voltage:

Voltage = Current x Resistance   or   Volts = Amperes x Ohms
$$\text{or} \quad E = I \times R$$

The voltage required to force a current through a circuit is equal to the resistance of the circuit multiplied by the current.

To find the resistance:

$$\text{Resistance} = \frac{\text{Voltage}}{\text{Current}} \quad \text{or} \quad \text{Ohms} = \frac{\text{Volts}}{\text{Amperes}} \quad \text{or} \quad R = \frac{E}{I}$$

The resistance of a circuit is equal to the voltage divided by the current flowing through that circuit.

Power Formulas

One horsepower = 746 watts   One kilowatt = 1000 watts

The power factor of electric motors varies from 80% to 90% in the larger size motors.

Single-Phase Alternating Current Circuits

Power in Watts = Volts x Amperes x Power Factor

To find current in amperes:

$$\text{Current} = \frac{\text{Watts}}{\text{Volts x Power Factor}} \quad \text{or}$$

$$\text{Amperes} = \frac{\text{Watts}}{\text{Volts x Power Factor}} \quad \text{or} \quad I = \frac{W}{E \times PF}$$

To find current of a motor, single phase:

$$\text{Current} = \frac{\text{Horsepower x 746}}{\text{Volts x Power Factor x Efficiency}} \quad \text{or}$$

$$I = \frac{HP \times 746}{E \times PF \times Eff.}$$

To find horsepower of a motor, single phase:

$$\text{Horsepower} = \frac{\text{Volts x Current x Power Factor x Efficiency}}{\text{746 Watts}}$$

$$HP = \frac{E \times I \times PF \times Eff.}{746}$$

To find power in watts of a motor, single phase:

Watts = Volts x Current x Power Factor x Efficiency   or
Watts = E x I x PF x Eff.

To find single phase kVA:

$$1 \text{ Phase kVA} = \frac{\text{Volts x Amps}}{1000}$$

Three-Phase Alternating Current Circuits

Power in Watts = Volts x Amperes x Power Factor x 1.73

To find current in amperes in each wire:

$$\text{Current} = \frac{\text{Watts}}{\text{Voltage x Power Factor x 1.73}} \quad \text{or}$$

$$\text{Amperes} = \frac{\text{Watts}}{\text{Volts x Power Factor x 1.73}} \quad \text{or} \quad I = \frac{W}{E \times PF \times 1.73}$$

To find current of a motor, 3 phase:

$$\text{Current} = \frac{\text{Horsepower x 746}}{\text{Volts x Power Factor x Efficiency x 1.73}} \quad \text{or}$$

$$I = \frac{HP \times 746}{E \times PF \times Eff. \times 1.73}$$

To find horsepower of a motor, 3 phase:

$$\text{Horsepower} = \frac{\text{Volts x Current x 1.73 x Power Factor}}{\text{746 Watts}}$$

$$HP = \frac{E \times I \times 1.73 \times PF}{746}$$

To find power in watts of a motor, 3 phase:

Watts = Volts x Current x 1.73 x Power Factor x Efficiency   or
Watts = E x I x 1.73 x PF x Eff.

To find 3 phase kVA:

$$3 \text{ phase kVA} = \frac{\text{Volts x Amps x 1.73}}{1000} \quad \text{or}$$

$$kVA = \frac{V \times A \times 1.73}{1000}$$

Power Factor (PF) is the percentage ratio of the measured watts (effective power) to the volt-amperes (apparent watts)

$$\text{Power Factor} = \frac{\text{Watts}}{\text{Volts x Amperes}} \times 100\%$$

A Conceptual Estimate of the costs for a building when final drawings are not available can be quickly figured by using **Table D5010-117 Cost Per S.F. for Electrical Systems for Various Building Types.** The following definitions apply to this table.

1. **Service and Distribution:** This system includes the incoming primary feeder from the power company, main building transformer, metering arrangement, switchboards, distribution panel boards, stepdown transformers, and power and lighting panels. Items marked (*) include the cost of the primary feeder and transformer. In all other projects the cost of the primary feeder and transformer is paid for by the local power company.

2. **Lighting:** Includes all interior fixtures for decor, illumination, exit and emergency lighting. Fixtures for exterior building lighting are included, but parking area lighting is not included unless mentioned. See also Section D5020 for detailed analysis of lighting requirements and costs.

3. **Devices:** Includes all outlet boxes, receptacles, switches for lighting control, dimmers and cover plates.

4. **Equipment Connections:** Includes all materials and equipment for making connections for heating, ventilating and air conditioning, food service and other motorized items requiring connections.

5. **Basic Materials:** This category includes all disconnect power switches not part of service equipment, raceways for wires, pull boxes, junction boxes, supports, fittings, grounding materials, wireways, busways, and wire and cable systems.

6. **Special Systems:** Includes installed equipment only for the particular system such as fire detection and alarm, sound, emergency generator and others as listed in the table.

Table D5010-117   Cost per S.F. for Electric Systems for Various Building Types

Type Construction	1. Service & Distrib.	2. Lighting	3. Devices	4. Equipment Connections	5. Basic Materials	6. Special Systems Fire Alarm & Detection	Lightning Protection	Master TV Antenna
Apartment, luxury high rise	$1.82	$1.13	$ .84	$1.12	$ 3.06	$ .54		$ .36
Apartment, low rise	1.06	.94	.77	.95	1.78	.46		
Auditorium	2.32	5.69	.69	1.60	3.66	.70		
Bank, branch office	2.78	6.32	1.13	1.64	3.47	2.00		
Bank, main office	2.09	3.44	.36	.69	3.76	1.03		
Church	1.45	3.49	.43	.42	1.80	1.03		
* College, science building	3.00	4.63	1.66	1.45	4.73	.90		
* College library	1.96	2.57	.33	.76	2.25	1.03		
* College, physical education center	3.08	3.58	.46	.62	1.75	.58		
Department store	1.02	2.47	.33	1.10	3.01	.46		
* Dormitory, college	1.39	3.18	.34	.69	2.96	.73		.47
Drive-in donut shop	3.87	9.50	1.62	1.65	4.84	—		
Garage, commercial	.52	1.28	.25	.51	1.04	—		
* Hospital, general	7.49	4.90	1.85	1.29	5.98	.63	$.21	
* Hospital, pediatric	6.56	7.29	1.60	4.75	11.11	.72		.56
* Hotel, airport	2.94	4.01	.34	.64	4.40	.59	.34	.52
Housing for the elderly	.83	.95	.45	1.25	3.75	.72		.46
Manufacturing, food processing	1.87	5.00	.31	2.36	4.09	.45		
Manufacturing, apparel	1.24	2.62	.37	.94	2.19	.40		
Manufacturing, tools	2.85	6.11	.35	1.11	3.66	.47		
Medical clinic	1.47	2.32	.56	1.85	3.18	.71		
Nursing home	1.94	3.95	.56	.50	3.66	.98		.38
Office Building	2.60	5.35	.31	.94	3.83	.52	.31	
Radio-TV studio	1.87	5.52	.84	1.73	4.53	.67		
Restaurant	6.97	5.24	1.01	2.62	5.44	.40		
Retail Store	1.47	2.78	.34	.64	1.71	—		
School, elementary	2.51	4.94	.68	.64	4.57	.62		.30
School, junior high	1.53	4.13	.34	1.16	3.63	.73		
* School, senior high	1.64	3.24	.59	1.51	4.04	.64		
Supermarket	1.68	2.79	.40	2.52	3.50	.32		
* Telephone Exchange	4.35	1.54	.38	1.41	2.72	1.20		
Theater	3.21	3.82	.70	2.19	3.55	.88		
Town Hall	1.97	3.02	.70	.79	4.70	.58		
* U.S. Post Office	5.76	3.88	.71	1.20	3.36	.58		
Warehouse, grocery	1.08	1.69	.26	.64	2.49	.37		

*Includes cost of primary feeder and transformer. Cont'd on next page.

COST ASSUMPTIONS:

Each of the projects analyzed in Table D5010-117 was bid within the last 10 years in the northeastern part of the United States. Bid prices have been adjusted to Jan. 1 levels. The list of projects is by no means all-inclusive, yet by carefully examining the various systems for a particular building type, certain cost relationships will emerge. The use of Square Foot/Cubic Costs, in the Reference section, along with the Project Size Modifier, should produce a budget S.F. cost for the electrical portion of a job that is consistent with the amount of design information normally available at the conceptual estimate stage.

Table D5010-117 Cost per S.F. for Electric Systems for Various Building Types (cont.)

Type Construction	6. Special Systems, (cont.)						
	Intercom Systems	Sound Systems	Closed Circuit TV	Snow Melting	Emergency Generator	Security	Master Clock Sys.
Apartment, luxury high rise	$ .72						
Apartment, low rise	.52						
Auditorium		$1.89	$ .87		$1.34		
Bank, branch office	1.00		1.99			$1.69	
Bank, main office	.56		.43		1.09	.92	$.40
Church	.70	.43				.28	
* College, science building	.73	.93			1.51	.48	.46
* College, library		.87			.70	.47	
* College, physical education center		.96				.49	
Department store					.33		
* Dormitory, college	.98						
Drive-in donut shop							.21
Garage, commercial							.18
Hospital, general	.74		.32		1.85		
* Hospital, pediatric	5.04	.51	.55		1.16		
* Hotel, airport	.73				.69		
Housing for the elderly	.88						
Manufacturing, food processing		.34			2.41		
Manufacturing apparel		.45					
Manufacturing, tools		.56		$.37			
Medical clinic							
Nursing home	1.67				.62		
Office Building		.30			.62	.33	.18
Radio-TV studio	.99				1.52		.70
Restaurant		.45					
Retail Store							
School, elementary		.32					.32
School, junior high		.81			.51		.55
* School, senior high	.67		.45		.70	.40	.41
Supermarket		.35			.65	.45	
* Telephone exchange					6.12	.25	
Theater		.65					
Town Hall							.32
* U.S. Post Office	.65		.17		.68		
Warehouse, grocery	.42						

*Includes cost of primary feeder and transformer. Cont'd on next page.

General: Variations in the following square foot costs are due to the type of structural systems of the buildings, geographical location, local electrical codes, designer's preference for specific materials and equipment, and the owner's particular requirements.

Table D5010-117 Cost per S.F. for Total Electric Systems for Various Building Types (cont.)

Type Construction	Basic Description	Total Floor Area in Square Feet	Total Cost per Square Foot for Total Electric Systems
Apartment building, luxury high rise	All electric, 18 floors, 86 1 B.R., 34 2 B.R.	115,000	$ 9.61
Apartment building, low rise	All electric, 2 floors, 44 units, 1 & 2 B.R.	40,200	6.47
Auditorium	All electric, 1200 person capacity	28,000	18.75
Bank, branch office	All electric, 1 floor	2,700	22.02
Bank, main office	All electric, 8 floors	54,900	14.78
Church	All electric, incl. Sunday school	17,700	10.03
*College, science building	All electric, 3-1/2 floors, 47 rooms	27,500	20.47
*College, library	All electric	33,500	10.93
*College, physical education center	All electric	22,000	11.52
Department store	Gas heat, 1 floor	85,800	8.73
*Dormitory, college	All electric, 125 rooms	63,000	10.73
Drive-in donut shop	Gas heat, incl. parking area lighting	1,500	21.70
Garage, commercial	All electric	52,300	3.78
*Hospital, general	Steam heat, 4 story garage, 300 beds	540,000	25.27
*Hospital, pediatric	Steam heat, 6 stories	278,000	39.85
Hotel, airport	All electric, 625 guest rooms	536,000	15.20
Housing for the elderly	All electric, 7 floors, 100 1 B.R. units	67,000	9.28
Manufacturing, food processing	Electric heat, 1 floor	9,600	16.83
Manufacturing, apparel	Electric heat, 1 floor	28,000	8.20
Manufacturing, tools	Electric heat, 2 floors	42,000	15.48
Medical clinic	Electric heat, 2 floors	22,700	10.09
Nursing home	Gas heat, 3 floors, 60 beds	21,000	14.27
Office building	All electric, 15 floors	311,200	15.29
Radio-TV studio	Electric heat, 3 floors	54,000	18.35
Restaurant	All electric	2,900	22.12
Retail store	All electric	3,000	6.94
School, elementary	All electric, 1 floor	39,500	14.89
School, junior high	All electric, 1 floor	49,500	13.38
*School, senior high	All electric, 1 floor	158,300	14.30
Supermarket	Gas heat	30,600	12.66
*Telephone exchange	Gas heat, 300 kW emergency generator	24,800	17.98
Theater	Electric heat, twin cinema	14,000	15.01
Town Hall	All electric	20,000	12.09
*U.S. Post Office	All electric	495,000	17.00
Warehouse, grocery	All electric	96,400	6.96

*Includes cost of primary feeder and transformer.

D5010 Electrical Service/Distribution

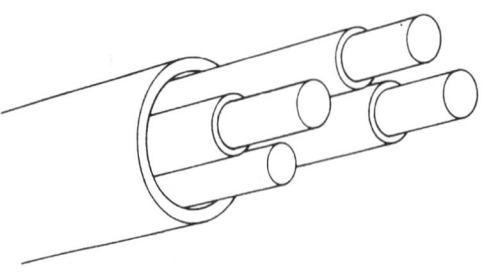

System Components	QUANTITY	UNIT	COST PER L.F.		
			MAT.	INST.	TOTAL
SYSTEM D5010 110 0200					
HIGH VOLTAGE CABLE, NEUTRAL AND CONDUIT INCLUDED, COPPER #2, 5 kV					
Shielded cable, no splice/termn, copper, XLP shielding, 5 kV, #2	.030	C.L.F.	10.50	7.83	18.33
Wire 600 volt, type THW, copper, stranded, #4	.010	C.L.F.	1.30	.99	2.29
Rigid galv steel conduit to 15' H, 2" diam, w/ term, ftng & support	1.000	L.F.	8.80	11.60	20.40
TOTAL			20.60	20.42	41.02

D5010 110	High Voltage Shielded Conductors	COST PER L.F.		
		MAT.	INST.	TOTAL
0200	High voltage cable, neutral & conduit included, copper #2, 5 kV	20.50	20.50	41
0240	Copper #1, 5 kV	23	20.50	43.50
0280	15 kV	36.50	30	66.50
0320	Copper 1/0, 5 kV	24	21	45
0360	15 kV	39.50	30.50	70
0400	25 kV	46	31	77
0440	35 kV	52.50	34.50	87
0480	Copper 2/0, 5 kV	35.50	25	60.50
0520	15 kV	42.50	31	73.50
0560	25 kV	54	34.50	88.50
0600	35 kV	61	37	98
0640	Copper 4/0, 5 kV	45.50	32.50	78
0680	15 kV	55	35.50	90.50
0720	25 kV	61.50	36.50	98
0760	35 kV	69.50	39	108.50
0800	Copper 250 kcmil, 5 kV	51.50	33.50	85
0840	15 kV	58.50	36.50	95
0880	25 kV	74.50	39.50	114
0920	35 kV	108	49	157
0960	Copper 350 kcmil, 5 kV	61	35.50	96.50
1000	15 kV	72	40.50	112.50
1040	25 kV	83	41.50	124.50
1080	35 kV	119	52	171
1120	Copper 500 kcmil, 5 kV	76.50	40	116.50
1160	15 kV	114	51	165
1200	25 kV	125	52	177
1240	35 kV	131	53.50	184.50

D5010 Electrical Service/Distribution

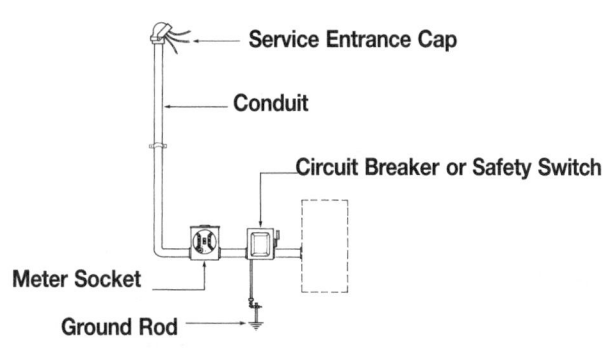

Service Entrance Cap

Conduit

Circuit Breaker or Safety Switch

Meter Socket

Ground Rod

System Components	QUANTITY	UNIT	COST EACH		
			MAT.	INST.	TOTAL
SYSTEM D5010 120 0220					
SERVICE INSTALLATION, INCLUDES BREAKERS, METERING, 20' CONDUIT & WIRE					
3 PHASE, 4 WIRE, 60 A					
Circuit breaker, enclosed (NEMA 1), 600 volt, 3 pole, 60 A	1.000	Ea.	585	187	772
Meter socket, single position, 4 terminal, 100 A	1.000	Ea.	41	163	204
Rigid galvanized steel conduit, 3/4", including fittings	20.000	L.F.	62.20	131	193.20
Wire, 600V type XHHW, copper stranded #6	.900	C.L.F.	88.65	72.45	161.10
Service entrance cap 3/4" diameter	1.000	Ea.	9.35	40	49.35
Conduit LB fitting with cover, 3/4" diameter	1.000	Ea.	13.65	40	53.65
Ground rod, copper clad, 8' long, 3/4" diameter	1.000	Ea.	32	98.50	130.50
Ground rod clamp, bronze, 3/4" diameter	1.000	Ea.	6.35	16.35	22.70
Ground wire, bare armored, #6-1 conductor	.200	C.L.F.	28.80	58	86.80
TOTAL			867	806.30	1,673.30

D5010 120	Electric Service, 3 Phase - 4 Wire	COST EACH		
		MAT.	INST.	TOTAL
0200	Service installation, includes breakers, metering, 20' conduit & wire			
0220	3 phase, 4 wire, 120/208 volts, 60 A	865	805	1,670
0240	100 A	1,100	970	2,070
0280	200 A	1,650	1,500	3,150
0320	400 A	3,875	2,750	6,625
0360	600 A	7,275	3,700	10,975
0400	800 A	9,650	4,450	14,100
0440	1000 A	11,900	5,150	17,050
0480	1200 A	14,700	5,250	19,950
0520	1600 A	26,700	7,550	34,250
0560	2000 A	29,800	8,600	38,400
0570	Add 25% for 277/480 volt			
0610	1 phase, 3 wire, 120/240 volts, 100 A	480	875	1,355
0620	200 A	1,000	1,275	2,275

D5010 Electrical Service/Distribution

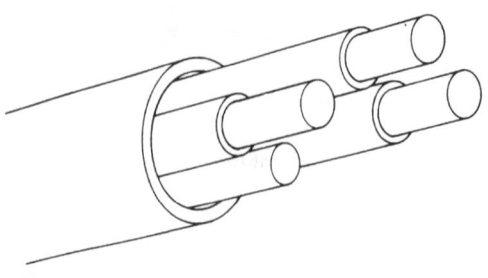

System Components	QUANTITY	UNIT	COST PER L.F.		
			MAT.	INST.	TOTAL
SYSTEM D5010 230 0200					
FEEDERS, INCLUDING STEEL CONDUIT & WIRE, 60 A					
Rigid galvanized steel conduit, 3/4", including fittings	1.000	L.F.	3.11	6.55	9.66
Wire 600 volt, type XHHW copper stranded #6	.040	C.L.F.	3.94	3.22	7.16
TOTAL			7.05	9.77	16.82

D5010 230	Feeder Installation	COST PER L.F.		
		MAT.	INST.	TOTAL
0200	Feeder installation 600 V, including RGS conduit and XHHW wire, 60 A	7.05	9.75	16.80
0240	100 A	12.45	12.90	25.35
0280	200 A	27	19.95	46.95
0320	400 A	54.50	40	94.50
0360	600 A	114	65	179
0400	800 A	161	78	239
0440	1000 A	186	100	286
0480	1200 A	231	102	333
0520	1600 A	325	156	481
0560	2000 A	370	200	570
1200	Branch installation 600 V, including EMT conduit and THW wire, 15 A	1.43	4.98	6.41
1240	20 A	1.43	4.98	6.41
1280	30 A	2.27	6.10	8.37
1320	50 A	3.70	7.05	10.75
1360	65 A	4.69	7.50	12.19
1400	85 A	7.45	8.85	16.30
1440	100 A	9.65	9.40	19.05
1480	130 A	12.40	10.55	22.95
1520	150 A	14.45	12.10	26.55
1560	200 A	19.50	13.65	33.15

D5010 Electrical Service/Distribution

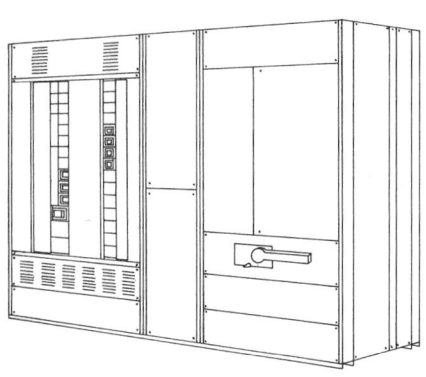

System Components	QUANTITY	UNIT	COST EACH		
			MAT.	INST.	TOTAL
SYSTEM D5010 240 0240					
SWITCHGEAR INSTALLATION, INCL SWBD, PANELS & CIRC BREAKERS, 600 A					
Panelboard, NQOD 225A 4W 120/208V main CB, w/20A bkrs 42 circ	1.000	Ea.	2,175	1,875	4,050
Switchboard, alum. bus bars, 120/208V, 4 wire, 600V	1.000	Ea.	4,025	1,050	5,075
Distribution sect., alum. bus bar, 120/208 or 277/480 V, 4 wire, 600A	1.000	Ea.	1,775	1,050	2,825
Feeder section circuit breakers, KA frame, 70 to 225 A	3.000	Ea.	3,300	489	3,789
TOTAL			11,275	4,464	15,739

D5010 240	Switchgear	COST EACH		
		MAT.	INST.	TOTAL
0200	Switchgear inst., incl. swbd., panels & circ bkr, 400 A, 120/208volt	3,975	3,275	7,250
0240	600 A	11,300	4,475	15,775
0280	800 A	14,200	6,375	20,575
0320	1200 A	17,500	9,750	27,250
0360	1600 A	23,700	13,700	37,400
0400	2000 A	29,900	17,400	47,300
0410	Add 20% for 277/480 volt			

D5020 Lighting and Branch Wiring

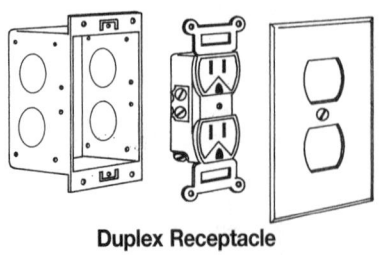

Duplex Receptacle

System Components	QUANTITY	UNIT	COST PER S.F.		
			MAT.	INST.	TOTAL
SYSTEM D5020 110 0200					
RECEPTACLES INCL. PLATE, BOX, CONDUIT, WIRE & TRANS. WHEN REQUIRED					
2.5 PER 1000 S.F., .3 WATTS PER S.F.					
Steel intermediate conduit, (IMC) 1/2" diam	167.000	L.F.	.35	.88	1.23
Wire 600V type THWN-THHN, copper solid #12	3.382	C.L.F.	.06	.16	.22
Wiring device, receptacle, duplex, 120V grounded, 15 amp	2.500	Ea.		.03	.03
Wall plate, 1 gang, brown plastic	2.500	Ea.		.02	.02
Steel outlet box 4" square	2.500	Ea.	.01	.07	.08
Steel outlet box 4" plaster rings	2.500	Ea.	.01	.02	.03
TOTAL			.43	1.18	1.61

D5020 110	Receptacle (by Wattage)	COST PER S.F.		
		MAT.	INST.	TOTAL
0190	Receptacles include plate, box, conduit, wire & transformer when required			
0200	2.5 per 1000 S.F., .3 watts per S.F.	.43	1.17	1.60
0240	With transformer	.48	1.23	1.71
0280	4 per 1000 S.F., .5 watts per S.F.	.48	1.36	1.84
0320	With transformer	.57	1.45	2.02
0360	5 per 1000 S.F., .6 watts per S.F.	.56	1.61	2.17
0400	With transformer	.68	1.72	2.40
0440	8 per 1000 S.F., .9 watts per S.F.	.59	1.78	2.37
0480	With transformer	.75	1.94	2.69
0520	10 per 1000 S.F., 1.2 watts per S.F.	.62	1.94	2.56
0560	With transformer	.88	2.20	3.08
0600	16.5 per 1000 S.F., 2.0 watts per S.F.	.72	2.42	3.14
0640	With transformer	1.16	2.87	4.03
0680	20 per 1000 S.F.,2.4 watts per S.F.	.75	2.63	3.38
0720	With transformer	1.27	3.15	4.42

D5020 Lighting and Branch Wiring

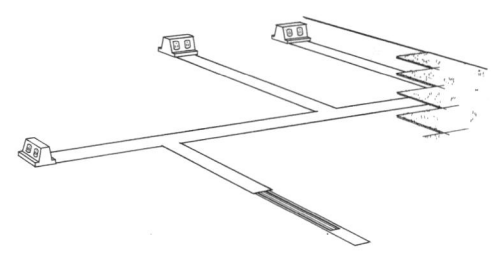

Underfloor Receptacle System

Description: Table D5020 115 includes installed costs of raceways and copper wire from panel to and including receptacle.

National Electrical Code prohibits use of undercarpet system in residential, school or hospital buildings. Can only be used with carpet squares.

Low density = (1) Outlet per 259 S.F. of floor area.

High density = (1) Outlet per 127 S.F. of floor area.

System Components	QUANTITY	UNIT	COST PER S.F.		
			MAT.	INST.	TOTAL
SYSTEM D5020 115 0200					
RECEPTACLE SYSTEMS, UNDERFLOOR DUCT, 5′ ON CENTER, LOW DENSITY					
Underfloor duct 3-1/8″ x 7/8″ w/insert 24″ on center	.190	L.F.	2.80	1.42	4.22
Underfloor duct junction box, single duct, 3-1/8″	.003	Ea.	.99	.39	1.38
Underfloor junction box carpet pan	.003	Ea.	.82	.02	.84
Underfloor duct outlet, high tension receptacle	.004	Ea.	.31	.26	.57
Wire 600V type THWN-THHN copper solid #12	.010	C.L.F.	.17	.48	.65
Vertical elbow for underfloor duct, 3-1/8″, included					
Underfloor duct conduit adapter, 2″ x 1-1/4″, included					
TOTAL			5.09	2.57	7.66

D5020 115	Receptacles, Floor	COST PER S.F.		
		MAT.	INST.	TOTAL
0200	Receptacle systems, underfloor duct, 5′ on center, low density	5.10	2.57	7.67
0240	High density	5.55	3.30	8.85
0280	7′ on center, low density	4.04	2.20	6.24
0320	High density	4.52	2.94	7.46
0400	Poke thru fittings, low density	1	1.18	2.18
0440	High density	1.98	2.32	4.30
0520	Telepoles, using Romex, low density	.97	.77	1.74
0560	High density	1.94	1.55	3.49
0600	Using EMT, low density	1.02	1.02	2.04
0640	High density	2.04	2.05	4.09
0720	Conduit system with floor boxes, low density	.93	.88	1.81
0760	High density	1.86	1.77	3.63
0840	Undercarpet power system, 3 conductor with 5 conductor feeder, low density	1.30	.32	1.62
0880	High density	2.54	.64	3.18

D5020 Lighting and Branch Wiring

Duplex Receptacle

Wall Switch

System Components	QUANTITY	UNIT	COST PER S.F.		
			MAT.	INST.	TOTAL
SYSTEM D5020 120 0520					
RECEPTACLES AND WALL SWITCHES					
4 RECEPTACLES PER 400 S.F.					
Steel intermediate conduit, (IMC), 1/2" diam	.220	L.F.	.46	1.16	1.62
Wire, 600 volt, type THWN-THHN, copper, solid #12	.005	C.L.F.	.08	.24	.32
Steel outlet box 4" square	.010	Ea.	.03	.26	.29
Steel outlet box, 4" square, plaster rings	.010	Ea.	.03	.08	.11
Receptacle, duplex, 120 volt grounded, 15 amp	.010	Ea.	.01	.13	.14
Wall plate, 1 gang, brown plastic	.010	Ea.		.07	.07
TOTAL			.61	1.94	2.55

D5020 120	Receptacles & Wall Switches	COST PER S.F.		
		MAT.	INST.	TOTAL
0520	Receptacles and wall switches, 400 S.F., 4 receptacles	.62	1.93	2.55
0560	6 receptacles	.67	2.25	2.92
0600	8 receptacles	.74	2.62	3.36
0640	1 switch	.14	.43	.57
0680	600 S.F., 6 receptacles	.62	1.93	2.55
0720	8 receptacles	.67	2.19	2.86
0760	10 receptacles	.72	2.45	3.17
0800	2 switches	.19	.57	.76
0840	1000 S.F., 10 receptacles	.61	1.93	2.54
0880	12 receptacles	.62	2.03	2.65
0920	14 receptacles	.66	2.19	2.85
0960	2 switches	.10	.31	.41
1000	1600 S.F., 12 receptacles	.56	1.71	2.27
1040	14 receptacles	.60	1.87	2.47
1080	16 receptacles	.61	1.93	2.54
1120	4 switches	.12	.37	.49
1160	2000 S.F., 14 receptacles	.59	1.76	2.35
1200	16 receptacles	.58	1.77	2.35
1240	18 receptacles	.60	1.87	2.47
1280	4 switches	.10	.31	.41
1320	3000 S.F., 12 receptacles	.47	1.35	1.82
1360	18 receptacles	.56	1.66	2.22
1400	24 receptacles	.58	1.77	2.35
1440	6 switches	.10	.31	.41
1480	3600 S.F., 20 receptacles	.54	1.61	2.15
1520	24 receptacles	.55	1.66	2.21
1560	28 receptacles	.57	1.71	2.28
1600	8 switches	.12	.36	.48
1640	4000 S.F., 16 receptacles	.47	1.35	1.82
1680	24 receptacles	.56	1.66	2.22

D50 Electrical

D5020 Lighting and Branch Wiring

D5020 120	Receptacles & Wall Switches	COST PER S.F.		
		MAT.	INST.	TOTAL
1720	30 receptacles	.56	1.71	2.27
1760	8 switches	.10	.31	.41
1800	5000 S.F., 20 receptacles	.47	1.35	1.82
1840	26 receptacles	.56	1.60	2.16
1880	30 receptacles	.56	1.66	2.22
1920	10 switches	.10	.31	.41

D5020 Lighting and Branch Wiring

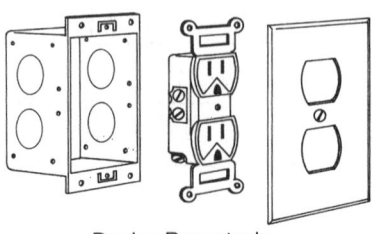

Duplex Receptacle

Wall Switch

System Components	QUANTITY	UNIT	COST PER EACH		
			MAT.	INST.	TOTAL
SYSTEM D5020 125 0520					
RECEPTACLES AND WALL SWITCHES, RECEPTICLE DUPLEX 120 V GROUNDED, 15 A					
Electric metallic tubing conduit, (EMT), 3/4" diam	22.000	L.F.	23.98	88.44	112.42
Wire, 600 volt, type THWN-THHN, copper, solid #12	.630	C.L.F.	10.52	29.93	40.45
Steel outlet box 4" square	1.000	Ea.	2.59	26	28.59
Steel outlet box, 4" square, plaster rings	1.000	Ea.	2.76	8.15	10.91
Receptacle, duplex, 120 volt grounded, 15 amp	1.000	Ea.	1.35	13.05	14.40
Wall plate, 1 gang, brown plastic	1.000	Ea.	.34	6.55	6.89
TOTAL			41.54	172.12	213.66

D5020 125	Receptacles & Switches by Each	COST PER EACH		
		MAT.	INST.	TOTAL
0460	Receptacles & Switches, with box, plate, 3/4" EMT conduit & wire			
0520	Receptacle duplex 120 V grounded, 15 A	41.50	172	213.50
0560	20 A	50.50	178	228.50
0600	Receptacle duplex ground fault interrupting, 15 A	72.50	178	250.50
0640	20 A	74.50	178	252.50
0680	Toggle switch single, 15 A	46	172	218
0720	20 A	48.50	178	226.50
0760	3 way switch, 15 A	48.50	182	230.50
0800	20 A	49.50	188	237.50
0840	4 way switch, 15 A	64	194	258
0880	20 A	80	207	287

346

D50 Electrical

D5020 Lighting and Branch Wiring

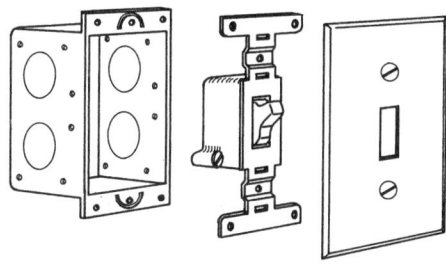

Description: Table D5020 130 includes the cost for switch, plate, box, conduit in slab or EMT exposed and copper wire. Add 20% for exposed conduit.

No power required for switches.

Federal energy guidelines recommend the maximum lighting area controlled per switch shall not exceed 1000 S.F. and that areas over 500 S.F. shall be so controlled that total illumination can be reduced by at least 50%.

System Components	QUANTITY	UNIT	COST PER S.F.		
			MAT.	INST.	TOTAL
SYSTEM D5020 130 0360					
WALL SWITCHES, 5.0 PER 1000 S.F.					
Steel, intermediate conduit (IMC), 1/2" diameter	88.000	L.F.	.19	.46	.65
Wire, 600V type THWN-THHN, copper solid #12	1.710	C.L.F.	.03	.08	.11
Toggle switch, single pole, 15 amp	5.000	Ea.	.03	.07	.10
Wall plate, 1 gang, brown plastic	5.000	Ea.		.03	.03
Steel outlet box 4" plaster rings	5.000	Ea.	.01	.13	.14
Plaster rings	5.000	Ea.	.01	.04	.05
TOTAL			.27	.81	1.08

D5020 130	Wall Switch by Sq. Ft.	COST PER S.F.		
		MAT.	INST.	TOTAL
0200	Wall switches, 1.0 per 1000 S.F.	.06	.19	.25
0240	1.2 per 1000 S.F.	.07	.21	.28
0280	2.0 per 1000 S.F.	.10	.30	.40
0320	2.5 per 1000 S.F.	.12	.38	.50
0360	5.0 per 1000 S.F.	.27	.81	1.08
0400	10.0 per 1000 S.F.	.55	1.64	2.19

D5020 Lighting and Branch Wiring

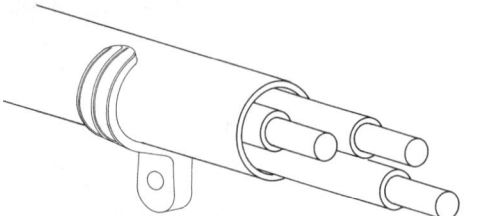

System D5020 135 includes all wiring and connections.

System Components	QUANTITY	UNIT	COST PER S.F.		
			MAT.	INST.	TOTAL
SYSTEM D5020 135 0200					
MISCELLANEOUS POWER, TO .5 WATTS					
Steel intermediate conduit, (IMC) 1/2" diam	15.000	L.F.	.03	.08	.11
Wire 600V type THWN-THHN, copper solid #12	.325	C.L.F.	.01	.02	.03
TOTAL			.04	.10	.14

D5020 135	Miscellaneous Power	COST PER S.F.		
		MAT.	INST.	TOTAL
0200	Miscellaneous power, to .5 watts	.04	.09	.13
0240	.8 watts	.05	.13	.18
0280	1 watt	.07	.17	.24
0320	1.2 watts	.08	.21	.29
0360	1.5 watts	.10	.25	.35
0400	1.8 watts	.11	.28	.39
0440	2 watts	.13	.34	.47
0480	2.5 watts	.16	.41	.57
0520	3 watts	.19	.49	.68

D5020 Lighting and Branch Wiring

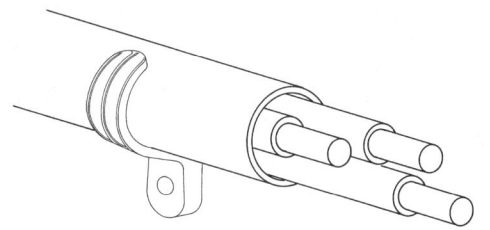

System D5020 140 includes all wiring and connections for central air conditioning units.

System Components	QUANTITY	UNIT	COST PER S.F.		
			MAT.	INST.	TOTAL
SYSTEM D5020 140 0200					
CENTRAL AIR CONDITIONING POWER, 1 WATT					
Steel intermediate conduit, 1/2″ diam.	.030	L.F.	.06	.16	.22
Wire 600V type THWN-THHN, copper solid #12	.001	C.L.F.	.02	.05	.07
TOTAL			.08	.21	.29

D5020 140	Central A. C. Power (by Wattage)	COST PER S.F.		
		MAT.	INST.	TOTAL
0200	Central air conditioning power, 1 watt	.08	.21	.29
0220	2 watts	.09	.24	.33
0240	3 watts	.13	.27	.40
0280	4 watts	.21	.35	.56
0320	6 watts	.42	.49	.91
0360	8 watts	.49	.51	1
0400	10 watts	.63	.59	1.22

D5020 Lighting and Branch Wiring

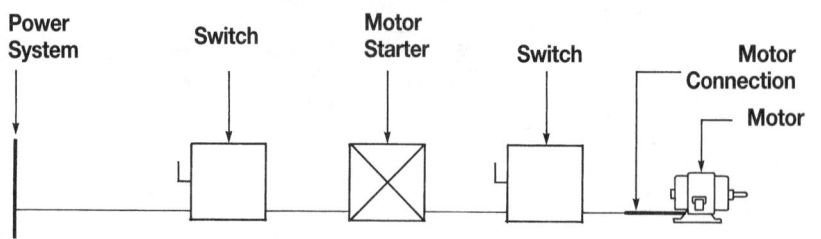

System D5020 145 installed cost of motor wiring as per Table D5010 170 using 50' of rigid conduit and copper wire. **Cost and setting of motor not included.**

System Components	QUANTITY	UNIT	COST EACH		
			MAT.	INST.	TOTAL
SYSTEM D5020 145 0200					
MOTOR INSTALLATION, SINGLE PHASE, 115V, TO AND INCLUDING 1/3 HP MOTOR SIZE					
Wire 600V type THWN-THHN, copper solid #12	1.250	C.L.F.	20.88	59.38	80.26
Steel intermediate conduit, (IMC) 1/2" diam.	50.000	L.F.	105.50	262.50	368
Magnetic FVNR, 115V, 1/3 HP, size 00 starter	1.000	Ea.	191	131	322
Safety switch, fused, heavy duty, 240V 2P 30 amp	1.000	Ea.	123	149	272
Safety switch, non fused, heavy duty, 600V, 3 phase, 30 A	1.000	Ea.	144	163	307
Flexible metallic conduit, Greenfield 1/2" diam.	1.500	L.F.	.62	3.92	4.54
Connectors for flexible metallic conduit Greenfield 1/2" diam.	1.000	Ea.	3.70	6.55	10.25
Coupling for Greenfield to conduit 1/2" diam flexible metalic conduit	1.000	Ea.	1.87	10.45	12.32
Fuse cartridge nonrenewable, 250V 30 amp	1.000	Ea.	1.78	10.45	12.23
TOTAL			592.35	796.25	1,388.60

D5020 145	Motor Installation	COST EACH		
		MAT.	INST.	TOTAL
0200	Motor installation, single phase, 115V, to and including 1/3 HP motor size	590	795	1,385
0240	To and incl. 1 HP motor size	615	795	1,410
0280	To and incl. 2 HP motor size	660	845	1,505
0320	To and incl. 3 HP motor size	750	860	1,610
0360	230V, to and including 1 HP motor size	595	805	1,400
0400	To and incl. 2 HP motor size	620	805	1,425
0440	To and incl. 3 HP motor size	695	865	1,560
0520	Three phase, 200V, to and including 1-1/2 HP motor size	695	885	1,580
0560	To and incl. 3 HP motor size	740	965	1,705
0600	To and incl. 5 HP motor size	795	1,075	1,870
0640	To and incl. 7-1/2 HP motor size	840	1,100	1,940
0680	To and incl. 10 HP motor size	1,325	1,375	2,700
0720	To and incl. 15 HP motor size	1,725	1,525	3,250
0760	To and incl. 20 HP motor size	2,200	1,750	3,950
0800	To and incl. 25 HP motor size	2,275	1,775	4,050
0840	To and incl. 30 HP motor size	3,475	2,075	5,550
0880	To and incl. 40 HP motor size	4,325	2,450	6,775
0920	To and incl. 50 HP motor size	7,275	2,850	10,125
0960	To and incl. 60 HP motor size	7,650	3,000	10,650
1000	To and incl. 75 HP motor size	10,200	3,450	13,650
1040	To and incl. 100 HP motor size	20,600	4,075	24,675
1080	To and incl. 125 HP motor size	21,600	4,475	26,075
1120	To and incl. 150 HP motor size	26,100	5,250	31,350
1160	To and incl. 200 HP motor size	30,500	6,425	36,925
1240	230V, to and including 1-1/2 HP motor size	665	875	1,540
1280	To and incl. 3 HP motor size	710	955	1,665
1320	To and incl. 5 HP motor size	765	1,050	1,815
1360	To and incl. 7-1/2 HP motor size	765	1,050	1,815
1400	To and incl. 10 HP motor size	1,175	1,300	2,475
1440	To and incl. 15 HP motor size	1,375	1,425	2,800
1480	To and incl. 20 HP motor size	2,050	1,700	3,750
1520	To and incl. 25 HP motor size	2,200	1,750	3,950

D5020 Lighting and Branch Wiring

D5020 145	Motor Installation	COST EACH		
		MAT.	INST.	TOTAL
1560	To and incl. 30 HP motor size	2,275	1,775	4,050
1600	To and incl. 40 HP motor size	4,175	2,400	6,575
1640	To and incl. 50 HP motor size	4,425	2,525	6,950
1680	To and incl. 60 HP motor size	7,300	2,850	10,150
1720	To and incl. 75 HP motor size	9,175	3,250	12,425
1760	To and incl. 100 HP motor size	10,600	3,625	14,225
1800	To and incl. 125 HP motor size	21,100	4,200	25,300
1840	To and incl. 150 HP motor size	23,100	4,775	27,875
1880	To and incl. 200 HP motor size	24,600	5,300	29,900
1960	460V, to and including 2 HP motor size	800	885	1,685
2000	To and incl. 5 HP motor size	845	960	1,805
2040	To and incl. 10 HP motor size	875	1,050	1,925
2080	To and incl. 15 HP motor size	1,150	1,225	2,375
2120	To and incl. 20 HP motor size	1,200	1,300	2,500
2160	To and incl. 25 HP motor size	1,325	1,375	2,700
2200	To and incl. 30 HP motor size	1,650	1,475	3,125
2240	To and incl. 40 HP motor size	2,150	1,575	3,725
2280	To and incl. 50 HP motor size	2,425	1,750	4,175
2320	To and incl. 60 HP motor size	3,575	2,050	5,625
2360	To and incl. 75 HP motor size	4,200	2,275	6,475
2400	To and incl. 100 HP motor size	4,650	2,525	7,175
2440	To and incl. 125 HP motor size	7,525	2,875	10,400
2480	To and incl. 150 HP motor size	9,375	3,225	12,600
2520	To and incl. 200 HP motor size	11,200	3,625	14,825
2600	575V, to and including 2 HP motor size	800	885	1,685
2640	To and incl. 5 HP motor size	845	960	1,805
2680	To and incl. 10 HP motor size	875	1,050	1,925
2720	To and incl. 20 HP motor size	1,150	1,225	2,375
2760	To and incl. 25 HP motor size	1,200	1,300	2,500
2800	To and incl. 30 HP motor size	1,650	1,475	3,125
2840	To and incl. 50 HP motor size	1,825	1,525	3,350
2880	To and incl. 60 HP motor size	3,500	2,050	5,550
2920	To and incl. 75 HP motor size	3,575	2,050	5,625
2960	To and incl. 100 HP motor size	4,200	2,275	6,475
3000	To and incl. 125 HP motor size	7,250	2,825	10,075
3040	To and incl. 150 HP motor size	7,525	2,875	10,400
3080	To and incl. 200 HP motor size	9,750	3,250	13,000

D5020 Lighting and Branch Wiring

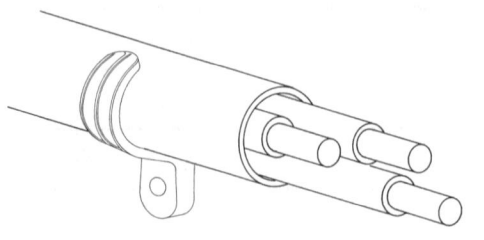

System Components			COST PER L.F.		
	QUANTITY	UNIT	MAT.	INST.	TOTAL
SYSTEM D5020 155 0200					
MOTOR FEEDER SYSTEMS, SINGLE PHASE, UP TO 115V, 1HP OR 230V, 2HP					
Steel intermediate conduit, (IMC) 1/2" diam	1.000	L.F.	2.11	5.25	7.36
Wire 600V type THWN-THHN, copper solid #12	.020	C.L.F.	.33	.95	1.28
TOTAL			2.44	6.20	8.64

D5020 155	Motor Feeder	COST PER L.F.		
		MAT.	INST.	TOTAL
0200	Motor feeder systems, single phase, feed up to 115V 1HP or 230V 2 HP	2.44	6.20	8.64
0240	115V 2HP, 230V 3HP	2.62	6.30	8.92
0280	115V 3HP	3.10	6.55	9.65
0360	Three phase, feed to 200V 3HP, 230V 5HP, 460V 10HP, 575V 10HP	2.61	6.70	9.31
0440	200V 5HP, 230V 7.5HP, 460V 15HP, 575V 20HP	2.88	6.85	9.73
0520	200V 10HP, 230V 10HP, 460V 30HP, 575V 30HP	3.60	7.20	10.80
0600	200V 15HP, 230V 15HP, 460V 40HP, 575V 50HP	5.10	8.20	13.30
0680	200V 20HP, 230V 25HP, 460V 50HP, 575V 60HP	7.55	10.40	17.95
0760	200V 25HP, 230V 30HP, 460V 60HP, 575V 75HP	8.50	10.60	19.10
0840	200V 30HP	10.75	10.95	21.70
0920	230V 40HP, 460V 75HP, 575V 100HP	12.60	12	24.60
1000	200V 40HP	14.75	12.80	27.55
1080	230V 50HP, 460V 100HP, 575V 125HP	16.65	14.10	30.75
1160	200V 50HP, 230V 60HP, 460V 125HP, 575V 150HP	21	14.95	35.95
1240	200V 60HP, 460V 150HP	24.50	17.60	42.10
1320	230V 75HP, 575V 200HP	31	18.30	49.30
1400	200V 75HP	35.50	18.70	54.20
1480	230V 100HP, 460V 200HP	42	22	64
1560	200V 100HP	57.50	27	84.50
1640	230V 125HP	66	29.50	95.50
1720	200V 125HP, 230V 150HP	74.50	34	108.50
1800	200V 150HP	78.50	37	115.50
1880	200V 200HP	115	54.50	169.50
1960	230V 200HP	105	40.50	145.50

D5020 Lighting and Branch Wiring

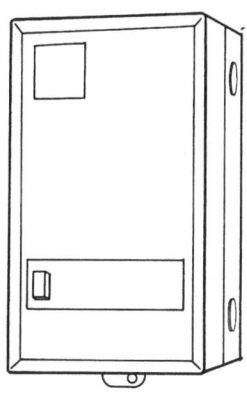

Starters are full voltage, type NEMA 1 for general purpose indoor application with motor overload protection and include mounting and wire connections.

System Components	QUANTITY	UNIT	COST EACH		
			MAT.	INST.	TOTAL
SYSTEM D5020 160 0200 MAGNETIC STARTER, SIZE 00 TO 1/3 HP, 1 PHASE 115V OR 1 HP 230V	1.000	Ea.	191	131	322
TOTAL			191	131	322

D5020 160	Magnetic Starter	COST EACH		
		MAT.	INST.	TOTAL
0200	Magnetic starter, size 00, to 1/3 HP, 1 phase, 115V or 1 HP 230V	191	131	322
0280	Size 00, to 1-1/2 HP, 3 phase, 200-230V or 2 HP 460-575V	210	149	359
0360	Size 0, to 1 HP, 1 phase, 115V or 2 HP 230V	213	131	344
0440	Size 0, to 3 HP, 3 phase, 200-230V or 5 HP 460-575V	254	227	481
0520	Size 1, to 2 HP, 1 phase, 115V or 3 HP 230V	245	174	419
0600	Size 1, to 7-1/2 HP, 3 phase, 200-230V or 10 HP 460-575V	285	325	610
0680	Size 2, to 10 HP, 3 phase, 200V, 15 HP-230V or 25 HP 460-575V	535	475	1,010
0760	Size 3, to 25 HP, 3 phase, 200V, 30 HP-230V or 50 HP 460-575V	875	580	1,455
0840	Size 4, to 40 HP, 3 phase, 200V, 50 HP-230V or 100 HP 460-575V	1,950	870	2,820
0920	Size 5, to 75 HP, 3 phase, 200V, 100 HP-230V or 200 HP 460-575V	4,550	1,150	5,700
1000	Size 6, to 150 HP, 3 phase, 200V, 200 HP-230V or 400 HP 460-575V	13,500	1,300	14,800

D5020 Lighting and Branch Wiring

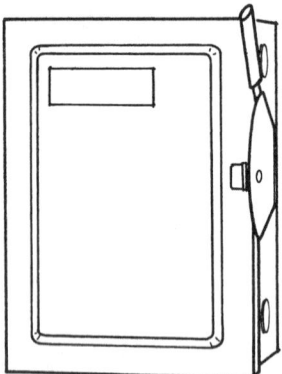

Safety switches are type NEMA 1 for general purpose indoor application, and include time delay fuses, insulation and wire terminations.

System Components	QUANTITY	UNIT	COST EACH		
			MAT.	INST.	TOTAL
SYSTEM D5020 165 0200					
SAFETY SWITCH, 30A FUSED, 1 PHASE, 115V OR 230V.					
Safety switch fused, hvy duty, 240V 2p 30 amp	1.000	Ea.	123	149	272
Fuse, dual element time delay 250V, 30 amp	2.000	Ea.	9.02	20.90	29.92
TOTAL			132.02	169.90	301.92

D5020 165	Safety Switches	COST EACH		
		MAT.	INST.	TOTAL
0200	Safety switch, 30 A fused, 1 phase, 2 HP 115 V or 3 HP, 230 V.	132	170	302
0280	3 phase, 5 HP, 200 V or 7 1/2 HP, 230 V	177	194	371
0360	15 HP, 460 V or 20 HP, 575 V	305	202	507
0440	60 A fused, 3 phase, 15 HP 200 V or 15 HP 230 V	300	258	558
0520	30 HP 460 V or 40 HP 575 V	385	266	651
0600	100 A fused, 3 phase, 20 HP 200 V or 25 HP 230 V	490	315	805
0680	50 HP 460 V or 60 HP 575 V	720	320	1,040
0760	200 A fused, 3 phase, 50 HP 200 V or 60 HP 230 V	875	445	1,320
0840	125 HP 460 V or 150 HP 575 V	1,100	450	1,550
0920	400 A fused, 3 phase, 100 HP 200 V or 125 HP 230 V	2,150	630	2,780
1000	250 HP 460 V or 350 HP 575 V	2,725	645	3,370
1020	600 A fused, 3 phase, 150 HP 200 V or 200 HP 230 V	3,650	935	4,585
1040	400 HP 460 V	4,475	950	5,425

D5020 Lighting and Branch Wiring

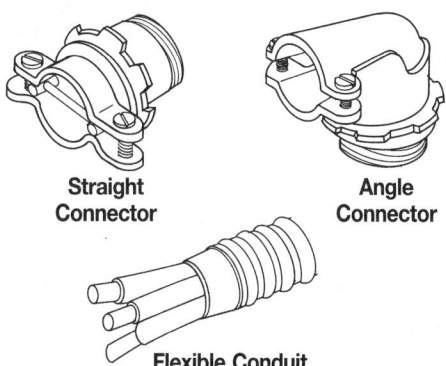

Straight Connector

Angle Connector

Flexible Conduit

Table below includes costs for the flexible conduit. Not included are wire terminations and testing motor for correct rotation.

System Components	QUANTITY	UNIT	COST EACH		
			MAT.	INST.	TOTAL
SYSTEM D5020 170 0200 **MOTOR CONNECTIONS, SINGLE PHASE, 115V/230V UP TO 1 HP** Motor connection, flexible conduit & fittings, 1 HP motor 115V	1.000	Ea.	9.12	62.23	71.35
TOTAL			9.12	62.23	71.35

D5020 170	Motor Connections	COST EACH		
		MAT.	INST.	TOTAL
0200	Motor connections, single phase, 115/230V, up to 1 HP	9.10	62	71.10
0240	Up to 3 HP	8.90	72.50	81.40
0280	Three phase, 200/230/460/575V, up to 3 HP	9.90	80.50	90.40
0320	Up to 5 HP	9.90	80.50	90.40
0360	Up to 7-1/2 HP	14.70	95	109.70
0400	Up to 10 HP	15.80	124	139.80
0440	Up to 15 HP	29	158	187
0480	Up to 25 HP	32.50	194	226.50
0520	Up to 50 HP	83.50	238	321.50
0560	Up to 100 HP	201	350	551

D5020 Lighting and Branch Wiring

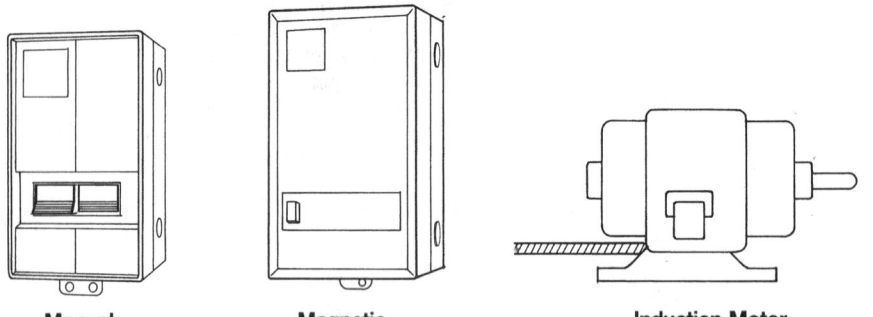

Manual Starter

Magnetic Starter

Induction Motor

For 230/460 Volt A.C., 3 phase, 60 cycle ball bearing squirrel cage induction motors, NEMA Class B standard line. Installation included.

No conduit, wire, or terminations included.

System Components	QUANTITY	UNIT	COST EACH		
			MAT.	INST.	TOTAL
SYSTEM D5020 175 0220					
MOTOR, DRIPPROOF CLASS B INSULATION, 1.15 SERVICE FACTOR, WITH STARTER					
1 H.P., 1200 RPM WITH MANUAL STARTER					
Motor, dripproof, class B insul, 1.15 serv fact, 1200 RPM, 1 HP	1.000	Ea.	268	116	384
Motor starter, manual, 3 phase, 1 HP motor	1.000	Ea.	192	149	341
TOTAL			460	265	725

D5020 175	Motor & Starter	COST EACH		
		MAT.	INST.	TOTAL
0190	Motor, dripproof, premium efficient, 1.15 service factor			
0200	1 HP, 1200 RPM, motor only	268	116	384
0220	With manual starter	460	265	725
0240	With magnetic starter	480	265	745
0260	1800 RPM, motor only	249	116	365
0280	With manual starter	440	265	705
0300	With magnetic starter	460	265	725
0320	2 HP, 1200 RPM, motor only	370	116	486
0340	With manual starter	560	265	825
0360	With magnetic starter	625	345	970
0380	1800 RPM, motor only	263	116	379
0400	With manual starter	455	265	720
0420	With magnetic starter	515	345	860
0440	3600 RPM, motor only	310	116	426
0460	With manual starter	500	265	765
0480	With magnetic starter	565	345	910
0500	3 HP, 1200 RPM, motor only	520	116	636
0520	With manual starter	710	265	975
0540	With magnetic starter	775	345	1,120
0560	1800 RPM, motor only	265	116	381
0580	With manual starter	455	265	720
0600	With magnetic starter	520	345	865
0620	3600 RPM, motor only	345	116	461
0640	With manual starter	535	265	800
0660	With magnetic starter	600	345	945
0680	5 HP, 1200 RPM, motor only	590	116	706
0700	With manual starter	820	375	1,195
0720	With magnetic starter	875	440	1,315
0740	1800 RPM, motor only	380	116	496
0760	With manual starter	610	375	985
0780	With magnetic starter	665	440	1,105
0800	3600 RPM, motor only	370	116	486

D5020 Lighting and Branch Wiring

D5020 175	Motor & Starter	COST EACH		
		MAT.	INST.	TOTAL
0820	With manual starter	600	375	975
0840	With magnetic starter	655	440	1,095
0860	7.5 HP, 1800 RPM, motor only	520	124	644
0880	With manual starter	750	385	1,135
0900	With magnetic starter	1,050	600	1,650
0920	10 HP, 1800 RPM, motor only	645	131	776
0940	With manual starter	875	390	1,265
0960	With magnetic starter	1,175	605	1,780
0980	15 HP, 1800 RPM, motor only	800	163	963
1000	With magnetic starter	1,325	640	1,965
1040	20 HP, 1800 RPM, motor only	1,075	201	1,276
1060	With magnetic starter	1,950	780	2,730
1100	25 HP, 1800 RPM, motor only	1,150	209	1,359
1120	With magnetic starter	2,025	790	2,815
1160	30 HP, 1800 RPM, motor only	1,450	218	1,668
1180	With magnetic starter	2,325	800	3,125
1220	40 HP, 1800 RPM, motor only	1,900	261	2,161
1240	With magnetic starter	3,850	1,125	4,975
1280	50 HP, 1800 RPM, motor only	2,125	325	2,450
1300	With magnetic starter	4,075	1,200	5,275
1340	60 HP, 1800 RPM, motor only	2,850	375	3,225
1360	With magnetic starter	7,400	1,525	8,925
1400	75 HP, 1800 RPM, motor only	3,100	435	3,535
1420	With magnetic starter	7,650	1,575	9,225
1460	100 HP, 1800 RPM, motor only	4,125	580	4,705
1480	With magnetic starter	8,675	1,725	10,400
1520	125 HP, 1800 RPM, motor only	4,775	745	5,520
1540	With magnetic starter	18,300	2,050	20,350
1580	150 HP, 1800 RPM, motor only	6,625	870	7,495
1600	With magnetic starter	20,100	2,175	22,275
1640	200 HP, 1800 RPM, motor only	8,475	1,050	9,525
1660	With magnetic starter	22,000	2,350	24,350
1680	Totally encl, premium efficient, 1.0 ser. fac., 1HP, 1200RPM, motor only	264	116	380
1700	With manual starter	455	265	720
1720	With magnetic starter	475	265	740
1740	1800 RPM, motor only	271	116	387
1760	With manual starter	465	265	730
1780	With magnetic starter	480	265	745
1800	2 HP, 1200 RPM, motor only	305	116	421
1820	With manual starter	495	265	760
1840	With magnetic starter	560	345	905
1860	1800 RPM, motor only	345	116	461
1880	With manual starter	535	265	800
1900	With magnetic starter	600	345	945
1920	3600 RPM, motor only	305	116	421
1940	With manual starter	495	265	760
1960	With magnetic starter	560	345	905
1980	3 HP, 1200 RPM, motor only	410	116	526
2000	With manual starter	600	265	865
2020	With magnetic starter	665	345	1,010
2040	1800 RPM, motor only	420	116	536
2060	With manual starter	610	265	875
2080	With magnetic starter	675	345	1,020
2100	3600 RPM, motor only	320	116	436
2120	With manual starter	510	265	775
2140	With magnetic starter	575	345	920
2160	5 HP, 1200 RPM, motor only	570	116	686
2180	With manual starter	800	375	1,175

D5020 Lighting and Branch Wiring

D5020 175	Motor & Starter	COST EACH		
		MAT.	INST.	TOTAL
2200	With magnetic starter	855	440	1,295
2220	1800 RPM, motor only	475	116	591
2240	With manual starter	705	375	1,080
2260	With magnetic starter	760	440	1,200
2280	3600 RPM, motor only	405	116	521
2300	With manual starter	635	375	1,010
2320	With magnetic starter	690	440	1,130
2340	7.5 HP, 1800 RPM, motor only	685	124	809
2360	With manual starter	915	385	1,300
2380	With magnetic starter	1,225	600	1,825
2400	10 HP, 1800 RPM, motor only	825	131	956
2420	With manual starter	1,050	390	1,440
2440	With magnetic starter	1,350	605	1,955
2460	15 HP, 1800 RPM, motor only	1,150	163	1,313
2480	With magnetic starter	1,675	640	2,315
2500	20 HP, 1800 RPM, motor only	1,350	201	1,551
2520	With magnetic starter	2,225	780	3,005
2540	25 HP, 1800 RPM, motor only	1,550	209	1,759
2560	With magnetic starter	2,425	790	3,215
2580	30 HP, 1800 RPM, motor only	1,850	218	2,068
2600	With magnetic starter	2,725	800	3,525
2620	40 HP, 1800 RPM, motor only	2,350	261	2,611
2640	With magnetic starter	4,300	1,125	5,425
2660	50 HP, 1800 RPM, motor only	2,900	325	3,225
2680	With magnetic starter	4,850	1,200	6,050
2700	60 HP, 1800 RPM, motor only	4,325	375	4,700
2720	With magnetic starter	8,875	1,525	10,400
2740	75 HP, 1800 RPM, motor only	5,575	435	6,010
2760	With magnetic starter	10,100	1,575	11,675
2780	100 HP, 1800 RPM, motor only	8,150	580	8,730
2800	With magnetic starter	12,700	1,725	14,425
2820	125 HP, 1800 RPM, motor only	10,700	745	11,445
2840	With magnetic starter	24,200	2,050	26,250
2860	150 HP, 1800 RPM, motor only	12,500	870	13,370
2880	With magnetic starter	26,000	2,175	28,175
2900	200 HP, 1800 RPM, motor only	15,200	1,050	16,250
2920	With magnetic starter	28,700	2,350	31,050

D50 Electrical

D5020 Lighting and Branch Wiring

General: The cost of the lighting portion of the electrical costs is dependent upon:
1. The footcandle requirement of the proposed building.
2. The type of fixtures required.
3. The ceiling heights of the building.
4. Reflectance value of ceilings, walls and floors.
5. Fixture efficiencies and spacing vs. mounting height ratios.

Footcandle Requirements: See Table D5020-204 for Footcandle and Watts per S.F. determination.

Table D5020-201 IESNA* Recommended Illumination Levels in Footcandles

Commercial Buildings			Industrial Buildings		
Type	Description	Footcandles	Type	Description	Footcandles
Bank	Lobby	50	Assembly Areas	Rough bench & machine work	50
	Customer Areas	70		Medium bench & machine work	100
	Teller Stations	150		Fine bench & machine work	500
	Accounting Areas	150	Inspection Areas	Ordinary	50
Offices	Routine Work	100		Difficult	100
	Accounting	150		Highly Difficult	200
	Drafting	200	Material Handling	Loading	20
	Corridors, Halls, Washrooms	30		Stock Picking	30
Schools	Reading or Writing	70		Packing, Wrapping	50
	Drafting, Labs, Shops	100	Stairways	Service Areas	20
	Libraries	70	Washrooms	Service Areas	20
	Auditoriums, Assembly	15	Storage Areas	Inactive	5
	Auditoriums, Exhibition	30		Active, Rough, Bulky	10
Stores	Circulation Areas	30		Active, Medium	20
	Stock Rooms	30		Active, Fine	50
	Merchandise Areas, Service	100	Garages	Active Traffic Areas	20
	Self-Service Areas	200		Service & Repair	100

*IESNA - Illuminating Engineering Society of North America

Table D5020-202 General Lighting Loads by Occupancies

Type of Occupancy	Unit Load per S.F. (Watts)
Armories and Auditoriums	1
Banks	5
Barber Shops and Beauty Parlors	3
Churches	1
Clubs	2
Court Rooms	2
*Dwelling Units	3
Garages — Commercial (storage)	½
Hospitals	2
*Hotels and Motels, including apartment houses without provisions for cooking by tenants	2
Industrial Commercial (Loft) Buildings	2
Lodge Rooms	1½
Office Buildings	5
Restaurants	2
Schools	3
Stores	3
Warehouses (storage)	¼
*In any of the above occupancies except one-family dwellings and individual dwelling units of multi-family dwellings:	
Assembly Halls and Auditoriums	1
Halls, Corridors, Closets	½
Storage Spaces	¼

Table D5020-203 Lighting Limit (Connected Load) for Listed Occupancies: New Building Proposed Energy Conservation Guideline

Type of Use	Maximum Watts per S.F.
Interior	3.00
Category A: Classrooms, office areas, automotive mechanical areas, museums, conference rooms, drafting rooms, clerical areas, laboratories, merchandising areas, kitchens, examining rooms, book stacks, athletic facilities.	
Category B: Auditoriums, waiting areas, spectator areas, restrooms, dining areas, transportation terminals, working corridors in prisons and hospitals, book storage areas, active inventory storage, hospital bedrooms, hotel and motel bedrooms, enclosed shopping mall concourse areas, stairways.	1.00
Category C: Corridors, lobbies, elevators, inactive storage areas.	0.50
Category D: Indoor parking.	0.25
Exterior	
Category E: Building perimeter: wall-wash, facade, canopy.	5.00 (per linear foot)
Category F: Outdoor parking.	0.10

Table D5020-204   Procedure for Calculating Footcandles and Watts Per Square Foot

1. Initial footcandles = No. of fixtures × lamps per fixture × lumens per lamp × coefficient of utilization ÷ square feet
2. Maintained footcandles = initial footcandles × maintenance factor
3. Watts per square foot = No. of fixtures × lamps × (lamp watts + ballast watts) ÷ square feet

Example: To find footcandles and watts per S.F. for an office 20′ x 20′ with 11 fluorescent fixtures each having 4–40 watt C.W. lamps.

Based on good reflectance and clean conditions:

Lumens per lamp = 40 watt cool white at 3150 lumens per lamp

Coefficient of utilization = .42 (varies from .62 for light colored areas to .27 for dark)

Maintenance factor = .75 (varies from .80 for clean areas with good maintenance to .50 for poor)

Ballast loss = 8 watts per lamp. (Varies with manufacturer. See manufacturers' catalog.)

1. Initial footcandles:

$$\frac{11 \times 4 \times 3150 \times .42}{400} = \frac{58,212}{400} = 145 \text{ footcandles}$$

2. Maintained footcandles:

$$145 \times .75 = 109 \text{ footcandles}$$

3. Watts per S.F.

$$\frac{11 \times 4 \, (40 + 8)}{400} = \frac{2,112}{400} = 5.3 \text{ watts per S.F.}$$

Table D5020-205   Approximate Watts Per Square Foot for Popular Fixture Types

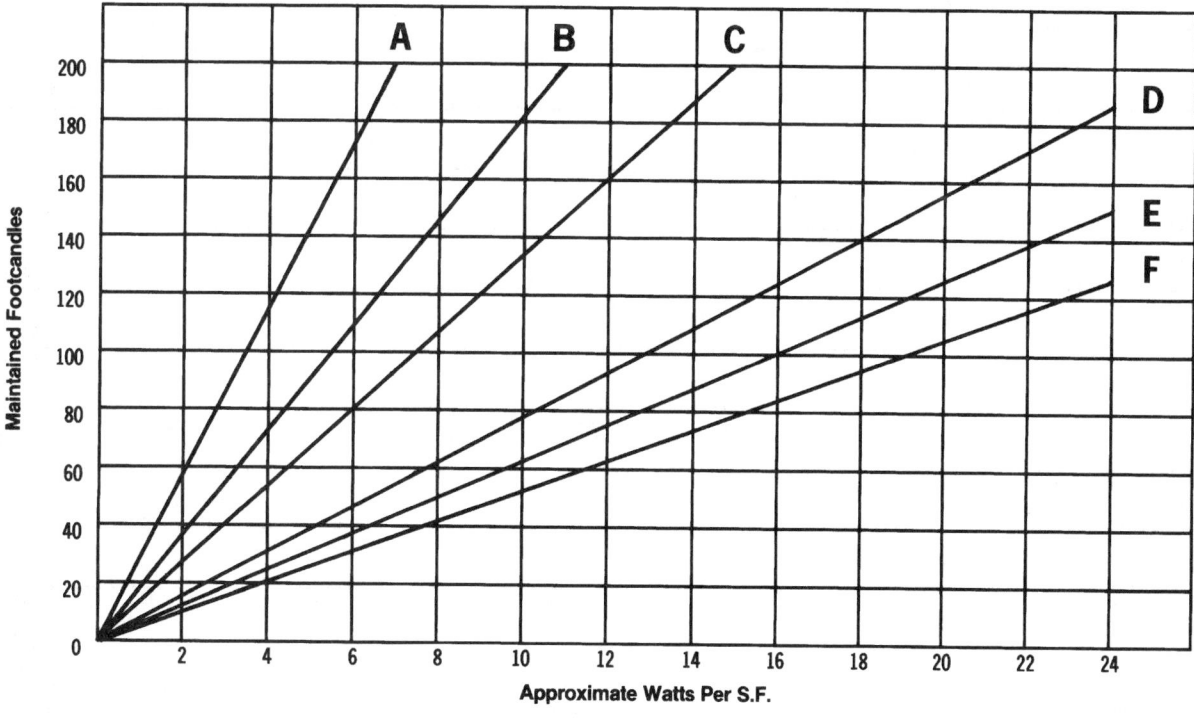

Due to the many variables involved, use for preliminary estimating only:
  a. Fluorescent – industrial System D5020 208
  b. Fluorescent – lens unit System D5020 208 Fixture types B & C
  c. Fluorescent – louvered unit
  d. Incandescent – open reflector System D5020 214, Type D
  e. Incandescent – lens unit System D5020 214, Type A
  f. Incandescent – down light System D5020 214, Type B

D5020 Lighting and Branch Wiring

A. Strip Fixture

B. Surface Mounted

C. Recessed

D. Pendent Mounted

Design Assumptions:

1. A 100 footcandle average maintained level of illumination.
2. Ceiling heights range from 9' to 11'.
3. Average reflectance values are assumed for ceilings, walls and floors.
4. Cool white (CW) fluorescent lamps with 3150 lumens for 40 watt lamps and 6300 lumens for 8' slimline lamps.
5. Four 40 watt lamps per 4' fixture and two 8' lamps per 8' fixture.
6. Average fixture efficiency values and spacing to mounting height ratios.
7. Installation labor is average U.S. rate as of January 1.

System Components	QUANTITY	UNIT	COST PER S.F.		
			MAT.	INST.	TOTAL
SYSTEM D5020 208 0520					
FLUORESCENT FIXTURES MOUNTED 9'-11" ABOVE FLOOR, 100 FC					
TYPE A, 8 FIXTURES PER 400 S.F.					
Steel intermediate conduit, (IMC) 1/2" diam.	.404	L.F.	.85	2.12	2.97
Wire, 600V, type THWN-THHN, copper, solid, #12	.008	C.L.F.	.13	.38	.51
Fluorescent strip fixture 8' long, surface mounted, two 75W SL	.020	Ea.	1.17	1.69	2.86
Steel outlet box 4" concrete	.020	Ea.	.29	.52	.81
Steel outlet box plate with stud, 4" concrete	.020	Ea.	.12	.13	.25
TOTAL			2.56	4.84	7.40

D5020 208	Fluorescent Fixtures (by Type)	COST PER S.F.		
		MAT.	INST.	TOTAL
0520	Fluorescent fixtures, type A, 8 fixtures per 400 S.F.	2.56	4.84	7.40
0560	11 fixtures per 600 S.F.	2.44	4.69	7.13
0600	17 fixtures per 1000 S.F.	2.37	4.62	6.99
0640	23 fixtures per 1600 S.F.	2.18	4.38	6.56
0680	28 fixtures per 2000 S.F.	2.18	4.38	6.56
0720	41 fixtures per 3000 S.F.	2.14	4.36	6.50
0800	53 fixtures per 4000 S.F.	2.10	4.26	6.36
0840	64 fixtures per 5000 S.F.	2.10	4.26	6.36
0880	Type B, 11 fixtures per 400 S.F.	4.79	7.10	11.89
0920	15 fixtures per 600 S.F.	4.47	6.80	11.27
0960	24 fixtures per 1000 S.F.	4.38	6.75	11.13
1000	35 fixtures per 1600 S.F.	4.11	6.45	10.56
1040	42 fixtures per 2000 S.F.	4.05	6.45	10.50
1080	61 fixtures per 3000 S.F.	4.03	6.25	10.28
1160	80 fixtures per 4000 S.F.	3.95	6.40	10.35
1200	98 fixtures per 5000 S.F.	3.93	6.35	10.28
1240	Type C, 11 fixtures per 400 S.F.	3.80	7.45	11.25
1280	14 fixtures per 600 S.F.	3.45	6.95	10.40
1320	23 fixtures per 1000 S.F.	3.43	6.90	10.33
1360	34 fixtures per 1600 S.F.	3.34	6.85	10.19
1400	43 fixtures per 2000 S.F.	3.34	6.75	10.09
1440	63 fixtures per 3000 S.F.	3.28	6.70	9.98
1520	81 fixtures per 4000 S.F.	3.21	6.60	9.81
1560	101 fixtures per 5000 S.F.	3.21	6.60	9.81
1600	Type D, 8 fixtures per 400 S.F.	3.51	5.80	9.31
1640	12 fixtures per 600 S.F.	3.50	5.75	9.25
1680	19 fixtures per 1000 S.F.	3.38	5.60	8.98
1720	27 fixtures per 1600 S.F.	3.20	5.45	8.65
1760	34 fixtures per 2000 S.F.	3.18	5.40	8.58
1800	48 fixtures per 3000 S.F.	3.06	5.25	8.31
1880	64 fixtures per 4000 S.F.	3.06	5.25	8.31
1920	79 fixtures per 5000 S.F.	3.06	5.25	8.31

D5020 Lighting and Branch Wiring

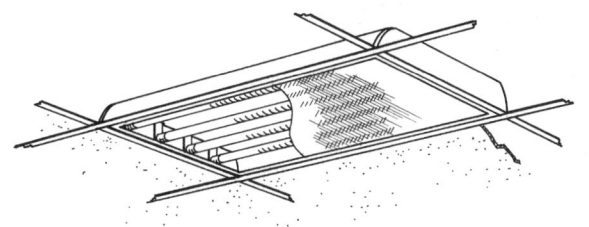

Type C. Recessed, mounted on grid ceiling suspension system, 2' x 4', four 40 watt lamps, acrylic prismatic diffusers.

5.3 watts per S.F. for 100 footcandles.

3 watts per S.F. for 57 footcandles.

System Components	QUANTITY	UNIT	COST PER S.F.		
			MAT.	INST.	TOTAL
SYSTEM D5020 210 0200					
FLUORESCENT FIXTURES RECESS MOUNTED IN CEILING					
1 WATT PER S.F., 20 FC, 5 FIXTURES PER 1000 S.F.					
Steel intermediate conduit, (IMC) 1/2" diam.	.128	L.F.	.27	.67	.94
Wire, 600 volt, type THW, copper, solid, #12	.003	C.L.F.	.05	.14	.19
Fluorescent fixture, recessed, 2'x 4', four 40W, w/ lens, for grid ceiling	.005	Ea.	.34	.56	.90
Steel outlet box 4" square	.005	Ea.	.07	.13	.20
Fixture whip, Greenfield w/#12 THHN wire	.005	Ea.	.03	.03	.06
TOTAL			.76	1.53	2.29

D5020 210	Fluorescent Fixtures (by Wattage)	COST PER S.F.		
		MAT.	INST.	TOTAL
0190	Fluorescent fixtures recess mounted in ceiling			
0200	1 watt per S.F., 20 FC, 5 fixtures per 1000 S.F.	.76	1.53	2.29
0240	2 watts per S.F., 40 FC, 10 fixtures per 1000 S.F.	1.50	3.02	4.52
0280	3 watts per S.F., 60 FC, 15 fixtures per 1000 S.F	2.25	4.55	6.80
0320	4 watts per S.F., 80 FC, 20 fixtures per 1000 S.F.	2.99	6.05	9.04
0400	5 watts per S.F., 100 FC, 25 fixtures per 1000 S.F.	3.75	7.55	11.30

D50 Electrical

D5020 Lighting and Branch Wiring

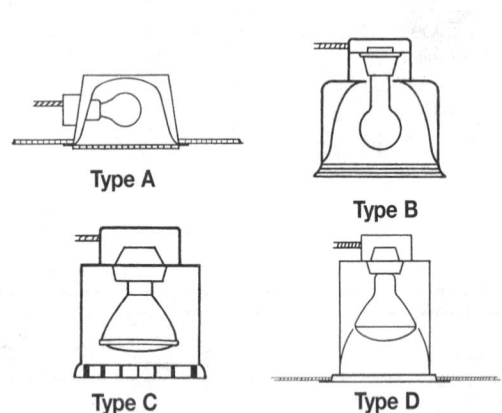

Type A. Recessed wide distribution reflector with flat glass lens 150 W.

Maximum spacing = 1.2 x mounting height.

13 watts per S.F. for 100 footcandles.

Type B. Recessed reflector down light with baffles 150 W.

Maximum spacing = 0.8 x mounting height.

18 watts per S.F. for 100 footcandles.

Type C. Recessed PAR–38 flood lamp with concentric louver 150 W.

Maximum spacing = 0.5 x mounting height.

19 watts per S.F. for 100 footcandles.

Type D. Recessed R–40 flood lamp with reflector skirt.

Maximum spacing = 0.7 x mounting height.

15 watts per S.F. for 100 footcandles.

System Components	QUANTITY	UNIT	COST PER S.F.		
			MAT.	INST.	TOTAL
SYSTEM D5020 214 0400					
INCANDESCENT FIXTURE RECESS MOUNTED, 100 FC					
TYPE A, 34 FIXTURES PER 400 S.F.					
Steel intermediate conduit, (IMC) 1/2" diam	1.060	L.F.	2.24	5.57	7.81
Wire, 600V, type THWN-THHN, copper, solid, #12	.033	C.L.F.	.55	1.57	2.12
Steel outlet box 4" square	.085	Ea.	1.22	2.21	3.43
Fixture whip, Greenfield w/#12 THHN wire	.085	Ea.	.50	.56	1.06
Incandescent fixture, recessed, w/lens, prewired, square trim, 200W	.085	Ea.	8.08	6.63	14.71
TOTAL			12.59	16.54	29.13

D5020 214	Incandescent Fixture (by Type)	COST PER S.F.		
		MAT.	INST.	TOTAL
0380	Incandescent fixture recess mounted, 100 FC			
0400	Type A, 34 fixtures per 400 S.F.	12.60	16.55	29.15
0440	49 fixtures per 600 S.F.	12.25	16.25	28.50
0480	63 fixtures per 800 S.F.	11.95	16.10	28.05
0520	90 fixtures per 1200 S.F.	11.55	15.75	27.30
0560	116 fixtures per 1600 S.F.	11.35	15.60	26.95
0600	143 fixtures per 2000 S.F.	11.25	15.55	26.80
0640	Type B, 47 fixtures per 400 S.F.	11.80	21	32.80
0680	66 fixtures per 600 S.F.	11.35	20.50	31.85
0720	88 fixtures per 800 S.F.	11.35	20.50	31.85
0760	127 fixtures per 1200 S.F.	11.15	20	31.15
0800	160 fixtures per 1600 S.F.	11	20	31
0840	206 fixtures per 2000 S.F.	11	19.95	30.95
0880	Type C, 51 fixtures per 400 S.F.	16.05	22	38.05
0920	74 fixtures per 600 S.F.	15.65	21.50	37.15
0960	97 fixtures per 800 S.F.	15.45	21.50	36.95
1000	142 fixtures per 1200 S.F.	15.20	21	36.20
1040	186 fixtures per 1600 S.F.	15.05	21	36.05
1080	230 fixtures per 2000 S.F.	14.95	21	35.95
1120	Type D, 39 fixtures per 400 S.F.	14.95	17.20	32.15
1160	57 fixtures per 600 S.F.	14.60	16.95	31.55
1200	75 fixtures per 800 S.F.	14.50	16.95	31.45
1240	109 fixtures per 1200 S.F.	14.20	16.80	31
1280	143 fixtures per 1600 S.F.	13.95	16.60	30.55
1320	176 fixtures per 2000 S.F.	13.85	16.55	30.40

D50 Electrical

D5020 Lighting and Branch Wiring

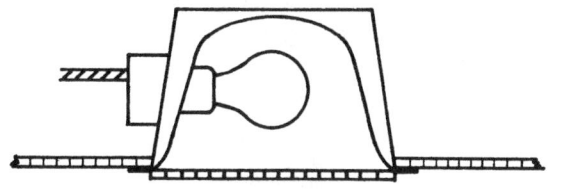

Type A. Recessed, wide distribution reflector with flat glass lens.

150 watt inside frost—2500 lumens per lamp.

PS–25 extended service lamp.

Maximum spacing = 1.2 x mounting height.

13 watts per S.F. for 100 footcandles.

System Components	QUANTITY	UNIT	COST PER S.F.		
			MAT.	INST.	TOTAL
SYSTEM D5020 216 0200					
INCANDESCENT FIXTURE RECESS MOUNTED, TYPE A					
1 WATT PER S.F., 8 FC, 6 FIXT PER 1000 S.F.					
Steel intermediate conduit, (IMC) 1/2″ diam	.091	L.F.	.19	.48	.67
Wire, 600V, type THWN-THHN, copper, solid, #12	.002	C.L.F.	.03	.10	.13
Incandescent fixture, recessed, w/lens, prewired, square trim, 200W	.006	Ea.	.57	.47	1.04
Steel outlet box 4″ square	.006	Ea.	.09	.16	.25
Fixture whip, Greenfield w/#12 THHN wire	.006	Ea.	.04	.04	.08
TOTAL			.92	1.25	2.17

D5020 216	Incandescent Fixture (by Wattage)	COST PER S.F.		
		MAT.	INST.	TOTAL
0190	Incandescent fixture recess mounted, type A			
0200	1 watt per S.F., 8 FC, 6 fixtures per 1000 S.F.	.92	1.24	2.16
0240	2 watt per S.F., 16 FC, 12 fixtures per 1000 S.F.	1.83	2.47	4.30
0280	3 watt per S.F., 24 FC, 18 fixtures, per 1000 S.F.	2.73	3.66	6.39
0320	4 watt per S.F., 32 FC, 24 fixtures per 1000 S.F.	3.65	4.90	8.55
0400	5 watt per S.F., 40 FC, 30 fixtures per 1000 S.F.	4.57	6.15	10.72

D50 Electrical

D5020 Lighting and Branch Wiring

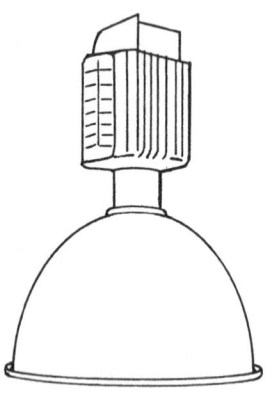

HIGH BAY FIXTURES

B. Metal halide 400 watt

C. High pressure sodium 400 watt

E. Metal halide 1000 watt

F. High pressure sodium 1000 watt

G. Metal halide 1000 watt
   125,000 lumen lamp

System Components	QUANTITY	UNIT	COST PER S.F.		
			MAT.	INST.	TOTAL
SYSTEM D5020 220 0880					
HIGH INTENSITY DISCHARGE FIXTURE, 8'-10' ABOVE WORK PLANE, 100 FC					
TYPE B, 12 FIXTURES PER 900 S.F.					
Steel intermediate conduit, (IMC) 1/2" diam	.460	L.F.	.97	2.42	3.39
Wire, 600V, type THWN-THHN, copper, solid, #10	.009	C.L.F.	.23	.47	.70
Steel outlet box 4" concrete	.009	Ea.	.13	.23	.36
Steel outlet box plate with stud 4" concrete	.009	Ea.	.05	.06	.11
Metal halide, hi bay, aluminum reflector, 400 W lamp	.009	Ea.	3.65	2.04	5.69
TOTAL			5.03	5.22	10.25

D5020 220	H.I.D. Fixture, High Bay, 8'-10' (by Type)	COST PER S.F.		
		MAT.	INST.	TOTAL
0500	High intensity discharge fixture, 8'-10' above work plane, 100 FC			
0880	Type B, 8 fixtures per 900 S.F.	5.05	5.20	10.25
0920	15 fixtures per 1800 S.F.	4.63	5	9.63
0960	24 fixtures per 3000 S.F.	4.63	5	9.63
1000	31 fixtures per 4000 S.F.	4.62	5	9.62
1040	38 fixtures per 5000 S.F.	4.62	5	9.62
1080	60 fixtures per 8000 S.F.	4.62	5	9.62
1120	72 fixtures per 10000 S.F.	4.16	4.65	8.81
1160	115 fixtures per 16000 S.F.	4.16	4.65	8.81
1200	230 fixtures per 32000 S.F.	4.16	4.65	8.81
1240	Type C, 4 fixtures per 900 S.F.	2.49	3.19	5.68
1280	8 fixtures per 1800 S.F.	2.49	3.19	5.68
1320	13 fixtures per 3000 S.F.	2.49	3.19	5.68
1360	17 fixtures per 4000 S.F.	2.49	3.19	5.68
1400	21 fixtures per 5000 S.F.	2.37	2.93	5.30
1440	33 fixtures per 8000 S.F.	2.34	2.88	5.22
1480	40 fixtures per 10000 S.F.	2.28	2.72	5
1520	63 fixtures per 16000 S.F.	2.28	2.72	5
1560	126 fixtures per 32000 S.F.	2.28	2.72	5

D5020 Lighting and Branch Wiring

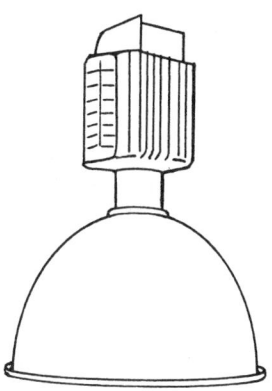

HIGH BAY FIXTURES

B. Metal halide 400 watt

C. High pressure sodium 400 watt

E. Metal halide 1000 watt

F. High pressure sodium 1000 watt

G. Metal halide 1000 watt
125,000 lumen lamp

System Components	QUANTITY	UNIT	COST PER S.F.		
			MAT.	INST.	TOTAL
SYSTEM D5020 222 0240					
HIGH INTENSITY DISCHARGE FIXTURE, 8'-10' ABOVE WORK PLANE					
1 WATT/S.F., TYPE B, 29 FC, 2 FIXTURES/1000 S.F.					
Steel intermediate conduit, (IMC) 1/2" diam	.100	L.F.	.21	.53	.74
Wire, 600V, type THWN-THHN, copper, solid, #10	.002	C.L.F.	.05	.11	.16
Steel outlet box 4" concrete	.002	Ea.	.03	.05	.08
Steel outlet box plate with stud, 4" concrete	.002	Ea.	.01	.01	.02
Metal halide, hi bay, aluminum reflector, 400 W lamp	.002	Ea.	.81	.45	1.26
TOTAL			1.11	1.15	2.26

D5020 222	H.I.D. Fixture, High Bay, 8'-10' (by Wattage)		COST PER S.F.		
			MAT.	INST.	TOTAL
0190	High intensity discharge fixture, 8'-10' above work plane				
0240	1 watt/S.F., type B, 29 FC, 2 fixtures/1000 S.F.		1.11	1.15	2.26
0280	Type C, 54 FC, 2 fixtures/1000 S.F.		.94	.87	1.81
0400	2 watt/S.F., type B, 59 FC, 4 fixtures/1000 S.F.		2.20	2.22	4.42
0440	Type C, 108 FC, 4 fixtures/1000 S.F.	R265723 -05	1.85	1.69	3.54
0560	3 watt/S.F., type B, 103 FC, 7 fixtures/1000 S.F.		3.70	3.55	7.25
0600	Type C, 189 FC, 6 fixtures/1000 S.F.		2.93	2.18	5.11
0720	4 watt/S.F., type B, 133 FC, 9 fixtures/1000 S.F.		4.79	4.66	9.45
0760	Type C, 243 FC, 9 fixtures/1000 S.F.		4.15	3.77	7.92
0880	5 watt/S.F., type B, 162 FC, 11 fixtures/1000 S.F.		5.90	5.75	11.65
0920	Type C, 297 FC, 11 fixtures/1000 S.F.		5.10	4.68	9.78

D5020 Lighting and Branch Wiring

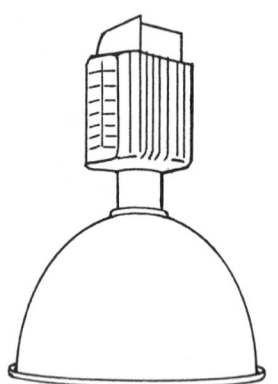

HIGH BAY FIXTURES

B. Metal halide 400 watt

C. High pressure sodium 400 watt

E. Metal halide 1000 watt

F. High pressure sodium 1000 watt

G. Metal halide 1000 watt
125,000 lumen lamp

System Components	QUANTITY	UNIT	COST PER S.F.		
			MAT.	INST.	TOTAL
SYSTEM D5020 224 1240					
HIGH INTENSITY DISCHARGE FIXTURE, 16' ABOVE WORK PLANE, 100 FC					
TYPE C, 5 FIXTURES PER 900 S.F.					
Steel intermediate conduit, (IMC) 1/2" diam	.260	L.F.	.55	1.37	1.92
Wire, 600V, type THWN-THHN, copper, solid, #10	.007	C.L.F.	.18	.37	.55
Steel outlet box 4" concrete	.006	Ea.	.09	.16	.25
Steel outlet box plate with stud, 4" concrete	.006	Ea.	.04	.04	.08
High pressure sodium, hi bay, aluminum reflector, 400 W lamp	.006	Ea.	2.25	1.36	3.61
TOTAL			3.11	3.30	6.41

D5020 224	H.I.D. Fixture, High Bay, 16' (by Type)	COST PER S.F.		
		MAT.	INST.	TOTAL
0510	High intensity discharge fixture, 16' above work plane, 100 FC			
1240	Type C, 5 fixtures per 900 S.F.	3.10	3.29	6.39
1280	9 fixtures per 1800 S.F.	2.88	3.45	6.33
1320	15 fixtures per 3000 S.F.	2.88	3.45	6.33
1360	18 fixtures per 4000 S.F.	2.73	3.08	5.81
1400	22 fixtures per 5000 S.F.	2.73	3.08	5.81
1440	36 fixtures per 8000 S.F.	2.73	3.08	5.81
1480	42 fixtures per 10,000 S.F.	2.48	3.19	5.67
1520	65 fixtures per 16,000 S.F.	2.48	3.19	5.67
1600	Type G, 4 fixtures per 900 S.F.	4.09	5.20	9.29
1640	6 fixtures per 1800 S.F.	3.47	4.87	8.34
1720	9 fixtures per 4000 S.F.	2.93	4.73	7.66
1760	11 fixtures per 5000 S.F.	2.93	4.73	7.66
1840	21 fixtures per 10,000 S.F.	2.74	4.26	7
1880	33 fixtures per 16,000 S.F.	2.74	4.26	7

D5020 Lighting and Branch Wiring

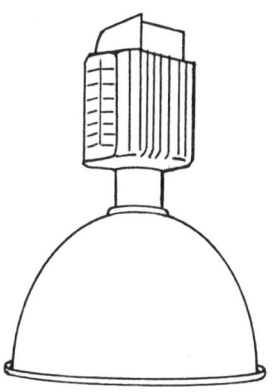

HIGH BAY FIXTURES

B. Metal halide 400 watt

C. High pressure sodium 400 watt

E. Metal halide 1000 watt

F. High pressure sodium 1000 watt

G. Metal halide 1000 watt
   125,000 lumen lamp

System Components	QUANTITY	UNIT	COST PER S.F.		
			MAT.	INST.	TOTAL
SYSTEM D5020 226 0240					
HIGH INTENSITY DISCHARGE FIXTURE, 16′ ABOVE WORK PLANE					
1 WATT/S.F., TYPE E, 42 FC, 1 FIXTURE/1000 S.F.					
Steel intermediate conduit, (IMC) 1/2″ diam	.160	L.F.	.34	.84	1.18
Wire, 600V, type THWN-THHN, copper, solid, #10	.003	C.L.F.	.08	.16	.24
Steel outlet box 4″ concrete	.001	Ea.	.01	.03	.04
Steel outlet box plate with stud, 4″ concrete	.001	Ea.	.01	.01	.02
Metal halide, hi bay, aluminum reflector, 1000 W lamp	.001	Ea.	.58	.26	.84
TOTAL			1.02	1.30	2.32

D5020 226	H.I.D. Fixture, High Bay, 16′ (by Wattage)	COST PER S.F.		
		MAT.	INST.	TOTAL
0190	High intensity discharge fixture, 16′ above work plane			
0240	1 watt/S.F., type E, 42 FC, 1 fixture/1000 S.F.	1.01	1.29	2.30
0280	Type G, 52 FC, 1 fixture/1000 S.F.	1.01	1.29	2.30
0320	Type C, 54 FC, 2 fixture/1000 S.F.	1.18	1.44	2.62
0440	2 watt/S.F., type E, 84 FC, 2 fixture/1000 S.F.	2.05	2.63	4.68
0480	Type G, 105 FC, 2 fixture/1000 S.F.	2.05	2.63	4.68
0520	Type C, 108 FC, 4 fixture/1000 S.F.	2.35	2.89	5.24
0640	3 watt/S.F., type E, 126 FC, 3 fixture/1000 S.F.	3.07	3.93	7
0680	Type G, 157 FC, 3 fixture/1000 S.F.	3.07	3.93	7
0720	Type C, 162 FC, 6 fixture/1000 S.F.	3.52	4.33	7.85
0840	4 watt/S.F., type E, 168 FC, 4 fixture/1000 S.F.	4.11	5.25	9.36
0880	Type G, 210 FC, 4 fixture/1000 S.F.	4.11	5.25	9.36
0920	Type C, 243 FC, 9 fixture/1000 S.F.	5.10	6.05	11.15
1040	5 watt/S.F., type E, 210 FC, 5 fixture/1000 S.F.	5.10	6.55	11.65
1080	Type G, 262 FC, 5 fixture/1000 S.F.	5.10	6.55	11.65
1120	Type C, 297 FC, 11 fixture/1000 S.F.	6.25	7.50	13.75

R265723
-05

D50 Electrical

D5020 Lighting and Branch Wiring

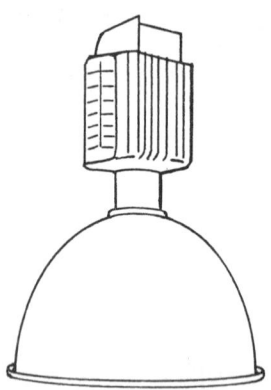

HIGH BAY FIXTURES

B. Metal halide 400 watt

C. High pressure sodium 400 watt

E. Metal halide 1000 watt

F. High pressure sodium 1000 watt

G. Metal halide 1000 watt
125,000 lumen lamp

System Components	QUANTITY	UNIT	COST PER S.F.		
			MAT.	INST.	TOTAL
SYSTEM D5020 228 1240					
HIGH INTENSITY DISCHARGE FIXTURE, 20′ ABOVE WORK PLANE, 100 FC					
TYPE C, 6 FIXTURES PER 900 S.F.					
Steel intermediate conduit, (IMC) 1/2″ diam	.350	L.F.	.74	1.84	2.58
Wire, 600V, type THWN-THHN, copper, solid, #10.	.011	C.L.F.	.28	.58	.86
Steel outlet box 4″ concrete	.007	Ea.	.10	.18	.28
Steel outlet box plate with stud, 4″ concrete	.007	Ea.	.04	.05	.09
High pressure sodium, hi bay, aluminum reflector, 400 W lamp	.007	Ea.	2.63	1.59	4.22
TOTAL			3.79	4.24	8.03

D5020 228	H.I.D. Fixture, High Bay, 20′ (by Type)		COST PER S.F.		
			MAT.	INST.	TOTAL
0510	High intensity discharge fixture 20′ above work plane, 100 FC				
1240	Type C, 6 fixtures per 900 S.F.		3.79	4.23	8.02
1280	10 fixtures per 1800 S.F.	R265723	3.39	3.97	7.36
1320	16 fixtures per 3000 S.F.	-05	3.04	3.82	6.86
1360	20 fixtures per 4000 S.F.		3.04	3.82	6.86
1400	24 fixtures per 5000 S.F.		3.04	3.82	6.86
1440	38 fixtures per 8000 S.F.		2.98	3.66	6.64
1520	68 fixtures per 16000 S.F.		2.83	4.03	6.86
1560	132 fixtures per 32000 S.F.		2.83	4.03	6.86
1600	Type G, 4 fixtures per 900 S.F.		4.06	5.10	9.16
1640	6 fixtures per 1800 S.F.		3.48	4.87	8.35
1680	7 fixtures per 3000 S.F.		3.42	4.71	8.13
1720	10 fixtures per 4000 S.F.		3.39	4.66	8.05
1760	11 fixtures per 5000 S.F.		3.05	5	8.05
1800	18 fixtures per 8000 S.F.		3.05	5	8.05
1840	22 fixtures per 10000 S.F.		3.05	5	8.05
1880	34 fixtures per 16000 S.F.		2.99	4.84	7.83
1920	66 fixtures per 32000 S.F.		2.82	4.42	7.24

D5020 Lighting and Branch Wiring

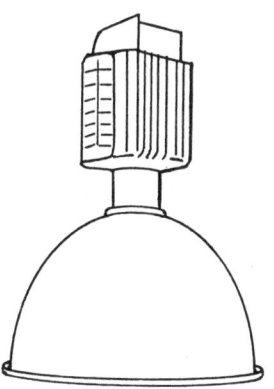

HIGH BAY FIXTURES

B. Metal halide 400 watt

C. High pressure sodium 400 watt

E. Metal halide 1000 watt

F. High pressure sodium 1000 watt

G. Metal halide 1000 watt
125,000 lumen lamp

System Components	QUANTITY	UNIT	COST PER S.F.		
			MAT.	INST.	TOTAL
SYSTEM D5020 230 0240					
HIGH INTENSITY DISCHARGE FIXTURE, 20' ABOVE WORK PLANE					
1 WATT/S.F., TYPE E, 40 FC, 1 FIXTURE 1000 S.F.					
Steel intermediate conduit, (IMC) 1/2" diam	.160	L.F.	.34	.84	1.18
Wire, 600V, type THWN-THHN, copper, solid, #10	.005	C.L.F.	.13	.26	.39
Steel outlet box 4" concrete	.001	Ea.	.01	.03	.04
Steel outlet box plate with stud, 4" concrete	.001	Ea.	.01	.01	.02
Metal halide, hi bay, aluminum reflector, 1000 W lamp	.001	Ea.	.58	.26	.84
TOTAL			1.07	1.40	2.47

D5020 230	H.I.D. Fixture, High Bay, 20' (by Wattage)		COST PER S.F.		
			MAT.	INST.	TOTAL
0190	High intensity discharge fixture, 20' above work plane				
0240	1 watt/S.F., type E, 40 FC, 1 fixture/1000 S.F.		1.07	1.40	2.47
0280	Type G, 50 FC, 1 fixture/1000 S.F.	R265723 -05	1.07	1.40	2.47
0320	Type C, 52 FC, 2 fixtures/1000 S.F.		1.25	1.61	2.86
0440	2 watt/S.F., type E, 81 FC, 2 fixtures/1000 S.F.		2.13	2.79	4.92
0480	Type G, 101 FC, 2 fixtures/1000 S.F.		2.13	2.79	4.92
0520	Type C, 104 FC, 4 fixtures/1000 S.F.		2.48	3.17	5.65
0640	3 watt/S.F., type E, 121 FC, 3 fixtures/1000 S.F.		3.20	4.19	7.39
0680	Type G, 151 FC, 3 fixtures/1000 S.F.		3.20	4.19	7.39
0720	Type C, 155 FC, 6 fixtures/1000 S.F.		3.73	4.78	8.51
0840	4 watt/S.F., type E, 161 FC, 4 fixtures/1000 S.F.		4.26	5.60	9.86
0880	Type G, 202 FC, 4 fixtures/1000 S.F.		4.26	5.60	9.86
0920	Type C, 233 FC, 9 fixtures/1000 S.F.		5.35	6.60	11.95
1040	5 watt/S.F., type E, 202 FC, 5 fixtures/1000 S.F.		5.35	7	12.35
1080	Type G, 252 FC, 5 fixtures/1000 S.F.		5.35	7	12.35
1120	Type C, 285 FC, 11 fixtures/1000 S.F.		6.60	8.20	14.80

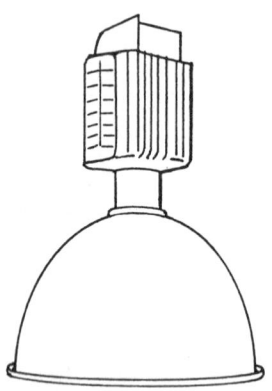

HIGH BAY FIXTURES

B. Metal halide 400 watt

C. High pressure sodium 400 watt

E. Metal halide 1000 watt

F. High pressure sodium 1000 watt

G. Metal halide 1000 watt
125,000 lumen lamp

System Components	QUANTITY	UNIT	COST PER S.F.		
			MAT.	INST.	TOTAL
SYSTEM D5020 232 1240					
HIGH INTENSITY DISCHARGE FIXTURE, 30' ABOVE WORK PLANE, 100 FC					
TYPE F, 4 FIXTURES PER 900 S.F.					
Steel intermediate conduit, (IMC) 1/2" diam	.580	L.F.	1.22	3.05	4.27
Wire, 600V, type THWN-THHN, copper, solid, #10	.018	C.L.F.	.46	.95	1.41
Steel outlet box 4" concrete	.004	Ea.	.06	.10	.16
Steel outlet box plate with stud, 4" concrete	.004	Ea.	.02	.03	.05
High pressure sodium, hi bay, aluminum, 1000 W lamp	.004	Ea.	2.16	1.04	3.20
TOTAL			3.92	5.17	9.09

D5020 232	H.I.D. Fixture, High Bay, 30' (by Type)		COST PER S.F.		
			MAT.	INST.	TOTAL
0510	High intensity discharge fixture, 30' above work plane, 100 FC				
1240	Type F, 4 fixtures per 900 S.F.		3.92	5.15	9.07
1280	6 fixtures per 1800 S.F.	R265723 -05	3.38	4.92	8.30
1320	8 fixtures per 3000 S.F.		3.38	4.92	8.30
1360	9 fixtures per 4000 S.F.		2.85	4.68	7.53
1400	10 fixtures per 5000 S.F.		2.85	4.68	7.53
1440	17 fixtures per 8000 S.F.		2.85	4.68	7.53
1480	18 fixtures per 10,000 S.F.		2.70	4.31	7.01
1520	27 fixtures per 16,000 S.F.		2.70	4.31	7.01
1560	52 fixtures per 32000 S.F.		2.67	4.26	6.93
1600	Type G, 4 fixtures per 900 S.F.		4.21	5.50	9.71
1640	6 fixtures per 1800 S.F.		3.52	4.98	8.50
1680	9 fixtures per 3000 S.F.		3.46	4.82	8.28
1720	11 fixtures per 4000 S.F.		3.39	4.66	8.05
1760	13 fixtures per 5000 S.F.		3.39	4.66	8.05
1800	21 fixtures per 8000 S.F.		3.39	4.66	8.05
1840	23 fixtures per 10,000 S.F.		3.07	4.94	8.01
1880	36 fixtures per 16,000 S.F.		3.07	4.94	8.01
1920	70 fixtures per 32,000 S.F.		3.07	4.94	8.01

D5020 Lighting and Branch Wiring

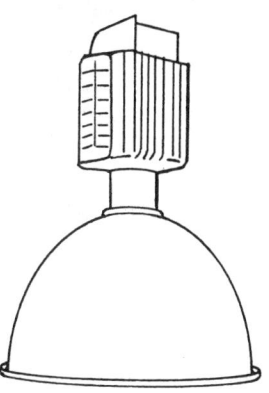

HIGH BAY FIXTURES

B. Metal halide 400 watt

C. High pressure sodium 400 watt

E. Metal halide 1000 watt

F. High pressure sodium 1000 watt

G. Metal halide 1000 watt
125,000 lumen lamp

System Components	QUANTITY	UNIT	COST PER S.F.		
			MAT.	INST.	TOTAL
SYSTEM D5020 234 0240					
HIGH INTENSITY DISCHARGE FIXTURE, 30' ABOVE WORK PLANE					
1 WATT/S.F., TYPE E, 37 FC, 1 FIXTURE/1000 S.F.					
Steel intermediate conduit, (IMC) 1/2" diam	.196	L.F.	.41	1.03	1.44
Wire, 600V type THWN-THHN, copper, solid, #10	.006	C.L.F.	.15	.32	.47
Steel outlet box 4" concrete	.001	Ea.	.01	.03	.04
Steel outlet box plate with stud, 4" concrete	.001	Ea.	.01	.01	.02
Metal halide, hi bay, aluminum reflector, 1000 W lamp	.001	Ea.	.58	.26	.84
TOTAL			1.16	1.65	2.81

D5020 234	H.I.D. Fixture, High Bay, 30' (by Wattage)		COST PER S.F.		
			MAT.	INST.	TOTAL
0190	High intensity discharge fixture, 30' above work plane				
0240	1 watt/S.F., type E, 37 FC, 1 fixture/1000 S.F.		1.17	1.64	2.81
0280	Type G, 45 FC., 1 fixture/1000 S.F.	R265723 -05	1.17	1.64	2.81
0320	Type F, 50 FC, 1 fixture/1000 S.F.		.97	1.28	2.25
0440	2 watt/S.F., type E, 74 FC, 2 fixtures/1000 S.F.		2.33	3.28	5.61
0480	Type G, 92 FC, 2 fixtures/1000 S.F.		2.33	3.28	5.61
0520	Type F, 100 FC, 2 fixtures/1000 S.F.		1.97	2.61	4.58
0640	3 watt/S.F., type E, 110 FC, 3 fixtures/1000 S.F.		3.53	4.97	8.50
0680	Type G, 138FC, 3 fixtures/1000 S.F.		3.53	4.97	8.50
0720	Type F, 150 FC, 3 fixtures/1000 S.F.		2.95	3.89	6.84
0840	4 watt/S.F., type E, 148 FC, 4 fixtures/1000 S.F.		4.69	6.60	11.29
0880	Type G, 185 FC, 4 fixtures/1000 S.F.		4.69	6.60	11.29
0920	Type F, 200 FC, 4 fixtures/1000 S.F.		3.95	5.25	9.20
1040	5 watt/S.F., type E, 185 FC, 5 fixtures/1000 S.F.		5.85	8.25	14.10
1080	Type G, 230 FC, 5 fixtures/1000 S.F.		5.85	8.25	14.10
1120	Type F, 250 FC, 5 fixtures/1000 S.F.		4.92	6.50	11.42

D5020 Lighting and Branch Wiring

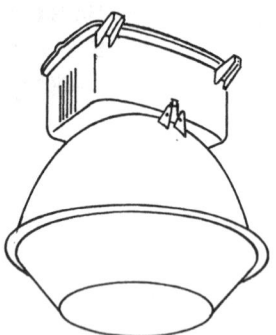

LOW BAY FIXTURES

J. Metal halide 250 watt

K. High pressure sodium 150 watt

System Components	QUANTITY	UNIT	COST PER S.F.		
			MAT.	INST.	TOTAL
SYSTEM D5020 236 0920					
HIGH INTENSITY DISCHARGE FIXTURE, 8'-10' ABOVE WORK PLANE, 50 FC					
TYPE J, 13 FIXTURES PER 1800 S.F.					
Steel intermediate conduit, (IMC) 1/2" diam	.550	L.F.	1.16	2.89	4.05
Wire, 600V, type THWN-THHN, copper, solid, #10	.012	C.L.F.	.31	.63	.94
Steel outlet box 4" concrete	.007	Ea.	.10	.18	.28
Steel outlet box plate with stud, 4" concrete	.007	Ea.	.04	.05	.09
Metal halide, lo bay, aluminum reflector, 250 W DX lamp	.007	Ea.	2.73	1.14	3.87
TOTAL			4.34	4.89	9.23

D5020 236	H.I.D. Fixture, Low Bay, 8'-10' (by Type)		COST PER S.F.		
			MAT.	INST.	TOTAL
0510	High intensity discharge fixture, 8'-10' above work plane, 50 FC				
0880	Type J, 7 fixtures per 900 S.F.		4.68	4.92	9.60
0920	13 fixtures per 1800 S.F.	R265723 -05	4.34	4.89	9.23
0960	21 fixtures per 3000 S.F.		4.34	4.89	9.23
1000	28 fixtures per 4000 S.F.		4.34	4.89	9.23
1040	35 fixtures per 5000 S.F.		4.34	4.89	9.23
1120	62 fixtures per 10,000 S.F.		3.93	4.69	8.62
1160	99 fixtures per 16,000 S.F.		3.93	4.69	8.62
1200	199 fixtures per 32,000 S.F.		3.93	4.69	8.62
1240	Type K, 9 fixtures per 900 S.F.		4.39	4.21	8.60
1280	16 fixtures per 1800 S.F.		4.07	4.07	8.14
1320	26 fixtures per 3000 S.F.		4.05	4.02	8.07
1360	31 fixtures per 4000 S.F.		3.75	3.93	7.68
1400	39 fixtures per 5000 S.F.		3.75	3.93	7.68
1440	62 fixtures per 8000 S.F.		3.75	3.93	7.68
1480	78 fixtures per 10,000 S.F.		3.75	3.93	7.68
1520	124 fixtures per 16,000 S.F.		3.70	3.82	7.52
1560	248 fixtures per 32,000 S.F.		3.62	4.26	7.88

D50 Electrical

D5020 Lighting and Branch Wiring

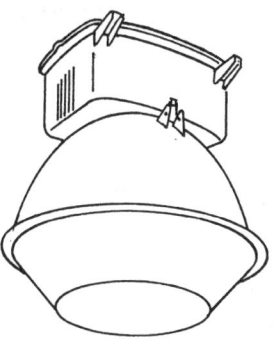

LOW BAY FIXTURES
J. Metal halide 250 watt
K. High pressure sodium 150 watt

System Components	QUANTITY	UNIT	COST PER S.F.		
			MAT.	INST.	TOTAL
SYSTEM D5020 238 0240					
HIGH INTENSITY DISCHARGE FIXTURE, 8'-10' ABOVE WORK PLANE					
1 WATT/S.F., TYPE J, 30 FC, 4 FIXTURES/1000 S.F.					
Steel intermediate conduit, (IMC) 1/2" diam	.280	L.F.	.59	1.47	2.06
Wire, 600V, type THWN-THHN, copper, solid, #10	.008	C.L.F.	.20	.42	.62
Steel outlet box 4" concrete	.004	Ea.	.06	.10	.16
Steel outlet box plate with stud, 4" concrete	.004	Ea.	.02	.03	.05
Metal halide, lo bay, aluminum reflector, 250 W DX lamp	.004	Ea.	1.56	.65	2.21
TOTAL			2.43	2.67	5.10

D5020 238	H.I.D. Fixture, Low Bay, 8'-10' (by Wattage)		COST PER S.F.		
			MAT.	INST.	TOTAL
0190	High intensity discharge fixture, 8'-10' above work plane				
0240	1 watt/S.F., type J, 30 FC, 4 fixtures/1000 S.F.		2.44	2.67	5.11
0280	Type K, 29 FC, 5 fixtures/1000 S.F.	R265723	2.32	2.40	4.72
0400	2 watt/S.F., type J, 52 FC, 7 fixtures/1000 S.F.	-05	4.36	4.90	9.26
0440	Type K, 63 FC, 11 fixtures/1000 S.F.		5	5.05	10.05
0560	3 watt/S.F., type J, 81 FC, 11 fixtures/1000 S.F.		6.70	7.40	14.10
0600	Type K, 92 FC, 16 fixtures/1000 S.F.		7.35	7.45	14.80
0720	4 watt/S.F., type J, 103 FC, 14 fixtures/1000 S.F.		8.70	9.75	18.45
0760	Type K, 127 FC, 22 fixtures/1000 S.F.		10.05	10.10	20.15
0880	5 watt/S.F., type J, 133 FC, 18 fixtures/1000 S.F.		11.10	12.30	23.40
0920	Type K, 155 FC, 27 fixtures/1000 S.F.		12.35	12.45	24.80

D50 Electrical

D5020 Lighting and Branch Wiring

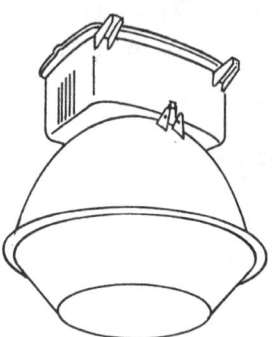

LOW BAY FIXTURES
J. Metal halide 250 watt
K. High pressure sodium 150 watt

System Components	QUANTITY	UNIT	COST PER S.F.		
			MAT.	INST.	TOTAL
SYSTEM D5020 240 0880					
HIGH INTENSITY DISCHARGE FIXTURE, 16' ABOVE WORK PLANE, 50 FC					
TYPE J, 9 FIXTURES PER 900 S.F.					
Steel intermediate conduit, (IMC) 1/2" diam	.630	L.F.	1.33	3.31	4.64
Wire, 600V type, THWN-THHN, copper, solid, #10	.012	C.L.F.	.31	.63	.94
Steel outlet box 4" concrete	.010	Ea.	.14	.26	.40
Steel outlet box plate with stud, 4" concrete	.010	Ea.	.06	.07	.13
Metal halide, lo bay, aluminum reflector, 250 W DX lamp	.010	Ea.	3.90	1.63	5.53
TOTAL			5.74	5.90	11.64

D5020 240	H.I.D. Fixture, Low Bay, 16' (by Type)	COST PER S.F.		
		MAT.	INST.	TOTAL
0510	High intensity discharge fixture, 16' above work plane, 50 FC			
0880	Type J, 9 fixtures per 900 S.F.	5.75	5.90	11.65
0920	14 fixtures per 1800 S.F.	4.94	5.55	10.49
0960	24 fixtures per 3000 S.F.	4.99	5.65	10.64
1000	32 fixtures per 4000 S.F.	4.99	5.65	10.64
1040	35 fixtures per 5000 S.F.	4.60	5.50	10.10
1080	56 fixtures per 8000 S.F.	4.60	5.50	10.10
1120	70 fixtures per 10,000 S.F.	4.60	5.50	10.10
1160	111 fixtures per 16,000 S.F.	4.60	5.50	10.10
1200	222 fixtures per 32,000 S.F.	4.60	5.50	10.10
1240	Type K, 11 fixtures per 900 S.F.	5.45	5.50	10.95
1280	20 fixtures per 1800 S.F.	5	5.10	10.10
1320	29 fixtures per 3000 S.F.	4.72	5	9.72
1360	39 fixtures per 4000 S.F.	4.70	4.95	9.65
1400	44 fixtures per 5000 S.F.	4.40	4.86	9.26
1440	62 fixtures per 8000 S.F.	4.40	4.86	9.26
1480	87 fixtures per 10,000 S.F.	4.40	4.86	9.26
1520	138 fixtures per 16,000 S.F.	4.40	4.86	9.26

R265723-05

D50 Electrical

D5020 Lighting and Branch Wiring

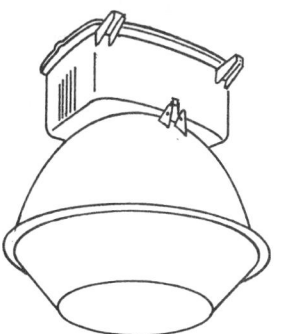

LOW BAY FIXTURES
J. Metal halide 250 watt
K. High pressure sodium 150 watt

System Components	QUANTITY	UNIT	COST PER S.F.		
			MAT.	INST.	TOTAL
SYSTEM D5020 242 0240					
HIGH INTENSITY DISCHARGE FIXTURE, 16′ ABOVE WORK PLANE					
1 WATT/S.F., TYPE J, 28 FC, 4 FIXTURES/1000 S.F.					
Steel intermediate conduit, (IMC) 1/2″ diam	.328	L.F.	.69	1.72	2.41
Wire, 600V, type THWN-THHN, copper, solid, #10	.010	C.L.F.	.26	.53	.79
Steel outlet box 4″ concrete	.004	Ea.	.06	.10	.16
Steel outlet box plate with stud, 4″ concrete	.004	Ea.	.02	.03	.05
Metal halide, lo bay, aluminum reflector, 250 W DX lamp	.004	Ea.	1.56	.65	2.21
TOTAL			2.59	3.03	5.62

D5020 242	H.I.D. Fixture, Low Bay, 16′ (by Wattage)		COST PER S.F.		
			MAT.	INST.	TOTAL
0190	High intensity discharge fixture, mounted 16′ above work plane				
0240	1 watt/S.F., type J, 28 FC, 4 fixt./1000 S.F.		2.59	3.03	5.62
0280	Type K, 27 FC, 5 fixt./1000 S.F.	R265723 -05	2.75	3.39	6.14
0400	2 watt/S.F., type J, 48 FC, 7 fixt/1000 S.F.		4.77	5.85	10.62
0440	Type K, 58 FC, 11 fixt/1000 S.F.		5.80	6.95	12.75
0560	3 watt/S.F., type J, 75 FC, 11 fixt/1000 S.F.		7.35	8.90	16.25
0600	Type K, 85 FC, 16 fixt/1000 S.F.		8.55	10.30	18.85
0720	4 watt/S.F., type J, 95 FC, 14 fixt/1000 S.F.		9.55	11.75	21.30
0760	Type K, 117 FC, 22 fixt/1000 S.F.		11.60	13.85	25.45
0880	5 watt/S.F., type J, 122 FC, 18 fixt/1000 S.F.		12.10	14.75	26.85
0920	Type K, 143 FC, 27 fixt/1000 S.F.		14.35	17.25	31.60

D5030 Communications and Security

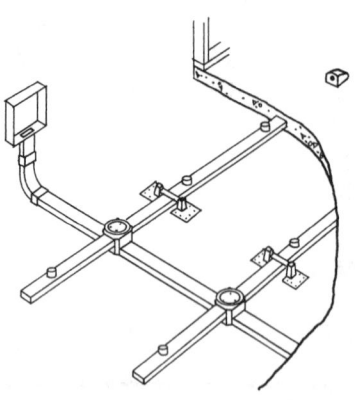

Description: System below includes telephone fitting installed. Does not include cable.

When poke thru fittings and telepoles are used for power, they can also be used for telephones at a negligible additional cost.

System Components	QUANTITY	UNIT	COST PER S.F.		
			MAT.	INST.	TOTAL
SYSTEM D5030 310 0200					
TELEPHONE SYSTEMS, UNDERFLOOR DUCT, 5' ON CENTER, LOW DENSITY					
Underfloor duct 7-1/4" w/insert 2' O.C. 1-3/8" x 7-1/4" super duct	.190	L.F.	4.85	1.99	6.84
Vertical elbow for underfloor superduct, 7-1/4", included					
Underfloor duct conduit adapter, 2" x 1-1/4", included					
Underfloor duct junction box, single duct, 7-1/4" x 3 1/8"	.003	Ea.	1.16	.39	1.55
Underfloor junction box carpet pan	.003	Ea.	.82	.02	.84
Underfloor duct outlet, low tension	.004	Ea.	.31	.26	.57
TOTAL			7.14	2.66	9.80

D5030 310	Telephone Systems	COST PER S.F.		
		MAT.	INST.	TOTAL
0200	Telephone systems, underfloor duct, 5' on center, low density	7.15	2.66	9.81
0240	5' on center, high density	7.45	2.92	10.37
0280	7' on center, low density	5.70	2.21	7.91
0320	7' on center, high density	6	2.47	8.47
0400	Poke thru fittings, low density	.92	.87	1.79
0440	High density	1.85	1.74	3.59
0520	Telepoles, low density	.91	.56	1.47
0560	High density	1.82	1.13	2.95
0640	Conduit system with floor boxes, low density	1.04	.93	1.97
0680	High density	2.06	1.84	3.90
1020	Telephone wiring for offices & laboratories, 8 jacks/MSF	.49	2.25	2.74

D50 Electrical

D5030 Communications and Security

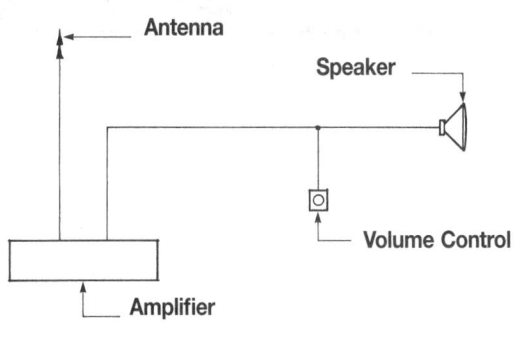

- Antenna
- Speaker
- Volume Control
- Amplifier

Sound System Includes AM–FM antenna, outlets, rigid conduit, and copper wire.
Fire Detection System Includes pull stations, signals, smoke and heat detectors, rigid conduit, and copper wire.
Intercom System Includes master and remote stations, rigid conduit, and copper wire.
Master Clock System Includes clocks, bells, rigid conduit, and copper wire.
Master TV Antenna Includes antenna, VHF–UHF reception and distribution, rigid conduit, and copper wire.

System Components	QUANTITY	UNIT	COST EACH		
			MAT.	INST.	TOTAL
SYSTEM D5030 910 0220					
SOUND SYSTEM, INCLUDES OUTLETS, BOXES, CONDUIT & WIRE					
Steel intermediate conduit, (IMC) 1/2" diam	1200.000	L.F.	2,532	6,300	8,832
Wire sound shielded w/drain, #22-2 conductor	15.500	C.L.F.	372	1,015.25	1,387.25
Sound system speakers ceiling or wall	12.000	Ea.	1,296	786	2,082
Sound system volume control	12.000	Ea.	966	786	1,752
Sound system amplifier, 250 Watts	1.000	Ea.	1,600	525	2,125
Sound system antenna, AM FM	1.000	Ea.	200	131	331
Sound system monitor panel	1.000	Ea.	360	131	491
Sound system cabinet	1.000	Ea.	780	525	1,305
Steel outlet box 4" square	12.000	Ea.	31.08	312	343.08
Steel outlet box 4" plaster rings	12.000	Ea.	33.12	97.80	130.92
TOTAL			8,170.20	10,609.05	18,779.25

D5030 910	Communication & Alarm Systems	COST EACH		
		MAT.	INST.	TOTAL
0200	Communication & alarm systems, includes outlets, boxes, conduit & wire			
0210	Sound system, 6 outlets	5,800	6,625	12,425
0220	12 outlets	8,175	10,600	18,775
0240	30 outlets	14,200	20,000	34,200
0280	100 outlets	43,300	67,000	110,300
0320	Fire detection systems, 12 detectors	2,850	5,500	8,350
0360	25 detectors	4,925	9,300	14,225
0400	50 detectors	9,550	18,300	27,850
0440	100 detectors	17,300	32,800	50,100
0480	Intercom systems, 6 stations	3,300	4,525	7,825
0520	12 stations	6,700	9,025	15,725
0560	25 stations	11,600	17,400	29,000
0600	50 stations	22,300	32,700	55,000
0640	100 stations	43,900	64,000	107,900
0680	Master clock systems, 6 rooms	4,500	7,375	11,875
0720	12 rooms	6,975	12,600	19,575
0760	20 rooms	9,500	17,800	27,300
0800	30 rooms	16,100	32,800	48,900
0840	50 rooms	26,200	55,500	81,700
0880	100 rooms	50,500	110,000	160,500
0920	Master TV antenna systems, 6 outlets	3,025	4,675	7,700
0960	12 outlets	5,625	8,700	14,325
1000	30 outlets	11,500	20,200	31,700
1040	100 outlets	38,000	66,000	104,000

D50 Electrical

D5030 Communications and Security

D5030 920	Data Communication	COST PER M.S.F.		
		MAT.	INST.	TOTAL
0100	Data communication system, incl data/voice outlets, boxes, conduit, & cable			
0102	Data and voice system, 2 data/voice outlets per 1000 S.F.	261	385	646
0104	4 data/voice outlets per 1000 S.F.	505	760	1,265
0106	6 data/voice outlets per 1000 S.F.	735	1,125	1,860
0110	8 data/voice outlets per 1000 S.F.	960	1,475	2,435

D5090 Other Electrical Systems

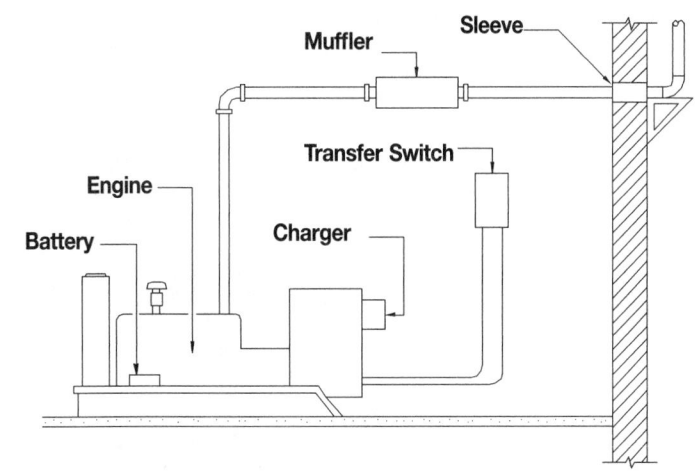

Description: System below tabulates the installed cost for generators by kW. Included in costs are battery, charger, muffler, and transfer switch.

No conduit, wire, or terminations included.

System Components	QUANTITY	UNIT	COST PER kW		
			MAT.	INST.	TOTAL
SYSTEM D5090 210 0200					
GENERATOR SET, INCL. BATTERY, CHARGER, MUFFLER & TRANSFER SWITCH					
GAS/GASOLINE OPER., 3 PHASE, 4 WIRE, 277/480V, 7.5 kW					
Generator set, gas or gasoline operated, 3 ph 4 W, 277/480 V, 7.5 kW	.133	Ea.	933.33	235.60	1,168.93
TOTAL			933.33	235.60	1,168.93

D5090 210	Generators (by kW)	COST PER kW		
		MAT.	INST.	TOTAL
0190	Generator sets, include battery, charger, muffler & transfer switch			
0200	Gas/gasoline operated, 3 phase, 4 wire, 277/480 volt, 7.5 kW	935	236	1,171
0240	11.5 kW	860	181	1,041
0280	20 kW	585	117	702
0320	35 kW	400	76.50	476.50
0360	80 kW	299	46	345
0400	100 kW	262	45	307
0440	125 kW	430	42	472
0480	185 kW	385	32	417
0560	Diesel engine with fuel tank, 30 kW	625	89.50	714.50
0600	50 kW	460	70	530
0720	125 kW	280	40.50	320.50
0760	150 kW	267	37.50	304.50
0800	175 kW	250	33.50	283.50
0840	200 kW	226	31	257
0880	250 kW	212	25.50	237.50
0920	300 kW	192	22.50	214.50
0960	350 kW	186	21	207
1000	400 kW	200	19.35	219.35
1040	500 kW	202	16.35	218.35
1200	750 kW	221	9.95	230.95
1400	1000 kW	206	10.55	216.55

D5090 Other Electrical Systems

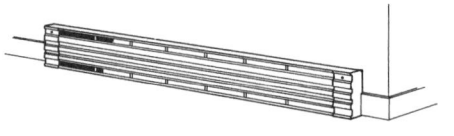

Low Density and Medium Density Baseboard Radiation

The costs shown in Table below are based on the following system considerations:

1. The heat loss per square foot is based on approximately 34 BTU/hr. per S.F. of floor or 10 watts per S.F. of floor.
2. Baseboard radiation is based on the low watt density type rated 187 watts per L.F. and the medium density type rated 250 watts per L.F.
3. Thermostat is not included.
4. Wiring costs include branch circuit wiring.

System Components	QUANTITY	UNIT	COST PER S.F.		
			MAT.	INST.	TOTAL
SYSTEM D5090 510 1000					
ELECTRIC BASEBOARD RADIATION, LOW DENSITY, 900 S.F., 31 MBH, 9 kW					
Electric baseboard radiator, 5' long 935 watt	.011	Ea.	.70	1.01	1.71
Steel intermediate conduit, (IMC) 1/2" diam	.170	L.F.	.36	.89	1.25
Wire 600 volt, type THW, copper, solid, #12	.005	C.L.F.	.08	.24	.32
TOTAL			1.14	2.14	3.28

D5090 510	Electric Baseboard Radiation (Low Density)	COST PER S.F.		
		MAT.	INST.	TOTAL
1000	Electric baseboard radiation, low density, 900 S.F., 31 MBH, 9 kW	1.15	2.14	3.29
1200	1500 S.F., 51 MBH, 15 kW	1.13	2.08	3.21
1400	2100 S.F., 72 MBH, 21 kW	1.01	1.87	2.88
1600	3000 S.F., 102 MBH, 30 kW	.92	1.70	2.62
2000	Medium density, 900 S.F., 31 MBH, 9 kW	1.02	1.95	2.97
2200	1500 S.F., 51 MBH, 15 kW	1	1.90	2.90
2400	2100 S.F., 72 MBH, 21 kW	.95	1.77	2.72
2600	3000 S.F., 102 MBH, 30 kW	.86	1.61	2.47

D5090 Other Electrical Systems

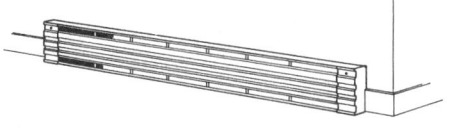

Commercial Duty Baseboard Radiation
The costs shown in Table below are based on the following system considerations:

1. The heat loss per square foot is based on approximately 41 BTU/hr. per S.F. of floor or 12 watts per S.F.
2. The baseboard radiation is of the commercial duty type rated 250 watts per L.F. served by 277 volt, single phase power.
3. Thermostat is not included.
4. Wiring costs include branch circuit wiring.

System Components	QUANTITY	UNIT	COST PER S.F.		
			MAT.	INST.	TOTAL
SYSTEM D5090 520 1000					
ELECTRIC BASEBOARD RADIATION, MEDIUM DENSITY, 1230 S.F., 51 MBH, 15 kW					
Electric baseboard radiator, 5' long	.013	Ea.	.83	1.19	2.02
Steel intermediate conduit, (IMC) 1/2" diam	.154	L.F.	.32	.81	1.13
Wire 600 volt, type THW, copper, solid, #12	.004	C.L.F.	.07	.19	.26
TOTAL			1.22	2.19	3.41

D5090 520	Electric Baseboard Radiation (Medium Density)	COST PER S.F.		
		MAT.	INST.	TOTAL
1000	Electric baseboard radiation, medium density, 1230 SF, 51 MBH, 15 kW	1.22	2.19	3.41
1200	2500 S.F. floor area, 106 MBH, 31 kW	1.14	2.05	3.19
1400	3700 S.F. floor area, 157 MBH, 46 kW	1.12	1.99	3.11
1600	4800 S.F. floor area, 201 MBH, 59 kW	1.03	1.84	2.87
1800	11,300 S.F. floor area, 464 MBH, 136 kW	.97	1.69	2.66
2000	30,000 S.F. floor area, 1229 MBH, 360 kW	.95	1.65	2.60

G1030   Site Earthwork

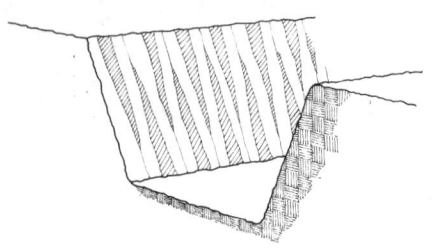

Trenching Systems are shown on a cost per linear foot basis. The systems include: excavation; backfill and removal of spoil; and compaction for various depths and trench bottom widths. The backfill has been reduced to accommodate a pipe of suitable diameter and bedding.

The slope for trench sides varies from none to 1:1.

The Expanded System Listing shows Trenching Systems that range from 2' to 12' in width. Depths range from 2' to 25'.

System Components		QUANTITY	UNIT	COST PER L.F.		
				EQUIP.	LABOR	TOTAL
SYSTEM G1030 805 1310						
TRENCHING COMMON EARTH, NO SLOPE, 2' WIDE, 2' DP, 3/8 C.Y. BUCKET						
Excavation, trench, hyd. backhoe, track mtd., 3/8 C.Y. bucket		.148	C.Y.	.26	.81	1.07
Backfill and load spoil, from stockpile		.153	C.Y.	.09	.24	.33
Compaction by vibrating plate, 6" lifts, 4 passes		.118	E.C.Y.	.03	.30	.33
Remove excess spoil, 6 C.Y. dump truck, 2 mile roundtrip		.040	C.Y.	.16	.15	.31
	TOTAL			.54	1.50	2.04

G1030 805	Trenching Common Earth	COST PER L.F.		
		EQUIP.	LABOR	TOTAL
1310	Trenching, common earth, no slope, 2' wide, 2' deep, 3/8 C.Y. bucket	.54	1.50	2.04
1320	3' deep, 3/8 C.Y. bucket	.74	2.25	2.99
1330	4' deep, 3/8 C.Y. bucket	.95	2.99	3.94
1340	6' deep, 3/8 C.Y. bucket	1.26	3.89	5.15
1350	8' deep, 1/2 C.Y. bucket	1.61	5.15	6.76
1360	10' deep, 1 C.Y. bucket	2.64	6.10	8.74
1400	4' wide, 2' deep, 3/8 C.Y. bucket	1.31	3	4.31
1410	3' deep, 3/8 C.Y. bucket	1.72	4.49	6.21
1420	4' deep, 1/2 C.Y. bucket	1.88	4.95	6.83
1430	6' deep, 1/2 C.Y. bucket	3	7.95	10.95
1440	8' deep, 1/2 C.Y. bucket	5.05	10.25	15.30
1450	10' deep, 1 C.Y. bucket	6	12.75	18.75
1460	12' deep, 1 C.Y. bucket	7.80	16.25	24.05
1470	15' deep, 1-1/2 C.Y. bucket	6.85	14.50	21.35
1480	18' deep, 2-1/2 C.Y. bucket	9.50	20.50	30
1520	6' wide, 6' deep, 5/8 C.Y. bucket w/trench box	7.05	12	19.05
1530	8' deep, 3/4 C.Y. bucket	8.80	15.50	24.30
1540	10' deep, 1 C.Y. bucket	8.50	16.15	24.65
1550	12' deep, 1-1/2 C.Y. bucket	9.15	17.35	26.50
1560	16' deep, 2-1/2 C.Y. bucket	12.40	21.50	33.90
1570	20' deep, 3-1/2 C.Y. bucket	17.80	26	43.80
1580	24' deep, 3-1/2 C.Y. bucket	21.50	31	52.50
1640	8' wide, 12' deep, 1-1/2 C.Y. bucket w/trench box	12.70	22	34.70
1650	15' deep, 1-1/2 C.Y. bucket	16.70	29	45.70
1660	18' deep, 2-1/2 C.Y. bucket	18	28.50	46.50
1680	24' deep, 3-1/2 C.Y. bucket	29	40	69
1730	10' wide, 20' deep, 3-1/2 C.Y. bucket w/trench box	22.50	38.50	61
1740	24' deep, 3-1/2 C.Y. bucket	35	46	81
1780	12' wide, 20' deep, 3-1/2 C.Y. bucket w/trench box	36.50	48.50	85
1790	25' deep, bucket	45.50	62	107.50
1800	1/2 to 1 slope, 2' wide, 2' deep, 3/8 C.Y. bucket	.74	2.25	2.99
1810	3' deep, 3/8 C.Y. bucket	1.20	3.93	5.13
1820	4' deep, 3/8 C.Y. bucket	1.76	6	7.76
1840	6' deep, 3/8 C.Y. bucket	2.96	9.70	12.66

G10 Site Preparation

G1030 Site Earthwork

G1030 805	Trenching Common Earth	COST PER L.F.		
		EQUIP.	LABOR	TOTAL
1860	8' deep, 1/2 C.Y. bucket	4.62	15.50	20.12
1880	10' deep, 1 C.Y. bucket	9.10	21.50	30.60
2300	4' wide, 2' deep, 3/8 C.Y. bucket	1.52	3.75	5.27
2310	3' deep, 3/8 C.Y. bucket	2.18	6.20	8.38
2320	4' deep, 1/2 C.Y. bucket	2.64	7.55	10.19
2340	6' deep, 1/2 C.Y. bucket	5	14.10	19.10
2360	8' deep, 1/2 C.Y. bucket	9.70	20.50	30.20
2380	10' deep, 1 C.Y. bucket	13.35	29	42.35
2400	12' deep, 1 C.Y. bucket	17.65	39.50	57.15
2430	15' deep, 1-1/2 C.Y. bucket	19.45	42.50	61.95
2460	18' deep, 2-1/2 C.Y. bucket	34	65	99
2840	6' wide, 6' deep, 5/8 C.Y. bucket w/trench box	9.80	17.55	27.35
2860	8' deep, 3/4 C.Y. bucket	14.15	26.50	40.65
2880	10' deep, 1 C.Y. bucket	13.75	27	40.75
2900	12' deep, 1-1/2 C.Y. bucket	17.50	35	52.50
2940	16' deep, 2-1/2 C.Y. bucket	28.50	50.50	79
2980	20' deep, 3-1/2 C.Y. bucket	44.50	69.50	114
3020	24' deep, 3-1/2 C.Y. bucket	63.50	95	158.50
3100	8' wide, 12' deep, 1-1/2 C.Y. bucket w/trench box	21.50	40	61.50
3120	15' deep, 1-1/2 C.Y. bucket	31.50	58	89.50
3140	18' deep, 2-1/2 C.Y. bucket	39	68	107
3180	24' deep, 3-1/2 C.Y. bucket	71	104	175
3270	10' wide, 20' deep, 3-1/2 C.Y. bucket w/trench box	44	80	124
3280	24' deep, 3-1/2 C.Y. bucket	78	114	192
3370	12' wide, 20' deep, 3-1/2 C.Y. bucket w/ trench box	66	93	159
3380	25' deep, 3-1/2 C.Y. bucket	91.50	133	224.50
3500	1 to 1 slope, 2' wide, 2' deep, 3/8 C.Y. bucket	.95	3	3.95
3520	3' deep, 3/8 C.Y. bucket	3.12	6.95	10.07
3540	4' deep, 3/8 C.Y. bucket	2.58	9	11.58
3560	6' deep, 3/8 C.Y. bucket	2.95	9.70	12.65
3580	8' deep, 1/2 C.Y. bucket	5.75	19.35	25.10
3600	10' deep, 1 C.Y. bucket	15.60	37	52.60
3800	4' wide, 2' deep, 3/8 C.Y. bucket	1.72	4.49	6.21
3820	3' deep, 3/8 C.Y. bucket	2.64	7.85	10.49
3840	4' deep, 1/2 C.Y. bucket	3.39	10.15	13.54
3860	6' deep, 1/2 C.Y. bucket	7	20.50	27.50
3880	8' deep, 1/2 C.Y. bucket	14.40	31	45.40
3900	10' deep, 1 C.Y. bucket	20.50	45.50	66
3920	12' deep, 1 C.Y. bucket	30.50	66	96.50
3940	15' deep, 1-1/2 C.Y. bucket	32	70.50	102.50
3960	18' deep, 2-1/2 C.Y. bucket	46.50	89.50	136
4030	6' wide, 6' deep, 5/8 C.Y. bucket w/trench box	12.85	23.50	36.35
4040	8' deep, 3/4 C.Y. bucket	18.50	32.50	51
4050	10' deep, 1 C.Y. bucket	19.85	39.50	59.35
4060	12' deep, 1-1/2 C.Y. bucket	26.50	53.50	80
4070	16' deep, 2-1/2 C.Y. bucket	44.50	80	124.50
4080	20' deep, 3-1/2 C.Y. bucket	72	113	185
4090	24' deep, 3-1/2 C.Y. bucket	105	159	264
4500	8' wide, 12' deep, 1-1/2 C.Y. bucket w/trench box	30.50	58	88.50
4550	15' deep, 1-1/2 C.Y. bucket	46	87.50	133.50
4600	18' deep, 2-1/2 C.Y. bucket	59.50	105	164.50
4650	24' deep, 3-1/2 C.Y. bucket	113	168	281
4800	10' wide, 20' deep, 3-1/2 C.Y. bucket w/trench box	65.50	122	187.50
4850	24' deep, 3-1/2 C.Y. bucket	120	178	298
4950	12' wide, 20' deep, 3-1/2 C.Y. bucket w/ trench box	95	137	232
4980	25' deep, 3-1/2 C.Y. bucket	136	201	337

G1030 Site Earthwork

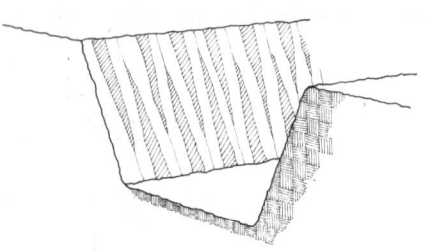

Trenching Systems are shown on a cost per linear foot basis. The systems include: excavation; backfill and removal of spoil; and compaction for various depths and trench bottom widths. The backfill has been reduced to accommodate a pipe of suitable diameter and bedding.

The slope for trench sides varies from none to 1:1.

The Expanded System Listing shows Trenching Systems that range from 2' to 12' in width. Depths range from 2' to 25'.

System Components			COST PER L.F.		
	QUANTITY	UNIT	EQUIP.	LABOR	TOTAL
SYSTEM G1030 806 1310		L.F.			
TRENCHING LOAM & SANDY CLAY, NO SLOPE, 2' WIDE, 2' DP, 3/8 C.Y. BUCKET					
Excavation, trench, hyd. backhoe, track mtd., 3/8 C.Y. bucket	.148	B.C.Y.	.24	.75	.99
Backfill and load spoil, from stockpile	.165	C.Y.	.09	.26	.35
Compaction by vibrating plate 18" wide, 6" lifts, 4 passes	.118	E.C.Y.	.03	.30	.33
Remove excess spoil, 6 C.Y. dump truck, 2 mile roundtrip	.042	C.Y.	.17	.15	.32
TOTAL			.54	1.50	2.04

G1030 806	Trenching Loam & Sandy Clay	COST PER L.F.		
		EQUIP.	LABOR	TOTAL
1310	Trenching, loam & sandy clay, no slope, 2' wide, 2' deep, 3/8 C.Y. bucket	.54	1.47	2
1320	3' deep, 3/8 C.Y. bucket	.80	2.40	3.20
1330	4' deep, 3/8 C.Y. bucket	.93	2.93	3.86
1340	6' deep, 3/8 C.Y. bucket	1.41	3.49	4.90
1350	8' deep, 1/2 C.Y. bucket	1.81	4.59	6.40
1360	10' deep, 1 C.Y. bucket	1.96	4.92	6.90
1400	4' wide, 2' deep, 3/8 C.Y. bucket	1.33	2.95	4.28
1410	3' deep, 3/8 C.Y. bucket	1.73	4.41	6.15
1420	4' deep, 1/2 C.Y. bucket	1.87	4.87	6.75
1430	6' deep, 1/2 C.Y. bucket	3.32	7.15	10.45
1440	8' deep, 1/2 C.Y. bucket	4.91	10.10	15
1450	10' deep, 1 C.Y. bucket	4.66	10.35	15
1460	12' deep, 1 C.Y. bucket	5.85	12.90	18.75
1470	15' deep, 1-1/2 C.Y. bucket	7.15	15	22
1480	18' deep, 2-1/2 C.Y. bucket	8.40	16.20	24.50
1520	6' wide, 6' deep, 5/8 C.Y. bucket w/trench box	6.80	11.80	18.60
1530	8' deep, 3/4 C.Y. bucket	8.50	15.20	23.50
1540	10' deep, 1 C.Y. bucket	7.75	15.50	23.50
1550	12' deep, 1-1/2 C.Y. bucket	8.90	17.35	26.50
1560	16' deep, 2-1/2 C.Y. bucket	12.05	21.50	33.50
1570	20' deep, 3-1/2 C.Y. bucket	16.25	26	42.50
1580	24' deep, 3-1/2 C.Y. bucket	20.50	31.50	52
1640	8' wide, 12' deep, 1-1/4 C.Y. bucket w/trench box	12.50	22	34.50
1650	15' deep, 1-1/2 C.Y. bucket	15.35	28	43.50
1660	18' deep, 2-1/2 C.Y. bucket	18.35	31	49.50
1680	24' deep, 3-1/2 C.Y. bucket	28	41	69
1730	10' wide, 20' deep, 3-1/2 C.Y. bucket w/trench box	28	41	69
1740	24' deep, 3-1/2 C.Y. bucket	35	50.50	85.50
1780	12' wide, 20' deep, 3-1/2 C.Y. bucket w/trench box	33.50	48.50	82
1790	25' deep, bucket	44	63	107
1800	1/2:1 slope, 2' wide, 2' deep, 3/8 C.Y. BK	.73	2.20	2.93
1810	3' deep, 3/8 C.Y. bucket	1.18	3.84	5
1820	4' deep, 3/8 C.Y. bucket	1.73	5.85	7.60
1840	6' deep, 3/8 C.Y. bucket	3.32	8.70	12

G10 Site Preparation

G1030 Site Earthwork

G1030 806	Trenching Loam & Sandy Clay	COST PER L.F.		
		EQUIP.	LABOR	TOTAL
1860	8' deep, 1/2 C.Y. bucket	5.20	13.90	19.10
1880	10' deep, 1 C.Y. bucket	6.70	17.40	24
2300	4' wide, 2' deep, 3/8 C.Y. bucket	1.53	3.68	5.20
2310	3' deep, 3/8 C.Y. bucket	2.17	6.05	8.20
2320	4' deep, 1/2 C.Y. bucket	2.61	7.40	10
2340	6' deep, 1/2 C.Y. bucket	5.55	12.70	18.25
2360	8' deep, 1/2 C.Y. bucket	9.45	20.50	30
2380	10' deep, 1 C.Y. bucket	10.25	24	34.50
2400	12' deep, 1 C.Y. bucket	17.25	39.50	57
2430	15' deep, 1-1/2 C.Y. bucket	20.50	44	64.50
2460	18' deep, 2-1/2 C.Y. bucket	33	66	99
2840	6' wide, 6' deep, 5/8 C.Y. bucket w/trench box	9.75	17.75	27.50
2860	8' deep, 3/4 C.Y. bucket	13.60	26	39.50
2880	10' deep, 1 C.Y. bucket	13.95	29	43
2900	12' deep, 1-1/2 C.Y. bucket	17.45	35.50	53
2940	16' deep, 2-1/2 C.Y. bucket	27.50	51	78.50
2980	20' deep, 3-1/2 C.Y. bucket	43	70.50	114
3020	24' deep, 3-1/2 C.Y. bucket	61	96.50	158
3100	8' wide, 12' deep, 1-1/2 C.Y. bucket w/trench box	21	40	61
3120	15' deep, 1-1/2 C.Y. bucket	28.50	56.50	85
3140	18' deep, 2-1/2 C.Y. bucket	38	68.50	107
3180	24' deep, 3-1/2 C.Y. bucket	68	106	174
3270	10' wide, 20' deep, 3-1/2 C.Y. bucket w/trench box	54.50	85.50	140
3280	24' deep, 3-1/2 C.Y. bucket	75.50	115	191
3320	12' wide, 20' deep, 3-1/2 C.Y. bucket w/trench box	60	93	153
3380	25' deep, 3-1/2 C.Y. bucket w/trench box	82.50	125	208
3500	1:1 slope, 2' wide, 2' deep, 3/8 C.Y. bucket	.93	2.93	3.86
3520	3' deep, 3/8 C.Y. bucket	1.63	5.50	7.15
3540	4' deep, 3/8 C.Y. bucket	2.52	8.75	11.25
3560	6' deep, 3/8 C.Y. bucket	3.32	8.70	12
3580	8' deep, 1/2 C.Y. bucket	8.60	23	31.50
3600	10' deep, 1 C.Y. bucket	11.40	30	41.50
3800	2' deep, 3/8 C.Y. bucket	1.73	4.41	6.15
3820	4' wide, 3' deep, 3/8 C.Y. bucket	2.62	7.70	10.30
3840	4' deep, 1/2 C.Y. bucket	3.34	9.95	13.30
3860	6' deep, 1/2 C.Y. bucket	7.75	18.25	26
3880	8' deep, 1/2 C.Y. bucket	14	31	45
3900	10' deep, 1 C.Y. bucket	15.90	37	53
3920	12' deep, 1 C.Y. bucket	23	52.50	75.50
3940	15' deep, 1-1/2 C.Y. bucket	33.50	72.50	106
3960	18' deep, 2-1/2 C.Y. bucket	45.50	90.50	136
4030	6' wide, 6' deep, 5/8 C.Y. bucket w/trench box	12.60	23.50	36
4040	8' deep, 3/4 C.Y. bucket	18.75	36.50	55.50
4050	10' deep, 1 C.Y. bucket	20	42.50	62.50
4060	12' deep, 1-1/2 C.Y. bucket	26	53.50	79.50
4080	20' deep, 3-1/2 C.Y. bucket	69.50	115	185
4090	24' deep, 3-1/2 C.Y. bucket	101	161	262
4500	8' wide, 12' deep, 1-1/4 C.Y. bucket w/trench box	29.50	58.50	88
4550	15' deep, 1-1/2 C.Y. bucket	42	85	127
4600	18' deep, 2-1/2 C.Y. bucket	57.50	106	164
4650	24' deep, 3-1/2 C.Y. bucket	109	171	280
4800	10' wide, 20' deep, 3-1/2 C.Y. bucket w/trench box	81	130	211
4850	24' deep, 3-1/2 C.Y. bucket	116	180	296
4950	12' wide, 20' deep, 3-1/2 C.Y. bucket w/trench box	87	138	225
4980	25' deep, bucket	131	204	335

G1030 Site Earthwork

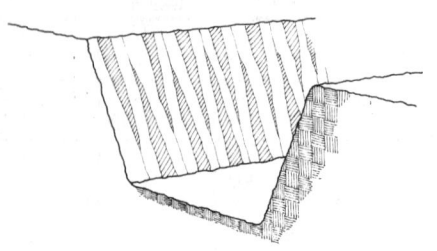

Trenching Systems are shown on a cost per linear foot basis. The systems include: excavation; backfill and removal of spoil; and compaction for various depths and trench bottom widths. The backfill has been reduced to accommodate a pipe of suitable diameter and bedding.

The slope for trench sides varies from none to 1:1.

The Expanded System Listing shows Trenching Systems that range from 2' to 12' in width. Depths range from 2' to 25'.

System Components	QUANTITY	UNIT	COST PER L.F.		
			EQUIP.	LABOR	TOTAL
SYSTEM G1030 807 1310		L.F.			
TRENCHING SAND & GRAVEL, NO SLOPE, 2' WIDE, 2' DEEP, 3/8 C.Y. BUCKET					
Excavation, trench, hyd. backhoe, track mtd., 3/8 C.Y. bucket	.148	B.C.Y.	.24	.74	.98
Backfill and load spoil, from stockpile	.118	C.Y.	.07	.19	.26
Compaction by vibrating plate 18" wide, 6" lifts, 4 passes	.118	E.C.Y.	.03	.30	.33
Remove excess spoil, 6 C.Y. dump truck, 2 mile roundtrip	.035	C.Y.	.14	.13	.27
TOTAL			.48	1.36	1.84

G1030 807	Trenching Sand & Gravel	COST PER L.F.		
		EQUIP.	LABOR	TOTAL
1310	Trenching, sand & gravel, no slope, 2' wide, 2' deep, 3/8 C.Y. bucket	.48	1.36	1.83
1320	3' deep, 3/8 C.Y. bucket	.75	2.32	3.07
1330	4' deep, 3/8 C.Y. bucket	.86	2.78	3.64
1340	6' deep, 3/8 C.Y. bucket	1.34	3.33	4.67
1350	8' deep, 1/2 C.Y. bucket	1.72	4.38	6.10
1360	10' deep, 1 C.Y. bucket	1.83	4.63	6.45
1400	4' wide, 2' deep, 3/8 C.Y. bucket	1.20	2.77	3.97
1410	3' deep, 3/8 C.Y. bucket	1.57	4.17	5.75
1420	4' deep, 1/2 C.Y. bucket	1.72	4.61	6.35
1430	6' deep, 1/2 C.Y. bucket	3.15	6.80	9.95
1440	8' deep, 1/2 C.Y. bucket	4.64	9.60	14.25
1450	10' deep, 1 C.Y. bucket	4.39	9.80	14.20
1460	12' deep, 1 C.Y. bucket	5.55	12.20	17.75
1470	15' deep, 1-1/2 C.Y. bucket	6.80	14.15	21
1480	18' deep, 2-1/2 C.Y. bucket	7.95	15.25	23
1520	6' wide, 6' deep, 5/8 C.Y. bucket w/trench box	6.40	11.15	17.55
1530	8' deep, 3/4 C.Y. bucket	8.05	14.45	22.50
1540	10' deep, 1 C.Y. bucket	7.35	14.65	22
1550	12' deep, 1-1/2 C.Y. bucket	8.45	16.40	25
1560	16' deep, 2 C.Y. bucket	11.55	20	31.50
1570	20' deep, 3-1/2 C.Y. bucket	15.45	24.50	40
1580	24' deep, 3-1/2 C.Y. bucket	19.55	29.50	49
1640	8' wide, 12' deep, 1-1/2 C.Y. bucket w/trench box	11.80	21	33
1650	15' deep, 1-1/2 C.Y. bucket	14.40	26.50	41
1660	18' deep, 2-1/2 C.Y. bucket	17.30	29	46.50
1680	24' deep, 3-1/2 C.Y. bucket	26.50	38	64.50
1730	10' wide, 20' deep, 3-1/2 C.Y. bucket w/trench box	26.50	38.50	65
1740	24' deep, 3-1/2 C.Y. bucket	33	47.50	80.50
1780	12' wide, 20' deep, 3-1/2 C.Y. bucket w/trench box	31.50	45.50	77
1790	25' deep, bucket	41.50	58.50	100
1800	1/2:1 slope, 2' wide, 2' deep, 3/8 CY bk	.68	2.09	2.77
1810	3' deep, 3/8 C.Y. bucket	1.10	3.66	4.76
1820	4' deep, 3/8 C.Y. bucket	1.61	5.60	7.20
1840	6' deep, 3/8 C.Y. bucket	3.17	8.30	11.45

G1030   Site Earthwork

G1030 807	Trenching Sand & Gravel	COST PER L.F.		
		EQUIP.	LABOR	TOTAL
1860	8' deep, 1/2 C.Y. bucket	4.98	13.25	18.25
1880	10' deep, 1 C.Y. bucket	6.30	16.40	22.50
2300	4' wide, 2' deep, 3/8 C.Y. bucket	1.38	3.47	4.85
2310	3' deep, 3/8 C.Y. bucket	1.99	5.75	7.75
2320	4' deep, 1/2 C.Y. bucket	2.41	7.05	9.45
2340	6' deep, 1/2 C.Y. bucket	5.30	12.15	17.45
2360	8' deep, 1/2 C.Y. bucket	8.95	19.45	28.50
2380	10' deep, 1 C.Y. bucket	9.70	22.50	32
2400	12' deep, 1 C.Y. bucket	16.35	37.50	54
2430	15' deep, 1-1/2 C.Y. bucket	19.25	41.50	61
2460	18' deep, 2-1/2 C.Y. bucket	31.50	62	93.50
2840	6' wide, 6' deep, 5/8 C.Y. bucket w/trench box	9.20	16.85	26
2860	8' deep, 3/4 C.Y. bucket	12.95	24.50	37.50
2880	10' deep, 1 C.Y. bucket	13.20	27.50	40.50
2900	12' deep, 1-1/2 C.Y. bucket	16.55	33.50	50
2940	16' deep, 2 C.Y. bucket	26.50	48	74.50
2980	20' deep, 3-1/2 C.Y. bucket	41	66	107
3020	24' deep, 3-1/2 C.Y. bucket	58	90.50	149
3100	8' wide, 12' deep, 1-1/4 C.Y. bucket w/trench box	20	38	58
3120	15' deep, 1-1/2 C.Y. bucket	27	53	80
3140	18' deep, 2-1/2 C.Y. bucket	36	64.50	101
3180	24' deep, 3-1/2 C.Y. bucket	65	99	164
3270	10' wide, 20' deep, 3-1/2 C.Y. bucket w/trench box	51.50	80	132
3280	24' deep, 3-1/2 C.Y. bucket	72	108	180
3370	12' wide, 20' deep, 3-1/2 C.Y. bucket w/trench box	57.50	89	147
3380	25' deep, bucket	83.50	125	209
3500	1:1 slope, 2' wide, 2' deep, 3/8 CY bk	1.83	3.57	5.41
3520	3' deep, 3/8 C.Y. bucket	1.52	5.25	6.75
3540	4' deep, 3/8 C.Y. bucket	2.36	8.35	10.70
3560	6' deep, 3/8 C.Y. bucket	3.17	8.35	11.50
3580	8' deep, 1/2 C.Y. bucket	8.25	22	30.50
3600	10' deep, 1 C.Y. bucket	10.75	28	39
3800	4' wide, 2' deep, 3/8 C.Y. bucket	1.57	4.17	5.75
3820	3' deep, 3/8 C.Y. bucket	2.41	7.30	9.70
3840	4' deep, 1/2 C.Y. bucket	3.10	9.45	12.55
3860	6' deep, 1/2 C.Y. bucket	7.45	17.50	25
3880	8' deep, 1/2 C.Y. bucket	13.30	29.50	43
3900	10' deep, 1 C.Y. bucket	15.05	35	50
3920	12' deep, 1 C.Y. bucket	22	50	72
3940	15' deep, 1-1/2 C.Y. bucket	31.50	68.50	100
3960	18' deep, 2-1/2 C.Y. bucket	43.50	85.50	129
4030	6' wide, 6' deep, 5/8 C.Y. bucket w/trench box	11.95	22.50	34.50
4040	8' deep, 3/4 C.Y. bucket	17.85	34.50	52.50
4050	10' deep, 1 C.Y. bucket	19.05	40.50	59.50
4060	12' deep, 1-1/2 C.Y. bucket	24.50	50.50	75
4070	16' deep, 2 C.Y. bucket	41.50	76	118
4080	20' deep, 3-1/2 C.Y. bucket	66	108	174
4090	24' deep, 3-1/2 C.Y. bucket	97	151	248
4500	8' wide, 12' deep, 1-1/2 C.Y. bucket w/trench box	28	55	83
4550	15' deep, 1-1/2 C.Y. bucket	39.50	80	120
4600	18' deep, 2-1/2 C.Y. bucket	55	99.50	155
4650	24' deep, 3-1/2 C.Y. bucket	104	160	264
4800	10' wide, 20' deep, 3-1/2 C.Y. bucket w/trench box	77	122	199
4850	24' deep, 3-1/2 C.Y. bucket	110	169	279
4950	12' wide, 20' deep, 3-1/2 C.Y. bucket w/trench box	82.50	129	212
4980	25' deep, bucket	125	191	315

G40 Site Electrical Utilities

G4020 Site Lighting

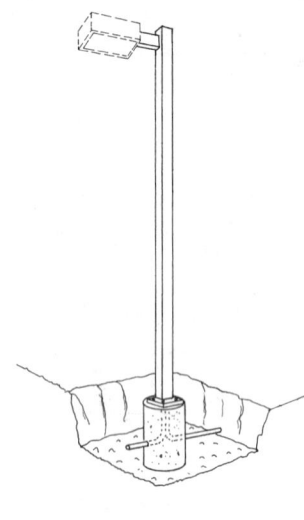

Table G4020 210 Procedure for Calculating Floodlights Required for Various Footcandles
Poles should not be spaced more than 4 times the fixture mounting height for good light distribution.

Estimating Chart
Select Lamp type.

Determine total square feet.

Chart will show quantity of fixtures to provide 1 footcandle initial, at intersection of lines. Multiply fixture quantity by desired footcandle level.

Chart based on use of wide beam luminaires in an area whose dimensions are large compared to mounting height and is approximate only.

To maintain 1 footcandle over a large area use these watts per square foot:

Incandescent	0.15
Metal Halide	0.032
Mercury Vapor	0.05
High Pressure Sodium	0.024

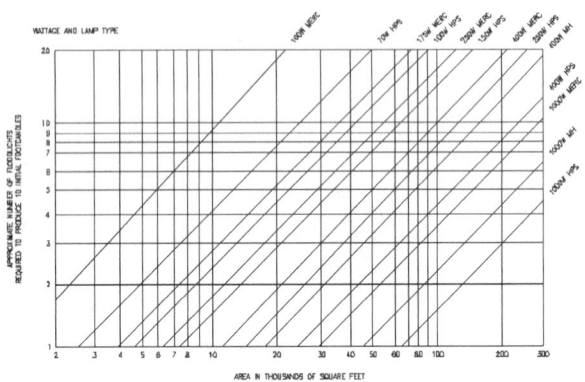

System Components	QUANTITY	UNIT	COST EACH		
			MAT.	INST.	TOTAL
SYSTEM G4020 210 0200					
LIGHT POLES, ALUMINUM, 20' HIGH, 1 ARM BRACKET					
Aluminum light pole, 20', no concrete base	1.000	Ea.	870	507	1,377
Bracket arm for Aluminum light pole	1.000	Ea.	114	65.50	179.50
Excavation by hand, pits to 6' deep, heavy soil or clay	2.368	C.Y.		211.94	211.94
Footing, concrete incl forms, reinforcing, spread, under 1 C.Y.	.465	C.Y.	73.47	74.36	147.83
Backfill by hand	1.903	C.Y.		61.85	61.85
Compaction vibrating plate	1.903	C.Y.		8.41	8.41
TOTAL			1,057.47	929.06	1,986.53

G4020 210	Light Pole (Installed)	COST EACH		
		MAT.	INST.	TOTAL
0200	Light pole, aluminum, 20' high, 1 arm bracket	1,050	930	1,980
0240	2 arm brackets	1,175	930	2,105
0280	3 arm brackets	1,300	965	2,265
0320	4 arm brackets	1,400	965	2,365
0360	30' high, 1 arm bracket	1,900	1,175	3,075
0400	2 arm brackets	2,025	1,175	3,200
0440	3 arm brackets	2,125	1,200	3,325
0480	4 arm brackets	2,250	1,200	3,450
0680	40' high, 1 arm bracket	2,325	1,575	3,900
0720	2 arm brackets	2,450	1,575	4,025
0760	3 arm brackets	2,550	1,600	4,150
0800	4 arm brackets	2,675	1,600	4,275
0840	Steel, 20' high, 1 arm bracket	1,275	985	2,260
0880	2 arm brackets	1,350	985	2,335
0920	3 arm brackets	1,375	1,025	2,400
0960	4 arm brackets	1,500	1,025	2,525
1000	30' high, 1 arm bracket	1,475	1,250	2,725
1040	2 arm brackets	1,575	1,250	2,825
1080	3 arm brackets	1,600	1,275	2,875
1120	4 arm brackets	1,700	1,275	2,975
1320	40' high, 1 arm bracket	1,925	1,675	3,600
1360	2 arm brackets	2,025	1,675	3,700

392

G40 Site Electrical Utilities

G4020 Site Lighting

G4020 210	Light Pole (Installed)	COST EACH		
		MAT.	INST.	TOTAL
1400	3 arm brackets	2,050	1,725	3,775
1440	4 arm brackets	2,150	1,725	3,875

Reference Section

All the reference information is in one section, making it easy to find what you need to know . . . and easy to use the book on a daily basis. This section is visually identified by a vertical gray bar on the page edges.

In this Reference Section, we've included Equipment Rental Costs, a listing of rental and operating costs; Crew Listings, a full listing of all crews, equipment, and their costs; Historical Cost Indexes for cost comparisons over time; City Cost Indexes and Location Factors for adjusting costs to the region you are in; Reference Tables, where you will find explanations, estimating information and procedures, or technical data; Square Foot Costs that allow you to make a rough estimate for the overall cost of a project; and an explanation of all the Abbreviations in the book.

Table of Contents

Estimating Tips

- This section contains the average costs to rent and operate hundreds of pieces of construction equipment. This is useful information when estimating the time and material requirements of any particular operation in order to establish a unit or total cost. Equipment costs include not only rental, but also operating costs for equipment under normal use.

Rental Costs

- Equipment rental rates are obtained from industry sources throughout North America–contractors, suppliers, dealers, manufacturers, and distributors.
- Rental rates vary throughout the country, with larger cities generally having lower rates. Lease plans for new equipment are available for periods in excess of six months, with a percentage of payments applying toward purchase.
- Monthly rental rates vary from 2% to 5% of the purchase price of the equipment depending on the anticipated life of the equipment and its wearing parts.
- Weekly rental rates are about 1/3 the monthly rates, and daily rental rates are about 1/3 the weekly rate.

Operating Costs

- The operating costs include parts and labor for routine servicing, such as repair and replacement of pumps, filters and worn lines. Normal operating expendables, such as fuel, lubricants, tires and electricity (where applicable), are also included.
- Extraordinary operating expendables with highly variable wear patterns, such as diamond bits and blades, are excluded. These costs can be found as material costs in the Unit Price section.
- The hourly operating costs listed do not include the operator's wages.

Crew Equipment Cost/Day

- Any power equipment required by a crew is shown in the Crew Listings with a daily cost.
- The daily cost of crew equipment is based on dividing the weekly rental rate by 5 (number of working days in the week), and then adding the hourly operating cost times 8 (the number of hours in a day). This "Crew Equipment Cost/Day" is shown in the far right column of the Equipment Rental pages.
- If equipment is needed for only one or two days, it is best to develop your own cost by including components for daily rent and hourly operating cost. This is important when the listed Crew for a task does not contain the equipment needed, such as a crane for lifting mechanical heating/cooling equipment up onto a roof.

Mobilization/ Demobilization

- The cost to move construction equipment from an equipment yard or rental company to the jobsite and back again is not included in equipment rental costs listed in the Reference section, nor in the bare equipment cost of any Unit Price line item, nor in any equipment costs shown in the Crew listings.
- Mobilization (to the site) and demobilization (from the site) costs can be found in the Unit Price section.
- If a piece of equipment is already at the jobsite, it is not appropriate to utilize mobil./demob. costs again in an estimate.

01 54 33 | Equipment Rental

			UNIT	HOURLY OPER. COST	RENT PER DAY	RENT PER WEEK	RENT PER MONTH	CREW EQUIPMENT COST/DAY	
10	0010	**CONCRETE EQUIPMENT RENTAL**, without operators	R015433 -10						10
	0150	For batch plant, see div. 01 54 33.50							
	0200	Bucket, concrete lightweight, 1/2 C.Y.	R033105 -70	Ea.	.60	17.35	52	156	15.20
	0300	1 C.Y.			.65	21	63	189	17.80
	0400	1-1/2 C.Y.			.80	28.50	85	255	23.40
	0500	2 C.Y.			.90	33.50	100	300	27.20
	0580	8 C.Y.			4.85	223	670	2,000	172.80
	0600	Cart, concrete, self propelled, operator walking, 10 C.F.			2.30	58.50	175	525	53.40
	0700	Operator riding, 18 C.F.			3.60	83.50	250	750	78.80
	0800	Conveyer for concrete, portable, gas, 16" wide, 26' long			8.05	118	355	1,075	135.40
	0900	46' long			8.40	143	430	1,300	153.20
	1000	56' long			8.55	153	460	1,375	160.40
	1100	Core drill, electric, 2-1/2 H.P., 1" to 8" bit diameter			2.20	89.50	268	805	71.20
	1150	11 HP, 8" to 18" cores			4.75	112	335	1,000	105
	1200	Finisher, concrete floor, gas, riding trowel, 48" diameter			5.55	91.50	275	825	99.40
	1300	Gas, manual, 3 blade, 36" trowel		Ea.	1.15	16.65	50	150	19.20
	1400	4 blade, 48" trowel			1.65	21	63	189	25.80
	1500	Float, hand-operated (Bull float) 48" wide			.08	13	39	117	8.45
	1570	Curb builder, 14 H.P., gas, single screw			10.45	228	685	2,050	220.60
	1590	Double screw			11.10	270	810	2,425	250.80
	1600	Grinder, concrete and terrazzo, electric, floor			2.92	127	380	1,150	99.35
	1700	Wall grinder			1.46	63.50	190	570	49.70
	1800	Mixer, powered, mortar and concrete, gas, 6 C.F., 18 H.P.			5.70	113	340	1,025	113.60
	1900	10 C.F., 25 H.P.			6.95	138	415	1,250	138.60
	2000	16 C.F.			7.30	162	485	1,450	155.40
	2100	Concrete, stationary, tilt drum, 2 C.Y.			5.75	223	670	2,000	180
	2120	Pump, concrete, truck mounted 4" line 80' boom			21.95	890	2,665	8,000	708.60
	2140	5" line, 110' boom			28.70	1,175	3,545	10,600	938.60
	2160	Mud jack, 50 C.F. per hr.			5.75	127	380	1,150	122
	2180	225 C.F. per hr.			7.50	143	430	1,300	146
	2190	Shotcrete pump rig, 12 CY/hr			13.70	235	705	2,125	250.60
	2600	Saw, concrete, manual, gas, 18 H.P.			4.00	36.50	110	330	54
	2650	Self-propelled, gas, 30 H.P.			7.80	93.50	280	840	118.40
	2700	Vibrators, concrete, electric, 60 cycle, 2 H.P.			.42	7.65	23	69	7.95
	2800	3 H.P.			.58	9	27	81	10.05
	2900	Gas engine, 5 H.P.			1.05	13.65	41	123	16.60
	3000	8 H.P.			1.55	15	45	135	21.40
	3050	Vibrating screed, gas engine, 8 H.P.			2.71	73.50	221	665	65.90
	3100	Concrete transit mixer, hydraulic drive							
	3120	6 x 4, 250 H.P., 8 C.Y., rear discharge			39.50	550	1,645	4,925	645
	3200	Front discharge			46.50	680	2,040	6,125	780
	3300	6 x 6, 285 H.P., 12 C.Y., rear discharge			45.60	640	1,920	5,750	748.80
	3400	Front discharge			47.65	690	2,065	6,200	794.20
20	0010	**EARTHWORK EQUIPMENT RENTAL**, without operators	R015433 -10						20
	0040	Aggregate spreader, push type 8' to 12' wide		Ea.	2.00	24.50	73	219	30.60
	0045	Tailgate type, 8' wide	R312323 -30	"	1.90	32.50	97	291	34.60
	0050	Augers for vertical drilling							
	0055	Earth auger, truck-mounted, for fence & sign posts	R312316 -40	Ea.	10.40	485	1,460	4,375	375.20
	0060	For borings and monitoring wells			35.95	645	1,935	5,800	674.60
	0070	Earth auger, portable, trailer mounted	R312316 -45		2.25	23	69	207	31.80
	0075	Earth auger, truck-mounted, for caissons, water wells, utility poles			174.00	3,500	10,535	31,600	3,499
	0080	Auger, horizontal boring machine, 12" to 36" diameter, 45 H.P.			19.00	192	575	1,725	267
	0090	12" to 48" diameter, 65 H.P.			27.05	340	1,025	3,075	421.40
	0095	Auger, for fence posts, gas engine, hand held			.40	4.67	14	42	6
	0100	Excavator, diesel hydraulic, crawler mounted, 1/2 C.Y. cap.			17.75	360	1,080	3,250	358
	0120	5/8 C.Y. capacity			21.40	485	1,455	4,375	462.20
	0140	3/4 C.Y. capacity			24.80	530	1,590	4,775	516.40

01 54 33 | Equipment Rental

			UNIT	HOURLY OPER. COST	RENT PER DAY	RENT PER WEEK	RENT PER MONTH	CREW EQUIPMENT COST/DAY	
20	0150	1 C.Y. capacity	Ea.	30.85	590	1,775	5,325	601.80	20
	0200	1-1/2 C.Y. capacity		37.95	785	2,360	7,075	775.60	
	0300	2 C.Y. capacity		49.45	1,000	2,995	8,975	994.60	
	0320	2-1/2 C.Y. capacity		65.60	1,350	4,040	12,100	1,333	
	0340	3-1/2 C.Y. capacity		106.15	2,200	6,585	19,800	2,166	
	0341	Attachments							
	0342	Bucket thumbs		2.55	212	635	1,900	147.40	
	0345	Grapples		2.35	182	545	1,625	127.80	
	0350	Gradall type, truck mounted, 3 ton @ 15' radius, 5/8 C.Y.		42.35	945	2,835	8,500	905.80	
	0370	1 C.Y. capacity		47.80	1,075	3,220	9,650	1,026	
	0400	Backhoe-loader, 40 to 45 H.P., 5/8 C.Y. capacity		9.85	197	590	1,775	196.80	
	0450	45 H.P. to 60 H.P., 3/4 C.Y. capacity		12.30	242	725	2,175	243.40	
	0460	80 H.P., 1-1/4 C.Y. capacity		15.05	275	825	2,475	285.40	
	0470	112 H.P., 1-1/2 C.Y. capacity		20.75	405	1,220	3,650	410	
	0480	Attachments							
	0482	Compactor, 20,000 lb		4.45	120	360	1,075	107.60	
	0485	Hydraulic hammer, 750 ft-lbs		2.05	70	210	630	58.40	
	0486	Hydraulic hammer, 1200 ft-lbs		4.30	135	405	1,225	115.40	
	0500	Brush chipper, gas engine, 6" cutter head, 35 H.P.		7.25	102	305	915	119	
	0550	12" cutter head, 130 H.P.		11.45	153	460	1,375	183.60	
	0600	15" cutter head, 165 H.P.		16.30	165	495	1,475	229.40	
	0750	Bucket, clamshell, general purpose, 3/8 C.Y.		1.05	35	105	315	29.40	
	0800	1/2 C.Y.		1.15	41.50	125	375	34.20	
	0850	3/4 C.Y.		1.30	51.50	155	465	41.40	
	0900	1 C.Y.		1.35	56.50	170	510	44.80	
	0950	1-1/2 C.Y.		2.10	76.50	230	690	62.80	
	1000	2 C.Y.		2.25	86.50	260	780	70	
	1010	Bucket, dragline, medium duty, 1/2 C.Y.		.60	22.50	67	201	18.20	
	1020	3/4 C.Y.		.65	23.50	71	213	19.40	
	1030	1 C.Y.		.65	25.50	77	231	20.60	
	1040	1-1/2 C.Y.		1.00	38.50	115	345	31	
	1050	2 C.Y.		1.05	43.50	130	390	34.40	
	1070	3 C.Y.		1.60	60	180	540	48.80	
	1200	Compactor, manually guided 2-drum vibratory smooth roller, 7.5 H.P.		5.35	152	455	1,375	133.80	
	1250	Rammer compactor, gas, 1000 lb. blow		2.05	36.50	110	330	38.40	
	1300	Vibratory plate, gas, 18" plate, 3000 lb. blow		2.00	22	66	198	29.20	
	1350	21" plate, 5000 lb. blow		2.45	27.50	83	249	36.20	
	1370	Curb builder/extruder, 14 H.P., gas, single screw		10.45	228	685	2,050	220.60	
	1390	Double screw		11.10	270	810	2,425	250.80	
	1500	Disc harrow attachment, for tractor		.38	63	189	565	40.85	
	1750	Extractor, piling, see lines 2500 to 2750							
	1810	Feller buncher, shearing & accumulating trees, 100 H.P.	Ea.	24.80	525	1,575	4,725	513.40	
	1860	Grader, self-propelled, 25,000 lb.		20.85	425	1,270	3,800	420.80	
	1910	30,000 lb.		25.05	510	1,525	4,575	505.40	
	1920	40,000 lb.		34.40	765	2,300	6,900	735.20	
	1930	55,000 lb.		45.75	1,100	3,275	9,825	1,021	
	1950	Hammer, pavement demo., hyd., gas, self-prop., 1000 to 1250 lb.		20.15	310	935	2,800	348.20	
	2000	Diesel 1300 to 1500 lb.		31.20	615	1,850	5,550	619.60	
	2050	Pile driving hammer, steam or air, 4150 ft.-lb. @ 225 BPM		6.75	275	825	2,475	219	
	2100	8750 ft.-lb. @ 145 BPM		8.65	445	1,340	4,025	337.20	
	2150	15,000 ft.-lb. @ 60 BPM		9.00	480	1,440	4,325	360	
	2200	24,450 ft.-lb. @ 111 BPM		11.95	530	1,590	4,775	413.60	
	2250	Leads, 15,000 ft.-lb. hammers	L.F.	.03	1.28	3.83	11.50	1	
	2300	24,450 ft.-lb. hammers and heavier	"	.05	2.33	7	21	1.80	
	2350	Diesel type hammer, 22,400 ft.-lb.	Ea.	28.40	615	1,840	5,525	595.20	
	2400	41,300 ft.-lb.		39.70	665	1,990	5,975	715.60	
	2450	141,000 ft.-lb.		64.00	1,150	3,415	10,200	1,195	
	2500	Vib. elec. hammer/extractor, 200 KW diesel generator, 34 H.P.		33.55	640	1,925	5,775	653.40	

R015433 -10

R312316 -30

R312316 -40

R312316 -45

01 54 33 | Equipment Rental

		UNIT	HOURLY OPER. COST	RENT PER DAY	RENT PER WEEK	RENT PER MONTH	CREW EQUIPMENT COST/DAY
2550	80 H.P.	Ea.	59.10	935	2,810	8,425	1,035
2600	150 H.P.		110.75	1,825	5,485	16,500	1,983
2700	Extractor, steam or air, 700 ft.-lb.		15.05	470	1,405	4,225	401.40
2750	1000 ft.-lb.		17.15	570	1,715	5,150	480.20
2800	Log chipper, up to 22" diam, 600 H.P.		32.75	430	1,290	3,875	520
2850	Logger, for skidding & stacking logs, 150 H.P.		39.55	835	2,500	7,500	816.40
2900	Rake, spring tooth, with tractor		8.04	219	658	1,975	195.90
3000	Roller, vibratory, tandem, smooth drum, 20 H.P.		6.15	122	365	1,100	122.20
3050	35 H.P.		8.75	235	705	2,125	211
3100	Towed type vibratory compactor, smooth drum, 50 H.P.		20.25	315	940	2,825	350
3150	Sheepsfoot, 50 H.P.		21.25	345	1,030	3,100	376
3170	Landfill compactor, 220 HP		62.00	1,225	3,650	11,000	1,226
3200	Pneumatic tire roller, 80 H.P.		12.20	330	990	2,975	295.60
3250	120 H.P.		18.75	565	1,690	5,075	488
3300	Sheepsfoot vibratory roller, 200 H.P.		50.15	925	2,775	8,325	956.20
3320	340 H.P.		69.85	1,375	4,090	12,300	1,377
3350	Smooth drum vibratory roller, 75 H.P.		18.30	505	1,510	4,525	448.40
3400	125 H.P.		23.95	625	1,870	5,600	565.60
3410	Rotary mower, brush, 60", with tractor		11.50	238	715	2,150	235
3420	Rototiller, 5 HP, walk-behind		1.89	55.50	166	500	48.30
3450	Scrapers, towed type, 9 to 12 C.Y. capacity		5.65	163	490	1,475	143.20
3500	12 to 17 C.Y. capacity		6.25	178	535	1,600	157
3550	Scrapers, self-propelled, 4 x 4 drive, 2 engine, 14 C.Y. capacity		98.80	1,550	4,635	13,900	1,717
3600	2 engine, 24 C.Y. capacity		137.50	2,425	7,245	21,700	2,549
3640	32 - 44 C.Y. capacity		163.05	2,825	8,465	25,400	2,997
3650	Self-loading, 11 C.Y. capacity		44.85	835	2,505	7,525	859.80
3700	22 C.Y. capacity		85.05	1,775	5,350	16,100	1,750
3710	Screening plant 110 H.P. w/ 5' x 10' screen		28.65	390	1,165	3,500	462.20
3720	5' x 16' screen		30.75	490	1,465	4,400	539
3850	Shovels, see Cranes division 01590-600						
3860	Shovel/backhoe bucket, 1/2 C.Y.	Ea.	1.90	56.50	170	510	49.20
3870	3/4 C.Y.		1.95	63.50	190	570	53.60
3880	1 C.Y.		2.05	73.50	220	660	60.40
3890	1-1/2 C.Y.		2.20	86.50	260	780	69.60
3910	3 C.Y.		2.50	122	365	1,100	93
3950	Stump chipper, 18" deep, 30 H.P.		5.54	103	310	930	106.30
4110	Tractor, crawler, with bulldozer, torque converter, diesel 80 H.P.		18.60	345	1,030	3,100	354.80
4150	105 H.P.		25.60	485	1,450	4,350	494.80
4200	140 H.P.		30.55	610	1,835	5,500	611.40
4260	200 H.P.		46.30	1,025	3,090	9,275	988.40
4310	300 H.P.		59.75	1,375	4,115	12,300	1,301
4360	410 H.P.		80.30	1,725	5,175	15,500	1,677
4370	500 H.P.		106.80	2,350	7,030	21,100	2,260
4380	700 H.P.		158.50	3,575	10,690	32,100	3,406
4400	Loader, crawler, torque conv., diesel, 1-1/2 C.Y., 80 H.P.		16.75	335	1,000	3,000	334
4450	1-1/2 to 1-3/4 C.Y., 95 H.P.		19.65	405	1,210	3,625	399.20
4510	1-3/4 to 2-1/4 C.Y., 130 H.P.		26.90	625	1,870	5,600	589.20
4530	2-1/2 to 3-1/4 C.Y., 190 H.P.		39.05	870	2,610	7,825	834.40
4560	3-1/2 to 5 C.Y., 275 H.P.		53.35	1,200	3,590	10,800	1,145
4610	Tractor loader, wheel, torque conv., 4 x 4, 1 to 1-1/4 C.Y., 65 H.P.		11.70	190	570	1,700	207.60
4620	1-1/2 to 1-3/4 C.Y., 80 H.P.		15.20	260	780	2,350	277.60
4650	1-3/4 to 2 C.Y., 100 H.P.		17.20	300	900	2,700	317.60
4710	2-1/2 to 3-1/2 C.Y., 130 H.P.		18.75	330	995	2,975	349
4730	3 to 4-1/2 C.Y., 170 H.P.		24.25	490	1,470	4,400	488
4760	5-1/4 to 5-3/4 C.Y., 270 H.P.		39.60	730	2,195	6,575	755.80
4810	7 to 8 C.Y., 375 H.P.		67.35	1,325	3,990	12,000	1,337
4870	12-1/2 C.Y., 690 H.P.		93.10	2,050	6,175	18,500	1,980
4880	Wheeled, skid steer, 10 C.F., 30 H.P. gas		6.50	117	350	1,050	122

Reference numbers: R015433-10, R312323-30, R312316-40, R312316-45

01 54 33 | Equipment Rental

			UNIT	HOURLY OPER. COST	RENT PER DAY	RENT PER WEEK	RENT PER MONTH	CREW EQUIPMENT COST/DAY	
20	4890	1 C.Y., 78 H.P., diesel	▼	12.15	208	625	1,875	222.20	**20**
	4891	Attachments for all skid steer loaders							
	4892	Auger	Ea.	.48	80.50	241	725	52.05	
	4893	Backhoe		.69	114	343	1,025	74.10	
	4894	Broom		.63	105	314	940	67.85	
	4895	Forks		.23	37.50	113	340	24.45	
	4896	Grapple		.56	93.50	281	845	60.70	
	4897	Concrete hammer		1.01	168	504	1,500	108.90	
	4898	Tree spade		1.04	173	519	1,550	112.10	
	4899	Trencher		.70	117	352	1,050	76	
	4900	Trencher, chain, boom type, gas, operator walking, 12 H.P.		3.00	45	135	405	51	
	4910	Operator riding, 40 H.P.		9.70	248	745	2,225	226.60	
	5000	Wheel type, diesel, 4' deep, 12" wide		54.15	765	2,290	6,875	891.20	
	5100	Diesel, 6' deep, 20" wide	▼	68.95	1,800	5,370	16,100	1,626	
	5150	Ladder type, diesel, 5' deep, 8" wide		37.95	860	2,585	7,750	820.60	
	5200	Diesel, 8' deep, 16" wide		63.00	1,825	5,510	16,500	1,606	
	5210	Tree spade, self-propelled	Ea.	12.85	267	800	2,400	262.80	
	5250	Truck, dump, tandem, 12 ton payload		24.80	283	850	2,550	368.40	
	5300	Three axle dump, 16 ton payload		33.50	435	1,310	3,925	530	
	5350	Dump trailer only, rear dump, 16-1/2 C.Y.	Ea.	4.35	122	365	1,100	107.80	
	5400	20 C.Y.		4.75	137	410	1,225	120	
	5450	Flatbed, single axle, 1-1/2 ton rating		13.30	58.50	175	525	141.40	
	5500	3 ton rating		16.90	83.50	250	750	185.20	
	5550	Off highway rear dump, 25 ton capacity		46.35	1,025	3,110	9,325	992.80	
	5600	35 ton capacity		47.25	1,050	3,180	9,550	1,014	
	5610	50 ton capacity		61.20	1,375	4,090	12,300	1,308	
	5620	65 ton capacity		65.65	1,450	4,340	13,000	1,393	
	5630	100 ton capacity		84.30	1,875	5,590	16,800	1,792	
	6000	Vibratory plow, 25 H.P., walking	▼	5.20	58.50	175	525	76.60	
40	0010	**GENERAL EQUIPMENT RENTAL**, without operators							**40**
	0150	Aerial lift, scissor type, to 15' high, 1000 lb. cap., electric	Ea.	2.40	45	135	405	46.20	
	0160	To 25' high, 2000 lb. capacity		2.80	63.50	190	570	60.40	
	0170	Telescoping boom to 40' high, 500 lb. capacity, gas		14.85	285	855	2,575	289.80	
	0180	To 45' high, 500 lb. capacity		15.80	325	980	2,950	322.40	
	0190	To 60' high, 600 lb. capacity		17.80	435	1,305	3,925	403.40	
	0195	Air compressor, portable, 6.5 CFM, electric		.42	10.35	31	93	9.55	
	0196	Gasoline		.67	15.65	47	141	14.75	
	0200	Air compressor, portable, gas engine, 60 C.F.M.		10.10	46.50	140	420	108.80	
	0300	160 C.F.M.		11.70	48.50	145	435	122.60	
	0400	Diesel engine, rotary screw, 250 C.F.M.		11.80	95	285	855	151.40	
	0500	365 C.F.M.		15.85	115	345	1,025	195.80	
	0550	450 C.F.M.		20.15	147	440	1,325	249.20	
	0600	600 C.F.M.		34.90	195	585	1,750	396.20	
	0700	750 C.F.M.		35.35	212	635	1,900	409.80	
	0800	For silenced models, small sizes, add							
	0900	Large sizes, add	▼						
	0920	Air tools and accessories							
	0930	Breaker, pavement, 60 lb.	Ea.	.40	8.65	26	78	8.40	
	0940	80 lb.		.40	9	27	81	8.60	
	0950	Drills, hand (jackhammer) 65 lb.		.50	15.35	46	138	13.20	
	0960	Track or wagon, swing boom, 4" drifter		46.05	700	2,095	6,275	787.40	
	0970	5" drifter		56.05	760	2,285	6,850	905.40	
	0975	Track mounted quarry drill, 6" diameter drill		77.10	1,150	3,445	10,300	1,306	
	0980	Dust control per drill		.91	17.35	52	156	17.70	
	0990	Hammer, chipping, 12 lb.		.40	20.50	62	186	15.60	
	1000	Hose, air with couplings, 50' long, 3/4" diameter		.04	6.65	20	60	4.30	
	1100	1" diameter	▼	.04	6.35	19	57	4.10	

Reference boxes:
R015433 -10
R312316 -40
R312316 -45
R312323 -30
R312323 -30
R314116 -40
R314116 -45

01 54 33 | Equipment Rental

		UNIT	HOURLY OPER. COST	RENT PER DAY	RENT PER WEEK	RENT PER MONTH	CREW EQUIPMENT COST/DAY
40	1200	1-1/2" diameter	Ea.	.05	9	27	81	5.80
	1300	2" diameter		.10	17	51	153	11
	1400	2-1/2" diameter		.12	19.35	58	174	12.55
	1410	3" diameter		.18	29.50	88	264	19.05
	1450	Drill, steel, 7/8" x 2'		.07	12.35	37	111	7.95
	1460	7/8" x 6'		.08	12.65	38	114	8.25
	1520	Moil points		.03	4.33	13	39	2.85
	1525	Pneumatic nailer w/accessories		.43	28.50	86	258	20.65
	1530	Sheeting driver for 60 lb. breaker		.04	6	18	54	3.90
	1540	For 90 lb. breaker		.12	8	24	72	5.75
	1550	Spade, 25 lb.		.35	6.65	20	60	6.80
	1560	Tamper, single, 35 lb.		.54	36	108	325	25.90
	1570	Triple, 140 lb.		.81	54	162	485	38.90
	1580	Wrenches, impact, air powered, up to 3/4" bolt		.30	8	24	72	7.20
	1590	Up to 1-1/4" bolt		.35	16.65	50	150	12.80
	1600	Barricades, barrels, reflectorized, 1 to 50 barrels		.02	3.33	10	30	2.15
	1610	100 to 200 barrels		.02	2.53	7.60	23	1.70
	1620	Barrels with flashers, 1 to 50 barrels		.02	4	12	36	2.55
	1630	100 to 200 barrels		.02	3.20	9.60	29	2.10
	1640	Barrels with steady burn type C lights		.03	5.35	16	48	3.45
	1650	Illuminated board, trailer mounted, with generator		.65	117	350	1,050	75.20
	1670	Portable barricade, stock, with flashers, 1 to 6 units		.02	4	12	36	2.55
	1680	25 to 50 units		.02	3.73	11.20	33.50	2.40
	1690	Butt fusion machine, electric		19.15	238	715	2,150	296.20
	1695	Electro fusion machine		14.15	102	305	915	174.20
	1700	Carts, brick, hand powered, 1000 lb. capacity		.25	41.50	124	370	26.80
	1800	Gas engine, 1500 lb., 7-1/2' lift		3.59	102	306	920	89.90
	1822	Dehumidifier, medium, 6 lb/hr, 150 CFM		.79	48	144	430	35.10
	1824	Large, 18 lb/hr, 600 CFM		1.56	95.50	286	860	69.70
	1830	Distributor, asphalt, trailer mtd, 2000 gal., 38 H.P. diesel		8.30	300	905	2,725	247.40
	1840	3000 gal., 38 H.P. diesel		9.50	330	985	2,950	273
	1850	Drill, rotary hammer, electric, 1-1/2" diameter		.79	25.50	76	228	21.50
	1860	Carbide bit for above		.04	6.65	20	60	4.30
	1865	Rotary, crawler, 250 H.P.		115.55	1,950	5,830	17,500	2,090
	1870	Emulsion sprayer, 65 gal., 5 H.P. gas engine		2.44	88	264	790	72.30
	1880	200 gal., 5 H.P. engine		6.10	147	440	1,325	136.80
	1920	Floodlight, mercury vapor, or quartz, on tripod						
	1930	1000 watt	Ea.	.31	12	36	108	9.70
	1940	2000 watt		.54	22	66	198	17.50
	1950	Floodlights, trailer mounted with generator, 1 - 300 watt light		2.70	66.50	200	600	61.60
	1960	2 - 1000 watt lights		3.80	112	335	1,000	97.40
	2000	4 - 300 watt lights		3.15	78.50	235	705	72.20
	2020	Forklift, wheeled, for brick, 18', 3000 lb., 2 wheel drive, gas		19.30	190	570	1,700	268.40
	2040	28', 4000 lb., 4 wheel drive, diesel		15.80	243	730	2,200	272.40
	2050	For rough terrain, 8000 lb., 16' lift, 68 H.P.		20.35	385	1,150	3,450	392.80
	2060	For plant, 4 T. capacity, 80 H.P., 2 wheel drive, gas		11.85	88.50	265	795	147.80
	2080	10 T. capacity, 120 H.P., 2 wheel drive, diesel		17.35	153	460	1,375	230.80
	2100	Generator, electric, gas engine, 1.5 KW to 3 KW		2.65	9.35	28	84	26.80
	2200	5 KW		3.40	13.65	41	123	35.40
	2300	10 KW		6.35	24.50	73	219	65.40
	2400	25 KW		8.25	55	165	495	99
	2500	Diesel engine, 20 KW		8.40	63.50	190	570	105.20
	2600	50 KW		15.20	86.50	260	780	173.60
	2700	100 KW		29.45	117	350	1,050	305.60
	2800	250 KW		58.15	220	660	1,975	597.20
	2850	Hammer, hydraulic, for mounting on boom, to 500 ft.-lb.		1.90	65	195	585	54.20
	2860	1000 ft.-lb.		3.35	103	310	930	88.80
	2900	Heaters, space, oil or electric, 50 MBH		1.64	7.65	23	69	17.70

Reference codes: R312323-30, R314116-40, R314116-45

01 54 33 | Equipment Rental

		UNIT	HOURLY OPER. COST	RENT PER DAY	RENT PER WEEK	RENT PER MONTH	CREW EQUIPMENT COST/DAY	
40	3000	100 MBH	Ea.	2.96	10.65	32	96	30.10	40
	3100	300 MBH		9.48	33.50	100	300	95.85	
	3150	500 MBH		19.15	50	150	450	183.20	
	3200	Hose, water, suction with coupling, 20' long, 2" diameter		.02	3	9	27	1.95	
	3210	3" diameter		.03	4.67	14	42	3.05	
	3220	4" diameter		.03	5	15	45	3.25	
	3230	6" diameter		.11	17.65	53	159	11.50	
	3240	8" diameter		.26	43.50	131	395	28.30	
	3250	Discharge hose with coupling, 50' long, 2" diameter		.01	1.33	4	12	.90	
	3260	3" diameter		.02	2.67	8	24	1.75	
	3270	4" diameter		.02	3.67	11	33	2.35	
	3280	6" diameter		.06	9.65	29	87	6.30	
	3290	8" diameter		.31	51.50	154	460	33.30	
	3295	Insulation blower		.25	7.35	22	66	6.40	
	3300	Ladders, extension type, 16' to 36' long		.19	31.50	95	285	20.50	
	3400	40' to 60' long		.28	47.50	142	425	30.65	
	3405	Lance for cutting concrete		2.50	79.50	239	715	67.80	
	3407	Lawn mower, rotary, 22", 5HP		1.63	43	129	385	38.85	
	3408	48" self propelled		3.92	131	394	1,175	110.15	
	3410	Level, laser type, for pipe and sewer leveling		1.21	80.50	242	725	58.10	
	3430	Electronic		.76	50.50	152	455	36.50	
	3440	Laser type, rotating beam for grade control		1.26	83.50	251	755	60.30	
	3460	Builders level with tripod and rod		.08	12.65	38	114	8.25	
	3500	Light towers, towable, with diesel generator, 2000 watt		3.15	78.50	235	705	72.20	
	3600	4000 watt		3.80	112	335	1,000	97.40	
	3700	Mixer, powered, plaster and mortar, 6 C.F., 7 H.P.		1.45	17.65	53	159	22.20	
	3800	10 C.F., 9 H.P.		1.75	29	87	261	31.40	
	3850	Nailer, pneumatic		.43	28.50	86	258	20.65	
	3900	Paint sprayers complete, 8 CFM		.72	48	144	430	34.55	
	4000	17 CFM		1.13	75.50	226	680	54.25	
	4020	Pavers, bituminous, rubber tires, 8' wide, 50 H.P., diesel		33.35	965	2,900	8,700	846.80	
	4030	10' wide, 150 H.P.		70.60	1,550	4,685	14,100	1,502	
	4050	Crawler, 8' wide, 100 H.P., diesel		70.50	1,625	4,905	14,700	1,545	
	4060	10' wide, 150 H.P.		79.35	1,975	5,900	17,700	1,815	
	4070	Concrete paver, 12' to 24' wide, 250 H.P.		70.40	1,525	4,555	13,700	1,474	
	4080	Placer-spreader-trimmer, 24' wide, 300 H.P.		101.85	2,525	7,545	22,600	2,324	
	4100	Pump, centrifugal gas pump, 1-1/2", 4 MGPH		3.05	40	120	360	48.40	
	4200	2", 8 MGPH		4.05	45	135	405	59.40	
	4300	3", 15 MGPH		4.30	48.50	145	435	63.40	
	4400	6", 90 MGPH		22.40	165	495	1,475	278.20	
	4500	Submersible electric pump, 1-1/4", 55 GPM		.37	15.35	46	138	12.15	
	4600	1-1/2", 83 GPM		.41	17.65	53	159	13.90	
	4700	2", 120 GPM		1.10	22	66	198	22	
	4800	3", 300 GPM		1.80	36.50	110	330	36.40	
	4900	4", 560 GPM		8.00	158	475	1,425	159	
	5000	6", 1590 GPM		11.75	215	645	1,925	223	
	5100	Diaphragm pump, gas, single, 1-1/2" diameter	Ea.	.98	41.50	124	370	32.65	
	5200	2" diameter		3.30	51.50	155	465	57.40	
	5300	3" diameter		3.30	51.50	155	465	57.40	
	5400	Double, 4" diameter		4.40	71.50	215	645	78.20	
	5500	Trash pump, self-priming, gas, 2" diameter		3.40	20.50	62	186	39.60	
	5600	Diesel, 4" diameter		8.85	56.50	170	510	104.80	
	5650	Diesel, 6" diameter		29.10	123	370	1,100	306.80	
	5655	Grout Pump		10.15	83.50	250	750	131.20	
	5660	Rollers, see division 01590-200							
	5700	Salamanders, L.P. gas fired, 100,000 BTU	Ea.	2.97	11.35	34	102	30.55	
	5705	50,000 BTU		2.22	8	24	72	22.55	
	5720	Sandblaster, portable, open top, 3 C.F. capacity		.40	20.50	62	186	15.60	

R312323 -30

R314116 -40

R314116 -45

01 54 33 | Equipment Rental

			UNIT	HOURLY OPER. COST	RENT PER DAY	RENT PER WEEK	RENT PER MONTH	CREW EQUIPMENT COST/DAY	
40	5730	6 C.F. capacity	R312323 -30	Ea.	.70	30	90	270	23.60	40
	5740	Accessories for above			.11	19	57	171	12.30	
	5750	Sander, floor	R314116 -40		.79	19	57	171	17.70	
	5760	Edger			.56	17.35	52	156	14.90	
	5800	Saw, chain, gas engine, 18" long	R314116 -45		1.65	16.35	49	147	23	
	5900	36" long			.55	50	150	450	34.40	
	5950	60" long			.55	50	150	450	34.40	
	6000	Masonry, table mounted, 14" diameter, 5 H.P.			1.28	54.50	164	490	43.05	
	6050	Portable cut-off, 8 H.P.			1.80	29.50	88	264	32	
	6100	Circular, hand held, electric, 7-1/4" diameter			.20	5	15	45	4.60	
	6200	12" diameter			.28	8.65	26	78	7.45	
	6250	Wall saw, w/hydraulic power, 10 H.P			5.65	55	165	495	78.20	
	6275	Shot blaster, walk behind, 20" wide			6.65	415	1,245	3,725	302.20	
	6280	Sidewalk broom, walk-behind			2.03	62	186	560	53.45	
	6300	Steam cleaner, 100 gallons per hour			2.80	63.50	190	570	60.40	
	6310	200 gallons per hour			3.90	78.50	235	705	78.20	
	6340	Tar Kettle/Pot, 400 gallon			3.99	51.50	155	465	62.90	
	6350	Torch, cutting, acetylene-oxygen, 150' hose			.50	21.50	65	195	17	
	6360	Hourly operating cost includes tips and gas			8.10				64.80	
	6410	Toilet, portable chemical			.11	18.35	55	165	11.90	
	6420	Recycle flush type			.13	22.50	67	201	14.45	
	6430	Toilet, fresh water flush, garden hose,			.15	25	75	225	16.20	
	6440	Hoisted, non-flush, for high rise			.13	22	66	198	14.25	
	6450	Toilet, trailers, minimum			.23	38	114	340	24.65	
	6460	Maximum			.69	115	344	1,025	74.30	
	6465	Tractor, farm with attachment			10.30	223	670	2,000	216.40	
	6470	Trailer, office, see division 01520-500								
	6500	Trailers, platform, flush deck, 2 axle, 25 ton capacity		Ea.	4.35	96.50	290	870	92.80	
	6600	40 ton capacity			5.60	135	405	1,225	125.80	
	6700	3 axle, 50 ton capacity			6.05	148	445	1,325	137.40	
	6800	75 ton capacity			7.55	195	585	1,750	177.40	
	6810	Trailer mounted cable reel for H.V. line work			4.66	222	665	2,000	170.30	
	6820	Trailer mounted cable tensioning rig			9.17	435	1,310	3,925	335.35	
	6830	Cable pulling rig			62.02	2,475	7,410	22,200	1,978	
	6850	Trailer, storage, see division 01520-500								
	6900	Water tank, engine driven discharge, 5000 gallons		Ea.	5.65	127	380	1,150	121.20	
	6925	10,000 gallons			7.80	177	530	1,600	168.40	
	6950	Water truck, off highway, 6000 gallons			56.50	735	2,210	6,625	894	
	7010	Tram car for H.V. line work, powered, 2 conductor			6.10	120	361	1,075	121	
	7020	Transit (builder's level) with tripod			.08	12.65	38	114	8.25	
	7030	Trench box, 3000 lbs. 6'x8'			.60	100	300	900	64.80	
	7040	7200 lbs. 6'x20'			.73	122	367	1,100	79.25	
	7050	8000 lbs., 8' x 16'			.93	154	463	1,400	100.05	
	7060	9500 lbs., 8'x20'			1.26	210	630	1,900	136.10	
	7065	11,000 lbs., 8'x24'			1.35	225	676	2,025	146	
	7070	12,000 lbs., 10' x 20'			1.50	250	750	2,250	162	
	7100	Truck, pickup, 3/4 ton, 2 wheel drive			6.35	56.50	170	510	84.80	
	7200	4 wheel drive			6.50	65	195	585	91	
	7250	Crew carrier, 9 passenger			8.65	78.50	235	705	116.20	
	7290	Tool van, 24,000 G.V.W.			10.85	108	325	975	151.80	
	7300	Tractor, 4 x 2, 30 ton capacity, 195 H.P.		Ea.	16.70	.170	510	1,525	235.60	
	7410	250 H.P.			21.65	232	695	2,075	312.20	
	7500	6 x 2, 40 ton capacity, 240 H.P.			20.90	282	845	2,525	336.20	
	7600	6 x 4, 45 ton capacity, 240 H.P.			26.80	300	905	2,725	395.40	
	7620	Vacuum truck, hazardous material, 2500 gallon			10.15	295	885	2,650	258.20	
	7625	5,000 gallon			13.03	415	1,240	3,725	352.25	
	7640	Tractor, with A frame, boom and winch, 225 H.P.			19.10	242	725	2,175	297.80	
	7650	Vacuum, H.E.P.A., 16 gal., wet/dry		Ea.	.92	22	66	198	20.55	

01 54 33 | Equipment Rental

		UNIT	HOURLY OPER. COST	RENT PER DAY	RENT PER WEEK	RENT PER MONTH	CREW EQUIPMENT COST/DAY	
40	7655	55 gal, wet/dry	Ea.	.89	33	99	297	26.90	**40**
	7660	Water tank, portable		.15	25.50	75.90	228	16.40	
	7690	Large production vacuum loader, 3150 CFM		17.37	620	1,860	5,575	510.95	
	7700	Welder, electric, 200 amp	Ea.	3.37	31.50	95	285	45.95	
	7800	300 amp		4.98	36	108	325	61.45	
	7900	Gas engine, 200 amp		10.90	23.50	71	213	101.40	
	8000	300 amp		12.50	25.50	76	228	115.20	
	8100	Wheelbarrow, any size		.07	11.65	35	105	7.55	
	8200	Wrecking ball, 4000 lb.		2.00	70	210	630	58	
50	0010	**HIGHWAY EQUIPMENT RENTAL**, without operators							**50**
	0050	Asphalt batch plant, portable drum mixer, 100 ton/hr.	Ea.	57.45	1,350	4,025	12,100	1,265	
	0060	200 ton/hr.		63.80	1,400	4,230	12,700	1,356	
	0070	300 ton/hr.		74.25	1,675	5,005	15,000	1,595	
	0100	Backhoe attachment, long stick, up to 185 HP, 10.5' long		.31	20.50	62	186	14.90	
	0140	Up to 250 HP, 12' long		.34	22.50	67	201	16.10	
	0180	Over 250 HP, 15' long		.45	29.50	89	267	21.40	
	0200	Special dipper arm, up to 100 HP, 32' long		.92	61	183	550	43.95	
	0240	Over 100 HP, 33' long		1.14	76	228	685	54.70	
	0300	Concrete batch plant, portable, electric, 200 CY/Hr		10.79	505	1,515	4,550	389.30	
	0500	Grader attachment, ripper/scarifier, rear mounted							
	0520	Up to 135 HP	Ea.	2.95	61.50	185	555	60.60	
	0540	Up to 180 HP		3.55	80	240	720	76.40	
	0580	Up to 250 HP		3.90	90	270	810	85.20	
	0700	Pvmt. removal bucket, for hyd. excavator, up to 90 HP		1.50	45	135	405	39	
	0740	Up to 200 HP		1.70	66.50	200	600	53.60	
	0780	Over 200 HP		1.80	80	240	720	62.40	
	0900	Aggregate spreader, self-propelled, 187 HP		42.35	800	2,400	7,200	818.80	
	1000	Chemical spreader, 3 C.Y.		2.40	41.50	125	375	44.20	
	1900	Hammermill, traveling, 250 HP		55.98	1,775	5,320	16,000	1,512	
	2000	Horizontal borer, 3" diam, 13 HP gas driven		4.90	53.50	160	480	71.20	
	2200	Hydromulchers, gas power, 3000 gal., for truck mounting		12.90	188	565	1,700	216.20	
	2400	Joint & crack cleaner, walk behind, 25 HP		2.40	48.50	145	435	48.20	
	2500	Filler, trailer mounted, 400 gal., 20 HP		6.55	183	550	1,650	162.40	
	3000	Paint striper, self propelled, double line, 30 HP		5.50	157	470	1,400	138	
	3200	Post drivers, 6" I-Beam frame, for truck mounting		10.30	420	1,255	3,775	333.40	
	3400	Road sweeper, self propelled, 8' wide, 90 HP		27.75	550	1,645	4,925	551	
	4000	Road mixer, self-propelled, 130 HP		35.40	685	2,050	6,150	693.20	
	4100	310 HP		67.30	2,125	6,405	19,200	1,819	
	4200	Cold mix paver, incl pug mill and bitumen tank,							
	4220	165 HP	Ea.	79.35	1,975	5,910	17,700	1,817	
	4250	Paver, asphalt, wheel or crawler, 130 H.P., diesel		78.05	1,875	5,630	16,900	1,750	
	4300	Paver, road widener, gas 1' to 6', 67 HP		37.40	780	2,345	7,025	768.20	
	4400	Diesel, 2' to 14', 88 HP		49.05	1,000	3,025	9,075	997.40	
	4600	Slipform pavers, curb and gutter, 2 track, 75 HP		31.95	720	2,160	6,475	687.60	
	4700	4 track, 165 HP		38.75	770	2,305	6,925	771	
	4800	Median barrier, 215 HP		39.35	800	2,395	7,175	793.80	
	4901	Trailer, low bed, 75 ton capacity		8.15	193	580	1,750	181.20	
	5000	Road planer, walk behind, 10" cutting width, 10 HP		2.30	28	84	252	35.20	
	5100	Self propelled, 12" cutting width, 64 HP		6.25	112	335	1,000	117	
	5200	Pavement profiler, 4' to 6' wide, 450 HP		197.95	3,250	9,760	29,300	3,536	
	5300	8' to 10' wide, 750 HP		313.05	4,450	13,360	40,100	5,176	
	5400	Roadway plate, steel, 1"x8'x20'		.06	10.65	32	96	6.90	
	5600	Stabilizer, self-propelled, 150 HP		36.55	580	1,735	5,200	639.40	
	5700	310 HP		62.05	1,275	3,845	11,500	1,265	
	5800	Striper, thermal, truck mounted 120 gal. paint, 150 H.P.		37.90	490	1,465	4,400	596.20	
	6000	Tar kettle, 330 gal., trailer mounted		3.67	36.50	110	330	51.35	
	7000	Tunnel locomotive, diesel, 8 to 12 ton		25.05	560	1,685	5,050	537.40	

R312323-30

R314116-40

R314116-45

01 54 33 | Equipment Rental

		UNIT	HOURLY OPER. COST	RENT PER DAY	RENT PER WEEK	RENT PER MONTH	CREW EQUIPMENT COST/DAY	
50	7005	Electric, 10 ton	Ea.	21.95	640	1,915	5,750	558.60	**50**
	7010	Muck cars, 1/2 C.Y. capacity		1.65	21.50	64	192	26	
	7020	1 C.Y. capacity		1.85	30	90	270	32.80	
	7030	2 C.Y. capacity		1.95	35	105	315	36.60	
	7040	Side dump, 2 C.Y. capacity		2.15	41.50	125	375	42.20	
	7050	3 C.Y. capacity		2.90	48.50	145	435	52.20	
	7060	5 C.Y. capacity		4.10	61.50	185	555	69.80	
	7100	Ventilating blower for tunnel, 7-1/2 H.P.		1.29	50	150	450	40.30	
	7110	10 H.P.		1.49	51.50	155	465	42.90	
	7120	20 H.P.		2.44	67.50	202	605	59.90	
	7140	40 H.P.		4.29	95	285	855	91.30	
	7160	60 H.P.		6.49	147	440	1,325	139.90	
	7175	75 H.P.		8.30	196	587	1,750	183.80	
	7180	200 H.P.		18.70	293	880	2,650	325.60	
	7800	Windrow loader, elevating		39.00	930	2,785	8,350	869	
60	0010	**LIFTING & HOISTING EQUIPMENT RENTAL**, without operators							**60**
	0120	Aerial lift truck, 2 person, to 80'	Ea.	22.35	615	1,840	5,525	546.80	
	0140	Boom work platform, 40' snorkel		13.25	223	670	2,000	240	
	0150	Crane, flatbed mntd, 3 ton cap.		17.70	217	650	1,950	271.60	
	0200	Crane, climbing, 106' jib, 6000 lb. capacity, 410 FPM		68.15	1,400	4,230	12,700	1,391	
	0300	101' jib, 10,250 lb. capacity, 270 FPM		73.80	1,775	5,360	16,100	1,662	
	0400	Tower, static, 130' high, 106' jib,							
	0500	6200 lb. capacity at 400 FPM	Ea.	71.45	1,625	4,890	14,700	1,550	
	0600	Crawler mounted, lattice boom, 1/2 C.Y., 15 tons at 12' radius		26.58	605	1,820	5,450	576.65	
	0700	3/4 C.Y., 20 tons at 12' radius		35.44	755	2,270	6,800	737.50	
	0800	1 C.Y., 25 tons at 12' radius		47.25	1,000	3,025	9,075	983	
	0900	1-1/2 C.Y., 40 tons at 12' radius		51.20	975	2,925	8,775	994.60	
	1000	2 C.Y., 50 tons at 12' radius		54.00	1,175	3,495	10,500	1,131	
	1100	3 C.Y., 75 tons at 12' radius		66.25	1,425	4,270	12,800	1,384	
	1200	100 ton capacity, 60' boom		76.70	1,850	5,515	16,500	1,717	
	1300	165 ton capacity, 60' boom		102.30	2,300	6,890	20,700	2,196	
	1400	200 ton capacity, 70' boom		130.85	2,675	8,030	24,100	2,653	
	1500	350 ton capacity, 80' boom		182.90	3,500	10,530	31,600	3,569	
	1600	Truck mounted, lattice boom, 6 x 4, 20 tons at 10' radius		33.59	1,175	3,540	10,600	976.70	
	1700	25 tons at 10' radius		36.28	1,275	3,840	11,500	1,058	
	1800	8 x 4, 30 tons at 10' radius		49.47	1,375	4,130	12,400	1,222	
	1900	40 tons at 12' radius		50.73	1,425	4,310	12,900	1,268	
	2000	8 x 4, 60 tons at 15' radius		52.34	1,525	4,540	13,600	1,327	
	2050	82 tons at 15' radius		54.44	1,625	4,840	14,500	1,404	
	2100	90 tons at 15' radius		60.67	1,775	5,310	15,900	1,547	
	2200	115 tons at 15' radius		74.05	1,975	5,900	17,700	1,772	
	2300	150 tons at 18' radius		74.51	2,100	6,280	18,800	1,852	
	2350	165 tons at 18' radius		87.35	2,200	6,615	19,800	2,022	
	2400	Truck mounted, hydraulic, 12 ton capacity		45.25	605	1,810	5,425	724	
	2500	25 ton capacity		45.45	625	1,880	5,650	739.60	
	2550	33 ton capacity		46.30	660	1,975	5,925	765.40	
	2560	40 ton capacity		44.70	630	1,895	5,675	736.60	
	2600	55 ton capacity		66.40	880	2,645	7,925	1,060	
	2700	80 ton capacity		78.75	975	2,920	8,750	1,214	
	2720	100 ton capacity		95.75	2,500	7,525	22,600	2,271	
	2740	120 ton capacity		93.21	2,750	8,280	24,800	2,402	
	2760	150 ton capacity		114.91	3,525	10,540	31,600	3,027	
	2800	Self-propelled, 4 x 4, with telescoping boom, 5 ton		19.10	290	870	2,600	326.80	
	2900	12-1/2 ton capacity		32.75	520	1,565	4,700	575	
	3000	15 ton capacity		35.00	615	1,850	5,550	650	
	3050	20 ton capacity		35.85	640	1,920	5,750	670.80	
	3100	25 ton capacity		47.95	820	2,460	7,375	875.60	

Reference boxes: R015433-10, R015433-15, R312316-45

01 54 33 | Equipment Rental

		UNIT	HOURLY OPER. COST	RENT PER DAY	RENT PER WEEK	RENT PER MONTH	CREW EQUIPMENT COST/DAY	
60	3150	40 ton capacity	Ea.	61.00	925	2,775	8,325	1,043	60
	3200	Derricks, guy, 20 ton capacity, 60' boom, 75' mast		19.00	345	1,036	3,100	359.20	
	3300	100' boom, 115' mast		30.02	590	1,770	5,300	594.15	
	3400	Stiffleg, 20 ton capacity, 70' boom, 37' mast		21.13	445	1,340	4,025	437.05	
	3500	100' boom, 47' mast		32.75	720	2,160	6,475	694	
	3550	Helicopter, small, lift to 1250 lbs. maximum, w/pilot		82.09	2,800	8,370	25,100	2,331	
	3600	Hoists, chain type, overhead, manual, 3/4 ton		.10	.67	2	6	1.20	
	3900	10 ton		.65	7.35	22	66	9.60	
	4000	Hoist and tower, 5000 lb. cap., portable electric, 40' high		4.42	199	597	1,800	154.75	
	4100	For each added 10' section, add		.09	15.65	47	141	10.10	
	4200	Hoist and single tubular tower, 5000 lb. electric, 100' high		5.95	278	833	2,500	214.20	
	4300	For each added 6'-6" section, add		.16	26.50	79	237	17.10	
	4400	Hoist and double tubular tower, 5000 lb., 100' high		6.38	305	918	2,750	234.65	
	4500	For each added 6'-6" section, add		.18	29.50	89	267	19.25	
	4550	Hoist and tower, mast type, 6000 lb., 100' high		6.91	315	952	2,850	245.70	
	4570	For each added 10' section, add		.11	19	57	171	12.30	
	4600	Hoist and tower, personnel, electric, 2000 lb., 100' @ 125 FPM		14.13	845	2,540	7,625	621.05	
	4700	3000 lb., 100' @ 200 FPM		16.14	955	2,870	8,600	703.10	
	4800	3000 lb., 150' @ 300 FPM		17.84	1,075	3,210	9,625	784.70	
	4900	4000 lb., 100' @ 300 FPM		18.50	1,100	3,270	9,800	802	
	5000	6000 lb., 100' @ 275 FPM		20.06	1,150	3,440	10,300	848.50	
	5100	For added heights up to 500', add	L.F.	.01	1.67	5	15	1.10	
	5200	Jacks, hydraulic, 20 ton	Ea.	.05	3.33	10	30	2.40	
	5500	100 ton	"	.30	9.65	29	87	8.20	
	6000	Jacks, hydraulic, climbing with 50' jackrods							
	6010	and control consoles, minimum 3 mo. rental							
	6100	30 ton capacity	Ea.	1.72	114	343	1,025	82.35	
	6150	For each added 10' jackrod section, add		.05	3.33	10	30	2.40	
	6300	50 ton capacity		2.76	184	552	1,650	132.50	
	6350	For each added 10' jackrod section, add		.06	4	12	36	2.90	
	6500	125 ton capacity		7.25	485	1,450	4,350	348	
	6550	For each added 10' jackrod section, add		.50	33	99	297	23.80	
	6600	Cable jack, 10 ton capacity with 200' cable		1.44	95.50	287	860	68.90	
	6650	For each added 50' of cable, add		.16	10.65	32	96	7.70	
70	0010	**WELLPOINT EQUIPMENT RENTAL**, without operators							70
	0020	Based on 2 months rental							
	0100	Combination jetting & wellpoint pump, 60 H.P. diesel	Ea.	13.00	283	850	2,550	274	
	0200	High pressure gas jet pump, 200 H.P., 300 psi	"	28.58	242	726	2,175	373.85	
	0300	Discharge pipe, 8" diameter	L.F.	.01	.46	1.38	4.14	.35	
	0350	12" diameter		.01	.68	2.04	6.10	.50	
	0400	Header pipe, flows up to 150 G.P.M., 4" diameter		.01	.41	1.24	3.72	.35	
	0500	400 G.P.M., 6" diameter		.01	.49	1.48	4.44	.40	
	0600	800 G.P.M., 8" diameter		.01	.68	2.04	6.10	.50	
	0700	1500 G.P.M., 10" diameter		.01	.71	2.14	6.40	.50	
	0800	2500 G.P.M., 12" diameter		.02	1.35	4.05	12.15	.95	
	0900	4500 G.P.M., 16" diameter		.03	1.73	5.18	15.55	1.30	
	0950	For quick coupling aluminum and plastic pipe, add		.03	1.79	5.36	16.10	1.30	
	1100	Wellpoint, 25' long, with fittings & riser pipe, 1-1/2" or 2" diameter	Ea.	.05	3.57	10.70	32	2.55	
	1200	Wellpoint pump, diesel powered, 4" diameter, 20 H.P.		5.78	163	490	1,475	144.25	
	1300	6" diameter, 30 H.P.		7.78	203	608	1,825	183.85	
	1400	8" suction, 40 H.P.		10.53	278	833	2,500	250.85	
	1500	10" suction, 75 H.P.		15.63	325	974	2,925	319.85	
	1600	12" suction, 100 H.P.		22.67	520	1,560	4,675	493.35	
	1700	12" suction, 175 H.P.		32.53	570	1,710	5,125	602.25	

Reference codes: R015433 -10, R015433 -15, R312316 -45, R312319 -90

			UNIT	HOURLY OPER. COST	RENT PER DAY	RENT PER WEEK	RENT PER MONTH	CREW EQUIPMENT COST/DAY	
80	0010	**MARINE EQUIPMENT RENTAL**, without operators							80
	0200	Barge, 400 Ton, 30' wide x 90' long	Ea.	17.65	255	765	2,300	294.20	
	0240	800 Ton, 45' wide x 90' long		29.15	365	1,090	3,275	451.20	
	2000	Tugboat, diesel, 100 HP		22.25	180	540	1,625	286	
	2040	250 HP		43.65	335	1,005	3,025	550.20	
	2080	380 HP	Ea.	93.95	990	2,970	8,900	1,346	

01 54 33 | Equipment Rental

Crews

Crew No.	Bare Costs Hr.	Daily	Incl. Subs O & P Hr.	Daily	Cost Per Labor-Hour Bare Costs	Incl. O&P
Crew A-1	Hr.	Daily	Hr.	Daily	Bare Costs	Incl. O&P
1 Building Laborer	$28.75	$230.00	$44.75	$358.00	$28.75	$44.75
1 Concrete saw, gas manual		54.00		59.40	6.75	7.42
8 L.H., Daily Totals		$284.00		$417.40	$35.50	$52.17
Crew A-1A	Hr.	Daily	Hr.	Daily	Bare Costs	Incl. O&P
1 Skilled Worker	$38.00	$304.00	$59.15	$473.20	$38.00	$59.15
1 Shot Blaster, 20"		302.20		332.42	37.77	41.55
8 L.H., Daily Totals		$606.20		$805.62	$75.78	$100.70
Crew A-1B	Hr.	Daily	Hr.	Daily	Bare Costs	Incl. O&P
1 Building Laborer	$28.75	$230.00	$44.75	$358.00	$28.75	$44.75
1 Concrete Saw		118.40		130.24	14.80	16.28
8 L.H., Daily Totals		$348.40		$488.24	$43.55	$61.03
Crew A-1C	Hr.	Daily	Hr.	Daily	Bare Costs	Incl. O&P
1 Building Laborer	$28.75	$230.00	$44.75	$358.00	$28.75	$44.75
1 Chain saw, gas, 18"		23.00		25.30	2.88	3.16
8 L.H., Daily Totals		$253.00		$383.30	$31.63	$47.91
Crew A-1D	Hr.	Daily	Hr.	Daily	Bare Costs	Incl. O&P
1 Building Laborer	$28.75	$230.00	$44.75	$358.00	$28.75	$44.75
1 Vibrating plate, gas, 18"		29.20		32.12	3.65	4.01
8 L.H., Daily Totals		$259.20		$390.12	$32.40	$48.77
Crew A-1E	Hr.	Daily	Hr.	Daily	Bare Costs	Incl. O&P
1 Building Laborer	$28.75	$230.00	$44.75	$358.00	$28.75	$44.75
1 Vibratory Plate, Gas, 21"		36.20		39.82	4.53	4.98
8 L.H., Daily Totals		$266.20		$397.82	$33.27	$49.73
Crew A-1F	Hr.	Daily	Hr.	Daily	Bare Costs	Incl. O&P
1 Building Laborer	$28.75	$230.00	$44.75	$358.00	$28.75	$44.75
1 Rammer/tamper, gas, 8"		38.40		42.24	4.80	5.28
8 L.H., Daily Totals		$268.40		$400.24	$33.55	$50.03
Crew A-1G	Hr.	Daily	Hr.	Daily	Bare Costs	Incl. O&P
1 Building Laborer	$28.75	$230.00	$44.75	$358.00	$28.75	$44.75
1 Rammer/tamper, gas, 8"		38.40		42.24	4.80	5.28
8 L.H., Daily Totals		$268.40		$400.24	$33.55	$50.03
Crew A-1H	Hr.	Daily	Hr.	Daily	Bare Costs	Incl. O&P
1 Building Laborer	$28.75	$230.00	$44.75	$358.00	$28.75	$44.75
1 Pressure washer		60.40		66.44	7.55	8.30
8 L.H., Daily Totals		$290.40		$424.44	$36.30	$53.06
Crew A-1J	Hr.	Daily	Hr.	Daily	Bare Costs	Incl. O&P
1 Building Laborer	$28.75	$230.00	$44.75	$358.00	$28.75	$44.75
1 Cultivator, Walk-Behind		48.30		53.13	6.04	6.64
8 L.H., Daily Totals		$278.30		$411.13	$34.79	$51.39
Crew A-1K	Hr.	Daily	Hr.	Daily	Bare Costs	Incl. O&P
1 Building Laborer	$28.75	$230.00	$44.75	$358.00	$28.75	$44.75
1 Cultivator, Walk-Behind		48.30		53.13	6.04	6.64
8 L.H., Daily Totals		$278.30		$411.13	$34.79	$51.39

Crew No.	Bare Costs Hr.	Daily	Incl. Subs O & P Hr.	Daily	Cost Per Labor-Hour Bare Costs	Incl. O&P
Crew A-1M	Hr.	Daily	Hr.	Daily	Bare Costs	Incl. O&P
1 Building Laborer	$28.75	$230.00	$44.75	$358.00	$28.75	$44.75
1 Snow Blower, Walk-Behind		53.45		58.80	6.68	7.35
8 L.H., Daily Totals		$283.45		$416.80	$35.43	$52.10
Crew A-2	Hr.	Daily	Hr.	Daily	Bare Costs	Incl. O&P
2 Laborers	$28.75	$460.00	$44.75	$716.00	$28.68	$44.53
1 Truck Driver (light)	28.55	228.40	44.10	352.80		
1 Flatbed Truck, gas, 1.5 Ton		141.40		155.54	5.89	6.48
24 L.H., Daily Totals		$829.80		$1224.34	$34.58	$51.01
Crew A-2A	Hr.	Daily	Hr.	Daily	Bare Costs	Incl. O&P
2 Laborers	$28.75	$460.00	$44.75	$716.00	$28.68	$44.53
1 Truck Driver (light)	28.55	228.40	44.10	352.80		
1 Flatbed Truck, gas, 1.5 Ton		141.40		155.54		
1 Concrete Saw		118.40		130.24	10.82	11.91
24 L.H., Daily Totals		$948.20		$1354.58	$39.51	$56.44
Crew A-3	Hr.	Daily	Hr.	Daily	Bare Costs	Incl. O&P
1 Truck Driver (heavy)	$29.55	$236.40	$45.65	$365.20	$29.55	$45.65
1 Dump Truck, 12 Ton, 8 C.Y.		368.40		405.24	46.05	50.66
8 L.H., Daily Totals		$604.80		$770.44	$75.60	$96.31
Crew A-3A	Hr.	Daily	Hr.	Daily	Bare Costs	Incl. O&P
1 Truck Driver (light)	$28.55	$228.40	$44.10	$352.80	$28.55	$44.10
1 Pickup truck, 4 x 4, 3/4 ton		91.00		100.10	11.38	12.51
8 L.H., Daily Totals		$319.40		$452.90	$39.92	$56.61
Crew A-3B	Hr.	Daily	Hr.	Daily	Bare Costs	Incl. O&P
1 Equip. Oper. (medium)	$38.40	$307.20	$57.85	$462.80	$33.98	$51.75
1 Truck Driver (heavy)	29.55	236.40	45.65	365.20		
1 Dump Truck, 16 Ton, 12 C.Y.		530.00		583.00		
1 F.E. Loader, W.M.,2.5 C.Y.		349.00		383.90	54.94	60.43
16 L.H., Daily Totals		$1422.60		$1794.90	$88.91	$112.18
Crew A-3C	Hr.	Daily	Hr.	Daily	Bare Costs	Incl. O&P
1 Equip. Oper. (light)	$36.85	$294.80	$55.55	$444.40	$36.85	$55.55
1 Loader, Skid Steer, 78 HP		222.20		244.42	27.77	30.55
8 L.H., Daily Totals		$517.00		$688.82	$64.63	$86.10
Crew A-3D	Hr.	Daily	Hr.	Daily	Bare Costs	Incl. O&P
1 Truck Driver, Light	$28.55	$228.40	$44.10	$352.80	$28.55	$44.10
1 Pickup truck, 4 x 4, 3/4 ton		91.00		100.10		
1 Flatbed Trailer, 25 Ton		92.80		102.08	22.98	25.27
8 L.H., Daily Totals		$412.20		$554.98	$51.52	$69.37
Crew A-3E	Hr.	Daily	Hr.	Daily	Bare Costs	Incl. O&P
1 Equip. Oper. (crane)	$39.80	$318.40	$60.00	$480.00	$34.67	$52.83
1 Truck Driver (heavy)	29.55	236.40	45.65	365.20		
1 Pickup truck, 4 x 4, 3/4 ton		91.00		100.10	5.69	6.26
16 L.H., Daily Totals		$645.80		$945.30	$40.36	$59.08
Crew A-3F	Hr.	Daily	Hr.	Daily	Bare Costs	Incl. O&P
1 Equip. Oper. (crane)	$39.80	$318.40	$60.00	$480.00	$34.67	$52.83
1 Truck Driver (heavy)	29.55	236.40	45.65	365.20		
1 Pickup truck, 4 x 4, 3/4 ton		91.00		100.10		
1 Truck Tractor, 240 H.P.		336.20		369.82		
1 Lowbed Trailer, 75 Ton		181.20		199.32	38.02	41.83
16 L.H., Daily Totals		$1163.20		$1514.44	$72.70	$94.65

409

Crew A-3G

Crew No.	Bare Costs Hr.	Daily	Incl. Subs O & P Hr.	Daily	Cost Per Labor-Hour Bare Costs	Incl. O&P
1 Equip. Oper. (crane)	$39.80	$318.40	$60.00	$480.00	$34.67	$52.83
1 Truck Driver (heavy)	29.55	236.40	45.65	365.20		
1 Pickup truck, 4 x 4, 3/4 ton		91.00		100.10		
1 Truck Tractor, 240 H.P.		395.40		434.94		
1 Lowbed Trailer, 75 Ton		181.20		199.32	41.73	45.90
16 L.H., Daily Totals		$1222.40		$1579.56	$76.40	$98.72

Crew A-4

Crew No.	Bare Costs Hr.	Daily	Incl. Subs O & P Hr.	Daily	Cost Per Labor-Hour Bare Costs	Incl. O&P
2 Carpenters	$36.70	$587.20	$57.15	$914.40	$35.37	$54.50
1 Painter, Ordinary	32.70	261.60	49.20	393.60		
24 L.H., Daily Totals		$848.80		$1308.00	$35.37	$54.50

Crew A-5

Crew No.	Bare Costs Hr.	Daily	Incl. Subs O & P Hr.	Daily	Cost Per Labor-Hour Bare Costs	Incl. O&P
2 Laborers	$28.75	$460.00	$44.75	$716.00	$28.73	$44.68
.25 Truck Driver (light)	28.55	57.10	44.10	88.20		
.25 Flatbed Truck, gas, 1.5 Ton		35.35		38.88	1.96	2.16
18 L.H., Daily Totals		$552.45		$843.09	$30.69	$46.84

Crew A-6

Crew No.	Bare Costs Hr.	Daily	Incl. Subs O & P Hr.	Daily	Cost Per Labor-Hour Bare Costs	Incl. O&P
1 Instrument Man	$38.00	$304.00	$59.15	$473.20	$37.02	$56.88
1 Rodman/Chainman	36.05	288.40	54.60	436.80		
1 Laser Transit/Level		58.10		63.91	3.63	3.99
16 L.H., Daily Totals		$650.50		$973.91	$40.66	$60.87

Crew A-7

Crew No.	Bare Costs Hr.	Daily	Incl. Subs O & P Hr.	Daily	Cost Per Labor-Hour Bare Costs	Incl. O&P
1 Chief Of Party	$47.15	$377.20	$73.10	$584.80	$40.40	$62.28
1 Instrument Man	38.00	304.00	59.15	473.20		
1 Rodman/Chainman	36.05	288.40	54.60	436.80		
1 Laser Transit/Level		58.10		63.91	2.42	2.66
24 L.H., Daily Totals		$1027.70		$1558.71	$42.82	$64.95

Crew A-8

Crew No.	Bare Costs Hr.	Daily	Incl. Subs O & P Hr.	Daily	Cost Per Labor-Hour Bare Costs	Incl. O&P
1 Chief of Party	$47.15	$377.20	$73.10	$584.80	$39.31	$60.36
1 Instrument Man	38.00	304.00	59.15	473.20		
2 Rodmen/Chainmen	36.05	576.80	54.60	873.60		
1 Laser Transit/Level		58.10		63.91	1.82	2.00
32 L.H., Daily Totals		$1316.10		$1995.51	$41.13	$62.36

Crew A-9

Crew No.	Bare Costs Hr.	Daily	Incl. Subs O & P Hr.	Daily	Cost Per Labor-Hour Bare Costs	Incl. O&P
1 Asbestos Foreman	$42.00	$336.00	$66.40	$531.20	$41.56	$65.70
7 Asbestos Workers	41.50	2324.00	65.60	3673.60		
64 L.H., Daily Totals		$2660.00		$4204.80	$41.56	$65.70

Crew A-10

Crew No.	Bare Costs Hr.	Daily	Incl. Subs O & P Hr.	Daily	Cost Per Labor-Hour Bare Costs	Incl. O&P
1 Asbestos Foreman	$42.00	$336.00	$66.40	$531.20	$41.56	$65.70
7 Asbestos Workers	41.50	2324.00	65.60	3673.60		
64 L.H., Daily Totals		$2660.00		$4204.80	$41.56	$65.70

Crew A-10A

Crew No.	Bare Costs Hr.	Daily	Incl. Subs O & P Hr.	Daily	Cost Per Labor-Hour Bare Costs	Incl. O&P
1 Asbestos Foreman	$42.00	$336.00	$66.40	$531.20	$41.67	$65.87
2 Asbestos Workers	41.50	664.00	65.60	1049.60		
24 L.H., Daily Totals		$1000.00		$1580.80	$41.67	$65.87

Crew A-10B

Crew No.	Bare Costs Hr.	Daily	Incl. Subs O & P Hr.	Daily	Cost Per Labor-Hour Bare Costs	Incl. O&P
1 Asbestos Foreman	$42.00	$336.00	$66.40	$531.20	$41.63	$65.80
3 Asbestos Workers	41.50	996.00	65.60	1574.40		
32 L.H., Daily Totals		$1332.00		$2105.60	$41.63	$65.80

Crew A-10C

Crew No.	Bare Costs Hr.	Daily	Incl. Subs O & P Hr.	Daily	Cost Per Labor-Hour Bare Costs	Incl. O&P
3 Asbestos Workers	$41.50	$996.00	$65.60	$1574.40	$41.50	$65.60
1 Flatbed Truck, gas, 1.5 Ton		141.40		155.54	5.89	6.48
24 L.H., Daily Totals		$1137.40		$1729.94	$47.39	$72.08

Crew A-10D

Crew No.	Bare Costs Hr.	Daily	Incl. Subs O & P Hr.	Daily	Cost Per Labor-Hour Bare Costs	Incl. O&P
2 Asbestos Workers	$41.50	$664.00	$65.60	$1049.60	$39.17	$60.58
1 Equip. Oper. (crane)	39.80	318.40	60.00	480.00		
1 Equip. Oper. Oiler	33.90	271.20	51.10	408.80		
1 Hydraulic Crane, 33 Ton		765.40		841.94	23.92	26.31
32 L.H., Daily Totals		$2019.00		$2780.34	$63.09	$86.89

Crew A-11

Crew No.	Bare Costs Hr.	Daily	Incl. Subs O & P Hr.	Daily	Cost Per Labor-Hour Bare Costs	Incl. O&P
1 Asbestos Foreman	$42.00	$336.00	$66.40	$531.20	$41.56	$65.70
7 Asbestos Workers	41.50	2324.00	65.60	3673.60		
2 Chipping Hammer, 12 Lb., Elec.		31.20		34.32	.49	.54
64 L.H., Daily Totals		$2691.20		$4239.12	$42.05	$66.24

Crew A-12

Crew No.	Bare Costs Hr.	Daily	Incl. Subs O & P Hr.	Daily	Cost Per Labor-Hour Bare Costs	Incl. O&P
1 Asbestos Foreman	$42.00	$336.00	$66.40	$531.20	$41.56	$65.70
7 Asbestos Workers	41.50	2324.00	65.60	3673.60		
1 Trk-mtd vac, 14 CY, 1500 Gal		510.95		562.04		
1 Flatbed Truck, 20,000 GVW		151.80		166.98	10.36	11.39
64 L.H., Daily Totals		$3322.75		$4933.82	$51.92	$77.09

Crew A-13

Crew No.	Bare Costs Hr.	Daily	Incl. Subs O & P Hr.	Daily	Cost Per Labor-Hour Bare Costs	Incl. O&P
1 Equip. Oper. (light)	$36.85	$294.80	$55.55	$444.40	$36.85	$55.55
1 Trk-mtd vac, 14 CY, 1500 Gal		510.95		562.04		
1 Flatbed Truck, 20,000 GVW		151.80		166.98	82.84	91.13
8 L.H., Daily Totals		$957.55		$1173.43	$119.69	$146.68

Crew B-1

Crew No.	Bare Costs Hr.	Daily	Incl. Subs O & P Hr.	Daily	Cost Per Labor-Hour Bare Costs	Incl. O&P
1 Labor Foreman (outside)	$30.75	$246.00	$47.90	$383.20	$29.42	$45.80
2 Laborers	28.75	460.00	44.75	716.00		
24 L.H., Daily Totals		$706.00		$1099.20	$29.42	$45.80

Crew B-1A

Crew No.	Bare Costs Hr.	Daily	Incl. Subs O & P Hr.	Daily	Cost Per Labor-Hour Bare Costs	Incl. O&P
1 Laborer Foreman	$30.75	$246.00	$47.90	$383.20	$29.42	$45.80
2 Laborers	28.75	460.00	44.75	716.00		
2 Cutting Torches		34.00		37.40		
2 Gases		129.60		142.56	6.82	7.50
24 L.H., Daily Totals		$869.60		$1279.16	$36.23	$53.30

Crew B-1B

Crew No.	Bare Costs Hr.	Daily	Incl. Subs O & P Hr.	Daily	Cost Per Labor-Hour Bare Costs	Incl. O&P
1 Laborer Foreman	$30.75	$246.00	$47.90	$383.20	$32.01	$49.35
2 Laborers	28.75	460.00	44.75	716.00		
1 Equip. Oper. (crane)	39.80	318.40	60.00	480.00		
2 Cutting Torches		34.00		37.40		
2 Gases		129.60		142.56		
1 Hyd. Crane, 12 Ton		724.00		796.40	27.74	30.51
32 L.H., Daily Totals		$1912.00		$2555.56	$59.75	$79.86

Crew B-2

Crew No.	Bare Costs Hr.	Daily	Incl. Subs O & P Hr.	Daily	Cost Per Labor-Hour Bare Costs	Incl. O&P
1 Labor Foreman (outside)	$30.75	$246.00	$47.90	$383.20	$29.15	$45.38
4 Laborers	28.75	920.00	44.75	1432.00		
40 L.H., Daily Totals		$1166.00		$1815.20	$29.15	$45.38

Crew B-3

Crew No.	Bare Costs Hr.	Bare Costs Daily	Incl. Subs O&P Hr.	Incl. Subs O&P Daily	Cost Per Labor-Hour Bare Costs	Cost Per Labor-Hour Incl. O&P
1 Labor Foreman (outside)	$30.75	$246.00	$47.90	$383.20	$30.96	$47.76
2 Laborers	28.75	460.00	44.75	716.00		
1 Equip. Oper. (med.)	38.40	307.20	57.85	462.80		
2 Truck Drivers (heavy)	29.55	472.80	45.65	730.40		
1 Crawler Loader, 3 C.Y.		834.40		917.84		
2 Dump Trucks, 16 Ton, 12 C.Y.		1060.00		1166.00	39.47	43.41
48 L.H., Daily Totals		$3380.40		$4376.24	$70.42	$91.17

Crew B-3A

Crew No.	Bare Costs Hr.	Bare Costs Daily	Incl. Subs O&P Hr.	Incl. Subs O&P Daily	Cost Per Labor-Hour Bare Costs	Cost Per Labor-Hour Incl. O&P
4 Laborers	$28.75	$920.00	$44.75	$1432.00	$30.68	$47.37
1 Equip. Oper. (med.)	38.40	307.20	57.85	462.80		
1 Hyd. Excavator, 1.5 C.Y.		775.60		853.16	19.39	21.33
40 L.H., Daily Totals		$2002.80		$2747.96	$50.07	$68.70

Crew B-3B

Crew No.	Bare Costs Hr.	Bare Costs Daily	Incl. Subs O&P Hr.	Incl. Subs O&P Daily	Cost Per Labor-Hour Bare Costs	Cost Per Labor-Hour Incl. O&P
2 Laborers	$28.75	$460.00	$44.75	$716.00	$31.36	$48.25
1 Equip. Oper. (med.)	38.40	307.20	57.85	462.80		
1 Truck Driver (heavy)	29.55	236.40	45.65	365.20		
1 Backhoe Loader, 80 H.P.		285.40		313.94		
1 Dump Truck, 16 Ton, 12 C.Y.		530.00		583.00	25.48	28.03
32 L.H., Daily Totals		$1819.00		$2440.94	$56.84	$76.28

Crew B-3C

Crew No.	Bare Costs Hr.	Bare Costs Daily	Incl. Subs O&P Hr.	Incl. Subs O&P Daily	Cost Per Labor-Hour Bare Costs	Cost Per Labor-Hour Incl. O&P
3 Laborers	$28.75	$690.00	$44.75	$1074.00	$31.16	$48.02
1 Equip. Oper. (med.)	38.40	307.20	57.85	462.80		
1 Crawler Loader, 4 C.Y.		1145.00		1259.50	35.78	39.36
32 L.H., Daily Totals		$2142.20		$2796.30	$66.94	$87.38

Crew B-4

Crew No.	Bare Costs Hr.	Bare Costs Daily	Incl. Subs O&P Hr.	Incl. Subs O&P Daily	Cost Per Labor-Hour Bare Costs	Cost Per Labor-Hour Incl. O&P
1 Labor Foreman (outside)	$30.75	$246.00	$47.90	$383.20	$29.22	$45.42
4 Laborers	28.75	920.00	44.75	1432.00		
1 Truck Driver (heavy)	29.55	236.40	45.65	365.20		
1 Truck Tractor, 195 H.P.		235.60		259.16		
1 Flatbed Trailer, 40 Ton		125.80		138.38	7.53	8.28
48 L.H., Daily Totals		$1763.80		$2577.94	$36.75	$53.71

Crew B-5

Crew No.	Bare Costs Hr.	Bare Costs Daily	Incl. Subs O&P Hr.	Incl. Subs O&P Daily	Cost Per Labor-Hour Bare Costs	Cost Per Labor-Hour Incl. O&P
1 Labor Foreman (outside)	$30.75	$246.00	$47.90	$383.20	$31.79	$48.94
4 Laborers	28.75	920.00	44.75	1432.00		
2 Equip. Oper. (med.)	38.40	614.40	57.85	925.60		
1 Air Compressor, 250 C.F.M.		151.40		166.54		
2 Breakers, Pavement, 60 lb.		16.80		18.48		
2 -50' Air Hoses, 1.5"		11.60		12.76		
1 Crawler Loader, 3 C.Y.		834.40		917.84	18.11	19.92
56 L.H., Daily Totals		$2794.60		$3856.42	$49.90	$68.86

Crew B-5A

Crew No.	Bare Costs Hr.	Bare Costs Daily	Incl. Subs O&P Hr.	Incl. Subs O&P Daily	Cost Per Labor-Hour Bare Costs	Cost Per Labor-Hour Incl. O&P
1 Foreman	$30.75	$246.00	$47.90	$383.20	$31.33	$48.25
6 Laborers	28.75	1380.00	44.75	2148.00		
2 Equip. Oper. (med.)	38.40	614.40	57.85	925.60		
1 Equip. Oper. (light)	36.85	294.80	55.55	444.40		
2 Truck Drivers (heavy)	29.55	472.80	45.65	730.40		
1 Air Compressor, 365 C.F.M.		195.80		215.38		
2 Breakers, Pavement, 60 lb.		16.80		18.48		
8 -50' Air Hoses, 1"		32.80		36.08		
2 Dump Trucks, 12 Ton, 8 C.Y.		736.80		810.48	10.23	11.25
96 L.H., Daily Totals		$3990.20		$5712.02	$41.56	$59.50

Crew B-5B

Crew No.	Bare Costs Hr.	Bare Costs Daily	Incl. Subs O&P Hr.	Incl. Subs O&P Daily	Cost Per Labor-Hour Bare Costs	Cost Per Labor-Hour Incl. O&P
1 Powderman	$38.00	$304.00	$59.15	$473.20	$33.91	$51.97
2 Equip. Oper. (med.)	38.40	614.40	57.85	925.60		
3 Truck Drivers (heavy)	29.55	709.20	45.65	1095.60		
1 F.E. Loader, W.M.,2.5 C.Y.		349.00		383.90		
3 Dump Trucks, 16 Ton, 12 C.Y.		1590.00		1749.00		
1 Air Compressor, 365 C.F.M.		195.80		215.38	44.48	48.92
48 L.H., Daily Totals		$3762.40		$4842.68	$78.38	$100.89

Crew B-5C

Crew No.	Bare Costs Hr.	Bare Costs Daily	Incl. Subs O&P Hr.	Incl. Subs O&P Daily	Cost Per Labor-Hour Bare Costs	Cost Per Labor-Hour Incl. O&P
3 Laborers	$28.75	$690.00	$44.75	$1074.00	$32.18	$49.31
1 Equip. Oper. (medium)	38.40	307.20	57.85	462.80		
2 Truck Drivers (heavy)	29.55	472.80	45.65	730.40		
1 Equip. Oper. (crane)	39.80	318.40	60.00	480.00		
1 Equip. Oper. Oiler	33.90	271.20	51.10	408.80		
2 Dump Trucks, 16 Ton, 12 C.Y.		1060.00		1166.00		
1 Crawler Loader, 4 C.Y.		1145.00		1259.50		
1 S.P. Crane, 4x4, 25 Ton		875.60		963.16	48.13	52.95
64 L.H., Daily Totals		$5140.20		$6544.66	$80.32	$102.26

Crew B-6

Crew No.	Bare Costs Hr.	Bare Costs Daily	Incl. Subs O&P Hr.	Incl. Subs O&P Daily	Cost Per Labor-Hour Bare Costs	Cost Per Labor-Hour Incl. O&P
2 Laborers	$28.75	$460.00	$44.75	$716.00	$31.45	$48.35
1 Equip. Oper. (light)	36.85	294.80	55.55	444.40		
1 Backhoe Loader, 48 H.P.		243.40		267.74	10.14	11.16
24 L.H., Daily Totals		$998.20		$1428.14	$41.59	$59.51

Crew B-6A

Crew No.	Bare Costs Hr.	Bare Costs Daily	Incl. Subs O&P Hr.	Incl. Subs O&P Daily	Cost Per Labor-Hour Bare Costs	Cost Per Labor-Hour Incl. O&P
.5 Labor Foreman (outside)	$30.75	$123.00	$47.90	$191.60	$33.01	$50.62
1 Laborer	28.75	230.00	44.75	358.00		
1 Equip. Oper. (med.)	38.40	307.20	57.85	462.80		
1 Vacuum Trk.,5000 Gal.		352.25		387.48	17.61	19.37
20 L.H., Daily Totals		$1012.45		$1399.88	$50.62	$69.99

Crew B-6B

Crew No.	Bare Costs Hr.	Bare Costs Daily	Incl. Subs O&P Hr.	Incl. Subs O&P Daily	Cost Per Labor-Hour Bare Costs	Cost Per Labor-Hour Incl. O&P
2 Labor Foremen (out)	$30.75	$492.00	$47.90	$766.40	$29.42	$45.80
4 Laborers	28.75	920.00	44.75	1432.00		
1 S.P. Crane, 4x4, 5 Ton		326.80		359.48		
1 Flatbed Truck, Gas, 1.5 Ton		141.40		155.54		
1 Butt Fusion Machine		296.20		325.82	15.93	17.52
48 L.H., Daily Totals		$2176.40		$3039.24	$45.34	$63.32

Crew B-7

Crew No.	Bare Costs Hr.	Bare Costs Daily	Incl. Subs O&P Hr.	Incl. Subs O&P Daily	Cost Per Labor-Hour Bare Costs	Cost Per Labor-Hour Incl. O&P
1 Labor Foreman (outside)	$30.75	$246.00	$47.90	$383.20	$30.69	$47.46
4 Laborers	28.75	920.00	44.75	1432.00		
1 Equip. Oper. (med.)	38.40	307.20	57.85	462.80		
1 Brush Chipper, 12", 130 H.P.		183.60		201.96		
1 Crawler Loader, 3 C.Y.		834.40		917.84		
2 Chainsaws, Gas, 36" Long		68.80		75.68	22.64	24.91
48 L.H., Daily Totals		$2560.00		$3473.48	$53.33	$72.36

Crew B-7A

Crew No.	Bare Costs Hr.	Bare Costs Daily	Incl. Subs O&P Hr.	Incl. Subs O&P Daily	Cost Per Labor-Hour Bare Costs	Cost Per Labor-Hour Incl. O&P
2 Laborers	$28.75	$460.00	$44.75	$716.00	$31.45	$48.35
1 Equip. Oper. (light)	36.85	294.80	55.55	444.40		
1 Rake w/Tractor		195.90		215.49		
2 Chain saws, gas, 18"		46.00		50.60	10.08	11.09
24 L.H., Daily Totals		$996.70		$1426.49	$41.53	$59.44

Crew No.	Bare Costs		Incl. Subs O & P		Cost Per Labor-Hour	

Crew B-8

Crew B-8	Hr.	Daily	Hr.	Daily	Bare Costs	Incl. O&P
1 Labor Foreman (outside)	$30.75	$246.00	$47.90	$383.20	$32.26	$49.44
2 Laborers	28.75	460.00	44.75	716.00		
2 Equip. Oper. (med.)	38.40	614.40	57.85	925.60		
1 Equip. Oper. Oiler	33.90	271.20	51.10	408.80		
2 Truck Drivers (heavy)	29.55	472.80	45.65	730.40		
1 Hyd. Crane, 25 Ton		739.60		813.56		
1 Crawler Loader, 3 C.Y.		834.40		917.84		
2 Dump Trucks, 16 Ton, 12 C.Y.		1060.00		1166.00	41.16	45.27
64 L.H., Daily Totals		$4698.40		$6061.40	$73.41	$94.71

Crew B-9

Crew B-9	Hr.	Daily	Hr.	Daily	Bare Costs	Incl. O&P
1 Labor Foreman (outside)	$30.75	$246.00	$47.90	$383.20	$29.15	$45.38
4 Laborers	28.75	920.00	44.75	1432.00		
1 Air Compressor, 250 C.F.M.		151.40		166.54		
2 Breakers, Pavement, 60 lb.		16.80		18.48		
2 -50' Air Hoses, 1.5"		11.60		12.76	4.50	4.94
40 L.H., Daily Totals		$1345.80		$2012.98	$33.65	$50.32

Crew B-9A

Crew B-9A	Hr.	Daily	Hr.	Daily	Bare Costs	Incl. O&P
2 Laborers	$28.75	$460.00	$44.75	$716.00	$29.02	$45.05
1 Truck Driver (heavy)	29.55	236.40	45.65	365.20		
1 Water Tanker, 5000 Gal.		121.20		133.32		
1 Truck Tractor, 195 H.P.		235.60		259.16		
2 -50' Discharge Hoses, 3"		3.50		3.85	15.01	16.51
24 L.H., Daily Totals		$1056.70		$1477.53	$44.03	$61.56

Crew B-9B

Crew B-9B	Hr.	Daily	Hr.	Daily	Bare Costs	Incl. O&P
2 Laborers	$28.75	$460.00	$44.75	$716.00	$29.02	$45.05
1 Truck Driver (heavy)	29.55	236.40	45.65	365.20		
2 -50' Discharge Hoses, 3"		3.50		3.85		
1 Water Tanker, 5000 Gal.		121.20		133.32		
1 Truck Tractor, 195 H.P.		235.60		259.16		
1 Pressure Washer		55.80		61.38	17.34	19.07
24 L.H., Daily Totals		$1112.50		$1538.91	$46.35	$64.12

Crew B-9C

Crew B-9C	Hr.	Daily	Hr.	Daily	Bare Costs	Incl. O&P
1 Labor Foreman (outside)	$30.75	$246.00	$47.90	$383.20	$29.15	$45.38
4 Laborers	28.75	920.00	44.75	1432.00		
1 Air Compressor, 250 C.F.M.		151.40		166.54		
2 -50' Air Hoses, 1.5"		11.60		12.76		
2 Breakers, Pavement, 60 lb.		16.80		18.48	4.50	4.94
40 L.H., Daily Totals		$1345.80		$2012.98	$33.65	$50.32

Crew B-9D

Crew B-9D	Hr.	Daily	Hr.	Daily	Bare Costs	Incl. O&P
1 Labor Foreman (Outside)	$30.75	$246.00	$47.90	$383.20	$29.15	$45.38
4 Common Laborers	28.75	920.00	44.75	1432.00		
1 Air Compressor, 250 C.F.M.		151.40		166.54		
2 -50' Air Hoses, 1.5"		11.60		12.76		
2 Air Powered Tampers		51.80		56.98	5.37	5.91
40 L.H., Daily Totals		$1380.80		$2051.48	$34.52	$51.29

Crew B-10

Crew B-10	Hr.	Daily	Hr.	Daily	Bare Costs	Incl. O&P
1 Equip. Oper. (med.)	$38.40	$307.20	$57.85	$462.80	$35.18	$53.48
.5 Laborer	28.75	115.00	44.75	179.00		
12 L.H., Daily Totals		$422.20		$641.80	$35.18	$53.48

Crew B-10A

Crew B-10A	Hr.	Daily	Hr.	Daily	Bare Costs	Incl. O&P
1 Equip. Oper. (med.)	$38.40	$307.20	$57.85	$462.80	$35.18	$53.48
.5 Laborer	28.75	115.00	44.75	179.00		
1 Roller, 2-Drum, W.B., 7.5 HP		133.80		147.18	11.15	12.27
12 L.H., Daily Totals		$556.00		$788.98	$46.33	$65.75

Crew B-10B

Crew B-10B	Hr.	Daily	Hr.	Daily	Bare Costs	Incl. O&P
1 Equip. Oper. (med.)	$38.40	$307.20	$57.85	$462.80	$35.18	$53.48
.5 Laborer	28.75	115.00	44.75	179.00		
1 Dozer, 200 H.P.		988.40		1087.24	82.37	90.60
12 L.H., Daily Totals		$1410.60		$1729.04	$117.55	$144.09

Crew B-10C

Crew B-10C	Hr.	Daily	Hr.	Daily	Bare Costs	Incl. O&P
1 Equip. Oper. (med.)	$38.40	$307.20	$57.85	$462.80	$35.18	$53.48
.5 Laborer	28.75	115.00	44.75	179.00		
1 Dozer, 200 H.P.		988.40		1087.24		
1 Vibratory Roller, Towed, 23 Ton		350.00		385.00	111.53	122.69
12 L.H., Daily Totals		$1760.60		$2114.04	$146.72	$176.17

Crew B-10D

Crew B-10D	Hr.	Daily	Hr.	Daily	Bare Costs	Incl. O&P
1 Equip. Oper. (med.)	$38.40	$307.20	$57.85	$462.80	$35.18	$53.48
.5 Laborer	28.75	115.00	44.75	179.00		
1 Dozer, 200 H.P.		988.40		1087.24		
1 Sheepsft. Roller, Towed		376.00		413.60	113.70	125.07
12 L.H., Daily Totals		$1786.60		$2142.64	$148.88	$178.55

Crew B-10E

Crew B-10E	Hr.	Daily	Hr.	Daily	Bare Costs	Incl. O&P
1 Equip. Oper. (med.)	$38.40	$307.20	$57.85	$462.80	$35.18	$53.48
.5 Laborer	28.75	115.00	44.75	179.00		
1 Tandem Roller, 5 Ton		122.20		134.42	10.18	11.20
12 L.H., Daily Totals		$544.40		$776.22	$45.37	$64.69

Crew B-10F

Crew B-10F	Hr.	Daily	Hr.	Daily	Bare Costs	Incl. O&P
1 Equip. Oper. (med.)	$38.40	$307.20	$57.85	$462.80	$35.18	$53.48
.5 Laborer	28.75	115.00	44.75	179.00		
1 Tandem Roller, 10 Ton		211.00		232.10	17.58	19.34
12 L.H., Daily Totals		$633.20		$873.90	$52.77	$72.83

Crew B-10G

Crew B-10G	Hr.	Daily	Hr.	Daily	Bare Costs	Incl. O&P
1 Equip. Oper. (med.)	$38.40	$307.20	$57.85	$462.80	$35.18	$53.48
.5 Laborer	28.75	115.00	44.75	179.00		
1 Sheepsft. Roll., 130 H.P.		956.20		1051.82	79.68	87.65
12 L.H., Daily Totals		$1378.40		$1693.62	$114.87	$141.13

Crew B-10H

Crew B-10H	Hr.	Daily	Hr.	Daily	Bare Costs	Incl. O&P
1 Equip. Oper. (med.)	$38.40	$307.20	$57.85	$462.80	$35.18	$53.48
.5 Laborer	28.75	115.00	44.75	179.00		
1 Diaphragm Water Pump, 2"		57.40		63.14		
1 -20' Suction Hose, 2"		1.95		2.15		
2 -50' Discharge Hoses, 2"		1.80		1.98	5.10	5.61
12 L.H., Daily Totals		$483.35		$709.07	$40.28	$59.09

Crew B-10I

Crew B-10I	Hr.	Daily	Hr.	Daily	Bare Costs	Incl. O&P
1 Equip. Oper. (med.)	$38.40	$307.20	$57.85	$462.80	$35.18	$53.48
.5 Laborer	28.75	115.00	44.75	179.00		
1 Diaphragm Water Pump, 4"		78.20		86.02		
1 -20' Suction Hose, 4"		3.25		3.58		
2 -50' Discharge Hoses, 4"		4.70		5.17	7.18	7.90
12 L.H., Daily Totals		$508.35		$736.57	$42.36	$61.38

Crew B-10J

Crew B-10J	Hr.	Daily	Hr.	Daily	Bare Costs	Incl. O&P
1 Equip. Oper. (med.)	$38.40	$307.20	$57.85	$462.80	$35.18	$53.48
.5 Laborer	28.75	115.00	44.75	179.00		
1 Centrifugal Water Pump, 3"		63.40		69.74		
1 -20' Suction Hose, 3"		3.05		3.36		
2 -50' Discharge Hoses, 3"		3.50		3.85	5.83	6.41
12 L.H., Daily Totals		$492.15		$718.75	$41.01	$59.90

Crew No.	Bare Costs		Incl. Subs O & P		Cost Per Labor-Hour	
Crew B-10K	**Hr.**	**Daily**	**Hr.**	**Daily**	**Bare Costs**	**Incl. O&P**
1 Equip. Oper. (med.)	$38.40	$307.20	$57.85	$462.80	$35.18	$53.48
.5 Laborer	28.75	115.00	44.75	179.00		
1 Centr. Water Pump, 6"		278.20		306.02		
1 -20' Suction Hose, 6"		11.50		12.65		
2 -50' Discharge Hoses, 6"		12.60		13.86	25.19	27.71
12 L.H., Daily Totals		$724.50		$974.33	$60.38	$81.19
Crew B-10L	**Hr.**	**Daily**	**Hr.**	**Daily**	**Bare Costs**	**Incl. O&P**
1 Equip. Oper. (med.)	$38.40	$307.20	$57.85	$462.80	$35.18	$53.48
.5 Laborer	28.75	115.00	44.75	179.00		
1 Dozer, 80 H.P.		354.80		390.28	29.57	32.52
12 L.H., Daily Totals		$777.00		$1032.08	$64.75	$86.01
Crew B-10M	**Hr.**	**Daily**	**Hr.**	**Daily**	**Bare Costs**	**Incl. O&P**
1 Equip. Oper. (med.)	$38.40	$307.20	$57.85	$462.80	$35.18	$53.48
.5 Laborer	28.75	115.00	44.75	179.00		
1 Dozer, 300 H.P.		1301.00		1431.10	108.42	119.26
12 L.H., Daily Totals		$1723.20		$2072.90	$143.60	$172.74
Crew B-10N	**Hr.**	**Daily**	**Hr.**	**Daily**	**Bare Costs**	**Incl. O&P**
1 Equip. Oper. (med.)	$38.40	$307.20	$57.85	$462.80	$35.18	$53.48
.5 Laborer	28.75	115.00	44.75	179.00		
1 F.E. Loader, T.M., 1.5 C.Y		334.00		367.40	27.83	30.62
12 L.H., Daily Totals		$756.20		$1009.20	$63.02	$84.10
Crew B-10O	**Hr.**	**Daily**	**Hr.**	**Daily**	**Bare Costs**	**Incl. O&P**
1 Equip. Oper. (med.)	$38.40	$307.20	$57.85	$462.80	$35.18	$53.48
.5 Laborer	28.75	115.00	44.75	179.00		
1 F.E. Loader, T.M., 2.25 C.Y.		589.20		648.12	49.10	54.01
12 L.H., Daily Totals		$1011.40		$1289.92	$84.28	$107.49
Crew B-10P	**Hr.**	**Daily**	**Hr.**	**Daily**	**Bare Costs**	**Incl. O&P**
1 Equip. Oper. (med.)	$38.40	$307.20	$57.85	$462.80	$35.18	$53.48
.5 Laborer	28.75	115.00	44.75	179.00		
1 Crawler Loader, 3 C.Y.		834.40		917.84	69.53	76.49
12 L.H., Daily Totals		$1256.60		$1559.64	$104.72	$129.97
Crew B-10Q	**Hr.**	**Daily**	**Hr.**	**Daily**	**Bare Costs**	**Incl. O&P**
1 Equip. Oper. (med.)	$38.40	$307.20	$57.85	$462.80	$35.18	$53.48
.5 Laborer	28.75	115.00	44.75	179.00		
1 Crawler Loader, 4 C.Y.		1145.00		1259.50	95.42	104.96
12 L.H., Daily Totals		$1567.20		$1901.30	$130.60	$158.44
Crew B-10R	**Hr.**	**Daily**	**Hr.**	**Daily**	**Bare Costs**	**Incl. O&P**
1 Equip. Oper. (med.)	$38.40	$307.20	$57.85	$462.80	$35.18	$53.48
.5 Laborer	28.75	115.00	44.75	179.00		
1 F.E. Loader, W.M., 1 C.Y.		207.60		228.36	17.30	19.03
12 L.H., Daily Totals		$629.80		$870.16	$52.48	$72.51
Crew B-10S	**Hr.**	**Daily**	**Hr.**	**Daily**	**Bare Costs**	**Incl. O&P**
1 Equip. Oper. (med.)	$38.40	$307.20	$57.85	$462.80	$35.18	$53.48
.5 Laborer	28.75	115.00	44.75	179.00		
1 F.E. Loader, W.M., 1.5 C.Y.		277.60		305.36	23.13	25.45
12 L.H., Daily Totals		$699.80		$947.16	$58.32	$78.93
Crew B-10T	**Hr.**	**Daily**	**Hr.**	**Daily**	**Bare Costs**	**Incl. O&P**
1 Equip. Oper. (med.)	$38.40	$307.20	$57.85	$462.80	$35.18	$53.48
.5 Laborer	28.75	115.00	44.75	179.00		
1 F.E. Loader, W.M.,2.5 C.Y.		349.00		383.90	29.08	31.99
12 L.H., Daily Totals		$771.20		$1025.70	$64.27	$85.47

Crew No.	Bare Costs		Incl. Subs O & P		Cost Per Labor-Hour	
Crew B-10U	**Hr.**	**Daily**	**Hr.**	**Daily**	**Bare Costs**	**Incl. O&P**
1 Equip. Oper. (med.)	$38.40	$307.20	$57.85	$462.80	$35.18	$53.48
.5 Laborer	28.75	115.00	44.75	179.00		
1 F.E. Loader, W.M., 5.5 C.Y.		755.80		831.38	62.98	69.28
12 L.H., Daily Totals		$1178.00		$1473.18	$98.17	$122.77
Crew B-10V	**Hr.**	**Daily**	**Hr.**	**Daily**	**Bare Costs**	**Incl. O&P**
1 Equip. Oper. (med.)	$38.40	$307.20	$57.85	$462.80	$35.18	$53.48
.5 Laborer	28.75	115.00	44.75	179.00		
1 Dozer, 700 H.P.		3406.00		3746.60	283.83	312.22
12 L.H., Daily Totals		$3828.20		$4388.40	$319.02	$365.70
Crew B-10W	**Hr.**	**Daily**	**Hr.**	**Daily**	**Bare Costs**	**Incl. O&P**
1 Equip. Oper. (med.)	$38.40	$307.20	$57.85	$462.80	$35.18	$53.48
.5 Laborer	28.75	115.00	44.75	179.00		
1 Dozer, 105 H.P.		494.80		544.28	41.23	45.36
12 L.H., Daily Totals		$917.00		$1186.08	$76.42	$98.84
Crew B-10X	**Hr.**	**Daily**	**Hr.**	**Daily**	**Bare Costs**	**Incl. O&P**
1 Equip. Oper. (med.)	$38.40	$307.20	$57.85	$462.80	$35.18	$53.48
.5 Laborer	28.75	115.00	44.75	179.00		
1 Dozer, 410 H.P.		1677.00		1844.70	139.75	153.72
12 L.H., Daily Totals		$2099.20		$2486.50	$174.93	$207.21
Crew B-10Y	**Hr.**	**Daily**	**Hr.**	**Daily**	**Bare Costs**	**Incl. O&P**
1 Equip. Oper. (med.)	$38.40	$307.20	$57.85	$462.80	$35.18	$53.48
.5 Laborer	28.75	115.00	44.75	179.00		
1 Vibr. Roller, Towed, 12 Ton		448.40		493.24	37.37	41.10
12 L.H., Daily Totals		$870.60		$1135.04	$72.55	$94.59
Crew B-11A	**Hr.**	**Daily**	**Hr.**	**Daily**	**Bare Costs**	**Incl. O&P**
1 Equipment Oper. (med.)	$38.40	$307.20	$57.85	$462.80	$33.58	$51.30
1 Laborer	28.75	230.00	44.75	358.00		
1 Dozer, 200 H.P.		988.40		1087.24	61.77	67.95
16 L.H., Daily Totals		$1525.60		$1908.04	$95.35	$119.25
Crew B-11B	**Hr.**	**Daily**	**Hr.**	**Daily**	**Bare Costs**	**Incl. O&P**
1 Equipment Oper. (light)	$36.85	$294.80	$55.55	$444.40	$32.80	$50.15
1 Laborer	28.75	230.00	44.75	358.00		
1 Air Powered Tamper		25.90		28.49		
1 Air Compressor, 365 C.F.M.		195.80		215.38		
2 -50' Air Hoses, 1.5"		11.60		12.76	14.58	16.04
16 L.H., Daily Totals		$758.10		$1059.03	$47.38	$66.19
Crew B-11C	**Hr.**	**Daily**	**Hr.**	**Daily**	**Bare Costs**	**Incl. O&P**
1 Equipment Oper. (med.)	$38.40	$307.20	$57.85	$462.80	$33.58	$51.30
1 Laborer	28.75	230.00	44.75	358.00		
1 Backhoe Loader, 48 H.P.		243.40		267.74	15.21	16.73
16 L.H., Daily Totals		$780.60		$1088.54	$48.79	$68.03
Crew B-11J	**Hr.**	**Daily**	**Hr.**	**Daily**	**Bare Costs**	**Incl. O&P**
1 Equipment Oper. (med.)	$38.40	$307.20	$57.85	$462.80	$33.58	$51.30
1 Laborer	28.75	230.00	44.75	358.00		
1 Grader, 30,000 Lbs.		505.40		555.94		
1 Ripper, beam & 1 shank		76.40		84.04	36.36	40.00
16 L.H., Daily Totals		$1119.00		$1460.78	$69.94	$91.30

413

Crews

Crew No.	Bare Costs		Incl. Subs O & P		Cost Per Labor-Hour	

Crew B-11K	Hr.	Daily	Hr.	Daily	Bare Costs	Incl. O&P
1 Equipment Oper. (med.)	$38.40	$307.20	$57.85	$462.80	$33.58	$51.30
1 Laborer	28.75	230.00	44.75	358.00		
1 Trencher, Chain Type, 8' D		1606.00		1766.60	100.38	110.41
16 L.H., Daily Totals		$2143.20		$2587.40	$133.95	$161.71

Crew B-11L	Hr.	Daily	Hr.	Daily	Bare Costs	Incl. O&P
1 Equipment Oper. (med.)	$38.40	$307.20	$57.85	$462.80	$33.58	$51.30
1 Laborer	28.75	230.00	44.75	358.00		
1 Grader, 30,000 Lbs.		505.40		555.94	31.59	34.75
16 L.H., Daily Totals		$1042.60		$1376.74	$65.16	$86.05

Crew B-11M	Hr.	Daily	Hr.	Daily	Bare Costs	Incl. O&P
1 Equipment Oper. (med.)	$38.40	$307.20	$57.85	$462.80	$33.58	$51.30
1 Laborer	28.75	230.00	44.75	358.00		
1 Backhoe Loader, 80 H.P.		285.40		313.94	17.84	19.62
16 L.H., Daily Totals		$822.60		$1134.74	$51.41	$70.92

Crew B-11N	Hr.	Daily	Hr.	Daily	Bare Costs	Incl. O&P
1 Labor Foreman	$30.75	$246.00	$47.90	$383.20	$31.65	$48.61
2 Equipment Operators (med.)	38.40	614.40	57.85	925.60		
6 Truck Drivers (hvy.)	29.55	1418.40	45.65	2191.20		
1 F.E. Loader, W.M., 5.5 C.Y.		755.80		831.38		
1 Dozer, 410 H.P.		1677.00		1844.70		
6 Dump Trucks, Off Hwy., 50 Ton		7848.00		8632.80	142.79	157.07
72 L.H., Daily Totals		$12559.60		$14808.88	$174.44	$205.68

Crew B-11Q	Hr.	Daily	Hr.	Daily	Bare Costs	Incl. O&P
1 Equipment Operator (med.)	$38.40	$307.20	$57.85	$462.80	$35.18	$53.48
.5 Laborer	28.75	115.00	44.75	179.00		
1 Dozer, 140 H.P.		611.40		672.54	50.95	56.05
12 L.H., Daily Totals		$1033.60		$1314.34	$86.13	$109.53

Crew B-11R	Hr.	Daily	Hr.	Daily	Bare Costs	Incl. O&P
1 Equipment Operator (med.)	$38.40	$307.20	$57.85	$462.80	$35.18	$53.48
.5 Laborer	28.75	115.00	44.75	179.00		
1 Dozer, 200 H.P.		988.40		1087.24	82.37	90.60
12 L.H., Daily Totals		$1410.60		$1729.04	$117.55	$144.09

Crew B-11S	Hr.	Daily	Hr.	Daily	Bare Costs	Incl. O&P
1 Equipment Operator (med.)	$38.40	$307.20	$57.85	$462.80	$35.18	$53.48
.5 Laborer	28.75	115.00	44.75	179.00		
1 Dozer, 300 H.P.		1301.00		1431.10		
1 Ripper, Beam & 1 Shank		76.40		84.04	114.78	126.26
12 L.H., Daily Totals		$1799.60		$2156.94	$149.97	$179.75

Crew B-11T	Hr.	Daily	Hr.	Daily	Bare Costs	Incl. O&P
1 Equipment Operator (med.)	$38.40	$307.20	$57.85	$462.80	$35.18	$53.48
.5 Laborer	28.75	115.00	44.75	179.00		
1 Dozer, 410 H.P.		1677.00		1844.70		
1 Ripper, Beam & 2 Shanks		85.20		93.72	146.85	161.54
12 L.H., Daily Totals		$2184.40		$2580.22	$182.03	$215.02

Crew B-11U	Hr.	Daily	Hr.	Daily	Bare Costs	Incl. O&P
1 Equipment Operator (med.)	$38.40	$307.20	$57.85	$462.80	$35.18	$53.48
.5 Laborer	28.75	115.00	44.75	179.00		
1 Dozer, 520 H.P.		2260.00		2486.00	188.33	207.17
12 L.H., Daily Totals		$2682.20		$3127.80	$223.52	$260.65

Crew B-11V	Hr.	Daily	Hr.	Daily	Bare Costs	Incl. O&P
3 Laborers	$28.75	$690.00	$44.75	$1074.00	$28.75	$44.75
1 Roller, 2-Drum, W.B., 7.5 HP		133.80		147.18	5.58	6.13
24 L.H., Daily Totals		$823.80		$1221.18	$34.33	$50.88

Crew B-11W	Hr.	Daily	Hr.	Daily	Bare Costs	Incl. O&P
1 Equipment Operator (med.)	$38.40	$307.20	$57.85	$462.80	$30.22	$46.59
1 Common Laborer	28.75	230.00	44.75	358.00		
10 Truck Drivers (hvy.)	29.55	2364.00	45.65	3652.00		
1 Dozer, 200 H.P.		988.40		1087.24		
1 Vibratory Roller, Towed, 23 Ton		350.00		385.00		
10 Dump Trucks, 12 Ton, 8 C.Y.		3684.00		4052.40	52.32	57.55
96 L.H., Daily Totals		$7923.60		$9997.44	$82.54	$104.14

Crew B-11Y	Hr.	Daily	Hr.	Daily	Bare Costs	Incl. O&P
1 Labor Foreman (Outside)	$30.75	$246.00	$47.90	$383.20	$32.19	$49.47
5 Common Laborers	28.75	1150.00	44.75	1790.00		
3 Equipment Operators (med.)	38.40	921.60	57.85	1388.40		
1 Dozer, 80 H.P.		354.80		390.28		
2 Roller, 2-Drum, W.B., 7.5 HP		267.60		294.36		
4 Vibratory Plates, Gas, 21"		144.80		159.28	10.66	11.72
72 L.H., Daily Totals		$3084.80		$4405.52	$42.84	$61.19

Crew B-12A	Hr.	Daily	Hr.	Daily	Bare Costs	Incl. O&P
1 Equip. Oper. (crane)	$39.80	$318.40	$60.00	$480.00	$34.27	$52.38
1 Laborer	28.75	230.00	44.75	358.00		
1 Hyd. Excavator, 1 C.Y.		601.80		661.98	37.61	41.37
16 L.H., Daily Totals		$1150.20		$1499.98	$71.89	$93.75

Crew B-12B	Hr.	Daily	Hr.	Daily	Bare Costs	Incl. O&P
1 Equip. Oper. (crane)	$39.80	$318.40	$60.00	$480.00	$34.27	$52.38
1 Laborer	28.75	230.00	44.75	358.00		
1 Hyd. Excavator, 1.5 C.Y.		775.60		853.16	48.48	53.32
16 L.H., Daily Totals		$1324.00		$1691.16	$82.75	$105.70

Crew B-12C	Hr.	Daily	Hr.	Daily	Bare Costs	Incl. O&P
1 Equip. Oper. (crane)	$39.80	$318.40	$60.00	$480.00	$34.27	$52.38
1 Laborer	28.75	230.00	44.75	358.00		
1 Hyd. Excavator, 2 C.Y.		994.60		1094.06	62.16	68.38
16 L.H., Daily Totals		$1543.00		$1932.06	$96.44	$120.75

Crew B-12D	Hr.	Daily	Hr.	Daily	Bare Costs	Incl. O&P
1 Equip. Oper. (crane)	$39.80	$318.40	$60.00	$480.00	$34.27	$52.38
1 Laborer	28.75	230.00	44.75	358.00		
1 Hyd. Excavator, 3.5 C.Y.		2166.00		2382.60	135.38	148.91
16 L.H., Daily Totals		$2714.40		$3220.60	$169.65	$201.29

Crew B-12E	Hr.	Daily	Hr.	Daily	Bare Costs	Incl. O&P
1 Equip. Oper. (crane)	$39.80	$318.40	$60.00	$480.00	$34.27	$52.38
1 Laborer	28.75	230.00	44.75	358.00		
1 Hyd. Excavator, .5 C.Y.		358.00		393.80	22.38	24.61
16 L.H., Daily Totals		$906.40		$1231.80	$56.65	$76.99

Crew B-12F	Hr.	Daily	Hr.	Daily	Bare Costs	Incl. O&P
1 Equip. Oper. (crane)	$39.80	$318.40	$60.00	$480.00	$34.27	$52.38
1 Laborer	28.75	230.00	44.75	358.00		
1 Hyd. Excavator, .75 C.Y.		516.40		568.04	32.27	35.50
16 L.H., Daily Totals		$1064.80		$1406.04	$66.55	$87.88

Crew No.	Bare Costs		Incl. Subs O & P		Cost Per Labor-Hour	
Crew B-12G	**Hr.**	**Daily**	**Hr.**	**Daily**	**Bare Costs**	**Incl. O&P**
1 Equip. Oper. (crane)	$39.80	$318.40	$60.00	$480.00	$34.27	$52.38
1 Laborer	28.75	230.00	44.75	358.00		
1 Crawler Crane, 15 Ton		576.65		634.32		
1 Clamshell Bucket, .5 C.Y.		34.20		37.62	38.18	42.00
16 L.H., Daily Totals		$1159.25		$1509.93	$72.45	$94.37
Crew B-12H	**Hr.**	**Daily**	**Hr.**	**Daily**	**Bare Costs**	**Incl. O&P**
1 Equip. Oper. (crane)	$39.80	$318.40	$60.00	$480.00	$34.27	$52.38
1 Laborer	28.75	230.00	44.75	358.00		
1 Crawler Crane, 25 Ton		983.00		1081.30		
1 Clamshell Bucket, 1 C.Y.		44.80		49.28	64.24	70.66
16 L.H., Daily Totals		$1576.20		$1968.58	$98.51	$123.04
Crew B-12I	**Hr.**	**Daily**	**Hr.**	**Daily**	**Bare Costs**	**Incl. O&P**
1 Equip. Oper. (crane)	$39.80	$318.40	$60.00	$480.00	$34.27	$52.38
1 Laborer	28.75	230.00	44.75	358.00		
1 Crawler Crane, 20 Ton		737.50		811.25		
1 Dragline Bucket, .75 C.Y.		19.40		21.34	47.31	52.04
16 L.H., Daily Totals		$1305.30		$1670.59	$81.58	$104.41
Crew B-12J	**Hr.**	**Daily**	**Hr.**	**Daily**	**Bare Costs**	**Incl. O&P**
1 Equip. Oper. (crane)	$39.80	$318.40	$60.00	$480.00	$34.27	$52.38
1 Laborer	28.75	230.00	44.75	358.00		
1 Gradall, 5/8 C.Y.		905.80		996.38	56.61	62.27
16 L.H., Daily Totals		$1454.20		$1834.38	$90.89	$114.65
Crew B-12K	**Hr.**	**Daily**	**Hr.**	**Daily**	**Bare Costs**	**Incl. O&P**
1 Equip. Oper. (crane)	$39.80	$318.40	$60.00	$480.00	$34.27	$52.38
1 Laborer	28.75	230.00	44.75	358.00		
1 Gradall, 3 Ton, 1 C.Y.		1026.00		1128.60	64.13	70.54
16 L.H., Daily Totals		$1574.40		$1966.60	$98.40	$122.91
Crew B-12L	**Hr.**	**Daily**	**Hr.**	**Daily**	**Bare Costs**	**Incl. O&P**
1 Equip. Oper. (crane)	$39.80	$318.40	$60.00	$480.00	$34.27	$52.38
1 Laborer	28.75	230.00	44.75	358.00		
1 Crawler Crane, 15 Ton		576.65		634.32		
1 F.E. Attachment, .5 C.Y.		49.20		54.12	39.12	43.03
16 L.H., Daily Totals		$1174.25		$1526.43	$73.39	$95.40
Crew B-12M	**Hr.**	**Daily**	**Hr.**	**Daily**	**Bare Costs**	**Incl. O&P**
1 Equip. Oper. (crane)	$39.80	$318.40	$60.00	$480.00	$34.27	$52.38
1 Laborer	28.75	230.00	44.75	358.00		
1 Crawler Crane, 20 Ton		737.50		811.25		
1 F.E. Attachment, .75 C.Y.		53.60		58.96	49.44	54.39
16 L.H., Daily Totals		$1339.50		$1708.21	$83.72	$106.76
Crew B-12N	**Hr.**	**Daily**	**Hr.**	**Daily**	**Bare Costs**	**Incl. O&P**
1 Equip. Oper. (crane)	$39.80	$318.40	$60.00	$480.00	$34.27	$52.38
1 Laborer	28.75	230.00	44.75	358.00		
1 Crawler Crane, 25 Ton		983.00		1081.30		
1 F.E. Attachment, 1 C.Y.		60.40		66.44	65.21	71.73
16 L.H., Daily Totals		$1591.80		$1985.74	$99.49	$124.11
Crew B-120	**Hr.**	**Daily**	**Hr.**	**Daily**	**Bare Costs**	**Incl. O&P**
1 Equip. Oper. (crane)	$39.80	$318.40	$60.00	$480.00	$34.27	$52.38
1 Laborer	28.75	230.00	44.75	358.00		
1 Crawler Crane, 40 Ton		994.60		1094.06		
1 F.E. Attachment, 1.5 C.Y.		69.60		76.56	66.51	73.16
16 L.H., Daily Totals		$1612.60		$2008.62	$100.79	$125.54

Crew No.	Bare Costs		Incl. Subs O & P		Cost Per Labor-Hour	
Crew B-12P	**Hr.**	**Daily**	**Hr.**	**Daily**	**Bare Costs**	**Incl. O&P**
1 Equip. Oper. (crane)	$39.80	$318.40	$60.00	$480.00	$34.27	$52.38
1 Laborer	28.75	230.00	44.75	358.00		
1 Crawler Crane, 40 Ton		994.60		1094.06		
1 Dragline Bucket, 1.5 C.Y.		31.00		34.10	64.10	70.51
16 L.H., Daily Totals		$1574.00		$1966.16	$98.38	$122.89
Crew B-12Q	**Hr.**	**Daily**	**Hr.**	**Daily**	**Bare Costs**	**Incl. O&P**
1 Equip. Oper. (crane)	$39.80	$318.40	$60.00	$480.00	$34.27	$52.38
1 Laborer	28.75	230.00	44.75	358.00		
1 Hyd. Excavator, 5/8 C.Y.		462.20		508.42	28.89	31.78
16 L.H., Daily Totals		$1010.60		$1346.42	$63.16	$84.15
Crew B-12R	**Hr.**	**Daily**	**Hr.**	**Daily**	**Bare Costs**	**Incl. O&P**
1 Equip. Oper. (crane)	$39.80	$318.40	$60.00	$480.00	$34.27	$52.38
1 Laborer	28.75	230.00	44.75	358.00		
1 Hyd. Excavator, 1.5 C.Y.		775.60		853.16	48.48	53.32
16 L.H., Daily Totals		$1324.00		$1691.16	$82.75	$105.70
Crew B-12S	**Hr.**	**Daily**	**Hr.**	**Daily**	**Bare Costs**	**Incl. O&P**
1 Equip. Oper. (crane)	$39.80	$318.40	$60.00	$480.00	$34.27	$52.38
1 Laborer	28.75	230.00	44.75	358.00		
1 Hyd. Excavator, 2.5 C.Y.		1333.00		1466.30	83.31	91.64
16 L.H., Daily Totals		$1881.40		$2304.30	$117.59	$144.02
Crew B-12T	**Hr.**	**Daily**	**Hr.**	**Daily**	**Bare Costs**	**Incl. O&P**
1 Equip. Oper. (crane)	$39.80	$318.40	$60.00	$480.00	$34.27	$52.38
1 Laborer	28.75	230.00	44.75	358.00		
1 Crawler Crane, 75 Ton		1384.00		1522.40		
1 F.E. Attachment, 3 C.Y.		93.00		102.30	92.31	101.54
16 L.H., Daily Totals		$2025.40		$2462.70	$126.59	$153.92
Crew B-12V	**Hr.**	**Daily**	**Hr.**	**Daily**	**Bare Costs**	**Incl. O&P**
1 Equip. Oper. (crane)	$39.80	$318.40	$60.00	$480.00	$34.27	$52.38
1 Laborer	28.75	230.00	44.75	358.00		
1 Crawler Crane, 75 Ton		1384.00		1522.40		
1 Dragline Bucket, 3 C.Y.		48.80		53.68	89.55	98.50
16 L.H., Daily Totals		$1981.20		$2414.08	$123.83	$150.88
Crew B-13	**Hr.**	**Daily**	**Hr.**	**Daily**	**Bare Costs**	**Incl. O&P**
1 Labor Foreman (outside)	$30.75	$246.00	$47.90	$383.20	$31.35	$48.29
4 Laborers	28.75	920.00	44.75	1432.00		
1 Equip. Oper. (crane)	39.80	318.40	60.00	480.00		
1 Equip. Oper. Oiler	33.90	271.20	51.10	408.80		
1 Hyd. Crane, 25 Ton		739.60		813.56	13.21	14.53
56 L.H., Daily Totals		$2495.20		$3517.56	$44.56	$62.81
Crew B-13A	**Hr.**	**Daily**	**Hr.**	**Daily**	**Bare Costs**	**Incl. O&P**
1 Foreman	$30.75	$246.00	$47.90	$383.20	$32.02	$49.20
2 Laborers	28.75	460.00	44.75	716.00		
2 Equipment Operators	38.40	614.40	57.85	925.60		
2 Truck Drivers (heavy)	29.55	472.80	45.65	730.40		
1 Crawler Crane, 75 Ton		1384.00		1522.40		
1 Crawler Loader, 4 C.Y.		1145.00		1259.50		
2 Dump Trucks, 12 Ton, 8 C.Y.		736.80		810.48	58.32	64.15
56 L.H., Daily Totals		$5059.00		$6347.58	$90.34	$113.35

Crews

Crew No.	Bare Costs		Incl. Subs O & P		Cost Per Labor-Hour	

Crew B-13B

Crew B-13B	Hr.	Daily	Hr.	Daily	Bare Costs	Incl. O&P
1 Labor Foreman (outside)	$30.75	$246.00	$47.90	$383.20	$31.35	$48.29
4 Laborers	28.75	920.00	44.75	1432.00		
1 Equip. Oper. (crane)	39.80	318.40	60.00	480.00		
1 Equip. Oper. Oiler	33.90	271.20	51.10	408.80		
1 Hyd. Crane, 55 Ton		1060.00		1166.00	18.93	20.82
56 L.H., Daily Totals		$2815.60		$3870.00	$50.28	$69.11

Crew B-13C	Hr.	Daily	Hr.	Daily	Bare Costs	Incl. O&P
1 Labor Foreman (outside)	$30.75	$246.00	$47.90	$383.20	$31.35	$48.29
4 Laborers	28.75	920.00	44.75	1432.00		
1 Equip. Oper. (crane)	39.80	318.40	60.00	480.00		
1 Equip. Oper. Oiler	33.90	271.20	51.10	408.80		
1 Crawler Crane, 100 Ton		1717.00		1888.70	30.66	33.73
56 L.H., Daily Totals		$3472.60		$4592.70	$62.01	$82.01

Crew B-13D	Hr.	Daily	Hr.	Daily	Bare Costs	Incl. O&P
1 Laborer	$28.75	$230.00	$44.75	$358.00	$34.27	$52.38
1 Equip. Oper. (crane)	39.80	318.40	60.00	480.00		
1 Hyd. Excavator, 1 C.Y.		601.80		661.98		
1 Trench Box		79.25		87.17	42.57	46.82
16 L.H., Daily Totals		$1229.45		$1587.16	$76.84	$99.20

Crew B-13E	Hr.	Daily	Hr.	Daily	Bare Costs	Incl. O&P
1 Laborer	$28.75	$230.00	$44.75	$358.00	$34.27	$52.38
1 Equip. Oper. (crane)	39.80	318.40	60.00	480.00		
1 Hyd. Excavator, 1.5 C.Y.		775.60		853.16		
1 Trench Box		79.25		87.17	53.43	58.77
16 L.H., Daily Totals		$1403.25		$1778.34	$87.70	$111.15

Crew B-13F	Hr.	Daily	Hr.	Daily	Bare Costs	Incl. O&P
1 Laborer	$28.75	$230.00	$44.75	$358.00	$34.27	$52.38
1 Equip. Oper. (crane)	39.80	318.40	60.00	480.00		
1 Hyd. Excavator, 3.5 C.Y.		2166.00		2382.60		
1 Trench Box		79.25		87.17	140.33	154.36
16 L.H., Daily Totals		$2793.65		$3307.78	$174.60	$206.74

Crew B-13G	Hr.	Daily	Hr.	Daily	Bare Costs	Incl. O&P
1 Laborer	$28.75	$230.00	$44.75	$358.00	$34.27	$52.38
1 Equip. Oper. (crane)	39.80	318.40	60.00	480.00		
1 Hyd. Excavator, .75 C.Y.		516.40		568.04		
1 Trench Box		79.25		87.17	37.23	40.95
16 L.H., Daily Totals		$1144.05		$1493.21	$71.50	$93.33

Crew B-13H	Hr.	Daily	Hr.	Daily	Bare Costs	Incl. O&P
1 Laborer	$28.75	$230.00	$44.75	$358.00	$34.27	$52.38
1 Equip. Oper. (crane)	39.80	318.40	60.00	480.00		
1 Gradall, 5/8 C.Y.		905.80		996.38		
1 Trench Box		79.25		87.17	61.57	67.72
16 L.H., Daily Totals		$1533.45		$1921.56	$95.84	$120.10

Crew B-13I	Hr.	Daily	Hr.	Daily	Bare Costs	Incl. O&P
1 Laborer	$28.75	$230.00	$44.75	$358.00	$34.27	$52.38
1 Equip. Oper. (crane)	39.80	318.40	60.00	480.00		
1 Gradall, 3 Ton, 1 C.Y.		1026.00		1128.60		
1 Trench Box		79.25		87.17	69.08	75.99
16 L.H., Daily Totals		$1653.65		$2053.78	$103.35	$128.36

Crew B-13J	Hr.	Daily	Hr.	Daily	Bare Costs	Incl. O&P
1 Laborer	$28.75	$230.00	$44.75	$358.00	$34.27	$52.38
1 Equip. Oper. (crane)	39.80	318.40	60.00	480.00		
1 Hyd. Excavator, 2.5 C.Y.		1333.00		1466.30		
1 Trench Box		79.25		87.17	88.27	97.09
16 L.H., Daily Totals		$1960.65		$2391.47	$122.54	$149.47

Crew B-14	Hr.	Daily	Hr.	Daily	Bare Costs	Incl. O&P
1 Labor Foreman (outside)	$30.75	$246.00	$47.90	$383.20	$30.43	$47.08
4 Laborers	28.75	920.00	44.75	1432.00		
1 Equip. Oper. (light)	36.85	294.80	55.55	444.40		
1 Backhoe Loader, 48 H.P.		243.40		267.74	5.07	5.58
48 L.H., Daily Totals		$1704.20		$2527.34	$35.50	$52.65

Crew B-15	Hr.	Daily	Hr.	Daily	Bare Costs	Incl. O&P
1 Equipment Oper. (med)	$38.40	$307.20	$57.85	$462.80	$31.96	$49.01
.5 Laborer	28.75	115.00	44.75	179.00		
2 Truck Drivers (heavy)	29.55	472.80	45.65	730.40		
2 Dump Trucks, 16 Ton, 12 C.Y.		1060.00		1166.00		
1 Dozer, 200 H.P.		988.40		1087.24	73.16	80.47
28 L.H., Daily Totals		$2943.40		$3625.44	$105.12	$129.48

Crew B-16	Hr.	Daily	Hr.	Daily	Bare Costs	Incl. O&P
1 Labor Foreman (outside)	$30.75	$246.00	$47.90	$383.20	$29.45	$45.76
2 Laborers	28.75	460.00	44.75	716.00		
1 Truck Driver (heavy)	29.55	236.40	45.65	365.20		
1 Dump Truck, 16 Ton, 12 C.Y.		530.00		583.00	16.56	18.22
32 L.H., Daily Totals		$1472.40		$2047.40	$46.01	$63.98

Crew B-17	Hr.	Daily	Hr.	Daily	Bare Costs	Incl. O&P
2 Laborers	$28.75	$460.00	$44.75	$716.00	$30.98	$47.67
1 Equip. Oper. (light)	36.85	294.80	55.55	444.40		
1 Truck Driver (heavy)	29.55	236.40	45.65	365.20		
1 Backhoe Loader, 48 H.P.		243.40		267.74		
1 Dump Truck, 12 Ton, 8 C.Y.		368.40		405.24	19.12	21.03
32 L.H., Daily Totals		$1603.00		$2198.58	$50.09	$68.71

Crew B-17A	Hr.	Daily	Hr.	Daily	Bare Costs	Incl. O&P
2 Laborer Foremen	$30.75	$492.00	$47.90	$766.40	$31.20	$48.57
6 Laborers	28.75	1380.00	44.75	2148.00		
1 Skilled Worker Foreman	40.00	320.00	62.25	498.00		
1 Skilled Worker	38.00	304.00	59.15	473.20		
80 L.H., Daily Totals		$2496.00		$3885.60	$31.20	$48.57

Crew B-18	Hr.	Daily	Hr.	Daily	Bare Costs	Incl. O&P
1 Labor Foreman (outside)	$30.75	$246.00	$47.90	$383.20	$29.42	$45.80
2 Laborers	28.75	460.00	44.75	716.00		
1 Vibratory Plate, gas, 21"		36.20		39.82	1.51	1.66
24 L.H., Daily Totals		$742.20		$1139.02	$30.93	$47.46

Crew B-19	Hr.	Daily	Hr.	Daily	Bare Costs	Incl. O&P
1 Pile Driver Foreman	$37.90	$303.20	$62.05	$496.40	$36.88	$58.52
4 Pile Drivers	35.90	1148.80	58.75	1880.00		
2 Equip. Oper. (crane)	39.80	636.80	60.00	960.00		
1 Equip. Oper. Oiler	33.90	271.20	51.10	408.80		
1 Crawler Crane, 40 Ton		994.60		1094.06		
60 L.F. of Leads, 15K Ft. Lbs.		60.00		66.00		
1 Hammer, Diesel, 22k Ft-Lb		595.20		654.72	25.78	28.36
64 L.H., Daily Totals		$4009.80		$5559.98	$62.65	$86.87

Crew B-19A

Crew No.	Bare Costs		Incl. Subs O & P		Cost Per Labor-Hour	
Crew B-19A	Hr.	Daily	Hr.	Daily	Bare Costs	Incl. O&P
1 Pile Driver Foreman	$37.90	$303.20	$62.05	$496.40	$36.88	$58.52
4 Pile Drivers	35.90	1148.80	58.75	1880.00		
2 Equip. Oper. (crane)	39.80	636.80	60.00	960.00		
1 Equip. Oper. Oiler	33.90	271.20	51.10	408.80		
1 Crawler Crane, 75 Ton		1384.00		1522.40		
60 L.F. Leads, 25K Ft. Lbs.		108.00		118.80		
1 Hammer, Diesel, 41k Ft-Lb		715.60		787.16	34.49	37.94
64 L.H., Daily Totals		$4567.60		$6173.56	$71.37	$96.46

Crew B-20

Crew B-20	Hr.	Daily	Hr.	Daily	Bare Costs	Incl. O&P
1 Labor Foreman (out)	$30.75	$246.00	$47.90	$383.20	$32.50	$50.60
1 Skilled Worker	38.00	304.00	59.15	473.20		
1 Laborer	28.75	230.00	44.75	358.00		
24 L.H., Daily Totals		$780.00		$1214.40	$32.50	$50.60

Crew B-20A

Crew B-20A	Hr.	Daily	Hr.	Daily	Bare Costs	Incl. O&P
1 Labor Foreman	$30.75	$246.00	$47.90	$383.20	$35.04	$53.49
1 Laborer	28.75	230.00	44.75	358.00		
1 Plumber	44.80	358.40	67.40	539.20		
1 Plumber Apprentice	35.85	286.80	53.90	431.20		
32 L.H., Daily Totals		$1121.20		$1711.60	$35.04	$53.49

Crew B-21

Crew B-21	Hr.	Daily	Hr.	Daily	Bare Costs	Incl. O&P
1 Labor Foreman (out)	$30.75	$246.00	$47.90	$383.20	$33.54	$51.94
1 Skilled Worker	38.00	304.00	59.15	473.20		
1 Laborer	28.75	230.00	44.75	358.00		
.5 Equip. Oper. (crane)	39.80	159.20	60.00	240.00		
.5 S.P. Crane, 4x4, 5 Ton		163.40		179.74	5.84	6.42
28 L.H., Daily Totals		$1102.60		$1634.14	$39.38	$58.36

Crew B-21A

Crew B-21A	Hr.	Daily	Hr.	Daily	Bare Costs	Incl. O&P
1 Labor Foreman	$30.75	$246.00	$47.90	$383.20	$35.99	$54.79
1 Laborer	28.75	230.00	44.75	358.00		
1 Plumber	44.80	358.40	67.40	539.20		
1 Plumber Apprentice	35.85	286.80	53.90	431.20		
1 Equip. Oper. (crane)	39.80	318.40	60.00	480.00		
1 S.P. Crane, 4x4, 12 Ton		575.00		632.50	14.38	15.81
40 L.H., Daily Totals		$2014.60		$2824.10	$50.37	$70.60

Crew B-21B

Crew B-21B	Hr.	Daily	Hr.	Daily	Bare Costs	Incl. O&P
1 Laborer Foreman	$30.75	$246.00	$47.90	$383.20	$31.36	$48.43
3 Laborers	28.75	690.00	44.75	1074.00		
1 Equip. Oper. (crane)	39.80	318.40	60.00	480.00		
1 Hyd. Crane, 12 Ton		724.00		796.40	18.10	19.91
40 L.H., Daily Totals		$1978.40		$2733.60	$49.46	$68.34

Crew B-21C

Crew B-21C	Hr.	Daily	Hr.	Daily	Bare Costs	Incl. O&P
1 Laborer Foreman	$30.75	$246.00	$47.90	$383.20	$31.35	$48.29
4 Laborers	28.75	920.00	44.75	1432.00		
1 Equip. Oper. (crane)	39.80	318.40	60.00	480.00		
1 Equip. Oper. Oiler	33.90	271.20	51.10	408.80		
2 Cutting Torches		34.00		37.40		
2 Gases		129.60		142.56		
1 Lattice Boom Crane, 90 Ton		1547.00		1701.70	30.55	33.60
56 L.H., Daily Totals		$3466.20		$4585.66	$61.90	$81.89

Crew B-22

Crew B-22	Hr.	Daily	Hr.	Daily	Bare Costs	Incl. O&P
1 Labor Foreman (out)	$30.75	$246.00	$47.90	$383.20	$33.96	$52.48
1 Skilled Worker	38.00	304.00	59.15	473.20		
1 Laborer	28.75	230.00	44.75	358.00		
.75 Equip. Oper. (crane)	39.80	238.80	60.00	360.00		
.75 S.P. Crane, 4x4, 5 Ton		245.10		269.61	8.17	8.99
30 L.H., Daily Totals		$1263.90		$1844.01	$42.13	$61.47

Crew B-22A

Crew B-22A	Hr.	Daily	Hr.	Daily	Bare Costs	Incl. O&P
1 Labor Foreman (out)	$30.75	$246.00	$47.90	$383.20	$32.86	$50.85
1 Skilled Worker	38.00	304.00	59.15	473.20		
2 Laborers	28.75	460.00	44.75	716.00		
.75 Equipment Oper. (crane)	39.80	238.80	60.00	360.00		
.75 S.P. Crane, 4x4, 5 Ton		245.10		269.61		
1 Generator, 5 KW		35.40		38.94		
1 Butt Fusion Machine		296.20		325.82	15.18	16.69
38 L.H., Daily Totals		$1825.50		$2566.77	$48.04	$67.55

Crew B-22B

Crew B-22B	Hr.	Daily	Hr.	Daily	Bare Costs	Incl. O&P
1 Skilled Worker	$38.00	$304.00	$59.15	$473.20	$33.38	$51.95
1 Laborer	28.75	230.00	44.75	358.00		
1 Electro Fusion Machine		174.20		191.62	10.89	11.98
16 L.H., Daily Totals		$708.20		$1022.82	$44.26	$63.93

Crew B-23

Crew B-23	Hr.	Daily	Hr.	Daily	Bare Costs	Incl. O&P
1 Labor Foreman (outside)	$30.75	$246.00	$47.90	$383.20	$29.15	$45.38
4 Laborers	28.75	920.00	44.75	1432.00		
1 Drill Rig, Truck-Mounted		3499.00		3848.90		
1 Flatbed Truck, gas, 3 Ton		185.20		203.72	92.11	101.32
40 L.H., Daily Totals		$4850.20		$5867.82	$121.26	$146.70

Crew B-23A

Crew B-23A	Hr.	Daily	Hr.	Daily	Bare Costs	Incl. O&P
1 Labor Foreman (outside)	$30.75	$246.00	$47.90	$383.20	$32.63	$50.17
1 Laborer	28.75	230.00	44.75	358.00		
1 Equip. Operator (medium)	38.40	307.20	57.85	462.80		
1 Drill Rig, Truck-Mounted		3499.00		3848.90		
1 Pickup Truck, 3/4 Ton		84.80		93.28	149.32	164.26
24 L.H., Daily Totals		$4367.00		$5146.18	$181.96	$214.42

Crew B-23B

Crew B-23B	Hr.	Daily	Hr.	Daily	Bare Costs	Incl. O&P
1 Labor Foreman (outside)	$30.75	$246.00	$47.90	$383.20	$32.63	$50.17
1 Laborer	28.75	230.00	44.75	358.00		
1 Equip. Operator (medium)	38.40	307.20	57.85	462.80		
1 Drill Rig, Truck-Mounted		3499.00		3848.90		
1 Pickup Truck, 3/4 Ton		84.80		93.28		
1 Centr. Water Pump, 6"		278.20		306.02	160.92	177.01
24 L.H., Daily Totals		$4645.20		$5452.20	$193.55	$227.18

Crew B-24

Crew B-24	Hr.	Daily	Hr.	Daily	Bare Costs	Incl. O&P
1 Cement Finisher	$35.55	$284.40	$52.20	$417.60	$33.67	$51.37
1 Laborer	28.75	230.00	44.75	358.00		
1 Carpenter	36.70	293.60	57.15	457.20		
24 L.H., Daily Totals		$808.00		$1232.80	$33.67	$51.37

Crew B-25

Crew B-25	Hr.	Daily	Hr.	Daily	Bare Costs	Incl. O&P
1 Labor Foreman	$30.75	$246.00	$47.90	$383.20	$31.56	$48.61
7 Laborers	28.75	1610.00	44.75	2506.00		
3 Equip. Oper. (med.)	38.40	921.60	57.85	1388.40		
1 Asphalt Paver, 130 H.P.		1750.00		1925.00		
1 Tandem Roller, 10 Ton		211.00		232.10		
1 Roller, Pneum. Whl, 12 Ton		295.60		325.16	25.64	28.21
88 L.H., Daily Totals		$5034.20		$6759.86	$57.21	$76.82

Crew No.	Bare Costs Hr.	Daily	Incl. Subs O & P Hr.	Daily	Cost Per Labor-Hour Bare Costs	Incl. O&P
Crew B-25B	Hr.	Daily	Hr.	Daily	Bare Costs	Incl. O&P
1 Labor Foreman	$30.75	$246.00	$47.90	$383.20	$32.13	$49.38
7 Laborers	28.75	1610.00	44.75	2506.00		
4 Equip. Oper. (medium)	38.40	1228.80	57.85	1851.20		
1 Asphalt Paver, 130 H.P.		1750.00		1925.00		
2 Tandem Rollers, 10 Ton		422.00		464.20		
1 Roller, Pneum. Whl, 12 Ton		295.60		325.16	25.70	28.27
96 L.H., Daily Totals		$5552.40		$7454.76	$57.84	$77.65
Crew B-25C	Hr.	Daily	Hr.	Daily	Bare Costs	Incl. O&P
1 Labor Foreman	$30.75	$246.00	$47.90	$383.20	$32.30	$49.64
3 Laborers	28.75	690.00	44.75	1074.00		
2 Equip. Oper. (medium)	38.40	614.40	57.85	925.60		
1 Asphalt Paver, 130 H.P.		1750.00		1925.00		
1 Tandem Roller, 10 Ton		211.00		232.10	40.85	44.94
48 L.H., Daily Totals		$3511.40		$4539.90	$73.15	$94.58
Crew B-26	Hr.	Daily	Hr.	Daily	Bare Costs	Incl. O&P
1 Labor Foreman (outside)	$30.75	$246.00	$47.90	$383.20	$32.45	$50.19
6 Laborers	28.75	1380.00	44.75	2148.00		
2 Equip. Oper. (med.)	38.40	614.40	57.85	925.60		
1 Rodman (reinf.)	41.30	330.40	67.75	542.00		
1 Cement Finisher	35.55	284.40	52.20	417.60		
1 Grader, 30,000 Lbs.		505.40		555.94		
1 Paving Mach. & Equip.		2324.00		2556.40	32.15	35.37
88 L.H., Daily Totals		$5684.60		$7528.74	$64.60	$85.55
Crew B-26A	Hr.	Daily	Hr.	Daily	Bare Costs	Incl. O&P
1 Labor Foreman (outside)	$30.75	$246.00	$47.90	$383.20	$32.45	$50.19
6 Laborers	28.75	1380.00	44.75	2148.00		
2 Equip. Oper. (med.)	38.40	614.40	57.85	925.60		
1 Rodman (reinf.)	41.30	330.40	67.75	542.00		
1 Cement Finisher	35.55	284.40	52.20	417.60		
1 Grader, 30,000 Lbs.		505.40		555.94		
1 Paving Mach. & Equip.		2324.00		2556.40		
1 Concrete Saw		118.40		130.24	33.50	36.85
88 L.H., Daily Totals		$5803.00		$7658.98	$65.94	$87.03
Crew B-26B	Hr.	Daily	Hr.	Daily	Bare Costs	Incl. O&P
1 Labor Foreman (outside)	$30.75	$246.00	$47.90	$383.20	$32.94	$50.83
6 Laborers	28.75	1380.00	44.75	2148.00		
3 Equip. Oper. (med.)	38.40	921.60	57.85	1388.40		
1 Rodman (reinf.)	41.30	330.40	67.75	542.00		
1 Cement Finisher	35.55	284.40	52.20	417.60		
1 Grader, 30,000 Lbs.		505.40		555.94		
1 Paving Mach. & Equip.		2324.00		2556.40		
1 Concrete Pump, 110' Boom		938.60		1032.46	39.25	43.17
96 L.H., Daily Totals		$6930.40		$9024.00	$72.19	$94.00
Crew B-27	Hr.	Daily	Hr.	Daily	Bare Costs	Incl. O&P
1 Labor Foreman (outside)	$30.75	$246.00	$47.90	$383.20	$29.25	$45.54
3 Laborers	28.75	690.00	44.75	1074.00		
1 Berm Machine		250.80		275.88	7.84	8.62
32 L.H., Daily Totals		$1186.80		$1733.08	$37.09	$54.16
Crew B-28	Hr.	Daily	Hr.	Daily	Bare Costs	Incl. O&P
2 Carpenters	$36.70	$587.20	$57.15	$914.40	$34.05	$53.02
1 Laborer	28.75	230.00	44.75	358.00		
24 L.H., Daily Totals		$817.20		$1272.40	$34.05	$53.02

Crew No.	Bare Costs Hr.	Daily	Incl. Subs O & P Hr.	Daily	Cost Per Labor-Hour Bare Costs	Incl. O&P
Crew B-29	Hr.	Daily	Hr.	Daily	Bare Costs	Incl. O&P
1 Labor Foreman (outside)	$30.75	$246.00	$47.90	$383.20	$31.35	$48.29
4 Laborers	28.75	920.00	44.75	1432.00		
1 Equip. Oper. (crane)	39.80	318.40	60.00	480.00		
1 Equip. Oper. Oiler	33.90	271.20	51.10	408.80		
1 Gradall, 5/8 C.Y.		905.80		996.38	16.18	17.79
56 L.H., Daily Totals		$2661.40		$3700.38	$47.52	$66.08
Crew B-30	Hr.	Daily	Hr.	Daily	Bare Costs	Incl. O&P
1 Equip. Oper. (med.)	$38.40	$307.20	$57.85	$462.80	$32.50	$49.72
2 Truck Drivers (heavy)	29.55	472.80	45.65	730.40		
1 Hyd. Excavator, 1.5 C.Y.		775.60		853.16		
2 Dump Trucks, 16 Ton, 12 C.Y.		1060.00		1166.00	76.48	84.13
24 L.H., Daily Totals		$2615.60		$3212.36	$108.98	$133.85
Crew B-31	Hr.	Daily	Hr.	Daily	Bare Costs	Incl. O&P
1 Labor Foreman (outside)	$30.75	$246.00	$47.90	$383.20	$30.74	$47.86
3 Laborers	28.75	690.00	44.75	1074.00		
1 Carpenter	36.70	293.60	57.15	457.20		
1 Air Compressor, 250 C.F.M.		151.40		166.54		
1 Sheeting Driver		5.75		6.33		
2 -50' Air Hoses, 1.5"		11.60		12.76	4.22	4.64
40 L.H., Daily Totals		$1398.35		$2100.03	$34.96	$52.50
Crew B-32	Hr.	Daily	Hr.	Daily	Bare Costs	Incl. O&P
1 Laborer	$28.75	$230.00	$44.75	$358.00	$35.99	$54.58
3 Equip. Oper. (med.)	38.40	921.60	57.85	1388.40		
1 Grader, 30,000 Lbs.		505.40		555.94		
1 Tandem Roller, 10 Ton		211.00		232.10		
1 Dozer, 200 H.P.		988.40		1087.24	53.27	58.60
32 L.H., Daily Totals		$2856.40		$3621.68	$89.26	$113.18
Crew B-32A	Hr.	Daily	Hr.	Daily	Bare Costs	Incl. O&P
1 Laborer	$28.75	$230.00	$44.75	$358.00	$35.18	$53.48
2 Equip. Oper. (medium)	38.40	614.40	57.85	925.60		
1 Grader, 30,000 Lbs.		505.40		555.94		
1 Roller, Vibratory, 25 Ton		565.60		622.16	44.63	49.09
24 L.H., Daily Totals		$1915.40		$2461.70	$79.81	$102.57
Crew B-32B	Hr.	Daily	Hr.	Daily	Bare Costs	Incl. O&P
1 Laborer	$28.75	$230.00	$44.75	$358.00	$35.18	$53.48
2 Equip. Oper. (medium)	38.40	614.40	57.85	925.60		
1 Dozer, 200 H.P.		988.40		1087.24		
1 Roller, Vibratory, 25 Ton		565.60		622.16	64.75	71.22
24 L.H., Daily Totals		$2398.40		$2993.00	$99.93	$124.71
Crew B-32C	Hr.	Daily	Hr.	Daily	Bare Costs	Incl. O&P
1 Labor Foreman	$30.75	$246.00	$47.90	$383.20	$33.91	$51.83
2 Laborers	28.75	460.00	44.75	716.00		
3 Equip. Oper. (medium)	38.40	921.60	57.85	1388.40		
1 Grader, 30,000 Lbs.		505.40		555.94		
1 Tandem Roller, 10 Ton		211.00		232.10		
1 Dozer, 200 H.P.		988.40		1087.24	35.52	39.07
48 L.H., Daily Totals		$3332.40		$4362.88	$69.42	$90.89
Crew B-33A	Hr.	Daily	Hr.	Daily	Bare Costs	Incl. O&P
1 Equip. Oper. (med.)	$38.40	$307.20	$57.85	$462.80	$35.64	$54.11
.5 Laborer	28.75	115.00	44.75	179.00		
.25 Equip. Oper. (med.)	38.40	76.80	57.85	115.70		
1 Scraper, Towed, 7 C.Y.		143.20		157.52		
1.25 Dozer, 300 H.P.		1626.25		1788.88	126.39	139.03
14 L.H., Daily Totals		$2268.45		$2703.90	$162.03	$193.14

Crews

Crew No.	Bare Costs		Incl. Subs O & P		Cost Per Labor-Hour	
Crew B-33B	**Hr.**	**Daily**	**Hr.**	**Daily**	**Bare Costs**	**Incl. O&P**
1 Equip. Oper. (med.)	$38.40	$307.20	$57.85	$462.80	$35.64	$54.11
.5 Laborer	28.75	115.00	44.75	179.00		
.25 Equip. Oper. (med.)	38.40	76.80	57.85	115.70		
1 Scraper, Towed, 10 C.Y.		157.00		172.70		
1.25 Dozer, 300 H.P.		1626.25		1788.88	127.38	140.11
14 L.H., Daily Totals		$2282.25		$2719.07	$163.02	$194.22

Crew B-33C	Hr.	Daily	Hr.	Daily	Bare Costs	Incl. O&P
1 Equip. Oper. (med.)	$38.40	$307.20	$57.85	$462.80	$35.64	$54.11
.5 Laborer	28.75	115.00	44.75	179.00		
.25 Equip. Oper. (med.)	38.40	76.80	57.85	115.70		
1 Scraper, Towed, 15 C.Y.		157.00		172.70		
1.25 Dozer, 300 H.P.		1626.25		1788.88	127.38	140.11
14 L.H., Daily Totals		$2282.25		$2719.07	$163.02	$194.22

Crew B-33D	Hr.	Daily	Hr.	Daily	Bare Costs	Incl. O&P
1 Equip. Oper. (med.)	$38.40	$307.20	$57.85	$462.80	$35.64	$54.11
.5 Laborer	28.75	115.00	44.75	179.00		
.25 Equip. Oper. (med.)	38.40	76.80	57.85	115.70		
1 S.P. Scraper, 14 C.Y.		1717.00		1888.70		
.25 Dozer, 300 H.P.		325.25		357.77	145.88	160.46
14 L.H., Daily Totals		$2541.25		$3003.97	$181.52	$214.57

Crew B-33E	Hr.	Daily	Hr.	Daily	Bare Costs	Incl. O&P
1 Equip. Oper. (med.)	$38.40	$307.20	$57.85	$462.80	$35.64	$54.11
.5 Laborer	28.75	115.00	44.75	179.00		
.25 Equip. Oper. (med.)	38.40	76.80	57.85	115.70		
1 S.P. Scraper, 24 C.Y.		2549.00		2803.90		
.25 Dozer, 300 H.P.		325.25		357.77	205.30	225.83
14 L.H., Daily Totals		$3373.25		$3919.18	$240.95	$279.94

Crew B-33F	Hr.	Daily	Hr.	Daily	Bare Costs	Incl. O&P
1 Equip. Oper. (med.)	$38.40	$307.20	$57.85	$462.80	$35.64	$54.11
.5 Laborer	28.75	115.00	44.75	179.00		
.25 Equip. Oper. (med.)	38.40	76.80	57.85	115.70		
1 Elev. Scraper, 11 C.Y.		859.80		945.78		
.25 Dozer, 300 H.P.		325.25		357.77	84.65	93.11
14 L.H., Daily Totals		$1684.05		$2061.05	$120.29	$147.22

Crew B-33G	Hr.	Daily	Hr.	Daily	Bare Costs	Incl. O&P
1 Equip. Oper. (med.)	$38.40	$307.20	$57.85	$462.80	$35.64	$54.11
.5 Laborer	28.75	115.00	44.75	179.00		
.25 Equip. Oper. (med.)	38.40	76.80	57.85	115.70		
1 Elev. Scraper, 20 C.Y.		1750.00		1925.00		
.25 Dozer, 300 H.P.		325.25		357.77	148.23	163.06
14 L.H., Daily Totals		$2574.25		$3040.28	$183.88	$217.16

Crew B-33H	Hr.	Daily	Hr.	Daily	Bare Costs	Incl. O&P
.25 Laborer	$28.75	$57.50	$44.75	$89.50	$36.74	$55.59
1 Equipment Operator (med.)	38.40	307.20	57.85	462.80		
.2 Equipment Operator (med.)	38.40	61.44	57.85	92.56		
1 Scraper, 32-44 C.Y.		2997.00		3296.70		
.2 Dozer, 410 H.P.		335.40		368.94	287.28	316.00
11.6 L.H., Daily Totals		$3758.54		$4310.50	$324.01	$371.59

Crew B-33J	Hr.	Daily	Hr.	Daily	Bare Costs	Incl. O&P
1 Equipment Operator (med.)	$38.40	$307.20	$57.85	$462.80	$38.40	$57.85
1 S.P. Scraper, 14 C.Y.		1717.00		1888.70	214.63	236.09
8 L.H., Daily Totals		$2024.20		$2351.50	$253.03	$293.94

Crew No.	Bare Costs		Incl. Subs O & P		Cost Per Labor-Hour	
Crew B-34A	**Hr.**	**Daily**	**Hr.**	**Daily**	**Bare Costs**	**Incl. O&P**
1 Truck Driver (heavy)	$29.55	$236.40	$45.65	$365.20	$29.55	$45.65
1 Dump Truck, 12 Ton, 8 C.Y.		368.40		405.24	46.05	50.66
8 L.H., Daily Totals		$604.80		$770.44	$75.60	$96.31

Crew B-34B	Hr.	Daily	Hr.	Daily	Bare Costs	Incl. O&P
1 Truck Driver (heavy)	$29.55	$236.40	$45.65	$365.20	$29.55	$45.65
1 Dump Truck, 16 Ton, 12 C.Y.		530.00		583.00	66.25	72.88
8 L.H., Daily Totals		$766.40		$948.20	$95.80	$118.53

Crew B-34C	Hr.	Daily	Hr.	Daily	Bare Costs	Incl. O&P
1 Truck Driver (heavy)	$29.55	$236.40	$45.65	$365.20	$29.55	$45.65
1 Truck Tractor, 240 H.P.		336.20		369.82		
1 Dump Trailer, 16.5 C.Y.		107.80		118.58	55.50	61.05
8 L.H., Daily Totals		$680.40		$853.60	$85.05	$106.70

Crew B-34D	Hr.	Daily	Hr.	Daily	Bare Costs	Incl. O&P
1 Truck Driver (heavy)	$29.55	$236.40	$45.65	$365.20	$29.55	$45.65
1 Truck Tractor, 240 H.P.		336.20		369.82		
1 Dump Trailer, 20 C.Y.		120.00		132.00	57.02	62.73
8 L.H., Daily Totals		$692.60		$867.02	$86.58	$108.38

Crew B-34E	Hr.	Daily	Hr.	Daily	Bare Costs	Incl. O&P
1 Truck Driver (heavy)	$29.55	$236.40	$45.65	$365.20	$29.55	$45.65
1 Dump Truck, Off Hwy., 25 Ton		992.80		1092.08	124.10	136.51
8 L.H., Daily Totals		$1229.20		$1457.28	$153.65	$182.16

Crew B-34F	Hr.	Daily	Hr.	Daily	Bare Costs	Incl. O&P
1 Truck Driver (heavy)	$29.55	$236.40	$45.65	$365.20	$29.55	$45.65
1 Dump Truck, Off Hwy., 35 Ton		1014.00		1115.40	126.75	139.43
8 L.H., Daily Totals		$1250.40		$1480.60	$156.30	$185.07

Crew B-34G	Hr.	Daily	Hr.	Daily	Bare Costs	Incl. O&P
1 Truck Driver (heavy)	$29.55	$236.40	$45.65	$365.20	$29.55	$45.65
1 Dump Truck, Off Hwy., 50 Ton		1308.00		1438.80	163.50	179.85
8 L.H., Daily Totals		$1544.40		$1804.00	$193.05	$225.50

Crew B-34H	Hr.	Daily	Hr.	Daily	Bare Costs	Incl. O&P
1 Truck Driver (heavy)	$29.55	$236.40	$45.65	$365.20	$29.55	$45.65
1 Dump Truck, Off Hwy., 65 Ton		1393.00		1532.30	174.13	191.54
8 L.H., Daily Totals		$1629.40		$1897.50	$203.68	$237.19

Crew B-34J	Hr.	Daily	Hr.	Daily	Bare Costs	Incl. O&P
1 Truck Driver (heavy)	$29.55	$236.40	$45.65	$365.20	$29.55	$45.65
1 Dump Truck, Off Hwy., 100 Ton		1792.00		1971.20	224.00	246.40
8 L.H., Daily Totals		$2028.40		$2336.40	$253.55	$292.05

Crew B-34K	Hr.	Daily	Hr.	Daily	Bare Costs	Incl. O&P
1 Truck Driver (heavy)	$29.55	$236.40	$45.65	$365.20	$29.55	$45.65
1 Truck Tractor, 240 H.P.		395.40		434.94		
1 Lowbed Trailer, 75 Ton		181.20		199.32	72.08	79.28
8 L.H., Daily Totals		$813.00		$999.46	$101.63	$124.93

Crew B-34N	Hr.	Daily	Hr.	Daily	Bare Costs	Incl. O&P
1 Truck Driver (heavy)	$29.55	$236.40	$45.65	$365.20	$29.55	$45.65
1 Dump Truck, 12 Ton, 8 C.Y.		368.40		405.24		
1 Flatbed Trailer, 40 Ton		125.80		138.38	61.77	67.95
8 L.H., Daily Totals		$730.60		$908.82	$91.33	$113.60

Crew B-34P

Crew No.	Bare Costs Hr.	Daily	Incl. Subs O & P Hr.	Daily	Cost Per Labor-Hour Bare Costs	Incl. O&P
1 Pipe Fitter	$45.20	$361.60	$68.00	$544.00	$37.38	$56.65
1 Truck Driver (light)	28.55	228.40	44.10	352.80		
1 Equip. Oper. (medium)	38.40	307.20	57.85	462.80		
1 Flatbed Truck, gas, 3 Ton		185.20		203.72		
1 Backhoe Loader, 48 H.P.		243.40		267.74	17.86	19.64
24 L.H., Daily Totals		$1325.80		$1831.06	$55.24	$76.29

Crew B-34Q

Crew No.	Bare Costs Hr.	Daily	Incl. Subs O & P Hr.	Daily	Cost Per Labor-Hour Bare Costs	Incl. O&P
1 Pipe Fitter	$45.20	$361.60	$68.00	$544.00	$37.85	$57.37
1 Truck Driver (light)	28.55	228.40	44.10	352.80		
1 Equip. Oper. (crane)	39.80	318.40	60.00	480.00		
1 Flatbed Trailer, 25 Ton		92.80		102.08		
1 Dump Truck, 12 Ton, 8 C.Y.		368.40		405.24		
1 Hyd. Crane, 25 Ton		739.60		813.56	50.03	55.04
24 L.H., Daily Totals		$2109.20		$2697.68	$87.88	$112.40

Crew B-34R

Crew No.	Bare Costs Hr.	Daily	Incl. Subs O & P Hr.	Daily	Cost Per Labor-Hour Bare Costs	Incl. O&P
1 Pipe Fitter	$45.20	$361.60	$68.00	$544.00	$37.85	$57.37
1 Truck Driver (light)	28.55	228.40	44.10	352.80		
1 Eqip. Oper. (crane)	39.80	318.40	60.00	480.00		
1 Flatbed Trailer, 25 Ton		92.80		102.08		
1 Dump Truck, 12 Ton, 8 C.Y.		368.40		405.24		
1 Hyd. Crane, 25 Ton		739.60		813.56		
1 Hyd. Excavator, 1 C.Y.		601.80		661.98	75.11	82.62
24 L.H., Daily Totals		$2711.00		$3359.66	$112.96	$139.99

Crew B-34S

Crew No.	Bare Costs Hr.	Daily	Incl. Subs O & P Hr.	Daily	Cost Per Labor-Hour Bare Costs	Incl. O&P
2 Pipe Fitters	$45.20	$723.20	$68.00	$1088.00	$39.94	$60.41
1 Truck Driver (heavy)	29.55	236.40	45.65	365.20		
1 Eqip. Oper. (crane)	39.80	318.40	60.00	480.00		
1 Flatbed Trailer, 40 Ton		125.80		138.38		
1 Truck Tractor, 240 H.P.		336.20		369.82		
1 Hyd. Crane, 80 Ton		1214.00		1335.40		
1 Hyd. Excavator, 2 C.Y.		994.60		1094.06	83.46	91.80
32 L.H., Daily Totals		$3948.60		$4870.86	$123.39	$152.21

Crew B-34T

Crew No.	Bare Costs Hr.	Daily	Incl. Subs O & P Hr.	Daily	Cost Per Labor-Hour Bare Costs	Incl. O&P
2 Pipe Fitters	$45.20	$723.20	$68.00	$1088.00	$39.94	$60.41
1 Truck Driver (heavy)	29.55	236.40	45.65	365.20		
1 Eqip. Oper. (crane)	39.80	318.40	60.00	480.00		
1 Flatbed Trailer, 40 Ton		125.80		138.38		
1 Truck Tractor, 240 H.P.		336.20		369.82		
1 Hyd. Crane, 80 Ton		1214.00		1335.40	52.38	57.61
32 L.H., Daily Totals		$2954.00		$3776.80	$92.31	$118.03

Crew B-35

Crew No.	Bare Costs Hr.	Daily	Incl. Subs O & P Hr.	Daily	Cost Per Labor-Hour Bare Costs	Incl. O&P
1 Laborer Foreman (out)	$30.75	$246.00	$47.90	$383.20	$36.00	$55.05
1 Skilled Worker	38.00	304.00	59.15	473.20		
1 Welder (plumber)	44.80	358.40	67.40	539.20		
1 Laborer	28.75	230.00	44.75	358.00		
1 Equip. Oper. (crane)	39.80	318.40	60.00	480.00		
1 Equip. Oper. Oiler	33.90	271.20	51.10	408.80		
1 Welder, electric, 300 amp		61.45		67.59		
1 Hyd. Excavator, .75 C.Y.		516.40		568.04	12.04	13.24
48 L.H., Daily Totals		$2305.85		$3278.03	$48.04	$68.29

Crew B-35A

Crew No.	Bare Costs Hr.	Daily	Incl. Subs O & P Hr.	Daily	Cost Per Labor-Hour Bare Costs	Incl. O&P
1 Laborer Foreman (out)	$30.75	$246.00	$47.90	$383.20	$34.96	$53.58
2 Laborers	28.75	460.00	44.75	716.00		
1 Skilled Worker	38.00	304.00	59.15	473.20		
1 Welder (plumber)	44.80	358.40	67.40	539.20		
1 Equip. Oper. (crane)	39.80	318.40	60.00	480.00		
1 Equip. Oper. Oiler	33.90	271.20	51.10	408.80		
1 Welder, gas engine, 300 amp		115.20		126.72		
1 Crawler Crane, 75 Ton		1384.00		1522.40	26.77	29.45
56 L.H., Daily Totals		$3457.20		$4649.52	$61.74	$83.03

Crew B-36

Crew No.	Bare Costs Hr.	Daily	Incl. Subs O & P Hr.	Daily	Cost Per Labor-Hour Bare Costs	Incl. O&P
1 Labor Foreman (outside)	$30.75	$246.00	$47.90	$383.20	$33.01	$50.62
2 Laborers	28.75	460.00	44.75	716.00		
2 Equip. Oper. (med.)	38.40	614.40	57.85	925.60		
1 Dozer, 200 H.P.		988.40		1087.24		
1 Aggregate Spreader		30.60		33.66		
1 Tandem Roller, 10 Ton		211.00		232.10	30.75	33.83
40 L.H., Daily Totals		$2550.40		$3377.80	$63.76	$84.44

Crew B-36A

Crew No.	Bare Costs Hr.	Daily	Incl. Subs O & P Hr.	Daily	Cost Per Labor-Hour Bare Costs	Incl. O&P
1 Labor Foreman (outside)	$30.75	$246.00	$47.90	$383.20	$34.55	$52.69
2 Laborers	28.75	460.00	44.75	716.00		
4 Equip. Oper. (med.)	38.40	1228.80	57.85	1851.20		
1 Dozer, 200 H.P.		988.40		1087.24		
1 Aggregate Spreader		30.60		33.66		
1 Tandem Roller, 10 Ton		211.00		232.10		
1 Roller, Pneum. Whl, 12 Ton		295.60		325.16	27.24	29.97
56 L.H., Daily Totals		$3460.40		$4628.56	$61.79	$82.65

Crew B-36B

Crew No.	Bare Costs Hr.	Daily	Incl. Subs O & P Hr.	Daily	Cost Per Labor-Hour Bare Costs	Incl. O&P
1 Labor Foreman (outside)	$30.75	$246.00	$47.90	$383.20	$33.92	$51.81
2 Laborers	28.75	460.00	44.75	716.00		
4 Equip. Oper. (medium)	38.40	1228.80	57.85	1851.20		
1 Truck Driver, Heavy	29.55	236.40	45.65	365.20		
1 Grader, 30,000 Lbs.		505.40		555.94		
1 F.E. Loader, crl, 1.5 C.Y.		399.20		439.12		
1 Dozer, 300 H.P.		1301.00		1431.10		
1 Roller, Vibratory, 25 Ton		565.60		622.16		
1 Truck Tractor, 240 H.P.		395.40		434.94		
1 Water Tanker, 5000 Gal.		121.20		133.32	51.37	56.51
64 L.H., Daily Totals		$5459.00		$6932.18	$85.30	$108.32

Crew B-36C

Crew No.	Bare Costs Hr.	Daily	Incl. Subs O & P Hr.	Daily	Cost Per Labor-Hour Bare Costs	Incl. O&P
1 Labor Foreman (outside)	$30.75	$246.00	$47.90	$383.20	$35.10	$53.42
3 Equip. Oper. (medium)	38.40	921.60	57.85	1388.40		
1 Truck Driver, Heavy	29.55	236.40	45.65	365.20		
1 Grader, 30,000 Lbs.		505.40		555.94		
1 Dozer, 300 H.P.		1301.00		1431.10		
1 Roller, Vibratory, 25 Ton		565.60		622.16		
1 Truck Tractor, 240 H.P.		395.40		434.94		
1 Water Tanker, 5000 Gal.		121.20		133.32	72.22	79.44
40 L.H., Daily Totals		$4292.60		$5314.26	$107.32	$132.86

Crew B-37

Crew No.	Bare Costs Hr.	Daily	Incl. Subs O & P Hr.	Daily	Cost Per Labor-Hour Bare Costs	Incl. O&P
1 Labor Foreman (outside)	$30.75	$246.00	$47.90	$383.20	$30.43	$47.08
4 Laborers	28.75	920.00	44.75	1432.00		
1 Equip. Oper. (light)	36.85	294.80	55.55	444.40		
1 Tandem Roller, 5 Ton		122.20		134.42	2.55	2.80
48 L.H., Daily Totals		$1583.00		$2394.02	$32.98	$49.88

Crew No.	Bare Costs Hr.	Daily	Incl. Subs O & P Hr.	Daily	Cost Per Labor-Hour Bare Costs	Incl. O&P
Crew B-38	**Hr.**	**Daily**	**Hr.**	**Daily**	**Bare Costs**	**Incl. O&P**
1 Labor Foreman (outside)	$30.75	$246.00	$47.90	$383.20	$32.70	$50.16
2 Laborers	28.75	460.00	44.75	716.00		
1 Equip. Oper. (light)	36.85	294.80	55.55	444.40		
1 Equip. Oper. (medium)	38.40	307.20	57.85	462.80		
1 Backhoe Loader, 48 H.P.		243.40		267.74		
1 Hyd.Hammer, (1200 lb)		115.40		126.94		
1 F.E. Loader, W.M., 4 C.Y.		488.00		536.80		
1 Pavt. Rem. Bucket		53.60		58.96	22.51	24.76
40 L.H., Daily Totals		$2208.40		$2996.84	$55.21	$74.92
Crew B-39	**Hr.**	**Daily**	**Hr.**	**Daily**	**Bare Costs**	**Incl. O&P**
1 Labor Foreman (outside)	$30.75	$246.00	$47.90	$383.20	$30.43	$47.08
4 Laborers	28.75	920.00	44.75	1432.00		
1 Equip. Oper. (light)	36.85	294.80	55.55	444.40		
1 Air Compressor, 250 C.F.M.		151.40		166.54		
2 Breakers, Pavement, 60 lb.		16.80		18.48		
2 -50' Air Hoses, 1.5"		11.60		12.76	3.75	4.12
48 L.H., Daily Totals		$1640.60		$2457.38	$34.18	$51.20
Crew B-40	**Hr.**	**Daily**	**Hr.**	**Daily**	**Bare Costs**	**Incl. O&P**
1 Pile Driver Foreman (out)	$37.90	$303.20	$62.05	$496.40	$36.88	$58.52
4 Pile Drivers	35.90	1148.80	58.75	1880.00		
2 Equip. Oper. (crane)	39.80	636.80	60.00	960.00		
1 Equip. Oper. Oiler	33.90	271.20	51.10	408.80		
1 Crawler Crane, 40 Ton		994.60		1094.06		
1 Vibratory Hammer & Gen.		1983.00		2181.30	46.52	51.18
64 L.H., Daily Totals		$5337.60		$7020.56	$83.40	$109.70
Crew B-40B	**Hr.**	**Daily**	**Hr.**	**Daily**	**Bare Costs**	**Incl. O&P**
1 Laborer Foreman	$30.75	$246.00	$47.90	$383.20	$31.78	$48.88
3 Laborers	28.75	690.00	44.75	1074.00		
1 Equip. Oper. (crane)	39.80	318.40	60.00	480.00		
1 Equip. Oper. Oiler	33.90	271.20	51.10	408.80		
1 Lattice Boom Crane, 40 Ton		1268.00		1394.80	26.42	29.06
48 L.H., Daily Totals		$2793.60		$3740.80	$58.20	$77.93
Crew B-41	**Hr.**	**Daily**	**Hr.**	**Daily**	**Bare Costs**	**Incl. O&P**
1 Labor Foreman (outside)	$30.75	$246.00	$47.90	$383.20	$29.85	$46.30
4 Laborers	28.75	920.00	44.75	1432.00		
.25 Equip. Oper. (crane)	39.80	79.60	60.00	120.00		
.25 Equip. Oper. Oiler	33.90	67.80	51.10	102.20		
.25 Crawler Crane, 40 Ton		248.65		273.51	5.65	6.22
44 L.H., Daily Totals		$1562.05		$2310.92	$35.50	$52.52
Crew B-42	**Hr.**	**Daily**	**Hr.**	**Daily**	**Bare Costs**	**Incl. O&P**
1 Labor Foreman (outside)	$30.75	$246.00	$47.90	$383.20	$32.60	$51.61
4 Laborers	28.75	920.00	44.75	1432.00		
1 Equip. Oper. (crane)	39.80	318.40	60.00	480.00		
1 Equip. Oper. Oiler	33.90	271.20	51.10	408.80		
1 Welder	41.35	330.80	74.90	599.20		
1 Hyd. Crane, 25 Ton		739.60		813.56		
1 Welder, gas engine, 300 amp		115.20		126.72		
1 Horz. Boring Csg. Mch.		421.40		463.54	19.94	21.93
64 L.H., Daily Totals		$3362.60		$4707.02	$52.54	$73.55

Crew No.	Bare Costs Hr.	Daily	Incl. Subs O & P Hr.	Daily	Cost Per Labor-Hour Bare Costs	Incl. O&P
Crew B-43	**Hr.**	**Daily**	**Hr.**	**Daily**	**Bare Costs**	**Incl. O&P**
1 Labor Foreman (outside)	$30.75	$246.00	$47.90	$383.20	$31.78	$48.88
3 Laborers	28.75	690.00	44.75	1074.00		
1 Equip. Oper. (crane)	39.80	318.40	60.00	480.00		
1 Equip. Oper. Oiler	33.90	271.20	51.10	408.80		
1 Drill Rig, Truck-Mounted		3499.00		3848.90	72.90	80.19
48 L.H., Daily Totals		$5024.60		$6194.90	$104.68	$129.06
Crew B-44	**Hr.**	**Daily**	**Hr.**	**Daily**	**Bare Costs**	**Incl. O&P**
1 Pile Driver Foreman	$37.90	$303.20	$62.05	$496.40	$36.23	$57.73
4 Pile Drivers	35.90	1148.80	58.75	1880.00		
2 Equip. Oper. (crane)	39.80	636.80	60.00	960.00		
1 Laborer	28.75	230.00	44.75	358.00		
1 Crawler Crane, 40 Ton		994.60		1094.06		
45 L.F. of Leads, 15K Ft. Lbs.		45.00		49.50	16.24	17.87
64 L.H., Daily Totals		$3358.40		$4837.96	$52.48	$75.59
Crew B-45	**Hr.**	**Daily**	**Hr.**	**Daily**	**Bare Costs**	**Incl. O&P**
1 Equip. Oper. (med.)	$38.40	$307.20	$57.85	$462.80	$33.98	$51.75
1 Truck Driver (heavy)	29.55	236.40	45.65	365.20		
1 Dist. Tanker, 3000 Gallon		273.00		300.30		
1 Tractor, 4 x 2, 250 H.P.		312.20		343.42	36.58	40.23
16 L.H., Daily Totals		$1128.80		$1471.72	$70.55	$91.98
Crew B-46	**Hr.**	**Daily**	**Hr.**	**Daily**	**Bare Costs**	**Incl. O&P**
1 Pile Driver Foreman	$37.90	$303.20	$62.05	$496.40	$32.66	$52.30
2 Pile Drivers	35.90	574.40	58.75	940.00		
3 Laborers	28.75	690.00	44.75	1074.00		
1 Chainsaw, gas, 36" Long		34.40		37.84	.72	.79
48 L.H., Daily Totals		$1602.00		$2548.24	$33.38	$53.09
Crew B-47	**Hr.**	**Daily**	**Hr.**	**Daily**	**Bare Costs**	**Incl. O&P**
1 Blast Foreman	$30.75	$246.00	$47.90	$383.20	$32.12	$49.40
1 Driller	28.75	230.00	44.75	358.00		
1 Equip. Oper. (light)	36.85	294.80	55.55	444.40		
1 Air Track Drill, 4"		787.40		866.14		
1 Air Compressor, 600 C.F.M.		396.20		435.82		
2 -50' Air Hoses, 3"		38.10		41.91	50.90	55.99
24 L.H., Daily Totals		$1992.50		$2529.47	$83.02	$105.39
Crew B-47A	**Hr.**	**Daily**	**Hr.**	**Daily**	**Bare Costs**	**Incl. O&P**
1 Drilling Foreman	$30.75	$246.00	$47.90	$383.20	$34.82	$53.00
1 Equip. Oper. (heavy)	39.80	318.40	60.00	480.00		
1 Oiler	33.90	271.20	51.10	408.80		
1 Air Track Drill, 5"		905.40		995.94	37.73	41.50
24 L.H., Daily Totals		$1741.00		$2267.94	$72.54	$94.50
Crew B-47C	**Hr.**	**Daily**	**Hr.**	**Daily**	**Bare Costs**	**Incl. O&P**
1 Laborer	$28.75	$230.00	$44.75	$358.00	$32.80	$50.15
1 Equip. Oper. (light)	36.85	294.80	55.55	444.40		
1 Air Compressor, 750 C.F.M.		409.80		450.78		
2 -50' Air Hoses, 3"		38.10		41.91		
1 Air Track Drill, 4"		787.40		866.14	77.21	84.93
16 L.H., Daily Totals		$1760.10		$2161.23	$110.01	$135.08
Crew B-47E	**Hr.**	**Daily**	**Hr.**	**Daily**	**Bare Costs**	**Incl. O&P**
1 Laborer Foreman	$30.75	$246.00	$47.90	$383.20	$29.25	$45.54
3 Laborers	28.75	690.00	44.75	1074.00		
1 Flatbed Truck, gas, 3 Ton		185.20		203.72	5.79	6.37
32 L.H., Daily Totals		$1121.20		$1660.92	$35.04	$51.90

Crew No.	Bare Costs Hr.	Daily	Incl. Subs O & P Hr.	Daily	Cost Per Labor-Hour Bare Costs	Incl. O&P
Crew B-47G						
1 Laborer Foreman	$30.75	$246.00	$47.90	$383.20	$31.27	$48.24
2 Laborers	28.75	460.00	44.75	716.00		
1 Equip. Oper. (light)	36.85	294.80	55.55	444.40		
1 Air Track Drill, 4"		787.40		866.14		
1 Air Compressor, 600 C.F.M.		396.20		435.82		
2 -50' Air Hoses, 3"		38.10		41.91		
1 Grout Pump		131.20		144.32	42.28	46.51
32 L.H., Daily Totals		$2353.70		$3031.79	$73.55	$94.74
Crew B-48						
1 Labor Foreman (outside)	$30.75	$246.00	$47.90	$383.20	$32.51	$49.83
3 Laborers	28.75	690.00	44.75	1074.00		
1 Equip. Oper. (crane)	39.80	318.40	60.00	480.00		
1 Equip. Oper. Oiler	33.90	271.20	51.10	408.80		
1 Equip. Oper. (light)	36.85	294.80	55.55	444.40		
1 Centr. Water Pump, 6"		278.20		306.02		
1 -20' Suction Hose, 6"		11.50		12.65		
1 -50' Discharge Hose, 6"		6.30		6.93		
1 Drill Rig, Truck-Mounted		3499.00		3848.90	67.77	74.54
56 L.H., Daily Totals		$5615.40		$6964.90	$100.28	$124.37
Crew B-49						
1 Labor Foreman (outside)	$30.75	$246.00	$47.90	$383.20	$33.91	$52.49
3 Laborers	28.75	690.00	44.75	1074.00		
2 Equip. Oper. (crane)	39.80	636.80	60.00	960.00		
2 Equip. Oper. Oilers	33.90	542.40	51.10	817.60		
1 Equip. Oper. (light)	36.85	294.80	55.55	444.40		
2 Pile Drivers	35.90	574.40	58.75	940.00		
1 Hyd. Crane, 25 Ton		739.60		813.56		
1 Centr. Water Pump, 6"		278.20		306.02		
1 -20' Suction Hose, 6"		11.50		12.65		
1 -50' Discharge Hose, 6"		6.30		6.93		
1 Drill Rig, Truck-Mounted		3499.00		3848.90	51.53	56.68
88 L.H., Daily Totals		$7519.00		$9607.26	$85.44	$109.17
Crew B-50						
2 Pile Driver Foremen	$37.90	$606.40	$62.05	$992.80	$35.07	$55.85
6 Pile Drivers	35.90	1723.20	58.75	2820.00		
2 Equip. Oper. (crane)	39.80	636.80	60.00	960.00		
1 Equip. Oper. Oiler	33.90	271.20	51.10	408.80		
3 Laborers	28.75	690.00	44.75	1074.00		
1 Crawler Crane, 40 Ton		994.60		1094.06		
60 L.F. of Leads, 15K Ft. Lbs.		60.00		66.00		
1 Hammer, 15K Ft. Lbs.		360.00		396.00		
1 Air Compressor, 600 C.F.M.		396.20		435.82		
2 -50' Air Hoses, 3"		38.10		41.91		
1 Chainsaw, gas, 36" Long		34.40		37.84	16.82	18.50
112 L.H., Daily Totals		$5810.90		$8327.23	$51.88	$74.35
Crew B-51						
1 Labor Foreman (outside)	$30.75	$246.00	$47.90	$383.20	$29.05	$45.17
4 Laborers	28.75	920.00	44.75	1432.00		
1 Truck Driver (light)	28.55	228.40	44.10	352.80		
1 Flatbed Truck, gas, 1.5 Ton		141.40		155.54	2.95	3.24
48 L.H., Daily Totals		$1535.80		$2323.54	$32.00	$48.41

Crew No.	Bare Costs Hr.	Daily	Incl. Subs O & P Hr.	Daily	Cost Per Labor-Hour Bare Costs	Incl. O&P
Crew B-52						
1 Carpenter Foreman	$38.70	$309.60	$60.25	$482.00	$33.86	$52.38
1 Carpenter	36.70	293.60	57.15	457.20		
3 Laborers	28.75	690.00	44.75	1074.00		
1 Cement Finisher	35.55	284.40	52.20	417.60		
.5 Rodman (reinf.)	41.30	165.20	67.75	271.00		
.5 Equip. Oper. (med.)	38.40	153.60	57.85	231.40		
.5 Crawler Loader, 3 C.Y.		417.20		458.92	7.45	8.20
56 L.H., Daily Totals		$2313.60		$3392.12	$41.31	$60.57
Crew B-53						
1 Equip. Oper. (light)	$36.85	$294.80	$55.55	$444.40	$36.85	$55.55
1 Trencher, Chain, 12 H.P.		51.00		56.10	6.38	7.01
8 L.H., Daily Totals		$345.80		$500.50	$43.23	$62.56
Crew B-54						
1 Equip. Oper. (light)	$36.85	$294.80	$55.55	$444.40	$36.85	$55.55
1 Trencher, Chain, 40 H.P.		226.60		249.26	28.32	31.16
8 L.H., Daily Totals		$521.40		$693.66	$65.17	$86.71
Crew B-54A						
.17 Labor Foreman (outside)	$30.75	$41.82	$47.90	$65.14	$37.29	$56.40
1 Equipment Operator (med.)	38.40	307.20	57.85	462.80		
1 Wheel Trencher, 67 H.P.		891.20		980.32	95.21	104.74
9.36 L.H., Daily Totals		$1240.22		$1508.26	$132.50	$161.14
Crew B-54B						
.25 Labor Foreman (outside)	$30.75	$61.50	$47.90	$95.80	$36.87	$55.86
1 Equipment Operator (med.)	38.40	307.20	57.85	462.80		
1 Wheel Trencher, 150 H.P.		1626.00		1788.60	162.60	178.86
10 L.H., Daily Totals		$1994.70		$2347.20	$199.47	$234.72
Crew B-55						
2 Laborers	$28.75	$460.00	$44.75	$716.00	$28.68	$44.53
1 Truck Driver (light)	28.55	228.40	44.10	352.80		
1 Truck-mounted earth auger		674.60		742.06		
1 Flatbed Truck, gas, 3 Ton		185.20		203.72	35.83	39.41
24 L.H., Daily Totals		$1548.20		$2014.58	$64.51	$83.94
Crew B-56						
1 Laborer	$28.75	$230.00	$44.75	$358.00	$32.80	$50.15
1 Equip. Oper. (light)	36.85	294.80	55.55	444.40		
1 Air Track Drill, 4"		787.40		866.14		
1 Air Compressor, 600 C.F.M.		396.20		435.82		
1 -50' Air Hose, 3"		19.05		20.95	75.17	82.68
16 L.H., Daily Totals		$1727.45		$2125.32	$107.97	$132.83
Crew B-57						
1 Labor Foreman (outside)	$30.75	$246.00	$47.90	$383.20	$33.13	$50.67
2 Laborers	28.75	460.00	44.75	716.00		
1 Equip. Oper. (crane)	39.80	318.40	60.00	480.00		
1 Equip. Oper. (light)	36.85	294.80	55.55	444.40		
1 Equip. Oper. Oiler	33.90	271.20	51.10	408.80		
1 Crawler Crane, 25 Ton		983.00		1081.30		
1 Clamshell Bucket, 1 C.Y.		44.80		49.28		
1 Centr. Water Pump, 6"		278.20		306.02		
1 -20' Suction Hose, 6"		11.50		12.65		
20 -50' Discharge Hoses, 6"		126.00		138.60	30.07	33.08
48 L.H., Daily Totals		$3033.90		$4020.25	$63.21	$83.76

Crews

Crew No.	Bare Costs		Incl. Subs O & P		Cost Per Labor-Hour	
Crew B-58	Hr.	Daily	Hr.	Daily	Bare Costs	Incl. O&P
2 Laborers	$28.75	$460.00	$44.75	$716.00	$31.45	$48.35
1 Equip. Oper. (light)	36.85	294.80	55.55	444.40		
1 Backhoe Loader, 48 H.P.		243.40		267.74		
1 Small Helicopter, w/pilot		2331.00		2564.10	107.27	117.99
24 L.H., Daily Totals		$3329.20		$3992.24	$138.72	$166.34
Crew B-59	Hr.	Daily	Hr.	Daily	Bare Costs	Incl. O&P
1 Truck Driver (heavy)	$29.55	$236.40	$45.65	$365.20	$29.55	$45.65
1 Truck Tractor, 195 H.P.		235.60		259.16		
1 Water Tanker, 5000 Gal.		121.20		133.32	44.60	49.06
8 L.H., Daily Totals		$593.20		$757.68	$74.15	$94.71
Crew B-59A	Hr.	Daily	Hr.	Daily	Bare Costs	Incl. O&P
2 Laborers	$28.75	$460.00	$44.75	$716.00	$29.02	$45.05
1 Truck Driver (heavy)	29.55	236.40	45.65	365.20		
1 Water Tanker, 5000 Gal.		121.20		133.32		
1 Truck Tractor, 195 H.P.		235.60		259.16	14.87	16.35
24 L.H., Daily Totals		$1053.20		$1473.68	$43.88	$61.40
Crew B-60	Hr.	Daily	Hr.	Daily	Bare Costs	Incl. O&P
1 Labor Foreman (outside)	$30.75	$246.00	$47.90	$383.20	$33.66	$51.37
2 Laborers	28.75	460.00	44.75	716.00		
1 Equip. Oper. (crane)	39.80	318.40	60.00	480.00		
2 Equip. Oper. (light)	36.85	589.60	55.55	888.80		
1 Equip. Oper. Oiler	33.90	271.20	51.10	408.80		
1 Crawler Crane, 40 Ton		994.60		1094.06		
45 L.F. of Leads, 15K Ft. Lbs.		45.00		49.50		
1 Backhoe Loader, 48 H.P.		243.40		267.74	22.91	25.20
56 L.H., Daily Totals		$3168.20		$4288.10	$56.58	$76.57
Crew B-61	Hr.	Daily	Hr.	Daily	Bare Costs	Incl. O&P
1 Labor Foreman (outside)	$30.75	$246.00	$47.90	$383.20	$30.77	$47.54
3 Laborers	28.75	690.00	44.75	1074.00		
1 Equip. Oper. (light)	36.85	294.80	55.55	444.40		
1 Cement Mixer, 2 C.Y.		180.00		198.00		
1 Air Compressor, 160 C.F.M.		122.60		134.86	7.57	8.32
40 L.H., Daily Totals		$1533.40		$2234.46	$38.34	$55.86
Crew B-62	Hr.	Daily	Hr.	Daily	Bare Costs	Incl. O&P
2 Laborers	$28.75	$460.00	$44.75	$716.00	$31.45	$48.35
1 Equip. Oper. (light)	36.85	294.80	55.55	444.40		
1 Loader, Skid Steer, 30 HP, gas		122.00		134.20	5.08	5.59
24 L.H., Daily Totals		$876.80		$1294.60	$36.53	$53.94
Crew B-63	Hr.	Daily	Hr.	Daily	Bare Costs	Incl. O&P
4 Laborers	$28.75	$920.00	$44.75	$1432.00	$30.37	$46.91
1 Equip. Oper. (light)	36.85	294.80	55.55	444.40		
1 Loader, Skid Steer, 30 HP, gas		122.00		134.20	3.05	3.36
40 L.H., Daily Totals		$1336.80		$2010.60	$33.42	$50.27
Crew B-64	Hr.	Daily	Hr.	Daily	Bare Costs	Incl. O&P
1 Laborer	$28.75	$230.00	$44.75	$358.00	$28.65	$44.42
1 Truck Driver (light)	28.55	228.40	44.10	352.80		
1 Power Mulcher (small)		119.00		130.90		
1 Flatbed Truck, gas, 1.5 Ton		141.40		155.54	16.27	17.90
16 L.H., Daily Totals		$718.80		$997.24	$44.92	$62.33

Crew No.	Bare Costs		Incl. Subs O & P		Cost Per Labor-Hour	
Crew B-65	Hr.	Daily	Hr.	Daily	Bare Costs	Incl. O&P
1 Laborer	$28.75	$230.00	$44.75	$358.00	$28.65	$44.42
1 Truck Driver (light)	28.55	228.40	44.10	352.80		
1 Power Mulcher (large)		204.20		224.62		
1 Flatbed Truck, gas, 1.5 Ton		141.40		155.54	21.60	23.76
16 L.H., Daily Totals		$804.00		$1090.96	$50.25	$68.19
Crew B-66	Hr.	Daily	Hr.	Daily	Bare Costs	Incl. O&P
1 Equip. Oper. (light)	$36.85	$294.80	$55.55	$444.40	$36.85	$55.55
1 Loader-Backhoe		196.80		216.48	24.60	27.06
8 L.H., Daily Totals		$491.60		$660.88	$61.45	$82.61
Crew B-67	Hr.	Daily	Hr.	Daily	Bare Costs	Incl. O&P
1 Millwright	$38.20	$305.60	$56.35	$450.80	$37.52	$55.95
1 Equip. Oper. (light)	36.85	294.80	55.55	444.40		
1 Forklift, R/T, 4,000 Lb.		272.40		299.64	17.02	18.73
16 L.H., Daily Totals		$872.80		$1194.84	$54.55	$74.68
Crew B-68	Hr.	Daily	Hr.	Daily	Bare Costs	Incl. O&P
2 Millwrights	$38.20	$611.20	$56.35	$901.60	$37.75	$56.08
1 Equip. Oper. (light)	36.85	294.80	55.55	444.40		
1 Forklift, R/T, 4,000 Lb.		272.40		299.64	11.35	12.48
24 L.H., Daily Totals		$1178.40		$1645.64	$49.10	$68.57
Crew B-69	Hr.	Daily	Hr.	Daily	Bare Costs	Incl. O&P
1 Labor Foreman (outside)	$30.75	$246.00	$47.90	$383.20	$31.78	$48.88
3 Laborers	28.75	690.00	44.75	1074.00		
1 Equip Oper. (crane)	39.80	318.40	60.00	480.00		
1 Equip Oper. Oiler	33.90	271.20	51.10	408.80		
1 Hyd. Crane, 80 Ton		1214.00		1335.40	25.29	27.82
48 L.H., Daily Totals		$2739.60		$3681.40	$57.08	$76.70
Crew B-69A	Hr.	Daily	Hr.	Daily	Bare Costs	Incl. O&P
1 Labor Foreman	$30.75	$246.00	$47.90	$383.20	$31.82	$48.70
3 Laborers	28.75	690.00	44.75	1074.00		
1 Equip. Oper. (medium)	38.40	307.20	57.85	462.80		
1 Concrete Finisher	35.55	284.40	52.20	417.60		
1 Curb/Gutter Paver, 2-Track		687.60		756.36	14.32	15.76
48 L.H., Daily Totals		$2215.20		$3093.96	$46.15	$64.46
Crew B-69B	Hr.	Daily	Hr.	Daily	Bare Costs	Incl. O&P
1 Labor Foreman	$30.75	$246.00	$47.90	$383.20	$31.82	$48.70
3 Laborers	28.75	690.00	44.75	1074.00		
1 Equip. Oper. (medium)	38.40	307.20	57.85	462.80		
1 Cement Finisher	35.55	284.40	52.20	417.60		
1 Curb/Gutter Paver, 4-Track		771.00		848.10	16.06	17.67
48 L.H., Daily Totals		$2298.60		$3185.70	$47.89	$66.37
Crew B-70	Hr.	Daily	Hr.	Daily	Bare Costs	Incl. O&P
1 Labor Foreman (outside)	$30.75	$246.00	$47.90	$383.20	$33.17	$50.81
3 Laborers	28.75	690.00	44.75	1074.00		
3 Equip. Oper. (med.)	38.40	921.60	57.85	1388.40		
1 Grader, 30,000 Lbs.		505.40		555.94		
1 Ripper, beam & 1 shank		76.40		84.04		
1 Road Sweeper, SP, 8' wide		551.00		606.10		
1 F.E. Loader, W.M., 1.5 C.Y.		277.60		305.36	25.19	27.70
56 L.H., Daily Totals		$3268.00		$4397.04	$58.36	$78.52

Crew No.	Bare Costs Hr.	Daily	Incl. Subs O&P Hr.	Daily	Cost Per Labor-Hour Bare Costs	Incl. O&P
Crew B-71	Hr.	Daily	Hr.	Daily	Bare Costs	Incl. O&P
1 Labor Foreman (outside)	$30.75	$246.00	$47.90	$383.20	$33.17	$50.81
3 Laborers	28.75	690.00	44.75	1074.00		
3 Equip. Oper. (med.)	38.40	921.60	57.85	1388.40		
1 Pvmt. Profiler, 750 H.P.		5176.00		5693.60		
1 Road Sweeper, SP, 8' wide		551.00		606.10		
1 F.E. Loader, W.M., 1.5 C.Y.		277.60		305.36	107.22	117.95
56 L.H., Daily Totals		$7862.20		$9450.66	$140.40	$168.76
Crew B-72	Hr.	Daily	Hr.	Daily	Bare Costs	Incl. O&P
1 Labor Foreman (outside)	$30.75	$246.00	$47.90	$383.20	$33.83	$51.69
3 Laborers	28.75	690.00	44.75	1074.00		
4 Equip. Oper. (med.)	38.40	1228.80	57.85	1851.20		
1 Pvmt. Profiler, 750 H.P.		5176.00		5693.60		
1 Hammermill, 250 H.P.		1512.00		1663.20		
1 Windrow Loader		869.00		955.90		
1 Mix Paver 165 H.P.		1817.00		1998.70		
1 Roller, Pneum. Whl, 12 Ton		295.60		325.16	151.09	166.20
64 L.H., Daily Totals		$11834.40		$13944.96	$184.91	$217.89
Crew B-73	Hr.	Daily	Hr.	Daily	Bare Costs	Incl. O&P
1 Labor Foreman (outside)	$30.75	$246.00	$47.90	$383.20	$35.03	$53.33
2 Laborers	28.75	460.00	44.75	716.00		
5 Equip. Oper. (med.)	38.40	1536.00	57.85	2314.00		
1 Road Mixer, 310 H.P.		1819.00		2000.90		
1 Tandem Roller, 10 Ton		211.00		232.10		
1 Hammermill, 250 H.P.		1512.00		1663.20		
1 Grader, 30,000 Lbs.		505.40		555.94		
.5 F.E. Loader, W.M., 1.5 C.Y.		138.80		152.68		
.5 Truck Tractor, 195 H.P.		117.80		129.58		
.5 Water Tanker, 5000 Gal.		60.60		66.66	68.20	75.02
64 L.H., Daily Totals		$6606.60		$8214.26	$103.23	$128.35
Crew B-74	Hr.	Daily	Hr.	Daily	Bare Costs	Incl. O&P
1 Labor Foreman (outside)	$30.75	$246.00	$47.90	$383.20	$34.02	$51.92
1 Laborer	28.75	230.00	44.75	358.00		
4 Equip. Oper. (med.)	38.40	1228.80	57.85	1851.20		
2 Truck Drivers (heavy)	29.55	472.80	45.65	730.40		
1 Grader, 30,000 Lbs.		505.40		555.94		
1 Ripper, beam & 1 shank		76.40		84.04		
2 Stabilizers, 310 H.P.		2530.00		2783.00		
1 Flatbed Truck, gas, 3 Ton		185.20		203.72		
1 Chem. Spreader, Towed		44.20		48.62		
1 Roller, Vibratory, 25 Ton		565.60		622.16		
1 Water Tanker, 5000 Gal.		121.20		133.32		
1 Truck Tractor, 195 H.P.		235.60		259.16	66.62	73.28
64 L.H., Daily Totals		$6441.20		$8012.76	$100.64	$125.20
Crew B-75	Hr.	Daily	Hr.	Daily	Bare Costs	Incl. O&P
1 Labor Foreman (outside)	$30.75	$246.00	$47.90	$383.20	$34.66	$52.81
1 Laborer	28.75	230.00	44.75	358.00		
4 Equip. Oper. (med.)	38.40	1228.80	57.85	1851.20		
1 Truck Driver (heavy)	29.55	236.40	45.65	365.20		
1 Grader, 30,000 Lbs.		505.40		555.94		
1 Ripper, beam & 1 shank		76.40		84.04		
2 Stabilizers, 310 H.P.		2530.00		2783.00		
1 Dist. Tanker, 3000 Gallon		273.00		300.30		
1 Truck Tractor, 240 H.P.		273.00		300.30		
1 Roller, Vibratory, 25 Ton		565.60		622.16	75.42	82.96
56 L.H., Daily Totals		$6164.60		$7603.34	$110.08	$135.77

Crew No.	Bare Costs Hr.	Daily	Incl. Subs O&P Hr.	Daily	Cost Per Labor-Hour Bare Costs	Incl. O&P
Crew B-76	Hr.	Daily	Hr.	Daily	Bare Costs	Incl. O&P
1 Dock Builder Foreman	$37.90	$303.20	$62.05	$496.40	$36.77	$58.54
5 Dock Builders	35.90	1436.00	58.75	2350.00		
2 Equip. Oper. (crane)	39.80	636.80	60.00	960.00		
1 Equip. Oper. Oiler	33.90	271.20	51.10	408.80		
1 Crawler Crane, 50 Ton		1131.00		1244.10		
1 Barge, 400 Ton		294.20		323.62		
1 Hammer, 15K Ft. Lbs.		360.00		396.00		
60 L.F. of Leads, 15K Ft. Lbs.		60.00		66.00		
1 Air Compressor, 600 C.F.M.		396.20		435.82		
2 -50' Air Hoses, 3"		38.10		41.91	31.66	34.83
72 L.H., Daily Totals		$4926.70		$6722.65	$68.43	$93.37
Crew B-76A	Hr.	Daily	Hr.	Daily	Bare Costs	Incl. O&P
1 Laborer Foreman	$30.75	$246.00	$47.90	$383.20	$31.02	$47.84
5 Laborers	28.75	1150.00	44.75	1790.00		
1 Equip. Oper. (crane)	39.80	318.40	60.00	480.00		
1 Equip. Oper. Oiler	33.90	271.20	51.10	408.80		
1 Crawler Crane, 50 Ton		1131.00		1244.10		
1 Barge, 400 Ton		294.20		323.62	22.27	24.50
64 L.H., Daily Totals		$3410.80		$4629.72	$53.29	$72.34
Crew B-77	Hr.	Daily	Hr.	Daily	Bare Costs	Incl. O&P
1 Labor Foreman	$30.75	$246.00	$47.90	$383.20	$29.11	$45.25
3 Laborers	28.75	690.00	44.75	1074.00		
1 Truck Driver (light)	28.55	228.40	44.10	352.80		
1 Crack Cleaner, 25 H.P.		48.20		53.02		
1 Crack Filler, Trailer Mtd.		162.40		178.64		
1 Flatbed Truck, gas, 3 Ton		185.20		203.72	9.89	10.88
40 L.H., Daily Totals		$1560.20		$2245.38	$39.01	$56.13
Crew B-78	Hr.	Daily	Hr.	Daily	Bare Costs	Incl. O&P
1 Labor Foreman	$30.75	$246.00	$47.90	$383.20	$29.05	$45.17
4 Laborers	28.75	920.00	44.75	1432.00		
1 Truck Driver (light)	28.55	228.40	44.10	352.80		
1 Paint Striper, S.P.		138.00		151.80		
1 Flatbed Truck, gas, 3 Ton		185.20		203.72		
1 Pickup Truck, 3/4 Ton		84.80		93.28	8.50	9.35
48 L.H., Daily Totals		$1802.40		$2616.80	$37.55	$54.52
Crew B-78A	Hr.	Daily	Hr.	Daily	Bare Costs	Incl. O&P
1 Equip. Oper. (light)	$36.85	$294.80	$55.55	$444.40	$36.85	$55.55
1 Line Rem. (metal balls) 115 HP		857.60		943.36	107.20	117.92
8 L.H., Daily Totals		$1152.40		$1387.76	$144.05	$173.47
Crew B-78B	Hr.	Daily	Hr.	Daily	Bare Costs	Incl. O&P
2 Laborers	$28.75	$460.00	$44.75	$716.00	$29.65	$45.95
.25 Equip. Oper. (light)	36.85	73.70	55.55	111.10		
1 Pickup Truck, 3/4 Ton		84.80		93.28		
1 Line Rem., 11 HP, walk behind		47.80		52.58		
.25 Road Sweeper, SP, 8' wide		137.75		151.53	15.02	16.52
18 L.H., Daily Totals		$804.05		$1124.48	$44.67	$62.47
Crew B-79	Hr.	Daily	Hr.	Daily	Bare Costs	Incl. O&P
1 Labor Foreman	$30.75	$246.00	$47.90	$383.20	$29.11	$45.25
3 Laborers	28.75	690.00	44.75	1074.00		
1 Truck Driver (light)	28.55	228.40	44.10	352.80		
1 Thermo. Striper, T.M.		596.20		655.82		
1 Flatbed Truck, Gas, 3 Ton		185.20		203.72		
2 Pickup Trucks, 3/4 Ton		169.60		186.56	23.77	26.15
40 L.H., Daily Totals		$2115.40		$2856.10	$52.88	$71.40

Means CostWorks®

Maximize Your Estimating & Budgeting Efforts with the RSMeans Online Construction Cost Estimator... or Means CostWorks 2007 CD-Rom

RSMeans Online Construction Cost Estimator: Estimates that are efficient, customized & online

- Create database-driven, customized estimates based on your specific building needs in a 24/7 secure online environment.

- Have immediate access to the entire RSMeans unit cost (70,000+ line items) & assemblies (19,000) databases, as well as referencing both union & open shop labor rates.

- Streamline your process with an optional "read-write" capability.

Means CostWorks 2007 CD-ROM The cost data titles you know and trust—delivered electronically in a quick & easy CD format.

- Purchase your titles separately or in value-priced packages & turn your desktop or laptop PC into a comprehensive cost library!

- Easily search & organize detailed cost data.

- Localize costs to your geographic area.

- Change & calculate results.

RSMeans

RSMeans Online Construction Cost Estimator

Estimates that are Efficient, Customized, and Online!

No More "Boilerplates" that Can Compromise Project Planning, Accurate Budgeting, or Your Bottom Line!

Your RSMeans Online Construction Cost Estimator Subscription Lets You:

- Select from the entire RSMeans database of authoritative and timely Unit and Assembly costs without referencing individual cost data titles. Your subscription includes over 70,000 unit cost lines and 19,000 assemblies.

- Choose from a comprehensive list of Facilities Repair & Remodeling costs – Easily quantify, localize and format your data.

- Effectively prepare or verify bids, establish budgets, and manage change orders.

- Allows individuals or large work groups to confidently manage construction cost information in a secure environment that is available 24/7 from any Web-enabled computer.

- Print your estimate directly or easily export to MS EXCEL and define user-selected report formats!

Create & Manage Your Own Unique Construction Cost Database With Our Optional Custom Database Utility!

- Adjust and customize RSMeans material cost data with read/write capabilities. Clone RSMeans lines or add your own cost lines!

- Easily reference both union and open shop wage rates for 46 labor trades.

An invaluable resource for Project Managers, Contractors, Facility Managers & Owners . . . or anyone who requires accurate *Customized* construction estimates.

Let Means CostWorks 2007 CD

take your estimating and budgeting to the next level.

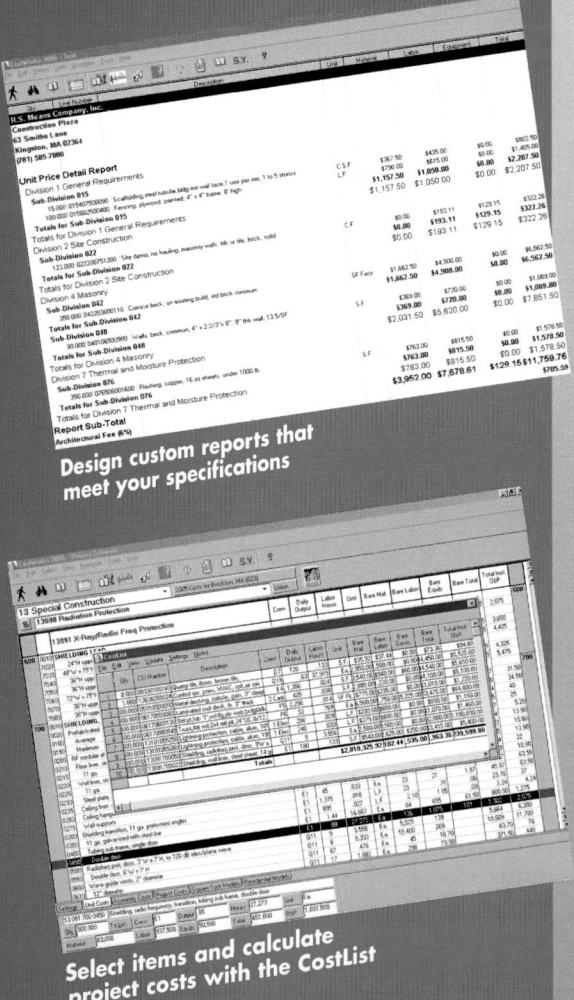

Design custom reports that meet your specifications

Select items and calculate project costs with the CostList

Means CostWorks 2007 subscriptions offer a choice of 17 detailed cost titles. Turn your desktop or laptop PC into a comprehensive cost library. *Purchase your titles separately or in value-priced packages.*

Your *Means CostWorks 2007* CD subscription now includes, Free:
- RSMeans Estimating HOTLINE
- RSMeans Quarterly Update Service
- CostWorks CD Estimator
  (if ordering a CD-Rom Package).

Cost Data delivered electronically in a CD format:
Localize costs to your area – determine best-cost solutions – change and calculate results automatically.

Save hundreds on specially priced CostWorks CD Package subscriptions!

Builder's Package
Includes:
- Residential Cost Data • Light Commercial Cost Data
- Site Work & Landscape Cost Data • Concrete & Masonry Cost Data
- Open Shop Building Construction Cost Data

Just **$469.95**
Over $300 off the price of buying each book separately!

Building Professional's Package
Includes:
- Building Construction Cost Data • Mechanical Cost Data
- Plumbing Cost Data • Electrical Cost Data • Square Foot Costs

Just **$549.95**
Over $250 off the price of buying each book separately!

Facility Manager's Package
Includes:
- Facilities Construction Cost Data • Facilities Maintenance & Repair Cost Data
- Repair & Remodeling Cost Data • Building Construction Cost Data

Just **$712.95**
Over $300 off the price of buying each book separately!

Design Professional's Package
Includes:
- Building Construction Cost Data • Square Foot Costs
- Interior Cost Data • Assemblies Cost Data

Just **$539.95**
Over $200 off the price of buying each book separately!

Quick and easy... Get this powerful tool before you estimate your next project!

Get started with *Means CostWorks* today!

Fax: 1-800-632-6732, phone: 1-800-334-3509, or at www.rsmeans.com/cwcw.asp
For information on network pricing call 1-800-334-3509.

Catalog No.	*Means CostWorks 2007* CD Subscription (12 Months)	Price	Cost
65067	Assemblies Cost Data	$241.95	
65017	Building Construction Cost Data	$169.95	
65317	Metric Construction Cost Data	$201.95	
65157	Open Shop Building Construction Cost Data	$169.95	
65117	Concrete & Masonry Cost Data	$169.95	
65037	Electrical Cost Data	$169.95	
65167	Heavy Construction Cost Data	$169.95	
65097	Interior Cost Data	$169.95	
65187	Light Commercial Cost Data	$210.95	
65027	Mechanical Cost Data	$169.95	
65217	Plumbing Cost Data	$169.95	
65287	Site Work & Landscape Cost Data	$169.95	
65057	Square Foot Costs	$239.95	
65607*	CostWorks CD Estimator	$299.95	
65207	Facilities Construction Cost Data	$392.95	
65047	Repair & Remodeling Cost Data	$169.95	
65177	Residential Cost Data	$195.95	
65307	Facilities Maintenance & Repair Cost Data	$392.95	

CD Package Subscription* (12 Months)

Catalog No.	Title	Price	Cost
65517	Builder's Package	$469.95	
65527	Building Professional's Package	$549.95	
65537	Facility Manager's Package	$712.95	
65547	Design Professional's Package	$539.95	

(Packages include FREE Means CD Estimator)
*All items shipped on a single CD with software code to unlock specific selections purchased.
Buy additional keys any time you need to unlock additional cost data books!*

RSMeans Online Construction Cost Estimator Subscription

Catalog No.	Title	Price	Cost
66100	1 Yr. Subscription – Standard Edition	Call for Pricing	
66300	1 Yr. Subscription – Customized Data Base Utility	Call for Pricing	

Name: _____
Company: _____
Street: _____
City: _____ State: _____ Zip: _____
Phone: (___) _____ Fax: (___) _____
E-mail: _____

☐ Bill me   ☐ Please charge my credit card
☐ Visa ☐ MasterCard ☐ American Express ☐ Discover

Card # _____ Expiration Date: _____
Cardholder Signature: _____

Subtotal
MA residents please add 5% sales tax
Shipping & Handling (CD) **$6.00**
Total

Reed Construction Data®

CWCW-2007

Crews

Crew B-79A

Crew B-79A	Bare Costs Hr.	Daily	Incl. Subs O&P Hr.	Daily	Cost Per Labor-Hour Bare Costs	Incl. O&P
1.5 Equip. Oper. (light)	$36.85	$442.20	$55.55	$666.60	$36.85	$55.55
.5 Line Remov. (grinder) 115 HP		411.20		452.32		
1 Line Rem. (metal balls) 115 HP		857.60		943.36	105.73	116.31
12 L.H., Daily Totals		$1711.00		$2062.28	$142.58	$171.86

Crew B-80

Crew B-80	Bare Costs Hr.	Daily	Incl. Subs O&P Hr.	Daily	Bare Costs	Incl. O&P
1 Labor Foreman	$30.75	$246.00	$47.90	$383.20	$31.23	$48.08
1 Laborer	28.75	230.00	44.75	358.00		
1 Truck Driver (light)	28.55	228.40	44.10	352.80		
1 Equip. Oper. (light)	36.85	294.80	55.55	444.40		
1 Flatbed Truck, gas, 3 Ton		185.20		203.72		
1 Fence Post Auger, T.M.		375.20		412.72	17.51	19.26
32 L.H., Daily Totals		$1559.60		$2154.84	$48.74	$67.34

Crew B-80A

Crew B-80A	Bare Costs Hr.	Daily	Incl. Subs O&P Hr.	Daily	Bare Costs	Incl. O&P
3 Laborers	$28.75	$690.00	$44.75	$1074.00	$28.75	$44.75
1 Flatbed Truck, gas, 3 Ton		185.20		203.72	7.72	8.49
24 L.H., Daily Totals		$875.20		$1277.72	$36.47	$53.24

Crew B-80B

Crew B-80B	Bare Costs Hr.	Daily	Incl. Subs O&P Hr.	Daily	Bare Costs	Incl. O&P
3 Laborers	$28.75	$690.00	$44.75	$1074.00	$30.77	$47.45
1 Equip. Oper. (light)	36.85	294.80	55.55	444.40		
1 Crane, Flatbed Mounted, 3 Ton		271.60		298.76	8.49	9.34
32 L.H., Daily Totals		$1256.40		$1817.16	$39.26	$56.79

Crew B-80C

Crew B-80C	Bare Costs Hr.	Daily	Incl. Subs O&P Hr.	Daily	Bare Costs	Incl. O&P
2 Laborers	$28.75	$460.00	$44.75	$716.00	$28.68	$44.53
1 Truck Driver (light)	28.55	228.40	44.10	352.80		
1 Flatbed Truck, gas, 1.5 Ton		141.40		155.54		
1 Manual fence post auger, gas		6.00		6.60	6.14	6.76
24 L.H., Daily Totals		$835.80		$1230.94	$34.83	$51.29

Crew B-81

Crew B-81	Bare Costs Hr.	Daily	Incl. Subs O&P Hr.	Daily	Bare Costs	Incl. O&P
1 Laborer	$28.75	$230.00	$44.75	$358.00	$32.23	$49.42
1 Equip. Oper. (med.)	38.40	307.20	57.85	462.80		
1 Truck Driver (heavy)	29.55	236.40	45.65	365.20		
1 Hydromulcher, T.M.		216.20		237.82		
1 Truck Tractor, 195 H.P.		235.60		259.16	18.82	20.71
24 L.H., Daily Totals		$1225.40		$1682.98	$51.06	$70.12

Crew B-82

Crew B-82	Bare Costs Hr.	Daily	Incl. Subs O&P Hr.	Daily	Bare Costs	Incl. O&P
1 Laborer	$28.75	$230.00	$44.75	$358.00	$32.80	$50.15
1 Equip. Oper. (light)	36.85	294.80	55.55	444.40		
1 Horiz. Borer, 6 H.P.		71.20		78.32	4.45	4.89
16 L.H., Daily Totals		$596.00		$880.72	$37.25	$55.05

Crew B-83

Crew B-83	Bare Costs Hr.	Daily	Incl. Subs O&P Hr.	Daily	Bare Costs	Incl. O&P
1 Tugboat Captain	$38.40	$307.20	$57.85	$462.80	$33.58	$51.30
1 Tugboat Hand	28.75	230.00	44.75	358.00		
1 Tugboat, 250 H.P.		550.20		605.22	34.39	37.83
16 L.H., Daily Totals		$1087.40		$1426.02	$67.96	$89.13

Crew B-84

Crew B-84	Bare Costs Hr.	Daily	Incl. Subs O&P Hr.	Daily	Bare Costs	Incl. O&P
1 Equip. Oper. (med.)	$38.40	$307.20	$57.85	$462.80	$38.40	$57.85
1 Rotary Mower/Tractor		235.00		258.50	29.38	32.31
8 L.H., Daily Totals		$542.20		$721.30	$67.78	$90.16

Crew B-85

Crew B-85	Bare Costs Hr.	Daily	Incl. Subs O&P Hr.	Daily	Cost Per Labor-Hour Bare Costs	Incl. O&P
3 Laborers	$28.75	$690.00	$44.75	$1074.00	$30.84	$47.55
1 Equip. Oper. (med.)	38.40	307.20	57.85	462.80		
1 Truck Driver (heavy)	29.55	236.40	45.65	365.20		
1 Aerial Lift Truck, 80'		546.80		601.48		
1 Brush Chipper, 12", 130 H.P.		183.60		201.96		
1 Pruning Saw, Rotary		7.45		8.20	18.45	20.29
40 L.H., Daily Totals		$1971.45		$2713.64	$49.29	$67.84

Crew B-86

Crew B-86	Bare Costs Hr.	Daily	Incl. Subs O&P Hr.	Daily	Bare Costs	Incl. O&P
1 Equip. Oper. (med.)	$38.40	$307.20	$57.85	$462.80	$38.40	$57.85
1 Stump Chipper, S.P.		106.30		116.93	13.29	14.62
8 L.H., Daily Totals		$413.50		$579.73	$51.69	$72.47

Crew B-86A

Crew B-86A	Bare Costs Hr.	Daily	Incl. Subs O&P Hr.	Daily	Bare Costs	Incl. O&P
1 Equip. Oper. (medium)	$38.40	$307.20	$57.85	$462.80	$38.40	$57.85
1 Grader, 30,000 Lbs.		505.40		555.94	63.17	69.49
8 L.H., Daily Totals		$812.60		$1018.74	$101.58	$127.34

Crew B-86B

Crew B-86B	Bare Costs Hr.	Daily	Incl. Subs O&P Hr.	Daily	Bare Costs	Incl. O&P
1 Equip. Oper. (medium)	$38.40	$307.20	$57.85	$462.80	$38.40	$57.85
1 Dozer, 200 H.P.		988.40		1087.24	123.55	135.91
8 L.H., Daily Totals		$1295.60		$1550.04	$161.95	$193.76

Crew B-87

Crew B-87	Bare Costs Hr.	Daily	Incl. Subs O&P Hr.	Daily	Bare Costs	Incl. O&P
1 Laborer	$28.75	$230.00	$44.75	$358.00	$36.47	$55.23
4 Equip. Oper. (med.)	38.40	1228.80	57.85	1851.20		
2 Feller Bunchers, 100 H.P.		1026.80		1129.48		
1 Log Chipper, 22" Tree		520.00		572.00		
1 Dozer, 105 H.P.		494.80		544.28		
1 Chainsaw, gas, 36" Long		34.40		37.84	51.90	57.09
40 L.H., Daily Totals		$3534.80		$4492.80	$88.37	$112.32

Crew B-88

Crew B-88	Bare Costs Hr.	Daily	Incl. Subs O&P Hr.	Daily	Bare Costs	Incl. O&P
1 Laborer	$28.75	$230.00	$44.75	$358.00	$37.02	$55.98
6 Equip. Oper. (med.)	38.40	1843.20	57.85	2776.80		
2 Feller Bunchers, 100 H.P.		1026.80		1129.48		
1 Log Chipper, 22" Tree		520.00		572.00		
2 Log Skidders, 50 H.P.		1632.80		1796.08		
1 Dozer, 105 H.P.		494.80		544.28		
1 Chainsaw, gas, 36" Long		34.40		37.84	66.23	72.85
56 L.H., Daily Totals		$5782.00		$7214.48	$103.25	$128.83

Crew B-89

Crew B-89	Bare Costs Hr.	Daily	Incl. Subs O&P Hr.	Daily	Bare Costs	Incl. O&P
1 Equip. Oper. (light)	$36.85	$294.80	$55.55	$444.40	$32.70	$49.83
1 Truck Driver (light)	28.55	228.40	44.10	352.80		
1 Flatbed Truck, gas, 3 Ton		185.20		203.72		
1 Concrete Saw		118.40		130.24		
1 Water Tank, 65 Gal.		16.40		18.04	20.00	22.00
16 L.H., Daily Totals		$843.20		$1149.20	$52.70	$71.83

Crew B-89A

Crew B-89A	Bare Costs Hr.	Daily	Incl. Subs O&P Hr.	Daily	Bare Costs	Incl. O&P
1 Skilled Worker	$38.00	$304.00	$59.15	$473.20	$33.38	$51.95
1 Laborer	28.75	230.00	44.75	358.00		
1 Core Drill (large)		105.00		115.50	6.56	7.22
16 L.H., Daily Totals		$639.00		$946.70	$39.94	$59.17

Crew No.	Bare Costs		Incl. Subs O & P		Cost Per Labor-Hour	
Crew B-89B	Hr.	Daily	Hr.	Daily	Bare Costs	Incl. O&P
1 Equip. Oper. (light)	$36.85	$294.80	$55.55	$444.40	$32.70	$49.83
1 Truck Driver, Light	28.55	228.40	44.10	352.80		
1 Wall Saw, Hydraulic, 10 H.P.		78.20		86.02		
1 Generator, Diesel, 100 KW		305.60		336.16		
1 Water Tank, 65 Gal.		16.40		18.04		
1 Flatbed Truck, gas, 3 Ton		185.20		203.72	36.59	40.25
16 L.H., Daily Totals		$1108.60		$1441.14	$69.29	$90.07

Crew No.	Bare Costs		Incl. Subs O & P		Cost Per Labor-Hour	
Crew B-90	Hr.	Daily	Hr.	Daily	Bare Costs	Incl. O&P
1 Labor Foreman (outside)	$30.75	$246.00	$47.90	$383.20	$31.23	$48.07
3 Laborers	28.75	690.00	44.75	1074.00		
2 Equip. Oper. (light)	36.85	589.60	55.55	888.80		
2 Truck Drivers (heavy)	29.55	472.80	45.65	730.40		
1 Road Mixer, 310 H.P.		1819.00		2000.90		
1 Dist. Truck, 2000 Gal.		247.40		272.14	32.29	35.52
64 L.H., Daily Totals		$4064.80		$5349.44	$63.51	$83.59

Crew No.	Bare Costs		Incl. Subs O & P		Cost Per Labor-Hour	
Crew B-90A	Hr.	Daily	Hr.	Daily	Bare Costs	Incl. O&P
1 Labor Foreman	$30.75	$246.00	$47.90	$383.20	$34.55	$52.69
2 Laborers	28.75	460.00	44.75	716.00		
4 Equip. Oper. (medium)	38.40	1228.80	57.85	1851.20		
2 Graders, 30,000 Lbs.		1010.80		1111.88		
1 Tandem Roller, 10 Ton		211.00		232.10		
1 Roller, Pneum. Whl, 12 Ton		295.60		325.16	27.10	29.81
56 L.H., Daily Totals		$3452.20		$4619.54	$61.65	$82.49

Crew No.	Bare Costs		Incl. Subs O & P		Cost Per Labor-Hour	
Crew B-90B	Hr.	Daily	Hr.	Daily	Bare Costs	Incl. O&P
1 Labor Foreman	$30.75	$246.00	$47.90	$383.20	$33.91	$51.83
2 Laborers	28.75	460.00	44.75	716.00		
3 Equip. Oper. (medium)	38.40	921.60	57.85	1388.40		
1 Tandem Roller, 10 Ton		211.00		232.10		
1 Roller, Pneum. Whl, 12 Ton		295.60		325.16		
1 Road Mixer, 310 H.P.		1819.00		2000.90	48.45	53.30
48 L.H., Daily Totals		$3953.20		$5045.76	$82.36	$105.12

Crew No.	Bare Costs		Incl. Subs O & P		Cost Per Labor-Hour	
Crew B-91	Hr.	Daily	Hr.	Daily	Bare Costs	Incl. O&P
1 Labor Foreman (outside)	$30.75	$246.00	$47.90	$383.20	$33.92	$51.81
2 Laborers	28.75	460.00	44.75	716.00		
4 Equip. Oper. (med.)	38.40	1228.80	57.85	1851.20		
1 Truck Driver (heavy)	29.55	236.40	45.65	365.20		
1 Dist. Tanker, 3000 Gallon		273.00		300.30		
1 Truck Tractor, 240 H.P.		273.00		300.30		
1 Aggreg. Spreader, S.P.		818.80		900.68		
1 Roller, Pneum. Whl, 12 Ton		295.60		325.16		
1 Tandem Roller, 10 Ton		211.00		232.10	29.24	32.16
64 L.H., Daily Totals		$4042.60		$5374.14	$63.17	$83.97

Crew No.	Bare Costs		Incl. Subs O & P		Cost Per Labor-Hour	
Crew B-92	Hr.	Daily	Hr.	Daily	Bare Costs	Incl. O&P
1 Labor Foreman (outside)	$30.75	$246.00	$47.90	$383.20	$29.25	$45.54
3 Laborers	28.75	690.00	44.75	1074.00		
1 Crack Cleaner, 25 H.P.		48.20		53.02		
1 Air Compressor, 60 C.F.M.		108.80		119.68		
1 Tar Kettle, T.M.		51.35		56.48		
1 Flatbed Truck, gas, 3 Ton		185.20		203.72	12.30	13.53
32 L.H., Daily Totals		$1329.55		$1890.11	$41.55	$59.07

Crew No.	Bare Costs		Incl. Subs O & P		Cost Per Labor-Hour	
Crew B-93	Hr.	Daily	Hr.	Daily	Bare Costs	Incl. O&P
1 Equip. Oper. (med.)	$38.40	$307.20	$57.85	$462.80	$38.40	$57.85
1 Feller Buncher, 100 H.P.		513.40		564.74	64.17	70.59
8 L.H., Daily Totals		$820.60		$1027.54	$102.58	$128.44

Crew No.	Bare Costs		Incl. Subs O & P		Cost Per Labor-Hour	
Crew B-94A	Hr.	Daily	Hr.	Daily	Bare Costs	Incl. O&P
1 Laborer	$28.75	$230.00	$44.75	$358.00	$28.75	$44.75
1 Diaphragm Water Pump, 2"		57.40		63.14		
1 -20' Suction Hose, 2"		1.95		2.15		
2 -50' Discharge Hoses, 2"		1.80		1.98	7.64	8.41
8 L.H., Daily Totals		$291.15		$425.26	$36.39	$53.16

Crew No.	Bare Costs		Incl. Subs O & P		Cost Per Labor-Hour	
Crew B-94B	Hr.	Daily	Hr.	Daily	Bare Costs	Incl. O&P
1 Laborer	$28.75	$230.00	$44.75	$358.00	$28.75	$44.75
1 Diaphragm Water Pump, 4"		78.20		86.02		
1 -20' Suction Hose, 4"		3.25		3.58		
2 -50' Discharge Hoses, 4"		4.70		5.17	10.77	11.85
8 L.H., Daily Totals		$316.15		$452.76	$39.52	$56.60

Crew No.	Bare Costs		Incl. Subs O & P		Cost Per Labor-Hour	
Crew B-94C	Hr.	Daily	Hr.	Daily	Bare Costs	Incl. O&P
1 Laborer	$28.75	$230.00	$44.75	$358.00	$28.75	$44.75
1 Centrifugal Water Pump, 3"		63.40		69.74		
1 -20' Suction Hose, 3"		3.05		3.36		
2 -50' Discharge Hoses, 3"		3.50		3.85	8.74	9.62
8 L.H., Daily Totals		$299.95		$434.94	$37.49	$54.37

Crew No.	Bare Costs		Incl. Subs O & P		Cost Per Labor-Hour	
Crew B-94D	Hr.	Daily	Hr.	Daily	Bare Costs	Incl. O&P
1 Laborer	$28.75	$230.00	$44.75	$358.00	$28.75	$44.75
1 Centr. Water Pump, 6"		278.20		306.02		
1 -20' Suction Hose, 6"		11.50		12.65		
2 -50' Discharge Hoses, 6"		12.60		13.86	37.79	41.57
8 L.H., Daily Totals		$532.30		$690.53	$66.54	$86.32

Crew No.	Bare Costs		Incl. Subs O & P		Cost Per Labor-Hour	
Crew B-95A	Hr.	Daily	Hr.	Daily	Bare Costs	Incl. O&P
1 Equip. Oper. (crane)	$39.80	$318.40	$60.00	$480.00	$34.27	$52.38
1 Laborer	28.75	230.00	44.75	358.00		
1 Hyd. Excavator, 5/8 C.Y.		462.20		508.42	28.89	31.78
16 L.H., Daily Totals		$1010.60		$1346.42	$63.16	$84.15

Crew No.	Bare Costs		Incl. Subs O & P		Cost Per Labor-Hour	
Crew B-95B	Hr.	Daily	Hr.	Daily	Bare Costs	Incl. O&P
1 Equip. Oper. (crane)	$39.80	$318.40	$60.00	$480.00	$34.27	$52.38
1 Laborer	28.75	230.00	44.75	358.00		
1 Hyd. Excavator, 1.5 C.Y.		775.60		853.16	48.48	53.32
16 L.H., Daily Totals		$1324.00		$1691.16	$82.75	$105.70

Crew No.	Bare Costs		Incl. Subs O & P		Cost Per Labor-Hour	
Crew B-95C	Hr.	Daily	Hr.	Daily	Bare Costs	Incl. O&P
1 Equip. Oper. (crane)	$39.80	$318.40	$60.00	$480.00	$34.27	$52.38
1 Laborer	28.75	230.00	44.75	358.00		
1 Hyd. Excavator, 2.5 C.Y.		1333.00		1466.30	83.31	91.64
16 L.H., Daily Totals		$1881.40		$2304.30	$117.59	$144.02

Crew No.	Bare Costs		Incl. Subs O & P		Cost Per Labor-Hour	
Crew C-1	Hr.	Daily	Hr.	Daily	Bare Costs	Incl. O&P
3 Carpenters	$36.70	$880.80	$57.15	$1371.60	$34.71	$54.05
1 Laborer	28.75	230.00	44.75	358.00		
32 L.H., Daily Totals		$1110.80		$1729.60	$34.71	$54.05

Crew No.	Bare Costs		Incl. Subs O & P		Cost Per Labor-Hour	
Crew C-2	Hr.	Daily	Hr.	Daily	Bare Costs	Incl. O&P
1 Carpenter Foreman (out)	$38.70	$309.60	$60.25	$482.00	$35.71	$55.60
4 Carpenters	36.70	1174.40	57.15	1828.80		
1 Laborer	28.75	230.00	44.75	358.00		
48 L.H., Daily Totals		$1714.00		$2668.80	$35.71	$55.60

Crews

Crew No.	Bare Costs Hr.	Daily	Incl. Subs O & P Hr.	Daily	Cost Per Labor-Hour Bare Costs	Incl. O&P
Crew C-2A	Hr.	Daily	Hr.	Daily	Bare Costs	Incl. O&P
1 Carpenter Foreman (out)	$38.70	$309.60	$60.25	$482.00	$35.52	$54.77
3 Carpenters	36.70	880.80	57.15	1371.60		
1 Cement Finisher	35.55	284.40	52.20	417.60		
1 Laborer	28.75	230.00	44.75	358.00		
48 L.H., Daily Totals		$1704.80		$2629.20	$35.52	$54.77
Crew C-3	Hr.	Daily	Hr.	Daily	Bare Costs	Incl. O&P
1 Rodman Foreman	$43.30	$346.40	$71.05	$568.40	$37.86	$60.89
4 Rodmen (reinf.)	41.30	1321.60	67.75	2168.00		
1 Equip. Oper. (light)	36.85	294.80	55.55	444.40		
2 Laborers	28.75	460.00	44.75	716.00		
3 Stressing Equipment		24.60		27.06		
.5 Grouting Equipment		73.00		80.30	1.52	1.68
64 L.H., Daily Totals		$2520.40		$4004.16	$39.38	$62.56
Crew C-4	Hr.	Daily	Hr.	Daily	Bare Costs	Incl. O&P
1 Rodman Foreman	$43.30	$346.40	$71.05	$568.40	$41.80	$68.58
3 Rodmen (reinf.)	41.30	991.20	67.75	1626.00		
3 Stressing Equipment		24.60		27.06	.77	.85
32 L.H., Daily Totals		$1362.20		$2221.46	$42.57	$69.42
Crew C-5	Hr.	Daily	Hr.	Daily	Bare Costs	Incl. O&P
1 Rodman Foreman	$43.30	$346.40	$71.05	$568.40	$40.31	$64.74
4 Rodmen (reinf.)	41.30	1321.60	67.75	2168.00		
1 Equip. Oper. (crane)	39.80	318.40	60.00	480.00		
1 Equip. Oper. Oiler	33.90	271.20	51.10	408.80		
1 Hyd. Crane, 25 Ton		739.60		813.56	13.21	14.53
56 L.H., Daily Totals		$2997.20		$4438.76	$53.52	$79.26
Crew C-6	Hr.	Daily	Hr.	Daily	Bare Costs	Incl. O&P
1 Labor Foreman (outside)	$30.75	$246.00	$47.90	$383.20	$30.22	$46.52
4 Laborers	28.75	920.00	44.75	1432.00		
1 Cement Finisher	35.55	284.40	52.20	417.60		
2 Gas Engine Vibrators		42.80		47.08	.89	.98
48 L.H., Daily Totals		$1493.20		$2279.88	$31.11	$47.50
Crew C-7	Hr.	Daily	Hr.	Daily	Bare Costs	Incl. O&P
1 Labor Foreman (outside)	$30.75	$246.00	$47.90	$383.20	$31.37	$48.09
5 Laborers	28.75	1150.00	44.75	1790.00		
1 Cement Finisher	35.55	284.40	52.20	417.60		
1 Equip. Oper. (med.)	38.40	307.20	57.85	462.80		
1 Equip. Oper. (oiler)	33.90	271.20	51.10	408.80		
2 Gas Engine Vibrators		42.80		47.08		
1 Concrete Bucket, 1 C.Y.		17.80		19.58		
1 Hyd. Crane, 55 Ton		1060.00		1166.00	15.56	17.12
72 L.H., Daily Totals		$3379.40		$4695.06	$46.94	$65.21
Crew C-7A	Hr.	Daily	Hr.	Daily	Bare Costs	Incl. O&P
1 Labor Foreman (outside)	$30.75	$246.00	$47.90	$383.20	$29.20	$45.37
5 Laborers	28.75	1150.00	44.75	1790.00		
2 Truck Drivers (Heavy)	29.55	472.80	45.65	730.40		
2 Conc. Transit Mixers		1588.40		1747.24	24.82	27.30
64 L.H., Daily Totals		$3457.20		$4650.84	$54.02	$72.67

Crew No.	Bare Costs Hr.	Daily	Incl. Subs O & P Hr.	Daily	Cost Per Labor-Hour Bare Costs	Incl. O&P
Crew C-7B	Hr.	Daily	Hr.	Daily	Bare Costs	Incl. O&P
1 Labor Foreman (outside)	$30.75	$246.00	$47.90	$383.20	$31.02	$47.84
5 Laborers	28.75	1150.00	44.75	1790.00		
1 Equipment Operator (heavy)	39.80	318.40	60.00	480.00		
1 Equipment Oiler	33.90	271.20	51.10	408.80		
1 Conc. Bucket, 2 C.Y.		27.20		29.92		
1 Lattice Boom Crane, 165 Ton		2022.00		2224.20	32.02	35.22
64 L.H., Daily Totals		$4034.80		$5316.12	$63.04	$83.06
Crew C-7C	Hr.	Daily	Hr.	Daily	Bare Costs	Incl. O&P
1 Labor Foreman (outside)	$30.75	$246.00	$47.90	$383.20	$31.41	$48.42
5 Laborers	28.75	1150.00	44.75	1790.00		
2 Equipment Operators (medium)	38.40	614.40	57.85	925.60		
2 F.E. Loaders, W.M., 4 C.Y.		976.00		1073.60	15.25	16.77
64 L.H., Daily Totals		$2986.40		$4172.40	$46.66	$65.19
Crew C-7D	Hr.	Daily	Hr.	Daily	Bare Costs	Incl. O&P
1 Labor Foreman (outside)	$30.75	$246.00	$47.90	$383.20	$30.41	$47.07
5 Laborers	28.75	1150.00	44.75	1790.00		
1 Equipment Operator (med.)	38.40	307.20	57.85	462.80		
1 Concrete Conveyer		160.40		176.44	2.86	3.15
56 L.H., Daily Totals		$1863.60		$2812.44	$33.28	$50.22
Crew C-8	Hr.	Daily	Hr.	Daily	Bare Costs	Incl. O&P
1 Labor Foreman (outside)	$30.75	$246.00	$47.90	$383.20	$32.36	$49.20
3 Laborers	28.75	690.00	44.75	1074.00		
2 Cement Finishers	35.55	568.80	52.20	835.20		
1 Equip. Oper. (med.)	38.40	307.20	57.85	462.80		
1 Concrete Pump (small)		708.60		779.46	12.65	13.92
56 L.H., Daily Totals		$2520.60		$3534.66	$45.01	$63.12
Crew C-8A	Hr.	Daily	Hr.	Daily	Bare Costs	Incl. O&P
1 Labor Foreman (outside)	$30.75	$246.00	$47.90	$383.20	$31.35	$47.76
3 Laborers	28.75	690.00	44.75	1074.00		
2 Cement Finishers	35.55	568.80	52.20	835.20		
48 L.H., Daily Totals		$1504.80		$2292.40	$31.35	$47.76
Crew C-8B	Hr.	Daily	Hr.	Daily	Bare Costs	Incl. O&P
1 Labor Foreman (outside)	$30.75	$246.00	$47.90	$383.20	$31.08	$48.00
3 Laborers	28.75	690.00	44.75	1074.00		
1 Equipment Operator	38.40	307.20	57.85	462.80		
1 Vibrating Screed		65.90		72.49		
1 Roller, Vibratory, 25 Ton		565.60		622.16		
1 Dozer, 200 H.P.		988.40		1087.24	40.50	44.55
40 L.H., Daily Totals		$2863.10		$3701.89	$71.58	$92.55
Crew C-8C	Hr.	Daily	Hr.	Daily	Bare Costs	Incl. O&P
1 Labor Foreman (outside)	$30.75	$246.00	$47.90	$383.20	$31.82	$48.70
3 Laborers	28.75	690.00	44.75	1074.00		
1 Cement Finisher	35.55	284.40	52.20	417.60		
1 Equipment Operator (med.)	38.40	307.20	57.85	462.80		
1 Shotcrete Rig, 12 CY/hr		250.60		275.66	5.22	5.74
48 L.H., Daily Totals		$1778.20		$2613.26	$37.05	$54.44
Crew C-8D	Hr.	Daily	Hr.	Daily	Bare Costs	Incl. O&P
1 Labor Foreman (outside)	$30.75	$246.00	$47.90	$383.20	$32.98	$50.10
1 Laborer	28.75	230.00	44.75	358.00		
1 Cement Finisher	35.55	284.40	52.20	417.60		
1 Equipment Operator (light)	36.85	294.80	55.55	444.40		
1 Air Compressor, 250 C.F.M.		151.40		166.54		
2 -50' Air Hoses, 1"		8.20		9.02	4.99	5.49
32 L.H., Daily Totals		$1214.80		$1778.76	$37.96	$55.59

Crew No.	Bare Costs Hr.	Bare Costs Daily	Incl. Subs O & P Hr.	Incl. Subs O & P Daily	Cost Per Labor-Hour Bare Costs	Cost Per Labor-Hour Incl. O&P
Crew C-8E	Hr.	Daily	Hr.	Daily	Bare Costs	Incl. O&P
1 Labor Foreman (outside)	$30.75	$246.00	$47.90	$383.20	$32.98	$50.10
1 Laborer	28.75	230.00	44.75	358.00		
1 Cement Finisher	35.55	284.40	52.20	417.60		
1 Equipment Operator (light)	36.85	294.80	55.55	444.40		
1 Air Compressor, 250 C.F.M.		151.40		166.54		
2 -50' Air Hoses, 1"		8.20		9.02		
1 Concrete Pump (small)		708.60		779.46	27.13	29.84
32 L.H., Daily Totals		$1923.40		$2558.22	$60.11	$79.94
Crew C-10	Hr.	Daily	Hr.	Daily	Bare Costs	Incl. O&P
1 Laborer	$28.75	$230.00	$44.75	$358.00	$33.28	$49.72
2 Cement Finishers	35.55	568.80	52.20	835.20		
24 L.H., Daily Totals		$798.80		$1193.20	$33.28	$49.72
Crew C-10B	Hr.	Daily	Hr.	Daily	Bare Costs	Incl. O&P
3 Laborers	$28.75	$690.00	$44.75	$1074.00	$31.47	$47.73
2 Cement Finishers	35.55	568.80	52.20	835.20		
1 Concrete Mixer, 10 CF		138.60		152.46		
2 Conc. Finishers, 46" Walk-Behind		51.60		56.76	4.75	5.23
40 L.H., Daily Totals		$1449.00		$2118.42	$36.23	$52.96
Crew C-11	Hr.	Daily	Hr.	Daily	Bare Costs	Incl. O&P
1 Struc. Steel Foreman	$43.35	$346.80	$78.50	$628.00	$40.57	$71.00
6 Struc. Steel Workers	41.35	1984.80	74.90	3595.20		
1 Equip. Oper. (crane)	39.80	318.40	60.00	480.00		
1 Equip. Oper. Oiler	33.90	271.20	51.10	408.80		
1 Lattice Boom Crane, 150 Ton		1852.00		2037.20	25.72	28.29
72 L.H., Daily Totals		$4773.20		$7149.20	$66.29	$99.29
Crew C-12	Hr.	Daily	Hr.	Daily	Bare Costs	Incl. O&P
1 Carpenter Foreman (out)	$38.70	$309.60	$60.25	$482.00	$36.23	$56.08
3 Carpenters	36.70	880.80	57.15	1371.60		
1 Laborer	28.75	230.00	44.75	358.00		
1 Equip. Oper. (crane)	39.80	318.40	60.00	480.00		
1 Hyd. Crane, 12 Ton		724.00		796.40	15.08	16.59
48 L.H., Daily Totals		$2462.80		$3488.00	$51.31	$72.67
Crew C-13	Hr.	Daily	Hr.	Daily	Bare Costs	Incl. O&P
1 Struc. Steel Worker	$41.35	$330.80	$74.90	$599.20	$39.80	$68.98
1 Welder	41.35	330.80	74.90	599.20		
1 Carpenter	36.70	293.60	57.15	457.20		
1 Welder, gas engine, 300 amp		115.20		126.72	4.80	5.28
24 L.H., Daily Totals		$1070.40		$1782.32	$44.60	$74.26
Crew C-14	Hr.	Daily	Hr.	Daily	Bare Costs	Incl. O&P
1 Carpenter Foreman (out)	$38.70	$309.60	$60.25	$482.00	$35.96	$56.19
5 Carpenters	36.70	1468.00	57.15	2286.00		
4 Laborers	28.75	920.00	44.75	1432.00		
4 Rodmen (reinf.)	41.30	1321.60	67.75	2168.00		
2 Cement Finishers	35.55	568.80	52.20	835.20		
1 Equip. Oper. (crane)	39.80	318.40	60.00	480.00		
1 Equip. Oper. Oiler	33.90	271.20	51.10	408.80		
1 Hyd. Crane, 80 Ton		1214.00		1335.40	8.43	9.27
144 L.H., Daily Totals		$6391.60		$9427.40	$44.39	$65.47

Crew No.	Bare Costs Hr.	Bare Costs Daily	Incl. Subs O & P Hr.	Incl. Subs O & P Daily	Cost Per Labor-Hour Bare Costs	Cost Per Labor-Hour Incl. O&P
Crew C-14A	Hr.	Daily	Hr.	Daily	Bare Costs	Incl. O&P
1 Carpenter Foreman (out)	$38.70	$309.60	$60.25	$482.00	$36.90	$57.81
16 Carpenters	36.70	4697.60	57.15	7315.20		
4 Rodmen (reinf.)	41.30	1321.60	67.75	2168.00		
2 Laborers	28.75	460.00	44.75	716.00		
1 Cement Finisher	35.55	284.40	52.20	417.60		
1 Equip. Oper. (med.)	38.40	307.20	57.85	462.80		
1 Gas Engine Vibrator		21.40		23.54		
1 Concrete Pump (small)		708.60		779.46	3.65	4.01
200 L.H., Daily Totals		$8110.40		$12364.60	$40.55	$61.82
Crew C-14B	Hr.	Daily	Hr.	Daily	Bare Costs	Incl. O&P
1 Carpenter Foreman (out)	$38.70	$309.60	$60.25	$482.00	$36.85	$57.59
16 Carpenters	36.70	4697.60	57.15	7315.20		
4 Rodmen (reinf.)	41.30	1321.60	67.75	2168.00		
2 Laborers	28.75	460.00	44.75	716.00		
2 Cement Finishers	35.55	568.80	52.20	835.20		
1 Equip. Oper. (med.)	38.40	307.20	57.85	462.80		
1 Gas Engine Vibrator		21.40		23.54		
1 Concrete Pump (small)		708.60		779.46	3.51	3.86
208 L.H., Daily Totals		$8394.80		$12782.20	$40.36	$61.45
Crew C-14C	Hr.	Daily	Hr.	Daily	Bare Costs	Incl. O&P
1 Carpenter Foreman (out)	$38.70	$309.60	$60.25	$482.00	$35.15	$54.99
6 Carpenters	36.70	1761.60	57.15	2743.20		
2 Rodmen (reinf.)	41.30	660.80	67.75	1084.00		
4 Laborers	28.75	920.00	44.75	1432.00		
1 Cement Finisher	35.55	284.40	52.20	417.60		
1 Gas Engine Vibrator		21.40		23.54	.19	.21
112 L.H., Daily Totals		$3957.80		$6182.34	$35.34	$55.20
Crew C-14D	Hr.	Daily	Hr.	Daily	Bare Costs	Incl. O&P
1 Carpenter Foreman (out)	$38.70	$309.60	$60.25	$482.00	$36.53	$56.96
18 Carpenters	36.70	5284.80	57.15	8229.60		
2 Rodmen (reinf.)	41.30	660.80	67.75	1084.00		
2 Laborers	28.75	460.00	44.75	716.00		
1 Cement Finisher	35.55	284.40	52.20	417.60		
1 Equip. Oper. (med.)	38.40	307.20	57.85	462.80		
1 Gas Engine Vibrator		21.40		23.54		
1 Concrete Pump (small)		708.60		779.46	3.65	4.01
200 L.H., Daily Totals		$8036.80		$12195.00	$40.18	$60.98
Crew C-14E	Hr.	Daily	Hr.	Daily	Bare Costs	Incl. O&P
1 Carpenter Foreman (out)	$38.70	$309.60	$60.25	$482.00	$36.28	$57.45
2 Carpenters	36.70	587.20	57.15	914.40		
4 Rodmen (reinf.)	41.30	1321.60	67.75	2168.00		
3 Laborers	28.75	690.00	44.75	1074.00		
1 Cement Finisher	35.55	284.40	52.20	417.60		
1 Gas Engine Vibrator		21.40		23.54	.24	.27
88 L.H., Daily Totals		$3214.20		$5079.54	$36.52	$57.72
Crew C-14F	Hr.	Daily	Hr.	Daily	Bare Costs	Incl. O&P
1 Laborer Foreman (out)	$30.75	$246.00	$47.90	$383.20	$33.51	$50.07
2 Laborers	28.75	460.00	44.75	716.00		
6 Cement Finishers	35.55	1706.40	52.20	2505.60		
1 Gas Engine Vibrator		21.40		23.54	.30	.33
72 L.H., Daily Totals		$2433.80		$3628.34	$33.80	$50.39

Crew C-14G	Hr.	Daily	Hr.	Daily	Bare Costs	Incl. O&P
1 Laborer Foreman (out)	$30.75	$246.00	$47.90	$383.20	$32.92	$49.46
2 Laborers	28.75	460.00	44.75	716.00		
4 Cement Finishers	35.55	1137.60	52.20	1670.40		
1 Gas Engine Vibrator		21.40		23.54	.38	.42
56 L.H., Daily Totals		$1865.00		$2793.14	$33.30	$49.88

Crew C-14H	Hr.	Daily	Hr.	Daily	Bare Costs	Incl. O&P
1 Carpenter Foreman (out)	$38.70	$309.60	$60.25	$482.00	$36.28	$56.54
2 Carpenters	36.70	587.20	57.15	914.40		
1 Rodman (reinf.)	41.30	330.40	67.75	542.00		
1 Laborer	28.75	230.00	44.75	358.00		
1 Cement Finisher	35.55	284.40	52.20	417.60		
1 Gas Engine Vibrator		21.40		23.54	.45	.49
48 L.H., Daily Totals		$1763.00		$2737.54	$36.73	$57.03

Crew C-15	Hr.	Daily	Hr.	Daily	Bare Costs	Incl. O&P
1 Carpenter Foreman (out)	$38.70	$309.60	$60.25	$482.00	$34.53	$53.44
2 Carpenters	36.70	587.20	57.15	914.40		
3 Laborers	28.75	690.00	44.75	1074.00		
2 Cement Finishers	35.55	568.80	52.20	835.20		
1 Rodman (reinf.)	41.30	330.40	67.75	542.00		
72 L.H., Daily Totals		$2486.00		$3847.60	$34.53	$53.44

Crew C-16	Hr.	Daily	Hr.	Daily	Bare Costs	Incl. O&P
1 Labor Foreman (outside)	$30.75	$246.00	$47.90	$383.20	$34.34	$53.32
3 Laborers	28.75	690.00	44.75	1074.00		
2 Cement Finishers	35.55	568.80	52.20	835.20		
1 Equip. Oper. (med.)	38.40	307.20	57.85	462.80		
2 Rodmen (reinf.)	41.30	660.80	67.75	1084.00		
1 Concrete Pump (small)		708.60		779.46	9.84	10.83
72 L.H., Daily Totals		$3181.40		$4618.66	$44.19	$64.15

Crew C-17	Hr.	Daily	Hr.	Daily	Bare Costs	Incl. O&P
2 Skilled Worker Foremen	$40.00	$640.00	$62.25	$996.00	$38.40	$59.77
8 Skilled Workers	38.00	2432.00	59.15	3785.60		
80 L.H., Daily Totals		$3072.00		$4781.60	$38.40	$59.77

Crew C-17A	Hr.	Daily	Hr.	Daily	Bare Costs	Incl. O&P
2 Skilled Worker Foremen	$40.00	$640.00	$62.25	$996.00	$38.42	$59.77
8 Skilled Workers	38.00	2432.00	59.15	3785.60		
.125 Equip. Oper. (crane)	39.80	39.80	60.00	60.00		
.125 Hyd. Crane, 80 Ton		151.75		166.93	1.87	2.06
81 L.H., Daily Totals		$3263.55		$5008.52	$40.29	$61.83

Crew C-17B	Hr.	Daily	Hr.	Daily	Bare Costs	Incl. O&P
2 Skilled Worker Foremen	$40.00	$640.00	$62.25	$996.00	$38.43	$59.78
8 Skilled Workers	38.00	2432.00	59.15	3785.60		
.25 Equip. Oper. (crane)	39.80	79.60	60.00	120.00		
.25 Hyd. Crane, 80 Ton		303.50		333.85		
.25 Conc. Finish.,46" Wlk-Behind		6.45		7.09	3.78	4.16
82 L.H., Daily Totals		$3461.55		$5242.55	$42.21	$63.93

Crew C-17C	Hr.	Daily	Hr.	Daily	Bare Costs	Incl. O&P
2 Skilled Worker Foremen	$40.00	$640.00	$62.25	$996.00	$38.45	$59.78
8 Skilled Workers	38.00	2432.00	59.15	3785.60		
.375 Equip. Oper. (crane)	39.80	119.40	60.00	180.00		
.375 Hyd. Crane, 80 Ton		455.25		500.77	5.48	6.03
83 L.H., Daily Totals		$3646.65		$5462.38	$43.94	$65.81

Crew C-17D	Hr.	Daily	Hr.	Daily	Bare Costs	Incl. O&P
2 Skilled Worker Foremen	$40.00	$640.00	$62.25	$996.00	$38.47	$59.78
8 Skilled Workers	38.00	2432.00	59.15	3785.60		
.5 Equip. Oper. (crane)	39.80	159.20	60.00	240.00		
.5 Hyd. Crane, 80 Ton		607.00		667.70	7.23	7.95
84 L.H., Daily Totals		$3838.20		$5689.30	$45.69	$67.73

Crew C-17E	Hr.	Daily	Hr.	Daily	Bare Costs	Incl. O&P
2 Skilled Worker Foremen	$40.00	$640.00	$62.25	$996.00	$38.40	$59.77
8 Skilled Workers	38.00	2432.00	59.15	3785.60		
1 Hyd. Jack with Rods		82.35		90.58	1.03	1.13
80 L.H., Daily Totals		$3154.35		$4872.19	$39.43	$60.90

Crew C-18	Hr.	Daily	Hr.	Daily	Bare Costs	Incl. O&P
.125 Labor Foreman (out)	$30.75	$30.75	$47.90	$47.90	$28.97	$45.10
1 Laborer	28.75	230.00	44.75	358.00		
1 Concrete Cart, 10 C.F.		53.40		58.74	5.93	6.53
9 L.H., Daily Totals		$314.15		$464.64	$34.91	$51.63

Crew C-19	Hr.	Daily	Hr.	Daily	Bare Costs	Incl. O&P
.125 Labor Foreman (out)	$30.75	$30.75	$47.90	$47.90	$28.97	$45.10
1 Laborer	28.75	230.00	44.75	358.00		
1 Concrete Cart, 18 C.F.		78.80		86.68	8.76	9.63
9 L.H., Daily Totals		$339.55		$492.58	$37.73	$54.73

Crew C-20	Hr.	Daily	Hr.	Daily	Bare Costs	Incl. O&P
1 Labor Foreman (outside)	$30.75	$246.00	$47.90	$383.20	$31.06	$47.71
5 Laborers	28.75	1150.00	44.75	1790.00		
1 Cement Finisher	35.55	284.40	52.20	417.60		
1 Equip. Oper. (med.)	38.40	307.20	57.85	462.80		
2 Gas Engine Vibrators		42.80		47.08		
1 Concrete Pump (small)		708.60		779.46	11.74	12.91
64 L.H., Daily Totals		$2739.00		$3880.14	$42.80	$60.63

Crew C-21	Hr.	Daily	Hr.	Daily	Bare Costs	Incl. O&P
1 Labor Foreman (outside)	$30.75	$246.00	$47.90	$383.20	$31.06	$47.71
5 Laborers	28.75	1150.00	44.75	1790.00		
1 Cement Finisher	35.55	284.40	52.20	417.60		
1 Equip. Oper. (med.)	38.40	307.20	57.85	462.80		
2 Gas Engine Vibrators		42.80		47.08		
1 Concrete Conveyer		160.40		176.44	3.17	3.49
64 L.H., Daily Totals		$2190.80		$3277.12	$34.23	$51.20

Crew C-22	Hr.	Daily	Hr.	Daily	Bare Costs	Incl. O&P
1 Rodman Foreman	$43.30	$346.40	$71.05	$568.40	$41.47	$67.80
4 Rodmen (reinf.)	41.30	1321.60	67.75	2168.00		
.125 Equip. Oper. (crane)	39.80	39.80	60.00	60.00		
.125 Equip. Oper. Oiler	33.90	33.90	51.10	51.10		
.125 Hyd. Crane, 25 Ton		92.45		101.69	2.20	2.42
42 L.H., Daily Totals		$1834.15		$2949.20	$43.67	$70.22

Crew C-23	Hr.	Daily	Hr.	Daily	Bare Costs	Incl. O&P
2 Skilled Worker Foremen	$40.00	$640.00	$62.25	$996.00	$38.17	$59.05
6 Skilled Workers	38.00	1824.00	59.15	2839.20		
1 Equip. Oper. (crane)	39.80	318.40	60.00	480.00		
1 Equip. Oper. Oiler	33.90	271.20	51.10	408.80		
1 Lattice Boom Crane, 90 Ton		1547.00		1701.70	19.34	21.27
80 L.H., Daily Totals		$4600.60		$6425.70	$57.51	$80.32

Crews

Crew C-23A

Crew No.	Bare Costs Hr.	Bare Costs Daily	Incl. Subs O&P Hr.	Incl. Subs O&P Daily	Cost Per Labor-Hour Bare Costs	Cost Per Labor-Hour Incl. O&P
1 Labor Foreman (outside)	$30.75	$246.00	$47.90	$383.20	$32.39	$49.70
2 Laborers	28.75	460.00	44.75	716.00		
1 Equip. Oper. (crane)	39.80	318.40	60.00	480.00		
1 Equip. Oper. Oiler	33.90	271.20	51.10	408.80		
1 Crawler Crane, 100 Ton		1717.00		1888.70		
3 Conc. buckets, 8 C.Y.		518.40		570.24	55.88	61.47
40 L.H., Daily Totals		$3531.00		$4446.94	$88.28	$111.17

Crew C-24

Crew No.	Bare Costs Hr.	Bare Costs Daily	Incl. Subs O&P Hr.	Incl. Subs O&P Daily	Cost Per Labor-Hour Bare Costs	Cost Per Labor-Hour Incl. O&P
2 Skilled Worker Foremen	$40.00	$640.00	$62.25	$996.00	$38.17	$59.05
6 Skilled Workers	38.00	1824.00	59.15	2839.20		
1 Equip. Oper. (crane)	39.80	318.40	60.00	480.00		
1 Equip. Oper. Oiler	33.90	271.20	51.10	408.80		
1 Lattice Boom Crane, 150 Ton		1852.00		2037.20	23.15	25.47
80 L.H., Daily Totals		$4905.60		$6761.20	$61.32	$84.52

Crew C-25

Crew No.	Bare Costs Hr.	Bare Costs Daily	Incl. Subs O&P Hr.	Incl. Subs O&P Daily	Cost Per Labor-Hour Bare Costs	Cost Per Labor-Hour Incl. O&P
2 Rodmen (reinf.)	$41.30	$660.80	$67.75	$1084.00	$32.33	$53.67
2 Rodmen Helpers	23.35	373.60	39.60	633.60		
32 L.H., Daily Totals		$1034.40		$1717.60	$32.33	$53.67

Crew C-27

Crew No.	Bare Costs Hr.	Bare Costs Daily	Incl. Subs O&P Hr.	Incl. Subs O&P Daily	Cost Per Labor-Hour Bare Costs	Cost Per Labor-Hour Incl. O&P
2 Cement Finishers	$35.55	$568.80	$52.20	$835.20	$35.55	$52.20
1 Concrete Saw		118.40		130.24	7.40	8.14
16 L.H., Daily Totals		$687.20		$965.44	$42.95	$60.34

Crew C-28

Crew No.	Bare Costs Hr.	Bare Costs Daily	Incl. Subs O&P Hr.	Incl. Subs O&P Daily	Cost Per Labor-Hour Bare Costs	Cost Per Labor-Hour Incl. O&P
1 Cement Finisher	$35.55	$284.40	$52.20	$417.60	$35.55	$52.20
1 Portable Air Compressor, gas		14.75		16.23	1.84	2.03
8 L.H., Daily Totals		$299.15		$433.82	$37.39	$54.23

Crew D-1

Crew No.	Bare Costs Hr.	Bare Costs Daily	Incl. Subs O&P Hr.	Incl. Subs O&P Daily	Cost Per Labor-Hour Bare Costs	Cost Per Labor-Hour Incl. O&P
1 Bricklayer	$38.05	$304.40	$57.90	$463.20	$33.35	$50.75
1 Bricklayer Helper	28.65	229.20	43.60	348.80		
16 L.H., Daily Totals		$533.60		$812.00	$33.35	$50.75

Crew D-2

Crew No.	Bare Costs Hr.	Bare Costs Daily	Incl. Subs O&P Hr.	Incl. Subs O&P Daily	Cost Per Labor-Hour Bare Costs	Cost Per Labor-Hour Incl. O&P
3 Bricklayers	$38.05	$913.20	$57.90	$1389.60	$34.51	$52.63
2 Bricklayer Helpers	28.65	458.40	43.60	697.60		
.5 Carpenter	36.70	146.80	57.15	228.60		
44 L.H., Daily Totals		$1518.40		$2315.80	$34.51	$52.63

Crew D-3

Crew No.	Bare Costs Hr.	Bare Costs Daily	Incl. Subs O&P Hr.	Incl. Subs O&P Daily	Cost Per Labor-Hour Bare Costs	Cost Per Labor-Hour Incl. O&P
3 Bricklayers	$38.05	$913.20	$57.90	$1389.60	$34.40	$52.42
2 Bricklayer Helpers	28.65	458.40	43.60	697.60		
.25 Carpenter	36.70	73.40	57.15	114.30		
42 L.H., Daily Totals		$1445.00		$2201.50	$34.40	$52.42

Crew D-4

Crew No.	Bare Costs Hr.	Bare Costs Daily	Incl. Subs O&P Hr.	Incl. Subs O&P Daily	Cost Per Labor-Hour Bare Costs	Cost Per Labor-Hour Incl. O&P
1 Bricklayer	$38.05	$304.40	$57.90	$463.20	$33.05	$50.16
2 Bricklayer Helpers	28.65	458.40	43.60	697.60		
1 Equip. Oper. (light)	36.85	294.80	55.55	444.40		
1 Grout Pump, 50 C.F./hr		122.00		134.20	3.81	4.19
32 L.H., Daily Totals		$1179.60		$1739.40	$36.86	$54.36

Crew D-5

Crew No.	Bare Costs Hr.	Bare Costs Daily	Incl. Subs O&P Hr.	Incl. Subs O&P Daily	Cost Per Labor-Hour Bare Costs	Cost Per Labor-Hour Incl. O&P
1 Bricklayer	$38.05	$304.40	$57.90	$463.20	$38.05	$57.90
8 L.H., Daily Totals		$304.40		$463.20	$38.05	$57.90

Crew D-6

Crew No.	Bare Costs Hr.	Bare Costs Daily	Incl. Subs O&P Hr.	Incl. Subs O&P Daily	Cost Per Labor-Hour Bare Costs	Cost Per Labor-Hour Incl. O&P
3 Bricklayers	$38.05	$913.20	$57.90	$1389.60	$33.48	$51.01
3 Bricklayer Helpers	28.65	687.60	43.60	1046.40		
.25 Carpenter	36.70	73.40	57.15	114.30		
50 L.H., Daily Totals		$1674.20		$2550.30	$33.48	$51.01

Crew D-7

Crew No.	Bare Costs Hr.	Bare Costs Daily	Incl. Subs O&P Hr.	Incl. Subs O&P Daily	Cost Per Labor-Hour Bare Costs	Cost Per Labor-Hour Incl. O&P
1 Tile Layer	$35.50	$284.00	$52.10	$416.80	$31.38	$46.05
1 Tile Layer Helper	27.25	218.00	40.00	320.00		
16 L.H., Daily Totals		$502.00		$736.80	$31.38	$46.05

Crew D-8

Crew No.	Bare Costs Hr.	Bare Costs Daily	Incl. Subs O&P Hr.	Incl. Subs O&P Daily	Cost Per Labor-Hour Bare Costs	Cost Per Labor-Hour Incl. O&P
3 Bricklayers	$38.05	$913.20	$57.90	$1389.60	$34.29	$52.18
2 Bricklayer Helpers	28.65	458.40	43.60	697.60		
40 L.H., Daily Totals		$1371.60		$2087.20	$34.29	$52.18

Crew D-9

Crew No.	Bare Costs Hr.	Bare Costs Daily	Incl. Subs O&P Hr.	Incl. Subs O&P Daily	Cost Per Labor-Hour Bare Costs	Cost Per Labor-Hour Incl. O&P
3 Bricklayers	$38.05	$913.20	$57.90	$1389.60	$33.35	$50.75
3 Bricklayer Helpers	28.65	687.60	43.60	1046.40		
48 L.H., Daily Totals		$1600.80		$2436.00	$33.35	$50.75

Crew D-10

Crew No.	Bare Costs Hr.	Bare Costs Daily	Incl. Subs O&P Hr.	Incl. Subs O&P Daily	Cost Per Labor-Hour Bare Costs	Cost Per Labor-Hour Incl. O&P
1 Bricklayer Foreman	$40.05	$320.40	$60.95	$487.60	$36.64	$55.61
1 Bricklayer	38.05	304.40	57.90	463.20		
1 Bricklayer Helper	28.65	229.20	43.60	348.80		
1 Equip. Oper. (crane)	39.80	318.40	60.00	480.00		
1 S.P. Crane, 4x4, 12 Ton		575.00		632.50	17.97	19.77
32 L.H., Daily Totals		$1747.40		$2412.10	$54.61	$75.38

Crew D-11

Crew No.	Bare Costs Hr.	Bare Costs Daily	Incl. Subs O&P Hr.	Incl. Subs O&P Daily	Cost Per Labor-Hour Bare Costs	Cost Per Labor-Hour Incl. O&P
1 Bricklayer Foreman	$40.05	$320.40	$60.95	$487.60	$35.58	$54.15
1 Bricklayer	38.05	304.40	57.90	463.20		
1 Bricklayer Helper	28.65	229.20	43.60	348.80		
24 L.H., Daily Totals		$854.00		$1299.60	$35.58	$54.15

Crew D-12

Crew No.	Bare Costs Hr.	Bare Costs Daily	Incl. Subs O&P Hr.	Incl. Subs O&P Daily	Cost Per Labor-Hour Bare Costs	Cost Per Labor-Hour Incl. O&P
1 Bricklayer Foreman	$40.05	$320.40	$60.95	$487.60	$33.85	$51.51
1 Bricklayer	38.05	304.40	57.90	463.20		
2 Bricklayer Helpers	28.65	458.40	43.60	697.60		
32 L.H., Daily Totals		$1083.20		$1648.40	$33.85	$51.51

Crew D-13

Crew No.	Bare Costs Hr.	Bare Costs Daily	Incl. Subs O&P Hr.	Incl. Subs O&P Daily	Cost Per Labor-Hour Bare Costs	Cost Per Labor-Hour Incl. O&P
1 Bricklayer Foreman	$40.05	$320.40	$60.95	$487.60	$35.32	$53.87
1 Bricklayer	38.05	304.40	57.90	463.20		
2 Bricklayer Helpers	28.65	458.40	43.60	697.60		
1 Carpenter	36.70	293.60	57.15	457.20		
1 Equip. Oper. (crane)	39.80	318.40	60.00	480.00		
1 S.P. Crane, 4x4, 12 Ton		575.00		632.50	11.98	13.18
48 L.H., Daily Totals		$2270.20		$3218.10	$47.30	$67.04

Crew E-1

Crew No.	Bare Costs Hr.	Bare Costs Daily	Incl. Subs O&P Hr.	Incl. Subs O&P Daily	Cost Per Labor-Hour Bare Costs	Cost Per Labor-Hour Incl. O&P
1 Welder Foreman	$43.35	$346.80	$78.50	$628.00	$40.52	$69.65
1 Welder	41.35	330.80	74.90	599.20		
1 Equip. Oper. (light)	36.85	294.80	55.55	444.40		
1 Welder, gas engine, 300 amp		115.20		126.72	4.80	5.28
24 L.H., Daily Totals		$1087.60		$1798.32	$45.32	$74.93

Crews

Crew E-2

Crew E-2	Hr.	Daily	Hr.	Daily	Bare Costs	Incl. O&P
1 Struc. Steel Foreman	$43.35	$346.80	$78.50	$628.00	$40.35	$69.89
4 Struc. Steel Workers	41.35	1323.20	74.90	2396.80		
1 Equip. Oper. (crane)	39.80	318.40	60.00	480.00		
1 Equip. Oper. Oiler	33.90	271.20	51.10	408.80		
1 Lattice Boom Crane, 90 Ton		1547.00		1701.70	27.63	30.39
56 L.H., Daily Totals		$3806.60		$5615.30	$67.97	$100.27

Crew E-3

Crew E-3	Hr.	Daily	Hr.	Daily	Bare Costs	Incl. O&P
1 Struc. Steel Foreman	$43.35	$346.80	$78.50	$628.00	$42.02	$76.10
1 Struc. Steel Worker	41.35	330.80	74.90	599.20		
1 Welder	41.35	330.80	74.90	599.20		
1 Welder, gas engine, 300 amp		115.20		126.72	4.80	5.28
24 L.H., Daily Totals		$1123.60		$1953.12	$46.82	$81.38

Crew E-4

Crew E-4	Hr.	Daily	Hr.	Daily	Bare Costs	Incl. O&P
1 Struc. Steel Foreman	$43.35	$346.80	$78.50	$628.00	$41.85	$75.80
3 Struc. Steel Workers	41.35	992.40	74.90	1797.60		
1 Welder, gas engine, 300 amp		115.20		126.72	3.60	3.96
32 L.H., Daily Totals		$1454.40		$2552.32	$45.45	$79.76

Crew E-5

Crew E-5	Hr.	Daily	Hr.	Daily	Bare Costs	Incl. O&P
2 Struc. Steel Foremen	$43.35	$693.60	$78.50	$1256.00	$40.85	$71.75
5 Struc. Steel Workers	41.35	1654.00	74.90	2996.00		
1 Equip. Oper. (crane)	39.80	318.40	60.00	480.00		
1 Welder	41.35	330.80	74.90	599.20		
1 Equip. Oper. Oiler	33.90	271.20	51.10	408.80		
1 Lattice Boom Crane, 90 Ton		1547.00		1701.70		
1 Welder, gas engine, 300 amp		115.20		126.72	20.78	22.86
80 L.H., Daily Totals		$4930.20		$7568.42	$61.63	$94.61

Crew E-6

Crew E-6	Hr.	Daily	Hr.	Daily	Bare Costs	Incl. O&P
3 Struc. Steel Foremen	$43.35	$1040.40	$78.50	$1884.00	$40.88	$71.95
9 Struc. Steel Workers	41.35	2977.20	74.90	5392.80		
1 Equip. Oper. (crane)	39.80	318.40	60.00	480.00		
1 Welder	41.35	330.80	74.90	599.20		
1 Equip. Oper. Oiler	33.90	271.20	51.10	408.80		
1 Equip. Oper. (light)	36.85	294.80	55.55	444.40		
1 Lattice Boom Crane, 90 Ton		1547.00		1701.70		
1 Welder, gas engine, 300 amp		115.20		126.72		
1 Air Compressor, 160 C.F.M.		122.60		134.86		
2 Impact Wrenches		25.60		28.16	14.14	15.56
128 L.H., Daily Totals		$7043.20		$11200.64	$55.03	$87.50

Crew E-7

Crew E-7	Hr.	Daily	Hr.	Daily	Bare Costs	Incl. O&P
1 Struc. Steel Foreman	$43.35	$346.80	$78.50	$628.00	$40.85	$71.75
4 Struc. Steel Workers	41.35	1323.20	74.90	2396.80		
1 Equip. Oper. (crane)	39.80	318.40	60.00	480.00		
1 Equip. Oper. Oiler	33.90	271.20	51.10	408.80		
1 Welder Foreman	43.35	346.80	78.50	628.00		
2 Welders	41.35	661.60	74.90	1198.40		
1 Lattice Boom Crane, 90 Ton		1547.00		1701.70		
2 Welder, gas engine, 300 amp		230.40		253.44	22.22	24.44
80 L.H., Daily Totals		$5045.40		$7695.14	$63.07	$96.19

Crew E-8

Crew E-8	Hr.	Daily	Hr.	Daily	Bare Costs	Incl. O&P
1 Struc. Steel Foreman	$43.35	$346.80	$78.50	$628.00	$40.62	$70.99
4 Struc. Steel Workers	41.35	1323.20	74.90	2396.80		
1 Welder Foreman	43.35	346.80	78.50	628.00		
4 Welders	41.35	1323.20	74.90	2396.80		
1 Equip. Oper. (crane)	39.80	318.40	60.00	480.00		
1 Equip. Oper. Oiler	33.90	271.20	51.10	408.80		
1 Equip. Oper. (light)	36.85	294.80	55.55	444.40		
1 Lattice Boom Crane, 90 Ton		1547.00		1701.70		
4 Welder, gas engine, 300 amp		460.80		506.88	19.31	21.24
104 L.H., Daily Totals		$6232.20		$9591.38	$59.92	$92.22

Crew E-9

Crew E-9	Hr.	Daily	Hr.	Daily	Bare Costs	Incl. O&P
2 Struc. Steel Foremen	$43.35	$693.60	$78.50	$1256.00	$40.88	$71.95
5 Struc. Steel Workers	41.35	1654.00	74.90	2996.00		
1 Welder Foreman	43.35	346.80	78.50	628.00		
5 Welders	41.35	1654.00	74.90	2996.00		
1 Equip. Oper. (crane)	39.80	318.40	60.00	480.00		
1 Equip. Oper. Oiler	33.90	271.20	51.10	408.80		
1 Equip. Oper. (light)	36.85	294.80	55.55	444.40		
1 Lattice Boom Crane, 90 Ton		1547.00		1701.70		
5 Welder, gas engines, 300 amp		576.00		633.60	16.59	18.24
128 L.H., Daily Totals		$7355.80		$11544.50	$57.47	$90.19

Crew E-10

Crew E-10	Hr.	Daily	Hr.	Daily	Bare Costs	Incl. O&P
1 Welder Foreman	$43.35	$346.80	$78.50	$628.00	$42.35	$76.70
1 Welder	41.35	330.80	74.90	599.20		
1 Welder, gas engine, 300 amp		115.20		126.72		
1 Flatbed Truck, gas, 3 Ton		185.20		203.72	18.77	20.65
16 L.H., Daily Totals		$978.00		$1557.64	$61.13	$97.35

Crew E-11

Crew E-11	Hr.	Daily	Hr.	Daily	Bare Costs	Incl. O&P
2 Painters, Struc. Steel	$33.50	$536.00	$61.15	$978.40	$33.15	$55.65
1 Building Laborer	28.75	230.00	44.75	358.00		
1 Equip. Oper. (light)	36.85	294.80	55.55	444.40		
1 Air Compressor, 250 C.F.M.		151.40		166.54		
1 Sandblaster, portable, 3 C.F.		15.60		17.16		
1 Sand Blasting Accessories		12.30		13.53	5.60	6.16
32 L.H., Daily Totals		$1240.10		$1978.03	$38.75	$61.81

Crew E-12

Crew E-12	Hr.	Daily	Hr.	Daily	Bare Costs	Incl. O&P
1 Welder Foreman	$43.35	$346.80	$78.50	$628.00	$40.10	$67.03
1 Equip. Oper. (light)	36.85	294.80	55.55	444.40		
1 Welder, gas engine, 300 amp		115.20		126.72	7.20	7.92
16 L.H., Daily Totals		$756.80		$1199.12	$47.30	$74.94

Crew E-13

Crew E-13	Hr.	Daily	Hr.	Daily	Bare Costs	Incl. O&P
1 Welder Foreman	$43.35	$346.80	$78.50	$628.00	$41.18	$70.85
.5 Equip. Oper. (light)	36.85	147.40	55.55	222.20		
1 Welder, gas engine, 300 amp		115.20		126.72	9.60	10.56
12 L.H., Daily Totals		$609.40		$976.92	$50.78	$81.41

Crew E-14

Crew E-14	Hr.	Daily	Hr.	Daily	Bare Costs	Incl. O&P
1 Welder Foreman	$43.35	$346.80	$78.50	$628.00	$43.35	$78.50
1 Welder, gas engine, 300 amp		115.20		126.72	14.40	15.84
8 L.H., Daily Totals		$462.00		$754.72	$57.75	$94.34

Crew E-16

Crew E-16	Hr.	Daily	Hr.	Daily	Bare Costs	Incl. O&P
1 Welder Foreman	$43.35	$346.80	$78.50	$628.00	$42.35	$76.70
1 Welder	41.35	330.80	74.90	599.20		
1 Welder, gas engine, 300 amp		115.20		126.72	7.20	7.92
16 L.H., Daily Totals		$792.80		$1353.92	$49.55	$84.62

Crew E-17

Crew No.	Bare Costs Hr.	Daily	Incl. Subs O & P Hr.	Daily	Cost Per Labor-Hour Bare Costs	Incl. O&P
1 Structural Steel Foreman	$43.35	$346.80	$78.50	$628.00	$42.35	$76.70
1 Structural Steel Worker	41.35	330.80	74.90	599.20		
16 L.H., Daily Totals		$677.60		$1227.20	$42.35	$76.70

Crew E-18

Crew No.	Bare Costs Hr.	Daily	Incl. Subs O & P Hr.	Daily	Cost Per Labor-Hour Bare Costs	Incl. O&P
1 Structural Steel Foreman	$43.35	$346.80	$78.50	$628.00	$41.16	$72.21
3 Structural Steel Workers	41.35	992.40	74.90	1797.60		
1 Equipment Operator (med.)	38.40	307.20	57.85	462.80		
1 Lattice Boom Crane, 20 Ton		976.70		1074.37	24.42	26.86
40 L.H., Daily Totals		$2623.10		$3962.77	$65.58	$99.07

Crew E-19

Crew No.	Bare Costs Hr.	Daily	Incl. Subs O & P Hr.	Daily	Cost Per Labor-Hour Bare Costs	Incl. O&P
1 Structural Steel Worker	$41.35	$330.80	$74.90	$599.20	$40.52	$69.65
1 Structural Steel Foreman	43.35	346.80	78.50	628.00		
1 Equip. Oper. (light)	36.85	294.80	55.55	444.40		
1 Lattice Boom Crane, 20 Ton		976.70		1074.37	40.70	44.77
24 L.H., Daily Totals		$1949.10		$2745.97	$81.21	$114.42

Crew E-20

Crew No.	Bare Costs Hr.	Daily	Incl. Subs O & P Hr.	Daily	Cost Per Labor-Hour Bare Costs	Incl. O&P
1 Structural Steel Foreman	$43.35	$346.80	$78.50	$628.00	$40.48	$70.51
5 Structural Steel Workers	41.35	1654.00	74.90	2996.00		
1 Equip. Oper. (crane)	39.80	318.40	60.00	480.00		
1 Oiler	33.90	271.20	51.10	408.80		
1 Lattice Boom Crane, 40 Ton		1268.00		1394.80	19.81	21.79
64 L.H., Daily Totals		$3858.40		$5907.60	$60.29	$92.31

Crew E-22

Crew No.	Bare Costs Hr.	Daily	Incl. Subs O & P Hr.	Daily	Cost Per Labor-Hour Bare Costs	Incl. O&P
1 Skilled Worker Foreman	$40.00	$320.00	$62.25	$498.00	$38.67	$60.18
2 Skilled Workers	38.00	608.00	59.15	946.40		
24 L.H., Daily Totals		$928.00		$1444.40	$38.67	$60.18

Crew E-24

Crew No.	Bare Costs Hr.	Daily	Incl. Subs O & P Hr.	Daily	Cost Per Labor-Hour Bare Costs	Incl. O&P
3 Structural Steel Workers	$41.35	$992.40	$74.90	$1797.60	$40.61	$70.64
1 Equipment Operator (medium)	38.40	307.20	57.85	462.80		
1 -25 Ton Crane		739.60		813.56	23.11	25.42
32 L.H., Daily Totals		$2039.20		$3073.96	$63.73	$96.06

Crew E-25

Crew No.	Bare Costs Hr.	Daily	Incl. Subs O & P Hr.	Daily	Cost Per Labor-Hour Bare Costs	Incl. O&P
1 Welder Foreman	$43.35	$346.80	$78.50	$628.00	$43.35	$78.50
1 Cutting Torch		17.00		18.70		
1 Gase		64.80		71.28	10.23	11.25
8 L.H., Daily Totals		$428.60		$717.98	$53.58	$89.75

Crew F-3

Crew No.	Bare Costs Hr.	Daily	Incl. Subs O & P Hr.	Daily	Cost Per Labor-Hour Bare Costs	Incl. O&P
4 Carpenters	$36.70	$1174.40	$57.15	$1828.80	$37.32	$57.72
1 Equip. Oper. (crane)	39.80	318.40	60.00	480.00		
1 Hyd. Crane, 12 Ton		724.00		796.40	18.10	19.91
40 L.H., Daily Totals		$2216.80		$3105.20	$55.42	$77.63

Crew F-4

Crew No.	Bare Costs Hr.	Daily	Incl. Subs O & P Hr.	Daily	Cost Per Labor-Hour Bare Costs	Incl. O&P
4 Carpenters	$36.70	$1174.40	$57.15	$1828.80	$36.75	$56.62
1 Equip. Oper. (crane)	39.80	318.40	60.00	480.00		
1 Equip. Oper. Oiler	33.90	271.20	51.10	408.80		
1 Hyd. Crane, 55 Ton		1060.00		1166.00	22.08	24.29
48 L.H., Daily Totals		$2824.00		$3883.60	$58.83	$80.91

Crew F-5

Crew No.	Bare Costs Hr.	Daily	Incl. Subs O & P Hr.	Daily	Cost Per Labor-Hour Bare Costs	Incl. O&P
1 Carpenter Foreman	$38.70	$309.60	$60.25	$482.00	$37.20	$57.92
3 Carpenters	36.70	880.80	57.15	1371.60		
32 L.H., Daily Totals		$1190.40		$1853.60	$37.20	$57.92

Crew F-6

Crew No.	Bare Costs Hr.	Daily	Incl. Subs O & P Hr.	Daily	Cost Per Labor-Hour Bare Costs	Incl. O&P
2 Carpenters	$36.70	$587.20	$57.15	$914.40	$34.14	$52.76
2 Building Laborers	28.75	460.00	44.75	716.00		
1 Equip. Oper. (crane)	39.80	318.40	60.00	480.00		
1 Hyd. Crane, 12 Ton		724.00		796.40	18.10	19.91
40 L.H., Daily Totals		$2089.60		$2906.80	$52.24	$72.67

Crew F-7

Crew No.	Bare Costs Hr.	Daily	Incl. Subs O & P Hr.	Daily	Cost Per Labor-Hour Bare Costs	Incl. O&P
2 Carpenters	$36.70	$587.20	$57.15	$914.40	$32.73	$50.95
2 Building Laborers	28.75	460.00	44.75	716.00		
32 L.H., Daily Totals		$1047.20		$1630.40	$32.73	$50.95

Crew G-1

Crew No.	Bare Costs Hr.	Daily	Incl. Subs O & P Hr.	Daily	Cost Per Labor-Hour Bare Costs	Incl. O&P
1 Roofer Foreman	$33.80	$270.40	$57.30	$458.40	$29.67	$50.33
4 Roofers, Composition	31.80	1017.60	53.95	1726.40		
2 Roofer Helpers	23.35	373.60	39.60	633.60		
1 Application Equipment		153.20		168.52		
1 Tar Kettle/Pot		62.90		69.19		
1 Crew Truck		116.20		127.82	5.93	6.53
56 L.H., Daily Totals		$1993.90		$3183.93	$35.61	$56.86

Crew G-2

Crew No.	Bare Costs Hr.	Daily	Incl. Subs O & P Hr.	Daily	Cost Per Labor-Hour Bare Costs	Incl. O&P
1 Plasterer	$33.55	$268.40	$50.75	$406.00	$30.35	$46.33
1 Plasterer Helper	28.75	230.00	43.50	348.00		
1 Building Laborer	28.75	230.00	44.75	358.00		
1 Grout Pump, 50 C.F./hr		122.00		134.20	5.08	5.59
24 L.H., Daily Totals		$850.40		$1246.20	$35.43	$51.92

Crew G-2A

Crew No.	Bare Costs Hr.	Daily	Incl. Subs O & P Hr.	Daily	Cost Per Labor-Hour Bare Costs	Incl. O&P
1 Roofer, composition	$31.80	$254.40	$53.95	$431.60	$27.97	$46.10
1 Roofer Helper	23.35	186.80	39.60	316.80		
1 Building Laborer	28.75	230.00	44.75	358.00		
1 Grout Pump, 50 C.F./hr		122.00		134.20	5.08	5.59
24 L.H., Daily Totals		$793.20		$1240.60	$33.05	$51.69

Crew G-3

Crew No.	Bare Costs Hr.	Daily	Incl. Subs O & P Hr.	Daily	Cost Per Labor-Hour Bare Costs	Incl. O&P
2 Sheet Metal Workers	$43.55	$696.80	$67.15	$1074.40	$36.15	$55.95
2 Building Laborers	28.75	460.00	44.75	716.00		
32 L.H., Daily Totals		$1156.80		$1790.40	$36.15	$55.95

Crew G-4

Crew No.	Bare Costs Hr.	Daily	Incl. Subs O & P Hr.	Daily	Cost Per Labor-Hour Bare Costs	Incl. O&P
1 Labor Foreman (outside)	$30.75	$246.00	$47.90	$383.20	$29.42	$45.80
2 Building Laborers	28.75	460.00	44.75	716.00		
1 Flatbed Truck, gas, 1.5 Ton		141.40		155.54		
1 Air Compressor, 160 C.F.M.		122.60		134.86	11.00	12.10
24 L.H., Daily Totals		$970.00		$1389.60	$40.42	$57.90

Crew G-5

Crew No.	Bare Costs Hr.	Daily	Incl. Subs O & P Hr.	Daily	Cost Per Labor-Hour Bare Costs	Incl. O&P
1 Roofer Foreman	$33.80	$270.40	$57.30	$458.40	$28.82	$48.88
2 Roofers, Composition	31.80	508.80	53.95	863.20		
2 Roofer Helpers	23.35	373.60	39.60	633.60		
1 Application Equipment		153.20		168.52	3.83	4.21
40 L.H., Daily Totals		$1306.00		$2123.72	$32.65	$53.09

Crew G-6A

Crew No.	Bare Costs Hr.	Daily	Incl. Subs O & P Hr.	Daily	Cost Per Labor-Hour Bare Costs	Incl. O&P
2 Roofers, Composition	$31.80	$508.80	$53.95	$863.20	$31.80	$53.95
1 Small Compressor, Electric		9.55		10.51		
2 Pneumatic Nailers		41.30		45.43	3.18	3.50
16 L.H., Daily Totals		$559.65		$919.13	$34.98	$57.45

Crew G-7

Crew No.	Bare Costs Hr.	Bare Costs Daily	Incl. Subs O&P Hr.	Incl. Subs O&P Daily	Cost Per Labor-Hour Bare Costs	Cost Per Labor-Hour Incl. O&P
1 Carpenter	$36.70	$293.60	$57.15	$457.20	$36.70	$57.15
1 Small Compressor, Electric		9.55		10.51		
1 Pneumatic Nailer		20.65		22.72	3.77	4.15
8 L.H., Daily Totals		$323.80		$490.42	$40.48	$61.30

Crew H-1

Crew No.	Bare Costs Hr.	Bare Costs Daily	Incl. Subs O&P Hr.	Incl. Subs O&P Daily	Cost Per Labor-Hour Bare Costs	Cost Per Labor-Hour Incl. O&P
2 Glaziers	$36.05	$576.80	$54.60	$873.60	$38.70	$64.75
2 Struc. Steel Workers	41.35	661.60	74.90	1198.40		
32 L.H., Daily Totals		$1238.40		$2072.00	$38.70	$64.75

Crew H-2

Crew No.	Bare Costs Hr.	Bare Costs Daily	Incl. Subs O&P Hr.	Incl. Subs O&P Daily	Cost Per Labor-Hour Bare Costs	Cost Per Labor-Hour Incl. O&P
2 Glaziers	$36.05	$576.80	$54.60	$873.60	$33.62	$51.32
1 Building Laborer	28.75	230.00	44.75	358.00		
24 L.H., Daily Totals		$806.80		$1231.60	$33.62	$51.32

Crew H-3

Crew No.	Bare Costs Hr.	Bare Costs Daily	Incl. Subs O&P Hr.	Incl. Subs O&P Daily	Cost Per Labor-Hour Bare Costs	Cost Per Labor-Hour Incl. O&P
1 Glazier	$36.05	$288.40	$54.60	$436.80	$31.70	$48.50
1 Helper	27.35	218.80	42.40	339.20		
16 L.H., Daily Totals		$507.20		$776.00	$31.70	$48.50

Crew J-1

Crew No.	Bare Costs Hr.	Bare Costs Daily	Incl. Subs O&P Hr.	Incl. Subs O&P Daily	Cost Per Labor-Hour Bare Costs	Cost Per Labor-Hour Incl. O&P
3 Plasterers	$33.55	$805.20	$50.75	$1218.00	$31.63	$47.85
2 Plasterer Helpers	28.75	460.00	43.50	696.00		
1 Mixing Machine, 6 C.F.		113.60		124.96	2.84	3.12
40 L.H., Daily Totals		$1378.80		$2038.96	$34.47	$50.97

Crew J-2

Crew No.	Bare Costs Hr.	Bare Costs Daily	Incl. Subs O&P Hr.	Incl. Subs O&P Daily	Cost Per Labor-Hour Bare Costs	Cost Per Labor-Hour Incl. O&P
3 Plasterers	$33.55	$805.20	$50.75	$1218.00	$31.98	$48.24
2 Plasterer Helpers	28.75	460.00	43.50	696.00		
1 Lather	33.70	269.60	50.20	401.60		
1 Mixing Machine, 6 C.F.		113.60		124.96	2.37	2.60
48 L.H., Daily Totals		$1648.40		$2440.56	$34.34	$50.84

Crew J-3

Crew No.	Bare Costs Hr.	Bare Costs Daily	Incl. Subs O&P Hr.	Incl. Subs O&P Daily	Cost Per Labor-Hour Bare Costs	Cost Per Labor-Hour Incl. O&P
1 Terrazzo Worker	$35.15	$281.20	$51.60	$412.80	$31.73	$46.58
1 Terrazzo Helper	28.30	226.40	41.55	332.40		
1 Terrazzo Grinder, Electric		99.35		109.29		
1 Terrazzo Mixer		155.40		170.94	15.92	17.51
16 L.H., Daily Totals		$762.35		$1025.43	$47.65	$64.09

Crew J-4

Crew No.	Bare Costs Hr.	Bare Costs Daily	Incl. Subs O&P Hr.	Incl. Subs O&P Daily	Cost Per Labor-Hour Bare Costs	Cost Per Labor-Hour Incl. O&P
1 Tile Layer	$35.50	$284.00	$52.10	$416.80	$31.38	$46.05
1 Tile Layer Helper	27.25	218.00	40.00	320.00		
16 L.H., Daily Totals		$502.00		$736.80	$31.38	$46.05

Crew K-1

Crew No.	Bare Costs Hr.	Bare Costs Daily	Incl. Subs O&P Hr.	Incl. Subs O&P Daily	Cost Per Labor-Hour Bare Costs	Cost Per Labor-Hour Incl. O&P
1 Carpenter	$36.70	$293.60	$57.15	$457.20	$32.63	$50.63
1 Truck Driver (light)	28.55	228.40	44.10	352.80		
1 Flatbed Truck, gas, 3 Ton		185.20		203.72	11.57	12.73
16 L.H., Daily Totals		$707.20		$1013.72	$44.20	$63.36

Crew K-2

Crew No.	Bare Costs Hr.	Bare Costs Daily	Incl. Subs O&P Hr.	Incl. Subs O&P Daily	Cost Per Labor-Hour Bare Costs	Cost Per Labor-Hour Incl. O&P
1 Struc. Steel Foreman	$43.35	$346.80	$78.50	$628.00	$37.75	$65.83
1 Struc. Steel Worker	41.35	330.80	74.90	599.20		
1 Truck Driver (light)	28.55	228.40	44.10	352.80		
1 Flatbed Truck, gas, 3 Ton		185.20		203.72	7.72	8.49
24 L.H., Daily Totals		$1091.20		$1783.72	$45.47	$74.32

Crew L-1

Crew No.	Bare Costs Hr.	Bare Costs Daily	Incl. Subs O&P Hr.	Incl. Subs O&P Daily	Cost Per Labor-Hour Bare Costs	Cost Per Labor-Hour Incl. O&P
1 Electrician	$43.90	$351.20	$65.35	$522.80	$44.35	$66.38
1 Plumber	44.80	358.40	67.40	539.20		
16 L.H., Daily Totals		$709.60		$1062.00	$44.35	$66.38

Crew L-2

Crew No.	Bare Costs Hr.	Bare Costs Daily	Incl. Subs O&P Hr.	Incl. Subs O&P Daily	Cost Per Labor-Hour Bare Costs	Cost Per Labor-Hour Incl. O&P
1 Carpenter	$36.70	$293.60	$57.15	$457.20	$32.02	$49.77
1 Carpenter Helper	27.35	218.80	42.40	339.20		
16 L.H., Daily Totals		$512.40		$796.40	$32.02	$49.77

Crew L-3

Crew No.	Bare Costs Hr.	Bare Costs Daily	Incl. Subs O&P Hr.	Incl. Subs O&P Daily	Cost Per Labor-Hour Bare Costs	Cost Per Labor-Hour Incl. O&P
1 Carpenter	$36.70	$293.60	$57.15	$457.20	$40.21	$61.70
.5 Electrician	43.90	175.60	65.35	261.40		
.5 Sheet Metal Worker	43.55	174.20	67.15	268.60		
16 L.H., Daily Totals		$643.40		$987.20	$40.21	$61.70

Crew L-3A

Crew No.	Bare Costs Hr.	Bare Costs Daily	Incl. Subs O&P Hr.	Incl. Subs O&P Daily	Cost Per Labor-Hour Bare Costs	Cost Per Labor-Hour Incl. O&P
1 Carpenter Foreman (outside)	$38.70	$309.60	$60.25	$482.00	$40.32	$62.55
.5 Sheet Metal Worker	43.55	174.20	67.15	268.60		
12 L.H., Daily Totals		$483.80		$750.60	$40.32	$62.55

Crew L-4

Crew No.	Bare Costs Hr.	Bare Costs Daily	Incl. Subs O&P Hr.	Incl. Subs O&P Daily	Cost Per Labor-Hour Bare Costs	Cost Per Labor-Hour Incl. O&P
2 Skilled Workers	$38.00	$608.00	$59.15	$946.40	$34.45	$53.57
1 Helper	27.35	218.80	42.40	339.20		
24 L.H., Daily Totals		$826.80		$1285.60	$34.45	$53.57

Crew L-5

Crew No.	Bare Costs Hr.	Bare Costs Daily	Incl. Subs O&P Hr.	Incl. Subs O&P Daily	Cost Per Labor-Hour Bare Costs	Cost Per Labor-Hour Incl. O&P
1 Struc. Steel Foreman	$43.35	$346.80	$78.50	$628.00	$41.41	$73.29
5 Struc. Steel Workers	41.35	1654.00	74.90	2996.00		
1 Equip. Oper. (crane)	39.80	318.40	60.00	480.00		
1 Hyd. Crane, 25 Ton		739.60		813.56	13.21	14.53
56 L.H., Daily Totals		$3058.80		$4917.56	$54.62	$87.81

Crew L-5A

Crew No.	Bare Costs Hr.	Bare Costs Daily	Incl. Subs O&P Hr.	Incl. Subs O&P Daily	Cost Per Labor-Hour Bare Costs	Cost Per Labor-Hour Incl. O&P
1 Structural Steel Foreman	$43.35	$346.80	$78.50	$628.00	$41.46	$72.08
2 Structural Steel Workers	41.35	661.60	74.90	1198.40		
1 Equip. Oper. (crane)	39.80	318.40	60.00	480.00		
1 S.P. Crane, 4x4, 25 Ton		875.60		963.16	27.36	30.10
32 L.H., Daily Totals		$2202.40		$3269.56	$68.83	$102.17

Crew L-6

Crew No.	Bare Costs Hr.	Bare Costs Daily	Incl. Subs O&P Hr.	Incl. Subs O&P Daily	Cost Per Labor-Hour Bare Costs	Cost Per Labor-Hour Incl. O&P
1 Plumber	$44.80	$358.40	$67.40	$539.20	$44.50	$66.72
.5 Electrician	43.90	175.60	65.35	261.40		
12 L.H., Daily Totals		$534.00		$800.60	$44.50	$66.72

Crew L-7

Crew No.	Bare Costs Hr.	Bare Costs Daily	Incl. Subs O&P Hr.	Incl. Subs O&P Daily	Cost Per Labor-Hour Bare Costs	Cost Per Labor-Hour Incl. O&P
2 Carpenters	$36.70	$587.20	$57.15	$914.40	$35.46	$54.78
1 Building Laborer	28.75	230.00	44.75	358.00		
.5 Electrician	43.90	175.60	65.35	261.40		
28 L.H., Daily Totals		$992.80		$1533.80	$35.46	$54.78

Crew L-8

Crew No.	Bare Costs Hr.	Bare Costs Daily	Incl. Subs O&P Hr.	Incl. Subs O&P Daily	Cost Per Labor-Hour Bare Costs	Cost Per Labor-Hour Incl. O&P
2 Carpenters	$36.70	$587.20	$57.15	$914.40	$38.32	$59.20
.5 Plumber	44.80	179.20	67.40	269.60		
20 L.H., Daily Totals		$766.40		$1184.00	$38.32	$59.20

Crews

Crew No.	Bare Costs		Incl. Subs O & P		Cost Per Labor-Hour	
Crew L-9	Hr.	Daily	Hr.	Daily	Bare Costs	Incl. O&P
1 Labor Foreman (inside)	$29.25	$234.00	$45.55	$364.40	$33.34	$53.92
2 Building Laborers	28.75	460.00	44.75	716.00		
1 Struc. Steel Worker	41.35	330.80	74.90	599.20		
.5 Electrician	43.90	175.60	65.35	261.40		
36 L.H., Daily Totals		$1200.40		$1941.00	$33.34	$53.92
Crew L-10	Hr.	Daily	Hr.	Daily	Bare Costs	Incl. O&P
1 Structural Steel Foreman	$43.35	$346.80	$78.50	$628.00	$41.50	$71.13
1 Structural Steel Worker	41.35	330.80	74.90	599.20		
1 Equip. Oper. (crane)	39.80	318.40	60.00	480.00		
1 Hyd. Crane, 12 Ton		724.00		796.40	30.17	33.18
24 L.H., Daily Totals		$1720.00		$2503.60	$71.67	$104.32
Crew L-11	Hr.	Daily	Hr.	Daily	Bare Costs	Incl. O&P
2 Wreckers	$28.75	$460.00	$50.70	$811.20	$33.54	$54.24
1 Equip. Oper. (crane)	39.80	318.40	60.00	480.00		
1 Equip. Oper. (light)	36.85	294.80	55.55	444.40		
1 Hyd. Excavator, 2.5 C.Y.		1333.00		1466.30		
1 Loader, Skid Steer, 78 HP		222.20		244.42	48.60	53.46
32 L.H., Daily Totals		$2628.40		$3446.32	$82.14	$107.70
Crew M-1	Hr.	Daily	Hr.	Daily	Bare Costs	Incl. O&P
3 Elevator Constructors	$53.40	$1281.60	$79.65	$1911.60	$50.73	$75.66
1 Elevator Apprentice	42.70	341.60	63.70	509.60		
5 Hand Tools		48.00		52.80	1.50	1.65
32 L.H., Daily Totals		$1671.20		$2474.00	$52.23	$77.31
Crew M-3	Hr.	Daily	Hr.	Daily	Bare Costs	Incl. O&P
1 Electrician Foreman (out)	$45.90	$367.20	$68.35	$546.80	$42.44	$63.74
1 Common Laborer	28.75	230.00	44.75	358.00		
.25 Equipment Operator, Medium	38.40	76.80	57.85	115.70		
1 Elevator Constructor	53.40	427.20	79.65	637.20		
1 Elevator Apprentice	42.70	341.60	63.70	509.60		
.25 S.P. Crane, 4x4, 20 Ton		167.70		184.47	4.93	5.43
34 L.H., Daily Totals		$1610.50		$2351.77	$47.37	$69.17
Crew M-4	Hr.	Daily	Hr.	Daily	Bare Costs	Incl. O&P
1 Electrician Foreman (out)	$45.90	$367.20	$68.35	$546.80	$42.04	$63.16
1 Common Laborer	28.75	230.00	44.75	358.00		
.25 Equipment Operator, Crane	39.80	79.60	60.00	120.00		
.25 Equipment Operator, Oiler	33.90	67.80	51.10	102.20		
1 Elevator Constructor	53.40	427.20	79.65	637.20		
1 Elevator Apprentice	42.70	341.60	63.70	509.60		
.25 S.P. Crane, 4x4, 40 Ton		260.75		286.82	7.24	7.97
36 L.H., Daily Totals		$1774.15		$2560.63	$49.28	$71.13
Crew Q-1	Hr.	Daily	Hr.	Daily	Bare Costs	Incl. O&P
1 Plumber	$44.80	$358.40	$67.40	$539.20	$40.33	$60.65
1 Plumber Apprentice	35.85	286.80	53.90	431.20		
16 L.H., Daily Totals		$645.20		$970.40	$40.33	$60.65
Crew Q-1C	Hr.	Daily	Hr.	Daily	Bare Costs	Incl. O&P
1 Plumber	$44.80	$358.40	$67.40	$539.20	$39.68	$59.72
1 Plumber Apprentice	35.85	286.80	53.90	431.20		
1 Equip. Oper. (medium)	38.40	307.20	57.85	462.80		
1 Trencher, Chain Type, 8' D		1606.00		1766.60	66.92	73.61
24 L.H., Daily Totals		$2558.40		$3199.80	$106.60	$133.32

Crew No.	Bare Costs		Incl. Subs O & P		Cost Per Labor-Hour	
Crew Q-2	Hr.	Daily	Hr.	Daily	Bare Costs	Incl. O&P
2 Plumbers	$44.80	$716.80	$67.40	$1078.40	$41.82	$62.90
1 Plumber Apprentice	35.85	286.80	53.90	431.20		
24 L.H., Daily Totals		$1003.60		$1509.60	$41.82	$62.90
Crew Q-3	Hr.	Daily	Hr.	Daily	Bare Costs	Incl. O&P
1 Plumber Foreman (inside)	$45.30	$362.40	$68.15	$545.20	$42.69	$64.21
2 Plumbers	44.80	716.80	67.40	1078.40		
1 Plumber Apprentice	35.85	286.80	53.90	431.20		
32 L.H., Daily Totals		$1366.00		$2054.80	$42.69	$64.21
Crew Q-4	Hr.	Daily	Hr.	Daily	Bare Costs	Incl. O&P
1 Plumber Foreman (inside)	$45.30	$362.40	$68.15	$545.20	$42.69	$64.21
1 Plumber	44.80	358.40	67.40	539.20		
1 Welder (plumber)	44.80	358.40	67.40	539.20		
1 Plumber Apprentice	35.85	286.80	53.90	431.20		
1 Welder, electric, 300 amp		61.45		67.59	1.92	2.11
32 L.H., Daily Totals		$1427.45		$2122.40	$44.61	$66.32
Crew Q-5	Hr.	Daily	Hr.	Daily	Bare Costs	Incl. O&P
1 Steamfitter	$45.20	$361.60	$68.00	$544.00	$40.67	$61.17
1 Steamfitter Apprentice	36.15	289.20	54.35	434.80		
16 L.H., Daily Totals		$650.80		$978.80	$40.67	$61.17
Crew Q-6	Hr.	Daily	Hr.	Daily	Bare Costs	Incl. O&P
2 Steamfitters	$45.20	$723.20	$68.00	$1088.00	$42.18	$63.45
1 Steamfitter Apprentice	36.15	289.20	54.35	434.80		
24 L.H., Daily Totals		$1012.40		$1522.80	$42.18	$63.45
Crew Q-7	Hr.	Daily	Hr.	Daily	Bare Costs	Incl. O&P
1 Steamfitter Foreman (inside)	$45.70	$365.60	$68.75	$550.00	$43.06	$64.78
2 Steamfitters	45.20	723.20	68.00	1088.00		
1 Steamfitter Apprentice	36.15	289.20	54.35	434.80		
32 L.H., Daily Totals		$1378.00		$2072.80	$43.06	$64.78
Crew Q-8	Hr.	Daily	Hr.	Daily	Bare Costs	Incl. O&P
1 Steamfitter Foreman (inside)	$45.70	$365.60	$68.75	$550.00	$43.06	$64.78
1 Steamfitter	45.20	361.60	68.00	544.00		
1 Welder (steamfitter)	45.20	361.60	68.00	544.00		
1 Steamfitter Apprentice	36.15	289.20	54.35	434.80		
1 Welder, electric, 300 amp		61.45		67.59	1.92	2.11
32 L.H., Daily Totals		$1439.45		$2140.40	$44.98	$66.89
Crew Q-9	Hr.	Daily	Hr.	Daily	Bare Costs	Incl. O&P
1 Sheet Metal Worker	$43.55	$348.40	$67.15	$537.20	$39.20	$60.45
1 Sheet Metal Apprentice	34.85	278.80	53.75	430.00		
16 L.H., Daily Totals		$627.20		$967.20	$39.20	$60.45
Crew Q-10	Hr.	Daily	Hr.	Daily	Bare Costs	Incl. O&P
2 Sheet Metal Workers	$43.55	$696.80	$67.15	$1074.40	$40.65	$62.68
1 Sheet Metal Apprentice	34.85	278.80	53.75	430.00		
24 L.H., Daily Totals		$975.60		$1504.40	$40.65	$62.68
Crew Q-11	Hr.	Daily	Hr.	Daily	Bare Costs	Incl. O&P
1 Sheet Metal Foreman (inside)	$44.05	$352.40	$67.95	$543.60	$41.50	$64.00
2 Sheet Metal Workers	43.55	696.80	67.15	1074.40		
1 Sheet Metal Apprentice	34.85	278.80	53.75	430.00		
32 L.H., Daily Totals		$1328.00		$2048.00	$41.50	$64.00

Crews

Crew Q-12

Crew No.	Bare Costs Hr.	Daily	Incl. Subs O & P Hr.	Daily	Cost Per Labor-Hour Bare Costs	Incl. O&P
1 Sprinkler Installer	$44.30	$354.40	$66.70	$533.60	$39.88	$60.05
1 Sprinkler Apprentice	35.45	283.60	53.40	427.20		
16 L.H., Daily Totals		$638.00		$960.80	$39.88	$60.05

Crew Q-13

Crew No.	Hr.	Daily	Hr.	Daily	Bare Costs	Incl. O&P
1 Sprinkler Foreman (inside)	$44.80	$358.40	$67.45	$539.60	$42.21	$63.56
2 Sprinkler Installers	44.30	708.80	66.70	1067.20		
1 Sprinkler Apprentice	35.45	283.60	53.40	427.20		
32 L.H., Daily Totals		$1350.80		$2034.00	$42.21	$63.56

Crew Q-14

Crew No.	Hr.	Daily	Hr.	Daily	Bare Costs	Incl. O&P
1 Asbestos Worker	$41.50	$332.00	$65.60	$524.80	$37.35	$59.05
1 Asbestos Apprentice	33.20	265.60	52.50	420.00		
16 L.H., Daily Totals		$597.60		$944.80	$37.35	$59.05

Crew Q-15

Crew No.	Hr.	Daily	Hr.	Daily	Bare Costs	Incl. O&P
1 Plumber	$44.80	$358.40	$67.40	$539.20	$40.33	$60.65
1 Plumber Apprentice	35.85	286.80	53.90	431.20		
1 Welder, electric, 300 amp		61.45		67.59	3.84	4.22
16 L.H., Daily Totals		$706.65		$1037.99	$44.17	$64.87

Crew Q-16

Crew No.	Hr.	Daily	Hr.	Daily	Bare Costs	Incl. O&P
2 Plumbers	$44.80	$716.80	$67.40	$1078.40	$41.82	$62.90
1 Plumber Apprentice	35.85	286.80	53.90	431.20		
1 Welder, electric, 300 amp		61.45		67.59	2.56	2.82
24 L.H., Daily Totals		$1065.05		$1577.19	$44.38	$65.72

Crew Q-17

Crew No.	Hr.	Daily	Hr.	Daily	Bare Costs	Incl. O&P
1 Steamfitter	$45.20	$361.60	$68.00	$544.00	$40.67	$61.17
1 Steamfitter Apprentice	36.15	289.20	54.35	434.80		
1 Welder, electric, 300 amp		61.45		67.59	3.84	4.22
16 L.H., Daily Totals		$712.25		$1046.40	$44.52	$65.40

Crew Q-17A

Crew No.	Hr.	Daily	Hr.	Daily	Bare Costs	Incl. O&P
1 Steamfitter	$45.20	$361.60	$68.00	$544.00	$40.38	$60.78
1 Steamfitter Apprentice	36.15	289.20	54.35	434.80		
1 Equip. Oper. (crane)	39.80	318.40	60.00	480.00		
1 Hyd. Crane, 12 Ton		724.00		796.40		
1 Welder, electric, 300 amp		61.45		67.59	32.73	36.00
24 L.H., Daily Totals		$1754.65		$2322.80	$73.11	$96.78

Crew Q-18

Crew No.	Hr.	Daily	Hr.	Daily	Bare Costs	Incl. O&P
2 Steamfitters	$45.20	$723.20	$68.00	$1088.00	$42.18	$63.45
1 Steamfitter Apprentice	36.15	289.20	54.35	434.80		
1 Welder, electric, 300 amp		61.45		67.59	2.56	2.82
24 L.H., Daily Totals		$1073.85		$1590.40	$44.74	$66.27

Crew Q-19

Crew No.	Hr.	Daily	Hr.	Daily	Bare Costs	Incl. O&P
1 Steamfitter	$45.20	$361.60	$68.00	$544.00	$41.75	$62.57
1 Steamfitter Apprentice	36.15	289.20	54.35	434.80		
1 Electrician	43.90	351.20	65.35	522.80		
24 L.H., Daily Totals		$1002.00		$1501.60	$41.75	$62.57

Crew Q-20

Crew No.	Hr.	Daily	Hr.	Daily	Bare Costs	Incl. O&P
1 Sheet Metal Worker	$43.55	$348.40	$67.15	$537.20	$40.14	$61.43
1 Sheet Metal Apprentice	34.85	278.80	53.75	430.00		
.5 Electrician	43.90	175.60	65.35	261.40		
20 L.H., Daily Totals		$802.80		$1228.60	$40.14	$61.43

Crew Q-21

Crew No.	Bare Costs Hr.	Daily	Incl. Subs O & P Hr.	Daily	Cost Per Labor-Hour Bare Costs	Incl. O&P
2 Steamfitters	$45.20	$723.20	$68.00	$1088.00	$42.61	$63.92
1 Steamfitter Apprentice	36.15	289.20	54.35	434.80		
1 Electrician	43.90	351.20	65.35	522.80		
32 L.H., Daily Totals		$1363.60		$2045.60	$42.61	$63.92

Crew Q-22

Crew No.	Hr.	Daily	Hr.	Daily	Bare Costs	Incl. O&P
1 Plumber	$44.80	$358.40	$67.40	$539.20	$40.33	$60.65
1 Plumber Apprentice	35.85	286.80	53.90	431.20		
1 Hyd. Crane, 12 Ton		724.00		796.40	45.25	49.77
16 L.H., Daily Totals		$1369.20		$1766.80	$85.58	$110.43

Crew Q-22A

Crew No.	Hr.	Daily	Hr.	Daily	Bare Costs	Incl. O&P
1 Plumber	$44.80	$358.40	$67.40	$539.20	$37.30	$56.51
1 Plumber Apprentice	35.85	286.80	53.90	431.20		
1 Laborer	28.75	230.00	44.75	358.00		
1 Equip. Oper. (crane)	39.80	318.40	60.00	480.00		
1 Hyd. Crane, 12 Ton		724.00		796.40	22.63	24.89
32 L.H., Daily Totals		$1917.60		$2604.80	$59.92	$81.40

Crew Q-23

Crew No.	Hr.	Daily	Hr.	Daily	Bare Costs	Incl. O&P
1 Plumber Foreman	$46.80	$374.40	$70.40	$563.20	$43.33	$65.22
1 Plumber	44.80	358.40	67.40	539.20		
1 Equip. Oper. (medium)	38.40	307.20	57.85	462.80		
1 Lattice Boom Crane, 20 Ton		976.70		1074.37	40.70	44.77
24 L.H., Daily Totals		$2016.70		$2639.57	$84.03	$109.98

Crew R-1

Crew No.	Hr.	Daily	Hr.	Daily	Bare Costs	Incl. O&P
1 Electrician Foreman	$44.40	$355.20	$66.10	$528.80	$38.47	$57.83
3 Electricians	43.90	1053.60	65.35	1568.40		
2 Helpers	27.35	437.60	42.40	678.40		
48 L.H., Daily Totals		$1846.40		$2775.60	$38.47	$57.83

Crew R-1A

Crew No.	Hr.	Daily	Hr.	Daily	Bare Costs	Incl. O&P
1 Electrician	$43.90	$351.20	$65.35	$522.80	$35.63	$53.88
1 Helper	27.35	218.80	42.40	339.20		
16 L.H., Daily Totals		$570.00		$862.00	$35.63	$53.88

Crew R-2

Crew No.	Hr.	Daily	Hr.	Daily	Bare Costs	Incl. O&P
1 Electrician Foreman	$44.40	$355.20	$66.10	$528.80	$38.66	$58.14
3 Electricians	43.90	1053.60	65.35	1568.40		
2 Helpers	27.35	437.60	42.40	678.40		
1 Equip. Oper. (crane)	39.80	318.40	60.00	480.00		
1 S.P. Crane, 4x4, 5 Ton		326.80		359.48	5.84	6.42
56 L.H., Daily Totals		$2491.60		$3615.08	$44.49	$64.56

Crew R-3

Crew No.	Hr.	Daily	Hr.	Daily	Bare Costs	Incl. O&P
1 Electrician Foreman	$44.40	$355.20	$66.10	$528.80	$43.28	$64.58
1 Electrician	43.90	351.20	65.35	522.80		
.5 Equip. Oper. (crane)	39.80	159.20	60.00	240.00		
.5 S.P. Crane, 4x4, 5 Ton		163.40		179.74	8.17	8.99
20 L.H., Daily Totals		$1029.00		$1471.34	$51.45	$73.57

Crew R-4

Crew No.	Hr.	Daily	Hr.	Daily	Bare Costs	Incl. O&P
1 Struc. Steel Foreman	$43.35	$346.80	$78.50	$628.00	$42.26	$73.71
3 Struc. Steel Workers	41.35	992.40	74.90	1797.60		
1 Electrician	43.90	351.20	65.35	522.80		
1 Welder, gas engine, 300 amp		115.20		126.72	2.88	3.17
40 L.H., Daily Totals		$1805.60		$3075.12	$45.14	$76.88

Crews

Crew R-5	Hr.	Daily	Hr.	Daily	Bare Costs	Incl. O&P
1 Electrician Foreman	$44.40	$355.20	$66.10	$528.80	$37.93	$57.07
4 Electrician Linemen	43.90	1404.80	65.35	2091.20		
2 Electrician Operators	43.90	702.40	65.35	1045.60		
4 Electrician Groundmen	27.35	875.20	42.40	1356.80		
1 Crew Truck		116.20		127.82		
1 Tool Van		151.80		166.98		
1 Pickup Truck, 3/4 Ton		84.80		93.28		
.2 Hyd. Crane, 55 Ton		212.00		233.20		
.2 Hyd. Crane, 12 Ton		144.80		159.28		
.2 Drill Rig, Truck-Mounted		699.80		769.78		
1 Tractor w/Winch		297.80		327.58	19.40	21.34
88 L.H., Daily Totals		$5044.80		$6900.32	$57.33	$78.41

Crew R-6	Hr.	Daily	Hr.	Daily	Bare Costs	Incl. O&P
1 Electrician Foreman	$44.40	$355.20	$66.10	$528.80	$37.93	$57.07
4 Electrician Linemen	43.90	1404.80	65.35	2091.20		
2 Electrician Operators	43.90	702.40	65.35	1045.60		
4 Electrician Groundmen	27.35	875.20	42.40	1356.80		
1 Crew Truck		116.20		127.82		
1 Tool Van		151.80		166.98		
1 Pickup Truck, 3/4 Ton		84.80		93.28		
.2 Hyd. Crane, 55 Ton		212.00		233.20		
.2 Hyd. Crane, 12 Ton		144.80		159.28		
.2 Drill Rig, Truck-Mounted		699.80		769.78		
1 Tractor w/Winch		297.80		327.58		
3 Cable Trailers		510.90		561.99		
.5 Tensioning Rig		167.68		184.44		
.5 Cable Pulling Rig		989.00		1087.90	38.35	42.18
88 L.H., Daily Totals		$6712.38		$8734.65	$76.28	$99.26

Crew R-7	Hr.	Daily	Hr.	Daily	Bare Costs	Incl. O&P
1 Electrician Foreman	$44.40	$355.20	$66.10	$528.80	$30.19	$46.35
5 Electrician Groundmen	27.35	1094.00	42.40	1696.00		
1 Crew Truck		116.20		127.82	2.42	2.66
48 L.H., Daily Totals		$1565.40		$2352.62	$32.61	$49.01

Crew R-8	Hr.	Daily	Hr.	Daily	Bare Costs	Incl. O&P
1 Electrician Foreman	$44.40	$355.20	$66.10	$528.80	$38.47	$57.83
3 Electrician Linemen	43.90	1053.60	65.35	1568.40		
2 Electrician Groundmen	27.35	437.60	42.40	678.40		
1 Pickup Truck, 3/4 Ton		84.80		93.28		
1 Crew Truck		116.20		127.82	4.19	4.61
48 L.H., Daily Totals		$2047.40		$2996.70	$42.65	$62.43

Crew R-9	Hr.	Daily	Hr.	Daily	Bare Costs	Incl. O&P
1 Electrician Foreman	$44.40	$355.20	$66.10	$528.80	$35.69	$53.97
1 Electrician Lineman	43.90	351.20	65.35	522.80		
2 Electrician Operators	43.90	702.40	65.35	1045.60		
4 Electrician Groundmen	27.35	875.20	42.40	1356.80		
1 Pickup Truck, 3/4 Ton		84.80		93.28		
1 Crew Truck		116.20		127.82	3.14	3.45
64 L.H., Daily Totals		$2485.00		$3675.10	$38.83	$57.42

Crew R-10	Hr.	Daily	Hr.	Daily	Bare Costs	Incl. O&P
1 Electrician Foreman	$44.40	$355.20	$66.10	$528.80	$41.23	$61.65
4 Electrician Linemen	43.90	1404.80	65.35	2091.20		
1 Electrician Groundman	27.35	218.80	42.40	339.20		
1 Crew Truck		116.20		127.82		
3 Tram Cars		363.00		399.30	9.98	10.98
48 L.H., Daily Totals		$2458.00		$3486.32	$51.21	$72.63

Crew R-11	Hr.	Daily	Hr.	Daily	Bare Costs	Incl. O&P
1 Electrician Foreman	$44.40	$355.20	$66.10	$528.80	$41.22	$61.75
4 Electricians	43.90	1404.80	65.35	2091.20		
1 Equip. Oper. (crane)	39.80	318.40	60.00	480.00		
1 Common Laborer	28.75	230.00	44.75	358.00		
1 Crew Truck		116.20		127.82		
1 Hyd. Crane, 12 Ton		724.00		796.40	15.00	16.50
56 L.H., Daily Totals		$3148.60		$4382.22	$56.23	$78.25

Crew R-12	Hr.	Daily	Hr.	Daily	Bare Costs	Incl. O&P
1 Carpenter Foreman	$37.20	$297.60	$57.90	$463.20	$34.43	$54.39
4 Carpenters	36.70	1174.40	57.15	1828.80		
4 Common Laborers	28.75	920.00	44.75	1432.00		
1 Equip. Oper. (med.)	38.40	307.20	57.85	462.80		
1 Steel Worker	41.35	330.80	74.90	599.20		
1 Dozer, 200 H.P.		988.40		1087.24		
1 Pickup Truck, 3/4 Ton		84.80		93.28	12.20	13.41
88 L.H., Daily Totals		$4103.20		$5966.52	$46.63	$67.80

Crew R-13	Hr.	Daily	Hr.	Daily	Bare Costs	Incl. O&P
1 Electrician Foreman	$44.40	$355.20	$66.10	$528.80	$41.90	$62.52
3 Electricians	43.90	1053.60	65.35	1568.40		
.25 Equip. Oper. (crane)	39.80	79.60	60.00	120.00		
1 Equipment Oiler	33.90	271.20	51.10	408.80		
.25 Hydraulic Crane, 33 Ton		191.35		210.49	4.56	5.01
42 L.H., Daily Totals		$1950.95		$2836.49	$46.45	$67.54

Crew R-15	Hr.	Daily	Hr.	Daily	Bare Costs	Incl. O&P
1 Electrician Foreman	$44.40	$355.20	$66.10	$528.80	$42.81	$63.84
4 Electricians	43.90	1404.80	65.35	2091.20		
1 Equipment Operator	36.85	294.80	55.55	444.40		
1 Aerial Lift Truck		289.80		318.78	6.04	6.64
48 L.H., Daily Totals		$2344.60		$3383.18	$48.85	$70.48

Crew R-18	Hr.	Daily	Hr.	Daily	Bare Costs	Incl. O&P
.25 Electrician Foreman	$44.40	$88.80	$66.10	$132.20	$33.75	$51.28
1 Electrician	43.90	351.20	65.35	522.80		
2 Helpers	27.35	437.60	42.40	678.40		
26 L.H., Daily Totals		$877.60		$1333.40	$33.75	$51.28

Crew R-19	Hr.	Daily	Hr.	Daily	Bare Costs	Incl. O&P
.5 Electrician Foreman	$44.40	$177.60	$66.10	$264.40	$44.00	$65.50
2 Electricians	43.90	702.40	65.35	1045.60		
20 L.H., Daily Totals		$880.00		$1310.00	$44.00	$65.50

Crew R-21	Hr.	Daily	Hr.	Daily	Bare Costs	Incl. O&P
1 Electrician Foreman	$44.40	$355.20	$66.10	$528.80	$43.89	$65.35
3 Electricians	43.90	1053.60	65.35	1568.40		
.1 Equip. Oper. (med.)	38.40	30.72	57.85	46.28		
.1 S.P. Crane, 4x4, 25 Ton		87.56		96.32	2.67	2.94
32.8 L.H., Daily Totals		$1527.08		$2239.80	$46.56	$68.29

Crew R-22	Hr.	Daily	Hr.	Daily	Bare Costs	Incl. O&P
.66 Electrician Foreman	$44.40	$234.43	$66.10	$349.01	$36.87	$55.61
2 Helpers	27.35	437.60	42.40	678.40		
2 Electricians	43.90	702.40	65.35	1045.60		
37.28 L.H., Daily Totals		$1374.43		$2073.01	$36.87	$55.61

Crews

Crew No.	Bare Costs		Incl. Sub O & P		Cost Per Labor-Hour	

Crew R-30	Hr.	Daily	Hr.	Daily	Bare Costs	Incl. O&P
.25 Electrician Foreman (out)	$45.90	$91.80	$68.35	$136.70	$34.73	$52.90
1 Electrician	43.90	351.20	65.35	522.80		
2 Laborers, (Semi-Skilled)	28.75	460.00	44.75	716.00		
26 L.H., Daily Totals		$903.00		$1375.50	$34.73	$52.90

Crew R-31	Hr.	Daily	Hr.	Daily	Bare Costs	Incl. O&P
1 Electrician	$43.90	$351.20	$65.35	$522.80	$43.90	$65.35
1 Core Drill, Elec, 2.5 HP		71.20		78.32	8.90	9.79
8 L.H., Daily Totals		$422.40		$601.12	$52.80	$75.14

Crew W-41E	Hr.	Daily	Hr.	Daily	Bare Costs	Incl. O&P
1 Laborers, (Semi-Skilled)	$28.75	$230.00	$44.75	$358.00	$38.78	$58.94
1 Plumber	44.80	358.40	67.40	539.20		
.5 Plumber	46.80	187.20	70.40	281.60		
20 L.H., Daily Totals		$775.60		$1178.80	$38.78	$58.94

Historical Cost Indexes

The table below lists both the RSMeans Historical Cost Index based on Jan. 1, 1993 = 100 as well as the computed value of an index based on Jan. 1, 2007 costs. Since the Jan. 1, 2007 figure is estimated, space is left to write in the actual index figures as they become available through either the quarterly "RSMeans Construction Cost Indexes" or as printed in the "Engineering News-Record." To compute the actual index based on Jan. 1, 2007 = 100, divide the Historical Cost Index for a particular year by the actual Jan. 1, 2007 Construction Cost Index. Space has been left to advance the index figures as the year progresses.

Year	Historical Cost Index Jan. 1, 1993 = 100		Current Index Based on Jan. 1, 2007 = 100		Year	Historical Cost Index Jan. 1, 1993 = 100	Current Index Based on Jan. 1, 2007 = 100		Year	Historical Cost Index Jan. 1, 1993 = 100	Current Index Based on Jan. 1, 2007 = 100	
	Est.	Actual	Est.	Actual		Actual	Est.	Actual		Actual	Est.	Actual
Oct 2007					July 1992	99.4	59.8		July 1974	41.4	24.9	
July 2007					1991	96.8	58.2		1973	37.7	22.7	
April 2007					1990	94.3	56.7		1972	34.8	20.9	
Jan 2007	166.3		100.0	100.0	1989	92.1	55.4		1971	32.1	19.3	
July 2006		162.0	97.4		1988	89.9	54.0		1970	28.7	17.3	
2005		151.6	91.2		1987	87.7	52.7		1969	26.9	16.2	
2004		143.7	86.4		1986	84.2	50.6		1968	24.9	15.0	
2003		132.0	79.4		1985	82.6	49.7		1967	23.5	14.1	
2002		128.7	77.4		1984	82.0	49.3		1966	22.7	13.7	
2001		125.1	75.2		1983	80.2	48.2		1965	21.7	13.0	
2000		120.9	72.7		1982	76.1	45.8		1964	21.2	12.7	
1999		117.6	70.7		1981	70.0	42.1		1963	20.7	12.4	
1998		115.1	69.2		1980	62.9	37.8		1962	20.2	12.1	
1997		112.8	67.8		1979	57.8	34.8		1961	19.8	11.9	
1996		110.2	66.3		1978	53.5	32.2		1960	19.7	11.8	
1995		107.6	64.7		1977	49.5	29.8		1959	19.3	11.6	
1994		104.4	62.8		1976	46.9	28.2		1958	18.8	11.3	
1993		101.7	61.2		1975	44.8	26.9		1957	18.4	11.1	

Adjustments to Costs

The Historical Cost Index can be used to convert National Average building costs at a particular time to the approximate building costs for some other time.

Example:

Estimate and compare construction costs for different years in the same city.

To estimate the National Average construction cost of a building in 1970, knowing that it cost $900,000 in 2007:

INDEX in 1970 = 28.7

INDEX in 2007 = 166.3

Note: The City Cost Indexes for Canada can be used to convert U.S. National averages to local costs in Canadian dollars.

Time Adjustment using the Historical Cost Indexes:

$$\frac{\text{Index for Year A}}{\text{Index for Year B}} \times \text{Cost in Year B} = \text{Cost in Year A}$$

$$\frac{\text{INDEX 1970}}{\text{INDEX 2007}} \times \text{Cost 2007} = \text{Cost 1970}$$

$$\frac{28.7}{166.3} \times \$900,000 = .173 \times \$900,000 = \$155,700$$

The construction cost of the building in 1970 is $155,700.

How to Use the City Cost Indexes

What you should know before you begin

RSMeans City Cost Indexes (CCI) are an extremely useful tool to use when you want to compare costs from city to city and region to region.

This publication contains average construction cost indexes for 316 U.S. and Canadian cities covering over 930 three-digit zip code locations, as listed directly under each city.

Keep in mind that a City Cost Index number is a percentage ratio of a specific city's cost to the national average cost of the same item at a stated time period.

In other words, these index figures represent relative construction factors (or, if you prefer, multipliers) for Material and Installation costs, as well as the weighted average for Total In Place costs for each CSI MasterFormat division. Installation costs include both labor and equipment rental costs. When estimating equipment rental rates only, for a specific location, use 01543 CONTRACTOR EQUIPMENT index.

The 30 City Average Index is the average of 30 major U.S. cities and serves as a National Average.

Index figures for both material and installation are based on the 30 major city average of 100 and represent the cost relationship as of July 1, 2006. The index for each division is computed from representative material and labor quantities for that division. The weighted average for each city is a weighted total of the components listed above it, but does not include relative productivity between trades or cities.

As changes occur in local material prices, labor rates and equipment rental rates, the impact of these changes should be accurately measured by the change in the City Cost Index for each particular city (as compared to the 30 City Average).

Therefore, if you know (or have estimated) building costs in one city today, you can easily convert those costs to expected building costs in another city.

In addition, by using the Historical Cost Index, you can easily convert National Average building costs at a particular time to the approximate building costs for some other time. The City Cost Indexes can then be applied to calculate the costs for a particular city.

Quick Calculations

Location Adjustment Using the City Cost Indexes:

$$\frac{\text{Index for City A}}{\text{Index for City B}} \times \text{Cost in City B} = \text{Cost in City A}$$

Time Adjustment for the National Average Using the Historical Cost Index:

$$\frac{\text{Index for Year A}}{\text{Index for Year B}} \times \text{Cost in Year B} = \text{Cost in Year A}$$

Adjustment from the National Average:

$$\frac{\text{Index for City A}}{100} \times \text{National Average Cost} = \text{Cost in City A}$$

Since each of the other RSMeans publications contains many different items, any *one* item multiplied by the particular city index may give incorrect results. However, the larger the number of items compiled, the closer the results should be to actual costs for that particular city.

The City Cost Indexes for Canadian cities are calculated using Canadian material and equipment prices and labor rates, in Canadian dollars. Therefore, indexes for Canadian cities can be used to convert U.S. National Average prices to local costs in Canadian dollars.

How to use this section

1. Compare costs from city to city.

In using the RSMeans Indexes, remember that an index number is not a fixed number but a ratio: It's a percentage ratio of a building component's cost at any stated time to the National Average cost of that same component at the same time period. Put in the form of an equation:

$$\frac{\text{Specific City Cost}}{\text{National Average Cost}} \times 100 = \text{City Index Number}$$

Therefore, when making cost comparisons between cities, do not subtract one city's index number from the index number of another city and read the result as a percentage difference. Instead, divide one city's index number by that of the other city. The resulting number may then be used as a multiplier to calculate cost differences from city to city.

The formula used to find cost differences between cities for the purpose of comparison is as follows:

$$\frac{\text{City A Index}}{\text{City B Index}} \times \text{City B Cost (Known)} = \text{City A Cost (Unknown)}$$

In addition, you can use RSMeans CCI to calculate and compare costs division by division between cities using the same basic formula. (Just be sure that you're comparing similar divisions.)

2. Compare a specific city's construction costs with the National Average.

When you're studying construction location feasibility, it's advisable to compare a prospective project's cost index with an index of the National Average cost.

For example, divide the weighted average index of construction costs of a specific city by that of the 30 City Average, which = 100.

$$\frac{\text{City Index}}{100} = \% \text{ of National Average}$$

As a result, you get a ratio that indicates the relative cost of construction in that city in comparison with the National Average.

3. Convert U.S. National Average to actual costs in Canadian City.

$$\frac{\text{Index for Canadian City}}{100} \times \text{National Average Cost} = \text{Cost in Canadian City in \$ CAN}$$

4. Adjust construction cost data based on a National Average.

When you use a source of construction cost data which is based on a National Average (such as RSMeans cost data publications), it is necessary to adjust those costs to a specific location.

$$\frac{\text{City Index}}{100} \quad \text{x} \quad \begin{array}{c} \text{``Book'' Cost Based on} \\ \text{National Average Costs} \end{array} \quad = \quad \begin{array}{c} \text{City Cost} \\ \text{(Unknown)} \end{array}$$

5. When applying the City Cost Indexes to demolition projects, use the appropriate division installation index. For example, for removal of existing doors and windows, use Division 8 (Openings) index.

What you might like to know about how we developed the Indexes

The information presented in the CCI is organized according to the Construction Specifications Institute (CSI) MasterFormat 2004.

To create a reliable index, RSMeans researched the building type most often constructed in the United States and Canada. Because it was concluded that no one type of building completely represented the building construction industry, nine different types of buildings were combined to create a composite model.

The exact material, labor and equipment quantities are based on detailed analysis of these nine building types, then each quantity is weighted in proportion to expected usage. These various material items, labor hours, and equipment rental rates are thus combined to form a composite building representing as closely as possible the actual usage of materials, labor and equipment used in the North American Building Construction Industry.

The following structures were chosen to make up that composite model:

1. Factory, 1 story
2. Office, 2–4 story
3. Store, Retail
4. Town Hall, 2–3 story
5. High School, 2–3 story
6. Hospital, 4–8 story
7. Garage, Parking
8. Apartment, 1–3 story
9. Hotel/Motel, 2–3 story

For the purposes of ensuring the timeliness of the data, the components of the index for the composite model have been streamlined. They currently consist of:

- specific quantities of 66 commonly used construction materials;
- specific labor-hours for 21 building construction trades; and
- specific days of equipment rental for 6 types of construction equipment (normally used to install the 66 material items by the 21 trades.)

A sophisticated computer program handles the updating of all costs for each city on a quarterly basis. Material and equipment price quotations are gathered quarterly from 316 cities in the United States and Canada. These prices and the latest negotiated labor wage rates for 21 different building trades are used to compile the quarterly update of the City Cost Index.

The 30 major U.S. cities used to calculate the National Average are:

Atlanta, GA	Memphis, TN
Baltimore, MD	Milwaukee, WI
Boston, MA	Minneapolis, MN
Buffalo, NY	Nashville, TN
Chicago, IL	New Orleans, LA
Cincinnati, OH	New York, NY
Cleveland, OH	Philadelphia, PA
Columbus, OH	Phoenix, AZ
Dallas, TX	Pittsburgh, PA
Denver, CO	St. Louis, MO
Detroit, MI	San Antonio, TX
Houston, TX	San Diego, CA
Indianapolis, IN	San Francisco, CA
Kansas City, MO	Seattle, WA
Los Angeles, CA	Washington, DC

What the CCI does not indicate

The weighted average for each city is a total of the divisional components weighted to reflect typical usage, but it does not include the productivity variations between trades or cities.

In addition, the CCI does not take into consideration factors such as the following:

- managerial efficiency
- competitive conditions
- automation
- restrictive union practices
- unique local requirements
- regional variations due to specific building codes

City Cost Indexes

		UNITED STATES			ALABAMA													
	DIVISION	30 CITY AVERAGE			BIRMINGHAM			HUNTSVILLE			MOBILE			MONTGOMERY			TUSCALOOSA		
		MAT.	INST.	TOTAL	MAT.	INST.	TOTAL	MAT.	INST.	TOTAL	MAT.	INST.	TOTAL	MAT.	INST.	TOTAL	MAT.	INST.	TOTAL
01590	EQUIPMENT RENTAL	.0	100.0	100.0	.0	101.5	101.5	.0	101.4	101.4	.0	97.9	97.9	.0	97.9	97.9	.0	101.4	101.4
02	SITE CONSTRUCTION	100.0	100.0	100.0	82.5	93.1	90.1	80.4	93.1	89.5	93.1	86.1	88.1	90.8	86.6	87.8	80.8	91.8	88.7
03100	CONCRETE FORMS & ACCESSORIES	100.0	100.0	100.0	94.4	71.1	74.4	96.4	64.6	69.1	96.6	49.8	56.4	95.4	45.3	52.4	96.4	37.4	45.8
03200	CONCRETE REINFORCEMENT	100.0	100.0	100.0	92.3	86.0	89.2	92.3	71.5	82.1	95.3	49.5	72.9	95.3	84.5	90.0	92.3	84.8	88.6
03300	CAST-IN-PLACE CONCRETE	100.0	100.0	100.0	92.7	64.1	81.9	85.9	62.7	77.2	90.7	51.6	76.0	87.3	46.6	72.0	89.5	44.5	72.6
03	CONCRETE	100.0	100.0	100.0	91.8	72.6	82.7	88.5	66.6	78.1	91.4	51.9	72.7	89.6	54.9	73.2	90.3	50.7	71.6
04	MASONRY	100.0	100.0	100.0	87.4	73.0	78.7	86.8	67.4	75.2	87.6	48.6	64.3	86.3	33.8	54.9	85.8	35.9	55.9
05	METALS	100.0	100.0	100.0	91.9	92.9	92.2	93.2	87.9	91.6	92.1	75.6	87.1	91.9	89.5	91.2	92.5	90.0	91.7
06	WOOD & PLASTICS	100.0	100.0	100.0	95.1	71.4	82.6	95.1	64.3	78.8	95.3	49.6	71.1	93.7	45.4	68.2	95.1	36.0	63.9
07	THERMAL & MOISTURE PROTECTION	100.0	100.0	100.0	98.3	82.7	92.0	96.1	78.3	88.9	96.1	71.4	86.1	96.1	65.3	83.7	96.4	64.1	83.4
08	DOORS & WINDOWS	100.0	100.0	100.0	97.5	74.3	91.5	97.8	61.0	88.3	97.8	49.6	85.3	97.8	55.3	86.8	97.8	55.2	86.8
09200	PLASTER & GYPSUM BOARD	100.0	100.0	100.0	105.2	71.2	85.0	101.0	63.9	78.9	103.1	48.7	70.7	103.1	44.5	68.2	103.1	34.9	62.5
095,098	CEILINGS & ACOUSTICAL TREATMENT	100.0	100.0	100.0	102.8	71.2	83.8	102.8	63.9	79.4	102.8	48.7	70.3	102.8	44.5	67.7	102.8	34.9	61.9
09600	FLOORING	100.0	100.0	100.0	105.1	46.6	89.5	105.1	50.0	90.4	115.6	50.5	98.2	113.9	26.1	90.5	105.1	38.3	87.3
097,099	WALL FINISHES, PAINTS & COATINGS	100.0	100.0	100.0	96.5	68.7	79.8	96.5	60.0	74.7	100.8	53.2	72.4	96.5	52.3	70.1	96.5	45.0	65.7
09	FINISHES	100.0	100.0	100.0	101.3	65.9	83.0	100.5	60.8	80.0	105.5	49.7	76.7	104.6	41.1	71.8	100.8	36.7	67.7
10 - 14	TOTAL DIV. 10000 - 14000	100.0	100.0	100.0	100.0	85.6	97.0	100.0	83.8	96.7	100.0	61.2	92.0	100.0	77.2	95.3	100.0	75.0	94.8
15	MECHANICAL	100.0	100.0	100.0	99.9	64.0	85.3	99.7	64.1	85.2	99.7	65.1	85.6	99.9	38.4	74.8	99.9	33.9	73.0
16	ELECTRICAL	100.0	100.0	100.0	98.0	64.2	81.8	97.9	68.3	83.7	97.9	47.9	73.9	97.2	64.6	81.5	97.4	64.2	81.5
01 - 16	WEIGHTED AVERAGE	100.0	100.0	100.0	96.2	73.1	86.2	95.8	70.4	84.8	96.8	59.3	80.6	96.2	55.6	78.7	95.8	54.0	77.8

		ALASKA									ARIZONA								
	DIVISION	ANCHORAGE			FAIRBANKS			JUNEAU			FLAGSTAFF			MESA/TEMPE			PHOENIX		
		MAT.	INST.	TOTAL	MAT.	INST.	TOTAL	MAT.	INST.	TOTAL	MAT.	INST.	TOTAL	MAT.	INST.	TOTAL	MAT.	INST.	TOTAL
01590	EQUIPMENT RENTAL	.0	118.7	118.7	.0	118.7	118.7	.0	118.7	118.7	.0	94.6	94.6	.0	98.2	98.2	.0	98.8	98.8
02	SITE CONSTRUCTION	145.7	133.7	137.1	128.9	133.7	132.4	141.0	133.7	135.8	81.9	99.9	94.8	86.2	103.2	98.4	86.6	104.0	99.1
03100	CONCRETE FORMS & ACCESSORIES	132.8	115.9	118.3	135.8	119.4	121.7	134.3	115.8	118.4	102.6	66.2	71.3	99.7	61.1	66.5	100.8	70.2	74.6
03200	CONCRETE REINFORCEMENT	142.6	106.8	125.1	145.7	106.9	126.7	109.3	106.8	108.1	105.1	73.2	89.5	110.6	73.2	92.3	108.9	73.3	91.5
03300	CAST-IN-PLACE CONCRETE	178.8	115.9	155.2	149.2	116.4	136.8	179.8	115.9	155.7	96.0	75.1	88.2	105.2	67.9	91.1	105.3	75.7	94.1
03	CONCRETE	146.2	113.6	130.8	127.9	115.4	122.0	141.5	113.6	128.3	123.2	70.5	98.3	112.0	65.8	90.2	111.6	72.6	93.2
04	MASONRY	216.0	120.7	159.0	218.1	120.7	159.8	206.3	120.7	155.1	98.3	63.9	77.7	110.5	50.3	74.5	98.1	64.9	78.2
05	METALS	125.4	102.3	118.4	125.5	102.6	118.5	125.4	102.2	118.4	97.7	69.2	89.1	100.6	69.9	91.3	102.1	71.0	92.7
06	WOOD & PLASTICS	119.4	114.2	116.7	119.7	118.5	119.1	119.4	114.2	116.7	107.3	66.1	85.5	99.6	66.3	82.0	100.6	71.3	85.1
07	THERMAL & MOISTURE PROTECTION	173.8	114.9	150.1	171.5	117.8	149.9	172.0	114.9	149.0	98.2	69.0	86.4	101.4	61.4	85.3	101.3	69.5	88.5
08	DOORS & WINDOWS	125.4	110.6	121.5	122.5	113.3	120.1	122.5	110.6	119.4	101.8	66.5	92.7	98.6	66.5	90.3	99.7	69.3	91.8
09200	PLASTER & GYPSUM BOARD	140.5	114.4	125.0	151.0	118.8	131.9	140.5	114.4	125.0	93.0	65.2	76.5	94.4	65.2	77.0	96.2	70.4	80.9
095,098	CEILINGS & ACOUSTICAL TREATMENT	130.7	114.4	120.9	130.7	118.8	123.5	130.7	114.4	120.9	105.1	65.2	81.1	95.1	65.2	77.1	101.9	70.4	82.9
09600	FLOORING	164.0	125.2	153.7	164.0	125.2	153.7	164.0	125.2	153.7	94.3	47.8	81.9	93.3	59.6	84.3	93.6	62.3	85.2
097,099	WALL FINISHES, PAINTS & COATINGS	161.6	115.4	134.0	161.6	124.9	139.7	161.6	115.4	134.0	91.3	56.0	70.2	95.2	56.0	71.8	95.2	56.5	72.1
09	FINISHES	154.3	117.9	135.5	153.9	121.5	137.1	153.0	117.9	134.9	97.1	61.1	78.5	95.0	59.7	76.8	97.1	66.9	81.5
10 - 14	TOTAL DIV. 10000 - 14000	100.0	110.8	102.2	100.0	111.3	102.3	100.0	110.8	102.2	100.0	82.9	96.5	100.0	78.1	95.5	100.0	83.9	96.7
15	MECHANICAL	100.4	107.7	103.4	100.4	116.3	106.9	100.4	101.7	100.9	100.1	77.1	90.7	100.2	66.8	86.6	100.2	77.1	90.8
16	ELECTRICAL	147.2	112.9	130.7	148.4	112.9	131.4	148.4	112.8	131.4	98.6	61.4	80.7	91.6	61.4	77.1	100.0	67.0	84.2
01 - 16	WEIGHTED AVERAGE	131.9	114.1	124.2	129.2	116.9	123.9	130.4	112.9	122.8	101.5	71.3	88.4	100.2	66.9	85.8	101.0	73.9	89.3

		ARIZONA						ARKANSAS											
	DIVISION	PRESCOTT			TUCSON			FORT SMITH			JONESBORO			LITTLE ROCK			PINE BLUFF		
		MAT.	INST.	TOTAL	MAT.	INST.	TOTAL	MAT.	INST.	TOTAL	MAT.	INST.	TOTAL	MAT.	INST.	TOTAL	MAT.	INST.	TOTAL
01590	EQUIPMENT RENTAL	.0	94.6	94.6	.0	98.2	98.2	.0	86.5	86.5	.0	108.2	108.2	.0	86.5	86.5	.0	86.5	86.5
02	SITE CONSTRUCTION	70.5	98.9	90.9	82.7	103.8	97.8	78.2	84.2	82.5	101.5	99.5	100.0	77.8	84.2	82.4	80.4	84.2	83.1
03100	CONCRETE FORMS & ACCESSORIES	98.3	52.9	59.3	100.2	69.8	74.1	100.5	43.5	51.5	87.3	60.0	63.9	94.7	73.8	76.8	80.0	73.5	74.4
03200	CONCRETE REINFORCEMENT	105.1	72.7	89.3	91.3	73.2	82.4	96.9	71.2	84.3	92.7	77.2	85.1	97.1	69.5	83.6	97.1	69.5	83.6
03300	CAST-IN-PLACE CONCRETE	95.9	60.7	82.7	108.3	75.5	95.9	90.6	70.9	83.2	86.3	64.5	78.1	88.6	72.1	82.1	83.1	71.1	78.6
03	CONCRETE	108.6	59.4	85.4	110.2	72.3	92.3	87.9	58.9	74.2	84.4	66.0	75.7	86.6	72.1	79.7	85.0	71.9	78.9
04	MASONRY	98.6	54.7	72.3	95.9	63.9	76.8	95.1	60.5	74.4	90.8	52.2	67.7	93.3	60.5	73.7	115.2	60.5	82.5
05	METALS	97.7	65.2	87.9	101.3	69.8	91.8	98.3	70.3	89.9	92.7	85.4	90.5	94.3	70.3	87.0	97.0	70.0	88.8
06	WOOD & PLASTICS	102.5	51.3	75.4	99.9	71.3	84.8	103.5	38.8	69.3	88.1	60.8	73.7	100.2	79.0	89.0	80.1	79.0	79.5
07	THERMAL & MOISTURE PROTECTION	97.0	58.4	81.5	102.9	66.1	88.1	99.9	55.8	82.1	102.6	59.9	85.4	98.7	60.0	83.1	98.5	60.0	83.0
08	DOORS & WINDOWS	101.9	55.2	89.8	94.8	69.3	88.2	97.0	46.9	84.0	98.8	65.0	90.0	97.0	69.7	89.9	92.3	69.7	86.5
09200	PLASTER & GYPSUM BOARD	90.7	50.0	66.5	99.1	70.4	82.0	88.6	37.7	58.3	100.5	59.9	76.3	88.6	79.1	82.9	82.6	79.1	80.5
095,098	CEILINGS & ACOUSTICAL TREATMENT	103.4	50.0	71.3	98.8	70.4	81.7	95.5	37.7	60.7	94.3	59.9	73.6	95.5	79.1	85.6	91.3	79.1	83.9
09600	FLOORING	92.6	47.6	80.6	94.9	47.8	82.4	113.5	71.8	102.4	75.8	64.3	72.8	115.0	71.8	103.5	102.8	71.8	94.5
097,099	WALL FINISHES, PAINTS & COATINGS	91.3	45.6	64.0	95.4	56.0	71.8	98.3	66.5	79.3	86.9	59.8	70.7	98.3	68.3	80.4	98.3	68.3	80.4
09	FINISHES	94.7	50.4	71.8	97.1	64.1	80.1	97.4	50.2	73.0	90.3	60.5	74.9	97.8	74.0	85.5	92.5	74.0	82.9
10 - 14	TOTAL DIV. 10000 - 14000	100.0	79.8	95.9	100.0	83.9	96.7	100.0	71.3	94.1	100.0	60.6	91.9	100.0	76.2	95.1	100.0	76.2	95.1
15	MECHANICAL	100.1	71.9	88.6	100.1	68.1	87.1	100.1	52.2	80.6	100.3	52.2	80.7	100.1	72.2	88.7	100.1	55.1	81.8
16	ELECTRICAL	98.3	61.3	80.6	93.6	59.3	77.2	96.7	80.5	88.9	105.9	55.8	81.8	96.8	81.8	89.6	96.1	81.8	89.2
01 - 16	WEIGHTED AVERAGE	99.2	64.9	84.4	99.4	70.2	86.8	96.7	62.3	81.9	96.4	63.8	82.3	95.8	73.2	86.0	96.1	69.6	84.6

441

City Cost Indexes

ARKANSAS / CALIFORNIA

DIVISION		TEXARKANA MAT.	INST.	TOTAL	ANAHEIM MAT.	INST.	TOTAL	BAKERSFIELD MAT.	INST.	TOTAL	FRESNO MAT.	INST.	TOTAL	LOS ANGELES MAT.	INST.	TOTAL	OAKLAND MAT.	INST.	TOTAL
01590	EQUIPMENT RENTAL	.0	87.2	87.2	.0	102.4	102.4	.0	99.8	99.8	.0	99.8	99.8	.0	98.9	98.9	.0	103.0	103.0
02	SITE CONSTRUCTION	96.5	84.8	88.1	104.2	109.2	107.8	109.7	106.4	107.3	110.9	106.1	107.5	105.0	108.2	107.3	152.6	103.8	117.6
03100	CONCRETE FORMS & ACCESSORIES	87.8	49.4	54.8	105.6	118.9	117.0	98.6	118.8	115.9	101.8	125.2	121.9	105.2	118.8	116.9	107.2	139.2	134.7
03200	CONCRETE REINFORCEMENT	96.6	75.0	86.0	97.4	116.6	106.8	109.7	116.5	113.0	93.0	116.7	104.6	120.0	116.9	118.4	104.0	117.4	110.6
03300	CAST-IN-PLACE CONCRETE	90.6	49.5	75.2	102.0	117.4	107.8	102.0	116.4	107.4	107.5	114.0	109.9	108.5	115.6	111.2	137.4	118.0	130.1
03	CONCRETE	85.0	54.9	70.8	106.3	117.1	111.4	105.6	116.7	110.8	108.2	118.7	113.2	115.3	116.5	115.8	124.0	126.6	125.3
04	MASONRY	95.1	35.7	59.5	89.5	111.2	102.5	111.5	110.5	110.9	113.9	110.5	111.9	100.2	116.7	110.1	162.8	120.6	137.6
05	METALS	89.7	69.0	83.4	107.5	102.4	106.0	102.5	101.9	102.3	107.8	102.0	106.0	108.6	101.3	106.4	99.8	106.9	101.9
06	WOOD & PLASTICS	90.2	52.7	70.4	96.2	118.2	107.9	86.7	118.3	103.4	100.2	126.9	114.3	90.6	118.0	105.1	103.1	143.1	124.0
07	THERMAL & MOISTURE PROTECTION	99.5	48.5	79.0	102.3	113.6	106.8	99.9	108.9	103.5	96.1	110.4	101.8	101.2	114.6	106.6	110.8	127.4	117.5
08	DOORS & WINDOWS	97.5	55.9	86.7	104.4	116.1	107.4	101.8	113.4	104.8	103.6	118.1	107.4	95.2	116.0	100.6	105.6	133.0	112.7
09200	PLASTER & GYPSUM BOARD	87.0	52.1	66.2	102.7	118.6	112.2	102.7	118.6	112.2	98.6	127.4	115.8	102.8	118.6	112.2	103.7	143.7	127.5
095,098	CEILINGS & ACOUSTICAL TREATMENT	98.0	52.1	70.4	117.8	118.6	118.3	119.7	118.6	119.1	120.6	127.4	124.7	123.8	118.6	120.7	122.9	143.7	135.4
09600	FLOORING	105.5	61.4	93.7	122.3	107.3	118.3	117.1	94.3	111.1	129.3	138.1	131.6	110.5	107.3	109.7	117.6	124.8	119.5
097,099	WALL FINISHES, PAINTS & COATINGS	98.3	35.1	60.5	108.7	109.8	109.4	108.8	95.4	100.8	129.5	96.8	109.9	97.5	109.8	104.9	111.5	137.9	127.2
09	FINISHES	96.1	49.9	72.3	113.7	115.7	114.8	114.6	112.0	113.3	119.2	125.6	122.5	110.5	115.6	113.1	118.7	137.7	128.6
10 - 14	TOTAL DIV. 10000 - 14000	100.0	43.1	88.3	100.0	112.4	102.5	100.0	120.9	104.3	100.0	121.4	104.4	100.0	111.8	102.4	100.0	125.1	105.2
15	MECHANICAL	100.1	38.5	75.0	100.1	107.6	103.2	100.2	104.0	101.8	100.3	115.8	106.6	100.1	107.5	103.1	100.4	136.3	115.0
16	ELECTRICAL	97.8	44.8	72.4	90.2	105.0	97.3	89.5	97.9	93.5	89.4	94.4	91.8	96.3	113.8	104.7	102.6	138.9	120.0
01 - 16	WEIGHTED AVERAGE	95.4	50.9	76.2	102.0	110.4	105.6	102.1	107.7	104.5	103.9	112.0	107.4	103.3	111.9	107.0	110.3	127.9	117.9

CALIFORNIA

DIVISION		OXNARD MAT.	INST.	TOTAL	REDDING MAT.	INST.	TOTAL	RIVERSIDE MAT.	INST.	TOTAL	SACRAMENTO MAT.	INST.	TOTAL	SAN DIEGO MAT.	INST.	TOTAL	SAN FRANCISCO MAT.	INST.	TOTAL
01590	EQUIPMENT RENTAL	.0	98.4	98.4	.0	99.4	99.4	.0	100.9	100.9	.0	102.6	102.6	.0	98.1	98.1	.0	108.3	108.3
02	SITE CONSTRUCTION	111.3	104.1	106.1	115.9	105.4	108.4	102.3	106.9	105.6	117.3	110.1	112.2	106.4	102.1	103.3	155.3	110.2	123.0
03100	CONCRETE FORMS & ACCESSORIES	104.2	119.0	116.9	103.9	125.2	122.2	105.9	118.9	117.1	105.8	125.6	122.8	106.1	109.9	109.3	107.6	140.0	135.4
03200	CONCRETE REINFORCEMENT	109.7	116.5	113.0	106.4	116.7	111.4	108.7	116.6	112.5	96.4	116.7	106.3	100.7	116.5	108.4	118.5	117.9	118.2
03300	CAST-IN-PLACE CONCRETE	108.3	116.6	111.4	122.7	113.9	119.4	105.8	117.4	110.1	116.1	114.7	115.6	109.8	107.3	108.8	140.9	119.6	132.9
03	CONCRETE	109.1	116.9	112.8	118.8	118.7	118.8	107.8	117.1	112.2	111.9	119.1	115.3	109.4	109.6	109.5	128.1	127.6	127.9
04	MASONRY	115.9	109.8	112.3	118.2	110.5	113.6	88.1	110.8	101.7	129.8	110.5	118.3	99.1	109.3	105.2	163.4	126.8	141.5
05	METALS	102.2	102.1	102.2	107.1	102.0	105.5	107.8	102.4	106.2	95.5	101.5	97.3	105.0	101.5	104.0	105.5	108.8	106.5
06	WOOD & PLASTICS	95.1	118.3	107.4	95.2	126.9	112.0	96.2	118.2	107.9	97.9	127.1	113.3	100.5	107.1	104.0	103.1	143.3	124.3
07	THERMAL & MOISTURE PROTECTION	105.5	111.4	107.8	105.9	112.2	108.5	102.6	113.5	107.0	113.0	113.4	113.2	109.2	104.2	107.2	114.0	131.3	121.0
08	DOORS & WINDOWS	100.7	116.2	104.7	103.3	120.1	107.7	102.8	116.1	106.3	118.9	120.2	119.2	103.3	108.7	104.7	110.2	133.1	116.1
09200	PLASTER & GYPSUM BOARD	102.7	118.6	112.2	101.3	127.4	116.8	102.1	118.6	111.9	99.5	127.4	116.1	107.8	107.1	107.4	106.2	143.7	128.5
095,098	CEILINGS & ACOUSTICAL TREATMENT	119.7	118.6	119.1	126.4	127.4	127.0	117.2	118.6	118.1	122.0	127.4	125.3	102.7	107.1	105.3	132.0	143.7	139.0
09600	FLOORING	115.6	107.3	113.4	116.4	117.7	116.7	121.5	107.3	117.7	121.1	117.7	120.2	108.3	107.3	108.1	117.6	125.9	119.8
097,099	WALL FINISHES, PAINTS & COATINGS	108.2	103.7	105.5	108.2	112.6	110.8	105.5	109.8	108.1	108.8	112.6	111.0	105.4	109.8	108.1	111.5	149.7	134.3
09	FINISHES	113.9	115.1	114.5	116.0	123.8	120.0	112.8	115.7	114.3	116.9	123.9	120.5	107.9	109.0	108.5	121.5	139.6	130.8
10 - 14	TOTAL DIV. 10000 - 14000	100.0	112.7	102.6	100.0	121.5	104.4	100.0	112.4	102.5	100.0	122.1	104.5	100.0	111.0	102.3	100.0	125.7	105.3
15	MECHANICAL	100.2	107.6	103.2	100.2	108.3	103.5	100.1	107.6	103.1	100.2	110.0	104.2	100.2	106.1	102.6	100.4	147.8	119.7
16	ELECTRICAL	95.7	106.3	100.8	98.4	104.8	101.5	90.4	102.9	96.4	97.3	104.8	100.9	100.0	97.8	99.0	102.7	151.3	126.0
01 - 16	WEIGHTED AVERAGE	103.5	109.8	106.2	106.4	111.8	108.7	101.9	109.8	105.3	106.0	112.6	108.9	103.4	105.6	104.3	112.6	133.8	121.8

CALIFORNIA / COLORADO

DIVISION		SAN JOSE MAT.	INST.	TOTAL	SANTA BARBARA MAT.	INST.	TOTAL	STOCKTON MAT.	INST.	TOTAL	VALLEJO MAT.	INST.	TOTAL	COLORADO SPRINGS MAT.	INST.	TOTAL	DENVER MAT.	INST.	TOTAL
01590	EQUIPMENT RENTAL	.0	100.2	100.2	.0	99.8	99.8	.0	99.4	99.4	.0	103.1	103.1	.0	94.3	94.3	.0	99.5	99.5
02	SITE CONSTRUCTION	150.9	99.7	114.2	111.2	106.4	107.7	108.9	105.5	106.4	115.6	110.2	111.7	93.0	95.1	94.5	92.1	104.3	100.9
03100	CONCRETE FORMS & ACCESSORIES	106.5	139.0	134.4	104.9	118.9	116.9	104.6	125.4	122.4	106.6	138.1	133.7	92.2	82.5	83.8	100.7	83.4	85.8
03200	CONCRETE REINFORCEMENT	97.0	117.4	107.0	109.7	116.5	113.0	110.3	116.7	113.4	103.6	117.5	110.4	105.9	85.5	95.9	105.9	85.6	96.0
03300	CAST-IN-PLACE CONCRETE	128.7	117.5	124.5	107.9	116.4	111.1	107.8	114.0	110.2	126.4	116.0	122.5	101.4	85.3	95.3	95.9	86.0	92.2
03	CONCRETE	117.6	126.4	121.8	109.0	116.8	112.7	109.0	118.8	113.7	118.4	125.3	121.6	108.4	84.1	96.9	102.9	84.8	94.4
04	MASONRY	149.5	120.7	132.3	112.0	110.4	111.0	115.5	110.5	112.5	92.2	122.6	110.3	101.3	77.7	87.2	103.4	81.8	90.5
05	METALS	101.6	108.3	103.6	102.7	102.0	102.5	104.1	102.3	103.5	98.6	103.6	100.1	99.6	87.8	96.1	102.0	88.1	97.8
06	WOOD & PLASTICS	102.0	142.8	123.6	95.1	118.3	107.4	97.0	126.9	112.8	96.3	142.8	120.9	91.5	85.2	88.1	100.7	85.0	92.4
07	THERMAL & MOISTURE PROTECTION	101.9	129.2	112.9	100.7	110.7	104.7	105.5	113.4	108.7	113.6	125.9	118.6	102.4	82.8	94.5	101.7	79.8	92.9
08	DOORS & WINDOWS	93.9	132.9	104.0	101.8	116.2	105.5	101.0	120.1	106.0	118.6	132.8	122.3	99.3	88.4	96.5	99.1	88.3	96.3
09200	PLASTER & GYPSUM BOARD	98.3	143.7	125.3	102.7	118.6	112.2	103.0	127.4	117.5	104.0	143.7	127.6	87.5	84.8	85.9	99.7	84.8	90.9
095,098	CEILINGS & ACOUSTICAL TREATMENT	110.6	143.7	130.5	119.7	118.6	119.1	119.7	127.4	124.3	132.2	143.7	139.1	102.9	84.8	92.0	101.1	84.8	91.3
09600	FLOORING	110.6	125.9	114.7	117.1	94.3	111.1	117.1	113.1	116.1	125.7	125.9	125.8	104.9	82.2	98.8	109.9	93.1	105.4
097,099	WALL FINISHES, PAINTS & COATINGS	110.1	137.9	126.7	108.2	103.7	105.5	108.2	113.0	111.1	108.3	137.9	126.0	108.2	50.1	73.5	108.5	75.9	89.0
09	FINISHES	111.4	137.8	125.0	114.6	112.9	113.7	114.5	123.0	118.9	119.1	137.5	128.6	98.1	79.0	88.2	101.0	84.5	92.5
10 - 14	TOTAL DIV. 10000 - 14000	100.0	124.5	105.0	100.0	112.7	102.6	100.0	121.6	104.4	100.0	123.8	104.9	100.0	89.6	97.9	100.0	89.9	97.9
15	MECHANICAL	100.2	135.2	114.5	100.2	107.6	103.2	100.2	108.4	103.5	100.3	123.4	109.7	100.1	78.1	91.1	100.0	85.3	94.0
16	ELECTRICAL	102.4	137.8	119.4	87.0	107.2	96.7	98.2	119.4	108.4	93.2	124.5	108.2	99.2	84.8	92.3	100.8	90.9	96.1
01 - 16	WEIGHTED AVERAGE	107.1	127.2	115.8	102.4	109.8	105.6	104.1	113.9	108.3	105.0	123.3	112.9	100.5	83.3	93.1	100.8	87.6	95.1

City Cost Indexes

	DIVISION	COLORADO												CONNECTICUT					
		FORT COLLINS			GRAND JUNCTION			GREELEY			PUEBLO			BRIDGEPORT			BRISTOL		
		MAT.	INST.	TOTAL	MAT.	INST.	TOTAL	MAT.	INST.	TOTAL	MAT.	INST.	TOTAL	MAT.	INST.	TOTAL	MAT.	INST.	TOTAL
01590	EQUIPMENT RENTAL	.0	95.9	95.9	.0	99.4	99.4	.0	95.9	95.9	.0	96.2	96.2	.0	100.7	100.7	.0	100.7	100.7
02	SITE CONSTRUCTION	102.6	97.6	99.0	121.3	100.5	106.4	90.2	96.3	94.6	113.3	94.8	100.1	96.8	104.2	102.1	95.9	104.2	101.8
03100	CONCRETE FORMS & ACCESSORIES	100.1	80.5	83.2	107.4	80.8	84.6	97.8	50.3	57.0	104.4	82.7	85.8	99.4	120.2	117.3	99.4	119.8	116.9
03200	CONCRETE REINFORCEMENT	106.9	80.4	93.9	113.0	80.3	97.0	106.8	79.8	93.6	108.9	85.5	97.5	104.3	125.7	114.7	104.3	125.6	114.7
03300	CAST-IN-PLACE CONCRETE	111.4	81.0	100.0	118.5	81.9	104.7	93.1	57.4	79.7	106.2	86.5	98.8	104.8	114.7	108.5	98.1	114.5	104.3
03	CONCRETE	115.7	80.7	99.2	122.4	81.1	102.9	100.3	59.2	80.9	111.6	84.7	98.9	109.1	119.2	113.9	105.7	118.9	112.0
04	MASONRY	113.7	56.6	79.6	141.5	58.9	92.1	108.0	37.5	65.8	103.2	77.7	87.9	107.1	127.1	119.1	99.2	127.1	115.9
05	METALS	97.9	81.4	92.9	100.9	80.5	94.8	97.9	79.7	92.4	102.4	88.9	98.3	96.9	124.1	105.2	96.9	123.7	105.0
06	WOOD & PLASTICS	100.4	85.0	92.3	105.0	85.2	94.5	97.6	50.2	72.6	101.8	85.5	93.2	99.1	118.5	109.4	99.1	118.5	109.4
07	THERMAL & MOISTURE PROTECTION	102.2	68.6	88.7	99.6	65.0	85.7	101.6	57.5	83.8	99.2	82.6	92.5	98.3	124.0	108.6	98.4	120.0	107.1
08	DOORS & WINDOWS	95.0	87.1	93.0	100.2	87.2	96.9	95.0	68.3	88.1	94.8	88.6	93.2	105.6	128.2	111.5	105.6	118.5	109.0
09200	PLASTER & GYPSUM BOARD	98.4	84.8	90.3	111.8	84.8	95.7	97.5	49.1	68.7	88.5	84.8	86.3	102.1	118.2	111.7	102.1	118.2	111.7
095,098	CEILINGS & ACOUSTICAL TREATMENT	95.2	84.8	89.0	99.2	84.8	90.6	95.2	49.1	67.5	111.0	84.8	95.3	96.2	118.2	109.4	96.2	118.2	109.4
09600	FLOORING	109.8	66.0	98.1	114.7	66.0	101.7	114.8	66.0	97.1	111.0	93.1	106.2	98.0	120.2	103.9	98.0	120.2	103.9
097,099	WALL FINISHES, PAINTS & COATINGS	108.5	49.7	73.3	111.6	72.7	88.3	108.5	31.0	62.1	111.6	46.4	72.6	91.0	114.8	105.3	91.0	114.8	105.3
09	FINISHES	99.9	74.8	86.9	107.0	78.1	92.1	98.7	49.8	73.5	104.5	81.0	92.4	100.0	119.5	110.1	100.0	119.5	110.1
10-14	TOTAL DIV. 10000 - 14000	100.0	88.2	97.6	100.0	89.4	97.8	100.0	80.2	95.9	100.0	90.5	98.0	100.0	110.4	102.1	100.0	110.4	102.1
15	MECHANICAL	100.0	82.6	92.9	99.9	69.4	87.5	100.0	77.3	90.7	99.9	70.3	87.9	100.0	111.9	104.8	100.0	111.8	104.8
16	ELECTRICAL	96.1	90.9	93.6	91.2	61.2	76.8	96.1	90.8	93.6	92.0	79.2	85.9	98.1	109.0	103.3	98.1	108.7	103.2
01-16	WEIGHTED AVERAGE	101.4	81.1	92.6	105.1	74.9	92.1	98.9	69.9	86.4	101.2	81.3	92.6	101.1	116.5	107.7	100.3	115.8	107.0

	DIVISION	CONNECTICUT																	
		HARTFORD			NEW BRITAIN			NEW HAVEN			NORWALK			STAMFORD			WATERBURY		
		MAT.	INST.	TOTAL	MAT.	INST.	TOTAL	MAT.	INST.	TOTAL	MAT.	INST.	TOTAL	MAT.	INST.	TOTAL	MAT.	INST.	TOTAL
01590	EQUIPMENT RENTAL	.0	100.7	100.7	.0	100.7	100.7	.0	101.3	101.3	.0	100.7	100.7	.0	100.7	100.7	.0	100.7	100.7
02	SITE CONSTRUCTION	95.4	104.2	101.7	96.1	104.2	101.9	96.0	105.0	102.5	96.6	104.2	102.0	97.2	104.2	102.2	96.4	104.2	102.0
03100	CONCRETE FORMS & ACCESSORIES	99.6	119.8	116.9	99.7	119.8	116.9	99.2	120.0	117.1	99.4	120.4	117.5	99.4	120.7	117.7	99.4	120.1	117.1
03200	CONCRETE REINFORCEMENT	104.3	125.6	114.7	104.3	125.6	114.7	104.3	125.6	114.7	104.3	125.8	114.8	104.3	125.8	114.8	104.3	125.6	114.7
03300	CAST-IN-PLACE CONCRETE	92.4	114.5	100.7	99.7	114.5	105.3	101.5	114.6	106.4	103.0	126.9	112.0	104.8	127.0	113.1	104.8	114.7	108.5
03	CONCRETE	102.9	118.9	110.4	106.6	118.9	112.4	121.3	119.1	120.3	108.2	123.5	115.4	109.1	123.7	116.0	109.1	119.1	113.8
04	MASONRY	98.7	127.1	115.7	101.4	127.1	116.7	99.4	127.1	116.0	98.9	128.9	116.8	99.7	128.9	117.2	99.7	127.1	116.1
05	METALS	101.7	123.7	108.3	93.6	123.7	102.7	93.8	124.0	102.9	96.9	124.5	105.3	96.9	124.9	105.4	96.9	124.0	105.1
06	WOOD & PLASTICS	100.7	118.5	110.1	99.1	118.5	109.4	99.1	118.5	109.4	99.1	118.5	109.4	99.1	118.5	109.4	99.1	118.5	109.4
07	THERMAL & MOISTURE PROTECTION	99.7	120.0	107.9	98.4	120.9	107.5	98.5	120.9	107.5	98.5	126.3	109.7	98.4	126.3	109.6	98.4	120.9	107.5
08	DOORS & WINDOWS	106.3	118.5	109.4	105.6	118.5	109.0	105.6	128.2	111.5	105.6	128.2	111.5	105.6	128.2	111.5	105.6	128.2	111.5
09200	PLASTER & GYPSUM BOARD	102.1	118.2	111.7	102.1	118.2	111.7	102.1	118.2	111.7	102.1	118.2	111.7	102.1	118.2	111.7	102.1	118.2	111.7
095,098	CEILINGS & ACOUSTICAL TREATMENT	96.2	118.2	109.4	96.2	118.2	109.4	96.2	118.2	109.4	96.2	118.2	109.4	96.2	118.2	109.4	96.2	118.2	109.4
09600	FLOORING	98.0	120.2	103.9	98.0	120.2	103.9	98.0	120.2	103.9	98.0	120.2	103.9	98.0	120.2	103.9	98.0	120.2	103.9
097,099	WALL FINISHES, PAINTS & COATINGS	91.0	114.8	105.3	91.0	114.8	105.3	91.0	114.8	105.3	91.0	114.8	105.3	91.0	114.8	105.3	91.0	118.0	107.1
09	FINISHES	100.0	119.5	110.1	100.0	119.5	110.1	100.1	119.5	110.1	100.0	119.5	110.1	100.1	119.5	110.1	99.9	119.9	110.2
10-14	TOTAL DIV. 10000 - 14000	100.0	110.4	102.1	100.0	110.4	102.1	100.0	110.4	102.1	100.0	110.4	102.1	100.0	110.6	102.2	100.0	110.4	102.1
15	MECHANICAL	100.0	111.8	104.8	100.0	111.8	104.8	100.0	111.9	104.8	100.0	111.9	104.9	100.0	111.9	104.9	100.0	111.9	104.8
16	ELECTRICAL	98.4	109.4	103.7	98.2	108.7	103.3	98.1	108.7	103.2	98.1	104.9	101.4	98.1	150.2	123.1	97.7	109.0	103.1
01-16	WEIGHTED AVERAGE	100.8	115.9	107.3	100.0	115.7	106.8	101.6	116.4	108.0	100.6	116.8	107.6	100.7	123.2	110.4	100.6	116.4	107.4

	DIVISION	D.C.			DELAWARE			FLORIDA											
		WASHINGTON			WILMINGTON			DAYTONA BEACH			FORT LAUDERDALE			JACKSONVILLE			MELBOURNE		
		MAT.	INST.	TOTAL	MAT.	INST.	TOTAL	MAT.	INST.	TOTAL	MAT.	INST.	TOTAL	MAT.	INST.	TOTAL	MAT.	INST.	TOTAL
01590	EQUIPMENT RENTAL	.0	103.5	103.5	.0	118.9	118.9	.0	97.9	97.9	.0	89.3	89.3	.0	97.9	97.9	.0	97.9	97.9
02	SITE CONSTRUCTION	109.6	91.5	96.6	89.0	112.6	105.9	112.4	86.7	94.0	98.6	73.9	80.9	112.5	87.3	94.4	120.6	87.4	96.8
03100	CONCRETE FORMS & ACCESSORIES	101.4	80.4	83.4	100.0	95.7	96.3	97.4	65.6	70.1	95.2	71.5	74.8	97.2	50.1	56.8	93.2	70.3	73.5
03200	CONCRETE REINFORCEMENT	110.8	88.7	100.0	98.6	91.3	95.1	95.3	82.5	89.0	95.3	66.9	81.4	95.3	47.0	71.7	96.4	82.6	89.6
03300	CAST-IN-PLACE CONCRETE	122.7	84.2	108.2	88.2	89.4	88.6	88.0	67.4	80.2	92.3	66.3	82.5	88.9	55.4	76.3	106.1	73.7	93.9
03	CONCRETE	117.3	84.6	101.9	99.8	93.8	96.9	90.2	70.6	81.0	92.2	70.0	81.7	90.6	53.0	72.9	101.6	74.9	89.0
04	MASONRY	100.7	79.7	88.1	110.8	88.1	97.2	86.2	61.1	71.2	86.5	64.5	73.3	86.1	47.8	63.2	85.4	71.9	77.3
05	METALS	103.5	108.0	104.8	100.8	112.0	104.2	93.6	92.9	93.4	93.5	85.5	91.1	92.2	76.4	87.4	101.9	93.5	99.3
06	WOOD & PLASTICS	97.3	81.0	88.7	99.9	96.9	98.3	96.3	68.2	81.4	91.8	70.8	80.7	96.3	49.0	71.3	91.6	68.2	79.2
07	THERMAL & MOISTURE PROTECTION	107.9	83.7	98.1	101.5	102.1	101.7	99.2	65.7	85.7	99.2	71.7	88.1	99.4	55.2	81.6	99.6	73.7	89.2
08	DOORS & WINDOWS	102.6	89.4	99.2	94.4	102.6	96.5	100.3	66.1	91.4	97.8	63.8	89.0	100.3	46.3	86.3	99.4	71.8	92.3
09200	PLASTER & GYPSUM BOARD	111.1	80.6	93.0	105.2	96.7	100.2	103.1	67.8	82.1	102.7	70.5	83.6	103.1	48.2	70.4	99.9	67.8	80.8
095,098	CEILINGS & ACOUSTICAL TREATMENT	108.6	80.6	91.7	101.3	96.7	98.6	102.8	67.8	81.8	102.8	70.5	83.4	102.8	48.2	69.9	98.6	67.8	80.1
09600	FLOORING	116.2	106.1	113.5	83.9	101.5	88.6	119.4	69.3	106.0	119.4	56.2	102.6	119.4	46.2	99.9	115.9	69.3	103.5
097,099	WALL FINISHES, PAINTS & COATINGS	121.1	83.6	98.6	94.5	99.9	97.7	111.7	67.4	85.2	107.4	46.6	71.1	111.7	45.0	71.8	111.7	92.9	100.4
09	FINISHES	108.2	85.2	96.4	100.7	97.0	98.8	108.9	66.5	87.1	106.9	65.9	85.7	109.0	48.2	77.6	106.9	72.2	89.0
10-14	TOTAL DIV. 10000 - 14000	100.0	96.5	99.3	100.0	102.7	100.6	100.0	80.6	96.0	100.0	88.5	97.6	100.0	77.3	95.3	100.0	84.7	96.9
15	MECHANICAL	100.2	90.7	96.3	100.3	108.4	103.6	99.9	67.1	86.5	99.9	70.9	88.1	99.9	47.2	78.4	99.9	73.2	89.0
16	ELECTRICAL	100.4	96.0	98.3	97.9	97.9	97.9	96.5	62.6	80.2	96.5	75.3	86.3	96.2	65.3	81.4	97.9	71.9	85.4
01-16	WEIGHTED AVERAGE	104.1	90.2	98.1	99.8	101.4	100.5	97.8	70.6	86.1	97.2	71.9	86.3	97.6	57.9	80.5	100.5	76.3	90.0

City Cost Indexes

FLORIDA

DIVISION		MIAMI			ORLANDO			PANAMA CITY			PENSACOLA			ST. PETERSBURG			TALLAHASSEE		
		MAT.	INST.	TOTAL	MAT.	INST.	TOTAL	MAT.	INST.	TOTAL	MAT.	INST.	TOTAL	MAT.	INST.	TOTAL	MAT.	INST.	TOTAL
01590	EQUIPMENT RENTAL	.0	89.3	89.3	.0	97.9	97.9	.0	97.9	97.9	.0	97.9	97.9	.0	97.9	97.9	.0	97.9	97.9
02	SITE CONSTRUCTION	97.7	73.9	80.6	112.9	86.5	93.9	126.8	83.9	96.1	126.5	86.4	97.8	113.3	85.8	93.6	113.7	85.9	93.7
03100	CONCRETE FORMS & ACCESSORIES	99.7	72.1	76.0	97.2	67.1	71.4	96.3	26.0	35.9	94.2	47.9	54.4	95.0	44.2	51.4	97.1	37.0	45.5
03200	CONCRETE REINFORCEMENT	95.3	67.0	81.4	95.3	79.4	87.5	99.5	45.9	73.3	102.0	46.3	74.8	98.7	52.9	76.3	95.3	46.5	71.5
03300	CAST-IN-PLACE CONCRETE	87.3	67.5	79.9	95.0	67.3	84.6	93.4	32.7	70.5	115.4	51.8	91.5	99.5	52.1	81.7	91.3	45.8	74.2
03	CONCRETE	90.0	70.6	80.9	95.1	70.7	83.6	98.6	33.8	68.0	109.3	50.6	81.6	96.5	50.3	74.7	91.8	43.8	69.1
04	MASONRY	84.9	66.6	74.0	89.9	61.1	72.7	90.7	25.6	51.8	109.3	46.4	71.6	134.9	45.8	81.5	88.9	36.1	57.3
05	METALS	98.4	85.6	94.6	99.1	91.4	96.7	93.4	62.2	84.0	94.4	75.6	88.7	96.6	76.4	90.5	87.1	74.9	83.4
06	WOOD & PLASTICS	97.2	70.8	83.2	96.3	70.6	82.7	95.1	26.2	58.7	93.5	48.3	69.6	93.5	44.3	67.5	94.7	35.1	63.2
07	THERMAL & MOISTURE PROTECTION	106.5	69.4	91.6	99.5	66.0	86.0	99.7	29.6	71.5	99.6	49.6	79.5	99.2	46.6	78.0	101.8	47.9	80.1
08	DOORS & WINDOWS	97.8	64.3	89.1	100.3	65.9	91.4	97.8	24.9	78.9	97.8	46.7	84.6	98.8	43.3	84.4	100.7	38.2	84.5
09200	PLASTER & GYPSUM BOARD	102.7	70.5	83.6	105.2	70.4	84.5	101.6	24.7	55.8	102.5	47.5	69.7	101.0	43.4	66.7	103.1	33.9	61.9
095,098	CEILINGS & ACOUSTICAL TREATMENT	102.8	70.5	83.4	102.8	70.4	83.3	97.7	24.7	53.8	98.6	47.5	67.9	97.7	43.4	65.1	102.8	33.9	61.4
09600	FLOORING	127.2	56.9	108.5	119.4	69.3	106.0	118.8	17.7	91.8	112.7	49.1	95.7	117.7	48.9	99.3	119.4	35.9	97.1
097,099	WALL FINISHES, PAINTS & COATINGS	107.4	46.6	71.1	111.7	51.6	75.8	111.7	23.0	58.6	111.7	52.2	76.1	111.7	43.5	70.9	111.7	37.0	67.0
09	FINISHES	109.2	66.5	87.1	109.3	66.2	87.1	108.5	23.7	64.7	106.7	48.4	76.6	106.9	44.9	74.9	109.1	35.7	71.3
10 - 14	TOTAL DIV. 10000 - 14000	100.0	89.1	97.8	100.0	80.8	96.1	100.0	40.9	87.8	100.0	45.8	88.8	100.0	47.8	89.3	100.0	69.4	93.7
15	MECHANICAL	99.9	72.2	88.6	99.9	62.8	84.8	99.9	23.1	68.6	99.9	46.0	77.9	99.9	45.8	77.8	99.9	36.7	74.1
16	ELECTRICAL	98.1	68.5	83.9	98.2	41.7	71.1	95.2	31.2	64.5	99.0	60.3	80.4	96.7	45.0	71.9	98.3	38.6	69.7
01 - 16	WEIGHTED AVERAGE	98.3	71.6	86.8	99.7	66.6	85.4	98.9	35.7	71.6	101.5	55.3	81.6	101.2	52.4	80.1	97.5	46.8	75.6

| DIVISION | | FLORIDA | | | GEORGIA | | | | | | | | | | | | | | |
		TAMPA			ALBANY			ATLANTA			AUGUSTA			COLUMBUS			MACON		
		MAT.	INST.	TOTAL	MAT.	INST.	TOTAL	MAT.	INST.	TOTAL	MAT.	INST.	TOTAL	MAT.	INST.	TOTAL	MAT.	INST.	TOTAL
01590	EQUIPMENT RENTAL	.0	97.9	97.9	.0	90.2	90.2	.0	92.2	92.2	.0	91.5	91.5	.0	90.2	90.2	.0	102.8	102.8
02	SITE CONSTRUCTION	113.6	86.5	94.2	98.9	77.2	83.3	100.6	94.4	96.1	97.0	91.8	93.3	98.8	77.6	83.6	99.2	94.6	95.9
03100	CONCRETE FORMS & ACCESSORIES	98.5	71.3	75.2	96.4	52.7	58.9	99.3	77.1	80.2	96.5	59.1	64.3	96.4	67.1	71.2	95.8	63.2	67.8
03200	CONCRETE REINFORCEMENT	95.3	87.1	91.3	94.9	89.9	92.5	95.2	91.8	93.5	96.2	71.7	84.2	95.3	91.1	93.2	96.5	90.2	93.4
03300	CAST-IN-PLACE CONCRETE	97.3	59.3	83.0	94.0	51.9	78.2	97.9	73.5	88.7	92.5	54.4	78.1	93.6	61.7	81.6	92.4	50.6	76.7
03	CONCRETE	95.0	71.2	83.8	93.0	60.8	77.8	99.4	78.6	89.6	94.9	60.2	78.6	92.9	70.9	82.5	92.4	65.2	79.6
04	MASONRY	87.4	76.0	80.6	89.4	48.8	65.1	87.3	77.0	81.1	87.5	48.5	64.1	88.5	68.7	76.6	101.9	46.3	68.6
05	METALS	95.6	92.9	94.8	92.9	89.0	91.7	88.5	81.0	86.2	87.4	69.3	81.9	92.5	93.0	92.6	88.5	91.4	89.3
06	WOOD & PLASTICS	97.8	72.7	84.5	95.1	47.6	70.0	96.9	78.8	87.3	93.8	59.7	75.8	95.1	65.7	79.6	102.7	64.5	82.5
07	THERMAL & MOISTURE PROTECTION	99.4	64.6	85.4	96.0	63.0	82.7	99.7	79.3	91.5	99.3	58.6	82.9	96.2	73.5	87.1	94.8	65.8	83.1
08	DOORS & WINDOWS	100.2	67.3	91.7	97.8	53.4	86.3	99.2	75.9	93.2	93.4	58.3	84.3	97.8	65.5	89.4	96.2	63.7	87.8
09200	PLASTER & GYPSUM BOARD	103.1	72.6	84.9	100.8	46.7	68.6	119.4	78.7	95.2	118.4	59.1	83.1	102.9	65.3	80.5	109.8	64.1	82.6
095,098	CEILINGS & ACOUSTICAL TREATMENT	102.8	72.6	84.6	102.0	46.7	68.7	111.1	78.7	91.6	112.1	59.1	80.2	102.0	65.3	79.9	98.1	64.1	77.6
09600	FLOORING	119.4	48.9	100.6	119.4	37.2	97.5	86.6	74.2	83.3	85.4	47.3	75.3	119.4	60.7	103.7	92.7	43.7	79.7
097,099	WALL FINISHES, PAINTS & COATINGS	111.7	43.5	70.9	107.4	50.0	73.1	91.4	85.9	88.1	91.4	45.2	63.8	107.4	56.5	76.9	109.4	55.7	77.3
09	FINISHES	109.0	64.4	86.0	106.3	47.9	76.2	99.3	77.4	88.0	98.8	54.9	76.1	106.6	63.9	84.6	95.3	58.1	76.1
10 - 14	TOTAL DIV. 10000 - 14000	100.0	83.8	96.7	100.0	81.0	96.1	100.0	84.3	96.8	100.0	63.7	92.5	100.0	83.2	96.5	100.0	80.7	96.0
15	MECHANICAL	99.9	73.0	88.9	99.7	65.3	85.7	100.1	78.3	91.2	100.1	79.9	91.9	99.9	60.5	83.8	99.9	66.3	86.2
16	ELECTRICAL	96.3	45.0	71.7	91.9	62.2	77.7	95.8	77.2	86.9	96.4	67.7	82.6	91.7	65.6	79.2	90.2	64.8	78.0
01 - 16	WEIGHTED AVERAGE	98.8	70.8	86.7	96.7	63.2	82.2	96.8	79.7	89.4	95.5	67.0	83.2	96.6	69.6	84.9	95.3	67.8	83.4

| DIVISION | | GEORGIA | | | | | | HAWAII | | | IDAHO | | | | | | | | |
		SAVANNAH			VALDOSTA			HONOLULU			BOISE			LEWISTON			POCATELLO		
		MAT.	INST.	TOTAL	MAT.	INST.	TOTAL	MAT.	INST.	TOTAL	MAT.	INST.	TOTAL	MAT.	INST.	TOTAL	MAT.	INST.	TOTAL
01590	EQUIPMENT RENTAL	.0	91.4	91.4	.0	90.2	90.2	.0	99.6	99.6	.0	101.1	101.1	.0	94.1	94.1	.0	101.1	101.1
02	SITE CONSTRUCTION	99.6	77.8	83.9	109.0	77.0	86.1	141.7	106.1	116.2	78.1	101.1	94.6	84.9	95.7	92.6	79.8	101.0	95.0
03100	CONCRETE FORMS & ACCESSORIES	96.6	57.8	63.3	84.4	49.8	54.7	111.0	141.5	137.2	98.5	78.6	81.4	113.7	66.1	72.8	98.6	78.2	81.1
03200	CONCRETE REINFORCEMENT	96.2	72.0	84.4	97.2	51.7	75.0	109.7	120.6	115.0	107.5	75.1	91.7	115.4	102.7	109.2	107.9	74.7	91.6
03300	CAST-IN-PLACE CONCRETE	92.7	52.1	77.4	92.0	54.7	78.0	190.6	126.5	166.5	96.8	85.6	92.6	108.8	91.0	102.1	99.6	85.4	94.2
03	CONCRETE	92.6	59.9	77.1	96.7	53.4	76.2	151.2	131.1	141.7	107.5	80.3	94.6	118.7	81.8	101.3	106.8	79.9	94.1
04	MASONRY	87.3	57.8	69.7	94.3	52.2	69.1	130.1	126.2	127.8	130.0	69.3	93.7	132.3	82.5	102.5	127.4	61.8	88.1
05	METALS	88.8	83.2	87.1	92.1	75.3	87.0	139.6	108.6	130.2	103.7	75.1	95.1	96.4	89.2	94.2	112.0	74.0	100.5
06	WOOD & PLASTICS	108.4	56.2	80.8	80.9	46.4	62.6	103.4	146.2	126.0	94.5	78.8	86.2	102.5	59.9	80.0	94.5	78.8	86.2
07	THERMAL & MOISTURE PROTECTION	96.6	59.5	81.7	96.4	64.1	83.4	108.2	126.9	115.7	92.4	75.4	85.6	137.9	77.4	113.6	92.5	66.5	82.0
08	DOORS & WINDOWS	98.9	53.8	87.2	93.2	43.7	80.4	105.9	136.5	113.8	93.6	74.4	88.6	113.8	67.7	101.9	94.3	68.1	87.5
09200	PLASTER & GYPSUM BOARD	103.1	55.6	74.8	95.8	45.5	65.9	130.1	147.2	140.3	89.0	78.1	82.5	152.7	58.8	96.8	86.9	78.1	81.7
095,098	CEILINGS & ACOUSTICAL TREATMENT	102.8	55.6	74.4	97.7	45.5	66.3	119.7	147.2	136.3	110.0	78.1	90.8	133.5	58.8	88.6	111.0	78.1	91.2
09600	FLOORING	120.4	55.8	103.2	110.9	46.6	93.8	168.6	130.9	158.5	97.2	51.9	85.1	136.2	89.1	123.6	99.0	51.9	86.4
097,099	WALL FINISHES, PAINTS & COATINGS	109.3	56.5	77.7	107.4	39.8	67.0	102.2	145.5	128.1	99.5	48.8	69.2	123.2	72.5	92.9	99.5	50.8	70.3
09	FINISHES	107.5	57.1	81.5	102.7	47.5	74.2	136.2	141.5	139.0	98.4	70.8	84.1	158.2	69.9	112.6	98.7	71.0	84.4
10 - 14	TOTAL DIV. 10000 - 14000	100.0	64.7	92.7	100.0	63.4	92.5	100.0	122.5	104.6	100.0	76.9	95.2	100.0	78.9	95.7	100.0	76.9	95.2
15	MECHANICAL	99.9	63.6	85.1	99.9	43.1	76.7	100.2	114.0	105.8	100.0	72.9	88.9	101.1	85.3	94.7	99.9	72.9	88.9
16	ELECTRICAL	94.4	59.3	77.6	90.0	38.4	65.3	111.1	120.4	115.6	95.1	73.8	84.9	81.2	81.6	81.4	91.4	75.0	83.5
01 - 16	WEIGHTED AVERAGE	96.6	63.5	82.3	96.5	52.4	77.4	120.2	122.7	121.3	100.8	76.4	90.3	108.7	81.9	97.1	101.6	75.1	90.2

444

DIVISION		IDAHO			ILLINOIS														
		TWIN FALLS			CHICAGO			DECATUR			EAST ST. LOUIS			JOLIET			PEORIA		
		MAT.	INST.	TOTAL	MAT.	INST.	TOTAL	MAT.	INST.	TOTAL	MAT.	INST.	TOTAL	MAT.	INST.	TOTAL	MAT.	INST.	TOTAL
01590	EQUIPMENT RENTAL	.0	101.1	101.1	.0	90.7	90.7	.0	102.2	102.2	.0	108.7	108.7	.0	88.5	88.5	.0	101.3	101.3
02	SITE CONSTRUCTION	86.2	99.1	95.4	97.9	90.7	92.7	87.5	95.9	93.6	104.2	96.4	98.6	97.7	88.8	91.4	96.5	94.8	95.3
03100	CONCRETE FORMS & ACCESSORIES	99.4	34.1	43.3	100.7	146.5	140.0	96.9	99.3	99.0	91.1	110.4	107.7	101.5	134.2	129.5	95.3	101.8	100.9
03200	CONCRETE REINFORCEMENT	109.9	74.3	92.5	97.3	148.9	122.5	95.3	96.2	95.7	97.2	105.2	101.1	97.3	129.7	113.1	95.3	100.2	97.7
03300	CAST-IN-PLACE CONCRETE	102.2	43.2	80.0	108.5	137.4	119.4	99.5	100.2	99.7	93.1	102.5	96.6	108.4	124.6	114.5	97.2	100.6	98.5
03	CONCRETE	114.8	45.9	82.2	106.4	142.8	123.6	96.6	99.5	98.0	88.1	107.6	97.3	106.4	129.2	117.2	95.4	101.5	98.3
04	MASONRY	130.4	36.3	74.0	96.0	138.4	121.4	70.0	103.8	90.2	73.7	109.4	95.1	99.0	122.2	112.9	111.0	109.7	110.2
05	METALS	112.1	71.7	99.8	98.3	127.7	107.2	99.0	104.6	100.7	96.7	120.4	103.9	96.1	115.9	102.1	99.0	107.5	101.5
06	WOOD & PLASTICS	95.1	33.1	62.3	102.8	146.7	126.0	99.4	95.9	97.6	95.1	111.4	103.7	104.4	137.1	121.7	99.4	98.0	98.7
07	THERMAL & MOISTURE PROTECTION	93.2	44.0	73.4	100.2	134.4	113.9	96.2	98.8	97.2	90.7	106.4	97.0	99.6	123.8	109.3	96.3	100.8	98.1
08	DOORS & WINDOWS	97.6	41.0	82.9	102.6	149.7	114.8	97.0	97.4	97.1	86.5	115.7	94.1	100.6	139.4	110.6	97.2	97.7	97.3
09200	PLASTER & GYPSUM BOARD	86.7	31.1	53.6	94.1	147.9	126.1	101.5	95.7	98.0	98.4	111.6	106.2	91.9	138.1	119.4	101.5	97.8	99.3
095,098	CEILINGS & ACOUSTICAL TREATMENT	104.2	31.1	60.2	100.0	147.9	128.8	94.8	95.7	95.3	88.1	111.6	102.2	100.0	138.1	122.9	94.8	97.8	96.6
09600	FLOORING	100.3	51.9	87.4	84.0	132.3	96.9	98.0	102.2	99.1	106.5	118.3	109.7	83.5	123.4	94.2	98.0	103.7	99.5
097,099	WALL FINISHES, PAINTS & COATINGS	99.6	29.9	57.9	83.0	141.5	118.0	91.5	98.5	95.7	99.3	98.7	98.9	81.0	127.9	109.0	91.5	90.1	90.7
09	FINISHES	98.0	36.0	66.0	90.6	144.0	118.1	95.6	99.9	97.8	94.4	109.4	102.1	90.0	132.6	112.0	95.7	100.9	98.4
10 - 14	TOTAL DIV. 10000 - 14000	100.0	43.3	88.3	100.0	124.6	105.1	100.0	84.7	96.9	100.0	102.1	100.4	100.0	119.2	104.0	100.0	98.5	99.7
15	MECHANICAL	99.9	60.2	83.7	99.9	129.7	112.0	100.0	102.0	100.8	99.9	96.7	98.6	100.0	113.2	105.4	100.0	94.2	97.6
16	ELECTRICAL	84.1	67.9	76.4	96.3	131.9	113.4	97.0	88.3	92.9	94.1	99.5	96.7	94.9	99.4	97.0	96.1	82.9	89.8
01 - 16	WEIGHTED AVERAGE	102.3	56.2	82.4	99.2	131.9	113.3	96.4	98.5	97.3	94.1	105.0	98.8	98.6	116.8	106.4	98.5	97.8	98.2

DIVISION		ILLINOIS						INDIANA											
		ROCKFORD			SPRINGFIELD			ANDERSON			BLOOMINGTON			EVANSVILLE			FORT WAYNE		
		MAT.	INST.	TOTAL	MAT.	INST.	TOTAL	MAT.	INST.	TOTAL	MAT.	INST.	TOTAL	MAT.	INST.	TOTAL	MAT.	INST.	TOTAL
01590	EQUIPMENT RENTAL	.0	101.3	101.3	.0	102.2	102.2	.0	96.3	96.3	.0	85.6	85.6	.0	120.9	120.9	.0	96.3	96.3
02	SITE CONSTRUCTION	95.6	95.7	95.7	93.2	95.8	95.1	85.3	94.7	92.0	74.5	94.7	89.0	79.1	128.8	114.8	86.1	94.7	92.3
03100	CONCRETE FORMS & ACCESSORIES	99.2	119.4	116.6	98.6	99.0	99.0	93.8	77.5	79.8	99.0	78.1	81.1	92.2	79.6	81.4	92.5	75.5	77.9
03200	CONCRETE REINFORCEMENT	90.2	119.7	104.6	95.3	92.4	93.9	96.6	77.1	87.1	86.7	82.3	84.5	95.0	78.9	87.1	96.6	78.9	88.0
03300	CAST-IN-PLACE CONCRETE	99.5	118.3	106.6	88.0	99.8	92.4	94.1	80.2	88.9	96.9	77.3	89.5	92.5	89.5	91.4	100.0	78.6	92.0
03	CONCRETE	96.0	119.2	106.9	90.9	98.5	94.5	91.7	78.9	85.6	99.4	78.4	89.5	99.5	83.1	91.8	94.6	77.8	86.7
04	MASONRY	86.2	108.7	99.7	74.0	103.1	91.4	89.6	81.4	84.7	88.0	77.4	81.7	84.1	82.9	83.4	93.2	82.6	86.9
05	METALS	99.0	118.9	105.0	101.5	102.6	101.8	92.4	87.1	90.8	95.4	76.3	89.6	88.2	85.2	87.3	92.4	87.6	90.9
06	WOOD & PLASTICS	99.4	118.3	109.4	102.0	95.9	98.8	112.1	76.8	93.5	117.0	77.3	96.0	96.4	77.4	86.4	111.9	73.9	91.8
07	THERMAL & MOISTURE PROTECTION	99.2	113.2	104.8	95.8	100.0	97.5	99.0	80.2	91.5	90.9	81.6	87.1	94.9	85.6	91.2	98.8	83.6	92.7
08	DOORS & WINDOWS	97.2	118.0	102.5	98.2	96.4	97.7	98.3	78.6	93.2	104.0	80.3	97.8	96.2	78.2	91.5	98.3	74.7	92.2
09200	PLASTER & GYPSUM BOARD	101.5	118.7	111.7	101.5	95.7	98.0	93.6	76.7	83.5	93.5	77.4	83.9	89.1	76.0	81.3	92.9	73.7	81.5
095,098	CEILINGS & ACOUSTICAL TREATMENT	94.8	118.7	109.2	94.8	95.7	95.3	88.3	76.7	81.3	80.3	77.4	78.6	84.8	76.0	79.5	88.3	73.7	79.5
09600	FLOORING	98.0	86.1	94.8	98.2	101.8	99.2	92.8	89.4	91.9	105.3	90.3	101.3	98.7	85.5	95.2	92.8	77.9	88.8
097,099	WALL FINISHES, PAINTS & COATINGS	91.5	117.1	106.9	91.5	98.5	95.7	95.5	74.6	83.0	92.1	86.3	88.6	98.7	88.1	92.4	95.5	77.6	84.8
09	FINISHES	95.7	113.0	104.6	95.7	99.7	97.8	90.1	79.3	84.5	94.0	81.1	87.4	92.4	81.3	86.7	89.8	76.0	82.7
10 - 14	TOTAL DIV. 10000 - 14000	100.0	108.8	101.8	100.0	84.5	96.8	100.0	90.5	98.1	100.0	85.8	97.1	100.0	90.7	98.1	100.0	85.2	97.0
15	MECHANICAL	100.0	111.1	104.5	100.0	97.2	98.8	99.9	75.2	89.8	99.7	81.9	92.4	99.9	83.7	93.3	99.9	77.6	90.8
16	ELECTRICAL	96.6	102.8	99.6	97.7	83.2	90.8	84.6	91.5	87.9	99.8	84.2	92.3	94.4	83.8	89.4	85.7	79.4	82.6
01 - 16	WEIGHTED AVERAGE	97.5	110.8	103.2	96.7	96.4	96.6	94.1	82.6	89.2	97.5	81.8	90.7	94.8	87.3	91.6	94.8	80.8	88.7

DIVISION		INDIANA															IOWA		
		GARY			INDIANAPOLIS			MUNCIE			SOUTH BEND			TERRE HAUTE			CEDAR RAPIDS		
		MAT.	INST.	TOTAL	MAT.	INST.	TOTAL	MAT.	INST.	TOTAL	MAT.	INST.	TOTAL	MAT.	INST.	TOTAL	MAT.	INST.	TOTAL
01590	EQUIPMENT RENTAL	.0	96.3	96.3	.0	91.3	91.3	.0	95.7	95.7	.0	106.2	106.2	.0	120.9	120.9	.0	96.0	96.0
02	SITE CONSTRUCTION	85.9	97.4	94.1	85.4	98.6	94.9	74.4	94.8	89.0	85.1	95.1	92.3	80.6	128.6	115.0	92.8	95.8	94.9
03100	CONCRETE FORMS & ACCESSORIES	93.9	102.9	101.6	94.3	87.4	88.4	90.2	77.2	79.0	97.9	79.0	81.7	93.6	79.6	81.6	101.3	81.1	84.0
03200	CONCRETE REINFORCEMENT	96.6	95.8	96.2	95.9	84.2	90.2	95.9	77.0	86.7	96.6	73.2	85.2	95.0	82.8	89.0	95.6	84.0	89.9
03300	CAST-IN-PLACE CONCRETE	98.3	108.3	102.0	90.5	88.0	89.6	101.8	79.2	93.3	89.4	83.2	87.1	89.6	88.8	89.3	104.1	77.7	94.2
03	CONCRETE	93.8	103.5	98.4	93.3	86.6	90.1	98.8	78.4	89.2	85.9	80.7	83.5	101.9	83.6	93.2	99.3	81.1	90.7
04	MASONRY	90.9	103.3	98.3	97.7	85.1	90.2	90.0	81.3	84.8	88.1	81.0	83.9	91.0	82.0	85.6	103.1	80.8	89.7
05	METALS	92.4	101.4	95.1	94.3	79.7	89.8	97.1	86.9	94.0	92.4	99.0	94.4	88.9	87.5	88.5	89.4	92.8	90.5
06	WOOD & PLASTICS	110.0	103.0	106.3	109.1	87.2	97.5	109.5	76.6	92.1	112.0	77.9	94.0	98.7	78.7	88.1	108.4	80.0	93.4
07	THERMAL & MOISTURE PROTECTION	98.1	102.7	100.0	97.2	86.9	93.0	93.2	80.2	88.0	96.7	84.5	91.8	95.0	82.4	89.9	99.3	81.5	92.1
08	DOORS & WINDOWS	98.3	100.8	99.0	106.1	86.1	100.9	98.9	78.5	93.6	91.7	77.9	88.1	96.7	81.1	92.7	98.6	84.0	94.8
09200	PLASTER & GYPSUM BOARD	89.1	103.5	97.7	92.1	87.1	89.1	88.9	76.7	81.7	93.2	77.8	84.1	89.1	77.4	82.2	99.2	79.8	87.7
095,098	CEILINGS & ACOUSTICAL TREATMENT	88.3	103.5	97.4	92.5	87.1	89.3	81.1	76.7	78.5	88.3	77.8	82.0	84.8	77.4	80.4	114.0	79.8	93.4
09600	FLOORING	92.8	115.6	98.9	92.6	92.0	92.4	97.8	89.4	95.6	91.4	82.3	88.9	98.7	93.7	97.3	125.6	53.6	106.4
097,099	WALL FINISHES, PAINTS & COATINGS	95.5	104.5	100.9	95.5	92.0	93.4	92.1	74.6	81.7	94.9	81.2	86.7	98.7	90.4	93.8	105.3	74.1	86.6
09	FINISHES	89.2	105.7	97.7	91.1	88.9	90.0	91.0	79.1	84.9	89.5	79.8	84.5	92.4	83.3	87.7	112.0	74.3	92.5
10 - 14	TOTAL DIV. 10000 - 14000	100.0	89.9	97.9	100.0	93.8	98.7	100.0	89.9	97.9	100.0	86.1	97.1	100.0	93.8	98.7	100.0	87.5	97.4
15	MECHANICAL	99.9	98.6	99.4	99.9	87.0	94.6	99.6	75.1	89.7	99.9	79.0	91.4	99.9	73.4	89.1	100.3	85.6	94.3
16	ELECTRICAL	98.0	99.7	98.8	101.6	91.4	96.7	89.7	82.2	86.1	96.7	84.9	91.0	92.9	84.3	88.8	95.5	83.9	89.9
01 - 16	WEIGHTED AVERAGE	95.9	100.9	98.1	97.7	88.1	93.6	95.9	81.2	89.5	94.1	83.9	89.7	95.5	85.8	91.3	98.8	84.2	92.5

IOWA

DIVISION		COUNCIL BLUFFS MAT.	INST.	TOTAL	DAVENPORT MAT.	INST.	TOTAL	DES MOINES MAT.	INST.	TOTAL	DUBUQUE MAT.	INST.	TOTAL	SIOUX CITY MAT.	INST.	TOTAL	WATERLOO MAT.	INST.	TOTAL
01590	EQUIPMENT RENTAL	.0	95.7	95.7	.0	100.1	100.1	.0	102.0	102.0	.0	94.6	94.6	.0	100.1	100.1	.0	100.1	100.1
02	SITE CONSTRUCTION	100.7	92.1	94.6	90.5	99.7	97.1	81.7	100.7	95.3	90.7	92.5	92.0	101.0	96.2	97.5	91.0	95.3	94.1
03100	CONCRETE FORMS & ACCESSORIES	81.1	55.2	58.9	101.0	92.6	93.8	102.9	78.3	81.8	82.7	73.9	75.1	101.3	66.8	71.6	101.8	46.6	54.4
03200	CONCRETE REINFORCEMENT	97.6	75.4	86.7	95.6	91.3	93.5	95.6	82.9	89.4	94.3	83.8	89.2	95.6	63.8	80.1	95.6	83.5	89.7
03300	CAST-IN-PLACE CONCRETE	108.3	70.1	93.9	100.2	96.7	98.9	97.8	98.4	98.0	101.9	94.4	99.1	103.2	55.0	85.1	104.1	45.9	82.2
03	CONCRETE	101.4	65.4	84.4	97.3	94.2	95.9	95.1	86.8	91.2	95.9	83.6	90.1	98.6	63.1	81.9	99.3	54.8	78.3
04	MASONRY	103.4	74.2	85.9	99.0	90.3	93.8	96.5	74.7	83.4	104.0	72.3	85.0	96.6	56.9	72.9	98.1	58.3	74.2
05	METALS	94.8	87.8	92.7	89.4	100.3	92.7	89.6	94.1	90.9	88.1	92.2	89.4	89.4	81.4	87.0	89.4	90.4	89.7
06	WOOD & PLASTICS	84.3	51.2	66.8	108.4	91.3	99.3	109.8	77.9	93.0	86.4	73.8	79.7	108.4	67.2	86.6	109.0	46.4	75.9
07	THERMAL & MOISTURE PROTECTION	98.7	64.7	85.0	98.8	91.1	95.7	100.2	79.5	91.9	98.9	70.5	87.5	98.8	55.4	81.3	98.6	50.9	79.4
08	DOORS & WINDOWS	97.6	58.2	87.4	98.6	92.3	97.0	98.6	83.1	94.6	97.7	81.1	93.4	98.6	64.1	89.7	94.3	59.6	85.3
09200	PLASTER & GYPSUM BOARD	90.7	50.2	66.6	99.2	91.1	94.4	96.7	77.4	85.2	91.0	73.5	80.6	99.2	66.4	79.7	99.2	45.0	66.9
095,098	CEILINGS & ACOUSTICAL TREATMENT	109.0	50.2	73.6	114.0	91.1	100.2	112.3	77.4	91.3	109.0	73.5	87.6	114.0	66.4	85.4	114.0	45.0	72.5
09600	FLOORING	100.3	43.8	85.2	111.5	100.0	108.4	111.3	40.0	92.3	114.7	37.3	94.1	112.0	56.5	97.2	113.3	56.8	98.2
097,099	WALL FINISHES, PAINTS & COATINGS	97.6	67.4	79.5	101.5	95.1	97.7	101.5	82.1	89.9	104.2	74.1	86.2	102.6	64.6	79.8	102.6	34.4	61.8
09	FINISHES	101.3	52.8	76.3	107.6	94.2	100.7	105.7	70.7	87.6	106.0	66.2	85.4	109.0	64.6	86.1	108.2	46.3	76.3
10-14	TOTAL DIV. 10000-14000	100.0	66.0	93.0	100.0	91.1	98.2	100.0	86.5	97.2	100.0	84.5	96.8	100.0	83.4	96.6	100.0	76.7	95.2
15	MECHANICAL	100.3	79.4	91.8	100.3	91.4	96.7	100.3	76.6	90.7	100.3	77.1	90.9	100.3	78.6	91.5	100.3	43.2	77.0
16	ELECTRICAL	100.0	84.9	92.8	92.2	89.1	90.7	96.1	77.8	87.3	98.8	78.7	89.2	95.5	76.4	86.3	95.5	66.7	81.7
01-16	WEIGHTED AVERAGE	99.3	74.2	88.5	97.5	93.2	95.7	97.2	81.5	90.5	97.7	79.2	89.7	98.3	72.7	87.3	97.7	60.7	81.7

KANSAS / KENTUCKY

DIVISION		DODGE CITY MAT.	INST.	TOTAL	KANSAS CITY MAT.	INST.	TOTAL	SALINA MAT.	INST.	TOTAL	TOPEKA MAT.	INST.	TOTAL	WICHITA MAT.	INST.	TOTAL	BOWLING GREEN MAT.	INST.	TOTAL
01590	EQUIPMENT RENTAL	.0	103.3	103.3	.0	99.6	99.6	.0	103.3	103.3	.0	101.3	101.3	.0	103.3	103.3	.0	95.5	95.5
02	SITE CONSTRUCTION	111.0	92.7	97.9	91.1	91.6	91.5	100.7	93.2	95.3	93.4	90.5	91.4	94.3	93.1	93.4	67.6	99.3	90.3
03100	CONCRETE FORMS & ACCESSORIES	93.3	65.6	69.5	98.7	103.1	102.4	89.4	46.3	52.4	98.7	49.2	56.2	96.9	54.5	60.5	83.6	84.9	84.7
03200	CONCRETE REINFORCEMENT	103.2	54.7	79.5	98.1	91.4	94.8	102.6	75.9	89.6	95.3	94.6	95.0	95.3	78.5	87.1	86.1	94.7	90.3
03300	CAST-IN-PLACE CONCRETE	115.4	54.6	92.6	90.5	95.5	92.4	100.1	46.8	80.0	91.6	51.7	76.6	85.6	56.1	74.5	84.5	112.9	95.2
03	CONCRETE	113.2	60.8	88.5	94.5	98.5	96.4	100.5	53.7	78.4	92.8	60.2	77.4	89.6	60.9	76.0	91.3	96.4	93.7
04	MASONRY	103.4	55.1	74.5	102.8	100.2	101.2	118.1	48.3	76.4	97.9	61.2	75.9	91.4	62.6	74.2	91.0	77.3	82.8
05	METALS	96.4	77.8	90.8	101.2	99.6	100.7	96.2	87.2	93.5	101.5	96.6	100.0	101.5	87.4	97.2	93.1	89.0	91.9
06	WOOD & PLASTICS	92.0	73.1	82.0	99.1	104.4	101.9	88.1	47.1	66.5	96.8	46.7	70.3	96.3	53.4	73.6	91.9	83.1	87.3
07	THERMAL & MOISTURE PROTECTION	96.8	60.7	82.3	94.9	102.0	97.8	96.4	55.2	79.8	97.2	68.3	85.6	96.5	62.8	82.9	84.6	90.6	87.1
08	DOORS & WINDOWS	96.9	61.2	87.7	95.9	96.5	96.1	96.8	51.1	85.0	97.2	62.2	88.1	97.2	60.1	87.6	95.8	83.4	92.6
09200	PLASTER & GYPSUM BOARD	98.6	72.3	82.9	95.5	104.3	100.8	98.0	45.6	66.8	99.8	45.1	67.2	99.8	52.0	71.3	85.3	83.0	83.9
095,098	CEILINGS & ACOUSTICAL TREATMENT	86.4	72.3	77.9	87.2	104.3	97.5	86.4	45.6	61.8	87.2	45.1	61.9	87.2	52.0	66.0	84.8	83.0	83.7
09600	FLOORING	97.3	47.2	84.0	88.3	102.0	91.9	95.2	34.5	79.0	99.0	43.8	84.3	98.0	75.8	92.1	94.3	91.3	93.5
097,099	WALL FINISHES, PAINTS & COATINGS	91.5	48.1	65.6	99.3	83.7	89.9	91.5	48.1	65.6	91.5	64.9	75.6	91.5	59.3	72.2	98.7	90.1	93.6
09	FINISHES	93.9	61.2	77.1	92.0	100.1	96.2	92.6	43.2	67.1	93.8	48.0	70.2	93.6	58.7	75.6	90.6	86.4	88.4
10-14	TOTAL DIV. 10000-14000	100.0	48.6	89.4	100.0	79.5	95.8	100.0	59.0	91.6	100.0	63.6	92.5	100.0	62.6	92.3	100.0	74.0	94.6
15	MECHANICAL	100.0	61.3	84.2	99.9	93.7	97.4	100.0	61.3	84.2	100.0	69.2	87.4	100.0	62.4	84.7	99.9	84.7	93.7
16	ELECTRICAL	97.8	74.3	86.5	102.6	103.7	103.1	97.6	69.1	83.9	102.5	77.1	90.3	100.3	69.1	85.3	93.7	82.4	88.3
01-16	WEIGHTED AVERAGE	100.1	66.4	85.6	98.5	97.7	98.1	99.0	62.0	83.0	98.5	69.2	85.8	97.5	67.4	84.5	94.1	86.9	91.0

KENTUCKY / LOUISIANA

DIVISION		LEXINGTON MAT.	INST.	TOTAL	LOUISVILLE MAT.	INST.	TOTAL	OWENSBORO MAT.	INST.	TOTAL	ALEXANDRIA MAT.	INST.	TOTAL	BATON ROUGE MAT.	INST.	TOTAL	LAKE CHARLES MAT.	INST.	TOTAL
01590	EQUIPMENT RENTAL	.0	103.1	103.1	.0	95.5	95.5	.0	120.9	120.9	.0	87.2	87.2	.0	87.5	87.5	.0	87.5	87.5
02	SITE CONSTRUCTION	74.6	101.9	94.2	65.5	99.0	89.5	79.1	129.3	115.1	104.2	85.4	90.7	109.4	85.4	92.2	111.4	85.1	92.6
03100	CONCRETE FORMS & ACCESSORIES	96.8	74.7	77.8	93.1	83.0	84.4	88.5	80.7	81.8	82.8	42.2	47.9	97.9	57.9	63.5	105.4	57.4	64.2
03200	CONCRETE REINFORCEMENT	95.0	94.7	94.9	95.0	94.9	94.9	86.2	92.1	89.1	98.5	63.2	81.2	101.2	62.8	82.4	101.2	62.8	82.4
03300	CAST-IN-PLACE CONCRETE	91.3	85.6	89.1	91.0	78.3	86.2	87.1	109.8	95.7	94.7	42.3	75.0	80.3	50.4	69.1	88.1	51.8	74.4
03	CONCRETE	94.0	82.5	88.6	93.7	83.7	89.0	99.7	92.9	96.5	89.6	47.3	69.7	89.6	56.8	74.2	94.1	57.1	76.6
04	MASONRY	88.0	73.3	79.2	89.7	80.8	84.4	87.8	79.8	83.0	112.3	50.2	75.2	102.9	51.6	72.2	100.5	52.6	71.8
05	METALS	94.6	90.3	93.3	101.9	89.0	98.0	85.1	89.6	86.5	88.4	73.3	83.9	95.4	70.4	87.8	90.2	70.4	84.2
06	WOOD & PLASTICS	103.0	71.9	86.6	103.5	83.1	92.7	92.3	77.0	84.2	84.0	41.3	61.4	103.8	61.3	81.3	110.7	60.6	84.2
07	THERMAL & MOISTURE PROTECTION	95.3	93.9	94.7	84.9	83.0	84.2	94.9	82.5	89.9	99.9	52.0	80.6	97.5	57.2	81.3	100.5	57.0	83.0
08	DOORS & WINDOWS	96.7	79.8	92.4	96.7	86.1	94.0	94.0	82.1	90.9	99.2	47.9	85.9	102.7	58.4	91.2	102.7	59.5	91.5
09200	PLASTER & GYPSUM BOARD	91.3	70.4	78.9	92.2	83.0	86.7	84.9	75.7	79.4	83.6	40.3	57.8	102.2	60.6	77.5	102.2	59.9	77.0
095,098	CEILINGS & ACOUSTICAL TREATMENT	89.0	70.4	77.8	89.0	83.0	85.4	74.7	75.7	75.3	94.7	40.3	62.0	98.7	60.6	75.8	98.7	59.9	75.4
09600	FLOORING	99.5	66.8	90.7	97.9	72.7	91.2	96.9	91.3	95.4	103.0	63.6	92.5	107.3	62.3	95.3	107.5	48.7	91.8
097,099	WALL FINISHES, PAINTS & COATINGS	98.7	77.6	86.1	98.7	79.0	86.9	98.7	99.7	99.3	98.3	39.4	63.1	102.9	45.1	68.3	102.9	41.6	66.2
09	FINISHES	94.1	72.3	82.9	93.5	80.4	86.7	88.8	84.2	86.4	94.5	45.0	69.0	103.3	57.7	79.8	103.3	54.4	78.1
10-14	TOTAL DIV. 10000-14000	100.0	96.4	99.3	100.0	95.7	99.1	100.0	97.4	99.5	100.0	63.7	92.5	100.0	73.4	94.5	100.0	73.4	94.5
15	MECHANICAL	99.9	74.8	89.7	99.9	82.6	92.9	99.9	79.1	91.4	100.1	59.6	83.6	100.0	55.4	81.8	100.0	60.6	84.0
16	ELECTRICAL	94.1	81.8	88.2	94.7	81.8	88.5	93.2	82.5	88.1	94.1	57.8	76.7	98.0	61.0	80.3	97.2	63.3	81.0
01-16	WEIGHTED AVERAGE	95.5	81.5	89.4	96.1	84.7	91.2	93.9	88.2	91.4	96.4	57.6	79.6	98.7	61.0	82.4	98.4	62.1	82.8

DIVISION		LOUISIANA MONROE MAT.	INST.	TOTAL	NEW ORLEANS MAT.	INST.	TOTAL	SHREVEPORT MAT.	INST.	TOTAL	MAINE AUGUSTA MAT.	INST.	TOTAL	BANGOR MAT.	INST.	TOTAL	LEWISTON MAT.	INST.	TOTAL
01590	EQUIPMENT RENTAL	.0	87.2	87.2	.0	88.7	88.7	.0	87.2	87.2	.0	100.7	100.7	.0	100.7	100.7	.0	100.7	100.7
02	SITE CONSTRUCTION	104.2	84.9	90.4	114.5	88.0	95.5	101.6	84.8	89.5	79.5	101.6	95.3	79.3	99.4	93.7	77.0	99.4	93.1
03100	CONCRETE FORMS & ACCESSORIES	82.1	42.3	47.9	104.8	66.1	71.6	101.7	45.1	53.1	98.5	61.3	66.6	92.7	86.9	87.7	98.3	86.9	88.5
03200	CONCRETE REINFORCEMENT	97.4	63.3	80.7	101.2	64.2	83.1	96.9	63.0	80.4	85.2	106.5	95.6	84.9	106.7	95.6	104.3	106.7	105.5
03300	CAST-IN-PLACE CONCRETE	94.7	45.7	76.3	87.2	74.9	82.6	84.5	49.3	71.2	81.7	59.9	73.5	81.1	61.0	73.6	82.8	61.0	74.6
03	CONCRETE	89.4	48.5	70.1	93.6	69.3	82.1	84.9	50.9	68.8	98.7	69.6	85.0	97.5	81.1	89.8	97.9	81.1	90.0
04	MASONRY	107.4	54.0	75.4	102.0	61.1	77.5	99.0	47.5	68.2	97.2	48.8	68.3	112.6	55.9	78.7	96.4	55.9	72.2
05	METALS	88.4	71.6	83.3	102.6	74.0	93.9	84.8	71.1	80.7	84.6	83.7	84.3	84.2	80.1	83.0	87.1	80.1	85.0
06	WOOD & PLASTICS	83.2	41.7	61.3	106.3	68.5	86.3	105.8	45.4	73.9	98.7	59.5	78.0	91.8	93.0	92.4	98.2	93.0	95.4
07	THERMAL & MOISTURE PROTECTION	99.9	54.6	81.7	101.3	62.4	85.7	98.9	53.3	80.6	98.7	52.2	80.0	98.6	57.2	81.9	98.4	57.2	81.8
08	DOORS & WINDOWS	99.1	52.6	87.1	103.4	68.2	94.3	97.0	50.2	84.9	102.5	58.4	91.1	102.4	78.1	96.1	105.6	78.1	98.5
09200	PLASTER & GYPSUM BOARD	83.3	40.7	58.0	101.3	68.0	81.5	88.6	44.5	62.4	100.9	57.6	75.1	97.9	92.0	94.4	102.8	92.0	96.4
095,098	CEILINGS & ACOUSTICAL TREATMENT	94.7	40.7	62.2	97.7	68.0	79.9	95.5	44.5	64.8	87.7	57.6	69.6	86.0	92.0	89.6	96.2	92.0	93.6
09600	FLOORING	102.6	40.2	86.0	107.9	47.8	91.9	113.3	59.5	98.9	98.0	40.5	82.7	95.6	50.5	83.6	98.8	50.5	85.9
097,099	WALL FINISHES, PAINTS & COATINGS	98.3	47.0	67.6	104.5	65.1	81.0	98.3	39.4	63.1	91.0	34.0	56.9	91.0	31.4	55.3	91.0	31.4	55.3
09	FINISHES	94.4	41.5	67.1	103.6	62.6	82.4	98.6	46.5	71.7	96.7	53.4	74.3	95.1	75.3	84.9	99.0	75.3	86.8
10-14	TOTAL DIV. 10000 - 14000	100.0	58.2	91.4	100.0	76.4	95.1	100.0	70.7	94.0	100.0	56.6	91.1	100.0	75.4	94.9	100.0	75.4	94.9
15	MECHANICAL	100.1	52.3	80.6	100.0	63.3	85.1	100.1	60.0	83.8	100.0	72.5	88.8	100.0	73.6	89.3	100.0	73.6	89.3
16	ELECTRICAL	95.8	58.7	78.0	98.2	67.8	83.7	95.0	69.1	82.6	98.1	84.1	91.4	96.6	82.4	89.8	98.3	82.4	90.7
01-16	WEIGHTED AVERAGE	96.3	56.2	79.0	100.7	68.2	86.6	94.9	59.7	79.7	96.4	70.7	85.3	96.6	77.1	88.2	97.1	77.1	88.5

DIVISION		MAINE PORTLAND MAT.	INST.	TOTAL	MARYLAND BALTIMORE MAT.	INST.	TOTAL	HAGERSTOWN MAT.	INST.	TOTAL	MASSACHUSETTS BOSTON MAT.	INST.	TOTAL	BROCKTON MAT.	INST.	TOTAL	FALL RIVER MAT.	INST.	TOTAL
01590	EQUIPMENT RENTAL	.0	100.7	100.7	.0	103.1	103.1	.0	98.9	98.9	.0	107.5	107.5	.0	102.8	102.8	.0	104.1	104.1
02	SITE CONSTRUCTION	80.1	99.4	94.0	94.7	93.8	94.0	84.3	89.6	88.1	84.7	108.4	101.7	82.3	104.4	98.1	81.4	104.5	97.9
03100	CONCRETE FORMS & ACCESSORIES	98.3	86.9	88.5	101.7	74.8	78.6	90.9	76.1	78.2	103.4	136.6	131.9	103.0	123.2	120.4	103.0	123.3	120.5
03200	CONCRETE REINFORCEMENT	104.3	106.7	105.5	101.7	88.7	95.3	89.9	72.5	81.4	103.1	138.0	120.6	104.3	138.5	121.0	104.3	127.7	115.7
03300	CAST-IN-PLACE CONCRETE	106.9	61.0	89.6	94.7	77.1	88.1	80.5	59.1	72.5	104.8	146.4	120.4	99.7	142.7	115.9	96.4	147.3	115.6
03	CONCRETE	110.1	81.1	96.4	105.3	79.3	93.0	90.2	70.6	80.9	111.9	139.6	125.0	109.1	132.1	120.0	107.5	131.7	118.9
04	MASONRY	96.6	55.9	72.2	98.9	71.2	82.3	103.1	74.9	86.2	115.7	151.9	137.4	111.9	143.2	130.6	111.8	147.5	133.2
05	METALS	88.4	80.1	85.9	97.4	99.0	97.9	93.9	90.3	92.8	98.9	125.5	106.9	95.4	122.2	103.8	95.9	118.2	102.6
06	WOOD & PLASTICS	99.3	93.0	96.0	99.1	76.1	87.0	86.7	75.9	81.0	102.1	136.2	120.1	101.1	121.4	111.8	101.1	121.7	112.0
07	THERMAL & MOISTURE PROTECTION	101.0	57.2	83.4	98.2	79.7	90.8	97.6	69.1	86.2	98.8	144.7	117.3	98.7	138.3	114.6	98.7	135.5	113.5
08	DOORS & WINDOWS	105.6	78.1	98.5	93.8	84.8	91.5	91.1	75.8	87.1	101.6	133.5	109.9	99.5	125.5	106.3	99.5	120.7	105.0
09200	PLASTER & GYPSUM BOARD	102.8	92.0	96.4	108.7	75.7	89.1	104.5	75.7	87.4	104.9	136.4	123.7	100.7	121.1	112.9	100.7	121.1	112.9
095,098	CEILINGS & ACOUSTICAL TREATMENT	96.2	92.0	93.6	98.8	75.7	84.9	99.7	75.7	85.3	95.2	136.4	120.0	97.1	121.1	111.6	97.1	121.1	111.6
09600	FLOORING	98.0	50.5	85.3	89.8	75.8	86.1	85.2	78.2	83.4	98.6	163.5	115.9	99.0	163.5	116.2	98.8	163.5	116.1
097,099	WALL FINISHES, PAINTS & COATINGS	91.0	31.4	55.3	94.9	84.7	88.8	94.9	38.4	61.1	93.8	155.8	130.9	93.3	129.6	115.0	93.3	129.6	115.0
09	FINISHES	98.8	75.3	86.7	95.2	75.3	85.0	92.8	72.5	82.3	99.0	144.2	122.3	98.9	131.5	115.7	98.9	131.7	115.8
10-14	TOTAL DIV. 10000 - 14000	100.0	75.4	94.9	100.0	86.1	97.1	100.0	88.6	97.7	100.0	120.3	104.2	100.0	116.9	103.5	100.0	117.7	103.6
15	MECHANICAL	100.0	73.6	89.3	99.9	83.1	93.0	99.9	85.9	94.2	100.0	128.8	111.7	100.0	107.3	103.0	100.0	107.3	103.0
16	ELECTRICAL	99.6	82.4	91.4	103.7	93.6	98.9	100.8	82.3	91.9	97.5	134.8	115.4	96.5	102.6	99.5	96.2	102.6	99.3
01-16	WEIGHTED AVERAGE	99.0	77.1	89.6	99.4	84.2	92.8	96.1	80.3	89.3	101.4	133.8	115.4	100.0	119.9	108.6	99.7	119.7	108.3

DIVISION		MASSACHUSETTS HYANNIS MAT.	INST.	TOTAL	LAWRENCE MAT.	INST.	TOTAL	LOWELL MAT.	INST.	TOTAL	NEW BEDFORD MAT.	INST.	TOTAL	PITTSFIELD MAT.	INST.	TOTAL	SPRINGFIELD MAT.	INST.	TOTAL
01590	EQUIPMENT RENTAL	.0	102.8	102.8	.0	102.8	102.8	.0	100.7	100.7	.0	104.1	104.1	.0	100.7	100.7	.0	100.7	100.7
02	SITE CONSTRUCTION	79.1	104.3	97.2	83.0	104.4	98.3	81.8	104.3	97.9	79.8	104.5	97.5	82.9	102.8	97.2	82.3	103.0	97.1
03100	CONCRETE FORMS & ACCESSORIES	94.1	122.9	118.8	102.8	123.3	120.4	99.7	123.5	120.1	103.0	123.3	120.5	99.7	96.7	97.2	99.9	107.3	106.2
03200	CONCRETE REINFORCEMENT	83.6	119.0	100.9	103.4	126.8	114.8	104.3	126.8	115.3	104.3	127.7	115.7	86.7	111.7	98.9	104.3	116.8	110.4
03300	CAST-IN-PLACE CONCRETE	91.0	146.6	111.9	100.5	142.7	116.4	91.4	142.8	110.7	84.9	147.3	108.4	99.7	117.4	106.4	95.1	119.5	104.3
03	CONCRETE	98.9	129.6	113.4	109.4	129.9	119.1	100.0	129.9	114.1	101.6	131.7	115.8	101.5	106.3	103.8	102.0	112.7	107.0
04	MASONRY	110.8	147.6	132.8	111.3	143.2	130.4	97.1	142.8	124.5	110.7	147.5	132.7	97.7	113.8	107.3	97.4	118.0	109.7
05	METALS	92.5	113.9	98.9	93.4	117.3	100.7	93.4	114.7	99.8	95.9	118.2	102.6	93.2	102.2	95.9	95.8	104.5	98.4
06	WOOD & PLASTICS	90.9	121.4	107.0	101.1	121.4	111.8	100.2	121.4	111.4	101.1	121.7	112.0	100.2	94.2	97.0	100.2	106.4	103.5
07	THERMAL & MOISTURE PROTECTION	98.3	136.4	113.6	98.7	138.3	114.6	98.4	138.1	114.4	98.7	135.5	113.5	98.5	109.9	103.1	98.4	112.9	104.3
08	DOORS & WINDOWS	95.7	118.1	101.5	99.5	122.3	105.4	105.6	122.3	109.9	99.5	120.7	105.0	105.6	102.0	104.7	105.6	109.9	106.7
09200	PLASTER & GYPSUM BOARD	94.2	121.1	110.2	102.8	121.1	113.7	102.8	121.1	113.7	100.7	121.1	112.9	102.8	93.3	97.1	102.8	105.8	104.6
095,098	CEILINGS & ACOUSTICAL TREATMENT	87.0	121.1	107.5	96.2	121.1	111.2	96.2	121.1	111.2	97.1	121.1	111.6	96.2	93.3	94.4	96.2	105.8	101.9
09600	FLOORING	95.1	163.5	113.3	98.0	163.5	115.5	98.0	163.5	115.5	98.8	163.5	116.1	98.2	128.3	106.2	97.9	129.3	106.3
097,099	WALL FINISHES, PAINTS & COATINGS	93.3	129.6	115.0	91.1	129.6	114.1	91.0	129.6	114.1	93.3	129.6	115.0	91.0	104.0	98.8	92.6	104.0	99.4
09	FINISHES	94.1	131.5	113.4	98.6	131.5	115.6	98.5	131.5	115.5	98.8	131.7	115.8	98.6	103.0	100.9	98.6	111.3	105.1
10-14	TOTAL DIV. 10000 - 14000	100.0	116.9	103.5	100.0	117.0	103.5	100.0	117.0	103.5	100.0	117.7	103.6	100.0	100.2	100.0	100.0	103.0	100.6
15	MECHANICAL	100.0	107.1	102.9	100.0	113.5	105.5	100.0	123.6	109.6	100.0	107.3	103.0	100.0	93.5	97.4	100.0	95.5	98.2
16	ELECTRICAL	94.9	102.6	98.6	97.8	115.3	106.2	98.2	115.3	106.4	97.5	102.6	100.0	98.2	91.4	94.9	98.2	91.4	95.0
01-16	WEIGHTED AVERAGE	97.1	118.8	106.5	99.7	122.1	109.4	98.5	123.9	109.4	99.1	119.7	108.0	98.7	100.8	99.6	99.1	104.4	101.4

DIVISION		MASSACHUSETTS			MICHIGAN														
		WORCESTER			ANN ARBOR			DEARBORN			DETROIT			FLINT			GRAND RAPIDS		
		MAT.	INST.	TOTAL	MAT.	INST.	TOTAL	MAT.	INST.	TOTAL	MAT.	INST.	TOTAL	MAT.	INST.	TOTAL	MAT.	INST.	TOTAL
01590	EQUIPMENT RENTAL	.0	100.7	100.7	.0	112.7	112.7	.0	112.7	112.7	.0	98.5	98.5	.0	112.7	112.7	.0	104.8	104.8
02	SITE CONSTRUCTION	82.2	104.3	98.0	72.8	96.3	89.6	72.5	96.4	89.6	85.5	98.2	94.6	64.1	95.2	86.4	79.4	86.8	84.7
03100	CONCRETE FORMS & ACCESSORIES	100.3	133.7	129.0	98.1	118.1	115.3	98.0	120.1	117.0	99.4	120.2	117.2	101.4	92.5	93.8	97.2	74.6	77.8
03200	CONCRETE REINFORCEMENT	104.3	137.6	120.6	96.1	128.5	112.0	96.1	128.8	112.1	95.4	128.7	111.7	96.1	127.8	111.6	95.0	90.1	92.6
03300	CAST-IN-PLACE CONCRETE	94.6	142.2	112.5	84.3	121.4	98.2	82.4	121.1	97.0	90.4	121.1	102.0	84.8	98.1	89.8	92.6	94.9	93.4
03	CONCRETE	101.7	136.2	118.0	88.9	121.4	104.3	87.9	122.3	104.1	92.0	120.9	105.7	89.4	101.9	95.3	94.7	84.1	89.7
04	MASONRY	96.8	142.8	124.4	96.6	116.1	108.3	96.5	120.5	110.9	95.6	120.5	110.5	96.7	91.1	93.3	92.0	52.2	68.2
05	METALS	95.8	118.5	102.7	97.5	124.9	105.8	97.6	125.7	106.1	101.8	104.6	102.7	97.6	122.0	105.0	94.9	82.0	91.0
06	WOOD & PLASTICS	100.6	136.3	119.5	100.2	118.9	110.1	100.2	120.2	110.8	101.1	120.2	111.2	103.7	93.4	98.2	97.9	75.0	85.8
07	THERMAL & MOISTURE PROTECTION	98.5	134.4	112.9	99.3	116.5	106.2	98.3	122.2	107.9	96.1	122.2	106.6	97.2	93.1	95.6	90.8	58.6	77.9
08	DOORS & WINDOWS	105.6	133.3	112.8	96.0	115.5	101.0	96.0	116.3	101.2	97.6	117.1	102.7	96.0	97.5	96.4	94.7	68.7	88.0
09200	PLASTER & GYPSUM BOARD	102.8	136.4	122.8	92.5	118.1	107.8	92.5	119.5	108.6	92.5	119.5	108.6	93.8	92.0	92.7	90.1	69.6	77.9
095,098	CEILINGS & ACOUSTICAL TREATMENT	96.2	136.4	120.4	89.0	118.1	106.5	89.0	119.5	107.3	90.0	119.5	107.7	89.0	92.0	90.8	89.0	69.6	77.3
09600	FLOORING	98.0	154.6	113.1	90.4	123.7	99.3	90.1	122.2	98.6	90.1	122.2	98.7	90.2	85.8	89.1	98.3	40.8	82.9
097,099	WALL FINISHES, PAINTS & COATINGS	91.0	129.6	114.1	89.2	106.4	99.5	89.2	114.2	104.1	90.8	114.2	104.8	89.2	95.5	93.0	98.7	39.9	63.5
09	FINISHES	98.5	138.4	119.1	91.5	118.4	105.4	91.4	120.2	106.3	92.4	120.2	106.7	91.1	90.9	91.0	93.5	64.2	78.4
10 - 14	TOTAL DIV. 10000 - 14000	100.0	109.0	101.9	100.0	110.5	102.2	100.0	111.4	102.3	100.0	111.4	102.3	100.0	98.2	99.6	100.0	102.5	100.5
15	MECHANICAL	100.0	105.3	102.2	99.9	106.8	102.7	99.9	114.7	105.9	99.9	114.7	105.9	99.9	95.4	98.1	99.9	52.6	80.6
16	ELECTRICAL	98.3	102.2	100.2	93.8	82.1	88.2	93.8	118.5	105.6	95.2	118.4	106.3	93.8	106.1	99.7	95.3	57.3	77.1
01 - 16	WEIGHTED AVERAGE	99.1	120.6	108.4	95.6	109.3	101.5	95.4	117.2	104.8	97.2	115.3	105.0	95.3	99.2	97.0	95.7	67.1	83.3

DIVISION		MICHIGAN												MINNESOTA					
		KALAMAZOO			LANSING			MUSKEGON			SAGINAW			DULUTH			MINNEAPOLIS		
		MAT.	INST.	TOTAL	MAT.	INST.	TOTAL	MAT.	INST.	TOTAL	MAT.	INST.	TOTAL	MAT.	INST.	TOTAL	MAT.	INST.	TOTAL
01590	EQUIPMENT RENTAL	.0	104.8	104.8	.0	112.7	112.7	.0	104.8	104.8	.0	112.7	112.7	.0	101.5	101.5	.0	105.9	105.9
02	SITE CONSTRUCTION	79.8	87.3	85.2	78.5	95.0	90.3	77.7	87.2	84.5	66.3	94.8	86.7	87.3	102.7	98.3	88.6	108.9	103.1
03100	CONCRETE FORMS & ACCESSORIES	96.1	88.0	89.1	101.5	95.3	96.1	96.8	87.1	88.4	98.1	93.8	94.4	99.3	119.6	116.7	100.0	139.0	133.5
03200	CONCRETE REINFORCEMENT	95.0	92.2	93.6	96.1	127.5	111.5	95.6	91.5	93.6	96.1	127.7	111.5	92.9	109.3	101.0	93.1	127.2	109.8
03300	CAST-IN-PLACE CONCRETE	93.0	109.6	99.2	84.3	97.8	89.4	90.9	104.1	95.9	83.2	99.6	89.4	102.0	109.0	104.6	110.8	126.3	116.6
03	CONCRETE	97.0	95.5	96.3	89.1	103.0	95.7	92.4	93.1	92.7	88.4	103.0	95.3	97.9	114.7	105.8	103.9	132.7	117.5
04	MASONRY	93.6	90.2	91.6	90.0	95.9	93.5	92.2	78.2	83.8	98.1	96.2	97.0	103.1	121.0	113.8	103.9	136.6	123.5
05	METALS	93.4	87.3	91.6	96.0	121.3	103.7	91.1	85.7	89.5	97.6	121.4	104.8	93.9	127.7	104.1	96.6	140.1	109.7
06	WOOD & PLASTICS	98.7	86.1	92.0	103.1	93.8	98.2	95.6	86.1	90.6	96.3	93.4	94.7	112.6	119.3	116.2	113.1	137.8	126.1
07	THERMAL & MOISTURE PROTECTION	90.6	90.5	90.5	97.9	96.0	97.2	89.8	79.0	85.4	97.7	94.4	96.4	98.3	121.3	107.6	98.3	138.7	114.6
08	DOORS & WINDOWS	91.2	83.9	89.3	96.0	97.8	96.4	90.5	82.2	88.3	93.8	97.5	94.8	94.7	121.5	101.6	97.9	144.6	110.0
09200	PLASTER & GYPSUM BOARD	90.1	81.1	84.7	94.6	92.4	93.3	78.6	81.1	80.1	92.5	92.0	92.2	98.5	120.3	111.5	98.7	139.1	122.8
095,098	CEILINGS & ACOUSTICAL TREATMENT	89.0	81.1	84.2	89.0	92.4	91.0	90.7	81.1	84.9	89.0	92.0	90.8	96.9	120.3	111.0	97.7	139.1	122.6
09600	FLOORING	98.3	76.2	92.4	100.0	94.4	98.5	97.4	75.1	91.5	90.4	85.8	89.2	106.1	127.2	111.7	103.5	126.5	109.6
097,099	WALL FINISHES, PAINTS & COATINGS	98.7	81.0	88.1	101.2	88.3	93.5	98.0	57.9	74.0	89.2	88.9	89.0	89.4	108.5	100.8	96.1	133.6	118.5
09	FINISHES	93.5	84.4	88.8	96.0	94.1	95.0	91.3	81.3	86.1	91.3	91.5	91.4	100.9	120.4	111.0	100.7	136.7	119.3
10 - 14	TOTAL DIV. 10000 - 14000	100.0	106.4	101.3	100.0	101.0	100.2	100.0	106.1	101.3	100.0	100.0	100.0	100.0	102.7	100.6	100.0	110.5	102.2
15	MECHANICAL	99.9	88.8	95.4	99.9	96.9	98.7	99.7	89.2	95.5	99.9	91.4	96.4	100.0	110.5	104.2	99.9	123.0	109.3
16	ELECTRICAL	93.5	84.9	89.4	92.6	82.4	87.7	94.1	79.0	86.9	91.1	77.8	84.7	102.5	96.4	99.5	103.9	118.7	111.0
01 - 16	WEIGHTED AVERAGE	95.3	88.8	92.5	95.4	97.3	96.2	94.0	85.5	90.3	94.8	95.1	94.9	98.5	112.9	104.7	100.2	128.3	112.3

DIVISION		MINNESOTA									MISSISSIPPI								
		ROCHESTER			SAINT PAUL			ST. CLOUD			BILOXI			GREENVILLE			JACKSON		
		MAT.	INST.	TOTAL	MAT.	INST.	TOTAL	MAT.	INST.	TOTAL	MAT.	INST.	TOTAL	MAT.	INST.	TOTAL	MAT.	INST.	TOTAL
01590	EQUIPMENT RENTAL	.0	101.5	101.5	.0	101.5	101.5	.0	102.0	102.0	.0	98.5	98.5	.0	98.5	98.5	.0	98.5	98.5
02	SITE CONSTRUCTION	87.6	101.8	97.8	90.9	103.4	99.9	80.7	106.5	99.2	100.9	86.3	90.4	104.7	86.0	91.3	96.2	86.0	88.9
03100	CONCRETE FORMS & ACCESSORIES	99.8	109.0	107.7	94.5	131.5	126.3	84.0	130.2	123.7	95.0	39.9	47.7	80.8	32.9	39.6	96.4	37.5	45.8
03200	CONCRETE REINFORCEMENT	92.9	126.7	109.4	89.7	127.1	108.0	96.6	126.7	111.3	95.3	61.7	78.9	103.3	41.2	72.9	95.3	44.7	70.6
03300	CAST-IN-PLACE CONCRETE	108.6	104.4	107.0	109.7	125.5	115.7	94.5	123.8	105.5	102.3	43.7	80.2	102.7	38.3	78.5	94.8	40.6	74.4
03	CONCRETE	101.3	111.8	106.2	104.6	129.1	116.2	89.9	127.8	107.8	97.1	47.2	73.6	100.2	38.3	71.0	93.5	41.9	69.1
04	MASONRY	103.2	119.5	113.0	113.1	136.6	127.1	105.7	128.7	119.5	88.6	35.4	56.8	131.2	36.7	74.7	90.2	36.7	58.2
05	METALS	93.7	137.6	107.0	93.3	139.5	107.3	92.3	137.0	105.8	89.7	80.9	87.0	88.0	71.1	82.9	89.6	72.8	84.5
06	WOOD & PLASTICS	113.1	106.8	109.7	106.7	128.1	118.0	90.8	128.0	110.4	95.1	39.8	65.9	77.6	32.3	53.7	97.2	38.4	66.1
07	THERMAL & MOISTURE PROTECTION	98.2	107.1	101.8	98.3	136.7	113.7	98.4	119.2	106.8	96.3	44.4	75.4	96.4	39.4	73.6	96.4	41.0	74.1
08	DOORS & WINDOWS	94.7	127.8	103.3	92.2	139.3	104.4	91.0	139.3	103.5	97.8	46.4	84.5	97.2	36.5	81.5	98.2	40.8	83.3
09200	PLASTER & GYPSUM BOARD	98.3	107.5	103.7	93.7	129.3	114.9	85.3	129.3	111.5	105.6	38.8	65.8	95.8	31.1	57.3	105.6	37.3	65.0
095,098	CEILINGS & ACOUSTICAL TREATMENT	96.1	107.5	102.9	94.4	129.3	115.4	70.3	129.3	105.8	102.8	38.8	64.3	97.7	31.1	57.6	102.8	37.3	63.4
09600	FLOORING	105.9	92.9	102.4	99.1	126.5	106.4	99.6	126.5	106.8	119.4	38.2	97.7	109.6	32.6	89.0	119.4	37.7	97.6
097,099	WALL FINISHES, PAINTS & COATINGS	92.0	104.9	99.7	96.1	123.9	112.7	102.0	133.6	120.9	107.4	34.1	63.6	107.4	33.5	63.2	107.4	33.5	63.2
09	FINISHES	100.7	105.9	103.4	98.1	129.9	114.5	90.7	130.4	111.2	107.3	38.5	71.8	101.9	32.7	66.2	107.3	37.1	71.2
10 - 14	TOTAL DIV. 10000 - 14000	100.0	101.0	100.2	100.0	108.9	101.8	100.0	106.9	101.4	100.0	49.0	89.5	100.0	47.5	89.2	100.0	48.2	89.3
15	MECHANICAL	99.9	103.0	101.2	99.9	121.1	108.5	99.6	121.6	108.6	99.9	51.4	80.1	99.9	32.7	72.5	99.9	32.4	72.4
16	ELECTRICAL	102.5	83.5	93.4	103.4	109.4	106.3	100.0	109.4	104.5	97.3	58.8	78.8	96.7	35.6	67.4	98.5	35.6	68.3
01 - 16	WEIGHTED AVERAGE	98.9	107.6	102.7	99.4	124.3	110.1	95.5	122.9	107.4	97.4	53.7	78.5	99.1	43.1	74.9	97.2	44.6	74.5

City Cost Indexes

MISSISSIPPI / MISSOURI

DIVISION		MERIDIAN MAT.	MERIDIAN INST.	MERIDIAN TOTAL	CAPE GIRARDEAU MAT.	CAPE GIRARDEAU INST.	CAPE GIRARDEAU TOTAL	COLUMBIA MAT.	COLUMBIA INST.	COLUMBIA TOTAL	JOPLIN MAT.	JOPLIN INST.	JOPLIN TOTAL	KANSAS CITY MAT.	KANSAS CITY INST.	KANSAS CITY TOTAL	SPRINGFIELD MAT.	SPRINGFIELD INST.	SPRINGFIELD TOTAL
01590	EQUIPMENT RENTAL	.0	98.5	98.5	.0	108.4	108.4	.0	108.7	108.7	.0	107.1	107.1	.0	104.6	104.6	.0	102.4	102.4
02	SITE CONSTRUCTION	96.1	86.0	88.9	92.9	93.8	93.5	102.2	95.0	97.0	102.4	99.0	99.9	95.6	98.0	97.3	95.1	93.6	94.0
03100	CONCRETE FORMS & ACCESSORIES	79.4	32.0	38.7	84.1	74.8	76.1	85.5	70.5	72.6	102.4	73.0	77.2	101.5	107.7	106.8	99.8	71.3	75.3
03200	CONCRETE REINFORCEMENT	102.0	44.2	73.7	94.9	86.2	90.6	99.0	112.8	105.8	108.6	72.3	90.8	103.1	113.8	108.3	95.3	112.4	103.7
03300	CAST-IN-PLACE CONCRETE	96.9	40.6	75.7	91.5	84.1	88.7	89.0	86.4	88.0	100.6	70.4	89.3	93.5	108.1	99.0	96.7	64.8	84.7
03	CONCRETE	93.1	39.3	67.7	89.0	81.8	85.6	85.3	85.8	85.5	97.9	73.0	86.1	93.7	109.3	101.1	95.4	77.9	87.1
04	MASONRY	88.3	27.4	51.8	112.4	78.3	92.0	130.1	83.9	102.5	92.2	59.1	72.4	97.0	106.5	102.7	86.2	79.4	82.1
05	METALS	88.0	71.9	83.1	96.9	109.2	100.6	93.9	121.8	102.3	96.3	87.6	93.7	103.7	115.1	107.2	97.7	106.6	100.4
06	WOOD & PLASTICS	76.3	31.0	52.4	77.9	72.2	74.9	88.8	64.2	75.8	105.6	73.6	88.7	104.9	107.5	106.3	99.7	71.0	84.5
07	THERMAL & MOISTURE PROTECTION	96.0	37.8	72.6	95.2	79.6	88.9	90.8	83.7	87.9	95.7	67.3	84.3	94.8	107.1	99.8	95.4	75.9	87.5
08	DOORS & WINDOWS	97.2	35.3	81.2	96.2	74.3	90.6	91.5	90.3	91.2	92.9	74.9	88.2	99.2	110.0	102.0	97.2	80.9	92.9
09200	PLASTER & GYPSUM BOARD	95.8	29.7	56.5	97.6	71.4	82.0	96.4	63.1	76.6	97.9	72.6	82.9	95.0	107.5	102.4	100.0	70.1	82.2
095,098	CEILINGS & ACOUSTICAL TREATMENT	97.7	29.7	56.8	84.8	71.4	76.7	88.1	63.1	73.1	91.1	72.6	80.0	97.0	107.5	103.3	88.1	70.1	77.2
09600	FLOORING	108.3	26.5	86.5	90.6	72.1	85.7	103.2	87.8	99.1	118.9	48.5	100.1	96.7	100.4	97.7	106.7	48.5	91.2
097,099	WALL FINISHES, PAINTS & COATINGS	107.4	33.1	63.0	99.5	75.8	85.3	99.3	77.9	86.5	91.6	43.3	62.7	96.4	115.4	107.8	93.7	62.0	74.8
09	FINISHES	101.0	30.5	64.6	90.0	73.2	81.4	93.0	71.1	81.7	101.1	65.5	82.7	96.6	107.1	102.0	96.5	65.5	80.5
10-14	TOTAL DIV. 10000 - 14000	100.0	47.4	89.2	100.0	63.5	92.5	100.0	92.1	98.4	100.0	73.7	94.6	100.0	97.8	99.6	100.0	87.7	97.5
15	MECHANICAL	99.9	32.4	72.4	100.0	96.4	98.5	99.9	97.9	99.1	100.1	54.8	81.7	100.0	109.3	103.8	100.0	66.6	86.4
16	ELECTRICAL	96.4	58.7	78.3	101.3	109.8	105.4	97.1	88.6	93.0	94.7	69.8	82.8	105.2	106.9	106.0	100.6	64.3	83.2
01-16	WEIGHTED AVERAGE	95.7	45.2	73.9	97.3	89.8	94.1	96.8	90.7	94.2	97.6	69.9	85.7	99.7	107.6	103.1	97.6	76.4	88.5

MISSOURI / MONTANA

DIVISION		ST. JOSEPH MAT.	ST. JOSEPH INST.	ST. JOSEPH TOTAL	ST. LOUIS MAT.	ST. LOUIS INST.	ST. LOUIS TOTAL	BILLINGS MAT.	BILLINGS INST.	BILLINGS TOTAL	BUTTE MAT.	BUTTE INST.	BUTTE TOTAL	GREAT FALLS MAT.	GREAT FALLS INST.	GREAT FALLS TOTAL	HELENA MAT.	HELENA INST.	HELENA TOTAL
01590	EQUIPMENT RENTAL	.0	103.1	103.1	.0	109.6	109.6	.0	99.0	99.0	.0	98.7	98.7	.0	98.7	98.7	.0	98.7	98.7
02	SITE CONSTRUCTION	97.1	93.5	94.5	92.6	97.5	96.1	88.7	96.6	94.4	98.4	95.2	96.1	102.2	96.0	97.7	103.3	95.9	98.0
03100	CONCRETE FORMS & ACCESSORIES	101.4	84.3	86.7	97.8	109.9	108.2	97.5	64.3	69.0	84.6	59.3	62.9	97.6	65.1	69.7	96.9	65.6	70.0
03200	CONCRETE REINFORCEMENT	101.9	98.7	100.3	86.8	113.7	99.9	95.6	72.4	84.2	103.7	72.5	88.4	95.6	72.3	84.2	99.0	60.5	80.2
03300	CAST-IN-PLACE CONCRETE	93.5	103.8	97.4	91.5	114.7	100.2	107.5	68.4	92.8	118.3	68.9	99.7	125.0	55.1	98.7	119.3	64.6	98.7
03	CONCRETE	93.5	94.4	93.9	88.6	113.1	100.2	101.0	68.2	85.5	104.7	66.1	86.5	109.6	63.9	88.1	107.2	65.1	87.3
04	MASONRY	96.6	91.9	93.8	93.8	112.3	104.9	121.5	69.7	90.5	117.2	70.2	89.1	121.4	72.4	92.1	115.7	63.9	84.7
05	METALS	99.8	104.2	101.2	101.6	125.6	108.9	107.9	84.2	100.8	100.8	84.0	95.7	104.6	83.7	98.3	103.7	77.1	95.7
06	WOOD & PLASTICS	105.6	80.7	92.4	93.6	107.7	101.0	101.9	65.0	82.4	88.0	59.8	73.1	103.3	65.0	83.1	102.5	65.0	82.7
07	THERMAL & MOISTURE PROTECTION	95.2	93.1	94.3	95.2	109.4	100.9	98.8	66.6	85.9	98.7	66.8	85.9	99.3	66.0	85.9	99.3	65.4	85.6
08	DOORS & WINDOWS	98.0	91.7	96.3	95.1	113.8	100.0	97.3	62.4	88.3	95.2	59.5	86.0	98.6	62.5	89.3	98.1	59.4	88.1
09200	PLASTER & GYPSUM BOARD	99.1	79.9	87.7	105.1	107.8	106.7	94.9	64.4	76.8	91.5	59.1	72.2	99.2	64.4	78.5	98.6	64.4	78.3
095,098	CEILINGS & ACOUSTICAL TREATMENT	96.2	79.9	86.4	89.9	107.8	100.6	112.1	64.4	83.4	111.5	59.1	79.9	114.0	64.4	84.2	111.5	64.4	83.2
09600	FLOORING	99.4	94.8	98.2	98.4	103.2	99.6	104.9	47.2	89.5	101.3	37.6	84.3	110.1	51.0	94.3	110.1	49.4	93.9
097,099	WALL FINISHES, PAINTS & COATINGS	92.1	91.8	91.9	99.5	109.6	105.5	97.2	57.8	73.6	97.6	39.4	62.8	97.6	44.3	65.8	97.6	49.2	68.7
09	FINISHES	97.6	86.3	91.8	94.9	107.8	101.5	104.1	60.2	81.4	102.2	52.4	76.5	106.9	60.1	82.7	106.3	60.8	82.8
10-14	TOTAL DIV. 10000 - 14000	100.0	92.5	98.5	100.0	102.7	100.6	100.0	64.4	92.7	100.0	62.9	92.4	100.0	65.5	92.9	100.0	54.1	90.6
15	MECHANICAL	100.1	94.8	98.0	100.0	112.5	105.1	100.2	75.1	90.0	100.3	69.7	87.9	100.3	74.6	89.9	100.3	74.1	89.7
16	ELECTRICAL	102.8	89.2	96.3	104.0	116.0	109.7	94.3	77.7	86.4	100.9	72.4	87.2	94.5	70.2	82.8	94.5	71.4	83.5
01-16	WEIGHTED AVERAGE	98.8	92.9	96.3	97.8	112.0	103.9	101.7	73.6	89.5	101.3	70.2	87.9	102.9	72.0	89.6	102.1	74.0	88.5

MONTANA / NEBRASKA / NEVADA

DIVISION		MISSOULA MAT.	MISSOULA INST.	MISSOULA TOTAL	GRAND ISLAND MAT.	GRAND ISLAND INST.	GRAND ISLAND TOTAL	LINCOLN MAT.	LINCOLN INST.	LINCOLN TOTAL	NORTH PLATTE MAT.	NORTH PLATTE INST.	NORTH PLATTE TOTAL	OMAHA MAT.	OMAHA INST.	OMAHA TOTAL	CARSON CITY MAT.	CARSON CITY INST.	CARSON CITY TOTAL
01590	EQUIPMENT RENTAL	.0	98.7	98.7	.0	101.3	101.3	.0	101.3	101.3	.0	101.3	101.3	.0	90.7	90.7	.0	101.1	101.1
02	SITE CONSTRUCTION	79.5	95.1	90.7	98.9	91.5	93.6	89.2	91.5	90.8	99.7	89.5	92.4	78.9	90.3	87.1	62.4	102.7	91.3
03100	CONCRETE FORMS & ACCESSORIES	88.0	55.6	60.1	96.3	48.0	54.9	101.5	44.4	52.4	96.2	44.3	51.6	95.8	71.2	74.7	98.8	97.6	97.8
03200	CONCRETE REINFORCEMENT	105.5	78.0	92.1	103.8	71.4	87.9	95.3	72.2	84.0	105.2	71.0	88.5	99.9	73.2	86.8	113.0	116.5	114.7
03300	CAST-IN-PLACE CONCRETE	87.1	61.2	77.4	116.2	55.7	93.4	99.9	58.1	84.2	116.2	49.3	91.0	103.5	71.9	91.6	108.9	94.9	103.6
03	CONCRETE	83.5	62.9	73.8	107.7	56.5	83.5	97.2	55.8	77.6	107.9	52.5	81.7	98.5	72.0	86.0	112.4	100.2	106.6
04	MASONRY	141.0	67.0	96.7	105.5	46.6	70.2	95.2	62.1	75.4	90.3	37.6	58.8	100.5	78.3	87.2	125.3	79.4	97.8
05	METALS	97.7	85.7	94.0	94.0	81.9	90.3	99.0	83.4	94.3	94.2	78.9	89.5	98.3	75.9	91.6	97.2	103.6	99.2
06	WOOD & PLASTICS	92.3	56.2	73.2	96.1	42.7	67.9	101.6	37.3	67.6	95.9	42.7	67.8	95.6	72.6	83.4	94.3	100.0	97.3
07	THERMAL & MOISTURE PROTECTION	98.0	69.1	86.4	95.9	55.8	79.8	95.0	59.4	80.7	95.9	44.5	75.2	91.8	70.4	83.2	98.9	87.4	94.3
08	DOORS & WINDOWS	95.2	59.8	86.1	90.7	47.3	79.4	96.3	46.9	83.5	89.9	49.5	79.5	99.6	65.1	90.7	94.3	107.0	97.5
09200	PLASTER & GYPSUM BOARD	92.8	55.4	70.5	98.0	41.1	64.1	101.5	35.6	62.2	98.4	41.1	64.3	103.5	72.4	84.9	85.4	99.8	94.0
095,098	CEILINGS & ACOUSTICAL TREATMENT	111.5	55.4	77.7	86.4	41.1	59.1	94.8	35.6	59.2	88.1	41.1	59.8	106.9	72.4	86.1	104.2	99.8	101.6
09600	FLOORING	103.5	58.8	91.6	96.2	33.5	79.4	98.0	40.3	82.6	96.1	34.8	79.8	124.2	44.6	103.0	100.0	61.6	89.7
097,099	WALL FINISHES, PAINTS & COATINGS	97.6	44.3	65.8	91.5	40.0	60.7	91.5	40.4	61.0	91.5	48.4	65.8	152.3	68.8	102.3	99.6	76.3	85.6
09	FINISHES	102.1	54.4	77.5	93.1	42.7	67.1	96.1	41.1	67.7	93.5	42.3	67.1	114.4	65.7	89.2	96.5	88.3	92.3
10-14	TOTAL DIV. 10000 - 14000	100.0	49.1	89.5	100.0	65.5	92.9	100.0	64.9	92.8	100.0	48.9	89.5	100.0	67.7	93.4	100.0	102.6	100.5
15	MECHANICAL	100.3	66.1	86.4	100.0	77.2	90.7	100.0	77.2	90.7	100.0	70.4	87.9	99.7	81.6	92.4	99.9	91.0	96.3
16	ELECTRICAL	99.0	76.7	88.3	90.8	67.3	79.5	96.6	67.3	82.5	92.5	41.2	67.9	90.4	85.8	88.2	91.2	106.2	98.4
01-16	WEIGHTED AVERAGE	98.9	69.3	86.1	97.5	64.6	83.3	97.7	66.0	84.0	96.9	56.8	79.6	98.8	77.3	89.5	99.3	96.0	97.9

DIVISION		NEVADA						NEW HAMPSHIRE									NEW JERSEY		
		LAS VEGAS			RENO			MANCHESTER			NASHUA			PORTSMOUTH			CAMDEN		
		MAT.	INST.	TOTAL	MAT.	INST.	TOTAL	MAT.	INST.	TOTAL	MAT.	INST.	TOTAL	MAT.	INST.	TOTAL	MAT.	INST.	TOTAL
01590	EQUIPMENT RENTAL	.0	101.1	101.1	.0	101.1	101.1	.0	100.7	100.7	.0	100.7	100.7	.0	100.7	100.7	.0	98.6	98.6
02	SITE CONSTRUCTION	63.5	105.3	93.5	62.9	102.7	91.4	83.3	99.0	94.6	83.8	99.0	94.7	78.2	98.0	92.4	86.4	104.5	99.3
03100	CONCRETE FORMS & ACCESSORIES	97.6	108.1	106.6	95.5	97.7	97.4	98.1	86.2	87.9	99.9	86.2	88.2	87.8	81.7	82.5	99.9	125.6	122.0
03200	CONCRETE REINFORCEMENT	107.4	123.4	115.2	107.4	122.8	114.9	104.3	88.1	96.4	104.3	88.1	96.4	83.4	88.0	85.6	104.3	109.9	107.0
03300	CAST-IN-PLACE CONCRETE	105.7	105.8	105.7	116.4	94.9	108.4	104.2	95.3	100.9	92.4	95.3	93.5	87.6	88.7	88.0	80.3	122.1	96.0
03	CONCRETE	109.8	110.0	109.9	115.1	101.4	108.6	108.7	89.4	99.6	102.8	89.4	96.5	94.0	85.0	89.8	96.7	120.3	107.9
04	MASONRY	119.3	96.2	105.4	125.1	79.4	97.7	97.8	90.9	93.7	97.9	90.9	93.7	93.4	80.1	85.4	91.4	124.1	111.0
05	METALS	97.8	109.5	101.3	97.6	106.4	100.3	95.8	84.9	92.5	95.8	84.9	92.5	92.4	83.0	89.6	95.7	100.7	97.2
06	WOOD & PLASTICS	94.0	108.2	101.5	89.9	100.0	95.3	98.5	87.0	92.4	100.2	87.0	93.2	85.8	87.0	86.4	100.2	127.0	114.4
07	THERMAL & MOISTURE PROTECTION	104.8	98.8	102.4	98.8	87.4	94.2	98.3	88.6	94.4	98.6	88.6	94.6	98.2	95.7	97.2	98.2	123.4	108.3
08	DOORS & WINDOWS	95.8	113.1	100.3	94.6	108.5	98.2	105.6	80.4	99.1	105.6	80.4	99.1	106.4	74.8	98.2	105.6	119.8	109.3
09200	PLASTER & GYPSUM BOARD	90.1	108.2	100.9	86.9	99.8	94.6	102.8	85.8	92.7	102.8	85.8	92.7	94.4	85.8	89.3	102.8	126.9	117.1
095,098	CEILINGS & ACOUSTICAL TREATMENT	111.2	108.2	109.4	111.0	99.8	104.3	96.2	85.8	89.9	96.2	85.8	89.9	86.0	85.8	85.9	96.2	126.9	114.7
09600	FLOORING	99.6	95.7	98.5	100.0	61.6	89.7	99.0	102.1	99.9	98.0	102.1	99.1	92.6	102.1	95.2	98.0	122.3	104.5
097,099	WALL FINISHES, PAINTS & COATINGS	101.7	113.5	108.8	99.6	76.3	85.6	91.0	97.6	94.9	91.0	97.6	94.9	91.0	37.7	59.1	91.0	129.6	114.1
09	FINISHES	99.0	106.4	102.8	98.4	88.3	93.2	99.0	91.0	94.8	98.9	91.0	94.8	93.2	81.5	87.1	99.2	125.9	113.0
10-14	TOTAL DIV. 10000-14000	100.0	98.7	99.7	100.0	102.6	100.5	100.0	75.4	94.9	100.0	75.4	94.9	100.0	71.3	94.1	100.0	114.1	102.9
15	MECHANICAL	99.9	102.3	100.9	99.9	91.1	96.3	100.0	86.8	94.6	100.0	86.8	94.6	100.0	81.0	92.3	100.0	111.5	104.7
16	ELECTRICAL	93.3	104.1	98.5	91.2	106.2	98.4	99.0	69.6	84.9	98.1	69.6	84.4	96.5	69.6	83.6	98.6	127.6	112.5
01-16	WEIGHTED AVERAGE	99.6	104.7	101.8	99.9	96.5	98.4	100.0	86.0	94.0	99.3	86.3	93.6	96.6	81.6	90.1	98.4	117.4	106.6

DIVISION		NEW JERSEY															NEW MEXICO		
		ELIZABETH			JERSEY CITY			NEWARK			PATERSON			TRENTON			ALBUQUERQUE		
		MAT.	INST.	TOTAL	MAT.	INST.	TOTAL	MAT.	INST.	TOTAL	MAT.	INST.	TOTAL	MAT.	INST.	TOTAL	MAT.	INST.	TOTAL
01590	EQUIPMENT RENTAL	.0	100.7	100.7	.0	98.6	98.6	.0	100.7	100.7	.0	100.7	100.7	.0	98.1	98.1	.0	115.6	115.6
02	SITE CONSTRUCTION	99.1	104.5	103.0	86.4	105.2	99.9	103.8	104.5	104.3	98.0	105.3	103.2	87.9	104.4	99.7	79.4	110.2	101.5
03100	CONCRETE FORMS & ACCESSORIES	109.4	126.3	123.9	99.9	126.6	122.8	97.9	126.4	122.4	98.7	126.4	122.5	98.1	125.9	122.0	97.0	69.1	73.1
03200	CONCRETE REINFORCEMENT	80.3	109.3	94.5	104.3	109.3	106.7	104.3	109.3	106.7	104.3	109.3	106.7	104.3	110.2	107.2	108.1	66.6	87.8
03300	CAST-IN-PLACE CONCRETE	88.4	127.3	103.1	80.3	127.4	98.0	97.8	127.3	108.9	102.4	127.4	111.8	92.4	124.8	104.6	108.9	74.9	96.1
03	CONCRETE	99.3	122.5	110.3	96.7	122.5	108.9	105.5	122.6	113.6	107.9	122.6	114.8	102.8	121.4	111.6	111.5	71.6	92.7
04	MASONRY	111.5	122.7	118.2	89.0	122.7	109.2	99.9	122.7	113.5	95.4	122.7	111.8	88.5	124.1	109.8	114.3	64.2	84.4
05	METALS	92.2	105.3	96.2	95.7	102.8	97.9	95.7	105.4	98.6	90.9	105.5	95.3	90.8	101.0	93.9	104.3	87.0	99.1
06	WOOD & PLASTICS	115.5	127.0	121.6	100.2	127.0	114.4	101.6	127.0	115.0	101.9	127.0	115.2	99.2	126.9	113.9	94.4	70.4	81.7
07	THERMAL & MOISTURE PROTECTION	98.9	128.7	110.9	98.1	128.7	110.4	98.3	128.7	110.5	98.8	122.3	108.2	97.0	121.8	107.0	98.5	72.5	88.0
08	DOORS & WINDOWS	106.7	119.6	110.1	105.6	119.6	109.2	111.7	119.6	113.7	111.7	119.6	113.7	106.2	118.9	109.5	94.4	71.6	88.5
09200	PLASTER & GYPSUM BOARD	106.0	126.9	118.4	102.8	126.9	117.1	102.8	126.9	117.1	102.8	126.9	117.1	102.8	126.9	117.1	90.8	69.1	77.9
095,098	CEILINGS & ACOUSTICAL TREATMENT	86.0	126.9	110.6	96.2	126.9	114.7	96.2	126.9	114.7	96.2	126.9	114.7	96.2	126.9	114.7	110.2	69.1	85.5
09600	FLOORING	102.6	122.3	107.8	98.0	122.3	104.5	98.2	122.3	104.6	98.0	122.3	104.5	98.2	122.3	104.6	101.6	70.0	93.2
097,099	WALL FINISHES, PAINTS & COATINGS	90.9	129.6	114.1	91.0	129.6	114.1	90.9	129.6	114.1	90.9	129.6	114.1	91.0	129.6	114.1	101.8	57.6	75.4
09	FINISHES	99.3	127.3	113.8	99.2	127.3	113.7	99.7	127.3	114.0	99.3	127.3	113.8	99.2	125.9	113.0	99.9	68.1	83.5
10-14	TOTAL DIV. 10000-14000	100.0	118.9	103.9	100.0	118.9	103.9	100.0	118.9	103.9	100.0	118.9	103.9	100.0	114.0	102.9	100.0	79.5	95.8
15	MECHANICAL	100.0	119.7	108.0	100.0	126.0	110.6	100.0	122.4	109.1	100.0	126.0	110.6	100.0	121.7	108.8	100.0	69.8	87.7
16	ELECTRICAL	96.6	138.1	116.5	100.0	141.9	120.1	99.9	141.9	120.0	100.0	138.1	118.3	98.7	142.2	119.6	84.9	76.3	80.8
01-16	WEIGHTED AVERAGE	99.5	121.7	109.1	98.4	123.3	109.2	101.0	122.8	110.4	100.2	122.9	110.0	98.2	121.7	108.3	99.8	75.7	89.4

DIVISION		NEW MEXICO												NEW YORK					
		FARMINGTON			LAS CRUCES			ROSWELL			SANTA FE			ALBANY			BINGHAMTON		
		MAT.	INST.	TOTAL	MAT.	INST.	TOTAL	MAT.	INST.	TOTAL	MAT.	INST.	TOTAL	MAT.	INST.	TOTAL	MAT.	INST.	TOTAL
01590	EQUIPMENT RENTAL	.0	115.6	115.6	.0	86.2	86.2	.0	115.6	115.6	.0	115.6	115.6	.0	115.3	115.3	.0	115.4	115.4
02	SITE CONSTRUCTION	85.2	110.2	103.1	89.6	85.9	86.9	89.4	110.2	104.3	78.7	110.2	101.3	72.7	106.4	96.9	94.3	90.2	91.3
03100	CONCRETE FORMS & ACCESSORIES	97.1	69.1	73.1	94.4	67.4	71.2	97.1	69.0	72.9	96.0	69.1	72.9	97.6	92.2	93.0	102.4	84.7	87.2
03200	CONCRETE REINFORCEMENT	118.1	66.6	92.9	111.7	54.4	83.7	117.2	54.6	86.6	115.9	66.6	91.8	97.0	94.6	95.8	96.1	91.5	93.8
03300	CAST-IN-PLACE CONCRETE	109.9	74.9	96.8	95.7	65.6	84.3	101.2	74.9	91.3	102.3	74.9	92.0	83.6	99.0	89.4	104.0	92.7	99.8
03	CONCRETE	115.6	71.6	94.8	92.4	64.9	79.4	115.6	69.2	93.7	109.3	71.6	91.5	95.5	95.8	95.6	98.3	90.5	94.6
04	MASONRY	121.3	64.2	87.1	107.3	63.8	81.3	122.4	64.2	87.6	115.1	64.2	84.7	90.2	95.2	93.2	107.7	84.6	93.9
05	METALS	101.9	87.0	97.4	101.8	72.8	93.0	101.9	80.9	95.5	101.9	87.0	97.4	96.0	106.4	99.1	92.6	116.1	99.7
06	WOOD & PLASTICS	94.5	70.4	81.8	86.4	69.0	77.2	94.5	70.4	81.8	93.1	70.4	81.1	97.3	91.5	94.2	106.3	86.2	95.7
07	THERMAL & MOISTURE PROTECTION	98.7	72.5	88.1	83.8	67.0	77.0	98.9	72.5	88.3	98.3	72.5	87.9	92.3	90.4	91.5	102.9	85.0	95.7
08	DOORS & WINDOWS	97.2	71.6	90.5	86.9	67.0	81.8	93.2	67.8	86.6	93.3	71.6	87.7	96.0	86.6	93.6	90.8	82.2	88.6
09200	PLASTER & GYPSUM BOARD	84.7	69.1	75.4	86.6	69.1	76.1	84.7	69.1	75.4	84.7	69.1	75.4	108.5	91.0	98.1	112.7	85.3	96.4
095,098	CEILINGS & ACOUSTICAL TREATMENT	100.9	69.1	81.7	96.1	69.1	79.8	100.9	69.1	81.7	100.9	69.1	81.7	97.3	91.0	93.5	97.3	85.3	90.1
09600	FLOORING	100.0	70.0	92.0	132.1	70.0	115.6	100.0	70.0	92.0	100.0	70.0	92.0	85.9	93.1	87.8	96.4	87.9	94.2
097,099	WALL FINISHES, PAINTS & COATINGS	99.6	57.6	74.5	92.7	57.6	71.7	99.6	57.6	74.5	99.6	57.6	74.5	83.5	75.9	79.0	88.4	86.9	87.5
09	FINISHES	96.3	68.1	81.7	109.6	67.0	87.6	96.7	68.1	81.9	96.1	68.1	81.6	96.2	90.6	93.3	97.6	85.6	91.4
10-14	TOTAL DIV. 10000-14000	100.0	79.5	95.8	100.0	75.8	95.0	100.0	79.5	95.8	100.0	79.5	95.8	100.0	92.1	98.4	100.0	89.4	97.8
15	MECHANICAL	99.9	69.8	87.6	100.2	69.2	87.6	99.9	69.6	87.6	99.9	69.8	87.6	100.3	88.8	95.6	100.5	83.1	93.4
16	ELECTRICAL	83.3	76.3	80.0	84.8	56.4	71.2	84.0	76.3	80.3	86.1	76.3	81.4	104.2	90.1	97.5	102.2	83.7	93.3
01-16	WEIGHTED AVERAGE	100.2	75.7	89.6	96.8	67.8	84.3	100.1	74.6	89.1	98.9	75.7	88.9	97.2	94.0	95.8	98.4	88.5	94.1

450

City Cost Indexes

	DIVISION	NEW YORK																
		BUFFALO			HICKSVILLE			NEW YORK			RIVERHEAD			ROCHESTER			SCHENECTADY		
		MAT.	INST.	TOTAL	MAT.	INST.	TOTAL	MAT.	INST.	TOTAL	MAT.	INST.	TOTAL	MAT.	INST.	TOTAL	MAT.	INST.	TOTAL
01590	EQUIPMENT RENTAL	.0	94.5	94.5	.0	116.9	116.9	.0	116.1	116.1	.0	116.9	116.9	.0	118.0	118.0	.0	115.3	115.3
02	SITE CONSTRUCTION	94.6	95.9	95.5	112.3	130.4	125.2	134.7	127.0	129.1	113.2	130.4	125.5	72.8	109.7	99.3	72.4	106.6	96.9
03100	CONCRETE FORMS & ACCESSORIES	98.4	115.1	112.8	90.0	152.5	143.7	108.5	180.3	170.1	95.3	152.5	144.5	100.2	99.4	99.5	102.0	92.9	94.2
03200	CONCRETE REINFORCEMENT	95.1	103.4	99.2	97.4	188.8	142.1	103.1	196.2	148.6	99.2	188.8	143.0	95.9	82.1	89.2	95.7	94.6	95.2
03300	CAST-IN-PLACE CONCRETE	114.9	118.9	116.4	97.2	155.0	119.0	108.5	162.3	128.7	98.9	155.0	120.0	105.9	100.7	104.0	95.5	100.1	97.2
03	CONCRETE	104.2	113.4	108.5	100.8	158.9	128.2	109.9	175.0	140.6	101.8	158.9	128.7	107.0	97.5	102.5	101.5	96.6	99.2
04	MASONRY	107.7	119.8	114.9	111.0	158.8	139.6	105.5	171.3	144.9	116.9	158.8	142.0	104.8	98.3	100.9	91.4	97.2	94.8
05	METALS	102.9	94.9	100.5	102.6	141.1	114.3	105.8	146.0	118.0	103.0	141.1	114.5	99.3	104.7	101.0	96.1	106.4	99.2
06	WOOD & PLASTICS	102.1	114.8	108.8	87.5	153.1	122.2	106.6	184.1	147.5	93.4	153.1	125.0	101.1	100.5	100.8	102.9	91.5	96.9
07	THERMAL & MOISTURE PROTECTION	99.0	110.1	103.5	108.5	146.1	123.7	110.6	163.0	131.7	109.0	146.1	124.0	95.5	96.2	95.8	92.4	91.3	91.9
08	DOORS & WINDOWS	94.1	103.4	96.5	88.6	155.5	105.9	96.9	176.5	117.5	88.6	155.5	105.9	97.8	89.8	95.8	96.0	86.6	93.6
09200	PLASTER & GYPSUM BOARD	97.3	114.9	107.8	101.4	154.5	133.0	115.8	185.9	157.5	103.7	154.5	133.9	99.0	100.6	99.9	108.5	91.0	98.1
095,098	CEILINGS & ACOUSTICAL TREATMENT	90.4	114.9	105.2	78.0	154.5	124.0	107.2	185.9	154.6	82.2	154.5	125.7	94.7	100.6	98.2	97.3	91.0	93.5
09600	FLOORING	98.7	119.2	104.2	95.9	83.1	92.5	95.5	171.8	115.9	97.2	83.1	93.5	87.9	104.9	92.4	85.9	93.1	87.8
097,099	WALL FINISHES, PAINTS & COATINGS	88.7	114.8	104.3	111.3	146.7	132.5	91.4	146.9	124.6	111.3	146.7	132.5	86.8	100.2	94.8	83.5	75.9	79.0
09	FINISHES	93.7	116.7	105.6	102.6	137.8	120.8	107.6	176.3	143.1	104.3	137.8	121.6	94.5	101.2	98.0	95.9	91.1	93.4
10 - 14	TOTAL DIV. 10000 - 14000	100.0	107.3	101.5	100.0	134.7	107.1	100.0	147.5	109.8	100.0	134.7	107.1	100.0	96.2	99.2	100.0	92.8	98.5
15	MECHANICAL	99.7	98.5	99.2	99.7	145.9	118.6	100.3	163.8	126.2	99.9	145.9	118.6	99.7	91.0	96.2	100.3	89.8	96.0
16	ELECTRICAL	100.4	98.0	99.3	103.1	153.8	127.4	110.4	173.6	140.7	104.6	153.8	128.1	104.6	92.1	98.6	102.2	90.1	96.4
01 - 16	WEIGHTED AVERAGE	100.1	105.3	102.4	101.1	147.5	121.1	105.3	164.5	130.9	102.0	147.5	121.6	99.9	97.3	98.8	97.8	94.6	96.4

	DIVISION	NEW YORK															NORTH CAROLINA		
		SYRACUSE			UTICA			WATERTOWN			WHITE PLAINS			YONKERS			ASHEVILLE		
		MAT.	INST.	TOTAL	MAT.	INST.	TOTAL	MAT.	INST.	TOTAL	MAT.	INST.	TOTAL	MAT.	INST.	TOTAL	MAT.	INST.	TOTAL
01590	EQUIPMENT RENTAL	.0	115.3	115.3	.0	115.3	115.3	.0	115.3	115.3	.0	115.7	115.7	.0	115.7	115.7	.0	93.6	93.6
02	SITE CONSTRUCTION	93.0	106.8	102.9	70.6	105.2	95.4	79.0	107.1	99.1	124.0	122.7	123.0	134.0	122.5	125.7	105.4	73.2	82.3
03100	CONCRETE FORMS & ACCESSORIES	101.3	87.2	89.2	102.5	81.4	84.4	85.6	87.6	87.3	108.1	138.7	134.3	108.3	138.8	134.5	96.3	44.1	51.5
03200	CONCRETE REINFORCEMENT	97.0	91.9	94.5	97.0	80.8	89.1	97.6	81.4	89.7	96.4	187.8	141.1	100.2	187.8	143.0	100.3	48.6	75.0
03300	CAST-IN-PLACE CONCRETE	96.5	97.8	97.0	88.3	92.7	89.9	102.8	80.7	94.5	95.6	131.3	109.0	107.0	131.3	116.2	100.7	49.9	81.6
03	CONCRETE	100.6	92.7	96.9	98.1	86.3	92.5	110.7	85.0	98.6	98.8	144.3	120.3	108.8	144.4	125.6	106.3	48.7	79.1
04	MASONRY	97.8	97.5	97.6	80.0	89.7	85.9	91.1	90.4	90.7	100.4	129.6	117.9	105.1	129.6	119.7	85.7	45.0	61.3
05	METALS	95.9	104.9	98.6	94.1	100.3	96.0	94.2	99.8	95.9	93.7	137.3	106.9	102.2	137.4	112.9	91.1	77.5	87.0
06	WOOD & PLASTICS	102.9	84.9	93.4	102.9	81.1	91.4	83.2	89.6	86.6	107.5	140.5	124.9	107.3	140.5	124.9	94.8	43.4	67.6
07	THERMAL & MOISTURE PROTECTION	101.5	94.7	98.8	92.3	91.1	91.8	92.5	89.8	91.4	111.1	137.4	121.7	111.4	137.4	121.9	104.0	47.8	81.4
08	DOORS & WINDOWS	93.9	81.3	90.6	96.0	76.7	91.0	96.0	78.2	91.4	91.6	148.7	106.4	95.0	151.4	109.6	92.2	43.4	79.5
09200	PLASTER & GYPSUM BOARD	108.5	84.2	94.1	108.5	80.3	91.7	101.8	89.1	94.2	110.0	141.2	128.6	115.4	141.2	130.7	109.8	41.8	69.3
095,098	CEILINGS & ACOUSTICAL TREATMENT	97.3	84.2	89.5	97.3	80.3	87.1	97.3	89.1	92.4	80.2	141.2	116.9	105.5	141.2	127.0	94.4	41.8	62.7
09600	FLOORING	87.6	92.4	88.9	85.9	87.4	86.3	78.0	87.4	80.5	92.0	167.0	112.0	91.5	167.0	111.6	101.3	50.2	87.6
097,099	WALL FINISHES, PAINTS & COATINGS	89.0	85.4	86.9	83.5	83.7	83.6	83.5	68.6	74.6	89.2	146.7	123.6	89.2	146.7	123.6	112.5	39.6	68.9
09	FINISHES	97.4	87.6	92.3	96.0	82.5	89.0	93.1	85.4	89.1	98.3	145.3	122.5	105.8	145.3	126.2	99.8	44.4	71.2
10 - 14	TOTAL DIV. 10000 - 14000	100.0	95.6	99.1	100.0	88.2	97.6	100.0	74.7	94.8	100.0	129.8	106.1	100.0	135.2	107.2	100.0	77.1	95.3
15	MECHANICAL	100.3	87.3	95.0	100.3	83.2	93.3	100.3	70.7	88.2	100.5	128.6	112.0	100.5	128.6	112.0	100.2	45.0	77.7
16	ELECTRICAL	102.2	93.0	97.8	99.4	87.9	93.9	102.2	87.5	95.2	100.4	146.9	122.7	107.6	146.9	126.4	98.2	42.9	71.7
01 - 16	WEIGHTED AVERAGE	98.7	93.3	96.4	96.6	88.4	93.1	98.3	85.9	92.9	99.2	137.1	115.6	104.0	137.3	118.4	97.9	51.4	77.8

	DIVISION	NORTH CAROLINA																	
		CHARLOTTE			DURHAM			FAYETTEVILLE			GREENSBORO			RALEIGH			WILMINGTON		
		MAT.	INST.	TOTAL	MAT.	INST.	TOTAL	MAT.	INST.	TOTAL	MAT.	INST.	TOTAL	MAT.	INST.	TOTAL	MAT.	INST.	TOTAL
01590	EQUIPMENT RENTAL	.0	93.6	93.6	.0	100.1	100.1	.0	100.1	100.1	.0	100.1	100.1	.0	100.1	100.1	.0	93.6	93.6
02	SITE CONSTRUCTION	106.3	73.3	82.6	105.6	83.0	89.4	104.4	83.2	89.2	105.4	83.2	89.5	106.7	83.2	89.9	106.5	73.2	82.6
03100	CONCRETE FORMS & ACCESSORIES	103.5	46.3	54.3	98.7	44.6	52.3	95.6	45.5	52.6	98.4	45.4	52.9	101.0	46.4	54.1	97.8	44.6	52.1
03200	CONCRETE REINFORCEMENT	100.7	45.3	73.6	101.9	56.8	79.8	104.2	56.9	81.1	100.7	56.7	79.2	100.7	56.9	79.3	101.1	54.9	78.5
03300	CAST-IN-PLACE CONCRETE	107.8	52.2	86.9	102.3	50.1	82.7	105.6	53.6	86.0	101.6	51.6	82.8	107.8	57.7	88.9	100.3	51.3	81.8
03	CONCRETE	109.7	49.9	81.4	106.8	50.6	80.3	108.2	52.2	81.8	106.2	51.4	80.4	109.5	54.0	83.3	106.2	50.6	80.0
04	MASONRY	92.5	49.3	66.6	91.5	40.5	61.0	88.6	46.1	63.2	88.8	41.2	60.3	91.3	47.1	64.9	75.2	42.7	55.7
05	METALS	95.4	76.4	89.6	105.1	81.2	97.9	110.0	81.3	101.3	98.5	80.8	93.1	95.9	81.3	91.5	90.6	80.1	87.5
06	WOOD & PLASTICS	104.0	45.9	73.3	97.3	45.1	69.7	93.3	45.1	67.8	97.0	45.0	69.5	100.4	45.9	71.6	96.6	44.3	69.0
07	THERMAL & MOISTURE PROTECTION	103.6	49.6	81.8	104.8	50.1	82.8	103.8	48.4	81.5	104.5	47.3	81.5	104.4	48.7	82.0	104.0	47.8	81.4
08	DOORS & WINDOWS	96.3	45.1	83.0	96.3	48.0	83.8	92.3	48.0	80.8	96.3	48.0	83.7	93.0	48.1	81.4	92.3	46.4	80.4
09200	PLASTER & GYPSUM BOARD	112.9	44.3	72.0	113.5	43.5	71.8	112.7	43.5	71.5	115.0	43.4	72.4	115.0	44.2	72.9	110.6	42.7	70.2
095,098	CEILINGS & ACOUSTICAL TREATMENT	100.3	44.3	66.6	100.3	43.5	66.1	95.2	43.5	64.1	100.3	43.4	66.1	100.3	44.2	66.6	95.2	42.7	63.6
09600	FLOORING	105.1	50.3	90.5	105.3	50.2	90.6	101.4	50.2	87.7	105.3	46.4	89.6	105.3	50.2	90.6	102.2	50.2	88.3
097,099	WALL FINISHES, PAINTS & COATINGS	112.5	45.2	72.2	112.5	41.5	70.0	112.5	39.6	68.9	112.5	38.5	68.2	112.5	43.6	71.3	112.5	38.4	68.2
09	FINISHES	102.8	46.7	73.9	103.0	45.0	73.1	100.5	45.4	72.0	103.3	44.6	73.0	103.4	46.6	74.1	100.4	44.7	71.6
10 - 14	TOTAL DIV. 10000 - 14000	100.0	77.7	95.4	100.0	75.8	95.0	100.0	76.8	95.2	100.0	77.3	95.3	100.0	77.3	95.3	100.0	76.5	95.2
15	MECHANICAL	100.1	45.2	77.7	100.3	43.5	77.2	100.1	44.9	77.6	100.2	45.0	77.7	100.2	42.9	76.9	100.2	44.6	77.6
16	ELECTRICAL	100.5	55.9	79.1	99.3	54.3	77.7	98.5	49.3	74.9	98.2	43.1	71.8	98.7	42.8	71.9	98.6	41.4	71.2
01 - 16	WEIGHTED AVERAGE	100.3	54.2	80.4	101.3	54.0	80.9	101.3	54.4	81.1	100.0	52.8	79.6	99.9	53.6	79.9	97.5	51.5	77.6

Section 1

DIVISION		NORTH CAROLINA WINSTON-SALEM			NORTH DAKOTA BISMARCK			NORTH DAKOTA FARGO			NORTH DAKOTA GRAND FORKS			NORTH DAKOTA MINOT			OHIO AKRON		
		MAT.	INST.	TOTAL	MAT.	INST.	TOTAL	MAT.	INST.	TOTAL	MAT.	INST.	TOTAL	MAT.	INST.	TOTAL	MAT.	INST.	TOTAL
01590	EQUIPMENT RENTAL	.0	100.1	100.1	.0	98.7	98.7	.0	98.7	98.7	.0	98.7	98.7	.0	98.7	98.7	.0	96.6	96.6
02	SITE CONSTRUCTION	105.8	83.1	89.5	93.2	97.1	96.0	92.1	97.1	95.7	102.8	94.6	96.9	100.1	97.1	97.9	96.0	105.2	102.6
03100	CONCRETE FORMS & ACCESSORIES	99.6	44.4	52.2	94.2	48.2	54.7	95.3	48.6	55.2	91.7	42.0	49.1	87.1	54.8	59.4	98.0	94.1	94.7
03200	CONCRETE REINFORCEMENT	100.7	45.0	73.5	103.7	74.9	89.6	95.6	74.0	85.1	102.2	75.2	89.0	105.7	75.2	90.8	96.2	94.8	95.5
03300	CAST-IN-PLACE CONCRETE	104.1	48.7	83.2	97.2	53.6	80.8	99.4	55.6	83.0	103.7	51.1	83.9	103.6	52.3	84.3	96.7	104.1	99.5
03	CONCRETE	107.6	47.7	79.3	96.6	56.3	77.6	105.3	57.0	82.5	104.4	52.6	80.0	103.2	58.8	82.3	96.7	96.8	96.8
04	MASONRY	89.0	42.5	61.1	101.3	62.3	78.0	103.4	46.6	69.4	103.2	67.4	81.8	103.5	68.2	82.4	88.8	99.7	95.3
05	METALS	95.9	75.8	89.8	94.3	81.1	90.3	96.6	79.8	91.5	94.3	77.4	89.2	94.6	81.7	90.7	91.5	81.0	88.3
06	WOOD & PLASTICS	97.0	44.3	69.1	87.0	43.4	64.0	87.1	44.0	64.3	84.1	39.9	60.7	79.1	52.1	64.9	91.3	91.9	91.6
07	THERMAL & MOISTURE PROTECTION	104.5	46.9	81.3	99.5	53.6	81.1	100.2	51.2	80.5	100.1	55.2	82.0	99.9	55.9	82.2	100.5	98.3	99.6
08	DOORS & WINDOWS	96.3	44.4	82.8	98.7	49.3	85.9	98.7	49.6	86.0	98.7	43.9	84.5	98.9	54.0	87.3	105.9	93.8	102.8
09200	PLASTER & GYPSUM BOARD	115.0	42.6	71.9	109.7	42.2	69.5	109.7	42.9	69.9	107.8	38.6	66.6	106.5	51.2	73.6	95.0	91.3	92.8
095,098	CEILINGS & ACOUSTICAL TREATMENT	100.3	42.6	65.6	139.5	42.2	81.0	139.5	42.9	81.4	139.5	38.6	78.8	139.5	51.2	86.4	94.5	91.3	92.6
09600	FLOORING	105.3	50.2	90.6	110.6	72.3	100.4	110.4	43.2	92.5	108.8	43.2	91.3	106.1	83.2	100.0	102.3	92.3	99.6
097,099	WALL FINISHES, PAINTS & COATINGS	112.5	40.4	69.3	97.6	36.4	61.0	97.6	70.8	81.6	97.6	27.8	55.9	97.6	30.4	57.5	107.8	114.8	112.0
09	FINISHES	103.3	44.8	73.1	115.0	50.3	81.6	114.9	48.5	80.6	114.8	39.9	76.1	113.6	57.0	84.4	99.4	95.8	97.5
10-14	TOTAL DIV. 10000 - 14000	100.0	76.9	95.3	100.0	65.8	93.0	100.0	65.9	93.0	100.0	42.9	88.2	100.0	66.9	93.2	100.0	95.3	99.0
15	MECHANICAL	100.2	43.4	77.1	100.5	64.1	85.7	100.5	68.0	87.3	100.5	43.1	77.1	100.5	62.4	85.0	100.1	98.0	99.2
16	ELECTRICAL	98.2	47.0	73.7	93.3	70.8	82.5	94.6	65.8	80.8	95.8	66.7	81.9	98.5	72.0	85.8	99.0	90.8	95.1
01-16	WEIGHTED AVERAGE	99.7	52.0	79.1	99.1	65.3	84.5	100.7	63.6	84.6	100.6	57.7	82.1	100.6	67.4	86.3	97.9	95.5	96.9

Section 2 — OHIO

DIVISION		CANTON			CINCINNATI			CLEVELAND			COLUMBUS			DAYTON			LORAIN		
		MAT.	INST.	TOTAL	MAT.	INST.	TOTAL	MAT.	INST.	TOTAL	MAT.	INST.	TOTAL	MAT.	INST.	TOTAL	MAT.	INST.	TOTAL
01590	EQUIPMENT RENTAL	.0	96.6	96.6	.0	101.9	101.9	.0	97.0	97.0	.0	95.4	95.4	.0	96.3	96.3	.0	96.6	96.6
02	SITE CONSTRUCTION	96.1	105.3	102.7	72.8	107.3	97.5	95.9	105.4	102.7	85.1	101.2	96.7	71.7	106.7	96.8	95.3	104.5	101.9
03100	CONCRETE FORMS & ACCESSORIES	98.0	85.4	87.2	95.9	85.9	87.3	98.1	105.4	104.3	96.9	84.5	86.3	95.8	76.7	79.4	98.0	100.9	100.5
03200	CONCRETE REINFORCEMENT	96.2	82.8	89.7	93.3	85.0	89.2	96.8	95.2	96.0	92.6	86.0	89.4	93.3	80.5	87.0	96.2	94.9	95.6
03300	CAST-IN-PLACE CONCRETE	97.7	100.9	98.9	83.1	86.7	84.4	94.9	112.4	101.5	90.6	89.2	90.1	77.2	86.9	80.9	92.0	104.9	96.9
03	CONCRETE	97.2	89.6	93.6	88.9	86.1	87.6	95.8	104.8	100.1	93.2	86.1	89.8	85.9	80.8	83.5	94.3	100.1	97.0
04	MASONRY	89.5	90.5	90.1	75.8	91.9	85.5	93.1	108.9	102.5	93.9	90.9	92.1	75.4	87.4	82.6	85.6	97.6	92.8
05	METALS	91.5	75.0	86.5	94.2	88.4	92.4	92.9	84.4	90.3	97.6	81.8	92.8	93.4	77.3	88.5	92.1	81.9	89.0
06	WOOD & PLASTICS	91.6	84.3	87.7	95.2	84.1	89.3	90.5	102.7	96.9	101.8	82.2	91.5	96.2	73.5	84.2	91.3	101.7	96.8
07	THERMAL & MOISTURE PROTECTION	101.2	94.2	98.4	95.3	94.2	94.9	99.3	112.6	104.7	98.0	94.2	96.5	100.6	88.5	95.7	101.1	104.0	102.6
08	DOORS & WINDOWS	100.1	76.7	94.1	97.3	86.8	94.6	96.4	100.2	97.4	101.1	82.7	96.3	97.7	76.3	92.1	100.1	99.7	100.0
09200	PLASTER & GYPSUM BOARD	95.6	83.4	88.4	95.1	83.9	88.5	94.3	102.3	99.1	91.6	81.7	85.7	95.1	73.1	82.0	95.0	101.4	98.8
095,098	CEILINGS & ACOUSTICAL TREATMENT	94.5	83.4	87.8	97.4	83.9	89.3	92.8	102.3	98.5	89.6	81.7	84.8	98.3	73.1	83.1	94.5	101.4	98.6
09600	FLOORING	102.5	83.7	97.5	111.3	97.9	107.7	102.1	108.2	103.7	94.1	88.1	92.5	114.1	85.5	106.5	102.5	104.2	102.9
097,099	WALL FINISHES, PAINTS & COATINGS	107.8	87.5	95.7	108.7	93.4	99.6	107.8	116.8	113.2	100.5	99.5	99.9	108.7	90.8	98.0	107.8	116.8	113.2
09	FINISHES	99.6	84.4	91.7	99.0	88.5	93.6	98.9	107.0	103.1	94.9	86.2	90.4	100.0	79.0	89.2	99.4	103.6	101.6
10-14	TOTAL DIV. 10000 - 14000	100.0	72.1	94.3	100.0	88.2	97.6	100.0	106.0	101.2	100.0	93.0	98.6	100.0	86.2	97.2	100.0	102.8	100.6
15	MECHANICAL	100.1	85.1	94.0	99.8	86.7	94.5	100.1	107.8	103.2	99.8	91.0	96.2	100.8	86.7	95.0	100.1	89.3	95.7
16	ELECTRICAL	98.4	93.5	96.1	98.6	78.1	88.7	98.8	108.2	103.3	96.7	84.6	90.9	96.6	84.6	90.8	98.6	84.9	92.0
01-16	WEIGHTED AVERAGE	97.5	87.8	93.3	95.1	88.3	92.2	97.3	104.9	100.6	97.3	88.6	93.5	94.9	85.0	90.6	97.0	94.9	96.1

Section 3 — OHIO / OKLAHOMA

DIVISION		OHIO SPRINGFIELD			OHIO TOLEDO			OHIO YOUNGSTOWN			OKLAHOMA ENID			OKLAHOMA LAWTON			OKLAHOMA MUSKOGEE		
		MAT.	INST.	TOTAL	MAT.	INST.	TOTAL	MAT.	INST.	TOTAL	MAT.	INST.	TOTAL	MAT.	INST.	TOTAL	MAT.	INST.	TOTAL
01590	EQUIPMENT RENTAL	.0	96.3	96.3	.0	98.3	98.3	.0	96.6	96.6	.0	77.7	77.7	.0	78.8	78.8	.0	87.2	87.2
02	SITE CONSTRUCTION	72.0	105.8	96.2	84.3	101.5	96.6	95.8	105.9	103.1	110.9	89.2	95.3	105.7	90.8	95.0	95.5	84.8	87.9
03100	CONCRETE FORMS & ACCESSORIES	95.8	82.7	84.6	96.9	94.0	94.4	98.0	90.1	91.2	96.6	38.5	46.7	100.4	52.5	59.3	101.9	34.7	44.2
03200	CONCRETE REINFORCEMENT	93.3	80.5	87.0	92.6	94.3	93.4	96.2	94.9	95.6	96.7	78.7	87.9	96.9	78.7	88.0	96.7	36.8	67.4
03300	CAST-IN-PLACE CONCRETE	79.3	86.7	82.1	90.6	109.3	97.6	95.8	104.9	99.2	93.4	49.5	76.9	90.4	49.5	75.0	84.0	38.0	66.7
03	CONCRETE	87.0	83.4	85.3	93.2	98.9	95.9	96.2	95.3	95.8	91.0	50.2	71.7	87.8	56.4	73.0	83.5	37.3	61.7
04	MASONRY	75.5	87.3	82.5	103.1	101.3	102.0	89.0	97.1	93.8	100.5	58.9	75.6	95.1	58.9	73.4	110.4	52.6	75.8
05	METALS	93.4	77.1	88.5	97.4	88.8	94.8	91.5	81.7	88.6	91.8	67.0	84.3	95.3	67.1	86.7	91.8	57.5	81.4
06	WOOD & PLASTICS	97.4	81.9	89.2	101.8	91.3	96.2	91.3	87.5	89.3	100.2	35.4	65.9	103.5	54.2	77.4	105.8	35.0	68.4
07	THERMAL & MOISTURE PROTECTION	100.5	89.3	96.0	99.9	106.3	102.5	101.3	99.1	100.4	100.2	60.0	84.0	100.0	62.0	84.7	99.8	46.2	78.2
08	DOORS & WINDOWS	95.6	78.3	91.1	99.1	92.4	97.4	100.1	92.3	98.1	95.4	48.7	83.3	97.0	58.9	87.1	95.4	34.2	79.5
09200	PLASTER & GYPSUM BOARD	95.1	81.7	87.1	91.6	91.0	91.2	95.0	86.8	90.1	86.6	34.4	55.5	88.6	53.7	67.9	89.9	33.8	56.5
095,098	CEILINGS & ACOUSTICAL TREATMENT	98.3	81.7	88.3	89.6	91.0	90.4	94.5	86.8	89.9	86.2	34.4	55.0	95.5	53.7	70.4	95.5	33.8	58.4
09600	FLOORING	114.1	85.5	106.5	93.3	93.0	93.2	102.5	93.8	100.2	110.7	49.5	94.4	113.5	49.5	96.4	114.7	39.9	94.8
097,099	WALL FINISHES, PAINTS & COATINGS	108.7	90.8	98.0	100.5	103.2	102.1	107.8	99.2	102.6	98.3	63.4	77.4	98.3	63.4	77.4	98.3	33.9	59.8
09	FINISHES	100.0	83.9	91.7	94.6	94.1	94.4	99.5	91.1	95.2	96.1	41.2	67.8	99.0	52.2	74.8	98.8	35.8	66.3
10-14	TOTAL DIV. 10000 - 14000	100.0	87.2	97.4	100.0	95.9	99.2	100.0	93.8	98.7	100.0	70.0	93.8	100.0	72.2	94.3	100.0	69.2	93.7
15	MECHANICAL	100.8	86.6	95.0	99.8	102.0	100.7	100.1	92.5	97.0	100.1	62.4	84.7	100.1	62.4	84.8	100.1	27.8	70.7
16	ELECTRICAL	96.6	83.3	90.2	96.5	101.2	98.8	98.6	90.1	94.5	96.5	68.5	83.0	98.0	68.6	83.9	96.1	32.8	65.8
01-16	WEIGHTED AVERAGE	94.8	85.9	91.0	97.5	98.6	98.0	97.3	93.2	95.6	96.9	60.5	81.2	97.2	63.7	82.8	96.3	43.0	73.3

City Cost Indexes

DIVISION		OKLAHOMA						OREGON											
		OKLAHOMA CITY			TULSA			EUGENE			MEDFORD			PORTLAND			SALEM		
		MAT.	INST.	TOTAL	MAT.	INST.	TOTAL	MAT.	INST.	TOTAL	MAT.	INST.	TOTAL	MAT.	INST.	TOTAL	MAT.	INST.	TOTAL
01590	EQUIPMENT RENTAL	.0	79.1	79.1	.0	87.2	87.2	.0	99.8	99.8	.0	99.8	99.8	.0	99.8	99.8	.0	99.8	99.8
02	SITE CONSTRUCTION	106.0	91.4	95.5	102.4	86.2	90.8	110.1	104.5	106.1	120.1	104.5	108.9	112.9	104.5	106.9	109.7	104.5	106.0
03100	CONCRETE FORMS & ACCESSORIES	101.6	45.3	53.2	101.5	45.6	53.5	104.1	105.1	104.9	99.3	104.8	104.1	105.7	105.2	105.3	105.1	105.1	105.1
03200	CONCRETE REINFORCEMENT	96.9	78.7	88.0	96.9	78.6	88.0	106.8	97.6	102.3	104.2	97.6	100.9	107.7	97.8	102.8	107.8	97.8	102.9
03300	CAST-IN-PLACE CONCRETE	88.1	52.6	74.8	91.6	48.3	75.3	105.7	104.8	105.3	109.1	104.7	107.5	108.6	104.9	107.2	109.6	104.8	107.8
03	CONCRETE	86.7	54.3	71.4	88.5	53.9	72.1	107.3	103.2	105.4	113.6	103.1	108.6	109.1	103.3	106.4	109.6	103.3	106.6
04	MASONRY	96.0	60.9	75.0	94.5	61.0	74.4	115.6	102.1	107.5	112.7	102.1	106.4	116.7	104.5	109.4	121.0	104.5	111.1
05	METALS	97.1	67.0	88.0	94.6	80.6	90.4	92.2	96.0	93.4	91.8	95.8	93.0	93.3	96.4	94.2	92.6	96.3	93.7
06	WOOD & PLASTICS	105.6	43.5	72.7	104.3	44.8	72.8	94.9	105.3	100.4	88.5	105.3	97.4	96.3	105.3	101.0	95.7	105.3	100.8
07	THERMAL & MOISTURE PROTECTION	98.3	61.7	83.6	99.8	58.8	83.3	105.1	95.2	101.1	105.7	92.7	100.5	104.8	100.1	102.9	105.0	97.4	101.9
08	DOORS & WINDOWS	97.0	53.1	85.6	97.0	53.4	85.7	98.9	105.6	100.6	101.6	105.6	102.6	96.5	105.6	98.8	98.3	105.6	100.2
09200	PLASTER & GYPSUM BOARD	88.6	42.7	61.3	88.6	43.9	62.0	100.0	105.2	103.1	98.2	105.2	102.4	102.3	105.2	104.0	98.1	105.2	102.3
095,098	CEILINGS & ACOUSTICAL TREATMENT	95.5	42.7	63.7	95.5	43.9	64.5	107.9	105.2	106.3	117.2	105.2	110.0	106.0	105.2	105.6	107.9	105.2	106.3
09600	FLOORING	113.5	49.5	96.4	113.3	51.6	96.8	115.4	97.9	110.7	112.9	97.9	108.9	114.0	97.9	109.7	115.4	97.9	110.7
097,099	WALL FINISHES, PAINTS & COATINGS	98.3	63.4	77.4	98.3	50.1	69.5	113.1	72.5	88.8	113.1	63.6	83.4	113.6	72.5	89.1	113.1	72.5	88.8
09	FINISHES	99.1	46.4	71.9	98.4	46.1	71.4	105.5	99.4	102.9	112.8	99.4	105.9	110.6	100.4	105.3	110.5	100.4	105.3
10 - 14	TOTAL DIV. 10000 - 14000	100.0	71.7	94.2	100.0	72.2	94.3	100.0	102.0	100.4	100.0	102.0	100.4	100.0	102.0	100.4	100.0	102.0	100.4
15	MECHANICAL	100.1	63.5	85.2	100.1	57.5	82.8	100.2	105.8	102.5	100.2	105.8	102.5	100.1	105.9	102.5	100.2	105.9	102.5
16	ELECTRICAL	97.5	68.5	83.6	98.0	42.8	71.5	98.3	96.4	97.4	101.5	86.3	94.2	98.8	100.9	99.8	98.6	96.4	97.5
01 - 16	WEIGHTED AVERAGE	97.3	62.8	82.4	97.0	58.6	80.4	101.6	101.7	101.6	103.1	100.0	101.7	101.8	102.7	102.2	102.1	102.0	102.0

DIVISION		PENNSYLVANIA																	
		ALLENTOWN			ALTOONA			ERIE			HARRISBURG			PHILADELPHIA			PITTSBURGH		
		MAT.	INST.	TOTAL	MAT.	INST.	TOTAL	MAT.	INST.	TOTAL	MAT.	INST.	TOTAL	MAT.	INST.	TOTAL	MAT.	INST.	TOTAL
01590	EQUIPMENT RENTAL	.0	115.3	115.3	.0	115.3	115.3	.0	115.3	115.3	.0	114.5	114.5	.0	95.2	95.2	.0	114.3	114.3
02	SITE CONSTRUCTION	91.8	106.6	102.4	96.5	106.4	103.6	93.0	107.0	103.1	82.5	105.1	98.7	103.1	96.5	98.3	98.8	107.7	105.2
03100	CONCRETE FORMS & ACCESSORIES	100.5	114.9	112.9	83.7	83.0	83.1	99.9	92.5	93.5	93.5	85.4	86.5	101.3	133.5	129.0	101.5	98.2	98.6
03200	CONCRETE REINFORCEMENT	97.0	107.7	102.3	94.1	89.5	91.9	96.1	86.8	91.5	97.0	94.0	95.5	100.3	137.3	118.4	95.6	105.1	100.3
03300	CAST-IN-PLACE CONCRETE	87.4	103.4	93.4	97.4	83.0	92.0	95.8	81.7	90.5	96.5	90.5	94.2	108.8	131.4	117.3	96.9	95.0	96.2
03	CONCRETE	95.0	110.4	102.3	91.5	85.8	88.8	90.6	89.0	89.8	97.6	90.3	94.1	110.5	133.1	121.2	93.4	99.6	96.3
04	MASONRY	95.1	101.9	99.1	97.4	80.2	87.1	87.1	91.8	89.9	96.0	85.7	89.8	99.1	136.0	121.2	88.4	97.7	94.0
05	METALS	96.6	123.9	104.8	90.4	112.7	97.1	90.5	111.7	96.9	98.4	116.3	103.8	103.3	129.3	111.2	91.9	122.2	101.1
06	WOOD & PLASTICS	102.2	118.5	110.8	79.4	85.0	82.3	98.7	92.2	95.2	95.8	84.4	89.8	100.4	132.2	117.2	100.7	98.3	99.4
07	THERMAL & MOISTURE PROTECTION	101.5	117.0	107.7	100.4	91.2	96.7	100.3	96.1	98.6	105.6	104.7	105.2	102.4	137.1	116.4	102.5	100.3	101.6
08	DOORS & WINDOWS	93.9	113.7	99.0	88.3	89.8	88.7	88.5	89.9	88.9	93.9	90.9	93.1	96.8	139.8	108.0	94.2	106.3	97.3
09200	PLASTER & GYPSUM BOARD	106.4	118.7	113.7	100.3	84.3	90.8	106.4	91.7	97.7	106.4	83.8	92.9	99.4	132.9	119.4	95.3	98.0	96.9
095,098	CEILINGS & ACOUSTICAL TREATMENT	88.1	118.7	106.5	92.3	84.3	87.5	88.1	91.7	90.3	88.1	83.8	85.5	99.7	132.9	119.7	89.8	98.0	94.7
09600	FLOORING	87.6	93.0	89.0	81.4	54.9	74.3	89.7	81.7	87.6	87.8	88.0	87.9	85.8	140.0	100.3	93.1	104.1	96.0
097,099	WALL FINISHES, PAINTS & COATINGS	89.0	87.2	87.9	84.6	106.0	97.4	95.6	88.4	91.3	89.0	86.6	87.5	96.1	141.7	123.4	98.3	110.0	105.3
09	FINISHES	94.8	107.8	101.5	93.2	79.4	86.1	96.3	89.8	93.0	93.7	85.6	89.5	99.8	135.5	118.2	94.6	100.0	97.4
10 - 14	TOTAL DIV. 10000 - 14000	100.0	103.1	100.6	100.0	97.5	99.5	100.0	102.4	100.5	100.0	95.2	99.0	100.0	123.5	104.8	100.0	102.8	100.6
15	MECHANICAL	100.3	110.8	104.6	99.7	86.3	94.3	99.7	92.7	96.9	100.3	91.5	96.7	100.1	135.5	114.5	99.9	99.4	99.7
16	ELECTRICAL	101.1	95.3	98.3	90.6	101.2	95.7	92.4	86.8	89.7	101.1	83.1	92.5	97.2	138.9	117.2	96.8	101.3	98.9
01 - 16	WEIGHTED AVERAGE	97.7	108.3	102.2	94.3	91.5	93.1	94.3	94.1	94.2	98.0	92.6	95.7	101.2	131.6	114.3	96.0	102.7	98.9

DIVISION		PENNSYLVANIA									PUERTO RICO			RHODE ISLAND			SOUTH CAROLINA		
		READING			SCRANTON			YORK			SAN JUAN			PROVIDENCE			CHARLESTON		
		MAT.	INST.	TOTAL	MAT.	INST.	TOTAL	MAT.	INST.	TOTAL	MAT.	INST.	TOTAL	MAT.	INST.	TOTAL	MAT.	INST.	TOTAL
01590	EQUIPMENT RENTAL	.0	118.7	118.7	.0	115.3	115.3	.0	114.5	114.5	.0	88.4	88.4	.0	102.5	102.5	.0	99.7	99.7
02	SITE CONSTRUCTION	100.0	112.0	108.6	92.3	106.5	102.4	81.5	105.1	98.4	115.5	90.9	97.8	77.7	103.1	95.9	101.0	81.8	87.3
03100	CONCRETE FORMS & ACCESSORIES	100.8	85.8	87.9	100.7	86.1	88.2	82.1	85.4	84.9	92.4	18.8	29.2	103.9	115.5	113.9	98.2	38.7	47.1
03200	CONCRETE REINFORCEMENT	98.6	95.1	96.9	97.0	112.3	104.5	96.1	94.0	95.1	187.3	11.5	101.3	104.3	118.7	111.3	100.7	64.4	83.0
03300	CAST-IN-PLACE CONCRETE	79.4	95.8	85.6	91.3	89.4	90.6	86.0	90.5	87.7	104.5	29.3	76.2	90.3	109.2	97.4	93.2	47.3	75.9
03	CONCRETE	95.4	92.5	94.0	97.0	93.7	95.5	94.9	90.3	92.7	113.6	21.7	70.2	104.4	113.5	108.7	102.0	48.3	76.6
04	MASONRY	100.2	88.2	93.0	95.4	93.3	94.2	95.4	85.7	89.6	86.6	16.1	44.4	107.0	123.0	116.6	96.9	36.2	60.6
05	METALS	100.0	117.4	105.3	98.4	125.1	106.5	96.1	116.3	102.2	109.8	29.6	85.6	95.8	111.4	100.5	91.5	79.4	87.8
06	WOOD & PLASTICS	100.5	83.6	91.6	102.2	83.0	92.0	86.4	84.4	85.3	90.0	19.1	52.6	103.5	114.9	109.5	97.0	38.0	65.8
07	THERMAL & MOISTURE PROTECTION	101.8	108.6	104.6	101.4	99.7	100.7	100.3	104.7	102.0	140.7	22.3	93.0	97.5	112.5	103.5	104.1	43.4	79.7
08	DOORS & WINDOWS	94.4	91.1	93.6	93.9	95.8	94.4	90.7	90.9	90.7	152.6	15.9	117.2	100.2	115.8	104.2	96.3	42.0	82.2
09200	PLASTER & GYPSUM BOARD	102.3	83.0	90.9	108.5	82.3	92.9	103.3	83.8	91.7	114.7	16.8	56.5	100.3	114.5	108.7	117.1	36.2	68.9
095,098	CEILINGS & ACOUSTICAL TREATMENT	88.7	83.0	85.3	97.3	82.3	88.3	89.8	83.8	86.1	199.8	16.8	89.8	95.4	114.5	106.9	100.3	36.2	61.7
09600	FLOORING	84.0	98.0	87.8	87.6	91.4	88.6	83.3	88.0	84.5	201.9	15.8	152.3	98.8	127.8	106.6	105.3	40.8	88.1
097,099	WALL FINISHES, PAINTS & COATINGS	94.5	100.0	97.8	89.0	98.8	94.9	89.0	86.6	87.5	198.0	14.8	88.4	93.3	118.7	108.5	112.5	39.9	69.1
09	FINISHES	97.1	88.5	92.7	97.4	87.5	92.3	92.4	85.6	88.9	192.8	18.1	102.6	98.3	118.7	108.8	103.9	38.7	70.3
10 - 14	TOTAL DIV. 10000 - 14000	100.0	95.5	99.1	100.0	96.8	99.4	100.0	95.2	99.0	100.0	18.6	83.3	100.0	105.0	101.0	100.0	62.3	92.2
15	MECHANICAL	100.3	106.3	102.7	100.3	95.3	98.3	100.3	91.5	96.7	95.9	14.6	62.8	100.0	104.7	101.9	100.2	44.1	77.4
16	ELECTRICAL	99.3	94.0	96.8	101.2	91.8	96.7	95.2	83.1	89.4	124.0	14.7	71.6	97.4	101.0	99.1	97.9	60.7	80.0
01 - 16	WEIGHTED AVERAGE	98.8	98.9	98.8	98.4	97.1	97.9	95.9	92.6	94.5	119.0	24.6	78.3	99.1	110.3	104.0	98.7	52.3	78.7

SOUTH CAROLINA / SOUTH DAKOTA

DIVISION		COLUMBIA MAT.	INST.	TOTAL	FLORENCE MAT.	INST.	TOTAL	GREENVILLE MAT.	INST.	TOTAL	SPARTANBURG MAT.	INST.	TOTAL	ABERDEEN MAT.	INST.	TOTAL	PIERRE MAT.	INST.	TOTAL
01590	EQUIPMENT RENTAL	.0	99.7	99.7	.0	99.7	99.7	.0	99.7	99.7	.0	99.7	99.7	.0	98.7	98.7	.0	98.7	98.7
02	SITE CONSTRUCTION	100.0	81.8	87.0	112.6	81.8	90.5	107.0	81.4	88.7	106.8	81.4	88.6	88.9	94.1	92.6	87.0	94.1	92.1
03100	CONCRETE FORMS & ACCESSORIES	102.7	40.9	49.6	84.4	40.8	47.0	98.0	40.7	48.8	101.6	40.7	49.3	95.3	37.6	45.8	93.8	39.1	46.9
03200	CONCRETE REINFORCEMENT	100.7	64.4	83.0	100.4	64.3	82.8	100.3	53.1	77.2	100.3	53.1	77.2	102.3	45.5	74.5	101.9	58.3	80.6
03300	CAST-IN-PLACE CONCRETE	87.0	46.9	71.9	79.4	47.1	67.3	79.4	47.1	67.2	79.4	47.1	67.2	98.8	48.1	79.7	96.0	44.1	76.5
03	CONCRETE	99.1	49.1	75.5	102.0	49.1	77.0	100.4	47.0	75.2	100.7	47.0	75.3	97.9	44.1	72.5	95.7	45.9	72.2
04	MASONRY	91.5	37.2	59.0	79.7	36.2	53.7	77.6	36.2	52.8	79.7	36.2	53.7	104.8	58.8	77.3	101.6	57.9	75.4
05	METALS	91.5	79.1	87.7	90.5	79.0	87.0	90.5	74.7	85.7	90.5	74.7	85.7	100.7	67.6	90.7	100.7	73.3	92.4
06	WOOD & PLASTICS	103.0	41.4	70.4	80.5	41.4	59.8	96.9	41.4	67.5	101.4	41.4	69.6	100.6	35.9	66.4	98.7	37.9	66.6
07	THERMAL & MOISTURE PROTECTION	103.7	43.3	79.4	104.4	43.7	80.0	104.4	43.7	80.0	104.4	43.7	80.0	98.6	50.1	79.1	98.7	48.2	78.4
08	DOORS & WINDOWS	96.3	43.8	82.7	92.3	43.8	79.7	92.2	41.2	79.0	92.2	41.2	79.0	94.4	38.3	79.9	97.6	43.1	83.5
09200	PLASTER & GYPSUM BOARD	115.0	39.6	70.1	105.8	39.6	66.4	111.1	39.6	68.5	112.7	39.6	69.2	100.5	34.5	61.2	99.2	36.6	61.9
095,098	CEILINGS & ACOUSTICAL TREATMENT	100.3	39.6	63.8	95.2	39.6	61.8	94.4	39.6	61.4	94.4	39.6	61.4	111.9	34.5	65.4	110.2	36.6	65.9
09600	FLOORING	103.2	40.8	86.6	95.4	40.8	80.8	102.6	41.6	86.4	104.3	41.6	87.6	110.9	54.1	95.7	110.1	38.5	91.0
097,099	WALL FINISHES, PAINTS & COATINGS	111.4	39.9	68.6	112.5	39.9	69.1	112.5	39.9	69.1	112.5	39.9	69.1	97.6	33.3	59.2	97.6	38.0	62.0
09	FINISHES	102.9	40.6	70.8	99.0	40.6	68.9	101.5	40.8	70.1	102.2	40.8	70.5	106.5	39.8	72.1	105.5	38.3	70.8
10 - 14	TOTAL DIV. 10000 - 14000	100.0	62.6	92.3	100.0	62.5	92.3	100.0	62.6	92.3	100.0	62.6	92.3	100.0	51.0	89.9	100.0	60.7	91.9
15	MECHANICAL	100.1	37.2	74.5	100.2	37.2	74.6	100.2	37.2	74.6	100.2	37.2	74.6	100.2	36.8	74.4	100.2	35.3	73.8
16	ELECTRICAL	98.4	35.3	68.2	96.0	20.2	59.7	97.9	32.1	66.3	97.9	32.1	66.3	96.2	56.8	77.3	92.8	52.4	73.4
01 - 16	WEIGHTED AVERAGE	98.0	47.8	76.4	96.8	45.6	74.7	96.9	46.5	75.2	97.2	46.5	75.3	99.5	51.8	78.9	98.8	51.8	78.5

SOUTH DAKOTA / TENNESSEE

DIVISION		RAPID CITY MAT.	INST.	TOTAL	SIOUX FALLS MAT.	INST.	TOTAL	CHATTANOOGA MAT.	INST.	TOTAL	JACKSON MAT.	INST.	TOTAL	JOHNSON CITY MAT.	INST.	TOTAL	KNOXVILLE MAT.	INST.	TOTAL
01590	EQUIPMENT RENTAL	.0	98.7	98.7	.0	99.9	99.9	.0	105.3	105.3	.0	105.2	105.2	.0	98.0	98.0	.0	98.0	98.0
02	SITE CONSTRUCTION	87.2	93.9	92.0	87.5	96.0	93.6	99.5	97.5	98.1	98.1	95.9	96.5	108.8	86.2	92.6	86.5	86.2	86.3
03100	CONCRETE FORMS & ACCESSORIES	102.3	35.0	44.5	94.5	40.4	48.0	96.1	44.8	52.1	88.7	36.0	43.5	82.1	45.5	50.7	95.1	45.5	52.5
03200	CONCRETE REINFORCEMENT	95.6	58.3	77.3	95.6	58.4	77.4	90.1	62.3	76.5	89.9	40.7	65.9	90.7	59.5	75.4	90.1	59.5	75.1
03300	CAST-IN-PLACE CONCRETE	95.3	41.8	75.2	93.5	47.0	76.0	98.4	46.8	79.0	98.4	40.3	76.5	79.1	51.2	68.6	92.4	46.9	75.3
03	CONCRETE	94.9	43.3	70.5	92.4	47.4	71.2	92.2	50.7	72.6	94.0	40.4	68.7	95.7	51.9	75.1	89.3	50.5	71.0
04	MASONRY	101.7	51.4	71.6	98.7	57.9	74.3	98.7	41.3	64.4	114.4	32.1	64.0	111.2	41.1	69.2	75.8	41.1	55.0
05	METALS	102.4	73.6	93.7	102.6	74.1	94.0	93.8	84.6	91.1	90.2	72.1	84.7	91.2	83.3	88.8	94.4	83.6	91.1
06	WOOD & PLASTICS	104.8	33.9	67.3	99.4	38.5	67.2	99.6	45.2	70.9	85.7	36.0	59.4	74.0	46.3	59.3	89.2	46.3	66.5
07	THERMAL & MOISTURE PROTECTION	99.0	50.1	79.3	100.0	53.7	81.4	98.8	57.5	82.2	95.2	38.2	72.3	92.8	58.1	78.8	90.9	57.6	77.5
08	DOORS & WINDOWS	98.6	40.9	83.7	98.9	43.4	84.5	100.7	51.4	88.0	101.4	37.8	85.0	96.5	51.9	85.0	93.2	51.9	82.5
09200	PLASTER & GYPSUM BOARD	100.5	32.5	60.0	100.5	37.2	62.8	85.0	44.3	60.8	94.1	34.8	58.8	96.0	45.4	65.9	102.4	45.4	68.5
095,098	CEILINGS & ACOUSTICAL TREATMENT	116.1	32.5	65.8	116.1	37.2	68.7	98.5	44.3	65.9	95.6	34.8	59.0	93.2	45.4	64.4	95.7	45.4	65.4
09600	FLOORING	110.1	73.1	100.2	110.1	76.8	101.2	104.0	47.6	89.0	93.5	23.0	74.7	97.3	48.9	84.4	103.1	48.9	88.7
097,099	WALL FINISHES, PAINTS & COATINGS	97.6	38.0	62.0	97.6	38.0	62.0	110.2	44.4	70.8	97.2	30.0	57.0	107.7	52.0	74.4	107.7	52.0	74.4
09	FINISHES	107.2	42.0	73.6	107.2	46.6	75.9	98.4	44.8	70.8	96.4	33.0	63.7	99.4	46.4	72.1	95.2	46.4	70.0
10 - 14	TOTAL DIV. 10000 - 14000	100.0	58.6	91.5	100.0	60.9	92.0	100.0	49.3	89.6	100.0	53.5	90.4	100.0	56.5	91.1	100.0	56.6	91.1
15	MECHANICAL	100.2	33.0	72.8	100.2	34.4	73.4	100.0	43.0	76.8	100.0	52.9	80.8	99.7	56.6	82.1	99.7	56.6	82.1
16	ELECTRICAL	93.3	52.4	73.7	92.8	73.8	83.7	104.3	73.9	89.7	102.2	48.9	76.7	92.8	55.7	75.0	97.9	62.2	80.8
01 - 16	WEIGHTED AVERAGE	99.3	50.6	78.3	98.9	56.3	80.5	98.4	57.9	80.9	98.1	50.2	77.4	97.2	57.7	80.1	94.5	58.4	78.9

TENNESSEE / TEXAS

DIVISION		MEMPHIS MAT.	INST.	TOTAL	NASHVILLE MAT.	INST.	TOTAL	ABILENE MAT.	INST.	TOTAL	AMARILLO MAT.	INST.	TOTAL	AUSTIN MAT.	INST.	TOTAL	BEAUMONT MAT.	INST.	TOTAL
01590	EQUIPMENT RENTAL	.0	102.9	102.9	.0	106.4	106.4	.0	87.2	87.2	.0	87.2	87.2	.0	86.8	86.8	.0	87.8	87.8
02	SITE CONSTRUCTION	89.6	92.8	91.9	93.7	100.1	98.3	102.8	85.1	90.1	103.4	85.9	90.9	93.0	85.5	87.6	98.7	85.4	89.2
03100	CONCRETE FORMS & ACCESSORIES	96.7	65.0	69.5	95.1	63.8	68.2	98.2	41.4	49.5	100.4	52.4	59.2	95.4	56.2	61.7	104.9	51.3	58.9
03200	CONCRETE REINFORCEMENT	95.1	67.9	81.8	98.4	66.9	83.0	94.1	50.3	72.7	94.1	51.6	73.3	88.0	49.6	69.3	94.3	42.2	68.8
03300	CAST-IN-PLACE CONCRETE	86.2	69.2	79.8	92.7	68.2	83.5	95.0	41.5	74.9	99.8	47.4	80.1	89.7	48.4	74.1	93.0	53.3	78.0
03	CONCRETE	85.7	68.5	77.6	92.6	67.3	80.7	89.5	44.2	68.1	92.1	51.3	72.9	79.8	52.9	67.1	92.0	51.1	72.7
04	MASONRY	89.2	73.1	79.6	83.5	64.4	72.1	98.0	53.5	71.4	99.8	49.1	69.5	88.8	52.4	67.0	100.9	57.7	75.1
05	METALS	94.7	93.0	94.2	95.8	90.5	94.2	98.3	65.3	88.3	98.3	65.8	88.5	97.4	63.7	87.2	104.0	62.9	91.6
06	WOOD & PLASTICS	94.6	65.8	79.4	96.2	64.2	79.3	99.9	40.8	68.7	101.3	55.5	77.1	96.5	60.1	77.2	113.5	51.8	80.9
07	THERMAL & MOISTURE PROTECTION	95.2	71.6	85.7	94.1	65.2	82.5	99.9	48.8	79.4	102.4	47.7	80.4	93.0	53.8	77.2	105.7	57.5	86.3
08	DOORS & WINDOWS	101.3	67.4	92.5	96.9	65.6	88.8	93.0	43.5	80.2	93.0	49.6	81.7	96.8	57.3	86.5	97.7	47.5	84.7
09200	PLASTER & GYPSUM BOARD	90.3	65.3	75.4	101.1	63.7	78.9	88.6	39.8	59.6	88.6	54.9	68.5	96.6	59.6	74.6	98.5	51.2	70.3
095,098	CEILINGS & ACOUSTICAL TREATMENT	102.8	65.3	80.3	96.9	63.7	77.0	95.5	39.8	62.0	95.5	54.9	71.1	91.6	59.6	72.4	104.4	51.2	72.4
09600	FLOORING	96.1	43.4	82.1	108.3	74.5	99.3	113.5	68.9	101.6	113.3	61.5	99.5	97.1	47.2	83.8	112.9	71.7	101.9
097,099	WALL FINISHES, PAINTS & COATINGS	101.4	58.8	75.9	111.0	66.0	84.1	97.1	52.9	70.7	97.1	36.7	60.9	93.9	41.2	62.4	95.9	49.4	68.1
09	FINISHES	94.8	59.7	76.7	105.2	65.9	84.9	98.4	47.5	72.1	98.5	52.6	74.8	93.4	52.9	72.5	97.5	54.7	75.4
10 - 14	TOTAL DIV. 10000 - 14000	100.0	80.7	96.0	100.0	80.2	95.9	100.0	74.0	94.7	100.0	71.6	94.2	100.0	72.0	94.2	100.0	77.8	95.4
15	MECHANICAL	99.9	72.0	88.5	99.8	81.3	92.3	100.1	43.1	76.9	100.1	51.3	80.2	99.9	56.8	82.3	100.1	63.6	85.2
16	ELECTRICAL	99.1	73.4	86.8	102.4	66.5	85.2	98.2	46.7	73.5	98.7	59.9	80.1	101.3	66.8	84.8	95.6	69.4	83.1
01 - 16	WEIGHTED AVERAGE	96.0	74.0	86.5	97.7	74.8	87.8	97.5	52.0	77.9	98.1	57.2	80.5	95.5	60.3	80.3	99.1	62.3	83.2

City Cost Indexes

TEXAS

DIVISION		CORPUS CHRISTI			DALLAS			EL PASO			FORT WORTH			HOUSTON			LAREDO		
		MAT.	INST.	TOTAL	MAT.	INST.	TOTAL	MAT.	INST.	TOTAL	MAT.	INST.	TOTAL	MAT.	INST.	TOTAL	MAT.	INST.	TOTAL
01590	EQUIPMENT RENTAL	.0	95.4	95.4	.0	98.2	98.2	.0	87.2	87.2	.0	87.2	87.2	.0	98.0	98.0	.0	86.8	86.8
02	SITE CONSTRUCTION	131.7	80.7	95.2	127.4	86.6	98.1	102.1	84.7	89.6	102.8	85.5	90.4	126.7	84.2	96.2	93.7	85.3	87.7
03100	CONCRETE FORMS & ACCESSORIES	98.9	37.6	46.3	94.0	57.9	63.0	97.7	45.8	53.1	95.6	57.2	62.7	93.4	66.5	70.3	93.1	37.9	45.7
03200	CONCRETE REINFORCEMENT	87.3	48.0	68.1	98.6	54.8	77.2	94.1	46.7	71.0	94.1	54.7	74.9	95.9	62.4	79.5	88.0	48.1	68.5
03300	CAST-IN-PLACE CONCRETE	107.1	44.9	83.7	95.8	54.9	80.4	88.3	36.5	68.8	93.4	50.6	77.3	95.9	67.9	85.3	82.4	60.1	74.0
03	CONCRETE	90.2	44.0	68.4	87.9	57.8	73.7	86.1	43.6	66.0	88.6	55.1	72.8	91.5	67.7	80.2	79.2	48.4	64.7
04	MASONRY	81.3	50.8	63.1	98.1	60.1	75.4	96.1	49.2	68.1	93.4	60.1	73.5	96.3	63.1	76.5	88.9	51.5	66.5
05	METALS	96.9	76.4	90.7	93.6	81.0	89.8	98.1	61.0	86.9	95.1	67.4	86.7	108.7	88.3	102.5	98.4	62.5	87.5
06	WOOD & PLASTICS	109.0	37.0	70.9	96.5	58.6	76.5	99.9	49.0	73.0	102.2	58.5	79.1	98.2	66.8	81.6	92.2	36.9	62.9
07	THERMAL & MOISTURE PROTECTION	95.7	46.6	75.9	96.9	61.6	82.7	99.4	52.3	80.5	100.1	53.5	81.4	99.0	65.8	85.7	92.0	50.8	75.4
08	DOORS & WINDOWS	103.8	38.6	86.9	103.9	54.2	91.0	93.0	44.3	80.4	87.3	54.1	78.7	107.2	65.5	96.4	97.4	39.2	82.3
09200	PLASTER & GYPSUM BOARD	96.6	35.8	60.4	91.6	57.9	71.5	88.6	48.2	64.6	88.6	57.9	70.4	97.0	66.4	78.8	94.0	35.8	59.4
095,098	CEILINGS & ACOUSTICAL TREATMENT	91.6	35.8	58.0	96.8	57.9	73.4	95.5	48.2	67.1	95.5	57.9	72.9	105.0	66.4	81.8	91.6	35.8	58.0
09600	FLOORING	109.7	46.6	92.9	101.6	57.2	89.7	113.5	67.0	101.1	146.9	45.7	119.9	100.2	63.1	90.3	93.8	46.6	81.2
097,099	WALL FINISHES, PAINTS & COATINGS	106.7	44.8	69.7	102.8	52.4	72.6	97.1	35.2	60.1	98.3	52.3	70.8	103.8	61.7	78.6	93.9	53.8	69.9
09	FINISHES	99.7	39.3	68.5	98.9	57.1	77.3	98.4	49.1	73.0	108.5	54.6	80.7	101.7	65.2	82.9	92.1	40.2	65.4
10 - 14	TOTAL DIV. 10000 - 14000	100.0	74.8	94.8	100.0	79.1	95.7	100.0	69.8	93.8	100.0	78.8	95.6	100.0	83.1	96.5	100.0	68.1	93.4
15	MECHANICAL	99.9	44.0	77.1	99.9	66.4	86.2	100.1	32.8	72.7	100.1	57.0	82.5	100.1	71.1	88.3	99.8	39.6	75.3
16	ELECTRICAL	99.4	52.0	76.6	96.8	62.7	80.4	97.2	53.8	76.4	97.0	62.6	80.5	98.0	69.5	84.3	100.8	65.0	83.6
01 - 16	WEIGHTED AVERAGE	98.4	52.0	78.3	98.0	65.6	84.0	96.9	50.3	76.8	97.0	61.4	81.6	101.4	71.4	88.5	95.4	52.8	77.0

DIVISION		TEXAS															UTAH		
		LUBBOCK			ODESSA			SAN ANTONIO			WACO			WICHITA FALLS			LOGAN		
		MAT.	INST.	TOTAL	MAT.	INST.	TOTAL	MAT.	INST.	TOTAL	MAT.	INST.	TOTAL	MAT.	INST.	TOTAL	MAT.	INST.	TOTAL
01590	EQUIPMENT RENTAL	.0	97.2	97.2	.0	87.2	87.2	.0	89.8	89.8	.0	87.2	87.2	.0	87.2	87.2	.0	100.2	100.2
02	SITE CONSTRUCTION	133.8	83.2	97.5	103.0	85.4	90.4	93.3	90.1	91.0	101.5	85.4	90.0	102.3	85.1	90.0	89.8	99.3	96.6
03100	CONCRETE FORMS & ACCESSORIES	97.8	40.7	48.7	98.1	38.7	47.1	93.1	57.1	62.2	98.9	40.2	48.5	98.9	41.8	49.9	102.0	57.7	63.9
03200	CONCRETE REINFORCEMENT	95.4	50.1	73.2	94.1	50.0	72.5	94.0	50.2	72.6	94.1	49.4	72.3	94.1	50.6	72.8	107.9	74.1	91.4
03300	CAST-IN-PLACE CONCRETE	95.2	47.5	77.2	95.0	43.5	75.6	80.8	68.0	76.0	85.7	52.2	73.1	91.6	47.1	74.8	92.3	68.6	83.4
03	CONCRETE	88.8	46.7	68.9	89.5	43.6	67.9	79.3	60.2	70.3	84.9	47.1	67.0	87.8	46.3	68.2	114.5	65.0	91.1
04	MASONRY	97.4	47.6	67.6	98.0	47.9	68.0	88.8	62.5	73.1	94.7	57.2	72.3	95.2	57.1	72.4	113.5	57.9	80.2
05	METALS	101.9	79.0	95.0	97.7	64.2	87.6	99.0	66.6	89.2	98.2	64.5	88.0	98.2	66.0	88.5	103.5	73.5	94.4
06	WOOD & PLASTICS	100.7	40.9	69.1	99.9	38.8	67.6	92.2	55.6	72.9	105.6	36.1	68.8	105.6	40.8	71.3	87.1	56.1	70.7
07	THERMAL & MOISTURE PROTECTION	91.4	49.5	74.6	99.9	44.9	77.8	92.0	64.0	80.7	100.4	49.1	79.7	100.4	52.6	81.1	98.2	65.5	85.0
08	DOORS & WINDOWS	103.9	41.8	87.8	93.0	40.3	79.3	99.3	54.2	87.6	87.3	37.4	74.4	87.3	43.4	75.9	89.1	57.6	81.0
09200	PLASTER & GYPSUM BOARD	89.2	39.8	59.8	88.6	37.8	58.4	94.0	55.0	70.8	88.6	34.9	56.7	88.6	39.8	59.6	84.9	54.8	66.9
095,098	CEILINGS & ACOUSTICAL TREATMENT	98.0	39.8	63.0	95.5	37.8	60.8	91.6	55.0	69.6	95.5	34.9	59.1	95.5	39.8	62.0	101.7	54.8	73.5
09600	FLOORING	105.1	38.8	87.4	113.5	38.3	93.4	93.8	67.3	86.7	147.1	35.6	117.4	148.3	75.0	128.8	100.0	57.5	88.6
097,099	WALL FINISHES, PAINTS & COATINGS	108.7	32.9	63.4	97.1	32.9	58.7	93.9	53.8	69.9	98.3	33.9	59.8	102.8	53.8	73.2	99.6	46.4	67.7
09	FINISHES	100.8	38.6	68.7	98.4	37.4	66.9	92.1	57.8	74.4	108.5	37.3	71.8	109.1	48.8	78.0	97.3	55.7	75.8
10 - 14	TOTAL DIV. 10000 - 14000	100.0	73.8	94.6	100.0	69.1	93.6	100.0	74.6	94.8	100.0	76.1	95.1	100.0	69.9	93.8	100.0	67.3	93.3
15	MECHANICAL	99.6	46.1	77.8	100.1	36.0	74.0	99.8	69.6	87.5	100.1	53.0	80.9	100.1	49.4	79.5	99.9	67.9	86.9
16	ELECTRICAL	96.6	43.9	71.3	98.3	40.9	70.8	101.3	65.0	83.9	98.3	76.8	88.0	100.8	64.8	83.6	91.2	71.4	81.7
01 - 16	WEIGHTED AVERAGE	99.4	51.9	78.9	97.5	47.4	75.8	95.7	66.1	83.0	97.2	57.5	80.1	98.0	56.8	80.2	100.2	68.0	86.3

DIVISION		UTAH									VERMONT						VIRGINIA		
		OGDEN			PROVO			SALT LAKE CITY			BURLINGTON			RUTLAND			ALEXANDRIA		
		MAT.	INST.	TOTAL	MAT.	INST.	TOTAL	MAT.	INST.	TOTAL	MAT.	INST.	TOTAL	MAT.	INST.	TOTAL	MAT.	INST.	TOTAL
01590	EQUIPMENT RENTAL	.0	100.2	100.2	.0	98.9	98.9	.0	100.1	100.1	.0	100.7	100.7	.0	100.7	100.7	.0	101.6	101.6
02	SITE CONSTRUCTION	78.7	99.3	93.5	86.4	97.4	94.3	78.6	99.2	93.4	76.3	97.9	91.8	76.2	97.9	91.7	115.0	87.1	95.0
03100	CONCRETE FORMS & ACCESSORIES	102.0	57.7	63.9	103.5	57.7	64.2	102.0	57.7	64.0	94.1	48.8	55.2	100.3	48.9	56.1	93.6	74.8	77.4
03200	CONCRETE REINFORCEMENT	107.5	74.1	91.2	116.7	74.1	95.9	109.9	74.1	92.4	104.3	51.4	78.4	104.3	51.4	78.4	89.6	82.1	85.9
03300	CAST-IN-PLACE CONCRETE	93.7	68.6	84.3	92.4	68.6	83.4	102.4	68.6	89.7	96.8	60.8	83.2	92.2	60.8	80.4	104.2	78.9	94.7
03	CONCRETE	104.0	65.0	85.6	114.4	65.0	91.1	124.1	65.0	96.2	104.7	54.0	80.8	102.8	54.0	79.8	108.8	78.7	94.6
04	MASONRY	107.2	57.9	77.7	119.1	57.9	82.5	121.5	57.9	83.4	105.3	57.6	76.7	86.1	57.6	69.1	92.3	71.3	79.7
05	METALS	103.9	73.5	94.7	101.8	73.6	93.3	108.2	73.6	97.7	97.4	64.3	87.4	95.8	64.4	86.3	100.0	96.4	98.9
06	WOOD & PLASTICS	87.1	56.1	70.7	88.6	56.1	71.4	89.1	56.1	71.7	92.6	47.0	68.5	100.6	47.0	72.3	96.0	75.5	85.2
07	THERMAL & MOISTURE PROTECTION	97.0	65.5	84.3	99.9	65.5	86.0	99.9	65.5	86.1	98.3	53.2	80.1	97.9	54.3	80.4	103.2	80.0	93.9
08	DOORS & WINDOWS	89.1	57.6	81.0	93.4	57.6	84.1	91.0	57.6	82.4	105.6	43.6	89.6	105.6	43.6	89.6	96.3	76.8	91.2
09200	PLASTER & GYPSUM BOARD	84.9	54.8	66.9	85.2	54.8	67.1	87.0	54.8	67.8	101.4	44.8	67.7	101.4	44.8	67.7	115.0	74.7	91.0
095,098	CEILINGS & ACOUSTICAL TREATMENT	101.7	54.8	73.5	101.7	54.8	73.5	100.8	54.8	73.1	90.3	44.8	62.9	90.3	44.8	62.9	100.3	74.7	84.9
09600	FLOORING	97.9	57.5	87.1	100.8	57.5	89.3	100.3	57.5	88.9	98.0	66.4	89.6	98.0	66.4	89.6	105.3	94.1	102.3
097,099	WALL FINISHES, PAINTS & COATINGS	99.6	46.4	67.7	99.6	59.2	75.4	102.1	59.2	76.4	91.0	38.4	59.5	91.0	38.4	59.5	124.7	83.0	99.7
09	FINISHES	95.7	55.7	75.1	98.0	57.2	76.9	97.1	57.2	76.5	96.4	50.3	72.6	96.3	50.3	72.6	104.0	79.0	91.1
10 - 14	TOTAL DIV. 10000 - 14000	100.0	67.3	93.3	100.0	67.3	93.3	100.0	67.3	93.3	100.0	87.4	97.4	100.0	87.4	97.4	100.0	86.9	97.3
15	MECHANICAL	99.9	67.9	86.9	99.9	67.9	86.9	100.1	67.9	87.0	100.0	63.1	85.0	100.0	63.1	85.0	100.2	85.7	94.3
16	ELECTRICAL	91.7	71.4	81.9	91.9	74.5	83.5	95.0	74.5	85.2	98.6	67.1	83.5	97.9	67.1	83.1	98.3	92.9	95.7
01 - 16	WEIGHTED AVERAGE	98.3	68.0	85.2	100.7	68.5	86.8	102.9	68.6	88.1	99.7	62.8	83.8	98.2	62.8	82.9	101.0	83.9	93.6

DIVISION		VIRGINIA																	
		ARLINGTON			NEWPORT NEWS			NORFOLK			PORTSMOUTH			RICHMOND			ROANOKE		
		MAT.	INST.	TOTAL	MAT.	INST.	TOTAL	MAT.	INST.	TOTAL	MAT.	INST.	TOTAL	MAT.	INST.	TOTAL	MAT.	INST.	TOTAL
01590	EQUIPMENT RENTAL	.0	100.1	100.1	.0	105.3	105.3	.0	106.0	106.0	.0	105.2	105.2	.0	105.2	105.2	.0	100.1	100.1
02	SITE CONSTRUCTION	126.6	84.4	96.3	109.7	86.5	93.0	109.1	87.6	93.7	108.3	85.9	92.2	109.6	87.0	93.4	107.2	83.8	90.4
03100	CONCRETE FORMS & ACCESSORIES	92.6	75.0	77.5	98.3	77.0	80.0	102.5	77.0	80.6	86.8	60.3	64.0	99.3	70.4	74.4	98.1	74.4	77.7
03200	CONCRETE REINFORCEMENT	101.0	84.6	93.0	100.7	71.6	86.5	100.7	71.6	86.5	100.4	71.5	86.3	100.7	71.0	86.2	100.7	70.6	86.0
03300	CAST-IN-PLACE CONCRETE	101.3	82.7	94.3	102.0	66.1	88.5	105.1	66.1	90.4	101.0	65.7	87.7	102.8	61.6	87.3	114.7	67.3	96.9
03	CONCRETE	113.9	80.1	97.9	106.5	73.3	90.8	108.3	73.3	91.7	105.1	65.6	86.5	106.9	68.7	88.9	112.9	72.3	93.7
04	MASONRY	105.3	73.3	86.2	97.9	60.0	75.2	104.3	60.0	77.8	102.8	58.2	76.1	96.2	65.4	77.8	98.3	65.7	78.8
05	METALS	98.8	88.0	95.5	100.0	91.7	97.5	99.1	91.7	96.8	99.1	89.9	96.3	102.0	92.2	99.0	99.8	89.4	96.7
06	WOOD & PLASTICS	92.5	75.5	83.5	97.0	83.6	89.9	102.6	83.6	92.6	83.4	61.4	71.7	98.2	73.8	85.3	97.0	79.1	87.6
07	THERMAL & MOISTURE PROTECTION	104.9	75.2	93.0	104.1	64.0	88.0	103.9	64.0	87.8	104.1	60.5	86.6	103.7	65.7	88.4	104.1	63.8	87.9
08	DOORS & WINDOWS	94.2	72.7	88.6	96.3	73.6	90.4	96.3	73.6	90.4	96.3	61.6	87.3	96.3	69.2	89.2	96.3	70.3	89.5
09200	PLASTER & GYPSUM BOARD	111.9	74.7	89.7	115.0	82.2	95.5	115.0	82.2	95.5	108.2	59.4	79.1	115.0	72.1	89.5	115.0	78.4	93.2
095,098	CEILINGS & ACOUSTICAL TREATMENT	95.2	74.7	82.9	100.3	82.2	89.4	100.3	82.2	89.4	100.3	59.4	75.7	100.3	72.1	83.3	100.3	78.4	87.1
09600	FLOORING	103.5	66.2	93.6	105.3	60.9	93.5	105.1	60.9	93.3	96.5	76.9	91.3	105.1	85.7	100.0	105.3	47.1	89.8
097,099	WALL FINISHES, PAINTS & COATINGS	124.7	83.0	99.7	112.5	52.0	76.3	112.5	62.8	82.8	112.5	62.8	82.8	112.5	72.7	88.7	112.5	51.4	75.9
09	FINISHES	102.8	74.8	88.4	103.4	72.2	87.3	103.3	73.4	87.9	99.7	63.2	80.9	103.2	74.0	88.2	103.2	67.7	84.9
10-14	TOTAL DIV. 10000 - 14000	100.0	86.9	97.3	100.0	82.7	96.4	100.0	82.7	96.4	100.0	76.3	95.1	100.0	82.3	96.4	100.0	77.7	95.4
15	MECHANICAL	100.2	86.6	94.7	100.2	63.2	85.2	100.2	63.4	85.2	100.2	63.4	85.2	100.2	67.0	86.7	100.2	57.7	82.9
16	ELECTRICAL	96.0	91.2	93.7	97.8	68.1	83.6	98.0	60.5	80.0	96.3	60.5	79.1	99.7	71.4	86.1	97.8	47.4	73.6
01-16	WEIGHTED AVERAGE	101.8	82.4	93.4	100.8	71.9	88.3	101.2	71.1	88.2	100.1	67.2	85.9	101.2	73.1	89.1	101.4	67.0	86.5

DIVISION		WASHINGTON																	
		EVERETT			RICHLAND			SEATTLE			SPOKANE			TACOMA			VANCOUVER		
		MAT.	INST.	TOTAL	MAT.	INST.	TOTAL	MAT.	INST.	TOTAL	MAT.	INST.	TOTAL	MAT.	INST.	TOTAL	MAT.	INST.	TOTAL
01590	EQUIPMENT RENTAL	.0	104.0	104.0	.0	90.2	90.2	.0	103.7	103.7	.0	90.2	90.2	.0	104.0	104.0	.0	97.2	97.2
02	SITE CONSTRUCTION	94.4	114.6	108.9	106.3	89.3	94.1	98.7	112.6	108.6	105.8	89.3	94.0	97.7	114.7	109.9	109.9	100.1	102.9
03100	CONCRETE FORMS & ACCESSORIES	110.0	101.0	102.3	114.2	79.1	84.1	100.6	101.8	101.6	120.7	79.1	84.9	100.6	101.5	101.4	101.4	96.2	96.9
03200	CONCRETE REINFORCEMENT	108.1	93.5	101.0	103.9	92.5	98.4	107.0	93.6	100.4	104.6	92.5	98.7	107.0	93.5	100.4	107.7	93.2	100.6
03300	CAST-IN-PLACE CONCRETE	97.6	105.7	100.6	112.1	82.2	100.8	102.4	108.0	104.5	116.4	82.2	103.5	100.3	107.9	103.2	112.1	99.8	107.4
03	CONCRETE	96.4	100.6	98.4	105.1	82.6	94.5	99.2	101.9	100.5	107.9	82.6	95.9	98.2	101.6	99.8	108.0	96.6	102.6
04	MASONRY	132.7	101.6	114.1	116.8	80.8	95.3	126.5	102.3	112.0	117.5	80.8	95.6	126.3	102.3	112.0	126.9	98.8	110.1
05	METALS	106.8	87.7	101.0	92.1	83.6	89.5	108.5	90.1	102.9	94.6	83.6	91.3	108.5	88.1	102.3	105.7	88.0	100.6
06	WOOD & PLASTICS	101.8	100.7	101.2	95.9	78.6	86.8	91.6	100.7	96.4	105.7	78.6	91.4	90.8	100.7	96.0	85.1	96.4	91.1
07	THERMAL & MOISTURE PROTECTION	103.0	97.4	100.8	144.1	80.0	118.3	103.0	99.6	101.6	141.4	80.1	116.7	102.7	98.1	100.8	104.6	90.1	98.8
08	DOORS & WINDOWS	100.9	98.1	100.2	111.1	76.3	102.1	103.1	98.1	101.8	111.7	76.3	102.5	101.6	98.1	100.7	99.1	94.8	98.0
09200	PLASTER & GYPSUM BOARD	107.3	100.7	103.3	136.3	77.9	101.5	102.0	100.7	101.2	129.9	77.9	99.0	104.0	100.7	102.0	102.4	96.6	98.9
095,098	CEILINGS & ACOUSTICAL TREATMENT	107.9	100.7	103.6	109.6	77.9	90.5	111.1	100.7	104.8	109.6	77.9	90.5	111.3	100.7	104.9	106.3	96.6	100.4
09600	FLOORING	122.9	105.2	118.2	110.0	41.7	91.8	115.2	105.2	112.5	110.8	72.5	100.6	115.8	105.2	113.0	121.7	76.7	109.7
097,099	WALL FINISHES, PAINTS & COATINGS	114.1	87.7	98.3	110.8	72.3	87.8	114.1	89.4	99.3	111.7	72.3	88.1	114.1	89.4	99.3	121.3	68.6	89.8
09	FINISHES	114.1	100.6	107.1	124.9	70.6	96.8	112.0	100.9	106.3	124.2	76.9	99.8	112.5	100.9	106.5	110.8	89.5	99.8
10-14	TOTAL DIV. 10000 - 14000	100.0	98.4	99.7	100.0	79.5	95.8	100.0	102.9	100.6	100.0	79.4	95.8	100.0	102.8	100.6	100.0	77.4	95.3
15	MECHANICAL	100.1	99.6	99.9	100.6	98.1	99.6	100.1	110.9	104.5	100.4	84.0	93.8	100.2	96.7	98.8	100.3	102.4	101.1
16	ELECTRICAL	105.8	94.3	100.3	95.3	94.1	94.7	105.7	102.7	104.2	94.7	79.9	87.6	105.7	99.0	102.5	113.1	98.7	106.2
01-16	WEIGHTED AVERAGE	104.4	99.4	102.2	104.6	86.0	96.6	104.7	103.5	104.1	105.3	81.8	95.1	104.4	99.9	102.5	106.0	96.2	101.8

DIVISION		WASHINGTON			WEST VIRGINIA												WISCONSIN		
		YAKIMA			CHARLESTON			HUNTINGTON			PARKERSBURG			WHEELING			EAU CLAIRE		
		MAT.	INST.	TOTAL	MAT.	INST.	TOTAL	MAT.	INST.	TOTAL	MAT.	INST.	TOTAL	MAT.	INST.	TOTAL	MAT.	INST.	TOTAL
01590	EQUIPMENT RENTAL	.0	104.0	104.0	.0	100.1	100.1	.0	100.1	100.1	.0	100.1	100.1	.0	100.1	100.1	.0	101.1	101.1
02	SITE CONSTRUCTION	100.7	113.1	109.6	104.9	86.8	91.9	106.3	87.6	92.9	111.4	86.8	93.7	112.0	86.4	93.7	89.1	103.7	99.6
03100	CONCRETE FORMS & ACCESSORIES	101.0	94.8	95.7	107.0	89.1	91.7	99.3	111.4	109.7	89.2	88.9	88.9	91.1	88.7	89.0	97.8	96.4	96.6
03200	CONCRETE REINFORCEMENT	107.4	92.1	99.9	100.7	83.4	92.3	100.7	87.1	94.1	99.3	85.2	92.4	98.6	88.6	93.7	94.2	104.2	99.1
03300	CAST-IN-PLACE CONCRETE	107.2	80.2	97.1	99.8	105.5	102.0	107.7	109.5	108.4	100.6	99.4	100.2	100.6	99.9	100.4	96.7	95.6	96.3
03	CONCRETE	103.0	88.8	96.3	105.9	94.3	100.4	109.4	106.3	107.9	109.3	92.5	101.4	109.3	93.2	101.7	96.0	97.9	96.9
04	MASONRY	118.6	68.1	88.4	95.4	91.2	92.9	97.7	90.4	93.3	82.1	87.4	85.3	106.5	86.6	94.6	89.8	99.6	95.7
05	METALS	106.7	82.2	99.3	100.0	97.7	99.3	100.1	99.4	99.9	98.7	98.2	98.5	98.8	99.6	99.1	93.7	105.0	97.1
06	WOOD & PLASTICS	91.1	100.7	96.2	107.5	87.8	97.0	97.0	116.6	107.4	85.8	87.9	86.9	87.7	87.9	87.8	111.5	95.4	103.0
07	THERMAL & MOISTURE PROTECTION	102.8	78.4	93.0	103.9	89.9	98.2	104.3	93.5	100.0	104.1	85.8	96.7	104.5	86.1	97.1	97.8	88.6	94.1
08	DOORS & WINDOWS	101.0	84.2	96.7	97.5	81.8	93.4	96.3	98.2	96.8	96.9	81.3	92.8	97.8	87.3	95.1	100.1	96.1	99.1
09200	PLASTER & GYPSUM BOARD	103.5	100.7	101.8	114.6	87.3	98.3	113.6	116.8	115.5	108.6	87.4	96.0	109.0	87.4	96.1	102.3	95.6	98.3
095,098	CEILINGS & ACOUSTICAL TREATMENT	106.2	100.7	102.9	98.6	87.3	91.8	94.4	116.8	107.9	93.5	87.4	89.8	93.5	87.4	89.8	98.8	95.6	96.9
09600	FLOORING	117.0	62.6	102.5	105.1	105.1	105.1	105.1	100.5	103.9	99.7	99.2	99.6	100.8	99.7	100.5	92.7	110.0	97.3
097,099	WALL FINISHES, PAINTS & COATINGS	114.1	72.3	89.1	112.5	92.9	100.8	112.5	92.3	100.4	112.5	93.6	101.2	112.5	98.1	103.9	88.3	91.2	90.1
09	FINISHES	111.7	87.0	98.9	102.9	92.7	97.6	101.5	108.9	105.3	99.4	91.4	95.3	99.8	91.7	95.6	97.7	98.5	98.1
10-14	TOTAL DIV. 10000 - 14000	100.0	98.0	99.6	100.0	90.7	98.1	100.0	94.8	98.9	100.0	90.5	98.0	100.0	97.5	99.5	100.0	94.2	98.8
15	MECHANICAL	100.2	97.1	98.9	100.2	81.5	92.6	100.2	88.0	95.2	100.1	88.4	95.4	100.2	90.3	96.2	100.2	86.9	94.8
16	ELECTRICAL	108.3	94.1	101.5	97.8	83.6	91.0	97.8	91.8	94.9	98.2	92.8	95.6	95.7	92.3	94.1	99.9	87.8	94.1
01-16	WEIGHTED AVERAGE	104.5	90.4	98.4	100.6	88.4	95.3	100.8	96.0	98.8	99.7	90.3	95.6	100.9	91.2	96.7	97.6	95.1	96.5

City Cost Indexes

		WISCONSIN																
DIVISION		GREEN BAY			KENOSHA			LA CROSSE			MADISON			MILWAUKEE			RACINE		
		MAT.	INST.	TOTAL	MAT.	INST.	TOTAL	MAT.	INST.	TOTAL	MAT.	INST.	TOTAL	MAT.	INST.	TOTAL	MAT.	INST.	TOTAL
01590	EQUIPMENT RENTAL	.0	98.7	98.7	.0	99.2	99.2	.0	101.1	101.1	.0	101.5	101.5	.0	87.6	87.6	.0	101.5	101.5
02	SITE CONSTRUCTION	92.3	99.6	97.5	95.7	104.1	101.8	82.7	103.7	97.7	89.0	107.6	102.3	90.7	95.8	94.4	89.9	108.1	103.0
03100	CONCRETE FORMS & ACCESSORIES	107.7	95.6	97.3	107.4	105.3	105.6	83.4	95.8	94.0	99.3	99.6	99.6	101.6	117.3	115.1	99.0	105.3	104.4
03200	CONCRETE REINFORCEMENT	92.4	89.2	90.9	96.3	97.0	96.7	93.9	88.2	91.1	96.5	88.4	92.5	96.5	97.3	96.9	96.5	97.0	96.7
03300	CAST-IN-PLACE CONCRETE	100.0	99.9	100.0	106.7	102.0	104.9	86.9	96.8	90.6	93.2	100.2	95.8	95.7	112.1	101.9	96.1	101.7	98.2
03	CONCRETE	97.1	96.2	96.7	100.7	102.6	101.6	87.6	95.0	91.1	93.4	97.8	95.4	95.0	110.8	102.5	94.8	102.5	98.4
04	MASONRY	120.3	98.7	107.4	98.4	109.2	104.9	89.0	99.6	95.4	101.7	102.3	102.1	101.4	118.1	111.4	101.4	109.2	106.1
05	METALS	95.7	98.1	96.4	98.0	101.1	99.0	93.6	97.4	94.7	100.0	95.9	98.8	100.7	92.9	98.3	99.0	101.1	99.7
06	WOOD & PLASTICS	117.4	95.4	105.7	114.3	105.0	109.4	94.2	95.4	94.8	108.6	99.5	103.8	111.6	117.5	114.7	109.2	105.0	107.0
07	THERMAL & MOISTURE PROTECTION	99.7	87.0	94.6	99.2	100.7	99.8	97.2	88.0	93.5	95.4	95.5	95.4	97.5	112.4	103.5	99.0	100.3	99.5
08	DOORS & WINDOWS	98.4	94.4	97.4	97.8	104.1	99.4	100.1	83.6	95.8	102.6	98.5	101.5	104.8	110.9	106.4	102.6	104.1	103.0
09200	PLASTER & GYPSUM BOARD	93.8	95.6	94.8	89.7	105.5	99.1	96.8	95.6	96.1	88.3	99.9	95.2	99.6	118.2	110.7	98.8	105.5	102.8
095,098	CEILINGS & ACOUSTICAL TREATMENT	92.2	95.6	94.2	80.2	105.5	95.4	97.2	95.6	96.2	87.6	99.9	95.0	89.1	118.2	106.6	85.7	105.5	97.6
09600	FLOORING	111.0	110.0	110.7	114.1	115.7	114.5	85.2	110.0	91.8	92.0	108.0	96.3	100.0	120.2	105.4	97.3	115.7	102.2
097,099	WALL FINISHES, PAINTS & COATINGS	97.6	77.3	85.5	104.1	104.5	104.3	88.3	69.3	77.0	94.3	97.0	95.9	96.5	111.8	105.6	94.3	104.5	100.4
09	FINISHES	101.2	96.7	98.9	98.0	107.7	103.0	93.8	96.0	95.0	92.3	101.5	97.1	97.1	118.1	108.0	95.2	107.7	101.7
10 - 14	TOTAL DIV. 10000 - 14000	100.0	93.7	98.7	100.0	100.1	100.0	100.0	94.2	98.8	100.0	94.3	98.8	100.0	104.2	100.9	100.0	100.1	100.0
15	MECHANICAL	100.5	86.0	94.6	100.2	93.8	97.6	100.2	86.5	94.6	99.8	89.0	95.4	100.0	101.8	100.7	100.0	93.8	97.5
16	ELECTRICAL	95.6	87.0	91.5	95.1	95.5	95.3	100.2	87.8	94.2	96.1	91.2	93.8	96.1	103.8	99.8	94.6	97.8	96.1
01 - 16	WEIGHTED AVERAGE	99.6	93.1	96.8	98.8	101.1	99.8	95.9	93.1	94.7	98.0	96.5	97.4	99.1	106.7	102.4	98.3	101.7	99.8

		WYOMING									CANADA								
DIVISION		CASPER			CHEYENNE			ROCK SPRINGS			CALGARY, ALBERTA			EDMONTON, ALBERTA			HALIFAX, NOVA SCOTIA		
		MAT.	INST.	TOTAL	MAT.	INST.	TOTAL	MAT.	INST.	TOTAL	MAT.	INST.	TOTAL	MAT.	INST.	TOTAL	MAT.	INST.	TOTAL
01590	EQUIPMENT RENTAL	.0	101.1	101.1	.0	101.1	101.1	.0	101.1	101.1	.0	107.0	107.0	.0	107.0	107.0	.0	98.6	98.6
02	SITE CONSTRUCTION	91.9	100.1	97.8	87.5	100.1	96.5	82.8	98.9	94.3	117.3	105.7	109.0	126.1	105.7	111.5	100.1	96.6	97.6
03100	CONCRETE FORMS & ACCESSORIES	100.1	48.0	55.4	103.6	65.8	71.1	99.2	42.9	50.9	126.8	94.8	99.3	124.7	94.8	99.0	92.7	78.6	80.6
03200	CONCRETE REINFORCEMENT	114.9	49.4	82.9	108.4	50.3	80.0	117.2	50.1	84.4	161.5	66.1	114.9	161.5	66.1	114.9	155.0	62.5	109.7
03300	CAST-IN-PLACE CONCRETE	97.9	78.9	90.8	98.0	79.1	90.9	99.0	58.2	83.7	184.0	104.6	154.1	198.7	104.6	163.3	171.3	76.1	135.5
03	CONCRETE	107.2	59.6	84.7	107.0	67.7	88.4	108.1	50.3	80.8	159.4	92.9	128.0	166.7	92.9	131.8	150.9	75.1	115.1
04	MASONRY	106.9	42.2	68.2	104.3	58.3	76.8	163.7	52.2	97.0	179.0	88.4	124.8	175.5	88.4	123.4	173.4	83.8	119.8
05	METALS	100.6	62.6	89.1	101.8	64.3	90.5	98.0	61.8	87.1	134.9	86.5	120.3	135.7	86.5	120.8	120.4	79.4	108.0
06	WOOD & PLASTICS	97.4	45.3	69.9	100.9	68.7	83.8	97.1	41.6	67.8	116.0	94.2	104.5	112.5	94.2	102.8	84.3	78.3	81.1
07	THERMAL & MOISTURE PROTECTION	98.5	55.1	81.1	97.9	61.8	83.4	98.8	52.9	80.3	124.8	90.7	111.1	123.4	90.7	110.3	104.9	77.8	94.0
08	DOORS & WINDOWS	93.6	44.9	81.0	95.7	57.5	85.8	100.1	43.0	85.3	92.3	84.2	90.2	92.3	84.2	90.2	82.9	72.5	80.2
09200	PLASTER & GYPSUM BOARD	96.2	43.6	64.9	89.8	67.6	76.6	91.6	39.9	60.8	140.8	93.6	112.7	146.9	93.6	115.1	165.1	77.7	113.1
095,098	CEILINGS & ACOUSTICAL TREATMENT	101.8	43.6	66.8	103.7	67.6	82.0	101.8	39.9	64.6	108.7	93.6	99.6	123.9	93.6	105.6	107.0	77.7	89.4
09600	FLOORING	99.1	39.7	83.3	104.1	63.8	93.4	102.6	56.1	90.2	134.9	95.4	124.3	134.9	95.4	124.3	109.5	69.3	98.8
097,099	WALL FINISHES, PAINTS & COATINGS	99.3	53.2	71.7	100.9	53.2	72.4	98.7	34.1	60.0	109.2	100.8	104.2	109.3	99.2	103.2	109.2	80.0	91.7
09	FINISHES	100.1	45.9	72.2	101.3	64.6	82.4	98.3	43.2	69.9	121.2	95.8	108.1	126.7	95.6	110.7	116.4	77.2	96.2
10 - 14	TOTAL DIV. 10000 - 14000	100.0	78.2	95.5	100.0	81.1	96.1	100.0	70.4	93.9	140.0	111.1	134.1	140.0	111.1	134.1	140.0	77.7	127.2
15	MECHANICAL	99.9	65.9	86.1	100.0	60.3	83.8	99.9	56.2	82.1	101.1	92.7	97.7	101.1	92.7	97.7	100.7	80.6	92.5
16	ELECTRICAL	94.4	63.3	79.5	94.4	72.2	83.7	90.1	62.9	77.1	119.8	90.5	105.8	120.0	90.5	105.9	130.5	79.6	106.1
01 - 16	WEIGHTED AVERAGE	99.8	61.4	83.2	100.0	67.8	86.1	102.1	58.1	83.1	124.8	93.1	111.1	126.3	93.1	111.9	119.8	80.4	102.8

		CANADA																	
DIVISION		HAMILTON, ONTARIO			KITCHENER, ONTARIO			LAVAL, QUEBEC			LONDON, ONTARIO			MONTREAL, QUEBEC			OSHAWA, ONTARIO		
		MAT.	INST.	TOTAL	MAT.	INST.	TOTAL	MAT.	INST.	TOTAL	MAT.	INST.	TOTAL	MAT.	INST.	TOTAL	MAT.	INST.	TOTAL
01590	EQUIPMENT RENTAL	.0	103.1	103.1	.0	103.0	103.0	.0	99.2	99.2	.0	103.2	103.2	.0	101.0	101.0	.0	103.0	103.0
02	SITE CONSTRUCTION	115.6	105.0	108.0	102.1	104.6	103.9	95.0	98.6	97.6	116.3	104.9	108.2	96.7	98.2	97.7	114.9	104.4	107.4
03100	CONCRETE FORMS & ACCESSORIES	121.7	99.5	102.6	116.3	91.4	94.9	129.7	92.9	98.1	127.5	94.0	98.7	130.0	99.1	103.5	122.9	94.6	98.6
03200	CONCRETE REINFORCEMENT	175.4	87.0	132.2	110.4	86.9	98.9	157.1	85.7	122.2	135.3	85.5	111.0	159.7	101.1	131.0	174.8	91.8	134.2
03300	CAST-IN-PLACE CONCRETE	154.0	99.7	133.6	140.8	84.1	119.5	134.5	100.5	121.7	150.8	98.4	131.1	132.0	105.5	122.0	162.9	90.8	135.8
03	CONCRETE	146.1	97.0	122.9	122.9	88.1	106.5	133.8	94.1	115.1	138.4	93.9	117.4	133.0	101.4	118.1	150.6	92.7	123.3
04	MASONRY	166.7	100.2	126.9	162.1	96.3	122.7	160.4	91.8	119.4	166.3	98.7	125.9	165.0	98.4	125.2	165.1	100.7	126.6
05	METALS	138.4	90.8	124.0	116.1	90.4	108.3	105.6	88.9	100.5	124.0	90.4	113.8	124.5	94.5	115.4	121.7	92.8	102.8
06	WOOD & PLASTICS	117.2	99.8	108.0	110.7	90.7	100.1	131.2	92.5	110.8	117.2	92.9	104.4	131.2	99.3	114.3	118.2	92.8	104.8
07	THERMAL & MOISTURE PROTECTION	110.1	95.3	104.1	108.8	91.8	102.0	104.5	94.2	100.4	112.7	93.2	104.9	105.1	101.5	103.6	109.8	92.3	102.7
08	DOORS & WINDOWS	92.3	95.6	93.1	83.5	88.9	84.9	92.3	80.4	89.2	93.2	90.7	92.6	92.3	86.8	90.8	91.2	92.6	91.6
09200	PLASTER & GYPSUM BOARD	184.9	99.9	134.3	151.8	90.5	115.3	142.1	92.3	112.4	185.2	92.8	130.2	144.7	98.9	117.4	153.8	92.7	117.4
095,098	CEILINGS & ACOUSTICAL TREATMENT	112.1	99.9	104.7	107.0	90.5	97.1	95.2	92.3	93.5	113.7	92.8	101.1	107.0	98.9	102.1	100.3	92.7	95.7
09600	FLOORING	132.5	101.8	124.3	128.1	101.8	121.1	132.5	102.5	124.5	135.0	101.8	126.2	132.5	102.5	124.5	132.5	104.4	125.0
097,099	WALL FINISHES, PAINTS & COATINGS	109.2	104.9	106.7	109.2	94.9	100.7	109.2	95.2	100.8	109.2	102.7	105.3	109.2	109.5	109.4	109.2	109.6	109.5
09	FINISHES	128.5	100.9	114.3	119.6	93.8	106.3	116.4	95.2	105.5	129.9	96.5	112.7	120.0	101.5	110.4	120.3	98.0	108.8
10 - 14	TOTAL DIV. 10000 - 14000	140.0	115.0	134.9	140.0	112.9	134.4	140.0	99.2	131.6	140.0	114.0	134.7	140.0	101.5	132.1	140.0	114.4	134.7
15	MECHANICAL	101.1	96.2	99.1	100.7	93.1	97.6	100.7	100.8	100.7	101.2	93.7	98.1	101.2	102.2	101.6	100.7	112.2	105.4
16	ELECTRICAL	132.2	100.0	116.8	126.0	97.2	112.2	125.8	79.5	103.6	132.8	97.2	115.7	133.2	89.8	112.4	127.1	99.0	113.6
01 - 16	WEIGHTED AVERAGE	124.8	98.7	113.5	115.5	94.5	106.4	115.4	92.8	105.6	122.0	96.2	110.9	119.8	98.1	110.5	118.8	100.7	111.0

457

	CANADA																
DIVISION	OTTAWA, ONTARIO			QUEBEC, QUEBEC			REGINA, SASKATCHEWAN			SASKATOON, SASKATCHEWAN			ST CATHARINES, ONTARIO			ST JOHNS, NEWFOUNDLAND		
	MAT.	INST.	TOTAL	MAT.	INST.	TOTAL	MAT.	INST.	TOTAL	MAT.	INST.	TOTAL	MAT.	INST.	TOTAL	MAT.	INST.	TOTAL
01590 EQUIPMENT RENTAL	.0	103.0	103.0	.0	101.5	101.5	.0	98.6	98.6	.0	98.6	98.6	.0	100.4	100.4	.0	98.6	98.6
02 SITE CONSTRUCTION	113.8	104.9	107.4	96.5	98.3	97.8	113.5	95.8	100.8	108.1	96.0	99.4	102.7	100.7	101.3	116.5	96.2	102.0
03100 CONCRETE FORMS & ACCESSORIES	120.4	96.7	100.1	129.9	99.4	103.7	103.5	63.9	69.5	103.5	63.8	69.4	114.0	96.6	99.1	101.1	68.9	73.4
03200 CONCRETE REINFORCEMENT	173.8	92.2	133.9	148.4	101.0	125.2	126.2	67.3	97.4	119.1	67.3	93.8	111.3	86.9	99.4	165.0	62.0	114.6
03300 CAST-IN-PLACE CONCRETE	155.1	103.0	135.5	145.5	105.9	130.6	156.3	72.8	124.9	142.1	72.7	116.0	134.5	99.5	121.3	172.4	79.8	137.6
03 CONCRETE	146.3	97.9	123.4	138.0	101.7	120.8	132.2	68.1	102.0	123.9	68.0	97.5	119.7	95.7	108.3	164.5	71.9	120.7
04 MASONRY	159.3	101.4	124.7	167.8	98.4	126.3	166.3	68.3	107.7	166.0	68.3	107.6	161.5	100.2	124.8	163.5	70.0	107.5
05 METALS	116.1	92.6	109.0	120.3	94.8	112.6	105.5	74.9	96.2	105.5	74.8	96.2	106.3	90.4	101.5	108.2	75.6	98.4
06 WOOD & PLASTICS	116.7	96.0	105.8	131.9	99.3	114.7	95.1	62.0	77.6	93.7	62.0	76.9	108.0	96.5	102.0	94.5	69.0	81.0
07 THERMAL & MOISTURE PROTECTION	110.2	96.1	104.6	104.7	101.7	103.5	105.3	67.5	90.1	104.1	66.4	88.9	108.8	94.6	103.1	108.3	67.2	91.8
08 DOORS & WINDOWS	92.3	93.5	92.6	92.3	94.8	92.9	87.1	59.7	80.0	86.1	59.7	79.3	83.0	92.4	85.5	98.6	64.6	89.8
09200 PLASTER & GYPSUM BOARD	229.4	96.0	150.0	190.1	98.9	135.8	165.2	61.0	103.2	144.1	61.0	94.7	136.4	96.5	112.6	170.4	68.2	109.6
095,098 CEILINGS & ACOUSTICAL TREATMENT	106.2	96.0	100.0	96.9	98.9	98.1	120.5	61.0	84.7	120.5	61.0	84.7	100.3	96.5	98.0	106.2	68.2	83.3
09600 FLOORING	132.5	100.3	123.9	132.5	102.5	124.5	119.8	65.1	105.2	119.8	65.1	105.2	126.2	101.8	119.7	114.4	58.9	99.6
097,099 WALL FINISHES, PAINTS & COATINGS	109.2	96.7	101.8	109.7	109.5	109.6	109.2	69.4	85.4	109.2	59.2	79.3	109.2	104.9	106.7	109.2	69.4	85.4
09 FINISHES	134.3	97.5	115.3	124.9	101.6	112.9	123.9	64.2	93.1	120.1	63.1	90.7	114.9	99.0	106.7	119.6	67.0	92.5
10-14 TOTAL DIV. 10000 - 14000	140.0	111.9	134.2	140.0	101.7	132.1	140.0	74.9	126.6	140.0	74.9	126.6	140.0	86.5	129.0	140.0	76.1	126.9
15 MECHANICAL	101.2	94.7	98.5	101.1	102.2	101.6	100.9	84.3	94.2	100.8	84.3	94.1	100.7	94.8	98.3	101.1	69.7	88.3
16 ELECTRICAL	125.8	98.2	112.6	129.9	89.8	110.6	129.3	68.7	100.2	129.6	68.7	100.4	128.5	98.3	114.0	125.1	70.3	98.8
01-16 WEIGHTED AVERAGE	120.7	97.8	110.8	119.9	98.5	110.7	116.3	73.9	98.0	114.8	73.7	97.0	113.4	96.3	106.0	120.7	72.5	99.9

	CANADA																	
DIVISION	THUNDER BAY, ONTARIO			TORONTO, ONTARIO			VANCOUVER, B C			WINDSOR, ONTARIO			WINNIPEG, MANITOBA					
	MAT.	INST.	TOTAL	MAT.	INST.	TOTAL	MAT.	INST.	TOTAL	MAT.	INST.	TOTAL	MAT.	INST.	TOTAL	MAT.	INST.	TOTAL
01590 EQUIPMENT RENTAL	.0	100.4	100.4	.0	103.2	103.2	.0	111.3	111.3	.0	100.4	100.4	.0	105.3	105.3	.0	.0	.0
02 SITE CONSTRUCTION	108.4	100.7	102.9	116.2	105.6	108.6	119.8	107.4	110.9	98.0	100.7	100.0	116.1	101.1	105.3	.0	.0	.0
03100 CONCRETE FORMS & ACCESSORIES	122.8	97.1	100.7	123.2	105.3	107.8	122.7	84.0	89.5	122.8	94.0	98.0	122.5	70.4	77.8	.0	.0	.0
03200 CONCRETE REINFORCEMENT	99.5	86.2	93.0	170.9	87.7	130.2	173.4	77.8	126.6	109.1	85.5	97.5	161.5	60.2	112.0	.0	.0	.0
03300 CAST-IN-PLACE CONCRETE	148.1	98.4	129.4	149.4	107.9	133.8	151.9	96.6	131.1	137.7	99.5	123.4	174.4	74.8	136.9	.0	.0	.0
03 CONCRETE	128.3	95.4	112.7	143.1	102.6	124.0	150.2	87.5	120.6	121.5	94.3	108.7	154.0	70.6	114.6	.0	.0	.0
04 MASONRY	162.3	100.2	125.1	175.1	105.2	133.3	167.8	87.2	119.6	161.7	98.6	123.9	178.2	66.8	111.5	.0	.0	.0
05 METALS	106.2	89.4	101.1	128.6	92.5	117.7	143.7	88.5	127.0	106.2	90.3	101.4	138.4	77.3	119.9	.0	.0	.0
06 WOOD & PLASTICS	118.2	96.8	106.9	118.2	105.4	111.4	113.5	80.9	96.3	118.2	93.3	105.0	115.9	71.0	92.2	.0	.0	.0
07 THERMAL & MOISTURE PROTECTION	109.0	94.9	103.4	110.7	100.1	106.5	133.3	87.5	114.9	108.9	93.1	102.5	105.1	71.6	91.6	.0	.0	.0
08 DOORS & WINDOWS	82.0	91.8	84.6	91.2	100.6	93.7	94.0	81.6	90.8	81.7	90.5	84.0	92.3	65.3	85.3	.0	.0	.0
09200 PLASTER & GYPSUM BOARD	167.5	96.8	125.4	166.6	105.6	130.3	144.0	79.9	105.9	156.9	93.1	119.0	148.6	69.7	101.6	.0	.0	.0
095,098 CEILINGS & ACOUSTICAL TREATMENT	95.2	96.8	96.2	119.6	105.6	111.2	109.5	79.9	91.7	95.2	93.1	94.0	107.0	69.7	84.6	.0	.0	.0
09600 FLOORING	132.5	58.6	112.8	132.5	107.9	125.9	134.9	98.8	125.2	132.5	102.5	124.5	132.5	72.3	116.4	.0	.0	.0
097,099 WALL FINISHES, PAINTS & COATINGS	109.2	98.7	102.9	111.0	109.6	110.2	109.2	95.4	100.9	109.2	98.2	102.6	109.3	51.7	74.8	.0	.0	.0
09 FINISHES	121.0	91.3	105.7	127.7	106.8	116.9	122.6	87.5	104.5	118.9	96.3	107.3	121.5	69.0	94.4	.0	.0	.0
10-14 TOTAL DIV. 10000 - 14000	140.0	87.0	129.1	140.0	117.0	135.3	140.0	108.7	133.6	140.0	85.9	128.9	140.0	77.1	127.1	.0	.0	.0
15 MECHANICAL	100.7	95.0	98.4	101.1	99.9	100.6	101.1	79.6	92.4	100.7	94.5	98.2	101.1	71.3	88.9	.0	.0	.0
16 ELECTRICAL	126.0	96.5	111.8	131.4	100.8	116.7	134.2	80.1	108.3	132.7	98.2	116.2	134.5	71.6	104.3	.0	.0	.0
01-16 WEIGHTED AVERAGE	114.8	95.0	106.3	123.1	102.2	114.1	126.9	86.9	109.6	114.3	95.4	106.1	125.8	73.5	103.2	.0	.0	.0

Location Factors

Costs shown in *RSMeans cost data publications* are based on National Averages for materials and installation. To adjust these costs to a specific location, simply multiply the base cost by the factor and divide by 100 for that city. The data is arranged alphabetically by state and postal zip code numbers. For a city not listed, use the factor for a nearby city with similar economic characteristics.

STATE/ZIP	CITY	MAT.	INST.	TOTAL
ALABAMA				
350-352	Birmingham	96.2	73.1	86.2
354	Tuscaloosa	95.8	54.0	77.8
355	Jasper	96.0	50.6	76.4
356	Decatur	95.7	54.1	77.7
357-358	Huntsville	95.8	70.4	84.8
359	Gadsden	95.7	57.3	79.1
360-361	Montgomery	96.2	55.6	78.7
362	Anniston	95.3	45.1	73.6
363	Dothan	96.0	47.4	75.0
364	Evergreen	95.3	48.6	75.2
365-366	Mobile	96.8	59.3	80.6
367	Selma	95.6	50.1	75.9
368	Phenix City	96.3	53.4	77.8
369	Butler	95.7	47.4	74.9
ALASKA				
995-996	Anchorage	131.9	114.1	124.2
997	Fairbanks	129.2	116.9	123.9
998	Juneau	130.4	112.9	122.8
999	Ketchikan	141.6	112.4	129.0
ARIZONA				
850,853	Phoenix	101.0	73.9	89.3
852	Mesa/Tempe	100.2	66.9	85.8
855	Globe	100.8	64.1	84.9
856-857	Tucson	99.4	70.2	86.8
859	Show Low	100.9	65.9	85.8
860	Flagstaff	101.5	71.3	88.4
863	Prescott	99.2	64.9	84.4
864	Kingman	97.8	67.5	84.7
865	Chambers	97.8	66.2	84.2
ARKANSAS				
716	Pine Bluff	96.1	69.6	84.6
717	Camden	94.1	46.0	73.3
718	Texarkana	95.4	50.9	76.2
719	Hot Springs	93.2	49.6	74.4
720-722	Little Rock	95.8	73.2	86.0
723	West Memphis	95.9	63.2	81.8
724	Jonesboro	96.4	63.8	82.3
725	Batesville	94.1	56.5	77.9
726	Harrison	95.5	57.4	79.0
727	Fayetteville	92.8	57.0	77.3
728	Russellville	94.1	58.1	78.5
729	Fort Smith	96.7	62.3	81.9
CALIFORNIA				
900-902	Los Angeles	103.3	111.9	107.0
903-905	Inglewood	99.0	110.8	104.1
906-908	Long Beach	100.8	110.8	105.1
910-912	Pasadena	99.9	111.0	104.7
913-916	Van Nuys	103.0	111.0	106.5
917-918	Alhambra	102.2	111.0	106.0
919-921	San Diego	103.4	105.6	104.3
922	Palm Springs	99.8	110.1	104.3
923-924	San Bernardino	97.6	109.7	102.8
925	Riverside	101.9	109.8	105.3
926-927	Santa Ana	99.6	110.0	104.1
928	Anaheim	102.0	110.4	105.6
930	Oxnard	103.5	109.8	106.2
931	Santa Barbara	102.4	109.8	105.6
932-933	Bakersfield	102.1	107.7	104.5
934	San Luis Obispo	103.4	108.6	105.6
935	Mojave	100.4	107.5	103.4
936-938	Fresno	103.9	112.0	107.4
939	Salinas	104.1	118.1	110.1
940-941	San Francisco	112.6	133.8	121.8
942,956-958	Sacramento	106.0	112.6	108.9
943	Palo Alto	105.2	127.4	114.8
944	San Mateo	108.3	128.5	117.0
945	Vallejo	105.0	123.3	112.9
946	Oakland	110.3	127.9	117.9
947	Berkeley	110.0	127.9	117.7
948	Richmond	109.3	126.1	116.6
949	San Rafael	110.2	127.1	117.5
950	Santa Cruz	109.2	118.1	113.1

STATE/ZIP	CITY	MAT.	INST.	TOTAL
CALIFORNIA (CONT'D)				
951	San Jose	107.1	127.2	115.8
952	Stockton	104.1	113.9	108.3
953	Modesto	104.0	113.4	108.1
954	Santa Rosa	104.1	126.7	113.9
955	Eureka	105.7	111.5	108.2
959	Marysville	104.9	113.1	108.5
960	Redding	106.4	111.8	108.7
961	Susanville	105.5	111.1	107.9
COLORADO				
800-802	Denver	100.8	87.6	95.1
803	Boulder	98.0	85.1	92.4
804	Golden	100.2	84.4	93.4
805	Fort Collins	101.4	81.1	92.6
806	Greeley	98.9	69.9	86.4
807	Fort Morgan	98.8	84.2	92.5
808-809	Colorado Springs	100.5	83.3	93.1
810	Pueblo	101.2	81.3	92.6
811	Alamosa	102.8	78.2	92.1
812	Salida	102.8	79.1	92.5
813	Durango	103.2	79.2	92.8
814	Montrose	101.7	77.8	91.4
815	Grand Junction	105.1	74.9	92.1
816	Glenwood Springs	102.7	81.3	93.4
CONNECTICUT				
060	New Britain	100.0	115.9	106.8
061	Hartford	100.8	115.9	107.3
062	Willimantic	100.5	115.8	107.1
063	New London	97.0	115.9	105.1
064	Meriden	99.0	116.4	106.5
065	New Haven	101.6	116.4	108.0
066	Bridgeport	101.1	116.5	107.7
067	Waterbury	100.6	116.4	107.4
068	Norwalk	100.6	116.8	107.6
069	Stamford	100.7	123.2	110.4
D.C.				
200-205	Washington	104.1	90.2	98.1
DELAWARE				
197	Newark	100.8	101.4	101.0
198	Wilmington	99.8	101.4	100.5
199	Dover	100.9	101.4	101.1
FLORIDA				
320,322	Jacksonville	97.6	57.9	80.5
321	Daytona Beach	97.8	70.6	86.1
323	Tallahassee	97.5	46.8	75.6
324	Panama City	98.9	35.7	71.6
325	Pensacola	101.5	55.3	81.6
326,344	Gainesville	99.3	60.3	82.4
327-328,347	Orlando	99.7	66.6	85.4
329	Melbourne	100.5	76.3	90.0
330-332,340	Miami	98.3	71.6	86.8
333	Fort Lauderdale	97.2	71.9	86.3
334,349	West Palm Beach	96.2	66.4	83.4
335-336,346	Tampa	98.8	70.8	86.7
337	St. Petersburg	101.2	52.4	80.1
338	Lakeland	98.0	70.5	86.1
339,341	Fort Myers	97.3	61.8	82.0
342	Sarasota	99.0	64.4	84.1
GEORGIA				
300-303,399	Atlanta	96.8	79.7	89.4
304	Statesboro	96.5	49.9	76.4
305	Gainesville	95.4	66.8	83.1
306	Athens	94.8	69.5	83.9
307	Dalton	97.0	54.9	78.9
308-309	Augusta	95.5	67.0	83.2
310-312	Macon	95.3	67.8	83.4
313-314	Savannah	96.6	63.5	82.3
315	Waycross	96.6	59.9	80.8
316	Valdosta	96.5	52.4	77.4
317,398	Albany	96.7	63.2	82.2
318-319	Columbus	96.6	69.6	84.9

459

Location Factors

STATE/ZIP	CITY	MAT.	INST.	TOTAL
HAWAII				
967	Hilo	115.4	122.7	118.5
968	Honolulu	120.2	122.7	121.3
STATES & POSS.				
969	Guam	134.9	52.6	99.4
IDAHO				
832	Pocatello	101.6	75.1	90.2
833	Twin Falls	102.3	56.2	82.4
834	Idaho Falls	99.6	62.3	83.5
835	Lewiston	108.7	81.9	97.1
836-837	Boise	100.8	76.4	90.3
838	Coeur d'Alene	107.9	79.9	95.8
ILLINOIS				
600-603	North Suburban	98.7	120.7	108.2
604	Joliet	98.6	116.8	106.4
605	South Suburban	98.7	120.7	108.2
606-608	Chicago	99.2	131.9	113.3
609	Kankakee	95.0	109.0	101.1
610-611	Rockford	97.5	110.8	103.2
612	Rock Island	95.3	97.5	96.2
613	La Salle	96.5	100.9	98.4
614	Galesburg	96.3	101.2	98.4
615-616	Peoria	98.5	97.8	98.2
617	Bloomington	95.6	103.4	99.0
618-619	Champaign	99.2	100.2	99.6
620-622	East St. Louis	94.1	105.0	98.8
623	Quincy	95.4	95.9	95.7
624	Effingham	94.8	95.6	95.1
625	Decatur	96.4	98.5	97.3
626-627	Springfield	96.7	96.4	96.6
628	Centralia	93.0	103.2	97.4
629	Carbondale	92.7	96.4	94.3
INDIANA				
460	Anderson	94.1	82.6	89.2
461-462	Indianapolis	97.7	88.1	93.6
463-464	Gary	95.9	100.9	98.1
465-466	South Bend	94.1	83.9	89.7
467-468	Fort Wayne	94.8	80.8	88.7
469	Kokomo	92.4	82.8	88.2
470	Lawrenceburg	91.7	79.6	86.5
471	New Albany	93.2	74.9	85.3
472	Columbus	95.6	81.1	89.3
473	Muncie	95.9	81.2	89.5
474	Bloomington	97.5	81.8	90.7
475	Washington	93.6	85.3	90.0
476-477	Evansville	94.8	87.3	91.6
478	Terre Haute	95.5	85.8	91.3
479	Lafayette	95.2	79.6	88.5
IOWA				
500-503,509	Des Moines	97.2	81.5	90.5
504	Mason City	96.0	64.1	82.2
505	Fort Dodge	96.1	60.1	80.6
506-507	Waterloo	97.7	60.7	81.7
508	Creston	96.4	64.3	82.6
510-511	Sioux City	98.3	72.7	87.3
512	Sibley	97.1	49.9	76.7
513	Spencer	98.8	48.3	77.0
514	Carroll	95.8	53.9	77.7
515	Council Bluffs	99.3	74.2	88.5
516	Shenandoah	96.4	51.2	76.9
520	Dubuque	97.7	79.2	89.7
521	Decorah	96.8	51.2	77.1
522-524	Cedar Rapids	98.8	84.2	92.5
525	Ottumwa	96.8	72.1	86.2
526	Burlington	96.0	74.1	86.6
527-528	Davenport	97.5	93.2	95.7
KANSAS				
660-662	Kansas City	98.5	97.7	98.1
664-666	Topeka	98.5	69.2	85.8
667	Fort Scott	97.3	70.1	85.6
668	Emporia	97.3	57.9	80.3
669	Belleville	99.0	61.9	83.0
670-672	Wichita	97.5	67.4	84.5
673	Independence	98.8	65.8	84.5
674	Salina	99.0	62.0	83.0
675	Hutchinson	94.3	59.3	79.2
676	Hays	98.2	65.2	83.9
677	Colby	99.0	64.0	83.9

STATE/ZIP	CITY	MAT.	INST.	TOTAL
KANSAS (CONT'D)				
678	Dodge City	100.1	66.4	85.6
679	Liberal	98.0	61.6	82.3
KENTUCKY				
400-402	Louisville	96.1	84.7	91.2
403-405	Lexington	95.5	81.5	89.4
406	Frankfort	95.6	85.4	91.2
407-409	Corbin	93.0	70.0	83.1
410	Covington	93.6	105.1	98.5
411-412	Ashland	92.2	103.4	97.0
413-414	Campton	93.9	70.4	83.7
415-416	Pikeville	94.6	86.8	91.2
417-418	Hazard	93.2	60.1	78.9
420	Paducah	91.8	90.1	91.0
421-422	Bowling Green	94.1	86.9	91.0
423	Owensboro	93.9	88.2	91.4
424	Henderson	91.6	90.0	90.9
425-426	Somerset	91.3	74.9	84.2
427	Elizabethtown	90.7	86.1	88.7
LOUISIANA				
700-701	New Orleans	100.7	68.2	86.6
703	Thibodaux	99.0	65.8	84.7
704	Hammond	96.1	60.8	80.9
705	Lafayette	98.3	60.5	82.0
706	Lake Charles	98.4	62.1	82.8
707-708	Baton Rouge	98.7	61.0	82.4
710-711	Shreveport	94.9	59.7	79.7
712	Monroe	96.3	56.2	79.0
713-714	Alexandria	96.4	57.6	79.6
MAINE				
039	Kittery	94.2	70.9	84.1
040-041	Portland	99.0	77.1	89.6
042	Lewiston	97.1	77.1	88.5
043	Augusta	96.4	70.7	85.3
044	Bangor	96.6	77.1	88.2
045	Bath	95.4	70.9	84.8
046	Machias	94.9	70.3	84.2
047	Houlton	95.0	73.7	85.8
048	Rockland	94.1	70.9	84.1
049	Waterville	95.3	70.8	84.7
MARYLAND				
206	Waldorf	100.3	71.8	88.0
207-208	College Park	100.3	80.2	91.6
209	Silver Spring	99.5	76.4	89.5
210-212	Baltimore	99.4	84.2	92.8
214	Annapolis	99.5	77.5	90.0
215	Cumberland	95.5	79.5	88.6
216	Easton	97.1	43.3	73.9
217	Hagerstown	96.1	80.3	89.3
218	Salisbury	97.5	51.7	77.7
219	Elkton	94.6	64.4	81.6
MASSACHUSETTS				
010-011	Springfield	99.1	104.4	101.4
012	Pittsfield	98.7	100.8	99.6
013	Greenfield	96.8	102.4	99.2
014	Fitchburg	95.5	120.6	106.3
015-016	Worcester	99.1	120.6	108.4
017	Framingham	95.1	124.8	107.9
018	Lowell	98.5	123.9	109.4
019	Lawrence	99.7	122.1	109.4
020-022, 024	Boston	101.4	133.8	115.4
023	Brockton	100.0	119.9	108.6
025	Buzzards Bay	94.4	118.8	105.0
026	Hyannis	97.1	118.8	106.5
027	New Bedford	99.1	119.7	108.0
MICHIGAN				
480,483	Royal Oak	93.6	108.7	100.1
481	Ann Arbor	95.6	109.3	101.5
482	Detroit	97.2	115.3	105.0
484-485	Flint	95.3	99.2	97.0
486	Saginaw	94.8	95.1	94.9
487	Bay City	95.0	95.2	95.1
488-489	Lansing	95.4	97.3	96.2
490	Battle Creek	95.0	91.2	93.4
491	Kalamazoo	95.3	88.8	92.5
492	Jackson	93.5	96.7	94.9
493,495	Grand Rapids	95.7	67.1	83.3
494	Muskegon	94.0	85.5	90.3

Location Factors

STATE/ZIP	CITY	MAT.	INST.	TOTAL
MICHIGAN (CONT'D)				
496	Traverse City	93.0	71.8	83.8
497	Gaylord	94.2	75.9	86.3
498-499	Iron Mountain	96.0	87.4	92.3
MINNESOTA				
550-551	Saint Paul	99.4	124.3	110.1
553-555	Minneapolis	100.2	128.3	112.3
556-558	Duluth	98.5	112.9	104.7
559	Rochester	98.9	107.6	102.7
560	Mankato	96.1	107.7	101.1
561	Windom	94.9	80.6	88.7
562	Willmar	94.4	86.7	91.1
563	St. Cloud	95.5	122.9	107.4
564	Brainerd	96.1	104.2	99.6
565	Detroit Lakes	97.9	99.2	98.5
566	Bemidji	97.3	101.1	98.9
567	Thief River Falls	96.9	96.2	96.6
MISSISSIPPI				
386	Clarksdale	95.6	29.9	67.2
387	Greenville	99.1	43.1	74.9
388	Tupelo	97.1	36.1	70.8
389	Greenwood	96.9	32.3	69.0
390-392	Jackson	97.2	44.6	74.5
393	Meridian	95.7	45.2	73.9
394	Laurel	97.0	32.4	69.1
395	Biloxi	97.4	53.7	78.5
396	McComb	95.5	51.5	76.5
397	Columbus	96.9	35.5	70.4
MISSOURI				
630-631	St. Louis	97.8	112.0	103.9
633	Bowling Green	96.6	93.6	95.3
634	Hannibal	95.5	84.4	90.7
635	Kirksville	97.8	77.4	89.0
636	Flat River	97.6	94.1	96.1
637	Cape Girardeau	97.3	89.8	94.1
638	Sikeston	95.4	81.1	89.2
639	Poplar Bluff	94.9	80.9	88.9
640-641	Kansas City	99.7	107.6	103.1
644-645	St. Joseph	98.8	92.9	96.3
646	Chillicothe	95.8	68.8	84.1
647	Harrisonville	95.5	98.1	96.6
648	Joplin	97.6	69.9	85.7
650-651	Jefferson City	95.6	88.7	92.6
652	Columbia	96.8	90.7	94.2
653	Sedalia	96.0	83.9	90.8
654-655	Rolla	94.8	76.3	86.8
656-658	Springfield	97.6	76.4	88.5
MONTANA				
590-591	Billings	101.7	73.6	89.5
592	Wolf Point	101.2	67.3	86.6
593	Miles City	99.1	67.8	85.6
594	Great Falls	102.9	72.0	89.6
595	Havre	100.1	67.5	86.0
596	Helena	102.1	70.4	88.5
597	Butte	101.3	70.2	87.9
598	Missoula	98.9	69.3	86.1
599	Kalispell	98.0	67.3	84.8
NEBRASKA				
680-681	Omaha	98.8	77.3	89.5
683-685	Lincoln	97.7	66.0	84.0
686	Columbus	96.1	49.2	75.8
687	Norfolk	97.8	59.8	81.4
688	Grand Island	97.5	64.6	83.3
689	Hastings	97.2	56.0	79.4
690	Mccook	96.9	46.6	75.2
691	North Platte	96.9	56.8	79.6
692	Valentine	99.0	37.3	72.4
693	Alliance	98.9	34.8	71.2
NEVADA				
889-891	Las Vegas	99.6	104.7	101.8
893	Ely	99.8	76.4	89.7
894-895	Reno	99.9	96.5	98.4
897	Carson City	99.3	96.0	97.9
898	Elko	98.5	82.5	91.6
NEW HAMPSHIRE				
030	Nashua	99.3	86.0	93.6
031	Manchester	100.0	86.0	94.0

STATE/ZIP	CITY	MAT.	INST.	TOTAL
NEW HAMPSHIRE (CONT'D)				
032-033	Concord	97.5	86.0	92.5
034	Keene	96.2	49.1	75.9
035	Littleton	96.2	59.6	80.4
036	Charleston	95.7	46.2	74.3
037	Claremont	94.8	46.2	73.8
038	Portsmouth	96.6	81.6	90.1
NEW JERSEY				
070-071	Newark	101.0	122.8	110.4
072	Elizabeth	99.5	121.7	109.1
073	Jersey City	98.4	123.3	109.2
074-075	Paterson	100.2	122.9	110.0
076	Hackensack	98.3	123.4	109.1
077	Long Branch	97.9	122.2	108.4
078	Dover	98.5	123.0	109.1
079	Summit	98.4	122.1	108.6
080,083	Vineland	96.4	118.6	106.0
081	Camden	98.4	117.4	106.6
082,084	Atlantic City	97.1	117.3	105.8
085-086	Trenton	98.2	121.7	108.3
087	Point Pleasant	98.4	120.7	108.0
088-089	New Brunswick	98.9	121.0	108.4
NEW MEXICO				
870-872	Albuquerque	99.8	75.7	89.4
873	Gallup	99.9	75.7	89.4
874	Farmington	100.2	75.7	89.6
875	Santa Fe	98.9	75.7	88.9
877	Las Vegas	98.4	75.7	88.6
878	Socorro	97.9	75.7	88.3
879	Truth/Consequences	98.1	70.6	86.3
880	Las Cruces	96.8	67.8	84.3
881	Clovis	98.5	74.5	88.1
882	Roswell	100.1	74.6	89.1
883	Carrizozo	100.6	75.7	89.8
884	Tucumcari	99.3	74.5	88.6
NEW YORK				
100-102	New York	105.3	164.5	130.9
103	Staten Island	101.3	157.6	125.6
104	Bronx	99.4	157.6	124.5
105	Mount Vernon	99.4	137.1	115.7
106	White Plains	99.2	137.1	115.6
107	Yonkers	104.0	137.3	118.4
108	New Rochelle	99.9	137.1	116.0
109	Suffern	99.6	123.7	110.0
110	Queens	101.2	157.4	125.4
111	Long Island City	102.8	157.4	126.4
112	Brooklyn	103.1	157.4	126.5
113	Flushing	103.3	157.4	126.6
114	Jamaica	101.5	157.4	125.6
115,117,118	Hicksville	101.1	147.5	121.1
116	Far Rockaway	103.4	157.4	126.7
119	Riverhead	102.0	147.5	121.6
120-122	Albany	97.2	94.0	95.8
123	Schenectady	97.8	94.6	96.4
124	Kingston	101.0	114.8	107.0
125-126	Poughkeepsie	100.2	116.1	107.1
127	Monticello	99.5	118.3	107.6
128	Glens Falls	92.6	91.1	92.0
129	Plattsburgh	96.7	85.4	91.8
130-132	Syracuse	98.7	93.3	96.4
133-135	Utica	96.6	88.4	93.1
136	Watertown	98.3	85.9	92.9
137-139	Binghamton	98.4	88.5	94.1
140-142	Buffalo	100.1	105.3	102.4
143	Niagara Falls	98.0	102.2	99.8
144-146	Rochester	99.9	97.3	98.8
147	Jamestown	96.9	85.8	92.1
148-149	Elmira	96.8	83.7	91.1
NORTH CAROLINA				
270,272-274	Greensboro	100.0	52.8	79.6
271	Winston-Salem	99.7	52.0	79.1
275-276	Raleigh	99.9	53.6	79.9
277	Durham	101.3	54.0	80.9
278	Rocky Mount	97.4	43.0	73.9
279	Elizabeth City	98.3	44.5	75.0
280	Gastonia	98.8	51.4	78.3
281-282	Charlotte	100.3	54.2	80.4
283	Fayetteville	101.3	54.4	81.1
284	Wilmington	97.5	51.5	77.6
285	Kinston	95.7	44.8	73.7

Location Factors

STATE/ZIP	CITY	MAT.	INST.	TOTAL
NORTH CAROLINA (CONT'D)				
286	Hickory	96.0	47.6	75.1
287-288	Asheville	97.9	51.4	77.8
289	Murphy	97.0	36.2	70.8
NORTH DAKOTA				
580-581	Fargo	100.7	63.6	84.6
582	Grand Forks	100.6	57.7	82.1
583	Devils Lake	100.2	59.6	82.7
584	Jamestown	100.2	50.9	78.9
585	Bismarck	99.1	65.3	84.5
586	Dickinson	100.9	59.6	83.1
587	Minot	100.6	67.4	86.3
588	Williston	99.4	59.5	82.2
OHIO				
430-432	Columbus	97.3	88.6	93.5
433	Marion	93.9	86.4	90.7
434-436	Toledo	97.5	98.6	98.0
437-438	Zanesville	94.4	84.1	90.0
439	Steubenville	95.4	93.5	94.6
440	Lorain	97.0	94.9	96.1
441	Cleveland	97.3	104.9	100.6
442-443	Akron	97.9	95.5	96.9
444-445	Youngstown	97.3	93.2	95.6
446-447	Canton	97.5	87.8	93.3
448-449	Mansfield	94.9	92.4	93.8
450	Hamilton	94.8	88.1	91.9
451-452	Cincinnati	95.1	88.3	92.2
453-454	Dayton	94.9	85.0	90.6
455	Springfield	94.8	85.9	91.0
456	Chillicothe	93.9	93.5	93.7
457	Athens	96.8	77.5	88.5
458	Lima	97.2	87.3	92.9
OKLAHOMA				
730-731	Oklahoma City	97.3	62.8	82.4
734	Ardmore	95.0	62.2	80.8
735	Lawton	97.2	63.7	82.8
736	Clinton	96.4	60.6	80.9
737	Enid	96.9	60.5	81.2
738	Woodward	95.1	60.6	80.2
739	Guymon	96.2	32.3	68.6
740-741	Tulsa	97.0	58.6	80.4
743	Miami	93.9	65.8	81.8
744	Muskogee	96.3	43.0	73.3
745	Mcalester	93.5	54.4	76.6
746	Ponca City	94.1	60.8	79.7
747	Durant	94.1	60.3	79.5
748	Shawnee	95.7	58.6	79.7
749	Poteau	93.1	63.2	80.2
OREGON				
970-972	Portland	101.8	102.7	102.2
973	Salem	102.1	102.0	102.0
974	Eugene	101.6	101.7	101.6
975	Medford	103.1	100.0	101.7
976	Klamath Falls	103.4	100.0	102.0
977	Bend	102.2	101.7	102.0
978	Pendleton	96.8	101.5	98.8
979	Vale	94.5	92.2	93.5
PENNSYLVANIA				
150-152	Pittsburgh	96.0	102.7	98.9
153	Washington	93.3	101.1	96.7
154	Uniontown	93.6	98.1	95.5
155	Bedford	94.5	91.4	93.1
156	Greensburg	94.6	99.0	96.5
157	Indiana	93.4	98.1	95.4
158	Dubois	94.7	94.6	94.7
159	Johnstown	94.4	94.9	94.6
160	Butler	92.3	99.9	95.6
161	New Castle	92.3	99.1	95.3
162	Kittanning	92.8	101.1	96.4
163	Oil City	92.3	96.3	94.0
164-165	Erie	94.3	94.1	94.2
166	Altoona	94.3	91.5	93.1
167	Bradford	95.6	94.3	95.0
168	State College	95.2	91.6	93.6
169	Wellsboro	96.2	91.7	94.3
170-171	Harrisburg	98.0	92.6	95.7
172	Chambersburg	95.7	89.5	93.0
173-174	York	95.9	92.6	94.5
175-176	Lancaster	94.5	87.9	91.7

STATE/ZIP	CITY	MAT.	INST.	TOTAL
PENNSYLVANIA (CONT'D)				
177	Williamsport	93.2	80.0	87.5
178	Sunbury	95.3	92.5	94.1
179	Pottsville	94.3	94.2	94.3
180	Lehigh Valley	95.8	112.1	102.8
181	Allentown	97.7	108.3	102.2
182	Hazleton	95.2	95.3	95.3
183	Stroudsburg	95.1	101.4	97.8
184-185	Scranton	98.4	97.1	97.9
186-187	Wilkes-Barre	95.1	95.6	95.3
188	Montrose	94.6	96.0	95.2
189	Doylestown	94.6	117.5	104.5
190-191	Philadelphia	101.2	131.6	114.3
193	Westchester	98.0	118.7	106.9
194	Norristown	96.9	126.7	109.7
195-196	Reading	98.8	98.9	98.8
PUERTO RICO				
009	San Juan	119.0	24.6	78.3
RHODE ISLAND				
028	Newport	98.4	110.3	103.5
029	Providence	99.1	110.3	104.0
SOUTH CAROLINA				
290-292	Columbia	98.0	47.8	76.4
293	Spartanburg	97.2	46.5	75.3
294	Charleston	98.7	52.3	78.7
295	Florence	96.8	45.6	74.7
296	Greenville	96.9	46.5	75.2
297	Rock Hill	96.8	33.3	69.4
298	Aiken	97.7	69.6	85.6
299	Beaufort	98.5	36.6	71.8
SOUTH DAKOTA				
570-571	Sioux Falls	98.9	56.3	80.5
572	Watertown	98.1	49.6	77.2
573	Mitchell	97.0	49.1	76.3
574	Aberdeen	99.5	51.8	78.9
575	Pierre	98.8	51.8	78.5
576	Mobridge	97.6	49.3	76.8
577	Rapid City	99.3	50.6	78.3
TENNESSEE				
370-372	Nashville	97.7	74.8	87.8
373-374	Chattanooga	98.4	57.9	80.9
375,380-381	Memphis	96.0	74.0	86.5
376	Johnson City	97.2	57.7	80.1
377-379	Knoxville	94.5	58.4	78.9
382	Mckenzie	96.2	57.4	79.4
383	Jackson	98.1	50.2	77.4
384	Columbia	94.7	59.2	79.4
385	Cookeville	96.0	61.0	80.9
TEXAS				
750	Mckinney	97.4	53.8	78.6
751	Waxahackie	97.4	54.6	78.9
752-753	Dallas	98.0	65.6	84.0
754	Greenville	97.5	39.5	72.5
755	Texarkana	96.4	52.4	77.4
756	Longview	96.8	41.2	72.8
757	Tyler	97.4	55.2	79.2
758	Palestine	93.5	41.6	71.1
759	Lufkin	94.4	45.6	73.3
760-761	Fort Worth	97.0	61.4	81.6
762	Denton	97.5	50.1	77.1
763	Wichita Falls	98.0	56.8	80.2
764	Eastland	96.7	40.7	72.5
765	Temple	95.5	50.8	76.2
766-767	Waco	97.2	57.5	80.1
768	Brownwood	97.7	38.3	72.0
769	San Angelo	97.3	46.5	75.4
770-772	Houston	101.4	71.4	88.5
773	Huntsville	100.0	38.9	73.6
774	Wharton	101.4	43.6	76.4
775	Galveston	99.3	70.4	86.9
776-777	Beaumont	99.1	62.3	83.2
778	Bryan	96.6	63.4	82.3
779	Victoria	101.5	46.0	77.5
780	Laredo	95.4	52.8	77.0
781-782	San Antonio	95.7	66.1	83.0
783-784	Corpus Christi	98.4	52.0	78.3
785	Mc Allen	98.5	46.4	76.0
786-787	Austin	95.5	60.3	80.3

Location Factors

STATE/ZIP	CITY	MAT.	INST.	TOTAL
TEXAS (CONT'D)				
788	Del Rio	97.7	32.3	69.5
789	Giddings	95.1	40.9	71.7
790-791	Amarillo	98.1	57.2	80.5
792	Childress	97.2	51.6	77.5
793-794	Lubbock	99.4	51.9	78.9
795-796	Abilene	97.5	52.0	77.9
797	Midland	99.7	48.7	77.7
798-799,885	El Paso	96.9	50.3	76.8
UTAH				
840-841	Salt Lake City	102.9	68.6	88.1
842,844	Ogden	98.3	68.0	85.2
843	Logan	100.2	68.0	86.3
845	Price	100.8	48.4	78.2
846-847	Provo	100.7	68.5	86.8
VERMONT				
050	White River Jct.	97.5	50.1	77.0
051	Bellows Falls	96.0	54.8	78.2
052	Bennington	96.4	70.8	85.3
053	Brattleboro	96.8	61.3	81.4
054	Burlington	99.7	62.8	83.8
056	Montpelier	96.4	62.8	81.9
057	Rutland	98.2	62.8	82.9
058	St. Johnsbury	97.7	59.7	81.3
059	Guildhall	96.3	59.6	80.4
VIRGINIA				
220-221	Fairfax	100.7	83.3	93.2
222	Arlington	101.8	82.4	93.4
223	Alexandria	101.0	83.9	93.6
224-225	Fredericksburg	99.4	76.0	89.3
226	Winchester	100.0	69.9	87.0
227	Culpeper	99.9	75.2	89.2
228	Harrisonburg	100.1	66.4	85.6
229	Charlottesville	100.6	69.2	87.0
230-232	Richmond	101.2	73.1	89.1
233-235	Norfolk	101.2	71.1	88.2
236	Newport News	100.8	71.9	88.3
237	Portsmouth	100.1	67.2	85.9
238	Petersburg	100.3	73.1	88.6
239	Farmville	99.7	59.8	82.4
240-241	Roanoke	101.4	67.0	86.5
242	Bristol	99.2	59.3	82.0
243	Pulaski	98.9	57.3	81.0
244	Staunton	99.9	66.3	85.4
245	Lynchburg	99.9	69.1	86.6
246	Grundy	99.3	56.6	80.9
WASHINGTON				
980-981,987	Seattle	104.7	103.5	104.1
982	Everett	104.4	99.4	102.2
983-984	Tacoma	104.4	99.9	102.5
985	Olympia	102.5	99.8	101.4
986	Vancouver	106.0	96.2	101.8
988	Wenatchee	104.5	84.9	96.0
989	Yakima	104.5	90.4	98.4
990-992	Spokane	105.3	81.8	95.1
993	Richland	104.6	86.0	96.6
994	Clarkston	103.9	83.3	95.0
WEST VIRGINIA				
247-248	Bluefield	98.0	77.1	89.0
249	Lewisburg	99.8	80.6	91.5
250-253	Charleston	100.6	88.4	95.3
254	Martinsburg	99.5	77.3	89.9
255-257	Huntington	100.8	96.0	98.8
258-259	Beckley	97.9	86.1	92.8
260	Wheeling	100.9	91.2	96.7
261	Parkersburg	99.7	90.3	95.6
262	Buckhannon	99.4	90.4	95.5
263-264	Clarksburg	99.9	90.5	95.8
265	Morgantown	100.0	90.6	95.9
266	Gassaway	99.3	89.3	95.0
267	Romney	99.3	84.0	92.7
268	Petersburg	99.1	86.6	93.7
WISCONSIN				
530,532	Milwaukee	99.1	106.7	102.4
531	Kenosha	98.8	101.1	99.8
534	Racine	98.3	101.7	99.8
535	Beloit	98.2	97.4	97.8
537	Madison	98.0	96.5	97.4

STATE/ZIP	CITY	MAT.	INST.	TOTAL
WISCONSIN (CONT'D)				
538	Lancaster	96.1	91.1	93.9
539	Portage	94.7	95.1	94.9
540	New Richmond	95.6	97.1	96.2
541-543	Green Bay	99.6	93.1	96.8
544	Wausau	94.9	92.2	93.7
545	Rhinelander	98.1	92.4	95.7
546	La Crosse	95.9	93.1	94.7
547	Eau Claire	97.6	95.1	96.5
548	Superior	95.4	100.6	97.6
549	Oshkosh	95.6	92.8	94.4
WYOMING				
820	Cheyenne	100.0	67.8	86.1
821	Yellowstone Nat'l Park	97.5	60.4	81.5
822	Wheatland	98.6	59.8	81.8
823	Rawlins	100.2	59.3	82.5
824	Worland	98.1	58.1	80.8
825	Riverton	99.1	58.2	81.4
826	Casper	99.8	61.4	83.2
827	Newcastle	97.9	59.3	81.2
828	Sheridan	100.7	62.1	84.0
829-831	Rock Springs	102.1	58.1	83.1
CANADIAN FACTORS (reflect Canadian currency)				
ALBERTA				
	Calgary	124.8	93.1	111.1
	Edmonton	126.3	93.1	111.9
	Fort McMurray	116.6	93.1	106.4
	Lethbridge	117.5	92.3	106.7
	Lloydminster	116.6	93.1	106.5
	Medicine Hat	116.7	92.3	106.2
	Red Deer	117.1	92.3	106.4
BRITISH COLUMBIA				
	Kamloops	117.8	96.4	108.6
	Prince George	118.9	96.4	109.2
	Vancouver	126.9	86.9	109.6
	Victoria	118.9	83.4	103.6
MANITOBA				
	Brandon	117.0	80.6	101.3
	Portage la Prairie	117.0	79.2	100.7
	Winnipeg	125.8	73.5	103.2
NEW BRUNSWICK				
	Bathurst	115.3	71.2	96.3
	Dalhousie	115.2	71.2	96.2
	Fredericton	116.9	76.2	99.3
	Moncton	115.6	72.0	96.8
	Newcastle	115.3	71.2	96.3
	Saint John	118.2	75.9	99.9
NEWFOUNDLAND				
	Corner Brook	120.2	71.4	99.1
	St. John's	120.7	72.5	99.9
NORTHWEST TERRITORIES				
	Yellowknife	121.1	90.8	108.0
NOVA SCOTIA				
	Bridgewater	116.7	79.1	100.5
	Dartmouth	118.1	79.1	101.3
	Halifax	119.8	80.4	102.8
	New Glasgow	116.1	79.1	100.1
	Sydney	113.5	79.1	98.7
	Truro	116.1	79.1	100.1
	Yarmouth	116.0	79.1	100.1
ONTARIO				
	Barrie	119.7	99.1	110.8
	Brantford	118.9	103.3	112.2
	Cornwall	119.3	99.5	110.7
	Hamilton	124.8	98.7	113.5
	Kingston	120.2	99.6	111.3
	Kitchener	115.5	94.5	106.4
	London	122.0	96.2	110.9
	North Bay	118.9	97.2	109.5
	Oshawa	118.8	100.7	111.0
	Ottawa	120.7	97.8	110.8
	Owen Sound	119.9	97.2	110.1
	Peterborough	118.9	99.3	110.4
	Sarnia	119.0	104.0	112.5

463

Location Factors

STATE/ZIP	CITY	MAT.	INST.	TOTAL
ONTARIO (CONT'D)				
	Sault Ste Marie	113.7	97.3	106.6
	St. Catharines	113.4	96.3	106.0
	Sudbury	113.4	94.8	105.4
	Thunder Bay	114.8	95.0	106.3
	Timmins	119.2	97.2	109.7
	Toronto	123.1	102.2	114.1
	Windsor	114.3	95.4	106.1
PRINCE EDWARD ISLAND				
	Charlottetown	118.4	66.4	95.9
	Summerside	117.7	66.4	95.5
QUEBEC				
	Cap-de-la-Madeleine	116.0	93.0	106.1
	Charlesbourg	116.0	93.0	106.1
	Chicoutimi	115.1	97.8	107.6
	Gatineau	115.3	92.8	105.6
	Granby	115.7	92.7	105.8
	Hull	115.6	92.8	105.7
	Joliette	116.3	93.0	106.3
	Laval	115.4	92.8	105.6
	Montreal	119.8	98.1	110.5
	Quebec	119.9	98.5	110.7
	Rimouski	115.5	97.8	107.8
	Rouyn-Noranda	115.3	92.8	105.6
	Saint Hyacinthe	115.0	92.8	105.4
	Sherbrooke	115.7	92.8	105.8
	Sorel	116.2	93.0	106.2
	St Jerome	115.4	92.8	105.6
	Trois Rivieres	116.2	93.0	106.2
SASKATCHEWAN				
	Moose Jaw	114.3	73.8	96.8
	Prince Albert	113.2	71.9	95.4
	Regina	116.3	73.9	98.0
	Saskatoon	114.8	73.7	97.0
YUKON				
	Whitehorse	113.9	72.6	96.1

R011105-05 Tips for Accurate Estimating

1. Use pre-printed or columnar forms for orderly sequence of dimensions and locations and for recording telephone quotations.

2. Use only the front side of each paper or form except for certain pre-printed summary forms.

3. Be consistent in listing dimensions: For example, length x width x height. This helps in rechecking to ensure that, the total length of partitions is appropriate for the building area.

4. Use printed (rather than measured) dimensions where given.

5. Add up multiple printed dimensions for a single entry where possible.

6. Measure all other dimensions carefully.

7. Use each set of dimensions to calculate multiple related quantities.

8. Convert foot and inch measurements to decimal feet when listing. Memorize decimal equivalents to .01 parts of a foot (1/8″ equals approximately .01′).

9. Do not "round off" quantities until the final summary.

10. Mark drawings with different colors as items are taken off.

11. Keep similar items together, different items separate.

12. Identify location and drawing numbers to aid in future checking for completeness.

13. Measure or list everything on the drawings or mentioned in the specifications.

14. It may be necessary to list items not called for to make the job complete.

15. Be alert for: Notes on plans such as N.T.S. (not to scale); changes in scale throughout the drawings; reduced size drawings; discrepancies between the specifications and the drawings.

16. Develop a consistent pattern of performing an estimate. For example:
    a. Start the quantity takeoff at the lower floor and move to the next higher floor.
    b. Proceed from the main section of the building to the wings.
    c. Proceed from south to north or vice versa, clockwise or counterclockwise.
    d. Take off floor plan quantities first, elevations next, then detail drawings.

17. List all gross dimensions that can be either used again for different quantities, or used as a rough check of other quantities for verification (exterior perimeter, gross floor area, individual floor areas, etc.).

18. Utilize design symmetry or repetition (repetitive floors, repetitive wings, symmetrical design around a center line, similar room layouts, etc.). Note: Extreme caution is needed here so as not to omit or duplicate an area.

19. Do not convert units until the final total is obtained. For instance, when estimating concrete work, keep all units to the nearest cubic foot, then summarize and convert to cubic yards.

20. When figuring alternatives, it is best to total all items involved in the basic system, then total all items involved in the alternates. Therefore you work with positive numbers in all cases. When adds and deducts are used, it is often confusing whether to add or subtract a portion of an item; especially on a complicated or involved alternate.

R011105-10 Unit Gross Area Requirements

The figures in the table below indicate typical ranges in square feet as a function of the "occupant" unit. This table is best used in the preliminary design stages to help determine the probable size requirement for the total project.

Building Type	Unit	Gross Area in S.F.		
		1/4	Median	3/4
Apartments	Unit	660	860	1,100
Auditorium & Play Theaters	Seat	18	25	38
Bowling Alleys	Lane		940	
Churches & Synagogues	Seat	20	28	39
Dormitories	Bed	200	230	275
Fraternity & Sorority Houses	Bed	220	315	370
Garages, Parking	Car	325	355	385
Hospitals	Bed	685	850	1,075
Hotels	Rental Unit	475	600	710
Housing for the elderly	Unit	515	635	755
Housing, Public	Unit	700	875	1,030
Ice Skating Rinks	Total	27,000	30,000	36,000
Motels	Rental Unit	360	465	620
Nursing Homes	Bed	290	350	450
Restaurants	Seat	23	29	39
Schools, Elementary	Pupil	65	77	90
Junior High & Middle		85	110	129
Senior High		102	130	145
Vocational		110	135	195
Shooting Ranges	Point		450	
Theaters & Movies	Seat		15	

General Requirements R0111 Summary of Work

R011110-30 Engineering Fees

Typical **Structural Engineering Fees** based on type of construction and total project size. These fees are included in Architectural Fees.

Type of Construction	Total Project Size (in thousands of dollars)			
	$500	$500-$1,000	$1,000-$5,000	Over $5000
Industrial buildings, factories & warehouses	Technical payroll times 2.0 to 2.5	1.60%	1.25%	1.00%
Hotels, apartments, offices, dormitories, hospitals, public buildings, food stores		2.00%	1.70%	1.20%
Museums, banks, churches and cathedrals		2.00%	1.75%	1.25%
Thin shells, prestressed concrete, earthquake resistive		2.00%	1.75%	1.50%
Parking ramps, auditoriums, stadiums, convention halls, hangars & boiler houses		2.50%	2.00%	1.75%
Special buildings, major alterations, underpinning & future expansion	↓	Add to above 0.5%	Add to above 0.5%	Add to above 0.5%

For complex reinforced concrete or unusually complicated structures, add 20% to 50%.

Typical **Mechanical and Electrical Engineering Fees** are based on the size of the subcontract. The fee structure for both are shown below. These fees are included in Architectural Fees.

Type of Construction	Subcontract Size							
	$25,000	$50,000	$100,000	$225,000	$350,000	$500,000	$750,000	$1,000,000
Simple structures	6.4%	5.7%	4.8%	4.5%	4.4%	4.3%	4.2%	4.1%
Intermediate structures	8.0	7.3	6.5	5.6	5.1	5.0	4.9	4.8
Complex structures	10.1	9.0	9.0	8.0	7.5	7.5	7.0	7.0

For renovations, add 15% to 25% to applicable fee.

General Requirements R0121 Allowances

R012153-10 Repair and Remodeling

Cost figures are based on new construction utilizing the most cost-effective combination of labor, equipment and material with the work scheduled in proper sequence to allow the various trades to accomplish their work in an efficient manner.

The costs for repair and remodeling work must be modified due to the following factors that may be present in any given repair and remodeling project.

1. Equipment usage curtailment due to the physical limitations of the project, with only hand-operated equipment being used.

2. Increased requirement for shoring and bracing to hold up the building while structural changes are being made and to allow for temporary storage of construction materials on above-grade floors.

3. Material handling becomes more costly due to having to move within the confines of an enclosed building. For multi-story construction, low capacity elevators and stairwells may be the only access to the upper floors.

4. Large amount of cutting and patching and attempting to match the existing construction is required. It is often more economical to remove entire walls rather than create many new door and window openings. This sort of trade-off has to be carefully analyzed.

5. Cost of protection of completed work is increased since the usual sequence of construction usually cannot be accomplished.

6. Economies of scale usually associated with new construction may not be present. If small quantities of components must be custom fabricated due to job requirements, unit costs will naturally increase. Also, if only small work areas are available at a given time, job scheduling between trades becomes difficult and subcontractor quotations may reflect the excessive start-up and shut-down phases of the job.

7. Work may have to be done on other than normal shifts and may have to be done around an existing production facility which has to stay in production during the course of the repair and remodeling.

8. Dust and noise protection of adjoining non-construction areas can involve substantial special protection and alter usual construction methods.

9. Job may be delayed due to unexpected conditions discovered during demolition or removal. These delays ultimately increase construction costs.

10. Piping and ductwork runs may not be as simple as for new construction. Wiring may have to be snaked through walls and floors.

11. Matching "existing construction" may be impossible because materials may no longer be manufactured. Substitutions may be expensive.

12. Weather protection of existing structure requires additional temporary structures to protect building at openings.

13. On small projects, because of local conditions, it may be necessary to pay a tradesman for a minimum of four hours for a task that is completed in one hour.

All of the above areas can contribute to increased costs for a repair and remodeling project. Each of the above factors should be considered in the planning, bidding and construction stage in order to minimize the increased costs associated with repair and remodeling jobs.

R012909-80   Sales Tax by State

State sales tax on materials is tabulated below (5 states have no sales tax). Many states allow local jurisdictions, such as a county or city, to levy additional sales tax.

Some projects may be sales tax exempt, particularly those constructed with public funds.

State	Tax (%)	State	Tax (%)	State	Tax (%)	State	Tax (%)
Alabama	4	Illinois	6.25	Montana	0	Rhode Island	7
Alaska	0	Indiana	6	Nebraska	5.5	South Carolina	5
Arizona	5.6	Iowa	5	Nevada	6.5	South Dakota	4
Arkansas	6	Kansas	5.3	New Hampshire	0	Tennessee	7
California	7.25	Kentucky	6	New Jersey	6	Texas	6.25
Colorado	2.9	Louisiana	4	New Mexico	5	Utah	4.75
Connecticut	6	Maine	5	New York	4	Vermont	6
Delaware	0	Maryland	5	North Carolina	4.5	Virginia	5
District of Columbia	5.75	Massachusetts	5	North Dakota	5	Washington	6.5
Florida	6	Michigan	6	Ohio	5.5	West Virginia	6
Georgia	4	Minnesota	6.5	Oklahoma	4.5	Wisconsin	5
Hawaii	4	Mississippi	7	Oregon	0	Wyoming	4
Idaho	5	Missouri	4.225	Pennsylvania	6	Average	4.84 %

Sales Tax by Province (Canada)

GST - a value-added tax, which the government imposes on most goods and services provided in or imported into Canada. PST - a retail sales tax, which five of the provinces impose on the price of most goods and some

services. QST - a value-added tax, similar to the federal GST, which Quebec imposes. HST - Three provinces have combined their retail sales tax with the federal GST into one harmonized tax.

Province	PST (%)	QST (%)	GST(%)	HST(%)
Alberta	0	0	6	0
British Columbia	7	0	6	0
Manitoba	7	0	6	0
New Brunswick	0	0	0	14
Newfoundland	0	0	0	14
Northwest Territories	0	0	6	0
Nova Scotia	0	0	0	14
Ontario	8	0	6	0
Prince Edward Island	10	0	6	0
Quebec	0	7.5	6	0
Saskatchewan	7	0	6	0
Yukon	0	0	6	0

R012909-85   Unemployment Taxes and Social Security Taxes

State Unemployment Tax rates vary not only from state to state, but also with the experience rating of the contractor. The Federal Unemployment Tax rate is 6.2% of the first $7,000 of wages. This is reduced by a credit of up to 5.4% for timely payment to the state. The minimum Federal Unemployment Tax is 0.8% after all credits.

Social Security (FICA) for 2007 is estimated at time of publication to be 7.65% of wages up to $94,200.

R012909-90   Overtime

One way to improve the completion date of a project or eliminate negative float from a schedule is to compress activity duration times. This can be achieved by increasing the crew size or working overtime with the proposed crew.

To determine the costs of working overtime to compress activity duration times, consider the following examples. Below is an overtime efficiency and cost chart based on a five, six, or seven day week with an eight through twelve hour day. Payroll percentage increases for time and one half and double time are shown for the various working days.

| Days per Week | Hours per Day | Production Efficiency | | | | | Payroll Cost Factors | |
		1st Week	2nd Week	3rd Week	4th Week	Average 4 Weeks	@ 1-1/2 Times	@ 2 Times
5	8	100%	100%	100%	100%	100 %	100 %	100 %
	9	100	100	95	90	96.25	105.6	111.1
	10	100	95	90	85	91.25	110.0	120.0
	11	95	90	75	65	81.25	113.6	127.3
	12	90	85	70	60	76.25	116.7	133.3
6	8	100	100	95	90	96.25	108.3	116.7
	9	100	95	90	85	92.50	113.0	125.9
	10	95	90	85	80	87.50	116.7	133.3
	11	95	85	70	65	78.75	119.7	139.4
	12	90	80	65	60	73.75	122.2	144.4
7	8	100	95	85	75	88.75	114.3	128.6
	9	95	90	80	70	83.75	118.3	136.5
	10	90	85	75	65	78.75	121.4	142.9
	11	85	80	65	60	72.50	124.0	148.1
	12	85	75	60	55	68.75	126.2	152.4

R013113-40   Builder's Risk Insurance

Builder's Risk Insurance is insurance on a building during construction. Premiums are paid by the owner or the contractor. Blasting, collapse and underground insurance would raise total insurance costs above those listed. Floater policy for materials delivered to the job runs $.75 to $1.25 per $100 value. Contractor equipment insurance runs $.50 to $1.50 per $100 value. Insurance for miscellaneous tools to $1,500 value runs from $3.00 to $7.50 per $100 value.

Tabulated below are New England Builder's Risk insurance rates in dollars per $100 value for $1,000 deductible. For $25,000 deductible, rates can be reduced 13% to 34%. On contracts over $1,000,000, rates may be lower than those tabulated. Policies are written annually for the total completed value in place. For "all risk" insurance (excluding flood, earthquake and certain other perils) add $.025 to total rates below.

| Coverage | Frame Construction (Class 1) | | | Brick Construction (Class 4) | | | Fire Resistive (Class 6) | | |
	Range		Average	Range		Average	Range		Average
Fire Insurance	$.350 to	$ .850	$.600	$.158 to	$.189	$.174	$.052 to	$.080	$.070
Extended Coverage	.115 to	.200	.158	.080 to	.105	.101	.081 to	.105	.100
Vandalism	.012 to	.016	.014	.008 to	.011	.011	.008 to	.011	.010
Total Annual Rate	$.477 to	$1.066	$.772	$.246 to	$.305	$.286	$.141 to	$.196	$.180

R013113-60   Workers' Compensation Insurance Rates by Trade

The table below tabulates the national averages for Workers' Compensation insurance rates by trade and type of building. The average "Insurance Rate" is multiplied by the "% of Building Cost" for each trade. This produces the "Workers' Compensation Cost" by % of total labor cost, to be added for each trade by building type to determine the weighted average Workers' Compensation rate for the building types analyzed.

Trade	Insurance Rate (% Labor Cost) Range	Average	% of Building Cost Office Bldgs.	Schools & Apts.	Mfg.	Workers' Compensation Office Bldgs.	Schools & Apts.	Mfg.
Excavation, Grading, etc.	4.5 % to 10.5%	10.4%	4.8%	4.9%	4.5%	.50%	.51%	.47%
Piles & Foundations	8.9 to 42.9	21.4	7.1	5.2	8.7	1.52	1.11	1.86
Concrete	5.1 to 29.8	15.5	5.0	14.8	3.7	.78	2.29	.57
Masonry	5.4 to 34.5	14.9	6.9	7.5	1.9	1.03	1.12	.28
Structural Steel	8.6 to 104.1	40.8	10.7	3.9	17.6	4.37	1.59	7.18
Miscellaneous & Ornamental Metals	5.2 to 22.6	11.9	2.8	4.0	3.6	.33	.48	.43
Carpentry & Millwork	7.6 to 53.2	18.4	3.7	4.0	0.5	.68	.74	.09
Metal or Composition Siding	6.8 to 34.1	16.7	2.3	0.3	4.3	.38	.05	.72
Roofing	8.4 to 77.1	32.3	2.3	2.6	3.1	.74	.84	1.00
Doors & Hardware	5 to 24.9	11.1	0.9	1.4	0.4	.10	.16	.04
Sash & Glazing	5.6 to 28.9	14.0	3.5	4.0	1.0	.49	.56	.14
Lath & Plaster	3.1 to 35.4	14.0	3.3	6.9	0.8	.46	.97	.11
Tile, Marble & Floors	2 to 29.4	9.5	2.6	3.0	0.5	.25	.29	.05
Acoustical Ceilings	2.3 to 50.2	11.6	2.4	0.2	0.3	.28	.02	.03
Painting	4.9 to 29.6	13.2	1.5	1.6	1.6	.20	.21	.21
Interior Partitions	7.6 to 53.2	18.4	3.9	4.3	4.4	.72	.79	.81
Miscellaneous Items	2.3 to 197.9	16.8	5.2	3.7	9.7	.88	.62	1.63
Elevators	2.9 to 13.7	6.9	2.1	1.1	2.2	.14	.08	.15
Sprinklers	2.8 to 15.3	8.3	0.5	—	2.0	.04	—	.17
Plumbing	2.9 to 14.2	8.1	4.9	7.2	5.2	.40	.58	.42
Heat., Vent., Air Conditioning	4.6 to 24.8	11.9	13.5	11.0	12.9	1.61	1.31	1.54
Electrical	2.7 to 13.6	6.6	10.1	8.4	11.1	.67	.55	.73
Total	2 % to 197.9%	—	100.0%	100.0%	100.0%	16.57%	14.87%	18.63%
			Overall Weighted Average	16.69%				

Workers' Compensation Insurance Rates by States

The table below lists the weighted average Workers' Compensation base rate for each state with a factor comparing this with the national average of 16.3%.

State	Weighted Average	Factor	State	Weighted Average	Factor	State	Weighted Average	Factor
Alabama	26.2%	161	Kentucky	19.6%	120	North Dakota	13.7%	84
Alaska	25.0	153	Louisiana	28.4	174	Ohio	14.1	87
Arizona	7.5	46	Maine	21.1	129	Oklahoma	14.6	90
Arkansas	14.0	86	Maryland	19.2	118	Oregon	13.3	82
California	18.9	116	Massachusetts	15.3	94	Pennsylvania	13.0	80
Colorado	13.3	82	Michigan	17.4	107	Rhode Island	21.3	131
Connecticut	21.3	131	Minnesota	25.7	158	South Carolina	17.0	104
Delaware	19.0	117	Mississippi	18.5	113	South Dakota	18.6	114
District of Columbia	15.9	98	Missouri	18.1	111	Tennessee	15.6	96
Florida	20.4	125	Montana	17.9	110	Texas	13.2	81
Georgia	22.8	140	Nebraska	24.4	150	Utah	12.4	76
Hawaii	17.0	104	Nevada	12.5	77	Vermont	24.3	149
Idaho	12.1	74	New Hampshire	21.5	132	Virginia	12.6	77
Illinois	18.8	115	New Jersey	12.9	79	Washington	11.0	67
Indiana	6.6	40	New Mexico	19.4	119	West Virginia	9.5	58
Iowa	12.0	74	New York	13.8	85	Wisconsin	14.8	91
Kansas	8.7	53	North Carolina	17.3	106	Wyoming	9.9	61
			Weighted Average for U.S. is	16.7% of payroll = 100%				

Rates in the following table are the base or manual costs per $100 of payroll for Workers' Compensation in each state. Rates are usually applied to straight time wages only and not to premium time wages and bonuses.

The weighted average skilled worker rate for 35 trades is 16.3%. For bidding purposes, apply the full value of Workers' Compensation directly to total labor costs, or if labor is 38%, materials 42% and overhead and profit 20% of total cost, carry 38/80 x 16.3% =7.7% of cost (before overhead and profit) into overhead. Rates vary not only from state to state but also with the experience rating of the contractor.

Rates are the most current available at the time of publication.

R013113-60 Workers' Compensation Insurance Rates by Trade and State (cont.)

State	Carpentry — 3 stories or less	Carpentry — interior cab. work	Carpentry — general	Concrete Work — NOC	Concrete Work — flat (flr., sdwk.)	Electrical Wiring — inside	Excavation — earth NOC	Excavation — rock	Glaziers	Insulation Work	Lathing	Masonry	Painting & Decorating	Pile Driving	Plastering	Plumbing	Roofing	Sheet Metal Work (HVAC)	Steel Erection — door & sash	Steel Erection — inter., ornam.	Steel Erection — structure	Steel Erection — NOC	Tile Work — (interior ceramic)	Waterproofing	Wrecking
	5651	5437	5403	5213	5221	5190	6217	6217	5462	5479	5443	5022	5474	6003	5480	5183	5551	5538	5102	5102	5040	5057	5348	9014	5701
AL	34.13	19.34	33.95	12.96	10.74	9.34	14.00	14.00	23.27	16.95	13.45	26.80	27.85	42.90	27.83	10.23	57.63	24.81	16.77	16.77	64.80	28.72	12.05	7.74	64.80
AK	20.75	14.96	16.46	15.76	14.11	13.57	20.48	20.48	28.87	34.86	12.30	34.35	24.83	41.80	33.32	11.83	46.92	12.89	13.20	13.20	53.01	25.10	9.07	8.50	53.01
AZ	6.88	5.05	11.62	5.56	3.49	3.85	4.74	4.74	7.40	7.69	4.27	5.37	4.87	9.83	5.20	3.90	12.19	5.65	9.50	9.50	18.98	8.11	2.04	2.34	50.31
AR	15.09	8.06	16.49	13.30	6.63	5.14	8.49	8.49	11.79	23.72	6.63	9.32	9.64	16.15	13.39	5.21	28.01	13.09	7.75	7.75	36.16	21.58	6.29	4.74	36.16
CA	28.28	9.07	28.28	13.18	13.18	9.93	7.88	7.88	16.37	23.34	10.90	14.55	19.45	21.18	18.12	11.59	39.79	15.33	13.55	13.55	23.93	21.91	7.74	19.45	21.91
CO	16.28	8.03	12.76	13.19	7.46	5.18	10.35	10.35	9.79	13.91	6.03	14.30	10.29	15.08	9.00	8.45	23.95	11.46	8.38	8.38	35.52	15.85	8.60	5.74	15.85
CT	18.32	16.06	25.93	24.48	12.29	8.04	11.27	11.27	19.79	16.24	22.62	25.73	16.24	24.89	18.63	10.52	37.59	15.83	16.42	16.42	50.14	26.51	9.01	6.09	56.71
DE	19.38	19.38	15.10	15.16	12.03	8.22	12.07	12.07	14.07	19.38	16.67	15.69	20.18	25.15	16.67	10.12	34.39	12.62	16.11	16.11	35.78	16.11	12.50	15.69	35.78
DC	10.42	10.52	11.61	12.06	13.54	6.02	10.60	10.60	20.19	9.97	11.19	17.94	9.87	13.76	12.24	14.19	18.73	8.23	13.03	13.03	45.04	17.90	29.37	4.20	45.04
FL	24.38	15.34	24.74	21.86	10.37	8.91	11.12	11.12	15.53	14.31	10.58	17.48	16.78	38.24	28.82	10.04	35.40	14.90	11.74	11.74	44.71	25.87	10.03	7.76	44.71
GA	32.85	16.50	24.47	14.19	10.63	9.63	16.68	16.68	15.64	22.01	14.48	19.28	18.57	26.27	18.16	10.69	43.72	20.35	15.97	15.97	51.69	40.46	9.92	8.47	51.69
HI	17.38	11.62	28.20	13.74	12.27	7.10	7.73	7.73	20.55	21.19	10.94	17.87	11.33	20.15	15.61	6.06	33.24	8.02	11.11	11.11	31.71	21.70	9.78	11.54	31.71
ID	11.92	6.31	13.47	10.64	5.99	5.25	6.90	6.90	9.58	8.14	6.31	9.91	8.88	15.41	9.45	5.99	29.54	9.11	7.94	7.94	33.81	14.13	13.67	4.09	33.81
IL	17.33	15.90	18.72	26.98	10.79	8.21	9.42	9.42	19.93	15.96	10.26	18.26	10.08	25.32	13.09	9.80	29.37	14.49	15.54	15.54	50.39	21.95	14.63	4.66	50.39
IN	6.77	5.04	7.58	5.08	3.27	2.66	4.64	4.64	5.63	5.55	2.25	5.64	5.54	9.05	4.04	2.94	11.82	4.60	5.20	5.20	21.57	8.02	3.19	2.68	21.57
IA	11.41	6.24	9.43	11.35	6.29	4.24	5.17	5.17	11.28	8.38	5.51	8.91	7.85	8.93	7.26	5.08	16.80	7.56	7.75	7.75	46.05	40.59	5.48	3.96	27.19
KS	8.44	8.42	10.36	7.95	5.18	3.44	4.51	4.51	8.63	8.03	4.28	7.83	6.15	11.55	6.35	5.42	8.44	7.02	7.09	7.09	25.33	12.78	5.36	3.20	27.57
KY	22.12	12.10	19.05	18.45	7.85	6.61	11.25	11.25	20.44	20.69	13.02	10.03	13.56	25.37	15.33	7.48	50.97	19.58	14.49	14.49	46.98	23.88	11.89	6.30	46.98
LA	23.14	24.93	53.17	26.43	15.61	9.88	17.49	17.49	20.14	21.43	24.51	28.33	29.58	31.25	22.37	8.64	77.12	21.20	18.98	18.98	51.76	24.21	13.77	13.78	66.41
ME	13.43	10.42	43.59	24.29	11.12	3.72	12.56	12.56	13.35	15.39	14.03	17.26	16.07	31.44	17.95	9.07	32.07	10.07	15.08	15.08	37.84	61.81	11.66	5.92	37.84
MD	13.71	5.95	10.55	18.82	8.70	5.15	12.91	12.91	25.52	20.07	11.37	13.31	9.37	26.22	14.21	8.58	48.72	12.74	13.66	13.66	56.25	37.81	8.03	7.14	26.80
MA	9.03	6.88	16.48	22.32	9.18	4.18	6.08	6.08	9.25	14.59	5.89	15.67	6.85	15.92	6.62	4.80	47.57	6.72	13.48	13.48	43.69	34.06	8.78	2.65	24.03
MI	18.40	11.39	20.64	19.45	8.95	4.72	10.20	10.20	12.41	14.75	13.59	15.54	13.64	40.27	15.18	6.84	30.67	10.00	10.16	10.16	40.27	22.21	11.11	4.87	39.06
MN	19.53	19.80	41.48	14.65	15.73	7.53	14.95	14.95	15.68	11.90	22.05	19.18	16.95	24.63	22.05	10.88	70.18	14.23	10.58	10.58	104.13	33.30	14.55	6.68	36.03
MS	15.56	12.84	23.21	11.49	8.98	6.57	12.44	12.44	15.13	18.03	7.40	9.64	14.52	24.87	35.36	8.27	35.09	19.67	13.57	13.57	35.55	29.83	10.22	5.63	35.55
MO	23.87	12.38	16.85	16.75	11.59	7.61	9.87	9.87	11.40	18.47	10.12	14.58	15.09	18.46	14.96	10.16	33.26	15.34	11.85	11.85	47.95	33.09	9.40	6.12	47.95
MT	16.07	12.88	21.13	12.84	11.28	7.97	14.85	14.85	11.94	30.09	15.86	13.71	9.70	33.39	11.78	10.03	39.71	11.20	9.70	9.70	36.35	18.79	7.76	6.04	36.35
NE	27.13	14.63	21.57	29.20	14.30	10.22	17.88	17.88	25.50	35.50	12.00	25.82	22.27	24.63	20.60	11.52	39.10	18.35	14.60	14.60	65.40	26.27	12.02	7.63	58.22
NV	16.61	7.10	11.57	9.28	6.51	5.39	9.50	9.50	12.80	10.21	8.27	8.90	9.60	14.19	8.05	6.78	14.14	19.04	10.96	10.96	23.34	26.59	6.18	5.18	33.18
NH	26.55	11.21	21.62	32.68	12.33	7.21	17.79	17.79	11.16	29.06	9.32	22.91	15.04	21.80	13.65	10.84	58.14	14.88	12.51	12.51	49.35	18.42	11.43	6.01	49.35
NJ	13.50	9.36	13.50	12.14	8.95	4.45	8.62	8.62	7.20	13.18	11.54	13.14	11.44	13.54	11.54	6.26	34.09	6.04	11.03	11.03	25.41	13.62	4.87	4.83	23.62
NM	24.50	8.22	21.85	20.69	11.35	7.41	11.51	11.51	19.30	14.97	10.20	16.38	16.09	24.75	12.34	9.96	33.44	12.58	18.13	18.13	52.69	29.16	7.79	7.63	52.69
NY	14.21	12.39	12.39	17.19	11.26	6.69	8.37	8.37	10.18	9.91	11.27	15.70	10.41	15.68	9.04	7.33	33.93	12.79	9.59	9.59	21.71	15.20	8.72	6.92	24.30
NC	15.87	13.33	16.07	15.61	7.64	10.21	10.51	10.51	11.41	13.79	14.12	11.11	12.95	17.63	13.75	9.20	26.70	13.53	9.22	9.22	79.26	20.69	6.50	5.68	79.26
ND	10.92	10.92	10.92	6.26	6.26	3.90	6.46	6.46	10.92	10.92	8.84	7.94	6.41	22.61	8.84	5.47	22.68	5.47	22.61	22.61	22.61	22.61	10.92	22.68	12.19
OH	7.70	7.61	9.64	11.57	10.29	5.45	8.10	8.10	7.09	17.31	50.15	11.49	11.69	28.24	3.08	6.58	24.34	9.30	8.14	8.14	26.41	10.66	9.21	5.70	26.41
OK	13.42	9.17	11.76	11.76	6.34	5.73	10.83	10.83	16.07	19.47	8.45	10.02	8.47	20.25	12.19	7.09	21.13	9.05	13.76	13.76	39.72	23.97	6.50	5.71	39.78
OR	16.00	8.81	15.78	13.64	8.79	4.26	9.88	9.88	14.41	10.15	7.41	14.86	10.88	15.81	11.82	5.87	23.65	9.60	8.95	8.95	28.75	14.60	11.37	4.49	28.75
PA	12.67	6.29	11.41	13.57	9.77	5.98	7.88	7.88	9.15	11.41	10.98	11.40	13.00	15.49	10.98	6.94	26.17	7.49	13.91	13.91	21.88	13.91	7.85	11.40	21.88
RI	19.53	11.65	18.07	18.23	16.24	4.43	10.38	10.38	12.85	22.78	11.97	25.11	24.13	37.66	17.25	8.52	33.92	10.29	14.07	14.07	59.49	37.50	14.36	7.70	78.79
SC	22.26	17.21	19.93	13.51	7.81	9.42	11.95	11.95	14.32	11.08	8.53	13.27	16.47	18.32	18.15	8.83	43.19	10.97	11.47	11.47	24.55	27.29	8.90	5.98	47.05
SD	22.94	7.74	24.42	28.24	8.33	5.60	12.27	12.27	11.43	12.46	8.11	9.46	10.34	21.36	12.74	12.97	19.88	12.48	8.97	8.97	84.39	33.82	8.80	4.25	84.39
TN	17.21	13.36	22.83	16.35	9.27	7.36	12.04	12.04	13.17	11.21	11.71	14.74	10.49	18.13	12.60	9.19	24.86	13.88	9.95	9.95	32.82	20.02	8.84	5.72	32.82
TX	13.44	9.34	13.44	11.97	7.27	7.70	8.65	8.65	9.83	12.10	7.83	12.38	9.08	20.20	9.08	6.92	19.68	16.57	12.11	12.11	30.40	12.18	6.82	6.89	13.00
UT	9.45	6.23	10.52	7.68	7.99	7.05	6.38	6.38	9.55	10.95	12.32	13.75	15.29	17.24	10.60	6.41	26.27	6.30	8.99	8.99	23.51	23.51	6.06	5.83	26.53
VT	23.50	10.31	19.77	29.84	12.44	8.34	14.86	14.86	24.74	25.93	10.12	20.66	14.15	21.89	15.99	10.47	45.95	14.75	14.56	14.56	74.50	62.00	10.22	10.24	74.50
VA	11.41	7.88	10.02	11.54	5.24	5.47	7.04	7.04	8.77	10.82	18.78	8.94	9.54	12.50	9.73	5.46	21.49	8.23	9.60	9.60	43.89	22.16	5.03	2.89	43.89
WA	9.30	9.30	9.30	8.12	8.12	3.35	8.04	8.04	12.92	10.56	9.30	10.21	9.88	17.82	11.50	4.64	18.57	4.61	8.65	8.65	8.65	8.65	10.40	18.57	7.87
WV	9.77	7.20	10.47	8.58	4.65	4.82	5.92	5.92	7.80	7.09	7.11	8.14	8.24	11.16	8.73	4.78	19.23	7.63	6.45	6.45	26.02	12.50	4.53	2.83	23.67
WI	12.40	11.36	16.86	12.29	9.90	4.24	7.44	7.44	13.31	11.62	8.09	18.99	14.32	13.64	12.11	5.42	34.72	7.16	9.18	9.18	37.88	24.06	13.99	5.52	37.88
WY	8.93	8.93	8.93	8.93	8.93	8.93	8.93	8.93	8.93	8.93	8.93	8.93	8.93	8.93	8.93	8.93	8.93	8.93	8.93	8.93	8.93	8.93	8.93	8.93	8.93
AVG.	16.75	11.08	18.39	15.53	9.55	6.59	10.39	10.39	14.05	15.81	11.61	14.92	13.19	21.39	14.04	8.10	32.30	11.89	11.86	11.86	40.80	23.82	9.53	7.16	39.10

R013113-60 Workers' Compensation (cont.) (Canada in Canadian dollars)

Province		Alberta	British Columbia	Manitoba	Ontario	New Brunswick	Newfndld. & Labrador	Northwest Territories	Nova Scotia	Prince Edward Island	Quebec	Saskat-chewan	Yukon
Carpentry—3 stories or less	Rate	8.17	5.91	4.48	4.58	3.94	8.84	4.16	7.67	5.73	15.41	7.33	4.79
	Code	42143	721028	40102	723	4226	4226	4-41	4226	401	80110	B1317	202
Carpentry—interior cab. work	Rate	2.56	6.13	4.48	4.58	4.67	4.92	4.16	5.74	3.98	15.41	3.68	4.79
	Code	42133	721021	40102	723	4279	4270	4-41	4274	402	80110	B11-27	202
CARPENTRY—general	Rate	8.17	5.91	4.48	4.58	3.94	4.92	4.16	7.67	5.73	15.41	7.33	4.79
	Code	42143	721028	40102	723	4226	4299	4-41	4226	401	80110	B1317	202
CONCRETE WORK—NOC	Rate	5.12	6.02	7.42	15.40	3.94	8.84	4.16	4.83	5.73	16.51	7.33	3.86
	Code	42104	721010	40110	748	4224	4224	4-41	4224	401	80100	B13-14	203
CONCRETE WORK—flat (flr. sidewalk)	Rate	5.12	6.02	7.42	15.40	3.94	8.84	4.16	4.83	5.73	16.51	7.33	3.86
	Code	42104	721010	40110	748	4224	4224	4-41	4224	401	80100	B13-14	203
ELECTRICAL Wiring—inside	Rate	2.22	2.18	2.56	3.25	2.45	2.92	3.91	2.23	3.98	7.64	3.68	3.86
	Code	42124	721019	40203	704	4261	4261	4-46	4261	402	80170	B11-05	206
EXCAVATION—earth NOC	Rate	2.95	4.20	3.78	4.55	2.88	4.29	4.04	4.11	4.23	8.36	4.37	3.86
	Code	40604	721031	40706	711	4214	4214	4-43	4214	404	80030	R11-06	207
EXCAVATION—rock	Rate	2.95	4.20	3.78	4.55	2.88	4.29	4.04	4.11	4.23	8.36	4.37	3.86
	Code	40604	721031	40706	711	4214	4214	4-43	4214	404	80030	R11-06	207
GLAZIERS	Rate	3.66	4.07	4.48	8.90	5.20	6.52	4.16	7.67	3.98	14.36	7.33	3.86
	Code	42121	715020	40109	751	4233	4233	4-41	4233	402	80150	B13-04	212
INSULATION WORK	Rate	3.03	7.02	4.48	8.90	5.20	6.52	4.16	7.67	5.73	15.41	6.15	4.79
	Code	42184	721029	40102	751	4234	4234	4-41	4234	401	80110	B12-07	202
LATHING	Rate	6.18	8.48	4.48	4.58	4.67	4.92	4.16	5.74	3.98	15.41	7.33	4.79
	Code	42135	721033	40102	723	4273	4279	4-41	4271	402	80110	B13-16	202
MASONRY	Rate	5.12	8.48	4.48	11.79	5.20	6.52	4.16	7.67	5.73	16.51	7.33	4.79
	Code	42102	721037	40102	741	4231	4231	4-41	4231	401	80100	B13-18	202
PAINTING & DECORATING	Rate	5.08	5.21	36.17	6.75	4.67	4.92	4.16	5.74	3.98	15.41	6.15	4.79
	Code	42111	721041	40105	719	4275	4275	4-41	4275	402	80110	B12-01	202
PILE DRIVING	Rate	5.12	5.52	3.78	6.26	3.94	9.78	4.04	4.83	5.73	8.36	7.33	4.79
	Code	42159	722004	40706	732	4221	4221	4-43	4221	401	80030	B13-10	202
PLASTERING	Rate	6.18	8.48	5.38	6.75	4.67	4.92	4.16	5.74	3.98	15.41	6.15	4.79
	Code	42135	721042	40108	719	4271	4271	4-41	4271	402	80110	B12-21	202
PLUMBING	Rate	2.22	3.89	2.92	4.02	2.89	3.24	3.91	2.23	3.98	7.61	3.68	3.27
	Code	42122	721043	40204	707	4241	4241	4-46	4241	402	80160	B11-01	214
ROOFING	Rate	9.20	10.86	7.12	12.53	8.10	8.84	4.16	9.49	5.73	22.49	7.33	4.79
	Code	42118	721036	40403	728	4236	4236	4-41	4236	401	80130	B13-20	202
SHEET METAL WORK (HVAC)	Rate	2.22	3.89	7.12	4.02	2.89	3.24	3.91	3.23	3.98	7.61	3.68	4.35
	Code	42117	721043	40402	707	4244	4244	4-46	4244	402	80160	B11-07	208
STEEL ERECTION—door & sash	Rate	3.03	16.01	12.96	15.40	3.94	8.84	4.16	7.67	5.73	29.92	7.33	4.79
	Code	42106	722005	40502	748	4227	4227	4-41	4227	401	80080	B13-22	202
STEEL ERECTION—inter., ornam.	Rate	3.03	16.01	12.96	15.40	3.94	8.84	4.16	7.67	5.73	29.92	7.33	4.79
	Code	42106	722005	40502	748	4227	4227	4-41	4227	401	80080	B13-22	202
STEEL ERECTION—structure	Rate	3.03	16.01	12.96	15.47	3.94	8.84	4.16	7.67	5.73	29.92	7.33	4.79
	Code	42106	722005	40502	748	4227	4227	4-41	4227	401	80080	B13-22	202
STEEL ERECTION—NOC	Rate	3.03	16.01	12.96	15.40	3.94	8.84	4.16	7.67	5.73	29.92	7.33	4.79
	Code	42106	722005	40502	748	4227	4227	4-41	4227	401	80080	B13-22	202
TILE WORK—inter. (ceramic)	Rate	4.22	6.58	2.05	6.75	4.67	4.92	4.16	5.74	3.98	15.41	7.33	4.79
	Code	42113	721054	40103	719	4276	4276	4-41	4276	402	80110	B13-01	202
WATERPROOFING	Rate	5.08	5.19	4.48	4.58	5.20	4.92	4.16	7.67	3.98	22.49	6.15	4.79
	Code	42139	721016	40102	723	4239	4299	4-41	4239	402	80130	B12-17	202
WRECKING	Rate	2.95	5.74	7.07	15.40	2.88	4.29	4.04	4.11	5.73	15.41	7.33	4.79
	Code	40604	721005	40106	748	4211	4211	4-43	4211	401	80110	B13-09	202

R013113-80   Performance Bond

This table shows the cost of a Performance Bond for a construction job scheduled to be completed in 12 months. Add 1% of the premium cost per month for jobs requiring more than 12 months to complete. The rates are "standard" rates offered to contractors that the bonding company considers financially sound and capable of doing the work. Preferred rates are offered by some bonding companies based upon financial strength of the contractor. Actual rates vary from contractor to contractor and from bonding company to bonding company. Contractors should prequalify through a bonding agency before submitting a bid on a contract that requires a bond.

Contract Amount	Building Construction Class B Projects			Highways & Bridges					
				Class A New Construction			Class A-1 Highway Resurfacing		
First $ 100,000 bid	$25.00 per M			$15.00 per M			$9.40 per M		
Next 400,000 bid	$ 2,500	plus $15.00	per M	$ 1,500	plus $10.00	per M	$ 940	plus $7.20	per M
Next 2,000,000 bid	8,500	plus 10.00	per M	5,500	plus 7.00	per M	3,820	plus 5.00	per M
Next 2,500,000 bid	28,500	plus 7.50	per M	19,500	plus 5.50	per M	15,820	plus 4.50	per M
Next 2,500,000 bid	47,250	plus 7.00	per M	33,250	plus 5.00	per M	28,320	plus 4.50	per M
Over 7,500,000 bid	64,750	plus 6.00	per M	45,750	plus 4.50	per M	39,570	plus 4.00	per M

R015113-65   Temporary Power Equipment

Cost data for the temporary equipment was developed utilizing the following information.

1) Re-usable material-services, transformers, equipment and cords are based on new purchase and prorated to three projects.
2) PVC feeder includes trench and backfill.
3) Connections include disconnects and fuses.
4) Labor units include an allowance for removal.
5) No utility company charges or fees are included.
6) Concrete pads or vaults are not included.
7) Utility company conduits not included.

R015423-10   Steel Tubular Scaffolding

On new construction, tubular scaffolding is efficient up to 60' high or five stories. Above this it is usually better to use a hung scaffolding if construction permits. Swing scaffolding operations may interfere with tenants. In this case, the tubular is more practical at all heights.

In repairing or cleaning the front of an existing building the cost of tubular scaffolding per S.F. of building front increases as the height increases above the first tier. The first tier cost is relatively high due to leveling and alignment.

The minimum efficient crew for erecting and dismantling is three workers. They can set up and remove 18 frame sections per day up to 5 stories high. For 6 to 12 stories high, a crew of four is most efficient. Use two or more on top and two on the bottom for handing up or hoisting. They can also set up and remove 18 frame sections per day. At 7' horizontal spacing, this will run about 800 SF per day of erecting and dismantling. Time for placing and removing planks must be added to the above. A crew of three can place and remove 72 planks per day up to 5 stories. For over 5 stories, a crew of four can place and remove 80 planks per day.

The table below shows the number of pieces required to erect tubular steel scaffolding for 1000 S.F. of building frontage. This area is made up of a scaffolding system that is 12 frames (11 bays) long by 2 frames high.

For jobs under twenty-five frames, add 50% to rental cost. Rental rates will be lower for jobs over three months duration. Large quantities for long periods can reduce rental rates by 20%.

Description of Component	CSI Line Item	Number of Pieces for 1000 S.F. of Building Front	Unit
5' Wide Standard Frame, 6'-4" High	01540-750-2200	24	Ea.
Leveling Jack & Plate	01540-750-2650	24	
Cross Brace	01540-750-2500	44	
Side Arm Bracket, 21"	01540-750-2700	12	
Guardrail Post	01540-750-2550	12	
Guardrail, 7' section	01540-750-2600	22	
Stairway Section	01540-750-2900	2	
Stairway Starter Bar	01540-750-2910	1	
Stairway Inside Handrail	01540-750-2920	2	
Stairway Outside Handrail	01540-750-2930	2	
Walk-Thru Frame Guardrail	01540-750-2940	2	

Scaffolding is often used as falsework over 15' high during construction of cast-in-place concrete beams and slabs. Two foot wide scaffolding is generally used for heavy beam construction. The span between frames depends upon the load to be carried with a maximum span of 5'.

Heavy duty shoring frames with a capacity of 10,000#/leg can be spaced up to 10' O.C. depending upon form support design and loading.

Scaffolding used as horizontal shoring requires less than half the material required with conventional shoring.

On new construction, erection is done by carpenters.

Rolling towers supporting horizontal shores can reduce labor and speed the job. For maintenance work, catwalks with spans up to 70' can be supported by the rolling towers.

R015433-10   Contractor Equipment

Rental Rates shown elsewhere in the book pertain to late model high quality machines in excellent working condition, rented from equipment dealers. Rental rates from contractors may be substantially lower than the rental rates from equipment dealers depending upon economic conditions; for older, less productive machines, reduce rates by a maximum of 15%. Any overtime must be added to the base rates. For shift work, rates are lower. Usual rule of thumb is 150% of one shift rate for two shifts; 200% for three shifts.

For periods of less than one week, operated equipment is usually more economical to rent than renting bare equipment and hiring an operator.

Costs to move equipment to a job site (mobilization) or from a job site (demobilization) are not included in rental rates, nor in any Equipment costs on any Unit Price line items or crew listings. These costs can be found elsewhere. If a piece of equipment is already at a job site, it is not appropriate to utilize mob/demob costs in an estimate again.

Rental rates vary throughout the country with larger cities generally having lower rates. Lease plans for new equipment are available for periods in excess of six months with a percentage of payments applying toward purchase.

Monthly rental rates vary from 2% to 5% of the cost of the equipment depending on the anticipated life of the equipment and its wearing parts. Weekly rates are about 1/3 the monthly rates and daily rental rates about 1/3 the weekly rate.

The hourly operating costs for each piece of equipment include costs to the user such as fuel, oil, lubrication, normal expendables for the equipment, and a percentage of mechanic's wages chargeable to maintenance. The hourly operating costs listed do not include the operator's wages.

The daily cost for equipment used in the standard crews is figured by dividing the weekly rate by five, then adding eight times the hourly operating cost to give the total daily equipment cost, not including the operator. This figure is in the right hand column of the Equipment listings under Crew Equipment Cost/Day.

Pile Driving rates shown for pile hammer and extractor do not include leads, crane, boiler or compressor. Vibratory pile driving requires an added field specialist during set-up and pile driving operation for the electric model. The hydraulic model requires a field specialist for set-up only. Up to 125 reuses of sheet piling are possible using vibratory drivers. For normal conditions, crane capacity for hammer type and size are as follows.

Crane Capacity	Hammer Type and Size		
	Air or Steam	Diesel	Vibratory
25 ton	to 8,750 ft.-lb.		70 H.P.
40 ton	15,000 ft.-lb.	to 32,000 ft.-lb.	170 H.P.
60 ton	25,000 ft.-lb.		300 H.P.
100 ton		112,000 ft.-lb.	

Cranes should be specified for the job by size, building and site characteristics, availability, performance characteristics, and duration of time required.

Backhoes & Shovels rent for about the same as equivalent size cranes but maintenance and operating expense is higher. Crane operators rate must be adjusted for high boom heights. Average adjustments: for 150' boom add 2% per hour; over 185', add 4% per hour; over 210', add 6% per hour; over 250', add 8% per hour and over 295', add 12% per hour.

Tower Cranes of the climbing or static type have jibs from 50' to 200' and capacities at maximum reach range from 4,000 to 14,000 pounds. Lifting capacities increase up to maximum load as the hook radius decreases.

Typical rental rates, based on purchase price are about 2% to 3% per month.

Erection and dismantling runs between 500 and 2000 labor hours. Climbing operation takes 10 labor hours per 20' climb. Crane dead time is about 5 hours per 40' climb. If crane is bolted to side of the building add cost of ties and extra mast sections. Climbing cranes have from 80' to 180' of mast while static cranes have 80' to 800' of mast.

Truck Cranes can be converted to tower cranes by using tower attachments. Mast heights over 400' have been used.

A single 100' high material **Hoist and Tower** can be erected and dismantled in about 400 labor hours; a double 100' high hoist and tower in about 600 labor hours. Erection times for additional heights are 3 and 4 labor hours per vertical foot respectively up to 150', and 4 to 5 labor hours per vertical foot over 150' high. A 40' high portable Buck hoist takes about 160 labor hours to erect and dismantle. Additional heights take 2 labor hours per vertical foot to 80' and 3 labor hours per vertical foot for the next 100'. Most material hoists do not meet local code requirements for carrying personnel.

A 150' high **Personnel Hoist** requires about 500 to 800 labor hours to erect and dismantle. Budget erection time at 5 labor hours per vertical foot for all trades. Local code requirements or labor scarcity requiring overtime can add up to 50% to any of the above erection costs.

Earthmoving Equipment: The selection of earthmoving equipment depends upon the type and quantity of material, moisture content, haul distance, haul road, time available, and equipment available. Short haul cut and fill operations may require dozers only, while another operation may require excavators, a fleet of trucks, and spreading and compaction equipment. Stockpiled material and granular material are easily excavated with front end loaders. Scrapers are most economically used with hauls between 300' and 1-1/2 miles if adequate haul roads can be maintained. Shovels are often used for blasted rock and any material where a vertical face of 8' or more can be excavated. Special conditions may dictate the use of draglines, clamshells, or backhoes. Spreading and compaction equipment must be matched to the soil characteristics, the compaction required and the rate the fill is being supplied.

R024119-10   Demolition Defined

Whole Building Demolition - Demolition of the whole building with no concern for any particular building element, component, or material type being demolished. This type of demolition is accomplished with large pieces of construction equipment that break up the structure, load it into trucks and haul it to a disposal site, but disposal or dump fees are not included. Demolition of below-grade foundation elements, such as footings, foundation walls, grade beams, slabs on grade, etc., is not included. Certain mechanical equipment containing flammable liquids or ozone-depleting refrigerants, electric lighting elements, communication equipment components, and other building elements may contain hazardous waste, and must be removed, either selectively or carefully, as hazardous waste before the building can be demolished.

Foundation Demolition - Demolition of below-grade foundation footings, foundation walls, grade beams, and slabs on grade. This type of demolition is accomplished by hand or pneumatic hand tools, and does not include saw cutting, or handling, loading, hauling, or disposal of the debris.

Gutting - Removal of building interior finishes and electrical/mechanical systems down to the load-bearing and sub-floor elements of the rough building frame, with no concern for any particular building element, component, or material type being demolished. This type of demolition is accomplished by hand or pneumatic hand tools, and includes loading into trucks, but not hauling, disposal or dump fees, scaffolding, or shoring. Certain mechanical equipment containing flammable liquids or ozone-depleting refrigerants, electric lighting elements, communication equipment components, and other building elements may contain hazardous waste, and must be removed, either selectively or carefully, as hazardous waste, before the building is gutted.

Selective Demolition - Demolition of a selected building element, component, or finish, with some concern for surrounding or adjacent elements, components, or finishes (see the first Subdivision (s) at the beginning of appropriate Divisions). This type of demolition is accomplished by hand or pneumatic hand tools, and does not include handling, loading,

storing, hauling, or disposal of the debris, scaffolding, or shoring. "Gutting" methods may be used in order to save time, but damage that is caused to surrounding or adjacent elements, components, or finishes may have to be repaired at a later time.

Careful Removal - Removal of a piece of service equipment, building element or component, or material type, with great concern for both the removed item and surrounding or adjacent elements, components or finishes. The purpose of careful removal may be to protect the removed item for later re-use, preserve a higher salvage value of the removed item, or replace an item while taking care to protect surrounding or adjacent elements, components, connections, or finishes from cosmetic and/or structural damage. An approximation of the time required to perform this type of removal is 1/3 to 1/2 the time it would take to install a new item of like kind (see Reference Number R260105-30). This type of removal is accomplished by hand or pneumatic hand tools, and does not include loading, hauling, or storing the removed item, scaffolding, shoring, or lifting equipment.

Cutout Demolition - Demolition of a small quantity of floor, wall, roof, or other assembly, with concern for the appearance and structural integrity of the surrounding materials. This type of demolition is accomplished by hand or pneumatic hand tools, and does not include saw cutting, handling, loading, hauling, or disposal of debris, scaffolding, or shoring.

Rubbish Handling - Work activities that involve handling, loading or hauling of debris. Generally, the cost of rubbish handling must be added to the cost of all types of demolition, with the exception of whole building demolition.

Minor Site Demolition - Demolition of site elements outside the footprint of a building. This type of demolition is accomplished by hand or pneumatic hand tools, or with larger pieces of construction equipment, and may include loading a removed item onto a truck (check the Crew for equipment used). It does not include saw cutting, hauling or disposal of debris, and, sometimes, handling or loading.

R053100-10 Decking Descriptions

General - All Deck Products

Steel deck is made by cold forming structural grade sheet steel into a repeating pattern of parallel ribs. The strength and stiffness of the panels are the result of the ribs and the material properties of the steel. Deck lengths can be varied to suit job conditions, but because of shipping considerations, are usually less than 40 feet. Standard deck width varies with the product used but full sheets are usually 12″, 18″, 24″, 30″, or 36″. Deck is typically furnished in a standard width with the ends cut square. Any cutting for width, such as at openings or for angular fit, is done at the job site.

Deck is typically attached to the building frame with arc puddle welds, self-drilling screws, or powder or pneumatically driven pins. Sheet to sheet fastening is done with screws, button punching (crimping), or welds.

Composite Floor Deck

After installation and adequate fastening, floor deck serves several purposes. It (a) acts as a working platform, (b) stabilizes the frame, (c) serves as a concrete form for the slab, and (d) reinforces the slab to carry the design loads applied during the life of the building. Composite decks are distinguished by the presence of shear connector devices as part of the deck. These devices are designed to mechanically lock the concrete and deck together so that the concrete and the deck work together to carry subsequent floor loads. These shear connector devices can be rolled-in embossments, lugs, holes, or wires welded to the panels. The deck profile can also be used to interlock concrete and steel.

Composite deck finishes are either galvanized (zinc coated) or phosphatized/painted. Galvanized deck has a zinc coating on both the top and bottom surfaces. The phosphatized/painted deck has a bare (phosphatized) top surface that will come into contact with the concrete. This bare top surface can be expected to develop rust before the concrete is placed. The bottom side of the deck has a primer coat of paint.

Composite floor deck is normally installed so the panel ends do not overlap on the supporting beams. Shear lugs or panel profile shape often prevent a tight metal to metal fit if the panel ends overlap; the air gap caused by overlapping will prevent proper fusion with the structural steel supports when the panel end laps are shear stud welded.

Adequate end bearing of the deck must be obtained as shown on the drawings. If bearing is actually less in the field than shown on the drawings, further investigation is required.

Roof Deck

Roof deck is not designed to act compositely with other materials. Roof deck acts alone in transferring horizontal and vertical loads into the building frame. Roof deck rib openings are usually narrower than floor deck rib openings. This provides adequate support of rigid thermal insulation board.

Roof deck is typically installed to endlap approximately 2″ over supports. However, it can be butted (or lapped more than 2″) to solve field fit problems. Since designers frequently use the installed deck system as part of the horizontal bracing system (the deck as a diaphragm), any fastening substitution or change should be approved by the designer. Continuous perimeter support of the deck is necessary to limit edge deflection in the finished roof and may be required for diaphragm shear transfer.

Standard roof deck finishes are galvanized or primer painted. The standard factory applied paint for roof deck is a primer paint and is not intended to weather for extended periods of time. Field painting or touching up of abrasions and deterioration of the primer coat or other protective finishes is the responsibility of the contractor.

Cellular Deck

Cellular deck is made by attaching a bottom steel sheet to a roof deck or composite floor deck panel. Cellular deck can be used in the same manner as floor deck. Electrical, telephone, and data wires are easily run through the chase created between the deck panel and the bottom sheet.

When used as part of the electrical distribution system, the cellular deck must be installed so that the ribs line up and create a smooth cell transition at abutting ends. The joint that occurs at butting cell ends must be taped or otherwise sealed to prevent wet concrete from seeping into the cell. Cell interiors must be free of welding burrs, or other sharp intrusions, to prevent damage to wires.

When used as a roof deck, the bottom flat plate is usually left exposed to view. Care must be maintained during erection to keep good alignment and prevent damage.

Cellular deck is sometimes used with the flat plate on the top side to provide a flat working surface. Installation of the deck for this purpose requires special methods for attachment to the frame because the flat plate, now on the top, can prevent direct access to the deck material that is bearing on the structural steel. It may be advisable to treat the flat top surface to prevent slipping.

Cellular deck is always furnished galvanized or painted over galvanized.

Form Deck

Form deck can be any floor or roof deck product used as a concrete form. Connections to the frame are by the same methods used to anchor floor and roof deck. Welding washers are recommended when welding deck that is less than 20 gauge thickness.

Form deck is furnished galvanized, prime painted, or uncoated. Galvanized deck must be used for those roof deck systems where form deck is used to carry a lightweight insulating concrete fill.

Thermal & Moist. Protec. | **R0784 Firestopping**

R078413-30 Firestopping

Firestopping is the sealing of structural, mechanical, electrical and other penetrations through fire-rated assemblies. The basic components of firestop systems are safing insulation and firestop sealant on both sides of wall penetrations and the top side of floor penetrations.

Pipe penetrations are assumed to be through concrete, grout, or joint compound and can be sleeved or unsleeved. Costs for the penetrations and sleeves are not included. An annular space of 1″ is assumed. Escutcheons are not included.

Metallic pipe is assumed to be copper, aluminum, cast iron or similar metallic material. Insulated metallic pipe is assumed to be covered with a thermal insulating jacket of varying thickness and materials.

Non-metallic pipe is assumed to be PVC, CPVC, FR Polypropylene or similar plastic piping material. Intumescent firestop sealant or wrap strips are included. Collars on both sides of wall penetrations and a sheet metal plate on the underside of floor penetrations are included.

Ductwork is assumed to be sheet metal, stainless steel or similar metallic material. Duct penetrations are assumed to be through concrete, grout or joint compound. Costs for penetrations and sleeves are not included. An annular space of 1/2″ is assumed.

Multi-trade openings include costs for sheet metal forms, firestop mortar, wrap strips, collars and sealants as necessary.

Structural penetrations joints are assumed to be 1/2″ or less. CMU walls are assumed to be within 1-1/2″ of metal deck. Drywall walls are assumed to be tight to the underside of metal decking.

Metal panel, glass or curtain wall systems include a spandrel area of 5′ filled with mineral wool foil-faced insulation. Fasteners and stiffeners are included.

R224000-10 Hot Water Consumption Rates

Type of Building	Size Factor	Maximum Hourly Demand	Average Day Demand
Apartment Dwellings	No. of Apartments: Up to 20 21 to 50 51 to 75 76 to 100 101 to 200 201 up	12.0 Gal. per apt. 10.0 Gal. per apt. 8.5 Gal. per apt. 7.0 Gal. per apt. 6.0 Gal. per apt. 5.0 Gal. per apt.	42.0 Gal. per apt. 40.0 Gal. per apt. 38.0 Gal. per apt. 37.0 Gal. per apt. 36.0 Gal. per apt. 35.0 Gal. per apt.
Dormitories	Men Women	3.8 Gal. per man 5.0 Gal. per woman	13.1 Gal. per man 12.3 Gal. per woman
Hospitals	Per bed	23.0 Gal. per patient	90.0 Gal. per patient
Hotels	Single room with bath Double room with bath	17.0 Gal. per unit 27.0 Gal. per unit	50.0 Gal. per unit 80.0 Gal. per unit
Motels	No. of units: Up to 20 21 to 100 101 Up	6.0 Gal. per unit 5.0 Gal. per unit 4.0 Gal. per unit	20.0 Gal. per unit 14.0 Gal. per unit 10.0 Gal. per unit
Nursing Homes		4.5 Gal. per bed	18.4 Gal. per bed
Office buildings		0.4 Gal. per person	1.0 Gal. per person
Restaurants	Full meal type Drive-in snack type	1.5 Gal./max. meals/hr. 0.7 Gal./max. meals/hr.	2.4 Gal. per meal 0.7 Gal. per meal
Schools	Elementary Secondary & High	0.6 Gal. per student 1.0 Gal. per student	0.6 Gal. per student 1.8 Gal. per student

For evaluation purposes, recovery rate and storage capacity are inversely proportional. Water heaters should be sized so that the maximum hourly demand anticipated can be met in addition to allowance for the heat loss from the pipes and storage tank.

R224000-20 Fixture Demands in Gallons Per Fixture Per Hour

Table below is based on 140°F final temperature except for dishwashers in public places (*) where 180°F water is mandatory.

Fixture	Apartment House	Club	Gym	Hospital	Hotel	Indust. Plant	Office	Private Home	School
Bathtubs	20	20	30	20	20			20	
Dishwashers, automatic	15	50-150*		50-150*	50-200*	20-100*		15	20-100*
Kitchen sink	10	20		20	30	20	20	10	20
Laundry, stationary tubs	20	28		28	28			20	
Laundry, automatic wash	75	75		100	150			75	
Private lavatory	2	2	2	2	2	2	2	2	2
Public lavatory	4	6	8	6	8	12	6		15
Showers	30	150	225	75	75	225	30	30	225
Service sink	20	20		20	30	20	20	15	20
Demand factor	0.30	0.30	0.40	0.25	0.25	0.40	0.30	0.30	0.40
Storage capacity factor	1.25	0.90	1.00	0.60	0.80	1.00	2.00	0.70	1.00

To obtain the probable maximum demand multiply the total demands for the fixtures (gal./fixture/hour) by the demand factor. The heater should have a heating capacity in gallons per hour equal to this maximum. The storage tank should have a capacity in gallons equal to the probable maximum demand multiplied by the storage capacity factor.

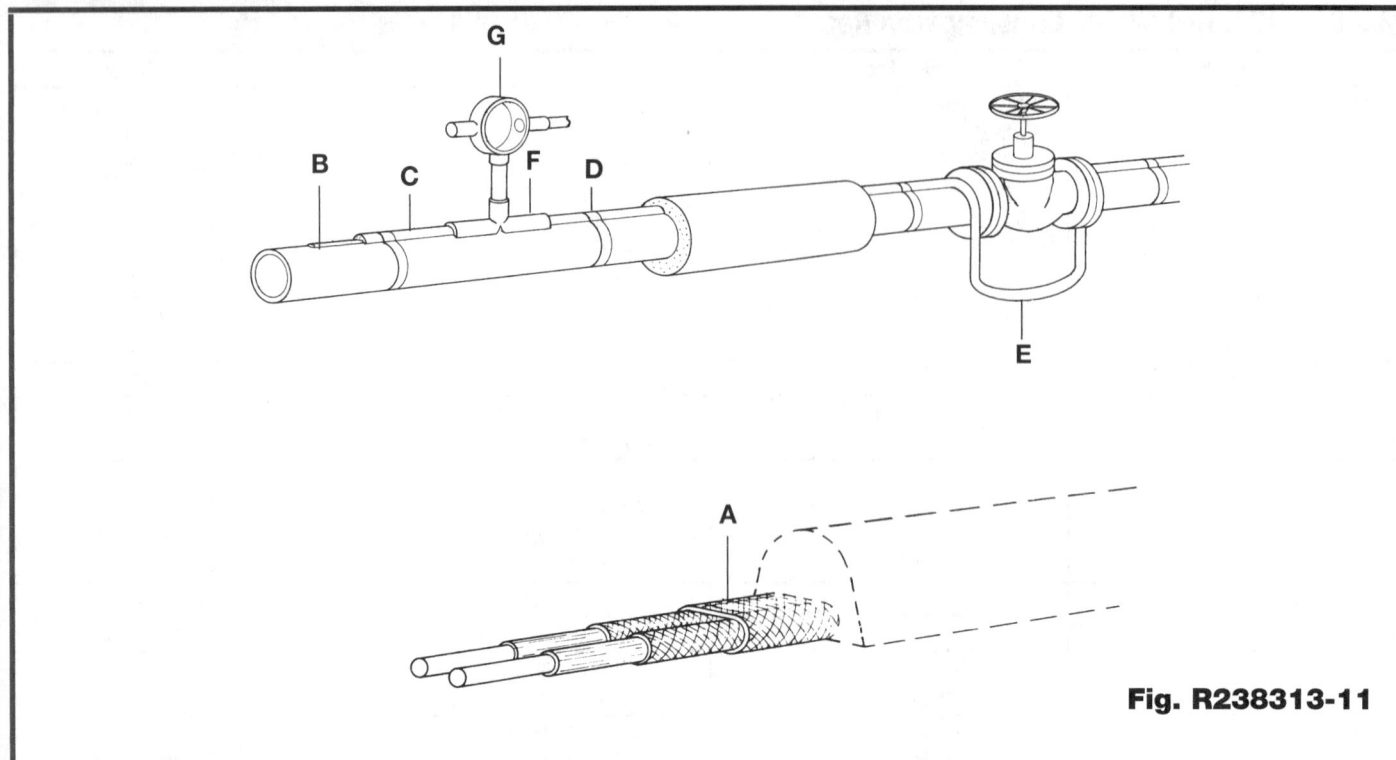

Fig. R238313-11

R238313-10   Heat Trace Systems

Before you can determine the cost of a HEAT TRACE installation the method of attachment must be established. There are (4) common methods:

1. Cable is simply attached to the pipe with polyester tape every 12′.
2. Cable is attached with a continuous cover of 2″ wide aluminum tape.
3. Cable is attached with factory extruded heat transfer cement and covered with metallic raceway with clips every 10′.
4. Cable is attached between layers of pipe insulation using either clips or polyester tape.

In all of the above methods each component of the system must be priced individually.

Example: Components for method 3 must include:

   A. Heat trace cable by voltage and watts per linear foot.
   B. Heat transfer cement, 1 gallon per 60 linear feet of cover.
   C. Metallic raceway by size and type.
   D. Raceway clips by size of pipe.

When taking off linear foot lengths of cable add the following for each valve in the system. (E)

SCREWED OR WELDED VALVE:			FLANGED VALVE:			BUTTERFLY VALVES:		
1/2″	=	6″	1/2″	=	1′ -0″	1/2″	=	0′
3/4″	=	9″	3/4″	=	1′ -6″	3/4″	=	0′
1″	=	1′ -0″	1″	=	2′ -0″	1″	=	1′ -0″
1-1/2″	=	1′ -6″	1-1/2″	=	2′ -6″	1-1/2″	=	1′ -6″
2″	=	2′	2″	=	2′ -6″	2″	=	2′ -0″
2-1/2″	=	2′ -6″	2-1/2″	=	3′ -0″	2-1/2″	=	2′ -6″
3″	=	2′ -6″	3″	=	3′ -6″	3″	=	2′ -6″
4″	=	4′ -0″	4″	=	4′ -0″	4″	=	3′ -0″
6″	=	7′ -0″	6″	=	8′ -0″	6″	=	3′ -6″
8″	=	9′ -6″	8″	=	11′ -0″	8″	=	4′ -0″
10″	=	12′ -6″	10″	=	14′ -0″	10″	=	4′ -0″
12″	=	15′ -0″	12″	=	16′ -6″	12″	=	5′ -0″
14″	=	18′ -0″	14″	=	19′ -6″	14″	=	5′ -6″
16″	=	21′ -6″	16″	=	23′ -0″	16″	=	6′ -0″
18″	=	25′ -6″	18″	=	27′ -0″	18″	=	6′ -6″
20″	=	28′ -6″	20″	=	30′ -0″	20″	=	7′ -0″
24″	=	34′ -0″	24″	=	36′ -0″	24″	=	8′ -0″
30″	=	40′ -0″	30″	=	42′ -0″	30″	=	10′ -0″

R238313-10 Heat Trace Systems (cont.)

Add the following quantities of heat transfer cement to linear foot totals for each valve:

Nominal Valve Size	Gallons of Cement per Valve
1/2"	0.14
3/4"	0.21
1"	0.29
1-1/2"	0.36
2"	0.43
2-1/2"	0.70
3"	0.71
4"	1.00
6"	1.43
8"	1.48
10"	1.50
12"	1.60
14"	1.75
16"	2.00
18"	2.25
20"	2.50
24"	3.00
30"	3.75

The following must be added to the list of components to accurately price HEAT TRACE systems:

1. Expediter fitting and clamp fasteners (F)
2. Junction box and nipple connected to expediter fitting (G)
3. Field installed terminal blocks within junction box
4. Ground lugs
5. Piping from power source to expediter fitting
6. Controls
7. Thermostats
8. Branch wiring
9. Cable splices
10. End of cable terminations
11. Branch piping fittings and boxes

Deduct the following percentages from labor if cable lengths in the same area exceed:

150' to 250'	10%	351' to 500'	20%
251' to 350'	15%	Over 500'	25%

Add the following percentages to labor for elevated installations:

15' to 20' high	10%	31' to 35' high	40%
21' to 25' high	20%	36' to 40' high	50%
26' to 30' high	30%	Over 40' high	60%

R238313-20 Spiral-Wrapped Heat Trace Cable (Pitch Table)

In order to increase the amount of heat, occasionally heat trace cable is wrapped in a spiral fashion around a pipe; increasing the number of feet of heater cable per linear foot of pipe.

Engineers first determine the heat loss per foot of pipe (based on the insulating material, its thickness, and the temperature differential across it). A ratio is then calculated by the formula:

$$\text{Feet of Heat Trace per Foot of Pipe} = \frac{\text{Watts/Foot of Heat Loss}}{\text{Watts/Foot of the Cable}}$$

The linear distance between wraps (pitch) is then taken from a chart or table. Generally, the pitch is listed on a drawing leaving the estimator to calculate the total length of heat tape required. An approximation may be taken from this table.

Feet of Heat Trace Per Foot of Pipe																
	Nominal Pipe Size in Inches															
Pitch In Inches	1	1¼	1½	2	2½	3	4	6	8	10	12	14	16	18	20	24
3.5	1.80															
4	1.65															
5	1.46	1.60	1.80													
6	1.34	1.45	1.55	1.75												
7	1.25	1.35	1.43	1.57	1.75											
8	1.20	1.28	1.34	1.45	1.60	1.80										
9	1.16	1.23	1.28	1.37	1.51	1.68										
10	1.13	1.19	1.24	1.32	1.44	1.57	1.82									
15	1.06	1.08	1.10	1.15	1.21	1.29	1.42	1.78								
20	1.04	1.05	1.06	1.08	1.13	1.17	1.25	1.49	1.73							
25		1.04	1.04	1.06	1.08	1.11	1.17	1.33	1.51	1.72						
30				1.04	1.05	1.07	1.12	1.24	1.37	1.54	1.70					
35					1.06	1.06	1.09	1.17	1.28	1.42	1.54	1.80				
40						1.05	1.07	1.14	1.22	1.33	1.44	1.64	1.78			
50							1.05	1.09	1.15	1.22	1.29	1.52	1.64	1.75		
60								1.06	1.11	1.16	1.21	1.35	1.44	1.53	1.64	1.83
70								1.05	1.08	1.12	1.17	1.25	1.31	1.39	1.46	1.62
80									1.06	1.09	1.13	1.19	1.24	1.30	1.35	1.47
90									1.04	1.06	1.10	1.15	1.19	1.24	1.28	1.38
100										1.05	1.08	1.10	1.13	1.16	1.23	1.32
														1.13	1.15	1.19	1.23

Note: Common practice would normally limit the lower end of the table to 5% of additional heat and above 80% an engineer would likely opt for two (2) parallel cables.

R260105-30 Electrical Demolition (Removal for Replacement)

The purpose of this reference number is to provide a guide to users for electrical "removal for replacement" by applying the rule of thumb: 1/3 of new installation time (typical range from 20% to 50%) for removal. Remember to use reasonable judgment when applying the suggested percentage factor. For example:

Contractors have been requested to remove an existing fluorescent lighting fixture and replace with a new fixture utilizing energy saver lamps and electronic ballast:

In order to fully understand the extent of the project, contractors should visit the job site and estimate the time to perform the renovation work in accordance with applicable national, state and local regulation and codes.

The contractor may need to add extra labor hours to his estimate if he discovers unknown concealed conditions such as: contaminated asbestos ceiling, broken acoustical ceiling tile and need to repair, patch and touch-up paint all the damaged or disturbed areas, tasks normally assigned to general contractors. In addition, the owner could request that the contractors salvage the materials removed and turn over the materials to the owner or dispose of the materials to a reclamation station. The normal removal item is 0.5 labor hour for a lighting fixture and 1.5 labor-hours for new installation time. Revise the estimate times from 2 labor-hours work up to a minimum 4 labor-hours work for just fluorescent lighting fixture.

For removal of large concentrations of lighting fixtures in the same area, apply an "economy of scale" to reduce estimating labor hours.

R260519-20 Armored Cable

Armored Cable – Quantities are taken off in the same manner as wire.

Bx Type Cable – Productivities are based on an average run of 50' before terminating at a box fixture, etc. Each 50' section includes field preparation of (2) ends with hacksaw, identification and tagging of wire. Set up is open coil type without reels attaching cable to snake and pulling across a suspended ceiling or open face wood or steel studding, price does not include drilling of studs.

Cable in Tray – Productivities are based on an average run of 100 L.F. with set up of pulling equipment for (2) 90° bends, attaching cable to pull-in means, identification and tagging of wires, set up of reels. Wire termination to breakers equipment, etc., are not included.

Job Conditions – Productivities are based on new construction to a height of 15' using rolling staging in an unobstructed area. Material staging is assumed to be within 100' of work being performed.

R260519-80 Undercarpet Systems

Takeoff Procedure for Power Systems: List components for each fitting type, tap, splice, and bend on your quantity takeoff sheet. Each component must be priced separately. Start at the power supply transition fittings and survey each circuit for the components needed. List the quantities of each component under a specific circuit number. Use the floor plan layout scale to get cable footage.

Reading across the list, combine the totals of each component in each circuit and list the total quantity in the last column. Calculate approximately 5% for scrap for items such as cable, top shield, tape, and spray adhesive. Also provide for final variations that may occur on-site.

Suggested guidelines are:
1. Equal amounts of cable and top shield should be priced.
2. For each roll of cable, price a set of cable splices.
3. For every 1 ft. of cable, price 2-1/2 ft. of hold-down tape.
4. For every 3 rolls of hold-down tape, price 1 can of spray adhesive.

Adjust final figures wherever possible to accommodate standard packaging of the product. This information is available from the distributor.

Each transition fitting requires:
1. 1 base
2. 1 cover
3. 1 transition block

Each floor fitting requires:
1. 1 frame/base kit
2. 1 transition block
3. 2 covers (duplex/blank)

Each tap requires:
1. 1 tap connector for each conductor
2. 1 pair insulating patches
3. 2 top shield connectors

Each splice requires:
1. 1 splice connector for each conductor
2. 1 pair insulating patches
3. 3 top shield connectors

Each cable bend requires:
1. 2 top shield connectors

Each cable dead end (outside of transition block) requires:
1. 1 pair insulating patches

Labor does not include:
1. Patching or leveling uneven floors.
2. Filling in holes or removing projections from concrete slabs.
3. Sealing porous floors.
4. Sweeping and vacuuming floors.
5. Removal of existing carpeting.
6. Carpet square cut-outs.
7. Installation of carpet squares.

Takeoff Procedures for Telephone Systems: After reviewing floor plans identify each transition. Number or letter each cable run from that fitting.

Start at the transition fitting and survey each circuit for the components needed. List the cable type, terminations, cable length, and floor fitting type under the specific circuit number. Use the floor plan layout scale to get the cable footage. Add some extra length (next higher increment of 5 feet) to preconnectorized cable.

Transition fittings require:
1. 1 base plate
2. 1 cover
3. 1 transition block

Floor fittings require:
1. 1 frame/base kit
2. 2 covers
3. Modular jacks

Reading across the list, combine the list of components in each circuit and list the total quantity in the last column. Calculate the necessary scrap factors for such items as tape, bottom shield and spray adhesive. Also provide for final variations that may occur on-site.

Adjust final figures whenever possible to accommodate standard packaging. Check that items such as transition fittings, floor boxes, and floor fittings that are to utilize both power and telephone have been priced as combination fittings, so as to avoid duplication.

Make sure to include marking of floors and drilling of fasteners if fittings specified are not the adhesive type.

Labor does not include:
1. Conduit or raceways before transition of floor boxes
2. Telephone cable before transition boxes
3. Terminations before transition boxes
4. Floor preparation as described in power section

Be sure to include all cable folds when pricing labor.

Takeoff Procedure for Data Systems: Start at the transition fittings and take off quantities in the same manner as the telephone system, keeping in mind that data cable does not require top or bottom shields.

The data cable is simply cross-taped on the cable run to the floor fitting.

Data cable can be purchased in either bulk form in which case coaxial connector material and labor must be priced, or in preconnectorized cut lengths.

Data cable cannot be folded and must be notched at 1 inch intervals. A count of all turns must be added to the labor portion of the estimate. (Note: Some manufacturers have prenotched cable.)

Notching required:
1. 90 degree turn requires 8 notches per side.
2. 180 degree turn requires 16 notches per side.

Floor boxes, transition boxes, and fittings are the same as described in the power and telephone procedures.

Since undercarpet systems require special hand tools, be sure to include this cost in proportion to number of crews involved in the installation.

Job Conditions: Productivity is based on new construction in an unobstructed area. Staging area is assumed to be within 200' of work being performed.

R260519-90 Wire

Wire quantities are taken off by either measuring each cable run or by extending the conduit and raceway quantities times the number of conductors in the raceway. Ten percent should be added for waste and tie-ins. Keep in mind that the unit of measure of wire is C.L.F. not L.F. as in raceways so the formula would read:

$$\frac{(\text{L.F. Raceway x No. of Conductors}) \times 1.10}{100} = \text{C.L.F.}$$

Price per C.L.F. of wire includes:
1. Set up wire coils or spools on racks
2. Attaching wire to pull in means
3. Measuring and cutting wire
4. Pulling wire into a raceway
5. Identifying and tagging

Price does not include:
1. Connections to breakers, panelboards or equipment
2. Splices

Job Conditions: Productivity is based on new construction to a height of 15' using rolling staging in an unobstructed area. Material staging is assumed to be within 100' of work being performed.

Economy of Scale: If more than three wires at a time are being pulled, deduct the following percentages from the labor of that grouping:

4-5 wires	25%
6-10 wires	30%
11-15 wires	35%
over 15	40%

If a wire pull is less than 100' in length and is interrupted several times by boxes, lighting outlets, etc., it may be necessary to add the following lengths to each wire being pulled:

Junction box to junction box	2 L.F.
Lighting panel to junction box	6 L.F.
Distribution panel to sub panel	8 L.F.
Switchboard to distribution panel	12 L.F.
Switchboard to motor control center	20 L.F.
Switchboard to cable tray	40 L.F.

Measure of Drops and Riser: It is important when taking off wire quantities to include the wire for drops to electrical equipment. If heights of electrical equipment are not clearly stated, use the following guide:

	Bottom A.F.F.	Top A.F.F.	Inside Cabinet
Safety switch to 100A	5'	6'	2'
Safety switch 400 to 600A	4'	6'	3'
100A panel 12 to 30 circuit	4'	6'	3'
42 circuit panel	3'	6'	4'
Switch box	3'	3'6"	1'
Switchgear	0'	8'	8'
Motor control centers	0'	8'	8'
Transformers - wall mount	4'	8'	2'
Transformers - floor mount	0'	12'	4'

R260519-91 Maximum Circuit Length (approximate) for Various Power Requirements Assuming THW, Copper Wire @ 75° C, Based Upon a 4% Voltage Drop

Maximum Circuit Length: Table R260519-91 indicates typical maximum installed length a circuit can have and still maintain an adequate voltage level at the point of use. The circuit length is similar to the conduit length.

If the circuit length for an ampere load and a copper wire size exceeds the length obtained from Table R260519-91, use the next largest wire size to compensate voltage drop.

Example: A 130 ampere load at 480 volts, 3 phase, 3 wire with No. 1 wire can be run a maximum of 555 L.F. and provide satisfactory operation. If the same load is to be wired at the end of a 625 L.F. circuit, then a larger wire must be used.

		Maximum Circuit Length in Feet				
	Wire	2 Wire, 1 Phase		3 Wire, 3 Phase		
Amperes	Size	120V	240V	240V	480V	600V
15	14*	50	105	120	240	300
	14	50	100	120	235	295
20	12*	60	125	145	290	360
	12	60	120	140	280	350
30	10*	65	130	155	305	380
	10	65	130	150	300	375
50	8	60	125	145	285	355
65	6	75	150	175	345	435
85	4	90	185	210	425	530
115	2	110	215	250	500	620
130	1	120	240	275	555	690
150	1/0	130	260	305	605	760
175	2/0	140	285	330	655	820
200	3/0	155	315	360	725	904
230	4/0	170	345	395	795	990
255	250	185	365	420	845	1055
285	300	195	395	455	910	1140
310	350	210	420	485	975	1220
380	500	245	490	565	1130	1415

*Solid Conductor

Note: The circuit length is the one-way distance between the origin and the load.

R260519-92 Minimum Copper and Aluminum Wire Size Allowed for Various Types of Insulation

Minimum Wire Sizes

Amperes	Copper THW THWN or XHHW	Copper THHN XHHW *	Aluminum THW XHHW	Aluminum THHN XHHW *	Amperes	Copper THW THWN or XHHW	Copper THHN XHHW *	Aluminum THW XHHW	Aluminum THHN XHHW *
15A	#14	#14	#12	#12	195	3/0	2/0	250kcmil	4/0
20	#12	#12	#10	#10	200	3/0	3/0	250kcmil	4/0
25	#10	#10	#10	#10	205	4/0	3/0	250kcmil	4/0
30	#10	#10	#8	#8	225	4/0	3/0	300kcmil	250kcmil
40	#8	#8	#8	#8	230	4/0	4/0	300kcmil	250kcmil
45	#8	#8	#6	#8	250	250kcmil	4/0	350kcmil	300kcmil
50	#8	#8	#6	#6	255	250kcmil	4/0	400kcmil	300kcmil
55	#6	#8	#4	#6	260	300kcmil	4/0	400kcmil	350kcmil
60	#6	#6	#4	#6	270	300kcmil	250kcmil	400kcmil	350kcmil
65	#6	#6	#4	#4	280	300kcmil	250kcmil	500kcmil	350kcmil
75	#4	#6	#3	#4	285	300kcmil	250kcmil	500kcmil	400kcmil
85	#4	#4	#2	#3	290	350kcmil	250kcmil	500kcmil	400kcmil
90	#3	#4	#2	#2	305	350kcmil	300kcmil	500kcmil	400kcmil
95	#3	#4	#1	#2	310	350kcmil	300kcmil	500kcmil	500kcmil
100	#3	#3	#1	#2	320	400kcmil	300kcmil	600kcmil	500kcmil
110	#2	#3	1/0	#1	335	400kcmil	350kcmil	600kcmil	500kcmil
115	#2	#2	1/0	#1	340	500kcmil	350kcmil	600kcmil	500kcmil
120	#1	#2	1/0	1/0	350	500kcmil	350kcmil	700kcmil	500kcmil
130	#1	#2	2/0	1/0	375	500kcmil	400kcmil	700kcmil	600kcmil
135	1/0	#1	2/0	1/0	380	500kcmil	400kcmil	750kcmil	600kcmil
150	1/0	#1	3/0	2/0	385	600kcmil	500kcmil	750kcmil	600kcmil
155	2/0	1/0	3/0	3/0	420	600kcmil	500kcmil		700kcmil
170	2/0	1/0	4/0	3/0	430		500kcmil		750kcmil
175	2/0	2/0	4/0	3/0	435		600kcmil		750kcmil
180	3/0	2/0	4/0	4/0	475		600kcmil		

*Dry Locations Only

Notes:

1. Size #14 to 4/0 is in AWG units (American Wire Gauge).
2. Size 250 to 750 is in kcmil units (Thousand Circular Mils).
3. Use next higher ampere value if exact value is not listed in table.
4. For loads that operate continuously increase ampere value by 25% to obtain proper wire size.
5. Refer to Table R260519-91 for the maximum circuit length for the various size wires.
6. Table R260519-92 has been written for estimating purpose only, based on ambient temperature of 30 °C (86°F); for ambient temperature other than 30 °C (86°F), ampacity correction factors will be applied.

R260519-93   Metric Equivalent, Wire

United States		European	
Size AWG or kcmil	Area Cir. Mils.(cmil) mm²	Size mm²	Area Cir. Mils.
18	1620/.82	.75	1480
16	2580/1.30	1.0	1974
14	4110/2.08	1.5	2961
12	6530/3.30	2.5	4935
10	10,380/5.25	4	7896
8	16,510/8.36	6	11,844
6	26,240/13.29	10	19,740
4	41,740/21.14	16	31,584
3	52,620/26.65	25	49,350
2	66,360/33.61	–	–
1	83,690/42.39	35	69,090
1/0	105,600/53.49	50	98,700
2/0	133,100/67.42	–	–
3/0	167,800/85.00	70	138,180
4/0	211,600/107.19	95	187,530
250	250,000/126.64	120	236,880
300	300,000/151.97	150	296,100
350	350,000/177.30	–	–
400	400,000/202.63	185	365,190
500	500,000/253.29	240	473,760
600	600,000/303.95	300	592,200
700	700,000/354.60	–	–
750	750,000/379.93	–	–

R260519-94   Size Required and Weight (Lbs./1000 L.F.) of Aluminum and Copper THW Wire by Ampere Load

Amperes	Copper Size	Aluminum Size	Copper Weight	Aluminum Weight
15	14	12	24	11
20	12	10	33	17
30	10	8	48	39
45	8	6	77	52
65	6	4	112	72
85	4	2	167	101
100	3	1	205	136
115	2	1/0	252	162
130	1	2/0	324	194
150	1/0	3/0	397	233
175	2/0	4/0	491	282
200	3/0	250	608	347
230	4/0	300	753	403
255	250	400	899	512
285	300	500	1068	620
310	350	500	1233	620
335	400	600	1396	772
380	500	750	1732	951

R260526-80   Grounding

Grounding When taking off grounding systems, identify separately the type and size of wire.

Example:

> Bare copper & size
>
> Bare aluminum & size
>
> Insulated copper & size
>
> Insulated aluminum & size

Count the number of ground rods and their size.

Example:

1.   8' grounding rod – 5/8" dia. 20 Ea.
2. 10' grounding rod – 5/8" dia. 12 Ea.
3. 15' grounding rod – 3/4" dia. 4 Ea.

Count the number of connections; the size of the largest wire will determine the productivity.

Example:

> Braze a #2 wire to a #4/0 cable
> The 4/0 cable will determine the L.H. and cost to be used.

Include individual connections to:

1. Ground rods
2. Building steel
3. Equipment
4. Raceways

Price does not include:

1. Excavation
2. Backfill
3. Sleeves or raceways used to protect grounding wires
4. Wall penetrations
5. Floor cutting
6. Core drilling

Job Conditions: Productivity is based on a ground floor area, using cable reels in an unobstructed area. Material staging area assumed to be within 100' of work being performed.

R260533-20   Conduit To 15' High

List conduit by quantity, size, and type. Do not deduct for lengths occupied by fittings, since this will be allowance for scrap. Example:

- A. Aluminum — size
- B. Rigid galvanized — size
- C. Steel intermediate (IMC) — size
- D. Rigid steel, plastic-coated 20 Mil. — size
- E. Rigid steel, plastic-coated 40 Mil. — size
- F. Electric metallic tubing (EMT) — size
- G. PVC Schedule 40 — size

Types (A) thru (E) listed above contain the following per 100 L.F.:

1. (11) Threaded couplings
2. (11) Beam-type hangers
3. (2) Factory sweeps
4. (2) Fiber bushings
5. (4) Locknuts
6. (2) Field threaded pipe terminations
7. (2) Removal of concentric knockouts

Type (F) contains per 100 L.F.:

1. (11) Set screw couplings
2. (11) Beam clamps
3. (2) Field bends on 1/2" and 3/4" diameter
4. (2) Factory sweeps for 1" and above
5. (2) Set screw steel connectors
6. (2) Removal of concentric knockouts

Type (G) contains per 100 L.F.:

1. (11) Field cemented couplings
2. (34) Beam clamps

3. (2) Factory sweeps
4. (2) Adapters
5. (2) Locknuts
6. (2) Removal of concentric knockouts

Labor-hours for all conduit to 15' include:

1. Unloading by hand
2. Hauling by hand to an area up to 200' from loading dock
3. Setup of rolling staging
4. Installation of conduit and fittings as described in Conduit models (A) thru (G)

Not included in the material and labor are:

1. Staging rental or purchase
2. Structural modifications
3. Wire
4. Junction boxes
5. Fittings in excess of those described in conduit models (A) thru (G)
6. Painting of conduit

Fittings

Only those fittings listed above are included in the linear foot totals, although they should be listed separately from conduit lengths, without prices, to ensure proper quantities for material procurement.

If the fittings required exceed the quantities included in the model conduit runs, then material and labor costs must be added to the difference. If actual needs per 100 L.F. of conduit are: (2) sweeps, (4) LBs and (1) field bend, then, (4) LBs and (1) field bend must be priced additionally.

R260533-21 Hangers

It is sometimes desirable to substitute an alternate style of hanger if the support being used is not the type described in the conduit models.

One approach is the substitution method:

1. Find the cost of the type hanger described in the conduit model.
2. Calculate the cost of the desired type hanger (it may be necessary to calculate individual components such as drilling, expansion shields, etc.).
3. Calculate the cost difference (delta) between the two types of hangers.
4. Multiply the cost delta by the number of hangers in the model.
5. Divide the total delta cost for hangers in the model by the length of the model to find the delta cost per L.F. for that model.
6. Modify the given unit costs per L.F. for the model by the delta cost per L.F.

Another approach to hanger configurations would be to start with the conduit only and add all the supports and any other items as separate lines. This procedure is most useful if the project involves racking many runs of conduit on a single hanger, for instance, a trapeze type hanger.

Example: Five (5) 2″ RGS conduits, 50 L.F. each, are to be run on trapeze hangers from one pull box to another. The run includes one 90° bend.

1. List the hangers' components to create an assembly cost for each 2′-wide trapeze.
2. List the components for the 50′ conduit run, noting that 6 trapeze supports will be required.

Job Conditions: Productivities are based on new construction to 15′ high, using scaffolding in an unobstructed area. Material storage is assumed to be within 100′ of work being performed.

Add to labor for elevated installations:

15′ to 20′ High–10%	30′ to 35 High–30%
20′ to 25′ High–20%	35′ to 40′ High–35%
25′ to 30′ High–25%	Over 40′ High–40%

Add these percentages to the L.F. labor cost, but not to fittings. Add these percentages only to quantities exceeding the different height levels, rather than the total conduit quantities.

Linear foot price for labor does not include penetrations in walls or floors and must be added to the estimate

R260533-22 Conductors in Conduit

Table below lists maximum number of conductors for various sized conduit using THW, TW or THWN insulations.

Copper Wire Size	1/2″ TW	1/2″ THW	1/2″ THWN	3/4″ TW	3/4″ THW	3/4″ THWN	1″ TW	1″ THW	1″ THWN	1-1/4″ TW	1-1/4″ THW	1-1/4″ THWN	1-1/2″ TW	1-1/2″ THW	1-1/2″ THWN	2″ TW	2″ THW	2″ THWN	2-1/2″ TW	2-1/2″ THW	2-1/2″ THWN	3″ THW	3″ THWN	3-1/2″ THW	3-1/2″ THWN	4″ THW	4″ THWN
#14	9	6	13	15	10	24	25	16	39	44	29	69	60	40	94	99	65	154	142	93		143		192			
#12	7	4	10	12	8	18	19	13	29	35	24	51	47	32	70	78	53	114	111	76	164	117		157			
#10	5	4	6	9	6	11	15	11	18	26	19	32	36	26	44	60	43	73	85	61	104	95	160	127		163	
#8	2	1	3	4	3	5	7	5	9	12	10	16	17	13	22	28	22	36	40	32	51	49	79	66	106	85	136
#6		1	1		2	4		4	6		7	11		10	15		16	26		23	37	36	57	48	76	62	98
#4		1	1		1	2		3	4		5	7		7	9		12	16		17	22	27	35	36	47	47	60
#3		1	1		1	1		2	3		4	6		6	8		10	13		15	19	23	29	31	39	40	51
#2		1	1		1	1		2	3		4	5		5	7		9	11		13	16	20	25	27	33	34	43
#1					1	1		1	1		3	3		4	5		6	8		9	12	14	18	19	25	25	32
1/0					1	1		1	1		2	3		3	4		5	7		8	10	12	15	16	21	21	27
2/0					1	1		1	1		1	2		3	3		5	6		7	8	10	13	14	17	18	22
3/0					1	1		1	1		1	1		2	3		4	5		6	7	9	11	12	14	15	18
4/0				1				1	1		1	1		1	2		3	4		5	6	7	9	10	12	13	15
250 kcmil								1	1		1	1		1	1		2	3		4	4	6	7	8	10	10	12
300								1	1		1	1		1	1		2	3		3	4	5	6	7	8	9	11
350									1		1	1		1	1		1	2		3	3	4	5	6	7	8	9
400											1	1		1	1		1	1		2	3	4	5	5	6	7	8
500											1	1		1	1		1	1		1	2	3	4	4	5	6	7
600												1		1	1		1	1		1	1	3	3	4	4	5	5
700														1	1		1	1		1	1	2	3	3	4	4	5
750														1	1		1	1		1	1	2	2	3	3	4	4

R260533-23 Metric Equivalent, Conduit

U.S. vs. European Conduit – Approximate Equivalents			
United States		European	
Trade Size	Inside Diameter Inch/mm	Trade Size	Inside Diameter mm
½	.622/15.8	11	16.4
¾	.824/20.9	16	19.9
1	1.049/26.6	21	25.5
1¼	1.380/35.0	29	34.2
1½	1.610/40.9	36	44.0
2	2.067/52.5	42	51.0
2½	2.469/62.7		
3	3.068/77.9		
3½	3.548/90.12		
4	4.026/102.3		
5	5.047/128.2		
6	6.065/154.1		

R260533-24 Conduit Weight Comparisons (Lbs. per 100 ft.) Empty

Type	1/2″	3/4″	1″	1-1/4″	1-1/2″	2″	2-1/2″	3″	3-1/2″	4″	5″	6″
Rigid Aluminum	28	37	55	72	89	119	188	246	296	350	479	630
Rigid Steel	79	105	153	201	249	332	527	683	831	972	1314	1745
Intermediate Steel (IMC)	60	82	116	150	182	242	401	493	573	638		
Electrical Metallic Tubing (EMT)	29	45	65	96	111	141	215	260	365	390		
Polyvinyl Chloride, Schedule 40	16	22	32	43	52	69	109	142	170	202	271	350
Polyvinyl Chloride Encased Burial						38		67	88	105	149	202
Fibre Duct Encased Burial						127		164	180	206	400	511
Fibre Duct Direct Burial						150		251	300	354		
Transite Encased Burial						160		240	290	330	450	550
Transite Direct Burial						220		310		400	540	640

R260533-25 Conduit Weight Comparisons (Lbs. per 100 ft.) with Maximum Cable Fill*

Type	1/2″	3/4″	1″	1-1/4″	1-1/2″	2″	2-1/2″	3″	3-1/2″	4″	5″	6″
Rigid Galvanized Steel (RGS)	104	140	235	358	455	721	1022	1451	1749	2148	3083	4343
Intermediate Steel (IMC)	84	113	186	293	379	611	883	1263	1501	1830		
Electrical Metallic Tubing (EMT)	54	116	183	296	368	445	641	930	1215	1540		

*Conduit & Heaviest Conductor Combination

R260533-60   Wireway

When "taking off" Wireway, list by size and type.

Example:
1. Screw cover, unflanged + size
2. Screw cover, flanged + size
3. Hinged cover, flanged + size
4. Hinged cover, unflanged + size

Each 10' length on Wireway contains:
1. 10' of cover either screw or hinged type
2. (1) Coupling or flange gasket
3. (1) Wall type mount

All fittings must be priced separately.

Substitution of hanger types is done the same as described in R260533-21, "HANGERS," keeping in mind that the wireway model is based on 10' sections instead of a 100' conduit run.

Labor-hours for wireway include:
1. Unloading by hand
2. Hauling by hand up to 100' from loading dock
3. Measuring and marking
4. Mounting wall bracket using (2) anchor type lead fasteners
5. Installing wireway on brackets, to 15' high (For higher elevations use factors in R260533-20)

Job Conditions: Productivity is based on new construction, to a height of 15' using rolling staging in an unobstructed area.

Material staging area is assumed to be within 100' of work being performed.

R260533-65   Outlet Boxes

Outlet boxes should be included on the same takeoff sheet as branch piping or devices to better explain what is included in each circuit.

Each unit price in this section is a stand alone item and contains no other component unless specified. For example, to estimate a duplex outlet, components that must be added are:
1. 4" square box
2. 4" plaster ring
3. Duplex receptacle
4. Device cover

The method of mounting outlet boxes is (2) plastic shield fasteners.

Outlet boxes plastic, labor-hours include:
1. Marking box location on wood studding
2. Mounting box

Economy of Scale – For large concentrations of plastic boxes in the same area deduct the following percentages from labor-hour totals:

1	to	10	0%
11	to	25	20%
26	to	50	25%
51	to	100	30%
	over	100	35%

Note: It is important to understand that these percentages are not used on the total job quantities, but only areas where concentrations exceed the levels specified.

R260533-70   Pull Boxes and Cabinets

List cabinets and pull boxes by NEMA type and size.

Example:	**TYPE**	**SIZE**
	NEMA 1	6"W x 6"H x 4"D
	NEMA 3R	6"W x 6"H x 4"D

Labor-hours for wall mount (indoor or outdoor) installations include:
1. Unloading and uncrating
2. Handling of enclosures up to 200' from loading dock using a dolly or pipe rollers
3. Measuring and marking
4. Drilling (4) anchor type lead fasteners using a hammer drill
5. Mounting and leveling boxes

Note: A plywood backboard is not included.

Labor-hours for ceiling mounting include:
1. Unloading and uncrating
2. Handling boxes up to 100' from loading dock

3. Measuring and marking
4. Drilling (4) anchor type lead fasteners using a hammer drill
5. Installing and leveling boxes to a height of 15' using rolling staging

Labor-hours for free standing cabinets include:
1. Unloading and uncrating
2. Handling of cabinets up to 200' from loading dock using a dolly or pipe rollers
3. Marking of floor
4. Drilling (4) anchor type lead fasteners using a hammer drill
5. Leveling and shimming

Labor-hours for telephone cabinets include:
1. Unloading and uncrating
2. Handling cabinets up to 200' using a dolly or pipe rollers
3. Measuring and marking
4. Mounting and leveling, using (4) lead anchor type fasteners

R260533-75   Weight Comparisons of Common Size Cast Boxes in Lbs.

Size NEMA 4 or 9	Cast Iron	Cast Aluminum	Size NEMA 7	Cast Iron	Cast Aluminum
6" x 6" x 6"	17	7	6" x 6" x 6"	40	15
8" x 6" x 6"	21	8	8" x 6" x 6"	50	19
10" x 6" x 6"	23	9	10" x 6" x 6"	55	21
12" x 12" x 6"	52	20	12" x 6" x 6"	100	37
16" x 16" x 6"	97	36	16" x 16" x 6"	140	52
20" x 20" x 6"	133	50	20" x 20" x 6"	180	67
24" x 18" x 8"	149	56	24" x 18" x 8"	250	93
24" x 24" x 10"	238	88	24" x 24" x 10"	358	133
30" x 24" x 12"	324	120	30" x 24" x 10"	475	176
36" x 36" x 12"	500	185	30" x 24" x 12"	510	189

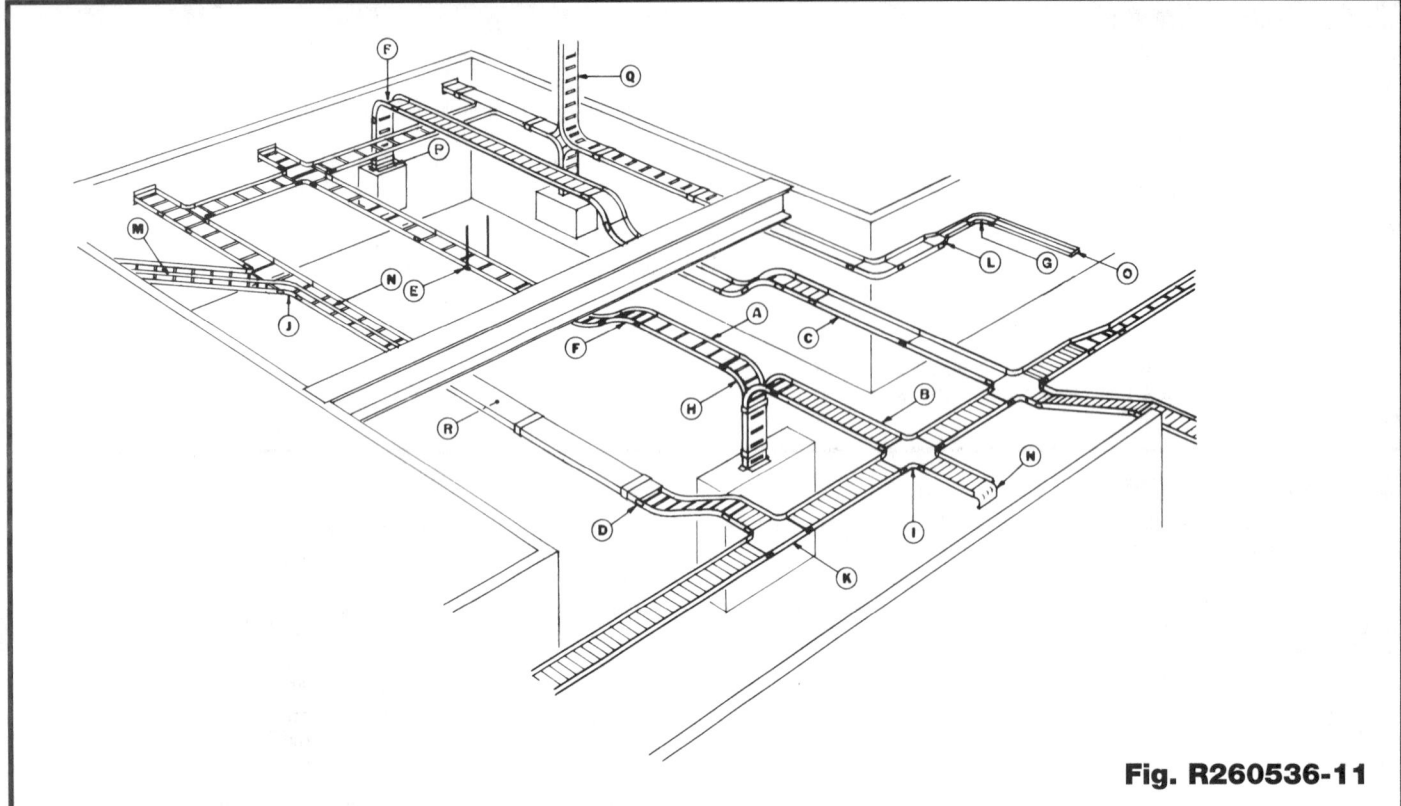

Fig. R260536-11

R260536-10   Cable Tray

Cable Tray - When taking off cable tray it is important to identify separately the different types and sizes involved in the system being estimated. (Fig. R260536-11)

  A. – Ladder Type, galvanized or aluminum
  B. – Trough Type, galvanized or aluminum
  C. – Solid Bottom, galvanized or aluminum

The unit of measure is calculated in linear feet; do not deduct from this footage any length occupied by fittings, this will be the only allowance for scrap. Be sure to include all vertical drops to panels, switch gear, etc.

Hangers – Included in the linear footage of cable tray is

  D. – 1 – Pair of connector plates per 12 L.F.

  E. – 1 – Pair clamp type hangers and 4' of 3/8" threaded rod per 12 L.F.

Not included are structural supports, which must be priced in addition to the hangers.

Fittings – Identify separately the different types of fittings
  1.) Ladder Type, galvanized or aluminum
  2.) Trough Type, galvanized or aluminum
  3.) Solid Bottom Type, galvanized or aluminum

The configuration, radius and rung spacing must also be listed. The unit of measure is "Ea."

  F. – Elbow, vertical Ea.

  G. – Elbow, horizontal Ea.

  H. – Tee, vertical Ea.

  I. – Cross, horizontal Ea.

  J. – Wye, horizontal Ea.

  K. – Tee, horizontal Ea.

  L. – Reducing fitting Ea.

Depending on the use of the system other examples of units which must be included are:

  M. – Divider strip L.F.

  N. – Drop-outs Ea.

  O. – End caps Ea.

  P. – Panel connectors Ea.

Wire and cable are not included and should be taken off separately, see Unit Price sections.

Job Conditions – Unit prices are based on a new installation to a work plane of 15' using rolling staging.

Add to labor for elevated installations:

15' to 20' High	10%
20' to 25' High	20%
25' to 30' High	25%
30' to 35' High	30%
35' to 40' High	35%
Over 40' High	40%

Add these percentages for L.F. totals but not to fittings. Add percentages to only those quantities that fall in the different elevations, in other words, if the total quantity of cable tray is 200' but only 75' is above 15' then the 10% is added to the 75' only.

Linear foot costs do not include penetrations through walls and floors which must be added to the estimate.

Cable Tray Covers

Covers – Cable tray covers are taken off in the same manner as the tray itself, making distinctions as to the type of cover. (Fig. R260536-11)

  Q. – Vented, galvanized or aluminum

  R. – Solid, galvanized or aluminum

Cover configurations are taken off separately noting type, specific radius and widths.

Note: Care should be taken to identify from plans and specifications exactly what is being covered. In many systems only vertical fittings are covered to retain wire and cable.

R260539-30 Conduit In Concrete Slab

List conduit by quantity, size and type.

Example:

A. Rigid galvanized steel + size

B. P.V.C. + size

Rigid galvanized steel (A) contains per 100 L.F.:

1. (20) Ties to slab reinforcing
2. (11) Threaded steel couplings
3. (2) Factory sweeps
4. (2) Field threaded conduit terminations

5. (2) Fiber bushings + locknuts
6. (2) Removal of concentric knockouts

P.V.C. (B) contains per 100 L.F.:

1. (20) Ties to slab reinforcing
2. (11) Field cemented couplings
3. (2) Factory sweeps
4. (2) Adapters
5. (2) Removal of concentric knockouts

R260539-40 Conduit In Trench

Conduit in trench is galvanized steel and contains per 100 L.F.:

1. (11) Threaded couplings
2. (2) Factory sweeps
3. (2) Fiber bushings + (4) locknuts
4. (2) Field threaded conduit terminations
5. (2) Removal of concentric knockouts

Note:

Conduit in Unit Price sections do not include:

1. Floor cutting
2. Excavation or backfill
3. Grouting or patching

Conduit fittings in excess of those listed in the above Conduit model must be added. (Refer to R260533-20 for Procedure example.)

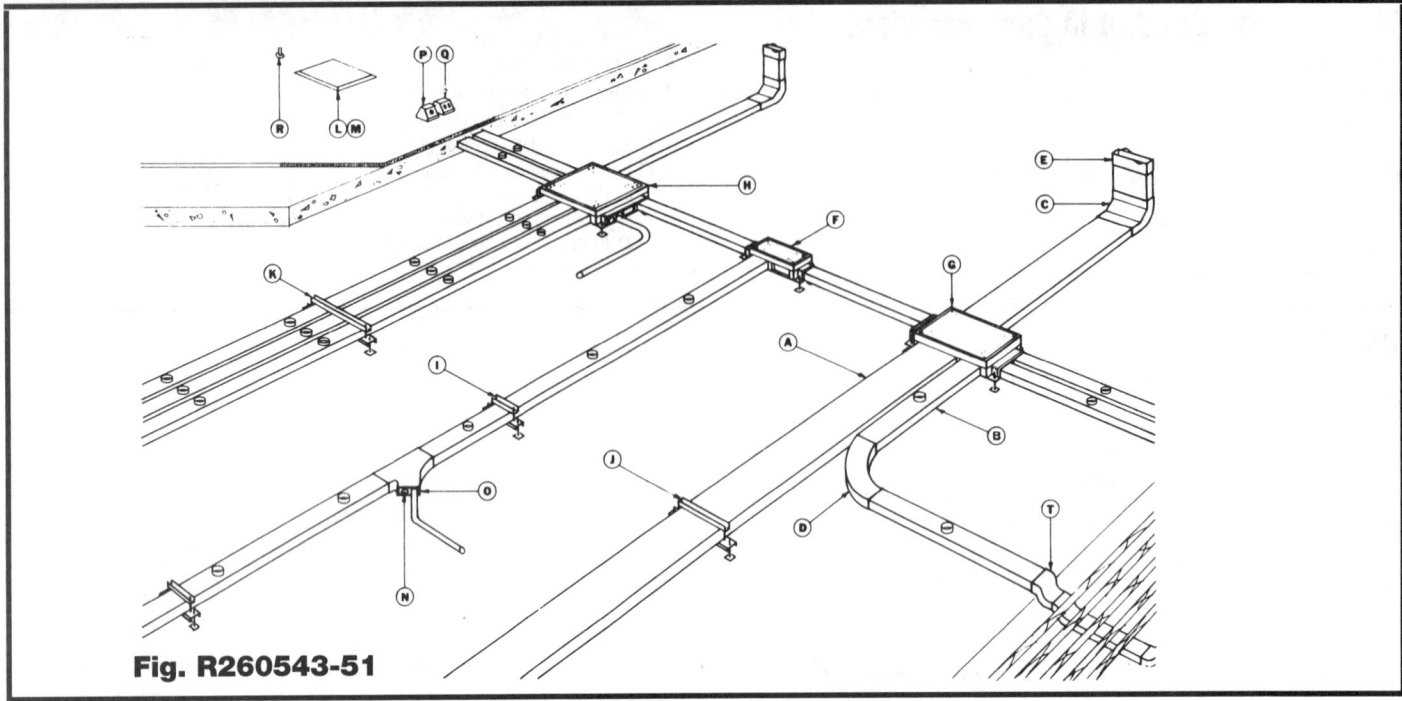

Fig. R260543-51

R260543-50   Underfloor Duct

When pricing Underfloor Duct it is important to identify and list each component, since costs vary significantly from one type of fitting to another. Do not deduct boxes or fittings from linear foot totals; this will be your allowance for scrap.

The first step is to identify the system as either:

    FIG. R260543-51 Single Level

    FIG. R260543-52 Dual Level

Single Level System

Include on your "takeoff sheet" the following unit price items, making sure to distinquish between Standard and Super duct:

  A. Feeder duct (blank) in L.F.
  B. Distribution duct (Inserts 2′ on center) in L.F.
  C. Elbows (Vertical) Ea.
  D. Elbows (Horizontal) Ea.
  E. Cabinet connector Ea.
  F. Single duct junction box Ea.
  G. Double duct junction box Ea.
  H. Triple duct junction box Ea.
   I. Support, single cell Ea.
   J. Support, double cell Ea.
  K. Support, triple cell Ea.
  L. Carpet pan Ea.
  M. Terrazzo pan Ea.
  N. Insert to conduit adapter Ea.
  O. Conduit adapter Ea.
  P. Low tension outlet Ea.
  Q. High tension outlet Ea.
  R. Galvanized nipple Ea.
  S. Wire per C.L.F.
  T. Offset (Duct type) Ea.

Dual Level System + Labor
see next page

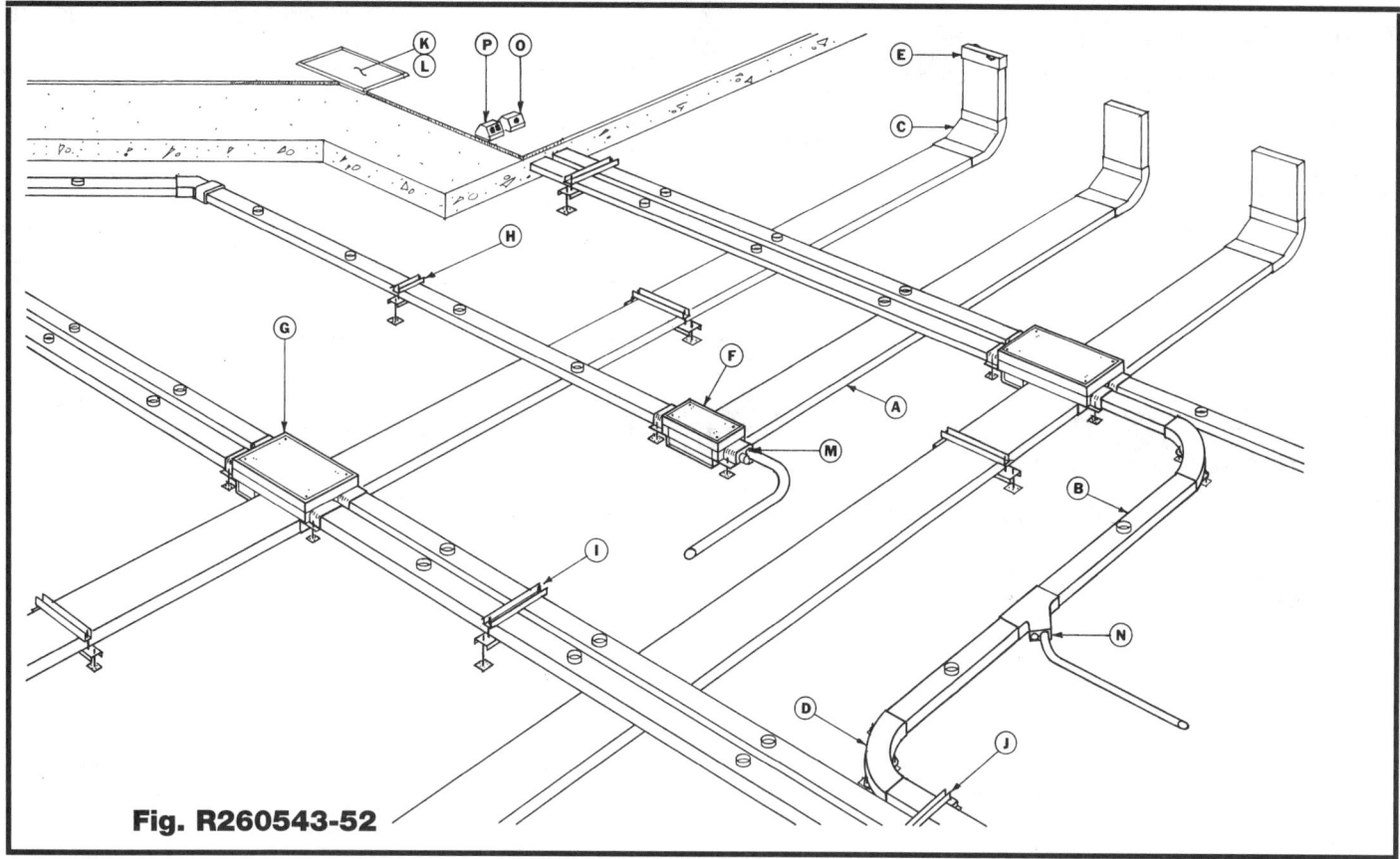

Fig. R260543-52

R260543-50 Underfloor Duct (cont.)

Dual Level

Include the following when "taking off" Dual Level systems:

Distinguish between Standard and Super duct.

  A. Feeder duct (blank) in L.F.
  B. Distribution duct (Inserts 2′ on center) in L.F.
  C. Elbows (Vertical) Ea.
  D. Elbows (Horizontal) Ea.
  E. Cabinet connector Ea.
  F. Single duct, 2 level, junction box Ea.
  G. Double duct, 2 level, junction box Ea.
  H. Support, single cell Ea.
  I. Support, double cell Ea.
  J. Support, triple cell Ea.
  K. Carpet pan Ea.
  L. Terrazzo pan Ea.
  M. Insert to conduit adapter Ea.
  N. Conduit adapter Ea.
  O. Low tension outlet Ea.
  P. High tension outlet Ea.
  Q. Wire per C.L.F.

Note: Make sure to include risers in linear foot totals. High tension outlets include box, receptacle, covers and related mounting hardware.

Labor-hours for both Single and Dual Level systems include:
  1. Unloading and uncrating
  2. Hauling up to 200′ from loading dock
  3. Measuring and marking
  4. Setting raceway and fittings in slab or on grade
  5. Leveling raceway and fittings

Labor-hours do not include:
  1. Floor cutting
  2. Excavation or backfill
  3. Concrete pour
  4. Grouting or patching
  5. Wire or wire pulls
  6. Additional outlets after concrete is poured
  7. Piping to or from Underfloor Duct

Note: Installation is based on installing up to 150′ of duct. If quantities exceed this, deduct the following percentages:
  1. 150′ to 250′ - 10%
  2. 250′ to 350′ - 15%
  3. 350′ to 500′ - 20%
  4. over 500′ - 25%

Deduct these percentages from labor only.

Deduct these percentages from straight sections only.

Do not deduct from fittings or junction boxes.

Job Conditions: Productivity is based on new construction.

Underfloor duct to be installed on first three floors.

Material staging area within 100′ of work being performed.

Area unobstructed and duct not subject to physical damage.

R260580-75   Motor Connections

Motor connections should be listed by size and type of motor.
Included in the material and labor cost is:

1. (2) Flex connectors
2. 18″ of flexible metallic wireway
3. Wire identification and termination
4. Test for rotation
5. (2) or (3) Conductors

Price does not include:

1. Mounting of motor
2. Disconnect Switch
3. Motor Starter
4. Controls
5. Conduit or wire ahead of flex

Note: When "Taking off" Motor connections, it is advisable to list connections on the same quantity sheet as Motors, Motor Starters and controls.

R260913-80   Switchboard Instruments

Switchboard instruments are added to the price of switchboards according to job specifications. This equipment is usually included when ordering "Gear" from the manufacturer and will arrive factory installed.

Included in the labor cost is:

1. Internal wiring connections
2. Wire identification
3. Wire tagging

Transition sections include:

1. Uncrating
2. Hauling sections up to 100′ from loading dock
3. Positioning sections
4. Leveling sections

5. Bolting enclosures
6. Bolting vertical bus bars

Price does not include:

1. Equipment pads
2. Steel channels embedded or grouted in concrete
3. Special knockouts
4. Rigging

Job Conditions: Productivity is based on new construction, equipment to be installed on the first floor within 200′ of the loading dock.

R262213-10   Electric Circuit Voltages

General: The following method provides the user with a simple non-technical means of obtaining comparative costs of wiring circuits. The circuits considered serve the electrical loads of motors, electric heating, lighting and transformers, for example, that require low voltage 60 Hertz alternating current.

The method used here is suitable only for obtaining estimated costs. It is **not** intended to be used as a substitute for electrical engineering design applications.

Conduit and wire circuits can represent from twenty to thirty percent of the total building electrical cost. By following the described steps and using the tables the user can translate the various types of electric circuits into estimated costs.

Wire Size: Wire size is a function of the electric load which is usually listed in one of the following units:

1. Amperes (A)
2. Watts (W)
3. Kilowatts (kW)
4. Volt amperes (VA)
5. Kilovolt amperes (kVA)
6. Horsepower (HP)

These units of electric load must be converted to amperes in order to obtain the size of wire necessary to carry the load. To convert electric load units to amperes one must have an understanding of the voltage classification of the power source and the voltage characteristics of the electrical equipment or load to be energized. The seven A.C. circuits commonly used are illustrated in Figures R262213-11 thru R262213-17 showing the transformer load voltage and the point of use voltage at the point on the circuit where the load is connected. The difference between the source and point of use voltages is attributed to the circuit voltage drop and is considered to be approximately 4%.

Motor Voltages: Motor voltages are listed by their point of use voltage and not the power source voltage.

For example: 460 volts instead of 480 volts
200 instead of 208 volts
115 volts instead of 120 volts

Lighting and Heating Voltages: Lighting and heating equipment voltages are listed by the power source voltage and not the point of wire voltage.

For example: 480, 277, 120 volt lighting
480 volt heating or air conditioning unit
208 volt heating unit

Transformer Voltages: Transformer primary (input) and secondary (output) voltages are listed by the power source voltage.

For example: Single phase 10 kVA
Primary 240/480 volts
Secondary 120/240 volts

In this case, the primary voltage may be 240 volts with a 120 volts secondary or may be 480 volts with either a 120V or a 240V secondary.

For example: Three phase 10 kVA
Primary 480 volts
Secondary 208Y/120 volts

In this case the transformer is suitable for connection to a circuit with a 3 phase 3 wire or 3 phase 4 wire circuit with a 480 voltage. This application will provide a secondary circuit of 3 phase 4 wire with 208 volts between phase wires and 120 volts between any phase wire and the neutral (white) wire.

R262213-11

3 Wire, 1 Phase, 120/240 Volt System

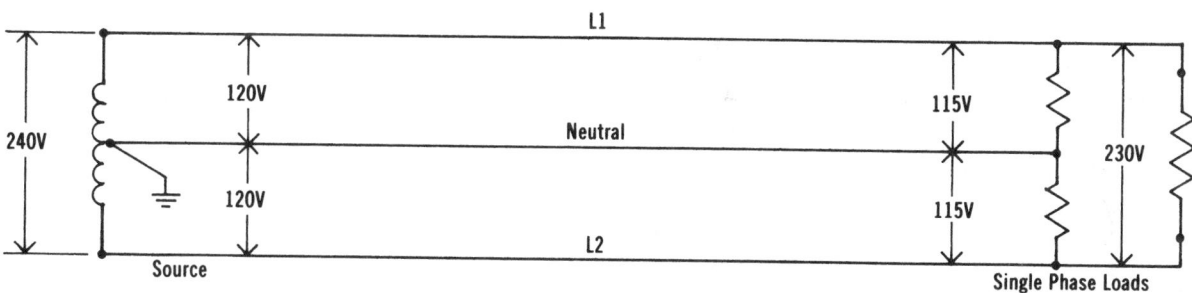

R262213-12

4 wire, 3 Phase, 208Y/120 Volt System

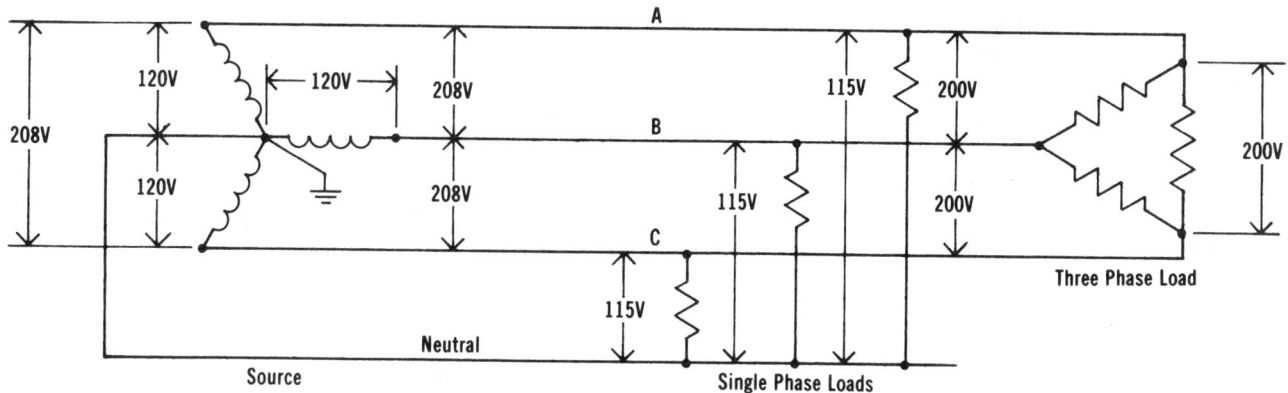

R262213-13 Electric Circuit Voltages (cont.)

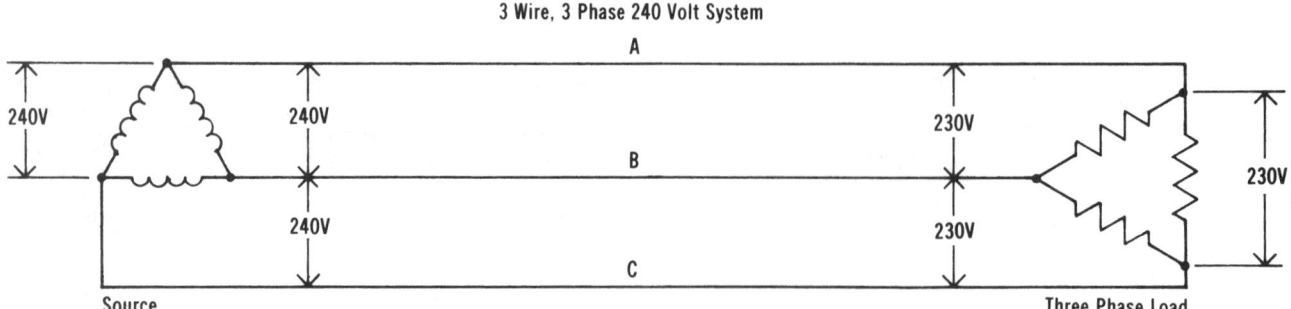

3 Wire, 3 Phase 240 Volt System

R262213-14

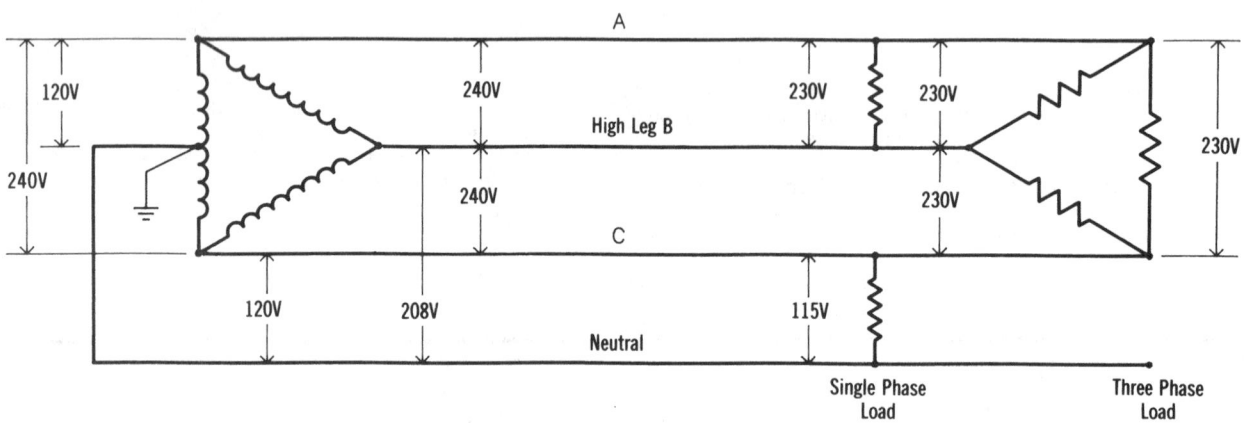

4 Wire, 3 Phase, 240/120 Volt System

R262213-15

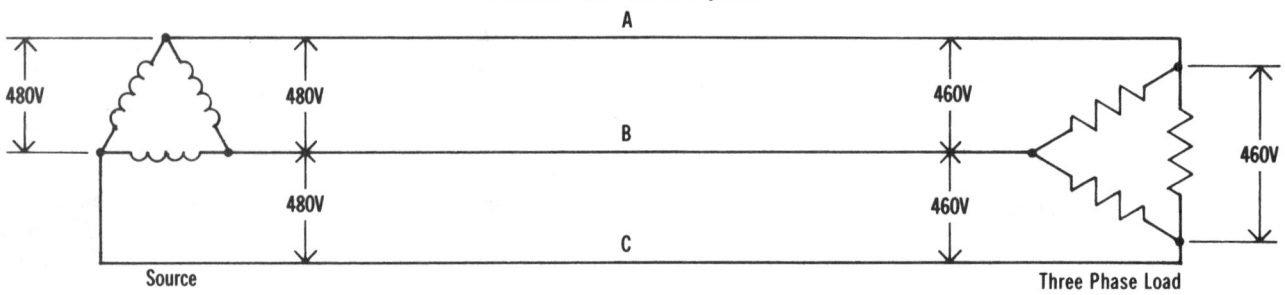

3 Wire, 3 Phase 480 Volt System

R262213-16

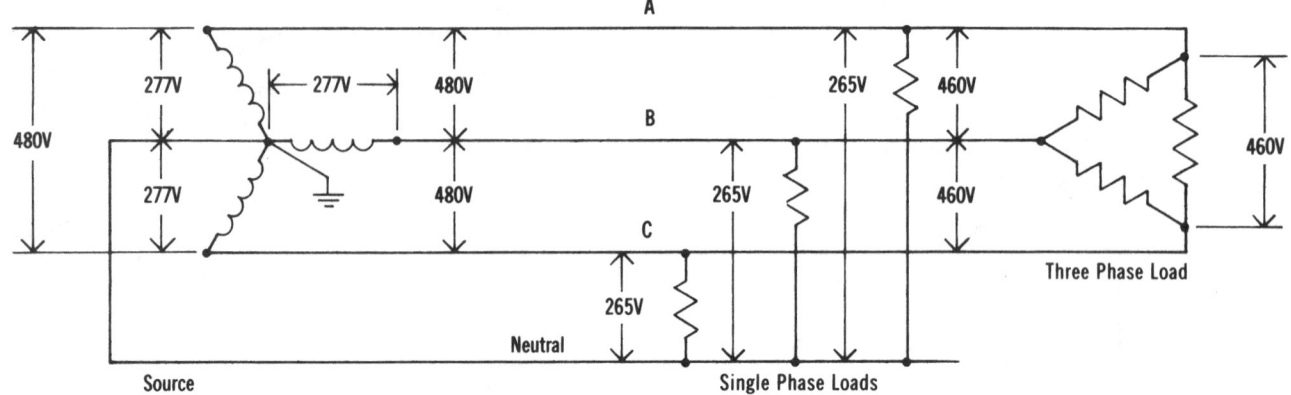

4 Wire, 3 Phase, 480Y/277 Volt System

R262213-17 Electric Circuit Voltages (cont.)

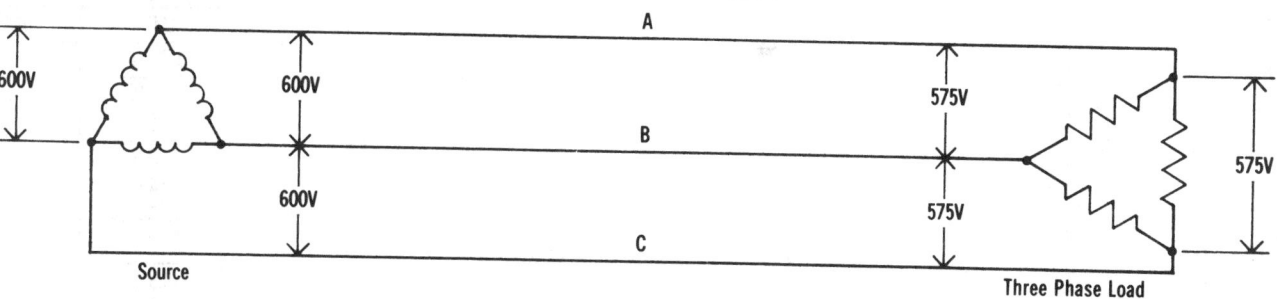

3 Wire, 3 Phase, 600 Volt System

R262213-20 kW Value/Cost Determination

General: Lighting and electric heating loads are expressed in watts and kilowatts.

Cost Determination:

The proper ampere values can be obtained as follows:
1. Convert watts to kilowatts
   (watts 1000 ÷ kilowatts)
2. Determine voltage rating of equipment.
3. Determine whether equipment is single phase or three phase.
4. Refer to Table R262213-21 to find ampere value from kW, Ton and Btu/hr. values.
5. Determine type of wire insulation – TW, THW, THWN.
6. Determine if wire is copper or aluminum.
7. Refer to Table R260519-92 to obtain copper or aluminum wire size from ampere values.
8. Next refer to Table R260533-22 for the proper conduit size to accommodate the number and size of wires in each particular case.
9. Next refer to unit cost data for the per linear foot cost of the conduit.
10. Next refer to unit cost data for the per linear foot cost of the wire. Multiply cost of wire per L.F. x number of wires in the circuits to obtain total wire cost per L.F.

11. Add values obtained in Step 9 and 10 for total cost per linear foot for conduit and wire x length of circuit = Total Cost.

Notes:
1. 1 Phase refers to single phase, 2 wire circuits.
2. 3 Phase refers to three phase, 3 wire circuits.
3. For circuits which operate continuously for 3 hours or more, multiply the ampere values by 1.25 for a given kw requirement.
4. For kW ratings not listed, add ampere values.

   For example: Find the ampere value of
   9 kW at 208 volt, single phase.

   $$\frac{\begin{array}{l}4 \text{ kW} = 19.2A\\5 \text{ kW} = 24.0A\end{array}}{9 \text{ kW} = 43.2A}$$

5. "Length of Circuit" refers to the one way distance of the run, not to the total sum of wire lengths.

R262213-21 Ampere Values as Determined by kW Requirements, BTU/HR or Ton, Voltage and Phase Values

			Ampere Values						
			120V	208V		240V		277V	480V
kW	Ton	BTU/HR	1 Phase	1 Phase	3 Phase	1 Phase	3 Phase	1 Phase	3 Phase
0.5	.1422	1,707	4.2A	2.4A	1.4A	2.1A	1.2A	1.8A	0.6A
0.75	.2133	2,560	6.2	3.6	2.1	3.1	1.9	2.7	.9
1.0	.2844	3,413	8.3	4.9	2.8	4.2	2.4	3.6	1.2
1.25	.3555	4,266	10.4	6.0	3.5	5.2	3.0	4.5	1.5
1.5	.4266	5,120	12.5	7.2	4.2	6.3	3.1	5.4	1.8
2.0	.5688	6,826	16.6	9.7	5.6	8.3	4.8	7.2	2.4
2.5	.7110	8,533	20.8	12.0	7.0	10.4	6.1	9.1	3.1
3.0	.8532	10,239	25.0	14.4	8.4	12.5	7.2	10.8	3.6
4.0	1.1376	13,652	33.4	19.2	11.1	16.7	9.6	14.4	4.8
5.0	1.4220	17,065	41.6	24.0	13.9	20.8	12.1	18.1	6.1
7.5	2.1331	25,598	62.4	36.0	20.8	31.2	18.8	27.0	9.0
10.0	2.8441	34,130	83.2	48.0	27.7	41.6	24.0	36.5	12.0
12.5	3.5552	42,663	104.2	60.1	35.0	52.1	30.0	45.1	15.0
15.0	4.2662	51,195	124.8	72.0	41.6	62.4	37.6	54.0	18.0
20.0	5.6883	68,260	166.4	96.0	55.4	83.2	48.0	73.0	24.0
25.0	7.1104	85,325	208.4	120.2	70.0	104.2	60.0	90.2	30.0
30.0	8.5325	102,390		144.0	83.2	124.8	75.2	108.0	36.0
35.0	9.9545	119,455		168.0	97.1	145.6	87.3	126.0	42.1
40.0	11.3766	136,520		192.0	110.8	166.4	96.0	146.0	48.0
45.0	12.7987	153,585			124.8	187.5	112.8	162.0	54.0
50.0	14.2208	170,650			140.0	208.4	120.0	180.4	60.0
60.0	17.0650	204,780			166.4		150.4	216.0	72.0
70.0	19.9091	238,910			194.2		174.6		84.2
80.0	22.7533	273,040			221.6		192.0		96.0
90.0	25.5975	307,170					225.6		108.0
100.0	28.4416	341,300							120.0

R262213-25   kVA Value/Cost Determination

General: Control transformers are listed in VA. Step-down and power transformers are listed in kVA.

Cost Determination:

1. Convert VA to kVA. Volt amperes (VA) ÷ 1000 = Kilovolt amperes (kVA).
2. Determine voltage rating of equipment.
3. Determine whether equipment is single phase or three phase.
4. Refer to Table R262213-26 to find ampere value from kVA value.
5. Determine type of wire insulation – TW, THW, THWN.
6. Determine if wire is copper or aluminum.
7. Refer to Table R260519-92 to obtain copper or aluminum wire size from ampere values.

Example: A transformer rated 10 kVA 480 volts primary, 240 volts secondary, 3 phase has the capacity to furnish the following:

1. Primary amperes = 10 kVA x 1.20 = 12 amperes (from Table R262213-26)
2. Secondary amperes = 10 kVA x 2.40 = 24 amperes (from Table R262213-26)

Note: Transformers can deliver generally 125% of their rated kVA. For instance, a 10 kVA rated transformer can safely deliver 12.5 kVA.

8. Next refer to Table R260533-22 for the proper conduit size to accommodate the number and size of wires in each particular case.
9. Next refer to unit price data for the per linear foot cost of the conduit.
10. Next refer to unit price data for the per linear foot cost of the wire. Multiply cost of wire per L.F. x number of wires in the circuits to obtain total wire cost.
11. Add values obtained in Step 9 and 10 for total cost per linear foot for conduit and wire x length of circuit = Total Cost.

R262213-26   Multiplier Values for kVA to Amperes Determined by Voltage and Phase Values

Volts	Multiplier for Circuits	
	2 Wire, 1 Phase	3 Wire, 3 Phase
115	8.70	
120	8.30	
230	4.30	2.51
240	4.16	2.40
200	5.00	2.89
208	4.80	2.77
265	3.77	2.18
277	3.60	2.08
460	2.17	1.26
480	2.08	1.20
575	1.74	1.00
600	1.66	0.96

R262213-27   Central Air Conditioning Watts per S.F., BTUs per Hour per S.F. of Floor Area and S.F. per Ton of Air Conditioning

Type Building	Watts per S.F.	BTUH per S.F.	S.F. per Ton	Type Building	Watts per S.F.	BTUH per S.F.	S.F. per Ton	Type Building	Watts per S.F.	BTUH per S.F.	S.F. per Ton
Apartments, Individual	3	26	450	Dormitory, Rooms	4.5	40	300	Libraries	5.7	50	240
Corridors	2.5	22	550	Corridors	3.4	30	400	Low Rise Office, Ext.	4.3	38	320
Auditoriums & Theaters	3.3	40	300/18*	Dress Shops	4.9	43	280	Interior	3.8	33	360
Banks	5.7	50	240	Drug Stores	9	80	150	Medical Centers	3.2	28	425
Barber Shops	5.5	48	250	Factories	4.5	40	300	Motels	3.2	28	425
Bars & Taverns	15	133	90	High Rise Off.-Ext. Rms.	5.2	46	263	Office (small suite)	4.9	43	280
Beauty Parlors	7.6	66	180	Interior Rooms	4.2	37	325	Post Office, Int. Office	4.9	42	285
Bowling Alleys	7.8	68	175	Hospitals, Core	4.9	43	280	Central Area	5.3	46	260
Churches	3.3	36	330/20*	Perimeter	5.3	46	260	Residences	2.3	20	600
Cocktail Lounges	7.8	68	175	Hotels, Guest Rooms	5	44	275	Restaurants	6.8	60	200
Computer Rooms	16	141	85	Public Spaces	6.2	55	220	Schools & Colleges	5.3	46	260
Dental Offices	6	52	230	Corridors	3.4	30	400	Shoe Stores	6.2	55	220
Dept. Stores, Basement	4	34	350	Industrial Plants, Offices	4.3	38	320	Shop'g. Ctrs., Sup. Mkts.	4	34	350
Main Floor	4.5	40	300	General Offices	4	34	350	Retail Stores	5.5	48	250
Upper Floor	3.4	30	400	Plant Areas	4.5	40	300	Specialty Shops	6.8	60	200

*Persons per ton   12,000 BTUH = 1 ton of air conditioning

R262213-60   Oil Filled Transformers

Transformers in this section include:
1. Rigging (as required)
2. Rental of crane and operator
3. Setting of oil filled transformer
4. (4) Anchor bolts, nuts and washers in concrete pad

Price does not include:
1. Primary and secondary terminations
2. Transformer pad
3. Equipment grounding
4. Cable
5. Conduit locknuts or bushings

Transformers in Unit Price sections for dry type, back-boost and isolating transformers include:
1. Unloading and uncrating
2. Hauling transformer to within 200' of loading dock

3. Setting in place
4. Wall mounting hardware
5. Testing

Price does not include:
1. Structural supports
2. Suspension systems
3. Welding or fabrication
4. Primary & secondary terminations

Add the following percentages to the labor for ceiling mounted transformers:

10' to 15'	= + 15%
15' to 25'	= + 30%
Over 25'	= + 35%

Job Conditions: Productivities are based on new construction. Installation is assumed to be on the first floor, in an obstructed area to a height of 10'. Material staging area is within 100' of final transformer location.

R262213-65 Transformer Weight (Lbs.) by kVA

Oil Filled 3 Phase 5/15 KV To 480/277			
kVA	Lbs.	kVA	Lbs.
150	1800	1000	6200
300	2900	1500	8400
500	4700	2000	9700
750	5300	3000	15000

Dry 240/480 To 120/240 Volt			
1 Phase		3 Phase	
kVA	Lbs.	kVA	Lbs.
1	23	3	90
2	36	6	135
3	59	9	170
5	73	15	220
7.5	131	30	310
10	149	45	400
15	205	75	600
25	255	112.5	950
37.5	295	150	1140
50	340	225	1575
75	550	300	1870
100	670	500	2850
167	900	750	4300

Electrical   R2624 Switchboards & Panelboards

R262416-50 Load Centers and Panelboards

When pricing Load Centers list panels by size and type. List Breakers in a separate column of the "Quantity Sheet," and define by phase and ampere rating.

Material and Labor prices include breakers; for example: a 100A, 3-Wire, 102/240V, 18 circuit panel w/ main breaker, as described in the unit cost section, contains 18 single pole 20A breakers.

If you do not choose to include a full panel of single pole breakers, use the following method to adjust material and labor costs.

Example: In an 18 circuit panel only 16 single pole breakers are desired, requiring that the cost of 2 breakers be subtracted from the panel cost.
1. Go to the appropriate unit cost section of the book to find the unit prices of the given circuit breaker type.
2. Modify those costs as follows: Bare material price x 0.50 Bare labor cost x 0.60.
3. Multiply those modified bare costs by 2 (breakers in this example).
4. Subtract the modified costs for the 2 breakers from the given cost of the panel.

Labor-hours for Load Center installation includes:
1. Unloading, uncrating, and handling enclosures 200' from unloading area
2. Measuring and marking

3. Drilling (4) lead anchor type fasteners using a hammer drill
4. Mounting and leveling panel to a height of 6'
5. Preparation and termination of feeder cable to lugs or main breaker
6. Branch circuit identification
7. Lacing using tie wraps
8. Testing and load balancing
9. Marking panel directory

Not included in the material and labor are:
1. Modifications to enclosure
2. Structural supports
3. Additional lugs
4. Plywood backboards
5. Painting or lettering

Note: Knockouts are included in the price of terminating pipe runs and need not be added to the Load Center costs.

Job Conditions: Productivity is based on new construction to a height of 6', in an unobstructed area. Material staging area is assumed to be within 100' of work being performed.

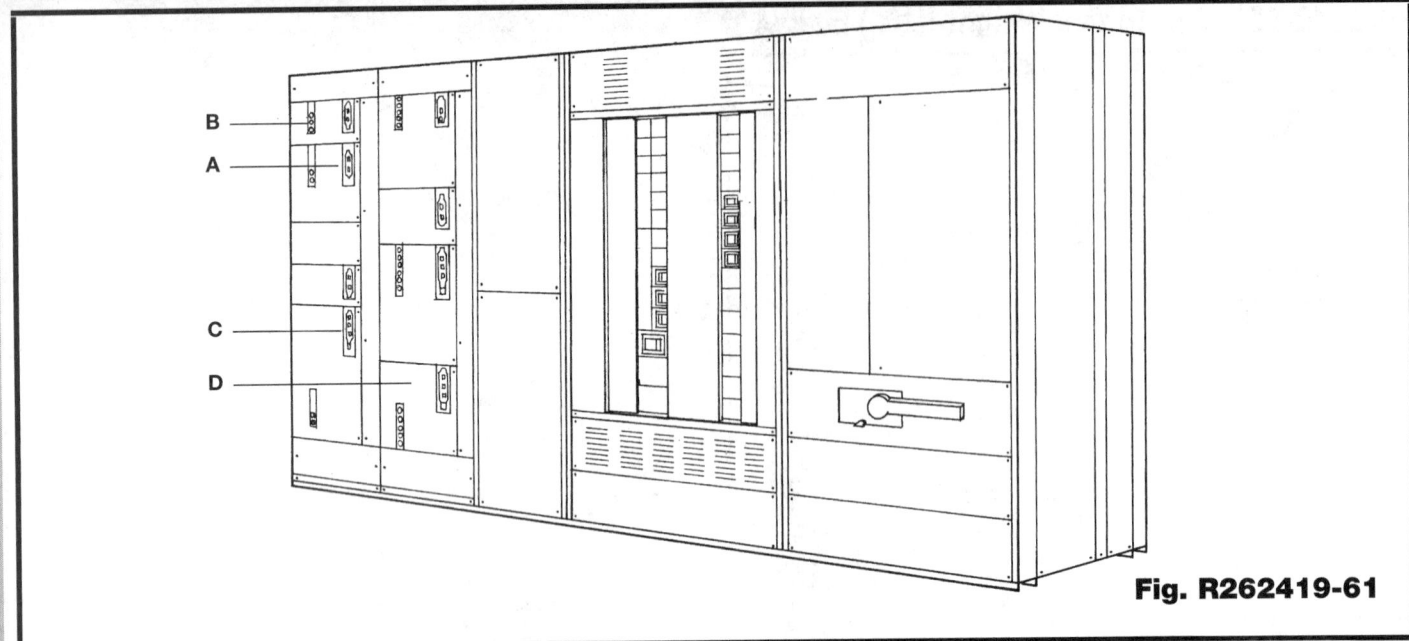

Fig. R262419-61

R262419-60   Motor Control Centers

When taking off Motor Control Centers, list the size, type and height of structures.

Example:
1. 600A, 22,000 RMS, 72″ high
2. 600A, back to back, 72″ high

Next take off individual starters; the number of structures can also be determined by adding the height in inches of starters divided by the height of the structure, and list on the same quantity sheet as the structures. Identify starters by Type, Horsepower rating, Size and Height in inches.

Example:
A. Class I, Type B, FVNR starter, 25 H.P., 18″ high.
   Add to the list with Starters, factory installed controls.

Example:
B. Pilot lights
C. Push buttons
D. Auxiliary contacts

Identify starters and structures as either copper or aluminum and by the NEMA type of enclosure.

When pricing starters and structures, be sure to add or deduct adjustments, using lines in the unit price section.

Included in the cost of Motor Control Structures are:
1. Uncrating
2. Hauling to location within 100′ of loading dock
3. Setting structures
4. Leveling
5. Aligning
6. Bolting together structure frames
7. Bolting horizontal bus bars

Labor-hours do not include:
1. Equipment pad
2. Steel channels embedded or grouted in concrete
3. Pull boxes
4. Special knockouts
5. Main switchboard section
6. Transition section
7. Instrumentation
8. External control wiring
9. Conduit or wire

Material for Starters includes:
1. Circuit breaker or fused disconnect
2. Magnetic motor starter
3. Control transformer
4. Control fuse and fuse block

Labor-hours for Starters include:
1. Handling
2. Installing starter within structure
3. Internal wiring connections
4. Lacing within enclosure
5. Testing
6. Phasing

Job Conditions: Productivity is based on new construction. Motor control location assumed to be on first floor in an unobstructed area. Material staging area within 100′ of final location.

Note: Additional labor-hours must be added if M.C.C. is to be installed on other than the first floor, or if rigging is required.

R262419-65 Motor Starters and Controls

Motor starters should be listed on the same "Quantity Sheet" as Motors and Motor Connections. Identify each starter by:

1. Size
2. Voltage
3. Type

Example:

  A. FVNR, 480V, 2HP, Size 00
  B. FVNR, 480V, 5HP, Size 0, Combination type
  C. FVR, 480V, Size 2

The NEMA type of enclosure should also be identified.

Included in the labor-hours are:

1. Unloading, uncrating and handling of starters up to 200' from loading area
2. Measuring and marking
3. Drilling (4) anchor type lead fasteners, using a hammer drill
4. Mounting and leveling starter
5. Connecting wire or cable to line and load sides of starter (when already lugged)
6. Installation of (3) thermal type heaters
7. Testing

The following is not included unless specified in the unit price description.

1. Control transformer
2. Controls, either factory or field installed
3. Conduit and wire to or from starter
4. Plywood backboard
5. Cable terminations

The following material and labor has been included for Combination type starters:

1. Unloading, uncrating and handling of starter up to 200' of loading dock
2. Measuring and marking
3. Drilling (4) anchor type lead fasteners using a hammer drill
4. Mounting and leveling
5. Connecting prepared cable conductors
6. Installation of (3) dual element cartridge type fuses
7. Installation of (3) thermal type heaters
8. Test for rotation

MOTOR CONTROLS

When pricing motor controls make sure you consider the type of control system being utilized. If the controls are factory installed and located in the enclosure itself, then you would add to your starter price the items 5200 thru 5800 in the Unit Price section. If control voltage is different from line voltage, add the material and labor cost of a control transformer.

For external control of starters include the following items:

1. Raceways
2. Wire & terminations
3. Control enclosures
4. Fittings
5. Push button stations
6. Indicators

Job Conditions: Productivity is based on new construction to a height of 10'.

Material staging area is assumed to be within 100' of work being performed.

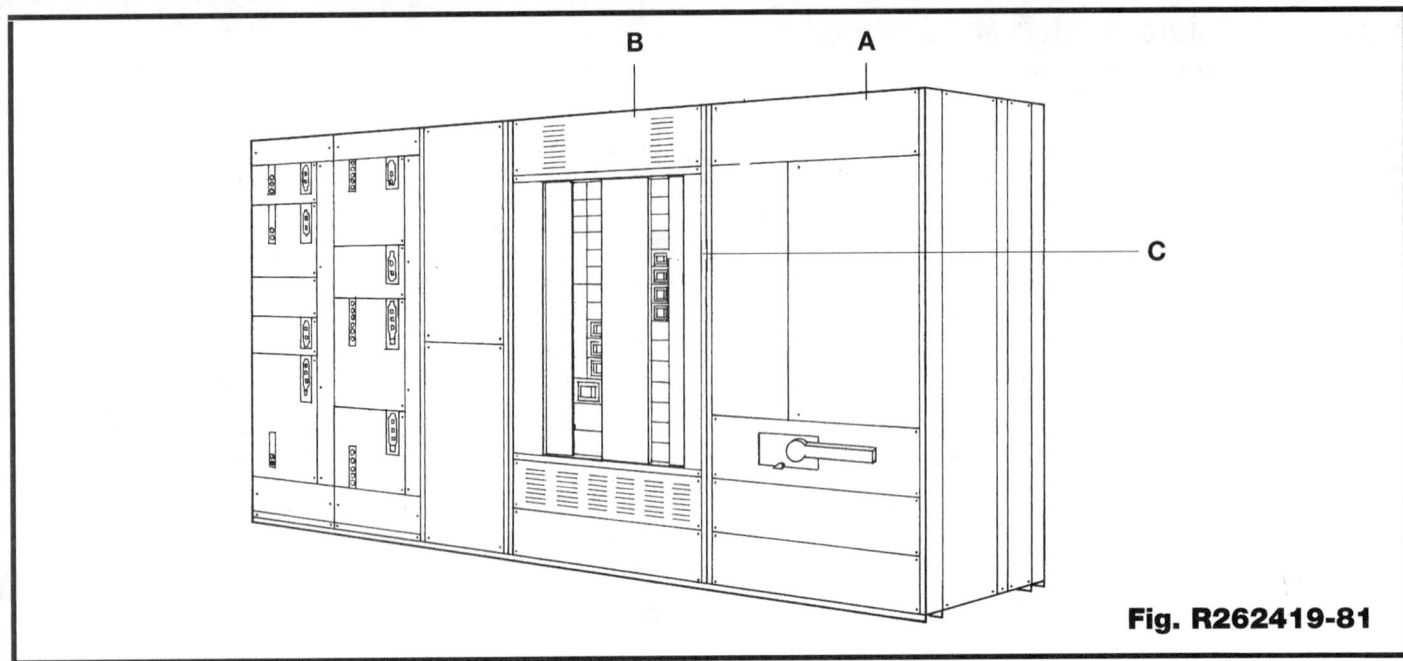

Fig. R262419-81

R262419-80   Distribution Section

After "Taking off" the Switchboard section, include on the same "Quantity sheet" the Distribution section; identify by:
1. Voltage
2. Ampere rating
3. Type

Example:

(Fig. R262419-81) (B) Distribution Section

Included in the labor costs of the Distribution section is:
1. Uncrating
2. Hauling to 200' of loading dock
3. Setting of distribution panel
4. Leveling & shimming

5. Anchoring of equipment to pad or floor
6. Bolting of horizontal bus bars between
7. Testing of equipment

Not included in the Distribution section is:
1. Breakers (C)
2. Equipment pads
3. Steel channels embedded or grouted in concrete
4. Pull boxes
5. Special knockouts
6. Transition section
7. Conduit or wire

R262419-82   Feeder Section

List quantities on the same sheet as the Distribution section. Identify breakers by: (Fig. R262419-81)
1. Frame type
2. Number of poles
3. Ampere rating

Installation includes:
1. Handling

2. Placing breakers in distribution panel
3. Preparing wire or cable
4. Lacing wire
5. Marking each phase with colored tape
6. Marking panel legend
7. Testing
8. Balancing

Reference Tables

R262419-84 Switchgear

It is recommended that "Switchgear" or those items contained in the following sections be quoted from equipment manufacturers as a package price.

Switchboard instruments

Switchboard distribution sections

Switchboard feeder sections

Switchboard Service Disc.

Switchboard In-plant Dist.

Included in these sections are the most common types and sizes of factory assembled equipment.

The recommended procedure for low voltage switchgear would be to price (Fig. R262419-81) (A) Main Switchboard

Identify by:
1. Voltage
2. Amperage
3. Type

Example:
1. 120/208V, 4-wire, 600A, nonfused
2. 277/480V, 4-wire, 600A, nonfused
3. 120/208V, 4-wire, 400A w/fused switch & CT compartment
4. 277/480V, 4-wire, 400A, w/fused switch & CT compartment
5. 120/208V, 4-wire, 800A, w/pressure switch & CT compartment
6. 277/480V, 4-wire, 800A, w/molded CB & CT compartment

Included in the labor costs for Switchboards are:
1. Uncrating
2. Hauling to 200' of loading dock
3. Setting equipment
4. Leveling and shimming
5. Anchoring
6. Cable identification
7. Testing of equipment

Not included in the Switchboard price is:
1. Rigging
2. Equipment pads
3. Steel channels embedded or grouted in concrete
4. Special knockouts
5. Transition or Auxiliary sections
6. Instrumentation
7. External control
8. Conduit and wire
9. Conductor terminations

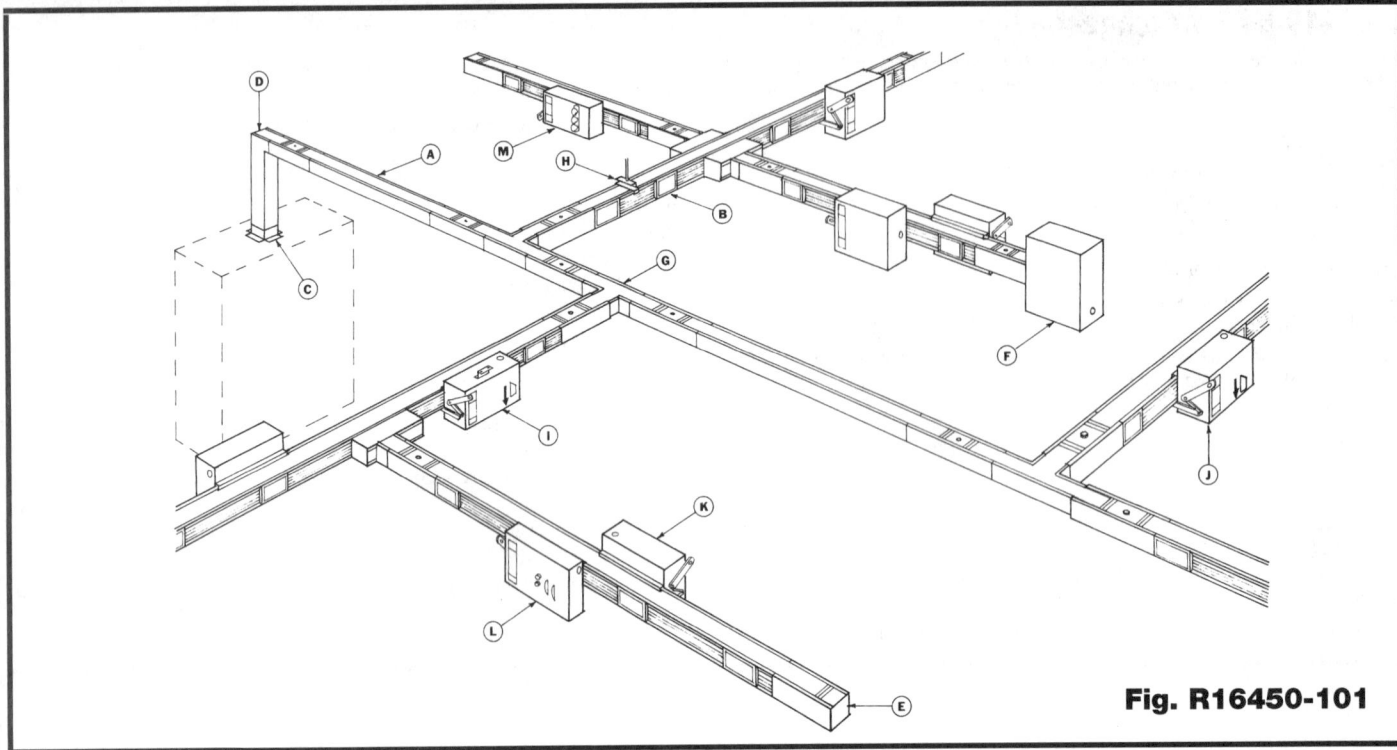

Fig. R16450-101

R262513-10 Aluminum Bus Duct

When taking off bus duct identify the system as either:
1. Aluminum
2. Copper

List straight lengths by type and size
 A. Plug-in — 800 A
 B. Feeder — 800 A

Do not measure thru fittings as you would on conduit, since there is no allowance for scrap in bus duct systems.

If upon taking off linear foot quantities of bus duct you find your quantities are not divisible evenly by 10 ft., then the remainder must be priced as a special item and quoted from the manufactuer. Do not use the bare material cost per L.F. for these special items. You can, however, safely use the bare labor cost per L.F. for the entire length.

Identify fittings by type and ampere rating.

Example:

C. Switchboard stub 800 A

D. Elbows 800 A

E. End box 800 A

F. Cable tap box 800 A

G. Tee Fittings 800 A

H. Hangers

Plug-in Units – List separately plug-in units and identify by type and ampere rating

I. Plug-in switches 600 Volt 3 phase 60 A

J. Plug-in molded case C.B. 60 A

K. Combination starter FVNR NEMA 1

L. Combination contactor & fused switch NEMA 1

M. Combination fusible switch & lighting control 60 A

Labor-hours for feeder and plug-in sections include:
1. Unloading and uncrating
2. Hauling up to 200 ft. from loading dock
3. Measuring and marking
4. Setup of rolling staging
5. Installing hangers
6. Hanging and bolting sections
7. Aligning and leveling
8. Testing

Labor-hours do not include:
1. Modifications to existing structure for hanger supports
2. Threaded rod in excess of 2 ft.
3. Welding
4. Penetrations thru walls
5. Staging rental

Deduct the following percentages from labor only:

  150 ft. to 250 ft. — 10%
  251 ft. to 350 ft. — 15%
  351 ft. to 500 ft. — 20%
     Over 500 ft. — 25%

Deduct percentage only if runs are contained in the same area.

Example: If the job entails running 100 ft. in 5 different locations do not deduct 20%, but if the duct is being run in 1 area and the quantity is 500 ft. then you would deduct 20%.

Deduct only from straight lengths, not fittings or plug-in units.

R262513-10 Aluminum Bus Duct (cont.)

Add to labor for elevated installations:

15 ft. to 20 ft. high	10%
21 ft. to 25 ft. high	20%
26 ft. to 30 ft. high	30%
31 ft. to 35 ft. high	40%
36 ft. to 40 ft. high	50%
Over 40 ft. high	60%

Bus Duct Fittings:

Labor-hours for fittings include:
1. Unloading and uncrating
2. Hauling up to 200 ft. from loading dock
3. Installing, fitting, and bolting all ends to in-place sections

Plug-in units include:
1. Unloading and uncrating
2. Hauling up to 200 ft. from loading dock
3. Installing plug-in into in-place duct
4. Setup of rolling staging
5. Connecting load wire to lugs
6. Marking wire
7. Checking phase rotation

Labor-hours for plug-ins do not include:
1. Conduit runs from plug-in
2. Wire from plug-in
3. Conduit termination

Economy of Scale – For large concentrations of plug-in units in the same area deduct the following percentages:

11	to	25	15%
26	to	50	20%
51	to	75	25%
76	to	100	30%
100	and	over	35%

Job Conditions: Productivities are based on new construction in an unobstructed first floor area to a height of 15 ft. using rolling staging.

Material staging area is within 100 ft. of work being performed.
Add to the duct fittings and hangers:
    Plug-in switches (fused)
    Plug-in breakers
    Combination starters
    Combination contactors
    Combination fusible switch and lighting control

Labor-hours for duct and fittings include:
1. Unloading and uncrating
2. Hauling up to 200 ft. from loading dock
3. Measuring and marking
4. Installing duct runs
5. Leveling
6. Sound testing

Labor-hours for plug-ins include:
1. Unloading and uncrating
2. Hauling up to 200 ft. from loading dock
3. Installing plug-ins
4. Preparing wire
5. Wire connections and marking

R262513-15 Weight (Lbs./L.F.) of 4 Pole Aluminum and Copper Bus Duct by Ampere Load

Amperes	Aluminum Feeder	Copper Feeder	Aluminum Plug–In	Copper Plug–In
225			7	7
400			8	13
600	10	10	11	14
800	10	19	13	18
1000	11	19	16	22
1350	14	24	20	30
1600	17	26	25	39
2000	19	30	29	46
2500	27	43	36	56
3000	30	48	42	73
4000	39	67		
5000		78		

R262716-40 Standard Electrical Enclosure Types

NEMA Enclosures

Electrical enclosures serve two basic purposes; they protect people from accidental contact with enclosed electrical devices and connections, and they protect the enclosed devices and connections from specified external conditions. The National Electrical Manufacturers Association (NEMA) has established the following standards. Because these descriptions are not intended to be complete representations of NEMA listings, consultation of NEMA literature is advised for detailed information.

The following definitions and descriptions pertain to NONHAZARDOUS locations.

NEMA Type 1: General purpose enclosures intended for use indoors, primarily to prevent accidental contact of personnel with the enclosed equipment in areas that do not involve unusual conditions.

NEMA Type 2: Dripproof indoor enclosures intended to protect the enclosed equipment against dripping noncorrosive liquids and falling dirt.

NEMA Type 3: Dustproof, raintight and sleet-resistant (ice-resistant) enclosures intended for use outdoors to protect the enclosed equipment against wind-blown dust, rain, sleet, and external ice formation.

NEMA Type 3R: Rainproof and sleet-resistant (ice-resistant) enclosures which are intended for use outdoors to protect the enclosed equipment against rain. These enclosures are constructed so that the accumulation and melting of sleet (ice) will not damage the enclosure and its internal mechanisms.

NEMA Type 3S: Enclosures intended for outdoor use to provide limited protection against wind-blown dust, rain, and sleet (ice) and to allow operation of external mechanisms when ice-laden.

NEMA Type 4: Watertight and dust-tight enclosures intended for use indoors and out – to protect the enclosed equipment against splashing water, see page of water, falling or hose-directed water, and severe external condensation.

NEMA Type 4X: Watertight, dust-tight, and corrosion-resistant indoor and outdoor enclosures featuring the same provisions as Type 4 enclosures, plus corrosion resistance.

NEMA Type 5: Indoor enclosures intended primarily to provide limited protection against dust and falling dirt.

NEMA Type 6: Enclosures intended for indoor and outdoor use – primarily to provide limited protection against the entry of water during occasional temporary submersion at a limited depth.

NEMA Type 6R: Enclosures intended for indoor and outdoor use – primarily to provide limited protection against the entry of water during prolonged submersion at a limited depth.

NEMA Type 11: Enclosures intended for indoor use – primarily to provide, by means of oil immersion, limited protection to enclosed equipment against the corrosive effects of liquids and gases.

NEMA Type 12: Dust-tight and driptight indoor enclosures intended for use indoors in industrial locations to protect the enclosed equipment against fibers, flyings, lint, dust, and dirt, as well as light splashing, see page, dripping, and external condensation of noncorrosive liquids.

NEMA Type 13: Oil-tight and dust-tight indoor enclosures intended primarily to house pilot devices, such as limit switches, foot switches, push buttons, selector switches, and pilot lights, and to protect these devices against lint and dust, see page, external condensation, and sprayed water, oil, and noncorrosive coolant.

The following definitions and descriptions pertain to HAZARDOUS, or CLASSIFIED, locations:

NEMA Type 7: Enclosures intended to use in indoor locations classified as Class 1, Groups A, B, C, or D, as defined in the National Electrical Code.

NEMA Type 9: Enclosures intended for use in indoor locations classified as Class 2, Groups E, F, or G, as defined in the National Electrical Code.

R262726-90 Wiring Devices

Wiring devices should be priced on a separate takeoff form which includes boxes, covers, conduit and wire.

Labor-hours for devices include:
1. Stripping of wire
2. Attaching wire to device using terminators on the device itself, lugs, set screws etc.
3. Mounting of device in box

Labor-hours do not include:
1. Conduit
2. Wire
3. Boxes
4. Plates

Economy of Scale – for large concentrations of devices in the same area deduct the following percentages from labor-hours:

1	to	10	0%
11	to	25	20%
26	to	50	25%
51	to	100	30%
	over	100	35%

R262726-90 Wiring Devices (cont.)

NEMA No.	15 R	20 R	30 R	50 R	60 R
1 125V 2 Pole, 2 Wire					
2 250V 2 Pole, 2 Wire					
5 125V 2 Pole, 3 Wire					
6 250V 2 Pole, 3 Wire					
7 277V, AC 2 Pole, 3 Wire					
10 125/250V 3 Pole, 3 Wire					
11 3 Phase 250V 3 Pole, 3 Wire					
14 125/250V 3 Pole, 4 Wire					
15 3 Phase 250V 3 Pole, 4 Wire					
18 3 Phase 208Y/120V 4 Pole, 4 Wire					

R262726-90   Wiring Devices (cont.)

NEMA No.	15 R	20 R	30 R	NEMA No.	15 R	20 R	30 R
L 1 125V 2 Pole, 2 Wire	●			**L 13** 3 Phase 600V 3 Pole, 3 Wire			●
L 2 250V 2 Pole, 2 Wire		● 15A		**L 14** 125/250V 3 Pole, 4 Wire		●	●
L 5 125 V 2 Pole, 3 Wire	●	●	●	**L 15** 3 Phase 250 V 3 Pole, 4 Wire		●	●
L 6 250 V 2 Pole, 3 Wire	●	●	●	**L 16** 3 Phase 480V 3 Pole, 4 Wire		●	●
L 7 227 V, AC 2 Pole, 3 Wire	●	●	●	**L 17** 3 Phase 600 V 3 Pole, 4 Wire			●
L 8 480 V 2 Pole, 3 Wire		●	●	**L 18** 3 Phase 208Y/120V 4 Pole, 4 Wire		●	●
L 9 600 V 2 Pole, 3 Wire		●	●	**L 19** 3 Phase 480Y/277V 4 Pole, 4 Wire		●	●
L 10 125 /250V 3 Pole, 3 Wire		●	●	**L 20** 3 Phase 600Y/347V 4 Pole, 4 Wire		●	●
L 11 3 Phase 250 V 3 Pole, 3 Wire		●	●	**L 21** 3 Phase 208Y/120V 4 Pole, 5 Wire		●	●
L 12 3 Phase 480 V 3 Pole, 3 Wire		●	●	**L 22** 3 Phase 480Y/277V 4 Pole, 5 Wire		●	●
				L 23 3 Phase 600Y/347V 4 Pole, 5 Wire		●	●

R262816-80   Safety Switches

List each Safety Switch by type, ampere rating, voltage, single or three phase, fused or nonfused.

Example:
  A. General duty, 240V, 3-pole, fused
  B. Heavy Duty, 600V, 3-pole, nonfused
  C. Heavy Duty, 240V, 2-pole, fused
  D. Heavy Duty, 600V, 3-pole, fused

Also include NEMA enclosure type and identify as:
  1. Indoor
  2. Weatherproof
  3. Explosionproof

Installation of Safety Switches includes:
  1. Unloading and uncrating
  2. Handling disconnects up to 200' from loading dock
  3. Measuring and marking location

  4. Drilling (4) anchor type lead fasteners, using a hammer drill
  5. Mounting and leveling Safety Switch
  6. Installing (3) fuses
  7. Phasing and tagging line and load wires

Price does not include:
  1. Modifications to enclosure
  2. Plywood backboard
  3. Conduit or wire
  4. Fuses
  5. Termination of wires

Job Conditions: Productivities are based on new construction to an installed height of 6' above finished floor.

Material staging area is assumed to be within 100' of work in progress.

R263213-45   Generator Weight (Lbs.) by kW

3 Phase 4 Wire /480 Volt			
Gas		Diesel	
kW	Lbs.	kW	Lbs.
7.5	600	30	1800
10	630	50	2230
15	960	75	2250
30	1500	100	3840
65	2350	125	4030
85	2570	150	5500
115	4310	175	5650
170	6530	200	5930
		250	6320
		300	7840
		350	8220
		400	10750
		500	11900

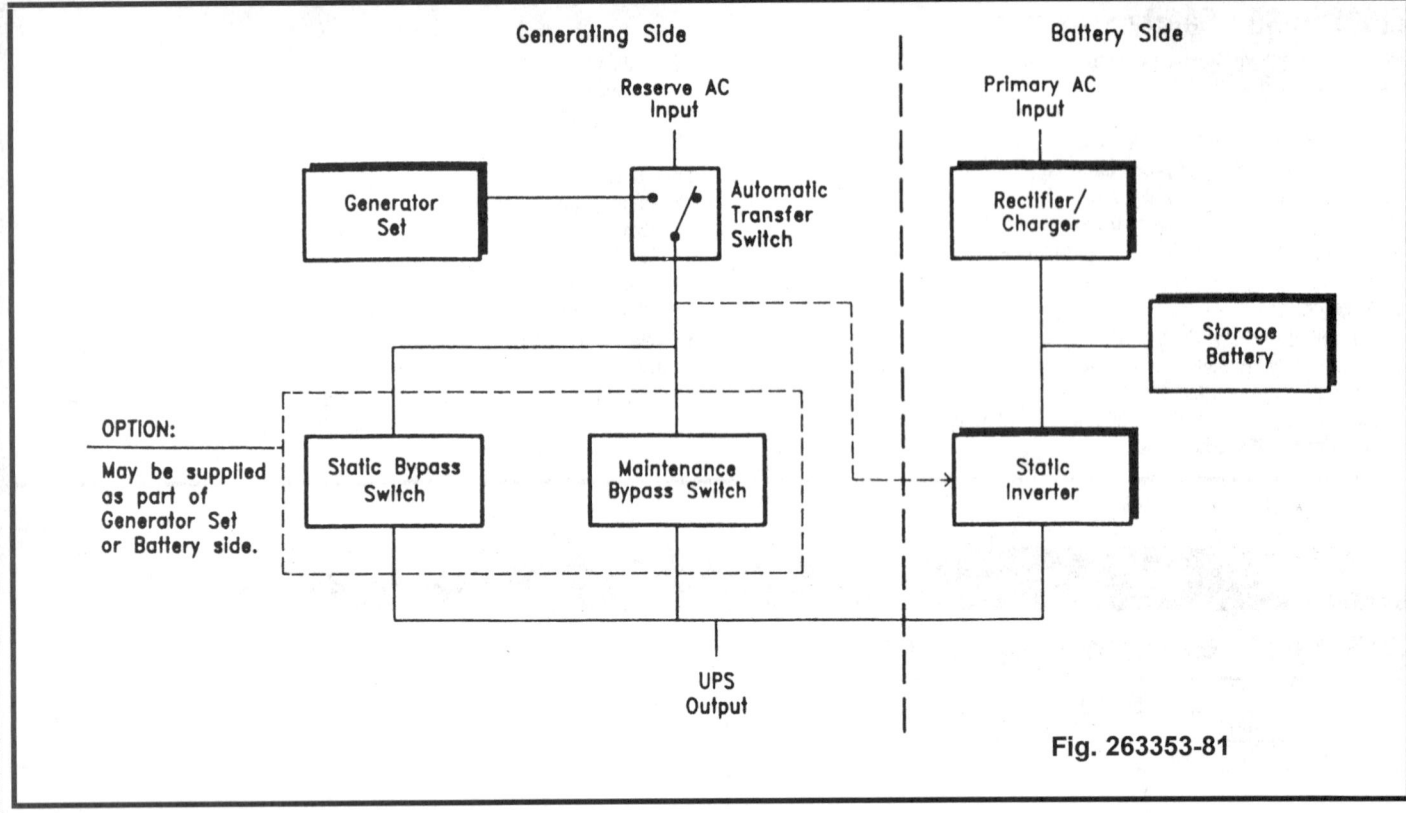

Generating Side — Battery Side

Reserve AC Input — Primary AC Input

Generator Set

Automatic Transfer Switch

Rectifier/ Charger

Storage Battery

OPTION:
May be supplied as part of Generator Set or Battery side.

Static Bypass Switch

Maintenance Bypass Switch

Static Inverter

UPS Output

Fig. 263353-81

R263353-80   Uninterruptible Power Supply Systems

General: Uninterruptible Power Supply (UPS) Systems are used to provide power for legally required standby systems. They are also designed to protect computers and provide optional additional coverage for any or all loads in a facility. Figure R263353-81 shows a typical configuration for a UPS System with generator set.

Cost Determination:

It is recommended that UPS System material costs be obtained from equipment manufacturers as a package. Installation costs should be obtained from a vendor or contractor.

The recommended procedure for pricing UPS Systems would be to identify:
1. Frequency – by Hertz (Hz.)
   For example: 50, 60, and 400/415 Hz.
2. Apparent power and real power
   For example: 200 kVA/170 kW
3. Input/Output voltage   For example: 120V, 208V or 240V

4. Phase – either single or three-phase
5. Options and accessories – For extended run time more batteries would be required. Other accessories include battery cabinets, battery racks, remote control panel, power distribution unit (PDU), and warranty enhancement plans.

Note:
1. Larger systems can be configured by paralleling two or more standard small size single modules.
   For example: two 15 kVA modules, combined together and configured to 30 kVA.
2. Maximum input current during battery recharge is typically 15% higher than normal current.
3. UPS Systems weights vary depending on options purchased and power rating in kVA.

R263413-30   HP Value/Cost Determination

General: Motors can be powered by any of the seven systems shown in Figure R262213-11 thru Figure R262213-17 provided the motor voltage characteristics are compatible with the power system characteristics.

Cost Determination:

Motor Amperes for the various size H.P. and voltage are listed in Table R263413-31. To find the amperes, locate the required H.P. rating and locate the amperes under the appropriate circuit characteristics.

For example:

   A. 100 H.P., 3 phase, 460 volt motor = 124 amperes (Table R263413-31)

   B. 10 H.P., 3 phase, 200 volt motor = 32.2 amperes (Table R263413-31)

Motor Wire Size: After the amperes are found in Table R263413-31 the amperes must be increased 25% to compensate for power losses. Next refer to Table R260519-92. Find the appropriate insulation column for copper or aluminum wire to determine the proper wire size.

For example:

   A. 100 H.P., 3 phase, 460 volt motor has an ampere value of 124 amperes from Table R263413-31

   B. 124A x 1.25 = 155 amperes

   C. Refer to Table R260519-92 for THW or THWN wire insulations to find the proper wire size. For a 155 ampere load using copper wire a size 2/0 wire is needed.

   D. For the 3 phase motor three wires of 2/0 size are required.

Conduit Size: To obtain the proper conduit size for the wires and type of insulation used, refer to Table R260533-22.

For example: For the 100 H.P., 460V, 3 phase motor, it was determined that three 2/0 wires are required. Assuming THWN insulated copper wire, use Table R260533-22 to determine that three 2/0 wires require 1-1/2" conduit.

Material Cost of the conduit and wire system depends on:

   1. Wire size required
   2. Copper or aluminum wire
   3. Wire insulation type selected
   4. Steel or plastic conduit
   5. Type of conduit raceway selected.

Labor Cost of the conduit and wire system depends on:

   1. Type and size of conduit
   2. Type and size of wires installed
   3. Location and height of installation in building or depth of trench
   4. Support system for conduit.

R263413-31   Ampere Values Determined by Horsepower, Voltage and Phase Values

H.P.	Amperes							
	Single Phase			Three Phase				
	115V	208V	230V	200V	208V	230V	460V	575V
1/6	4.4A	2.4A	2.2A					
1/4	5.8	3.2	2.9					
1/3	7.2	4.0	3.6					
1/2	9.8	5.4	4.9	2.5A	2.4A	2.2A	1.1A	0.9A
3/4	13.8	7.6	6.9	3.7	3.5	3.2	1.6	1.3
1	16	8.8	8	4.8	4.6	4.2	2.1	1.7
1-1/2	20	11	10	6.9	6.6	6.0	3.0	2.4
2	24	13.2	12	7.8	7.5	6.8	3.4	2.7
3	34	18.7	17	11.0	10.6	9.6	4.8	3.9
5	56	30.8	28	17.5	16.7	15.2	7.6	6.1
7-1/2	80	44	40	25.3	24.2	22	11	9
10	100	55	50	32.2	30.8	28	14	11
15				48.3	46.2	42	21	17
20				62.1	59.4	54	27	22
25				78.2	74.8	68	34	27
30				92	88	80	40	32
40				120	114	104	52	41
50				150	143	130	65	52
60				177	169	154	77	62
75				221	211	192	96	77
100				285	273	248	124	99
125				359	343	312	156	125
150				414	396	360	180	144
200				552	528	480	240	192
250							302	242
300							361	289
350							414	336
400							477	382

R263413-30 Cost Determination (cont.)

Magnetic starters, switches, and motor connection:

To complete the cost picture from H.P. to Costs additional items must be added to the cost of the conduit and wire system to arrive at a total cost.

1. Assembly Table D5020-160 Magnetic Starters Installed Cost lists the various size starters for single phase and three phase motors.
2. Assembly Table D5020-165 Heavy Duty Safety Switches Installed Cost lists safety switches required at the beginning of a motor circuit and also one required in the vicinity of the motor location.
3. Assembly Table D5020-170 Motor Connection lists the various costs for single and three phase motors.

Worksheet to obtain total motor wiring costs:

It is assumed that the motors or motor driven equipment are furnished and installed under other sections for this estimate and the following work is done under this section:

1. Conduit
2. Wire (add 10% for additional wire beyond conduit ends for connections to switches, boxes, starters, etc.)
3. Starters
4. Safety switches
5. Motor connections

Figure R263413-32

				Cost	
Item	Type	Size	Quantity	Unit	Total
Wire					
Conduit					
Switch					
Starter					
Switch					
Motor Connection					
Other					
Total Cost					

Worksheet for Motor Circuits

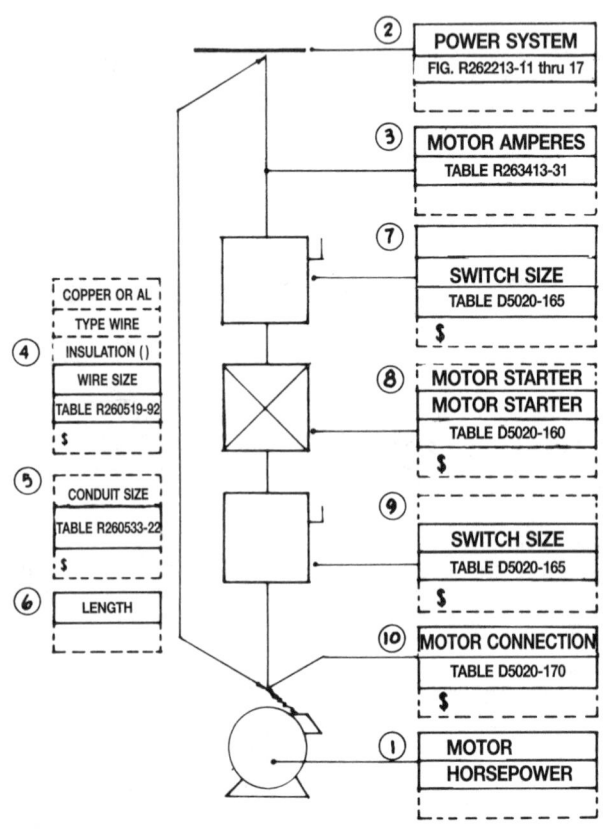

R263413-33 Maximum Horsepower for Starter Size by Voltage

Starter Size	Maximum HP (3φ)			
	208V	240V	480V	600V
00	1½	1½	2	2
0	3	3	5	5
1	7½	7½	10	10
2	10	15	25	25
3	25	30	50	50
4	40	50	100	100
5		100	200	200
6		200	300	300
7		300	600	600
8		450	900	900
8L		700	1500	1500

R263623-60   Automatic Transfer Switches

When taking off Automatic transfer switches identify by voltage, amperage and number of poles.

Example: Automatic transfer switch, 480V, 3 phase, 30A, NEMA 1 enclosure

Labor-hours for transfer switches include:
1. Unloading, uncrating and handling switches up to 200′ from loading dock
2. Measuring and marking
3. Drilling (4) lead type anchors, using a hammer drill
4. Mounting and leveling
5. Circuit identification
6. Testing and load balancing

Labor-hours do not include:
1. Modifications in enclosure
2. Structural supports
3. Additional lugs
4. Plywood backboard
5. Painting or lettering
6. Conduit runs to or from transfer switch
7. Wire
8. Termination of wires

R265113-40   Interior Lighting Fixtures

When taking off interior lighting fixtures, it is advisable to set up your quantity work sheet to conform to the lighting schedule as it appears on the print. Include the alpha-numeric code plus the symbol on your work sheet.

Take off a particular section or floor of the building and count each type of fixture before going on to another type. It would also be advantageous to include on the same work sheet the pipe, wire, fittings and circuit number associated with each type of lighting fixture. This will help you identify the costs associated with any particular lighting system and in turn make material purchases more specific as to when and how much to order under the classification of lighting.

By taking off lighting first you can get a complete "WALK THRU" of the job. This will become helpful when doing other phases of the project.

Materials for a recessed fixture include:
1. Fixture
2. Lamps
3. 6′ of jack chain
4. (2) S hooks
5. (2) Wire nuts

Labor for interior recessed fixtures include:
1. Unloading by hand
2. Hauling by hand to an area up to 200′ from loading dock
3. Uncrating
4. Layout
5. Installing fixture
6. Attaching jack chain & S hooks
7. Connecting circuit power
8. Reassembling fixture
9. Installing lamps
10. Testing

Material for surface mounted fixtures includes:
1. Fixture
2. Lamps
3. Either (4) lead type anchors, (4) toggle bolts, or (4) ceiling grid clips
4. (2) Wire nuts

Material for pendent mounted fixtures includes:
1. Fixture
2. Lamps
3. (2) Wire nuts

4. Rigid pendents as required by type of fixtures
5. Canopies as required by type of fixture

Labor hours include the following for both surface and pendent fixtures:
1. Unloading by hand
2. Hauling by hand to an area up to 200′ from loading dock
3. Uncrating
4. Layout and marking
5. Drilling (4) holes for either lead anchors or toggle bolts using a hammer drill
6. Installing fixture
7. Leveling fixture
8. Connecting circuit power
9. Installing lamps
10. Testing

Labor for surface or pendent fixtures does not include:
1. Conduit
2. Boxes or covers
3. Connectors
4. Fixture whips
5. Special support
6. Switching
7. Wire

Economy of Scale: For large concentrations of lighting fixtures in the same area deduct the following percentages from labor:

25	to	50	fixtures	15%
51	to	75	fixtures	20%
76	to	100	fixtures	25%
101 and over				30%

Job Conditions: Productivity is based on new construction in a unobstructed first floor location, using rolling staging to 15′ high.

Material staging is assumed to be within 100′ of work being performed.

Add the following percentages to labor for elevated installations:

15′	to	20′	high	10%
21′	to	25′	high	20%
26′	to	30′	high	30%
31′	to	35′	high	40%
36′	to	40′	high	50%
41′ and over				60%

R265723-05   Comparison - Operation of High Intensity Discharge Lamps

Lamp Type	Wattage	Life (Hours)	1 Circuit Wattage	Average Initial Lumens	2 L.L.D.	3 Mean Lumens
M.V.	100 DX	24,000	125	4,000	61%	2,440
L.P.S.	SOX-35	18,000	65	4,800	100%	4,800
H.P.S.	LU-70	12,000	84	5,800	90%	5,220
M.H.	No Equivalent					
M.V.	175 DX	24,000	210	8,500	66%	5,676
M.V.	250 DX	24,000	295	13,000	66%	7,986
L.P.S.	SOX-55	18,000	82	8,000	100%	8,000
H.P.S.	LU-100	12,000	120	9,500	90%	8,550
M.H.	No Equivalent					
M.V.	400 DX	24,000	465	24,000	64%	14,400
L.P.S.	SOX-90	18,000	141	13,500	100%	13,500
H.P.S.	LU-150	16,000	188	16,000	90%	14,400
M.H.	MH-175	7,500	210	14,000	73%	10,200
M.V.	No Equivalent					
L.P.S.	SOX-135	18,000	147	22,500	100%	22,500
H.P.S.	LU-250	20,000	310	25,500	92%	23,205
M.H.	MH-250	7,500	295	20,500	78%	16,000
M.V.	1000 DX	24,000	1,085	63,000	61%	37,820
L.P.S.	SOX-180	18,000	248	33,000	100%	33,000
H.P.S.	LU-400	20,000	480	50,000	90%	45,000
M.H.	MH-400	15,000	465	34,000	72%	24,600
H.P.S.	LU-1000	15,000	1,100	140,000	91%	127,400

1. Includes ballast losses and average lamp watts
2. Lamp lumen depreciation (% of initial light output at 70% rated life)
3. Lamp lumen output at 70% rated life (L.L.D. x initial)
4. Cost of yearly operation = Cicuit Wattage/1000 x annual operating hours x average cost per KWH

M.V. = Mercury Vapor
L.P.S. = Low pressure sodium
H.P.S. = High pressure sodium
L.H. = Metal halide

R265723-10   For Other than Regular Cool White (CW) Lamps

Multiply Material Costs as Follows:					
Regular Lamps	Cool white deluxe (CWX)	x 1.35	Energy Saving Lamps	Cool white (CW/ES)	x 1.35
	Warm white deluxe (WWX)	x 1.35		Cool white deluxe (CWX/ES)	x 1.65
	Warm white (WW)	x 1.30		Warm white (WW/ES)	x 1.55
	Natural (N)	x 2.05		Warm white deluxe (WWX/ES)	x 1.65

R265723-20 Lamp Comparison Chart with Enclosed Floodlight, Ballast, & Lamp for Pole Mounting

Type	Watts	Initial Lumens	Lumens per Watt	Lumens @ 40% Life	Life (Hours)
Incandescent	150	2,880	19	85	750
	300	6,360	21	84	750
	500	10,850	22	80	1,000
	1,000	23,740	24	80	1,000
	1,500	34,400	23	80	1,000
Tungsten	500	10,950	22	97	2,000
Halogen	1,500	35,800	24	97	2,000
Fluorescent	40	3,150	79	88	20,000
Cool	110	9,200	84	87	12,000
White	215	16,000	74	81	12,000
Deluxe	250	12,100	48	86	24,000
Mercury	400	22,500	56	85	24,000
	1,000	63,000	63	75	24,000
Metal	175	14,000	80	77	7,500
Halide	400	34,000	85	75	15,000
	1,000	100,000	100	83	10,000
	1,500	155,000	103	92	1,500
High	70	5,800	83	90	20,000
Pressure	100	9,500	95	90	20,000
Sodium	150	16,000	107	90	24,000
	400	50,000	125	90	24,000
	1,000	140,000	140	90	24,000
Low	55	4,600	131	98	18,000
Pressure	90	12,750	142	98	18,000
Sodium	180	33,000	183	98	18,000

Color: High Pressure Sodium — Slightly Yellow
Low Pressure Sodium — Yellow
Mercury Vapor — Green-Blue
Metal Halide — Blue White

Note: Pole not included.

R265723-25 Energy Efficiency Rating for Luminaires

The energy efficiency program for luminaires recommends the use of a metric called the Luminaire Efficacy Rating (LER). The LER value expresses the total luminaires generated by a lamp, to the watts consumed by the lamp.

Lamp Type	Lumens per Watt
Incandescent	17
Tungsten Halogen	14 - 20
Fluorescent	50 - 104
Metal Halide	64 - 96
High Pressure Sodium	76 - 116

R271323-40 Fiber Optics

Fiber optic systems use optical fiber such as plastic, glass, or fused silica, a transparent material, to transmit radiant power (i.e., light) for control, communication, and signaling applications. The types of fiber optic cables can be nonconductive, conductive, or composite. The composite cables contain fiber optics and current-carrying electrical conductors. The configuration for one of the fiber optic systems is as follows:

The transceiver module acts as transmitting and receiving equipment in a common house, which converts electrical energy to light energy or vice versa.

Pricing the fiber optic system is not an easy task. The performance of the whole system will affect the cost significantly. New specialized tools and techniques decrease the installing cost tremendously. In the fiber optic section of Means Electrical Cost Data, a benchmark for labor-hours and material costs is set up so that users can adjust their costs according to unique project conditions.

Units for Measure: Fiber optic cable is measured in hundred linear feet (C.L.F.) or industry units of measure - meter (m) or kilometer (km). The connectors are counted as units (EA.)

Material Units: Generally, the material costs include only the cable. All the accessories shall be priced separately.

Labor Units: The following procedures are generally included for the installation of fiber optic cables:

- Receiving
- Material handling
- Setting up pulling equipment
- Measuring and cutting cable
- Pulling cable

These additional items are listed and extended: Terminations

Takeoff Procedure: Cable should be taken off by type, size, number of fibers, and number of terminations required. List the lengths of each type of cable on the takeoff sheets. Total and add 10% for waste. Transfer the figures to a cost analysis sheet and extend.

R271513-75 High Performance Cable

There are several categories used to describe high performance cable. The following information includes a description of categories CAT 3, 5, 5e, 6, and 7, and details classifications of frequency and specific standards. The category standards have evolved under the sponsorship of organizations such as the Telecommunication Industry Association (TIA), the Electronic Industries Alliance (EIA), the American National Standards Institute (ANSI), the International Organization for Standardization (ISO), and the International Electrotechnical Commission (IEC), all of which have catered to the increasing complexities of modern network technology. For network cabling, users must comply with national or international standards. A breakdown of these categories is as follows:

Category 3: Designed to handle frequencies up to 16 MHz.

Category 5: (TIA/EIA 568A) Designed to handle frequencies up to 100 MHz.

Category 5e: Additional transmission performance to exceed Category 5.

Category 6 (draft): Development by TIA and other international groups to handle frequencies of 250 MHz.

Category 7 (draft): Under development to handle a frequency range from 1 to 600 MHz.

R337116-60 Average Transmission Line Material Requirements (Per Mile)

Terrain:		Flat				Rolling				Mountain			
Item		69kV	161kV	161kV	500kV	69kV	161kV	161kV	500kV	69kV	161kV	161kV	500kV
Pole Type	Unit	Wood	Wood	Steel	Steel	Wood	Wood	Steel	Steel	Wood	Wood	Steel	Steel
Conductor: 397,500 – Cir. Mil., 26/7–ACSR													
Structures	Ea.	12[1]				9[2]				7[2]			
Poles	Ea.	12				18				14			
Crossarms	Ea.	24[3]				9[4]				7[4]			
Conductor	Ft.	15,990				15,990				15,990			
Insulators[5]	Ea.	180				135				105			
Ground Wire	Ft.	5,330				10,660				10,660			
Conductor: 636,000 – Cir. Mil., 26/7–ACSR													
Structures	Ea.	13[1]	11[6]			10[2]	9[2]	6[9]		8[2]	8[2]	6[9]	
Excavation	C.Y.	–	–			–	–	120		–	–	120	
Concrete	C.Y.	–	–			–	–	10		–	–	10	
Steel Towers	Tons	–	–			–	–	32		–	–	32	
Poles	Ea.	13	11			20	18	–		16	16	–	
Crossarms	Ea.	26[3]	33[7]			10[4]	9[8]	–		8[4]	8[8]	–	
Conductor	Ft.	15,990	15,990			15,990	15,990	15,990		15,990	15,990	15,990	
Insulators[5]	Ea.	195	165			150	297	297		120	264	297	
Ground Wire	Ft.	5330	5330			10,660	10,660	10,660		10,660	10,660	10,660	
Conductor: 954,000 – Cir. Mil., 45/7–ACSR													
Structures	Ea.	14[1]	12[6]		4[11]	10[2]	9[2]	6[9]	4[11]	8[2]	8[2]	6[9]	4[13]
Excavation	C.Y.	–	–		200	–	–	125	214	–	–	125	233
Concrete	C.Y.	–	–		20	–	–	10	21	–	–	10	21
Steel Towers	Tons	–	–		57	–	–	33	57	–	–	33	63
Poles	Ea.	14	12		–	20	18	–	–	16	16	–	–
Crossarms	Ea.	28[10]	36[7]		–	10[4]	9[8]	–	–	8[4]	8[8]	–	–
Conductor	Ft.	15,990	15,990		47,970[12]	15,990	15,990	15,990	47,970[12]	15,990	15,990	15,990	47,970[12]
Insulators[5]	Ea.	210	180		288	150	297	297	288	120	264	297	576
Ground Wire	Ft.	5330	5330		10,660	10,660	10,660	10,660	10,660	10,660	10,660	10,660	10,660
Conductor: 1,351,500 – Cir. Mil., 45/7–ACSR													
Structures	Ea.	8[14]				8[14]							
Excavation	C.Y.	220				220							
Concrete	C.Y.	28				28							
Steel Towers	Ton	46				46							
Conductor	Ft.	31,680[15]				31,680[15]							
Insulators	Ea.	528				528							
Ground Wire	Ft.	10,660				10,660							

1. Single pole two-arm suspension type construction
2. Two-pole wood H-frame construction
3. 4¾" x 5¾" x 8' and 4¾" x 5¾" x 10' wood crossarm
4. 6" x 8" x 26'-0" wood crossarm
5. 5¾" x 10" disc insulator
6. Single pole construction with 3 fiberglass crossarms (5 fog- type insulators per phase)
7. 7'-0" fiberglass crossarms
8. 6" x 10" x 35'-0" wood crossarm
9. Laced steel tower, single circuit construction
10. 5" x 7" x 8' and 5" x 7" x 10' wood crossarm
11. Laced steel tower, single circuit 500-kV construction
12. Bundled conductor (3 sub-conductors per phase)
13. Laced steel tower, single circuit restrained phases (500-kV)
14. Laced steel tower, double circuit construction
15. Both sides of double circuit strung

Note: To allow for sagging, a mile (5280 Ft.) of transmission line uses 5330 Ft. of conductor per wire (called a wire mile).

R337119-30 Concrete for Conduit Encasement

Table below lists C.Y. of concrete for 100 L.F. of trench. Conduits separation center to center should meet 7.5″ (N.E.C.).

Number of Conduits	1	2	3	4	6	8	9	Number of Conduits
Trench Dimension	11.5″ x 11.5″	11.5″ x 19″	11.5″ x 27″	19″ x 19″	19″ x 27″	19″ x 38″	27″ x 27″	Trench Dimension
Conduit Diameter 2.0″	3.29	5.39	7.64	8.83	12.51	17.66	17.72	Conduit Diameter 2.0″
2.5″	3.23	5.29	7.49	8.62	12.19	17.23	17.25	2.5″
3.0″	3.15	5.13	7.24	8.29	11.71	16.59	16.52	3.0″
3.5″	3.08	4.97	7.02	7.99	11.26	15.98	15.84	3.5″
4.0″	2.99	4.80	6.76	7.65	10.74	15.30	15.07	4.0″
5.0″	2.78	4.37	6.11	6.78	9.44	13.57	13.12	5.0″
6.0″	2.52	3.84	5.33	5.74	7.87	11.48	10.77	6.0″

Estimating Tips

- The cost figures in this Square Foot Cost section were derived from approximately 11,200 projects contained in the RSMeans database of completed construction projects. They include the contractor's overhead and profit, but do not generally include architectural fees or land costs. The figures have been adjusted to January of the current year. New projects are added to our files each year, and outdated projects are discarded. For this reason, certain costs may not show a uniform annual progression. In no case are all subdivisions of a project listed.

- These projects were located throughout the U.S. and reflect a tremendous variation in square foot (S.F.) and cubic foot (C.F.) costs. This is due to differences, not only in labor and material costs, but also in individual owners' requirements. For instance, a bank in a large city would have different features than one in a rural area. This is true of all the different types of buildings analyzed. Therefore, caution should be exercised when using these Square Foot costs. For example, for court houses, costs in the database are local court house costs and will not apply to the larger, more elaborate federal court houses. As a general rule, the projects in the 1/4 column do not include any site work or equipment, while the projects in the 3/4 column may include both equipment and site work. The median figures do not generally include site work.

- None of the figures "go with" any others. All individual cost items were computed and tabulated separately. Thus, the sum of the median figures for Plumbing, HVAC and Electrical will not normally total up to the total Mechanical and Electrical costs arrived at by separate analysis and tabulation of the projects.

- Each building was analyzed as to total and component costs and percentages. The figures were arranged in ascending order with the results tabulated as shown. The 1/4 column shows that 25% of the projects had lower costs and 75% had higher. The 3/4 column shows that 75% of the projects had lower costs and 25% had higher. The median column shows that 50% of the projects had lower costs and 50% had higher.

- There are two times when square foot costs are useful. The first is in the conceptual stage when no details are available. Then square foot costs make a useful starting point. The second is after the bids are in and the costs can be worked back into their appropriate units for information purposes. As soon as details become available in the project design, the square foot approach should be discontinued and the project priced as to its particular components. When more precision is required, or for estimating the replacement cost of specific buildings, the current edition of *RSMeans Square Foot Costs* should be used.

- In using the figures in this section, it is recommended that the median column be used for preliminary figures if no additional information is available. The median figures, when multiplied by the total city construction cost index figures (see City Cost Indexes) and then multiplied by the project size modifier at the end of this section, should present a fairly accurate base figure, which would then have to be adjusted in view of the estimator's experience, local economic conditions, code requirements, and the owner's particular requirements. There is no need to factor the percentage figures, as these should remain constant from city to city. All tabulations mentioning air conditioning had at least partial air conditioning.

- The editors of this book would greatly appreciate receiving cost figures on one or more of your recent projects, which would then be included in the averages for next year. All cost figures received will be kept confidential, except that they will be averaged with other similar projects to arrive at Square Foot cost figures for next year's book. See the last page of the book for details and the discount available for submitting one or more of your projects.

50 17 | Square Foot Costs

		50 17 00 | S.F. Costs	UNIT	UNIT COSTS 1/4	MEDIAN	3/4	% OF TOTAL 1/4	MEDIAN	3/4	
01	0010	APARTMENTS Low Rise (1 to 3 story)	S.F.	60	75.50	101				01
	0020	Total project cost	C.F.	5.40	7.15	8.80				
	0100	Site work	S.F.	5.15	7	12.30	6.05%	10.55%	14.05%	
	0500	Masonry		1.18	2.75	4.75	1.54%	3.67%	6.35%	
	1500	Finishes		6.35	8.75	10.80	9.05%	10.75%	12.85%	
	1800	Equipment		1.96	2.97	4.42	2.73%	4.03%	5.95%	
	2720	Plumbing		4.68	6	7.65	6.65%	8.95%	10.05%	
	2770	Heating, ventilating, air conditioning		2.98	3.67	5.40	4.20%	5.60%	7.60%	
	2900	Electrical		3.47	4.61	6.20	5.20%	6.65%	8.40%	
	3100	Total: Mechanical & Electrical		12.05	15.30	19.10	15.90%	18.05%	23%	
	9000	Per apartment unit, total cost	Apt.	56,000	85,500	126,000				
	9500	Total: Mechanical & Electrical	"	10,600	16,600	21,700				
02	0010	APARTMENTS Mid Rise (4 to 7 story)	S.F.	79.50	96	119				02
	0020	Total project costs	C.F.	6.20	8.55	11.70				
	0100	Site work	S.F.	3.18	6.30	11.35	5.25%	6.70%	9.15%	
	0500	Masonry		5.30	7.30	10.40	5.10%	7.25%	10.50%	
	1500	Finishes		10	13.95	16.45	10.55%	13.45%	17.70%	
	1800	Equipment		2.43	3.73	4.75	2.54%	3.48%	4.31%	
	2500	Conveying equipment		1.80	2.22	2.65	1.94%	2.27%	2.69%	
	2720	Plumbing		4.67	7.45	8.30	5.70%	7.20%	8.95%	
	2900	Electrical		5.25	7.50	8.75	6.65%	7.20%	8.95%	
	3100	Total: Mechanical & Electrical		16.80	21	25.50	18.50%	21%	23%	
	9000	Per apartment unit, total cost	Apt.	90,000	106,000	176,000				
	9500	Total: Mechanical & Electrical	"	17,000	19,700	24,800				
03	0010	APARTMENTS High Rise (8 to 24 story)	S.F.	90	109	133				03
	0020	Total project costs	C.F.	8.75	10.70	13				
	0100	Site work	S.F.	3.27	5.30	7.40	2.58%	4.84%	6.15%	
	0500	Masonry		5.20	9.50	11.80	4.74%	9.65%	11.05%	
	1500	Finishes		10	12.50	14.80	9.75%	11.80%	13.70%	
	1800	Equipment		2.90	3.57	4.73	2.78%	3.49%	4.35%	
	2500	Conveying equipment		2.05	3.11	4.23	2.23%	2.78%	3.37%	
	2720	Plumbing		6.65	7.85	11	6.80%	7.20%	10.45%	
	2900	Electrical		6.20	7.85	10.60	6.45%	7.65%	8.80%	
	3100	Total: Mechanical & Electrical		18.55	23.50	28.50	17.95%	22.50%	24.50%	
	9000	Per apartment unit, total cost	Apt.	94,000	103,500	143,500				
	9500	Total: Mechanical & Electrical	"	20,300	23,200	24,500				
04	0010	AUDITORIUMS	S.F.	93	126	182				04
	0020	Total project costs	C.F.	5.85	8.15	11.70				
	2720	Plumbing	S.F.	5.95	8.20	10.40	5.85%	7.20%	8.70%	
	2900	Electrical		7.25	10.50	14.10	6.80%	8.95%	11.30%	
	3100	Total: Mechanical & Electrical		14.40	20.50	41.50	24.50%	30.50%	31.50%	
05	0010	AUTOMOTIVE SALES	S.F.	67.50	94	115				05
	0020	Total project costs	C.F.	4.57	5.50	7.10				
	2720	Plumbing	S.F.	3.16	5.50	6	2.89%	6.05%	6.50%	
	2770	Heating, ventilating, air conditioning		4.88	7.45	8.05	4.61%	10%	10.35%	
	2900	Electrical		5.60	8.60	11.90	7.40%	9.95%	12.40%	
	3100	Total: Mechanical & Electrical		15.60	22	28	19.15%	20.50%	26%	
06	0010	BANKS	S.F.	135	169	213				06
	0020	Total project costs	C.F.	9.70	13.20	17.45				
	0100	Site work	S.F.	15.55	23.50	34	7.85%	12.95%	17%	
	0500	Masonry		7.05	13.25	24.50	3.36%	6.95%	10.35%	
	1500	Finishes		12.05	16.45	21	5.85%	8.45%	11.25%	
	1800	Equipment		5.45	11.25	24	1.34%	5.95%	10.65%	
	2720	Plumbing		4.26	6.10	8.90	2.82%	3.90%	4.93%	
	2770	Heating, ventilating, air conditioning		8.10	10.80	14.40	4.86%	7.15%	8.50%	
	2900	Electrical		12.85	17.15	22.50	8.20%	10.20%	12.20%	
	3100	Total: Mechanical & Electrical		30	40.50	49	16.55%	19.45%	23%	
	3500	See also division 11020 & 11030 (MF2004 11 16 00 & 11 17 00)								

522

50 17 00 | S.F. Costs

		UNIT	UNIT COSTS			% OF TOTAL		
			1/4	MEDIAN	3/4	1/4	MEDIAN	3/4
13	0010	**CHURCHES**	S.F.	91.50	116	150			
	0020	Total project costs	C.F.	5.70	7.15	9.45			
	1800	Equipment	S.F.	1.10	2.62	5.55	.95%	2.24%	4.50%
	2720	Plumbing		3.57	4.99	7.35	3.51%	4.96%	6.25%
	2770	Heating, ventilating, air conditioning		8.35	10.85	15.40	7.50%	10%	12%
	2900	Electrical		7.70	10.55	14.15	7.30%	8.75%	10.95%
	3100	Total: Mechanical & Electrical	↓	23.50	30.50	41.50	18.25%	22%	24.50%
	3500	See also division 11040 (MF2004 11 91 00)							
15	0010	**CLUBS, COUNTRY**	S.F.	98.50	118	149			
	0020	Total project costs	C.F.	7.90	9.65	13.35			
	2720	Plumbing	S.F.	6.30	8.80	20	5.60%	7.90%	10%
	2900	Electrical		7.75	10.60	13.80	7%	8.95%	11%
	3100	Total: Mechanical & Electrical	↓	23.50	41	51.50	19%	26.50%	29.50%
17	0010	**CLUBS, SOCIAL Fraternal**	S.F.	78.50	113	151			
	0020	Total project costs	C.F.	4.90	7.45	8.85			
	2720	Plumbing	S.F.	4.93	6.15	9.30	5.60%	6.90%	8.55%
	2770	Heating, ventilating, air conditioning		7.10	8.60	11.05	8.20%	9.25%	14.40%
	2900	Electrical		6.25	9.70	11.75	6.50%	9.50%	10.55%
	3100	Total: Mechanical & Electrical	↓	17.40	33	42	21%	23%	23.50%
18	0010	**CLUBS, Y.M.C.A.**	S.F.	98.50	133	163			
	0020	Total project costs	C.F.	4.54	7.60	11.30			
	2720	Plumbing	S.F.	6.20	12.40	13.90	5.65%	7.60%	10.85%
	2900	Electrical		7.90	10.20	14.50	6.25%	7.80%	10.20%
	3100	Total: Mechanical & Electrical	↓	28	33	37.50	18.40%	22.50%	28.50%
19	0010	**COLLEGES Classrooms & Administration**	S.F.	109	144	196			
	0020	Total project costs	C.F.	7.95	11.30	17.80			
	0500	Masonry	S.F.	7.30	13.75	16.90	5.65%	8.25%	10.50%
	2720	Plumbing		5.40	10.55	19.35	5.10%	6.60%	8.95%
	2900	Electrical		8.95	13.65	17.15	7.70%	9.85%	12%
	3100	Total: Mechanical & Electrical	↓	33	46	55	24%	28%	31.50%
21	0010	**COLLEGES Science, Engineering, Laboratories**	S.F.	185	216	263			
	0020	Total project costs	C.F.	10.60	15.50	17.55			
	1800	Equipment	S.F.	10.30	23.50	25.50	2%	6.45%	12.65%
	2900	Electrical		15.25	21	33.50	7.10%	9.40%	12.10%
	3100	Total: Mechanical & Electrical	↓	56.50	67	104	28.50%	31.50%	41%
	3500	See also division 11600 (MF2004 11 53 00)							
23	0010	**COLLEGES Student Unions**	S.F.	118	165	194			
	0020	Total project costs	C.F.	6.60	8.65	11.10			
	3100	Total: Mechanical & Electrical	S.F.	31	48	57	23.50%	26%	29%
25	0010	**COMMUNITY CENTERS**	"	96.50	120	160			
	0020	Total project costs	C.F.	6.35	9.10	11.75			
	1800	Equipment	S.F.	2.45	4.17	6.55	1.69%	3.12%	5.60%
	2720	Plumbing		4.73	8.05	11.75	4.85%	7%	8.95%
	2770	Heating, ventilating, air conditioning		7.80	11.20	15.40	6.80%	10.35%	12.90%
	2900	Electrical		8.25	10.65	15.45	7.35%	9.10%	10.85%
	3100	Total: Mechanical & Electrical	↓	29	34.50	49.50	20.50%	25.50%	32.50%
28	0010	**COURT HOUSES**	S.F.	140	161	188			
	0020	Total project costs	C.F.	10.75	12.90	16.25			
	2720	Plumbing	S.F.	6.70	9.35	13.40	5.95%	7.45%	8.20%
	2900	Electrical		13.75	16.25	19.60	8.55%	9.95%	11.55%
	3100	Total: Mechanical & Electrical	↓	38	52	57	22.50%	29.50%	30.50%
30	0010	**DEPARTMENT STORES**	S.F.	52	70.50	88.50			
	0020	Total project costs	C.F.	2.78	3.61	4.90			
	2720	Plumbing	S.F.	1.62	2.04	3.10	1.82%	4.21%	5.90%
	2770	Heating, ventilating, air conditioning	↓	4.72	7.30	10.95	8.20%	9.10%	14.80%

		50 17 00 | S.F. Costs		UNIT COSTS			% OF TOTAL			
			UNIT	1/4	MEDIAN	3/4	1/4	MEDIAN	3/4	
30	2900	Electrical	S.F.	5.95	8.20	9.65	9.05%	12.15%	14.95%	30
	3100	Total: Mechanical & Electrical	↓	10.50	13.40	23.50	13.20%	21.50%	50%	
31	0010	**DORMITORIES Low Rise (1 to 3 story)**	S.F.	97.50	125	156				31
	0020	Total project costs	C.F.	5.90	9	13.50				
	2720	Plumbing	S.F.	5.95	7.95	10	8.05%	9%	9.65%	
	2770	Heating, ventilating, air conditioning		6.25	7.50	10	4.61%	8.05%	10%	
	2900	Electrical		6.40	9.55	13.30	6.55%	8.90%	9.55%	
	3100	Total: Mechanical & Electrical	↓	33	36	56.50	22.50%	26%	29%	
	9000	Per bed, total cost	Bed	41,800	46,300	99,500				
32	0010	**DORMITORIES Mid Rise (4 to 8 story)**	S.F.	121	158	193				32
	0020	Total project costs	C.F.	13.35	14.65	17.55				
	2900	Electrical	S.F.	12.85	14.60	19	8.20%	10.20%	11.10%	
	3100	Total: Mechanical & Electrical	"	33.50	37.50	72.50	19.50%	30.50%	37.50%	
	9000	Per bed, total cost	Bed	17,200	39,300	81,200				
34	0010	**FACTORIES**	S.F.	45.50	68	105				34
	0020	Total project costs	C.F.	2.94	4.38	7.25				
	0100	Site work	S.F.	5.25	9.55	15.10	6.95%	11.45%	17.95%	
	2720	Plumbing		2.47	4.59	7.60	3.73%	6.05%	8.10%	
	2770	Heating, ventilating, air conditioning		4.81	6.90	9.30	5.25%	8.45%	11.35%	
	2900	Electrical		5.70	9	13.75	8.10%	10.50%	14.20%	
	3100	Total: Mechanical & Electrical	↓	16.25	21.50	33	21%	28.50%	35.50%	
36	0010	**FIRE STATIONS**	S.F.	90.50	124	162				36
	0020	Total project costs	C.F.	5.30	7.30	9.70				
	0500	Masonry	S.F.	12.80	24	31.50	8.60%	12.40%	16.45%	
	1140	Roofing		2.96	8	9.10	1.90%	4.94%	5.05%	
	1580	Painting		2.43	3.43	3.51	1.37%	1.57%	2.07%	
	1800	Equipment		1.92	2.72	6.55	.81%	2.25%	3.69%	
	2720	Plumbing		5.75	8.50	12.45	5.85%	7.35%	9.50%	
	2770	Heating, ventilating, air conditioning		4.95	7.85	12.30	4.86%	7.25%	9.25%	
	2900	Electrical		6.45	11.10	14.80	6.80%	8.75%	10.60%	
	3100	Total: Mechanical & Electrical	↓	32	38.50	44.50	19.60%	23%	26%	
37	0010	**FRATERNITY HOUSES and Sorority Houses**	S.F.	91	117	160				37
	0020	Total project costs	C.F.	9.05	9.45	12.10				
	2720	Plumbing	S.F.	6.85	7.85	14.40	6.80%	8%	10.85%	
	2900	Electrical		6	12.95	15.85	6.60%	9.90%	10.65%	
	3100	Total: Mechanical & Electrical	↓	16	23	27.75		15.10%	15.90%	
38	0010	**FUNERAL HOMES**	S.F.	96	131	237				38
	0020	Total project costs	C.F.	9.80	10.90	21				
	2900	Electrical	S.F.	4.23	7.75	8.50	3.58%	4.44%	5.95%	
	3100	Total: Mechanical & Electrical	"	15	22.50	30.50	12.90%	12.90%	12.90%	
39	0010	**GARAGES, COMMERCIAL (Service)**	S.F.	56	84	116				39
	0020	Total project costs	C.F.	3.65	5.25	7.65				
	1800	Equipment	S.F.	3.06	6.90	10.70	2.69%	4.62%	6.80%	
	2720	Plumbing		3.76	5.80	10.55	5.45%	7.85%	10.65%	
	2730	Heating & ventilating		4.94	6.75	9.30	5.25%	6.85%	8.20%	
	2900	Electrical		5.25	8	11.40	7.15%	9.25%	10.85%	
	3100	Total: Mechanical & Electrical	↓	11.70	21.50	32.50	13.60%	17.40%	27%	
40	0010	**GARAGES, MUNICIPAL (Repair)**	S.F.	79.50	106	150				40
	0020	Total project costs	C.F.	4.98	6.30	10.85				
	0500	Masonry	S.F.	7.50	14.65	22.50	5.60%	9.15%	12.50%	
	2720	Plumbing		3.58	6.85	12.95	3.59%	6.70%	7.95%	
	2730	Heating & ventilating		6.10	8.85	17.10	6.15%	7.45%	13.50%	
	2900	Electrical		5.90	9.25	13.35	6.65%	8.15%	11.15%	
	3100	Total: Mechanical & Electrical	↓	18.95	38	55	21.50%	25.50%	28.50%	

50 17 00 | S.F. Costs

			UNIT COSTS			% OF TOTAL		
		UNIT	1/4	MEDIAN	3/4	1/4	MEDIAN	3/4
41 0010	**GARAGES, PARKING**	S.F.	31	45.50	78			
0020	Total project costs	C.F.	2.92	3.96	5.75			
2720	Plumbing	S.F.	.88	1.36	2.10	1.72%	2.70%	3.85%
2900	Electrical	↓	1.70	2.09	3.28	4.33%	5.20%	6.30%
3100	Total: Mechanical & Electrical	↓	3.48	4.85	6.05	7%	8.90%	11.05%
3200								
9000	Per car, total cost	Car	13,100	16,400	21,000			
43 0010	**GYMNASIUMS**	S.F.	86.50	115	147			
0020	Total project costs	C.F.	4.30	5.85	7.15			
1800	Equipment	S.F.	2.05	3.85	7.40	2.03%	3.30%	5.20%
2720	Plumbing		5.45	6.75	8.35	4.95%	6.75%	7.75%
2770	Heating, ventilating, air conditioning		5.85	8.95	17.95	5.80%	9.80%	11.10%
2900	Electrical		6.55	8.85	11.05	6.60%	8.50%	10.30%
3100	Total: Mechanical & Electrical	↓	23.50	33	39	20.50%	24%	29%
3500	See also division 11480 (MF2004 11 67 00)							
46 0010	**HOSPITALS**	S.F.	164	203	300			
0020	Total project costs	C.F.	12.55	15.60	22.50			
1800	Equipment	S.F.	4.20	8.10	13.95	1.10%	2.68%	5%
2720	Plumbing		14.25	19.95	25.50	7.60%	9.10%	10.85%
2770	Heating, ventilating, air conditioning		21	27	36	7.80%	12.95%	16.65%
2900	Electrical		18.05	23.50	36.50	9.85%	11.55%	13.90%
3100	Total: Mechanical & Electrical	↓	51	68.50	110	27%	33.50%	37%
9000	Per bed or person, total cost	Bed	135,500	217,000	290,000			
9900	See also division 11700 (MF2004 11 71 00)							
48 0010	**HOUSING For the Elderly**	S.F.	81.50	103	127			
0020	Total project costs	C.F.	5.80	8.05	10.30			
0100	Site work	S.F.	6.05	8.95	12.90	5.05%	7.90%	12.10%
0500	Masonry		2.48	9.25	13.55	1.30%	6.05%	11%
1800	Equipment		1.97	2.71	4.32	1.88%	3.23%	4.43%
2510	Conveying systems		1.98	2.66	3.61	1.78%	2.20%	2.81%
2720	Plumbing		6.05	7.70	9.70	8.15%	9.55%	10.50%
2730	Heating, ventilating, air conditioning		3.10	4.40	6.55	3.30%	5.60%	7.25%
2900	Electrical		6.05	8.25	10.55	7.30%	8.50%	10.25%
3100	Total: Mechanical & Electrical	↓	21	25	33	18.10%	22.50%	29%
9000	Per rental unit, total cost	Unit	75,500	88,500	98,500			
9500	Total: Mechanical & Electrical	"	16,900	19,400	22,600			
50 0010	**HOUSING Public (Low Rise)**	S.F.	68.50	95	124			
0020	Total project costs	C.F.	6.10	7.60	9.45			
0100	Site work	S.F.	8.70	12.55	20.50	8.35%	11.75%	16.50%
1800	Equipment		1.86	3.04	4.63	2.26%	3.03%	4.24%
2720	Plumbing		4.94	6.50	8.25	7.15%	9.05%	11.60%
2730	Heating, ventilating, air conditioning		2.48	4.81	5.25	4.26%	6.05%	6.45%
2900	Electrical		4.14	6.15	8.55	5.10%	6.55%	8.25%
3100	Total: Mechanical & Electrical	↓	19.65	25.50	28.50	14.50%	17.55%	26.50%
9000	Per apartment, total cost	Apt.	75,000	85,500	107,500			
9500	Total: Mechanical & Electrical	"	16,000	19,800	21,900			
51 0010	**ICE SKATING RINKS**	S.F.	58.50	137	150			
0020	Total project costs	C.F.	4.30	4.40	5.05			
2720	Plumbing	S.F.	2.19	4.10	4.19	3.12%	3.23%	5.65%
2900	Electrical		6.25	9.60	10.15	6.30%	10.15%	15.05%
3100	Total: Mechanical & Electrical	↓	10.45	14.75	18.45	18.95%	18.95%	18.95%
52 0010	**JAILS**	S.F.	178	230	297			
0020	Total project costs	C.F.	16.05	22.50	27.50			
1800	Equipment	S.F.	6.95	20.50	35	2.80%	5.55%	11.90%
2720	Plumbing		18.15	23	30.50	7%	8.90%	13.35%
2770	Heating, ventilating, air conditioning		16.05	21.50	41.50	7.50%	9.45%	17.75%
2900	Electrical	↓	18.55	24.50	30.50	8.20%	11.55%	14.70%

50 17 | Square Foot Costs

50 17 00 | S.F. Costs		UNIT	UNIT COSTS			% OF TOTAL		
			1/4	MEDIAN	3/4	1/4	MEDIAN	3/4
52 3100	Total: Mechanical & Electrical	S.F.	47.50	89	105	27.50%	30%	34%	**52**
53 0010	**LIBRARIES**	S.F.	113	141	186				**53**
0020	Total project costs	C.F.	7.70	9.65	12.30			
0500	Masonry	S.F.	8.80	15.60	26	5.80%	7.80%	12.35%
1800	Equipment		1.54	4.14	6.25	.41%	1.50%	4.16%
2720	Plumbing		4.16	6.05	8.20	3.38%	4.60%	5.70%
2770	Heating, ventilating, air conditioning		9.20	15.60	20.50	7.80%	10.95%	12.80%
2900	Electrical		11.30	14.75	18.65	8.30%	10.25%	11.95%
3100	Total: Mechanical & Electrical		34	43	53.50	19.65%	22.50%	26.50%
54 0010	**LIVING, ASSISTED**	S.F.	104	123	145				**54**
0020	Total project costs	C.F.	8.75	10.25	11.65			
0500	Masonry	S.F.	3.06	3.65	4.29	2.37%	3.16%	3.86%
1800	Equipment		2.37	2.75	3.53	2.12%	2.45%	2.66%
2720	Plumbing		8.75	11.70	12.10	6.05%	8.15%	10.60%
2770	Heating, ventilating, air conditioning		10.35	10.85	11.85	7.95%	9.35%	9.70%
2900	Electrical		10.20	11.25	13.05	9%	10%	10.70%
3100	Total: Mechanical & Electrical		28.50	33.50	38.50	26%	29%	31.50%
55 0010	**MEDICAL CLINICS**	S.F.	106	131	166				**55**
0020	Total project costs	C.F.	7.75	10.05	13.35			
1800	Equipment	S.F.	2.87	6	9.35	1.05%	2.94%	6.35%
2720	Plumbing		7.05	9.90	13.25	6.15%	8.40%	10.10%
2770	Heating, ventilating, air conditioning		8.40	11	16.20	6.65%	8.85%	11.35%
2900	Electrical		9.10	12.95	16.90	8.10%	10%	12.25%
3100	Total: Mechanical & Electrical		29	39.50	54	22.50%	27%	33.50%
3500	See also division 11700 (MF2004 11 71 00)							
57 0010	**MEDICAL OFFICES**	S.F.	100	123	151				**57**
0020	Total project costs	C.F.	7.45	10.10	13.65			
1800	Equipment	S.F.	3.46	6.50	9.25	.98%	5.10%	7.05%
2720	Plumbing		5.50	8.50	11.45	5.60%	6.80%	8.50%
2770	Heating, ventilating, air conditioning		6.65	9.80	12.70	6.15%	8.05%	9.70%
2900	Electrical		8	11.60	16.20	7.60%	9.80%	11.70%
3100	Total: Mechanical & Electrical		21.50	31	47	19.35%	23%	30.50%
59 0010	**MOTELS**	S.F.	63	91	119				**59**
0020	Total project costs	C.F.	5.60	7.50	12.25			
2720	Plumbing	S.F.	6.40	8.10	9.70	9.45%	10.60%	12.55%
2770	Heating, ventilating, air conditioning		3.89	5.80	10.40	5.60%	5.60%	10%
2900	Electrical		5.95	7.50	9.35	7.45%	9.05%	10.45%
3100	Total: Mechanical & Electrical		20	25.50	43.50	18.50%	24%	25.50%
5000								
9000	Per rental unit, total cost	Unit	32,000	61,000	66,000			
9500	Total: Mechanical & Electrical	"	6,250	9,450	11,000			
60 0010	**NURSING HOMES**	S.F.	98.50	127	156				**60**
0020	Total project costs	C.F.	7.75	9.70	13.25			
1800	Equipment	S.F.	3.11	4.12	6.85	2.02%	3.62%	4.99%
2720	Plumbing		8.45	12.80	15.45	8.75%	10.10%	12.70%
2770	Heating, ventilating, air conditioning		8.90	13.50	17.90	9.70%	11.45%	11.80%
2900	Electrical		9.75	12.20	16.60	9.40%	10.55%	12.45%
3100	Total: Mechanical & Electrical		23.50	32.50	54.50	26%	29.50%	30.50%
9000	Per bed or person, total cost	Bed	43,800	55,000	70,500			
61 0010	**OFFICES Low Rise (1 to 4 story)**	S.F.	83	107	139				**61**
0020	Total project costs	C.F.	5.95	8.20	10.75			
0100	Site work	S.F.	6.45	11.05	16.40	5.90%	9.70%	13.55%
0500	Masonry	"	2.86	6.35	11.75	2.62%	5.45%	8.20%
1800	Equipment	S.F.	.88	1.73	4.71	.73%	1.50%	3.66%
2720	Plumbing		2.95	4.59	6.70	3.66%	4.50%	6.10%
2770	Heating, ventilating, air conditioning		6.60	9.15	13.40	7.20%	10.30%	11.70%
2900	Electrical		6.75	9.65	13.60	7.45%	9.65%	11.40%

50 17 | Square Foot Costs

50 17 00 | S.F. Costs

			UNIT	UNIT COSTS			% OF TOTAL			
				1/4	MEDIAN	3/4	1/4	MEDIAN	3/4	
61	3100	Total: Mechanical & Electrical	S.F.	17.60	25	37	18%	22.50%	27%	61
62	0010	**OFFICES Mid Rise (5 to 10 story)**	S.F.	88	107	141				62
	0020	Total project costs	C.F.	6.25	7.95	11.30				
	2720	Plumbing	S.F.	2.66	4.12	5.90	2.83%	3.74%	4.50%	
	2770	Heating, ventilating, air conditioning		6.70	9.55	15.25	7.65%	9.40%	11%	
	2900	Electrical		6.55	8.35	11.60	6.35%	7.80%	10%	
	3100	Total: Mechanical & Electrical	↓	16.70	21.50	41.50	18.95%	21%	27.50%	
63	0010	**OFFICES High Rise (11 to 20 story)**	S.F.	108	136	168				63
	0020	Total project costs	C.F.	7.55	9.45	13.55				
	2900	Electrical	S.F.	6.55	8	11.90	5.80%	7.85%	10.50%	
	3100	Total: Mechanical & Electrical	"	21	28.50	48	16.90%	23.50%	34%	
64	0010	**POLICE STATIONS**	S.F.	130	166	212				64
	0020	Total project costs	C.F.	10.30	12.65	17.30				
	0500	Masonry	S.F.	12.20	21.50	27	6.70%	10.55%	11.35%	
	1800	Equipment		2.06	9.05	14.35	1.43%	4.07%	6.70%	
	2720	Plumbing		7.25	14.45	18	5.65%	6.90%	10.75%	
	2770	Heating, ventilating, air conditioning		11.30	15.05	20.50	5.85%	10.55%	11.70%	
	2900	Electrical		14.10	20.50	27	9.80%	11.85%	14.80%	
	3100	Total: Mechanical & Electrical	↓	46.50	56	75.50	28.50%	32%	32.50%	
65	0010	**POST OFFICES**	S.F.	102	126	161				65
	0020	Total project costs	C.F.	6.15	7.80	8.85				
	2720	Plumbing	S.F.	4.61	5.70	7.20	4.24%	5.30%	5.60%	
	2770	Heating, ventilating, air conditioning		7.20	8.90	9.90	6.65%	7.15%	9.35%	
	2900	Electrical		8.45	11.90	14.10	7.25%	9%	11%	
	3100	Total: Mechanical & Electrical	↓	24.50	32	36	16.25%	18.80%	22%	
66	0010	**POWER PLANTS**	S.F.	710	940	1,725				66
	0020	Total project costs	C.F.	19.55	42.50	91				
	2900	Electrical	S.F.	50	106	159	9.30%	12.75%	21.50%	
	8100	Total: Mechanical & Electrical	"	125	405	910	32.50%	32.50%	52.50%	
67	0010	**RELIGIOUS EDUCATION**	S.F.	82.50	108	132				67
	0020	Total project costs	C.F.	4.58	6.55	8.20				
	2720	Plumbing	S.F.	3.45	4.89	6.85	4.40%	5.30%	7.10%	
	2770	Heating, ventilating, air conditioning		8.70	9.85	13.95	10.05%	11.45%	12.35%	
	2900	Electrical		6.55	9.25	12.20	7.60%	9.05%	10.35%	
	3100	Total: Mechanical & Electrical	↓	26.50	35	42.50	22%	23%	27%	
69	0010	**RESEARCH Laboratories and Facilities**	S.F.	123	177	258				69
	0020	Total project costs	C.F.	9.65	18.55	22				
	1800	Equipment	S.F.	5.50	10.80	26.50	.90%	4.58%	8.80%	
	2720	Plumbing		12.40	15.85	25.50	6.15%	8.30%	10.80%	
	2770	Heating, ventilating, air conditioning		11.10	37.50	44	7.25%	16.50%	17.50%	
	2900	Electrical		14.20	24	40.50	9.45%	11.15%	15.40%	
	3100	Total: Mechanical & Electrical	↓	43.50	83.50	118	29.50%	37%	45.50%	
70	0010	**RESTAURANTS**	S.F.	119	153	200				70
	0020	Total project costs	C.F.	10.05	13.20	17.30				
	1800	Equipment	S.F.	7.75	19	28.50	6.10%	13%	15.65%	
	2720	Plumbing		9.50	11.50	15.10	6.10%	8.15%	9%	
	2770	Heating, ventilating, air conditioning		12.05	16.65	22	9.20%	12%	12.40%	
	2900	Electrical		12.65	15.60	20.50	8.35%	10.55%	11.55%	
	3100	Total: Mechanical & Electrical	↓	39	41.50	53	19.25%	24%	29.50%	
	9000	Per seat unit, total cost	Seat	4,400	5,850	6,925				
	9500	Total: Mechanical & Electrical	"	1,100	1,450	1,725				
72	0010	**RETAIL STORES**	S.F.	55.50	75	99				72
	0020	Total project costs	C.F.	3.79	5.40	7.50				
	2720	Plumbing	S.F.	2.03	3.38	5.75	3.26%	4.60%	6.80%	
	2770	Heating, ventilating, air conditioning		4.38	6	9	6.75%	8.75%	10.15%	
	2900	Electrical		5	6.85	9.95	7.25%	9.90%	11.65%	
	3100	Total: Mechanical & Electrical	↓	12.85	17.15	23	17.05%	21%	23.50%	

		50 17 00 | S.F. Costs	UNIT	UNIT COSTS			% OF TOTAL			
				1/4	MEDIAN	3/4	1/4	MEDIAN	3/4	
74	0010	**SCHOOLS Elementary**	S.F.	90	111	137				74
	0020	Total project costs	C.F.	5.95	7.65	9.85				
	0500	Masonry	S.F.	8.15	13.50	20.50	5.80%	10.65%	14.95%	
	1800	Equipment		2.65	4.41	8.30	1.90%	3.38%	4.98%	
	2720	Plumbing		5.25	7.40	9.90	5.70%	7.15%	9.35%	
	2730	Heating, ventilating, air conditioning		7.85	12.50	17.45	8.15%	10.80%	14.90%	
	2900	Electrical		8.50	11.15	14	8.40%	10%	11.70%	
	3100	Total: Mechanical & Electrical	↓	30	38	45.50	24.50%	27.50%	30%	
	9000	Per pupil, total cost	Ea.	10,500	15,600	46,200				
	9500	Total: Mechanical & Electrical	"	2,950	3,750	13,400				
76	0010	**SCHOOLS Junior High & Middle**	S.F.	92.50	115	137				76
	0020	Total project costs	C.F.	5.95	7.70	8.65				
	0500	Masonry	S.F.	11.70	15.10	17.65	8%	11.60%	14.30%	
	1800	Equipment		3	4.84	7.45	1.81%	3.26%	4.96%	
	2720	Plumbing		5.95	6.75	8.75	5.40%	6.80%	7.25%	
	2770	Heating, ventilating, air conditioning		10.95	13.30	23.50	9%	11.80%	17.45%	
	2900	Electrical		9.20	11.10	14.25	7.90%	9.30%	10.60%	
	3100	Total: Mechanical & Electrical	↓	28.50	38	47.50	23%	26.50%	29.50%	
	9000	Per pupil, total cost	Ea.	11,900	15,600	21,000				
78	0010	**SCHOOLS Senior High**	S.F.	96	118	149				78
	0020	Total project costs	C.F.	6.10	8.70	14.40				
	1800	Equipment	S.F.	2.57	6.25	8.95	1.86%	3.22%	4.80%	
	2720	Plumbing		5.60	8.30	15.35	5.70%	7%	8.35%	
	2770	Heating, ventilating, air conditioning		11.20	12.85	24.50	8.95%	11.60%	15%	
	2900	Electrical		9.80	12.50	19.30	8.45%	10.05%	11.95%	
	3100	Total: Mechanical & Electrical	↓	32.50	37.50	64	23%	26.50%	28.50%	
	9000	Per pupil, total cost	Ea.	9,225	18,800	23,500				
80	0010	**SCHOOLS Vocational**	S.F.	79	112	140				80
	0020	Total project costs	C.F.	4.92	7.05	9.75				
	0500	Masonry	S.F.	4.65	11.50	17.55	3.53%	4.61%	10.95%	
	1800	Equipment	"	2.37	3.26	8.50	1.24%	3.13%	4.68%	
	2720	Plumbing	S.F.	5.05	7.55	11.10	5.40%	6.90%	8.55%	
	2770	Heating, ventilating, air conditioning		7.10	13.20	22	8.60%	11.90%	14.65%	
	2900	Electrical		8.25	10.80	15.40	8.45%	10.95%	13.20%	
	3100	Total: Mechanical & Electrical	↓	28.50	31.50	54.50	23.50%	29.50%	31%	
	9000	Per pupil, total cost	Ea.	11,000	29,500	44,000				
83	0010	**SPORTS ARENAS**	S.F.	69	92.50	142				83
	0020	Total project costs	C.F.	3.76	6.75	8.70				
	2720	Plumbing	S.F.	4.01	6.10	12.85	4.35%	6.35%	9.40%	
	2770	Heating, ventilating, air conditioning		8.65	10.20	14.20	8.80%	10.20%	13.55%	
	2900	Electrical		7.20	9.80	12.65	8.60%	9.90%	12.25%	
	3100	Total: Mechanical & Electrical	↓	17.95	32	42	21.50%	25%	27.50%	
85	0010	**SUPERMARKETS**	S.F.	64	74	87				85
	0020	Total project costs	C.F.	3.56	4.30	6.50				
	2720	Plumbing	S.F.	3.57	4.50	5.25	5.40%	6%	7.45%	
	2770	Heating, ventilating, air conditioning		5.25	7	8.50	8.60%	8.65%	9.60%	
	2900	Electrical		8	9.20	10.90	10.40%	12.45%	13.60%	
	3100	Total: Mechanical & Electrical	↓	20.50	22.50	31	20.50%	26.50%	31%	
86	0010	**SWIMMING POOLS**	S.F.	103	173	370				86
	0020	Total project costs	C.F.	8.30	10.35	11.30				
	2720	Plumbing	S.F.	9.55	10.95	15	4.80%	9.70%	20.50%	
	2900	Electrical		7.80	12.60	18.35	6.50%	7.25%	7.60%	
	3100	Total: Mechanical & Electrical	↓	18.95	48	66	11.15%	14.10%	23.50%	
87	0010	**TELEPHONE EXCHANGES**	S.F.	137	201	255				87
	0020	Total project costs	C.F.	8.55	13.75	18.85				
	2720	Plumbing	S.F.	5.80	9.20	13.10	4.52%	5.80%	6.90%	
	2770	Heating, ventilating, air conditioning	↓	13.45	27	33.50	11.80%	16.05%	18.40%	

		50 17 00 | S.F. Costs	UNIT	UNIT COSTS			% OF TOTAL		
				1/4	MEDIAN	3/4	1/4	MEDIAN	3/4
87	2900	Electrical	S.F.	14	22	39	10.90%	14%	17.85%	87
	3100	Total: Mechanical & Electrical	↓	41.50	78.50	111	29.50%	33.50%	44.50%	
91	0010	**THEATERS**	S.F.	86.50	107	163				91
	0020	Total project costs	C.F.	3.99	5.90	8.70				
	2720	Plumbing	S.F.	2.88	3.12	12.75	2.92%	4.70%	6.80%	
	2770	Heating, ventilating, air conditioning		8.40	10.15	12.60	8%	12.25%	13.40%	
	2900	Electrical		7.55	10.20	21	8.05%	9.95%	12.25%	
	3100	Total: Mechanical & Electrical	↓	19.45	29	59.50	23%	26.50%	27.50%	
94	0010	**TOWN HALLS City Halls & Municipal Buildings**	S.F.	100	127	159				94
	0020	Total project costs	C.F.	8.80	10.75	14.85				
	2720	Plumbing	S.F.	4.03	7.55	13.85	4.31%	5.95%	7.95%	
	2770	Heating, ventilating, air conditioning		7.30	14.45	21	7.05%	9.05%	13.45%	
	2900	Electrical		9.15	13.40	17.85	8.05%	9.50%	12.05%	
	3100	Total: Mechanical & Electrical	↓	31	36	48	22%	26.50%	31%	
97	0010	**WAREHOUSES & Storage Buildings**	S.F.	37.50	54	77				97
	0020	Total project costs	C.F.	1.95	2.95	4.88				
	0100	Site work	S.F.	3.72	7.40	11.15	6.05%	12.95%	19.85%	
	0500	Masonry		2.25	5.10	11.05	3.73%	7.40%	12.30%	
	1800	Equipment		.58	1.25	7	.91%	1.82%	5.55%	
	2720	Plumbing		1.20	2.15	4.03	2.90%	4.80%	6.55%	
	2730	Heating, ventilating, air conditioning		1.37	3.86	5.20	2.41%	5%	8.90%	
	2900	Electrical		2.13	4.01	6.65	5.15%	7.20%	10.10%	
	3100	Total: Mechanical & Electrical	↓	5.95	9.10	19.95	12.75%	18.90%	26%	
99	0010	**WAREHOUSE & OFFICES Combination**	S.F.	44	59	79.50				99
	0020	Total project costs	C.F.	2.27	3.29	4.86				
	1800	Equipment	S.F.	.77	1.48	2.21	.52%	1.21%	2.40%	
	2720	Plumbing		1.71	3.03	4.44	3.74%	4.76%	6.30%	
	2770	Heating, ventilating, air conditioning		2.70	4.22	5.90	5%	5.65%	10.05%	
	2900	Electrical		2.97	4.40	6.95	5.85%	8%	10%	
	3100	Total: Mechanical & Electrical	↓	8.25	12.60	20	14.40%	19.95%	24.50%	

Square Foot Project Size Modifier

One factor that affects the S.F. cost of a particular building is the size. In general, for buildings built to the same specifications in the same locality, the larger building will have the lower S.F. cost. This is due mainly to the decreasing contribution of the exterior walls plus the economy of scale usually achievable in larger buildings. The Area Conversion Scale shown below will give a factor to convert costs for the typical size building to an adjusted cost for the particular project.

The Square Foot Base Size lists the median costs, most typical project size in our accumulated data, and the range in size of the projects.

The Size Factor for your project is determined by dividing your project area in S.F. by the typical project size for the particular Building Type. With this factor, enter the Area Conversion Scale at the appropriate Size Factor and determine the appropriate cost multiplier for your building size.

Example: Determine the cost per S.F. for a 100,000 S.F. Mid-rise apartment building.

$$\frac{\text{Proposed building area} = 100,000 \text{ S.F.}}{\text{Typical size from below} = 50,000 \text{ S.F.}} = 2.00$$

Enter Area Conversion scale at 2.0, intersect curve, read horizontally the appropriate cost multiplier of .94. Size adjusted cost becomes .94 x $96.00 = $90.25 based on national average costs.

Note: For Size Factors less than .50, the Cost Multiplier is 1.1
For Size Factors greater than 3.5, the Cost Multiplier is .90

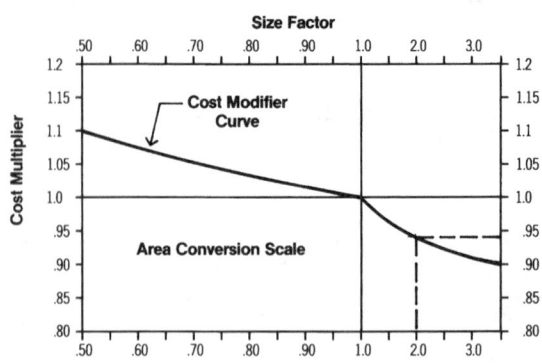

Building Type	Median Cost per S.F.	Typical Size Gross S.F.	Typical Range Gross S.F.		Building Type	Median Cost per S.F.	Typical Size Gross S.F.	Typical Range Gross S.F.	
Apartments, Low Rise	$ 75.50	21,000	9,700 -	37,200	Jails	$230.00	40,000	5,500 -	145,000
Apartments, Mid Rise	96.00	50,000	32,000 -	100,000	Libraries	141.00	12,000	7,000 -	31,000
Apartments, High Rise	109.00	145,000	95,000 -	600,000	Living, Assisted	123.00	32,300	23,500 -	50,300
Auditoriums	126.00	25,000	7,600 -	39,000	Medical Clinics	131.00	7,200	4,200 -	15,700
Auto Sales	94.00	20,000	10,800 -	28,600	Medical Offices	123.00	6,000	4,000 -	15,000
Banks	169.00	4,200	2,500 -	7,500	Motels	91.00	40,000	15,800 -	120,000
Churches	116.00	17,000	2,000 -	42,000	Nursing Homes	127.00	23,000	15,000 -	37,000
Clubs, Country	118.00	6,500	4,500 -	15,000	Offices, Low Rise	107.00	20,000	5,000 -	80,000
Clubs, Social	113.00	10,000	6,000 -	13,500	Offices, Mid Rise	107.00	120,000	20,000 -	300,000
Clubs, YMCA	133.00	28,300	12,800 -	39,400	Offices, High Rise	136.00	260,000	120,000 -	800,000
Colleges (Class)	144.00	50,000	15,000 -	150,000	Police Stations	166.00	10,500	4,000 -	19,000
Colleges (Science Lab)	216.00	45,600	16,600 -	80,000	Post Offices	126.00	12,400	6,800 -	30,000
College (Student Union)	165.00	33,400	16,000 -	85,000	Power Plants	940.00	7,500	1,000 -	20,000
Community Center	120.00	9,400	5,300 -	16,700	Religious Education	108.00	9,000	6,000 -	12,000
Court Houses	161.00	32,400	17,800 -	106,000	Research	177.00	19,000	6,300 -	45,000
Dept. Stores	70.50	90,000	44,000 -	122,000	Restaurants	153.00	4,400	2,800 -	6,000
Dormitories, Low Rise	125.00	25,000	10,000 -	95,000	Retail Stores	75.00	7,200	4,000 -	17,600
Dormitories, Mid Rise	158.00	85,000	20,000 -	200,000	Schools, Elementary	111.00	41,000	24,500 -	55,000
Factories	68.00	26,400	12,900 -	50,000	Schools, Jr. High	115.00	92,000	52,000 -	119,000
Fire Stations	124.00	5,800	4,000 -	8,700	Schools, Sr. High	118.00	101,000	50,500 -	175,000
Fraternity Houses	117.00	12,500	8,200 -	14,800	Schools, Vocational	112.00	37,000	20,500 -	82,000
Funeral Homes	131.00	10,000	4,000 -	20,000	Sports Arenas	92.50	15,000	5,000 -	40,000
Garages, Commercial	84.00	9,300	5,000 -	13,600	Supermarkets	74.00	44,000	12,000 -	60,000
Garages, Municipal	106.00	8,300	4,500 -	12,600	Swimming Pools	173.00	20,000	10,000 -	32,000
Garages, Parking	45.50	163,000	76,400 -	225,300	Telephone Exchange	201.00	4,500	1,200 -	10,600
Gymnasiums	115.00	19,200	11,600 -	41,000	Theaters	107.00	10,500	8,800 -	17,500
Hospitals	203.00	55,000	27,200 -	125,000	Town Halls	127.00	10,800	4,800 -	23,400
House (Elderly)	103.00	37,000	21,000 -	66,000	Warehouses	54.00	25,000	8,000 -	72,000
Housing (Public)	95.00	36,000	14,400 -	74,400	Warehouse & Office	59.00	25,000	8,000 -	72,000
Ice Rinks	137.00	29,000	27,200 -	33,600					

A	Area Square Feet; Ampere
ABS	Acrylonitrile Butadiene Stryrene; Asbestos Bonded Steel
A.C.	Alternating Current; Air-Conditioning; Asbestos Cement; Plywood Grade A & C
A.C.I.	American Concrete Institute
AD	Plywood, Grade A & D
Addit.	Additional
Adj.	Adjustable
af	Audio-frequency
A.G.A.	American Gas Association
Agg.	Aggregate
A.H.	Ampere Hours
A hr.	Ampere-hour
A.H.U.	Air Handling Unit
A.I.A.	American Institute of Architects
AIC	Ampere Interrupting Capacity
Allow.	Allowance
alt.	Altitude
Alum.	Aluminum
a.m.	Ante Meridiem
Amp.	Ampere
Anod.	Anodized
Approx.	Approximate
Apt.	Apartment
Asb.	Asbestos
A.S.B.C.	American Standard Building Code
Asbe.	Asbestos Worker
A.S.H.R.A.E.	American Society of Heating, Refrig. & AC Engineers
A.S.M.E.	American Society of Mechanical Engineers
A.S.T.M.	American Society for Testing and Materials
Attchmt.	Attachment
Avg.	Average
A.W.G.	American Wire Gauge
AWWA	American Water Works Assoc.
Bbl.	Barrel
B&B	Grade B and Better; Balled & Burlapped
B.&S.	Bell and Spigot
B.&W.	Black and White
b.c.c.	Body-centered Cubic
B.C.Y.	Bank Cubic Yards
BE	Bevel End
B.F.	Board Feet
Bg. cem.	Bag of Cement
BHP	Boiler Horsepower; Brake Horsepower
B.I.	Black Iron
Bit.; Bitum.	Bituminous
Bk.	Backed
Bkrs.	Breakers
Bldg.	Building
Blk.	Block
Bm.	Beam
Boil.	Boilermaker
B.P.M.	Blows per Minute
BR	Bedroom
Brg.	Bearing
Brhe.	Bricklayer Helper
Bric.	Bricklayer
Brk.	Brick
Brng.	Bearing
Brs.	Brass
Brz.	Bronze
Bsn.	Basin
Btr.	Better
BTU	British Thermal Unit
BTUH	BTU per Hour
B.U.R.	Built-up Roofing
BX	Interlocked Armored Cable
c	Conductivity, Copper Sweat
C	Hundred; Centigrade
C/C	Center to Center, Cedar on Cedar
Cab.	Cabinet
Cair.	Air Tool Laborer
Calc	Calculated
Cap.	Capacity
Carp.	Carpenter
C.B.	Circuit Breaker
C.C.A.	Chromate Copper Arsenate
C.C.F.	Hundred Cubic Feet
cd	Candela
cd/sf	Candela per Square Foot
CD	Grade of Plywood Face & Back
CDX	Plywood, Grade C & D, exterior glue
Cefi.	Cement Finisher
Cem.	Cement
CF	Hundred Feet
C.F.	Cubic Feet
CFM	Cubic Feet per Minute
c.g.	Center of Gravity
CHW	Chilled Water; Commercial Hot Water
C.I.	Cast Iron
C.I.P.	Cast in Place
Circ.	Circuit
C.L.	Carload Lot
Clab.	Common Laborer
Clam	Common maintenance laborer
C.L.F.	Hundred Linear Feet
CLF	Current Limiting Fuse
CLP	Cross Linked Polyethylene
cm	Centimeter
CMP	Corr. Metal Pipe
C.M.U.	Concrete Masonry Unit
CN	Change Notice
Col.	Column
CO_2	Carbon Dioxide
Comb.	Combination
Compr.	Compressor
Conc.	Concrete
Cont.	Continuous; Continued
Corr.	Corrugated
Cos	Cosine
Cot	Cotangent
Cov.	Cover
C/P	Cedar on Paneling
CPA	Control Point Adjustment
Cplg.	Coupling
C.P.M.	Critical Path Method
CPVC	Chlorinated Polyvinyl Chloride
C.Pr.	Hundred Pair
CRC	Cold Rolled Channel
Creos.	Creosote
Crpt.	Carpet & Linoleum Layer
CRT	Cathode-ray Tube
CS	Carbon Steel, Constant Shear Bar Joist
Csc	Cosecant
C.S.F.	Hundred Square Feet
CSI	Construction Specifications Institute
C.T.	Current Transformer
CTS	Copper Tube Size
Cu	Copper, Cubic
Cu. Ft.	Cubic Foot
cw	Continuous Wave
C.W.	Cool White; Cold Water
Cwt.	100 Pounds
C.W.X.	Cool White Deluxe
C.Y.	Cubic Yard (27 cubic feet)
C.Y./Hr.	Cubic Yard per Hour
Cyl.	Cylinder
d	Penny (nail size)
D	Deep; Depth; Discharge
Dis.;Disch.	Discharge
Db.	Decibel
Dbl.	Double
DC	Direct Current
DDC	Direct Digital Control
Demob.	Demobilization
d.f.u.	Drainage Fixture Units
D.H.	Double Hung
DHW	Domestic Hot Water
Diag.	Diagonal
Diam.	Diameter
Distrib.	Distribution
Dk.	Deck
D.L.	Dead Load; Diesel
DLH	Deep Long Span Bar Joist
Do.	Ditto
Dp.	Depth
D.P.S.T.	Double Pole, Single Throw
Dr.	Driver
Drink.	Drinking
D.S.	Double Strength
D.S.A.	Double Strength A Grade
D.S.B.	Double Strength B Grade
Dty.	Duty
DWV	Drain Waste Vent
DX	Deluxe White, Direct Expansion
dyn	Dyne
e	Eccentricity
E	Equipment Only; East
Ea.	Each
E.B.	Encased Burial
Econ.	Economy
E.C.Y	Embankment Cubic Yards
EDP	Electronic Data Processing
EIFS	Exterior Insulation Finish System
E.D.R.	Equiv. Direct Radiation
Eq.	Equation
Elec.	Electrician; Electrical
Elev.	Elevator; Elevating
EMT	Electrical Metallic Conduit; Thin Wall Conduit
Eng.	Engine, Engineered
EPDM	Ethylene Propylene Diene Monomer
EPS	Expanded Polystyrene
Eqhv.	Equip. Oper., Heavy
Eqlt.	Equip. Oper., Light
Eqmd.	Equip. Oper., Medium
Eqmm.	Equip. Oper., Master Mechanic
Eqol.	Equip. Oper., Oilers
Equip.	Equipment
ERW	Electric Resistance Welded
E.S.	Energy Saver
Est.	Estimated
esu	Electrostatic Units
E.W.	Each Way
EWT	Entering Water Temperature
Excav.	Excavation
Exp.	Expansion, Exposure
Ext.	Exterior
Extru.	Extrusion
f.	Fiber stress
F	Fahrenheit; Female; Fill
Fab.	Fabricated
FBGS	Fiberglass
F.C.	Footcandles
f.c.c.	Face-centered Cubic
$f'c.$	Compressive Stress in Concrete; Extreme Compressive Stress
F.E.	Front End
FEP	Fluorinated Ethylene Propylene (Teflon)
F.G.	Flat Grain
F.H.A.	Federal Housing Administration
Fig.	Figure
Fin.	Finished
Fixt.	Fixture
Fl. Oz.	Fluid Ounces
Flr.	Floor
F.M.	Frequency Modulation; Factory Mutual
Fmg.	Framing
Fndtn.	Foundation

Fori.	Foreman, Inside	I.W.	Indirect Waste	M.C.F.	Thousand Cubic Feet
Foro.	Foreman, Outside	J	Joule	M.C.F.M.	Thousand Cubic Feet per Minute
Fount.	Fountain	J.I.C.	Joint Industrial Council	M.C.M.	Thousand Circular Mils
FPM	Feet per Minute	K	Thousand; Thousand Pounds;	M.C.P.	Motor Circuit Protector
FPT	Female Pipe Thread		Heavy Wall Copper Tubing, Kelvin	MD	Medium Duty
Fr.	Frame	K.A.H.	Thousand Amp. Hours	M.D.O.	Medium Density Overlaid
F.R.	Fire Rating	KCMIL	Thousand Circular Mils	Med.	Medium
FRK	Foil Reinforced Kraft	KD	Knock Down	MF	Thousand Feet
FRP	Fiberglass Reinforced Plastic	K.D.A.T.	Kiln Dried After Treatment	M.F.B.M.	Thousand Feet Board Measure
FS	Forged Steel	kg	Kilogram	Mfg.	Manufacturing
FSC	Cast Body; Cast Switch Box	kG	Kilogauss	Mfrs.	Manufacturers
Ft.	Foot; Feet	kgf	Kilogram Force	mg	Milligram
Ftng.	Fitting	kHz	Kilohertz	MGD	Million Gallons per Day
Ftg.	Footing	Kip.	1000 Pounds	MGPH	Thousand Gallons per Hour
Ft. Lb.	Foot Pound	KJ	Kiljoule	MH, M.H.	Manhole; Metal Halide; Man-Hour
Furn.	Furniture	K.L.	Effective Length Factor	MHz	Megahertz
FVNR	Full Voltage Non-Reversing	K.L.F.	Kips per Linear Foot	Mi.	Mile
FXM	Female by Male	Km	Kilometer	MI	Malleable Iron; Mineral Insulated
Fy.	Minimum Yield Stress of Steel	K.S.F.	Kips per Square Foot	mm	Millimeter
g	Gram	K.S.I.	Kips per Square Inch	Mill.	Millwright
G	Gauss	kV	Kilovolt	Min., min.	Minimum, minute
Ga.	Gauge	kVA	Kilovolt Ampere	Misc.	Miscellaneous
Gal.	Gallon	K.V.A.R.	Kilovar (Reactance)	ml	Milliliter, Mainline
Gal./Min.	Gallon per Minute	KW	Kilowatt	M.L.F.	Thousand Linear Feet
Galv.	Galvanized	KWh	Kilowatt-hour	Mo.	Month
Gen.	General	L	Labor Only; Length; Long;	Mobil.	Mobilization
G.F.I.	Ground Fault Interrupter		Medium Wall Copper Tubing	Mog.	Mogul Base
Glaz.	Glazier	Lab.	Labor	MPH	Miles per Hour
GPD	Gallons per Day	lat	Latitude	MPT	Male Pipe Thread
GPH	Gallons per Hour	Lath.	Lather	MRT	Mile Round Trip
GPM	Gallons per Minute	Lav.	Lavatory	ms	Millisecond
GR	Grade	lb.; #	Pound	M.S.F.	Thousand Square Feet
Gran.	Granular	L.B.	Load Bearing; L Conduit Body	Mstz.	Mosaic & Terrazzo Worker
Grnd.	Ground	L. & E.	Labor & Equipment	M.S.Y.	Thousand Square Yards
H	High; High Strength Bar Joist;	lb./hr.	Pounds per Hour	Mtd.	Mounted
	Henry	lb./L.F.	Pounds per Linear Foot	Mthe.	Mosaic & Terrazzo Helper
H.C.	High Capacity	lbf/sq.in.	Pound-force per Square Inch	Mtng.	Mounting
H.D.	Heavy Duty; High Density	L.C.L.	Less than Carload Lot	Mult.	Multi; Multiply
H.D.O.	High Density Overlaid	L.C.Y.	Loose Cubic Yard	M.V.A.	Million Volt Amperes
Hdr.	Header	Ld.	Load	M.V.A.R.	Million Volt Amperes Reactance
Hdwe.	Hardware	LE	Lead Equivalent	MV	Megavolt
Help.	Helper Average	LED	Light Emitting Diode	MW	Megawatt
HEPA	High Efficiency Particulate Air	L.F.	Linear Foot	MXM	Male by Male
	Filter	Lg.	Long; Length; Large	MYD	Thousand Yards
Hg	Mercury	L & H	Light and Heat	N	Natural; North
HIC	High Interrupting Capacity	LH	Long Span Bar Joist	nA	Nanoampere
HM	Hollow Metal	L.H.	Labor Hours	NA	Not Available; Not Applicable
H.O.	High Output	L.L.	Live Load	N.B.C.	National Building Code
Horiz.	Horizontal	L.L.D.	Lamp Lumen Depreciation	NC	Normally Closed
H.P.	Horsepower; High Pressure	lm	Lumen	N.E.M.A.	National Electrical Manufacturers
H.P.F.	High Power Factor	lm/sf	Lumen per Square Foot		Assoc.
Hr.	Hour	lm/W	Lumen per Watt	NEHB	Bolted Circuit Breaker to 600V.
Hrs./Day	Hours per Day	L.O.A.	Length Over All	N.L.B.	Non-Load-Bearing
HSC	High Short Circuit	log	Logarithm	NM	Non-Metallic Cable
Ht.	Height	L-O-L	Lateralolet	nm	Nanometer
Htg.	Heating	L.P.	Liquefied Petroleum; Low Pressure	No.	Number
Htrs.	Heaters	L.P.F.	Low Power Factor	NO	Normally Open
HVAC	Heating, Ventilation & Air-	LR	Long Radius	N.O.C.	Not Otherwise Classified
	Conditioning	L.S.	Lump Sum	Nose.	Nosing
Hvy.	Heavy	Lt.	Light	N.P.T.	National Pipe Thread
HW	Hot Water	Lt. Ga.	Light Gauge	NQOD	Combination Plug-on/Bolt on
Hyd.;Hydr.	Hydraulic	L.T.L.	Less than Truckload Lot		Circuit Breaker to 240V.
Hz.	Hertz (cycles)	Lt. Wt.	Lightweight	N.R.C.	Noise Reduction Coefficient
I.	Moment of Inertia	L.V.	Low Voltage	N.R.S.	Non Rising Stem
I.C.	Interrupting Capacity	M	Thousand; Material; Male;	ns	Nanosecond
ID	Inside Diameter		Light Wall Copper Tubing	nW	Nanowatt
I.D.	Inside Dimension; Identification	M²CA	Meters Squared Contact Area	OB	Opposing Blade
I.F.	Inside Frosted	m/hr; M.H.	Man-hour	OC	On Center
I.M.C.	Intermediate Metal Conduit	mA	Milliampere	OD	Outside Diameter
In.	Inch	Mach.	Machine	O.D.	Outside Dimension
Incan.	Incandescent	Mag. Str.	Magnetic Starter	ODS	Overhead Distribution System
Incl.	Included; Including	Maint.	Maintenance	O.G.	Ogee
Int.	Interior	Marb.	Marble Setter	O.H.	Overhead
Inst.	Installation	Mat; Mat'l.	Material	O&P	Overhead and Profit
Insul.	Insulation/Insulated	Max.	Maximum	Oper.	Operator
I.P.	Iron Pipe	MBF	Thousand Board Feet	Opng.	Opening
I.P.S.	Iron Pipe Size	MBH	Thousand BTU's per hr.	Orna.	Ornamental
I.P.T.	Iron Pipe Threaded	MC	Metal Clad Cable	OSB	Oriented Strand Board

Abbreviations

O.S.&Y.	Outside Screw and Yoke	Rsr	Riser	Th.;Thk.	Thick
Ovhd.	Overhead	RT	Round Trip	Thn.	Thin
OWG	Oil, Water or Gas	S.	Suction; Single Entrance; South	Thrded	Threaded
Oz.	Ounce	SC	Screw Cover	Tilf.	Tile Layer, Floor
P.	Pole; Applied Load; Projection	SCFM	Standard Cubic Feet per Minute	Tilh.	Tile Layer, Helper
p.	Page	Scaf.	Scaffold	THHN	Nylon Jacketed Wire
Pape.	Paperhanger	Sch.; Sched.	Schedule	THW.	Insulated Strand Wire
P.A.P.R.	Powered Air Purifying Respirator	S.C.R.	Modular Brick	THWN;	Nylon Jacketed Wire
PAR	Parabolic Reflector	S.D.	Sound Deadening	T.L.	Truckload
Pc., Pcs.	Piece, Pieces	S.D.R.	Standard Dimension Ratio	T.M.	Track Mounted
P.C.	Portland Cement; Power Connector	S.E.	Surfaced Edge	Tot.	Total
P.C.F.	Pounds per Cubic Foot	Sel.	Select	T-O-L	Threadolet
P.C.M.	Phase Contrast Microscopy	S.E.R.; S.E.U.	Service Entrance Cable	T.S.	Trigger Start
P.E.	Professional Engineer;	S.F.	Square Foot	Tr.	Trade
	Porcelain Enamel;	S.F.C.A.	Square Foot Contact Area	Transf.	Transformer
	Polyethylene; Plain End	S.F. Flr.	Square Foot of Floor	Trhv.	Truck Driver, Heavy
Perf.	Perforated	S.F.G.	Square Foot of Ground	Trlr	Trailer
Ph.	Phase	S.F. Hor.	Square Foot Horizontal	Trlt.	Truck Driver, Light
P.I.	Pressure Injected	S.F.R.	Square Feet of Radiation	TTY	Teletypewriter
Pile.	Pile Driver	S.F. Shlf.	Square Foot of Shelf	TV	Television
Pkg.	Package	S4S	Surface 4 Sides	T.W.	Thermoplastic Water Resistant
Pl.	Plate	Shee.	Sheet Metal Worker		Wire
Plah.	Plasterer Helper	Sin.	Sine	UCI	Uniform Construction Index
Plas.	Plasterer	Skwk.	Skilled Worker	UF	Underground Feeder
Pluh.	Plumbers Helper	SL	Saran Lined	UGND	Underground Feeder
Plum.	Plumber	S.L.	Slimline	U.H.F.	Ultra High Frequency
Ply.	Plywood	Sldr.	Solder	U.L.	Underwriters Laboratory
p.m.	Post Meridiem	SLH	Super Long Span Bar Joist	Unfin.	Unfinished
Pntd.	Painted	S.N.	Solid Neutral	URD	Underground Residential
Pord.	Painter, Ordinary	S-O-L	Socketolet		Distribution
pp	Pages	sp	Standpipe	US	United States
PP; PPL	Polypropylene	S.P.	Static Pressure; Single Pole; Self-	USP	United States Primed
P.P.M.	Parts per Million		Propelled	UTP	Unshielded Twisted Pair
Pr.	Pair	Spri.	Sprinkler Installer	V	Volt
P.E.S.B.	Pre-engineered Steel Building	spwg	Static Pressure Water Gauge	V.A.	Volt Amperes
Prefab.	Prefabricated	S.P.D.T.	Single Pole, Double Throw	V.C.T.	Vinyl Composition Tile
Prefin.	Prefinished	SPF	Spruce Pine Fir	VAV	Variable Air Volume
Prop.	Propelled	S.P.S.T.	Single Pole, Single Throw	VC	Veneer Core
PSF; psf	Pounds per Square Foot	SPT	Standard Pipe Thread	Vent.	Ventilation
PSI; psi	Pounds per Square Inch	Sq.	Square; 100 Square Feet	Vert.	Vertical
PSIG	Pounds per Square Inch Gauge	Sq. Hd.	Square Head	V.F.	Vinyl Faced
PSP	Plastic Sewer Pipe	Sq. In.	Square Inch	V.G.	Vertical Grain
Pspr.	Painter, Spray	S.S.	Single Strength; Stainless Steel	V.H.F.	Very High Frequency
Psst.	Painter, Structural Steel	S.S.B.	Single Strength B Grade	VHO	Very High Output
P.T.	Potential Transformer	sst	Stainless Steel	Vib.	Vibrating
P. & T.	Pressure & Temperature	Sswk.	Structural Steel Worker	V.L.F.	Vertical Linear Foot
Ptd.	Painted	Sswl.	Structural Steel Welder	Vol.	Volume
Ptns.	Partitions	St.;Stl.	Steel	VRP	Vinyl Reinforced Polyester
Pu	Ultimate Load	S.T.C.	Sound Transmission Coefficient	W	Wire; Watt; Wide; West
PVC	Polyvinyl Chloride	Std.	Standard	w/	With
Pvmt.	Pavement	STK	Select Tight Knot	W.C.	Water Column; Water Closet
Pwr.	Power	STP	Standard Temperature & Pressure	W.F.	Wide Flange
Q	Quantity Heat Flow	Stpi.	Steamfitter, Pipefitter	W.G.	Water Gauge
Quan.;Qty.	Quantity	Str.	Strength; Starter; Straight	Wldg.	Welding
Q.C.	Quick Coupling	Strd.	Stranded	W. Mile	Wire Mile
r	Radius of Gyration	Struct.	Structural	W-O-L	Weldolet
R	Resistance	Sty.	Story	W.R.	Water Resistant
R.C.P.	Reinforced Concrete Pipe	Subj.	Subject	Wrck.	Wrecker
Rect.	Rectangle	Subs.	Subcontractors	W.S.P.	Water, Steam, Petroleum
Reg.	Regular	Surf.	Surface	WT., Wt.	Weight
Reinf.	Reinforced	Sw.	Switch	WWF	Welded Wire Fabric
Req'd.	Required	Swbd.	Switchboard	XFER	Transfer
Res.	Resistant	S.Y.	Square Yard	XFMR	Transformer
Resi.	Residential	Syn.	Synthetic	XHD	Extra Heavy Duty
Rgh.	Rough	S.Y.P.	Southern Yellow Pine	XHHW; XLPE	Cross-Linked Polyethylene Wire
RGS	Rigid Galvanized Steel	Sys.	System		Insulation
R.H.W.	Rubber, Heat & Water Resistant;	t.	Thickness	XLP	Cross-linked Polyethylene
	Residential Hot Water	T	Temperature; Ton	Y	Wye
rms	Root Mean Square	Tan	Tangent	yd	Yard
Rnd.	Round	T.C.	Terra Cotta	yr	Year
Rodm.	Rodman	T & C	Threaded and Coupled	Δ	Delta
Rofc.	Roofer, Composition	T.D.	Temperature Difference	%	Percent
Rofp.	Roofer, Precast	Tdd	Telecommunications Device for	~	Approximately
Rohe.	Roofer Helpers (Composition)		the Deaf	Ø	Phase
Rots.	Roofer, Tile & Slate	T.E.M.	Transmission Electron Microscopy	@	At
R.O.W.	Right of Way	TFE	Tetrafluoroethylene (Teflon)	#	Pound; Number
RPM	Revolutions per Minute	T. & G.	Tongue & Groove;	<	Less Than
R.S.	Rapid Start		Tar & Gravel	>	Greater Than

Index

Index

Index

541

Notes

545

547

Reed Construction Data/RSMeans...
a tradition of excellence in Construction Cost Information and Services since 1942.

For more information visit RSMeans Web site at www.rsmeans.com

Book Selection Guide

The following table provides definitive information on the content of each cost data publication. The number of lines of data provided in each unit price or assemblies division, as well as the number of reference tables and crews, is listed for each book. The presence of other elements such as equipment rental costs, historical cost indexes, city cost indexes, square foot models, or cross-referenced index is also indicated. You can use the table to help select the RSMeans book that has the quantity and type of information you most need in your work.

Unit Cost Divisions	Building Construction Costs	Mechanical	Electrical	Repair & Remodel	Square Foot	Site Work Landsc.	Assemblies	Interior	Concrete Masonry	Open Shop	Heavy Construc.	Light Commercial	Facil. Construc.	Plumbing	Western Construction Costs	Residential
1	548	283	368	477		490		304	455	547	499	209	976	317	546	160
2	540	237	58	528		768		331	189	539	665	410	1088	260	540	218
3	1469	179	165	797		1336		261	1852	1463	1478	256	1392	148	1464	288
4	847	18	0	649		660		561	1060	823	582	426	1098	0	833	354
5	1802	205	159	917		749		954	686	1770	989	801	1781	282	1742	725
6	1398	78	78	1358		487		1127	317	1376	645	1437	1356	22	1735	1591
7	1275	167	71	1284		476		493	430	1270	0	969	1329	176	1274	759
8	1906	61	10	1936		299		1670	671	1868	0	1325	2107	0	1865	1270
9	1650	70	23	1474		237		1743	370	1614	0	1373	1866	55	1641	1253
10	877	17	10	512		223		777	156	879	36	396	998	233	877	212
11	944	213	169	488		124		796	28	929	0	216	1053	176	925	104
12	529	0	2	273		267		1681	31	520	19	339	1734	27	520	312
13	700	119	111	233		385		245	69	692	287	71	707	62	610	68
14	287	36	0	233		29		264	0	286	0	12	304	21	285	6
21	78	0	16	25		0		244	0	78	0	59	362	367	77	0
22	1024	6968	135	1039		1450		598	20	1032	1655	728	6407	8663	1057	531
23	1093	7172	606	765		159		661	38	1094	110	568	5021	1851	1071	326
26	1226	493	9755	985		715		1040	55	1221	527	1046	9702	438	1162	556
27	75	0	195	18		9		67	0	75	35	48	197	0	66	4
28	87	59	93	65		0		69	0	72	0	42	108	45	87	27
31	853	199	108	311		1375		7	657	826	1470	90	973	127	854	97
32	706	76	0	575		3802		360	251	680	1242	344	1331	188	692	412
33	517	1030	430	199		1893		0	225	520	1852	95	1561	1191	514	93
34	99	0	20	4		134		0	31	55	110	0	119	0	99	0
35	18	0	0	0		166		0	0	18	166	0	83	0	18	0
41	58	0	0	24		7		28	0	58	30	0	65	14	58	0
44	98	82	0	0		29		0	0	23	29	0	98	105	23	0
Totals	20704	17711	12582	15169		16269		14281	7591	20328	12426	11260	43735	14768	20635	9366

Assembly Divisions	Building Construction Costs	Mechanical	Electrical	Repair & Remodel	Square Foot	Site Work Landsc.	Assemblies	Interior	Concrete Masonry	Open Shop	Heavy Construc.	Light Commercial	Facil. Construc.	Plumbing	Western Construction Costs	Asm Div	Residential
A		19	0	192	150	540	612	0	550		580	153	24	0		1	374
B		0	0	809	2479	0	5588	332	1914		368	2022	144	0		2	217
C		0	0	635	850	0	1206	1550	131		0	753	238	0		3	588
D		1014	789	693	1783	0	2401	753	0		0	1262	1011	890		4	867
E		0	0	85	257	0	294	5	0		0	257	5	0		5	393
F		0	0	0	123	0	124	0	0		0	123	3	0		6	358
G		465	172	331	111	1857	584	0	482		910	110	113	559		7	300
																8	760
																9	80
																10	0
																11	0
																12	0
Totals		1498	961	2745	5753	2397	10809	2640	3077		1858	4680	1538	1449			3937

Reference Section	Building Construction Costs	Mechanical	Electrical	Repair & Remodel	Square Foot	Site Work Landsc.	Assemblies	Interior	Concrete Masonry	Open Shop	Heavy Construc.	Light Commercial	Facil. Construc.	Plumbing	Western Construction Costs	Residential
Reference Tables	yes	yes	yes	yes	no	yes	yes	yes	yes	yes	yes	yes	yes	yes	yes	yes
Models					105							46				32
Crews	461	461	461	443		461		461	461	440	461	440	443	461	461	461
Equipment Rental Costs	yes	yes	yes	yes		yes		yes	yes	yes	yes	yes	yes	yes	yes	yes
Historical Cost Indexes	yes	yes	yes	yes	yes	yes	yes	yes	yes	yes	yes	yes	yes	yes	yes	no
City Cost Indexes	yes	yes	yes	yes	yes	yes	yes	yes	yes	yes	yes	yes	yes	yes	yes	yes

Annual Cost Guides

For more information
visit RSMeans Web site
at www.rsmeans.com

RSMeans Building Construction Cost Data 2007

Available in Both Softbound and Looseleaf Editions

Many customers enjoy the convenience and flexibility of the looseleaf binder, which increases the usefulness of *RSMeans Building Construction Cost Data 2007* by making it easy to add and remove pages. You can insert your own cost information pages, so everything is in one place. Copying pages for faxing is easier also. Whichever edition you prefer, softbound or the convenient looseleaf edition, you'll be eligible to receive *RSMeans Quarterly Update Service* FREE. Current subscribers can receive *RSMeans Quarterly Update Service* via e-mail.

Unit Prices Now Updated to MasterFormat 2004!

Now Available in Spanish!

$169.95 per copy, Looseleaf
Available Oct. 2006
Catalog No. 61017

RSMeans Building Construction Cost Data 2007

Now Available in Spanish!

Offers you unchallenged unit price reliability in an easy-to-use arrangement. Whether used for complete, finished estimates or for periodic checks, it supplies more cost facts better and faster than any comparable source. Over 20,000 unit prices for 2007. The City Cost Indexes and Location Factors cover over 930 areas, for indexing to any project location in North America. Order and get *RSMeans Quarterly Update Service* FREE. You'll have year-long access to the RSMeans Estimating **HOTLINE** FREE with your subscription. Expert assistance when using RSMeans data is just a phone call away.

$136.95 per copy (English or Spanish)
Catalog No. 60017 (English) Available Oct. 2006
Catalog No. 60717 (Spanish) Available Jan. 2007

Unit Prices Now Updated to MasterFormat 2004!

RSMeans Metric Construction Cost Data 2007

A massive compendium of all the data from both the 2007 *RSMeans Building Construction Cost Data* AND *Heavy Construction Cost Data*, in **metric** format! Access all of this vital information from one complete source. It contains more than 600 pages of unit costs and 40 pages of assemblies costs. The Reference Section contains over 200 pages of tables, charts and other estimating aids. A great way to stay in step with today's construction trends and rapidly changing costs.

$162.95 per copy
Available Dec. 2006
Catalog No. 63017

RSMeans Mechanical Cost Data 2007

- **HVAC**
- **Controls**

Total unit and systems price guidance for mechanical construction... materials, parts, fittings, and complete labor cost information. Includes prices for piping, heating, air conditioning, ventilation, and all related construction.

Plus new 2007 unit costs for:

- Over 2500 installed HVAC/controls assemblies
- "On Site" Location Factors for over 930 cities and towns in the U.S. and Canada
- Crews, labor, and equipment

$136.95 per copy
Available Oct. 2006
Catalog No. 60027

Unit Prices Now Updated to MasterFormat 2004!

RSMeans Plumbing Cost Data 2007

Comprehensive unit prices and assemblies for plumbing, irrigation systems, commercial and residential fire protection, point-of-use water heaters, and the latest approved materials. This publication and its companion, *RSMeans Mechanical Cost Data*, provide full-range cost estimating coverage for all the mechanical trades.

New for '07: More lines of no-hub CI soil pipe fittings, more flange-type escutcheons, fiberglass pipe insulation in a full range of sizes for 2-1/2" and 3" wall thicknesses, 220 lines of grease duct, and much more.

$136.95 per copy
Available Oct. 2006
Catalog No. 60217

RSMeans Electrical Cost Data 2007

Pricing information for every part of electrical cost planning. More than 13,000 unit and systems costs with design tables; clear specifications and drawings; engineering guides; illustrated estimating procedures; complete labor-hour and materials costs for better scheduling and procurement; and the latest electrical products and construction methods.

- A variety of special electrical systems including cathodic protection
- Costs for maintenance, demolition, HVAC/ mechanical, specialties, equipment, and more

$136.95 per copy
Available Oct. 2006
Catalog No. 60037

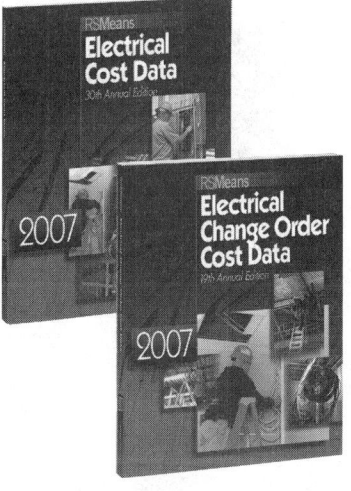

Unit Prices Now Updated to MasterFormat 2004!

RSMeans Electrical Change Order Cost Data 2007

RSMeans Electrical Change Order Cost Data 2007 provides you with electrical unit prices exclusively for pricing change orders—based on the recent, direct experience of contractors and suppliers. Analyze and check your own change order estimates against the experience others have had doing the same work. It also covers productivity analysis and change order cost justifications. With useful information for calculating the effects of change orders and dealing with their administration.

$136.95 per copy
Available Nov. 2006
Catalog No. 60237

RSMeans Square Foot Costs 2007

It's Accurate and Easy To Use!

- **Updated 2007 price information**, based on nationwide figures from suppliers, estimators, labor experts, and contractors
- "How-to-Use" sections, with **clear examples** of commercial, residential, industrial, and institutional structures
- Realistic graphics, offering true-to-life illustrations of building projects
- Extensive information on using square foot cost data, including sample estimates and alternate pricing methods

$148.95 per copy
Over 450 pages, illustrated, available Nov. 2006
Catalog No. 60057

RSMeans Repair & Remodeling Cost Data 2007

Commercial/Residential

Use this valuable tool to estimate commercial and residential renovation and remodeling.

Includes: New costs for hundreds of unique methods, materials, and conditions that only come up in repair and remodeling, PLUS:

- Unit costs for over 15,000 construction components
- Installed costs for over 90 assemblies
- Over 930 "On-Site" localization factors for the U.S. and Canada.

Unit Prices Now Updated to MasterFormat 2004!
$116.95 per copy
Available Nov. 2006
Catalog No. 60047

Annual Cost Guides

For more information
visit RSMeans Web site
at www.rsmeans.com

RSMeans Facilities Construction Cost Data 2007

For the maintenance and construction of commercial, industrial, municipal, and institutional properties. Costs are shown for new and remodeling construction and are broken down into materials, labor, equipment, overhead, and profit. Special emphasis is given to sections on mechanical, electrical, furnishings, site work, building maintenance, finish work, and demolition.

More than 43,000 unit costs, plus assemblies costs and a comprehensive Reference Section are included.

$323.95 per copy
Available Dec. 2006
Catalog No. 60207

**Unit Prices Now Updated to
MasterFormat 2004!**

RSMeans Light Commercial Cost Data 2007

Specifically addresses the light commercial market, which is a specialized niche in the construction industry. Aids you, the owner/designer/contractor, in preparing all types of estimates—from budgets to detailed bids. Includes new advances in methods and materials.

Assemblies Section allows you to evaluate alternatives in the early stages of design/planning.

Over 11,000 unit costs for 2007 ensure you have the prices you need... when you need them.

$116.95 per copy
Available Dec. 2006
Catalog No. 60187

RSMeans Residential Cost Data 2007

Contains square foot costs for 30 basic home models with the look of today, plus hundreds of custom additions and modifications you can quote right off the page. With costs for the 100 residential systems you're most likely to use in the year ahead. Complete with blank estimating forms, sample estimates, and step-by-step instructions.

Now contains line items for cultured stone and brick, PVC trim lumber, and TPO roofing.

$116.95 per copy
Available Oct. 2006
Catalog No. 60177

**Unit Prices Now Updated to
MasterFormat 2004!**

RSMeans Site Work & Landscape Cost Data 2007

Includes unit and assemblies costs for earthwork, sewerage, piped utilities, site improvements, drainage, paving, trees & shrubs, street openings/repairs, underground tanks, and more. Contains 57 tables of Assemblies Costs for accurate conceptual estimates.

2007 update includes:
- Estimating for infrastructure improvements
- Environmentally-oriented construction
- ADA-mandated handicapped access
- Hazardous waste line items

$136.95 per copy
Available Nov. 2006
Catalog No. 60287

RSMeans Assemblies Cost Data 2007

RSMeans Assemblies Cost Data 2007 takes the guesswork out of preliminary or conceptual estimates. Now you don't have to try to calculate the assembled cost by working up individual component costs. We've done all the work for you.

Presents detailed illustrations, descriptions, specifications, and costs for every conceivable building assembly—240 types in all—arranged in the easy-to-use UNIFORMAT II system. Each illustrated "assembled" cost includes a complete grouping of materials and associated installation costs, including the installing contractor's overhead and profit.

$223.95 per copy
Available Oct. 2006
Catalog No. 60067

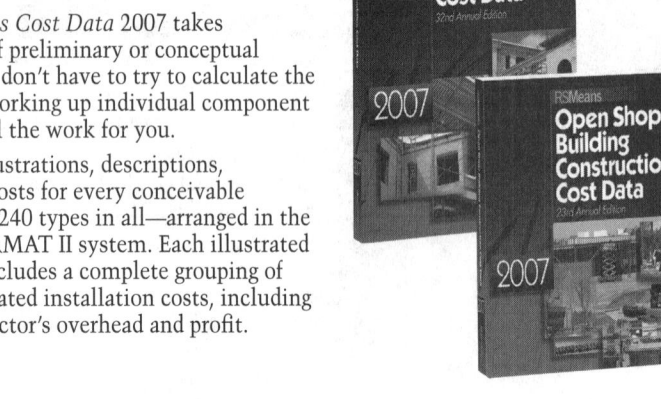

RSMeans Open Shop Building Construction Cost Data 2007

The latest costs for accurate budgeting and estimating of new commercial and residential construction... renovation work... change orders... cost engineering.

RSMeans Open Shop "BCCD" will assist you to:
- Develop benchmark prices for change orders
- Plug gaps in preliminary estimates and budgets
- Estimate complex projects
- Substantiate invoices on contracts
- Price ADA-related renovations

Unit Prices Now Updated to MasterFormat 2004!

$136.95 per copy
Available Dec. 2006
Catalog No. 60157

For more information
visit RSMeans Web site
at www.rsmeans.com

Annual Cost Guides

RSMeans Building Construction Cost Data 2007
Western Edition

This regional edition provides more precise cost information for western North America. Labor rates are based on union rates from 13 western states and western Canada. Included are western practices and materials not found in our national edition: tilt-up concrete walls, glu-lam structural systems, specialized timber construction, seismic restraints, and landscape and irrigation systems.

$136.95 per copy
Available Dec. 2006
Catalog No. 60227

Unit Prices Now Updated to MasterFormat 2004!

RSMeans Heavy Construction Cost Data 2007

A comprehensive guide to heavy construction costs. Includes costs for highly specialized projects such as tunnels, dams, highways, airports, and waterways. Information on labor rates, equipment, and material costs is included. Features unit price costs, systems costs, and numerous reference tables for costs and design.

$136.95 per copy
Available Dec. 2006
Catalog No. 60167

RSMeans Construction Cost Indexes 2007

Who knows what 2007 holds? What materials and labor costs will change unexpectedly? By how much?

- Breakdowns for 316 major cities
- National averages for 30 key cities
- Expanded five major city indexes
- Historical construction cost indexes

$294.00 per year (subscription)
$73.50 individual quarters
Catalog No. 60147 A,B,C,D

RSMeans Interior Cost Data 2007

Provides you with prices and guidance needed to make accurate interior work estimates. Contains costs on materials, equipment, hardware, custom installations, furnishings, and labor costs... for new and remodel commercial and industrial interior construction, including updated information on office furnishings, and reference information.

Unit Prices Now Updated to MasterFormat 2004!

$136.95 per copy
Available Nov. 2006
Catalog No. 60097

RSMeans Concrete & Masonry Cost Data 2007

Provides you with cost facts for virtually all concrete/masonry estimating needs, from complicated formwork to various sizes and face finishes of brick and block—all in great detail. The comprehensive unit cost section contains more than 7,500 selected entries. Also contains an Assemblies Cost section, and a detailed Reference section that supplements the cost data.

$124.95 per copy
Available Dec. 2006
Catalog No. 60117

Unit Prices Now Updated to MasterFormat 2004!

RSMeans Labor Rates for the Construction Industry 2007

Complete information for estimating labor costs, making comparisons, and negotiating wage rates by trade for over 300 U.S. and Canadian cities. With 46 construction trades listed by local union number in each city, and historical wage rates included for comparison. Each city chart lists the county and is alphabetically arranged with handy visual flip tabs for quick reference.

$296.95 per copy
Available Dec. 2006
Catalog No. 60127

RSMeans Facilities Maintenance & Repair Cost Data 2007

RSMeans Facilities Maintenance & Repair Cost Data gives you a complete system to manage and plan your facility repair and maintenance costs and budget efficiently. Guidelines for auditing a facility and developing an annual maintenance plan. Budgeting is included, along with reference tables on cost and management, and information on frequency and productivity of maintenance operations.

The only nationally recognized source of maintenance and repair costs. Developed in cooperation with the Civil Engineering Research Laboratory (CERL) of the Army Corps of Engineers.

$296.95 per copy
Available Dec. 2006
Catalog No. 60307

Reference Books

For more information
visit RSMeans Web site
at www.rsmeans.com

Home Addition & Renovation Project Costs

This essential home remodeling reference gives you 35 project estimates, and guidance for some of the most popular home renovation and addition projects... from opening up a simple interior wall to adding an entire second story. Each estimate includes a floor plan, color photos, and detailed costs. Use the project estimates as backup for pricing, to check your own estimates, or as a cost reference for preliminary discussion with homeowners.

Includes:

- Case studies—with creative solutions and design ideas.
- Alternate materials costs—so you can match the estimates to the particulars of your projects.
- Location Factors—easy multipliers to adjust the book's costs to your own location.

$29.95 per copy
Over 200 pages, illustrated, Softcover
Catalog No. 67349

Kitchen & Bath Project Costs:
Planning & Estimating
Successful Projects

Project estimates for 35 of the most popular kitchen and bath renovations... from replacing a single fixture to whole-room remodels. Each estimate includes:

- All materials needed for the project
- Labor-hours to install (and demolish/remove) each item
- Subcontractor costs for certain trades and services
- An allocation for overhead and profit

PLUS! Takeoff and pricing worksheets—forms you can photocopy or access electronically from the book's Web site; alternate materials—unit costs for different finishes and fixtures; location factors—easy multipliers to adjust the costs to your location; and expert guidance on estimating methods, project design, contracts, marketing, working with homeowners, and tips for each of the estimated projects.

$29.95 per copy
Over 175 pages
Catalog No. 67347

Residential & Light Commercial Construction Standards 2nd Edition
By RSMeans and Contributing Authors

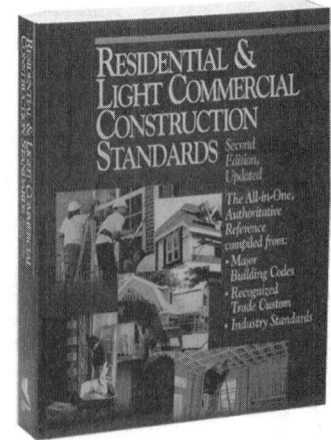

For contractors, subcontractors, owners, developers, architects, engineers, attorneys, and insurance personnel, this book provides authoritative requirements and recommendations compiled from the nation's leading professional associations, industry publications, and building code organizations.

It's an all-in-one reference for establishing a standard for workmanship, quickly resolving disputes, and avoiding defect claims. Includes practical guidance from professionals who are well-known in their respective fields for quality design and construction.

$59.95 per copy
600 pages, illustrated, Softcover
Catalog No. 67322A

Value Engineering: Practical Applications
By Alphonse Dell'Isola, PE

For Design, Construction, Maintenance & Operations

A tool for immediate application—for engineers, architects, facility managers, owners, and contractors. Includes: making the case for VE—the management briefing, integrating VE into planning, budgeting, and design, conducting life cycle costing, using VE methodology in design review and consultant selection, case studies, VE workbook, and a life cycle costing program on disk.

$79.95 per copy
Over 450 pages, illustrated, Softcover
Catalog No. 67319

Facilities Operations & Engineering Reference
By the Association for Facilities Engineering and RSMeans

An all-in-one technical reference for planning and managing facility projects and solving day-to-day operations problems. Selected as the official Certified Plant Engineer reference, this handbook covers financial analysis, maintenance, HVAC and energy efficiency, and more.

$54.98 per copy
Over 700 pages, illustrated, Hardcover
Catalog No. 67318

The Building Professional's Guide to Contract Documents
3rd Edition
By Waller S. Poage, AIA, CSI, CVS

A comprehensive reference for owners, design professionals, contractors, and students.

- Structure your documents for maximum efficiency.
- Effectively communicate construction requirements.
- Understand the roles and responsibilities of construction professionals.
- Improve methods of project delivery.

$32.48 per copy, 400 pages
Diagrams and construction forms, Hardcover
Catalog No. 67261A

Building Security: Strategies & Costs
By David Owen

This comprehensive resource will help you evaluate your facility's security needs, and design and budget for the materials and devices needed to fulfill them.

Includes over 130 pages of RSMeans cost data for installation of security systems and materials, plus a review of more than 50 security devices and construction solutions.

$44.98 per copy
350 pages, illustrated, Hardcover
Catalog No. 67339

Cost Planning & Estimating for Facilities Maintenance

In this unique book, a team of facilities management authorities shares their expertise on:

- Evaluating and budgeting maintenance operations
- Maintaining and repairing key building components
- Applying *RSMeans Facilities Maintenance & Repair Cost Data* to your estimating

Covers special maintenance requirements of the ten major building types.

$89.95 per copy
Over 475 pages, Hardcover
Catalog No. 67314

Life Cycle Costing for Facilities
By Alphonse Dell'Isola and Dr. Steven Kirk

Guidance for achieving higher quality design and construction projects at lower costs! Cost-cutting efforts often sacrifice quality to yield the cheapest product. Life cycle costing enables building designers and owners to achieve both. The authors of this book show how LCC can work for a variety of projects — from roads to HVAC upgrades to different types of buildings.

$99.95 per copy
450 pages, Hardcover
Catalog No. 67341

Planning & Managing Interior Projects 2nd Edition
By Carol E. Farren, CFM
Expert guidance on managing renovation & relocation projects.

This book guides you through every step in relocating to a new space or renovating an old one. From initial meeting through design and construction, to post-project administration, it helps you get the most for your company or client. Includes sample forms, spec lists, agreements, drawings, and much more!

$69.95 per copy
200 pages, Softcover
Catalog No. 67245A

Builder's Essentials: Best Business Practices for Builders & Remodelers
An Easy-to-Use Checklist System
By Thomas N. Frisby

A comprehensive guide covering all aspects of running a construction business, with more than 40 user-friendly checklists. Provides expert guidance on: increasing your revenue and keeping more of your profit, planning for long-term growth, keeping good employees, and managing subcontractors.

$29.95 per copy
Over 220 pages, Softcover
Catalog No. 67329

Reference Books

For more information
visit RSMeans Web site
at www.rsmeans.com

Interior Home Improvement
Costs New 9th Edition

Updated estimates for the most popular remodeling and repair projects—from small, do-it-yourself jobs to major renovations and new construction. Includes: Kitchens & Baths; New Living Space from your Attic, Basement, or Garage; New Floors, Paint, and Wallpaper; Tearing Out or Building New Walls; Closets, Stairs, and Fireplaces; New Energy-Saving Improvements, Home Theaters, and More!

$24.95 per copy
250 pages, illustrated, Softcover
Catalog No. 67308E

Exterior Home Improvement
Costs New 9th Edition

Updated estimates for the most popular remodeling and repair projects—from small, do-it-yourself jobs, to major renovations and new construction. Includes: Curb Appeal Projects—Landscaping, Patios, Porches, Driveways, and Walkways; New Windows and Doors; Decks, Greenhouses, and Sunrooms; Room Additions and Garages; Roofing, Siding, and Painting; "Green" Improvements to Save Energy & Water.

$24.95 per copy
Over 275 pages, illustrated, Softcover
Catalog No. 67309E

Builder's Essentials: Plan Reading & Material Takeoff
By Wayne J. DelPico

For Residential and Light Commercial Construction

A valuable tool for understanding plans and specs, and accurately calculating material quantities. Step-by-step instructions and takeoff procedures based on a full set of working drawings.

$35.95 per copy
Over 420 pages, Softcover
Catalog No. 67307

Builder's Essentials: Framing & Rough Carpentry 2nd Edition
By Scot Simpson

Develop and improve your skills with easy-to-follow instructions and illustrations. Learn proven techniques for framing walls, floors, roofs, stairs, doors, and windows. Updated guidance on standards, building codes, safety requirements, and more. Also available in Spanish!

$24.95 per copy
Over 150 pages, Softcover
Catalog No. 67298A
Spanish Catalog No. 67298AS

Concrete Repair and Maintenance Illustrated
By Peter Emmons

Hundreds of illustrations show users how to analyze, repair, clean, and maintain concrete structures for optimal performance and cost effectiveness. From parking garages to roads and bridges to structural concrete, this comprehensive book describes the causes, effects, and remedies for concrete wear and failure. Invaluable for planning jobs, selecting materials, and training employees, this book is a must-have for concrete specialists, general contractors, facility managers, civil and structural engineers, and architects.

$34.98 per copy
300 pages, illustrated, Softcover
Catalog No. 67146

Means Unit Price Estimating Methods
New 3rd Edition

This new edition includes up-to-date cost data and estimating examples, updated to reflect changes to the CSI numbering system and new features of RSMeans cost data. It describes the most productive, universally accepted ways to estimate, and uses checklists and forms to illustrate shortcuts and timesavers. A model estimate demonstrates procedures. A new chapter explores computer estimating alternatives.

$29.98 per copy
Over 350 pages, illustrated, Hardcover
Catalog No. 67303A

Total Productive Facilities Management
By Richard W. Sievert, Jr.

Today, facilities are viewed as strategic resources... elevating the facility manager to the role of asset manager supporting the organization's overall business goals. Now, Richard Sievert Jr., in this well-articulated guidebook, sets forth a new operational standard for the facility manager's emerging role... a comprehensive program for managing facilities as a true profit center.

$29.98 per copy
275 pages, Softcover
Catalog No. 67321

Means Environmental Remediation Estimating Methods 2nd Edition
By Richard R. Rast

Guidelines for estimating 50 standard remediation technologies. Use it to prepare preliminary budgets, develop estimates, compare costs and solutions, estimate liability, review quotes, negotiate settlements.

$49.98 per copy
Over 750 pages, illustrated, Hardcover
Catalog No. 64777A

For more information
visit RSMeans Web site
at www.rsmeans.com

Reference Books

Means Illustrated Construction
Dictionary Condensed, 2nd Edition
Recognized in the industry as the best resource of its kind.

This essential tool has been further enhanced with updates to existing terms and the addition of hundreds of new terms and illustrations—in keeping with recent developments. For contractors, architects, insurance and real estate personnel, homeowners, and anyone who needs quick, clear definitions for construction terms.

$59.95 per copy
Over 500 pages, Softcover
Catalog No. 67282A

Means Repair & Remodeling
Estimating New 4th Edition
By Edward B. Wetherill & RSMeans

This important reference focuses on the unique problems of estimating renovations of existing structures, and helps you determine the true costs of remodeling through careful evaluation of architectural details and a site visit.

New section on disaster restoration costs.

$69.95 per copy
Over 450 pages, illustrated, Hardcover
Catalog No. 67265B

Facilities Planning &
Relocation
New, lower price and user-friendly format.
By David D. Owen

A complete system for planning space needs and managing relocations. Includes step-by-step manual, over 50 forms, and extensive reference section on materials and furnishings.

$89.95 per copy
Over 450 pages, Softcover
Catalog No. 67301

Means Square Foot &
Assemblies Estimating
Methods 3rd Edition

Develop realistic square foot and assemblies costs for budgeting and construction funding. The new edition features updated guidance on square foot and assemblies estimating using UNIFORMAT II. An essential reference for anyone who performs conceptual estimates.

$34.98 per copy
Over 300 pages, illustrated, Hardcover
Catalog No. 67145B

Means Electrical Estimating
Methods 3rd Edition

Expanded edition includes sample estimates and cost information in keeping with the latest version of the CSI MasterFormat and UNIFORMAT II. Complete coverage of fiber optic and uninterruptible power supply electrical systems, broken down by components, and explained in detail. Includes a new chapter on computerized estimating methods. A practical companion to *RSMeans Electrical Cost Data.*

$64.95 per copy
Over 325 pages, Hardcover
Catalog No. 67230B

Means Mechanical
Estimating Methods 3rd Edition

This guide assists you in making a review of plans, specs, and bid packages, with suggestions for takeoff procedures, listings, substitutions, and pre-bid scheduling. Includes suggestions for budgeting labor and equipment usage. Compares materials and construction methods to allow you to select the best option.

$64.95 per copy
Over 350 pages, illustrated, Hardcover
Catalog No. 67294A

Means ADA Compliance Pricing
Guide New Second Edition
By Adaptive Environments and RSMeans

Completely updated and revised to the new 2004 *Americans with Disabilities Act Accessibility Guidelines,* this book features more than 70 of the most commonly needed modifications for ADA compliance. Projects range from installing ramps and walkways, widening doorways and entryways, and installing and refitting elevators, to relocating light switches and signage.

$79.95 per copy
Over 350 pages, illustrated, Softcover
Catalog No. 67310A

Project Scheduling &
Management for Construction
New 3rd Edition
By David R. Pierce, Jr.

A comprehensive yet easy-to-follow guide to construction project scheduling and control—from vital project management principles through the latest scheduling, tracking, and controlling techniques. The author is a leading authority on scheduling, with years of field and teaching experience at leading academic institutions. Spend a few hours with this book and come away with a solid understanding of this essential management topic.

$64.95 per copy
Over 300 pages, illustrated, Hardcover
Catalog No. 67247B

Reference Books

For more information
visit RSMeans Web site
at www.rsmeans.com

The Practice of Cost Segregation Analysis

by Bruce A. Desrosiers and Wayne J. DelPico

This expert guide walks you through the practice of cost segregation analysis, which enables property owners to defer taxes and benefit from "accelerated cost recovery" through depreciation deductions on assets that are properly identified and classified.

With a glossary of terms, sample cost segregation estimates for various building types, key information resources, and updates via a dedicated Web site, this book is a critical resource for anyone involved in cost segregation analysis.

$99.95 per copy
Over 225 pages
Catalog No. 67345

Preventive Maintenance for Multi-Family Housing

by John C. Maciha

Prepared by one of the nation's leading experts on multi-family housing.

This complete PM system for apartment and condominium communities features expert guidance, checklists for buildings and grounds maintenance tasks and their frequencies, a reusable wall chart to track maintenance, and a dedicated Web site featuring customizable electronic forms. A must-have for anyone involved with multi-family housing maintenance and upkeep.

$89.95 per copy
225 pages
Catalog No. 67346

Means Landscape Estimating Methods

4th Edition

By Sylvia H. Fee

This revised edition offers expert guidance for preparing accurate estimates for new landscape construction and grounds maintenance. Includes a complete project estimate featuring the latest equipment and methods, and chapters on Life Cycle Costing and Landscape Maintenance Estimating.

$62.95 per copy
Over 300 pages, illustrated, Hardcover
Catalog No. 67295B

Job Order Contracting

Expediting Construction Project Delivery

by Allen Henderson

Expert guidance to help you implement JOC—fast becoming the preferred project delivery method for repair and renovation, minor new construction, and maintenance projects in the public sector and in many states and municipalities. The author, a leading JOC expert and practitioner, shows how to:

- Establish a JOC program
- Evaluate proposals and award contracts
- Handle general requirements and estimating
- Partner for maximum benefits

$89.95 per copy
192 pages, illustrated, Hardcover
Catalog No. 67348

Builder's Essentials: Estimating Building Costs

For the Residential & Light Commercial Contractor

By Wayne J. DelPico

Step-by-step estimating methods for residential and light commercial contractors. Includes a detailed look at every construction specialty—explaining all the components, takeoff units, and labor needed for well-organized, complete estimates. Covers correctly interpreting plans and specifications, and developing accurate and complete labor and material costs.

$29.95 per copy
Over 400 pages, illustrated, Softcover
Catalog No. 67343

Building & Renovating Schools

This all-inclusive guide covers every step of the school construction process—from initial planning, needs assessment, and design, right through moving into the new facility. A must-have resource for anyone concerned with new school construction or renovation. With square foot cost models for elementary, middle, and high school facilities, and real-life case studies of recently completed school projects.

The contributors to this book—architects, construction project managers, contractors, and estimators who specialize in school construction—provide start-to-finish, expert guidance on the process.

$99.95 per copy
Over 425 pages, Hardcover
Catalog No. 67342

For more information
visit RSMeans Web site
at www.rsmeans.com

Reference Books

Historic Preservation: Project Planning & Estimating

By Swanke Hayden Connell Architects

Expert guidance on managing historic restoration, rehabilitation, and preservation building projects and determining and controlling their costs. Includes:

- How to determine whether a structure qualifies as historic
- Where to obtain funding and other assistance
- How to evaluate and repair more than 75 historic building materials

$49.98 per copy
Over 675 pages, Hardcover
Catalog No. 67323

Means Illustrated Construction Dictionary

Unabridged 3rd Edition, with CD-ROM

Long regarded as the industry's finest, *Means Illustrated Construction Dictionary* is now even better. With the addition of over 1,000 new terms and hundreds of new illustrations, it is the clear choice for the most comprehensive and current information. The companion CD-ROM that comes with this new edition adds many extra features: larger graphics, expanded definitions, and links to both CSI MasterFormat numbers and product information.

$99.95 per copy
Over 790 pages, illustrated, Hardcover
Catalog No. 67292A

Designing & Building with the IBC 2nd Edition

By Rolf Jensen & Associates, Inc.

This updated, comprehensive guide helps building professionals make the transition to the 2003 International Building Code®. Includes a side-by-side code comparison of the IBC 2003 to the IBC 2000 and the three primary model codes, a quick-find index, and professional code commentary. With illustrations, abbreviations key, and an extensive Resource section.

$99.95 per copy
Over 875 pages, Softcover
Catalog No. 67328A

Means Plumbing Estimating Methods 3rd Edition

By Joseph Galeno and Sheldon Greene

Updated and revised! This practical guide walks you through a plumbing estimate, from basic materials and installation methods through change order analysis. *Plumbing Estimating Methods* covers residential, commercial, industrial, and medical systems, and features sample takeoff and estimate forms and detailed illustrations of systems and components.

$29.98 per copy
330+ pages, Softcover
Catalog No. 67283B

Builder's Essentials: Advanced Framing Methods

By Scot Simpson

A highly illustrated, "framer-friendly" approach to advanced framing elements. Provides expert, but easy to interpret, instruction for laying out and framing complex walls, roofs, and stairs, and special requirements for earthquake and hurricane protection. Also helps bring framers up to date on the latest building code changes, and provides tips on the lead framer's role and responsibilities, how to prepare for a job, and how to get the crew started.

$24.95 per copy
250 pages, illustrated, Softcover
Catalog No. 67330

Means Estimating Handbook

2nd Edition

Updated Second Edition answers virtually any estimating technical question—all organized by CSI MasterFormat. This comprehensive reference covers the full spectrum of technical data required to estimate construction costs. The book includes information on sizing, productivity, equipment requirements, code-mandated specifications, design standards, and engineering factors.

$99.95 per copy
Over 900 pages, Hardcover
Catalog No. 67276A

Preventive Maintenance Guidelines for School Facilities

By John C. Maciha

A complete PM program for K-12 schools that ensures sustained security, safety, property integrity, user satisfaction, and reasonable ongoing expenditures.

Includes schedules for weekly, monthly, semiannual, and annual maintenance in hard copy and electronic format.

$149.95 per copy
Over 225 pages, Hardcover
Catalog No. 67326

Preventive Maintenance for Higher Education Facilities

By Applied Management Engineering, Inc.

An easy-to-use system to help facilities professionals establish the value of PM, and to develop and budget for an appropriate PM program for their college or university. Features interactive campus building models typical of those found in different-sized higher education facilities, and PM checklists linked to each piece of equipment or system in hard copy and electronic format.

$149.95 per copy
150 pages, Hardcover
Catalog No. 67337

For more information
visit RSMeans Web site
at www.rsmeans.com

Seminars

Means CostWorks® Training

This one-day seminar has been designed with the intention of assisting both new and existing users to become more familiar with the *Means CostWorks* program. The class is broken into two unique sections: (1) A one-half day presentation on the function of each icon; and each student will be shown how to use the software to develop a cost estimate. (2) Hands-on estimating exercises that will ensure that each student thoroughly understands how to use *CostWorks*. You must bring your own laptop computer to this course.

Means CostWorks Benefits/Features:
- Estimate in your own spreadsheet format
- Power of RSMeans National Database
- Database automatically regionalized
- Save time with keyword searches
- Save time by establishing common estimate items in "Bookmark" files
- Customize your spreadsheet template
- Hot Key to Product Manufacturers' listings and specs
- Merge capability for networking environments
- View crews and assembly components
- AutoSave capability
- Enhanced sorting capability

Unit Price Estimating

This interactive two-day seminar teaches attendees how to interpret project information and process it into final, detailed estimates with the greatest accuracy level.

The single most important credential an estimator can take to the job is the ability to visualize construction in the mind's eye, and thereby estimate accurately.

Some Of What You'll Learn:
- Interpreting the design in terms of cost
- The most detailed, time-tested methodology for accurate "pricing"
- Key cost drivers—material, labor, equipment, staging, and subcontracts
- Understanding direct and indirect costs for accurate job cost accounting and change order management

Who Should Attend: Corporate and government estimators and purchasers, architects, engineers… and others needing to produce accurate project estimates.

Square Foot and Assemblies Estimating

This two-day course teaches attendees how to quickly deliver accurate square foot estimates using limited budget and design information.

Some Of What You'll Learn:
- How square foot costing gets the estimate done faster
- Taking advantage of a "systems" or "assemblies" format
- The RSMeans "building assemblies/square foot cost approach"
- How to create a very reliable preliminary and systems estimate using bare-bones design information

Who Should Attend: Facilities managers, facilities engineers, estimators, planners, developers, construction finance professionals… and others needing to make quick, accurate construction cost estimates at commercial, government, educational, and medical facilities.

Repair and Remodeling Estimating

This two-day seminar emphasizes all the underlying considerations unique to repair/remodeling estimating and presents the correct methods for generating accurate, reliable R&R project costs using the unit price and assemblies methods.

Some Of What You'll Learn:
- Estimating considerations—like labor-hours, building code compliance, working within existing structures, purchasing materials in smaller quantities, unforeseen deficiencies
- Identifying problems and providing solutions to estimating building alterations
- Rules for factoring in minimum labor costs, accurate productivity estimates, and allowances for project contingencies
- R&R estimating examples calculated using unit price and assemblies data

Who Should Attend: Facilities managers, plant engineers, architects, contractors, estimators, builders… and others who are concerned with the proper preparation and/or evaluation of repair and remodeling estimates.

Mechanical and Electrical Estimating

This two-day course teaches attendees how to prepare more accurate and complete mechanical/electrical estimates, avoiding the pitfalls of omission and double-counting, while understanding the composition and rationale within the RSMeans Mechanical/Electrical database.

Some Of What You'll Learn:
- The unique way mechanical and electrical systems are interrelated
- M&E estimates–conceptual, planning, budgeting, and bidding stages
- Order of magnitude, square foot, assemblies, and unit price estimating
- Comparative cost analysis of equipment and design alternatives

Who Should Attend: Architects, engineers, facilities managers, mechanical and electrical contractors… and others needing a highly reliable method for developing, understanding, and evaluating mechanical and electrical contracts.

Plan Reading and Material Takeoff

This two-day program teaches attendees to read and understand construction documents and to use them in the preparation of material takeoffs.

Some of What You'll Learn:
- Skills necessary to read and understand typical contract documents—blueprints and specifications
- Details and symbols used by architects and engineers
- Construction specifications' importance in conjunction with blueprints
- Accurate takeoff of construction materials and industry-accepted takeoff methods

Who Should Attend: Facilities managers, construction supervisors, office managers… and others responsible for the execution and administration of a construction project, including government, medical, commercial, educational, or retail facilities.

Facilities Maintenance and Repair Estimating

This two-day course teaches attendees how to plan, budget, and estimate the cost of ongoing and preventive maintenance and repair for existing buildings and grounds.

Some Of What You'll Learn:
- The most financially favorable maintenance, repair, and replacement scheduling and estimating
- Auditing and value engineering facilities
- Preventive planning and facilities upgrading
- Determining both in-house and contract-out service costs
- Annual, asset-protecting M&R plan

Who Should Attend: Facility managers, maintenance supervisors, buildings and grounds superintendents, plant managers, planners, estimators… and others involved in facilities planning and budgeting.

Scheduling and Project Management

This two-day course teaches attendees the most current and proven scheduling and management techniques needed to bring projects in on time and on budget.

Some Of What You'll Learn:
- Crucial phases of planning and scheduling
- How to establish project priorities and develop realistic schedules and management techniques
- Critical Path and Precedence Methods
- Special emphasis on cost control

Who Should Attend: Construction project managers, supervisors, engineers, estimators, contractors… and others who want to improve their project planning, scheduling, and management skills.

Assessing Scope of Work for Facility Construction Estimating

This two-day course is a practical training program that addresses the vital importance of understanding the SCOPE of projects in order to produce accurate cost estimates in a facility repair and remodeling environment.

Some Of What You'll Learn:
- Discussions on site visits, plans/specs, record drawings of facilities, and site-specific lists
- Review of CSI divisions, including means, methods, materials, and the challenges of scoping each topic
- Exercises in SCOPE identification and SCOPE writing for accurate estimating of projects
- Hands-on exercises that require SCOPE, take-off, and pricing

Who Should Attend: Corporate and government estimators, planners, facility managers… and others needing to produce accurate project estimates.

562

Seminars

2007 RSMeans Seminar Schedule

Location	Dates
Las Vegas, NV	March
Washington, DC	April
Phoenix, AZ	April
Denver, CO	May
San Francisco, CA	June
Philadelphia, PA	June
Washington, DC	September
Dallas, TX	September
Las Vegas, NV	October
Orlando, FL	November
Atlantic City, NJ	November
San Diego, CA	December

Note: Call for exact dates and details.

Registration Information

Register Early... Save up to $100! Register 30 days before the start date of a seminar and save $100 off your total fee. *Note: This discount can be applied only once per order. It cannot be applied to team discount registrations or any other special offer.*

How to Register Register by phone today! RSMeans' toll-free number for making reservations is: **1-800-334-3509.**

Individual Seminar Registration Fee $935. *Means CostWorks®* **Training Registration Fee $375.** To register by mail, complete the registration form and return with your full fee to: Seminar Division, Reed Construction Data, RSMeans Seminars, 63 Smiths Lane, Kingston, MA 02364.

Federal Government Pricing All federal government employees save 25% off regular seminar price. Other promotional discounts cannot be combined with Federal Government discount.

Team Discount Program Two to four seminar registrations, call for pricing: 1-800-334-3509, Ext. 5115

Multiple Course Discounts When signing up for two or more courses, call for pricing.

Refund Policy Cancellations will be accepted up to ten days prior to the seminar start. There are no refunds for cancellations received later than ten working days prior to the first day of the seminar. A $150 processing fee will be applied for all cancellations. Written notice of cancellation is required. Substitutions can be made at any time before the session starts. **No-shows are subject to the full seminar fee.**

AACE Approved Courses Many seminars described and offered here have been approved for 14 hours (1.4 recertification credits) of credit by the AACE International Certification Board toward meeting the continuing education requirements for recertification as a Certified Cost Engineer/Certified Cost Consultant.

AIA Continuing Education We are registered with the AIA Continuing Education System (AIA/CES) and are committed to developing quality learning activities in accordance with the CES criteria. Many seminars meet the AIA/CES criteria for Quality Level 2. AIA members may receive (14) learning units (LUs) for each two-day RSMeans course.

NASBA CPE Sponsor Credits We are part of the National Registry of CPE Sponsors. Attendees may be eligible for (16) CPE credits.

Daily Course Schedule The first day of each seminar session begins at 8:30 A.M. and ends at 4:30 P.M. The second day is 8:00 A.M.–4:00 P.M. Participants are urged to bring a hand-held calculator since many actual problems will be worked out in each session.

Continental Breakfast Your registration includes the cost of a continental breakfast, a morning coffee break, and an afternoon break. These informal segments will allow you to discuss topics of mutual interest with other members of the seminar. (You are free to make your own lunch and dinner arrangements.)

Hotel/Transportation Arrangements RSMeans has arranged to hold a block of rooms at most host hotels. To take advantage of special group rates when making your reservation, be sure to mention that you are attending the RSMeans Seminar. You are, of course, free to stay at the lodging place of your choice. (**Hotel reservations and transportation arrangements should be made directly by seminar attendees.**)

Important Class sizes are limited, so please register as soon as possible.

Note: Pricing subject to change.

Registration Form Call 1-800-334-3509 to register or FAX 1-800-632-6732. Visit our Web site: www.rsmeans.com

Please register the following people for the RSMeans Construction Seminars as shown here. We understand that we must make our own hotel reservations if overnight stays are necessary.

❏ Full payment of $_____ enclosed.

❏ Bill me

Name of Registrant(s)
(To appear on certificate of completion)
_____
_____
_____

P.O. #: _____

GOVERNMENT AGENCIES MUST SUPPLY PURCHASE ORDER NUMBER OR TRAINING FORM.

Firm Name _____

Address _____

City/State/Zip _____

Telephone No. _____ Fax No. _____

E-mail Address _____

Charge our registration(s) to: ☐ MasterCard ☐ VISA ☐ American Express ☐ Discover

Account No. _____ Exp. Date _____

Cardholder's Signature _____

Seminar Name	City	Dates

Please mail check to: Seminar Division, Reed Construction Data, RSMeans Seminars, 63 Smiths Lane, P.O. Box 800, Kingston, MA 02364 USA

MeansData™

CONSTRUCTION COSTS FOR SOFTWARE APPLICATIONS

Your construction estimating software is only as good as your cost data.

A proven construction cost database is a mandatory part of any estimating package. The following list of software providers can offer you MeansData™ as an added feature for their estimating systems. See the table below for what types of products and services they offer (match their numbers). Visit online at **www.rsmeans.com/demosource/** for more information and free demos. Or call their numbers listed below.

1. **3D International**
   713-871-7000
   venegas@3di.com

2. **4Clicks-Solutions, LLC**
   719-574-7721
   mbrown@4clicks-solutions.com

3. **Aepco, Inc.**
   301-670-4642
   blueworks@aepco.com

4. **Applied Flow Technology**
   800-589-4943
   info@aft.com

5. **ArenaSoft Estimating**
   888-370-8806
   info@arenasoft.com

6. **Ares Corporation**
   925-299-6700
   sales@arescorporation.com

7. **Beck Technology**
   214-303-6293
   stewartcarroll@beckgroup.com

8. **BSD - Building Systems Design, Inc.**
   888-273-7638
   bsd@bsdsoftlink.com

9. **CMS - Computerized Micro Solutions**
   800-255-7407
   cms@proest.com

10. **Corecon Technologies, Inc.**
    714-895-7222
    sales@corecon.com

11. **CorVet Systems**
    301-622-9069
    sales@corvetsys.com

12. **Earth Tech**
    303-771-3103
    kyle.knudson@earthtech.com

13. **Estimating Systems, Inc.**
    800-967-8572
    esipulsar@adelphia.net

14. **HCSS**
    800-683-3196
    info@hcss.com

15. **MC² - Management Computer**
    800-225-5622
    vkeys@mc2-ice.com

16. **Maximus Asset Solutions**
    800-659-9001
    assetsolutions@maximus.com

17. **Sage Timberline Office**
    800-628-6583
    productinfo.timberline@sage.com

18. **Shaw Beneco Enterprises, Inc.**
    877-719-4748
    inquire@beneco.com

19. **US Cost, Inc.**
    800-372-4003
    sales@uscost.com

20. **Vanderweil Facility Advisors**
    617-451-5100
    info@VFA.com

21. **WinEstimator, Inc.**
    800-950-2374
    sales@winest.com

TYPE	1	2	3	4	5	6	7	8	9	10	11	12	13	14	15	16	17	18	19	20	21
BID					•			•	•		•			•	•		•				•
Estimating		•			•	•	•	•	•	•	•	•	•	•	•		•	•	•		•
DOC/JOC/SABER		•			•			•			•		•			•	•	•			•
ID/IQ		•									•		•			•	•	•			•
Asset Mgmt.																•				•	•
Facility Mgmt.	•		•													•				•	
Project Mgmt.	•	•					•			•	•			•			•	•			
TAKE-OFF					•				•	•	•	•		•	•		•		•		
EARTHWORK											•	•		•	•						
Pipe Flow				•										•							
HVAC/Plumbing					•				•		•										
Roofing					•				•												
Design	•				•		•									•	•				•
Other Offers/Links:																					
Accounting/HR		•			•										•		•	•			
Scheduling					•	•											•		•		•
CAD							•										•				•
PDA																	•				•
Lt. Versions		•							•								•		•		
Consulting	•	•			•		•		•		•	•				•	•	•	•	•	•
Training		•			•		•	•	•	•	•	•	•	•	•	•	•	•	•	•	•

Qualified re-seller applications now being accepted. Call Carol Polio, Ext. 5107.

FOR MORE INFORMATION
CALL 1-800-448-8182, EXT. 5107 OR FAX 1-800-632-6732

For more information
visit RSMeans Web site
at www.rsmeans.com

New Titles

Understanding & Negotiating Construction Contracts

By Kit Werremeyer

Take advantage of the author's 30 years' experience in small-to-large (including international) construction projects. Learn how to identify, understand, and evaluate high risk terms and conditions typically found in all construction contracts—then negotiate to lower or eliminate the risk, improve terms of payment, and reduce exposure to claims and disputes. The author avoids "legalese" and gives real-life examples from actual projects.

$69.95 per copy
300 pages, Softcover
Catalog No. 67350

Green Building: Project Planning & Cost Estimating, 2nd Edition

This new edition has been completely updated with the latest in green building technologies, design concepts, standards, and costs. Now includes a 2007 Green Building *CostWorks* CD with more than 300 green building assemblies and over 5,000 unit price line items for sustainable building. The new edition is also full-color with all new case studies—plus a new chapter on deconstruction, a key aspect of green building.

$129.95 per copy
350 pages, Softcover
Catalog No. 67338A

How to Estimate with Means Data & CostWorks

New 3rd Edition

By RSMeans and Saleh A. Mubarak, Ph.D.

New 3rd Edition—fully updated with new chapters, plus new CD with updated *CostWorks* cost data and MasterFormat organization. Includes all major construction items—with more than 300 exercises and two sets of plans that show how to estimate for a broad range of construction items and systems—including general conditions and equipment costs.

$59.95 per copy
272 pages, Softcover
Includes CostWorks CD
Catalog No. 67324B

Means Spanish/English Construction Dictionary

2nd Edition

By RSMeans and the International Code Council

This expanded edition features thousands of the most common words and useful phrases in the construction industry with easy-to-follow pronunciations in both Spanish and English. Over 800 new terms, phrases, and illustrations have been added. It also features a new stand-alone "Safety & Emergencies" section, with colored pages for quick access. Unique to this dictionary are the systems illustrations showing the relationship of components in the most common building systems for all major trades.

$23.95 per copy
Over 400 pages
Catalog No. 67327A

Construction Business Management

By Nick Ganaway

Only 43% of construction firms stay in business after four years. Make sure your company thrives with valuable guidance from a pro with 25 years of success as a commercial contractor. Find out what it takes to build all aspects of a business that is profitable, enjoyable, and enduring. With a bonus chapter on retail construction.

$49.95 per copy
200 pages, Softcover
Catalog No. 67352

The Homeowner's Guide to Mold

Expert guidance to protect your health and your home.

Mold, whether caused by leaks, humidity or flooding, is a real health and financial issue—for homeowners and contractors. This full-color book explains:

- Construction and maintenance practices to prevent mold
- How to inspect for and remove mold
- Mold remediation procedures and costs
- What to do after a flood
- How to deal with insurance companies if you're thinking of submitting a mold damages claim

$21.95 per copy
144 pages, Softcover
Catalog No. 67344

Qty.	Book No.	COST ESTIMATING BOOKS	Unit Price	Total	Qty.	Book No.	REFERENCE BOOKS (Cont.)	Unit Price	Total
	60067	Assemblies Cost Data 2007	$223.95			67276A	Estimating Handbook, 2nd Ed.	$99.95	
	60017	Building Construction Cost Data 2007	136.95			67318	Facilities Operations & Engineering Reference	54.98	
	61017	Building Const. Cost Data–Looseleaf Ed. 2007	169.95			67301	Facilities Planning & Relocation	89.95	
	60717	Building Const. Cost Data–Spanish 2007	136.95			67338A	Green Building: Proj. Planning & Cost Est., 2nd Ed.	129.95	
	60227	Building Const. Cost Data–Western Ed. 2007	136.95			67323	Historic Preservation: Proj. Planning & Est.	49.98	
	60117	Concrete & Masonry Cost Data 2007	124.95			67349	Home Addition & Renovation Project Costs	29.95	
	50147	Construction Cost Indexes 2007 (subscription)	294.00			67308E	Home Improvement Costs–Int. Projects, 9th Ed.	24.95	
	60147A	Construction Cost Index–January 2007	73.50			67309E	Home Improvement Costs–Ext. Projects, 9th Ed.	24.95	
	60147B	Construction Cost Index–April 2007	73.50			67344	Homeowner's Guide to Mold	21.95	
	60147C	Construction Cost Index–July 2007	73.50			67324B	How to Est.w/Means Data & CostWorks, 3rd Ed.	59.95	
	60147D	Construction Cost Index–October 2007	73.50			67282A	Illustrated Const. Dictionary, Condensed, 2nd Ed.	59.95	
	60347	Contr. Pricing Guide: Resid. R & R Costs 2007	39.95			67292A	Illustrated Const. Dictionary, w/CD-ROM, 3rd Ed.	99.95	
	60337	Contr. Pricing Guide: Resid. Detailed 2007	39.95			67348	Job Order Contracting	89.95	
	60327	Contr. Pricing Guide: Resid. Sq. Ft. 2007	39.95			67347	Kitchen & Bath Project Costs	29.95	
	60237	Electrical Change Order Cost Data 2007	136.95			67295B	Landscape Estimating Methods, 4th Ed.	62.95	
	60037	Electrical Cost Data 2007	136.95			67341	Life Cycle Costing for Facilities	99.95	
	60207	Facilities Construction Cost Data 2007	323.95			67294A	Mechanical Estimating Methods, 3rd Ed.	64.95	
	60307	Facilities Maintenance & Repair Cost Data 2007	296.95			67245A	Planning & Managing Interior Projects, 2nd Ed.	69.95	
	60167	Heavy Construction Cost Data 2007	136.95			67283B	Plumbing Estimating Methods, 3rd Ed.	29.98	
	60097	Interior Cost Data 2007	136.95			67345	Practice of Cost Segregation Analysis	99.95	
	60127	Labor Rates for the Const. Industry 2007	296.95			67337	Preventive Maint. for Higher Education Facilities	149.95	
	60187	Light Commercial Cost Data 2007	116.95			67346	Preventive Maint. for Multi-Family Housing	89.95	
	60027	Mechanical Cost Data 2007	136.95			67326	Preventive Maint. Guidelines for School Facil.	149.95	
	63017	Metric Construction Cost Data 2007	162.95			67247B	Project Scheduling & Management for Constr. 3rd Ed.	64.95	
	60157	Open Shop Building Const. Cost Data 2007	136.95			67265B	Repair & Remodeling Estimating Methods, 4th Ed.	69.95	
	60217	Plumbing Cost Data 2007	136.95			67322A	Resi. & Light Commercial Const. Stds., 2nd Ed.	59.95	
	60047	Repair and Remodeling Cost Data 2007	116.95			67327A	Spanish/English Construction Dictionary, 2nd Ed.	23.95	
	60177	Residential Cost Data 2007	116.95			67145B	Sq. Ft. & Assem. Estimating Methods, 3rd Ed.	34.98	
	60287	Site Work & Landscape Cost Data 2007	136.95			67321	Total Productive Facilities Management	29.98	
	60057	Square Foot Costs 2007	148.95			67350	Understanding and Negotiating Const. Contracts	69.95	
	62017	Yardsticks for Costing (2007)	136.95			67303A	Unit Price Estimating Methods, 3rd Ed.	29.98	
	62016	Yardsticks for Costing (2006)	126.95			67319	Value Engineering: Practical Applications	79.95	
		REFERENCE BOOKS							
	67310A	ADA Compliance Pricing Guide, 2nd Ed.	79.95						
	67330	Bldrs Essentials: Adv. Framing Methods	24.95						
	67329	Bldrs Essentials: Best Bus. Practices for Bldrs	29.95						
	67298A	Bldrs Essentials: Framing/Carpentry 2nd Ed.	24.95						
	67298AS	Bldrs Essentials: Framing/Carpentry Spanish	24.95						
	67307	Bldrs Essentials: Plan Reading & Takeoff	35.95						
	67342	Building & Renovating Schools	99.95						
	67261A	Bldg. Prof. Guide to Contract Documents, 3rd Ed.	32.48						
	67339	Building Security: Strategies & Costs	44.98						
	67146	Concrete Repair & Maintenance Illustrated	34.98						
	67352	Construction Business Management	49.95						
	67314	Cost Planning & Est. for Facil. Maint.	89.95						
	67328A	Designing & Building with the IBC, 2nd Ed.	99.95						
	67230B	Electrical Estimating Methods, 3rd Ed.	64.95						
	64777A	Environmental Remediation Est. Methods, 2nd Ed.	49.98						
	67343	Estimating Bldg. Costs for Resi. & Lt. Comm.	29.95						

MA residents add 5% state sales tax

Shipping & Handling**

Total (U.S. Funds)*

Prices are subject to change and are for U.S. delivery only. *Canadian customers may call for current prices. **Shipping & handling charges: Add 7% of total order for check and credit card payments. Add 9% of total order for invoiced orders.

Send Order To: **ADDV-1000**

Name (Please Print) _____

Company _____

☐ **Company**

☐ **Home Address** _____

_____

City/State/Zip _____

Phone # _____ P.O. # _____

(Must accompany all orders being billed)

Reed Construction Data, Inc.

Reed Construction Data, Inc. is a leading worldwide provider of total construction information solutions. The company's portfolio of information products and services is designed specifically to help construction industry professionals advance their businesses with timely, accurate, and actionable project, product, and cost data. Each of these groups offers a variety of innovative products and services created for the full spectrum of design, construction, distribution, and manufacturing professionals. Reed Construction Data is a division of Reed Business Information, a member of the Reed Elsevier plc group of companies.

Cost Information

RSMeans, the undisputed market leader and authority on construction costs, publishes current cost and estimating information in annual cost books, on *Means CostWorks*® CD, and on the Web. RSMeans furnishes the construction industry with a rich library of corresponding reference books and a series of professional seminars that are designed to sharpen personal skills and maximize the effective use of cost estimating and management tools. RSMeans also provides construction cost consulting for owners, manufacturers, designers, and contractors.

Project Data

Reed Construction Data maintains a significant role in the overall construction process by facilitating the assembly of public and private project data and delivering this information to contractors, distributors, and building product manufacturers in a secure, accurate, and timely manner. Acting on behalf of architects, engineers, owner/developers, and government agencies, Reed Construction Data is the construction community's premier resource for project leads and bid documents. Our wide variety of products and services provides the most up-to-date, enhanced details on many of the country's largest public sector, commercial, industrial, and multi-family residential projects.

Reed Bulletin and Reed CONNECT™ provide project leads and project data through all stages of construction. Customers are supplied industry data through leads, project reports, contact lists, plans, specifications, and addenda—either online, via e-mail, or in paper format.

Building Product Information

The Reed First Source suite of products is the only integrated building product information system offered to the commercial construction industry for searching, selecting, and specifying building products. These online and print resources include the *First Source* catalog, SPEC-DATA™, MANU-SPEC ™, First Source CAD, and manufacturer's catalogs. Written by industry professionals and organized using CSI MasterFormat 2004 criteria, the design and construction community uses this information to make better, more informed design decisions.

Research & Analytics

Reed Construction Data's forecasting tools cover most aspects of the construction business. Our vast network of resources makes us uniquely qualified to give you the information you need to keep your business profitable.

Associated Construction Publications (ACP)

Reed Construction Data's regional construction magazines cover the nation through a network of 14 regional magazines. Serving the construction market for more than 100 years, our magazines are a trusted source of news and information in the local and national construction communities.

International

Reed Construction Data Canada is the "voice of construction" in Canada, and leading provider of project information and construction market intelligence. Products and services include: *Building Reports*, Canadian ICI project leads tracking construction projects through life cycles—from concept to construction. *Building Reports* can be customized and delivered through CONNECT™, RCD's web-based sales and contact management system, or by fax or email; *Daily Commercial News*, Ontario's leading construction industry daily newspaper, contains ICI project leads, Building Reports, Bidders' Register, news and trends, construction tenders, certificates of substantial performance, and CanaData Economic Snapshot; *Journal of Commerce*, Alberta and British Columbia's twice-weekly construction industry newspaper, is a one-stop source for private and public projects, tenders and expressions of interest, top industry news and construction trends, and more; *CanaData*, a variety of products tracking market performance, economic analysis, forecasts, and trends in the construction market; *Buildcore*, organized by MasterFormat™, is Canada's most comprehensive listing of commercial building products and manufacturers active and available in Canada—in print and online at www.buildcore.com.

Reed Business Information – Scandinavia, with offices in Denmark, Norway, Finland, and Sweden, is the construction industry's source for project information in the Nordic region.

Reed Construction Data—Australia (formerly Cordell Building Information Services) is Australia's leading provider and respected industry authority on building and construction information.

For more information, please visit our website at www. reedconstructiondata.com

Reed Construction Data, Inc.
30 Technology Parkway South
Norcross, GA 30092-2912
(800) 322-6996
(800) 895-8661 (fax)
Email: info@reedbusiness.com

RSMeans Project Cost Report

By filling out this report, your project data will contribute to the database that supports the RSMeans Project Cost Square Foot Data. When you fill out this form, RSMeans will provide a $30 discount off one of the RSMeans products advertised in the preceding pages. Please complete the form including all items where you have cost data, and all the items marked (✔).

$30.00 Discount per product for each report you submit.

Project Description (No remodeling projects, please.)

✔ Building Use (Office, School...) _____

✔ Address (City, State) _____

_____

✔ Frame (Wood, Steel...) _____

✔ Exterior Wall (Brick, Tilt-up...) _____

✔ Basement: (check one)   ☐ Full ☐ Partial ☐ None

✔ Number Stories _____

✔ Floor-to-Floor Height _____

% Air Conditioned _____ Tons _____

Comments _____

Total Project Cost $ _____

Owner _____

Architect _____

General Contractor _____

✔ Bid Date _____

Typical Bay Size _____

✔ Labor Force: _____ % Union _____ % Non-Union

✔ Project Description (Circle one number in each line)

1. Economy 2. Average   3. Custom 4. Luxury
1. Square   2. Rectangular 3. Irregular 4. Very Irregular

A	✔ General Conditions	$
B	✔ Site Work	$
C	✔ Concrete	$
D	✔ Masonry	$
E	✔ Metals	$
F	✔ Wood & Plastics	$
G	✔ Thermal & Moisture Protection	
GR	Roofing & Flashing	$
H	✔ Doors and Windows	$
J	✔ Finishes	$
JP	Painting & Wall Covering	$

K	✔ Specialties	$
L	✔ Equipment	$
M	✔ Furnishings	$
N	✔ Special Construction	$
P	✔ Conveying Systems	$
Q	✔ Mechanical	$
QP	Plumbing	$
QB	HVAC	$
R	✔ Electrical	$
S	✔ Mech./Elec. Combined	$

Please specify the RSMeans product you wish to receive. Complete the address information.

Product Name _____

Product Number _____

Your Name _____

Title _____

Company _____

☐ Company
☐ Home   Street Address _____

City, State, Zip _____

Method of Payment:

Credit Card # _____

Expiration Date _____

Check _____

Purchase Order _____

Reed Construction Data/RSMeans
Square Foot Costs Department
P.O. Box 800
Kingston, MA 02364-9988

Return by mail or Fax 888-492-6770.